Separations Division 2015

Core Programming Area at the 2015 AIChE Annual Meeting

Salt Lake City, Utah, USA
8-13 November 2015

Volume 1 of 2

ISBN: 978-1-5108-1873-6

Printed from e-media with permission by:

Curran Associates, Inc.
57 Morehouse Lane
Red Hook, NY 12571

Printed by Curran Associates, Inc. (2016)

For permission requests, please contact AIChE
at the address below.

AIChE
120 Wall Street, FL 23
New York, NY 10005-4020

Phone: (800) 242-4363
Fax: (203) 775-5177

www.aiche.org

Additional copies of this publication are available from:

Curran Associates, Inc.
57 Morehouse Lane
Red Hook, NY 12571 USA
Phone: 845-758-0400
Fax: 845-758-2634
Email: curran@proceedings.com
Web: www.proceedings.com

TABLE OF CONTENTS

VOLUME 1

VOLUME 2

408192 White Graphene As a Novel Type of Adsorbent

Abstract:

Boron nitride nanosheets (BNNSs) are 2D flat nanosheets consisting of an equimolar ratio of sp^2-bonded B and N atoms. Despite the structural similarity to graphene (hence the name "white graphene"), BNNSs exhibit unique properties, such as an enhanced thermal and chemical robustness. It is a highly porous material, with a theoretical surface area of ~2640 $m^2\ g^{-1}$ for one sheet, which makes it an excellent candidate for adsorption applications. However, controlling the synthesis of porous BNNSs is not trivial and the synthesis mechanisms remain unclear. If these were unveiled, one could tune the structure of BNNSs to target specific separation applications.

This study investigates the synthesis mechanisms of BNNSs via a bottom-up approach (i.e. synthesis from chemical precursors rather than via exfoliation). Specifically, the effects of the reaction parameters on the chemical and physical features of the resulting BNNSs are analyzed using a range of analytical techniques (e.g. N_2 sorption, XRD, FT-IR, Raman, NMR, thermogravimetric analysis). The findings highlight the role of the reaction intermediate on the resulting porous BNNSs (S_{BET} > 1,000 $m^2\ g^{-1}$) as well the key parameters controlling its structure and chemistry. Preliminary gas adsorption tests (including CO_2 capture) indicate the promising role of BNNSs as a robust adsorbent.

428530 The Application of Reactive Dyes to Functionalize Fibers with MOFs, Quantum Dots, and Gold Nanoparticles

Abstract:

Reactive dyes conventionally used to chemically bind chromophores to fabrics have been used to develop a platform technology that can modify commercially available fibers with nanoscale structures. The reactive dye method was used to attach Cu-BTC to both cotton and nylon. The cotton and nylon Cu-BTC modified fabrics have surface areas of approximately 976 and 680 m^2/g, respectively, X-ray diffraction patterns consistent with Cu-BTC, and FESEM images that show crystals decorating the fabric surfaces. To demonstrate the ability of the fibers to perform separations, ethane and ethylene single component gas adsorption isotherms were measured for both nylon and cotton MOF modified fibers. Both of the MOF modified fibers show preferential adsorption of ethylene over ethane. The MOF modified fabrics also have the ability to adsorb ammonia with the cotton sample adsorbing 2 mol/kg of ammonia from a 25 ^{o}C 50% relative humidity air stream. The results are consistent with bulk powder Cu-BTC MOF showing that the attachment of the adsorbent to the fiber does not alter the properties of the adsorbent. In addition to MOFs, quantum dots and gold nanoparticles have also been bound to cellulose and nylon. The results illustrate how the reactive dye method provides a platform to modify fibers to perform separations.

426991 Preparation of a Hybrid Biosorbent Using Self-Immobilizing Mycelia and Magnetic Nanoparticles

Abstract:

Many porous materials have heavy metal ion or dye adsorption abilities. Natural materials such as clay, ash and various biomasses have been investigated as cheap adsorbents for water treatments, and the adsorption is mainly based on the affinity between biosorbents and adsorptive pollutants, which is usually due to the functional groups such as sulfonate, phosphonate and amine existed on the surface of biomass materials. Meanwhile, many biosorbents used are in suspended forms, which may be not effective and make the post separation of these biomasses from treated water extremely difficult. Moreover, those biomasses left in the water may cause secondary pollution.

In order to overcome this drawback, research used immobilized microorganisms as biosorbents for easy separation. Magnetic collection is also an easy and efficient way to collect adsorbents. In addition, the combination of magnetic nanoparticles with various immobilized biosorbents can form hybrid adsorbents with magnetic induction heatable properties. This temperature remote controlling feature is useful for pollutant adsorption processes. In this study, a marine-derived fungus was used for dye adsorption from aqueous solutions. This fungus had strong pellet forming ability even with high concentrations of particles in culture media, which is not possible for many other fungi reported. Magnetic mycelial pellets were prepared by adding Fe_3O_4 nanoparticles in culture media, which can be imbedded into the mycelial pellets during hyphal growth. Properties of the self-immobilized biosorbent were characterized, and its dye adsorption capacity was investigated with adsorption kinetics studied under magnetic induction heating. The results showed that the solution heated by magnetic induction had faster adsorption rates and less energy consumption.

420216 In-Pore Synthesis of Reactive Sites in Carbon Silica Composites for Gas-Phase Adsorption of Toxic Industrial Chemicals

Abstract:

Adsorbent technology is critical in maintaining breathable air environments for diverse civilian and military applications. Gas masks for military personnel and first responders are examples of such technologies used to capture toxic industrial chemicals (TICs) from air. Our approach involves the synthesis of composite adsorbents with various phases each capable of targeting a variety of adsorbates. This contribution describes the development of carbon-silica composites (CSCs) containing reactive sites created via in-pore reactions. These sites provide targeted removal capabilities for basic and acid-forming TICs such as ammonia and sulfur dioxide. Breakthrough experiments are conducted on synthesized materials using NH_3 (1500 ppm) and SO_2 (500 ppm) to determine their adsorption capacities. Adsorbents are also characterized by porosimetry, crystallography, and microscopy techniques.

The base composite material consists of a nonpolar carbon phase and a polar silaceous phase made from polymerizing furfuryl alcohol within the pores of MCM-41 to form the microporous structure of the CSC. The composite is further functionalized by introducing various metal salts as reactive sites for redox reactions to enhance the removal capacities for TICs. A unique advantage of this functionalization is the capability of performing precipitation reactions between combinations of chemical moieties within the pores of the substrate. The mechanism of in-pore synthesis of otherwise non-soluble functionalities incorporated into well-defined porous adsorbent structures may be appealing for many catalysis and separation processes.

430997 Partial Etching of Titanium Carbide-Derived Carbon for VOC Adsorption

Abstract:

Volatile organic compounds (VOCs) are produced in many chemical processes and can cause harm to human health and the environment. Specifically, methanol may adversely react with other VOCs to form smog. Carbide-derived carbons (CDCs) offer one solution to remove methanol from waste gas streams via adsorption. Control over pore size and distribution, chemical selectivity, and chemical and thermal stability make CDCs ideal for this process. CDCs are formed selectively by removing the metal atoms within a metal carbide using a high temperature etching process. This selective removal of metal atoms with chlorine leaves behind a narrow distribution of pore sizes in a carbon complex. Additionally, partial etching of the carbide yields a carbon complex with residual metal species that add active sorption and catalytic sites. Partial etching and post-etching modification negates the need for metal doping seen in activated carbons because the metal is present in the precursor. Titanium has been shown to catalyze methanol in the reduced metal state, while titanium dioxide has been shown to photocatalytically degrade it. Based upon these metal-to-VOC interactions, titanium carbide-derived carbon (TiCDC) has excellent potential to be tuned for methanol adsorption. In this work, we use a horizontal bed reactor to partially etch TiC to create TiCDC with various amounts of residual metal. However, partial etching in a horizontal bed reactor shows a significant downside: the development of two layers in the bed which consist of completely etched TiCDC and unreacted TiC. Methanol adsorption studies show that the TiCDC samples with various residual metal contents are surface area dependent, rather than dependent on the amount of residual metal. This suggests that much of the residual metal remains inaccessible for use as active sorption sites. Further control over metal extraction in CDCs is necessary to optimize active sorption sites.

(7f) Batch Adsorption of Heavy Metal Ions Using Biomass-Based Activated Carbon Sorbents

Alhwaige, A. A., Case Western Reserve Univesity
Al-Shareef, A., Al-Mergib University
Sowedan, A. M., Al-Mergib University
Ebshish, A. S., Al-Mergib University
Bshish, A. M., Al-Mergib University
Ghareba, S. M., Al-Mergib University
Originally presented on: 11/8/2015 17:10:00 - 17:30:00

Water is the most important and essential component on the earth for vital activities of living beings. Unfortunately, water pollution has become a serious issue in the present century because affecting all living organisms and human health [1]. The selection of efficient removal technologies for heavy metals has been given much attention due to their economic and energy impacts. There are numerous techniques have been applied to remove metal ions from contaminated water and many physico-chemical methods have been proposed for simulated wastewater treatment including sorption, precipitation, membrane, filtration, and ion exchange. Among the available technologies, adsorption is considered an effective purification and separation technique used in industry especially in water and wastewater treatments [2].

Currently extensive research is focused on developing and designing new porous materials for water purification such as adsorption of heavy metal ions. Porous carbon materials, which have received much interest because they are cheap, thermally and chemically stable, and highly pores. In addition, the removal of heavy metal ions using nitrogen-doped porous carbon (N-PC) has been shown to be efficient, recyclable, and selective with extraordinary adsorption capacity. Furthermore, these sorbents exhibit fast adsorption-desorption kinetics and outstanding adsorption capacity, as well as good stability over multiple adsorption-desorption cycles. [3]. Biomass has attracted multidisciplinary study because of its unique physicochemical properties. Biomass-derived porous carbon sorbents have interesting features, including eco-friendly materials, control of sorbent design, unique structure, and ease of handling [4,5].

This study reports for preparation of activated porous carbon sorbents from waste-tea leaves, an environmentally-benign biomass. The sorption performance of the developed activated carbon sorbents for metal ions removal is studied. Theoretical adsorption isotherms have been calculated and compared to the experimental results. The isosteric heat of adsorption is obtained and the results show the endothermic nature of adsorption. The effects of pH and temperature on the adsorption performance are also investigated. The obtained adsorbents show good adsorption capacities for the removal of metal ions from simulated wastewater systems, suggesting excellent potential for metals ions adsorption.

References

[1] I. Ali, Chem. Rev. 2012, 112, 5073–5091.

[2] G. Zhao, J. Li, X. Ren, C. Chen, and X. Wang, Environ. Sci. Technol. 2011, 45, 10454–10462.

[3] L. Ye, H. You, J. Yao, H Su, Desalination 298 (2012) 1–12.

[4] M. Kazmi, A. R. Saleemi, N. Feroze, A. Yaqoob, S. W. Ahmad, Pol. J. Chem. Tech., 15, 1–6.

[5] A. A. Alhwaige, T. Agag, H. Ishida and S. Qutubuddin, RSC Adv., 2013, 3, 16011–16020.

409064 Hybrid Anion Exchange/Activated Carbon for Cr(VI) Removal in Drinking Water

Abstract:

A novel hybrid anion exchangeable activated carbons (HACAX) is synthesized through chemical modification of porous carbon surface. Ion exchange capacity, kinetics and regeneration tests suggest that HACAX contains stable positive charged functional groups. The surface characterization results suggest HACAX not only contains stable positive charged functional groups, but also a broad spectrum of surface oxygen functional groups. In the study of adsorption of hexavalent chromium, HACAX shows excellent kinetics, selectivity and regeneration efficiency towards Cr (VI). Combined with rich surface oxygen functional groups, this material provides a pathway to adsorb and convert Cr(VI) into Cr(III) at natural pH. The results suggest HACAX could potentially provide an effective point-of-use (POU) solution for hexavalent chromium removal from drinking water.

409065 2. Activated Carbon Fiber Composites for Ammonia Removal in Air

Abstract:

Chemically Activated Carbon Fibers (CAFs) were synthesized for ammonia adsorption. NH_3 adsorption performance was measured in flowthrough configuration at both 0% and 50% relative humidity conditions. The surface of these materials was characterized using SEM, BET, elemental analysis, FTIR, XPS, and Boehm titration. High surface area >1000 m^2/g of all ACFCs are essential to providing more accessible surface functional groups and thus adsorption efficiency. Surface chemistry analysis results showed that adsorption performance is correlated with surface acidic oxygen groups.

417491 Mesoporous MgO for Carbon Dioxide Capture at an Ambient Condition and the Effect of Calcination Steps to Sorption Performances

Abstract:

To cope with the climate change caused by the emission of CO_2 gas, much research has been done to develop high performance materials for CO_2 capture. In the study, mesoporous MgO sorbents were synthesized by using an aerogel method for CO_2 sorption at ambient temperatures. At the last step of the synthesis, a calcination method was developed: 1-step calcined MgO and 4-step calcined MgO. The effect of the calcination condition and method on CO_2 sorption capacity was evaluated.

The sorption capacity of as-synthesized MgO samples at 30-350°C was measured and CO_2 sorption/desorption cyclic test was carried out by using Thermogravimetric analysis(TGA). The 4-step calcined MgO has the sorption capacity of 10.6wt% and 70% of saturation was reached within 5 minutes. On the other hand, the 1-step calcined MgO showed better stability than the 4-step one in the cyclic test. Visual images and crystalline data of sorbents were confirmed by SEM, TEM and XRD. In addition, textual properties (surface area, pore size and pore volumes) of the sorbents before and after the cyclic test were compared by measuring the N_2 isotherm analysis. It was confirmed that the CO_2 sorption capacity depended on the textual properties controlled by the calcination method.

419776 Enhancement of CO2 Storage on Mesoporous Carbon in the Presence of Water

Abstract:

Enhancement of CO_2 Storage on Mesoporous Carbons in the Presence of Water

Jia Zhou[a], Wei Su[ab], Yan Sun[c], Xiaojing Wang[a], Shuguang Deng[d]

aSchool of Chemical Engineering and Technology, Tianjin University, Tianjin 300072, China

bTianjin Key Laboratory of Membrane and Desalination Technology,Tianjin 300072,China

cDepartment of Chemistry, School of Science, Tianjin University, Tianjin 300072, China

d Department of Chemical Engineering and Materials New Mexico State University, Las Cruces, NM, 88003, USA

Carbon dioxide is a primary contributor of greenhouse gas in atmosphere. Hydrate-forming method is a promising solution to capture and store CO_2. The equilibrium isotherms of CO_2 on mesoporous carbon CMK-3 with different water contents were collected at 275K. When the water weight ratio was 0.55, a rapid increase at about 2.6MPa was observed in the isotherm, for the CO_2 hydrate was formed in the carbon pores. However, there were two sharp increases in the isotherms at the pressure of 2.1MPa and 2.6MPa respectively, as the water weight ratio was at 1.55, 1.75 and 2.35. At high water weight ratio, after the pores were fully filled with water, there still existed extra water outside the pores. As a result, CO_2 hydrate was formed in the space among carbon particles. Compared with dry CMK-3, the CO_2 adsorption capacity wet CMK-3 was improved as much as 122% at the pressure of 3.5MPa, and the highest adsorption capacity is about 41mmol/g.

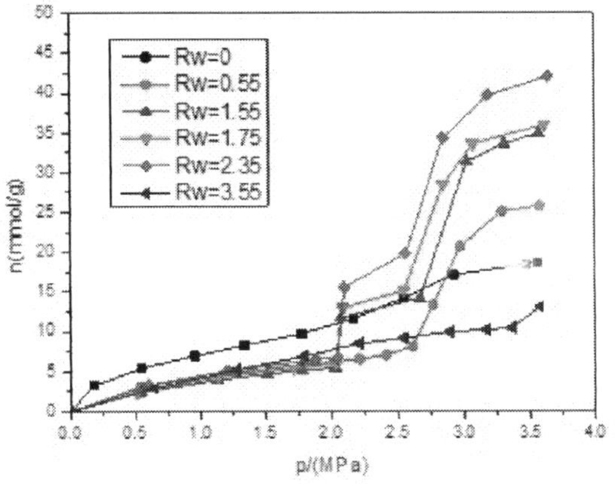

Fig 1. CO_2 isotherms of on mesoporous carbon CMK-3 with different water contents at 275 K

416586 Carbon Spheres for Heavy Metal Removal from Refinery Water

Abstract:

An environmentally-friendly hydrothermal procedure was employed to synthesize carbon spheres (CS) from glucose solution at 170^{0}C. To study the effect of reaction time, four CS samples were synthesized at 12h, 14h, 16h and 19h, respectively. A portion of the samples were surface activated by the treatment with 0.5M NaOH, the other portion treated with 0.5M CH_3COOH, and the remaining samples were left untreated (native). All samples were characterized using various methods including SEM, XRD, Raman spectroscopy, N_2 adsorption and FT-IR. Batch adsorption experiments were carried out at room temperature to remove heavy metal ions (Ag^+ and Cu^{2+}) from aqueous solution . It was found that increasing reaction time resulted in CS with a lower amorphous content and lowered adsorption capacity, NaOH-treated CS had the highest adsorption capacity of 454 mg Ag^+/g and 172 mg Cu^{2+}/g, while CH_3COOH-treated CS had a lower Ag^+ and Cu^{2+} uptake than untreated CS. The adsorption isotherms were well fit by the Langmuir isotherm equation (for base-treated CS) and the Freundlich isotherm equation (for acid-treated CS). The CS showed a high potential for the removal of heavy-metals from aqueous solutions.

421209 Preparation and Characterizations of Activated Carbon from Char Produced from Woody Biomass Gasification and Its Application for Dye Removal

Abstract:

Apart from syngas, woody biomass gasification process also generates significant amount of carbon residues (char) and ashes as by-products, which usually contain high concentration of harmful compounds, hence causing health and environmental problems during disposal. In view of this, it is essential to develop beneficial uses of these solid residual wastes. In this work, carbon residue obtained from wood gasification was employed as a precursor for preparing adsorbent via physical activation. The effect of activating agents and temperature was examined in order to obtain the optimal conditions for the activation stage. The obtained activated carbon was characterized and tested as an adsorbent for dye (Rhodamine B) removal, and the influence of adsorption time, initial concentration and pH were also studied. Kinetics and isotherm analysis were also performed.

From BET analysis, the surface area of unprocessed char was less than 200 m^2 per gram and it was greatly increased to 776.46 m^2 per gram after steam activation at 900°C. Moreover, the FTIR spectra also showed the abundance of hydroxyl (-OH) and carboxyl (-COOH) groups on the surface of activated carbon obtained from steam activation. The activated carbon produced at the optimal conditions was used for dye adsorption and it exhibited high dye adsorption capability. This can be attributed to the increase in surface area as well as the abundance of hydroxyl and carboxyl groups on the activated carbon surface. Moreover, it was found that the dye adsorption capability was significantly increased at basic condition. This can be attributed to the increased electrostatic interaction between the deprotonated (negatively charged) activated carbon and dye molecules. Furthermore, kinetics showed that the dye adsorption data follow the pseudo-second-order adsorption model, whereas the isotherm analysis indicated that the dye adsorption data can be represented by the Langmuir model. Our results clearly showed that the activated carbon prepared from steam activation of carbon residue from wood gasification can be effectively used as an adsorbent for dye removal.

The use of carbon residues produced from wood gasification as a source of absorbent materials not only provides a cost-effective and environmental friendly way of recycling those residual wastes, reducing its negative environmental effects, but also produces high-value products, offering the economic benefits for woody biomass gasification process.

421753 A Rapid Pressure Swing Adsorption Process for Oxygen Generation

Abstract:

Rapid pressure swing adsorption is commonly used in the design of portable oxygen concentrators, which serve as a viable alternative to oxygen cylinders in the medical field. Rapid pressure swing adsorption processes present challenges in process design that slower PSA processes do not possess. These mainly deal with how to efficiently operate a PSA process on a small scale while dealing with an increase in mass and heat transfer resistance caused by higher gas velocities. This study illustrates how to deal with these challenges through the design of a rapid PVSA process using a rate enhanced adsorbent that provides a high purity product (97% oxygen using argon free air) and extremely favorable recoveries (50-60% at pressure ratios between 4-5) with a BSF under 100 lbs. /TPD oxygen.

Using a commercial Li-X zeolite, a lab-scale two column PVSA system was used to design a process utilizing rapid PVSA. The design included first examining the effect of cycle time on the oxygen recovery at a range of pressure ratios (varying from 2.5-7 by changing the desorption pressure). The adsorption/desorption pressure, amount of gas used for the feed and purge steps, and product purity were maintained as the cycle time was reduced. This provided a method in which any change in oxygen recovery was almost solely assessed to the reduction in cycle time and not variations in other process parameters. Additionally, a range of adsorption pressures were examined to determine an ideal operating range using rapid PVSA for this adsorbent.

425034 Enhance the Water Stability in Cr-BTC Metal-Organic Frameworks

Abstract:

Intense interest in the porous materials for selective adsorption processes. Metal organic frameworks (MOFs) attracting attention over the recent years, which can be manipulated to target specific adsorption behavior by introducing open metal sites and functional groups into the structure. Nowdays, MOF behavior in the presence of water is a topic of significant importance when considering these materials for adsorption applications. Among the water sensitive MOFs, Cr-BTC (Cr_3(1,3,5-benzenetricarboxylate)$_2$, with a same topological structure as HKUST-1) is a very typical material which can highly-selective and reversible binding O_2 by the open Cr^{2+}sites, however, in our study, which cannot survival in the open air even if only trace of water molecules exist. How to prevent the invasion of water molecules, and keep the structure of the Cr-BTC is the focus of our research.

In this work, the original samples Cr-BTC was synthesized from literature report: $Cr(CO)_6$, 1,3,5-benzenetricarboxylate acid and anhydrous, air-free *N,N*-dimethylformamide (DMF) under a dry nitrogen atmosphere. O_2 adsorption volume reach to 14 wt% (very close to the theoretical value 14.4%), but we cannot found any diffraction peaks from the XRD pattern of binding O_2 Cr-BTC sample exposed to the open air, that means the structure was destroyed in the air condition. We suspect the water molecular in the air is the only killer, so the water vapor adsorption was tested under 298 K and 3.16 KPa, and then exposed the sample in the oxygen environment for 1 h. The experiment result showed BET surface of Cr-BTC was decreased dramatically from 1369 m^2/g to 51 m^2/g, and there is no diffraction peaks from the XRD pattern.

Surprisingly, when the no O_2 binding Cr-BTC was placed in saturation ethanol (C_2H_5OH) and ethyl bromide (C_2H_5Br) vapor environments separately for 12 h, we can found the XRD diffraction peaks after the two samples exposed in the open air for 1 h, and then they could keep high BET surface area. We also verified this result by DFT and GCMC simulation.

426528 Lead Removal from Aqueous Solution By a Zirconium-Phosphate Impregnated PVDF Membrane

Abstract:

Abstract:

Lead contamination in water has become a serious issue due to its high toxicity and easy accumulation in human body. Excess uptake of lead can damage the nervous system, kidneys and cardiovascular system. United States Environmental Protection Agency has regulated that the maximum lead concentration is drinking water is 15 g/L.

In this study, a novel zirconium phosphate impregnated PVDF membrane was developed for lead removal from aqueous solution. The zirconium ions were firstly impregnated within the PVDF membrane by coordinated with ether and hydrogen bond, which were produced by crosslinking of PVA (polyvinyl alcohol) and glutaraldehyde. The resulting membrane was then placed in the phosphate solution for the further surface modification.

The membrane could be used for lead removal. The adsorption kinetics study showed that the adsorption equilibrium was established within 10 h. The adsorption increased with an increase in solution pH; the optimal pH for adsorption was 5.5. The experimental data were better described by Langmuir equation than Freundlich equation; the maximum adsorption capacity of 121.2 mg-Pb/g was achieved at the optimal pH. The adsorption of lead was mainly due to ion exchange between hydrogen of lead ions.

Key words: PVDF membrane, zirconium phosphate, PVA coating, lead, adsorption

427173 Optimization of PSA Plant for CBM Purification

Abstract:

This work presents a dynamic simulation and optimization model of a pressure swing adsorption (PSA) process to purify coal bed methane (CBM). The developed simulation model is well validated by comparing experimental results. Based on the simulation model, the optimization is carried out to scale up the PSA process design from a laboratory scale to a commercial size plant via a pilot plant scale. The scale up optimization results are verified by PSA vendor company in Korea. The PSA optimization methodology (Ko et al., 2005) is implemented in this study within the gPROMS modeling system. As a result, PSA pilot plant has been constructed and will be operated at CBM site.

Preparation of bi-polar membranes and their application to sodium hypochlorite generation

K. Y. Kim[1] , H. S. Shin[2], C. S. Kim[2], J. W. Rhim[1*]

[1]*Department of Chemical Engineering, Hannam University, Daejeon, Korea*
[2] *Techwin Co. Ltd., Cheongju-si, Chungcheongbuk-do, Korea; (jwrhim@hnu.kr)*

Introduction

Motivation

- Ion exchange membranes are classified into anion exchange membranes and cation exchange membranes relying on the type of ionic groups attached to the membrane surface. They are used in a number of industrial application areas owing to their ecological safeness and comparative low energy consumption.
- In this study, polysulfone(PSf) and polyehterimide(PEI) were aminated to synthesize anion exchange polymers at different amination ratios and polyether ether ketone (PEEK) was sulfonated to make cation exchange polymers. Two bipolar membranes such as APSf/SPEEK and APEI/SPEEK were prepared by using synthesized polymers and hypochlorite production test was conducted.

Direct fluorination

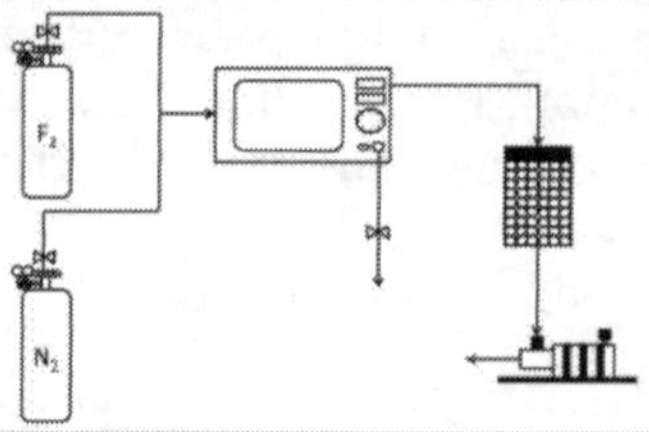

- Reaction to insert or replace the fluorine element in the surface of the material through the process of pouring directly the fluorine gas to the surface of the material
- To improve durability via the surface modification

Experimental

Production process of hypochlorite

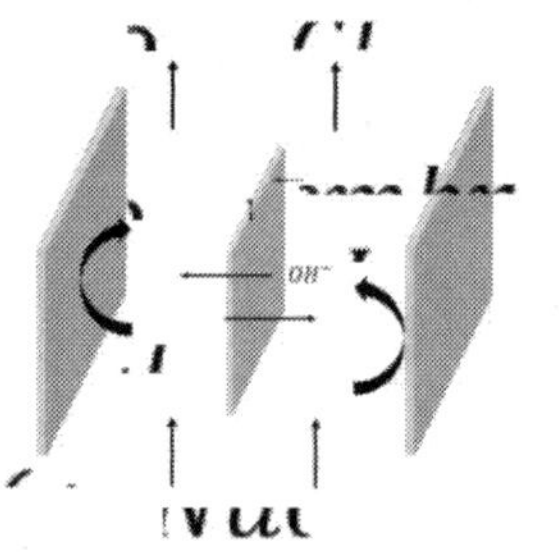

Manufacturing process of AEM and CEM

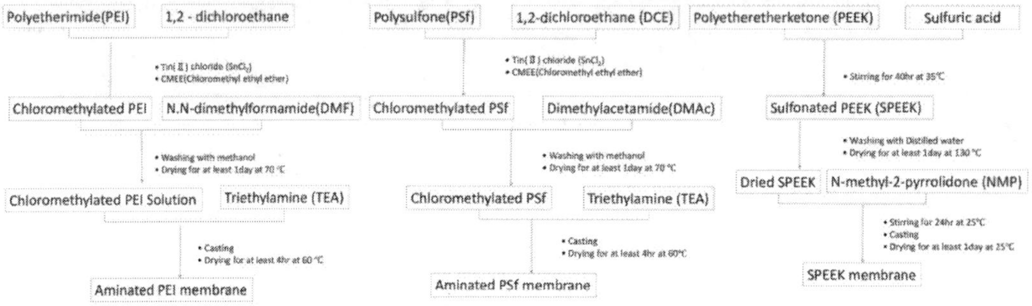

Results

- **Ion exchange capacity**

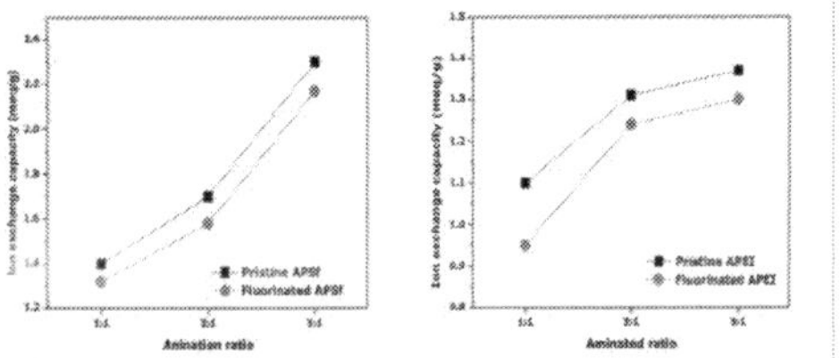

- **Ion conductivity**

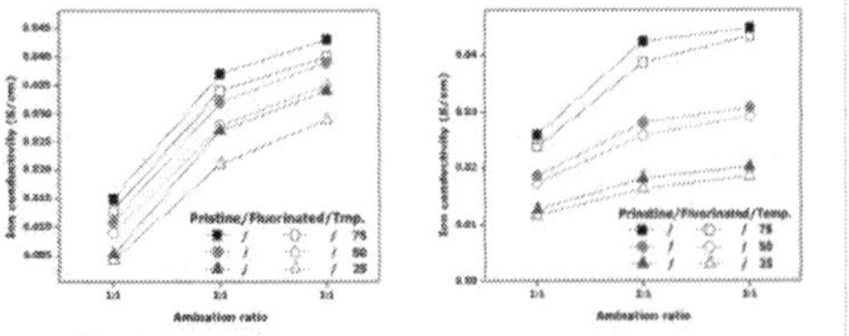

- **Contact angle**

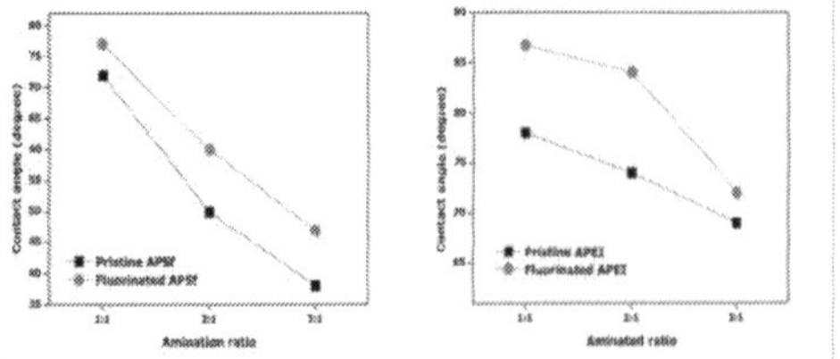

- **Feed : 0.5wt% NaCl**
- **Feed flow rate : 30ml/min**
- **Current density : 5 mA/cm²**
- **Membrane area : 25cm²**
- **Operating temperature : 25℃**
- **Fluorination time : 2hr**

- **APSf 1:1, 2:1, 3:1/SPEEK bi-polar**

Pristined bi-polar

Sample	Anode pH	Cathode pH	Voltage (V)	Generation (ppm)	Test time (min)
SPEEK/APSf 1:1	3.77	11.52	3	54.24	250
SPEEK/APSf 2:1	3.49	11.66	4	54.36	235
SPEEK/APSf 3:1	3.31	12.18	6	60.97	220

Fluorinated bi-polar

Sample	Anode pH	Cathode pH	Voltage (V)	Generation (ppm)	Test time (min)
SPEEK/APSf 1:1	3.77	11.26	4	48.46	600
SPEEK/APSf 2:1	3.47	11.64	5	50.13	580
SPEEK/APSf 3:1	3	12.28	5	53.18	568

- **APEI 1:1, 2:1, 3:1/SPEEK bi-polar**

Pristined bi-polar

Sample	Anode pH	Cathode pH	Voltage (V)	Generation (ppm)	Test time (min)
SPEEK/APEI1:1	3.27	11.58	3	36.87	170
SPEEK/APEI 2:1	3.19	11.64	4	52.47	162
SPEEK/APEI 3:1	3.05	11.84	5	59.56	150

Fluorinated bi-polar

Sample	Anode pH	Cathode pH	Voltage (V)	Generation (ppm)	Test time (min)
SPEEK/APEI1:1	3.27	11.6	3	29.78	468
SPEEK/APEI 2:1	3.07	11.72	4	50.52	456
SPEEK/APEI 3:1	2.96	12.01	5	58.27	442

Conclusions

- Polysulfone(PSf) and polyetherimide(PEI) were both aminated and polyether ether ketone was sulfonated. Bi-polar membranes were characterized by measurement of the degree of swelling, ion exchange capacity (IEC), and ion conductivity.
- Membranes were modified by surface fluorination using 2000 ppm F_2 gas against N_2 gas for 2 h at room temperature. The surface fluorinated APSf and APEI membranes were characterized again to determine any differences between the pristine and fluorinated membranes.
- The hypochlorite concentration generated by using the surface fluorinated bi-polar membranes was dependent on the IEC of the APSf and ranged from 48.5 to 53.2 ppm. For APEI, the generated hypochlorite concentration ranged from 29.8 to 58.3 ppm.

423898 Design of Resin Polymers for the Adsorption of Metal Ions Using Computer Aided Molecular Design (CAMD)

Abstract:

Design of resin polymers for the adsorption of metal ions using computer aided molecular design (CAMD)

AIChE Annual Meeting, Salt Lake City, Utah, Nov. 8-13, 2015

For submission to "Adsorbent Materials"

Rajib Mukherjee and Urmila M. Diwekar

Vishwamitra Research Institute, Center for Uncertain Systems:

Tools for Optimization and Management (VRI-CUSTOM),

2714 Crystal Way, Crystal Lake, IL 60012

Abstract:

The advancement and economic viability of vertical drilling technology has lead to the expansion of gas production with hydraulic fracturing of sedimentary rocks known as shale formations. A huge amount of water is used in the process. The flowback water from the process of hydraulic fracturing may contain a variety of formation materials, including brines, heavy metals, natural occurring radioactive material (NORM), and organics, which can make wastewater treatment difficult and expensive. One of the easiest and common practices for the removal of heavy metals and radionuclides from water is the use of ion exchange resin polymers.

<> Commonly occurring NORM in hydraulic fracking includes radium and barium. NORM and other heavy metals can be removed using cation exchange resins by adsorption. The adsorption mechanism depends on the difference in activities of NORM and other heavy metals in adsorbents and the bulk fluid phase. The activity coefficients can be obtained using the group contribution methods. In the present work, the activity coefficient in the adsorbent resin polymers is obtained using group contribution methods. For example, polystyrene sulfonate is a cation exchange resin. It contains sulphonic acid functional group as cation exchanger. Other functional groups can also act as cation exchangers includes carboxylic acid, phosphonic acid, phosphinic acid, phenolic, arsonic acid, silinonic acid etc. that can act as a strong acid or weak acid cation exchangers. Keeping the same polymer backbone, different functional group for cation exchange shows different adsorption capacity of the resin polymer. For the backbone polymer, other than polystyrene, different poly venyl aryl compounds can also be used. The monomer units of the polymers can also be divided into structural and functional groups with varied interaction parameters with NORM and other heavy metals. Thus, the activity coefficient in a complex adsorbent can be obtained using group contribution method.

The UNIFAC group contribution is generally used for different compounds. In case of polymer, a free volume of polymer has been added. The modified UNIFAC method is used for estimating activities of adsorbents in the polymer phase. In our research, the interaction parameters of NORM and other heavy metals with structural and functional groups obtained from different resin polymers used as cation exchange resins are obtained using the modified UNIFAC method. The interaction parameters are found by solving an optimization problem where the goal is to minimize the percentage error in the theoretical estimation of adsorption with the experimental results obtained from adsorption isotherm.

The interaction parameters and other group properties obtained with modified UNIFAC group contribution method can be used in a computer-aided molecular design (CAMD) methodology for optimal design of novel resin polymers for NORM and other heavy metals removal. In this work, optimal design of novel polymer is solved as a mixed integer nonlinear programming problem. A new algorithm called efficient ant colony optimization (EACO) is used to solve the problem. The algorithm for CAMD of adsorption resin polymer design maximizes the adsorption capacity of the adsorbent subject to structural feasibility, thermodynamic property correlations, process conditions and process constraints.

In this work we have developed a technology for the design of novel resin polymers useful for the removal of heavy metals and NORM from water. Polymers can be divided into structural and functional groups. The structural and functional property of a polymer depends on these groups. We have used the adsorption theory and develop new group contribution methods (GCMs) to predict interaction properties of each group present in the adsorbents based on their thermodynamics. We have also used a novel CAMD framework to design adsorbent polymers with improved properties using properties of the structural and functional groups that have been developed by the group contribution methods.

428438 Ligand-Assisted Displacement Chromatography for Lanthanide Separations

Abstract:

Lanthanides (Ln's) are the major components of rare earth elements, which are critical components of many high-valued products. The ions of Ln's have the same valence and little differences in size and other physical and chemical properties. They cannot be separated using conventional ion exchange or chromatography processes. Multiple sequential and parallel solvent extraction steps and large amounts of solvents have been used for producing high-purity Ln's, resulting in grave impacts on the environment.

The feasibility of ligand-assisted displacement chromatography using polymeric ion exchangers was demonstrated for Ln separation by Spedding and Powell in the 1950's and 1960's. Three lanthanides were recovered with relatively high purity (~99%) and with yields from 83 to 93%. However, each separation run took more than three weeks, resulting in very low adsorbent productivity. More importantly, the complex mechanisms in this system were not well understood, and no systematic design/optimization method or scaling rules have been reported.

In this study, intrinsic adsorption, reaction, and mass transfer parameters for three lanthanides were estimated from frontal experiments. Rate model simulations were developed and verified using frontal and ligand-assisted displacement chromatography data. A verified model and model parameters were used to optimize the displacement process. Experimental results show the average yield of high-purity (99%) products is more than 97%, and sorbent productivity is an order of magnitude higher than that of Spedding and Powell.

Dynamic simulation studies show that the mechanisms and the criteria for separation in ligand-assisted displacement processes are quite different from those of conventional displacement chromatography. Displacement and migration of solute bands is driven by the reactions of the solutes with the ligand in the mobile phase. The complexation reactions help the solute with higher affinity for the ligand migrate downstream from a solute with a lower affinity for the ligand. If the column is sufficiently long, the different solutes eventually are separated to form pure bands with sharp boundaries. Both boundaries migrate at the same speed, resulting in a fully developed displacement train. The migration speed of the train is controlled by the interstitial velocity, the ligand concentration, the complexation equilibrium constant of the ligand with the presaturant, and the adsorption equilibrium constant of the presaturant with the sorbent. The solute with the highest affinity for the ligand will elute first. By contrast, in conventional displacement chromatography, the solute with the lowest affinity for the sorbent elutes first.

429525 Development of a New Lithium Sorbent

Abstract:

With the rapid development of lithium ion battery industry, the markets demand for lithium resources are booming across the world. Therefore, the interest in extracting lithium from brines including sea water has been significantly increasing because the amount of lithium in sea water is as high as 2.6×10^{11} tons. In this research, lithium manganese oxides (LMO) with spinel structure have been developed for lithium recovery from sea water or salt lake brines because of their unique advantages: high lithium sorption selectivity, relative low cost and environmentally friendly process for lithium adsorption/desorption. In the past few decades, the Li^+ recovery capabilities of various kinds of LMO such as $LiMn_2O_4$, $Li_4Mn_5O_{12}$, $Li_{1.6}Mn_{1.6}O_4$ have been evaluated. However, the Li^+ adsorption amounts of these adsorbents were low and reached about 50% theoretical adsorption value. In addition, the regeneration performances of the sorbents were poor. In this study, the porous LMO with 3D nano structure were prepared by using polystyrene template method. The test results showed the sorbent has a high Li^+ adsorption capacity (higher than 55 mg/g in simulated lithium-rich sea water), an excellent stability, and a high lithium sorption selectivity. Moreover, the electrochemical systems for lithium recovery from aqueous system of the sorbent were also proposed to improve the lithium extraction efficiency.

430260 Role of Nanoporous Carbons in Controlled Release of Drugs

Abstract:

Controlled and sustained release of bioactive agents, including drug and related materials are one of the major challenges of human health care and medical sciences. A well-designed and novel passive drug delivery system can be described as a smart medium that maintains the therapeutic level of drug concentration in the specific targets of human physiology for a sustained period of time. Passive drug delivery is controlled by the molecular diffusion of a confined drug from its carrier media to its specific neighboring physiological environment by concentration gradient, response of immune system or by the presence of some external stimuli. Porous materials can serve as an excellent passive drug delivery media and several successful researches on the role of porous media have widened its scope to be a universal drug carrier. It can confine the poorly soluble drug via high specific area within its porous moiety and well control the release of desired drug in the therapeutic level for a sustained interval of time with the help of diffusion barrier. This process leads to zero premature release and eliminates the possibility of saw tooth profile of drug concentration or toxic level of prolonged heavy drug exposure. Mesoporous silica materials were most widely employed as drug delivery media for multitude of drugs, however, owing to several drawbacks of silica based materials; there is a certain need of other porous matrices that can perform well as a universal drug carrier. In this research, several micro- and mesoporous carbons were examined for the controlled release of different kinds of model drugs. All the carbon materials were also examined for their biocompatibility. The results of these experiments suggest that the nanoporous carbon materials are biocompatible and can well control the release of the drugs and thereby making them suitable candidates for controlled release purposes.

428458 Modeling Ion Exchange Kinetics for Molecular Imprinted Polymer

Abstract:

Molecular imprinted polymers (MIPs) have the potential to reduce the process steps needed to extract gold, and should demonstrate that the same core bead technology could be extended to other metals and toxic substances that are extracted as byproducts of gold mining. These extensions should show significant reductions in tailing waste products and a significant reduction in environmental impact of mining. Ion exchange resins or activated charcoal are the most commonly used media for separation and sequestration of metal ions from solution. Molecularly imprinted polymer (MIP) beads have been shown to improve rates of separation and selectivity when compared to existing ion exchange resin solutions because they are made of more stable polymers and can be used multiple times. Integrating mathematical models with experimental data can aid in predicting the separation efficiency of an individual MIP bead and optimize the extraction process. Here we will present a mathematical model that describes limiting cases for the kinetics of a particle diffusion controlled ion exchange process, and compare it to elution data by varying the mobilities, (ratio of the diffusivities between counter ions), between spherical ion exchanger. The calculations are based on the nonlinear Nernst-Planck equations of ionic motion, which take into account the effect of the electric forces within the system. Solutions to the equations were obtained using finite difference techniques, and compared with elution data and found that using a mobility ratio of 1 gave the best fit to the data, which implies that the diffusivities of species A and B are roughly equal to one another.

433619 Synthesis, Characterization and Application of Metal Organic Framework with One or Two Metal Sites for CO2 Adsorption

Abstract:

Crystalline metal-organic frameworks (MOFs) have recently emerged as potential candidate adsorbent for CO_2 capture applications mainly due to their high surface area and tailorable structure. The presence of open metal sites in some MOFs strengthens their adsorption sites, which are favorable for CO_2 capture and CO_2 separation. The objective of this study is to compare the performance of five different MOFs for CO_2 capture. Specifically, single metal (Mg, Co, and Zn-MOF-74 and Cu-MOF-199) and two metal (Mg-Co-MOF-74) MOFs were synthesized and characterized using X-ray diffraction (XRD) patterns, surface and bulk elemental analysis (XPS and CHN), scanning electron microscopy (SEM), CO_2 adsorption isotherm. XRD and XPS showed that the structures of the samples were in a good agreement with previous reports and all metal cations were incorporated within the MOF structure. The XRD pattern for Mg-Co-MOF-74 was identical to that of single metal MOF74. Higher affinity was found for Mg-Co-MOF-74 compared to other MOFs, based on Henry's law constant. CO_2 adsorption capacity of 5.6, 4.0, 2.2, 4.9, and 6.4 mmol/g were obtained for Mg, Co, and Zn-MOF-74, Cu-MOF-199, and Mg-Co-MOF-74, respectively, indicating higher adsorption capacity of two metal MOF towards CO_2 compared to single metal MOFs. The higher affinity and adsorption capacity of Mg-Co-MOF-74 suggests that it is the most favorable MOF, among the five tested MOFs, for CO_2 capture application.

436249 Effect of Activation Condition for Production of Activated Carbon from Fly Ash

Abstract:

Activated carbon (AC) was produced from Waste oil fly ash (OFA) by physico-chemical treatment method. The OFA was first treated with different types of oxidizing agent followed by steam gasification at different activation temperature. The produced AC was functionalized with hydroxyl and amine functional groups for removal of acidic gases from downstream. Several techniques such as FTIR, SEM and BET were used to characterize the amine functionalized OFA. Kinetic parameters for adsorption of H2S were studied to obtain the rate of adsorption. The results revealed that the surface morphology was quite different as a result of the treatment of OFA. The BET surface area increased from 59 m^2/g for untreated ash to 318 m^2/g after chemical activation. The maximum adsorption capacity of the treated ash was found to be 240 mg/g. The adsorption rate increased with the increase in flow rate and concentration. The results indicated the potential of using waste OFA in H2S removal.

408523 Fast and Accurate Analysis of Extended Natural Gas Composition and Physical Properties Using a Temperature Programmable Micro GC Fusion

Abstract:

Precise and fast measurement of natural gas composition is critical for producers, gatherers, and natural gas appliance and engine manufacturers. Physical property calculations, such as heating value (measured in BTU), provide valuable information to the end user. Due to variations in natural gas composition, gas samples must be analyzed frequently, as small changes in physical properties can have a significant financial impact.

INFICON Micro GC Fusion builds upon proven MEMS based technology and introduces temperature programming, which allows for faster, more accurate analyses. Compared to isothermal operation, temperature programming provides sharper peaks and extends application capability. With a single column, C1-C8+ components can be analyzed within 4 minutes. For extended analysis up to C12, a second module may be utilized. Micro GC Fusion's small size and transportability makes it an ideal instrument either in the field or in the lab.

408643 Removal of Copper Ions from Acidified Dilute Solution By Using Multi-Rotating Disc Electrochemical Reactor

Abstract:

Abstract

Heavy metal removal from waste water has received a great concern due to their disobedience and persistence in the environment. There are various methods have been extensively used for this purpose such as adsorption, chemical precipitation, cementation, ion exchange, etc. In this study a multi-rotating disc electrochemical reactor with a grooved and smooth surface has been used to remove copper ions from acidified dilute solutions of copper sulfate at the limiting current (Constant Current Electrolysis). The experimental module consists of cell and electrical circuit, the cell consists of Plexiglas cylinder of 15.5 cm and 30 cm height. The electrical circuit consisted of insoluble lead anode and a cathode of multi-rotating rough/smooth discs. The discs are made of pure copper with diameter of 7 cm (five discs), thickness of parallel grooves is 2 mm, the distance between discs was fixed at 1.5 cm, and distance between parallel grooves is fixed at 3 mm. The work was carried out at different speed or rotation (50, 100,and 150 RPM) and different initial concentration of copper sulfate (0.008M CuSo4 + 1.5 M H2So4, 0.008M CuSo4 + 1.5 M H2So4, and 0.008M CuSo4 + 1.5 M H2So4) at constant temperature. The results revealed that the percent of copper removal in the rough discs is ranging from 39.4 to 78.77% and for smooth discs is ranging from 34.3% to 68.9 % depending on the speed of discs and concentration of copper ions. In general, the rough discs have a higher percent of copper removal than the smooth discs with a factor ranging from 1.09 to 1.22 depending on the concentration of copper ions and disc rotation speed, due to high exposed area in rough than smooth discs.

Keywords:

Rotating disc, Electrochemical reactor; Copper ions; Grooved surface, Smooth surface

412909 A New Step in Process Integration: The Enzymatic Catalyzed Reactive Dividing Wall Column

Abstract:

Abstract to AICHE 2015

Topic : **Poster Session: General Topics on Separations**

Title: **A new Step in Process Integration: The Enzymatic Catalyzed Reactive Dividing Wall Column**

Abstract

Increasing global competition and environmental legislation require the chemical sector to produce more cost efficient and reduce emissions. Process integration is an efficient way to achieve both goals, by reducing the energy consumption of a process as well as the investment costs. The concepts of reactive distillation and the dividing wall column have been subject of intensive research and have already been implemented on an industrial scale. The reactive dividing wall column (RDWC) combines the two concepts, creating a whole new piece of highly integrated equipment. Reactive dividing wall columns allow the simultaneous execution of a chemical reaction and distillative separation into three or more fractions. The use of an enzymatic catalyst combines the advantages of high selectivity with low temperature levels. The high level of integration leads to a complex design process that requires suitable and reliable simulation models.

Our research aims at reducing uncertainties about designing and operating reactive dividing wall columns. By modelling, simulating and operating a reactive dividing wall column we create the basis for designing RDWCs on an industrial scale and generate valuable operational experience. The approach is based on a model system: the heterogeneously catalyzed trans-esterification of butyl acetate and hexanol. After successfully developing and validating a rigorous mathematical simulation model for chemical catalyst trans-esterification, based on the concept of equilibrium stages (using experiments with our own pilot plant scale RDWC), we take RDWC research a new step forward. The existing rigorous mathematical simulation model was extended for an enzyme catalyzed reactive dividing wall column. The model can be used for simulating reactive as well as non-reactive dividing wall columns with chemical or enzymatic catalyst. This allows an easy comparison between different types of catalyst. Additionally important input parameters for the simulation have been determined by our own extensive experiments. The calculation of the separation is based on experimental measurements of the vapor liquid equilibrium of the system for the pressure range from 30 to 900 mbar. The calculation of the reaction rates are based on measured chemical kinetics of the enzyme catalyzed reaction. For the model validation an existing pilot plant scale reactive dividing wall column will be refitted for the enzymatic catalyzed process to validate the simulation results and generate operational experience with the enzymatic catalyzed process.

The poster presentation will give a short introduction into the concept of our research on the enzyme catalyzed reactive wall column. The mathematical model will be explained to give an insight into modelling reactive dividing wall columns. Furthermore it will be shown which experimental data are necessary to model the separation process and the trans-esterification and the results of these experiments will be presented. Finally the pilot plant will be described and a choice of simulation results will be shown.

416555 Enhancing Cross-Linked PEO Membrane Properties By a Highly-Efficient Post-Treatment Strategy

Abstract:

Herein, we first report an extremely versatile strategy to significantly improve gas transport properties of PEO-containing cross-linked membranes based on their specific swelling properties. Low-molecular-weight poly (ethylene glycol) (PEG) was deliberately incorporated into synthesized swellable pure PEO cross-linked membrane via immersing PEO membrane into PEG-water solution and membranes with various PEG loadings are obtained. The imbedded PEG molecules can be located between polymer chains to improve chain mobility and increase chain distance, resulting in the increase of fractional free volume (FFV), which is in favor of gas diffusion. On the other hand, the increment of CO_2-philic building units resulting from the imbedded PEG molecules leads to higher CO_2 solubility. Finally, the simple treated membranes exhibit significant increments in both CO_2 permeability (up to 762%) and CO_2/H_2 selectivity (up to 86%). The separation performance was correlated nicely with the microstructure of the membranes. This study may provide useful insights about the formation and mass transport behaviour of highly-efficient polymeric membrane applicable to clean energy purification and CO_2 capture, and possibly bridge the material-induced technology gap between academia and industry. Most importantly, our discovered strategy has great potential for treating composite membranes and regenerating "aged" membranes to realize clean energy purification and CO_2 capture by membrane technology.

419015 Novel Composite Ultrafiltration Hollow Fbier Membranesï¼šFabrication and Characterization

Abstract:

Reported herein is the fabrication of a novel superhydrophilic composite hollow fiber ultrafiltration (UF) membrane. Base material for both separating and supporting layers are PVDF. Graft copolymers of poly (vinylidene fluoride) with poly (oxyethylene methacrylate) (PVDF-g-POEM) is used as hydrophilic modifier in the selective layer. It is synthesised via atom transfer radical polymerisation (ATRP) and characterized by FTIR, XRD etc. The characterisations indicate that the PVDF-g-POEM with high purity can be delivered using water as a precipitator and a purifier. In phase inversion to form the membrane, two types of water soluble salts (i.e. organic and inorganic salts) are employed to induce the tuning of the surface structure and hydrophilicity critical to separation efficiency. The resultant UF membranes are characterized by solute transport experiments to reveal the pore size and the pore size distribution characteristics. The surface composition by XPS and the contact angle test show that the surface hydrophilicity of the new type of membrane is significantly improved compared to the neat PVDF membrane. In addition, the chemical stability and anti-fouling properties of the membrane are also investigated.

419304 Designing Reactive Dividing-Wall Columns: Leveraging Optimization Algorithms By Fundamental Process Understanding

Abstract:

Process intensification has become an important concept to meet the challenges of an increasingly cost driven chemical industry. The integration of two or more unit operations into a single apparatus can lead to significantly lower cost of investment as well as cost of operation. While in dividing-wall columns two distillation columns are integrated into a single shell, reactive distillation columns combine reaction and distillation in just one apparatus. Furthermore, these two integrated processes can be combined to the reactive dividing-wall column (RDWC).

The design of reactive dividing-wall columns has to handle more degrees of freedom compared to less integrated process alternatives. This is due to the liquid split and the vapor split, which distribute the liquid flow at the top and the vapor flow at the bottom on both sides of the dividing-wall. In total, this highly integrated process shows a strongly non-linear behavior, making its design complex. Many local optima exist and various design variables proof to be highly sensitive on each other.

Therefore, an effective approach to determine an (e.g. cost) optimal column design is the application of suitable optimization algorithms. But also these algorithms cannot guarantee to find the best possible design. Especially the combination of local optima and high dependency of design variables complicate the search in the solution space.

To overcome these limitations, we will show in this work how fundamental process understanding can be used to transform the solution space and thereby facilitate the search for the global (cost) optimal column design. By combining fundamental process understanding with optimization algorithms the process design becomes not only more effective, but also more efficient. Furthermore, the application of this method to a wide variety of process alternatives and reaction systems allows developing heuristics regarding the optimal process integration level, and hence, to speed up the conceptual design phase drastically.

In the presentation we will deduce firstly fundamental process understanding of the reactive dividing-wall column by extensive and systematic simulation studies. Furthermore, we will explain the optimization approach, and finally how its performance can be leveraged by fundamental process understanding.

419554 Hollow Fiber Membrane Dehumidification Device for Air Conditioning System

Abstract:

In order to provide a comfortable living and working environment indoor in humid tropical countries, the outdoor air must be cooled and dehumidified before it is circulated into the rooms. The current technologies used for removing moisture from the air are either chiller or desiccants. In the air handling unit (AHU), the chiller cools the humid air to its saturation temperature and condenses the excessive moisture in the humid air. The moisture goes through the phase transition process. Using desiccant is a simple method for removing moisture. However the process to regenerate the desiccants is energy intensive and it has the risk of cross-contamination. Membrane technology for air dehumidification is based on the solution diffusion mechanism. The water molecules will be preferentially penetrating through the membranes due to its smaller kinetic diameter than the other counterparts. There is no phase transition involved neither the contacting between the fresh air stream and the polluted streams, hence it will require minimal energy consumption and it can avoid the cross-contamination problems.

A pilot scale air dehumidification system is built up with 9 pieces one-inch PAN/PDMS composite hollow fiber modules. PAN/PDMS composite hollow fiber was chosen due to its high gas permeability at low operation pressures. The hollow fiber spinning and coating technology, module design and system integration for membrane dehumidification were studied. A 150 hour long-term performance test shows that our membrane system has a stably good performance. In our test, the inlet fresh air has an average RH of 76.6±8.1% and temperature of 27.8±1.6¢ªC. After dehumidification, the dry air has an average RH of 47.3±6.5% and temperature of 31.8±2.3¢ªC. The calculation shows that using our membrane dehumidification together with air conditioning can save the energy up to 27% compared to conventional air conditioning process.

423408 Design and Prepare Hyperbranched Polymer Composite Membrane from an Industrial Material-Boltorn W3000

Abstract:

Design and Prepare Hyperbranched Polymer Composite Membrane from an Industrial Material-Boltorn W3000

Lin Wang, Naixin Wang*, Tong Wu, Yue Zhang, Cui Zhao, Shulan Ji*

Beijing Key Laboratory for Green Catalysis and Separation and Department of Chemistry and Chemical Engineering, Beijing University of Technology, Beijing 100124, P. R. China.

Nowadays, hyperbranched polymers (HBPs) have been widely investigated in membrane fabrication process, due to its unique physical and chemical properties, such as low viscosity, small molecular weight and abounding of active terminated groups [1-3]. Herein, a commercially available HBP, Boltorn W3000, was used as the membrane material, which contains hyperbranched polyester as the core, grafted by hydrophobic fatty acid chains (see Fig. 1) [4]. On the basis of the distinctive molecular structure, two simple but effective approaches were utilized to obtain composite membranes for nanofiltration (NF) and pervaporation (PV).

Firstly, NF membranes were prepared on the PAN flat substrate from aqueous W3000 emulsion via a novel one-step self-assembly method (see Fig. 2). The obtained membrane presents an excellent performance in dye desalination. The dye (methyl blue) and salt retentions can reach 97.0% and less than 12.0%, respectively. Meanwhile, the flux of the composite membrane for dye desalination can reach 55.0 $L \cdot m^{-2} \cdot h^{-1}$ under the operation pressure of 0.5 MPa. In addition, the composite membrane maintained stable separation performance under different operating conditions and a long-time running test.

Beside, a simple self-crosslinking strategy was employed to prepare composite membranes for PV separation of toluene/*n*-heptane mixtures. After thermal crosslinking, the hydroxyl and carboxyl groups on Boltorn W3000 molecules could react at intramolecular and intermolecular to form a stable dense layer. The separation factor and permeation flux of the "pore-filling" membrane could respectively reach 5.1 and 63.1 $g/(m^2h)$ (40 ^oC). With respect to the "non-pore-filling" membrane, the composite membrane with "pore-filling" structure performed considerably stable separation properties.

Moreover, as the materials are commercially available and the fabrication process are facile, the membrane obtained via these two approaches would have great potential in industrial application.

Keywords: Hyperbranched polymer; Self-emulsify; Self-crosslinking; Nanofiltration; Pervaporation.

References

[1] L. Wang, S. Ji, N. Wang, R. Zhang, G. Zhang and J. Li, One-step self-assembly fabrication of amphiphilic hyperbranched polymer composite membrane from aqueous emulsion for dye desalination. *J. Membr. Sci.*, 2014, 452: 143-152.

[2] N. Wang, L. Wang, R. Zhang, J. Li, C. Zhao, T. Wu and S. Ji, Highly stable "pore-filling" tubular composite membrane by self-crosslinkable hyperbranched polymers for toluene/n-heptane separation. *J. Membr. Sci.*, 2015, 474: 263-272.

[3] C. Gao, D. Yan, Hyperbranched polymers: from synthesis to applications. *Prog. Polym. Sci.*, 2004, 29: 183-275.

[4] U. Doma¨½ska, Z. Żołek-Tryznowska, Solubility of hyperbranched polymer, Boltorn W-3000, in alcohols, ethers and hydrocarbons, J. Chem. Thermodynamics 42(2010) 1304-1309.

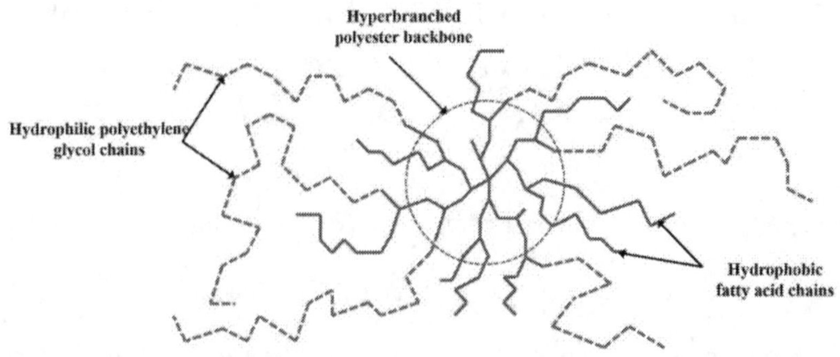

Fig. 1 Schematic representation of the amphiphilic HBP Boltorn W3000 molecular structure.

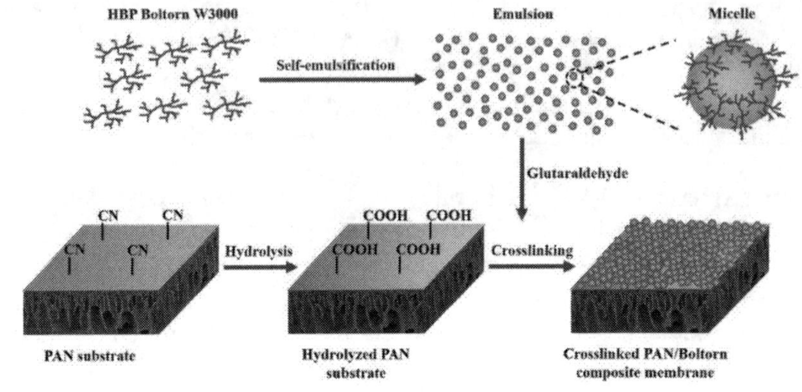

Fig. 2 Schematic illustration of the preparation process of cross-linked PAN/Boltorn composite membrane from aqueous emulsion.

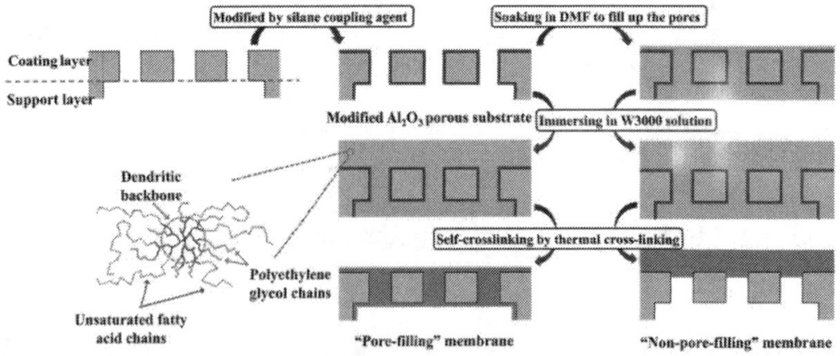

Fig. 3 Schematic illustration of the preparation process of Boltorn W3000 composite membrane with simple self-crosslinking strategy.

424200 Diffusion of CH4, CO and CH4/CO Mixture Under the Single-File Conditions By Pulsed Field Gradient NMR

Abstract:

Among the many technologies used to separate gases, membrane-based separations are the most energetically-efficient option. Significant progress has been made recently in the development of highly efficient porous membranes to tackle challenging gas separations, such as CH_4/CO_2, CH_4/CO, and olefin/paraffin separations. Isolating greenhouse gases such as CH_4 and CO_2 is essential in the fight against global warming. Many industrial applications, therefore, require membranes with even greater selectivity. A novel way to achieve an order of magnitude improvement in membrane selectivity is to exploit an anomalous molecular transport mode known as single-file diffusion (SFD), i.e., diffusion in narrow one-dimensional channels where gas molecules cannot pass each other in the channels. Recent results of molecular dynamics (MD) simulations indicate that the transport rate of a selected component in the nanochannels can be reduced by many orders of magnitude if the diffusion mechanism of this component is changed from normal diffusion to SFD.

To achieve this it is essential to have an understanding of mechanism of SFD in gas mixtures. However, currently there are no experimental studies that explore molecular SFD of gas mixtures. Here, we report the observation of single-file diffusion of pure CO and CH_4 as well as their mixture in Ala-L-Val (AV) nanochannels by pulsed field gradient (PFG) NMR. ^{13}C PFG NMR was applied to study diffusion of these ^{13}C-labeled molecules. High magnetic field gradients up to 23T/m were used to study diffusion for root MSD as small as one micrometer. The reported PFG NMR data will be discussed in the context of relevant molecular simulation studies that have investigated SFD of mixtures.

424903 Influence of Solvent on the Structure and Performance of Poly(L-lactic acid) Membrane

Abstract:

Poly(L-lactic acid)(PLLA) is a biodegradable polymer derived from renewable resources and has extensive biomedical applications. In recent years, more and more researchers have prepared successfully PLLA membrane for separation and purification using different methods.

In this study, we fabricated PLLA membranes through non-solvent induced phase separation technique. The binary solvent mixture consisting of 1,4-Dioxane and N-methyl-2-pyrrolidone was used as solvent and water was used as non-solvent. The influence of solvent on the structure and performance of PLLA membrane was investigated using various instruments and experiments. The morphology of the fabricated membranes was observed by using a scanning electron microscope. The viscosity of the casting solution was measured by a rotation viscometer. Contact angle was measured on the membrane surface using a video contact angle system . Filtration experiments were carried out at room temperature using a dead-end ultrafiltration cell with a magnetic stirrer.

The membrane morphology was tuned from closed sponge-like structure to paralleled finger-like structure through the systematical variation of solvent composition. Filtration experiments indicated that the pure water flux reached a maximum and the membrane hydraulic resistance reached a minimum when the solvent composition was fixed at a critical value. The membrane prepared from the binary solvent mixture had larger pore size and higher flux than that prepared from single solvent due to the change of the solubility parameter difference between the solvent and PLLA. In addition, the effect of the concentration of PLLA in casting solution and the wet membrane exposure tine in air on the structure and performance of PLLA membrane were also studied respectively. This work demonstrated that a series of membranes with different structure and performance could be prepared by adjusting the solvent ratio of the binary solvent mixture.

426957 'Effect of Nucleation Agents on Spherical Agglomeration in the Cooling Crystallization of L-Methionine

Abstract:

L-methionine, the sulfur-containing essential amino acid, is generally achieved to thin and plate-like crystals with {001} planes by a batch cooling crystallization. This morphology of crystals has an adverse effect on powder flowability. However, when a small amount of nucleation agents was added, spherical agglomeration of L-methionine having better flow characteristics was produced in the cooling crystallization. To understand the effect of these additives, the measurement of induction times and statistical approach on the induction time distribution were implemented. It was confirmed that additives have roles as nucleation promoters and nucleation rate kinetics was evaluated in terms of supersaturation. Using the thermodynamic parameters of this model, the values of the effective interfacial energy for heterogeneous nucleation was estimated. Surface energy for the crystal-solution interface was changed according to the kinds of additives due to the hydrophobic-hydrophobic interactions on {001} planes between sulfur atoms of L-methionine and additive molecules. Also effects of process variables, cooling rates, kinds and concentrations of additives, on the size distribution of spherical agglomerates which achieve the increase on powder flowability were investigated

Pervaporation separation of isopropyl alcohol-water mixtures using hollow fiber composite membranes

B. S. Cheon [2], Chung-Seop LEE [1], J. W. Rhim [2]

[1] Airrane Co.Ltd., Deajeon 305-805, korea

[2] Department of Chemical Engineering, Hannam University, Daejeon 305-811, Korea

Research motivation

- High concentration of IPA in water is widely used as cleaning agent in modern chemical and electronic industries. After cleaning operations the waste IPA is required to recover and recycle from environmental as well as economic point of view.
- However, IPA forms an azeotrope with water (at 87.8 wt.% IPA, atmospheric pressure) and thus, it is difficult to separate by conventional separation processes like distillation.
- Thus in recent times various membranes have been tried for pervaporative (PV) separation of IPA-water mixtures. IPA generation in Korea : 14-15% IPA, 50 Ton/day/every site. At present, a number of site is 6 sites (will 18 in a next 2 yrs.)

Preliminary test results

- For 14 wt.% IPA of 'A' Semiconductor Co. in Korea, Tested using the following the modules.

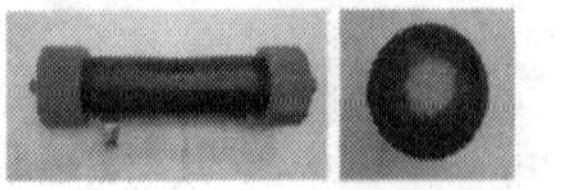

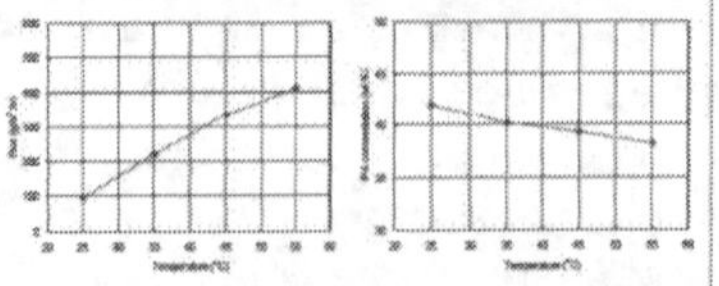

Development of polyetherimide supporter membrane

➤ Polyetherimide (PEI)

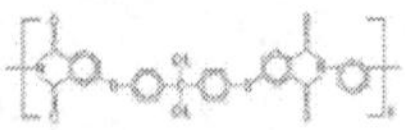

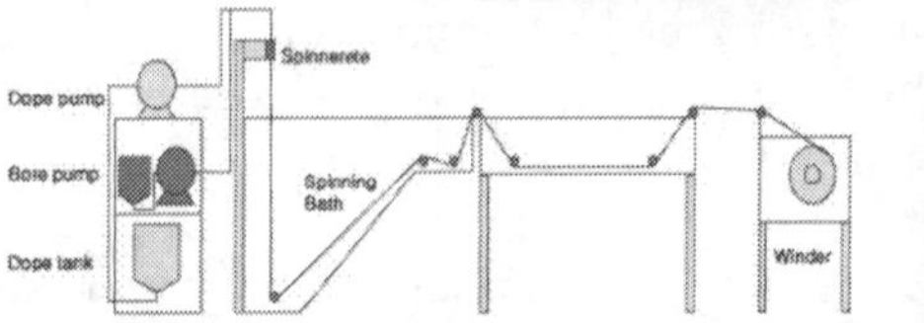

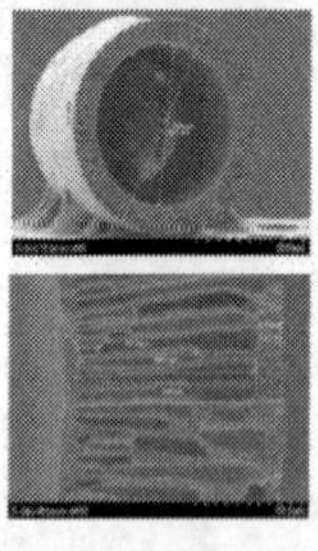

Experimental

Part. 1 : Lab test

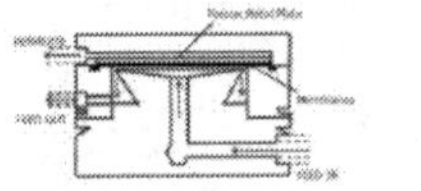

- The pervaporation separation characteristics were investigated for the iso-propyl alcohol (IPA)-water system using PDMS dense membranes.
- The ratio, 9:1 and 10:1, of the basic material based on RTV-655 and crosslinking agent were used to prepare dense membranes at the reaction temperatures, 40, 60, and 80°C.
- And these resulting membranes were characterized by pervaporation technique in terms of permeabilities and separation factors for the feed composition of IPA 85 wt% at the operating temperatures, 25, 35, 45, and 55°C.

Part. 2 : Pilot test

Effective membrane area : 1 m^2

Effective membrane area : 0.5 m^2

- CARBONE Deutsche Carbone Membrane Trennverfahren GFT
- Friedrichsthaler strabe 19, D 66540, Neunkirchen/Heinitz
  Tel. 068 21/792-0. Telefax 06821/792 50, Telex 444 147
- SERIAL No. : 2442, Year Built 1994

Part. 3 : Application to on-site

Results

Part 1

➤ Using hollow fiber composite membranes [Lab test]

- **Feed** : 15 wt.% aqueous IPA solution / **Operating temperature** : 25, 35, 45, 55 ℃ / **Reacting temperature** : 40℃, 60 ℃, 80 ℃

Polydimethylsiloxane (PDMS) - RTV-655 (MOMENTIVE) / A : B = 9 : 1 Polydimethylsiloxane (PDMS) - RTV-655 (MOMENTIVE) / A : B = 9 : 1

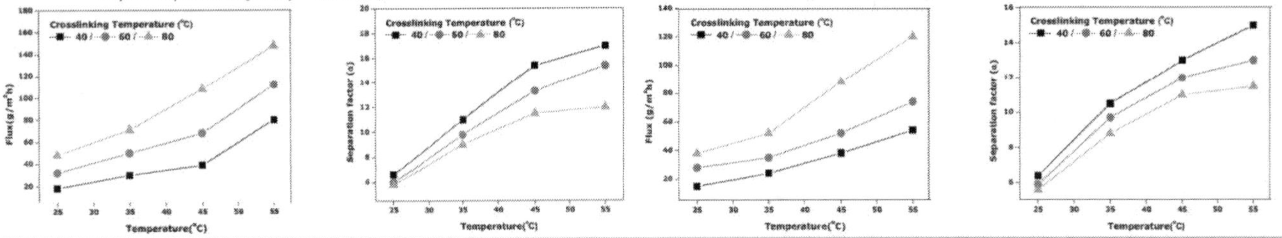

Part 2

➤ Using hollow fiber pilot module [Pilot test]

- **Feed** : 15 wt.% aqueous IPA solution
- **Operating temperature** : 25, 35, 45, 55 ℃ / 55 ℃

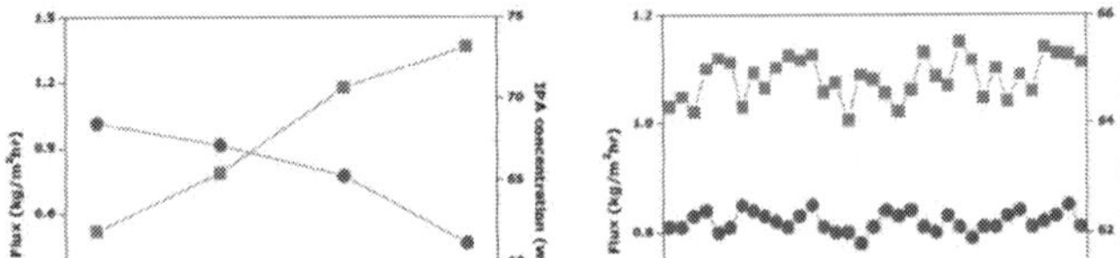

- **Feed** : 15 wt.% aqueous IPA solution / **Module :** Airrane , Nagassep
- **Operating temperature** : 25, 35, 45, 55 ℃

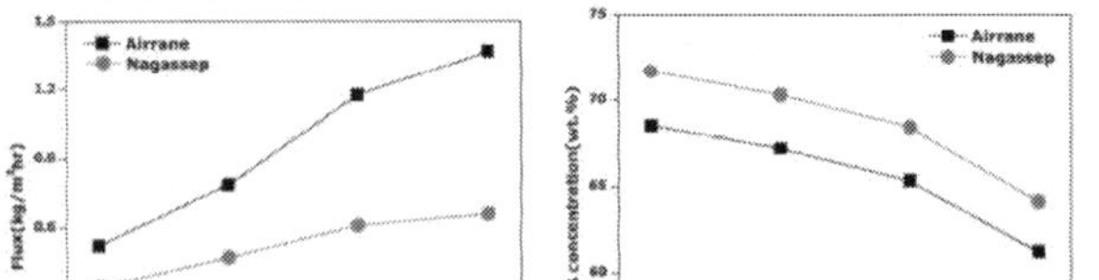

Part 3

➤ Using hollow fiber module [Application to on-site]

- **Feed** : Wastewater containing 15% IPA provided by A Company
- **Operating temperature** : 25, 35, 45, 55 ℃ / 55 ℃

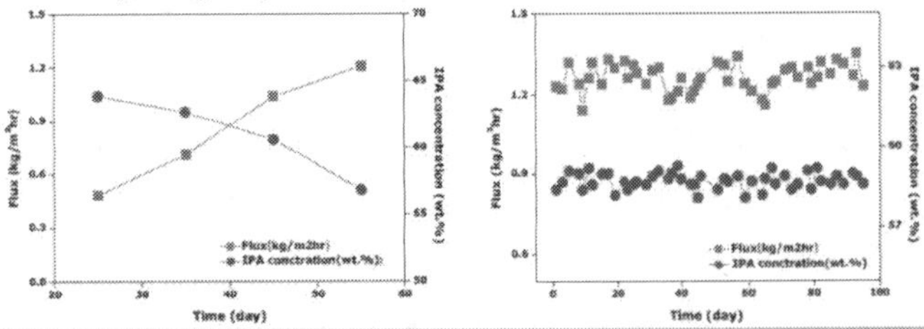

Conclusion

[Part.1]

As the reaction temperatures increased, there as an increase in both the flux and separation factor.

In case the crosslinking agent was 10% rather than 20% of the PDMS quantity at a reaction temperature of 60 ℃, the flux and separation factor were 116 g/m2/hr and 74%, respectively.

[Part.2]

When the Nagasep hollow fiber membrane was used with 15 wt% IPA feed solution, the flux was two times higher than that with the composite membranes of Airrane Co., while the permeate concentration was similar.

When the pilot test was carried out using the GFT apparatus at 55 ℃ for the feed solution of 15 wt% IPA, the flux was shown to be 1~1.5 kg/m2/hr, while the permeate concentration was observed as 61.9~62.5 wt%.

[Part.3]

One hundred on-site scale modules were installed at A company.

The IPA concentration in the waste stream varied from 5~10 wt%.

The flux ranged from 0.48 kg/m2/hr to 1.21 kg/m2/hr, while the concentrated IPA was 57 to 70 wt%. The modules have been running stably for more than 1 month.

Preparation of hollow fiber nanofiltration(NF) membranes by PSP method and their performance studies.

Y. S. Jeon, J. W. Rhim*

Department of Chemical Engineering, Hannam University, Daejeon, Korea (jwrhim@hnu.kr)

Introduction

Motivation

- Environmental issues and shortages in energy resources due to abrupt urbanization, population increases, and weather changes have led to increased interest about the specifically in economic, social and ecological areas in the water industry caused by the unbalance between demand and supply.
- Various researches for the development of separation membranes and related materials, which are key materials in the water industry, suggest that multi-layer membrane technology is the most preferable method.

What is Salting - out effect?

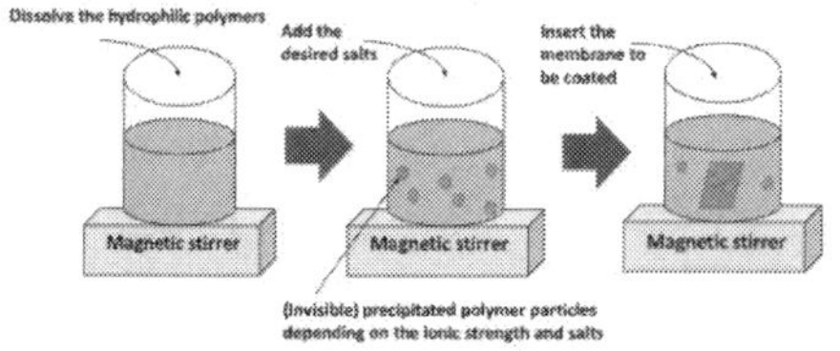

Experimental

SEM images of PVDF membranes

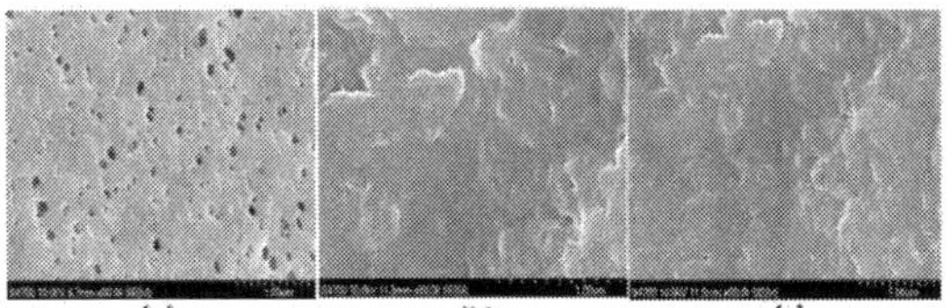

(a) (b) (c)

- SEM image of PVDF membrane surface coated in accordance with the conditions (a)pristine PVDF membrane, (b) PSSA2% for 2min at 4atm, (c) PEI 3% coating onto PSSA 2% coating layer for 1min at 4atm

Performance test apparatus

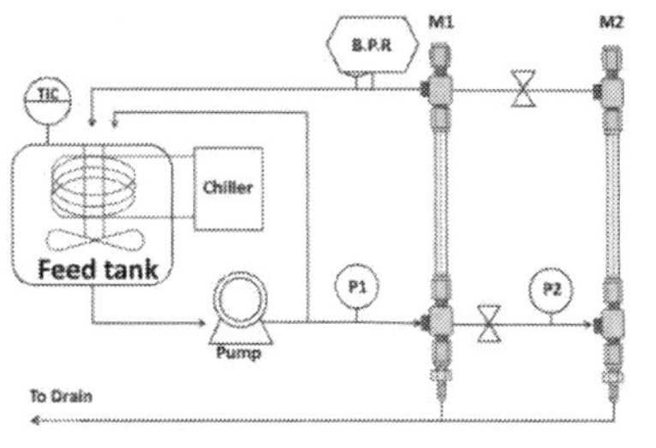

Calculation method

Permeate flux (L/m^2hr)

$$F(LMH) = Q/(A*X)$$

Q = weight of permeate
A = effective membrane area
T = operating time

Rejection (%)

$$R=(C_f-C_p)/C_f*100$$

C_f = feed concentration
C_p = permeate concentration

Results

➢ **Comparison of** Dip coating **and** PSP coating
1st coating : PSSA_MA 2wt% , IS : 0.3, Coating pressure : 4atm, Coating time 2, 4 , 8 min
2nd coating : PEI 3wt% , IS : 0.1, Coating pressure : 4atm, Coating time : 1min

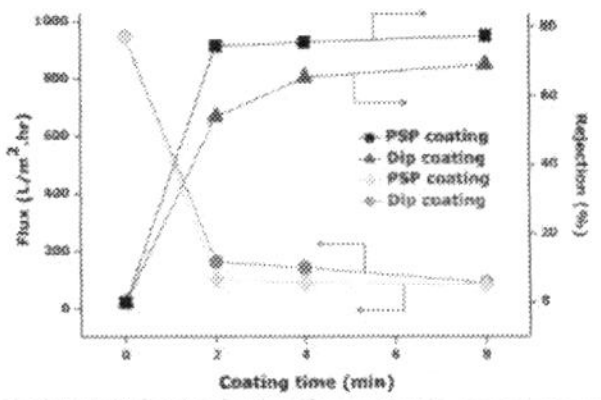

➢ **Effect of** PEI Coating time
1st coating : PSSA_MA 2wt% , IS : 0.3, Coating pressure : 4atm, Coating time 2, 4, 8 min
2nd coating : PEI 3wt% , IS : 0.1, Coating pressure : 4atm, Coating time : 1min

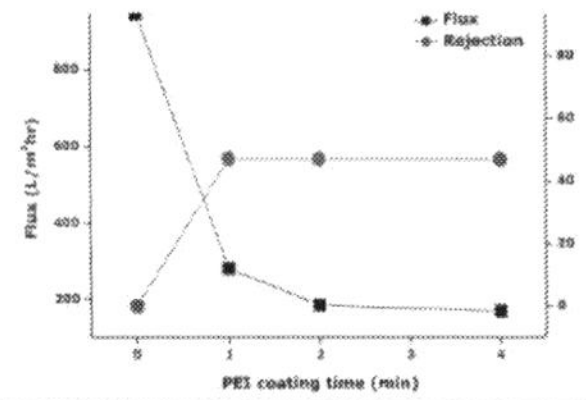

➢ **Effect of** cross-linking time
1st coating : PSSA 2wt% , IS : 0.3, Coating pressure : 4atm, Coating time : 4 min ,Drying temp. and time : 60℃, 10min
2nd coating : PEI 3wt% , IS : 0.1, Coating pressure : 4atm, Coating time : 2min ,Drying temp. and time : 60℃, 30min

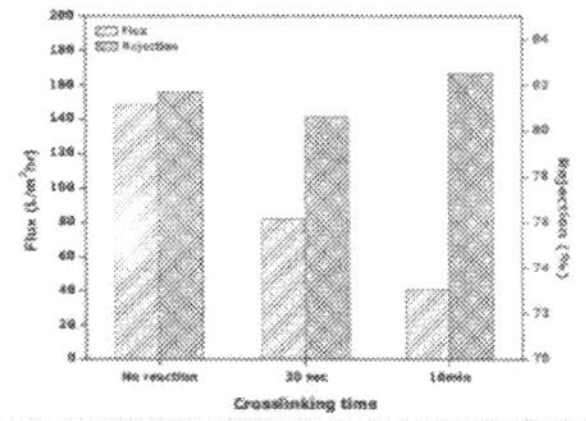

➢ **Effect of** 0.3% IPC (cross-linking agent)
1st coating : PSSA 2wt% , IS : 0.3, Coating pressure : 4atm, Coating time : 4 min ,Drying temp. and time : 60℃, 10min
2nd coating : PEI 3wt% , IS : 0.1, Coating pressure : 4atm, Coating time : 2min ,Drying temp. and time : 60℃, 30min

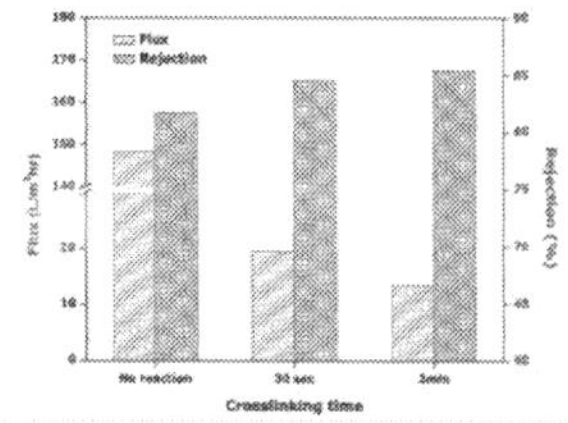

➢ **Effect of** feed concentration
1st coating : PEI 5%, Coating pressure : 6atm, Coating time : 1, 5min , Heat treatment : 90℃, 5min, PSP method
2nd coating : PVSA 0.5% , Coating pressure : 4atm, Coating time 5min, PSP method

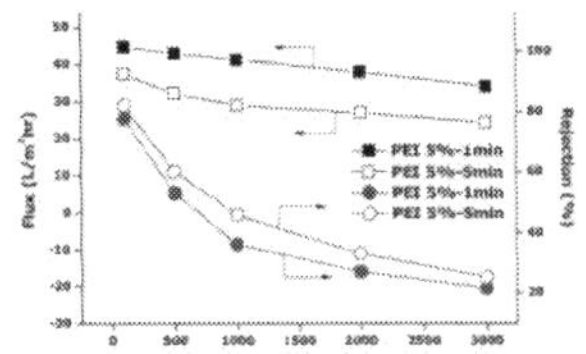

➢ **Durability Test**
PEI (2%) / PVSA(1%) Double coating
PSP method
Coating pressure : 4atm
Coating time : 5min

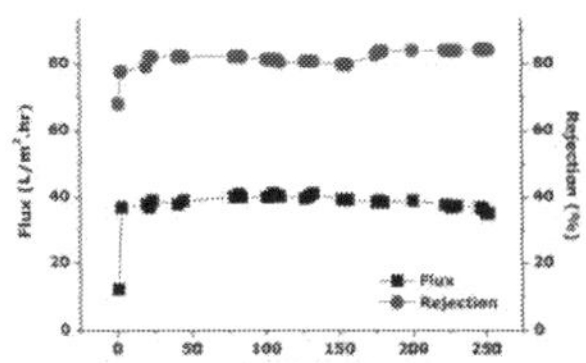

➢ **Different** concentration of IPC
1st coating : PSSA 2wt% , IS : 0.3, Coating pressure : 4atm, Coating time : 4 min ,Drying temp. and time : 60℃, 10min
2nd coating : PEI 3wt% , IS : 0.1, Coating pressure : 4atm, Coating time : 2min ,Drying temp. and time : 60℃, 30min

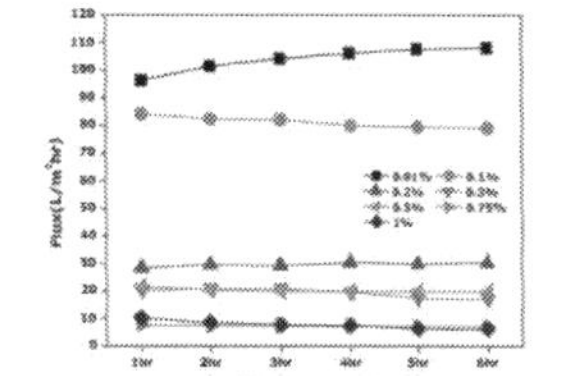

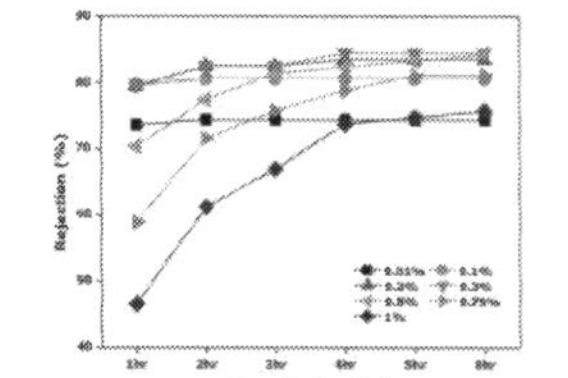

Conclusions

- ➢ The resulting membranes were characterized in terms of the flux and rejection for 100 ppm NaCl at 4 atm to determine their suitability for application in a household water purifier.
- ➢ The double layer composite membrane prepared by 148.3 LMH and 81.7% for 100 ppm NaCl.
- ➢ The cross-linking by isophthaloylchloride (IPC) caused the flux to be very low to almost half-time, but the rejection was not affected. The low IPC concentration up to 0.3% showed a better performance with flux of 20 LMH and rejection of 84.5%.
- ➢ A combination of PVSA and PEI polyelectrolyte complex produced by the PSP method showed the best performance with flux of 43 LMH and salt rejection rate of 83%, and this performance was maintained without loss of flux or rejection rate in a durability test carried out for 10 days.

416162 Construction of Vapor-Induced Covalently Interconnecting Polymeric Network (CIPN): A Potential Method to Mitigate Non-Selective Interfacial Voids in Mixed Matrix Membranes with Enhanced Molecular-Sieving Properties

Abstract:

Interfacial void induced low selectivity is a critical drawback for mixed matrix membranes (MMMs) applied to gas separation. In this study, post-synthetic functionalization of fabricated MMMs by TAEA vapor-induced covalently interconnecting polymeric networks (CIPNs) technology has been employed to molecularly tune the 6FDA-Durene-ZIF-71 MMMs from CO_2-selective into H_2-selective via enhancement of the molecular-sieving properties. The new selective layer comprise of polyamide-based matrix and ZIF-71, interfacial void was refilled by vaporous TAEA treatment.

Both X-ray Photoelectron Spectroscopy (XPS) and Doppler Broadening Energy Spectroscopy (DBES) results reveal that non-selective interfacial voids between particle-particle and particle-polymer have been mitigated by construction of vapor-induced based CIPNs inside the ZIF-71 based MMMs. TGA-IR demonstrated the high thermal stability of resultant membranes and the feasibility of applying the fabricated membranes in a high temperature H_2/CO_2 separation application. Embedding 10 wt% of ZIF-71 nano-particles with size of less than 100 nm inside the TAEA-treated 6FDA-Durene membrane further enhances the pure H_2 permeability by 80% (from 300 Barrer to 540 Barrer) and the ideal H_2/CO_2 selectivity by 34% (from 46.8 to 62.5). The newly developed technology of tripodal amine-based vapor induced CIPNs may solve the most challenging issue in MMMs fabrication and may molecularly tune the MMMs to fit in a specific industrial gas separation application.

428503 Optimization of Membrane Properties for CO2 Removal from Biogas

Abstract:

Biomethane from biogas is a valuable renewable energy source. Biogas contains approximately 60% CH_4 and 40% CO_2 and other contaminants, such as water. The CO_2 must be removed below 4% before biomethane can be used as fuel for transportation or delivered to pipelines. Membrane technology is an attractive approach for CO_2 removal, due to its inherent advantages, such as compactness, small footprint and simple operation. The objective of this work is to evaluate a two step membrane process for CO_2 removal and identify the optimal membrane properties for low capital cost and energy consumption. A biogas separation and optimization model for a binary CO_2/CH_4 mixture is created, which is incorporated with the calculation for power, capital and operating cost. Membranes with separation properties on the Robeson's upper bound were evaluated. In disagreement with the conventional approach utilizing higher membrane selectivity to lower costs, it was found that the membrane system achieves best economics when the membrane has high permeance and moderate selectivity. For example, membranes with a CO_2/CH_4 selectivity of 10-20 provided the lowest operating cost. This presentation will also describe the effect of the biomethane purity and methane recovery on the economics of the CO_2 removal.

430029 An Investigation of Thermal Stability of Pd-Ag Composite Membranes Fabricated By Surfactant Induced Electroless Plating (SIEP) Method for Hydrogen Separation

Abstract:

Hydrogen is considered as an alternative, clean and sustainable energy carrier for reducing the usage of global fossil fuel in large-scale process industries. However, the existing large-scale separation and purification of H_2 for these industries requires large capital investment and production costs. In this regard, Pd- and its alloy based membranes have the potential to play an important role in separation and purification of hydrogen. The dense Pd-Ag composite membranes are particularly attractive for the superior performance in terms of permselectivity and stability compared to pure Pd membranes of similar thickness. In this work, dense Pd-Ag composite membrane on microporous stainless steel (MPSS) substrates were fabricated by sequential deposition of Pd and Ag using surfactant induced electroless plating (SIEP) process, a modified electroless plating process (EP). A cationic surfactant, dodecyl trimethyl ammonium bromide (DTAB), is used to provide a controlled grain-size distribution and agglomeration process. In this work, 4xCMC and 1xCMC of DTAB were used in Pd- and Ag- bath, respectively. Prior to the metal deposition, the MPSS substrates were oxidized at 600 °C for 18 hrs to create an oxide layer as intermetallic diffusion barrier. The membranes fabricated by SIEP method were tested for long term (> 950 hrs) hydrogen permeation measurements at temperatures cycled between 623 K and 723 K. These membranes were characterized by SEM, XRD, EDS and AFM to study physical and morphological features before and after stability tests. The results of these studies will be presented.

431499 Urea-Functionalized Silica on Top of Graded Structured Gamma-Alumina Membranes

Abstract:

A functionalized silica membrane is synthesized on top of γ-alumina intermediate layer which has graded structure by controlling particle size of boehmite sol. Three successive layers of γ-alumina with decreasing particle sizes are coated on an α-alumina substrate and functionalized by co-condensation of 1-[3-(Trimethoxysilyl) propyl]urea (TMPU) and TEOS with different ratio and is used for selective separation of CO_2.

The graded γ-alumina membrane was characterized as the following: DLS from boehmite sol, BET from unsupported γ-alumina membrane, permporometry and SEM from γ-alumina membrane were performed. The urea functionalized membranes was characterized by gas separation measurements, Si NMR, C NMR and TG analysis. The γ-alumina intermediate layers were made by dip-coating and calcination of boehmite (AlOOH) sols with different particle size mixed with PVA solution on top of a macroporous α-alumina support sequentially to give a graded structure substantially free of defects.

The size of the sol particles was tuned carefully by controlling synthesis parameters including quantity of acid and hydrolysis time. Different hydrolysis time for aluminum isopropoxide in presence of water and the subsequent acid peptization using different quantity of acetic acid was used.The size distribution of boehmite sols was measured by Dynamic Light Scattering (DLS). The boehmite sizes were controlled by adjusting H^+/alkoxide molar ratios of 0.04, 0.07 and 0.15 and hydrolysis times of 24, 24 and 0.5 h, respectively. The boehmite average sizes obtained were 160, 60 and 6 nm, correspondingly. The DLS results showed that higher acid concentration and shorter hydrolysis time led to smaller particle size. In accordance with the DLS results, BET surface area for the unsupported γ-alumina of 160, 60, 6 nm boehmite average sizes were measured to be 214, 236 and 326 m^2/g, respectively.

n-hexane/helium permporometry experiment for the γ-alumina membrane after several coating steps with boehmite sols were carried out. The permporometry data of γ-alumina with graded structure showed that the most permeance contribution of pores are less than Kelvin diameter of 4 nm while the dry He permeance for γ-alumina was about 1.4 $\times 10^{-5}$ mol.m^{-2} Pa^{-1} s^{-1}. Cross-sectional scanning electron microscopy images of the intermediate layers of γ-alumina were recorded. The images showed that we could prepare a membrane with a graded structure. The final thickness was about 6 μm. No infiltration of γ-alumina particles into the α-alumina macroporous support was indicated due to the first coating of larger boehmite particles. It was also observed that the top γ-alumina layer had a finer structure as compared to the layers underneath.

Urea-functionalized silica sols were synthesized from X TMPU and 1 TEOS in 22 EtOH, 5 H_2O and 0.4 HNO_3, in which three X's of 0.05, 0.2 and 0.4 were used for the co-condensation. 10% CO_2 in either CH_4 or N_2gas mixture was separated on the membranes at 100°C and feed pressure of 150 kPa. CO_2/N_2 and CO_2/CH_4 measurements were obtained for the γ-alumina and the functionalized silica membrane. As the γ-alumina is functionalized with urea group, the CO_2 separation factor increases up to 4 times. The separation factor increases with the quantity of the functional groups, indicating a proper interaction between the urea groups and CO_2molecules. CO_2/CH_4 separation factor of 2.5, and CO_2/N_2 separation factor of 3 were observed for the urea functionalized membrane, as compared to the γ-alumina membrane with separation factor of 0.6 and 0.8 for CO_2/CH_4 and CO_2/N_2, respectively, according to Knudsen separation factor. CO_2 permeance decreased from 10×10^{-8} to 0.7×10^{-8} mol.m^{-2} Pa^{-1} s^{-1} where X increased from 0.05 to 0.4. This decrease in the permeance may be attributed to an increase in amount of the adsorbed water.

The successful incorporation of functional groups in the membrane structure was confirmed by FTIR methodology, Si MAS NMR and C MAS NMR. The FTIR spectra of the functionalized silica are mainly characterized by bands at 3150–3600 cm^{-1} (N–H stretching vibrations), 2800–3000 cm^{-1} (C–H asymmetric and symmetric stretching vibrations), 1600–1800 cm^{-1} (C=O vibrations), 1400-1500 cm^{-1} (NH_2 bending vibration),1200 cm^{-1} (Si-C vibrations), 400,800,1078 cm^{-1}(Si-O-Si vibrations).

Covalent bond between organic functional groups and the silica network was confirmed by assigning T site in the Si MAS NMR results for the functionalized silica. In the functionalized samples, the resonances at around −64 ppm represent T3(RSi(OSi)3) silicon site, where R represents the functional groups. To find a semi quantitative analysis for functional group incorporation in silica network, curve deconvolution of Si MAS NMR spectra of functionalized samples were performed to determine the relative concentration of T and Q resonances. The relative concentration of T site to the total sites was 16 and 28% for X=0.2 and 0.4, respectively. This result indicates that almost all of the functional groups are covalently attached to the silica network and the degree of functional group incorporation in the silica network is in accordance with the amount of functional groups in the starting silica sol as it increases from X=0.2 to 0.4.

The carbon atoms present in the functional groups were detected in C MAS NMR; especially carbonyl group were observed as a strong signal at 160 ppm. For measuring the functional groups thermal stability as well as real degree of functionality for the functionalized silica, thermal gravimetric analysis was conducted under Argon flow. The TG results showed that the functional groups were stable under 200°C. The real functionality of the functionalized silica for X= 0.2 and 0.4 was found to be 20 and 26% respectively, which is in accordance with the weight percent of the functional groups in the starting sol.

(41j) Improving the Chlorine Resistance of RO Membrane By Using a New Monomer of 1,3,5-Triazine-2,4,6-Tricarbonyl Trichloride

Zeng, Y., Zhejiang University
Chen, J., Zhejiang University
Zhang, L., Zhejiang University
Li, J., Zhejiang University
Hou, L., Zhejiang University, Hangzhou, China

A acyl chloride incorporation of triazine ring, 1,3,5-triazine- 2,4,6-tricarbonyl trichloride, was synthesized to fabricate a novel polyamide thin film composite(TFC) membranes via interfacial polymerization with m-phenylenediamine (MPD) on porous polysulfone supports. Compared with cyanuric chloride, the triazine with acyl chloride group monomer exhibits higher reactivity towards MPD. The prepared membranes were characterized by ATR-IR spectroscopy, elemental analysis, SEM, AFM, contact angle and zeta potential measurements, and the chloride resistance was studied by the evaluation of membrane performance before and after chloride exposure. The results reveal that the chloride stability of TFC membranes had been improved in comparison to the membrane prepared from MPD and TMC.

404455 Purification of Menthol Enantiomers By a New Technique Combining Distillation and Crystallization

Abstract:

Stripping crystallization (SC) is a new separation technique which combines distillation and crystallization. Basically, SC is operated at a triple-point condition, in which the liquid is simultaneously vaporized and crystallized due to the three-phase equilibrium. In essence, the process is continued until the liquid phase is completely eliminated and only the pure solid crystals remain in the feed. Thus, crystal washing is not required since only pure solid crystals remain in the feed and no impurities are adhered on the crystal surfaces at the end of SC. In the previous research, SC has been successfully applied to separate p-xylene from the mixed xylenes by Shiau et al. (2005, 2008).

Menthol is widely used in perfumery, cigarette, cough drops, and nasal inhalers for its cooling effect and refreshing flavor. Only L-menthol has been reported to have analgesic effect whereas D-menthol is lack of analgesic and flavor properties. Although natural L-menthol is obtained by distillation from the leaves of various subspecies of mint, menthol can also be synthesized via a number of routes, by which a racemic mixture is obtained. Thus, separation and purification of L-menthol from the racemic mixture has been an important research issue. A thermodynamic model is developed in this work to simulate the three-phase equilibrium during the SC operation and to direct the batch SC experiments. The experiments show that no crystallization occurs during SC without stirring due to subcooling of the liquid enantiomers. Stirring can eliminate subcooling during SC, leading to the L-menthol crystals gradually formed. The experimental results, including the final enantiomeric purity and recovery ratio of L-menthol, are consistent with the simulation results predicted by the model.

1. Shiau, L. D., Wen, C. C. and Lin, B. S. Separation and purification of p-xylene from the mixture of m-xylene and p-xylene by distillative freezing, *Ind. Eng. Chem. Res*, **44(7)**, 2258-2265, 2005.

2. Shiau, L. D., Wen, C. C. and Lin, B. S. Separation of p-xylene from the multicomponent xylene system by stripping crystallization, *AIChE Journal*, **54(1)**, 337-342, 2008.

(42ae) Intermediate Temperature Ceramic Proton Conductor for Hydrogen Separation

Elangovan, E., Ceramatec, Inc
Frost, L., Ceramatec, Inc.
Hartvigsen, J. J., Ceramatec, Inc.

Current hydrogen membranes make use of precious metals that are expensive and can be adversely affected by the presence of contaminants. Substantial work has been done on proton conducting high temperature ceramics for use in hydrogen membrane separation systems but physical and electronic degradation have made the use of these membranes questionable.

Ceramatec initiated work on a new ceramic material set under funding provided by the US Department of Energy. After encouraging results under the DOE funded program, Ceramatec was able to obtain funding from a private firm and continue research on an intermediate temperature proton conductor for hydrogen separation. This paper will discuss the technology basis established in literature, the approach initiated by Ceramatec, and results obtained in the laboratory experiments to date. The discussion will also include an overview of on-going efforts and planned experiments.

409289 Electrophoresis of Two Spheres Along Their Line of Centers: Influence of Particle-Particle Interaction Forces

Abstract:

Electrophoresis of two spheres along their line of centers: influence of particle-particle interaction forces

Shiojenn Tseng,[1] Chih-Hua Huang,[2] Jyh-Ping Hsu[2,*]

[1]Department of Mathematics, Tamkang University, Tamsui, Taipei, Taiwan 25137

[2]Department of Chemical Engineering, National Taiwan University, Taipei, Taiwan 10617 Tel: 886-2-23637448, Fax: 886-2-23623040, E-mail: jphsu@ntu.edu.tw

ABSTRACT

Considering recent applications of electrophoresis conduced in nanoscaled devices, where particle-particle interaction can play a role, we studied for the first time the electrophoresis of two rigid spheres along their center line, taking account of the hydrodynamic, electric, and van der Waals interactions between them. Under the conditions of constant surface potential and surface charge density, the influences of the level of surface potential/charge density, the bulk salt concentration, and the particle-particle distance on their electrokinetic behaviors are examined. Numerical simulation reveals that these behaviors are much more complicated and interesting than those of isolated particles. In particular, we show that care must be taken in choosing an appropriate particle concentration in relevant experiment to avoid obtaining unreliable mobility data.

Keywords: electrophoresis; two spheres; electric double layer force; van der Waals force.

* Corresponding author e-mail: jphsu@ntu.edu.tw (J.-P. Hsu)

409292 Diffusiophoresis of a pH-Regulated Polyelectrolyte in a Nanopore of Nonunifrom Cross Section

Abstract:

Diffusiophoresis of a pH-Regulated Polyelectrolyte in a Nanopore of Nonunifrom Cross Section

Chun-Yuan Su,[1] Jyh-Ping Hsu,[1,*] Shiojenn Tseng[2,*]

[1]Department of Chemical Engineering, National Taiwan University, Taipei, Taiwan 10617 [2]Department of Mathematics, Tamkang University, Tamsui, Taipei, Taiwan 25137

Tel: 886-2-23637448, Fax: 886-2-23623040, E-mail: jphsu@ntu.edu.tw

Abstract

The diffusiophoresis of a pH-regulated polyelectrolyte (PE) along the axis of a charged, nanopore having nonuniform cross section is modeled, aimed with improving the performance of a nanopore-based sensing device for single biomolecules such as DNAs and proteins. Previous analysis based on a rigid particle of constant charge density in a solution containing binary ionic species is extended to a porous, pH-regulated PE in a solution containing multiple ionic species so that the conditions considered are much closer to reality. The behavior of the PE under various conditions is simulated by varying its charge, relative size, and position in the nanopore, the solution pH, and the background salt concentration. We show that the PE translocation velocity can be regulated by the pore width and its charged conditions. In particular, both the magnitude and the direction of the translocation velocity can be adjusted.

* Corresponding authors' e-mail: jphsu@ntu.edu.tw (J.P. Hsu), topology@mail.tku.edu.tw (S. Tseng)

Modeling and Simulation of Diffusion Effect on the Glycol Dehydration Unit of a Natural Gas Plant

M. Wigwe

Department of Petroleum Engineering, Texas Tech University, Lubbock, TX 79409

J. G Akpa and E. N Wami

Chemical Engineering Dept., Rivers State University of Science & Technology, Port Harcourt.

ABSTRACT

Mathematical models of the absorber of a glycol dehydration facility was developed using the principles of conservation of mass and energy. A model that predicts the variation of the water content of gas in mole fraction across the column was developed. Similar models were developed for the variation of the gas and liquid (TEG) temperatures across the packing height. The models developed contain contributions from bulk and diffusion flows. The effect of diffusion on the process occurring in the absorber was studied in this work. The models were validated using the initial conditions in the plant data from a gas processing plant to ascertain if the outlet conditions predicted by the models meet the plant data outlet requirements. The results obtained showed a good match due to the effect of diffusion observed between z=0 and z=0.004m. The results obtained showed that the effect of diffusion was noticed between z=0 and z=0.004m. A deviation from plant data of 0% was observed for the gas water content at a residence time of 20 seconds, at z=0.004m. Similarly, deviations of 1.584% and 2.844% were observed for the gas and TEG temperatures. A reasonable conclusion to draw from this study is that the effect of diffusion on absorption of a flow process occurs over small average height in the absorber.
Key words: separations, absorption, simulation, dehydration, water content, Triethylene glycol.

Introduction

Natural gas used by consumers is composed almost entirely of methane. However, natural gas found at the wellhead, although still composed primary of methane, is by no means pure. Raw natural gas comes from three types of wells: oil wells, gas wells and condensate wells. Natural gas that comes from oil wells is termed "associated gas" and can exist separate from the crude oil (as free gas) or dissolved in the oil (as dissolved gas). Those from gas or condensate wells are termed "non-associated gas", since they are found with little or no crude oil. Whatever the source of the natural gas, once separated from crude oil (if present), it commonly exists in mixture with other hydrocarbons, principally ethane, propane, butane, and pentanes. In addition, raw natural gas contains water vapor, hydrogen sulfide, carbon dioxide, helium, nitrogen and other trace compounds. Water vapor is the most common undesirable impurity found in natural gas. By its source, natural gas is almost always associated with water vapor, usually in the range of 400-600lbH_20/MMSCF of gas [2]. This water must be removed in order to protect the system through which it would be transported or processed from corrosion and hydrate formation which lead to severe operational problems.

To be able to overcome these problems, we need a well designed dehydration unit or plant which would restrict the water content to less than 4-7lbH_20/MMSCF of gas according to pipeline specification [2,3]. The principle of gas absorption is employed using a suitable solvent (triethylene glycol (TEG) in this case) to selectively absorb water vapor from the gaseous mixture when counter-currently brought into intimate contact with the gaseous mixture in an absorption column. The aim of this project is to develop and simulate an unsteady state non-isothermal mathematical model of the absorber for the glycol dehydration of a natural gas. To achieve these, we shall develop an unsteady state non-isothermal mathematical model that will predict the variation of gas water content across the absorber, develop an unsteady state non-isothermal mathematical model that will predict the variation of gas and TEG temperatures across the absorber, study the effect of diffusion on the process, solve the resulting model equations analytically using the Laplace transform method and validate models by comparing results with available plant data.

Gomez et al [4] simulated a reactive absorption system for the absorption of CO_2 in a packed column using methyl diethanolamine (MDEA) as the solvent. Steady state conditions and plug flow were assumed for the gas phase, leading to a set of ordinary differential equations. In John [5], a mathematical model for the wet scrubbing of CO_2 using chilled ammonia was studied. Diffusion and conduction terms were included in the development of the unsteady state models of the system. These models predict the variation of the concentration of the reactants and products with time across the packed height, as well as the variation of the temperature of the system with time across the packed height. The partial differential equations developed were solved using the numerical technique of Matlab by applying the Robin boundary condition (BC), the Neumann BC, and the Dirichlet BC. A similar study on CO_2 absorption was carried out by Ahmed et al [6] using a highly concentrated monoethanolamine solutions (MEA). The model developed is equivalent to eqn. (8) with the inclusion of the reaction term. Most studies on gas dehydration using TEG were simulated using special packages like HYSYS used by Peyghambarzadeh and Jafarpour [7].

Model Development

Mathematical models that could be used to predict the performance of the absorber are developed. In the absorber, the model would monitor the amount of water vapor removed from the gas stream, as well as the temperature variation (for non-isothermal process) along the packing height of the absorber for both gas and liquid (TEG) phases. The aim is to reduce the water content of the gas to the standard pipeline specification of less than 7 1b/MMscf of gas [2,3].

Model Assumptions

The following assumptions were made in the dynamic modeling of the TEG gas absorber:

1. Packed column is used [8].
2. The absorber is well-lagged, hence heat losses are negligible.
3. There is no reaction occurring, as a result, the reaction term would not be considered in the model development.
4. The effect of change in total flow rate in the column is ignored, and an average value is assumed constant. [9].
5. Since the water vapor is the only diffusing component, no diffusion term would be considered for the liquid phase.

Material Balance

Fig.1 (a) shows the simplified TEG absorber with the inflow and outflow streams and their compositions, whereas in fig. 1(b), an elemental packed height is shown, with its associated inflow and outflow streams. In fig. 1(a), G is the molar flow rate of the gas stream, y_0 its initial water content, C_{Aog}, the inlet water vapor concentration, G, y and C_{AG} are respectively the dry gas parameters. For the TEG stream, L, x_0 and C_{A0L} represent the inlet flow rate, initial water content and water concentration respectively, similarly, L, x and C_{AL} are their respective rich TEG parameters. In fig. 1(b), we make a material balance on the elemental packing height (ΔZ). The component balance for component **A** (water vapor) is taken across the elemental packing height (ΔZ). The inflow parameter can be split into two types: flow due to bulk flow and diffusion. Diffusion can occur because of the concentration gradient in the axial direction. Since water vapor is the diffusing component, we would include this diffusion term for only the gas stream. If a component **A** is diffusing through another component **B**, in a binary mixture, then the diffusion flux N_A of **A** (in moles of **A** per unit time per unit area) is given by Fick's law as:

$$N_A = -D_A \frac{dC_A}{dX} \quad (1)$$

Where D_A is the diffusion coefficient due to diffusion and turbulence (called the eddy diffusivity), in length^2 per unit time) [11]. On substitution into the material balance equation, the concentration of the gas water content in mole fraction becomes:

$$\frac{\partial y_A}{\partial t} = -V_g \frac{\partial y_A}{\partial z} + D_A \frac{\partial^2 y_A}{\partial z^2} \quad (2)$$

Where V_g is the gas velocity in m/s.

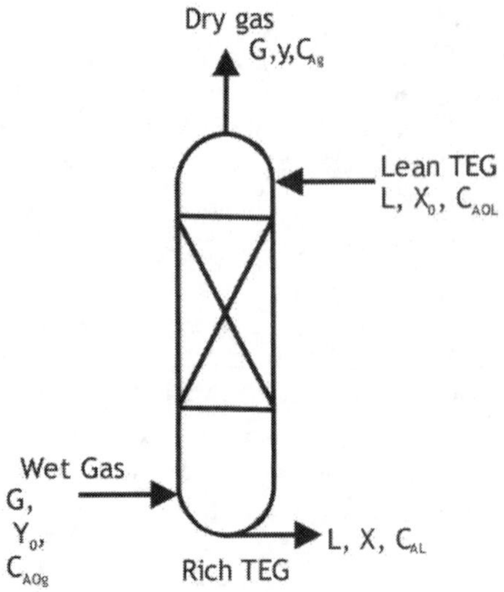

Fig. 1a: Absorber streams

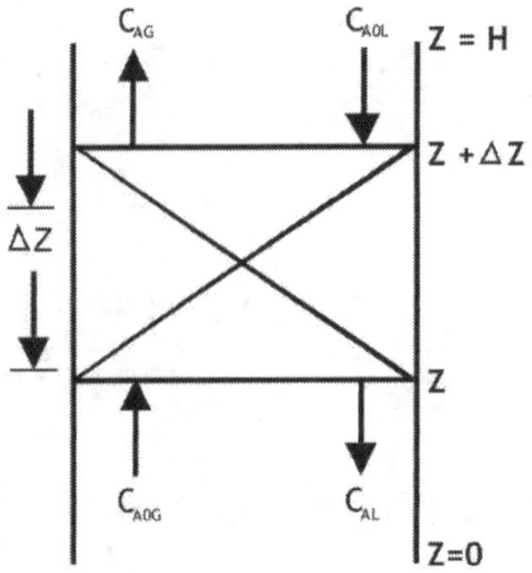

Fig.1b: Material balance on element ΔZ

Energy Balance

Energy balance is carried out for both gas and liquid (glycol) phases. The glycol which enters the column at a higher temperature leaves at a lower temperature (Plant data). As a result, the glycol loses some amount of heat (Q) to the gas phase. This heat shall be included in the energy balance for the gas phase.

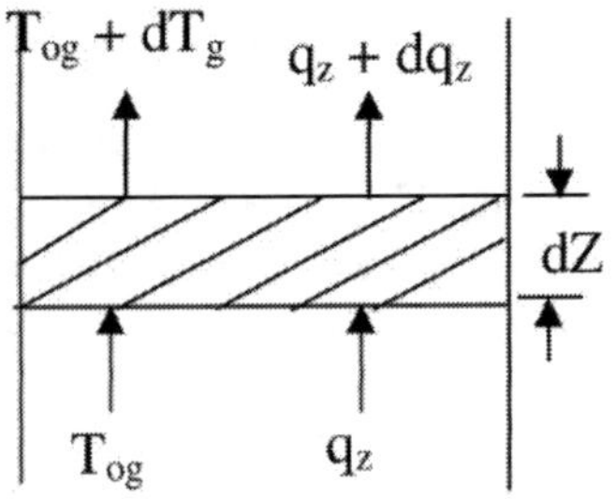

Fig. 2: Energy balance for gas phase

Where q_z is a heat flux in the z-direction due to conduction, given by Fourier's law of conduction of heat as

$$q_z = -K_A \frac{dT}{dz} \tag{3}$$

where K_A is the effective thermal conductivity of **A** (water vapor) in KW/(m^{2}K). The material balance for the liquid phase becomes:

$$\frac{\partial T_g}{\partial t} = -\frac{G}{AC_{Ag}}\frac{\partial T_g}{\partial z} + K_T\frac{\partial^2 T_g}{\partial z^2} + \frac{LC_{pl}}{AC_{Ag}C_{pg}}\frac{dT_L}{dz}$$

$$\frac{\partial T_g}{\partial t} = \alpha\frac{\partial T_g}{\partial z} + K_T\frac{\partial^2 T_g}{\partial z^2} + \beta \tag{4}$$

Where $\alpha = -\frac{G}{AC_{Ag}}$, $\beta = \frac{LC_{pl}}{AC_{Ag}C_{pg}}\frac{dT_L}{dz}$

C_{pg} is the specific heat of the water vapor in the gas stream. T_{og} and T_g are the inlet and outlet temperatures of the gas stream, A is the cross-sectional area of the packed section, C_{pl} is heat capacity of the liquid stream, dT_L is the temperature difference of the liquid stream and K_T is the thermal diffusivity of water vapor.

The energy balance for the liquid phase is made from the top of the column.

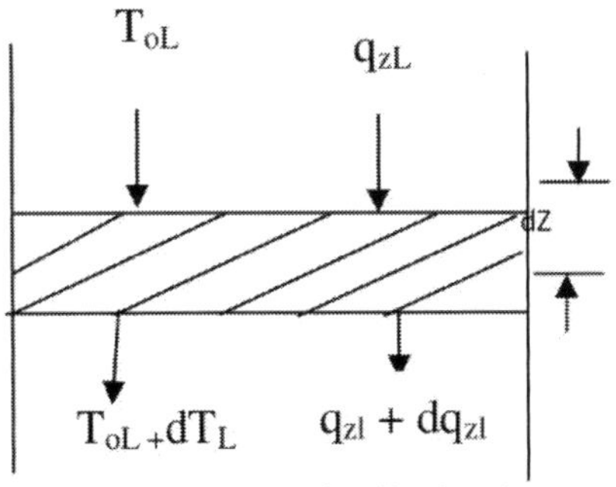

Fig. 3: Energy balance for liquid phase

The model equation for the liquid temperature is given as:

$$\frac{\partial T_L}{\partial t} = -\frac{L}{AC_{AL}}\frac{\partial T_L}{\partial z} + K_G\frac{\partial^2 T_L}{\partial z^2} \qquad (5)$$

Where K_G is the thermal diffusivity of the liquid in m^2/s

Solution for the Effect of Diffusion

It can be seen that the equations developed contain effects due to bulk flow and diffusion flow. The effect of diffusion is the focus of this work and will be studied further. The effect of diffusion on the gas dehydration process can be studied by neglecting the contribution of bulk flow. As a result, eqns. (2), (4) and (5) become:

$$Gas\ water\ content:\ \frac{\partial y_A}{\partial t} = D_A\frac{\partial^2 y_A}{\partial z^2} \qquad (6)$$

$$Gas\ temperature\ variation:\ \frac{\partial T_g}{\partial t} = K_T\frac{\partial^2 T_g}{\partial z^2} + \beta \qquad (7)$$

$$TEG\ temperature\ variation:\ \frac{\partial T_L}{\partial t} = K_G\frac{\partial^2 T_L}{\partial z^2} \qquad (8)$$

Since both z and t have infinite values, we may take the Laplace transform with respect to either variable. It has been proven that taking the Laplace transform with respect to *z* would make the solution path complex and difficult [12]. A simplified solution can be obtained when the Laplace transform is taken with respect to time variable, *t*. We solve the resulting equation using Laplace Transform, expand the result in Taylor's series and transform the final solution to its exponential equivalent; we obtain the final solution as eqn. 9. Eqns. 10 and 11 are solutions to eqns. 8 and 7 respectively.

$$y_A(z,t) = y_{Ao}e^{-\frac{z}{\sqrt{\Pi D_A t}}} \qquad (9)$$

$$T_L(z,t) = T_{oL}e^{-\frac{z}{\sqrt{\Pi K_G t}}} \qquad (10)$$

$$T_g(z,t) = T_{og}e^{-\frac{z}{\sqrt{\Pi K_T t}}} + [T_{oL} - T_L(z,t)] \qquad (11)$$

Operating Parameters

The mass diffusivity of water vapor in triethylene glycol (TEG) is given by Perry et al [14] in eqn. (2.7) as:

$$D_A = D_{12} = 1.1728 x 10^{-16} \frac{T(x_2 M_2)^{1/2}}{\mu_2 V_1^{0.6}} \quad (12)$$

subcript 1 *represents water vapour and subscript* 2 *is for TEG.*

Where the solvent viscosity at temperature T is μ_2 in Pa sec, the solute molar volume at the normal boiling point V_1 in m^3/kmole, and the solvent association parameter χ_2 multiplied by the solvent molecular weight are required input parameters. Data were taken at gas inlet conditions (Company W Plant data, 2007)

$T = 50°C = 323.15K$

$At\ T = 50°C,\ \mu_2 = 0.01355515\, Pa.\sec,$

solvent association parameter, $\chi_2 = 1$ (*for TEG*), *Perry et al*,[14]

$V_1 = 0.0183 m^3 / kmol$, *Perry et al*,[14]

$M_2 = 150.17\ kg / kmol$

On substituti on of these parameters int *o eqn*.(12), *we obtain* :

$D_A = 3.80 \times 10^{-10}\ m^2 / s$

The thermal diffusivity of water vapor can be obtained from Perry et al, [14], to be $K_T = 2.338 * 10^{-5}\ m^2/s$. The inlet gas and glycol temperatures were obtained from the plant data as:

$T_{og} = 50°C$

$T_{OL} = 55°C$

Inlet gas and glycol water content (in weight %) were obtained from plant data and were converted to mole % (assuming a binary mixture) using the relations:

$$y_1 = \frac{x_1 / M_1}{x_1 / M_1 + x_2 / M_2} \quad (13)$$

$$y_2 = 1 - y_1 \quad (14)$$

Where x_1 and x_2 are concentrations of the gas and glycol in weight %, y_1 and y_2 are their respective mole %, M_1 and M_2 are their respective molecular weights. The results for the input and output streams of the gas and glycol streams are tabulated in Table 1 and Table 2.

Table 1: Plant Data showing inlet conditions (Company W data, 2007)

COMPONENT	INPUT STREAM			
	GAS STREAM		GLYCOL STREAM	
	Weight %	*Mole%*	*Weight %*	*Mole%*
TEG	-	-	99.51	96.054
H_2O	0.17	0.187	0.49	3.946
GAS	99.83	99.813	-	-
TOTAL	100.00	100.000	100.00	100.000
TEMPERATURE, °C	50		55	

Table 2: Plant Data showing outlet conditions (Company W data, 2007)

COMPONENT	OUTPUT STREAM			
	GAS STREAM		GLYCOL STREAM	
	Weight %	*Mole%*	*Weight %*	*Mole%*
TEG	-	-	95.36	71.127
H_2O	0.01	0.011	4.64	28.873
GAS	99.99	99.989	-	-
TOTAL	100.00	100.000	100.00	100.000
TEMPERATURE, °C	51.3		51	

The physical properties of the wet gas and glycol such as density, molecular weight, and molar volume, mass and thermal diffusivities were obtained as shown above and from Perry et al [14]. These physical properties are tabulated as shown in Table 3.

Table 3: Physical properties of components [14]

Property	TEG	H2O	Gas
Molar mass	150.17	18.02	19.83
Molar volume, m^3/kmol		0.01813	
Mass diffusivity, m^2/s		$2.338x10^{-5}$	
Thermal diffusivity, m^2/s		$3.80x10^{-10}$	
Density, kg/m^3	1125	1000	

Results and Discussion

The results obtained (equations 13-15) were simulated across the packed height and varying residence times. The mass and thermal diffusivities were also varied to see the effects they will have on the process. Also, the gas inlet water content was also varied to know how fast the model will reach the stipulated standard gas pipeline specification given by the Gas Processors and Suppliers Association [3].

Table 4: Comparison of model result and plant data at z=0.04m and t=20seconds

Parameter	Plant data	Model result	% Deviation
Gas water content, mole	0.00011	$1.0717x10^{-14}$	0
Gas temperature, °C	51.3	50.4874	1.584
TEG temperature, °C	51	49.5497	2.844

The results in Table 4 show that the effect of diffusion was noticed between z=0 and z=0.004m. A deviation from plant data of 0% was observed for the gas water content at a residence time of 20 seconds, at z=0.004m. Similarly, deviations of 1.584% and 2.844% were observed for the gas and TEG temperatures.

Variation of water content of gas with column height at different residence time

From fig. 4 shown below, it is observed that the water content of the gas (in mole fraction) reduces as the gas moves from the bottom of the column to the top. This is in agreement with plant data (Table 2). Also, it is observed that the shorter the residence time of the gas in the column the higher the rate of removal of the water vapor from the gas. This is because a shorter time is needed to establish equilibrium between the water vapor in the gas and that in the liquid phase. So, as the residence time increases, water vapour returns to the vapor phase. This is in agreement to the analysis made by Chapra and Canale [15] on a similar model.

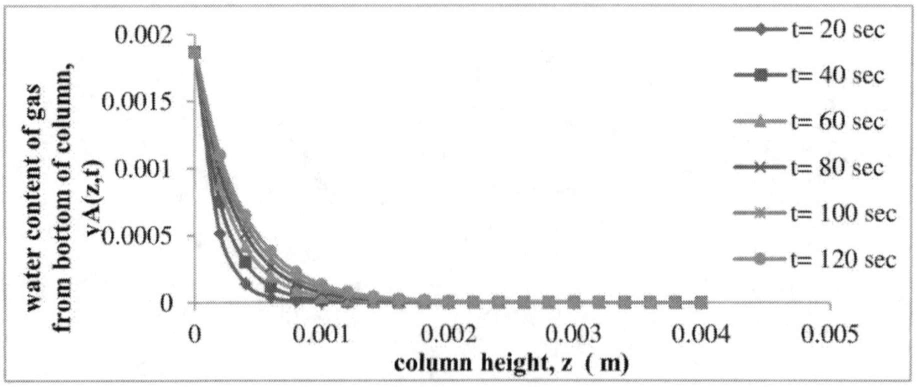

Fig. 4: Variation of water content of gas up the column in mole fraction

Fig. 5 shown below is similar to Fig. 4., except that the water content of the gas has been expressed in terms of pounds of water per million standard cubic feet of gas (*lb H_2O/MMscf* of gas) where 1 *lb H_2O/MMscf* of gas is equivalent to 0.002098951 mole%, to enable us have a quick view of when the maximum standard gas pipeline specification of 7 *lb H_2O/MMscf* given by Ikoku [2] and the GPSA [3] has been attained. This value was achieved by the model eqn. (9). Thus at z = 0.0004m and a residence time of 20 seconds we have a value of 6.695 *lb H_2O/MMscf of gas*. If the residence time is increased to 40 seconds, a value of 5.722 *lb H_2O/MMscf of gas* is obtained at z = 0.0006m. In all the cases considered, that is from t = 20 seconds to 120 seconds, it is observed that the outlet gas water content meets the standard pipeline requirement given in Ikoku [2] and the GPSA [3].

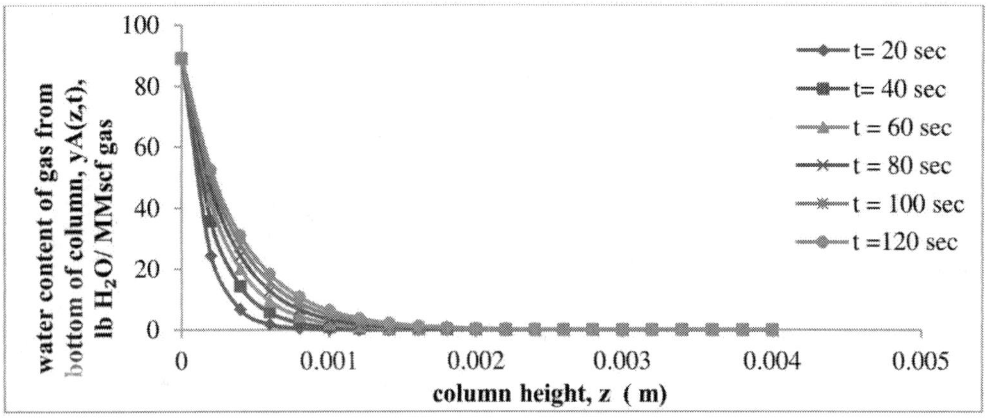

Fig. 5: Variation of water content of gas up the column in lb H2O/MMscf

Variation of temperature of Triethylene glycol across the column height at different residence times

In fig. 6, the temperature of the absorbing solvent, triethylene glycol (TEG), reduces gradually as it moves from the top of the column to the bottom in accordance with plant data requirement. This reduction in temperature becomes slower at higher residence time, obviously because more water vapor has been absorbed at higher residence time. Triethylene glycol enters the column at a temperature of 55°C and leaves at 51°C (Tables 1 & 2). An outlet temperature of 49.55°C is predicted by the model eqn. (10) at t = 20 seconds and at z = 0.004m. This value increases to 51.08°C when t = 40 seconds and z = 0.004m.

Variation of Temperature of Gas with Column Height at Different Residence Times

In the gas absorber being modeled, the gas enters the column at a temperature of 50°C and increases very slowly until it leaves the column at a slightly higher temperature of 51.3°C. This is close to the value predicted by model equation (41), requiring the gas to leave the column at 50.5°C at a residence time t = 20 seconds and at z = 0.004m. It can be noticed in Fig. 7 that this value becomes lower at higher residence time. This implies that if we reduce the residence time further, the required outlet gas temperature would be achieved. It should be noted that the increase in residence time results from the heat transfer from the liquid stream to the gas stream.

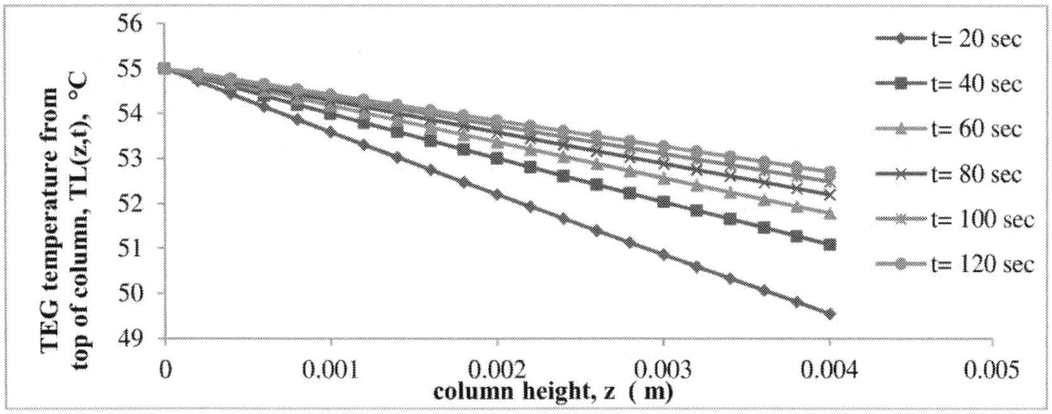

Fig. 6: Variation of temperature of triethylene glycol down the column

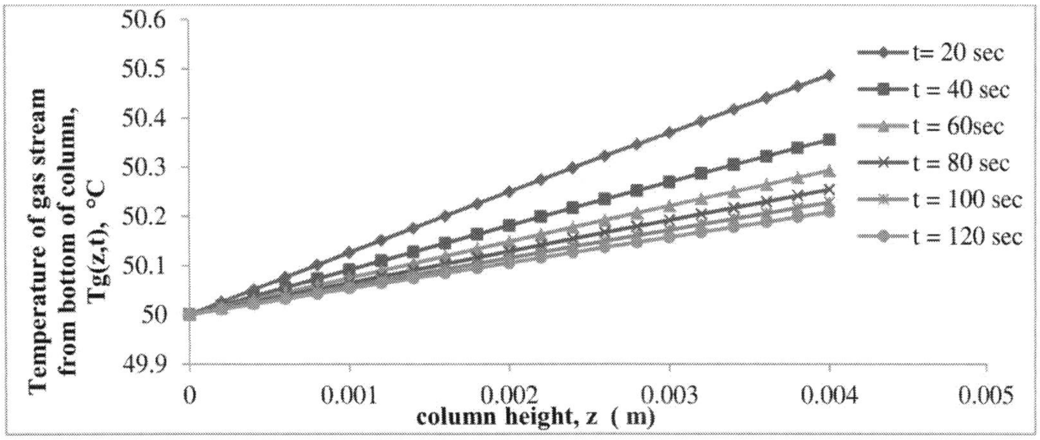

Fig. 7: Variation of the temperature of the gas stream from column bottom

Conclusion

Mathematical models of the absorber of a glycol dehydration facility was developed using the principles of conservation of mass and energy. A model that predicts the variation of the water content of gas in mole fraction across the column was developed. Similar models were developed for the variation of the gas and liquid (TEG) temperatures across the packing height.

The models developed contain contributions from bulk and diffusion flows. The effect of diffusion on the process occurring in the absorber was studied in this work. The models were validated using the initial conditions in the plant data from a company W TEG unit in Nigeria to ascertain if the outlet conditions predicted by the models meet the plant data outlet requirements. The results obtained showed that the effect of diffusion was noticed between z=0 and z=0.004m. A deviation from plant data of 0% was observed for the gas water content at a residence time of 20 seconds, at z=0.004m. Similarly, deviations of 1.584% and 2.844% were observed for the gas and TEG temperatures.

References

1. Ikoku C. U, *Natural Gas Production Engineering,* Reprint ed., Kreiger Publishing Company, Malabar, Florida, USA, (1992), pp 20, 133-171.
2. Gas Processors Suppliers Association, *Engineering Data Book,* (11th ed., Electronic), Gas Processors Suppliers Association, Tusla, Oklahoma, USA, (GPSA, 1998), Chapter 20.
3. Gómez C., Borio D. O, and Schbib N. S, *Simulation of an Industrial Packed Column for Reactive Absorption of* CO_2, Latin American Applied Research, Vol. 33, Argentina, (2003), pp 201-205.
4. John N. *Development of a Model for Wet Scrubbing of Carbon Dioxide by Chilled Ammonia*, Department of Chemical Engineering, Lund University, Lund, Sweden, (2009).
5. Ahmed A., Paitoon T. and Raphael I., applications of new absorption kinetics and vapor/liquid equilibrium models to simulation of a pilot plant for carbon dioxide absorption into high co2-loaded, concentrated monoethanolamine solutions, *chemindix*, ccu/09, International Test Centre for Carbon Dioxide Capture (ITC), Faculty of Engineering, University of Regina, Regina, SK, S4S 0A2, Canada, (2007).
6. Peyghambarzadeh, S.M and Jafarpour M., *Impact of Thermodynamic Model on the Simulation of Natural Gas Dehydration Unit,* 6th National-Student Chemical Engineering Congress, University of Isfaham, (2006) pp 1-10.
7. Max S. P, Klaus D. T and Ronald E. W, *Plant Design and Economics for Chemical Engineers,* 5th ed., McGraw Hill International Edition, New York, (2004), pp 797-80.
8. McCabe W. L., Smith J. and Harriott P., *Unit Operations of Chemical Engineering,* 7th ed., McGraw Hill International Edition, New York, (2005), pp 576-578.
9. Richardson, J.F , Harker J. H., with Backhurst J. R.: *Coulson & Richardson's Chemical Engineering: Particle Technology and Separation Process,* Vol.2, 5th ed., Elsevier, New Delhi, India, (2002), pp 542-717.
10. Luyben, W.L., *Process modeling, simulation, and control for chemical engineers*, 2nd ed., McGraw-Hill Education, Columbus, OH, USA, (1990), pp. 22-24.
11. Stephenson G., Partial Differential Equations for Scientists and Engineers, 3rd ed., Longman Inc., New York, (1986), pp 89-134.
12. Kreyszig, E., *Advanced Engineering Mathematics*, 8th ed., John Wiley and Sons Inc., Singapore, (2005), pp 2-138, 582, and 755.
13. Perry, R.H., and Green D. W, *Perry's Chemical Engineers' Handbook*, 7th ed., McGraw Hill companies Inc, New York, (1999), sections 1, 2 and 14.
14. Chapra S. C and Canale R. P, *Numerical Method for Engineers*, 5th ed., Tata McGraw-Hill Publishing Company Ltd, New Delhi, India, (2007), pp. 722-727.

410124 Fundamental Study of Washed Cell-Free and Concentrated Ascites Reinfusion Therapy (w-CART)

Abstract:

Cell-free and concentrated ascites reinfusion therapy (CART) is treatment for refractory ascites by ultrafiltration and reinfusion of their ascites. Recently, CART is frequently performed as palliative care. However, adverse events such as more than 37.5°C fever were often reported. In order to prevent fever, high pre-dose of adrenal corticosteroid and antipyretics are frequently prescribed. The one of the cause of the fever may be concentrated cytokine that is included in the original ascites. In this study, we examined new CART method incorporating the washing process to remove cytokine fractions and purify albumin.

We performed in vitro normal CART and w-CART using 2700 mL of pseudo-ascites (Total protein was 3.0 g/dL), prepared using human plasma discarded by plasma exchange treatment, and concentrated spent dialysate containing maker solutes such as b2 and a1-microglobulin (MG). Molecular size of b2-MG (MW: 12kDa) and a1-MG (30kDa) is equivalent to the cytokine fraction. All experiments were performed under the approval of Ethics Committee of the University with respect to the collection of dialysate drainage. First, 300mL of concentrated ascites was obtained by normal CART operation. In addition, we mixed 2700 mL saline to a concentrated ascites and concentrated it again by filtration (washing process).

While maintaining the approximately 90% of albumin, not only urea but also larger solutes such as b2-MG, myoglobin (18kDa), prolactin (22kDa) and a1-MG were more removed by w-CART compared with conventional CART. These results indicated that w-CART could selectively remove cytokines without excessive loss of albumin compared with conventional CART. Newly proposed w-CART would be effective for preventing fever without pre-dose of adrenal corticosteroid and antipyretics.

Washed Cell-free and concentrated Ascites Reinfusion Therapy (w-CART) may be effective for removal of cytokines compared with conventional CART.

415947 A Novel Mussel-Inspired Superhydrophobic Cloth Boat for Self-Driven Crude Oil Spill Cleanup

Abstract:

Superhydrophobic fabric was simply prepared by a novel mussel-inspired strategy at a much lower concentration of dopamine with the aid of folic acid (FA), avoiding the use of any additional nanoparticles. Intriguingly, different from the additional nanoparticles, FA takes the role of structure directing agents for preparation rough polydopamine (pDA) coatings, and such hierarchical structures can be readily controlled by adjusting the FA concentrations or the coating duration. Higher FA concentration is adverse to formation of obvious hierarchical structure because of the inhibition effect of FA on the pDA formation, while appropriate duration of Stage I is indispensable for construction of rough pDA coatings. After octadecylamine (ODA) chemical manipulation, the as-prepared cloth boat with hierarchical structure exhibits excellent superhydrophobicity and the water contact angle and rolling off angle were about 162° and 7°, respectively. Importantly, the superhydrophobic fabric can withstand continuous and drastic 3.5 wt.% NaCl solution rinse and repeated tear with adhesive tape more than 30 times, exhibiting the excellent durability. Moreover, an energy-saving and highly efficient mini boat fabricated by our novel superhydrophobic fabric was innovatively utilized for self-driven oil spill cleanup, which can automatically recycle crude oil spill while float freely on water with the cleanup rate of crude oil spill up to 97.1%, demonstrating great potential in environmental remediation. The novel strategy designed in this study can inspire the fast development of mussel-inspired superhydrophobic materials for applications in various fascinating fields.

416191 Temperature Effects on Heat Transfer Fouling of Thin Stillage from Ethanol Production

Abstract:

Heat transfer fouling is the accumulation and formation of unwanted materials on heat transfer surfaces which leads to a decrease in the overall heat transfer coefficient. Fouling of heat transfer equipment increases energy consumption and maintenance costs and thus decreases processing efficiency. In the fuel ethanol industry, evaporator fouling occurs when thin stillage is concentrated. Fouling affects the efficiency and environmental footprint of more than 200 biorefineries in the US. Thin stillage is the liquid fraction of unfermented materials from fermentation and is composed of carbohydrate, protein, fat and ash.

Research on thin stillage fouling has focused on effects of corn oil, pH, Reynolds number, solids concentration and carbohydrates (Singh et al 1999, Wilkins et al 2006ab, Arora et al 2010, Challa et al 2014). However, temperature effects on fouling rates have not been studied. The objectives were to investigate the influence of bulk fluid temperature, initial probe temperature and their temperature difference on thin stillage fouling characteristics. Experiments were conducted using model thin stillage (1% starch solution) and commercial thin stillage with varying temperature conditions. Bulk temperatures varied from 60 to 80°C and initial probe temperatures varied from 100 to 120°C; thus temperature differences varied from 20 to 60°C. Fouling resistances were measured using an annular probe with a 7 L batch system. Mean fouling rate, maximum fouling resistance and induction period were used to characterize fouling behavior. For commercial thin stillage, fouling rate and maximum fouling resistance increased as the initial probe temperature increased from 100 to 120°C. At 120°C initial probe temperature, fouling rate increased with bulk temperature. For the model thin stillage, similar results were observed. Induction periods were longer than 5 hr for an initial probe temperature of 100°C. At an initial probe temperature of 120°C, longer induction periods, lower fouling rate and maximum fouling resistance were observed at lower bulk temperature. Bulk temperature and initial probe temperature had an effect on thin stillage fouling characteristics. Comparing with commercial thin stillage, bulk temperature had larger effects on model thin stillage. Using higher initial probe temperature, such as 120°C, and fluid bulk temperature of 80°C would provide repeatable and more rapid (within 5 hr) characterization of fouling behavior.

419386 Energy-Saving Design and Control of a Methyl Methacrylate Separation Process

Abstract:

In the industrial process to produce methyl methacrylate (MMA), a mixture including MMA, methanol (MeOH) and water (H2O) needs to be separated. The MMA and water are designed to be drawn out of this separation process as two product streams while methanol as an excess reactant should be recycled back to the esterification reactor. Wu et al. (2011) devised a design flowsheet including a distillation column with a bottom decanter and another stripper. The bottom composition of the distillation column was designed to be near MMA/H2O azeotrope so that a decanter can be designed to obtain aqueous and organic streams. The aqueous stream is pure enough to draw-out of the system while another small stripper was needed to purify the organic stream to MMA product specification. This design flowsheet was further simplified by combining two columns into one column with a middle decanter.

The drawback of the above design is that the bottom composition of the distillation column needs to be placed near a saddle point (MMA/H2O azeotrope) of this ternary system. Thus, larger reboiler duty was required at the distillation column to satisfy this bottom purity requirement. In this paper, an alternative design using the same process units (a distillation column, a decanter and another stripper) is developed. In this alternative design, the fresh feed is first mixed with some water and condensed stripper top stream so that the composition of the mixture is pulled within the LLE envelope. After liquid-liquid separation in the decanter, the aqueous stream can further be separated in a distillation column with top product of mostly MeOH (at MeOH/MMA azeotrope) and the bottom product of pure water. The MMA product can be obtained from further purify the organic stream in a small stripper. Note that in this alternative design, top and products of the distillation column are designed to be at unstable node and stable node, respectively. Thus, signification saving in the steam cost (16.3%) can be realized by using this alternative design with the same process units. Another benefit of this alternative design is that the MMA product loss through the water outlet stream is also less than the design in Wu et al. (2011). This represents another 9.6% savings of the operating cost for this alternative design.

Furthermore, in this research, a suitable and simple control structure for this alternative design is also devised based on both open loop and closed loop sensitivity tests. By controlling temperature difference at two trays of the distillation column and another tray temperature at stripper, both MMA and water products can be maintained at high-purities despite large variations in feed flow rate and feed composition changes.

Reference:

Wu, Y. C.; Hsu, C. S.; Huang, H. P.; Chien, I. L.; Design and Control of a Methyl Methacrylate Separation Process with a Middle Decanter, *Ind. Eng. Chem. Res.*, **2011**, 50, 4595-4607.

420388 Three-Dimensional CFD Model of a Multitube Pd/Au Membrane Module for Hydrogen Purification

Abstract:

Three-dimensional CFD Model of a Multitube Pd/Au Membrane Module for Hydrogen Purification

Rui Ma, Bernardo Castro-Dominguez, Anthony G. Dixon, Yi Hua Ma

Department of Chemical Engineering, Worcester Polytechnic Institute, 100 Institute Road, 01609 Worcester, MA, USA.

Hydrogen is an important chemical product due to its broad use in the generation of ammonia and methanol as well as a clean energy carrier. The U.S Energy Information Administration estimated that in the U.S., energy demand grows 0.4% per year. Coal gasification is a potential method to solve this problem since it generates electricity and H_2 simultaneously. Nevertheless, this method also produces syngas, which is composed of methane, carbon dioxide, carbon monoxide and other impurities. Consequently, a separation process is required for the purification of H_2. Compared to traditional separation methods, palladium (Pd) membrane separation technology has the advantage of high H_2 selectivity, continuous operation and economic manufacturing costs. The objective of this work is to model accurately the performance of a multitube Pd membrane module and gain insight on the effects of different operating conditions.

A seven-tube membrane module has been designed and manufactured as shown in Fig. 1, capable of separating pure H_2 from syngas mixtures. In this module, H_2 enriched coal-derived syngas is introduced into the system as the feed; H_2 permeates across the membrane and is collected at the inner tube side, while the unwanted gases are washed-out at the retentate side. Due to the broad variety of operating conditions, the membrane performance is hard to predict; thus, a computational fluid dynamics (CFD) simulation is used to optimize the operating conditions of this process. Compared to previously reported simulations, the module in this work takes heat transfer into consideration, making this framework more realistic since the temperature influences considerably the permeability and stability of the membranes. The performance of the multitube membrane module is accurately simulated, by generating a 3D model using COMSOL Multiphysics 5.0 solving simultaneously the continuity and Navier-Stokes equations as well as conservation of energy. Sieverts' law was used to define the permeation rate of H_2 across each membrane, as shown in equation 1.

$$F_{H_2} = \frac{Q}{\delta}(P^{0.5}_{r,H_2} - P^{0.5}_{p,H_2}) \qquad (1)$$

Where F_{H_2} is H_2 flux across the membrane, Q is the permeability and δ is the thickness of membrane. $P^{0.5}_{r,H_2}$ and $P^{0.5}_{p,H_2}$ are the H_2 partial pressures at retentate side and permeate side, respectively. Equation 1 represents the solution-diffusion mechanism for H_2 transport; the permeability (Q) is highly dependent on temperature. The overall effect of temperature on permeability is introduced using the Arrhenius correlation as shown in equation 2.

$$Q = Q_0 e^{-E_p/(RT)} \qquad (2)$$

Where Q_0 is permeability constant and E_p is activation energy. Higher temperatures lead to higher permeabilities and thus higher H_2 diffusion rates through the membrane.

The simulation was carried out under different Reynolds numbers (Re) within the laminar regime and various feed temperatures (523K-723K); the influences on the Péclet number and mass transfer resistances are clearly described under these circumstances. The depletion of H_2 in the proximity of the membrane surface induces the formation of a gas boundary layer often called concentration polarization. The H_2 partial pressure adjacent to the surface of the membranes is reduced due to the mass transfer resistance generated by concentration polarization. Furthermore, the simulation shows this phenomenon accurately by depicting a cross-sectional concentration profile of the shell side, as shown in Fig. 2. The results show that higher Re reduces concentration polarization by reducing the mass transfer boundary layer.

The performance of the system is analyzed based on the total H_2 recovered on the tube side from the feed under the different operation conditions. Axial H_2 concentration profiles (Fig. 3) show the H_2 distribution, implying better H_2 recovery at low Re. On the other hand, operating at high Re number or high Péclet numbers minimizes the effect of diffusion by markedly enhancing the convective forces. Although low feed flow rates lead to a high H_2 recovery, they diminish the usage of the membrane. This is caused by early H_2 depletion at the beginning of the membrane and a noteworthy axial H_2 partial pressure gradient.

Furthermore, temperature distribution profiles show the influence of convective heat transfer of the gases to the membranes. Permeability distribution profiles along the membranes display the influence of temperature on H_2 permeance. Operating at higher feed temperatures, close to the membrane temperature, increases the H_2 recovery of the module; nevertheless lower temperatures display lower efficiencies due to the reduction of temperature at the surface of the membranes.

Based on the results obtained, the optimum operating conditions maximize H_2 recovery; typically enhanced by low-moderate feed rates and high feed temperatures. This heat and mass transfer model is validated by comparing the results obtained in this study with previously reported experimental results of a single tube membrane module. The evaluation of the actual industrial model will be performed afterwards at the National Carbon Capture Center in Alabama.

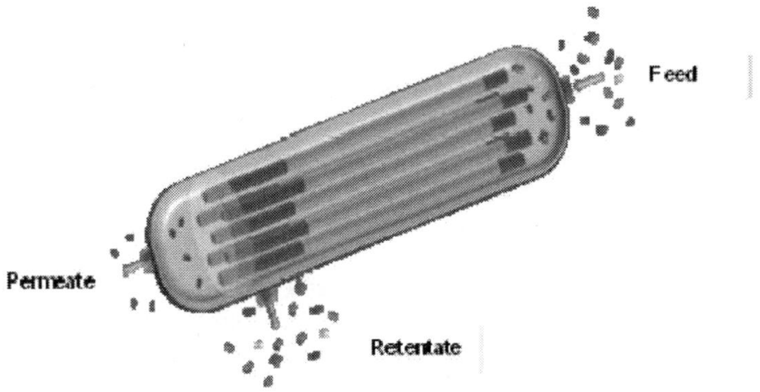

Fig. 1. Membrane module set-up

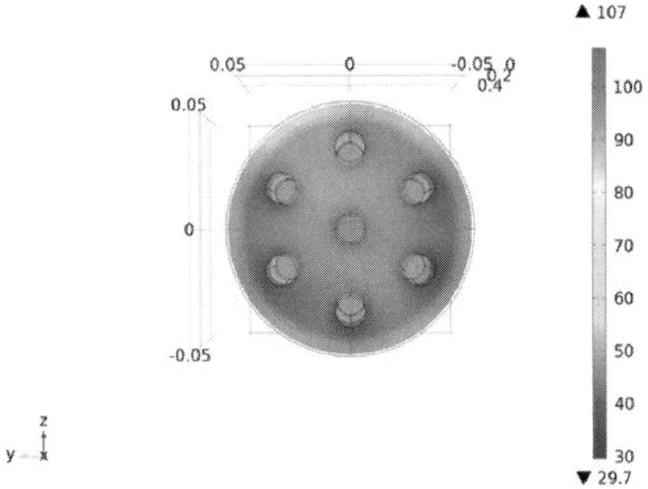

Fig. 2. Cross-section concentration profile

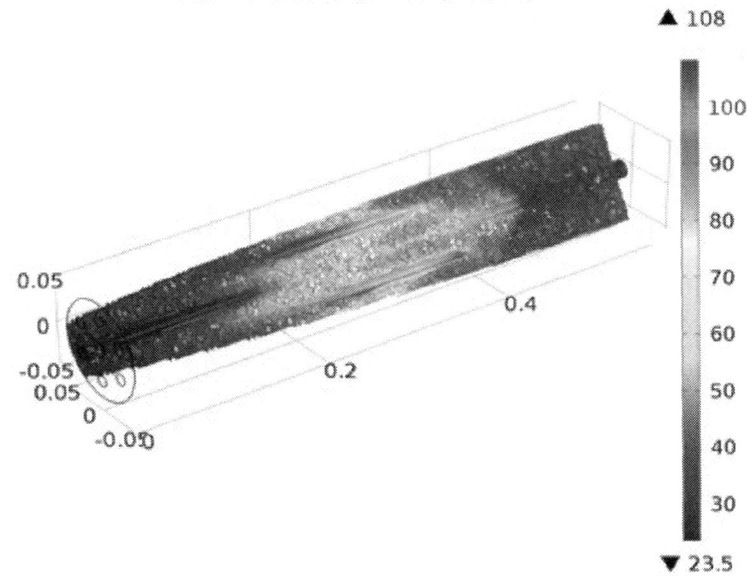

Fig. 3. Hydrogen concentration profile across the retentate side of the multitube membrane module

421743 A Parametric Study for Oxygen Plant Based on the Claude Cycle

Abstract:

The air liquefaction system for oxygen plants in gas industries is based on the Claude cycle. The effect of process parameters on the overall system is difficult to analyse via manual calculations and this provides the motivation to use process simulators for understanding the steady state and dynamic behaviour of such systems. In this paper, a mathematical model for an industrial oxygen plant that produces 80 m3/hr of oxygen at Mitt Oxygen Pvt. Ltd. Bhavnagar was developed and analyzed. The effect of cycle pressure ratio, heat exchanger effectiveness, inlet air temperature, flow fraction and turbo-expander efficiency on overall performance were studied and optimal operating parameters were determined. It was found that with increase in pressure ratio the losses associated with the upper turbo-expander stage increase and this eventually reduces the excess refrigeration available in that stage. The excess refrigeration, which is passed to the final throttle stage for liquefaction, decreases the liquid yield. When forward stream pressure is increased, at a certain pressure ratio the gain in refrigeration by increase in enthalpy drop across the expanders is balanced by the loss refrigeration due to increase in expander losses. The study on the effect of heat exchanger performance on the cycle reveals that, compared to the rest of the heat exchangers, the inlet temperature to the expanders have more pronounced effect on the liquid yield of the cycle. For expander it was found that the effect of variation of isentropic efficiency is more sensitive when the flow through expander is at its optimum value, which yields maximum liquefaction.The results of this study are utilized to develop guidelines for improving the plant performance of the oxygen plant.

422143 Development and Characterization of Carbon Nanotube Membranes for Membrane Distillation

Abstract:

Membrane distillation (MD) is an attractive separation process that uses hydrophobic membranes and is driven by the vapor pressure difference across the membrane. Membrane distillation is an emerging technology for water desalination, industrial wastewater treatment, radioactive waste treatment, removal of volatile organic compounds, and concentrating organic solutions. Further development of MD technology depends on the development of cost-effective, robust, and scalable membranes with the desired characteristics, including optimized porosity, high hydrophobicity, the proper surface chemistry, high thermal stability, and good mechanical properties.

This work was aimed at the development of carbon nanotube (CNT) composite membranes that can be utilized for both conventional MD and high-temperature (e.g., >120 °C) separation applications. Composite membranes were prepared by growing CNT, by the chemical vapor deposition method, on various metallic and ceramic substrates, including stainless steel and Hastelloy C alloy meshes, nickel foam, and quartz filter paper. Substrates were subjected to different pretreatments before the carbon deposition stage to create active sites or adjust pore openings. Pretreatments included acid treatment using concentrated HCl at 25 °C, air oxidation at 500-800 °C, surface treatment by Argon plasma, iron or nickel catalyst deposition by solution coating or sputtering, and nickel coating by electroless nickel solution. Membranes were characterized by scanning electron microscopy, Raman spectroscopy, water contact angle measurement, and other characterization methods. A commercially available Teflon membrane and a CNT buckypaper were characterized and tested for comparison purposes. A direct-contact MD configuration was used for the desalination tests. Characterization results indicated that prepared composite CNT membranes are highly hydrophobic and composed of CNT and other carbon materials grown on the substrate materials. Work is in progress to evaluate the performance of developed membranes for MD application. Properly optimized CNT composite membranes can potentially be used for water desalination and other separation applications, particularly at elevated temperatures, where existing polymer-based membranes may fail.

422170 Feasibility Analysis of Precombustion CO2 Capture in IGCC Power Plants Using MgO-Based Solid Sorbent

Abstract:

Climate Change, which has been associated with increasing concentration of greenhouse gases is regarded as one of the key environmental issues in the 21st century. Carbon dioxide (CO_2) is the primary greenhouse gas emitted through human activities. Coal-fired power plants still generate a major portion of electricity in the U.S. and globally.and national and global reliance on coal for electricity generation is expected to continue to be significant in the next several decades. Advanced coal based power generation technologies, such as Integrated Gasification Combined Cycles (IGCC) processes, are expected to be among the leading contenders for power generation because of their higher efficiencies, flexible products and potential environmental advantages compared to conventional coal combustion (pulverized coal) processes. The three general categories of CO_2 capture technologies that can be applied to coal-based power plants are pre-combustion, post combustion, and oxy-combustion. Pre-combustion technologies in IGCC processes involve separating CO_2 before the fuel is combusted. The high temperature of syngas in the outlet of gasifier suggests that MgO-based sorbents can be a good candidate for Carbon capture processes because its temperature-pressure equilibrium is in the operating range of industrial power plants. The syngas leaving the gasifier is cleaned up of other contaminants (Sulfur, particulates etc.) and then enters the carbon capture unit before entering to the turbines. This paper addresses the process design and economic feasibility of a Carbon Capture and Storage (CCS) process involving regenerable MgO-based sorbents for an IGCC power plant. The process consists of several circulating fluidized beds (CFB) operating at elevated temperature and pressure, to capture CO_2 in a carbonator reactor and regenerate the sorbent in a fluidized bed regenerator. A make-up stream also enters the system to compensate the sorbent degradation as it goes through carbonation/regeneration cycles. Cost analysis demonstrate the make-up flow rate is the major source of operating cost in such a system.

424098 Heat Transfer Effects of Phytic Acid Concentration and Phytase Addition on Fouling Behavior of Steepwater

Sunday, November 8, 2015
Exhibit Hall 1 (Salt Palace Convention Center)
Ju Tian, Agricultural and Biological Engineering, Univeristy of Illinois at Urbana-Champaign, Urbana, IL

Heat Transfer Effects of Phytic Acid Concentration and Phytase Addition on Fouling Beh avior of Steepwater

Ju Tian1, David B. Johnston2, Nicki J. Engeseth1, Vijay Singh1, M. E. Tumbleson1 and Kent D. Rausch1,*

1Univeristy of Illinois at Urbana-Champaign, Urbana, IL

2Eastern Regional Research Center, ARS, USDA, Wyndmoor, PA

*(217-265-0697) krausch@illinois.edu

Heat transfer fouling describes the phenomenon of unwanted materials forming and accumulating on heat transfer surfaces. This can lead to a decrease in process efficiency, as fouling of heat transfer equipment increases energy consumption and maintenance costs. In the corn wet milling industry, evaporator fouling takes place when steepwater is concentrated. Steepwater is the solution resulting from the corn steeping process and is composed of solubilized kernel compounds, microbes (principally lactobacilli) and solid from recycled process streams.

Research on corn processing fouling has focused on the effects of steepwater solid, corn oil, pH, Reynolds number, solids concentration and carbohydrates (Agbisit et al 2003, Singh et al 1999, Wilkins et al 2006ab, Arora et al 2010, Challa et al 2014). However, effects of phytic acid concentration or phytase addition on the fouling characteristics is not known. There are reports from industry suggesting phytic acid content is positively correlated with fouling rate, and adding phytase to the process stream may help reduce fouling. Research also has shown the solubility of certain phytic acid metal complexes, the product of chelation reaction between metal ions available in corn process streams (Mg, Ca, K) and phytic acid, can be influenced by adding Phytase (Ekholm et al 2003). A study on one of these phytic acid metal complexes (Ca) concluded the concentration was positively correlated with fouling in the dairy industry at high temperatures (Bansal et al 2006). The purpose of this study was to evaluate effects of phytic acid concentration and phytase addition on steepwater fouling behavior. Experiments were conducted using commercial steepwater with different phytic acid concentrations. With phytic acid addition, the phytic acid concentrations of the samples were adjusted to vary from 25 mg/g sample to 75 mg/g sample. Fouling resistances were measured using an annular probe with a 7 L batch system. Mean fouling rate, maximum fouling resistance and induction period characterized fouling behavior. The results will provide a better understanding of phytic acid effects on wet milling fouling and provide possible solutions to fouling mitigation in the wet milling process.

LITERATURE CITED

Agbisit, R. M., Singh, V., Valenti, J. J., Kakleas, M., Tumbleson, M. E., & Rausch, K. D. 2003. Technique to measure surface-fouling tendencies of steepwater from corn wet milling. Cereal Chemistry, 80(1), 84-86.

Arora, A., Dien, B. S., Belyea, R. L., Singh, V., Tumbleson, M. E. and Rausch, K. D. 2010.

Heat transfer fouling characteristics of microfiltered thin stillage from the dry grind process. Bioresource Technology 101:6521-6527.

Bansal, B. and Chen, X. D. 2006. A critical review of milk fouling in heat exchangers. Comprehensive Reviews in Food Science and Food Safety, 5:27–33.

Challa, R. K., Johnston, D. B., Singh, V., Tumbleson, M. E. and Rausch K. D. 2014. Fouling characteristics of model carbohydrate mixtures and their interaction effect. Food and Bioproducts Processing, 93:197-204.

Ekholm, P., Virkki, L., Ylinen, M., Johansson, L., 2003. The effect of phytic acid and some natural chelating agents on the solubility of mineral elements in oat bran. Food Chemistry 80, no. 2: 165.

Singh, V., Panchal, C. B. and Eckhoff, S. R. 1999. Effect of corn oil on thin stillage evaporators. Cereal Chemistry 76:846-849.

Wilkins, M. R., Belyea, R. L., Singh, V., Buriak, P., Wallig, M. A., Tumbleson, M. E. and Rausch, K. D. 2006a. Analysis of heat transfer fouling by dry-grind maize thin stillage using an annular fouling apparatus. Cereal Chemistry 83:121-126.

Wilkins, M. R., Singh, V., Belyea, R. L., Buriak, P., Wallig, M. A., Tumbleson, M. E. and Rausch, K. D. 2006b. Effect of pH on fouling characteristics and deposit compositions in dry-grind thin stillage. Cereal Chemistry 83:311-314.

425035 Preparation and Stability of Imidazolium-Functionalized Alkaline Electrolyte Membrane Incorporating ZrO2 Nanoparticles

Abstract:

As a potential next-generation clean energy technology, alkaline anion exchange membrane fuel cells (AAEMFC) has attracted enormous attention in the past few decades. They have several advantanges over proton exchange membrane fuel cells(PEMFCs) and conventional alkaline fuel cells, including fast oxygen reduction reaction, free of carbonate crystallization at the electrode, more options to use inexpensive catalysts such as nickel, and low fuel crossover. Anion exchange membrane (AEM), one of the core components of AAEMFCs, is a key factor that determines the fuel cells' power density and life. The desirable AEM should possess high ionic conductivity, good mechanical and thermal stability, and long-term alkaline stability. So far, commercial AEMs still cannot meet the requirements of AAEMFCs due to their low ionic conductivity and poor chemical stability.

The reason for poor alkaline stability of AAEM lies in two aspects: vulnerableness of the ion exchange groups and the strong attack of hydroxide ions under fuel cell operating conditions. In this work, we address the low stability issue by alleviating the intensity of hydroxide attack.

We will make hybrid AAEMs with poly(ether ether ketone) (PEEK) and zirconium dioxide (ZrO_2) nanoparticles, and investigate how the nanoparticle content influence the membranes' water uptake, water retention, alkaline stability and other properties. PEEK is a semicrystalline high-performance polymer with excellent chemical stability, thermal stability and mechanical property. It has been widely used for making ion exchange membranes. ZrO_2 nanoparticle has large specific surface area, good hydrophilicity and mechanical properties. We synthesize imidazolium-functionalized PEEK first, and then disperse nano-ZrO_2 fillers into it to make hybrid AAEMs.

The introduction of hydrophilic nano-ZrO_2 fillers may probably produce a solvation effect on the OH^- ions, thus enhancing the alkaline stabilities. The possible mechanism is that the nano-ZrO_2 enhances water uptake of the membrane, thus "diluting" OH^- ions inside the membrane. In other words, there will be a lower OH^- concentration at the vicinity of the imiazolium groups and the OH^- ion is surrounded by large amount of water. The high water uptake also may also have a protection effect to the ion exchange group, thus weakening the OH^- ion's attack and improving the alkaline stabilities. The main purpose of this work is verify the above hypotheses.

Using the concentrated sulphuric acid as solvent and catalyst, chloromethyl octyl ether (CMOE) as chloromethylation reagent, we successfully synthesized CMPEEK. A low reaction temperature, e.g. -10 °C, is required for the good controllability of chloromethylation degree (DC) of CMPEEK. By finely tuning reactant ratio and reaction time, CMPEEKs with different DC were successfully obtained, and the solubility test manifested that all the CMPEEKs have good solubility in high-boiling-point membrane casting solvents.

The Im-PEEK was synthesized by reacting CMPEEK with 1,2- dimethylimidazole. Differed amounts (2.5%, 5%, 7.5% and 10%) of nano-ZrO_2 was then incorporated into Im-PEEK by a simple blending method, and the resultant membraes are designated as CMPEEK-x%ZrO_2. A good compatibility between ImPEEK and the nanoparticle was comfirmed by the uniform morphology of the composite membranes as seen from a scanning electron microscopy study. The X-ray diffraction was used to determine the crystallinity of the composite membranes. It is seen in the TGA curves that the incorporation of ZrO_2 improves the thermal stability of the membrane. More importantly, the water uptake of the composite membranes increases with the amount of ZrO_2. The composite membrane containing 7.5% ZrO_2has the highest water uptake (25.27% at 20°C), and also exhibites the highest conductivity (53.96 ms/cm at 80°C), but with the lowest swelling ratio(18.52%, 20°C) and the best alkaline stability (3M NaOH treated 24h, the conductivity decreased by 4.27% only).

This work provides some new insight into the AAEM's stability issue. It may open a new possibility of and a simple route toward making high-performance AAEM with good alkaline stability and mechanical strength.

424890 Preparing Alkaline Anion Exchange Membrane with Enhanced Hydroxide Conductivity Via Blending Imidazolium-Functionalized and Sulfonated Poly(ether ether ketone)

Abstract:

The development of alkaline anion exchange membrane (AEM) with both high ion conductivity and stabilities is of great significance for fuel cell applications. The ion conductivity of AEM is usually governed by three factors: 1) the amount of charge carriers in unit volume (IEC); 2) the ionic mobility of charge carriers; and 3) the morphology of ion channels. The low ion conductivity of AEM comes from the low inherent hydroxide ion mobility ratio, which is far lower than that of proton. Increasing the IEC of membrane has been demonstrated as a feasible way to enhance the hydroxide conductivity of AEMs. When utilized as the matrix of AEMs, most of functional polymers exhibited a positive correlation between IEC and hydroxide conductivity. In this study, a facile acid-base blending method is designed to improve AEM performances *via* increasing the IEC of membranes. Basic imidazolium-functionalized poly (ether ether ketone) with a high functionalization degree is employed as polymer matrix to pursue IEC as well as high hydroxide conductivity, meanwhile acidic sulfonated poly (ether ether ketone) (SPEEK) is employed as the cross-linking agent to enhance the stabilities of the blend membranes. Particularly, an *in-situ* Menshutkin/crosslinking method is exploited to prevent the flocculation in the preparation process of blend membranes. As a result, dense and defect-free blend membranes are obtained. The blend membranes exhibit high level of IEC up to 3.15 mmol g^{-1}, and consequently possess elevated hydroxide conductivity up to 31.59 mS cm^{-1} at 30 °C. In addition, benefiting from the strong electrostatic interaction introduced by the acid-base blending, the stabilities and methanol resistance of blend membranes are enhanced.

425362 Validation of an Exergy Analisys Using Aspen PLUS for the Methanol/Water System – a Case of Study

Abstract:

Between all industrial separation processes, the distillation is one of the largest energy consumers. So is justified the search for methods of energy optimization for distillation columns. The optimization of parameters such feed stage, thermal condition of the feed, use of intermediate reboilers can be obtained analysing the *exergy* profile of the column, which in turn depends on the determination of consistent thermodynamic properties of all feeds inside the collumn. This work aims to analyze and validate a determination of the *exergy* loss profile along a distillation column using the software *Aspen Plus* for the binary system Methanol / Water with a rigorous design method, comparing the simulation results with the results obtained by *Faria* in his doctoral thesis (Unicamp - 1996) applying the appropriate thermodynamic models and performing the material and energy balances for each stage of the column, then identifying models that result in consistent *exergy* balances for the column. The results show that appropriate models allow to obtain consistent balances of *exergy* only in a limited set of cases, and then, to correct this, it is necessary to use models that consider the the values of enthalpy and entropy with excess. The validation of the method allowed us to search optimized operational conditions for the distillation column, which aims to minimize the *exergy* loss in each separation stage and makes it more efficient thermodynamically.

425637 Divalent Vs Monovalent Ion Sorption and Transport in Ion Exchange Polymers Based on Cross-Linked Polystyrene with Fixed Sulfonate Anions

Abstract:

Divalent vs monovalent ion sorption and transport in ion exchange polymers based on cross-linked polystyrene with fixed sulfonate anions

Michele Galizia, Francesco M. Benedetti, Donald R. Paul, Benny D. Freeman

J.McKetta Department of Chemical Engineering, 200 E. Dean Keeton Street, 78712 Austin,

and Center for Energy and Environmental Resources, 10100 Burnet Road, 78758 Austin,

The University of Texas at Austin, Texas (USA)

Sorption and transport of alkali-chlorides (NaCl, KCl, LiCl) and alkaline earth-chlorides ($CaCl_2$, $MgCl_2$) in a commercial cation exchange membrane based on cross-linked polystyrene with fixed sulfonate ions was investigated. Salt transport parameters are correlated to the ion size and valency, as well to the membrane water content. It is found that divalent salts show higher sorption and permeability coefficients than monovalent salts. These differences are due to the stronger interaction between divalent counter-ions and the polymer's sulfonate ions, which significantly reduces the actual fixed charge of the polymer. Due to this partial neutralization of fixed charges, the Donnan exclusion [1] becomes less effective in the case of divalent salts, thus determining the observed increases in permeability and sorption coefficients. Moreover, water content in membranes in Ca^{++} of Mg^{++} form is significantly less than in membranes in Na^{+}, K^{+} or Li^{+} form. Divalent ions tend to more strongly screen the polymer fixed charges, thus reducing the hydrophilicity of the membrane.

Thermodynamics of sorption was also investigated by calculating the salt activity coefficients in membrane, by using a Donnan-type treatment. The results confirm that divalent ions can strongly interact with fixed sulfonate ions relative to monovalent ions.

[1] F.G. Donnan, Chem. Rev. 1924 1 73-90

426975 Constructing Ultrathin, Methanol Sieving Graphene Oxide Layer on Nafion Membrane for Direct Methanol Fuel Cells

Abstract:

Perfluorosulfonic polymers such as Nafion are the most common membrane electrolytes used in polymer electrolyte and direct methanol fuel cells (PEFCs and DMFCs) owing to their high proton conductivity and good chemical and thermal stability. However, methanol readily migrates from the anode, through the Nafion membrane, to the cathode, considerably reducing the open-circuit potential and poisoning the electrocatalysts at the cathode. Sieving methanol with proton by constructing a thin separation layer on Nafion is an idea strategy to solve this proplem. In this study, we aim to precisely sieving methanol/proton by constructing an ultrathin sulfonated graphene oxide (GO) membrane on Nafion. Through adjusting graphene oxide spacing, an exceptionally high selectivity is obtained. This high selectivity is due to the GO membrane having selective nanochannels which can block the transport of methanol and permit the transport of proton.

425104 Enhancing CO2 Separation Performance of Composite Membrane By Incorporation of Amino Acid-Functionalized Graphene Oxide

Abstract:

Composite membranes composed of polymer phase and filler phase have been extensively studied for energy-efficient gas separation, especially for CO_2 separation, gasoline desulfurization, and alcohol dehydration, *etc*. The multiphase characteristics of composite membranes provide more degree of freedom to manipulate multiple interactions, tailor multiscale structures and integrate multiple functionalities in comparison with unfilled membrane. As the basic unit of protein, amino acid molecules are important in biology for promotion CO_2 transport. The composite membranes are fabricated by incorporating amino acid-functionalized graphene oxide nanosheets into sulfonated poly (ether ether ketone) (SPEEK) polymer matrix. Graphene oxide (GO) nanosheets are functionalized with amino acids through a facile two-step method by using dopamine and cysteine in succession. CO_2 separation performance of the as-prepared membranes is evaluated for CO_2/CH_4 and CO_2/N_2 systems. GO nanosheets increase more tortuous paths for the larger molecules, enhancing diffusivity selectivity. Amino acid with the carboxylic acid groups and primary amine groups simultaneously enhances solubility selectivity and reactivity selectivity. Accordingly, CO_2 molecules can transport quickly with the enhanced selectivity. When the content is 8 wt %, the optimum separation performance is achieved with selectivities of 82 and 115 for CO_2/CH_4 and CO_2/N_2, respectively, and CO_2 permeability of 1247 Barrer, significantly surpassing the Robeson upper bound reported in 2008. Besides, the mechanical and thermal stabilities of the composite membranes are also improved compared with pristine SPEEK membrane.

427136 Effect of Synthesis and Template Removal Condition on the Quality of Mesoporous Silica Membrane

Abstract:

To establish how to prepare high quality mesoporous silica membrane, the effect of synthesis condition and template removal condition on the membrane quality was examined. Ordered mesoporous silica membrane was prepared by spin-coating on porous alumina disc using tetraethoxysilane (TEOS) as a silica source and cetyltrimethylammonium bromide (CTAB) as a template surfactant. Since the substrate had relatively large pores, the pores were filled with paraffin to prevent a mesoporous silica preparation solution from immersing. After forming mesoporous silica membrane on the alumina substrate, the paraffin was removed by toluene. It is also known that mesoporous silica membrane experiences the shrinkage of silica framework, which causes defect, during removing process of template surfactant in the case of using calcinations for removal. To reduce the shrinkage, template removal by solvent extraction method was tried instead of calcination. Shrinkage of silica framework was greatly reduced by solvent extraction method using formic acid solution, but the extracted membrane showed structure collapse to some extent after removing surfactant. Therefore, pre-treatment of as-synthesized mesoporous silica membrane by tetramethoxysilane (TMOS) vapor was performed to increase silica density before template removal. As a result, the TMOS-treated membrane using solvent extraction showed lowest shrinkage ratio and maintained mesoporous silica structure even after surfactant removal.

Synthesis of ion exchange membranes and their application to MCDI

S. H. Kwon[1], H. S. Shin[2], C. S. Kim[2], J. W. Rhim[1*]

[1]Department of Chemical Engineering, Hannam University, Daejeon, Korea
[2]Techwin Co. Ltd., Cheongju-si, Chungcheongbuk-do, Korea; (jwrhim@hnu.kr)

Introduction

Why desalination?

➢ Why desalination ?

- Drinking water
- Industrial water supply
- Health & environmental problems
- Ultra-pure water
- Purification of products

➢ Type of desalination processes

- Distillation
- Membrane processes
- Reverse osmosis (RO)
- Electro-membrane process
- Ion exchange resins
- Capacitive deionization (CDI)

Carbon materials for CDI applications

➢ Properties of electrode materials

- High specific surface area
- High electronic conductivity
- Fast response of the entire surface area
- Chemical and electrochemical stability
- Easily shaped

» **Carbon materials**

➢ Carbon Materials used

- Activated carbon powder
- Activated carbon cloth
- Carbon felt
- Carbon nanotube
- Carbon aerogels

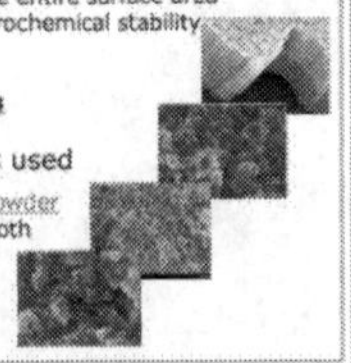

Membrane capacitive deionization (MCDI)

➢ Structure of MCDI cell

➢ Composite carbon electrode coated with ion-exchange polymer

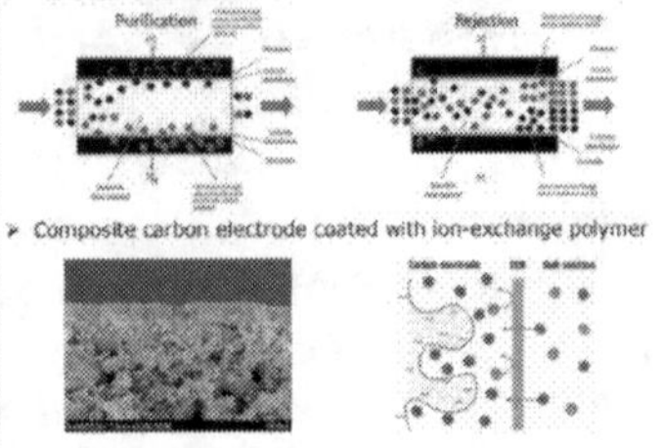

MCDI unit cell and setup for experiment

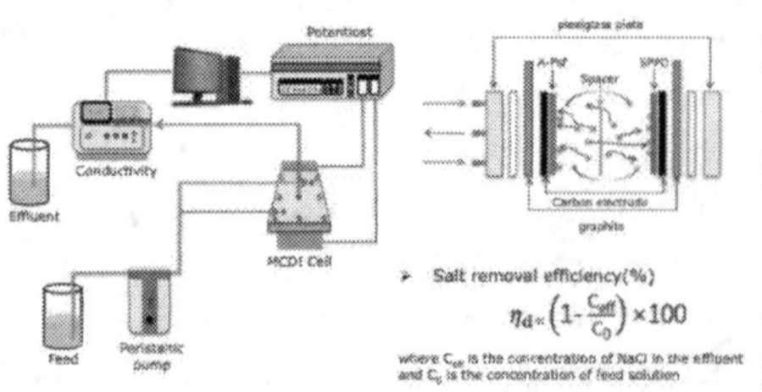

➢ Salt removal efficiency(%)

$$\eta_d = \left(1 - \frac{C_{eff}}{C_0}\right) \times 100$$

where C_{eff} is the concentration of NaCl in the effluent and C_0 is the concentration of feed solution

Preparation of carbon electrodes loaded with ion exchange polymers

➢ Synthesis of anion exchange polymer

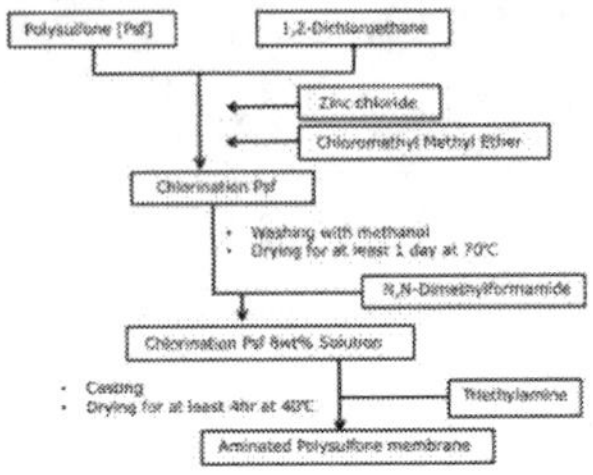

➢ Synthesis of cation exchange polymer

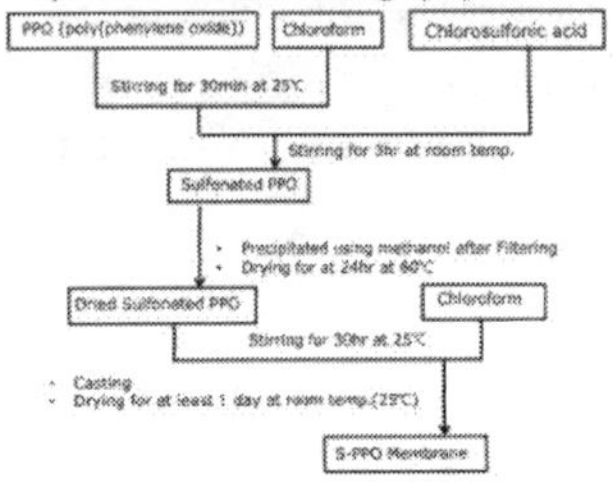

➢ Ion exchange capacity

anion

cation

Physicochemical characteristics of A-PSF membranes

- FT-IR

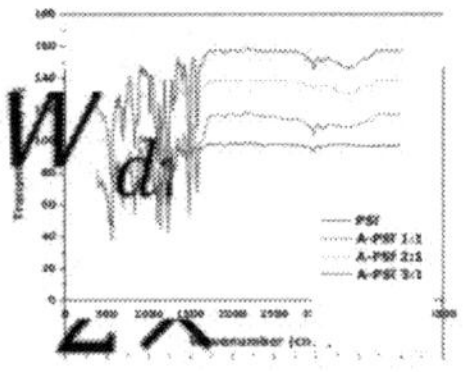

- Water content

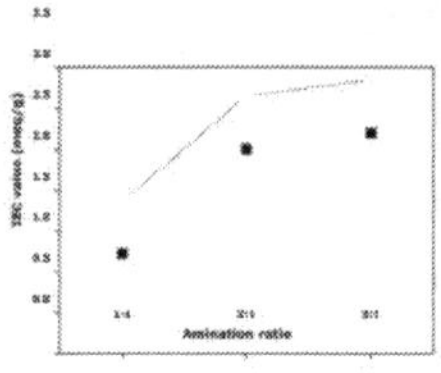

- Ion exchange capacity

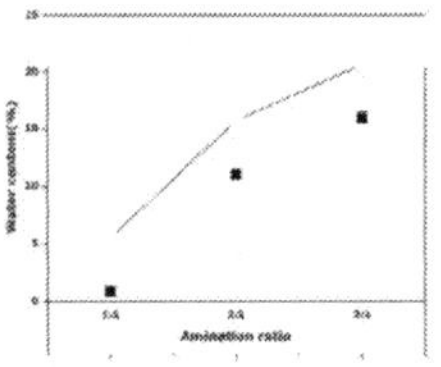

- Proton conductivity

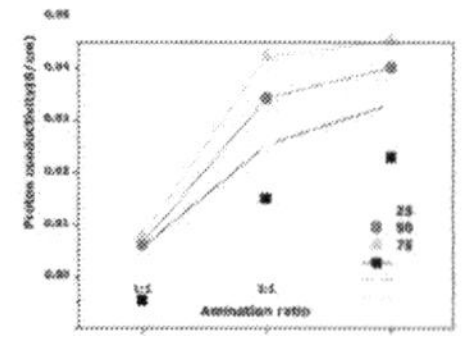

Morphologies of the electrodes

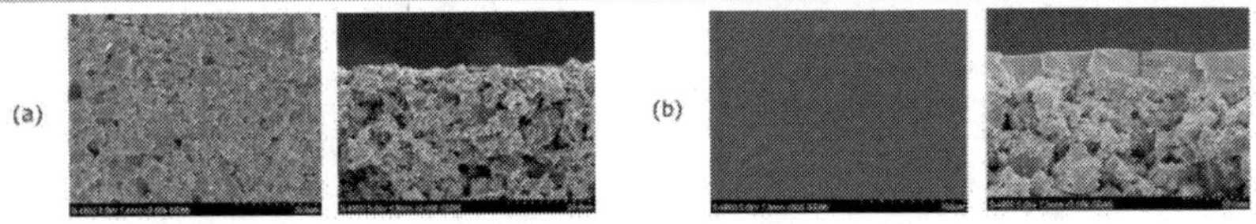

SEM mages of surface (top) and cross-section (bottom) of (a) carbon electrode (b) coated electrode and

Results

➢ Operation conditions

- MCDI cell: CE coated with ion-exchange polymer / Cathode: A-PSf (2:1) = 2.5 (meq/g) / Anode: SPPO = 2.5 (meq/g)
- Adsorption voltage : 1V – desorption voltage : 0V (except types of feed concentration)

- Types of adsorption time

Feed solution: 100 mg/L NaCl
Flow rate : 25 (ml/min)
adsorption time : 120, 180, 300s
desorption time 60s

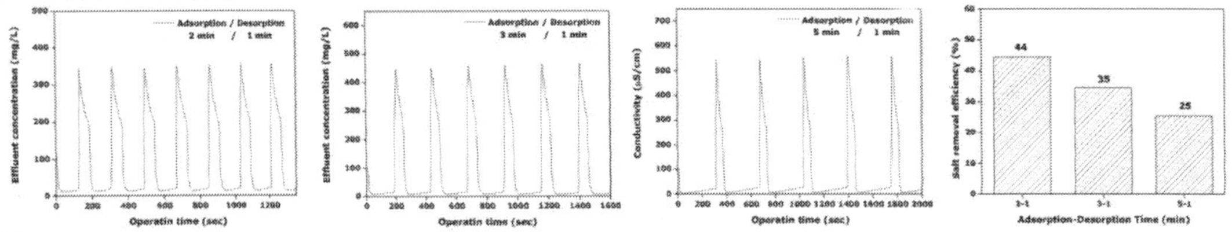

- Types of feed concentration

Feed solution: 100, 200, 300, 500 mg/L NaCl / Flow rate : 15 (ml/min)
Adsorption time : 180s - desorption time for 60s (* desorption voltage : -0.3V)

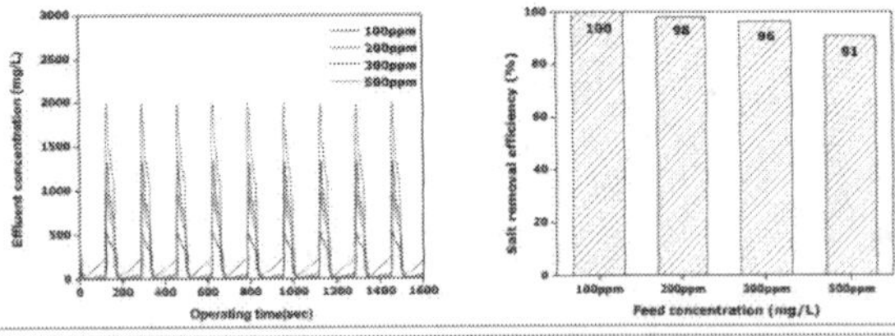

- Types of flow rate

Feed solution: 100 mg/L NaCl / Flow rate : 10, 25, 30 (ml/min) / adsorption 300s – desorption 60s

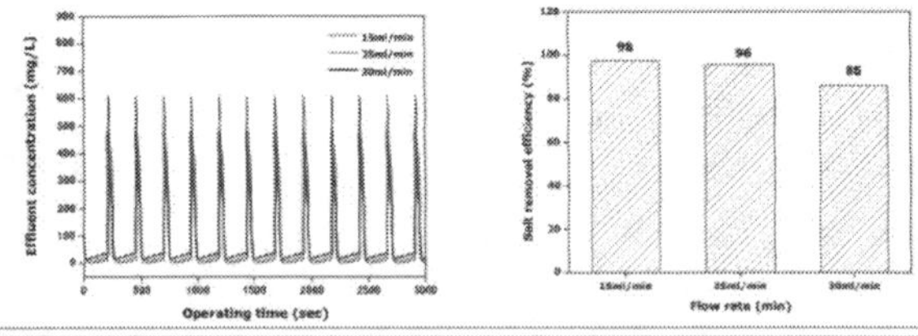

Conclusions

- The aminated polysulfone and sulfonated poly(phenylene oxide) were synthesized as the anionic exchange polymer and the cationic exchange polymer respectively, and these polymers were coated onto the carbon electrodes.
- In case the IEC of S-PPO is similar to that of 2:1 A-PSf, the salt removal ratio is higher than in all other cases where the IEC is different from that of S-PPO. It appears that if there are any differences between the IEC of the cation and anion exchange polymers, comparable results cannot be obtained.
- In case the feed flow rate is lower, i.e., 15ml/min rather than 25 or 30ml/min, the salt removal efficiency is higher because the residence time in the cell is longer.
- As the feed concentration is increased from 100 ppm NaCl to 200, 300, and 500 ppm, the salt removal efficiency decreases because of existing ions in the feed solution.
- This electrostatic interaction became weak as a result of the low voltage applied to the electrodes; hence, salt removal was low.

427186 Novel Solvent Recovery Method for Solvent Deasphalting Process

Abstract:

Recently, demand for crude oils has been increased but their reserves have been diminishing, leading to more attention toward unconventional oils. Unconventional oils account for 70% of total oil reserves but they cannot be easily utilized because of highly contained heteroatoms such as metal (V, Ni), nitrogen and sulfur, and their high viscosity causes transportation problem. Therefore, heavy oil upgrading processes are required for unconventional oil to be converted into light oils, and these processes can be categorized as carbon rejection and hydrogen addition. Solvent deasphalting (SDA) process is a carbon rejection process for heavy oil upgrading in which solvents (C_3-C_7) extract asphaltene-removed light components that are called deasphalted oil (DAO) and alkane-insoluble component, asphaltene, is enriched in the bottom of extractor. In SDA process, required amount of solvent is normally four to ten times higher than that of heavy oil feed, which makes solvent recovery process important to save operating cost. Conventional solvent recovery process is operated in supercritical conditions of solvent, obtaining phase separation between DAO and solvent, and high temperature and pressure are required, resulting in high energy consumption. In this study, novel solvent recovery process is proposed, which can separate solvent selectively from DAO using supercritical CO_2 (Sc-CO_2). The results show that at moderate pressure (<100 bar) and temperature (<60 °C), solvent recovery reaches over 80%, indicating the possibility of alternative solvent recovery process.

427262 Enhanced Water Retention and Low-Humidity Proton Conductivity of Sulfonated Poly(ether ether ketone) Hybrid Membrane By Incorporating Ellipsoidal Microcapsules

Abstract:

Two kinds of polymeric ellipsoidal microcapsules (EMCs) with different functional groups (phosphoric acid groups and imidazole groups) were synthesized *via* precipitation polymerization and incorporated into sulfonated poly(ether ether ketone) (SPEEK) matrix to prepare hybrid membranes. The structure, thermal stability and composition of EMCs were characterized by transmission electron microscopy (TEM), Fourier transform infrared spectroscopy (FTIR), and thermogravimetric analysis (TGA). The quasi-one dimensional structure endowed EMCs with high water uptake and water retention. Especially, for the hybrid membranes filled with 5 wt.% phosphoric acid-functionalized polymeric EMCs, water uptake was increased from 19.0% for pristine SPEEK membranes to 58.2%, and the water retention was 15.1% after 180 min testing at 40 ^{o}C and 20% RH, which is threefold and ninefold higher than those of pristine SPEEK membranes, respectively. Particularly, the proton conductivity at low RH was still up to 3.51×10^{-3}S/cm after 60 min testing. The results manifested that the ellipsoidal microcapsules, as a new kind of filler, had great potential in enhancing the water retention and low-humidity proton conductivity of proton exchange membranes.

427275 Graphitic Carbon Nitride Nanosheets/Speek Nanocomposite Membrane for Direct Methanol Fuel Cells

Abstract:

As a novel 2D nanomaterial, graphitic carbon nitride (g-C_3N_4) has been actively explored in photocatalysis and heterogeneous chemical catalysis, however, its application in proton exchange membranes (PEMs) remained unexplored. Different from other 2D nanomaterials, g-C_3N_4possesses periodic vacancies in the lattice which could allow hydrated proton to selectively pass through. This molecular sieving effect could lead to a decrease in methanol crossover and more direct proton transfer pathways.

In this study, g-C_3N_4 nanosheets were incorporated into sulfonated poly(ether ether ketone) (SPEEK) for the first time to prepare nanocomposite membranes. The resultant nanocomposite membrane showed increased proton conductivity and decreased methanol crossover, which indicated that g-C_3N_4 nanosheets could serve as efficient methanol barriers; meanwhile, incorporation of g-C_3N_4 nanosheets in PEMs led to better connectivity of the ionic groups, triggered the acid-base pair effect and provided more continuous pathways for protons. In addition, the ultimate tensile strength of the control SPEEK membrane was increased by 68 %, arisen from the superior intrinsic mechanical stability of the g-C_3N_4 nanosheets, as well as the interfacial interactions between the g-C_3N_4 nanosheets and the SPEEK matrix. This study suggested the application potential of g-C_3N_4nanosheets as novel and versatile fillers to prepare high performance nanocomposite membranes for direct methanol fuel cells.

427452 Antibody Purification from Cell Culture Supernatant Using Overload and Elute Chromatography with New Polymeric Multimodal Membranes As Stationary Phase

Abstract:

This presentation will describe our latest research efforts on devising improved methods to capture and purify antibodies from cell culture supernatant using overload and elute chromatography with newly designed multimodal membranes (MMMs). The membranes incorporate ligands that can bind proteins through a combination of Coulombic interactions, hydrophobic interactions, and/or hydrogen bonding, thus widening the operation ranges for separation and improving selectivity.

Protein binding experiments with IgG show the MMM has a high equilibrium binding capacity (180 mg IgG/ml) and excellent salt tolerance. The multimodal membranes maintain binding capacities in excess of 100 mg IgG/ml at salt concentrations that are typical for elution buffers used in multi-stage chromatographic bioseparation processes. Protein dynamic binding capacities in excess of 50 mg IgG/ml were determined from breakthrough curve analysis.

The MMMs were used as the stationary phase following a Protein A purification step in the purification human IgG1 from Chinese hamster ovary (CHO) cell culture supernatant. The MMM stationary phase greatly enhanced load productivity compared to resin-based ion-exchange media. Further, no buffer exchange step was required between the Protein A and MMM chromatography purification steps.

In a second study, the MMMs were used as the stationary phase for the primary capture step, in place of Protein A resin chromatography. Using an overload and elute chromatography process, the MMMs demonstrated a high selectivity for capturing human IgG from CHO cell culture supernatant. A high quality protein product was obtained from the MMM chromatography purification step.

This work shows that the new MMM chromatography operation can effectively decrease the number of purification steps and shorten the purification time and/or increase load productivity in downstream processing of biologics.

COMPUTATIONAL FLUID DYNAMICS SIMULATION OF BAFFLE EFFECTS IN NON-CIRCULAR-CHANNELED MICROFILTRATION MEMBRANES

Sophon Leechasan and Pimporn Ponpesh[*]
Chemical Engineering Research Unit for Value Adding of Bioresources and Computational Process Engineering Research Unit, Chulalongkorn University, Bangkok, Thailand
**pimporn.p@chula.ac.th*

Abstract

Non-circular-channeled microfiltration membranes have been successfully applied commercially. Studies show that irregular-shaped channels, such as star-shaped channels, can not only improve the critical flux by reducing membrane fouling, but also require less energy to operate. However, it was found that stagnant areas could occur in the non-circular channels and a careful design of baffle insertion might be needed. This study investigated flow behavior and membrane fouling as affected by non-circular-shaped channels and baffle integration in microfiltration membrane in detail using computational fluid dynamic (CFD) modeling. The CFD model was validated with the published empirical data. The results will be used to guide the design and operation of baffles and membrane modules to achieve optimal permeation flux and reduce energy consumption.
Keywords: Separations

Introduction

Membrane microfiltration has been widely used in pharmaceutical, water purification, mineral separation, food and beverage processing. However, fouling has been a persisting problem for membrane separation. The substances can be collected on the membrane surface and over a period of time cause fouling which obstructs the flux and degrades membrane's performance. It also increases energy consumption for the operation. The impact of hydrodynamics induced by turbulent flow has been applied to reduce membrane fouling [1]. This includes microfiltration with special geometries such as star-shaped microfilter. It was shown to be able to improve the flux while requiring less pump input power [2, 3]. However, Chiu et al. (2005) claimed that stagnant zones might be presented near the tips of the star channels, causing the accumulation of the substances. Later, Chiu and James (2006) found that baffle insertion in star-shaped microfiltration membrane could improve the flux and reduce energy consumption. However, details of the transport phenomena in the star-shaped microfiltration membranes are still limited.

Computational Fluid Dynamic (CFD) modeling has been applied in this study to solve the transport phenomena occurred in a star-shaped microfiltration membrane. CFD modeling can provide the quantitative and qualitative details of the systems with a short turnaround time and less or no experimental resources required. It has been successfully applied to investigate flow behaviors in membrane filtration systems. Xin et al. (2011) applied CFD modeling to investigate the effect of square bar and cylinder turbulence promoters on flow pattern in a regular ultrafiltration membrane channel. They found that the geometries, which enhance flow instability, yield greater permeation velocity and improve mass transfer. Liu et al. (2012) applied RNG k-epsilon model to simulate the effect of helical screw insert in a tubular microfiltration membrane. They found that a helical screw insert, which was thought to generate the secondary flow and enhance permeate flux, in fact suppressed the secondary flow. However, the permeate flux was still improved compared to the case in an empty tube due to surface fluctuation and wall shear stress.

Various geometries and arrangements of the filtration channel and baffle insertion can create complex hydrodynamics in the membrane modules. Therefore, thorough investigation using CFD modeling is helpful in providing detailed information and better understanding of the phenomena. Although CFD simulation of membrane filtration processes is widely recognized, there is still lack of elucidation by CFD models in flow behaviors and particle deposition in star-shaped microfiltration membranes. This study aimed to provide novel and comprehensive information about membrane filtration and fouling in a microfiltration membrane module with star-shaped channels. The results will be used to guide the design and operation of membrane microfiltration to achieve optimal permeate flux with high energy efficiency.

Materials and methods

1. System description and computational domain

The computational domain of the star-shaped microfiltration membrane in this study was based on the system applied by Chiu and James (2006). It included a microfiltration membrane module which was consisted of seven star-shaped channels. The schematic diagram and dimension of the system are shown in Figure 1. Membrane resistant and pore size were $2x10^{-11}$ m^{-1} and $2x10^{-7}$ m in diameter, respectively. The system was operated with 3.3 m/s inlet flow velocity at 25 °C.

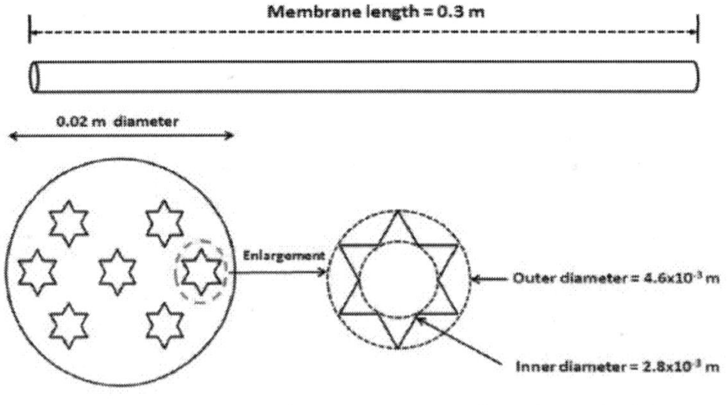

Figure 1. Computational geometry of membrane microfilter with star-shaped channels [1].

2. Governing equations

Due to the effect of swirling flow induced by baffle insertion in the membrane channel, the RNG k-ε model was applied [7]. The model is composed of the following transport equations

$$\frac{\partial}{\partial t}(\rho k)+\frac{\partial}{\partial x_i}(\rho k u_i)=\frac{\partial}{\partial x_j}\left(\alpha_k \mu_{eff}\frac{\partial k}{\partial x_j}\right)+G_k+G_b-\rho\varepsilon \tag{1}$$

and

$$\frac{\partial(\rho\varepsilon)}{\partial t}+\frac{\partial(\rho\varepsilon u_i)}{\partial x_i}=\frac{\partial}{\partial x_j}\left(\alpha_\varepsilon \mu_{eff}\frac{\partial \varepsilon}{\partial x_j}\right)+C_{1\varepsilon}\frac{\varepsilon}{k}(G_k+C_{3\varepsilon}G_b)-C_{2\varepsilon}\rho\frac{\varepsilon^2}{k}-R_\varepsilon \tag{2}$$

In addition, the continuity equation for incompressible fluid was also applied. It can be expressed as follows.

$$\frac{\partial\rho}{\partial t}+\nabla\cdot(\rho\mathbf{u})=0 \tag{3}$$

The velocity vector ($\mathbf{u}$) of fluid is defined by $\mathbf{u}=u\mathbf{i}+v\mathbf{j}+w\mathbf{k}$. Additional source term (eq. 4) for flow in homogeneous porous media was also applied in the RNG k-ε model.

$$S_i=-\left(\frac{\mu}{\alpha}u_i+C_2\frac{1}{2}\rho|u|u_i\right) \tag{4}$$

Furthermore, discrete phase model with Saffman's lift force (eq.5) was used to simulate particle trajectory and predict surface deposition.

$$\frac{\partial u_p}{\partial t}=F_D(u-u_p)+\frac{g_i(\rho_p-\rho)}{\rho_p}+\frac{2\kappa\upsilon^{\frac{1}{2}}\rho d_{ij}}{\rho_p d_p(d_{lk}d_{kl})^{\frac{1}{4}}}(u-u_p) \tag{5}$$

Results and discussion

To validate the CFD model, the simulated result of permeate flux was compared with the experimental result by Gésan-Guiziou et al. (2002). It is shown in Figure 2 that the simulation agrees reasonable well with the experiment with maximum difference of about 15 percent. The deviation might be due to the estimation of some unknown parameters. Thus, further analysis is required. However, the validation result show that the CFD model can be applied to predict the real system under acceptable accuracy. In addition, it can be estimated from the results that the critical flux, when membrane fouling becomes significant, is at about $1.2x10^{-5}$ – $1.4x10^{-5}$ m/s. The CFD model was also validated by comparing the simulated results with the experimental data of transmembrane pressure and permeate flux from Chiu and James (2006). The study investigated the effect of rod baffle insertion in a star-shaped microfiltration membrane. It was found that the predicted result from the CFD model was in good agreement with the experimental result [8].

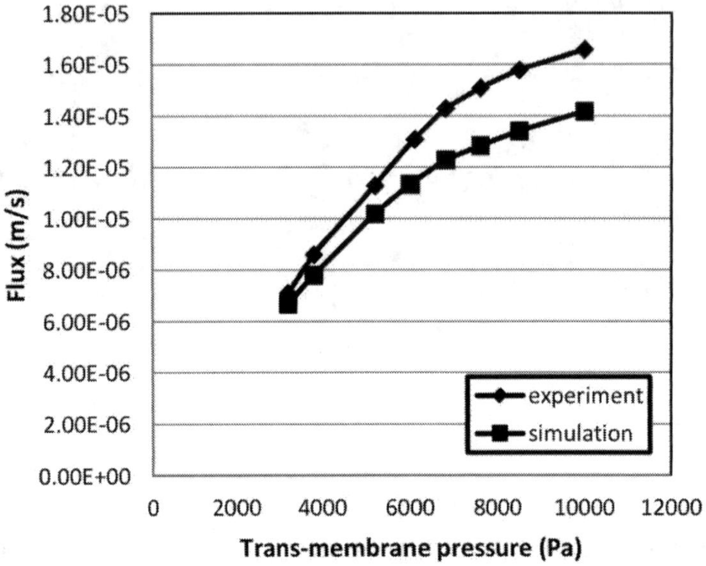

Figure 2. Comparison of permeate flux from the simulation and the experimental measurement [9].

As a result, the CFD model was applied to further investigate the system and to acquire detailed information of the flow profile and particle motion in the star-shaped microfiltration membrane with various baffle insertion. It was found that membrane fouling is not uniformly occurred along the microfiltration channels and thus the installation of baffle throughout membrane modules is not necessary. Furthermore, the insertion of a geometry, which causes an abrupt change in fluid velocity, can enhance wall shear stress and particle lift force which help reduce membrane fouling and improve the critical flux. The publication of these results is under progressed.

Conclusions

CFD simulation was applied to investigate the transport phenomena occurred in a microfiltration membrane with star-shaped channels. It was found that RNG k-ε model and particle discrete phase model could predict the permeate flux and the effect of membrane fouling relatively well. Further investigation of the system using the validated CFD model has provided a guideline for membrane module design and operation for optimal permeate flux and efficient energy consumption.

Nomenclature

α	=	Membrane permeability
α_k, α_ε	=	Inverse effective Prandtl number for k and ε respectively
ε	=	Dissipation rate of turbulent kinetic energy
κ	=	Saffman's lift force constant
μ_{eff}	=	Effective dynamic viscosity coefficient
ν	=	Kinematic viscosity
ρ, ρ_p	=	Fluid and particle density respectively
σ_k, σ_ε	=	Turbulent Prandtl number of k and ε respectively
$C_{1\varepsilon}$, $C_{2\varepsilon}$, $C_{3\varepsilon}$	=	Constants
C_2	=	Inertial resistance
d_{ij}	=	Deformation tensor
F_D	=	Drag force
G_b	=	Generation of turbulent kinetic energy due to buoyancy
G_k	=	Generation of turbulent kinetic energy due to the mean velocity gradient
k	=	Turbulent kinetic energy
R_ε	=	Additional term of ε equation
u_i, u_p	=	Fluid and particle velocity respectively

Acknowledgements

This research has been supported by New Researcher Scholarship of CSTS, MOST-STDA 2014 (SCH-NR2014-097), Ratchadaphiseksomphot Endowment Fund 2014, and Computational Process Engineering Research Unit of Chulalongkorn University.

References

1. Chiu, T.Y., and A.E. James (2006), "Effects of axial baffles in non-circular multi channel ceramic membranes using organic feed," *Sep. Purif. Technol.*, 51, pp. 233-239.
2. Mantec Technical Ceramics Ltd. (2013), "Case study: Star-Sep™ ceramic membranes in the pharmaceuticals industry," available at www.mantectechnicalceramics.com, accessed on 24.03.2014.
3. Chiu, T.Y. and A.E. James (2005), "Critical flux determination of non-circular multi-channel ceramic membranes using TiO_2 suspensions," *J. Membr. Sci.*, 254(1-2), pp. 295-301.

4. Chiu, T.Y., M.V. Lara Dominguez, and A.E. James (2005), "Critical flux and rejection behavior of non-circular-channelled membranes influence of some operating conditions," *Sep. Purif. Technol.*, 50, pp. 212-219.
5. Xin L., L. Xinping, P. Jiong, and S. Yi (2011), "Comparison of turbulence promoter geometry on flow pattern from view point of field synergy principle," *Procedia Environ. Sci.*, 11, pp. 1566-1573.
6. Liu, Y., G. He, L. Ding, H. Dou, J. Ju, and B. Li (2012), " Experiment and CFD studies on the performance of microfiltration enhanced by a turbulence promoter," *Chin. J. Chem. Eng.*, 20(4), pp. 617-624.
7. ANSYS Inc, (2013), ANSYS Fluent 14.5 Theory guide.
8. Leechasan, S. and P. Ponpesh (2014), "Computacional fluid dynamics modeling of effect of baffles in star shaped microfiltration membranes for water treatment," PSU-USM-NSTRU international conference on arts and sciences 2014, June 2-3, 2014, Songkla, Thailand.
9. Gésan-Guiziou, G., R.J. Wakeman, and G. Daufin (2002), "Stability of latex crossflow filtration: cake properties and critical conditions of deposition," *Chem. Eng. J.*, 85, pp. 27-34.

428947 Predicting Ion Transport in Ion-Exchange Membranes

Abstract:

Fundamental understanding of ion sorption and transport in ion-exchange polymers is critical for development of high performance membranes. Despite the long history of literature on the subject, this topic is currently not well understood, making it difficult to build the appropriate connections between polymer structure and transport properties. Experimental techniques for characterizing ion sorption and transport in charged membranes have been established, however, a simple theoretical framework to describe the experimental findings is currently lacking. In this work, a model to predict ion sorption and permeability in ion-exchange membranes has been formulated and tested against experimental data. The model requires basic membrane information (chemical structure, fixed charge density, and water content). For the membranes used in this study, the model was completely predictive and no adjustable parameters were required.

Quantitative description of ion sorption in charged membranes equilibrated in salt solutions typically relies on a rigorous thermodynamic formalism (i.e., Donnan-type analysis). Previous studies have often overlooked the importance of including ion activity coefficients in such treatments, mainly because no reliable models or experimental techniques were available for obtaining these values in the polymer phase. In this work, Manning's counter-ion condensation theory, originally developed for polyelectrolytes, was extended to ion-exchange membranes and used to estimate membrane ion activity coefficients. Solution activity coefficients were obtained by the Pitzer model. The results of the model were compared to ion sorption data obtained experimentally for various commercially available ion-exchange membranes. Good agreement was found between the predicted values for membrane co-ion concentration and those obtained experimentally when solution and membrane non-idealities were included in the model. The generality and shortcomings of the model, as well as the applicability of polyelectrolyte theories to charged membrane systems, are discussed.

The model was extended to predict salt permeability coefficients in charged membranes within the framework of the solution-diffusion mechanism. According to this model, the permeability of a penetrant across dense polymers is comprised of the product between partition and diffusion coefficients. Ion diffusion coefficients in the membrane were estimated using the Mackie and Meares model. This model has no adjustable parameters and requires knowledge of polymer volume fraction and solution diffusion coefficient of the penetrant. Ion sorption coefficients were estimated by the model presented in this work. Thus, predicted values for permeability coefficients were obtained, and these values were compared to those determined experimentally. Good agreement between model and experiment was found without use of adjustable parameters. General applicability and limitation of this treatment are discussed.

429266 Continuous Roll-to-Roll Fabrication of Amine-Containing Membranes for CO2 Separation and Capture

Abstract:

Continuous Roll-to-Roll Fabrication of Amine-Containing Membranes for CO_2 Separation and Capture

Witopo Salim[1], Varun Vakharia[1], Dongzhu Wu[1], Yuanxin Chen[1], Lin Zhao[1], Yang Han[1], and W.S. Winston Ho[1,2]

[1]William G. Lowrie Department of Chemical and Biomolecular Engineering, The Ohio State University, 151 West Woodruff Avenue, Columbus, OH 43210-1180, USA

[2]Department of Materials Science and Engineering, The Ohio State University, 2041 College Road, Columbus, OH 43210-1178, USA

Abstract

Amine-containing membranes for CO_2 capture and separation were scaled up to fabricate 14-inch wide membranes by using a continuous roll-to-roll machine equipped with a thin-film-coating assembly. Aided by a material balance equation, three variables, including the coating knife gap setting, substrate rolling speed, and coating solution concentration, were identified as the critical factors to control the membrane selective layer thickness. For the CO_2 separation from a synthesis gas stream containing CO_2 and H_2S, i.e., pre-combustion CO_2 capture in IGCC (integrated gasification combined cycle) technology or hydrogen purification for fuel cells, the thicker amine-containing membranes with around 15 ± 2 microns of selective layer thicknesses were fabricated by using a gap setting of 17 mils, a rolling speed of 0.5 ft/min (set by the drying capacity of the drier, i.e., its length), and an amine coating solution of 18 wt%. For the post-combustion CO_2 capture from flue gas, the same thin-film-coating assembly machine was used to fabricate thinner amine containing membranes with $< 250 \pm 20$ nm of selective layer thicknesses by using various rolling speeds and amine-containing solution concentrations at a fixed gap setting of 0.45 mils. The fabrication of thinner membranes was improved further by using an insert in the coating solution trough to control the coating solution delivery rate and a smooth coating knife base to reduce the thickness variation. From the gas transport measurements, the flat-sheet samples of the scale-up membranes for both applications exhibited similar performances compared to the membranes synthesized in lab scale. CO_2 permeance of approximately 200 GPU and CO_2/H_2 selectivity > 180 for the thicker membranes for the purification of hydrogen and synthesis gas along with CO_2 capture (e.g., pre-combustion capture), and CO_2 permeance of about 800 GPU and CO_2/N_2 selectivity > 140 for the thinner membranes for CO_2 capture from flue gas. Techno-economic analysis of the post-combustion CO_2 capture membrane process using the thin membrane resulted in about $41/tonne CO_2 captured (based on 2007 dollar). The scale-up membranes can be used to prepare pilot-scale membrane modules, either as spiral-wound or plate-and-frame configuration, for testing with the acid gas mixtures or flue gas.

432775 Gas-Diffusion Cathode Expedites Ammonia Removal from Aqueous Phase

Abstract:

Ammonia is a pollutant in many types of aqueous streams such as industrial wastewater from coking, chemical fertilizer, coal gasification, petroleum refining, pharmaceutical and catalyst factory. It is also a byproduct in processes such as microbial cultivation, fish aquaculture, and livestock/poultry feed digestion. Ammonia is toxic to fishes and microorganisms in bioreactors, and has a tendency of causing eutrophication and depletion of dissolved oxygen due to nitrification. It is critical to protect water-borne species and aquatic environments by removing ammonia from entering aquatic environments. The reactors designed and manufactured in this study aimed to expedite the efficiency of ammonia removal from aqueous phase by utilizing the principle of membrane contactor and modifying the cathode microcosm environment via electrochemical or microbial electrochemical reduction of oxygen. Our experiment evaluated cathode materials and structures that favored the catalytic reduction of oxygen as well as the improved ammonia volatilization via hydrophobic gas-diffusion layers. The volatized ammonia gas was eventually collected and concentrated in sulfuric acid solution.

Several designs of cathode materials and structures were tested. The cathode reaction, oxygen reduction reaction (ORR), was tested under three types of catalysts, i.e., Pt/carbon black, activated carbon, and biofilm. The gas diffusion layers were prepared in laboratory from different materials, i.e., poly (tetrafluoroethylene) (PTFE), poly (dimethylsiloxane)(PDMS), poly (vinylidene fluoride)(PVDF) by natural drying or phase inversion. The structures of cathode were optimized by studying the impact on ammonia removal of omitting the macro-porous substratum, the micro-porous layers, the position of chemical catalyst layer, and the application of conductive polymer layer (polyaniline). Different cathode materials and structures were evaluated by linear scanning voltammetry (LSV) for the catalytic activity of ORR, under both dissolved oxygen saturated and depleted conditions. Ammonia removal was further studied in the cathodes that showed high ORR catalytic effects, at initial ammonia concentrations of from 200 to 1000 mg/L with an increment of 200 mg/L, pH of 6, 6.5, 7, 7.5 and 8, and under different cathode potentials of -200 mV, -100 mV, 0 mV, +100 mV, and +200 mV vs. Ag/AgCl (3 M NaCl). Further experiments were run by sustainably providing energy that pushed ORR through the operation of microbial fuel cell mode and cultivation of biofilm cathode (biocathode). This design effectively removed ammonium/ammonia nitrogen from contaminated aqueous phase with a benefit of pure ammonium enrichment, and will have potential applications in real wastewater operated in an cost-effective and energy-sustainable way.

431844 Nanostructured Membranes for CO2 Separation and Other Applications

Abstract:

Nanostructured Membranes for CO2 Separation and Other Applications

Witopo Salim[1] and W.S. Winston Ho[1,2]

[1]William G. Lowrie Department of Chemical and Biomolecular Engineering, The Ohio State University, 151 West Woodruff Avenue, Columbus, OH 43210-1350, USA

[2]Department of Materials Science and Engineering, The Ohio State University, 2041 College Road, Columbus, OH 43210-1178, USA

Abstract

Advances in nanotechnology in recent years enhance the R&D in nanostructured polymer-based membranes. This work reports on the recent developments of the nanostructured polymer-based membranes, formed by incorporating nanomaterials and/or porous membranes with nanoscale pores, for gas separations, reverse osmosis, ultrafiltration, and other potential applications. Significant progress has been made on the development of nanostructured membranes for gas separations, particularly for CO_2 separations from H_2, N_2 and CH_4. A polymer/zeolite composite nanostructured membrane is presented as an example of the potential of nanostructured polymer-based membrane for the gas separation application of CO_2 capture from flue gas in power plants. Continued improvements on the nanostructured reverse osmosis polyamide membranes via interfacial polymerization have been made along with the efforts on developing novel nanostructured membranes for desalination. The modification of both the surface and bulk properties of the ultrafiltration membranes by nanomaterials can improve the membrane performance and open up the possibility of new applications in addition to those for the current ultrafiltration membrane processes. Amine-containing membranes composed of a selective layer on nanoporous polysulfone support are presented as an example of the application of ultrafiltration membrane for the CO_2 separation from a synthesis gas stream containing CO_2 and H_2S, i.e., for hydrogen purification for fuel cells or for pre-combustion CO_2 capture in IGCC (integrated gasification combined cycle) technology. Similarly, amine-containing membranes composed of a selective layer on nanoporous polyethersulfone support are another example of the application of ultrafiltration membrane for CO_2 capture from flue gas.

433649 Crystallization Process Development of a Chemically Modified Small Protein

Abstract:

Crystallization of proteins has been proven as a processing step for protein purification in addition to the conventional application of Protein structure analysis. Crystallization increases the advantage provided to the formulation for solid crystalline protein compared to protein in solution. A crystalline drug substance for a chemically modified small protein process was developed to encompass the two advantages of crystallization - stability and purity. The challenges encountered during the development were due to the high solubility of the molecule, shear sensitivity, and lack of a representative scale down model. Through cross functional collaboration, an adequate scale down model was developed that led to crystallization process conditions and investigation into multiple factors found to affect crystal morphology, kinetics, and yield. Use of PAT tools for reaction monitoring such as FBRM and PVM aided in process understanding and in setting of scale-up parameters. In conclusion, a crystallization process has the potential to be included in downstream processing as a low cost and high efficiency step.

433692 Use of Forward Osmosis- Membrane Distillation Process in Treatment of Hyper-Saline Produced Water

Abstract:

Oilfield produced water (PW) can be considered as a potential water resource if treated efficiently and economically. However, its high salinity and complex chemistry can limit treatment options. In this work, the feasibility of using a hybrid forward osmosis (FO) - membrane distillation (MD) process for treating of hyper saline produced water with total dissolved salts (TDS) of about 240,000 ppm was studied. MD is applied as a separation method for the draw solution of the forward osmosis process. A TFC membrane from Hydration Technologies Innovation (HTI) was used in the FO experiment and a polypropylene (PP) membrane from 3M with pore size of 0.45 micron was used in the MD test. Three draw solutions (i.e. NaCl, KCl and $MgCl_2$) at concentrations near to their saturation limits (5, 4, and 4.8 molar respectively) were considered. Results showed that each draw solution behaved quite differently in both FO and MD due to substantial differences in diffusivity, solubility, activity, and solution viscosity. $MgCl_2$ solutions gave the highest water fluxes in FO, while NaCl and KCl showed much higher fluxes in MD. FO water flux appears to be the limiting factor of design, especially with high salinity feeds, and we found that $MgCl_2$ represent the best draw solution for the treatment of hyper saline PW using FO-MD process. Stable fluxes of about 4 l/m^2.hr in the FO test and 3 kg/m^2.hr in the MD test were achieved over 20 hr. of operation. Adjusting the temperature of the draw solution was found to equalize fluxes. We also found that no oil contaminated the draw solution, assuring that the MD process would be protected from oil fouling.

434556 Comparison of an Extractive Dividing Wall Column Operating at Vacuum and a Conventional Extractive Distillation Sequence

Abstract:

Bioethanol has been considered one of the most promising substitutes for fossil fuels. In order to be used in engines available in the market, bioethanol needs to have a maximum water content of 1.0 vol % according to the international standard ASTM D 4806, USA. Ethanol-water mixtures are highly nonideal with a binary azeotrope at 95.6 wt. %. In order to achieve the separation, different kinds of technologies have been developed, such as pervaporation, adsorption with molecular sieves, azeotropic distillation, pressure swing distillation and extractive distillation, among others. In the specific case of extractive distillation, a solvent is used to modify the phase equilibrium and allows the separation of ethanol at high purity. The inclusion of an additional component results in higher capital and operation costs due to the need of an extra column for the recovery of the solvent.

Some studies have worked on the design, optimization and control of dividing wall columns for the separation of azeotropic mixtures ethanol-water. They have reported some advantages in terms of energy savings and investment costs (Bravo-Bravo et al., 2010) (Kiss & Suszwalak, 2012) (Segovia-Hernandez et al., 2014) (Segovia-Hernandez et al., 2014) (Kiss & Ignat, 2012). Most of them have compared the results of the dividing wall column with a conventional sequence of distillation columns working at the same pressure of 1 atm and with reboiler temperatures around 200°C using ethylene glycol as solvent. According to (Meirelles, Weiss, & Herfurth, 2007), the recovery column in the real distillation sequence should work at lower temperature (less than 150°C) in order to avoid degradation of the solvent. This is achieved by operating the recovery column at lower pressure, which is usually less than the pressure in the extractive column. Similar requirement was reported by (Garcia-Herreros, Gomez, Gil, & Rodriguez, 2011) with the use of glycerol. The difference of pressures between the columns in the conventional extractive distillation sequence is a limiting factor in the use of dividing wall columns as an equivalent process. In dividing wall columns, once the position of the wall is fixed the design should ensure the same pressure in both sides of the wall.

In this work, a comparison of a conventional sequence of two distillation columns working at different pressure with an equivalent separation using a dividing wall column working at one single pressure is made. In the conventional separation, the first column (extractive) works at 1 atm and the second column (recovery) is designed with the aim to ensure a reboiler temperature lower than 150°C. In the case of the dividing wall column, the extractive zone and the recovery zone are in the same shell, so that the pressure at both sides of the wall is the same. Modeling and simulation are done using Aspen Plus®, where NRTL was selected as thermodynamic model and RadFrac as solution module. In order to achieve the minimal energy consumption for the dividing wall column, the optimal operating vacuum pressure is found using as constraint the degradation temperature of the solvent. This energy consumption is compared with the result obtained for the conventional sequence. In addition, two different solvents, ethylene glycol and glycerol, are used for comparison purposes. Results of this work validate the use of dividing wall columns in systems with common operative constraints.

433515 Efficient Separation-Process Synthesis

Abstract:

Separation processes are integral to chemical, petrochemical, pharmaceutical and gas separation industries. These processes usually account for close-to 40-70% of chemical plant costs. Despite the huge costs involved, industrially, separation sequences are built using the experience and intuition of a few engineers. This has created a need for the development of tools that systematically identify attractive candidates for separation, and we address this need in this research work, focusing on distillation. Our research work is divided into two parts. In the first part, we expand and generate the search space of possible distillation configurations by introducing hitherto unknown configurations such as basic configurations, heat and mass integrated configurations, dividing wall columns, etc. In the second part, we globally optimize the entire generated search space of configurations to determine their minimum heat duty, overall costs, exergy loss, etc. This systematic procedure allows an industrial practitioner to screen unwanted configurations, and concentrate only on a handful of useful configurations for further design considerations.

434507 A Hybrid Dead-End Cross-Flow Forward Osmosis System for Evaluation of High Recovery Osmotic Flux Performance with Produced Water

Abstract:

Water management during the extraction of petroleum products from earth is a major issue in oil and gas production industries. Many of these waters are challenging to manage because of oil content, high salinity, fouling and scaling potential, and overall toxicity. Forward osmosis (FO), has recently been considered as a means of treating or concentrating this water because of its low fouling propensity and relative independence from osmotic pressure limitations. In this work, we propose a new method for testing the efficacy of forward osmosis in treating produced waters using a hybrid dead-end cross-flow FO system that enables all of the conventional testing metrics to be measured (water and salt flux), but do so at high recoveries and with small volumes of feed. We demonstrate the usefulness of this system at testing osmotic flux at high recoveries using a real produced water provided by Chevron. The unprocessed produced water contains approximately 7,500 TDS and osmotic recovery of 90% is demonstrated using the hybrid dead-end cross-flow FO system.

435407 Tuning Al Spatial Distribution in ZSM-5 Membranes: A New Strategy to Improve Separation Performance and Stability of Zeolite Membranes for Dehydration of Acetic Acid

Sunday, November 8, 2015
Exhibit Hall 1 (Salt Palace Convention Center)

Liangqing Li[1], Zan Chen[1], Jianhua Yang[2,3], Jinqu Wang[1], Jinming Lu[1], Yan Zhang[1] and Hongcheng Guo[2], (1)Chemical Engineering, Dalian University of Technology, Dalian, China, (2)Dalian University of Technology, Dalian, China, (3)Chemical Engineering of Material Science, University of Minnesota, Minneapolis, MN

Tuning Al spatial distribution in ZSM-5 membranes: a new strategy to improve separation performance and stability of zeolite membranes for dehydration of acetic acid

Liangqing Li [a], Zan Chen[a], Jianhua Yang [a,b]*, Wanze Li [a], Jinqu Wang [a], Jinming Lu [a] ,Yan Zhang[a], Hongchen Guo[a]

a:State Key Laboratory of Fine Chemicals, School of Chemical Engineering, Dalian University of Technology, Dalian, Liaoning,

116024, China.

b: Department of Chemical Engineering and Materials Science, University of Minnesota, MN 55455;

*corresponding authors: Prof. Jianhua Yang, E-mail add.: yjianhua@dlut.edu.cn;

Abstract

Dehydration of acetic acid (AcOH) by the conventional distillation is particularly energy-intensive process. Membrane pervaporation is attracting great interest for this purpose because of the greatly reduced energy requirement. However, the dehydration of AcOH is very challenging because of the demanded high acid resistance and the membrane materials are very limited. So far the water flux, selectivity and acid resistance of the both of polymer and zeolite membranes for the dehydration of AcOH are low.

The fabrication of a high performance zeolite membrane significantly relies on the control of essential membrane microstructures including the membrane thickness[1], preferential grain orientation[2] and grain boundary defects[3]. In the present work, a new strategy through tuning the Al spatial distribution of ZSM-5 membrane was proposed to improve the acid resistance of zeolite membrane. The organic template-free synthesis route was designed to tune the spatial distribution of ZSM-5 membrane. A novel ZSM-5 membrane with a low Si/Al ratio and homogeneous aluminum spatial distribution was achieved from the designed synthesis route and the obtained ZSM-5 membrane high selectivity and flux and improved stability for dehydration of acetic acid in a wide AcOH content range. The work firstly demonstrates that the Al spatial distribution is a key factor for tuning the separation performance of the zeolite membranes[4].

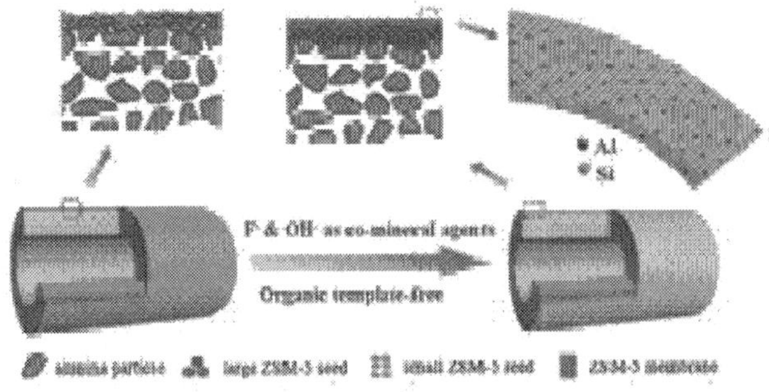

Fig. 1 Schematic illustration of preparation of the ZSM-5 membrane preparation with uniform spatial distribution

Reference:

1. (a) P. Lee, X. Zhang, J. A. Stoeger, A. Malek, W. Fan, S. Kumar, W. C. Yoo, S. A. Hashimi, R. L. Penn, A. Stein and M. Tsapatsis, J. Am. Chem. Soc. 2011; 133: 493; (b) J.Hedlund, F. Jareman, A. Bons and M. Anthonis, *J. Membr. Sci.* 2003; 222:163.

2. Z. Lai, B. Griselda, D. Isabel, J. Nery, K. Sujaoti, M. Amat, E. Kokkoli, O. Terasaki, R. W. Thompson, M. Tsapatsis and D. G. Vlachos, *Science*, 2003; 300:457.

3. J. Choi, H. Jeong, M. A. Snyder, J. A. Stoeger, R. I. Masel and M. Tsapatsis, Science, 2009; 31:590.

4. Jianhua Yang, Liangqing Li, Wanze Li, Jinqu Wang,a Zan Chen, Dehong Yin, Jinming Lu, Yan Zhanga and Hongchen Guo, *ChemCommu.* 2014; 50:14657

437761 Filling of MCM-41 Silica Nanoparticles in Thin Film Nanocomposite Membranes to Purify Brackish Water

Abstract:

Reverse osmosis (RO) is a widely applied process for saline water purification. Thin film composite membranes (TMC) are the most common type of membranes for this purpose. In this work, a group of TFC membranes were prepared by the in situ interfacial polymerization (IP) reaction of m-phenylenediamine (MPD) solution containing triethylamine/ camphorsulfonic acid salt (TEA/CSA salt) and calcium chloride with trimesoyl chloride (TMC) dissolved in isooctane. The prepared membranes were fabricated on polysulfone (PSU) support sheets. The PSU support sheets were prepared by dissolving PSU in N,N-dimethylformamide (DMF) and then casting them on a glass plate. The MCM-41 silica nanoparticles were mixed in both MPD and TMC solutions in an increasing amount (0.01%, 0.015%, 0.02%, 0.025%, 0.03% and 0.04%) during the synthesis of the thin-film nanocomposite. The performance testing demonstrated that the water flux was increased for all membranes with the silica nanoparticles though the salt rejection was decreased slightly. It was noted that the addition of silica nanoparticles to MPD solution leading to slightly better and more stable results than adding them to TMC solution. With the addition of 0.02% silica nanoparticles to MPD solution, the water flux was increased from 44 to 64.1 L/m^2hr while the rejection of NaCl maintained essentially the same at 95%.

435313 Development of Lab Scale Membrane Process for CO2 Capture Using Polymeric Membranes

Abstract:

In order to apply membrane process to the CCS processes, permeation and separation characteristics of each membrane should be evaluated with appropriate working conditions and rigorous regulations. 0.2 Nm^3/hr and 1.0 Nm^3/hr scale test equipments and 0.2 Nm^3/hr multi stage modules were built at KIER combustion-gas processing complex laboratory building. Those test equipments can control temperature, pressure, flow rate, steam amount, gas composition, stage cut, and also real flue gas from 10MW_{th} scale CFB(circulating fluidized bed) boiler can be supplied. In this study, hollow fiber type membranes and their multi stage configurations were developed for establishing standardization of membrane processes for CCS and certificating membrane performance such as permeance, selectivity, chemical-physical resistance and long term operability.

437672 Moving Towards a Sterile Chromatography Platform for Virus Purification

Abstract:

Process development of live virus vaccines presents many unique challenges owing to their size, molecular complexity, and structural heterogeneity. For downstream purification, one of the main challenges is the inability to sterile filter the virus at the end of the process. Because of this limitation, all components of the purification process have to be sterilized and all operations must be performed under aseptic conditions. This need for sterile operation limits the purification techniques that can be used to purify a virus. While many biologics use chromatography as the work horse of the purification process, these chromatography steps are usually not operated aseptically. Membrane chromatography offers one route to sterile operation; however, the number of functional chemistries are limited. Therefore, this work examines the potential for sterile operation of conventional particle-based chromatography for live-virus vaccine purification by comparing batch and column modes as well as the impact of autoclave and gamma sterilization processes on resin functionality.

438058 Development of Improved Polysulfone Based Hemodialysis Membranes by Using Temperature and pH Responsive Pentablock Copolymers

Abstract:

Introduction:
Polymeric membranes intended for hemodialysis applications have been attracting major attention in the medical field and have a huge place in the market. Polymer based hemodialysis membranes are expected to show outstanding blood compatibility, oxidative, thermal and hydrolytic stability, and high permeability for low molecular weight proteins as well as good mechanical and film-forming properties for an efficient operation and the comfort of the patient. However, the inadequate blood compatibility of current membranes entails the injections of anticoagulants during hemodialysis. In addition, the significant amount of reactive oxygen species, generated during hemodialysis, react with the proteins and lipids in blood causing oxidative stress which leads to cardiovascular, cancer, and age-associated neurodegenerative diseases. As a result, antioxidants are administrated to hemodialysis patients orally or intravenously [1, 2]. Therefore, the motivation for the present work is a need for the development of functional hemodialysis membranes possessing improved blood compatibility and suppressed oxidative stress. For this purpose, we propose the use of temperature and pH responsive amphiphilic pentablock copolymer family blended in with polysulfone (PSF) membranes. The temperature responsive Pluronic F127, consisting of polyethylene oxide - polypropylene oxide - polyethlylene oxide (PEO-PPO-PEO) blocks, constitutes the middle block while the pH responsive 2-N,N-(Diisopropylamino)ethyl methacrylate (DiPAEM), tertbutylaminoethylmethacrylate (tBAEM), N,N-(Dimethylamino)ethyl methacrylate (DMAEM), or N,N-(diethylamino)ethyl methacrylate (DEAEM) blocks form the end blocks for each member of pentablock copolymer family. The pentablock copolymers are blended in PSF matrix and alpha-lipoic acid (ALA), an antioxidant used for prevention of various chronic diseases associated with oxidative stress, is electrostatically immobilized on both surface and cross section of the membrane. These pentablock copolymers are being investigated in hemodialysis applications for the first time.
Experimental:
The PDEAEM, PDMAEM, PtBAEM and PDiPAEM pentablock copolymers were synthesized via atom transfer radical polymerization (ATRP) with the Pluronic macrointiator as described in our previous work [3]. PSF membranes were obtained by dry wet phase inversion technique and characterized by scanning electron microscopy (SEM). The antioxidant activities of immobilized ALA are measured using the free-radical scavenging activity of 2,2-diphenyl-1-picrylhydrazyl (DPPH) radicals. The blood compatibilities of the membranes are evaluated in terms of inhibition of reactive oxygen species in blood plasma, amount of adsorbed plasma proteins, platelet adhesion, activation and cytotoxicity on blood cells. The transport and structural properties of the membranes are also characterized.
Results and Discussion:
The NMR results depicted that pentablock copolymers were successfully synthesized. Our results indicated that the blending of pentablock copolymers did not block the pores and did not cause a significant difference in the classical finger-like structure of the PSF membranes which is preferable for maintaining the transport properties of the membranes. The protein adsorption results indicated that as the pentablock copolymer percentage in the casting solution increased the adsorbed protein amount on the membranes decreased. It was anticipated that the Pluronic F127 middle block prevents the protein adsorption through the temperature responsive micellization on the membrane surface and cross section.
Conclusion:
The use of proposed pentablock copolymers along with ALA in PSF hemodialysis membrane applications appears promising

References:
1. Ran, F., et al., Biocompatibility of modified polyethersulfone membranes by blending an amphiphilic triblock co-polymer of poly(vinyl pyrrolidone)–b-poly(methyl methacrylate)–b-poly(vinyl pyrrolidone). Acta Biomaterialia, 2011. 7(9): p. 3370-3381.
2. Mahlicli, F.Y. and S.A. Altinkaya, Immobilization of alpha lipoic acid onto polysulfone membranes to suppress hemodialysis induced oxidative stress. Journal of Membrane Science, 2014. 449(0): p. 27-37.
3. Determan, M.D., et al., Synthesis and characterization of temperature and pH-responsive pentablock copolymers. Polymer, 2005. 46(18): p. 6933-6946.

425103 A Rapid Pressure Swing Adsorption Process for the Improvement of Portable Oxygen Concentrators

Abstract:

With the market for portable oxygen concentrators (POC) expected to rise to over $2 billion dollars by 2021, the demand for lighter, more efficient units is going to increase rapidly. While POC units on the market today have been recently improved with respect to weight and battery life, the technology is still relatively new and there remains significant room for development. The heart of these units is a rapid pressure swing adsorption (PSA) process that separates oxygen and nitrogen in air to deliver a stream of at least 90% oxygen to patients who require this level of oxygen purity to conduct daily activities. The proper design of this PSA process is one of the keys to improving these units in the future. Through this study, a lab-scale rapid PSA process has been developed with a commercial adsorbent that would significantly improve the efficiency of a POC. The process developed, when properly employed in a POC, has the potential to reduce the weight of a POC in half and double its battery life, both critical selling factors to potential consumers. The process also demonstrated a high level of efficiency at a wide range of operating conditions, which will provide flexibility for implementation of the process in a larger scale POC.

426137 Novel Pyridinium-Based Ionic Liquids for Efficient and Reversible Capture of SO2

Abstract:

Novel Pyridinium-based Ionic Liquids for Efficient and Reversible Capture of SO_2

Shaojuan Zeng [a, b], Hongshuai Gao [a], Xiangping Zhang [a,*], Suojiang Zhang [a,*]

[a]*Beijing Key Laboratory of Ionic Liquids Clean Process, State Key Laboratory of Multiphase Complex System, Institute of Process Engineering, Chinese Academy of Sciences, 100190, Beijing, China*

[b]*College of Chemical and Engineering, University of Chinese Academy of Sciences, 100049, Beijing, China*

**Corresponding author: sjzhang@ipe.ac.cn; xpzhang@ipe.ac.cn*

The reduction of SO_2 emission from flue gases has become one of the most urgent environmental issues for a sustainable society. Although some traditional methods for flue gas desulfurization (FGD) have been commercialized, there are still some inherent drawbacks. For example, limestone scrubbing produces a large amount of solid wastes and waste water, which may result in secondary pollution to the environment[1]. Therefore, the development of a proper absorbent for efficient and reversible capture of SO_2 is of critical importance.

Recently, ionic liquids (ILs) have drawn much attention as potential absorbents for SO_2 capture due to their unique properties[2]. To our knowledge, very few studies on pyridinium-based ILs for SO_2 removal were reported, although they have higher thermal stability, lower cost and higher biodegradability than imidazolium-based ILs[3]. In this work, the interaction between various cations, e.g. different alkyl chain length and functionalized substituent groups on the pyridinium ring, and anions of ILs and SO_2 was firstly studied by Quantum Chemical calculation and Molecular Dynamic simulation. Based on the calculation results, a series of pyridinium-based ILs were designed and synthesized for SO_2 capture. The results indicated that the gravimetric capacity of SO_2 decreases with increase of the length of alkyl chain. Comparing with cation of IL, anion plays a dominant role in SO_2 absorption, which is consistent with the calculation resluts. Among the investigted anions, the conventional IL with [SCN] shows the highest SO_2 capacity, and good selectivity for SO_2 to other gases and excellent reversibility[4]. Meanwhile, the incorporation of functionalized substituent groups on the cation, such as tertiary amino group, ether group and nitrile group plays a key role in improving SO_2 capture, which exhibited much better absorption performances of SO_2 than the conventional IL with [SCN]. This work will provide some useful information to design more competitive absorbents for SO_2 capture.

REFERENCES

1. X. X. Ma, T. Kaneko, T. Tashimo, T. Yoshida and K. Kato, *Chem Eng Sci*, 2000, **55**, 4643-4652.

2. S. Y. Hong, J. Im, J. Palgunadi, S. D. Lee, J. S. Lee, H. S. Kim, M. Cheong and K.-D. Jung, *Energy Environ Sci*, 2011, **4**, 1802-1806.

3. K. M. Docherty, J. K. Dixon and C. F. Kulpa Jr, *Biodegradation*, 2006, **18**, 481-493.

4. S. Zeng, H. Gao, X. Zhang, H. Dong, X. Zhang and S. Zhang, *Chem Eng J*, 2014, **251**, 248-256.

EXPERIMENTAL DETERMINATION OF LIQUID-VAPOR EQUILIBRIUM OF THE PSEUDO-BINARY MIXTURE OF BIODIESEL AND ETHANOL

Author(s): Savio L. Bertoli[1], Pollyana M. S. Kalvelage[1], Edésio L. Simionatto[2], Antônio A. C. Barros[3], Juscelino de Almeida Junior[1], Juliano de Almeida[1], Luiza F. Peyerl[1] and Vinicyus R. Wiggers[1]

[1] Department of Chemical Engineering, Regional University of Blumenau - FURB, Blumenau, Brazil;
[2] Department of Chemistry , Regional University of Blumenau, Blumenau, Brazil;
[3] ISPTEC- Instituto Superior Politécnico de Tecnologias e Ciências; DET- Departamento de Engenharia e Tecnologia, Av. Luanda Sul, Rua Lateral Via S10, Talatona – Luanda Sul, República de Angola.

Abstract

The quality of biofuels is related to the detailed research to the several stages of production processes, mainly those related to the purification ones. In order to improve the performance of these processes, it is required the knowledge of the vapor-liquid equilibrium of the mixtures. Phase equilibrium data allows establishing the appropriate conditions of temperature and pressure of the separation processes performed by distillation, related to the recovering of methanol or ethanol in the biodiesel purification. In this paper, we present experimental results obtained by ebulliometry techniques of the liquid-vapor equilibrium of the pseudo-binary mixture of biodiesel-ethanol.

Keywords: Vapor-liquid equilibrium, pseudo binary mixture, transesterification, biodiesel.

1. Introduction

Currently the debates regarding environmental problems occupy a featured place in all parts of the world. Under this perspective several sectors of modern society are concerned and aware of the importance in preserving natural resources, especially the fossil fuels, which are nonrenewable and cause environmental problems (Kalvelage, 2014).

Biofuels are an alternative to the fossil fuels since its use promotes the reduction of environmental impacts. The most well-known biofuels used today are biodiesel and bioethanol (Oliveira, 2003), and in the Brazilian territory the use of them is regulated (ANP, 2014).

The biodiesel has been studied and proposed as one of the alternatives to fossil fuels; it is renewable and biodegradable, with gaseous emissions free of sulfur compounds, toxic and carcinogenic substances and emits 90% less smoke compared to the mineral diesel (Rocha et al., 2008).

In order to improve the performance of the biodiesel production and purification processes, the determination of the equilibrium conditions in the separations units is required; specially the vapor-liquid equilibrium (VLE) present in the recovery of ethanol and biodiesel purification, carried out by distillation operations (Coelho, 2011).

The determination of the phase equilibrium and the thermodynamic modelling consolidates the knowledge of chemical processes involving industrial separation operations such as distillation, absorption and extraction. Based on conservation principles and in the phase equilibrium it is possible to predict and evaluate the process equipment performance relative to the heat and mass transfer efficiency and in determining the amount of utilities for the units of production and purification (Mehl, 2009).

From this perspective, it is possible to obtain VLE data for the pseudo-binary mixture ethanol and biodiesel, according to the method described below.

2. Materials and Methods

2.1. Materials

To execute of this study, the following materials were used: a) biodiesel (obtained from virgin soybean oil and frying oil); b) PA anhydrous ethyl alcohol; c) distilled water; d) soy oil, and e) frying oil used in food processing.

2.2 Equipment

The equipment used is as follows:

- Pilot-scale plant: The biodiesel used in this study was produced using the pilot-scale plant designed to biodiesel production, built and installed in the Chemical Engineering Department at the Regional University of Blumenau (FURB). The pilot plant is a compact unit that incorporates various unit operations as shown in Figure 1.

The pilot scale plant was operated according the following steps:

a) Initially the vegetable oil or animal fat is introduced into the tank (1) and reagent (alcohol) in the tank (2);

b) The pump (4) then is turned on for the transport of the fluid contained in these tanks to tank (5);

c) Once transport is completed, the electrical resistances installed on tube surface (7) are ignited to start the heating of the mixture in the tank (5);

d) The pump (4) is turned on for the heating and recirculation of the mixture contained in tank (5); the recirculation is carried out over the tank (5), the pump (4) and the tube (7). The system is kept open due to the presence of the condenser (6) installed on the tank (5), which ensures condensation and returning of the vaporized reactant;

e) When the reaction temperature is reached, the pre-heated catalyst is added in the tank (5), starting out the transesterification reaction;

f) When the reaction is complete, the temperature increases, to recover the reagent (alcohol) used in excess to ensure the reversibility of the reaction, thus opening the alcohol condenser outlet valve to the tank (3);

g) After the recovery of the excess alcohol, the remaining mixture is pumped into the tank (11) where the phase separation occurs. Then, using the same pump, the biodiesel, light phase, is directed to the tank (13), and glycerol, heavy phase, is directed to the tank (12);

h) The glycerol and the biodiesel present in the tanks (12) and (13) respectively, go through purification processes to ensure the commercial characteristics;

i) In the tank (3), the excess of alcohol recovered can also be purified further using the distillation column (9) coupled to the reboiler (8). The products of this stage are collected in the tank (10) and redirected to the tank (2), returning to the process.

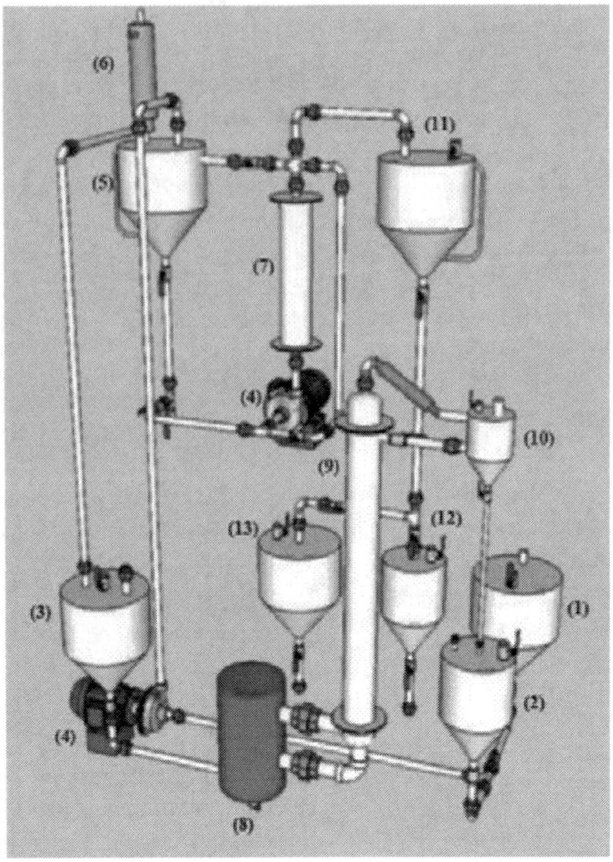

Figure 1 - Micro plant designed, built and installed in the Department of Chemical Engineering (FURB) used in the production of biodiesel.

- Fischer Ebulliometer: Figure 2 shows the Model 602, Fischer ebulliometer consisting of a dynamic measuring cell for the vapor-liquid equilibrium data whose phases, liquid and vapor, are circulating. This device is suitable for no electrolyte systems and allows the study of substances with a high boiling point, if it is provided together with the appropriate safety equipment. The Fischer ebulliometer is based on the liquid and vapor phases recirculating, in contact to reach equilibrium and comprises the following parts: (1) Bottle mixture; (2) chamber? (3) tube of the ascent of the mixture or "Cottrell pump"; (4) thermocouple; (5)condenser; (6) withdrawal port of the samples of liquid phase and vapor phase; (7) valves, each connected for ebulliometry system.

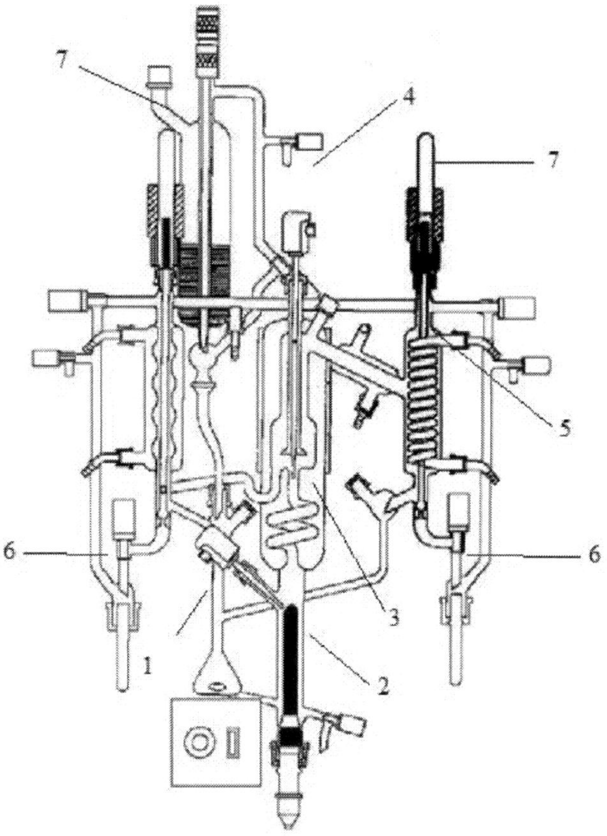

Figure 2 - Model 602 Fischer Ebulliometer. ILUDEST (2011)

- Gas Chromatography: The esters analysis is performed by gas chromatography. This work was performed using a gas chromatograph with ionization detector GC-2010 model Shimadzu, with auto-injector AOC-5000 headspace model.
- Refractometer: in order to determine the refractive index of the ethanol / water mixture an Abbe refractometer was used. The index information was read using a graduated scale, through an optical system. In this case, the measured value is obtained by comparing it with the values on a table provided by the instrument manufacturer.

3. Results and Discussion

3.1 Analysis of the vapor-liquid equilibrium of the pseudo binary mixture ethanol/biodiesel

Figure 3 depicts the behavior T-x1 of the pseudo-binary mixture of biodiesel and ethanol for biodiesel obtained from virgin soybean oil. The experimental data of liquid-vapor equilibrium were determined operating the system at atmospheric pressure in all tests.

The experimental data were evaluated and compared with the literature data (GUO et al., 2007) by graphical depiction of the T-x1. However, it is essential to note that the literature data (GUO et al., 2007) were obtained experimentally from pressures below atmospheric pressure, unlike the pressure used in this work. Secondly, the biodiesel used in the work of Guo et al. (2007) was obtained from sunflower seed oil.

The comparison between the literature data and those from this study shows similarity in the behavior of these data mainly for ethanol concentrations greater than 50%. This behavior is possible considering the proximity of pressures used in both cases referenced herein. It can be seen that the data obtained virtually overlaps the data of Guo et al. (2007), despite of the 6.32 kPa pressure difference.

The Figure 4 presents the T-x1 data obtained in this work at a pressure of 101.32 kPa and the data from Guo et al. (2007) at a pressure of 24 kPa, for the mixture of ethanol and biodiesel obtained from virgin soybean oil and sunflower seed oil, respectively. Despite the significant difference in the operating pressure, it can be observed a similar qualitative behavior between the data.

Figure 3 and 5 also show a similar behavior data from this work and the literature, despite the different pressures used and the biodiesel obtained from different vegetable oils and produced at different operation conditions. GUO et al. (2007) used the biodiesel obtained from sunflower seeds at a 95 kPa pressure, whereas this study used the biodiesel from virgin soybean oil at 101, 32 kPa.

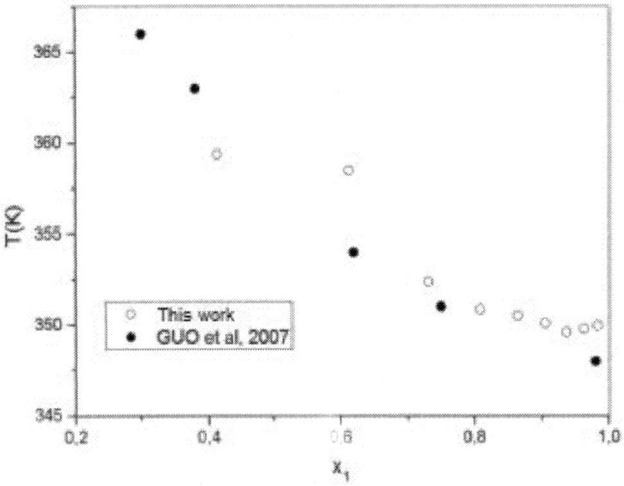

Figure 3 - Isobaric VLE data at 95 kPa (GUO et al., 2007) and 101,32 kPa (used in this work) for the pseudo-binary system ethanol (1) /biodiesel (2) obtained from the virgin soybean oil (used in this work) and sunflower seed oil (GUO et al., 2007).

Figures 4 e 6 also shows similarity, but the working pressure data from the literature is 24 kPa. The biodiesel used by Guo *et al.* (2007) is from sunflower seeds while the present work uses biodiesel from soybean oil at 101.32 kPa.

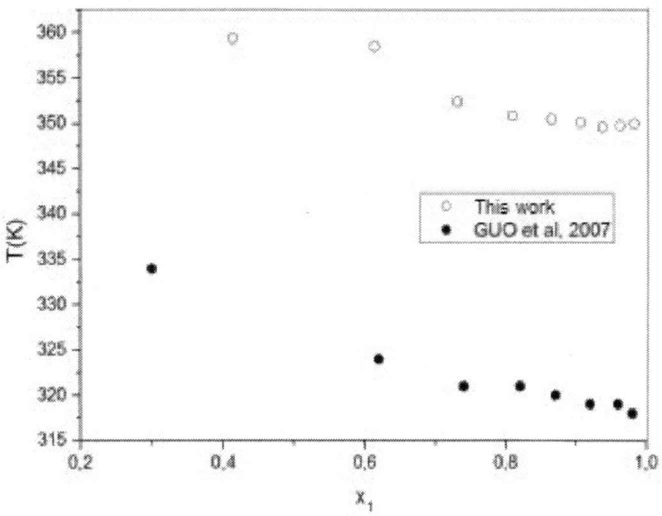

Figure 4 - Isobaric VLE data at 24 kPa (GUO et al., 2007) and 101,32kPa (used in this work) for the pseudo-binary system ethanol (1) /biodiesel (2) obtained from virgin soybean oil (for this work) and sunflower seed oil (GUO et al., 2007).

Figure 5 and 6 shows a comparison between the data at 101,32kPa obtained from biodiesel produced with frying oil (this work) with the literature data for biodiesel produced from sunflower seed oil (GUO et al., 2007) at pressures of 95 kPa and 24 kPa, Fig. 5 and 6 respectively. Thus, due the similarity presented, it can be inferred that the virgin soybean oil, frying oil, and the sunflower seed oil present a formation of fatty acid esters with similar physical and chemical characteristics.

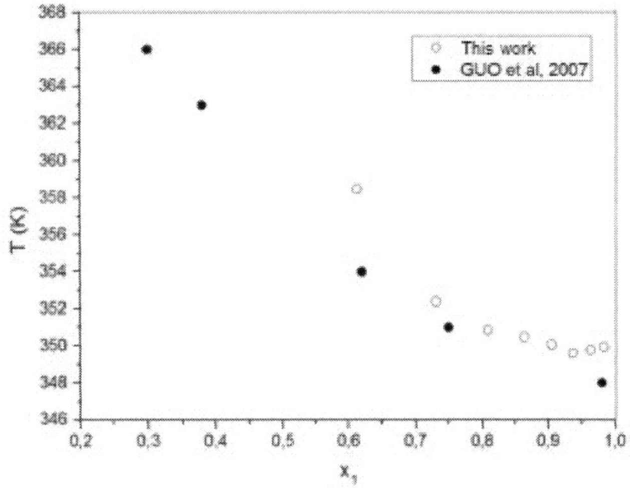

Figure 5 - Isobaric VLE data at 95 kPa (GUO et al., 2007) and 101.32 kPa (this work) for the pseudo/binary system ethanol (1) /biodiesel (2) obtained from frying oil (this work) and sunflower seed oil (GUO et al., 2007)

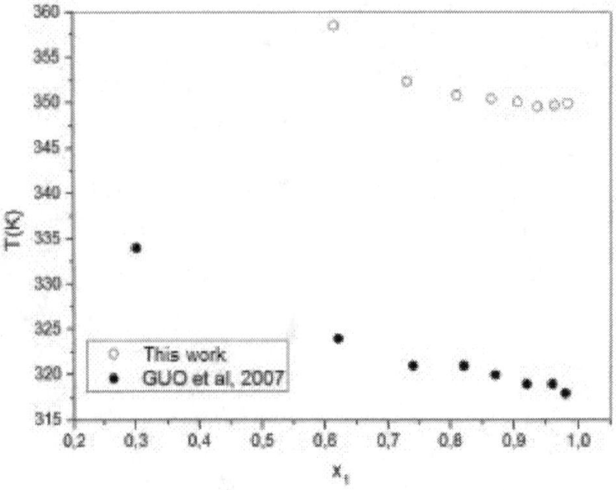

Figure 6 - Isobaric VLE data at 24 kPa (GUO et al., 2007) and 101.32 kPa (this work) for the pseudo-binary system ethanol (1) /biodiesel (2) obtained from frying oil (this work) and sunflower seed oil (GUO et al., 2007).

As expected, as the biodiesel fraction increases, there is an increase in the equilibrium temperature. This data set presented difficulty in regression due to the limited set of points and the large difference in the boiling point temperatures between the biodiesel and ethanol.

4. Conclusion

Aiming to improve the performance of the biodiesel production and purification processes, the determination of phase equilibria is required, especially the vapor-liquid equilibrium (VLE) of ethanol/biodiesel. In this paper, experimental vapor-liquid equilibrium (VLE) was obtained, data with experiments conducted at atmospheric pressure, in pseudo-binary systems (ethanol + biodiesel). Furthermore, it was possible to observe a qualitative and quantitative similarity in the results when compared to experiments found in literature [GUO et al., 2007] utilizing sunflower seed oil under distinct pressures. The data suggest the possibility of representing the thermodynamic behavior of biodiesel from the esters that compose it, which is an important fact that facilitates the study of such systems in experimental bench.

Finally, the results obtained in this study allows the improvement of thermodynamic data, in order to predict the behavior of the phenomena involved consistently and accurately in process control and design.

5. References

1. ANP – Agencia Nacional de Petróleo, Gás Natural e Biocombustíveis. *Resolução ANP N° 45 Agosto 2014*. Brazil.

2. Coelho, A.C. (2011), "Equilíbrio Líquido-Vapor de Sistemas Binários Envolvendo Ésteres Etílicos do Biodiesel (Glicerol ou Água) + Etanol: Dados Experimentais e Modelagem Termodinâmica." (Liquid-Vapor Equilibrium of Binary Systems Involving Ethylic Esters of Biodiesel (Glycerol or Water) + Ethanol: Experimental

Data and Thermodynamic Modeling). *MSc Thesis. Setor de Tecnologia, Programa de Pós Graduação em Tecnologia de Alimentos* , Universidade Federal do Paraná, Brazil.

3. Guo, Y. ; Zhong, J. ; Xing, Y.; LI, D.; Lin. (2007), "Volatility of blended fuel of Biodiesel and Ethanol". *Energy & Fuels* 21, 1188 – 1192.

4. Iludest Destillatiosanlagem G.M.B.H. (2011), "Installation Manual for Equilibrium Apparatus Fischer Labodest VLE 602". *Waldbuttelbrunn*. Germany.

5. Mehl, A. (2009), "Estudo do Equilíbrio de Fases de Hidrocarbonetos e CO2 Supercrítico" (Study of Phase Equilibria of Hydrocarbon and Supercritical CO2). *Doctoral Thesis in Chemical and Biochemical Process Technology*. Universidade Federal do Rio de Janeiro- RJ, Brazil.

6. Kalvelage, P. M. S. (2014), "Determinação Experimental do Equilíbrio Líquido-Vapor da Mistura Pseudo- Binária Biodiesel e Etanol" (Experimental Determination of the Vapor-Liquid Equilibrium of the Pseudo-binary Mixture of Biodiesel and Ethanol). *MSc Thesis in Chemical Engineering*. Universidade Regional de Blumenau-SC, Brazil.

7. Oliveira, H. N. (2003), "Determinação de dados de Equilíbrio Líquido-Vapor para Sistemas Hidrocarbonetos e Desenvolvimento de uma nova Célula Dinâmica" (Determination of Vapor Liquid Equilibrium Data for Hydrocarbon Systems and Development of a New Dynamic Cell). *Doctoral Thesis in Chemical Engineering*. Universidade Federal do Rio Grande do Norte. Natal-RN, Brazil.

8. Rocha, D. Q .; Barros, D. K .; Costa, E. J. C .; Passos, P. R .; Veiga, V. F. V .; Chaar, J.S. (2008), "Determinação da matéria-prima utilizada na produção do Biodiesel mineral através de monitoramento seletivo de íons". (Determination of raw material used in the production of mineral biodiesel through selective ion monitoring). *Química Nova*, 31, 1062-1066.

434052 Extraction of Antioxidants and Tannins from Walnut Husks Using Supercritical Carbon Dioxide with an Ethanol Modifier

Abstract:

During the processing of nuts, a considerable amount of material such as husks and shells are produced and often discarded as compost or landfill. In producing walnuts for retail sale, the husks are often discarded. Walnut husks contain tannins and polyphenolic compounds and the husk may be used for treatment of intestinal parasites and fungal infections. Supercritical fluids, due to their high diffusivity are excellent solvents for extracting chemical species from plant matter. Supercritical carbon dioxide, by itself has a relatively low dielectric constant, and ethanol can be added as a modifier to help improve the extraction of polyphenolic and tannin compounds. The effects of temperature and ethanol content with constant density supercritical carbon dioxide with ethanol upon the extraction of ground walnut husks was evaluated using a factorial design of experiments. Extractions were performed in a 22 mL stainless reactor. Temperature was varied between 50 and 70 °C and ethanol mole fraction between 10 and 20%. Both dried and damp walnut husks were evaluated. The antioxidant potential was evaluated using the total phenolic content (TPC) assay, ferric reducing ability of plasma (FRAP) assay and DPPH assay. DNA protection abilities of the extracts, as well as chemical composition, will also be presented.

434444 Energy Efficient Bio-Butanol Recovery By Gel Stripping

Abstract:

Design of alternative, energy-efficient processes for recovery of n-butanol from dilute (10 to 20 g/L) fermentation effluents is vital because of the high energy requirements of multi-stage distillation. We are investigating a new, energy-efficient process called "gel stripping" for recovery of n-butanol from fermentations via a column-based absorption process. The raw fermentation product is passed through a vertical column containing particles of a selectively absorbent polymeric gel material, which absorbs the alcohol component preferentially while draining excess water and other waste components from the system. The concentrated alcohol is subsequently recovered from the gel by one of several options without the need to feed large quantities of water to a multi-stage distillation unit.

One major challenge in gel stripping is to design an absorbent gel material that is both reasonably selective for the alcohol and able to swell to a high degree in dilute alcohol-water mixtures. There is usually a trade-off between selectivity and equilibrium swelling, however. In addition, the entropy of mixing of alcohol and water in the external solution is favorable enough to discourage the alcohol from entering the polymeric gel, even when the gel has a high affinity for pure alcohol. This problem is analogous to the problem faced in liquid-liquid extraction (LLE) operations in which the system must offer attractive selectivity without compromising the distribution coefficient for alcohol.

Groot et al. previously reported that Castor oil provides an attractive combination of high distribution coefficient for butanol (2.6), low distribution coefficient for water (0.00955), low toxicity, and low price ($1.2 per kg).[1] However, Castor oil is an ineffective solvent for LLE because it is prone to forming stubborn emulsions and is subject to fouling by proteins. However, by forming a crosslinked, polymeric gel material from Castor oil, we find that its desirable separation characteristics are retained, while the resulting rubber-like material is unable to form an emulsion. A novel gel based on Castor oil was tested in the gel stripping process and permitted 95%+ removal of butanol from dilute (20 g/L) aqueous solutions, though additional work is needed to increase the absorption capacity of the material.
The second stage of gel stripping is the energy-efficient removal of the absorbed butanol from the gel. The most promising options are 1) extraction into a low-boiling liquid such as diethyl ether, 2) direct vacuum distillation from the gel, and 3) gas stripping. Current work in our group focuses on evaluating the strengths and weaknesses of these approaches and on evaluating the energy costs of each approach. Calculations suggest that the energy cost of recovering butanol by gel stripping might be reduced by more than 50 % compared to simple multi-stage distillation.

Without requiring sophisticated techniques or processing equipment, gel stripping potentially provides high separation efficiency with low energy cost. Advantages and limitations of gel stripping compared to other processes such as extractive fermentation, LLE, and pervaporation will be discussed.

[1] Groot, W. J.; Soedjak, H. S.; Donck, P. B.; Van der Lans, R. G. J. M.; Luyben, K. C. A.; Timmer, J. M. K. Butanol recovery from fermentations by liquid-liquid extraction and membrane solvent extraction. *Bioprocess engineering*, **1990**, *5*(5), 203-216.

435595 Corresponding States Theory of Adsorbed FILMS: Influence of Wetting and Wall Roughness

Monday, November 9, 2015: 8:30 AM
255D (Salt Palace Convention Center)
Keith E. Gubbins, Chemical & Biomolecular Engineering, North Carolina State University, Raleigh, NC

CORRESPONDING STATES THEORY OF ADSORBED FILMS: INFLUENCE OF WETTING AND WALL ROUGHNESS

Yun Long[a], An Rong[b], Małgorzata Śliwinska-Bartkowiak[c], Matthias Thommes[d] and Keith E. Gubbins[b]

[a] *Department of Chemical& Biomolecular Engineering, National University of Singapore, Singapore 117585*

[b] *Department of Chemical& Biomolecular Engineering, North Carolina State University,*

Raleigh, NC 27695, USA

[c] *Faculty of Physics, Adam Mickiewicz University, Poznan 61-614, Poland*

[d] *Quantachrome Instruments, 1900 Corporate Drive, Boynton Beach, FL 33426, USA*
A statistical mechanical analysis of adsorption shows that the equilibrium properties of the adsorbed film (isotherms, isosteric heats, phase change conditions, etc.) are determined by the dimensionless pore width, *H*, and pore geometry, and by a *microscopic wetting parameter*, α_w, that is defined in terms of intermolecular forces and solid structure; α_w is a measure of wetting that applies at all scales and for any kind of adsorbed film (gas, liquid or solid) [1,2].

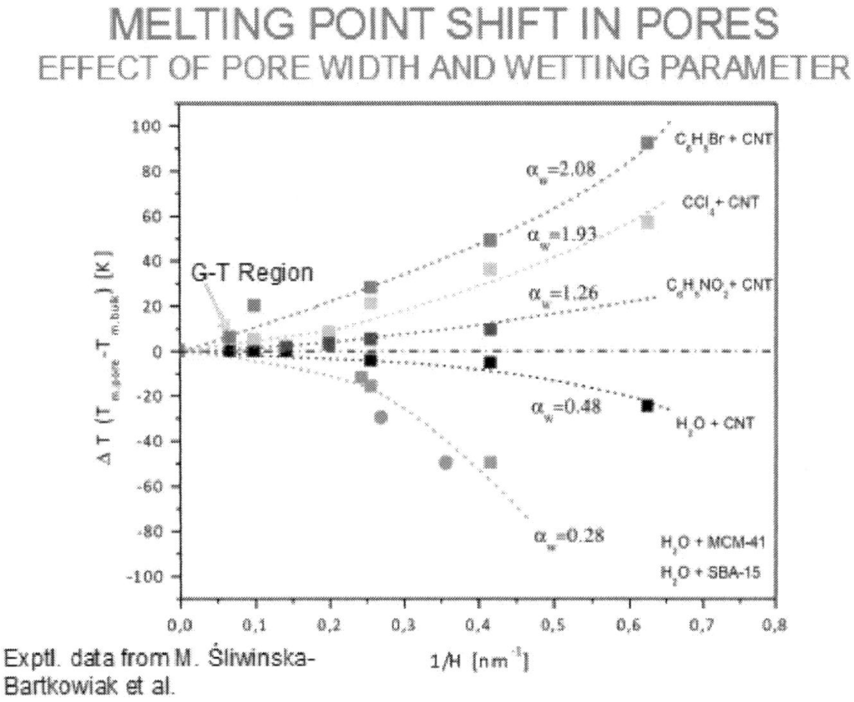

Figure 1.
Experimental results for the effect of confinement in cylindrical pores on the melting point of various adsorbates. Here CNT = carbon nanotube and α_w is the microscopic wetting parameter; high values of α_w indicate strong wetting. For large pore widths the temperature shift is approximately linear, as required by the Gibbs-Thomson equation, but departures occur for smaller widths.

We illustrate the usefulness of this approach using examples drawn from both experimental and molecular simulation studies of gas-liquid and liquid-solid phase separations [3,4], and pressure enhancement in nanopores [5-7], with emphasis on simple pore geometries. These examples illustrate the central role played by wetting, and also the breakdown of some concepts and macroscopic laws, such as the equations of Kelvin, Laplace and Gibbs-Thomson, for nano-phases confined within small pores. They also suggest consistency tests for experimental data that should be applied in the event of capillary condensation or other phase changes in mesoporous materials.

The approach can be extended to account for geometric or energetic wall roughness [8], and to adsorbate molecules that interact with site-site potentials having both Lennard-Jones and point charge sites.

1. R .Radhakrishnan, K.E. Gubbins and M. Śliwinska-Bartkowiak, "Global Phase Diagrams for Freezing in Porous Media", *Journal of Chemical Physics*, **116**, 1147-1155 (2002).

2. Keith E. Gubbins, Yun Long and Malgorzata Sliwinska-Bartkowiak, "Thermodynamics of Confined Nano-Phases", *Journal of Chemical Thermodynamics*, **74**, 169-183 (2014).

3. L.D. Gelb, K.E. Gubbins, R. Radhakrishnan and M. Sliwinska-Bartkowiak, "Phase Separation in Confined Systems", *Reports on Progress in Physics*, **62**, 1573-1659 (1999).

4. C. Alba-Simionesco, B. Coasne, G. Dosseh, G. Dudziak, K.E. Gubbins, R. Radhakrishnan and M. Śliwinska-Bartkowiak, "Effects of Confinement on Freezing and Melting", *Journal of Physics: Condensed Matter*, **18**, R15-R68 (2006).

5. Yun Long, Jeremy C. Palmer, Benoit Coasne, Małgorzata Śliwinska-Bartkowiak and Keith E. Gubbins, "Pressure enhancement in carbon nanopores: A major confinement effect", *Physical Chemistry Chemical Physics*, **13**, 17163-17170 (2011).

6. H. Drozdowski, M. Kempinski, M. Śliwinska-Bartkowiak, M. Jazdzewska, Y. Long, J.C. Palmer and K.E. Gubbins, „Structural Analysis of the Behavior of Water Adsorbed in Activated Carbon Fibers", *Physical Chemistry Chemical Physics*, **14**, 7145-7153 (2012).

7. Yun Long, Jeremy C. Palmer, Benoit Coasne, Małgorzata Śliwinska-Bartkowiak, George Jackson, Erich A. Müller and Keith E. Gubbins, "On the Molecular Origin of High Pressure Effects in Nanoconfinement: Effects of Surface Chemistry and Roughness", *Journal of Chemical Physics*, **139**, 144701 (2013).

8. M. Śliwinska-Bartkowiak, A. Sterczyńska, Y. Long and K.E. Gubbins, "Influence of Microroughness on the Wetting Properties of Nano-Porous Silica Matrices", *Molecular Physics*, **112**, 2365-2371 (2014).

426374 First-Principles Modeling of Carbon Dioxide Capture and Conversion in Metal Organic Frameworks

Abstract:

The development of novel porous materials for CO2 capture from flue gas and other waste streams coupled with the conversion of CO2 into useful materials constitutes a grand challenge problem in today's world. Capture of CO2 is critical to achieve in the short-term because fossil fuel use is expected to remain high for the next several decades. However, capturing CO2 is only half of the problem. What to do with the enormous quantities of CO2 generated from coal-fired power plants alone is a major challenge. The only alternative to sequestration is conversion of CO2 into fuel through hydrogenation. Ideally, one would have a technology that used a renewable energy source, such as solar or wind, to provide the reducing agent (e.g., hydrogen) to drive the conversion of CO2 into hydrogen-rich fuels.

In this work we provide examples of how electronic structure density functional theory (DFT) can be used to provide insight into the problems of CO2 capture and subsequent reduction through the addition of hydrogen to create value-added products. We report calculations on adsorption of CO2 and other small molecules in metal organic framework (MOF) materials using both classical and quantum mechanical simulations. We show that quantum mechanical DFT calculations provide important information about spin transitions and geometric relaxations in the MOFs that affect adsorption. We also demonstrate the utility of DFT methods for designing catalytically active MOFs for the reduction of CO2.

433659 Superadaptability of Water Molecules to Hydrophobic Nanoenvironments

Abstract:

Superadaptability of Water Molecules to Hydrophobic Nanoenvironments

Katsumi Kaneko

Center for Energy and Environmental Science, Shinshu University, Nagano, 380-8553, Japan

The interaction of water with solid surfaces is one of importance research issues from fundamental science and technology, although it has been studied for a long time. When we drop a water droplet on the compressed disk plate of hydrophobic carbons having nanoscale pores, the water droplet forms a stable hemisphere. This indicates that the porous carbon-packed surface is rather hydrophobic. It is well-known that water vapor begins to be adsorbed suddenly on nanoporous carbons of fewer amounts of surface oxygen groups and hydrophobic zeolites such as $AlPO_4$-5. The behavior can be phenomenologically described by the transformation of hydrophobic surface to hydrophilic surface[1]; we need to understand that water molecules can adapt to chemically unfavorable nanoenvironments through varying the intermolecular structure biomimically[1-3]. However, this quasi transformation does not occur on carbon aerogels having large mesopores[4-6]. The dependence of this quasi transformation on the pore size may be studied. Also we need to understand the contribution of edge carbons to water adsorption[7].

References

[1] Miyawaki, J., Kanda, T., Kaneko, K. 2001 *Langmuir*, 17 (2001) 664.

[2] Ohba, T., Kanoh, H., Kaneko, K. 2004 *J. Am. Chem. Soc*. 126, 1560.

[3] Kaneko,K. 2015, *Nature Chem*. **7**, 194.

[4] Hanzawa, Y., Kaneko, K. 1997 *Langmuir* 13 (1997) 5802.

[5] Thommes, M., Morell, J., Cychosz, K.A., Froba, M., 2013 *Langmuir*29, 14893.

[6] Zoraida Pina-Salazar, E., Kaneko, K. Colloid Interface Sci. Comm. To be submitted.

[7] Asai, M., Ohba, T., Iwanaga, T., Kanoh, H., Endo, M., Campos-Delgado, J., Terrones, M., Nakai, K.,Kaneko, K. 2011 *J. Am. Chem. Soc.* 133,14880.

431349 Advanced Physical Adsorption Characterization of Nanoporous Materials with Hierarchical Pore Structure

Abstract:

A comprehensive textural characterization of nanoporous materials has become more important than ever for the optimization of novel systems used in many important existing and potentially new applications. The most popular method to obtain surface area, pore size, pore size distribution and porosity information from powders and porous solids is gas adsorption. Within the last two decades major progress has been achieved in understanding the adsorption and phase behavior of fluids in ordered nanoporous materials and in the development of advanced approaches based on statistical mechanics such as molecular simulation and density functional theory (DFT) of inhomogeneous fluids. This progress, coupled with the availability of high resolution experimental procedures for the adsorption of various subcritical fluids, has led to major advances in the structural characterization by physical adsorption.

Recently, new challenges emerged concerning the surface and structural characterization of nanoporous materials with more complex pore structures such as hierarchically structured porous materials (e.g., micro-mesoporous zeolites and novel nanoporous carbons). Within this context we will focus on some selected examples to discuss recent advances, but will also reflect briefly on the existing challenges in physical adsorption characterization.

424045 Impact of Microfluidic Device Material on Cell Separation

Abstract:

Microfluidic devices are capable of performing a complete blood analysis in which the first step is the separation of white blood cells (WBCs) from red blood cells (RBCs). Due to the difference in the size of the WBCs and RBCs, this cell separation can be performed by size exclusion using cross-flow filtration in the microfluidic device. The properties of the materials used in fabrication of the microfluidic device, however, may affect the separation of the WBCs and RBCs. The current work is an evaluation of polymeric material on the performance of the microfluidic device. Microfluidic devices with 10 μm square pillars were successfully fabricated from polydimethylsiloxane, polystyrene, polymethylmethacrylate, and polycarbonate. Device characteristics, including polarity, surface roughness, porosity, feature dimensions, and cell separation efficiency are compared for the different materials. The efficiency of separating WBCs while discarding RBCs is evaluated using video microscopy of colored microspheres.

High-resolution protein separation using a laterally-fed membrane chromatography device

Pedram Madadkar, Raja Ghosh*

Department of Chemical Engineering

McMaster University

Canada

*rghosh@mcmaster.ca

Improvements in upstream processing and bio-reactor technologies have resulted in dramatic enhancements in titres of biopharmaceuticals such as monoclonal antibodies. Downstream processing which still relies primarily on packed-bed chromatography is widely identified as the bottleneck in bio-manufacturing. The speed of separation in packed-bed chromatography is limited by slow diffusion-based mass transport [1]. Over the past decade or so, membrane chromatography has been proposed as a cost-effective alternative separation technique where diffusion-linked limitations are addressed by using micro-porous adsorptive membranes with flow-through pores, instead of resin beads [2, 3]. The predominance of convection transport of target species to and from their binding sites makes membrane chromatography a very fast separation technique. In addition, the efficiency of separation is largely independent of the superficial velocity, even for large macromolecules, making process design and scale-up much easier and more flexible. The high speed of separation translated to a high production rate, and low product degradation. Other advantages include lower buffer usage, lower pressure drop, and lower capital cost of equipment. In addition, membrane devices being disposable, eliminates expensive and time-consuming protocols such as cleaning and validation [4, 5].

Membrane chromatography has two efficiency determining components: the membrane and the device. Recent studies have shown that the device design is just as important as the membrane [6]. The device affects factors such as flow distribution, dead volume, and variability in flow path lengths, all of which are critically important. Currently used membrane devices are available in three formats: stacked-disc, radial-flow, and hollow fiber, the first two being more popular [7, 8]. Stacked-discs resemble syringe-type filters whereas radial-flow devices are made by wrapping or pleating flat-sheet membrane over cylindrical support structures (Fig. 1). Stacked discs are based on a simple design but are not suitable for large-scale applications. While radial-flow devices are suitable for large-scale processes, their use is largely restricted to polishing application. Radial-flow devices are now widely used in the flow-through mode for removing impurities such as host cell proteins (HCPs), viruses, endotoxins and aggregates. However, these devices are not suitable for bind-and-elute separations, particularly where the multi-component separation of proteins is required [9, 10].

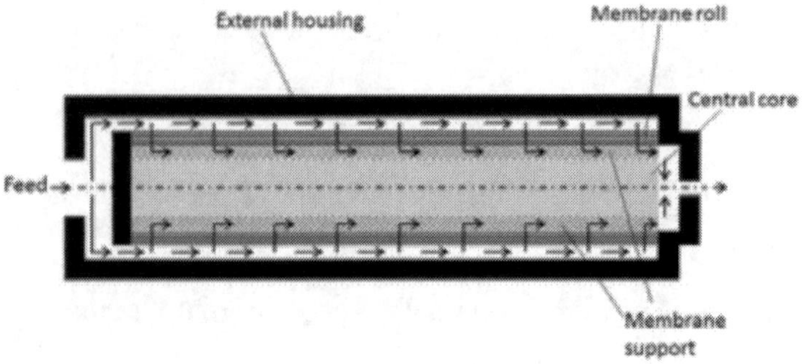

Figure 1 Radial-flow membrane chromatography device.

The current study focuses on a laterally-fed flat-sheet membrane chromatography (LFMC) device suitable for stacking rectangular adsorptive membranes as conceptualized in Fig. 2 [11, 12]. The working device comprises three components (see Fig. 3): the top and bottom plates which contain the inlets and outlet, as well as feed and permeate channels; and the central plate within which the membrane stack is glued in place. The two most critical design criteria for the LFMC device are: a. the hydraulic resistance of the membrane stack should be greater than the resistance in the feed or the permeate channels, b. the lateral resistances in the feed and permeate channels should be similar [13].

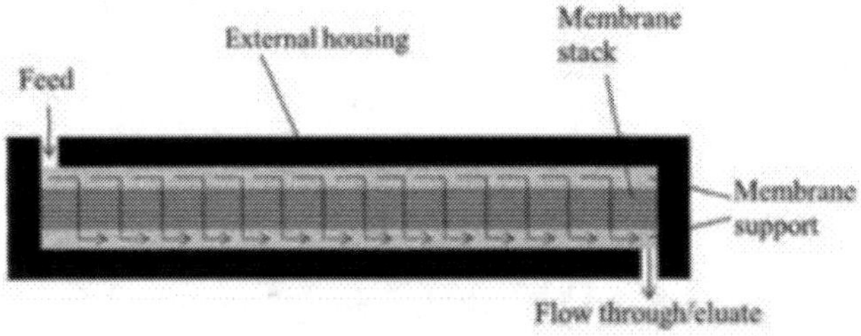

Figure 2 Laterally-fed membrane chromatography device.

LFMC device was fabricated and a stack of commercially sources ion-exchange membranes (Sartobind S, Sartorius Stedim Biotech, Gottingen, Germany) was housed in it (see

Fig 3). The effective bed volume of the device was 7 mL. The performance of the device was compared with an equivalent radial-flow device.

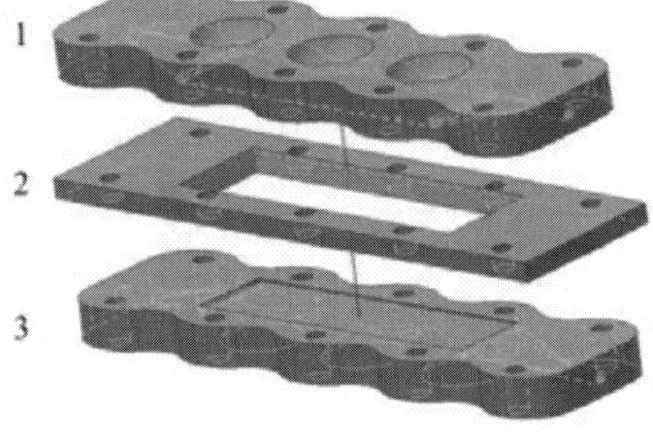

Figure 3 Diagram of the prototype laterally-fed membrane chromatography device used in the current study (1: acrylic top plate; 2: Delrin frame; 3: acrylic bottom plate).

Bind-and-elute experiments were carried out using lysozyme as the model protein. The results are summarized in Fig 4. The LFMC device gave sharper and significantly more symmetrical peaks. The superior results obtained using the LFMC device could be attributed to a variety of factors such as good flow distribution, lower dead volume and greater uniformity in solute flow path length. These results indicate that the LFMC device would suitable for carrying out multi-component protein separation in bind-and-elute mode.

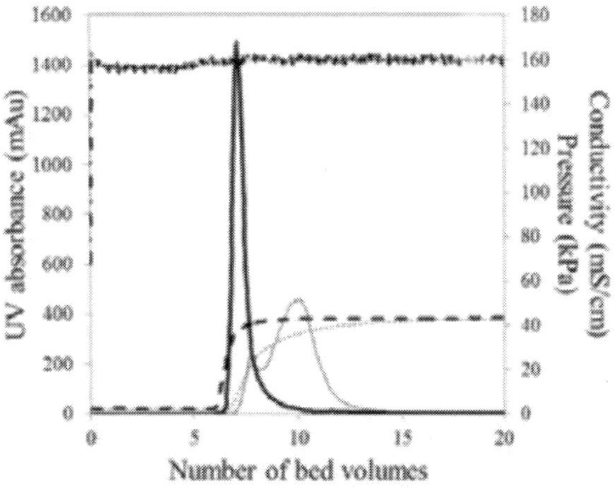

Figure 4 Lysozyme elution peaks obtained with the radial-flow (thin lines) and the laterally-fed (thick lines) membrane chromatography devices (membrane: Sartobind S; membrane bed volume: 7 mL; feed concentration: 8 mg/mL; volume injected: 2 mL; binding buffer: 20 mM sodium phosphate, pH = 7.0; eluting buffer: binding buffer + 0.5 M NaCl; flow rate: 10 mL/min; UV absorbance: solid lines; conductivity: dashed lines; pressure: dotted lines).

Multi-component separation experiments were carried using three proteins: ovalbumin, conalbumin, and lysozyme. The experimental conditions were designed such that ovalbumin was obtained as flow-through, while conalbumin and lysozyme were obtained as eluted peaks. The bound proteins were eluted using linear salt gradient. The results obtained with the radial-flow and LFMC devices are shown in Figs 5 and 6 respectively. The eluted peaks obtained with the LFMC device were very sharp and symmetrical at all three gradients examined. The sharpness of the eluted peaks decreased slightly, while their resolution increased as the gradient length was increased. Three model proteins could be fractionated in about 10 bed volumes using the laterally-fed device as opposed to 25 bed volumes with the equivalent radial-flow device. The design of the laterally-fed device is simple and could potentially be mass-produced by polymer molding. The flat shape of the laterally-fed device gives it significantly lower footprint and offers additional advantages such as stackability and ease of multiplexing.

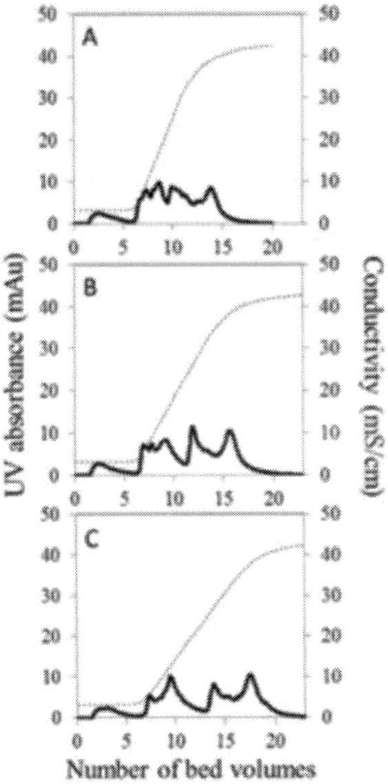

Figure 5 Multi-component separation peaks obtained with the radial-flow device using 40 mL (A), 60mL (B) and 80 mL (C) linear gradients (membrane: Sartobind S; membrane bed volume: 7 mL; ovalbumin concentration: 0.25 mg/mL, conalbumin concentration: 0.5 mg/mL; lysozyme concentration: 0.25 mg/mL; loop size: 2 mL; binding buffer: 20 mM citrate, pH = 5.5; eluting buffer: binding buffer + 0.5 M NaCl; flow rate: 10 mL/min; solid line: UV absorbance; dashed line: conductivity).

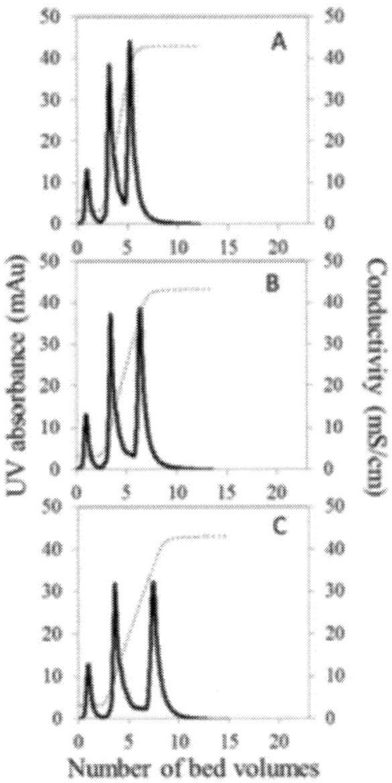

Figure 6 Multi-component separation peaks obtained with the laterally device using 20 mL (A), 30mL (B) and 40 mL (C) linear gradients (membrane: Sartobind S; membrane bed volume: 7 mL; ovalbumin concentration: 0.25 mg/mL, conalbumin concentration: 0.5 mg/mL; lysozyme concentration: 0.25 mg/mL; loop size: 2 mL; binding buffer: 20 mM citrate, pH = 5.5; eluting buffer: binding buffer + 0.5 M NaCl; flow rate: 10 mL/min; solid line: UV absorbance; dashed line: conductivity)

References

1 U. Gottschalk, K. Brorson, A. A. Shukla, The need for innovation in biomanufacturing, Nat. Biotechnol, 30 (2012) 489-492.

2 R. Ghosh, Protein separation using membrane chromatography: opportunities and challenges, J. Chromatogr. A, 952 (2002) 13–27.

3 D. K. Roper, E. N. Lightfoot, Separation of biomolecules using adsorptive membranes, J. Chromatogr. A, 702 (1995) 3–26.

4 V. Orr, L. Zhong, M. Moo-Young, C. P. Chou, Recent advances in bioprocessing application of membrane chromatography, Biotechnol. Adv, 31 (2013) 450–65.

5 R. Van Reis, A. Zydney, Membrane separations in biotechnology, Curr. Opin. Biotechnol, 12 (2001) 208–11.

6 R. Ghosh, T. Wong, Effect of module design on the efficiency of membrane chromatographic separation processes, J. Membr. Sci, 281 (2006) 532–540.

7 K. D. Program, Affinity adsorption devices prepared from microporous poly (amide) hollow fibers and sheet membranes, J. Membr. Sci, 129 (1997) 31–46.

8 U. Klein, E. Yeager, D. Seshadri, R. Baurmeister, Affinity adsorption devices, J. Membr. Sci, 129 (1997) 31–46.\

9 J. Weaver, S. M. Husson, L. Murphy, A. R. Wickramasinghe, Anion-exchange membrane adsorbers for flow-through polishing steps: part I. Clearance of minute virus of mice, Biotechnol. Bioeng, 110 (2013) 491-499.

10 J. H. Vogel, H. Nguyen, R. Giovannini, J. Ignowski, S. Garger, A. Salgotra, J. Tom, A new large-scale manufacturing platform complex for biopharmaceuticals, Biotechnol. Bioeng, 109 (2012) 3049-3058.

11 R. Ghosh, Laterally-fed membrane module, US Patent Application No. 62074858 filing date November 4, 2014.

12 P. Madadkar, Q. Wu, R. Ghosh, A laterally-fed membrane chromatography module, J. Membr. Sci, 487 (2015) 173–179.

13 P. Madadkar, Raja Ghosh, High-resolution protein separation using a laterally-fed membrane chromatography device, submitted to J. Membr. Sci.

430797 Purification and Recovery of Infectious Virus Particles Using Osmolyte Flocculation

Abstract:

Over the last two centuries, vaccines have played an important role in global public health by saving millions of lives every year. The market is demanding less expensive vaccine manufacturing operations due to the extraordinary growth in vaccine manufacturing in India, China and African union countries. To curb the cost of vaccines, these countries are looking for innovative manufacturing techniques that reduce the cost of virus purification and concentration. Various methods, such as ultrafiltration, nanofiltration, and chromatographic techniques, are available for virus purification. However, novel unit operations are needed to reduce manufacturing costs. A low cost vaccine is essential for the equitable distribution of vaccines among developing countries.

In an effect to provide a lower cost option for vaccine purification, we have developed a novel flocculation process using osmolytes, followed by microfiltration. Osmolytes are naturally occurring compounds found in the cells of many organisms, and their main function is to stabilize intracellular proteins against environmental stresses, such as high osmotic pressure. This is accomplished by changing the water content of the cells. Osmolytes also have the advantage of being common excipients in therapeutic products; therefore removal of the flocculant before formulation is less of a concern. We have demonstrated that osmolyte flocculation is specific to an enveloped and nonenveloped virus, as compared to model proteins. We are using diafiltration for recovery and purification of model virus particles. During diafiltration, addition of flocculating agent improves the purification with successive diavolumes. The flocculated virus particles are easy to filter using a 0.1 micron filter. To date, we have achieved >50% recovery of a non-enveloped virus and >30% recovery of an enveloped virus using 1M mannitol as the flocculating agent. We are exploring filter pore size and flocculation time to optimize the recovery and purity of infectious virus particles. This flocculation process shows promise as a future platform approach for vaccine purification and can be applied to a wide variety of virus particles.

430919 Adsorption Kinetics of Proteins for Increased Understanding of Immobilisation Behaviour

Abstract:

Adsorption Kinetics of Proteins for Increased Understanding of Immobilisation Behaviour

S. Hedberg*, J.Y.Y Heng*, D.R. Williams* and J. Liddell**

*Surface and Particle Engineering Laboratory

Department of Chemical Engineering

Imperial College London, SW7 2BY,

United Kingdom

**Research and Development

Fujifilm Diosynth Biotechnologies

Billingham, TS23 1LH, United Kingdom

ABSTRACT

Immobilisation of proteins and enzymes onto a solid phase has shown to be very important process for both commercial and research uses. A key application for the immobilising of a protein with a solid phase is that it allows the study of protein-protein interactions between a 'free' protein and an immobilised protein.

Protein-protein molecular interactions are known to be involved in protein solution aggregation behaviour which can lead to better understanding of the mechanisms behind protein aggregation and aggregation kinetics. This work includes a number of experimental studies that includes both model proteins and therapeutic mAbs, of different sizes being immobilised onto different standard affinity resins of with various active surface groups.

The maximum amount of protein that can be adsorbed onto the resin and the kinetics of the immobilisation procedure have been investigated. Also the amount of protein has been reduced and the resin type changed to observe the difference in kinetics observed. After the immobilisation procedure has finished the resins with immobilised proteins are washed a number of times with different solution conditions, including high salt buffers, to evaluate the strength of the covalent immobilisation linkage of the proteins. This study links the experimental results to the theory of the kinetics of adsorption. It will also be able to determine the precise amount of protein needed for a certain surface coverage of the resin when there is a shortage of proteins available to improve the economics of the process. These results are useful for the application of Self-Interaction Chromatography that has become an increasingly popular method of determining the osmotic second virial coefficient of proteins as well as the cross virial coefficient between two proteins.

The results have shown that many proteins with amine surface groups used have very high rate of adsorption with at least around 75% of protein immobilised on the resin with aldehyde surface groups with continuous mixing. The amount will differ for individual proteins, with almost 80% immobilised for the therapeutic mAb. However, resins with different surface groups and different solution conditions can easily and significantly affect the extent of immobilisation.

In summary these results obtained will contribute better understanding of adsorption kinetics and equilibria of a number of model and therapeutic proteins.

REFERENCES

1. Quigley, A. (2012) *Protein Aggregation Behaviour and the Second Virial Coefficient,* Ph.D. Thesis, Imperial College London

2. Roberts, C.J., Das, T.K., and Sahin, E. (2011) *Predicting Solution Aggregation Rates for Therapeutic Proteins: Approaches and Challenges.* International Journal of Pharmaceutics. **418**(2): p. 318-333.

416956 Novel Dendron-Based Chromatographic Adsorbents for HIC

Abstract:

The work describes the preparation of a novel dendronized chromatographic adsorbent for hydrophobic interaction chromatography (HIC). This new matrix was chemically synthesized by covalently attaching polyester dendrons with hydroxyl group (-OH) ends to NHS-activated chromatographic media. The dendronized resin was then functionalized through organic carboxylic acid incorporation by reaction with the available –OH groups. This allowed the creation of a dendronized resin with hydrophobic ligands on the periphery. The obtained branched architecture of the dendron allows the attachment of a higher amount of ligands to the resin and generating hydrophobic clusters throughout it. This results in a higher interaction site number and a stronger hydrophobic interaction between proteins and the chromatographic adsorbent. To the best of our knowledge, this is the first work describing a support of such nature.

The purpose of this work was to evaluate the performance of this new HIC support in the separation of proteins of distinct hydrophobicity. In doing so, five protein models (i.e. ribonuclease A (RNasa A), lysozyme, β-lactoglobulin, α-quimotripsyn and bovine serum albumin (BSA)) were loaded separately into the column. The proteins were eluted by applying a linear decreasing gradient of 1.5 or 2M ammonium sulfate (AS)/20 mM Tris-HCl pH7 buffer. Results showed that at 2M AS concentrations, all proteins were adsorbed onto the support and eluted almost at the end of the gradient, which suggests that all proteins were strongly adsorbed onto the support. When 1.5 M AS was used, only RNase A did not adsorb to the support and was eluted in the follow-through. The other sample proteins eluted earlier in the gradient indicating a weaker binding onto the resin. This innovative dendronized support opens a window to a new chromatographic support generation to develop novel downstream processing strategies.

438795 Crystallization Science and Technology: Current Status and Future Directions

Abstract:

Crystallization from solution is a major separation and purification processes used in the chemical , pharmaceutical and food industry. Recent years have seen major advances in our understanding of the crystallization process through improvements in instrumentation and through molecular modeling studies. This lecture will discuss the current state of the art as well as a future research directions.

438796 Lessons Learned from the Use of Unconventional Materials for CO_2 Capture

Abstract:

Having worked on several approaches to CO2 capture since about 2005, I have explored a number of physical and chemical solvents as well as polymer membranes. Initially, a great deal of these of materials were based upon ionic liquids (ILs). Due to a number of challenges encountered in the practical application of ILs due to the demanding requirements in CO2 separation processes, I started to reconsider how best to use ILs to address CO2 capture. Ultimately, more promising and robust materials will not come from ILs themselves, but from retrosynthetic analysis and a reconsideration of which structural variables and properties are (and are not) important. The hybridization of the constituent parts into entirely new, yet seemingly familiar substances with improved properties and economics. This talk will highlight lessons learned from ILs and how these have influenced recent work from my group in the development of new amine solvents and polymer materials that better address the demanding process conditions and requirements of CO2 capture and related separations.

438797 Combining Crystallization and Chemical Reactions in Separation and Purification Processes

Abstract:

Separation and purification processes are created to recover specific compounds from mixtures found in nature or those resulting from chemical transformations. The process selected often must address unique challenges presented by the product of interest. In some cases, integration of reaction and separation can be effective in meeting such challenges. Here, for example, we are concerned with two challenges: (1) overcoming limitations associated with solid-liquid equilibrium in recovering an enantiomerically pure compound (EPC) and (2) enhancing the yield of an intermediate in a complex reaction network. In both of these instances, we show how crystallization and tailored reaction conditions can be coupled to achieve enhanced purity and greater yield. Moreover, since such attributes are achieved using a single stage, the reactive separations offer operational advantages such as reducing the cost, time, and energy consumption of the process.

Separation of EPCs is challenging because such compounds have similar chemical and physical properties, with their structures differing only in their spatial orientation. Crystallization is one of the few techniques that has been successful in resolving a significant number of racemic systems, and here we exploit reactive crystallization to recover enantiomerically pure amino acids from racemic systems by overcoming thermodynamic limitations.

β-lactam antibiotics can be produced by means of an enzymatically catalyzed reaction, but the catalyst also hydrolyzes the antibiotics into undesired by-products. This challenge corresponds to a classic kinetically controlled system where the desired product is an intermediate of the overall reaction. We show how a reactive crystallization scheme that selectively removes the desired product from solution, before it is consumed by undesired reactions, can significantly enhance the yield of the product.

438798 Ionic Liquid Composite Membranes for Separation of CO_2/Light Gas Mixtures

Abstract:

A novel approach for performing these separations is to use room temperature ionic liquids (RTILs) in various morphologies as membranes. In comparison to conventional polymers, they perform gas separations due to solubility differences. They can be converted to polymerizable molecules. This allows promising RTILs to be prepared as membranes. Thus, they can be converted to various morphologies as membranes while maintaining the inherent selectivity of the material. They can be prepared as polymer films, composite structures with ionic liquid within the structure and gels. Polymer films have been tested for gas separations up to 40 bar feed pressure. The conditioning or morphology change due to the incorporation of gases such as CO_2 at these high pressures is reversible. This is an important advantage in comparison to conventional polymers. The wide range of morphologies and chemical structures allow for finding materials that not only have the desired physicochemical properties but also the mechanical properties needed to produce viable membranes.

Mixed matrix membranes (MMMs) look to be a very promising candidate for commercialization and large scale implementation. These MMMs consist of three individual components: polymerizable room temperature ionic liquids (poly(RTILs), normal RTILs, and zeolite particles. These components synergize to produce enhanced gas transport. The polymer matrix allows for facile and economical fabrication, while the zeolite provides excellent separation performance. The RTIL prevents defect formation within the membrane by acting as a wetting agent to help the polymer matrix interface aptly with the zeolite. We have investigated how varying the three components in our MMMs affects membrane permeability and selectivity. This initial study identified the optimal type of zeolite particles. More in depth analysis of the ionic liquid revealed the significance of molar volume in the RTIL and free volume in the poly(RTIL) on the membrane's gas separation ability. Previous work established the correlation between increasing RTIL molar volume and decreasing CO_2 solubility selectivity. Current work shows how the polymerized analogues (poly(RTILs) display a similar trend with free volume. This resulted in the creation of MMMs made with smaller molar and free volume RTILs and poly(RTILs) that performed above Robeson's 2008 upper bound for CO_2/CH_4. With this knowledge we aim to devise a systematic design approach for creating state of the art MMMs.

438799 SAPO-34 Membranes for N2/CH4 Separation

Abstract:

In this presentation the development of SAPO-34 membranes for N2/CH4 separation is discussed. SAPO-34 membranes were synthesized by several routes towards N2/CH4 separation. Membrane synthesis parameters including water content in the gel, crystallization time, support pore size, and aluminum source were investigated. These membranes separated N2 from CH4 with N2 permeance as high as 500 GPU with separation selectivity of 8 at room temperature for a 50/50 N2/CH4 mixture. To our best knowledge, this separation performance is superior to those of the state-of-the-art membranes. For small nitrogen-contaminated gas wells, these SAPO-34 membranes have the potential to compete with the benchmark technology cryogenic distillation for N2 rejection.

434922 Cation Exchange in Forward Osmosis: Elucidating Impacts of Membrane and Solution Chemistry

Abstract:

Thin film composite (TFC) membranes designed for use in forward osmosis (FO) have recently reached the ubiquity previously occupied by the cellulose triacetate membranes. FO processes make use of an osmotic pressure gradient to drive water flux across a semi-permeable membrane. The solution having a higher osmotic pressure, or draw solution, can make use of a myriad of different solutes from nanoparticles, to organic solvents, to electrolytes. Electrolyte draw solutions are commonly used in many large scale applications of forward osmosis processes.

A unique interaction between draw and feed solution containing differing electrolytes solutions and the thin film composite membrane chemistry has been noted in recent studies. This unique interaction is the exchange of cations across the membrane, which is present as high forward and reverse permeation of cations between the draw and feed solutions. It has been hypothesized that this behavior originates for carboxylic acid functional groups forming part of the chemical structure of the polyamide selective layer of TFC membranes. This functional group is shared by weak acid cation exchange resins which are characteristically sensitive to pH. This study seeks to further investigate the cation exchange behavior in TFC membranes with focus upon two primary goals:

1. To understand the importance of pH in affecting the bi-directional transport of cations across the selective layer of a commercial TFC FO membrane.
2. To form a TFC membrane possessing has superior resilience to cation exchange.

In this study potassium chloride has been selected as a model draw solute serving as an analog for the draw solutions envisaged for forward osmosis desalination processes. This selection eliminates the use of a draw solution which contains weak acids or bases, and the use of draw feed solutes having a metal cation allow for easy quantification of ion concentrations using atomic absorption spectroscopy. Water flux and ion fluxes were observed for a feed and draw solution kept at identical pHs, buffered at pHs of 2, 4, 6, 8, and 10. It was found that cation flux dropped significantly at pHs lower than 6.

New TFC membranes tailored for FO were prepared of differing selective layer chemistries to measure the rate of cation transport between the previously mentioned draw solution of potassium chloride and feed solution of sodium chloride. The selective layers of these membranes were formed on a composite support layer prepared from polysulfone cast over electrospun polyvinylidiene fluoride fibers. Membranes having this support layer were found to have low structural parameters and give adequate performance in FO. For the membranes prepared in this portion of the study differences in both water and ion flux were observed.

416070 Carboxyethyl Amine Sodium Salts As Novel Draw Solutes for Forward Osmosis Applications

Abstract:

For the successful implement of forward osmosis (FO) technology, development of suitable draw solutes is a critical issue, which however still faces problems of severe leakage and energy-intensive recovery. In this work, a series of novel carboxyethyl amine sodium salts (CASSs) with different carboxyl group numbers, i.e., diethylenetriamine pentapropionic acid sodium (DTPP-Na), triethylenetetramine hexapropionic acid sodium (TTHP-Na), and tetraethylenepentamine heptapropionic acid sodium (TPHP-Na), are synthesized and employed as draw solutes for FO application. Chemical structures of the synthesized CASSs are examined by ^{1}HNMR and HRMS. Their FO performances are investigated and compared in terms of different physicochemical properties. The effects of the CASS concentration on the osmotic pressure and viscosity of the draw solutions, as well as the resulted FO performance are also systematically investigated. A high water flux of 23.07 LMH and an acceptable reverse salt flux of 0.75 gMH can be achieved with 0.5 g/mL TTHP-Na draw solution under PRO mode, which is superior to most other draw solutes reported in previous literatures. TTHP-Na draw solution is further evaluated to recycle the Congo red solution to examine its applicability for waste water treatment. This study provides useful insights for the molecular design of novel high-performance draw solutes for future FO applications.

414952 Nanometric Graphene Oxide Framework Membrane: Enhancing Heavy Metal Removal Via Nanofiltration

Abstract:

GO framework membranes were developed for effective heave metal removal via nanofiltration. A thin GO framework layer was firstly constructed by the crosslinking with ethylenediamine (EDA) and subsequently assembled on a polycarbonate substrate via a pressure-assisted filtration process. The resultant membrane was further post-treated by hyperbranched polyethyleneimine (HPEI) to alter the charge property on the membrane surface. The EDA crosslinking was found not only to enhance the membrane stability, but also to enlarge the nanochannels between the laminated GO nanosheets for higher water permeability. Different post-treatment agents were compared and HPEI 60K was observed to provide a higher surface charge while lowering the mass transport resistance. In addition, all materials were prepared in aqueous solutions and the as-fabricated GO framework layer was only 70 nm thick, indicating a green and cost effective fabrication process. The newly developed membrane exhibited a high pure water permeability of 5.01 L m^{-2} h^{-1} bar^{-1} and comparably high rejection towards Mg^{2+}, Pb^{2+}, Ni^{2+}, Cd^{2+} and Zn^{2+}. These results have demonstrated the great potential of GO framework materials in waste water treatment and may provide insights for the design and fabrication of the next generation 2D-based nanofiltration membrane.

419495 Highly Permeable Dual Selective Thin Film Composite Membranes for Forward Osmosis with Viscous Draw Solutes

Abstract:

Recently, a wide range of large draw solutes have been successfully developed and employed in the forward osmosis (FO) process with the advantages of high water flux, minimum reverse flux and easy recovery. However, the high viscosity of these large draw solutes tends to suffer from more severe internal concentration polarization (ICP) effects and internal fouling of the draw solutes in the membrane sublayers. To solve this problem, a dual selective flat sheet membrane has been designed in this study in order to prevent the viscous draw solutes from entering the porous support layer and minimize the ICP effects. Dense and thin polyamide layers have been fabricated on both the top and bottom surfaces of highly porous cellulose ester support through interfacial polymerization (IP). The dual selective thin film composite (TFC) membrane demonstrated reasonably high water flux and low salt reverse flux using sodium chloride as draw solutions at the concentrations ranging from 0.25M to 1M. When tested with viscous draw solutes including sucrose, ferric citric acid complex (Fe-CA) and polyethylene glycol monolaurate (PEG 640ML), significantly mitigated ICP effects and reduced fouling propensity were observed compared to single selective TFC membranes. This study proves the great potential for the combined use of the newly developed dual selective membrane with viscous draw solutes for various FO applications.

419344 Membrane Based Separation in Metallurgy: Current Status and Perspective

Abstract:

Metallurgy is concerning metal extraction and purification. It is a very important or pillar industry in many countries worldwide. Nevertheless, in its traditional manner it has significantly negative impacts on environment. In the past decade of years, the body of literature on membrane application in metallurgy to improve its environmental friendliness and production efficiency has expanded rapidly and the new trends observed are listed below:

1) The previous ideas have been studied intensively by researchers from other countries with the advantages and disadvantages being analysed critically.

2) Incremental innovations either by modification of original technologies or transferring the technologies to the process concerning other groups of metals are emerging.

3) Various novel processes that combine membrane separation with metallurgy are proposed and investigated.

4) More and more attention is also given to tackling metallurgical waste using membrane process.

5) A distinct branch which is called "Special Industrial Membrane" mainly targeting metallurgy is growing rapidly. In addition to special polymeric membranes, inorganic and metal membranes attract great attention in recent years and are on the way of being commercialized.

With this presentation, the latest development of sustainability of metallurgy approached within the framework of membrane separation technologies will be summarized. We will goes through the above five aspects, trying to understand the strength and identify the future of membrane-based separation in this area, by describing and analyzing the real-world and potential design and applications.

419457 Nanofiltration Membrane Consisting of a Sulfonated Pentablock Copolymer Selective Layer for the Removal of Heavy Metals

Abstract:

Heavy metal pollution has become a major problem in numerous countries due to the indiscriminate discharge of untreated wastewater. The situation is expected to be further aggravated in the near future by rapid population growth and economic development. As heavy metals are toxic and non-biodegradable, their bio-accumulation in human body will cause the life dysfunction. As such, there is an urgent need to address the current situation to prevent further damage to the environment and human health. Among the methods available for wastewater treatment, nanofiltration (NF) is a feasible and promising option. However, the development of new membrane materials is constantly required for the advancement of this technology. In this study, we attempt to develop a composite NF membrane consisting of a molecularly designed pentablock copolymer rejection layer for the removal of heavy metal ions. The resultant NF membrane has a reasonably high pure water permeability (*A*) of 2.4 LMH/bar but a relatively small mean effective pore diameter of 0.50 nm, and a molecular weight cut-off of 255 Da. . The newly developed NF membrane can effectively remove heavy metal cations such as Pb^{2+}, Cd^{2+}, Zn^{2+} and Ni^{2+} with a rejection greater than 98.0%. In addition, the membrane also shows reasonably high rejections towards anions such as $HAsO_4^{2-}$ (99.9%) and $HCrO_4^-$ (92.3%). This performance can be attributed to (1) the pentablock copolymer's unique ability to form a continuous water transport passageway with a defined pore size and (2) the incorporation of polyethyleneimine (PEI) as a gutter layer between the selective layer and the substrate. To the best of our knowledge, this is the first reported NF membrane comprising this pentablock copolymer as the selective material. The promising preliminary results achieved in the current study may provide a useful platform for the development of new NF membranes for heavy metal removal.

435602 An Economical and UV Assisted Crumb Rubber Device for Harnessing Drinking Water from River

Abstract:

Drinking water shortage is becoming a very challenging problem for future due to the impact of calamities such as climate change, storms, burst of population and sinking water tables. In the last decade new techniques such as nano filtration, reverse osmosis, and solar distillation have been developed. Many developing countries find the new techniques expensive. In the work conducted at COMSATS Lahore, use has been made of recycled crumb rubber which is originally a source of breeding of mosquitoes. Crumb rubber fortunately acts as a very good filtration media because of its elasticity and porosity. In the experimental work, sizes of crumb rubber ranging from 1.2 – 4.2mm and 5.0 – 6.5mm were used to create columns for filtration. The average porosity was 0.50. The pilot plant consisted of a rectangular glass chamber 1.0m height and 0.62m width with a gravel support 0.3m. Glass tubes were inserted in the column to count the head loss. A UV reactor with immobilized nano-TiO_2 was attached for intake of the effluent after filtration. River water with turbidity of 238(NTU) and coliform count of 238(MPN/100ml) was used as effluent. After filtration it was found that the crumb rubber size 1.2 – 4.2mm was found to be highly effective in reducing the turbidity from 238(NTU) to 4.2(NTU) with 90% efficiency. The head loss was found to be minimum at a depth of 0.3m compared to other media depths. The turbidity decreased with increasing filtration rate of $12m^3/hm^3$. On final treatment by the photocatalytic reactor no presence of coliform was observed. The drinking water quality was in accordance with WHO specifications. The above technique has been found to be cost effective and it can be scaled up for large scale applications.

428788 Absolute and Net Adsorption in the Ideal Adsorbed Solution Theory

Abstract:

At the last AIChE meeting we saw contributions based on two approaches that applied the Ideal Adsorbed Solution Theory. The first is based on the definition of the Gibbs adsorption isotherm from absolute adsorption and corresponds to the original approach of Myers and Prausnitz (1965). The second is based on the derivation from the net adsorption framework of Talu (2013) and leads to a series of equations which are similar but not the same. Given the fundamental importance of the IAST in the treatment of multicomponent adsorption equilibria we have investigated the two approaches in detail and are able to show that in fact the original formulation based on the absolute adsorption is correct. We will identify the problem in Talu's derivation and show that a correct derivation even starting from the Net adsorption framework should lead to the same set of equations for the IAST. Furthermore we will show that the Net-IAST approach of Talu leads to numerical complications and physically impossible solutions when implementing it to mixtures were a strongly adsorbed species is combined with a weakly adsorbing one.

431057 High Throughput Separation By Rapid Pressure Swing Adsorption: Bench Scale Demonstration of Model Binary Gas Mixture

Abstract:

Pressure swing adsorption (PSA) is a cyclic gas separation technique that is just over 50 years old. It has many successful commercial applications including gas dehydration, air separation, hydrogen purification, and solvent vapor recovery. More applications are also on the horizon, including CO_2 capture from flue gas

Traditional, commercial PSA systems utilize cycle times on the order of minutes to tens of minutes, with individual step times no shorter than 0.5 min (30 s) or so. However, since the mid 80's there has been a great deal of interest in trying to reduce the cycle time to make the columns smaller and thus increase the feed throughput, which is the amount of gas processed per unit time per unit mass of adsorbent. These so called rapid PSA processes have met with limited success, however, until only recently.

Within the past 10 years, SeQual Technologies in San Diego and QuestAir in Vancouver (among less notable others) developed the rapid PSA concept into viable, commercial processes using rotary valve and rotary bed technology, respectively. They both have commercial processes that operate with cycle times on the order of a few seconds, with individual step times of less than 1 s. SeQual has been focusing on medical oxygen technology, while QuestAir has been focusing on hydrogen technology.

However, a paucity of work has been done on other rapid PSA applications; and overall, the understanding of the rapid PSA concept is quite limited with little fundamental information available. This leaves considerable room not only for the development of new theories and mechanisms underlying the rapid PSA concept, but also for the development of new applications for rapid PSA processes. The goal of this project has been to shed some light on both of these notions, theoretically, experimentally and via process modeling.

To this end, a 1-bed rapid PSA system has been constructed and is currently being utilized to separate a model binary gas mixture consisting of 15 vol% CO_2 in N_2 using beaded 13X zeolite over a wide range of cycle times and process conditions. Step times as short as 0.25 s are possible in this 1-bed system under both positive and negative gauge pressures and at different temperatures. A variety of PSA cycle schedules are being studied, including light reflux, heavy reflux and/or dual reflux steps, and including equalization steps with tanks. The ultimate goal is to demonstrate the separation of this model binary gas mixture at exceptionally high feed throughputs. This presentation will discuss the latest results obtained with this unique 1-bed rapid PSA system.

421184 Integrated Adsorbent-Process Optimization for Carbon Capture and Concentration Using Vacuum Swing Adsorption Cycles

Abstract:

A large number of adsorbents are being regularly reported in the literature claiming their suitability for carbon capture and concentration (CCC) from flue gas. One thing common in all these adsorbents is they all show higher selectivity of carbon dioxide over nitrogen. Adsorbents that have been tested for the transport mechanism of the diffusing gases in the pores show macropore molecular diffusion as the controlling step. As such, it is possible to simulate a CCC process for each adsorbent using only equilibrium isotherm information, a suitable mixture equilibrium model based on single component parameters and molecular diffusivity as inputs, and determine the optimum operating conditions as decision variables under which the process will give minimum energy or maximum productivity while meeting the purity-recovery constraints imposed by the regulatory agencies. We have developed such a simulation and optimization platform for CCC using Vacuum Swing Adsorption (VSA). This platform provides a reliable tool to evaluate adsorbents for their suitability for CCC. It has been applied to many adsorbents reported in the literature.

Here we propose a novel approach for integrated adsorbent and process design. The scope of our aforementioned simulation and optimization platform is expanded to include adsorbent isotherm characteristics, in addition to the process operating conditions, as decision variables. By including adsorbent isotherm characteristics as decision variables, we are now able to ask what is the minimum energy or maximum productivity possible in a VSA process for CCC and what are the shapes of the corresponding isotherms? We have answered these questions for two VSA cycles, namely a 4-step process with light product pressurization (LPP) and a 6-step process with light and heavy reflux in addition to LPP. We have established the minimum energy and maximum productivity bounds for these two cycles as functions of the evacuation pressure. While the 4-step VSA cycle enjoys lower energy penalty advantage, it requires a very low evacuation pressure and fails to achieve the purity-recovery constraints if the evacuation pressure is increased beyond 0.04 bar. In contrast, the 6-step cycle enjoys a wider operating window for the low evacuation pressure and higher productivity, but these come at the expense of increased energy penalty.

The integrated adsorbent-process optimization adopted here directly gives the shape of the isotherm necessary to achieve a certain target for a chosen VSA cycle along with the necessary operating conditions. It may be viewed as adsorbent design by process inversion. The results obtained set benchmarks for the development of new adsorbent and VSA cycles for CCC.

425232 Unusual Nitrogen-Oxygen Adsorption Properties of Large Pore Titanium Silicate (ETS-10)

Abstract:

ETS-10 contains structural titanium atoms cross-linked by silica tetrahedral. Almost straight-linear (like rods) octahedrally coordinated titania chains extend in two different crystallographic directions displaced by ½ u.c. Each titanium imparts negative two charge to the structure which are balanced by exchangeable cations typically as Na+ and K+ during synthesis. The structure results in a 3D channel system prescribed by 10- and 12-membered oxygen rings in two directions, therefore ETS-10 can be classified as a large-pore molecular sieve zeolite. Unlike its cousin ETS-4, ETS-10 structure is very robust and can withstand high temperatures up to 450 C and a wide range of pH. Although ETS-4 was heavily studied and even resulted in some applications (i.e. Molecular Gates) due to fine tunable pore size within 0.1 A, information about ETS-10 adsorption almost does not exist in literature. In this work, we have looked at the oxygen and nitrogen adsorption properties of ETS-10 as pure and binary isotherms. The data show several unexpected phenomena quite different than other zeolites with respect to pure component heat of adsorption, dependency of nitrogen selectivity on coverage, and impact of ion exchange. This presentation will attempt to explain these unusual observations based on structure information and try to postulate if and how these behavior can be exploited in PSA applications.

428191 Speedy Standing Wave Design of Size-Exclusion Simulated Moving Bed: Solvent Consumption and Sorbent Productivity Related to Material Properties and Design Parameters

Abstract:

Size-exclusion simulated moving beds (SEC-SMB) have been used for large scale separations of linear alkanes from branched alkanes. While SEC-SMBs are orders of magnitude more efficient than batch chromatography, they are not widely used. One key barrier is the complexity in design and optimization. A four-zone SEC-SMB for a binary separation has seven material parameters and 14 design parameters (two yields, five operating parameters, and seven equipment parameters). Previous optimization studies using numerical methods do not guarantee global optima or explicitly express solvent consumption (*D/F*) or sorbent productivity (P_R) as functions of the material properties and design parameters.

The Standing Wave concept is used to develop analytical expressions for *D/F* and P_R as functions of 14 dimensionless groups, which consist of 21 material and design parameters. The resulting Speedy Standing Wave Design (SSWD) solutions are simplified for two limiting cases: diffusion or dispersion controlled. An example of SEC-SMB for insulin purification is used to illustrate how *D/F* and P_R change with the dimensionless groups. The results show that maximum P_R for both diffusion and dispersion controlled systems is mainly determined by yields, equipment parameters, material properties, and two key dimensionless groups: (1) the ratio of step time to diffusion time and (2) the ratio of diffusion time to pressure-limited convection time. A sharp trade off of *D/F* and P_R occurs when the yield is greater than 99%. The column configuration for maximum P_R is analytically related to the diffusivity ratio and the selectivity. To achieve maximum sorbent productivity, one should match step time, diffusion time, and pressure-limited convection time for diffusion controlled systems. For dispersion controlled systems, the axial dispersion time should be about 10 times the step time and about 50 times the pressure-limited convection time. Its value can be estimated from given yields, material properties, and column configuration. Particle size and 14 design parameters can be optimized for minimum *D/F*, maximum P_R, or minimum cost within one minute on a laptop computer.

EXPERIENCES IN INORGANIC FOULING REMOVAL FROM CONDENSATE POLISHING RESIN USING WEAK-ACID BASED CLEANER IN BADAK LNG

Ertanto Vetra, Badak LNG, Bontang, Indonesia
Inta Wiyatmoko, GE Water & Process Technologies, Jakarta, Indonesia

Introduction

Badak LNG (PTB) is a natural gas liquefaction plant situated in Bontang, East Kalimantan, Indonesia. Operating since 1977, PTB currently has eight processing trains, namely Train-A through Train-H, with a maximum sustainable capacity of 22.1 MTPA (Million Tonnes per Annum). The process of liquefying natural gas requires a refrigeration cycle to reduce the natural gas temperature and eventually change the phase of natural gas to liquid. In PTB, the refrigeration process is powered by three large centrifugal compressors which are driven by high pressure steam (62 kg/cm^2 g). The high pressure steam is also utilized in a turbine power generator to generate electricity for plant and community use.

To generate the above-mentioned high pressure steam, a total of twenty one boilers are employed with a total steam generating capacity of 61 MTPA. Deep water well is used as the raw water source for the boiler feed water which is treated through a combination of aeration-filtration, ion exchange demineralization, and de-aeration process. The simplified flow diagram of the steam-condensate cycle in PTB can be seen in **Figure-1**.

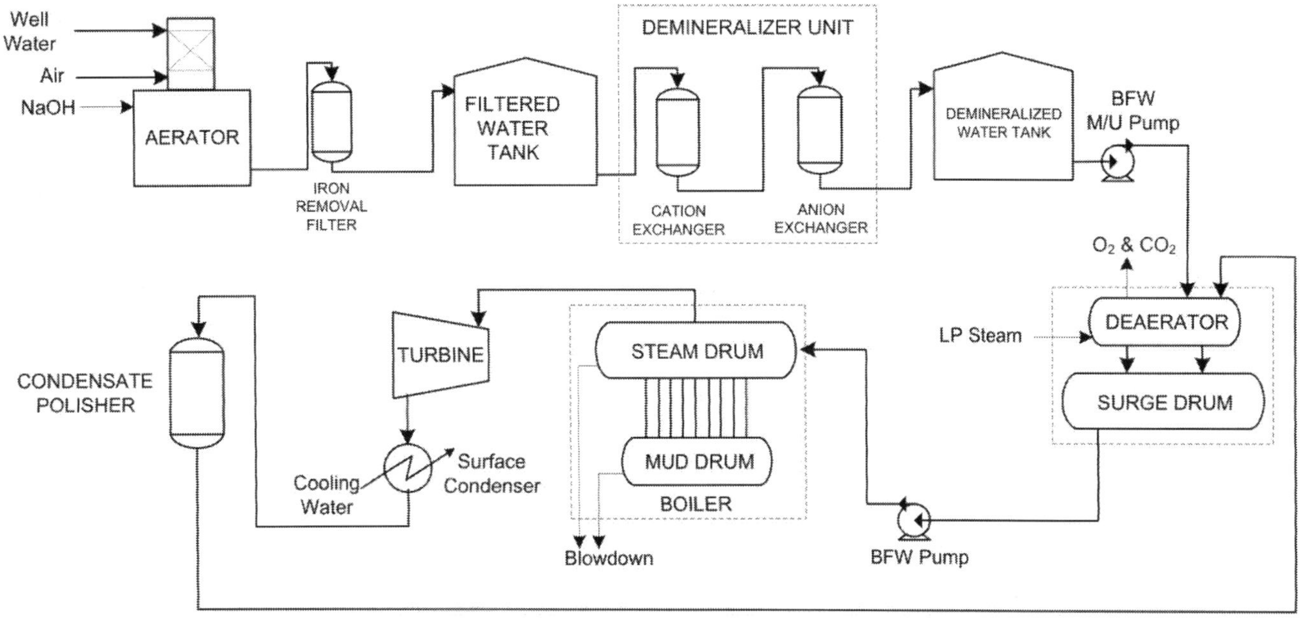

Figure-1. Simplified Steam-Condensate Cycle in Badak LNG

Around 5% of the generated steam is loss throughout the system via boiler blow down, steam trap discharges, and various leakages. The remaining 95% can be recovered as condensate and treated in Condensate Polishers. A total of 15 (fifteen) condensate polishers, filled with macro-porous strong acid cation resin, are employed at the condensate return system to remove corrosion product, mainly iron and copper, from condensate stream before it is sent back to the deaerator.

Problem Background

Condensate polisher treats condensate water for the removal of ions that have been picked up during the water cycle. Irreversible sorption or the precipitation of a foulant within resin particles can cause deterioration of resin performance. If the resin becomes fouled with iron or copper oxides, they will inhibit proper exchange by coating the resin beads. The beads will also weaken physically. During internal inspection of one the condensate polisher (36V-7B), signs of inorganic fouling was visually observed. Further resin analysis has shown that build up of inorganic foulant has occurred inside the resin. Initial effort to remove the foulant was done through regular regeneration; however this only removes around 40% of inorganic foulant. The next effort is to do resin cleaning to remove the inorganic fouling material.

There are several methods that are typically recommended from the manufacturer to remove inorganic, specifically iron, fouling from the resin, i.e.: Air brushing, HCl treatment, and treatment by reducing agent such as $Na_2S_2O_4$ (Sodium dithionite). Air brushing is basically a physical process which involves agitating the resin by injection of compressed air at the bottom of the bed, followed by backwash. However, this method is only effective for removing insoluble, loose and easy to loosen particulate iron (crud). The next method, HCl treatment, is used to remove crud and soluble contaminants. However, it possesses certain safety and health risk especially during material handling and the cleaning process itself. In addition, although the polisher vessel has internal lining, but the pump, valve, and piping system is not designed to withstand strong acid service, thereby adding the risk of material damage from acidic corrosion. The last method is treatment by reducing agent such as Sodium dithionite. Sodium dithionite is a powerful reducing agent and when applied to an iron fouled resin bed will reduce any ferric iron present to the soluble ferrous form. However, this also has a significant health and safety risk from the formation of strong fumes of an obnoxious nature during the mixing process. It is also classified as a flammable solid and is shipped under the appropriate caution label. The solution itself is relative unstable especially when in contact with air which results in oxidation of sodium dithionite. Thereby mixing or agitation using air is not allowed.

An alternative to the abovementioned methods is cleaning using citric acid based cleaner (Aquamax IEC-2). It is a blend of an acid chelant, an organic phosphonate and a corrosion inhibitor specially formulated to assist in restoration of ion exchange resins. It is designed to solubilise and disperses many metallic and inorganic foulants, including iron, copper, aluminium and hardness salts. Major constituent of Aquamax IEC-2 is citric acid (30-60 %-wt), which is less corrosive and less hazardous than hydrochloric acid. Citric acid also present in lower concentration (10-30 %-wt) in the Steamate Regen-3 which is regularly used in PTB for Polisher regeneration. This ensures the compatibility of the cleaner with the existing steam and condensate system. This resin cleaner is intended to solubilise and disperse the inorganic foulants from resin beads.

Evaluation

Cleaning Procedure

The cleaning process is conducted offline and onsite, which means after the polisher is taken out of service, the resin does not need to be unloaded from the vessel, thereby saving downtime and maintenance work. After the vessel is taken offline, the following steps are conducted:

1. Open upper manway of the vessel to visually check the internal condition of the resin,
2. Conduct overflow by using backwash stream so that any debris/foams that may form at the water surface can be removed through the upper manway,
3. After the water surface is visually clean (or can be measured by turbidity level less than 30 NTU), reduce the water level by draining the vessel until the level is approximately 4-6 inches above the resin layer. Ensure that the resin is still submerged in water to prevent resin dry out,
4. Take resin sample for analysis, (pre-cleaning sample),
5. Add the Resin cleaner into the cation polisher, the recommended dosage of this cleaner is 40-80 L-cleaner/m^3-resin, depending on the severity of the fouling,
6. Conduct air bubbling for approximately 2 hours to slowly mix the resin bed with the cleaner solution, this can be done through two methods:
   a. Using external air sparger/lance (**Figure-2a**). For this method, to ensure good mixing, change the lance position every 15 minutes to cover all area inside the vessel.
   b. Using internal lateral distributor (**Figure-2b**),

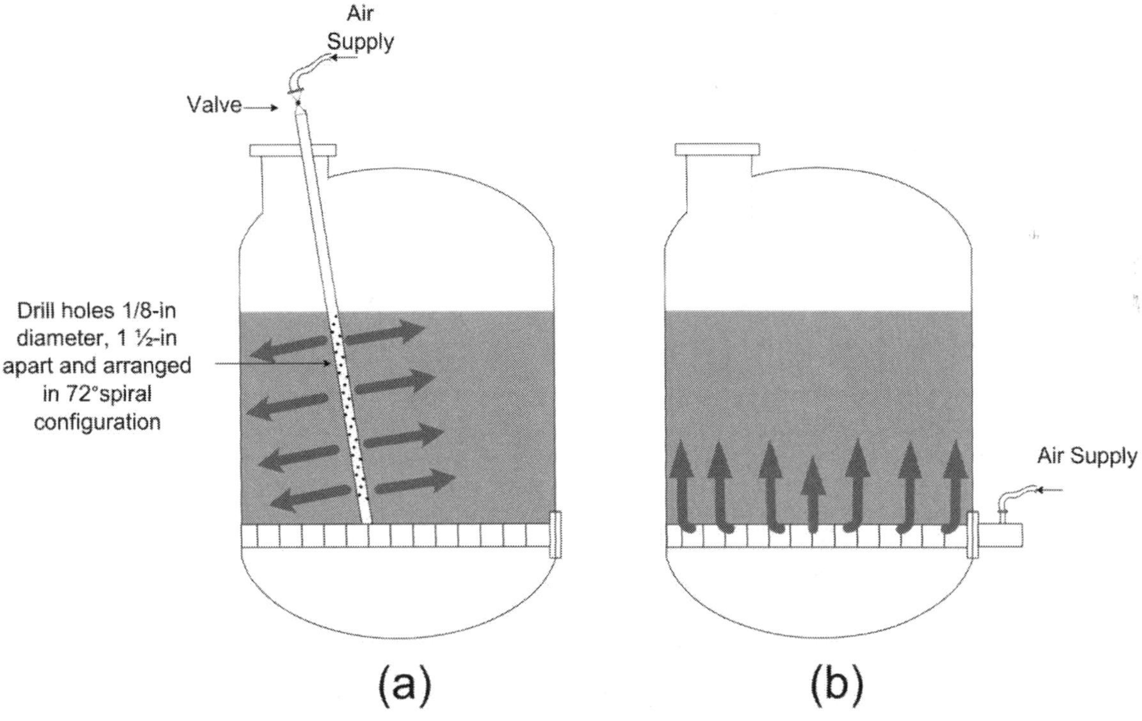

Figure-2 Air Bubbling Method (a) Using external air lance/sparger (b) Using internal lateral distributor

7. Stop the air bubbling and allow the cleaner to soak for 2 hours,
8. Take resin sample for analysis, (post-cleaning sample),
   a. If the visual inspection shows that the resin is relatively clean, proceed to step No. 9 and monitor the turbidity in rinse water until less than 10 NTU.
   b. If the visual inspection shows that the resin is still fouled, repeat the cleaning procedure (step No. 5-6). To initiate a repeat cleaning, drain and refill the vessel with fresh condensate water before adding the cleaner.
9. Close the upper manway,
10. Rinse the resin bed until the pH at the outlet stream to sewer is stable,
11. Conduct double regeneration (2x normal regeneration dosage)
12. Reopen the upper manway and take resin sample (post-regeneration sample),
13. Close the upper manway and vessel is ready to be put back into service

Results and Analysis

Until 2015, six polishers (**Table-1**) have been cleaned using citric acid based resin cleaner. The first two vessels to be cleaned (36V-7A & 36C-7F) was chosen as a trial test before plant-wide implementation takes place. The selection was made based on consideration of regeneration schedule, which is conducted every 2 years in PTB, since the cleaning process includes double regeneration process afterwards. The remaining four vessels are also chosen based on the regeneration schedule. Typically to determine whether it's time to clean or replace the resin, the following standard guidelines are usually used:

1. Cleaning of resin should be considered when total metallic fouling exceeds approximately 3,500 ppm
2. Replacement of resin should be considered when: capacity is less than 75% of new resin guidelines or moisture content exceeds 115-125% of new resin guidelines

Table-1 Summary Information for Condensate Polisher Cleaning

Vessel Tag	Date	Resin Type	Resin Age [years]	Resin Volume [m^3]	Dosage [L/m^3]	Repetition [Times]	Bubbling method
36V-7A	Dec-13	Amberlite IR200C	30	6.8	40	2	Air Lancer
36V-7E	Dec-14	Amberlite IR200C	30	8.0	40	2	Air Lancer
36V-7F	Dec-14	Amberlite IR200C	30	8.0	40	2	Air Lancer
36C-7A	Dec-14	Mitsubishi PK228	25	10.7	40	2	Internal Distributor
36C-7C	Dec-14	Mitsubishi PK228	25	10.7	40	2	Air Lancer
36C-7F	Dec-13	Amberlite IR200C	16	10.7	65	1	Air Lancer

Comparison of inorganic fouling content of the resin before the cleaning, after the cleaning, and after regeneration is given in **Figure-7/8/9.** In overall, the reduction of inorganic fouling content is up to 38% after cleaning process and up to 72% after regeneration process. It can be seen that after regeneration, the inorganic content of all the six vessels has been reduced to below the recommended guidelines of 3,500 ppm as $CaCO_3$.

There are two kinds of resin fouling, external and internal fouling. Both of this fouling is usually analyzed and reported in resin analysis. External fouling (mud, silt, iron, oil, etc) usually covers the surface of ion-exchange resin particles, thereby blinding the exchange sites on the surface as well as the pores leading to the internal exchange sites. Early stage of external fouling can be mitigated by conducting periodic sub-surface backwash. However, for heavily fouled resin bead, backwashing alone will not be effective and the resin must undergo a cleaning process.

Internal fouling (iron, copper, manganese, etc) occurs inside the resin bead on the internal exchange sites. This internal fouling is more difficult to remove than external ones. If resin bead has been covered by external and internal fouling, resin regeneration alone will not be effective, because the regenerant will try to remove the external fouling first before it can penetrate deeper into the internal exchange sites. In most cases, the regenerant dosage is simply insufficient to remove all the fouling material, which eventually leads to inability to regenerate all of the exhausted resin.

PTB uses amine based regenerant for this Cation Polisher, which contains a small amount of citric acid that helps remove a small amount of inorganic fouling. However, as explained earlier, if the resin bead has been covered by external and internal fouling, normal resin regeneration alone will be ineffective. The presence of resin cleaner will assist the overall removal process, since the cleaner will remove the external fouling so the succeeding regenerant can penetrate deeper into the resin pores and remove the internal fouling effectively. **Figure-5** also shows that the regeneration step helps reduce the inorganic content after the cleaning is conducted.

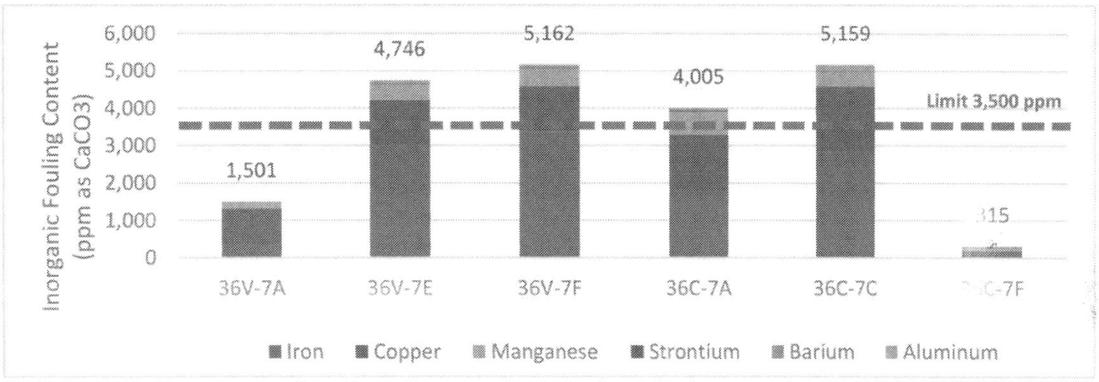

Figure-3 Inorganic Fouling Content of Cation Polisher Resin before Cleaning

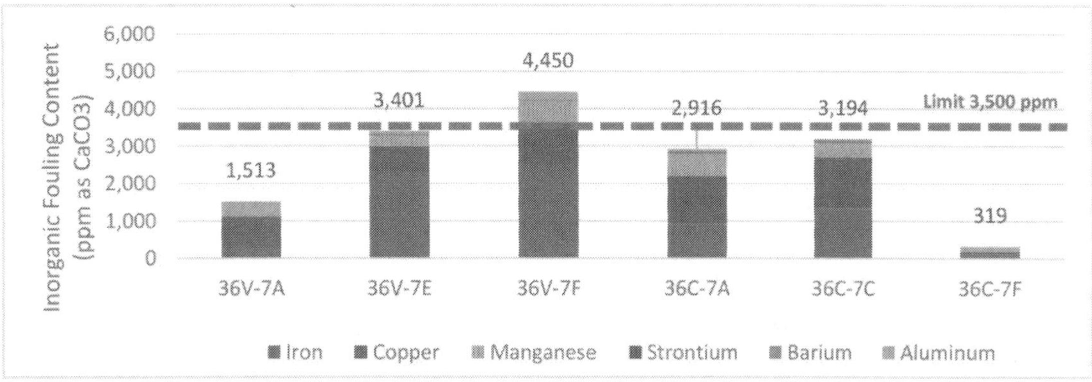

Figure-4 Inorganic Fouling Content of Cation Polisher Resin after Cleaning

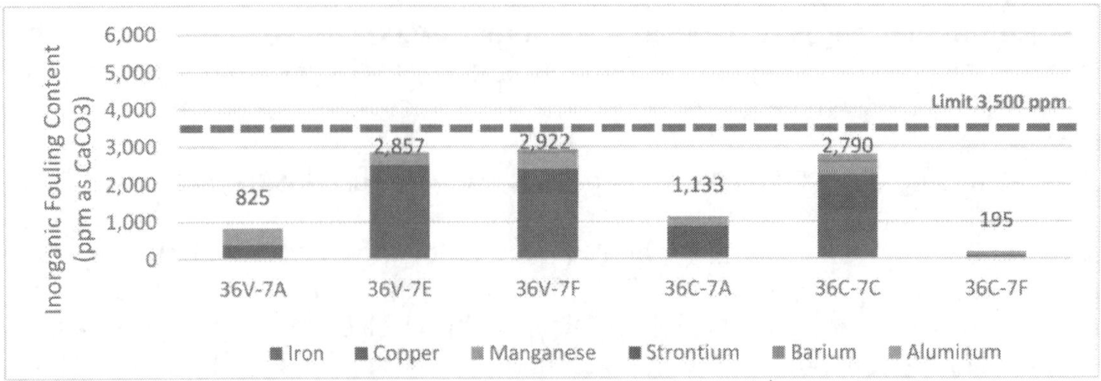

Figure-5 Inorganic Fouling Content of Cation Polisher Resin after Regeneration

Previous figures also show the typical distribution of inorganic content inside the resin. Majority of the foulant is in form of iron and copper, which came from corrosion products. Iron mainly came from the corrosion of condensate piping while copper came from corrosion of copper-nickel tubes in the turbine surface condensers. A closer look on the degree of removal from those two metallic compound during cleaning and regeneration is laid out in **Table-2** and **Table-3**.

Table-2 Concentration and Degree of Removal for Iron Fouling

Vessel Tag	Total Fe Fouling			% Removal	
	Before Cleaning	After Cleaning	After Regeneration	After Cleaning	After Regeneration
	[ppm as $CaCO_3$]				
36V-7A	340	286	48	16%	86%
36V-7E	3,036	2,321	1,964	24%	35%
36V-7F	3,393	2,500	1,714	26%	49%
36C-7A	1,775	915	70	48%	96%
36C-7C	2,857	1,339	1,286	53%	55%
36C-7F	120	127	15	0%	87%

Table-3 Concentration and Degree of Removal for Copper Fouling

Vessel Tag	Total Cu Fouling			% Removal	
	Before Cleaning	After Cleaning	After Regeneration	After Cleaning	After Regeneration
	[ppm as $CaCO_3$]				
36V-7A	973	832	345	15%	65%
36V-7E	1,180	661	566	44%	52%
36V-7F	1,196	1,133	708	5%	41%
36C-7A	1,523	1,287	816	15%	46%
36C-7C	1,731	1,353	960	22%	45%
36C-7F	86	63	82	27%	5%

It can be seen from above tables that the cleaner has successfully removed inorganic fouling; with removal degree varies between 35-96% for iron and 5-65% for copper. For 36V-

7A and 36C-7F, some anomaly is observed, with very high iron removal rate (86-87%) for both of the vessels, but very low copper removal rate for 36V-7F. This anomaly can be caused by the relatively low initial inorganic content as shown in **Figure-3**. This may contributes to sampling error caused by uneven sample distribution inside the vessel.

Another important thing to notice between 36V-7A and 36C-7F is the different dosage and cleaning repetition applied (**Table-1**). This was done to evaluate which dosage and repetition was more effective. After the trial, it was concluded that lower dosage with more repetition give better result compared with higher dosage with no repetition. One possible explanation is the repetition process increases the contact time between the cleaning agent and the resin, hence better removal. Therefore, for the subsequent cleaning program, the selected dosage was 40 L/m^3 with two cleaning repetition.

For the heavily fouled polishers (Fe>1000 ppm as $CaCO_3$), 36C-7A shows higher removal percentage (96%) compared with other vessel. One possible contributing factor is the different in bubbling method during the cleaning process (see **Table-1**). Bubbling in other vessels is done using an external air lancer. Whereas in 36C-7A, the bubbling is done through existing pressure gauge connection at the bottom of the vessel. This method of bubbling takes use of existing lateral distributor inside the polisher vessel, thereby ensuring more uniform and vigorous mixing of air which eventually resulted in a better contact between the resin and the cleaner.

The visual comparison of resin condition prior and after the cleaning process is given in **Figure-6.a** and **Figure-6.b**. It can be seen that prior to cleaning, reddish-brown colored spots are evident at the surface of the resin. This indicates the occurrence of inorganic fouling on the resin, particularly iron fouling. After the cleaning, the amount of reddish-brown stain is reduced. The regeneration step at the end of cleaning also improved the fouling removal as evident by the visual appearance of the resin shown in **Figure-6.c**. A closer look at individual resin shows the difference between heavily iron-fouled resin (**Figure-7.a**) and clean resin (**Figure-7.b**)

(a) **(b)** **(c)**

Figure-6 Visual Sample of Cation Polisher Resin (a) Before cleaning (b) After cleaning (c) After regeneration

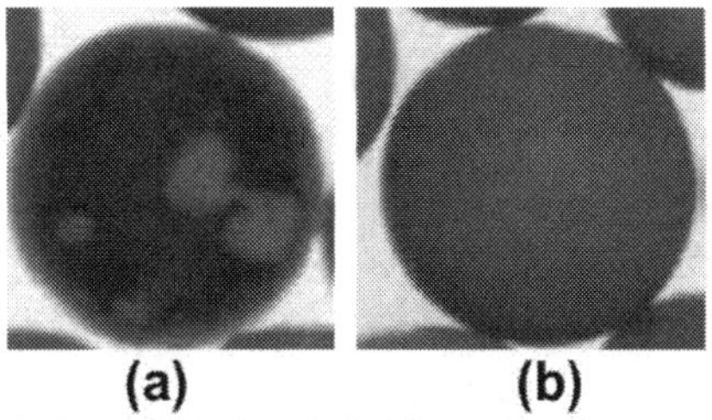

(a) **(b)**

Figure-7 Cation Polisher Resin (a) Heavily fouled with Iron (b) Clean

Potential Economic Benefits

To analyze the benefit of conducting resin cleaning, comparison was made with the total cost required for resin replacement. Resin replacement costs consist of the cost for the resin itself and the cost of initial regeneration. The initial regeneration is typically between 3-4 times the dosages of normal regeneration if the resin came in H-form. However, usually new resin is shipped in Na-form which is more stable but requires additional step of conversion from Na-form to H-form using acidic solution such as citric acid, HCl, or H_2SO_4. In this evaluation, it is conservatively assumed that the resin is already in H-form and it only requires triple dosage for the initial regeneration. Meanwhile, the cost of resin cleaning consists of the cost of cleaner itself and the cost of double regeneration. The cost comparison for various resin volume is given in **Table-4.** It can be concluded that resin cleaning only cost around a third of the cost for total resin replacement, with potential saving varies between $43,700 and $65,850 per vessel.

Table-4 Cost Comparison between Resin Replacement and Resin Cleaning

Resin Volume	Replacement Cost		Cleaning Cost		Total Cost		Potential Saving
[m^3]	Resin Cost	Initial Regeneration Cost	Cleaner Cost	Regeneration Cost	Resin Replacement	Resin Cleaning	
6.8	$40,800	$36,000	$9,100	$24,000	$76,800	$33,100	$43,700
8	$48,000	$48,000	$10,850	$32,000	$96,000	$42,850	$53,150
10.7	$64,200	$48,000	$14,350	$32,000	$112,200	$46,350	$65,850

Conclusions

1. Citric acid based cleaner has been proven to successfully remove inorganic fouling from cation polisher resin with degree of removal varies between 35-96% for iron and 5-65% for copper.
2. Citric acid based cleaner offers good cleaning performance with less safety and health hazards compared with other conventional cleaning method such as cleaning with Hydrochloric Acid.
3. Resin cleaning can potentially save up to $65,850 per vessel compared with total resin replacement cost.

References

1. Dow Water Solutions, "Dowex Ion Exchange Resins: Water Conditioning Manual", pp. 57
2. GE Water & Process Technologies (2007), "Aquamax IEC-2 Ion Exchange Resin Cleaner Fact Sheet", pp. 1
3. Purolite Service Bulletin (1999), "The Fouling of Ion Exchange Resins and Methods of Cleaning", pp. 6-7

433294 Dynamic Adsorption Process Simulator: Evolution of the Energy Balance on the Approach to Periodicity

Abstract:

The dynamic adsorption process simulator (DAPS) developed over the past fifteen years has slowly transformed into a very rigorous, very robust and very useful adsorption process simulator. It has been used and is being used to develop and study countless cyclic adsorption processes including but not limited to air separation, air purification, air revitalization, hydrogen purification, solvent vapor recovery, gasoline vapor recovery, natural gas purification, refinery gas upgrading, ammonia purification, isotope separation, etc. DAPS is capable of simulating the most complex PSA cycle schedules including the 12-bed 13-step hydrogen purification process, which utilizes 3 to 4 layers of different adsorbents in a column. It can handle multiple input and output flows along the column, not just at the ends. DAPS can also equalize bed-to-bed or bed-to-tank-to-bed, depending on the demands of the cycle schedule. Finally, it can be coupled with a reactor, either as a layered or mixed bed of adsorbent and catalyst.

DAPS accounts for mass, energy and momentum changes throughout every step of every cycle of a pressure swing adsorption (PSA), temperature swing adsorption (TSA) or concentration swing adsorption (CSA) process until the periodic state is reached. It utilizes many assumptions, like all process simulators; but, DAPS still accounts for velocity changes due to adsorption, temperature, and pressure changes along the column. It accounts for mass transfer effects via a liner driving force expression and heat transfer effects via an overall heat transfer coefficient. To ensure accuracy, the mass and energy balances are checked for closure from cycle to cycle during the approach to periodic behavior.

The energy balance closure looks at the energy in the input and output flows, the energy accumulated in the gas and adsorbed phases, the energy gains and losses due to adsorption and desorption, and the energy losses and gains through the wall of the bed. This cycle to cycle energy balance closure reveals some surprising dynamics that appear in the first cycle and the then they change from cycle to cycle until they finally disappear at the periodic state after many ensuing cycles. Because these energy balance closure dynamics are real and thus exist in actual cyclic adsorption processes, they reveal very interesting features about the energy changes that take place within the column during the approach to periodicity. They also are a much more stringent check on the energy balance closure within DAPS because much more information is available to analyze compared to that at the periodic state. This presentation will provide an overview of DAPS, with particular emphasis on the cycle to cycle energy balance closure dynamics.

424769 Transdermal Cancer Vaccine Using Antigen-Coating Nano Drug Carrier

Abstract:

Transcutaneous immunization using the skin's immune system is increasingly attracting attention. In contrast to conventional immunization by injection, transcutaneous immunization requires only topical application of antigens to the skin and is a simple, non-invasive immunization method that does not require medical personnel. Furthermore, the epidermal and dermal layers of skin contain an abundance of antigen-presenting cells (APCs) such as Langerhans cells (LCs) and dermal dendritic cells (dDCs). These APCs capture antigens, migrate to the lymph node from the skin, and then induce antigen-specific immunity by antigen presentation. Using the skin's immune system, transcutaneous immunization can offer a more effective method than subcutaneous (s.c.) or intramuscular injection. However, the major problem in transcutaneous administration is that the stratum corneum (SC), the outermost layer of the skin, serves as a hydrophobic barrier at the skin surface that hinders the penetration of hydrophilic biomolecules ($M_w > 500$ Da). Because most antigens are hydrophilic molecules, such as proteins and peptides, it is difficult to deliver antigens to APCs through the skin.

To overcome the SC barrier and induce immunity by topical application of antigens, various antigen formulations have been developed, including hydrogel patches, poly(lactic acid) nanoparticles, chitosan nanoparticles, and flexible liposomes. Our approach is to coat the hydrophilic antigens with hydrophobic surfactant molecules using water-in-oil (W/O) emulsification, followed by freeze-drying to produce oil-dispersible antigen–surfactant complexes. Because hydrophobic materials are more permeable through the hydrophobic SC than hydrophilic materials, oil dispersions of the complexes, namely solid-in-oil (S/O) nanodispersions, can improve the delivery of antigens into the skin. Using the S/O technique, we previously demonstrated antigen-specific antibody production by transcutaneous immunization, without physical enhancement or pre-treatment of skin. Furthermore, co-encapsulation of polyarginine as a skin penetration enhancer or CpG oligodeoxynucleotide as an immune adjuvant in S/O nanodispersions induced antibody production more effectively.

<>In this study, we applied the S/O nanodispersion technique to the induction of cancer immunity by transdermal delivery of cancer antigen. Cancer immunotherapy by activation of immune system against cancer has recently received attention for its feasibility in the treatment of malignancies with little toxicity. The key step for inducing the cancer immunity is the delivery of cancer-specific antigens to DCs, followed by the antigen-presenting by the DCs. We prepared S/O nanodispersions bearing ovalbumin (OVA) as a model cancer-antigen for efficient transdermal OVA delivery. Our investigation revealed the ability of this approach to induce antigen-specific cellular immune responses against cancer by evaluating the growth of OVA-bearing tumors and the production of cytokines from splenocytes. Inhibition of OVA-bearing tumor growth was achieved, demonstrating the applicability of S/O nano carriers to the induction of cancer immunity.

425662 Isolation and Characterization of Material-Binding Peptides for Site-Specific Immobilization of Proteins

Abstract:

Immobilization technologies for proteins are important in the fields of bioseparation and bioanalysis. Especially, oriented immobilization of antibodies and/or their fragments onto the surfaces of solid supports such as plastic plates and chromatography resins are highly requisite for improvement of binding capacity. Affinity peptide tags such as 6xHis-tag and FLAG-tag might be useful to immobilize target proteins site-specifically onto the certain surfaces that ligand molecules are immobilized. Use of the conventional affinity peptide tags are limited because ligand molecules for capturing affinity peptide tags are sometimes not efficiently introduced to the target surfaces.

Here, we developed material-binding peptides that directly recognize the surface structures of solid materials. By introducing material-binding peptides to target proteins, they may be site-specifically immobilized with improvement of density, orientation and remaining activity after immobilization. We developed a screening method of material-binding peptides from *E. coli* soluble proteins. First, protein solution extracted from *E. coli* cells was adsorbed onto the surface of solid materials, and the proteins that were strongly interacted with target surfaces were recovered. "The affinity proteins" that had strong affinity toward the surface of target materials were individually separated by 2D electrophoresis (2DE) and identified by peptide mass fingerprinting method utilizing MALDI-TOF MS. The affinity proteins with high purity were prepared by recombinant DNA technologies, followed by digestion with 3 types of proteases (Trypsin, Chymotrypsin and V8 protease). Binding characteristics of peptide fragments produced by proteases were analyzed by reverse-phase HPLC, and the binding peptide candidates are identified by MALDI-TOF MS and/or protein sequencer. The identified peptides were introduced to C-terminus of glutathione S-transferase as a model protein and then adsorption characteristics of peptide-tagged GSTs are further evaluated. So far, the material-binding peptides toward polystyrene (PS), polycarbonate (PC), poly(methyl methacrylate) (PMMA), silicone nitride (SiN) and other materials were successfully isolated. By introducing these peptides to antibody fragments such as single-chain Fv (scFv) antibodies, their binding capacities on the target surfaces became higher than those without peptides. Thus, the method for screening of material-binding peptides developed in this study will be useful, and other material-binding peptides will be also isolated by this method.

427255 Metal Affinity Immobilized Liposome Chromatography for Recognition of Peptides

Abstract:

Metal affinity-immobilized liposome chromatography (MA-ILC) was newly developed as a chromatographic technique to separate and analyze peptides. The MA-ILC matrix gel was first prepared by immobilizing liposomes modified with functional ligands. The functional ligand used to adsorb metal ions was N-hexadecyl iminodiacetic acid (HIDA), which is obtained by attaching a long alkyl chain to an iminodiacetic acid (IDA). Cu(II) ion was first adsorbed on the gel matrix through its complex formation with the HIDA on the surface of the immobilized liposome. Synthetic peptides of various types ranging in size from 5 to 40 residues were then used, and their retention properties on the MA-ILC were evaluated. The retention property of peptides on the MA-ILC by using a usual imidazole elution was compared with the retention property in the case of the immobilized metal affinity chromatography (IMAC) and an immobilized liposome chromatography (ILC). It was found that the retention property of peptides on the MA-ILC has the features of both the IMAC and the ILC; the retention ability of peptides depends on both the number of histidine residues in peptides and the liposome membrane affinity of the peptides. Histidine and tryptophan residues among amino acid residues in peptides indicated a high contribution coefficient for the peptide retention on the MA-ILC, probably due to their metal ion and membrane interaction properties, respectively.

429252 Monitoring of Pegylation Reaction By Monolith Chromatography

Abstract:

PEGylation has been used for making a smart protein drug as it prolongs the residence time in the body and makes proteins more stable and soluble. However, PEGylation reaction is so complicated that undesired PEGylated protein isoforms are also produced unless the process is carefully controlled. It is therefore needed to develop a very fast and high-resolution analytical method for monitoring PEGylation reaction.

Separation of randomly PEGylated protein (lysozyme) isoforms by monolith disk ion exchange chromatography (IEC) with a CIM SO3 strong cation-exchange monolith disk was optimized based on the number of resolved peaks, peak resolution, analysis time and pressure drop. In order to increase the retention of mono- and di-PEGylated protein isomers the mobile phase was decreased to pH 4.5, where a large number of mono- and di-PEGylated isomers were resolved within a few minutes. Based on the linear gradient elution (LGE) optimization model, the following values were determined; gradient slope =0.016 M/mL, disk thickness = 3 mm (single disk) and flow rate = 10 mL/min. Under these optimal conditions, the analysis was completed within 4 minutes while the pressure drop was below 1 MPa. We compared the performance of this method with pH gradient elution IEC. The number of peaks retained and resolved in the method developed (salt LGE) was much larger that in pH gradient elution. This method was successfully applied to monitoring mono and di-PEGylated positional isoforms in the reaction mixture of random PEGylation of lysozyme.

429398 Improving the Separation Performance of Modified Proteins By Electrostatic Interaction Chromatography

Abstract:

When native proteins are modified (intendedly or unintendedly), modified forms must be separated from the native protein and other unwanted contaminants. A well-known intended modification is PEGylation, where a PEG molecule is attached to a protein. Unintended modified proteins are dimer and aggregates, which must be removed by chromatography during downstream processes of antibody drugs.

Electrostatic interaction based chromatography (IEC) is known to be most efficient for PEGylated protein separations. PEGylated proteins are much more weakly retained compared with the native form in IEC. PEGylated protein positional isomers can be separated by IEC and the binding sites for mono-PEGylated proteins are similar to those for native proteins. Weak retention of PEGylated proteins is likely to be due to a steric hindrance between the ion-exchange ligand and charges of PEGylated proteins.

Since dimer and aggregates are more hydrophobic and have more charges, their retention volumes are larger in IEC. We investigated the number of binding sites $\boldsymbol{B}$ for a model protein monomer and dimer. As the mobile phase pH increased both the B values and the difference between the B values of monomer and dimer (aggregates) increased for an-ion exchange chromatography. Consequently, the removal of dimer became better with increasing pH. Another efficient method for improving the separation performance is to add polyethylene glycol (PEG) into the mobile phase. As PEG increased the viscosity, the pore diffusion coefficient decreased, which resulted in peak broadening. We determined the pore diffusion coefficient in the presence of PEG in the mobile phase of IEC. Based on the pore diffusion coefficient and the mobile phase viscosity the optimum conditions were detetemined for the monomer separation.

416646 Comparison of Large-Pore Media and Monoliths for Protein and VLP Chromatography

Abstract:

In this paper we examine the performance of large-pore particles with that of monoliths for the cation exchange chromatography of proteins and VLPs. Chromatographic experiments in bind-and-elute mode are conducted with POROS HS 50 and HS 20, having mean particle size in the 50 and 20 um range, respectively, from Applied Biosystems, and with radial flow monoliths, 1 mL in volume, from BIA Separations. Structural characteristics of both particles and monoliths are determined by scanning and transmission electron microscopy as well as by inverse size exclusion chromatography. The particles are found to have much higher equilbrium binding capacities than the monoliths for proteins, including IgG and thyroglobulin. The dynamic binding capacities are also higher for particle based columns except at extremely high flow rates exceeding 1000 cm/h. Protein recoveries are high for both particles and monoliths. In the case of VLPs, binding is limited to a thin layer on the outside surface of the particles, so that the capacity advantage of the POROS resins is greatly reduced compared to the monoliths. However, VLP recovery from the monoliths is low and is dependent on the direction of flow. Closer microscopic examination, indicates that physical heterogenities in the monolith structure may result in physical trapping of the VLPs.

409853 Metal-Ferrocyanide Functionalized Magnetic Adsorbents for the Removal of Radioactive Cesium in Contaminated Water

Abstract:

The nuclear accident at the Fukushima Daiichi nuclear power station in 2011 released a huge quantity of radioactive contaminants into the environment [1]. Among these, ^{137}Cs is the most problematic contaminant due to its long half-life (30.2 years), high solubility in water, and high-energy gamma ray emissions. Various types of Cs adsorbents such as zeolites, calixarene, aluminum molybdophosphate, and metal ferrocyanide have been investigated. [2]. However, they cannot be easily used to remove Cs in an open environment because there is no easy way to collect the used adsorbents after applying them into the environment. Magnetic adsorbents have attracted significant attention for environmental remediation application, as they display a high surface area and provide unique magnetic separation property.

In this study, metal (cupper or iron) ferrocyanide-functionalized magnetic adsorbents were fabricated for the highly efficient removal of radioactive cesium from contaminated water. The hydrophilic polymer such as polyethyleneimine, polyvinylpyrrolidone were coated on the surface of magnetic nanocluster to stabilize the magnetic adsorbent in water and used as coating materials to immobilize the copper and ferrocyanide on the surface of magnetic nanoclusters. Furthermore, Iron ferrocyanide was directly coated onto the surface of magnetic nanoparticles without template such as hydrophilic polymer for the enhanced magnetic property and Cs removal performance. The resulting metal-ferrocyanide functionalized magnetic adsorbents having supeparamagnetic property could be easily separated from water using an external magnet, and also showed excellent removal efficiency of radioactive cesium from contaminated water due to the ion exchange capacity of the metal-ferrocyanide.

References

1. M. Manolopoulou, E. Vagena, S. Stoulos, A. Ioannidou, C. Papastefanou, J. Environ. Radioact. 102, 796 (2011).

2. P. A. Haas, Sep. Sci. Technol. 28, 2479 (1993).

411131 Confined Swelling Pressure Tests for Spherical Resorcinol-Formaldehyde Resin

Abstract:

Confined Swelling Pressure Tests for Spherical Resorcinol-Formaldehyde Resin

Joseph Walker, Paul Taylor, and Trent Walker

Oak Ridge National Laboratory

An ion exchange process using spherical resorcinol-formaldehyde (sRF) resin from Microbeads AS (Skedsmokorset, Norway) is the baseline process for removing cesium from the approximately 53 million gallons of highly radioactive nuclear waste currently stored in underground tanks at the Department of Energy (DOE) Hanford site. Since the selected sRF resin is known to swell during processing as it is converted from a hydrogen form to a sodium form, there is a concern about the pressure the resin could exert if it were to become trapped in process piping.

A test system (see Fig. 1) was fabricated and operated at Oak Ridge National Laboratory to measure the swelling pressure of various batches of sRF resin. The test vessel consists of a 6 inch long section of 2 inch, schedule 80, 304L stainless steel pipe welded to 2 inch, class 600 pipe flanges at the top and bottom. With the flanges, the total depth of the resin bed within the test vessel is 6.5 inches. The inlet and outlet filter assemblies, which are modified Swagelok ½-in. stainless inline filters, are welded to the sides of the vessel. Pressure elements/transmitters are bolted to the top and bottom flanges of the test vessel to monitor the pressures during testing. The top pressure assembly is a 0–750 psi, WIKA general industrial transmitter (WIKA Instrument, LP, Lawrenceville, GA) with a flush diaphragm seal, and the bottom pressure assembly is a 0–1500 psi assembly of the same design.

Fig. 1. Photograph of the test system before instruments were connected to the data acquisition system.

Six batches of the sRF resin were tested:

1. Blue drum (batch # 5J-370/686)
2. Black drum (batch # BSC-3380-3-0200)
3. Finncont-1 (batch # BSC-3380-2P-0201)
4. Finncont-2 (batch # 6C-370/745)
5. Finncont-3 (batch # 0L-370-1312)
6. Microbeads (batch # 1f-370-1392)

For each test, the test vessel was filled to the top flange (see Fig. 2) with preconditioned, hydrogen-form sRF resin in water, and then the top flange/pressure transducer was connected, air was vented from the lines, and dilute sodium hydroxide was pumped through the resin to convert it to the hydroxide form. The pressures at the top and bottom of the test vessel were measured as the resin swelled inside the test vessel. When the pressures had stabilized, the flow of sodium hydroxide was stopped so that the static pressures inside the test vessel could be measured. A small amount of deionized water was then pumped through the test vessel to displace the sodium hydroxide, and the top flange was removed, allowing the resin to expand (see Fig. 3).

Fig. 2. H-form sRF resin loaded in test vessel

Fig. 3. sRF resin expansion after top flange was removed

One run each was made with the first five resin batches listed above, and five runs were made with the sixth batch. After each test, samples of resin were collected from different locations within the column, titrated to determine the extent of conversion to the sodium form, and examined microscopically to look for bead breakage and changes in the bead sizes and shapes.

426160 Analysis of Mobile Helium Cluster Reactions Near Surfaces and Grain Boundaries of Plasma-Exposed Tungsten

Abstract:

The evolution of the surface morphology and the near-surface structure of plasma facing components (PFCs) in nuclear fusion reactors is impacted significantly by the implantation of helium (He) atoms. Tungsten (W) is an important PFC material due to its thermomechanical properties. In tungsten, such interstitial He atoms are very mobile and aggregate to form clusters of different sizes. The smallest of these helium clusters, containing n helium atoms with n = 1-7, also are mobile and their diffusional transport mediates the evolution of surface morphology and the sub-surface helium gas bubble structure and dynamics.

In this presentation, we report the results of a systematic atomic-scale analysis of the reactions of small mobile helium clusters near low-Miller-index tungsten surfaces, aiming at a fundamental understanding of the near-surface dynamics of helium-carrying species in plasma-exposed tungsten. These small mobile helium clusters are attracted to the surface and migrate to the surface by Fickian diffusion and drift due to the thermodynamic driving force for surface segregation. As the clusters migrate toward the surface, trap mutation (TM) and cluster dissociation reactions are activated at rates higher than in the bulk. TM produces W adatoms and immobile complexes of helium clusters surrounding W vacancies located within the lattice planes at a short distance from the surface. These reactions are identified and characterized in detail based on analysis of a large number of molecular-dynamics (MD) trajectories for each such mobile cluster near W(100), W(110), and W(111) surfaces. TM is found to be the dominant cluster reaction for all cluster and surface combinations, except for the clusters with n = 4 and n = 5 near W(100) where cluster partial dissociation following TM dominates. We find that there exists a critical cluster size, n = 4 near W(100) and W(111) and n = 5 near W(110), beyond which formation of multiple W adatoms and vacancies in TM reactions is observed. The identified cluster reactions are responsible for important structural, morphological, and compositional features in plasma-exposed tungsten, including surface adatom populations, near-surface immobile helium-vacancy complexes, and retained helium content, which are expected to influence the amount of hydrogen re-cycling and tritium retention in fusion tokamaks.

We have also carried out a systematic atomic-scale analysis of reactions of small mobile helium clusters near a symmetric tilt grain boundary (GB) based on MD simulations. We found that TM reactions are dominant for clusters with n = 4-7, and only a single vacancy is formed in each of the TM reactions identified regardless of cluster size n. The displaced W atom generated from the TM reaction near the GB forms an extended W interstitial configuration on the GB. We also found that, for n = 1-4, the helium clusters (including single He interstitial atoms) that segregate on the GB are rendered practically immobile. However, the mobility of the extended W interstitial configuration on the GB depends on the exact location where the TM reactions take place, with the W interstitial being highly mobile if the TM reaction occurs on the GB. Although this study is limited to a simple, prototypical model GB structure, the insights obtained are very useful for understanding surface morphology and helium retention in plasma-exposed polycrystalline tungsten.

432458 Membrane Distillation for Nutrient and Water Recovery from Thermally Treated Dairy Manure

Abstract:

Manure management is a critical aspect of modern dairy operations. Mismanagement can result in contamination to local and regional watersheds that may have adverse environmental impacts. Because fresh manure has a high water content (~85%), direct use is typically limited to local direct land application as a carbon source for soil amendment and as low-quality fertilizer. However, this practice may not be a viable option for all operations, so the manure is often transported to treatment facilities at a considerable cost. This project focuses on the development of an on-site treatment process for the dairy manure that can produce a solid fuel source, liquid fertilizer, and high-quality reuse water, all of which can be used on-site or sold for profit.

This presentation will focus on the use of direct contact membrane distillation (DCMD) for the treatment of the aqueous byproduct of thermally treated dairy manure to concentrate nutrients and produce high-quality reuse water. Hydrothermal carbonization (HTC) was applied to wet dairy manure with hot compressed water to produce a solid, carbon-rich hydrochar and an aqueous, nutrient-rich byproduct. Once the solid fuel products have been removed, the hot aqueous stream serves as a feed to DCMD, where soluble nutrients (nitrogen and phosphorus) are concentrated for use as a liquid fertilizer and the distillate water can serve as high-quality reuse water. No heating is required for DCMD operation as the HTC liquid byproduct is received at an elevated temperature, thereby making the treatment of the liquid phase a very low-energy process. Preliminary results indicate >99% rejection of TOC and total nitrogen (TN) in the feed solution and >95% rejection of phosphorus, with concentration factors greater than 6 achievable. The coupling of DCMD with HTC offers a great low-energy opportunity for nutrient and water recovery from a traditional waste source.

434015 Free Surface Electrospun Polyvinylidene Fluoride Membranes for Direct Contact Membrane Distillation

Abstract:

The earth consists of 1% total potable water accessible to humans. The use of current technologies such as osmosis processes, filtration, and distillation support the production of purified water, however, direct contact membrane distillation (DCMD) has the potential to be a competitive process to these conventional methods. The current process for manufacturing DCMD membranes is inefficient, expensive, and poorly optimized; resulting in limited industrialization. Here, we consider membranes produced by free surface electrospinning, which produces DCMD membranes with desirable properties such as large pore size (1 -3 μm) and low membrane thicknesses (50- 400 μm). In this study, we produced DCMD membranes from a polymeric solution containing 22 % w/w Polyvinylidene Fluoride (PVdF) in Dimethylacetamide (DMAc). The free surface electrospinning setup consisted of a solution bath with a partially submerged rotating wire spindle. A high voltage, in excess of 30 kV, was applied to the wire spindle. The induced electric field causes charge to accumulate on the surface of the entrained solution, leading to the formation of electrohydrodynamic jets. Electrospinning experiments were performed in a controlled environment with relative humidity ranging from 40-80RH%. In the presence of high humidity (> 50%), water absorbs into the polymeric jet causing the PVDF to phase separate. The gelled fibers were collected on a grounded rotating drum and left at room conditions to solidify over a period of 24 hours. The final morphology of the membrane was examined using scanning electron microscopy and the performance of the produced membranes was evaluated with an in-house built DCMD apparatus.

Two Dimensional Dynamic Modeling and Identification of a Lab-Scale Direct Contact Membrane Distillation Plant

Bader Bin Ashoor[1,2], Shadi Wajih Hasan[1], Wolfgang Marquardt[2] , Adel Mhamdi[2]

[1]Institute Center for Water and Environment (iWATER),
Department of Chemical and Environmental Engineering, Masdar Institute of Science and Technology,
P.O. Box 54224, Abu Dhabi, United Arab Emirates
[2]AVT-Process Systems Engineering, RWTH Aachen University, D-52056 Aachen, Germany

Session: **Membrane Distillation and Membrane Contactors**
Date: **Monday, November 9, 2015**

The 2015 AIChE Annual Meeting Salt Lake City USA

Presentation Outline

Masdar INSTITUTE

- Introduction
- Motivation
- Methodology
  - Development of 2D dynamic model of DCMD.
    - Mass balance
    - Energy balances
    - gProms ModelBuilder
  - Experimentation
  - Parameter estimation and model validation using gProms
- Results and Discussion
- Cunclusion
- Acknowledgments
- Questions

AACHENER VERFAHRENSTECHNIK

Introduction

Masdar INSTITUTE مصدر معهد

- Sourcing of non-potable water from desalinated and groundwater

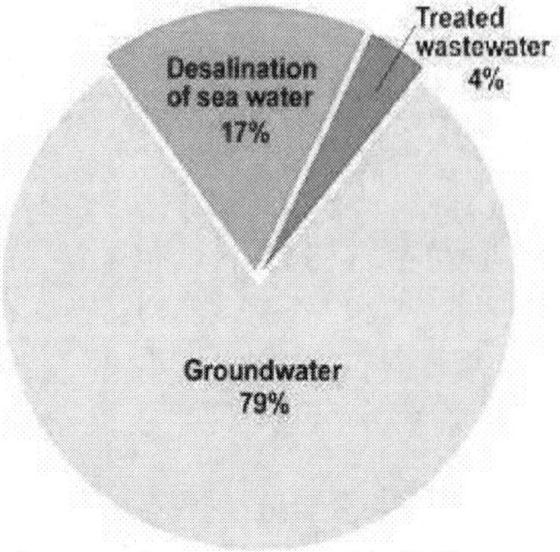

Sources of water consumed in UAE

Source: Environmental Agency Abu Dhabi, 2014

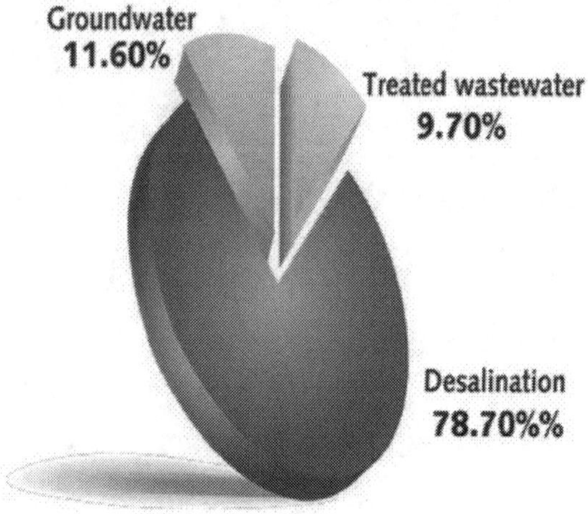

Sources of water consumed in Abu Dhabi

Source: WaterWorld, 2014

Introduction

Masdar INSTITUTE مصدر معهد

- Desalination production capacity and population in UAE

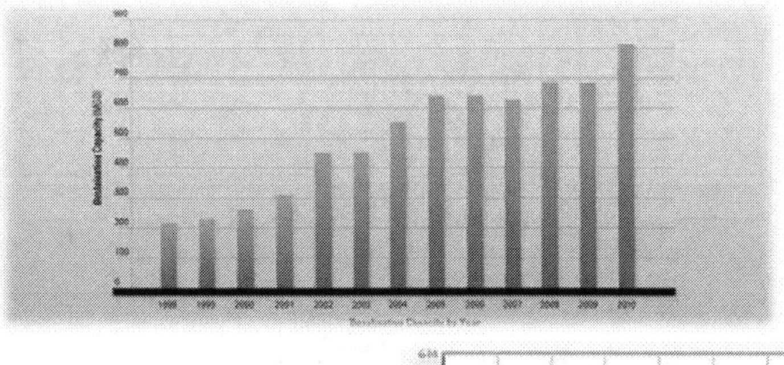

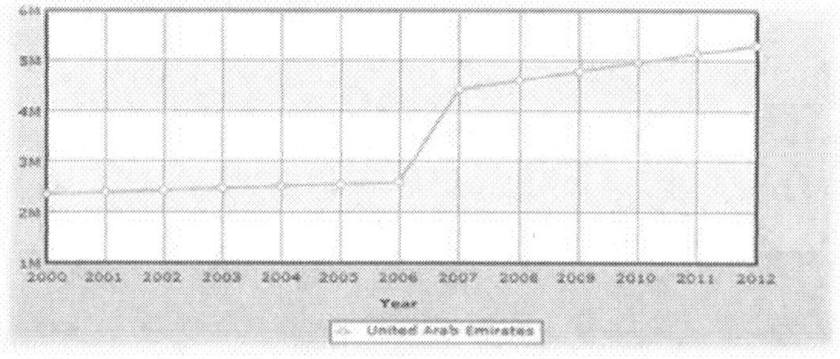

Introduction

Masdar INSTITUTE

➢About 98% of water in the world cannot be use directly due to the water salinity

UAE Desalination Share In the World 2010

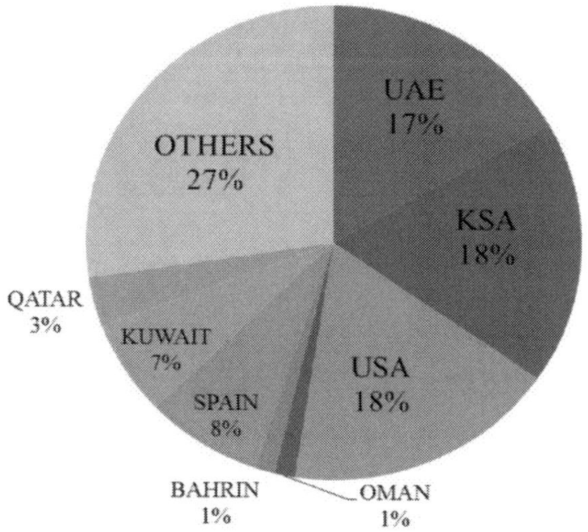

AACHENER VERFAHRENSTECHNIK

Introduction

Masdar INSTITUTE

Forecast capital expenditure on new desalination plants

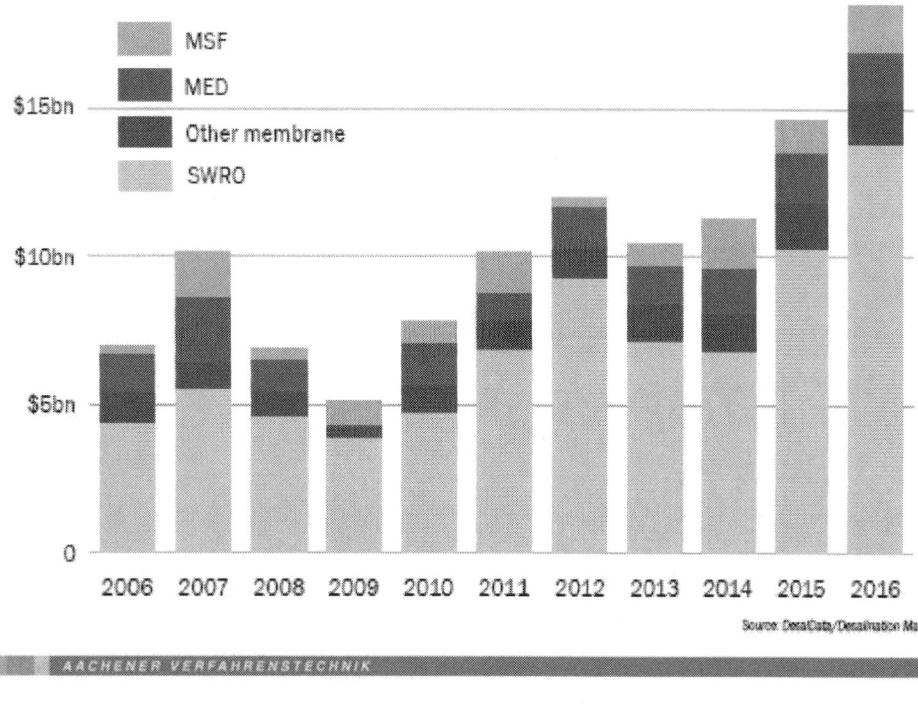

Source: DesalData/Desalination Markets 2010

AACHENER VERFAHRENSTECHNIK

Motivation

Masdar INSTITUTE معهد مصدر

- Membrane distillation (MD) known for not more than 50 years.
- Different configurations of MD: DC, V, AG, ...
- MD is non-isothermal process
- Temperature and concentration polarization
- Use MD with waste heat energy or renewable energy
  → transient (dynamic) operation
- Only stationary models are available in literature

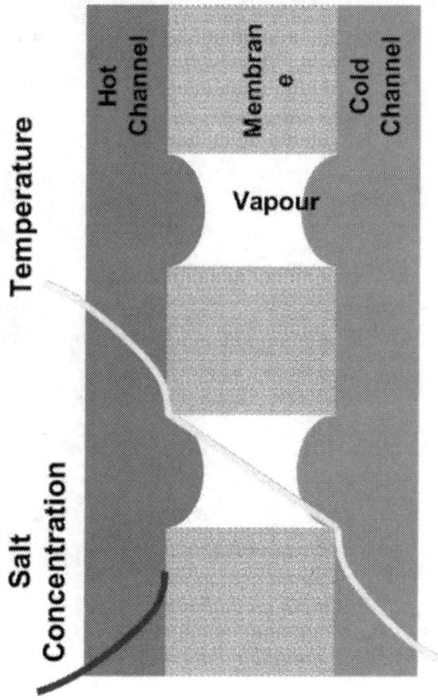

AACHENER VERFAHRENSTECHNIK

Motivation - Schematic diagram of DCMD

Masdar INSTITUTE معهد مصدر

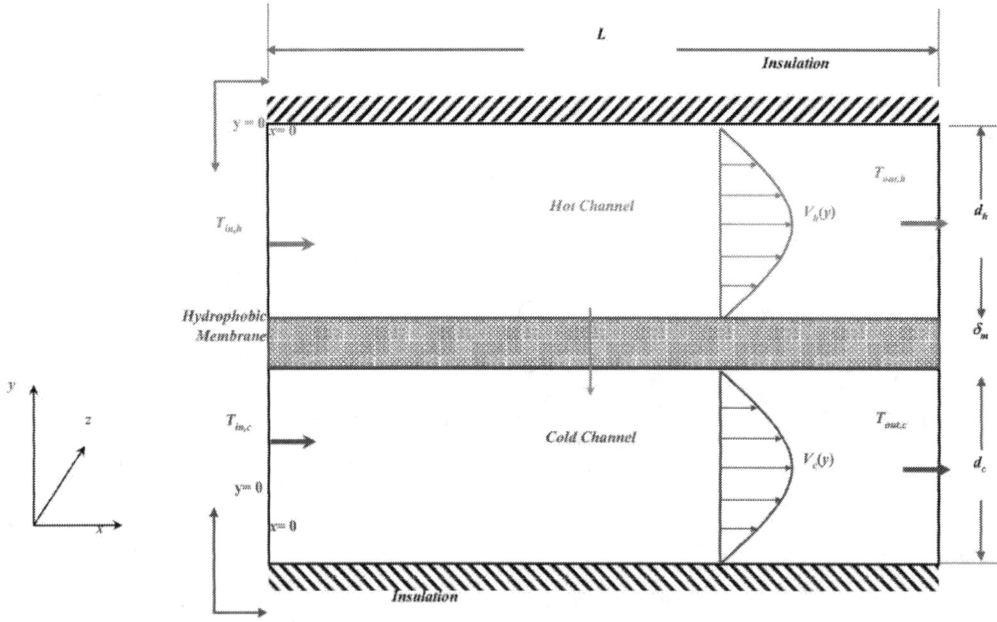

AACHENER VERFAHRENSTECHNIK

Methodology – 2D Dynamic Model of DCMD

Masdar INSTITUTE معهد مصدر

Channels

- Mass and energy balances

$$\rho C_p \frac{\partial T(x,z)}{\partial t} + \rho C_p u(x) \frac{\partial T(x,z)}{\partial z} = k \left[\frac{\partial^2 T(x,z)}{\partial x^2} + \frac{\partial^2 T(x,z)}{\partial z^2} \right]$$

$$\rho \frac{\partial C_s(x,z)}{\partial t} + \rho u(x) \frac{\partial C_s(x,z)}{\partial t} = D \left[\frac{\partial^2 C_s(x,z)}{\partial x^2} + \frac{\partial^2 C_s(x,z)}{\partial z^2} \right]$$

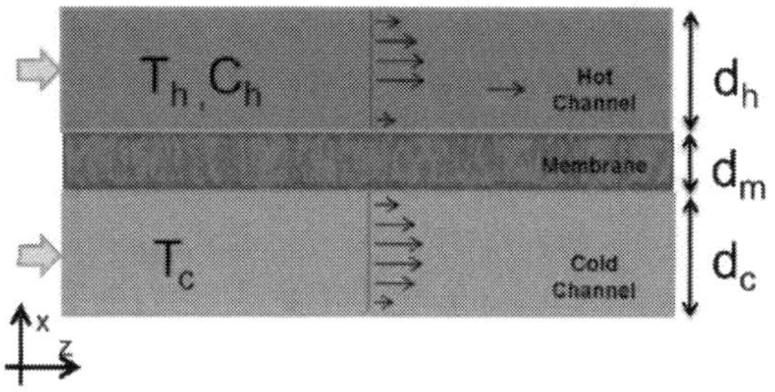

Membrane

- The vapor flux density j across the hydrophobic membrane

$$j = C_m \Delta P^{sat} = C_m (\gamma(X) P_h^{sat} - P_c^{sat})$$

- The membrane coefficient C_m : a weighted sum of the Knudsen diffusion and the Poiseuille flow

Interface (boundary conditions)

- mass and heat balance at the interface between liquid and vapor:

Implemetation

- Model implemented and simulated in gProms

AACHENER VERFAHRENSTECHNIK

Methodology – Experimental Pilot Plant

Masdar INSTITUTE معهد مصدر

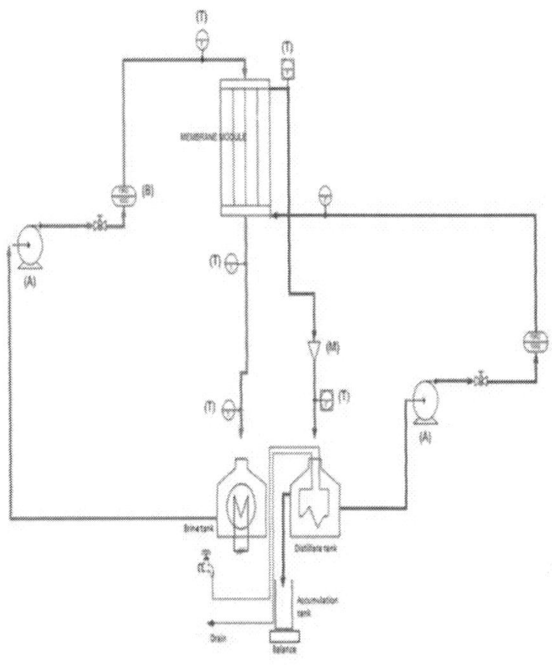

AACHENER VERFAHRENSTECHNIK

Methodology - Parameter Estimation and Model Validation

Masdar INSTITUTE معهد مصدر

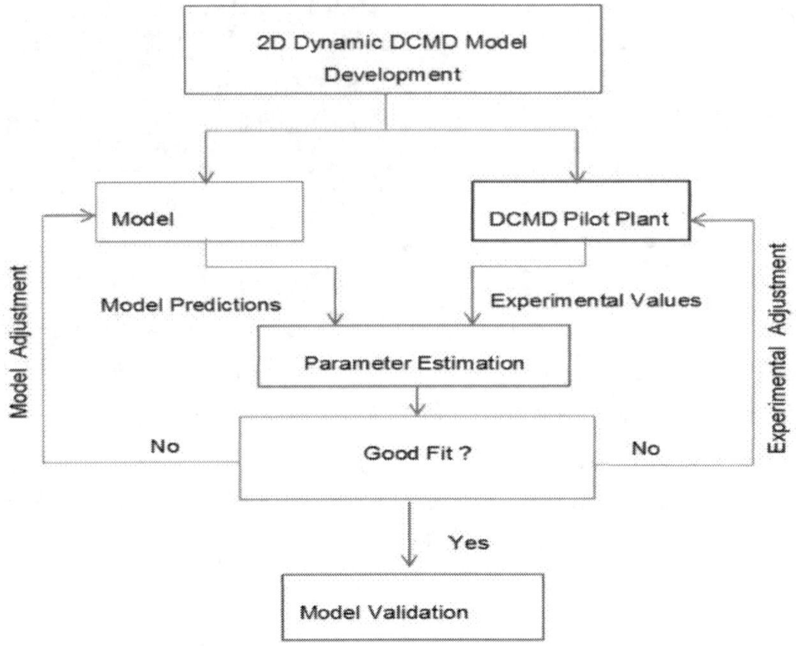

AACHENER VERFAHRENSTECHNIK

Results and Discussion – Case Study

Masdar INSTITUTE معهد مصدر

1. Geometry, physical properties and operating conditions → case study from the literature
2. The computed results are in good agreement with the experimental data from the literature.
3. Numerical simulations for different operational parametrs at the module inlet, e.g. permeate temperature, feed and permeate flow rate, etc.
4. The increase between the feed and the permeate temperatures causes monotone increase of MD flux

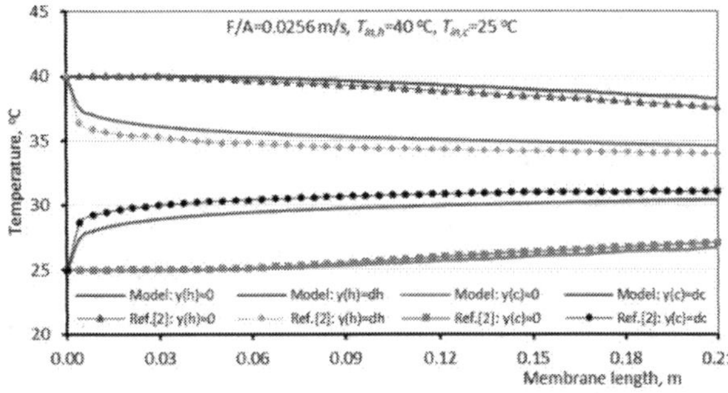

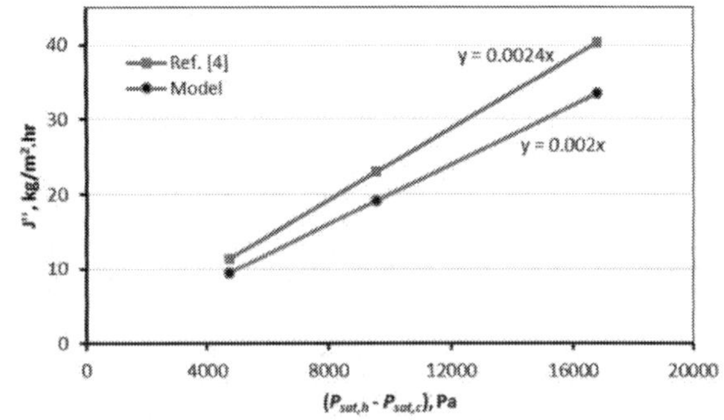

AACHENER VERFAHRENSTECHNIK

Results and Discussion – 2D Dynamic Model Simulation

Masdar INSTITUTE

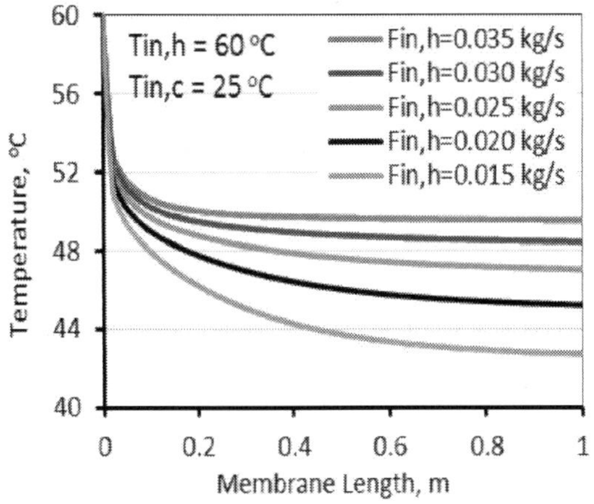

Variation of Feed temperature

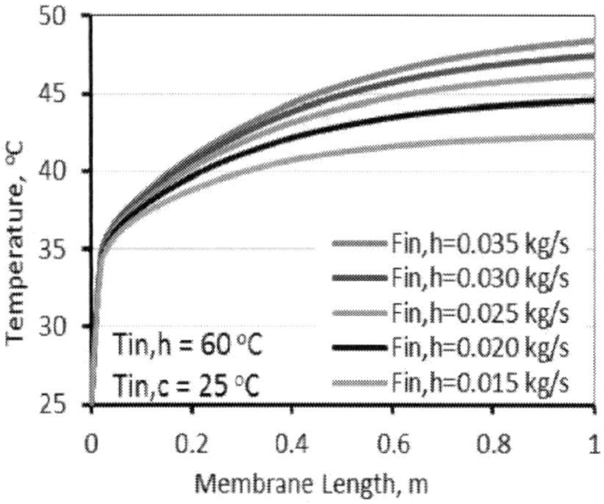

Variation of Permeate temperature

AACHENER VERFAHRENSTECHNIK

Results and Discussion – 2D Dynamic Model Simulation

Masdar INSTITUTE

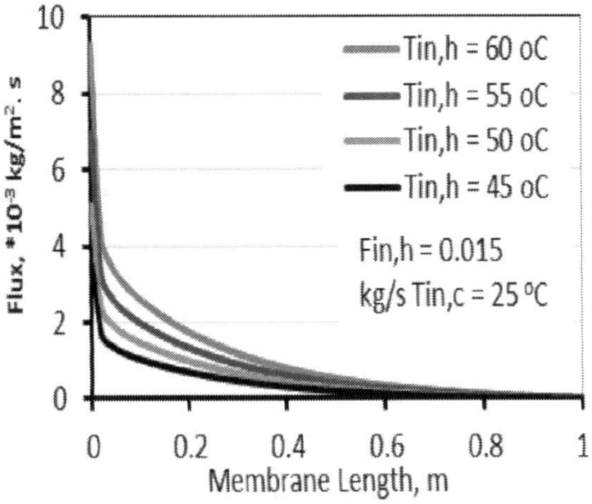

Flux transportation across the membrane surface

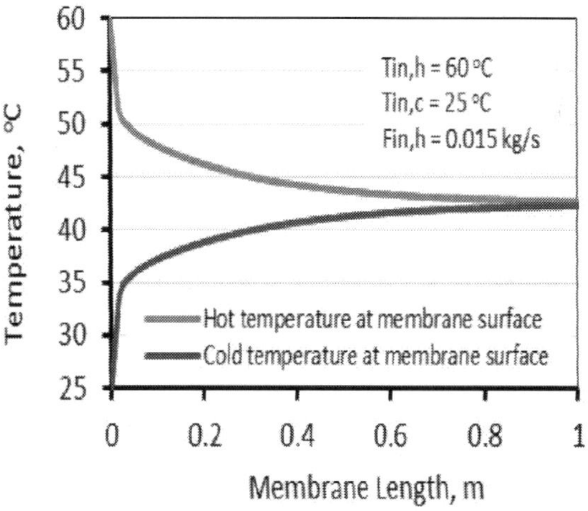

Temperature profile along the membrane surface

AACHENER VERFAHRENSTECHNIK

Results and Discussion – 2D Dynamic Model Simulation

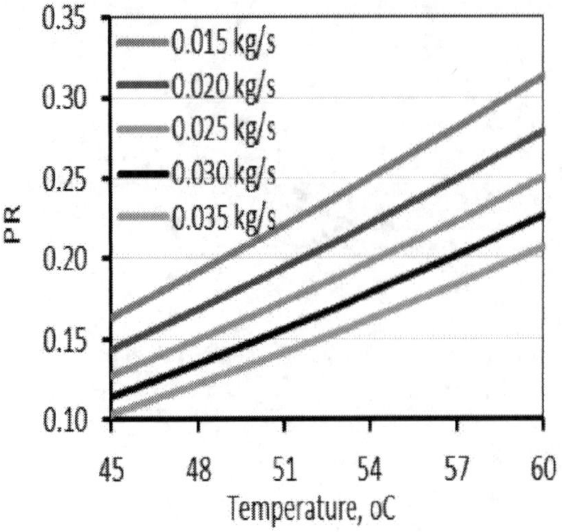

Performance Ratio as function of hot feedwater flow rate and temperature

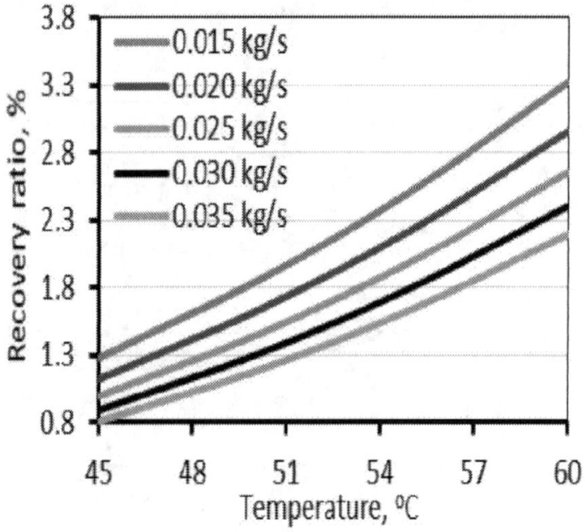

Recovery Ratio as function of hot feedwater flow rate and temperature

AACHENER VERFAHRENSTECHNIK

Results and Discussion – 2D Dynamic Model Simulation

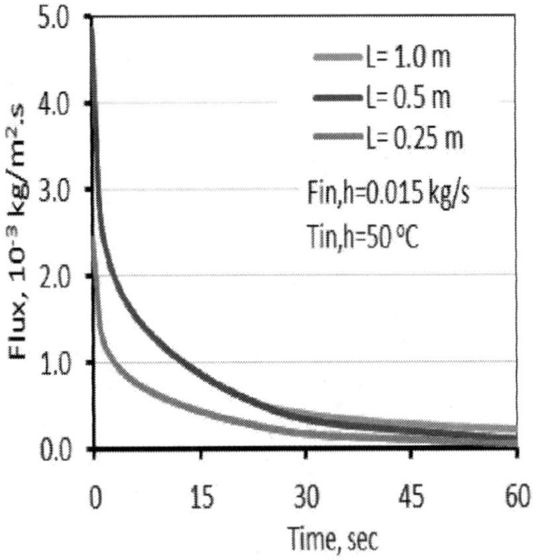

Transient flux at different location for counter flow

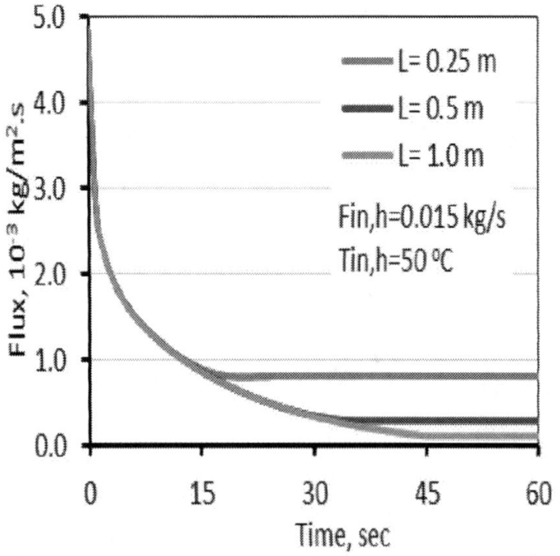

Transient flux at different location for parallel flow

AACHENER VERFAHRENSTECHNIK

Results - Parameter Estimation

Masdar INSTITUTE معهد مصدر

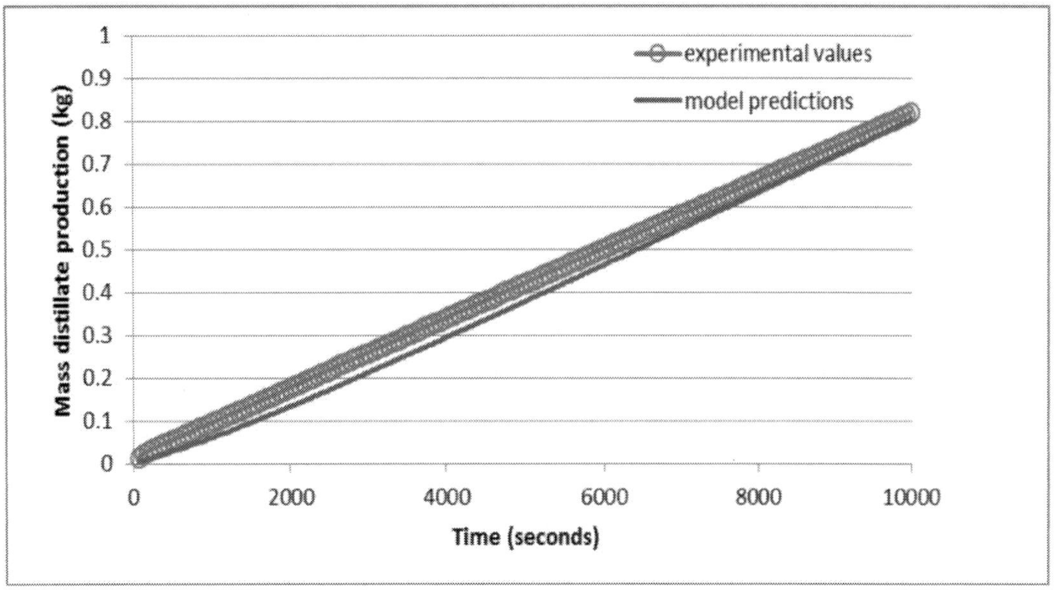

Model prediction and experimental values for mass distillate production

AACHENER VERFAHRENSTECHNIK

Results - Parameter Estimation

Masdar INSTITUTE معهد مصدر

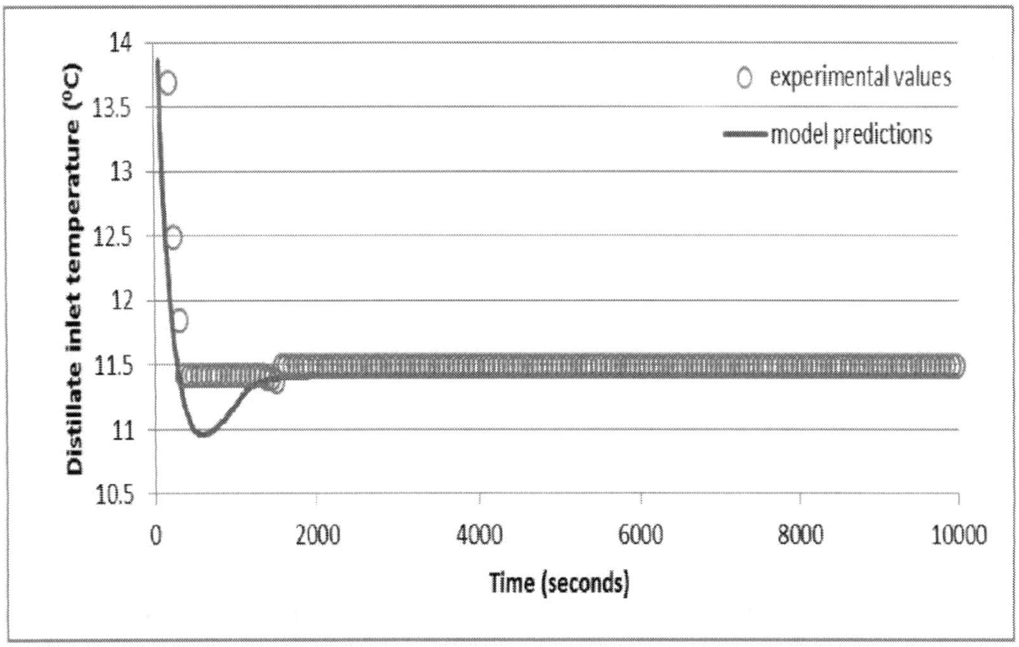

Model prediction and experimental values for distillate inlet temperature

AACHENER VERFAHRENSTECHNIK

Results - Parameter Estimation

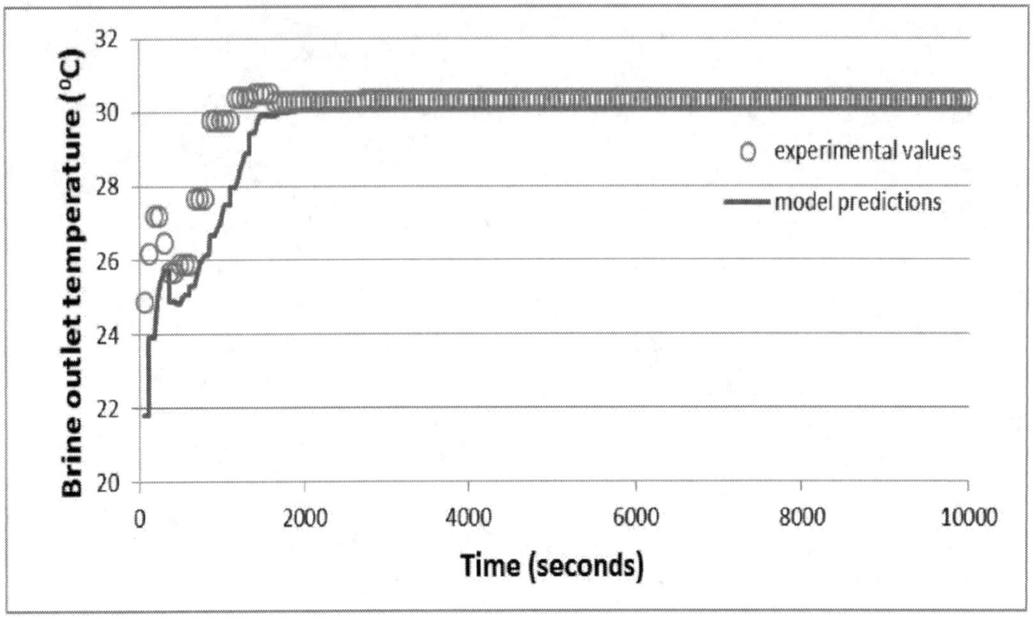

Model prediction and experimental values for brine outlet temperature

AACHENER VERFAHRENSTECHNIK

Results - Parameter Estimation

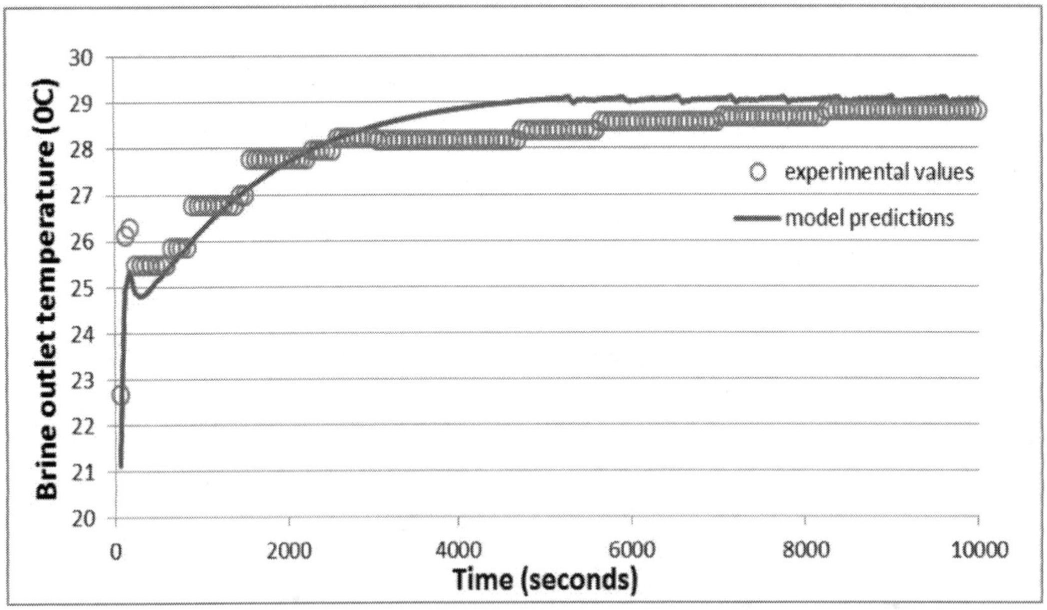

Model prediction and experimental values for brine outlet temperature

AACHENER VERFAHRENSTECHNIK

Results - Parameter Estimation

Masdar INSTITUTE معهد مصدر

Table 1. Estimated heat transfer coefficients

Parameter	Optimal Estimation	95% t-Value	Standard Deviation
$\alpha_{Heatloss}$	**2.7913E+00**	**67.97**	**1.866E-2**
α_{Coil}	**1.0964E+2**	**8.865**	**5.468E+00**
			Reference t-Value: 1.8335

AACHENER VERFAHRENSTECHNIK

Results - Parameter Estimation

Masdar INSTITUTE معهد مصدر

Table 2. Estimated membrane coefficient and conductivity of water liquid in the feed and permeate channels

Parameter	Optimal Estimation	95% t-Value	Standard Deviation
k_{ff}	1.2649 E+00	1.938	6.278 E-1
k_{fp}	4.39987 E-2	3447	6.569 E-6
Cm	3.73686 E-5	58.27	3.271 E-7
			Reference t-Value: 1.64502

AACHENER VERFAHRENSTECHNIK

Results - Model Validation

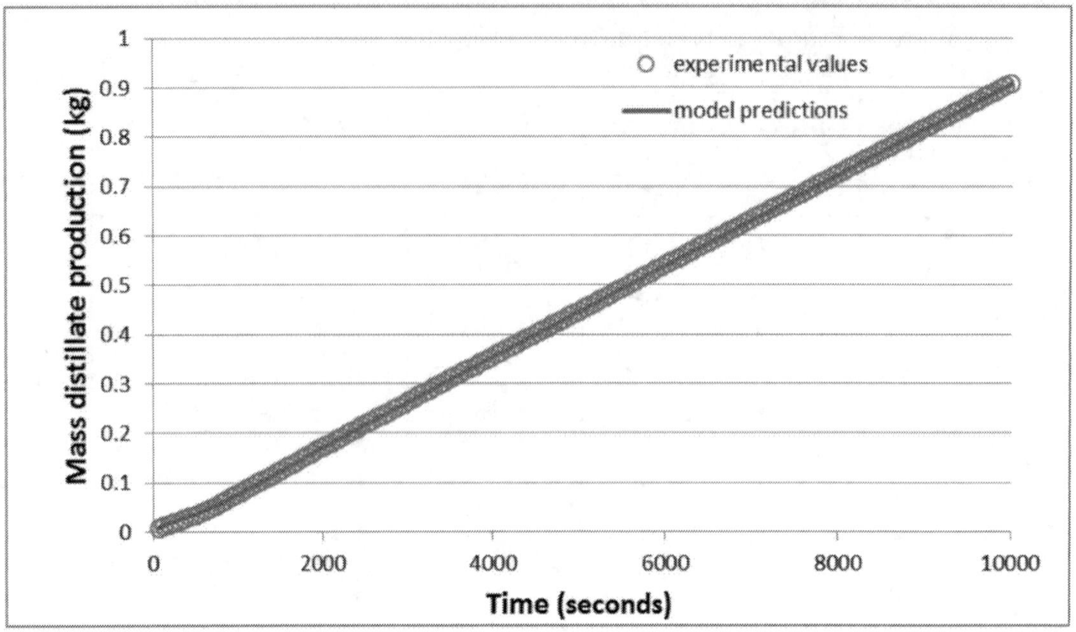

Model prediction and experimental values for mass distillate production

AACHENER VERFAHRENSTECHNIK

Results - Model Validation

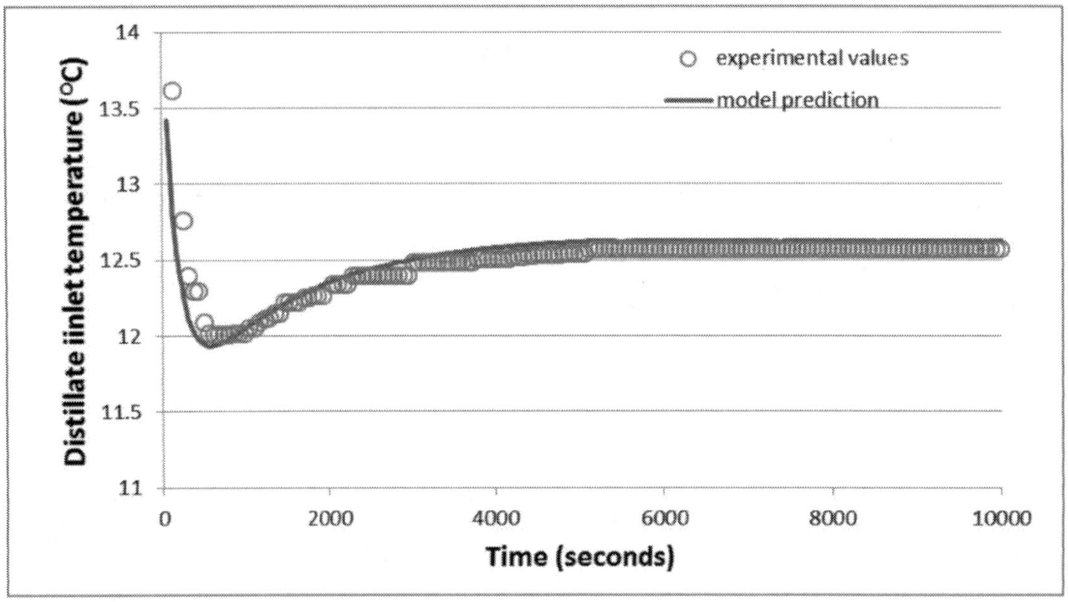

Model prediction and experimental values for distillate inlet temperature

AACHENER VERFAHRENSTECHNIK

Results - Model Validation

Masdar INSTITUTE معهد مصدر

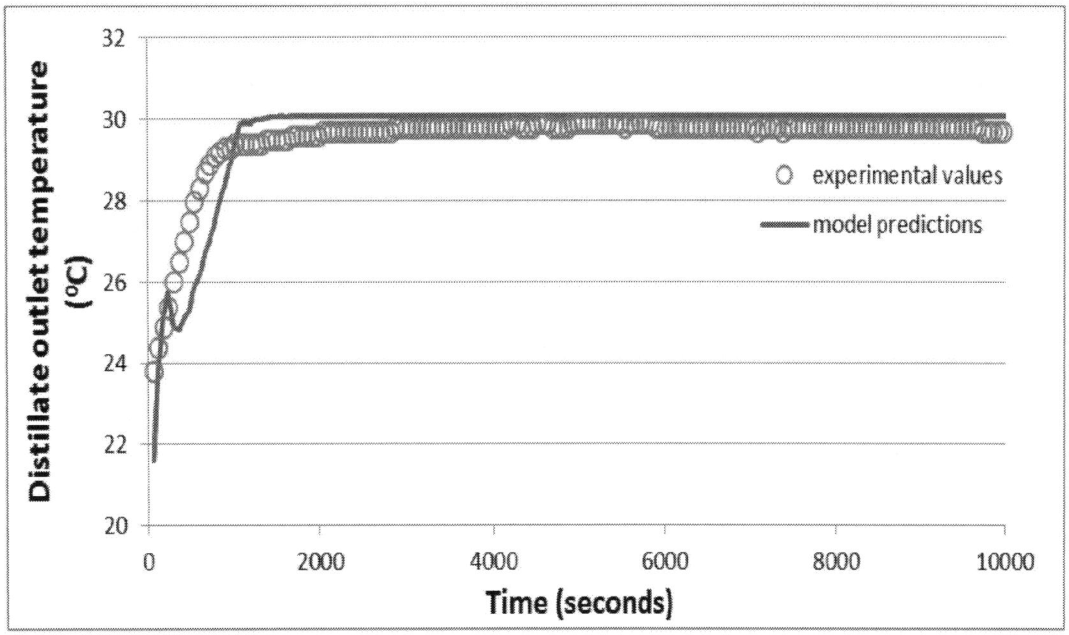

Model prediction and experimental values for distillate outlet temperature

Results - Model Validation

Masdar INSTITUTE معهد مصدر

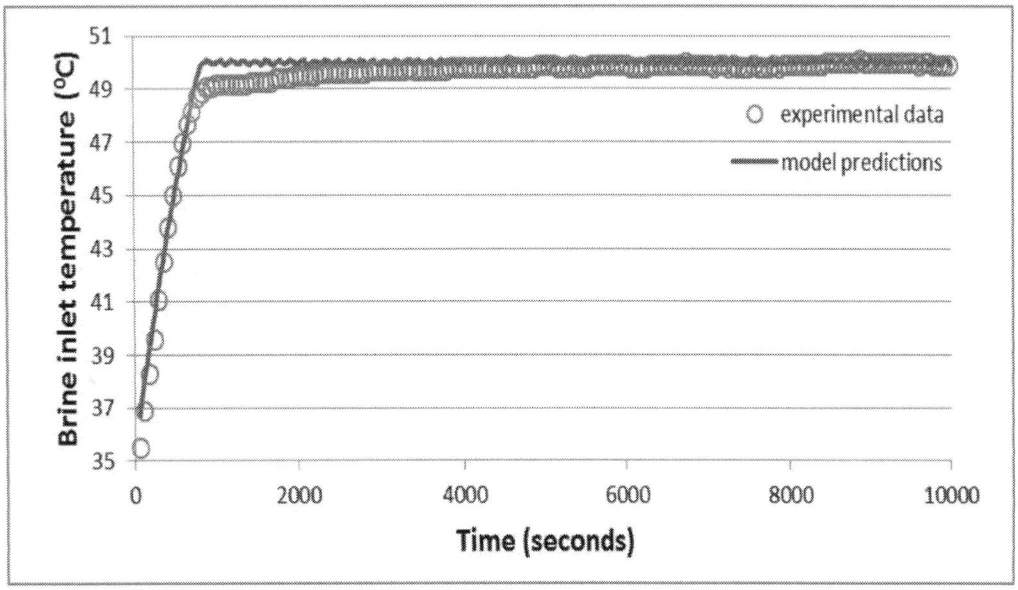

Model prediction and experimental values for brine inlet temperature

Results - Model Validation

Masdar INSTITUTE

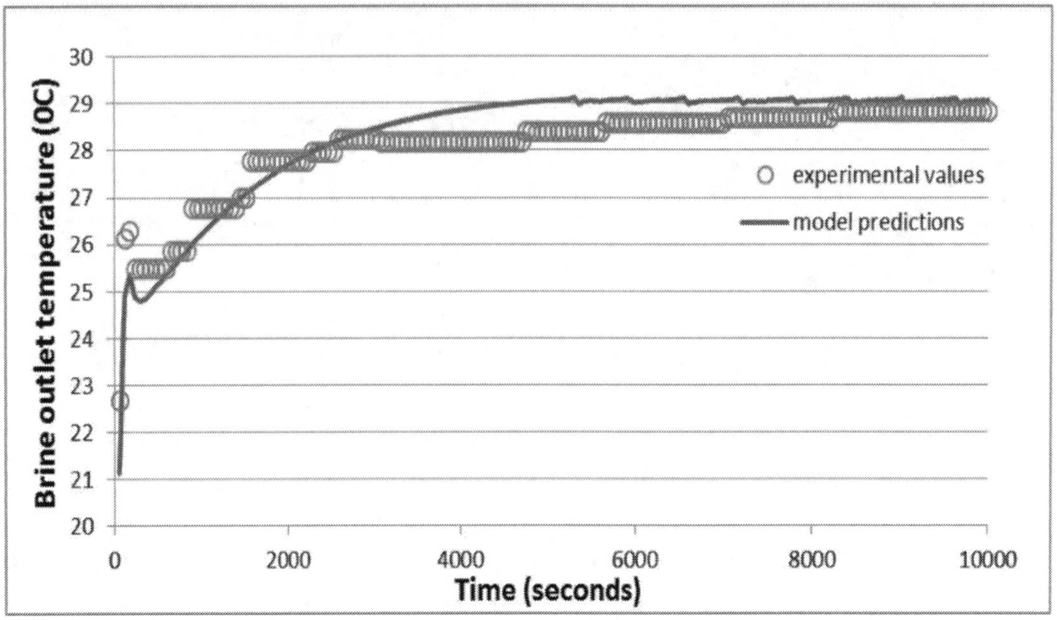

Model prediction and experimental values for brine outlet temperature

AACHENER VERFAHRENSTECHNIK

Conclusions

Masdar INSTITUTE

- 2-Dimensional mathematical model was developed which is in agreement with published studies.
- Dedicated experiments w ere conducted to estimate $\alpha_{Heatloss}$, α_{Coil}, k_{ff} and k_{fp}.
- The deviation model predictions and experimental data were minimized by gPROMS objective function (weighted least squares formulations).
- Accurate tests have been performed for all the parameters estimated.
- After optimal values of all parameters have been determined by parameter estimation, they were used in the DCMD 2-D model to perform model validation.
- The model validation shows that the DCMD 2-D model is a good representation of the DCMD pilot plant.

AACHENER VERFAHRENSTECHNIK

Acknowledgements

Masdar INSTITUTE

- Professor Marquardt, Dr. Adel Mhamdi and my colleagues at AVT-Process System Engineering Aachen University for their help and support
- Dr. Shadi Hasan and his team at iWATER Masdar Institute of Science and Technology for their help and support
- AquaStill Sittard Netherlands for fabricating the DCMD pilot plant

AACHENER VERFAHRENSTECHNIK

Masdar INSTITUTE

Thank you

Questions?

AACHENER VERFAHRENSTECHNIK

MD cost analysis

Masdar INSTITUTE معهد مصدر

1. Capital cost (CC):

   cost of equipment: MD membrane module, auxiliary equipment, installation charges.

2. Annual operating costs:

   Amortization or fixed charges (A_{fixed})

$$a = \frac{i(1+i)^n}{(1+i)^n - 1}$$

$$A_{fixed} = a \times CC$$

   O&M Costs ($A_{O\&M}$) = 0.2X A_{fixed}

   Membrane replacement cost (A_{MR}) =

   20% of Module cost (MC)

   Electric cost ($A_{electric}$) = $c\ w\ M\ f\ 365$

c electric cost, w specific consumption (kWh/m3), M plant capacity (m3/day), f plant availability

AACHENER VERFAHRENSTECHNIK

Cost analysis

Masdar INSTITUTE معهد مصدر

- Total annual cost (A_{total}):

$$A_{total} = A_{fixed} + A_{O\&M} + A_{MR} + A_{electric}$$

3. Water production cost (*WPC):*

$$WPC = \frac{A_{fixed}}{f\ M\ 365}$$

AACHENER VERFAHRENSTECHNIK

Dynamic Optimization

Masdar INSTITUTE معهد مصدر

- Using the 2-D dynamic model to optimize the system that leads to the lowest unit cost for a specified water production rate.
- The optimization problem is defined as:

$$\min(C_w) = f(T_f, T_p, F_f, F_p, C_s)$$

Subject to:

- The desired distillate water production rate (F_{dw}).
- The maximum temperature of the brine ($T_b \leq 95°C$)

AACHENER VERFAHRENSTECHNIK

433191 Application of Direct Contact Membrane Distillation for Treating High Salinity Solutions: Impact of Membrane Structure and Chemistry

Abstract:

Direct contact membrane distillation (DCMD) is a thermally driven membrane process in which water vapor transport occurs through a non-wetted porous hydrophobic membrane. In DCMD, a hot feed solution is flowing on one side of the membrane while a cold distillate solution is flowing on the other side creating a driving force of vapor pressure difference across the membrane. Highly concentrated solutions are required for the forward osmosis as a draw solution for the process. In this work, DCMD is investigated as a separation method for the draw solution of the forward osmosis process. Three different types of membranes with the same pore size of 0.45 μm (i.e. PVDF, PP, and ECTFE) were used in this research: polyvinylidene fluoride (PVDF) from Millipore, poly propylene (PP), and ethylene chloro trifluoro ethylene (ECTFE) from 3M. Also PP-3M membranes with different pore sizes of 0.45, 0.2, and 0.1 μm were used to study the effect of the pore size of the membrane on the process efficiency. Three highly concentrated solutions were used (i.e. $MgCl_2$, KCl and NaCl) as feed solutions for the MD process. The feed solution and the distillate were kept at 50 ^{0}C and 20 ^{0}C respectively. The results showed that the ECTFE membrane gives the highest water flux, likely because it is thinner (about half than the thickness of the PVDF and the PP membranes) and therefore exhibits less mass flux resistance. The PVDF membrane showed the lowest flux as it has the lowest porosity (66%) compared to 85% and 81% for the PP and the ECTFE membranes respectively. Membranes with larger pore sizes exhibited slightly higher flux. Feed chemistry also impacted the flux, with KCl solutions having higher flux than either NaCl or $MgCl_2$. These differences were largely driven by solution vapour pressure differences. Excitingly, all salt rejection measurements were in excess of 99.5%.

420106 Influence of Microporous Membrane Properties on DCMD

Abstract:

It would very useful if one can predict the performance of a microporous hydrophobic membrane in Direct Contact Membrane Distillation (DCMD) just from the nominal properties of the membrane if the boundary layer heat transfer coefficients are known. To evaluate the influence of membrane properties on their performance in desalination by the DCMD process, a variety of microporous hydrophobic flat sheet membranes of polyvinylidene fluoride (PVDF) and polytetrafluoroethylene (PTFE) were studied over a range of hot brine temperatures, 65-85deg C. The membranes had a wide range of values of the pore size and thickness. Experiments were done using two different flat membrane test cells, one made out of stainless steel and the other from chlorinated polyvinyl chloride (CPVC). Membrane properties, such as the maximum pore size were experimentally determined. Mass transfer modeling involved both the Knudsen and the transition regimes. The boundary layer heat transfer resistances in the membrane cells and the membrane surface temperatures were determined from experimental data via Wilson plots. Good agreements of membrane mass transfer coefficients and water vapor fluxes were found between the model simulations and the experimental results.

(152f) Lignocellulose to Biofuels: Membranes for Process Intensification

Wickramasinghe, S. R., University of Arkansas
Qian, X., University of Arkansas
Originally presented on: 11/9/2015 14:20:00 - 14:42:00

Lignocellulose to biofuels: membranes for process intensification

Membrane based processes are often attractive as they can frequently lead to tremendous levels of process intensification. This is particularly important in the development of sustainable manufacturing processes. This presentation focuses on the conversion of lingnocellulosic biomass to biofuels and chemicals, where development of cost efficient manufacturing processes are essential. The figure below gives the glucose concentration as a function of time for a both a batch and continuous process (left hand side y-axis), as well as the rate of glucose production (right hand side y-axis) for enzymatic hydrolysis of cellulose. As can be seen in the continuous process the rate of glucose production is several times higher than for the batch process but the glucose concentration is almost10 times lower. This is due to product inhibition of the enzyme

A membrane reactor with an appropriate pore size could lead to continuous removal of glucose in the permeate as it is produced. Thus the rate of glucose production will be maximized. However it will be essential to concentrate the glucose prior to further biochemical or thermochemical, processing. Here results are presented for a submerged membrane reactor that enables continuous enzymatic hydrolysis. Development of such a process can lead to more rapid hydrolysis and smaller reactor volumes.

Finally we have developed a unique catalytic membrane that not only enables removal of glucose as it is produced, but also catalyzes the hydrolysis of cellulose in one step. As cellulase enzyme continues to represent one of the most expensive components in the production of biofuels and chemicals, replacement of the enzyme is highly desirable. Using our results for the hydrolysis of cellulose, the tremendous potential of membranes for process intensification will be discussed.

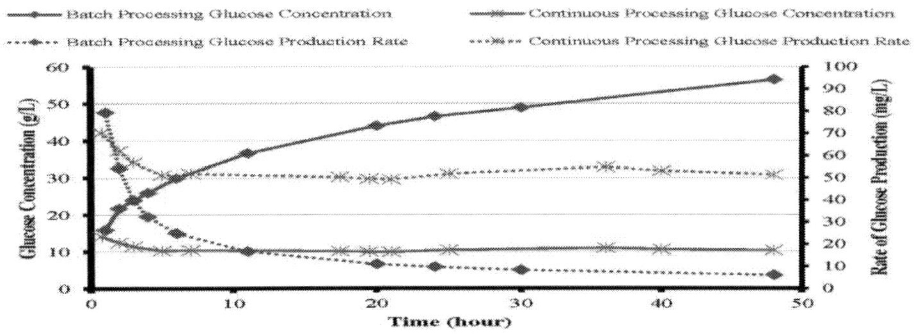

Variation of glucose concentration and glucose production rate as a function of time for batch and continuous processing.

425529 Effect of Membrane Structure on Membranes Performance during Membrane Distillation

Abstract:

Membrane Distillation (MD) has received increasing interest as an emerging desalination technology. Although significant research has been dedicated to advance this emerging technology the transition from research to commercialization has remained slow. One of the major barriers for this slow progress is the absence of commercially available membranes specifically made for MD. Current membranes employed in MD have been developed for microfiltration applications. Low permeate flux, membrane pore wetting, low mechanical stability, scaling, fouling and absence of membrane modules are amongst the major drawbacks that limit the commercialization of MD.

Here we focus on studying MD performance of commercially available microfiltration membranes. We relate performance to membrane structural parameters. There are numerous structural parameters studied in the literature, including porosity, tortuosity, thickness, polymer material, hydrophobicity, pore size, pore size distribution etc. Researchers have tried to couple these parameters to the MD performance. However, there is not yet a systematic study that shows how these structural parameters correlate with membrane performance in MD applications. Most studies have focused on a single membrane property and related it to membrane performance. Yet MD performance depends on the combined effect of a number of membrane properties such as hydrophobicity, tortuosity, pore size etc. In addition, most of the literature is limited to reporting MD fluxes and none of the previous literature investigates the interplay between the structural parameters of the membranes with their overall performance, such as permeate flux, fouling and scaling. There is also some difficulty associated with studying the structural parameters due to the fact that they are intertwined (like porosity and tortuosity). This creates significant difficulties when discussing the effect of structural parameters.

In this study, we have established a direct contact MD setup to investigate the performance of 20 hydrophobic and superhydrophobic membranes. We have also designed a unique membrane module which led to significant increase in flux while keeping the pressure drop within the module very low. We have characterized the membranes using well-established methods such as bubble point, liquid entry pressure, atomic force microscopy, scanning electron microscopy, contact angle, surface energy, porosity, and air permeability. Our results indicate the strong interplay between the membrane structural parameters and membrane performance.

424575 Self-Assembly for Membrane Formation: Influence of Solvent Quality, Hydrogen Bonds and Coulombic Interactions

Abstract:

Self-assembly for membrane formation: influence of solvent quality, hydrogen bonds and coulombic interactions

S. P. Nunes[a], P. Madhavan[a], K. V. Peinemann[b]

[a]Biological and Environmental Science and Engineering Division, Water Desalination and Reuse Center,

[b]Physical Science and Engineering Division, Advanced Membrane and Porous Materials Center,

King Abdullah University of Science and Technology, 23955-69 Thuwal, Saudi Arabia

The combination of regular phase inversion process for membrane preparation and the self-assembly of block copolymer in solution as a strategy to achieve isoporosity and high flux was demonstrated by Peinemann et al. [1] in 2007 using polystyrene-b-poly(4-vinyl pyridine) copolymer casting solutions. However the development of this technology could only progress years later[2, 3], after understanding the mechanism of pore formation. Based on evidences of electron microscopy (cryo transmission[2] and scanning[3] electron microscopy and focus ion beam), we proposed the key steps are the formation of micelles in solution, their supramolecular assembly and the order immobilization on the top layer of the incipient membrane, during the phase inversion process[2]. Far from the water-copolymer solution interface a rather disordered gradient of larger pores continues by macroscopic phase separation, following spinodal decomposition or nucleation and growth mechanisms. Since then the process has been explored by different groups using different approaches with analogous copolymer systems[4-6].

The thermodynamic interaction between blocks and the solvent mixture guides the assembly and the membrane formation. The assembly order is then kinetically trapped when solvent-non-solvent exchange takes place during phase inversion. A rough estimation of the expected order in solution can be obtained by using dissipative particle dynamics, DPD, and values of Hansen solubility parameters, taking in account van der Waals, polarity and hydrogen bond contributions. By using small angle x-ray scattering, supported by DPD modeling, we demonstrated for polystyrene-b-poly(4-vinylpyridine) how the predominant order varies with the copolymer concentration and the interaction between blocks and solvents, for instance from hexagonal micelle assembly to lamellar[6]. The order observed in the final membrane structure remarkably reflects the solution assembly. In this way diblock and triblock copolymers solutions with hexagonal and bcc micelle orders led to the corresponding pore arrangement in the final membrane[7]. The structuration leading to the final membrane morphology could be also followed by time-resolved cryo microscopy and graze incident small angle x-ray scattering[8].

A simple way to guide different morphologies is the use of small molecular weight additives in the membrane casting solutions. We have investigated the effect of complexing metal ions [1,2], molecules with hydroxyl and carboxylic groups, able to form hydrogen bonds[9] and more recently ionic liquids[10].

Block copolymer membranes have been manufactured as flat-sheet, hollow fibers[11] and spherical particles[12]. Their sharp pore size distribution and gate response to pH has allowed their effective application for protein separations with similar molecular weights[12, 13].

References

1. K. V. Peinemann, V. Abetz, P. F. W. Simon, Asymmetric superstructure formed in a

block copolymer via phase separation, Nature Materials 6 (2007) 992-996.

2. S. P. Nunes, R. Sougrat, B. Hooghan, D. H. Anjum, A. R. Behzad, L. Zhao, N. Pradeep, I. Pinnau, U. Vainio, and K.-V. Peinemann, Ultraporous Films with Uniform Nanochannels by Block Copolymer Micelles Assembly, *Macromolecules 43* (2010) 8079–8085.

3. S. P. Nunes, A. R. Behzad, B. Hooghan, Rachid Sougrat, Madhavan Karunakaran, Neelakanda Pradeep, Ulla Vainio, and Klaus-Viktor Peinemann. Switchable pH-Responsive Polymeric Membranes Prepared *via* Block Copolymer Micelle Assembly, *ACS Nano* 5 (2011) 3516-3522.

4. W. A. Phillip, R. M. Dorin, J. Werner, E. M. V. Hoek, U. Wiesner, M. Elimelech,

Tuning structure and properties of graded triblock terpolymer-based mesoporous and

hybrid films, Nano Lett. 11(2011), 2892–2900

5. S. P. Nunes, M. Karunakaran, N. Pradeep, A. R. Behzad, B. Hooghan, R. Sougrat, H. He, and K.-V. Peinemann, From Micelle Supramolecular Assemblies in Selective Solvents to Isoporous Membranes, *Langmuir* 27 (2011) 10184-10190.

6. D Marques, Ulla Vainio, Nicolas Moreno Chaparro, Victor Manuel Calo, Ali Reza Bezahd, Jed Pitera, Klaus-Viktor Peinemann and Suzana P. Nunes, Self-assembly in casting solutions of block copolymer membranes, *Soft Matter* 9 (23) (2013), 5557-5564.

7. R. M. Dorin, D. S. Marques, H. Sai, U. Vainio, W. A. Phillip, K.-V. Peinemann, S. P. Nunes and U. Wiesner, Solution Small-Angle X-ray Scattering as a Screening and Predictive Tool in the Fabrication of Asymmetric Block Copolymer Membranes, *ACS Macro Lett.* (2012) *1,* 614−617.

8. D. S. Marques, R. M. Dorin, U. Wiesner, D. M. Smilgies, A. R. Behzad, U. Vainio, K. V. Peinemann, S. P. Nunes*, Time-resolved GISAXS and cryo-microscopy characterization of block copolymer membrane formation, *Polymer 55* (2014) 1327-1

9. P. Madhavan, K. V. Peinemann, S. P. Nunes, Complexation-tailored morphology of asymmetric block copolymer membranes, *ACS Applied Materials and Interfaces* 5 (2013) 7152-7159.

10. P. Madhavan, R. Sougrat, A. R. Behzad, K. V. Peinemann, S. P. Nunes, Ionic Liquids as Self-Assembly Guide for the Formation of Nanostructured Block Copolymer Membranes, J. Membrane Sci. 2015.

11. R. Hilke, N. Pradeep, P. Madhavan, U. Vainio, A. R. Behzad, R. Sougrat, S. P. Nunes, K. V. Peinemann*, Block copolymer hollow fiber membranes with catalytic activity and pH-response, *ACS Applied Materials and Interfaces*, 5 (2013) 7001-7006

12. Haizhou Yu, Xiaoyan Qiu, Suzana P. Nunes, Klaus-Viktor Peinemann, Biomimetic block copolymer particles with gated nanopores and ultrahigh protein sorption capacity, *Nature Communications*, 5 (2014) 4110.

13. X. Qiu, H. Yu, M. Karunakaran, N. Pradeep, S. P. Nunes, K. V. Peinemann, Selective separation of similarly sized proteins with tunable nanoporous block copolymer membranes, *ACS Nano* 7 (2013) 768-776.

422431 Transport Channels in Polymer-Grafted Nanocomposite Membranes for Vapor Separations

Abstract:

Inspired by the increasing need for polymeric membranes with tunable separation ability, here we focus on the mass transport of a small molecule penetrant in matrix-free thin films of PMA-grafted, nonporous silica nanoparticles. Preliminary permeability data measured with the quartz crystal microbalance technique are larger than the neat polymer and directly contrast conventional composite theory. We postulate that the self-assembly of the particles grafted with moderate to high grafting densities of moderate length polymer chains into ordered arrays creates interstitial spaces which the polymer chains can only fill by stretching. Solute particles are thus driven to diffuse through these channels as their placement in these regions serve to reduce the unfavorable entropy of the stretched chains. This, in turn, increases the available "free volume" within the film and creates channels, similar in spirit to transport channels in cell membranes, which facilitate penetrant diffusion. Importantly, we expect that these channels will disappear in the limit where the grafts become infinitely long: in this limit the medium should transport solute exactly as a normal polymeric membrane. The potential tunability of these transport channels could be a paradigm shifting discovery for this class of membrane materials.

417349 Block Copolymer Pervaporation Membrane for Organics-Water Separation

Abstract:

Biobutanol can be produced via acetone-butanol-ethanol, or ABE, fermentation. One of the challenges in industrializing this process is product inhibition. Organic products of the fermentation deter productivity of the cells above certain concentrations of products. These concentrations are typically very low (< 2%), thus resulting in low productivity and restricting fermentations to batch processes. Many researches have been conducted to address this problem by in situ product recovery. Of the various approaches, pervaporation has advantages in the fact that the separated permeate is very concentrated and that it is non-invasive to the cell culture. However, the pervaporation technology is currently limited by the effectiveness of the membranes. An optimal pervaporation membrane would have a high permeability and selectivity for the fermentation products. Simultaneously, it would have to have enough mechanical stability to be produced in a thin film. This is to minimize the time that it takes for the molecules to diffuse through the membrane.

We try to address these two design aspects by using a block copolymer membrane. We covalently joined polydimethylsiloxane(PDMS), which is the most widely used membrane material for this application of pervaporation, with two polystyrene(PS) blocks on either end, forming a PS-b-PDMS-b-PS(PSDS) tri block copolymer. The PS blocks are expected to provide the polymer with enhanced mechanical properties, replacing the chemical cross-links which is normally used as a way of tethering the PDMS. The PDMS blocks provide the transporting phase through which molecules selectively diffuse. We fabricated PSDS block copolymer membranes with comparable thicknesses to those of commercially available cross-linked PDMS membranes. Our binary butanol-water pervaporation experiments show that the PSDS membrane has both the higher butanol selectivity and higher butanol permeability than those of the commercial PDMS membrane.

We were also able test the PSDS and the commercial PDMS membranes in a pervaporation-incorporated ABE fermentation system. In our study, we show that the pervaporative-fermentations were able to deter product inhibition, resulting in the increase of ABE production. Also, PSDS membranes had higher permeabilities and selectivities for the products than those of the commercial PDMS membranes, which resulted in a further increase in ABE production for the PSDS pervaporative-fermentation.

429223 Block Polymer Membrane Adsorbers Chemically-Tailored for Selenium Sorption

Abstract:

Abstract: Traditionally, the targeted capture of analytes relies on packed beds for high-recovery and high-capacity sorption. Membrane adsorbers provide an alternative technological pathway that replaces microporous resins with high-flux membranes. However, these devices are stymied by low sorption capacities, complex processing methodologies, and have limited tunability. Here, we demonstrate the use of the self-assembly and non-solvent induced phase separation (SNIPS) technique to generate nanoporous thin films with tunable pore wall chemistries; in this way, chemically-selective membranes can be generated in a straightforward manner. Specifically, the A-B-C triblock polymer polyisoprene-*b*-polystyrene-*b*-poly(*N*,*N*-dimethylacrylamide) (PI-PS-PDMA) is processed from solution to yield a membrane with a pore density of $\sim 10^{13}$ pores m^{-2}. The PDMA moiety that lines the pore walls is converted, quantitatively, to thiol group chemical functionalities in the solid state using facile reaction conditions. These thiol-functionalized membranes act as high capacity, high flux adsorbers for the removal of selenium salts from an aqueous solution. This scalable system demonstrates a platform for high-flux chemically-selective purification for use in next-generation industrial water treatment and reuse applications.

430617 Tunable-Porosity Membranes for Water Treatment By Discrete Nanoparticle Assembly

Monday, November 9, 2015: 1:54 PM
155B (Salt Palace Convention Center)

Patrizia Marchetti[1], Martin Mechelhoff[2] and Andrew G. Livingston[1], (1)Chemical Engineering, Imperial College London, London, United Kingdom, (2)Lanxess Deutschland GmbH, Leverkusen, Germany

Tunable-Porosity Membranes For Water Treatment By

Discrete Nanoparticle Assembly

Patrizia Marchetti [1], Martin Mechelhoff [2], Andrew G. Livingston [1,*]

[1]*Department of Chemical Engineering, Imperial College London, South Kensington Campus, London SW7 2AZ, UK*

[2]*Lanxess Deutschland GmbH ,Group Function Innovation & Technology, 51369 Leverkusen, Germany*

* a.livingston@imperial.ac.uk

Thin film composite (TFC) membranes are an important class of membranes, mainly used for reverse osmosis and nanofiltration (NF) [1]. They consist of a thin (submicron) separating "barrier" on top of a chemically different porous support. For aqueous applications, they are usually prepared by interfacial polymerization of polyamide on a support-layer consisting of, for example, polysulfone. The two main challenges with this procedure are the need for organic solvents, such as n-hexane, and the difficulty in a precise control of the pore size at nanometer scale. Colloidal assembly of polymeric nanoparticles represents an interesting alternative to conventional fabrication procedures for TFC membranes [2]. This approach permits the organization of nanomaterials, acting as pre-structured building blocks, to create larger entities with new properties. The production process is also more environmentally friendly, as crosslinked organic nanoparticles can be easily handled in an aqueous solution without the need for an organic solvent.

In this work, a novel NF membrane was prepared by a facile single-step coating procedure of highly crosslinked poly(styrene-co-butadiene) polymer nanoparticles on top of a hydrophilic ultrafiltration support. The nanoparticles employed were characterised by high chemical and mechanical stability in extreme conditions, such as concentrated acid and base solutions and a wide range of organic solvents. During membrane fabrication, the nanoparticle surface loading was optimized, in order to avoid cracking during the drying step and obtain a thin, defect-free membrane separation layer of 144 nm (Figure 1).

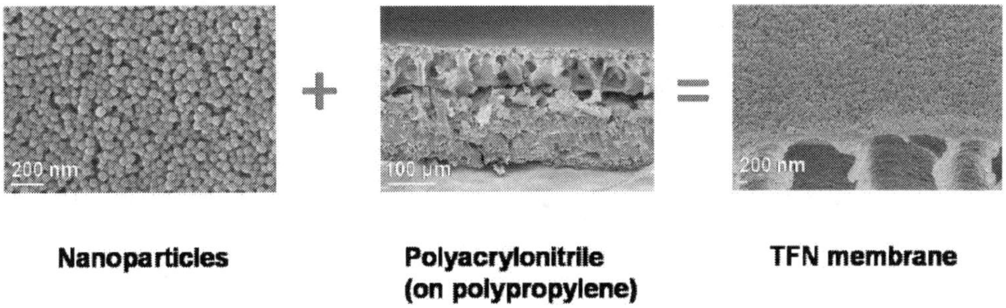

Figure 1. Thin Film Nanocomposite membrane by nanoparticle assembly on top of polyacrylonitrile ultrafiltration support

The membrane exhibited a significant charge rejection mechanism and low fouling tendency, due to the high density of hydroxyl group functionalization on the nanoparticle surface. Interestingly, upon heating above the glass transition temperature (or minimum film formation temperature) the particles coalesced, forming a rubber-like dense film. This was a consequence of deformation and interpenetration of the nanoparticles, which in turn comprise a highly crosslinked particle core and a more flexible and less dense outer shell. The novel membrane formed by nanoparticle annealing showed a molecular weight cut-off below 500 g mol^{-1} and a viable water permeance of about 40 L m^{-2} h^{-1} bar^{-1}, unlike membranes formed from conventional linear rubber-like polymers, which usually have insignificant water permeance of less than 0.4 L m^{-2} h^{-1} bar^{-1}. The permeation pathways were found to be proportional to, and therefore tunable with, the annealing temperature.

References

[1] K. Peng Lee, T. C. Arnot, D. Mattia, *J. Membr. Sci.* 370 (2011) 1.

[2] Q. Zhang, S. Ghosh, S. Samitsu, X. Peng, I. Ichinose. *J. Mater. Chem.* 21 (2011) 1684.

431610 Multiscale Ionomeric Block Polymer Assemblies for Membrane Applications

Abstract:

Polymeric membranes for filtration and desalination applications play a critical role in producing the billions of gallons of water required daily. Current polymers used for water desalination membranes suffer from coupled water sorption and salt selectivity, which leads to membranes with either (1) high permeation rates and low selectivity, or (2) highly selective membranes with poor permeation rates. Decoupling and optimizing water and ion transport in polymeric membranes is critical for reducing the energy costs of desalination. Ionomeric block polymers are a tremendously powerful platform for separations membranes and offer the potential to decouple moisture transport and salt rejection from mechanical integrity. These exciting materials synergistically incorporate electrostatic interactions to modulate transport performance with microphase separation to customize the mechanical integrity. Thus, we will discuss the development of structure-property-diffusion relationships that have been developed in our lab that contribute toward a better understanding of the interplay between polymer morphology, water and salt sorption and transport characteristics, and thermomechanical properties. In addition to variations in nanoscale structure and functionality, our lab is focused on analyzing changes in the macroscopic structure of the membrane. For example, filtration processes that remove large particles and biomass demand a high water flux, which make fibrous nonwoven mats desirable. Electrospinning non-woven polymeric mats comprised of ionomeric block polymers offers a strategy to generate large surface area filtration substrates with tailorable surface functionality. Other membrane architectures of interest include dense membranes and hollow fiber membranes, and our lab is interested in correlating the properties of ionomeric block polymers to the membrane architecture.

416018 Designing Clay-Polyelectrolyte Hybrid Membranes for Effluent Treatment

Abstract:

In this study, clay-polyelectrolyte composite membranes were employed for reducing the COD level of an electrocoagulation—treated high strength effluent. A few recent studies have shown the incorporation of Montmorrillonite (MMT) clay nanoplatelets within polyelectrolyte multilayers for developing desalination membranes with high permeability. However, most of these studies involved lab-based synthetic solutions constituting mainly the target ion, for testing these membranes. These solutions fail to emulate the complexities involved in actual wastewater solutions. In our research we showed, for the first time, the application of these composite membranes to treating real effluents. This involved the selection of the appropriate substrate as well as the appropriate polyelectrolyte combination. The number of bilayers deposited and the pH of the polyelectrolytes used were also tweaked in order to get the best combination of high permeability as well as high COD removal. This complex effluent system also presented an opportunity to test the fouling propensity of these hybrid systems in presence of the various types of natural organic matters present in the solution. The performance of these membranes were compared with commercial nanofiltration and reverse osmosis membranes as well as pure polyelectrolyte-based membranes in terms of permeability, removal and anti-fouling property. Our results show that these hybrid membranes show lower flux decline over a given period of time than the remaining membranes, rendering them highly permeable as well as fouling resistant.

438527 Controlled Release from Polyelectrolyte Complex Drug Carriers

Abstract:

Comparison studies of release of a model drug, Rhodamine B, from two different types of polyelectrolyte complex beads (PECs) made of poly(styrenesulfonate, PSS)-poly(allylamine, PAA) and chitosan-alginate were investigated. The release was carried out by using a modified Franz diffusion cell into deionized water at 37°C. The long term sustained release (> 7 days) of Rhodamine B was obtained with PSS-PAA PECs due to their tighter networks. By tuning the compositions of PSS-PAA, the time needed to completely release the drug could be varied from two days to 14 days. For the case of chitosan-alginate PECs, 100% accumulated release occurred within 24 hours. With the incorporation of a thermo-responsive polymer, e.g., poly(N-isopropylacrylamine), the release could be extended to two or more days for the chitosan-alginate PECs. By drying the beads, the release of Rhodamine B was found to be further extended.

438450 3D Printed Microfluidic Device for Dynamic Investigation of the Blood Brain Barrier

Abstract:

The "Blood Brain Barrier" (BBB), surrounding the capillaries in the Central Nervous System, is a tissue composed of endothelial cells joined to each other by tight junctions. It serves to impede the passage of any impurities from the blood to the brain and is essential to brain homeostasis, disruption of which can result in a myriad of brain disorders like Alzheimer's, Multiple Sclerosis, and Parkinson's disease. Current in-vitro models of the BBB include a seeded Transwell ® Membrane under static conditions, where the apical and basal side fluids do not exhibit dynamic flow and apply shear stress on the membrane. Being a simplified model of the tissue, it does not adequately model the tissue as it is in-vivo, where there is dynamic flow of both basal (cerebrospinal fluid) and apical (blood) fluids. Consequently, we have designed a microfluidics device, the μTRANS chip, to model the BBB under dynamic conditions, allowing us to study not only the tissue structure of the BBB but also the cellular level dynamics of it under different environmental conditions. Our initial results, revolving around the construction of the device, its characterization, and the biochemical/mechanical modulations of the tissue will be presented at the conference. In terms of construction, the device was built using a seeded Transwell ® membrane (consisting of a layer of human brain endothelial cells on the apical side and a layer of astrocytes on the basal side) inserted between two glass slides. Each of the glass slides served as a distinct chamber, containing two 3D printed micro-channels, interconnected by smaller micro-channels. The smaller channels were designed to allow for fluid exchange between the larger micro-channels. Gold electrodes were also built into the device to analyze the permeability of the BBB using electrochemical impedance spectroscopy (EIS). The channels, glass substrate layers, and Transwell ® Membrane were all placed into a 3D printed vacuum chamber as the final step in construction, in order to maintain contact between the different layers of the device during experimental modulations. The permeability characteristics of the BBB were studied using fluorescence and EIS, specifically in the form of transendothelial electrical resistance tests (TEER), under different biochemical and mechanical conditions, including varying levels of pH, shear stress, and nutrients. We believe that our ability to study the BBB in a dynamic environment allows us to not only accurately model the tissue under physiological conditions but to control its characteristics using biochemical and mechanical means.

438448 Self-Assembly Simulations of Polymer Functionalized Virus Capsids

Abstract:

Viral capsids have been the subject of a significant amount of research in a variety of fields. Their potential relevance to biomedical science arises from their ability to encapsulate therapeutic agents and deliver them to targeted cellular sites. The capsids can be decorated with polymers to target specific sites or promote controlled aggregation for creating novel materials. We used coarse-grained molecular dynamics to simulate the aggregation dynamics of amphiphilic functionalized icosahedral capsids, representing the Cowpea Mosaic Virus capsid, when grafted with diblock copolymers of polyethylene glycol and polylactic acid. Additional variations of the system includes changing the polymer grafting density and capsid volume fraction to show how these changes affect the self-assembled structures. Our observations suggest that (a) the excluded volume of the icosahedral and (b) the polymer grafting density play the largest role on aggregation dynamics and morphologies when compared to the other variants.

438410 Investigation of the Interaction Between a Novel Drug Delivery System and an Epithelial Cell Layer

Abstract:

Surgery and other invasive processes are the current treatment options for patients with diabetic retinopathy, glaucoma, and age-related macular degeneration, which are eye diseases that affect at least nine million Americans. We are developing a targeted drug-delivery system, which would provide a non-invasive option for treatment. The system features a collagen membrane with drug-loaded poly(lactic-co-glycolic acid) (PLGA) nanoparticles incorporated into a thin membrane. This membrane can be worn by a patient in conjunction with a contact lens, and provides for continuous drug release. The purpose of this project is to study the interactions of lidocaine-loaded nanoparticles in a collagen membrane with a layer of corneal epithelial cells. Lidocaine-loaded PLGA nanoparticles were made and characterized by measuring size and drug content. Nanoparticles were mixed into a collagen solution and then gelled to form a thin membrane. Corneal epithelial cells were cultured on semipermeable well inserts to form a confluent layer of cells. The cell layer was exposed to the collagen membrane containing the lidocaine-loaded nanoparticles. Measurements were taken to determine cell viability and permeability, as well as the amount of lidocaine that crosses the cell layer. Results showed that lidocaine was released from the system and delivered across the cell layer. The release profile shows an initial burst release, followed by steady drug release over the testing period. Cell viability decreased with the application of the membranes, compared to a control sample of cells only. Further work is being done to determine the cause of cell death and to test the system for longer periods of time

438407 Quantitative Analysis of Fundus Images for Grading of Vitreous Haze

Abstract:

Purpose: Intraocular inflammatory processes lead to progressive accumulation of cells and protein exudate in the vitreous, which can be detected by ophthalmoloscopic inspection of the eye as decreased visibility of the retinal vasculature, optic nerve head, and other fundus details. The objective of this work was to develop a quantitative method for grading the blurriness of ocular fundus images in order to provide automated grading of the severity of uveitis.

Methods: A computer algorithm was developed to quantify image blurriness through application of entropy filtering and Fourier spatial-frequency analysis. The algorithm was refined using a set of 8 reference images that were created by optically filtering a single standard fundus picture to varying degrees that simulate increasing vitreous haze. It was then applied without modification to a dataset of 120 digital fundus images collected from patients with different grades of uveitis. The algorithm performance was evaluated against the acutance method, which is commonly used to quantify the overall sharpness of edges in an image. Computed scores of vitreous haze were also compared for both methods in a masked fashion with the subjective readings of three expert clinicians.

Results: The Pearson correlation coefficient between computed automated grades of the algorithm and clinician grades of blurriness was 0.81-0.88 (R^2 = 0.66-0.78). Acutance measurements calculated from the raw images and from images with background fluctuations subtracted out did not perform as well. The Pearson correlation coefficient between acutance and clinician grades was found to be -0.05 (R = 0.003) and 0.33 (R=0.109), respectively, which is quite weak.

Conclusions: We have developed a computer algorithm for grading vitreous haze in an unbiased and quantitative manner that correlates strongly with the subjective readings of expert clinicians. The algorithm performance with fundus images is much better than that of standard acutance measures.

McCormick

Northwestern Engineering

Department of Chemical and Biological Engineering

Developing a Strategy for Constructing Modular Biosensors

Andrew K.D. Younger[1], **Neil C. Dalvie[2]**, Austin G. Rottinghaus[3], and Joshua N. Leonard[2,4,5,*]

[1] Interdisciplinary Biological Sciences Graduate Program, Northwestern University, Evanston, IL, 60208, USA
[2] Department of Chemical and Biological Engineering, Northwestern University, Evanston, IL, 60208, USA
[3] Department of Chemical and Biological Engineering, University of Wisconsin-Madison, Madison, WI, 53706, USA
[4] Chemistry of Life Processes Institute, Northwestern University, Evanston, IL, 60208, USA
[5] Member, Robert H. Lurie Comprehensive Cancer Center, Northwestern University, Evanston, IL, 60208, USA

1

McCormick

Northwestern Engineering

Department of Chemical and Biological Engineering

Motivation: Quantifying metabolism

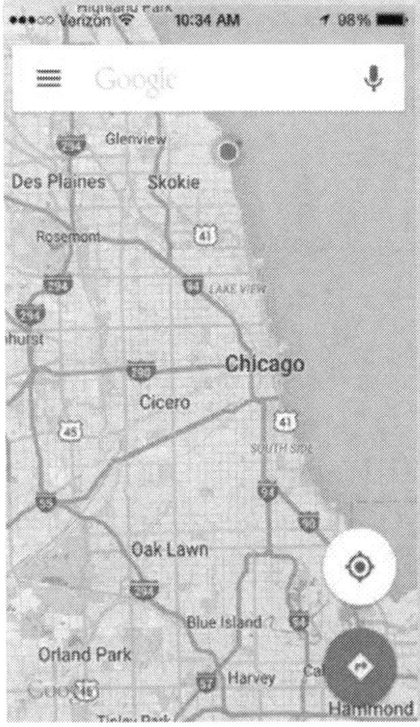

2

McCormick

Northwestern Engineering

Department of Chemical and Biological Engineering

Motivation: Quantifying metabolism

3

McCormick

Northwestern Engineering

Department of Chemical and Biological Engineering

Motivation: Quantifying metabolism

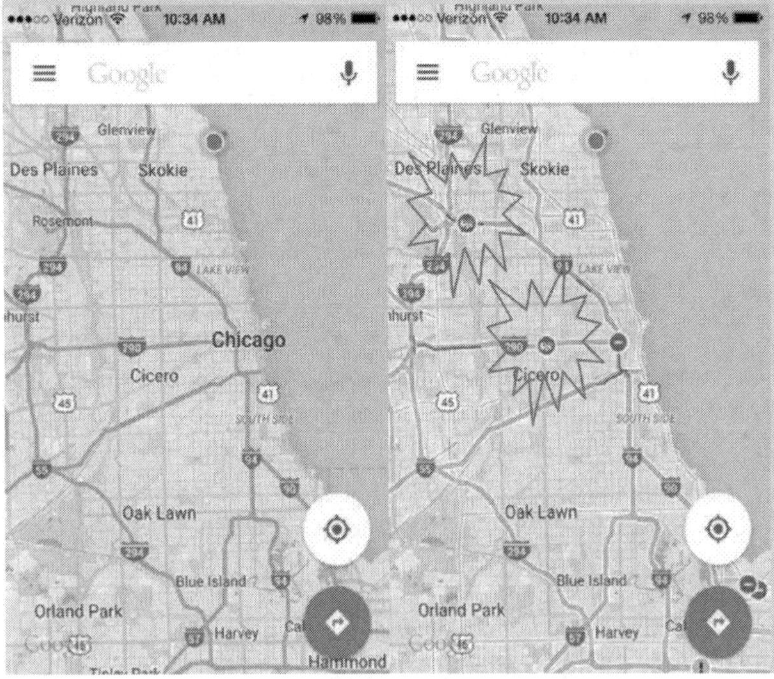

4

McCormick

Northwestern Engineering

Department of Chemical and Biological Engineering

Motivation: Quantifying metabolism

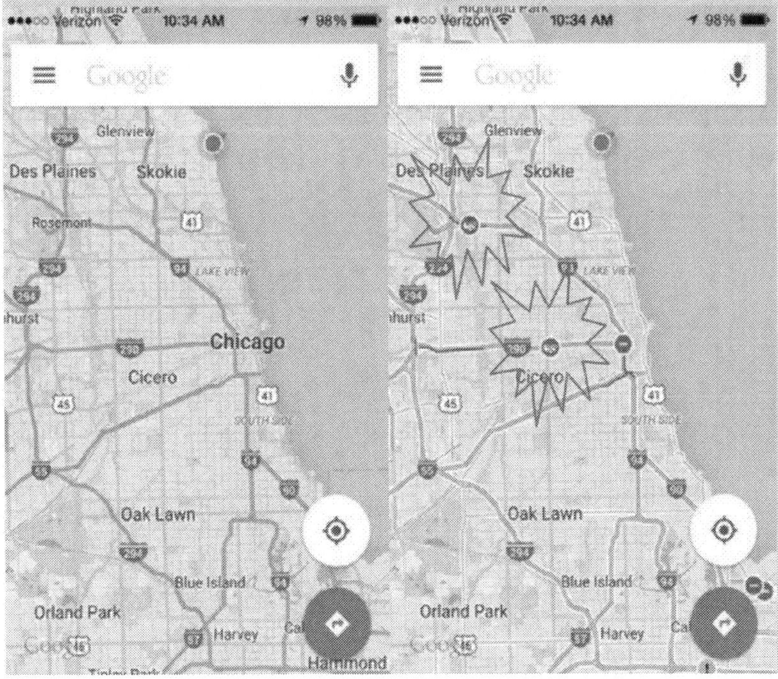

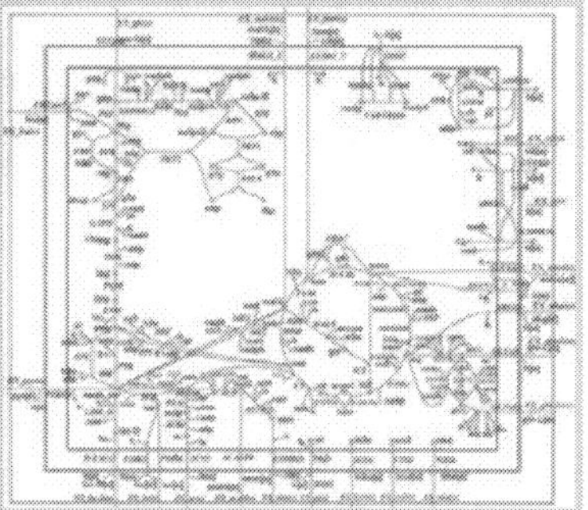

5

Orth *et al.* 2011

McCormick

Northwestern Engineering

Department of Chemical and Biological Engineering

Motivation: Quantifying metabolism

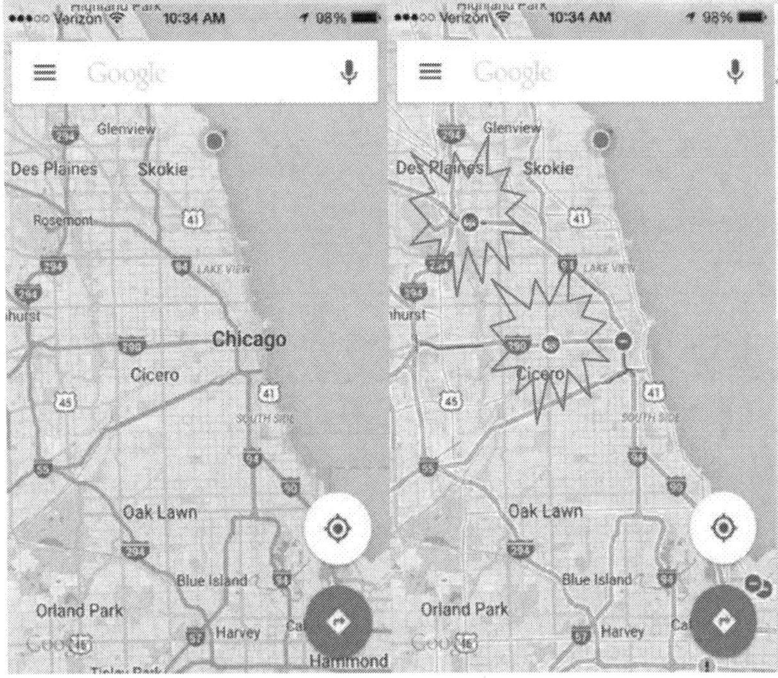

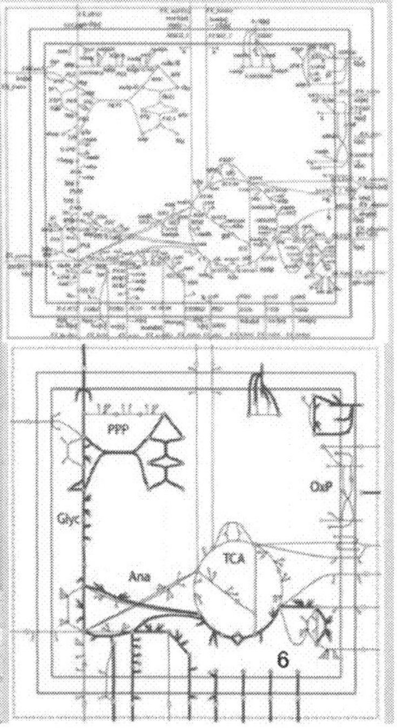

6

McCormick
Northwestern Engineering
Department of Chemical and Biological Engineering

Useful functions of metabolite biosensors

High Throughput Library Screening

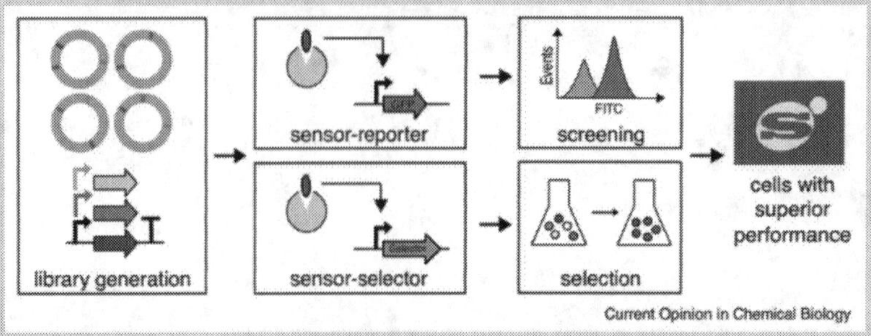

7

Zhang *et al*. 2015

McCormick
Northwestern Engineering
Department of Chemical and Biological Engineering

Useful functions of metabolite biosensors

High Throughput Library Screening

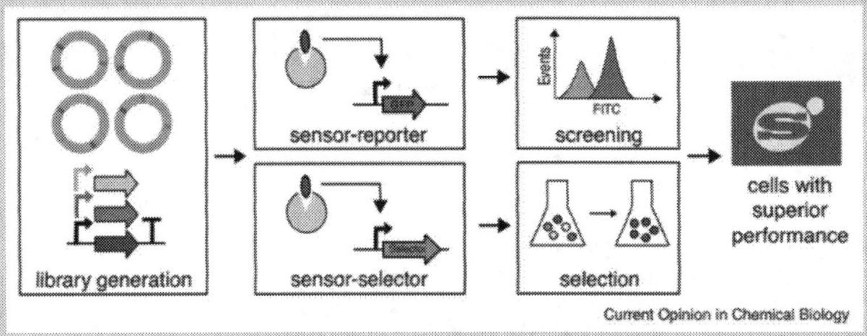

Feedback control

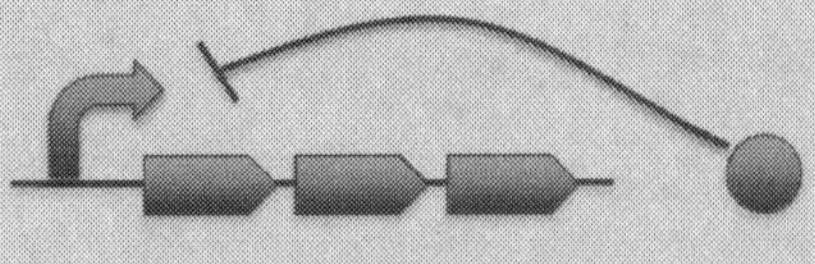

8

Zhang *et al*. 2015

McCormick
Northwestern Engineering
Department of Chemical and Biological Engineering

Useful functions of metabolite biosensors

High Throughput Library Screening

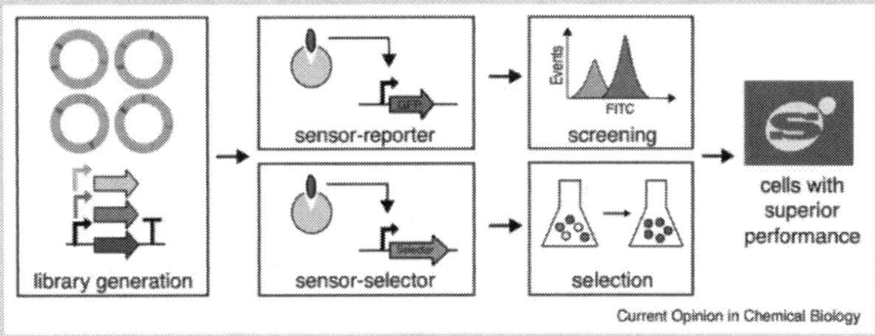

Feedback control

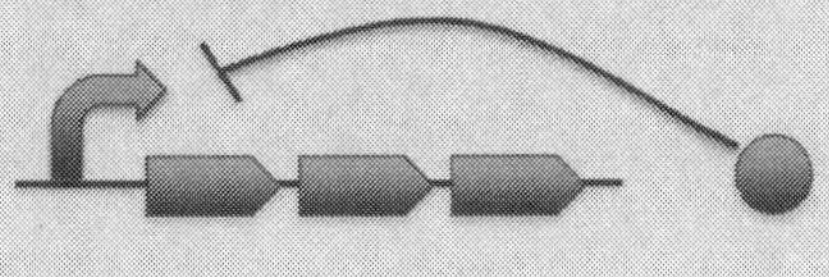

Limited natural options

9

Zhang *et al.* 2015

McCormick
Northwestern Engineering
Department of Chemical and Biological Engineering

Metabolite responsive transcriptional regulators

<u>Goal:</u>

Ligand responsive transcriptional regulator

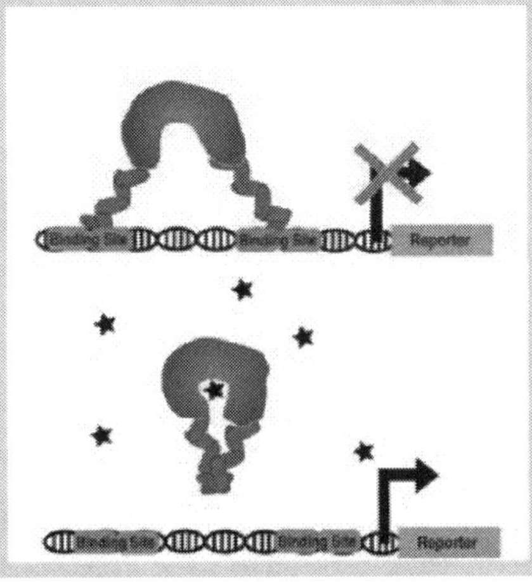

<u>Approach:</u>

Modular design

-Convert ligand binding protein to biosensor

-DNA binding domain

10

McCormick

Northwestern Engineering

Department of Chemical and Biological Engineering

Metabolite responsive transcriptional regulators

Goal:

Ligand responsive transcriptional regulator

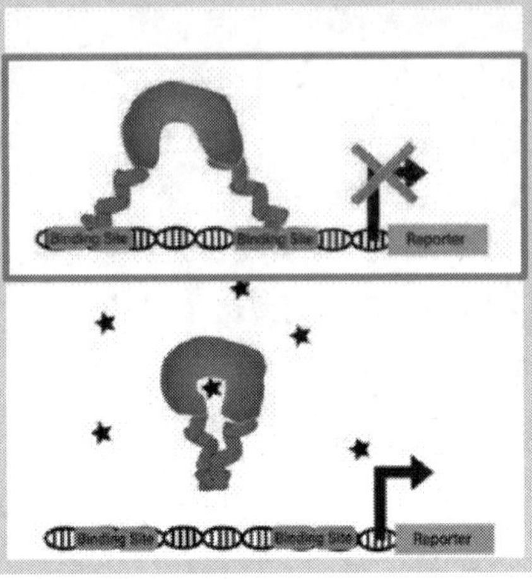

Approach:

Modular design

-Convert ligand binding protein to biosensor

-DNA binding domain

11

McCormick

Northwestern Engineering

Department of Chemical and Biological Engineering

DNA binding domain: BCR1 zinc finger

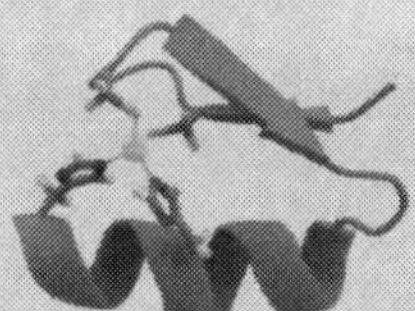

12

PDB #1A1L Thomas Splettstoesser

McCormick

Northwestern Engineering

Department of Chemical and Biological Engineering

DNA binding domain: BCR1 zinc finger

9 base pair binding site

13

PDB #1A1L Thomas Splettstoesser

McCormick

Northwestern Engineering

Department of Chemical and Biological Engineering

DNA binding domain: BCR1 zinc finger

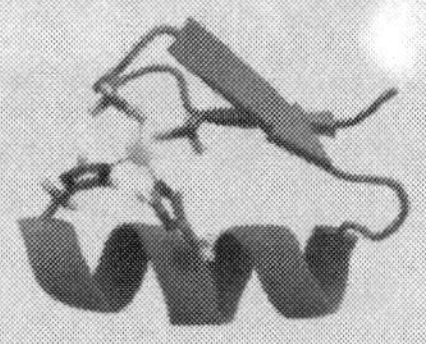

9 base pair binding site

Constitutive GFP reporter

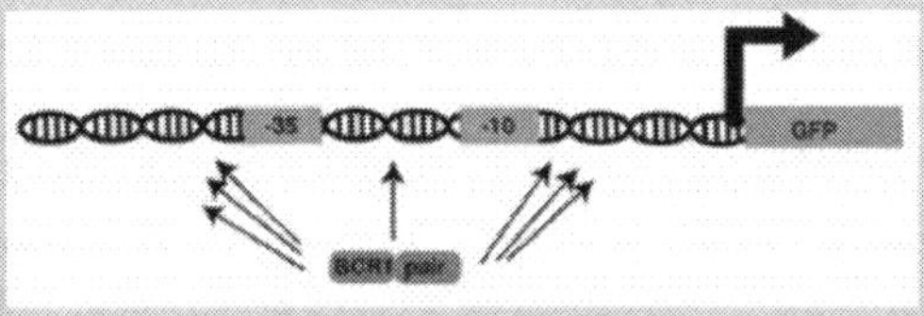

14

PDB #1A1L Thomas Splettstoesser

McCormick
Northwestern Engineering
Department of Chemical and Biological Engineering

DNA binding domain: BCR1 zinc finger

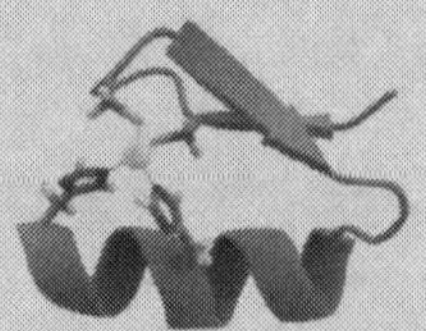

9 base pair binding site

Constitutive GFP reporter

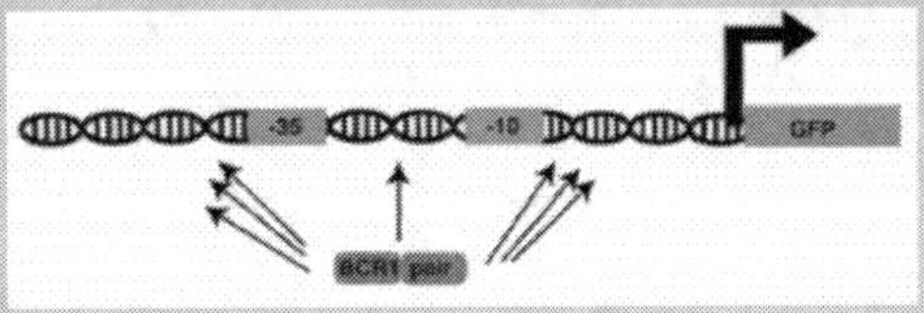

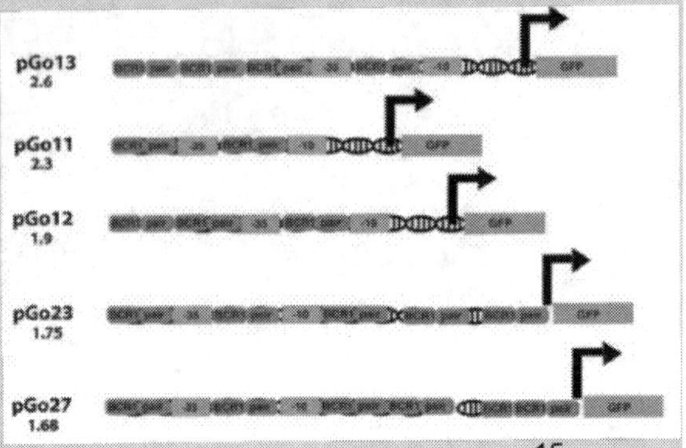

15

PDB #1A1L Thomas Splettstoesser

McCormick
Northwestern Engineering
Department of Chemical and Biological Engineering

Investigate expression of promoter library

McCormick

Northwestern Engineering

Department of Chemical and Biological Engineering

Investigate expression of promoter library

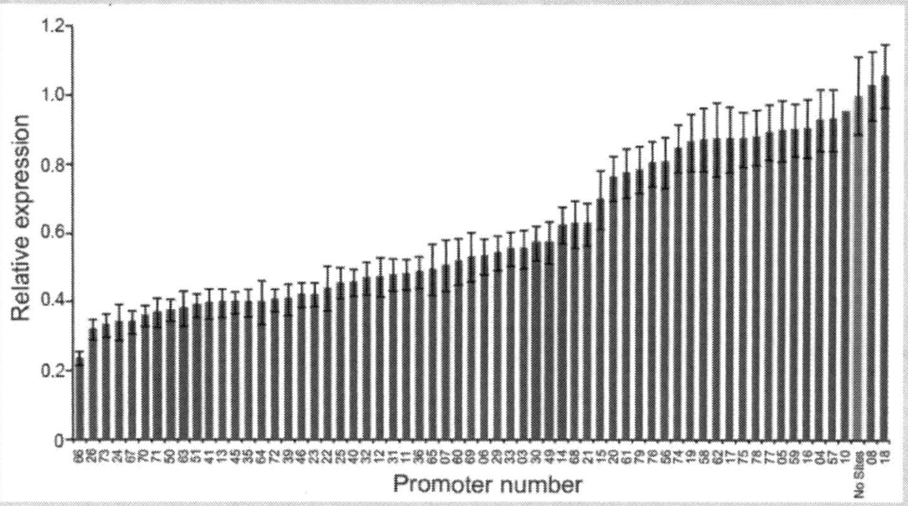

McCormick

Northwestern Engineering

Department of Chemical and Biological Engineering

Investigate expression of promoter library

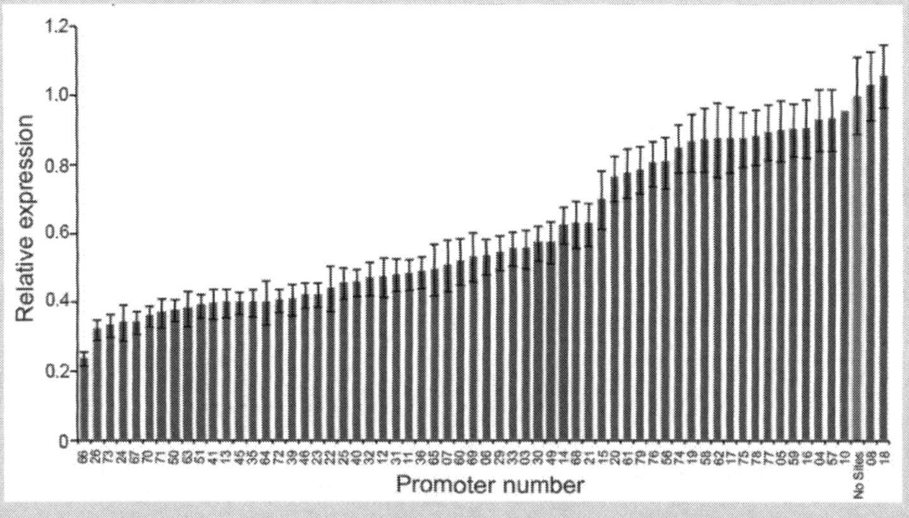

Wide range of repression – Why?

McCormick
Northwestern Engineering
Department of Chemical and Biological Engineering

Method for quantitatively describing the library

Goal: Select a set of descriptors that describe promoters.

Risk: Too many descriptors will over fit the system.

19

McCormick
Northwestern Engineering
Department of Chemical and Biological Engineering

Method for quantitatively describing the library

Goal: Select a set of descriptors that describe promoters.

Risk: Too many descriptors will over fit the system.

Assumption: The only thing that affects reporter expression is the binding of the biosensor

Features only describe the location of zinc finger binding sites

20

McCormick
Northwestern Engineering
Department of Chemical and Biological Engineering

Number	Shorthand	Description of Feature
1 ★	*BS in core*	Number of binding sites in the core
2 ★	*pairs in core*	Number of pairs in the core
3	*BS upstream*	Number of binding sites upstream of the -35 region
4	*pairs upstream*	Number of pairs upstream of the -35 region
5	*BS downstream*	Number of binding sites downstream of the -10 box
6	*pairs downstream*	Number of pairs downstream of the -10 box
7	*length*	Length of the promoter in base pairs
8 ★	*spacer -10*	Base pairs downstream of the -10 box, before the first site
9 ★	*spacer -35*	Base pairs upstream of the -35 region, before the first site
10 ★	*spacer inside -10*	Base pairs upstream of the -10 box, before the first site
11 ★	*spacer insider -35*	Base pairs downstream of the -35 region, before the first site
12	*core middle space*	Non-binding site base pairs in the core
13	*ave downstream spacing*	Average number of base pairs between sites downstream
14	*ave upstream spacing*	Average number of base pairs between sites upstream
15	*down length*	Base pairs downstream of the -10 box
16	*up length*	base pairs upstream of the -35 box
17	*lopsidedness*	Absolute value of (sites upstream) - (sites downstream)

21

McCormick
Northwestern Engineering
Department of Chemical and Biological Engineering

Determining important features

Define output:
"repressibility" = (-1)*expression

Questions:

What is the relative importance of features in determining repressibility?

How many of these features are actually significant?

22

McCormick
Northwestern Engineering
Department of Chemical and Biological Engineering

Approach: Three independent feature selection methods

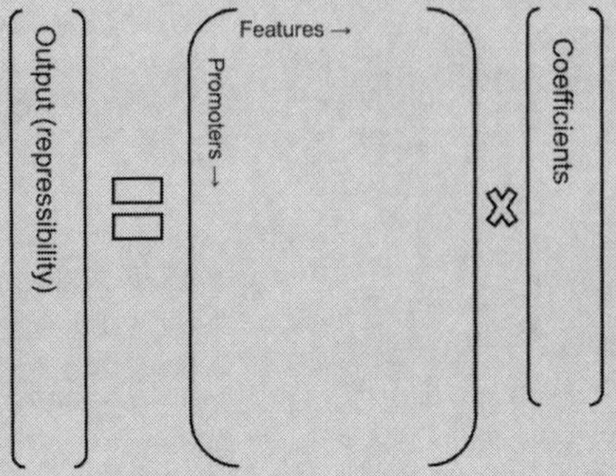

- Partial least squares regression
- Random forest
- Lasso regression

Iterative

Regularized

23

McCormick
Northwestern Engineering
Department of Chemical and Biological Engineering

How many features are significant?

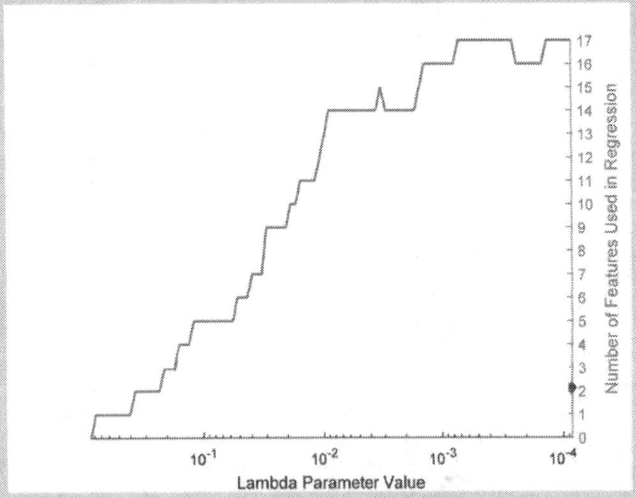

Decreasing penalty on coefficients

Lasso regression shrinks the coefficients of unimportant features to 0

After a certain number of features, the fits stops improving

24

McCormick

Northwestern Engineering

Department of Chemical and Biological Engineering

How many features are significant?

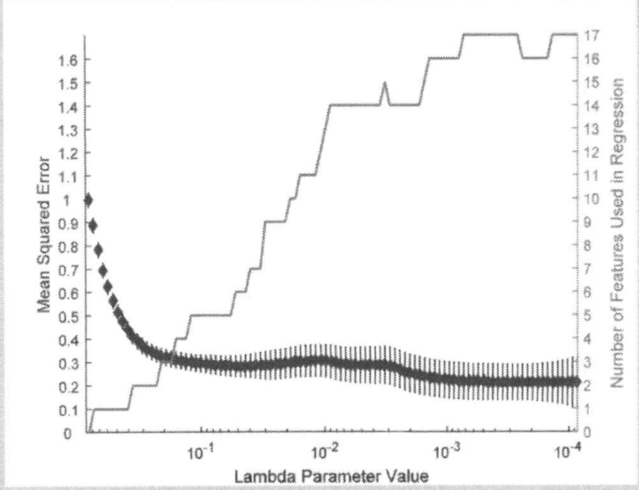

Decreasing penalty on coefficients

Lasso regression shrinks the coefficients of unimportant features to 0

After a certain number of features, the fits stops improving

25

McCormick

Northwestern Engineering

Department of Chemical and Biological Engineering

How many features are significant?

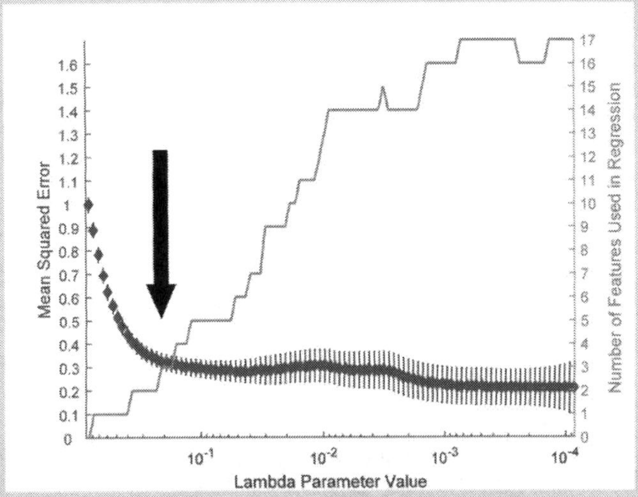

Decreasing penalty on coefficients

Lasso regression shrinks the coefficients of unimportant features to 0

After a certain number of features, the fits stops improving

26

McCormick

Northwestern Engineering

Department of Chemical and Biological Engineering

How many features are significant?

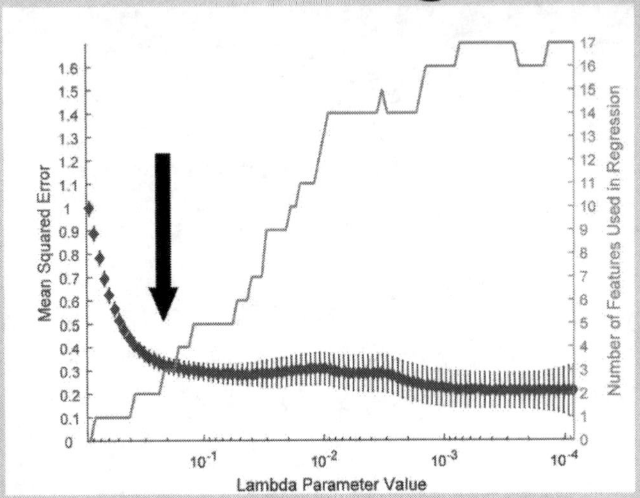

Decreasing penalty on coefficients

Lasso regression shrinks the coefficients of unimportant features to 0

After a certain number of features, the fits stops improving

~3 important features

27

McCormick

Northwestern Engineering

Department of Chemical and Biological Engineering

Comparison of all three methods

Green = positive regression coefficient (positively correlated with repressibility)

Pink = negative regression coefficient (negatively correlated with repressibility)

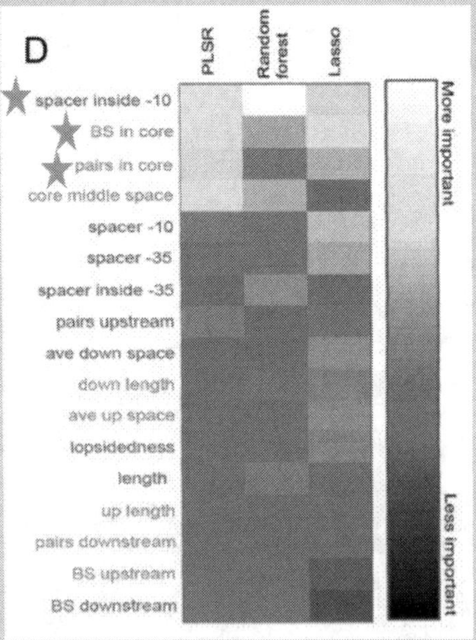

28

McCormick

Northwestern Engineering

Department of Chemical and Biological Engineering

Rules for zinc finger repression

Crowding promoter boxes, especially -10, tends to create strong repression

Binding site quantity does not matter
- No titration effect

Regions away from the boxes are not important (upstream, downstream)

Proposed mechanism: The zinc finger blocks the sigma factor when near the -10 box

29

McCormick

Northwestern Engineering

Department of Chemical and Biological Engineering

Application to biosensors

Promoter library analysis revealed elegant design rules for synthetic promoters

Useful feature selection methods for describing transcription factor binding

Application of this DNA binding domain in biosensors

McCormick
Northwestern Engineering
Department of Chemical and Biological Engineering

Application to biosensors

Promoter library analysis revealed elegant design rules for synthetic promoters

Useful feature selection methods for describing transcription factor binding

Application of this DNA binding domain in biosensors

Development of a small molecule responsive transcriptional regulator platform.
Andrew Younger

Session 734: Emerging Tools for Synthetic Biology

Thursday 3:15, Salt Palace Convention Center 150G

McCormick
Northwestern Engineering
Department of Chemical and Biological Engineering

Acknowledgements

Josh Leonard
Andrew Younger
Austin Rottinghaus
Rachel Dudek
Andy Scarpelli
Michelle Hung
Kelly Schwarz
Taylor Dolberg
Alia Zander
Danny Wells

Collaborators
Declan Bates (U. Warrick)
Paul Freemont (Imperial College London)
Jay Keasling (UC Berkeley)
Anne Osbourn (John Innes Center)
Susan Rosser (U. Edinburgh)

Chemistry of Life Processes Institute

Office of Undergraduate Research
UR@NU

NSF
EPSRC
Pioneering research and skills

32

McCormick
Northwestern Engineering
Department of Chemical and Biological Engineering

Decision tree regression: Random forest

Uses decision trees on subsets of features

Quantify how much accuracy is lost when feature is not present

Creates a ranked list of importance

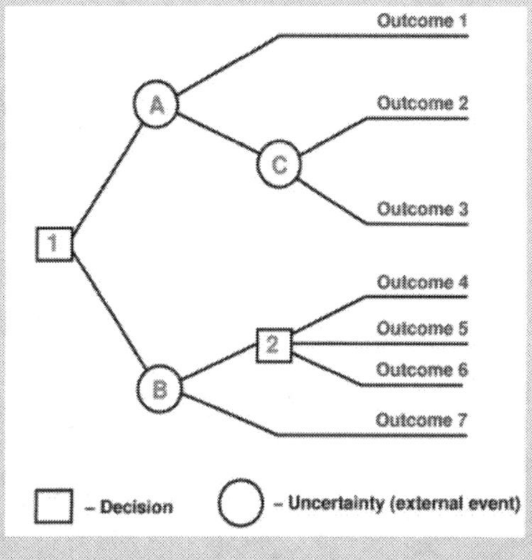

34

McCormick
Northwestern Engineering
Department of Chemical and Biological Engineering

Partial Least Squares Regression: Permutation Test

Shuffle output randomly

Obtain a variance for each feature coefficient

Compare the real output to the random variance

Again, creates a ranked list

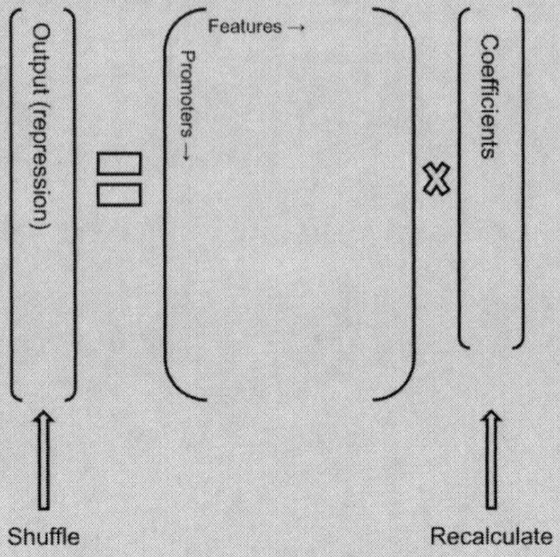

35

McCormick
Northwestern Engineering
Department of Chemical and Biological Engineering

Regularized regression: Lasso

Constrains $\sum|\beta|$
(sum of absolute coefficients)

penalty

$$\sum_{i=1}^{N}(y_i - \sum_j \beta_j x_{ij})^2 + \lambda \sum_j |\beta_j|$$

Eliminates features by shrinking unimportant coefficients to 0

Creates a ranked list of importance (again)

36

438528 Immunomodulatory Amphiphilic Polyanhydride Microparticles for Peripheral Nerve Regeneration

Abstract:

: Peripheral nerve repair using implanted biomaterial conduits seeded with cells face many challenges, including the host immune response. Tissue engineering approaches often employ the use of drug delivery vehicles to supply sustained release of trophic factors which support the differentiation and continued growth of the implanted cells. However, host immune cells responding to inflammatory cytokines released by the damaged nerve may inhibit this growth. This project focuses on the fabrication of amphiphilic degradable polyanhydride micro particles that can release surface adsorbed Interleukin-4 and encapsulated nerve growth factor simultaneously. Interleukin-4, an anti-inflammatory cytokine, serves to promote an M2-like phenotype in invading macrophages as a method to support wound healing. These M2-like macrophages may reduce the inflammatory response, thereby generating an environment which supports cell growth and tissue regeneration, thus allowing implanted cells to benefit from the sustained release of neurotrophic factors. The release of nerve growth factor helps stimulate the peripheral nerve regeneration. We have focused on the fabrication and release kinetics of the two macromolecules from the polyanhydride microparticles.

437145 Graphene Oxide Hollow Fibre Membranes: Stability, Modifications, and Challenges

Abstract:

Graphene, a two dimensional single atomic carbon layer, is an attractive material for thin film and membrane applications. Graphene oxide (GO) synthesized from chemical exfoliation is a commonly used precursor for graphene based materials. GO membranes made by assembling small GO flakes have demonstrated great potential in gas separation and liquid filtration. For upscale applications, GO membranes in hollow fibre geometry are of particular interest due to the high surface area per volume and easy assembly features at module level.

In our research group, inorganic hollow fibres have been developed for various membrane processes. Among all types of hollow fibre substrates, ceramic hollow fibre is an excellent candidate for GO membranes due to its low mass transfer resistance, robust mechanical property and inert chemical nature, which enable it to work under extreme conditions. However, our recent study revealed that GO membranes on ceramic hollow fibre are unstable in dry state, mainly due to drying-related shrinkage [1]. A simple way to tackle the stability problem is to reserve GO hollow fibre membranes in water after initial drying to avoid further shrinkage and the formation of fatal defects. Though such a strategy worked well to preserve the microstructure of GO membranes, it limits the applications of GO membranes to wet processes. Modifications of the membranes at dry state will also be difficult. In another approach, GO hollow fibre membranes were stabilized by using a porous poly(methyl methacrylate) (PMMA) sacrificial layer, which created a space between the hollow fibre substrate and the GO membrane that allows stress-free shrinkage. Defect free GO hollow fibre membrane was successfully determined and the membrane was stable in a long term (1200 hours) gastight stability test.

Post treatment of the stabilized GO membranes was also successfully carried out using UV radiation. The UV treatment changed the surface of GO membranes to superhydrophilic, and induced mild reduction that could have led to some controlled defect formations. All these modifications have enhanced the performance of GO hollow fibre membranes in both aqueous and solvent nanofiltration, while still maintaining the molecular sieve property of GO membranes. The pure water and acetone flux increased from 0.05 to 2.8 LMH, and from 0.14 to 7.5 LMH, respectively, with a molecular weight cut-off lower than 250 Da.

[1] Aba, N.F.D., et al., Graphene oxide membranes on ceramic hollow fibers – Microstructural stability and nanofiltration performance. Journal of Membrane Science, 2015. 484(0): p. 87-94

414949 Nanometric Graphene Oxide Framework Membranes with Enhanced Heavy Metal Removal Via Nanofiltration

Abstract:

A novel dual-modification strategy, including (1) the crosslinking and construction of a GO framework by ethylenediamine (EDA) and (2) the amine-enrichment modification by hyperbranched polyethyleneimine (HPEI), has been proposed to design stable and highly charged GO framework membranes with the GO selective layer thickness of 70 nm for effective heave metal removal via nanofiltration (NF). Results from sonication experiments and positron annihilation spectroscopy confirmed that EDA crosslinking not only enhanced structural stability, but also enlarged the nanochannels among the laminated GO nanosheets for higher water permeability. HPEI 60K was found to be the most effective post-treatment agent that resulted in GO framework membranes with a higher surface charge and lower transport resistance. The newly developed membrane exhibited a high pure water permeability of 5.01 L m^{-2} h^{-1} bar^{-1} and comparably high rejections towards Mg^{2+}, Pb^{2+}, Ni^{2+}, Cd^{2+} and Zn^{2+}. These results have demonstrated the great potential of GO framework materials in wastewater treatment and may provide insights for the design and fabrication of the next generation 2D-based NF membranes.

(195c) Freestanding Ultrathin Graphene-Based Membranes for Water Purification

Liu, H., Monash University
Zhang, X., Monash University

Originally presented on: 11/9/2015 16:05:00 - 16:30:00

Graphene-based membranes, consisting of two-dimensional (2D) graphene oxide (GO) or reduced graphene oxide (rGO) nanosheets, have recently attracted increasing attention in separation of liquid mixtures. The nanochannels of these membranes, formed between 2D GO or rGO nanosheets, allow gases and ions with smaller sizes than those of the channels to permeate, while blocking all other larger species. However, hydration of GO in aqueous solution enlarges the channels, making it more challenging to be used in various exciting applications, such as desalination, energy production and hydrofracking water treatment. Although the hydration effect can be inhibited by reduction, a substantial loss in membrane permeability inevitably occurs. In this research, we aim to develop a next generation of high-flux and energy-efficient membranes based on graphene materials, and at the same time maintain their superior selectivity for precise ionic and molecular sieving in aqueous solution. Preliminary work has demonstrated the fabrication of freestanding ultrathin rGO membranes which are more selective and permeable than the state-of-the-art membranes in forward osmosis process. Future work will include further enhancement of the performance of graphene-based membranes in terms of permeability and selectivity, exploration of the uses of graphene-based membranes in other water treatment processes and understanding the mechanisms of mass transport in graphene-based membranes.

426311 Tailoring Inter-Layer Nanospace of Graphene Oxide Laminates for High-Efficient Membrane Gas Separation

Abstract:

As a well-known two-dimensional material, graphene exhibits preeminent electrical, mechanical and thermal properties owing to its unique one-atom-thick structure. Graphene and its derivatives (e.g., graphene oxide) become emerging nano-building blocks for separation membranes featuring distinct laminar structure and tunable physicochemical properties. Extraordinary molecular separation properties for purifying water and gases have been demonstrated by graphene-based membranes, which attract a huge surge of interest during last a few years [1]. The pioneering work of Geim and co-workers found that laminar graphene oxide (GO) membranes are impermeable to gases and liquids but solely allow unimpeded evaporation of water. Thus, GO membranes have been reported for application in water-purification and solvent-dehydration processes [2]. Meanwhile, recent works found that GO membranes can also exhibit gas-separation characteristics if their stacking structure is care-fully controlled [3]. As the prominent characteristic of GO laminates, inter-layer space has been proved to play a significantly important role in molecular transport.

We are trying to tailor the nanospace between graphene oxide sheets for fast and selective transport of gases. Two approaches were demonstrated here. First, membranes with fast and selective CO_2 transport channels of GO laminates were proposed based on the construction of hydrogen bonding between GO and polymer. Our explorations (e.g., TEM, positron annihilation) found that the formed molecular-sieving interlayer spacing and straight diffusion pathways of the GO laminates play critical roles in determining gas transport properties. Second, molecular-intercalated GO membranes were developed by a facile spin-casting assisted assembly technique. With a well-ordered stacking structure and enlarged interlayer spacing, the intercalated GO membranes show excellent molecular sieving properties. Our prepared GO membranes exhibited preferential CO_2 (1st approach) or H_2 (2^{nd}approach) permeation performance for gas mixtures that transcends the upper bounds of state-of-the-art membranes. Moreover, we firstly realized GO-based membranes having extraordinary operational stability, which is believed to be a notable step toward pushing this membrane into industrialized implementation.

Keywords: Graphene oxide, inter layer space, membrane, gas separation

Acknowledgements

The work was supported by the National Natural Science Foundation of China (Nos. 21406107, 2149580015, 21476107), Innovative Research Team Program by the Ministry of Education of China (No. IRT13070), Natural Science Foundation of Jiangsu Province (No. BK20140930).

References

1. Liu G. P., Jin W. Q. & Xu N. P. Graphene-based membranes. *Chem. Soc. Rev.* DOI: 10.1039/c4cs00423j.
2. Shen J., Liu G. P., Huang K., Jin W. Q., Lee K.-R. & Xu N. P. Membranes with fast and selective gas-transport channels of laminar graphene oxide for efficient CO2 capture. *Angew. Chem. Int. Ed.* **54**, 578-582 (2015).
3. Huang K., Liu G. P., Lou Y. Y., Dong Z. Y., Shen J. & Jin W. Q. A graphene oxide membrane with highly selective molecular separation of aqueous organic solution. *Angew. Chem. Int. Ed.* **53**, 6929-6932 (2014).

431010 Scalable and Ultra-Selective Carbon Molecular Sieve (CMS) Membranes for Challenging Gas Separations

Abstract:

Scalable and Ultra-Selective Carbon Molecular Sieve (CMS) Membranes for Challenging Gas Separations

Chen Zhang, William J. Koros

School of Chemical and Biomolecular Engineering, Georgia Institute of Technology, Atlanta, GA

Carbon molecular sieve (CMS) membranes have received increasing attention in the past years for advanced gas separations. CMS membranes can be formed into asymmetric hollow fibers, by controlled pyrolysis of polymeric precursor hollow fiber membranes, and are capable of delivering simultaneously attractive productivity and separation efficiency without compromising scalability. Pyrolysis temperature is among the parameters that control CMS membrane's ultramicropore size distribution and therefore, gas separation performance. In general, more densely packed sp^2-hybridized graphene-like sheets with lower permeability and higher selectivity are obtained with increasing pyrolysis temperature. For example, previous studies[1] showed that CO_2/CH_4 selectivity of Matrimid®-derived CMS membranes was enhanced by 200% as pyrolysis temperature increased from 650 °C to 800 °C.

Nevertheless, formation of CMS membranes at pyrolysis temperatures above 800 °C has been rarely reported. This was partially due to challenges involved with processing brittle CMS dense films. We have discovered that this challenge can be successfully overcome by using hollow fiber precursors instead of dense film precursors. This talk will present CMS hollow fiber membranes formed at pyrolysis temperatures above 800 °C. Gas permeation measurements suggested that these CMS hollow fiber membranes were capable of delivering superior selectivities that are well-above the polymer upper bound.

References:

1. (a) Bhuwania, N.; Labreche, Y.; Achoundong, C. S. K.; Baltazar, J.; Burgess, S. K.; Karwa, S.; Xu, L.; Henderson, C. L.; Williams, P. J.; Koros, W. J., Engineering substructure morphology of asymmetric carbon molecular sieve hollow fiber membranes. *Carbon* **2014,** *76* (0), 417-434; (b) Ning, X.; Koros, W. J., Carbon molecular sieve membranes derived from Matrimid® polyimide for nitrogen/methane separation. *Carbon* **2014,** *66* (0), 511-522.

433115 Vertically-Aligned Carbon Nanotube Membranes for Breathable and Protective Fabrics

Abstract:

Vertically-aligned Carbon Nanotube Membranes for Breathable and Protective Fabrics

N. Bui, S. Kim, E. Meshot, J. Pena, S. Guo, K. J. Wu, F. Fornasiero*

Lawrence Livermore National Laboratory, Livermore, CA

P. Gibson

U.S. Army Natick Soldier Research, Development and Engineering Center, Natick, MA

Membrane materials with high breathability, high degree of protection from chemical and biological (CB) agents, and rapid response to CB threats are highly desirable for protective fabrics. At LLNL, we recently proposed a CB threat responsive fabric platform based on a carbon nanotube membrane. The membrane consists of an array of a-few-nm-wide vertically-aligned carbon nanotubes (VACNTs) as through pores and a flexible, impermeable parylene-N matrix filling the inter-tube spacing. It promises to provide high water-vapor transport rate and an effective barrier function to CB agents by best leveraging the outstanding mass transport properties of narrow CNT channels. Furthermore, this membrane platform is ideally suited to fundamentally understand the mechanisms of mass transport through CNT pores.

In this work, we empirically investigated the transport properties through the well-aligned, narrow CNT channels at varied length scale. Results showed that these VACNTs membranes sustained an ultrahigh liquid and gas transport rates when compared with conventional theory predictions. Also, they performed outstanding moisture vapor transport rates (> 4000 $g.m^{-2}.day^{-1}$ at all relative humidities), comparable to or exceeding state-of-the-art breathable fabrics. Selectivity study showed that these membranes completely excluded 5-7nm-Au nanoparticles and charged 3-nm dye molecules during dead-end filtration tests. Furthermore, our membranes demonstrated bioprotection capability through size-exclusion mechanism when Dengue virus was used as a surrogate for a biological threat agent. These findings suggest that, with appropriate functionalization at the CNTs pore entrances, VACNT-polymer composite membranes may provide excellent candidates for next generation of breathable fabrics with active protection from CB threats.

This work was performed under the auspices of the U.S. Department of Energy by Lawrence Livermore National Laboratory under Contract DE-AC52-07NA27344.

422390 Absolute, Excess and Net Adsorption and Helium Adsorption

Abstract:

The aim of this contribution is to highlight some inconsistencies often encountered in the literature. We will discuss the relative advantages and disadvantages of reporting absolute, excess and net adsorption with particular reference to the determination of adsorption of helium on HISIV 3000. Simple relationships will be derived to show the limiting behaviour of the three approaches at low pressures and in the saturation limit. The discussion will highlight the difficulties that lead to issues with the definition of the excess adsorbed amount and the approaches of Sircar (2001) and Gumma and Talu (2003) to determine the skeletal density of microporous solids. Classical thermodynamic approaches and molecular simulations will be used to identify the correct representation of the temperature dependence of the Henry law constant. This will be used to identify the correct temperature at which to carry out helium pycnometry experiments.

427521 Assessment of the O2/N2 Separation Ability of Carbon Molecular Sieves By Spontaneous Liquid-Gas Imbibition

Monday, November 9, 2015: 3:40 PM
255D (Salt Palace Convention Center)

Xinyu Jia, Environmental and Biological Science and Technology, Dalian University of Technology, Dalian, China

Assessment of the O_2/N_2 separation ability of carbon molecular sieves by spontaneous liquid-gas imbibition

Xinyu Jia,Yuanyuan Cong,Shaoping Xu*

State Key Laboratory of Fine Chemicals, Faculty of Chemical, Environmental and Biological Science and Technology, Dalian University of Technology, Dalian 116024, P.R.China

(*Corresponding Author's E-mail: huizixu@hotmail.com)

ABSTRACT

Objective:

The kinetics of spontaneous water-oxygen or nitrogen imbibition at 303.2K on carbon molecular sieves (CMSs) were carried out to establish an assessment method for the O_2/N_2 separation performance of the CMSs in pressure swing adsorption (PSA).

Method:

The spontaneous water-gas imbibition was conducted this way: in the sample holder with a constant volume, CMS particles saturated with gas were immersed into water at a given temperature, and the pressure of the gas with time on stream, which is a measure of the volume of the gas recovered by water, is recorded. The results of the spontaneous water-oxygen imbibition on eight CMSs are shown in Fig.1.

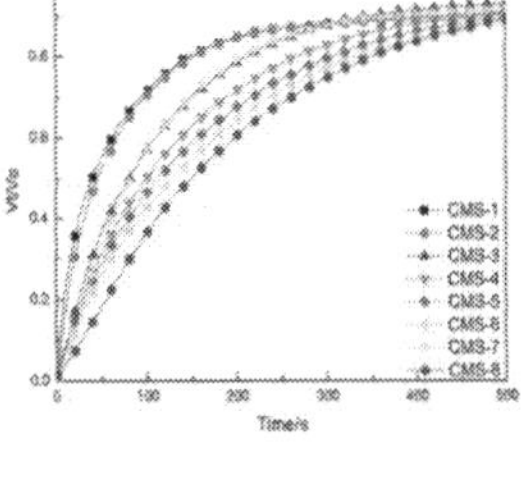

Fig. 1. Kinetics of the spontaneous water-oxygen imbibition at 303.2K on CMS-1°«CMS-8.

Results and discussion:

The spontaneous water-gas imbibition on CMSs involves the surface adsorption of water, the diffusion of the gas along the micropores, and the diffusion of the gas through the barrier at the micropore entrance of the CMS. To elucidate the rate-controlling step of the process, the pseudo-second-order (PSO) kinetic model and the linear driving force (LDF) model has been used to describe the spontaneous water-gas imbibition.

For CMS-1°«CMS-6, the kinetics of the water-oxygen imbibition obeys the PSO model, indicating the surface adsorption is the rate-controlling step. On the contrary, for CMS-8 the diffusion of oxygen through the barrier at the micropore mouth is the rate-controlling step with the kinetics following the LDF model. For CMS-7, up to 200s, the dynamics obeys the LDF model, and over 200s it follows the PSO model. Hence, the size distribution of the micropore mouths for CMS-7 is far less uniform than that of other CMSs. The Kinetic parameters of the two models demonstrate that the average sizes of the micropore mouths are in the order of CMS-1>CMS-2>CMS-3>CMS-7>CMS-4>CMS-5> CMS-6>CMS-8.

The kinetics of the water-nitrogen imbibition on the CMSs indicate that CMS-1 and CMS-2 fit with the PSD model and the physical adsorption process is the rate-controlling step. For CMS-3°«CMS-5, the rate-controlling step is mainly the diffusion of nitrogen through the micropore mouth. For CMS-6 and CMS-8, their micropore sizes are small enough to make the kinetics of the process controlled by not only the barrier resistance at the micropore mouth but also the diffusion in the micropore interior. And for CMS-7, the kinetics is controlled by all three steps mentioned above. In addition, the same order of the average sizes of the micropore mouths of the CMSs are obtained from the kinetics of the water-nitrogen imbibition as that of the water-oxygen imbibition.

An assessment method for the O_2/N_2 separation ability of CMSs has been established, which has been verified by the PSA of air:

The micropore mouths of CMSs such as CMS-1 are too large to separate O_2/N_2 sufficiently on which the rate-controlling steps of the water-O_2 or N_2 imbibition are both the surface adsorption.

The micropore mouths of CMSs are too small to adsorb O_2 within a short time in SPA cycles leading to lower selectivity of O_2/N_2. Since the diffusion in the micropore interior is one of the rate-controlling steps in the water-nitrogen imbibition on them.

CMSs whose micropore mouths dimensions are not uniform as CMS-7 are unsuitable for the kinetic separation of O_2/N_2.

CMSs owning proper and uniform micropore mouths dimensions such as CMS-3°«CMS-5 are favorite adsorbents for the O_2/N_2 separation by PSA.

420130 Investigation of CO_2-H_2o Co-Adsorption on Amine Functionalized Cellulose in a Closed-Loop Differential Bed

Abstract:

The greenhouse effect is one of the most destructive global climate change phenomena. The CO_2 released into the atmosphere via the combustion of fossil fuels presents the largest portion of the total greenhouse gas emission. Therefore, in spite of technical challenges associated with it, "Direct Air Capture" (DAC), i.e. CO_2 capture from ambient air, has been considered an option to mitigate CO_2 emissions from the transportation sector. In addition, by turning the air-captured CO_2 into liquid fuel via a solar thermochemical redox cycle [1], the carbon footprint of the conventional transportation fuels can be considerably lowered. To achieve these goals, we consider a DAC technology based on amine functionalized nanofibrillated cellulose (NFC) sorbent subjected to cyclic temperature-vacuum swing (TVS) adsorption/desorption. The simultaneous amine-based CO_2 and H_2O chemisorption from air offers two advantages for the coupling of DAC technology with solar thermochemical fuel production: (i) simultaneous adsorption of H_2O from air, which is another feedstock for the solar thermochemical fuel production process, and (ii) the enhancement of CO_2 adsorption with the presence of humidity. Substantial work on the characterization the adsorbent material and the TVS process have been performed[2-6]. However, the adsorption kinetics, which is the key to the design and modeling of the reactor and the process, has not been resolved yet. This work focuses on development of an experimental setup for determination of the co-adsorption kinetics of CO_2 and H_2O on the amine-functionalized NFC. The approach combines the concepts of a differential packed bed and a volumetric adsorption system. The use of a differential packed bed allows a simpler mathematical treatment of the experimental data and isothermal adsorption condition. The difficulty of quantifying the small sorbate concentration difference between the upstream and the downstream of the differential packed bed is compensated by placing it in a closed system, where a fixed volume of gas is constantly circulated through the packed bed. Subsequently, the adsorption kinetics is extracted from the transient sorbate concentration profile in the system. The characterization of the experimental setup will be presented along with the preliminary adsorption kinetics of the CO_2-H_2O co-adsorption on amine functionalized NFC.

References:

1. Chueh WC, Falter C, Abbott M, et al. High-Flux Solar-Driven Thermochemical Dissociation of CO2 and H2O Using Nonstoichiometric Ceria. *Science.* December 24, 2010 2010;330(6012):1797-1801.

2. Gebald C, Wurzbacher JA, Borgschulte A, Zimmermann T, Steinfeld A. Single-Component and Binary CO2 and H2O Adsorption of Amine-Functionalized Cellulose. *Environmental Science & Technology.* 2014.

3. Gebald C, Wurzbacher JA, Tingaut P, Steinfeld A. Stability of Amine-Functionalized Cellulose during Temperature-Vacuum-Swing Cycling for CO2 Capture from Air. *Environmental science & technology.* 2013;47(17):10063-10070.

4. Gebald C, Wurzbacher JA, Tingaut P, Zimmermann T, Steinfeld A. Amine-based nanofibrillated cellulose as adsorbent for CO2 capture from air. *Environmental science & technology.* 2011;45(20):9101-9108.

5. Wurzbacher JA, Gebald C, Piatkowski N, Steinfeld A. Concurrent Separation of CO2 and H2O from Air by a Temperature-Vacuum Swing Adsorption/Desorption Cycle. *Environmental science & technology.* 2012;46(16):9191-9198.

6. Wurzbacher JA, Gebald C, Steinfeld A. Separation of CO2 from air by temperature-vacuum swing adsorption using diamine-functionalized silica gel. *Energy & Environmental Science.* 2011;4(9):3584-3592.

417352 Characterizing Mass Transfer Properties in Porous Materials with Transient Data

Abstract:

Obtaining adsorption equilibria and kinetic data in porous materials is imperative for the proper design of adsorption units. Generating the equilibria information is relatively simple; whereas, kinetic measurements face experimental challenges such as eliminating heat and mass transfer impact. Additionally, these measurements are affected by the accuracy of mathematical model used to obtain the mass transfer parameters. The macroscopic methods used in mass transfer studies consist of measuring either the uptake into or release from the material. Subsequently, the experimental data is matched with an appropriate mass transfer model to extract the kinetic parameters. Consequently, the accuracy of kinetic parameter obtained strongly depends on how well the model resembles the experimental conditions. Gravimetric methods are commonly used and consist of recording the weight gain (or loss) in a sample as a function of time. For such methods, it is typically desirable to have a step change in pressure or concentration. A sudden increase in pressure, however, may cause the weight balance to oscillate. Moreover, the conventional method omits a significant amount of data measured before the pressure stabilizes, and hence is unable to utilize the information during the transient period which can be is critical for materials with complex pore structure.

A new methodology that represents an improved experimental design has been developed. It makes use of the entire dataset including the transient data. The method is based on a mathematical model that mimics the transient experimental conditions. More accurate determination of the kinetic parameters is thus possible. For complex materials, additional insights into the mass transfer behavior can be determined. The new method and its applications to several materials including activated carbon, zeolite, and shale rock are presented.

411477 Adsorbed Bubble Phase Transport for Removal of Pollutants from Water

Abstract:

Adsorbed Phase Transport for Removal of Pollutants from Water

The process of adsorbed phase transport of pollutants in a bubble column was investigated in this paper. The process of the removal of 2,4-Dichloro phenoxy acetic acid from water was used as an example . A lab scale unit was set up and various experimental runs were carried out to study the adsorption process. The experimental findings show that the adsorption method is very effective (>90% removal) in removing traces (ppb level) of the pollutants which is not easily removable by simple air stripping. In addition a mathematical model was developed to describe the process. Some parameters of the model were measured or calculated while others such as the aqueous mass transfer coefficient and the solute partition coefficient were adjusted to fit the experimental data. The calibration of the model was carried out using the experimental results of change in gas flow rate (the easiest parameter to vary). A numerical sensitivity analysis was carried out using the calibrated model to study the effect of various parameters such as the bubble radius, aqueous phase drag-up by air, column radius and ratio of organic to aqueous volume phases.

Keywords

- Adsorbed phase transport;
- Mathematical model;
- 2, 4 D pesticide;

408035 On the Choice of Batch or Fixed-Bed Adsorption Processes for Wastewater Treatment

Abstract:

Newly emerged nanomaterials, such as graphene or carbon nanotubes, are of increasingly widespread importance in environmental remediation, especially in their ability to remove undesirable chemicals from hydrological systems. Despite recent advances, the high cost of these materials and the relative inability to regenerate them hinders their use at larger scales. While it is commonly held that adsorption in fixed bed processes is more efficient than in batch processes, the definition of efficiency needs to be examined closely. When the goal is to remove all contaminants from a solution, fixed bed adsorption is the only process that may be used. However, the situation is distinctly different when there is a threshold level of effluent contamination—in such cases, a process may be chosen that maximizes the quantity of contaminant adsorbed per unit mass of adsorbent. In this paper, we show that there exists a critical effluent concentration above which batch processes are more efficient than fixed bed processes from this perspective. The present analysis is supported by results from previous literature for which comparisons are possible between fixed bed and batch process adsorption. This work proposes a simple method to choose either batch or fixed bed processes when the goal is to minimize the quantity of adsorbent needed to remove contaminants to below a specified threshold level. It also provides a systematic set of self-consistency checks on experimental measurements that may be used for comparative studies of these methods.

437385 Virus Retention by Viresolve® NFP Filters: Effects of Solution Conditions

Abstract:

Virus filtration is a critical step in the overall viral clearance strategy for the production of biotherapeutics. Virus filtration membranes provide a robust, primarily size-based, removal mechanism that compliments both adsorptive and inactivation steps. However, there is limited data suggesting that the actual degree of virus removal during virus filtration is also a function of solution conditions, suggesting that electrostatic interactions may also play a role in determining the effectiveness of virus capture. The objective of this work was to examine the effects of solution pH and ionic strength on virus retention by the Viresolve® NFP membrane.

Experiments were performed using the bacteriophage fX174 as a model virus, with the virus retention data complemented by the use of confocal microscopy to directly visualize the capture of fluorescently-labeled fX174 within the filter. Data were obtained during both constant pressure filtration and after a process disruption, in this case a short pressure release as might occur during typical process operations. Virus retention was highest at low pH and low ionic strength, conditions under which there was an attractive electrostatic interaction between the negatively-charged membrane and the positively-charged phage. In addition, the transient increase in virus transmission seen in response to a pressure disruption at pH 7.8 and 10 was completely absent at pH 4.9, suggesting that the trapped virus are unable to overcome the electrostatic attraction and diffuse out of the pores when the pressure is released.

This physical picture was further confirmed by the acquired confocal images. A two-fluorescent dye based technique was used to image and differentiate between virus particles captured before and after the pressure release. Images obtained at pH 10 showed the migration of previously captured phage; this phenomenon was absent at pH 4.9. These results provide important new insights into the factors governing virus retention using virus filtration membranes.

437386 Impact of virus filtration load handling on virus filter performance

Abstract:

Virus-retentive filtration (VF) is routinely incorporated into monoclonal antibody downstream purification processes as a dedicated virus removal unit operation to provide robust and reliable virus clearance capacity. VF is applicable to a wide range of viruses since the primary mechanism is size based. Performing experiments at bench scale introduces the possibilities that the feed and material handling used for validation studies are not representative of those at manufacturing scale. In this work, investigations into the filter flux decay during a small scale mock filtration run before viral clearance study were conducted. A comparison of the performance of two filter designs was completed, along with biochemical and biophysical characterization of the feed streams. It was determined that both virus filter load handling (freezing and thawing) and preparation (pre-filtration) impacted virus filtration performance without and with virus spiking.

422667 Beyond Air-Water Diffusion: Improving Integrity Test Accuracy for Sterilizing and Virus Membrane Filters By Measurement of Permeate Gas Composition

Abstract:

Reliability of retention performance is of paramount importance for membrane filters designed for sterile and virus filtration. To achieve dependable retention, filter suppliers employ multiple quality assurance measures, including an integrity test to ensure the absence of oversize pores or defects that can compromise the retention capability of the filter. Integrity testing is also routinely practiced immediately prior to and after use of the filter to detect any defects that may have been incurred during installation or use of the filter. Probably the most commonly applied non-destructive integrity test for membrane filters - especially sterilizing grade and virus filters - is the gas-liquid diffusion test, with air and water often used as the gas-liquid pair. In a gas-liquid diffusion test, the measured diffusive flow rate through a wetted membrane is compared to a theoretical or empirically established flow rate for an integral membrane. A measured flow rate value that exceeds the specification for an integral filter signals the presence of one or more defects. However, the sensitivity of the air-water diffusion test is limited by the fact that the diffusive flow rate for an integral membrane can span a range that is large compared to the flow contributed by a defect, resulting in a relatively poor signal-to-noise ratio.

In this talk, a novel non-destructive air-water integrity test for microporous and nanoporous membranes is described. The test is based on the principle of differing rates of gas diffusion of different gases through the liquid layer of a wetted membrane. Oxygen permeates through water faster that nitrogen, so with air as the challenge gas and water as the wetting fluid, the permeate stream will be enriched in oxygen. The permeate oxygen concentration is predictable, accurately measurable, and within a narrow and repeatable range for an integral membrane. A leak through the membrane will result in a deviation from the integral permeate concentration, signaling a defect. Compared to the conventional air-water diffusion test, this air binary gas (i.e., O_2 and N_2) test in which the permeate gas composition is measured (in addition to the diffusive flow rate) has a superior signal-to-noise ratio and was demonstrated to provide a significantly higher level of retention assurance for both sterilizing grade and virus filters. Since air and water are used as the gas-liquid pair, the air binary gas test also maintains the convenience, safety, and environmentally friendly aspects of the air-water diffusion test.

414034 Use of Preconditioning to Control Membrane Fouling and Enhance Performance during Ultrafiltration of Plasmid DNA

Abstract:

Several recent studies have demonstrated that small pore size ultrafiltration membranes can be used for purification of supercoiled plasmid DNA for therapeutic applications, but the performance of these membrane systems is severely limited by membrane fouling and low selectivity. The objective of this work was to examine the potential of using "pre-conditioning", in this case accomplished by pre-elongating the DNA by passage through a region with large pore size, to minimize fouling and enhance DNA separations.

Data were obtained using both asymmetric hollow fiber ultrafiltration membranes, with flow in either the normal or reverse orientation, and with composite membrane structures made by placing a larger pore size flat sheet microfiltration membrane in series with an ultrafiltration membrane. In all cases, flow through the larger pore size region pre-stretched the plasmid, leading to an increase in plasmid transmission and a significant reduction in fouling. This pre-conditioning also provided a significant increase in selectivity for separation of the linear and supercoiled isoforms. The effectiveness of the pre-conditioning was a function of the pore size and morphology of the top layer in the composite membrane structure, with the best performance obtained using a pre-conditioning layer that had a pore size similar to the radius of gyration of the plasmid. These results clearly demonstrate the potential for dramatically increasing the performance of membrane systems for plasmid DNA separations by controlling the pore morphology to effectively pre-stretch the DNA before passing through the narrow pores of the selective ultrafiltration membrane.

426346 Antibody Purification from Cell Culture Supernatant Using Overload and Elute Chromatography with New Multimodal Membranes As Stationary Phase

Abstract:

This presentation will describe our latest research efforts on devising improved methods to capture and purify antibodies from cell culture supernatant using overload and elute chromatography with newly designed multimodal membranes (MMMs). The membranes incorporate ligands that can bind proteins through a combination of Coulombic interactions, hydrophobic interactions, and/or hydrogen bonding, thus widening the operation ranges for separation and improving selectivity.

Protein binding experiments with IgG show the MMM has a high equilibrium binding capacity (180 mg IgG/ml) and excellent salt tolerance. The multimodal membranes maintain binding capacities in excess of 100 mg IgG/ml at salt concentrations that are typical for elution buffers used in multi-stage chromatographic bioseparation processes. Protein dynamic binding capacities in excess of 50 mg IgG/ml were determined from breakthrough curve analysis.

The MMMs were used as the stationary phase following a Protein A purification step in the purification human IgG1 from Chinese hamster ovary (CHO) cell culture supernatant. The MMM stationary phase greatly enhanced load productivity compared to resin-based ion-exchange media. Further, no buffer exchange step was required between the Protein A and MMM chromatography purification steps.

In a second study, the MMMs were used as the stationary phase for the primary capture step, in place of Protein A resin chromatography. Using an overload and elute chromatography process, the MMMs demonstrated a high selectivity for capturing human IgG from CHO cell culture supernatant. A high quality protein product was obtained from the MMM chromatography purification step.

This work shows that the new MMM chromatography operation can effectively decrease the number of purification steps and shorten the purification time and/or increase load productivity in downstream processing of biologics.

432536 Permeability and Selectivity of Gases in Glassy Polymeric Membranes

Abstract:

The transport of gases and vapors in polymeric systems is of great interest for various applications, among the others for membranes technology. Therefore, solubility and the permeability of low molecular weight species in glassy and rubbery polymers are of great importance, and their prediction based on reliable models are highly needed.

In this work, use is made of a simple and fundamental approach recently proposed for the description of gas and vapor solubility and permeability, which considers the diffusion coefficient as the product of two contributions, the penetrant mobility (a purely kinetic parameter), and the thermodynamic factor (accounting for the dependence of the penetrant chemical potential on its concentration) [1].

The nonequilibrium thermodynamic approach (NET-GP) provides the description of the thermodynamic properties of the solute/polymer mixtures in the glassy state, and in particular of the solubility coefficient as well as of the thermodynamic factor for diffusion [2]; the penetrant mobility, on the other hand, is endowed with a simple exponential dependence on penetrant concentration. The simple and general expression derived for penetrant diffusivity and permeability requires two adjustable parameters only: the penetrant infinite dilution mobility L_0, and the plasticization factor b, accounting for its dependence on concentration.

The resulting model is able to describe successfully all the different experimental behaviors of gas and vapor permeability in glassy polymers as a function of upstream pressure and even the non-monotonous behavior known as plasticization effect [3-5], and can also be employed for the a priori estimation of the permeability and selectivity properties, on the basis of the pure polymer and penetrant properties.
Indeed, the NET-GP model can be used predictively for the solubility coefficient, while the two mobility parameters of the model are endowed with a precise physical meaning which allow to relate them with pure polymer and penetrant properties through simple and effective correlations. In fact, the infinite dilution mobility depends on the penetrant size (e.g. critical volume), and on the polymer fractional free volume, while the plasticization factor is related to the swelling induced by the diffusant in the matrix, readily estimated by the NET-GP. On the other hand, the membrane sieving ability (diffusion selectivity) is strictly related to its cohesive energy, and the thermodynamic model readily provides the solubility selectivity contribution.

Thus, the model can be applied to obtained to a simple method for the calculation of permeability and selectivity of various gas pairs in glassy polymers. This can be then applied to the analysis of the upper bound correlations provided by Robeson [6], often employed in the evaluation of the membrane separation performance.

Noteworthy, the same approach can provide on one hand the detailed description of the permeability behaviors with pressure or temperature with a remarkable accuracy and, on the other hand, it also offers an estimation a priori of the permeability and selectivity of the various penetrants of interest, on the basis of simple physical correlations.

References:

[1] Minelli M, Sarti, J Membr Sci. 2013, 435, 176–185.

[2] Doghieri F, Sarti GC, Macromolecules. 1996, 29, 7885–7896.

[3] Minelli M, Sarti GC, J Membr Sci. 2013, 444, 429–439.

[4] Minelli M, J Membr Sci. 2014, 451, 305–311.

[5] Minelli M, Sarti GC. J Membr Sci. 2015;473:137–145.

[6] Robeson LM, J Membr Sci. 2008, 320, 390-400.

415748 Development of Chiral Membranes for Purification of Proteins and Small Molecules

Abstract:

There is a great need to improve synthetic membrane filtration performance (increased permeation flow rates and higher selectivity) for pressure-driven membrane systems for water, energy, pharmaceutical, and biotechnology applications like desalination, fuel cells, small molecule purification, and protein separation, respectively. Recently, we have developed a set of chiral membranes that allow for selective protein separation and reduced fouling. Two chiral surfaces of opposite chirality (S and R) were synthesized in order to determine if they had an effect on fouling and protein separations, since almost all amino acids of proteins are of one specific stereoisomeric form, *i.e.,* the (S) form. Two proteins that have been tested, BSA and ovalbumin, resulted in the same preliminary qualitative result. That is, when the surface of the membrane was the same chirality as the protein feed, (S), minimal fouling occurred; however, when the chirality of the surface was the opposite to that of the foulant, (R), the membrane fouled significantly. These represent a new class of chiral membranes using covalently attached chiral monomers to the surface of poly(ether sulfone) (PES) membranes. These membranes will allow for low fouling separation of proteins and small molecules based on chirality, in addition to the classical separation parameters of membranes such as pore size, charge, trans-membrane pressure and surface wettability.

Using our high-throughput platform, a vast number of model surfaces can be produced using a library of chiral monomers and grafted from PES membrane surfaces. A chiral library will be synthesized using the 20 natural amino acids and a common linker containing a methacrylate and various different repeated PEG units (n = 1-6) to graft the amino acids to the PES surface. We utilize a high-throughput platform with an Atmospheric Pressure Plasma (APP) polymerization system to activate irradiation-sensitive commercial PES membranes. The synthesized chiral monomers can then be grafted from the membrane surface and evaluated. These chiral surfaces will be tested using various proteins foulants of different molecular weights, such as bovine serum albumin (67 kDa) and ovalbumin (45 kDa). The selected winners are determined by measuring fouling index, selectivity, and permeation flux or inverse resistance. We believe these results could lead to a fifth fundamental rule to use as a guideline when choosing anti-protein fouling surfaces. That is, chirality can be important when separating chiral solutions. Developing a chiral library will allow us to better understand what is necessary for the chiral effects to occur. Steric effects in addition to chain length and functionality should all be critical when the chiral surface interacts with the foulant. We speculate that chiral membranes could have drastic implications in the pharmaceutical industry, when only one stereoisomer is desired from a racemic mixture.

435898 Validation of Flow Simulations in Spacer Filled Channels with Particle Image Velocimetry

Abstract:

Membranes in the form of sheets are used to form spiral wound membrane modules for a wide variety of separations ranging from seawater desalination to carbon dioxide capture from coal-fired power plants. Spacers are a critical component of these modules. Spacers create and maintain uniform flow channels. Spacers also are used to mix fluid within the channel to enhance mass transfer. Unfortunately, spacers increase pressure drop. This increase depends strongly on spacer geometry.

Computational fluid dynamics (CFD) has been used extensively to investigate flows within spacer filled channels. This body of work has examined the dependence of velocity field and pressure drop on spacer geometry variables.

Simulation results typically are validated by comparison to experimental measurements of pressure drop. Limited point-wise, volumetric comparisons of velocity fields with experimental measurements are available.

Detailed comparisons of experimental velocity fields, obtained using Particle Image Velocimetry (PIV), with simulation results are reported here. PIV captures the movement of seed particles in the flowing fluid to determine the local fluid velocity. Measurement of the two velocity components in the interrogation plane are presented for three types of spacers: 1) symmetric, 2) asymmetric, and 3) static mixing. The results indicate CFD simulations are in good agreement with experimental results. The greatest experimental challenges are uniform particle distribution throughout the flow channel and precise orientation of the flow cell.

Stereographic measurements of all three velocity components in the interrogation plane also are reported. The out-of-plane velocity component is small for the symmetric and asymmetric spacers and values are near the resolution level of the experiment. However, good agreement is found for the static mixing spacer which possesses much larger out-of-plane velocities.

The results suggest CFD simulations provide good predictions of spacer performance. This further validates the use of CFD to optimize spacer geometry.

432783 Tunability and Pollutant Degradation By Pnipam Functionalized MF Membranes: Modeling, Simulation, and Experimental Results

Abstract:

Tunability and Pollutant Degradation by PNIPAm Functionalized MF Membranes: Modeling, Simulation, and Experimental Results

Anthony Saad (speaker), Li Xiao, Minghui Gui, Dibakar Bhattacharyya

Department of Chemical and Materials Engineering, University of Kentucky, Lexington, KY 40506

Abstract

This study is aimed at creating predictive models to estimate the temperature responsive behavior of poly-N-isopropylacrylamide (PNIPAm) as it relates to water flux behavior and partitioning of toxic pollutants, specifically PCBs. Flow through PNIPAm functionalized PVDF MF membranes is modeled mathematically to predict effective pore size based on the swelling extent of the temperature-responsive PNIPAm polymer around its LCST. PNIPAm is known to show a sharp transition to a hydrophobic state at its lower critical solution temperature (LCST) at around 32°C, making PNIPAM attractive for biomedical applications. A model is developed to predict the effect of changing temperature about the LCST of NIPAm on water flux through the membrane, caused by a change in effective pore size. Swelling ratios of PNIPAm polymeric networks have been reported in water, and are utilized to estimate changes in both flux and pore diameter. Water permeation is tested at various temperatures to verify model predictions.

Furthermore, the entrapment of reactive Fe/Pd nanoparticles in the temperature responsive PNIPAm polymer network for dechlorination and contaminant degradation of PCBs has been reported. This study also aims to model PCB degradation by reactive immobilized nanoparticles in a PNIPAm functionalized PVDF MF membrane. The model accounts for diffusion through the membrane, as well as reactivity of the nanoparticles. Solute concentration is predicted as a function of length through the membrane, and also as a function of time. Changing the temperature of the reaction affects inter-particle spacing and solute adsorption because of the changing PNIPAm conformation, as well as the intrinsic rate constant for the reaction. These effects are predicted, and verified by experimental degradation results. This research is supported by the NIEHS-SRP grant P42ES007380, and by the NSF KY EPSCOR program. Full-scale PAA functionalized PVDF membranes were developed through collaborative work with Nanostone-Sepro (Oceanside, CA, USA).

429420 The Effects of Salt on the Free Energy of Thermo-Responsive PNIPAM Transition

Abstract:

Thermo-responsive polymers exhibit globule-to-coil transition at their lower critical solution temperatures (LCST). The thermo-responsive poly (N-isopropylacrylamide) (PNIPAM) and its copolymers have been often grafted on membrane substrate as stimuli-responsive membranes. However, the driving force for the folding and unfolding of PNIPAM, particularly at the molecular level remains controversial and unresolved. Here classical molecular dynamics simulations coupled with metadynamics technique was used to investigate the free energies associated with globule-to-coil transition for PNIPAM in water and NaCl salt solution at their respective transition temperatures. Our results indicate that globule-to-coil transition is mainly entropically driven. Further unfolding process appears to have a higher barrier and is energetically less favored compared to the folding process at the transition temperature. The presence of salt ions alters the free energy landscape and stabilizes the folded structure.

425897 Hydrogen Solubility Predictions Via First-Principles Calculations: Investigations on Deviated Predictions from Experimental Solubility

Abstract:

Hydrogen solubility is an important parameter that affects hydrogen-selective membrane properties because the hydrogen permeability through a metallic membrane is half the product of the solubility and diffusivity of atomic hydrogen in the bulk metal. Accurate solubility predictions is crucial for the prediction of overall membrane properties. Our study may provide useful information in the search for new materials for hydrogen-selective membranes and hydrogen storage by identifying a proper alloy for the Pd-based and non-Pd-based membranes that have enhanced hydrogen solubility. In the last year, we have reported hydrogen solubility in ten transition metals (V, Nb, Ta, W, Ni, Pd, Pt, Cu, Ag, and Au) predicted by first principles based on density functional theory (DFT) combined with chemical potential equilibrium between hydrogen in the gas and solid-solution phases. This year, we have extended our work to further understand the deviations in the predictions of late transition metals. The solubility predictions in this study match experimental data within a factor of two in the cases of V, Nb, Ta and W, and within a factor of three in the cases of Ni, Cu, and Ag. Order of magnitude deviations are obtained in the cases of Pd, Pt, and Au. In the case of Pd, the deviation in solubility prediction is mainly attributed to the errors involved in the calculated vibrational frequencies of dissolved hydrogen. In Pt and Au, hydrogen in the octahedral interstitial site is less stable than in the tetrahedral site, contradicting the predictions based on the hard-sphere model. Potential energy surface analysis reveals a slightly downward concavity near the center of the octahedral sites in Pt and Au, which may explain the calculated imaginary vibrational frequencies in these sites and lead to unreliable solubility predictions.

426310 Effect of Pegylated Gold Nanoparticle and Nanorod Permeation on Lipid Bilayer Membranes

Abstract:

In targeted drug delivery systems, nanoparticle carriers are often functionalized with polyethylene glycol (PEG) polymers to improve solubility and in-vivo stability in biological media. Gold nanoparticles decorated with ligands are excellent candidates in particular for delivery of therapeutic agents. This is due to their distinctive physical and chemical characteristics, which include the ability to be synthesized in a wide range of shapes and ease of functionalization of the nanoparticle surface. In this study, we have carried out a series of coarse-grained molecular dynamics simulations to investigate the permeation of polyethylene oxide (methyl-terminated PEG or PEO) functionalized gold spherical nanoparticles and nanorods across a protein free phospholipid bilayer membrane. Investigation of permeation of functionalized nanoparticles under various conditions is important because it can aid in developing strategies for using nanoparticle as drug carriers while suppressing cytotoxic effects as they permeate cell membranes in-vivo. Previously we have shown that ligand-coated gold nanoparticle permeation across lipid bilayer membranes induces water and ion penetration as well as incidence of lipid flip flop and lipid displacement.

In our studies we have studied the effect of ion concentration gradients, nanoparticle shapes, nanoparticle permeation velocities, length of PEO ligands, and chain density of grafted PEO ligands on water and ion leakage, lipid flip-flop, and lipid loss from the lipid bilayer membrane. Additionally, the mechanism of penetration of the PEO functionalized nanorods and spherical nanoparticles into the bilayer and the nature of the interactions between the PEO ligands and the lipids of the bilayer are discussed. The results of this work will be of interest to experimentalists who engineer nanoparticles with surface modifications for biomedical applications.

426986 Prediction of Membrane Plasticization Via Molecular Simulation

Abstract:

Polyimides are attractive membrane materials for gas separation applications such as CO2 removal from natural gas or flue gas treatment processes, due to their good permeability, selectivity, and mechanical resistance. The principle issue in these applications is the plasticization of the polyimide membranes at high partial pressures of CO_2, which can lead to reduced membrane selectivity and unpredictable membrane properties.

In this study we investigated structure-plasticization relationships of three sulfonated copolyimides: 6FDA/BTDA-pBAPS, 6FDA-pBAPS/DABA, and 6FDA-pBAPS/mPDA. These polymers were selected based on our previous studies in which the permeability coefficients of H2, O2, He, CO2, N2, and CH4 gases for more than 2200 possible co-polyimide structures were estimated by the group contribution method. These selected copolyimides were then synthesized and characteristics of the membranes were determined experimentally up to 10 bar. However, these studies were unable to deterimine the plasticization resistance of these materials, since the plasticization phenomenon occurs usually well above 10 bar. In this work we predicted the plasticization behavior of these materials up to 40 bar using molecular simulation methods.

We used Molecular Dynamic (MD) and Monte Carlo (MC) simulations to model the relationship between structural properties and CO2-induced plasticization behavior of copolyimides in CO2/CH4 separation. The physical characteristics of simulated membranes matched well with the experimental data Simulated sorption coefficients agreed well with the experimental results obtained from gravimetric sorption (IGA) measurements. Sorption simulations up to 40 bars were performed and fractional free volume increases of copolyimides were analyzed. Calculation of diffusion coefficients were carried out at the sorption pressures in order to calculate permeabilities. Finally we estimated the plasticization resistance of each copolyimide based on the variation of permeabilities as a function of CO2 pressure. Finally, plasticization pressures of polyimides were reported and compared.

Abbreviations: 6FDA: 4,4-hexafluoro isopropylidene diphthalicanhydride; DABA: 3,5-diamino benzoic acid; BTDA: 3,3-4,4-benzophenone tetracarboxyclic dianhydride; pBAPS: bis [4-(4-aminophenoxy) phenyl] sulfone; mPDA: 1,3-phenylenediamine.

Acknowledgment: This work was supported by Turkish Scientific and Technological Research Council (TUBITAK) trough grant no. 113M336.

414742 Efficient Calibration for in-Situ Measuring L-Glutamic Acid Solution Concentration Via ATR-FTIR

Abstract:

In this presentation, an efficient calibration method is proposed for in-situ measuring the concentration of L-glutamic acid (LGA) solution via ATR-FTIR. We clarify that using the unsaturated zone spectra as commonly practiced for establishing a concentration calibration model may lead to evidently biased measurement of the solution concentration. By comparison on using the spectra of unsaturated and metastable zones for calibrating the concentration of LGA aqueous solution, it is demonstrated that the use of metastable zone spectra to establish a calibration model can guarantee prediction accuracy for the solution concentration via real-time ATR-FTIR monitoring. Experiments under different concentrations and temperatures are performed to verify the efficiency of the proposed calibration method. In addition, it is shown that the proposed calibration model can also be used for nucleation detection, therefore facilitating real-time monitoring of batch cooling crystallization in practice.

415213 Optimization of a Reactive Crystallization Using a Pharmaceutical Engineering Approach

Abstract:

Reactive crystallizations are complex and difficult to control due to the number of competing parameters influencing the chemical transformation and particle growth. Here we will discuss a salt-to-salt transformation in the presence of multiple reactive chemical species. The crystallization is further burdened by the potential to create a poorly soluble free form that poses a risk to both product quality and yield. The presentation will highlight the use of FBRM in combination with other analytical techniques to identify and elucidate the side reactions responsible for observed variability in product potency. Detailed understanding of the upstream composition and the seed point proved vital to delivering a robust process. To offer reasonably broad parameter ranges to manufacturing and further mitigate risk, the refined process incorporates a crude crystallization focused upon yield and a re-crystallization that assures high quality. A fundamental understanding of the interactions of the chemical species and the root causes of variability were needed to deliver a suitable process for commercial manufacturing.

417528 Monitoring Co-Crystal Purity with on-Line Raman Spectrascopy

Abstract:

Pharmaceutical co-crystals are defined as stoichiometric molecular complexes that contain an active pharmaceutical ingredient (API) and a coformer in crystal lattice via noncovalent interactions, predominantly hydrogen bonds. Compared with APIs themselves, pharmaceutical co-crystals have advantages in solid state properties, such as solubility, melting point, chemical interaction, stability and bioavailability. Solution crystallization is an important method to produce co-crystals. However, crystals of constituent components, either API or co-former, may crystallize out concomitantly due to operational fluctuations in starting composition or temperature, compromising the purity of co-crystals. Therefore, it is desirable to detect crystals of single components when they appear during co-crystallization, whereby corrective operation can be taken to ensure co-crystal purity. In this study, a methodology is developed to monitor co-crystal purity using Raman spectroscopy with caffeine-glutaric acid-acetonitrile as the model system. Caffeine and glutaric acid can form co-crystal from solution in acetonitrile at the appropriate conditions. Raman spectra were collected in three classes of suspensions with a solid mixture of caffeine crystals and co-crystals, pure co-crystals, and a mixture of glutaric acid crystals and co-crystals, respectively at different temperatures. Principal component analysis (PCA) was performed for these spectra and the resulting first few principal components were put to discriminant analysis. It was found that Raman spectroscopy is able to distinguish the three classes of suspensions clearly.

Key Words, co-crystal, PAT, Raman spectroscopy, crystallization, principal component analysis

425352 Automated Multi-Loop Direct Nucleation Control Using Focused Beam Reflectance Measurement in Multistage Continuous Mixed Suspension Mixed Product Removal Crystallizer

Abstract:

Batch crystallization is an economical and widely used separation and purification unit operation especially in pharmaceutical industry. It has been extensively studied in literature and very well understood due to the development of various process analytical technology (PAT) tools like focused beam reflectance measurement (FRBM), particle video microscope (PVM), UV/Vis spectroscopy, IR spectroscopy and Raman spectroscopy. In addition, some of these tools have been successfully applied in feedback control frameworks in order to control the batch crystallization process in real time. Direct nucleation control (DNC) is a methodology that uses the chord count information from in situ FBRM to actuate adaptive heating or cooling rate to have controlled dissolution in the batch cooling crystallization process. It has been proved being able to produce much larger and more uniform crystal as compared with simple temperature control (T-control) approach in batch process.[1-2] Supersaturation control (SSC) and active polymorphic feedback control (APFC) apply UV/Vis or/and Raman spectroscopy in similar feedback control loops to maintain controlled supersaturation level or polymorphic transformation during batch cooling process.[3-4] These model-free feedback control approaches are proved very robust to automatically reject disturbances or to adjust the process when the required target product quality is changed. They can provide automated fast recipe design and only require simple mathematical tools without any modeling effort or much system knowledge.[5]

Continuous Crystallization is of great interest in recent decades due to consistent product quality, high process and equipment efficiency and reduced operation cost compared with batch crystallization. Mixed suspension mixed product removal (MSMPR) crystallizer is one of the most important continuous crystallization systems due to its simple operation and capability for long residence time. There are some published papers that investigate the recipe design for continuous MSMPR crystallization systems via modelling or experiment.[6-9] However, according to the authors' best knowledge, currently there are no published works that study the application of previously mentioned feedback control approaches (i.e. DNC, SSC, APFC) in any type of continuous crystallization systems.

In this work, a multi-loop DNC approach is experimentally investigated in a multistage MSMPR cooling crystallization process in order to automatically control the nucleation rate at each MSMPR stage. The feed solution is free of crystals. Each MSMPR stage is equipped with a heating circulator, a temperature sensor and a FBRM probe. A Labview based software Crystallization Monitoring and Control (CryMOCO) is installed on the supervisory computer to collect the chord count information from each individual FBRM probe. The information is then compared with the desired set-point of the corresponding MSMPR stage in CryMOCO, which remotely controls each temperature circulator and actuates a predefined adaptive heating or cooling rate in the corresponding MSMPR stage. In each DNC loop, the total chord count is targeted as the controlled variable and the set-point since it indicates information about nucleation, generation of fine particles as well as the average particle size. For example, in a certain stage the adaptive heating rate is actuated automatically once the measured FBRM total count is larger than the set-point of the corresponding stage, and vice versa. Paracetamol and ethanol are used as the model compound and solvent, respectively. The proposed multi-loop DNC approach is found being able to reach desired steady state and produce crystals of consistent desired size distribution after a few heating/cooling cycles with or without disturbances. The proposed approach has the following advantages: (1) It can automatically help avoid nucleation in the second stage and after when fine particles are not desired in the final product. This is achieved by designing a proper set-point for each MSMPR stage so that limited amount of nucleation or breakage is allowed. (2) It can automatically achieve optimal operating recipes to produce desired crystals of various sizes (i.e. either small or big). (3) It can automatically reject disturbances such as accidental seeding or nucleation. It is faster than simply washing out without any control. (4) It can reduce startup duration time and startup waste as compared with simple T-control that uses constant temperatures. Simulation work is also carried out in the group to validate and explain some of the experimental observations, and therefore to optimize the control performance.

References

[1] Saleemi, A.; Rielly, C.; Nagy, Z. K. Automated direct nucleation control for in situ dynamic fines removal in batch cooling crystallization. *CrystEngComm* **2012**, 14(6), 2196-2203.

[2] Abu Bakar, M. R.; Nagy, Z. K.; Saleemi, A. N.; Rielly, C. D. The impact of direct nucleation control on crystal size distribution in pharmaceutical crystallization processes. *Cryst. Growth Des.* **2009**, 9(3), 1378-1384.

[3] Saleemi, A. N.; Rielly, C. D.; Nagy, Z. K. Comparative investigation of supersaturation and automated direct nucleation control of crystal size distribution using ATR-UV/vis spectroscopy and FBRM. *Cryst. Grwoth Des.* **2012**, 12(4), 1792-1807.

[4] Simone, E.; Saleemi, A. N.; Tonnon, N.; Nagy, Z. K. Active polymorphic feedback control of crystallization processes using a combined Raman and ATR-UV/Vis spectroscopy approach. *Cryst. Growth Des.* **2014**, 14(4), 1839-1850.

[5] Nagy, Z. K.; Fujiwara, M.; Braatz, R. D. Modelling and control of combined cooling and antisolvent crystallization processes. *J Process Control* **2008**, 18, 856-864.

[6] Zhang, H.; Quon, J.; Alvarez, A. J.; Evans, J.; Myerson, A. S.; Trout, B. Development of continuous anti-solvent/cooling crystallization process using cascaded mixed suspension, mixed product removal crystallizers. *Org. Process Res. Dev.* **2012**, 16, 915-924.

[7] Vetter, T.; Burcham, C. L.; Doherty, M. F. Regions of attainable particle size in continuous and batch crystallization processes. *Chem. Eng. Sci.* **2014**, 106, 167-180.

[8] Yang, Y.; Nagy, Z. K. Advanced control approaches for combined cooling/antisolvent crystallization in continuous mixed suspension mixed product removal cascade crystallizers. *Chem. Eng. Sci.* **2015**, 127, 362-373.

[9] Yang, Y.; Nagy, Z. K. Combined cooling and antisolvent crystallization in continuous mixed suspension mixed product removal cascade: steady state and startup optimization. *Ind. Eng. Chem. Res.* **2015**, DOI: 10.1021/ie5034254.

423519 Digital Phase Diagram of K2SO4-KOH-H2O and Crystal Growth of K2SO4 in the Strong Alkali Solution

Abstract:

Potassium sulfate has always been an important and superior potash fertilizer. In this paper, we investigate the crystallization process of potassium sulfate from the strong alkali solution that is involved in the clean production process of raw alunite mineral. The research focuses on the digital phase diagram for a ternary system of K_2SO_4-KOH-H_2O and the crystal growth of K_2SO_4 in the strong alkali solution.

Firstly, the phase equilibrium data for the ternary system of K_2SO_4-KOH-H_2O are measured from 40°C to 80°C, where the contents of potassium and sulfur in the aqueous solutions are detected by ICP¨OES and the composition of solid phases are tested by X-ray diffraction, respectively. Based on the measured phase equilibrium data, origin is used to present the 3D colormap and ternary contour drawings, as shown in Figure 1. With the interpolation and fitting results obtained from the software, the saturation temperature of a given K_2SO_4-KOH-H_2O content can be predicted from the smooth surface of the ternary phase diagram. It is found that the K_2SO_4 solubility decreases with the increase of the KOH concentration in the aqueous solution, and the K_2SO_4 solubility becomes smaller in the strong alkali solution which is beneficial to separate K_2SO_4 from the KOH solution.

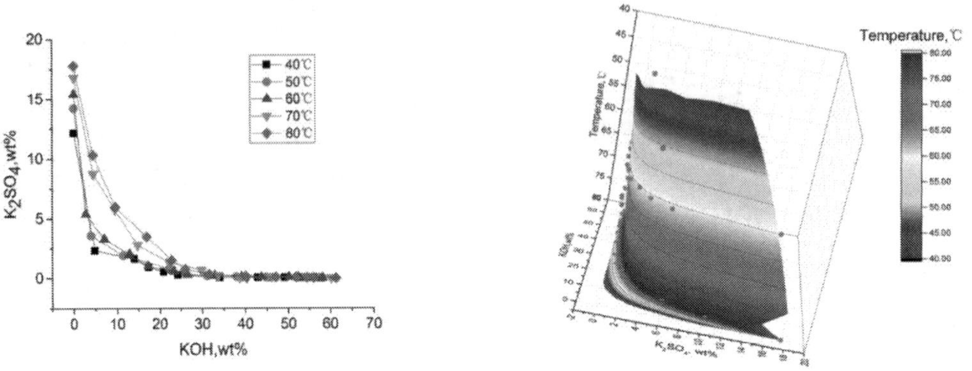

Figure 1. Solubility of K_2SO_4 in KOH solution and digital phase diagram of K_2SO_4-KOH-H_2O at different temperatures

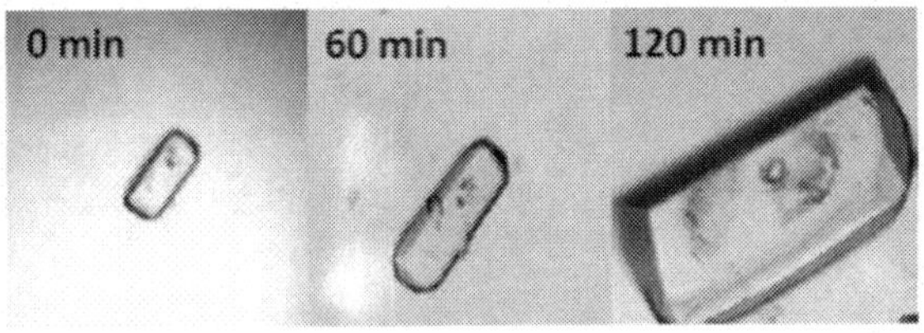

Figure 2. Crystal growth of K_2SO_4 in the strong KOH solution measured using a reflection microscopic technique at 313K

The supersolubility and metastable zone width of K_2SO_4 in the concentrated KOH solution are investigated by focused beam reflectance measurement (FBRM) and ultrasonic sensor online monitoring for different operating conditions, which will provide the important information for the design and optimization of K_2SO_4 crystallization process. For the system of K_2SO_4-KOH-H_2O, the FBRM and the ultrasonic sensor can provide a high sensitivity for nucleation detection, which detects changes related to the amount of grown nuclei and changes in solution concentration, respectively. Furthermore, the crystal growth rates of K_2SO_4 in the strong KOH solution are measured using a reflection microscopic technique, where the saturated solution is poured into a thermo-controlled microscopic cell with a volume of 5 mL and a diameter of 3.5cm. Experiments are conducted by decreasing the solution to different temperatures to create different supersaturation levels, and then the crystal growth rates of K_2SO_4 are calculated by measuring the changes in the dimensions of the crystals with the analysis Image Processing Program (Olympus). Figure 2 shows the crystal growth process of potassium sulfate in the strong KOH solution measured under the microscope at 313K. The crystal shape of K_2SO_4 is a typical rectangular shape, which is convenient for easy identification and measurements of specific crystallographic faces. The influence of impurities from the alkaline solution on the crystal shape will be investigated using the reflection microscopic technique. Moreover, the two-dimensional population balance equation is proposed for the calculation of the crystal growth rate of K_2SO_4 with an acceptable accuracy. From this study of crystal growth, a given content of solution can be designed with a desired growth mechanism in order to effectively separate K_2SO_4 from the strong KOH solution by crystallization.

435128 Controlling Crystal Size, Shape and Purity during Crystallization By Using Additives

Abstract:

Crystallization which is a purification and separation unit operation plays an important role in the industrial production of pharmaceuticals. The properties of the produced crystals such as purity or size and shape distribution have strong impact on the downstream processes as well as on the end usage. It is known, that chemicals are generally contaminated by different amounts of impurities, which can come from different sources, such as from upstream processes or chemical side reactions [1]. In most of the industrial processes, the effect of impurities may result e.g. in needle shape crystals which impeding manufacturability of the solid dosage form. While crystallization is often used as a purification process, trace amount of impurities generally exist in the product. While in some cases impurities of various amounts are acceptable for certain highly toxic impurities the acceptable limits are very low and difficult to achieve using a single crystallization step. For example genotoxic impurities carry significant health risk and strict regulations are applied on the concentration of these materials [6,8,9]. Impurities, even if it is in trace amounts, can influence the crystal properties like crystal shape. In this case, impurities are called as crystal growth modifiers (CGMs) [2-6]. In this work, a method is presented, which reduces the possible undesired effect of the CGMs during crystallization by using a control additive (CA). The control method is based on a chemical reaction between the control additive and the growth modifier, which results in a non-adsorptive reaction product. Thus, the concentration of CGM in the solution and thus in the solid phase is the function of the additive concentration in the solution. It is also known that the concentration of the CGMs in the solution affects the rate of the change of crystal growth.

This work deals with the investigation and control of batch cooling crystallization in the presence of crystal growth modifiers. The objectives of the control method are to maintain the purity of the crystal product and also to control the crystal size and shape distribution. A mathematical model including nucleation and growth of two characteristic crystal sizes as well as impurity adsorption and chemical reaction model is developed in order to investigate and control the purity and the crystal shape. The growth modifiers influence directly the growth kinetics. This inhibitory effect on the crystal surfaces leads to changed crystal size and shape distribution [2-6]. On the other hand, growth modifiers are built in the crystal structure according to surface adsorption mechanism and so CGMs also change the purity of the crystal product [7]. The proposed purity control method is based on chemical reaction between the undesired impurity and the control additives which provides an alternative transformation route for the compound against the surface adsorption. Hence, the chemical reaction inhibits the adsorption rate by lowering the concentration in the solution.

Utilizing the benefits of the competitive adsorption and reaction mechanisms, model based control for crystal shape and purity is developed. The simulator is suitable to calculate the optimal concentration of control additive in order to achieve desired crystal shape and/or reducing the impurity concentration below the desired limits. The results demonstrate the efficiency of the proposed approach for shape and purity control.

Acknowledgement: Financial support provided by the European Research Council grant no.[280106-CrySys] is acknowledged.

References:

[1] N., Variankaval, A.S., Cote, M.F., Docherty, 2008. From form to function: Crystallization of active pharmaceutical ingredients. AIChE Journal, 54(7), 1682-1688.

[2] A., Majumder, Z.K., Nagy, 2013. Prediction and control of crystal shape distribution in the presence of crystal growth modifiers. Chemical Engineering Science, 101, 598-602.

[3] R.J., Davey, 1976. The effect of impurity adsorption ion the kinetics of crystal growth from solution. Journal of Crystal Growth, 34, 109-119.

[4] N., Kubota, J.W., Mullin, 1995. A kinetic model for crystal growth from aqueous solution in the presence of impurity. Journal of Crystal Growth, 152(3), 203-208.

[5] N., Kubota, S., Sasaki, N., Doki, N., Minamikawa, M., Yokota, 2004. Absorption of an Al(III) impurity onto the (100) face of a growing KDP crystal in supersaturated solution. Crystal Growth and Design, 4(3), 533-537.

[6] A., Borsos, A., Majumder, Z. K., Nagy, 2014. Model development and experimental validation for crystal shape control by using tailored mixtures of growth modifiers. Computer Aided Chemical Engineering, 33, 781-786.

[7] K., Maeda, R., Tabuchi, Y., Asakuma, K., Fukui, 2006. Distribution of metallic ions in single KDP crystal grown from aqueous solution. Cryst. Res. Technol., 41(10), 955-960.

[8] A., Garcia-Arieta, 214. Interactions between active pharmaceutical ingredients and excipients affecting bioavailability: impact on bioequivalence. European Journal of Pharmaceutical Sciences, 65, 89-97.

[9] S., Gorog, 2006. The importance and the challenges of impurity profiling in modern pharmaceutical analysis. Trends in Analytical Chemistry, 25 (8), 755-757.

409298 Recent Advances in Continuous Slug-Flow Crystallization

Abstract:

Continuous slug-flow crystallization is an approach for the manufacture of pharmaceutical crystals with improved control of product characteristics, simplified post-crystallization operations, higher production rate flexibility, and reduced capital costs and footprint. Explanations are provided for the reported order-of-magnitude reductions in secondary nucleation and production times compared to batch, semibatch, and mixed-suspension mixed-product-removal crystallizers. After reviewing the literature on slug/segmented-flow crystallization, recent advances are described in equipment design/configurations. Pros and cons of different process intensification strategies are discussed, such as (1) the spontaneous generation of hydrodynamically stable slug flow; (2) indirect vs. direct and spatially localized vs. spatially distributed ultrasonication, (3) dual-impinging jet, coaxial, and radial micromixers, and (4) operating micromixers under laminar vs. turbulent flow conditions. Guidance is provided for quickly assessing whether slug-flow crystallization is the appropriate technology for a particular solute-solvent(s) system and, if so, systematically select equipment configurations, design parameters, and control system designs.

437611 Microfluidic Platforms for Solid Form Screening

Abstract:

This talk will summarize the achievements that resulted from a collaborative project between AbbVie and UIUC to develop microfluidic tools to identify solid forms of candidate pharmaceuticals, with the notion that microfluidic approaches would enable screening for, and down selection of, suitable solid forms earlier in the drug development process. Microfluidic platforms will be presented that allow for different crystallization methods (free interface diffusion, evaporation, antisolvent addition) as well as on-chip analysis of the solid forms through Raman Spectroscopy and X-ray diffraction. These platforms were validated through screening for polymorphs, salts and cocrystals of different compounds.

437031 Unsolved Problems in Crystal Growth: Challenges for the Future

Abstract:

In the last six decades revolutionary changes have occurred in the field of crystal growth. The mechanisms by which crystals grow at low supersaturations (spiral growth from screw dislocations), and high supersaturations (2 dimensional nucleation of various types) are understood. Surface characterization techniques have improved to molecular resolution. However, progress has been unexpectedly slow on several fronts, including models that are capable of predicting absolute growth rates (not just relative growth rates of crystal faces), the precise effects of solution environment at the solvent-surface interface on growth, and most perplexing of all, the mechanisms by which polar crystals grow. In this presentation I will focus on solution effects, which seem within reach, and on polar crystals, which do not.

410668 In-Vitro Dewetting Dynamics on Silicone Hydrogel Contact Lenses

Abstract:

The tear film on our eyes provides a smooth optical interface as well as protects the eyes against environmental challenges. Of particular interest in this presentation is the meibomian lipid layer, secreted by glands lining our eyelids, that has been shown to possess remarkable interfacial viscoelastic properties. By using a newly developed instrument called the Interfacial Dewetting and Drainage Optical Platform(i-DDrOP) developed in the Fuller laboratory at Stanford University, we investigate the *in-vitro* dewetting dynamics of thin aqueous films laden with meibomian lipid layer on silicone hydrogel contact lenses at physiological conditions. Our *in-vitro* measurements show dewetting times that are comparable to the *in-vivo* tear break up times. This highlights the utility of the i-DDrOP as a tool to study biomimetic tear film solutions and their interactions with contact lens surfaces.

438452 Prediction of Vapor/Liquid Characteristics By Quantum Mechanics Data

Abstract:

A large number of chemical processes involve the purification of mixtures in industry, such as liquid extraction, distillation, and recrystallization. Infinite dilution coefficients can be used to characterize the vapor-liquid, solid liquid, and liquid-liquid phase behavior that underlies these unit operations. The largest deviation from ideality occurs in the dilute solution system. The ability to predict infinite dilution coefficients based on simple guidelines of hydrogen bonding could provide valuable design formulations and useful thermodynamic insights. The MOSCED model is sufficiently simple to fill this need. Unfortunately, a lack of characterized parameters for many compounds limits the capabilities of the MOSCED model. The goal of this paper is to develop a simple method to estimate the missing parameters based on readily available quantum mechanical results.

Sigma profiles have been calculated using quantum mechanics and tabulated at a website hosted by Virginia Tech. The charge density of the molecule over the molecular surface is described by the values where the projected surface charge is opposite of that of the molecule. Hydrogen bonding can occur when the charge density exceeds a threshold value. These observations can be used to calculate acidity and basicity parameters for the MOSCED model. Given that the bonding starts at a specific point it was assumed the rest of the contributions could be interpreted as another parameter, polarizability factor.

The acidity and basicity parameters were characterized by fitting the data from the area beneath the sigma profile at a certain distance and correlating the values to pre-determined values from Lazzaroni et al. The fit for the polarizability factor was determined by using a weighted function to give greater emphasis on values further from the center of the sigma profile.

Correlations of acidity and basicity were characterized based on experimental data for roughly 4500 binary solutions. A comparison of the results for the fit vs. the MOSCED with Lazzaroni parameter values and UNIFAC model for the solutions was used to determine the accuracy of the model. Several specific solutions were poorly characterized in comparison to the MOSCED model, but the correlation provides qualitative estimates of activity coefficients with in roughly 0.280 log units of the experimental data, compared to 0.106 log units for the original MOSCED method and 0.183 log units for UNIFAC.

Effect of Electrical Field on the Effective Parameters of Non-Newtonian Fluid

11/9/15

STEPHEN DUECK
STEFFANO OYANADER
DR. MARIO OYANADER

cbu | Gordon and Jill Bourns COLLEGE OF ENGINEERING AIChE

Outline

- Physics of the Problem & Applications
- Description of System
- Analytical Model
- Computational Model
- Graphical Illustrations
- Conclusions
- Future Work
- Acknowledgements

cbu | Gordon and Jill Bourns COLLEGE OF ENGINEERING AIChE

Physics of the Problem & Applications

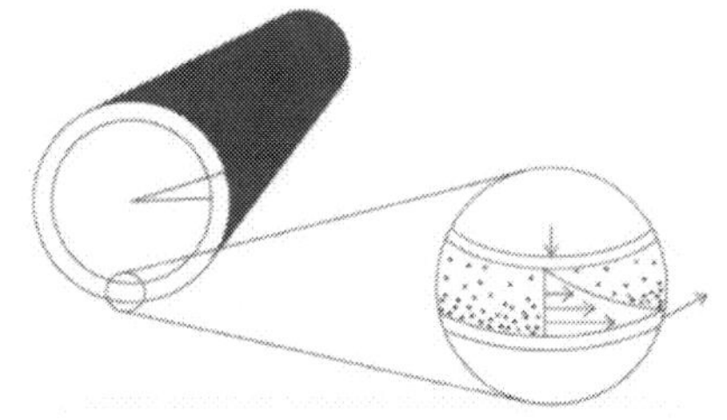

Applications:

- Environmental and toxicological investigations of nanomaterials
- Determination of size and shape of environmental and biological particles
- Separations
  - high and ultra high molar mass polymers
  - Macromolecules
  - Particulates
  - Nanoparticles
  - Gels
- Replacement of centrifugation
- Blood Cleansing (dialysis)
- Separation of pharmaceuticals and metabolites for treatment and diagnosis from biosamples

cbu | Gordon and Jill Bourns COLLEGE OF ENGINEERING

AIChE

Rectangular Simplification Does not Capture Non-Newtonian Velocity Field "n"

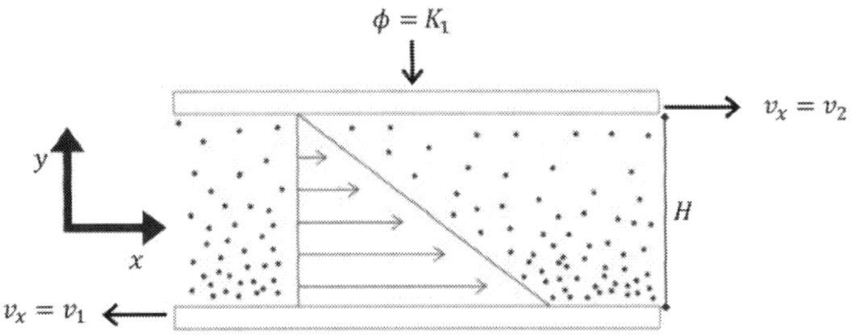

Governing Equation:

$$0 = \frac{\partial}{\partial y} \mathrm{T}_{y_x}$$

Boundary Conditions:

$$v_x = v_1 \ @ \ y = 0$$
$$v_x = v_2 \ @ \ y = H$$

Shear Stress Equation:

$$C_1 = \mathrm{T}_{y_x} = \mu \left| \frac{\partial v_x}{\partial y} \right|^n$$

Fluid Velocity Equation:

$$v_x(y) = \left(\frac{C_1}{\mu_0}\right)^{\frac{1}{n}} y + C_2 \rightarrow v_x(y) = \frac{v_2 - v_1}{H} y + v_1$$

cbu | Gordon and Jill Bourns COLLEGE OF ENGINEERING

AIChE

Cylindrical Captures "n" Parameter

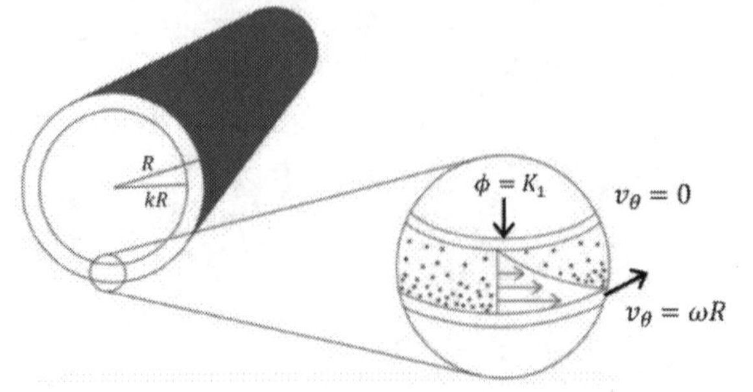

Governing Equation:

$$\frac{-1}{r^2}\frac{\partial}{\partial y}(r^2 T_{r\theta}) = 0$$

Boundary Conditions:

$$v_\theta = 0 \ @\ r = kR\ (inner)$$
$$v_\theta = \omega R \ @\ r = R\ (outer)$$

Shear Stress Equation:

$$T_{r\theta} = -m\left(r\frac{\partial}{\partial y}\left(\frac{v_\theta}{r}\right)\right)^n$$

Fluid Velocity Equation:

$$v_\theta = -\frac{n}{2} r^{1-\frac{2}{n}} C_1^{\frac{1}{n}} + rC_2 \rightarrow \frac{v_\theta}{\omega r} = \frac{1-\left(\frac{kR}{r}\right)^{\frac{2}{n}}}{\left(1-k^{\frac{2}{n}}\right)} = v_\theta^*$$

cbu | Gordon and Jill Bourns COLLEGE OF ENGINEERING

AIChE The Global Home of Chemical Engineers

Cylindrical Captures "n" Parameter

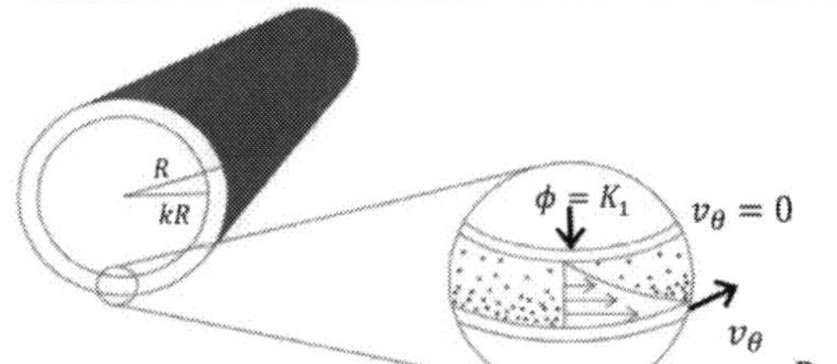

Governing Equation:

$$\frac{-1}{r^2}\frac{\partial}{\partial y}(r^2 T_{r\theta}) = 0$$

Boundary Conditions:

$$v_\theta = 0 \ @\ r = kR\ (inner)$$
$$v_\theta = \omega R \ @\ r = R\ (outer)$$

Shear Stress Equation:

$$T_{r\theta} = -m\left(r\frac{\partial}{\partial y}\left(\frac{v_\theta}{r}\right)\right)^n$$

Fluid Velocity Equation:

$$\frac{v_\theta}{\omega r} = \frac{1-\left(\frac{kR}{r}\right)^{\frac{2}{n}}}{\left(1-k^{\frac{2}{n}}\right)} = v_\theta^*$$

$$\langle v_\theta^* \rangle = \frac{\int_{kR}^{R} v_\theta^*(r) r dr}{\int_{kR}^{R} r dr}$$

$$\langle v_\theta^* \rangle = \omega \left\{ \frac{1-k^2-nk^{\frac{2}{n}}+2nk^2-n}{\left(1-k^{\frac{2}{n}}\right)(n-1)(k^2-1)} \right\}$$

cbu | Gordon and Jill Bourns COLLEGE OF ENGINEERING

AIChE The Global Home of Chemical Engineers

Electrical Field Governing Equation

$$\frac{1}{r}\frac{\partial}{\partial r}\left(r\frac{\partial \phi}{\partial r}\right) = 0$$

Boundary Conditions:

$\phi = 0 \quad @\, r = kR$

$\phi = K_1 \quad @\, r = R$

$$\phi = K_1\left(1 - \frac{\ln\left(\frac{r}{R}\right)}{\ln(K)}\right)$$

$$\langle\phi\rangle = \frac{\int_{kR}^{R}\phi(r)r\,dr}{\int_{kR}^{R}r\,dr} = K_1\left\{\frac{1}{1-k^2} + \frac{1}{lnk^2}\right\}$$

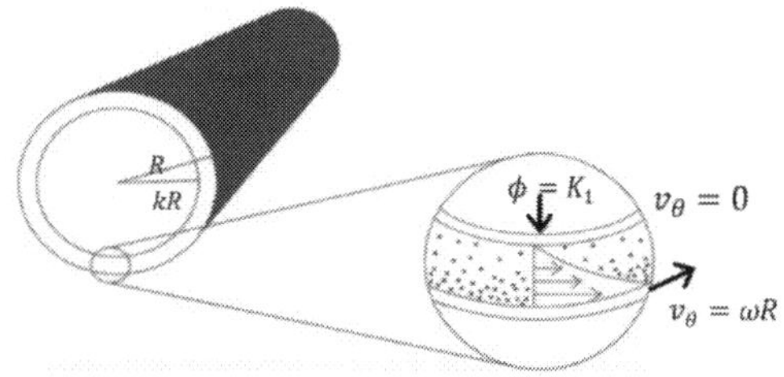

cbu | Gordon and Jill Bourns COLLEGE OF ENGINEERING AIChE

Molar Species Continuity Governing Equation

$$\frac{\partial C}{\partial t} + \frac{v_\theta}{r}\frac{\partial}{\partial \theta} = D\left[\frac{1}{r}\frac{\partial}{\partial r}\left(r\frac{\partial C}{\partial r}\right) + \frac{1}{r^2}\frac{\partial^2 C}{\partial \theta^2}\right] + \mu^*\frac{1}{r}\frac{\partial}{\partial r}\left(rC\frac{\partial \phi}{\partial r}\right)$$

Boundary Conditions:

$$D\frac{\partial C}{\partial r} + \mu^*(C * \frac{\partial \phi}{\partial r}) = 0 \; @\, r = kR$$

$$D\frac{\partial C}{\partial r} + \mu^*(C * \frac{\partial \phi}{\partial r}) = 0 \; @\, r = R$$

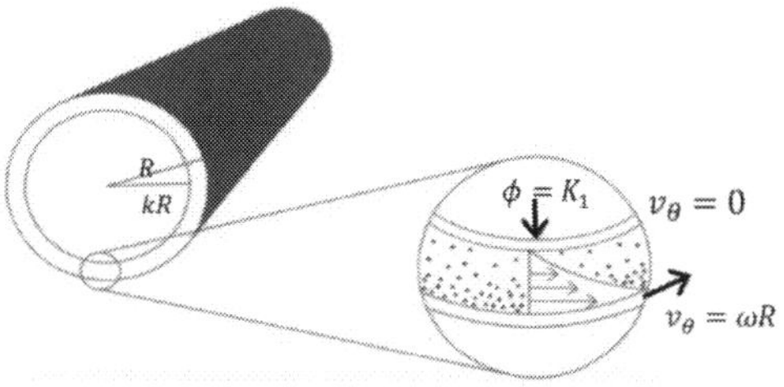

cbu | Gordon and Jill Bourns COLLEGE OF ENGINEERING AIChE

Modeling Approach

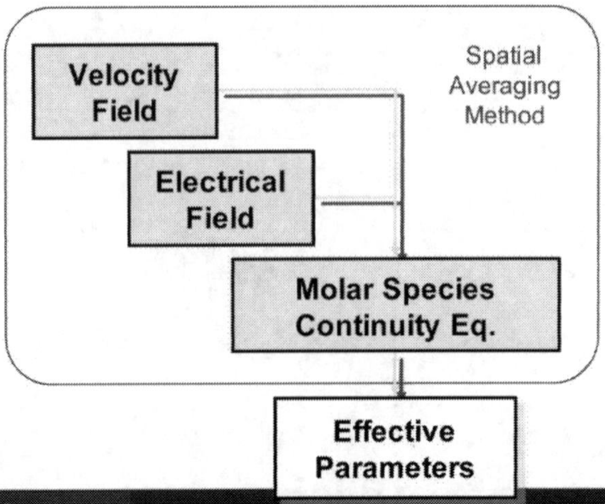

$$\frac{\partial C}{\partial t}+\frac{v_\theta}{r}\frac{\partial}{\partial \theta}=D\left[\frac{1}{r}\frac{\partial}{\partial r}\left(r\frac{\partial C}{\partial r}\right)+\frac{1}{r^2}\frac{\partial^2 C}{\partial \theta^2}\right]+\mu^*\frac{1}{r}\frac{\partial}{\partial r}\left(rC\frac{\partial \phi}{\partial r}\right)$$

Effective Parameters Using the Area Averaging Method

$$\hat{v}_{eff}=\frac{v_{eff}}{\langle v_\theta^* \rangle}=1+\langle \tilde{v}_\theta * f(r)\rangle$$

$$\Theta=\frac{\langle \tilde{v}_\theta^* * g(r)\rangle}{\langle v_\theta^* \rangle R^2}=(1-\hat{D}_{eff}/\mathrm{D})/Pe^2$$

f(r) and g(r) from Cwirko & Carbonell (1989)

Effective Parameters Using the Area Averaging Method

Effective Velocity

$$\hat{v}_{eff} = 1 + A_0 + A_1 * f$$

$$\Omega \equiv \frac{\mu * K_1}{D}$$

Effective Dispersion Coefficient

$$\Theta = \frac{A_0}{2(2-\Omega)}\left\{\frac{(k^4 - k^4)}{(k^2 - k^4)}\frac{(A-2)}{(A-4)}A_0 - \frac{1}{2}(k^2+1)A_1 f\right\} - \left\{\frac{A_0}{2\Omega} + \frac{A_1}{(2-\Omega)\Omega}\right\}\{A_0 + A_1 f\}$$
$$- \frac{A_1}{(2-A)(2-A-\Omega)}\left\{\frac{(k^{2-A}+1)A_0}{2} - \frac{2(k^{4-A}-1)}{(k^2-1)(A-4)}A_1 f\right\}$$

cbu | Gordon and Jill Bourns COLLEGE OF ENGINEERING — AIChE The Global Home of Chemical Engineers

Effective Parameters Using the Area Averaging Method

$$A_0 = \frac{2k^2 - 2k^{\frac{2}{n}}}{\frac{2}{n} - 2 - \frac{2}{n}k^2 + 2k^{\frac{2}{n}}}$$

$$A_1 = \frac{\left(\frac{2}{n} - 2\right)(k^2 - 1)k^{\frac{2}{n}}}{\frac{2}{n} - 2 - \frac{2}{n}k^2 + 2k^{\frac{2}{n}}}$$

$$f = \frac{\left(k^{2+\frac{\Omega}{\ln(k)}} - k^{\frac{2}{n}}\right)\left(2 + \frac{\Omega}{\ln(k)}\right)}{k^{\frac{2}{n}}\left(k^{2+\frac{\Omega}{\ln(k)}} - 1\right)\left(2 + \frac{\Omega}{\ln(k)} - \frac{2}{n}\right)}$$

$$\Omega \equiv \frac{\mu * K_1}{D}$$

cbu | Gordon and Jill Bourns COLLEGE OF ENGINEERING — AIChE The Global Home of Chemical Engineers

V_{eff} Analysis

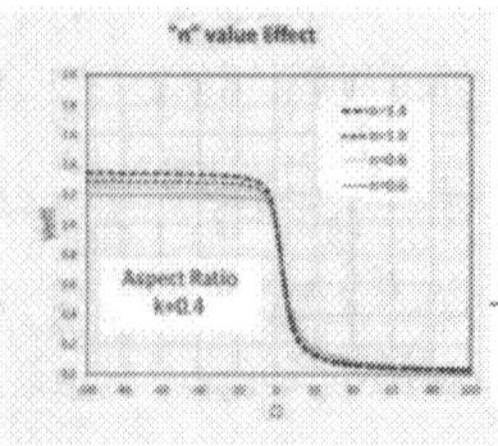

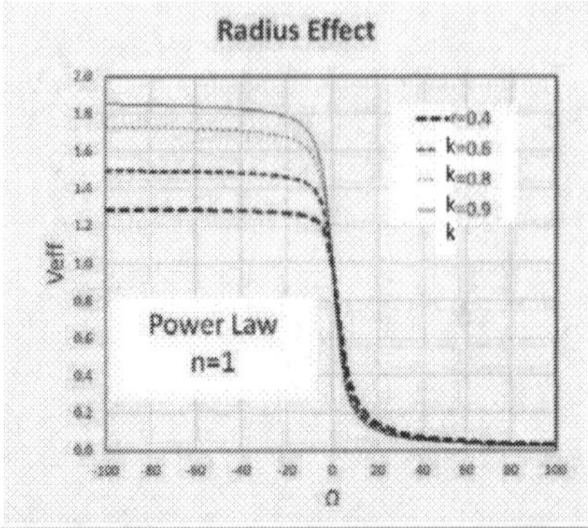

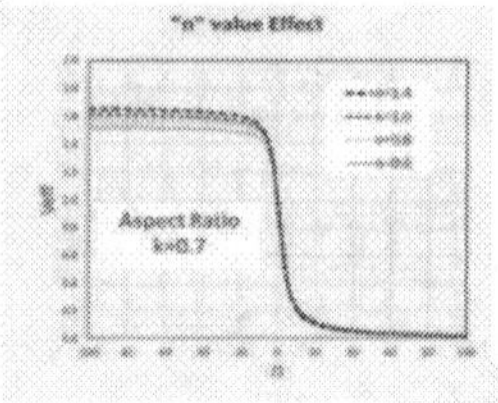

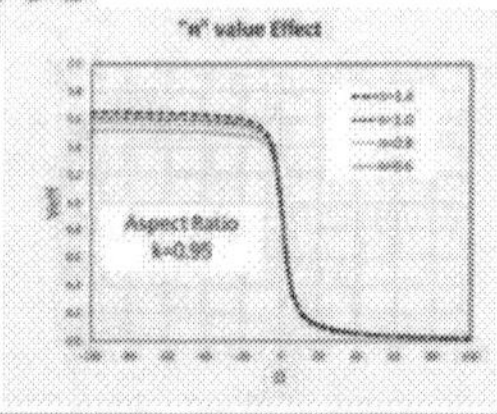

cbu | Gordon and Jill Bourns COLLEGE OF ENGINEERING

AIChE The Global Home of Chemical Engineers

D_{eff} Analysis

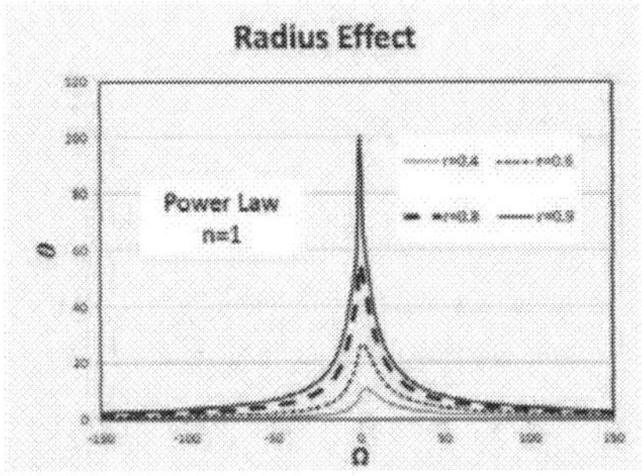

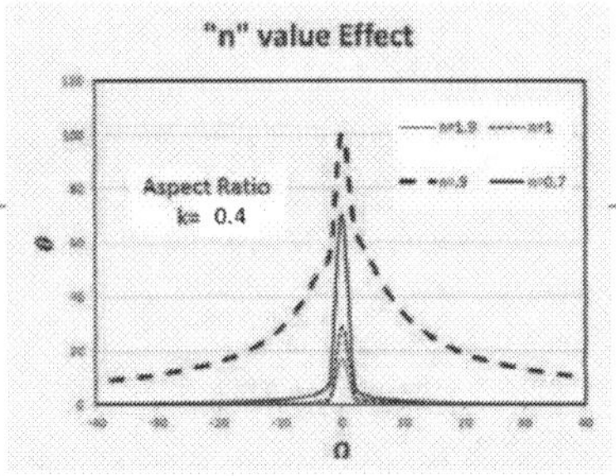

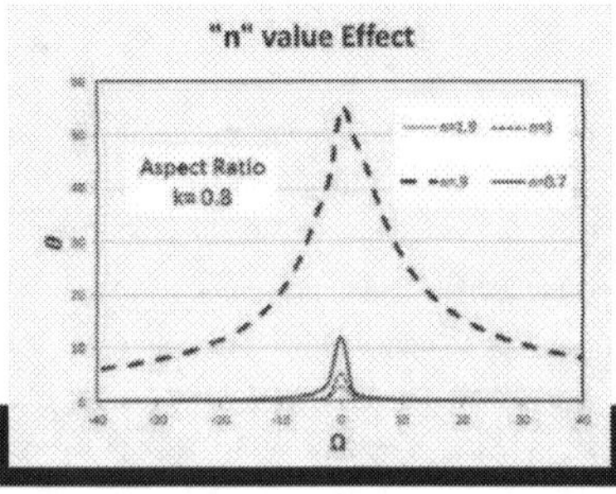

cbu | Gordon and Jill Bourns COLLEGE OF ENGINEERING

AIChE The Global Home of Chemical Engineers

General Conclusions

- Effective Dispersion Coefficient values decrease as the Aspect Ratio parameter (k) increases.
- Effective Velocity values increase as the Aspect Ratio parameter (k) increases.
- Values of the Power Law model parameter "n" produces lower Effective velocity values as it decreases.
- Inversion of the electrical Field Direction could bring Effective Velocity from maximum values to a minimum of zero.

Future Work

- Computation of Optimal Time of Separation
- New Non-Newtonian Fluid Model
- Collaborative Paper Writing

Optimal Separation Time

$$\tau_{op} = \left[\frac{\sqrt{\frac{D_{effA}}{D_A}} + \sqrt{\frac{D_{effB}}{D_B}}}{\hat{v}_{effA} - \hat{v}_{effB}} \right]^2$$

cbu | Gordon and Jill Bourns COLLEGE OF ENGINEERING AIChE

Acknowledgements

- California Baptist University, College of Engineering
- Department of Chemical Engineering

cbu | Gordon and Jill Bourns COLLEGE OF ENGINEERING

AIChE The Global Home of Chemical Engineers

438449 Degradation of a Thermoplastic Polyether Ester Elastomer in Downhole Conditions

Abstract:

Thermoplastic elastomers (TPEs) combine the strength of thermoplastics with the flexibility of elastomers, a characteristic potentially useful in downhole applications. We studied the degradation process of a particular TPE of interest from both a mechanical and chemical perspective, and have determined the Fickian diffusion constant, the chemical degradation rate, the change in crystallinity, and the storage modulus with respect to varying temperature and time. Techniques used to determine such rates include: dynamic mechanical Aanalysis (DMA), differential scanning calorimetry (DSC), and end group analysis. Applying Arrhenius' empirical relationship to the determined rates gives rise to a temperature-dependent model that predicts the degradation behaviour of the TPE outside of the experimental temperature range. Our results indicate that the chemical degradation via hydrolysis follows pseudo first order kinetics, and the degradation results in an increase of crystallinity and thus storage modulus and a decrease of tensile strength and strain. The polymer eventually deteriorates due to brittleness.

438453 Alternative Sampling Densities in Non-Uniform Sampling

Abstract:

Non-uniform sampling (NUS) is more widely practiced, but many implementations either do not weight the distribution of samples, or follow exponential sampling densities. Weighting the sampling density in proportion to the signal envelope provides intrinsic sensitivity enhancements (MRC, V49, 483-491, 2011). However sinusoidal weighting is a compelling alternative to exponential weighting matched to the signal decay (JBNMR, V58: 303-314, 2014). We now describe families of alternative densities as candidates to replace common exponential NUS weightings. Yielding equal intrinsic sensitivity to their exponential counterparts, these alternatives possess improved constraints on line shapes by distributing more samples beyond 1xT2. An algorithm to generate properly gapped sampling densities (JBNMR, V58: 303-314, 2014) is extended, and is also contrasted with deterministic NUS schedules.

438447 Dynamic Light Scattering Study of Sulfonated Polystyrene Ionomers in Benzene

Abstract:

From golf ball coatings (Surlyn®) to fuel cell membranes (Nafion®), ionic polymers (ionomers) are widely used. Ionomers are molecular compounds that contain long hydrocarbon chains with ions randomly distributed along the chain. These polymers form ionic aggregates when ionic groups attract each other in nonpolar environment, and these interactions dictate the properties exhibited. We studied the aggregation of sulfonated polystyrene (SPS) ionomers of varying molecular weights and ion contents in benzene. Dynamic light scattering (DLS) experiments determine the effective diameter and the polydispersity of the aggregates and molecules in the solution. To do this, we created solutions of the SPS and its acidic equivalent in benzene at different concentrations and analyzed the scattering. We also studied the effects of an ionic liquid, BMIM-PF_6 (1-butyl-3-methylimidazolium hexafluorophosphate) on SPS ionomers of higher ion contents. It was found that the ionic liquid weakens the ionic bonds and partly destroys ionic aggregates. We will study the properties of ionomers with high ion contents with other ionic liquids and with toluene as a solvent, for comparisons.

438443 A DFT Study of Methanol Reforming on C- and N-Doped TiO2 (110) Rutile Surfaces

Abstract:

: Biomass reforming is a sustainable method for hydrogen gas production. Hydrogen gas is an increasingly desirable clean burning fuel source that is commercially produced by methane reforming, which is a high energy process. Methanol reforming, a model for biomass reforming, over TiO_2 offers a lower energy alternative with a greater hydrogen gas yield; however, the process is not well understood, and thus it is not economically viable with current technologies. Nonmetal dopants on TiO_2 have been shown to reduce band gaps of the TiO_2 for visible light photocatalysis, but the mechanism on these altered catalysts are not well understood. We apply periodic DFT+U to gain a better mechanistic understanding of methanol reforming over C- and N-doped TiO_2 in order to inform future catalyst design. We compare binding energies and reaction barriers for both lattice and interstitial doping on rutile TiO_2 (110).

438313 Thermochemical Analysis of Intermolecular Vs. Intramolecular Reactions in Iridium Complexes

Abstract:

Transition metals are commonly used in catalytic reactions in academic and industrial settings. Transition metals have a variety of oxidation states and can bind ligands in a variety of geometries, including octahedral, tetrahedral, or square-planar coordination to name a few. The Cp^*IrPPh_3 complex (**1**) was studied for its ability to activate C-H bonds under mild conditions. As observed in experimental work, complex **1** in benzene has the ability to undergo both intermolecular and intramolecular reactions in a nearly 1:1 ratio. Literature suggests that the intramolecular reaction should be favored. Computational methods were used to investigate the details of the intermolecular and intramolecular reaction mechanisms. With the detailed knowledge of the reaction mechanism, we can determine if the 1:1 ratio is a kinetic or thermodynamic effect.

437858 Quantitative and Contextual Characterization of Plasmid Copy Number

Abstract:

Intracellular plasmid copy numbers directly affect the dosage of genetic constructs expressed from extrachromosomal replicons. This is particularly relevant for complex, multigene systems that require precise gene expression levels to be functional (e.g., genetic circuits, metabolic pathways). Furthermore, mathematical models of plasmid-based gene expression require more than a qualitative or semi-quantitative description of plasmid copy number to be accurate or meaningful. We have developed a straightforward method for determining plasmid copy number using a simple mass balance-based model of gene expression with two measurable inputs, steady-state fluorescence and growth rate. Using our approach, we have investigated various genetic, cellular and environmental context dependencies of plasmid copy number in commonly used laboratory strains of *Escherichia coli*.

(237j) A Novel Approach to Data Processing in High Throughput Material Science Experimentation

Albing, M.,

Originally presented on: 11/9/2015 17:30:00 - 17:45:00

Advancements in materials modeling techniques lead to rapid development of new materials. To test the properties of these modeled materials, a more rapid approach to material science research is required. High throughput experiments (HTE) in material science research accomplishes this challenge by providing a methodology for streamlining and optimizing the efficiency of experiments. In current HTE experiments large amounts of data are collected from a range of material compositions. While in the past HTE was limited by data collection, advances in computing have helped overcome these issues. To address the burgeoning big data in materials science problem we have been working to develop novel data processing algorithms. A typical experiment requires that a sample be scanned with a synchrotron light source, such as the one at the Stanford Linear Accelerator Center (SLAC) to gather information about the underlying crystal structure and composition. The resulting large volume of data would take a long time to process manually. An algorithm can process the data in seconds. A typical algorithm begins with a background subtraction. This is done quickly with a cubic spline fit. After the background is subtracted, peaks are fit with a pseudo-Voigt profile. Using the full width half maximum, peak location and peak height for each sample, the features across a sample can be analyzed quickly[J1] [s2] . This type of signal processing is widely used and greatly increases the data interpretation step of an experiment. We have been developing assisted machine learning algorithms to more [s3] quickly analyze crystal phase with respect to sample composition and other properties. The algorithm first separates data into many groups, called clusters. This initial clustering procedure does not require human input and forms clusters based on common traits in the XRD spectra. The next part of the algorithm requires that a human identify features on characteristic data picked by the algorithm from each cluster. The algorithm uses human input to create standard criteria for crystal phase identification in the rest of the data. This greatly reduces the time from data collection to sample characterization, and ultimately new material discovery. By developing new data processing techniques, experiments are optimized for efficiency and precision.Advancements in materials modeling techniques lead to rapid development of new materials. To test the properties of these modeled materials, a more rapid approach to material science research is required. High throughput experiments (HTE) in material science research accomplishes this challenge by providing a methodology for streamlining and optimizing the efficiency of experiments. In current HTE experiments large amounts of data are collected from a range of material compositions. While in the past HTE was limited by data collection, advances in computing have helped overcome these issues. To address the burgeoning big data in materials science problem we have been working to develop novel data processing algorithms. A typical experiment requires that a sample be scanned with a synchrotron light source, such as the one at the Stanford Linear Accelerator Center (SLAC) to gather information about the underlying crystal structure and composition. The resulting large volume of data would take a long time to process manually. An algorithm can process the data in seconds. A typical algorithm begins with a background subtraction. This is done quickly with a cubic spline fit. After the background is subtracted, peaks are fit with a pseudo-Voigt profile. Using the full width half maximum, peak location and peak height for each sample, the features across a sample can be analyzed quickly[J1] [s2] . This type of signal processing is widely used and greatly increases the data interpretation step of an experiment. We have been developing assisted machine learning algorithms to more [s3] quickly analyze crystal phase with respect to sample composition and other properties. The algorithm first separates data into many groups, called clusters. This initial clustering procedure does not require human input and forms clusters based on common traits in the XRD spectra. The next part of the algorithm requires that a human identify features on characteristic data picked by the algorithm from each cluster. The algorithm uses human input to create standard criteria for crystal phase identification in the rest of the data. This greatly reduces the time from data collection to sample characterization, and ultimately new material discovery. By developing new data processing techniques, experiments are optimized for efficiency and precision.

428724 Investigation of Titanium and Cerium MOFs for Improved Acid Gas Separation

Abstract:

Developing sorbents for the adsorption of toxic chemicals has garnered much research attention in the field of porous materials. Impregnated activated carbons[1-3] have been widely explored for acid gas separation applications; however, metal-organic frameworks (MOFs) have emerged as a promising class of materials for acid gas separations. MOFs are characterized by metal clusters and organic linkers, large surface areas, and tunable chemical properties that make them excellent candidates for gas adsorption.[4-8] Several MOFs using titanium for the metal source have shown promising adsorption properties for CO_2, water, and H_2S.[9, 10] In an effort to determine the adsorption and separation performance of these materials, dynamic breakthrough studies with a wide range of acid gases were performed on titanium and cerium MOFs. In addition, recent work has identified MOFs as excellent templates for producing MOF-derived structured metal oxides.[11-13]The dynamic performance of these oxides was examined in comparison with the parent MOFs as well as bulk commercial oxides to determine advantages of using structured material derivatives. In addition to dynamic breakthrough performance, the effect of acid gas exposure was also explored through water exposure experiments, as well as structure analysis before and after breakthrough measurements.

(1) Smith, J. W. H.; Westreich, P.; Croll, L. M.; Reynolds, J. H.; Dahn, J. R. *J. Colloid Interface Sci.* **2009,** *337*, 313.

(2) Petit, C.; Karwacki, C.; Peterson, G.; Bandosz, T. J. *The Journal of Physical Chemistry C* **2007,** *111*, 12705.

(3) Mahle, J. J.; Peterson, G. W.; Schindler, B. J.; Smith, P. B.; Rossin, J. A.; Wagner, G. W. *J. Phys. Chem. C* **2010,** *114*, 20083.

(4) Czaja, A. U.; Trukhan, N.; Müller, U. *Chem. Soc. Rev.* **2009,** *38*, 1284.

(5) Li, J. R.; Kuppler, R. J.; Zhou, H. C. *Chem. Soc. Rev.* **2009,** *38*, 1477.

(6) Keskin, S.; van Heest, T. M.; Sholl, D. S. *ChemSusChem* **2010,** *3*, 879.

(7) Yilmaz, B.; Trukhan, N.; Müller, U. *Chinese Journal of Catalysis* **2012,** *33*, 3.

(8) Sumida, K.; Rogow, D. L.; Mason, J. A.; McDonald, T. M.; Bloch, E. D.; Herm, Z. R.; Bae, T. H.; Long, J. R. *Chem. Rev.* **2012,** *112*, 724.

(9) Im, J. H.; Ko, N.; Yang, S. J.; Park, H. J.; Kim, J.; Park, C. R. *New J. Chem.* **2014,** *38*, 2752.

(10) Vaesen, S.; Guillerm, V.; Yang, Q.; Wiersum, A. D.; Marszalek, B.; Gil, B.; Vimont, A.; Daturi, M.; Devic, T.; Llewellyn, P. L.; Serre, C.; Maurin, G.; De Weireld, G. *Chem Commun (Camb)* **2013,** *49*, 10082.

(11) Liu, D. D.; Dai, F. N.; Li, X. H.; Liang, J. L.; Liu, Y. Q.; Liu, C. G. *Rsc Advances* **2015,** *5*, 15182.

(12) Im, J. H.; Kang, E.; Yang, S. J.; Park, H. J.; Kim, J.; Park, C. R. *Bull. Korean Chem. Soc.* **2014,** *35*, 2477.

(13) Wang, Z.; Li, X.; Xu, H.; Yang, Y.; Cui, Y.; Pan, H.; Wang, Z.; Chen, B.; Qian, G. *Journal of Materials Chemistry A* **2014,** *2*, 12571.

429302 Encapsulation of Gold Nanoparticles in Uio-66 for CO Oxidation

Abstract:

Gold nanoparticles (AuNPs) are incorporated in the zirconium-based metal-organic framework, UiO-66, via encapsulation, or the formation of crystals around pre-formed nanoparticles. Encapsulation is advantageous over conventional impregnation techniques because (1) the nanoparticles can be designed for specific target applications, (2) the spatial location can be controlled, (3) there is limited pore blockage, and (4) the complete MOF database can be used as supports. The greatest limitation of encapsulation is nanoparticle stability; they cannot react adversely with any of the precursors, intermediates, or products. The conventional UiO-66 synthesis generates hydrochloric acid during the reaction which often reacts with or dissolves nanoparticles, including AuNPs. Therefore, the HCl-sensitive AuNPs require an alternative UiO-66 synthesis that does not generate HCl. The use of zirconium propoxide (Zr(O*n*Pr)) in place of zirconium chloride ($ZrCl_4$) leads to an alternative synthesis route for producing high-quality crystals of UiO-66 with no generation of HCl. This new method enables the inclusion of HCl-sensitive AuNPs into the mother solution for encapsulation by UiO-66, yielding a composite with 8nm AuNPs randomly distributed throughout high-quality UiO-66 crystals. The resulting Au@UiO-66 exhibits enhanced catalytic activity for carbon monoxide oxidation compared to the parent material. It also demonstrates improved catalytic activity over physical mixtures of AuNPs and UiO-66, suggesting a synergistic effect that developed during encapsulation. This alternative synthesis lays the ground-work for the encapsulation of a broad base of HCl-sensitive nanoparticles in chemically, mechanically, and thermally stable UiO-66 for designed applications.

429895 Rapid Room-Temperature Synthesis of HKUST-1 Powder, Patterns and Coatings Using ZnO

Abstract:

Conventional solvothermal synthesis to produce MOFs typically involves harsh conditions and consumes hours to days for a batch process. Consequently, these disadvantages hinder MOF industrial implementation and expansion of potential MOF applications. Rapid synthesis methods at mild conditions are therefore highly desired for MOFs.

We find ZnO can accelerate the synthesis of HKUST-1 ($Cu_3(BTC)_2$) at room temperature. HKUST-1 space-time-yield reaches up to 2.9×10^4 $kg{\cdot}m^{-3}{\cdot}d^{-1}$, which is over 10× higher than previous reports. Powder XRD and BET analysis confirm the formation of crystalline HKUST-1 with surface area of 1874±73 m^2/g. We investigated the reaction mechanism using STEM-EDX, XRD and FTIR, and the results suggest hydroxyl double salt ($(Zn,Cu)(OH)_3NO_3$) is an important intermediate to drive the rapid formation of HKUST-1. ICP-OES shows the concentration of Zn residue in the MOF product is generally less than 0.70 wt%, and that n(Zn):n(Cu) in the filtrate collected after the rapid synthesis is larger than the initial molar ratio in the mixed reactants. These results all indicate ZnO is consumed and converted into soluble Zn-containing species during the rapid synthesis.

Using the same synthesis strategy, we deposited ZnO thin films via atomic layer deposition (ALD) onto polymer spheres, silicon wafers and fibrous substrates for growing HKUST-1 coatings. MOF patterns were also obtained on pre-patterned ZnO surfaces. We will discuss how MOF-functionalized fibrous materials can be used for hazardous gas mitigation. This rapid room-temperature approach is also promising for applications such as membrane separation and chemical sensing.

432797 The Evaluation of the Temperature Dependence of Gate-Opening Phenomena in ZIF-8

Abstract:

ZIF materials have attracted significant attention due to their exceptional chemical and thermal stability, as well as the documented flexibility of their pore apertures. While it's been made clear that some degree of linker flexibility plays a major role in understanding the diffusion of certain species into these frameworks, the gate-opening phenomena exhibited in many isotherms has not been well studied. Specifically, the understanding of how framework deformation occurs as a function of temperature and adsorbate is not yet well understood in ZIFs. Many adsorbate/adsorbent pairs exhibit a sharp increase in uptake capacity at certain pressures, generally attributed to the structural deformation of pore opening. It is clear that this material property could have significant effects on its adsorptive behavior, especially during operation of a packed bed column. Investigations of various adsorbate isotherms will help elucidate the nature of gate-opening phenomena and its potential effect on future adsorptive processes.

In this work, the onset of gate-opening behavior in ZIF-8 is evaluated through a detailed gas adsorption study. Increases in temperature show a shift in the gate-opening pressure of the adsorbent, in some instances disappearing at higher temperatures. The adsorbate pressures and loadings at gate-opening are evaluated as a function of adsorbate, and the disappearance of the phenomena is related to material properties. Various gas isotherms are presented in a wide temperature range to investigate the thermodynamics of adsorption and structural deformation of ZIF-8. The fundamental understanding of the conditions under which this material deformation occurs is important to the successful application of ZIFs in industrial processes.

432343 Particle Size Analysis of the Sonochemically Synthesized HKUST-1: Towards Scale-up to High Throughput Synthesis of Metal Organic Frameworks

Abstract:

Metal organic framework (MOF) crystals have shown promise in many different applications such as separations, adsorption, and catalysis due in part to their highly tuneable functional groups. The initial particle size of crystals plays a vital role in overall performance for these applications, for example, emphasis on seed crystal size in gas separations has been is highly studied for growing membranes, and in catalysis many reactions are shown to occur at the crystal surface since diffusion into the pores is a rate-limiting step. High throughput methods to produce small MOF crystals of nearly uniform size are a necessary step for any of these processes to become commercially feasible. Sonochemical synthesis is a potential technique that may meet each of these demands due to the relatively small particle sizes, relatively high crystallinity, reaction speed. Thus, a study has been performed to investigate the effects that amplitude, solvent concentrations, time, and sonicator tip size have on the particle size of the metal organic framework HKUST-1 during sonochemical synthesis.

427238 Polymer Threaded MOF As an Effective Ion Exchange Material

Abstract:

We extend the concept of "*in situ* polymerization in MOF" from Kitigawa and co-workers [1], to synthesize ion-exchange polymers into MOFs . We report a porous metal−organic framework composite with flexible anion-exchange polymers threaded within the host cavity.[2] Polyvinyl benzyl trimethylammonium hydroxide (PVBTAH) threaded in ZIF-8 (PVBTAH~ZIF-8) is synthesized in steps of chloro-monomer impregnation, *in situ* polymerization, amination, and alkaline ion exchange. The synthesized non-cross-linked PVBTAH~ZIF-8 material exhibits superior ion-exchange kinetics compared to conventional ion-exchange resins. The open porous structure of MOF allows the physically trapped polymers to contact solvent freely and exchange ions efficiently, as oppose to the conventional ion-exchange resins in which the binding sites are hidden within the polymeric beads, preventing full utilization of ion exchange sites. It has good potential to be applied as a hybrid polymeric electrolyte in advanced fuel cells and flow batteries which demands high ionic selectivity.

[1] T.Uemura, K.Kitagawa, S. Horike, T. Kawamura, S. Kitagawa, M. Mizuno, and K. Endo, "Radical polymerisation of styrene in porous coordination polymers", *Chem. Commun.*, 48 (2005) 5968-70.
[2] L. Gao, C.Y. V. Li, K.Y. Chan, and Z.N. Chen, "Metal-Organic Framework Threaded with Aminated Polymer Formed in Situ for Fast and Reversible Ion Exchange", *J. Am. Chem. Soc.*, 136 (2014) 7209-7212.

Highly Stable Tunable Rare Earth fcu-MOF Platform. Access to Adsorption Kinetics and Molecular Exclusion Driven Gas/Vapor Separations via Pore Size Contraction

Youssef Belmabkhout, Dong-u Xue, Ayalew Assen, Karim Adil, Prashant Bhatt and Mohamed Eddaoudi

Functional Materials Design, Discovery and Development (FMD³)
King Abdullah University of Science and Technology KAUST (KSA)

AIChE Meeting Salt Lake City, Nov, 10th 2015

FMD³ Functional Materials Design, Discovery and Development
King Abdullah University of Science and Technology KAUST

Motivation:
Separation of molecules with small difference in physical/chemical properties

≈ 0.38 nm

- Ethylene/Ethane
- Ethane/propane
- Propane/Butane
- Propylene/propane
- Butane/iso-butane
- Pentane/iso-pentane

≈ 0.5 nm

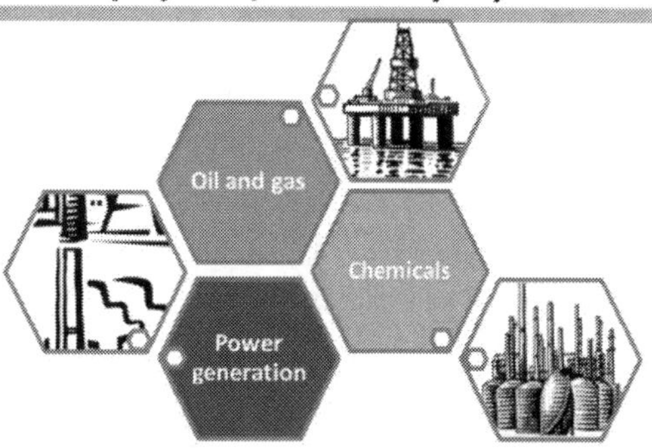

Huge potential - a lot of problems

Breakdown of the use of distillation energy (Total: 2400 TByu/yr (53%))

Feed	Typical Components Light/Heavy Key	Distillation Energy Consumption
Petroleum	gasoline/naphta	26%
Crude oil	Light naphta/heavy naphta/light distillate	22%
Liquefied petroleum gas (LPG)	Ethane/propane/butane	11%
Olefins	Ethylene/ethane, propylene/propane	6%
Miscellaneous hydrocarbons	Cumene/phenol, acetone/acrylonitrile	5%
Water-oxygenated hydrocarbons	Methanol/water, water/acetic acid	5%
Aromatics	Ethylbenzene/styrene, benzene/toluene	4%
Water-inorganics	Ammonia/water	3%
Air	Nitrogen/oxygen	1%
Water-hydrocarbons	p-xylene/water	1%
Other		16%

Relative energy use by various separation technologies

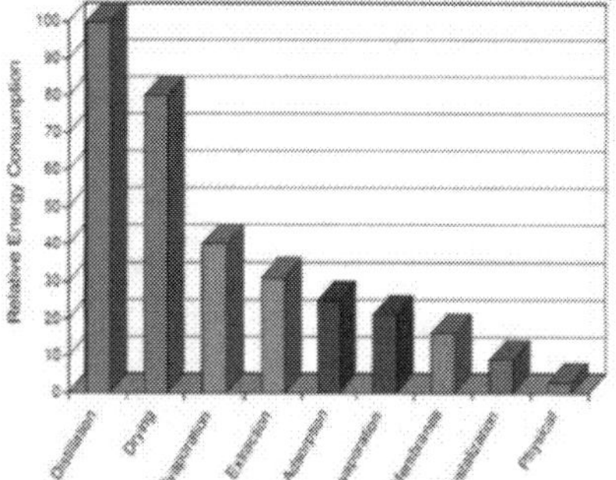

Source: US Department of Energy Report. Materials for separation technology: energy and emission reduction opportunities. http://www1.eere.energy.gov/manufacturing/industries_technologies/imf/pdfs/separationsreport.pdf; 2004. Humphrey and Keller, Separation Process Technology,

Intrinsic material properties vs. process performance relationship

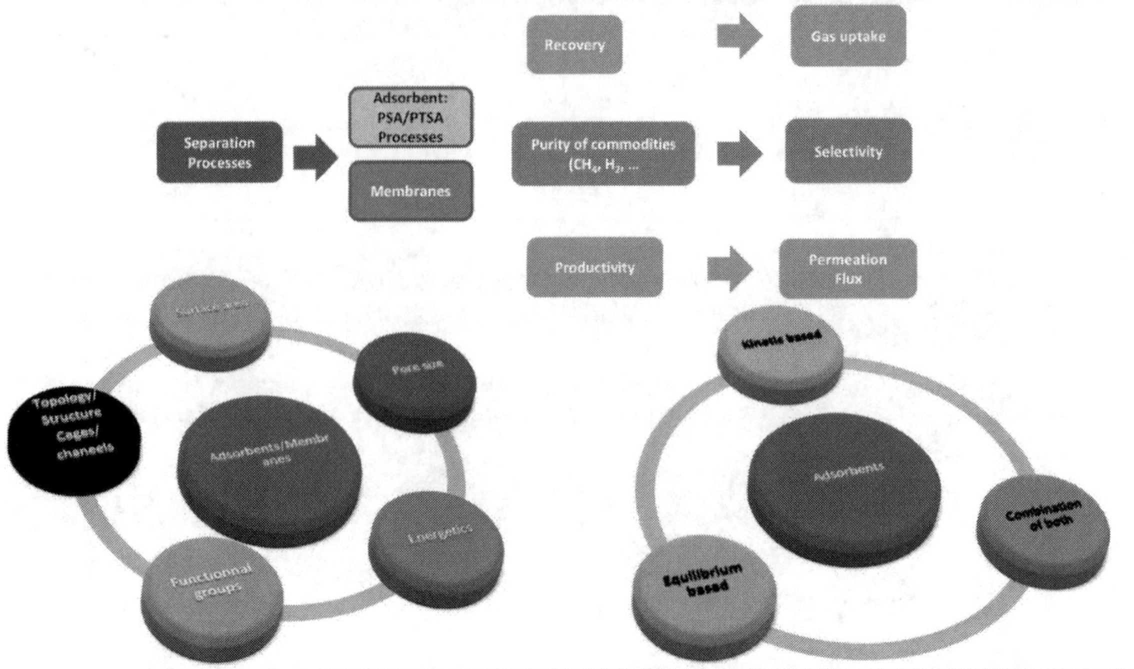

FMD³ Functional Materials Design, Discovery and Development
King Abdullah University of Science and Technology KAUST

Isoreticular MOFs

IRMOFs

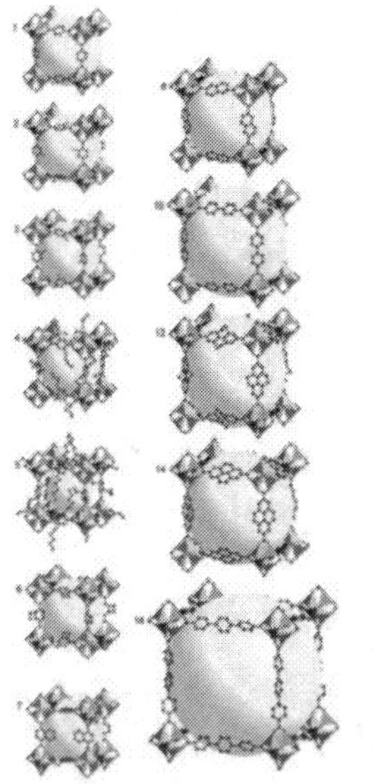

ZIFs

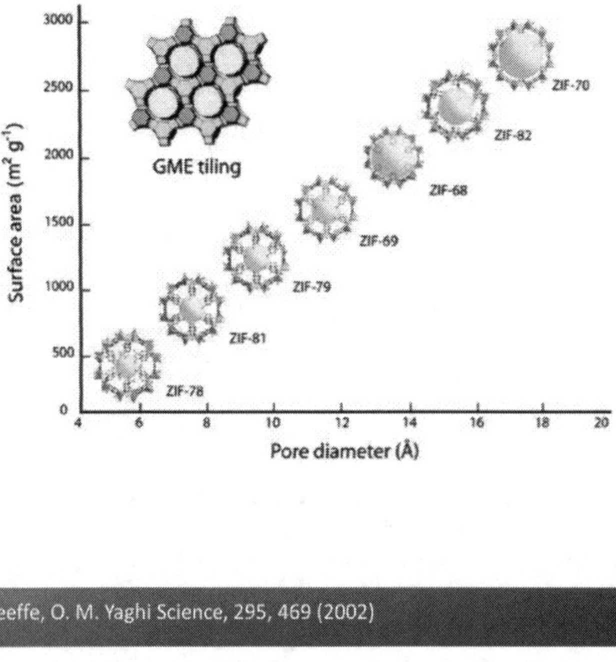

M. Eddaoudi, J. Kim. D. Vodak, J. Wachter, M. O'Keeffe, O. M. Yaghi Science, 295, 469 (2002)

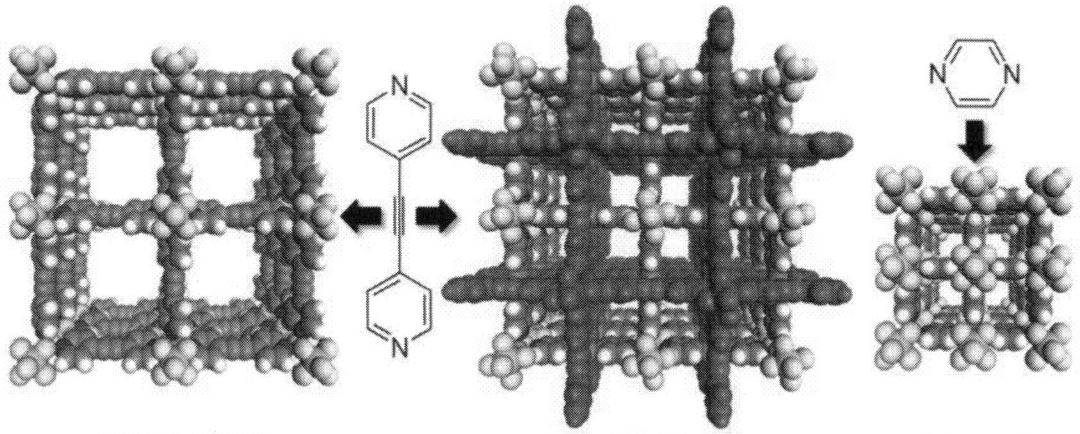

SIFSIX-2-Cu
10.5 Å pore
3140 m^2/g (N_2)

SIFSIX-2-Cu-i
4.9 Å pore
735 m^2/g (N_2)

SIFSIX-3-Zn
3.9 Å pore
250 m^2/g (CO_2)

SIFSIX Platform:
CO_2 capture and separation

Strategies for pore size tuning

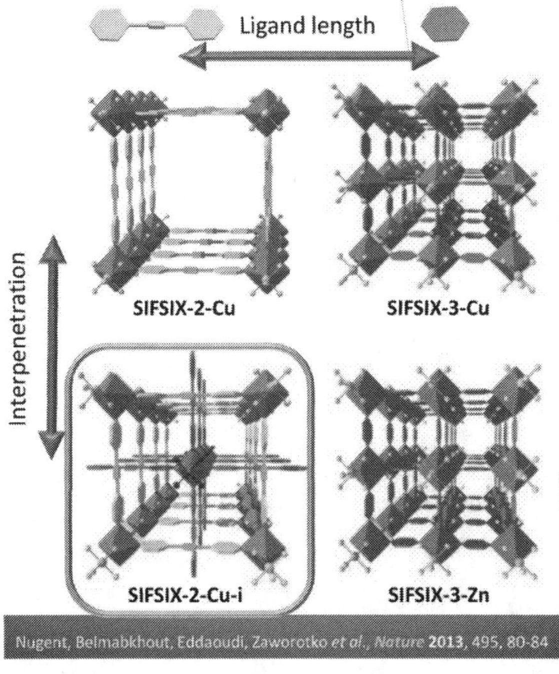

SIFSIX-3-Cu
Best material to date
Steady heats of adsorption
for CO_2 [illegible] confined space
no loss of affinity

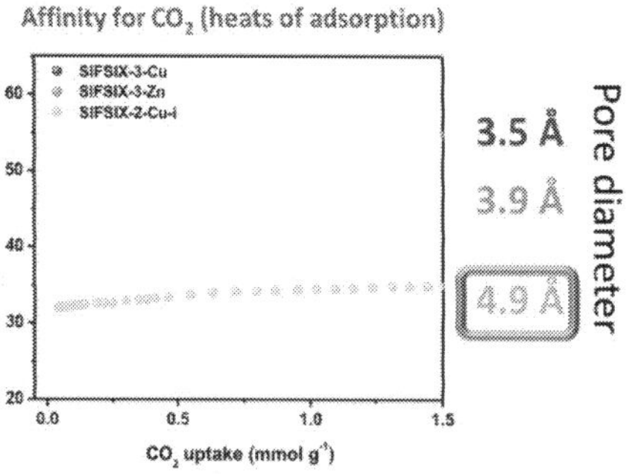

SIFSIX-3-Zn
Best material to date
for natural gas upgrading
Enhanced charge density by
pore size contraction

JACS
JOURNAL OF THE AMERICAN CHEMICAL SOCIETY
Article
pubs.acs.org/JACS

Tunable Rare-Earth fcu-MOFs: A Platform for Systematic Enhancement of CO_2 Adsorption Energetics and Uptake

Dong-Xu Xue,[†] Amy J. Cairns,[†] Youssef Belmabkhout,[†] Lukasz Wojtas,[‡] Yunling Liu,[†] Mohamed H. Alkordi,[†] and Mohamed Eddaoudi*[,†]

[†]Functional Materials Design, Discovery & Development Research Group (FMD³), Advanced Membranes & Porous Materials Center, Division of Physical Sciences and Engineering, 4700 King Abdullah University of Science and Technology (KAUST), Thuwal 23955-6900, Kingdom of Saudi Arabia

[‡]Department of Chemistry, University of South Florida, 4202 East Fowler Avenue, Tampa, Florida 33620, United States

Supporting Information

ABSTRACT: A series of fcu-MOFs based on rare-earth (RE) metals and linear fluorinated/nonfluorinated, homo/hetero-functional ligands were targeted and synthesized. This particular fcu-MOF platform was selected because of its unique structural characteristics combined with the ability/potential to dictate and regulate its chemical properties (e.g., tuning of the electron-rich RE metal ions and high localized charge density, a property arising from the proximal positioning of polarizing tetrazolate moieties and fluoro-groups that decorate the exposed inner surfaces of the confined

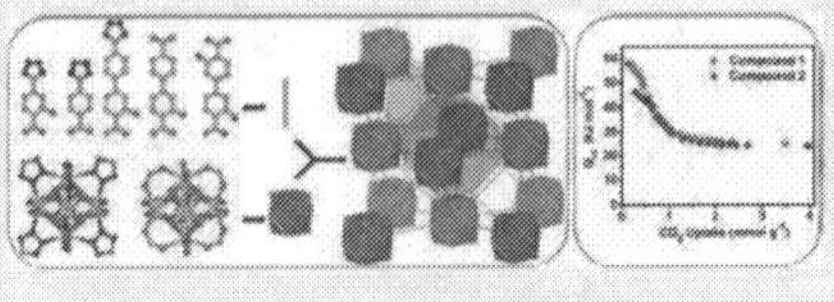

Tunable Rare-Earth fcu-MOFs Platform

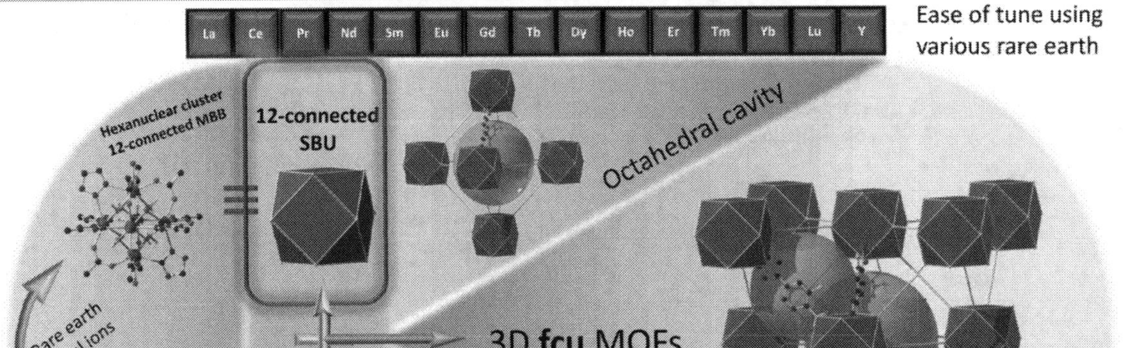

Xue, D. –X. et. al. *J. Am. Chem. Soc.* 2013, *135*, 7660-7667.

Tunable Rare-Earth fcu-MOF platform:

Equilibrium based adsorbents

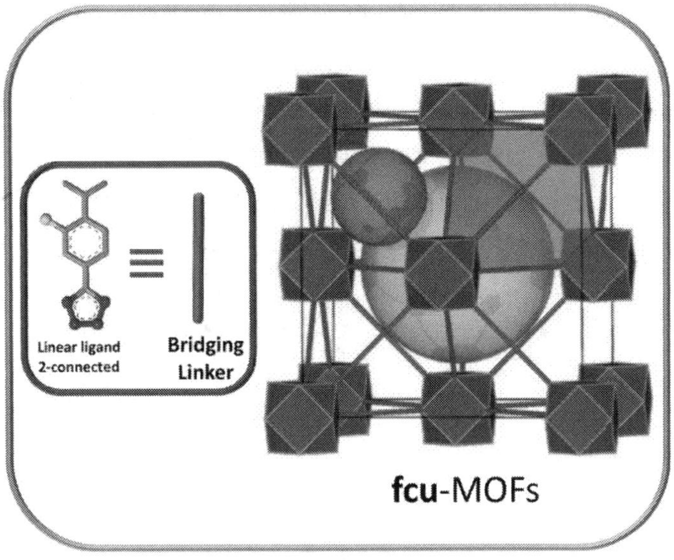

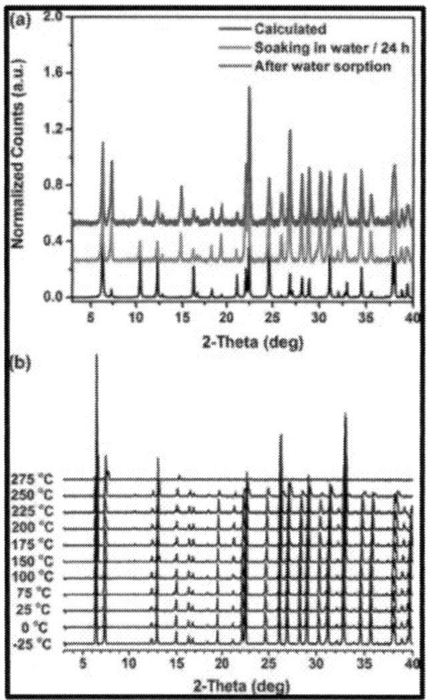

Chemical and Thermal stability

D.-X. Xue, A.J. Cairns, Y. Belmabkhout, L. Wojtas, Y. Liu, M.H. Alkordi, and M. Eddaoudi, *J. Am. Chem. Soc.* **2013**, 135, 7660-7667. Patent issued

Porosity and CO_2 Adsorptions

Compound (Metal)	1 (Tb)	2 (Y)	3 (Tb)	4 (Tb)	5 (Y)	6 (Tb)	7 (Tb)
Ligand	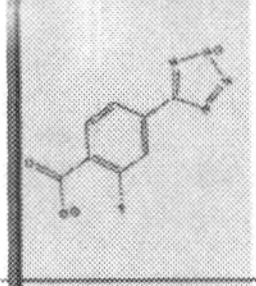	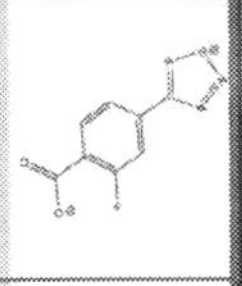	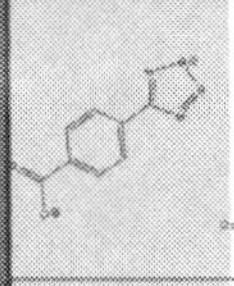		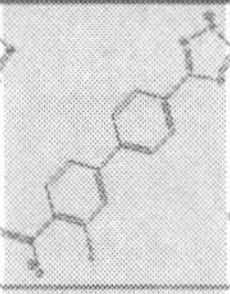	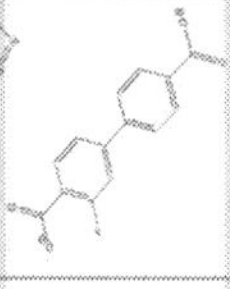	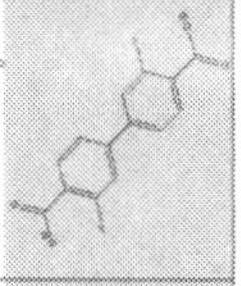
BET ($m^2 g^{-1}$)	1220	1310	904	2200	**2410**	1940	1854
P.V. ($cm^3 g^{-1}$)	0.51	0.56	0.39	0.86	0.94	0.78	0.72
CO_2 Uptake (mmol g^{-1})	7.53, 5.86 and 3.54	8.33, 6.46 and 4.12	5.96, 4.53 and 2.50	3.93, 2.75 and 1.64	4.01, 2.81 and 1.59	3.50, 2.40 and 1.37	3.91, 2.63 and 1.36
Q_{st} for CO_2 (kJ mol^{-1})	**58.1**-25.0	46.1-24.0	46.6-23.8	36.7-20.3	27.2-19.5	39.1-18.5	41.6-20.7

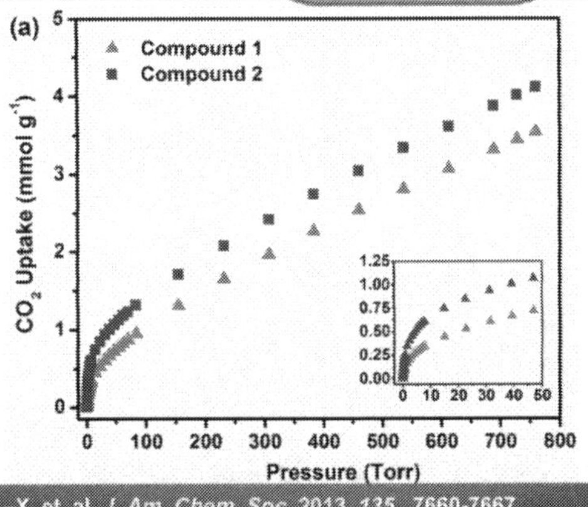

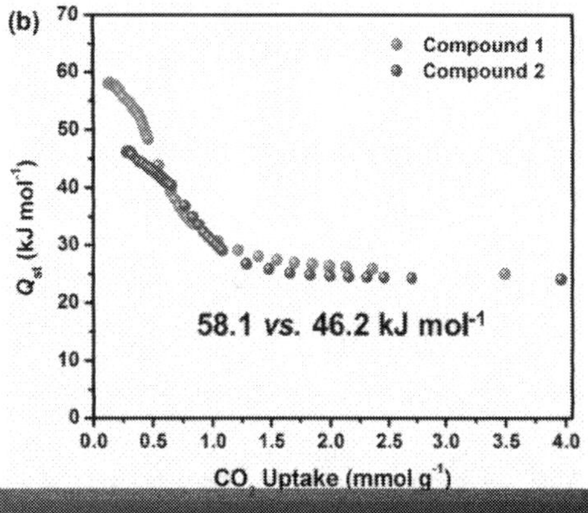

Xue, D. –X. et. al. *J. Am. Chem. Soc.* 2013, *135*, 7660-7667.

Traces and Low CO_2 concentration removal. Column breakthrough

Conical pockets decorated with –F, -OH moieties and tetrazolate groups (proximal to Oms)

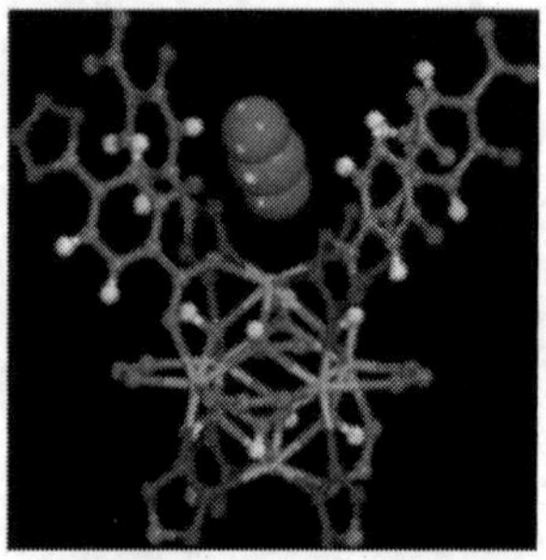

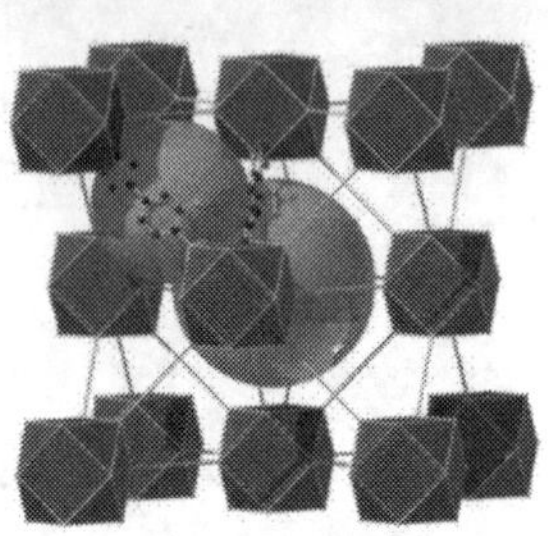

Column breakthrough tests using CO_2/N_2: 0.1/99.9% mixture

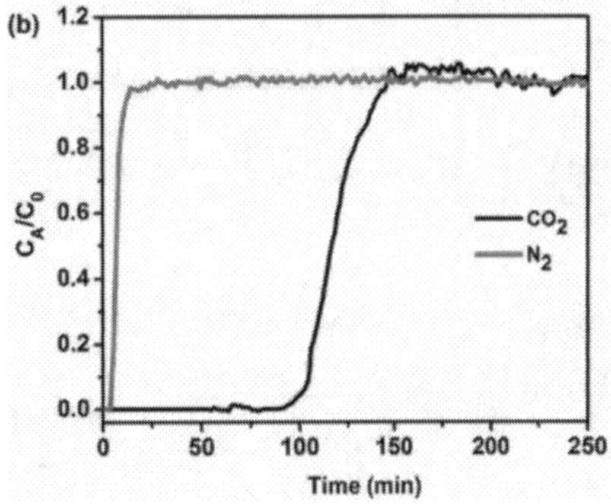

CO_2/N_2 (0.1/99.9%): ca. 1051

Xue, D. –X. et. al. *J. Am. Chem. Soc.* 2013, *135*, 7660-7667. 11

FMD³ Functional Materials Design, Discovery and Development
King Abdullah University of Science and Technology KAUST

JACS JOURNAL OF THE AMERICAN CHEMICAL SOCIETY
Article
pubs.acs.org/JACS

Tunable Rare Earth fcu-MOF Platform: Access to Adsorption Kinetics Driven Gas/Vapor Separations via Pore Size Contraction

Dong-Xu Xue, Youssef Belmabkhout, Osama Shekhah, Hao Jiang, Karim Adil, Amy J. Cairns, and Mohamed Eddaoudi*

Functional Materials Design, Discovery and Development Research Group (FMD³), Advanced Membranes and Porous Materials Center (AMPM), Division of Physical Sciences and Engineering, King Abdullah University of Science and Technology (KAUST), Thuwal 23955-6900, Kingdom of Saudi Arabia

Supporting Information

ABSTRACT: Reticular chemistry approach was successfully employed to deliberately construct new rare-earth (RE, i.e., Eu^{3+}, Tb^{3+}, and Y^{3+}) **fcu** metal–organic frameworks (MOFs) with restricted window apertures. Controlled and selective access to the resultant contracted **fcu**-MOF pores permits the achievement of the requisite sorbate cutoff, ideal for selective adsorption kinetics based separation and/or molecular sieving of gases and vapors. Predetermined reaction conditions that permitted the formation *in situ* of the 12-connected RE hexanuclear molecular building block (MBB) and the establishment of the first RE-**fcu**-MOF platform, especially in the presence of 2-fluorobenzoic acid (2-FBA) as a modulator and a structure directing agent, were used to synthesize isostructural RE-1,4-NDC-**fcu**-MOFs based on a relatively bulkier 2-connected bridging ligand, namely 1,4-naphthalenedicarboxylate (1,4-NDC). The subsequent RE-1,4-NDC-**fcu**-MOF structural features, contracted windows/pores and high concentration of open metal sites combined with exceptional hydrothermal and chemical stabilities, yielded notable gas/solvent separation properties, driven mostly by adsorption kinetics as exemplified in this work for *n*-butane/methane, butanol/methanol, and butanol/water pair systems.

12

Tunable Rare-Earth fcu-MOF platform:
Naphthalene analogue: Highly stable MOF

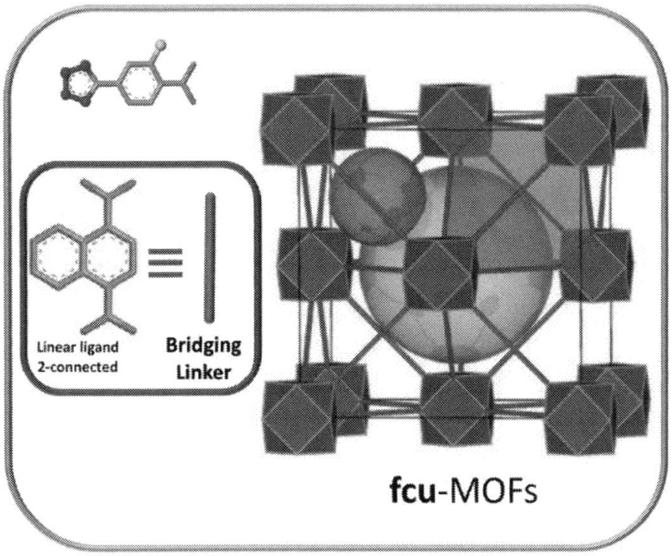

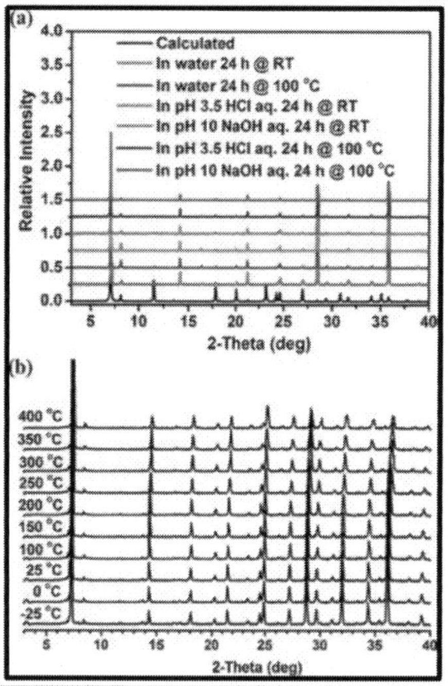

Chemical and Thermal stability

D.-X. Xue, Y. Belmabkhout, O. Shekhah, H. Jiang, K. Adil, A.J. Cairns, and M. Eddaoudi, *J. Am. Chem. Soc.* **2015**, 137, 5034-5040.

FMD[3] Functional Materials Design, Discovery and Development
King Abdullah University of Science and Technology KAUST

Tunable Rare-Earth fcu-MOF platform:
Naphthalene analogue. Single gas adsorption

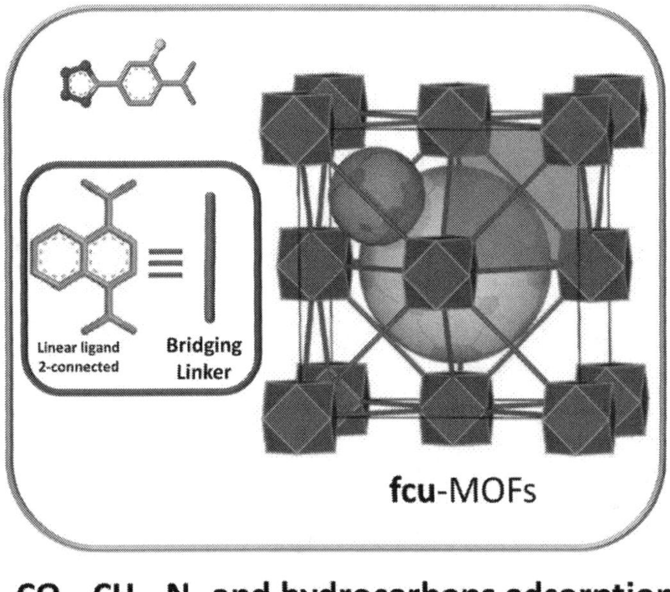

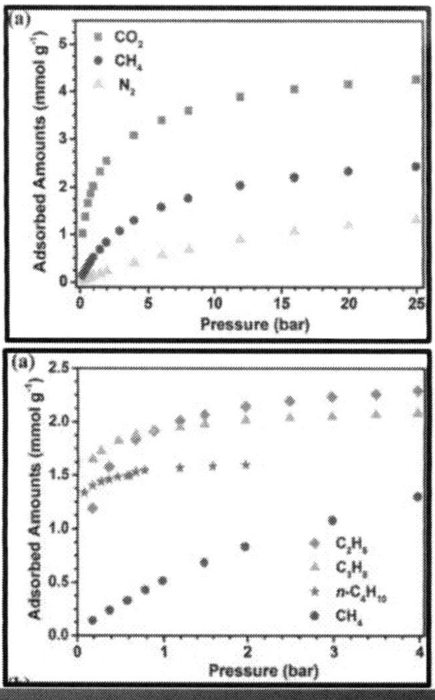

CO_2, CH_4, N_2 and hydrocarbons adsorption

D.-X. Xue, Y. Belmabkhout, O. Shekhah, H. Jiang, K. Adil, A.J. Cairns, and M. Eddaoudi, *J. Am. Chem. Soc.* **2015**, 137, 5034-5040

Tunable Rare-Earth fcu-MOF platform:

Equilibrium based separation dominate for C_3/C_1 and C_4/C_1

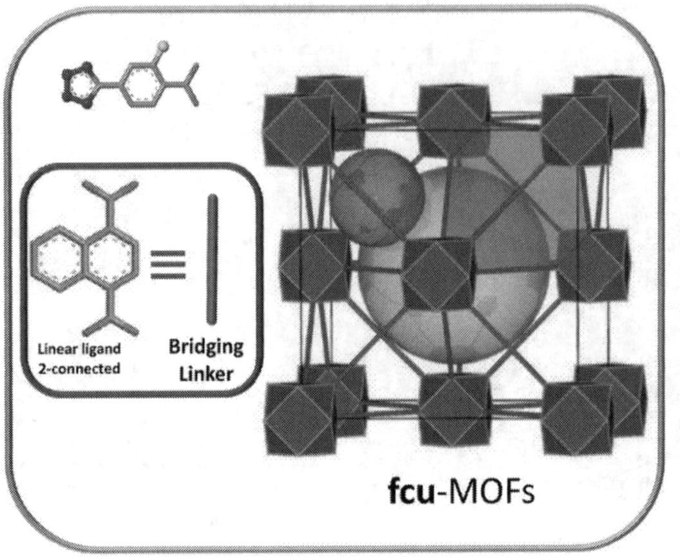

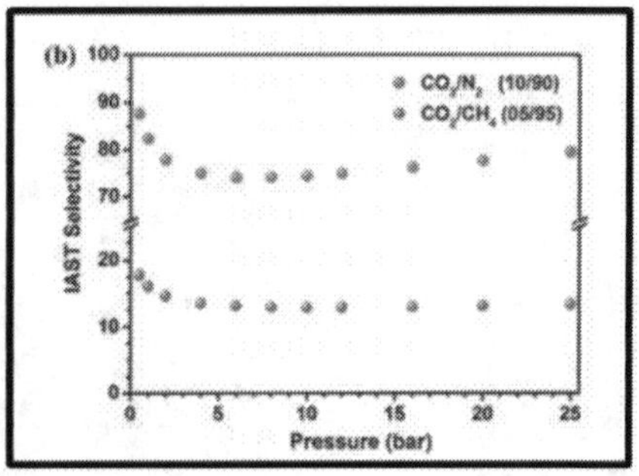

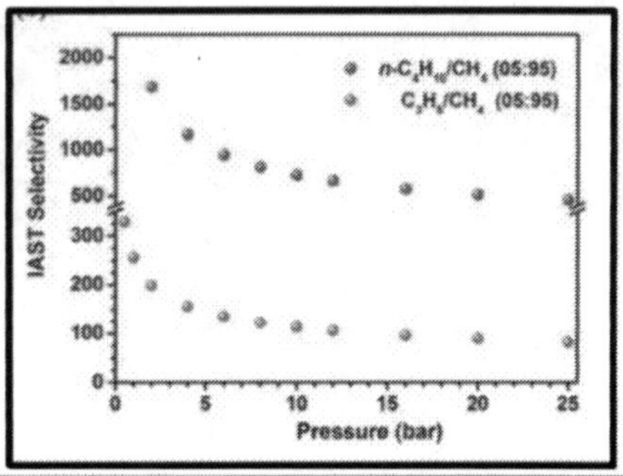

Potential for separation of hydrocarbons

D.-X. Xue, Y. Belmabkhout, O. Shekhah, H. Jiang, K. Adil, A.J. Cairns, and M. Eddaoudi , *J. Am. Chem. Soc.* **2015**, 137, 5034-5040

FMD[3] Functional Materials Design, Discovery and Development
King Abdullah University of Science and Technology KAUST

Tunable Rare-Earth fcu-MOF platform:

Kinetics based adsorbents

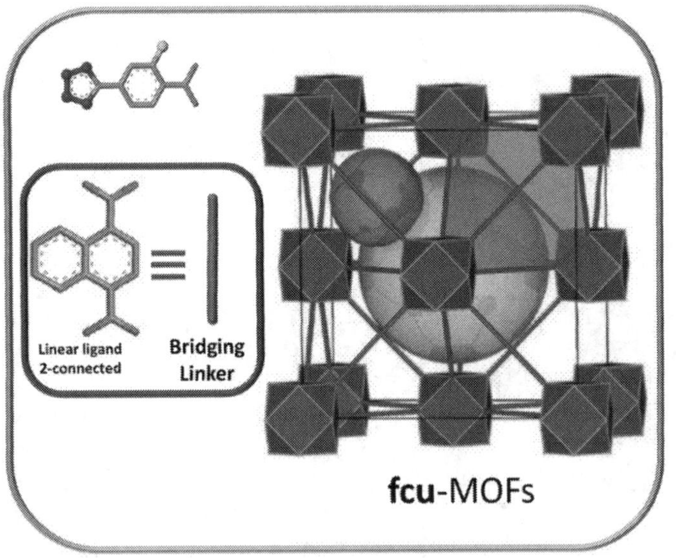

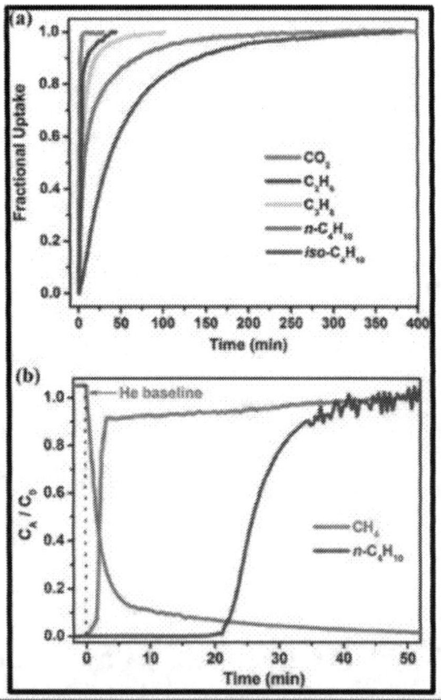

n-C_4H_{10}/ CH_4 mixed gas adsorption (2/98)

D.-X. Xue, Y. Belmabkhout, O. Shekhah, H. Jiang, K. Adil, A.J. Cairns, and M. Eddaoudi , *J. Am. Chem. Soc.* **2015**, 137, 5034-5040

FMD³ Functional Materials Design, Discovery and Development

King Abdullah University of Science and Technology KAUST

Tunable Rare-Earth fcu-MOF platform:

Full sieving of butanol-Fcu-MOF thin Film

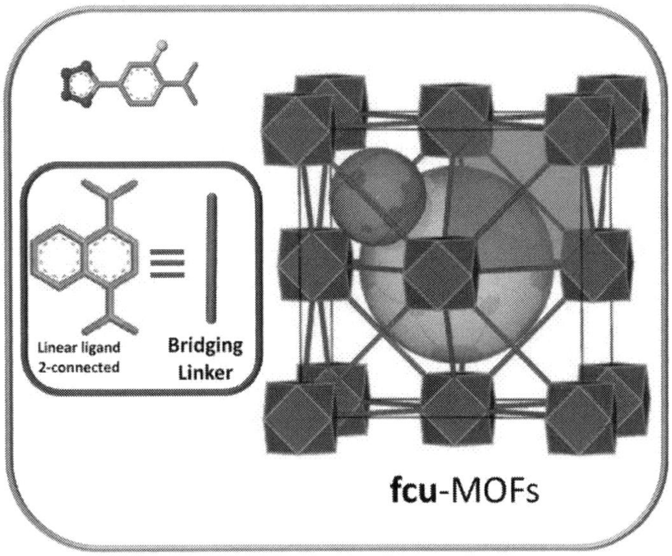

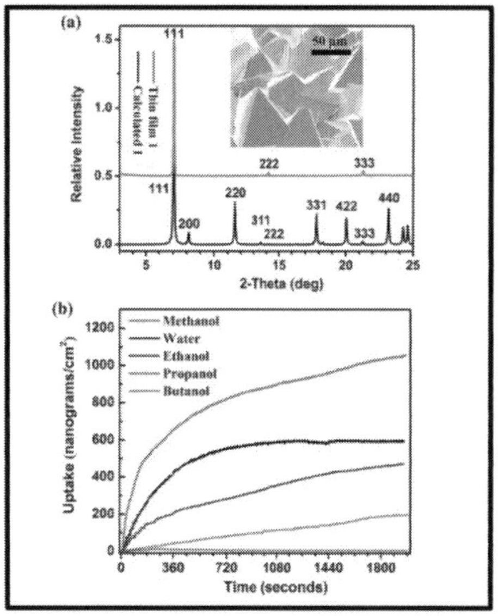

D.-X. Xue, Y. Belmabkhout, O. Shekhah, H. Jiang, K. Adil, A.J. Cairns, and M. Eddaoudi , *J. Am. Chem. Soc.* **2015**, 137, 5034-5040

FMD³ Functional Materials Design, Discovery and Development

King Abdullah University of Science and Technology KAUST

Molecular Exclusion of Branched Paraffins

Angewandte Chemie

Paraffin Separation

International Edition: DOI: 10.1002/anie.201506345
German Edition: DOI: 10.1002/ange.201506345

Ultra-Tuning of the Rare-Earth fcu-MOF Aperture Size for Selective Molecular Exclusion of Branched Paraffins

*Ayalew H. Assen, Youssef Belmabkhout, Karim Adil, Prashant M. Bhatt, Dong-Xu Xue, Hao Jiang, and Mohamed Eddaoudi**

Abstract: *Using isoreticular chemistry allows the design and construction of a new rare-earth metal (RE) fcu-MOF with a suitable aperture size for practical steric adsorptive separations. The judicious choice of a relatively short organic building block, namely fumarate, to bridge the 12-connected RE hexanuclear clusters has afforded the contraction of the well-defined RE-fcu-MOF triangular window aperture, the sole access to the two interconnected octahedral and tetrahedral cages. The newly constructed RE (Y^{3+} and Tb^{3+}) fcu-MOF analogues display unprecedented total exclusion of branched paraffins from normal paraffins. The resultant window aperture size of about 4.7 Å, regarded as a sorbate-size cut-off, enabled a complete sieving of branched paraffins from normal paraffins. The results are supported by collective single gas and mixed gas/vapor adsorption and calorimetric studies.*

chemical and physical properties (boiling point, vapor pressure, etc), the conventional low-temperature fractional distillation technique is the currently used separation technology for the separation of normal paraffins from branched paraffins; however, this technology is recognized to be energy intensive and not cost effective.[3]

Accordingly, deployment of alternative cost-effective and energy efficient separation approaches is of prime industrial importance. Principally, adsorptive separation based on porous materials as separation agents is regarded as a prospective economically viable technology and a suitable alternate to the energy-intensive distillation-based separation. Nevertheless, the practicality of the adsorptive separation technology and its associated efficiency are strongly dependent and directly correlated to the intrinsic properties of the

Advanced Membranes & Porous Materials Center (AMPM) 18

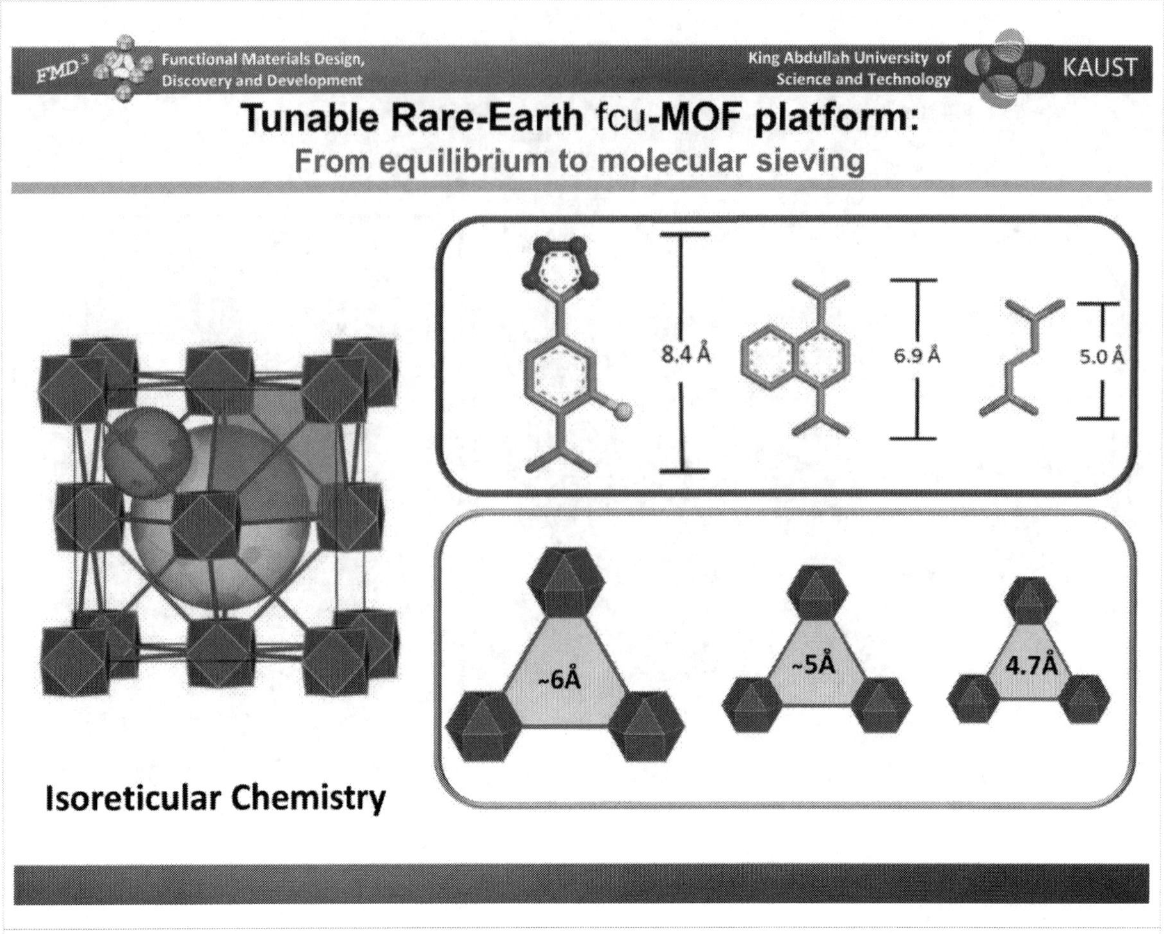
FMD³
Functional Materials Design, Discovery and Development
King Abdullah University of Science and Technology
KAUST
Tunable Rare-Earth fcu-MOF platform:
From equilibrium to molecular sieving
8.4 Å
6.9 Å
5.0 Å
~6Å
~5Å
4.7Å
Isoreticular Chemistry

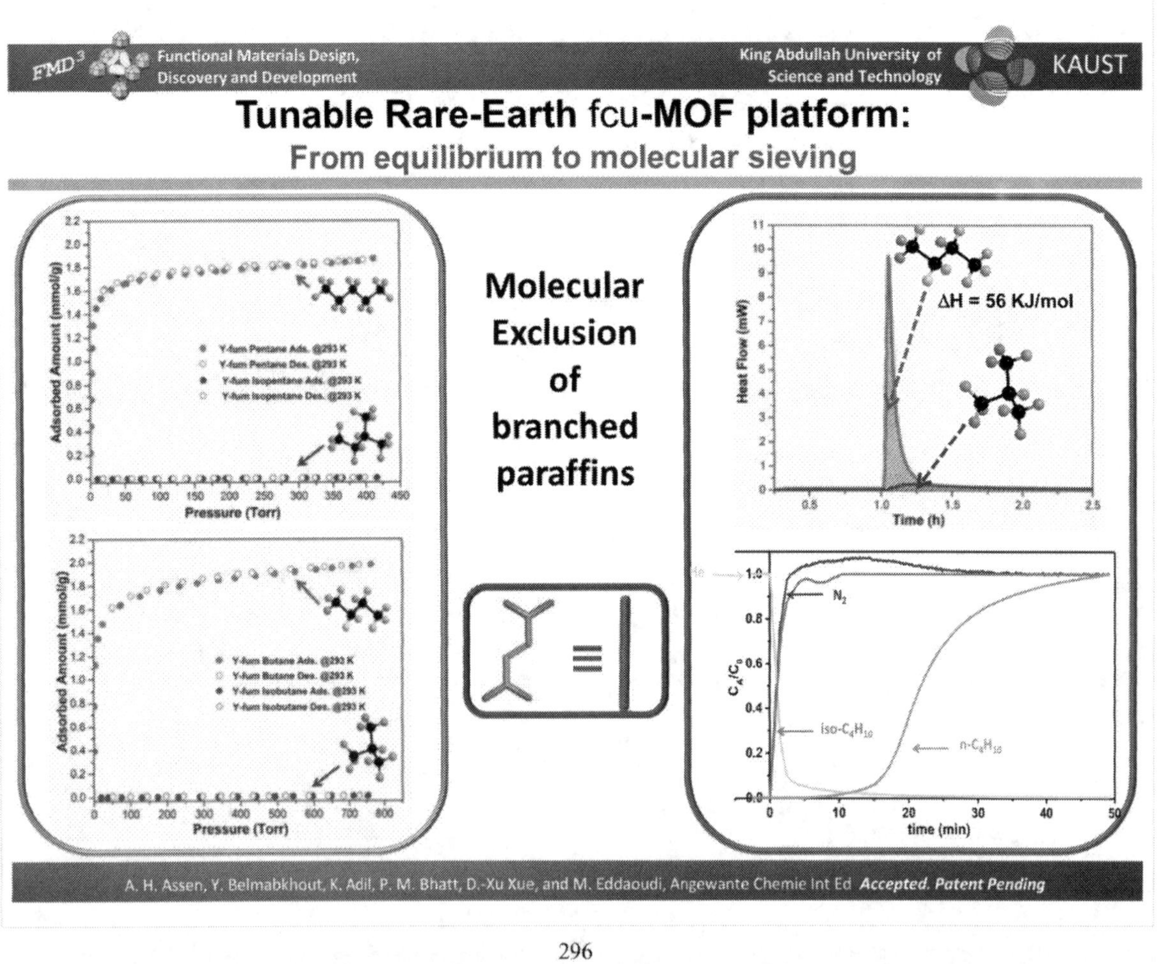
FMD³
Functional Materials Design, Discovery and Development
King Abdullah University of Science and Technology
KAUST
Tunable Rare-Earth fcu-MOF platform:
From equilibrium to molecular sieving
Adsorbed Amount (mmol/g)
Pressure (Torr)
Y-fum Pentane Ads. @293 K
Y-fum Pentane Des. @293 K
Y-fum Isopentane Ads. @293 K
Y-fum Isopentane Des. @293 K
Y-fum Butane Ads. @293 K
Y-fum Butane Des. @293 K
Y-fum Isobutane Ads. @293 K
Y-fum Isobutane Des. @293 K
Molecular Exclusion of branched paraffins
Heat Flow (mW)
Time (h)
ΔH = 56 KJ/mol
C_A/C_0
time (min)
N_2
iso-C_4H_{10}
n-C_4H_{10}
A. H. Assen, Y. Belmabkhout, K. Adil, P. M. Bhatt, D.-Xu Xue, and M. Eddaoudi, Angewante Chemie Int Ed *Accepted. Patent Pending*

FMD[3] Functional Materials Design, Discovery and Development

King Abdullah University of Science and Technology KAUST

Tunable Rare-Earth fcu-MOF platform:

From equilibrium to molecular sieving

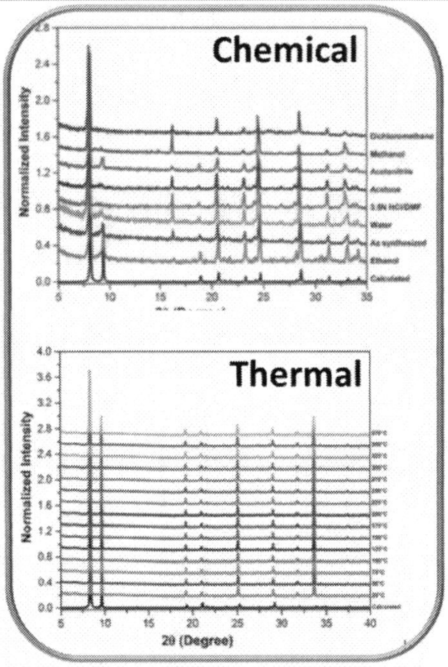

Stability

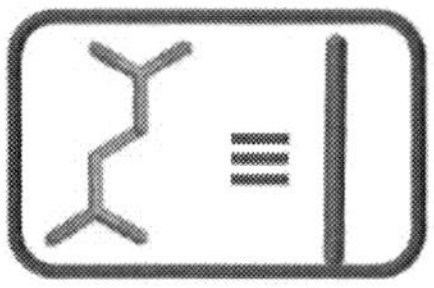

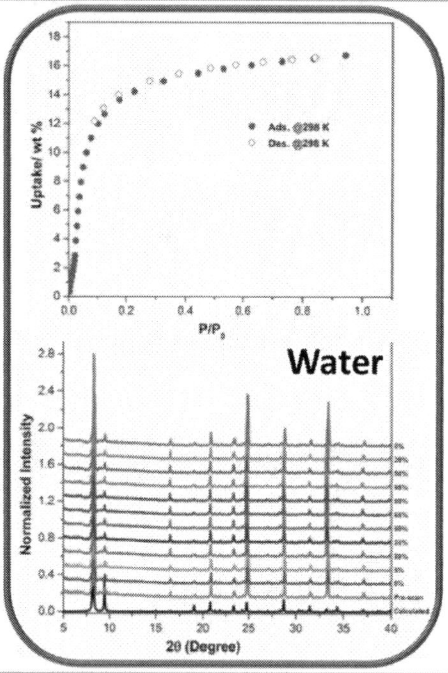

A. H. Assen, Y. Belmabkhout, K. Adil, P. M. Bhatt, D.-Xu Xue, and M. Eddaoudi, Angewante Chemie Int Ed *Accepted. Patent Pending*

FMD[3] Functional Materials Design, Discovery and Development

King Abdullah University of Science and Technology KAUST

Tunable Rare-Earth fcu-MOF platform:

Summary

fcu-MOF

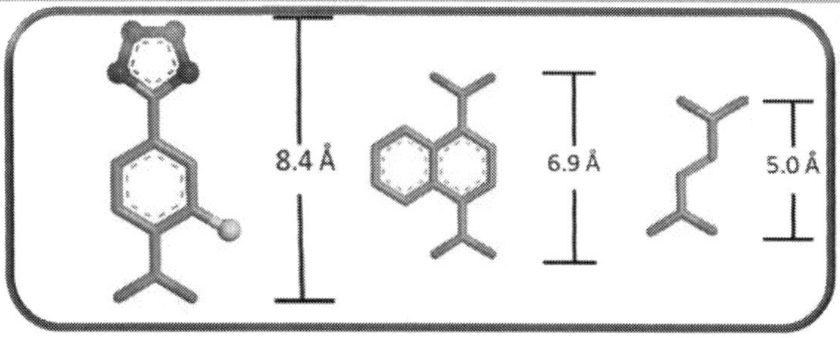

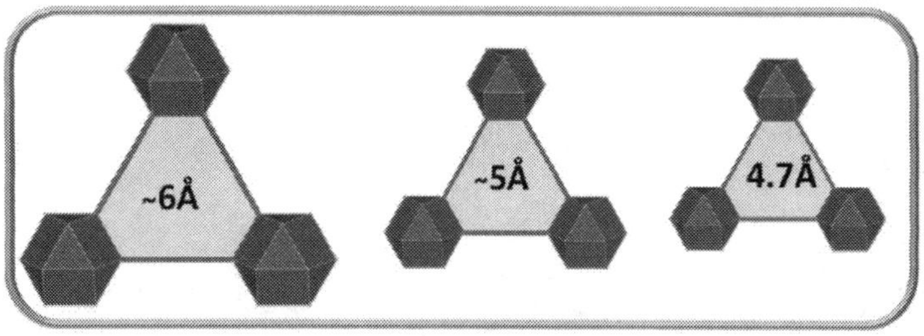

Equilibrium based adsorption

Kinetically driven separation

Molecular Exclusion

A. H. Assen, Y. Belmabkhout, K. Adil, P. M. Bhatt, D.-Xu Xue, and M. Eddaoudi, Angewante Chemie Int Ed *Accepted. Patent Pending*

FMD³ Functional Materials Design, Discovery and Development — King Abdullah University of Science and Technology KAUST

Nothing would get done without them

FMD³ members

Acknowledgment

23

FMD³ Functional Materials Design, Discovery and Development — King Abdullah University of Science and Technology KAUST

24

432103 Techno-Economic Feasibility of Biomass Hydrolysate Desalination By Membrane Capacitive Deionization

Abstract:

The bio-based industry is putting significant effort into research and development to replace refined sugars by much cheaper non-food renewable feed stocks for the production of bio-fuels and chemicals. Although this obviously increases sustainability, it also brings about a number of technological challenges due to the higher complexity of these secondary substrates. Biomass hydrolysates, for example, can contain high concentrations of sodium and potassium that inhibit fermentation and hence need to be reduced. The commonly used ion-exchange processes carry high operational costs and generate a waste stream through the use of chemicals for regeneration. Electrochemical pretreatment of biomass hydrolysates with membrane capacitive deionization or MCDI could provide a lower cost alternative with minimal waste generation.

This study considers the electrochemical treatment of biomass hydrolysates by membrane capacitive deionization (MCDI) as an novel chemical free, lower cost alternative in comparison to the commonly used ion-exchange processes. Model experiments with model solutions and a commercial bench-scale MCDI set-up indicated that none of the most abundant hydrolysate components (sugars, organic acids and furans) prohibited MCDI implementation for this application. Compared to water desalination performance was somewhat lowered by the competition for electro-sorption between the protons originating from organic acid dissociation and the cations. This effect was not observed during MCDI treatment of a real biomass hydrolysate sample. Instead, the achieved sodium (Na) and potassium (K) removal and energy usage were comparable to that for a model solution with equal conductivity and sugar concentration.

Based on the experimentally obtained desalination efficiencies, the economic viability of MCDI was investigated for a production capacity of 500 ton sugar/day and a target Na removal from 3 to 0.1 g/kg hydrolysate. Although capital costs were higher for MCDI than for IEX due to the expensive MCDI cells and power supplies, operating costs were lower because less water and chemicals are used and less wastewater is generated. Cost calculations for different initial feed concentrations indicated that IEX was only preferential over MCDI when the feed Na^+ concentration was below 0.4 g/kg hydrolysate. Then the higher chemical, water and wastewater treatment costs for IEX no longer outweighed the higher cost of MCDI cells compared to IEX resins.

In conclusion, this study clearly demonstrates the technical and economic potential of MCDI for process streams such as biomass hydrolysates, hereby considerably broadening its potential application field.

424162 Modeling Dynamic Gas Evolution Behavior in Hydrocarbons for Subsea Separator Applications

Abstract:

Offshore oil and gas production requires a minimal facilities footprint, especially regarding the equipment that is used for gas/liquid separation. As a result, compact subsea separators are becoming a more prevalent option in these production environments. Produced hydrocarbons will experience less residence time in these compact separators, thereby introducing the potential for gas carry under events that can damage downstream equipment. The gas evolution rate from the produced liquids is a key parameter that will dictate the ultimate performance of these separators. Surprisingly, the fundamental nature of gas evolution from liquid streams in these production scenarios is poorly understood, especially at high pressures in the presence of shear. This work seeks to address this gap by focusing on the numerical simulation of dynamic gas/liquid interaction and gas disengagement behavior from a liquid under flowing conditions at high pressures. The impact of various parameters on gas evolution will be explored in the context of numerical simulations supported by experimental data. Remarks on the potential applications of this model to aid in the design and development of subsea separators with a focus on high pressure applications will be made.

420511 CFD Investigations and Optimization of a Novel Dynamic Oil Water Separator

Abstract:

In the petroleum industry, economical separation of oil and water has been a major challenge. Typically produced water accounts for an average of 75% of extracted fluid; oil water separation totally costs $40B/yr. worldwide to the industry. For separation of oil and water, industry traditionally relies on conventional gravity based vessels that are bulky, heavy and expensive. More importantly, separation by only gravity suffers from low efficiency for oil water emulsion as extracted fluid.

In order to improve the split ratio and the separation efficiency of traditional API gravity separation, a dynamic separator with a helical impeller and draft tube (KDS) has been proposed by the Dow Chemical Company. In such device, the centrally-located impeller produces up-pumping flow inside cylindrical draft tube and downward flow between draft tube and container. The flow circulation inside container facilitates phase separation by increasing probability of oil droplet collision. In addition, with slow rotational speed, the shear created by helical impeller inside draft tube may further enhance droplet coalesce by deformation of oil droplet. The novel dynamic separator was found promising with about 8 times faster than gravity separation in preliminary experiments. There is still huge space for improvement with better designs.

Herein, Computational Fluid Dynamics analyses have been used to understand working principles as well as provide optimization for the preliminary design of the dynamic oil-water coalescer. Firstly, the flow pattern of fluid inside has been studied with single phase CFD simulations with SST transition turbulent model. Since a higher shear will lead to droplet breakup rather than coalesce, CFD investigations have been adopted in this paper to identify the critical shear rate and different zones for droplet breakup or coalesce respectively. With the objectives of enhancing droplet coalesce while suppressing breakup, an optimized design has been found by parametric study with different parameters, e.g., impeller twisting angle, draft tube diameter, impeller rotational speed. In addition, multi-phase simulation with Eulerian Model and population Balance Model has also been adopted to studies oil-water separation process and oil droplet diameter evolution. CFD results showed increasing twisting angle of impeller is helpful to develop a more homogeneous flow field inside draft tube and thus reduce high shear region for breakup. The optimized draft tube diameter has also been discussed for highest global circulate flow rate.

436124 Assessment of Polysulfone Membranes Embedded with Carbon Nanotubes and Polyester Non-Woven Fabric Support for the Treatment of Oil and Gas Produced Wastewater

Abstract:

D. N. B. Nkazi, Kwame O-A., Jean Mulopo, S. E. Iyuke,

School of Chemical and Metallurgical Engineering University of the Witwatersrand, PO Box 3, Johannesburg, Wits 2050, South Africa

Abstract

Keywords: Produced oil wastewater; Phase Inversion (PI) membranes; Carbon nanotubes; polyester non-woven fabric; Characterization; permeate flux, oil rejection

With industrial development, there is increase in the amount of oil used, but various technical and management developments lag behind other reasons that are not perfect and make a lot of oil into the water, forming pollution. The use of modified Polysulfone membranes for the treatment of wastewater produced in the oil and gas industry were studied. The micro-porous anisotropic membranes were prepared using the Phase Inversion (PI) method and embedded with carbon nanotubes (CNTs) and/or not supported on a polyester non-woven fabric. The produced membranes were characterized using the Scanning Electron Microscopy (SEM), atomic force microscopy (AFM), Fourier Transform Infrared Spectroscopy (FTIR), and tensile strength analysis.

The multi-walled carbon nanotubes MWCNTs/PSF membrane with 0.4 wt% MWCNTs has the roughest surface and the largest pore size. For membranes containing more than 0.4 wt% MWCNTs, the surfaces become smoother and the pore sizes of cross-section become smaller due to the to the increasing density of functionalized MWCNTs and van der Waals forces; which cause the steric hindrance between the functionalized MWCNTs to agglomerate inside the polymer matrix during the phase inversion. The pore sizes ranged from 1.598μm for the membranes with the fabric support to 0.191mm for the CNT imbedded membranes without a fabric support. The results showed that the functionalized CNTs interacted with the hydrophobic membranes to enhance its physical, chemical and mechanical properties. The addition of a fabric support decreased the pore sizes of the PI membranes. The membranes were tested for performance and it was shown that increasing pressure increased permeates flux and CNTs increased permeate flux while controlling fouling via pore blockage. The fabric influenced the flux decline by providing further resistance to the flow of permeates through the membrane. Permeates were collected and tested for oil concentrations. The results showed oil rejections ranging between 78% and 90% with the mixed matrix phase inversion membrane supported on a polyester non-woven fabric having the highest oil rejections.

The performance studies with addition of different fractions of MWCNTs on the polymeric membranes produced with different solvents (DMF, $CHCL_3$ and THF) show an appreciable 48.6 % increase in water flux with DMF as solvent. At values above 0.4 wt% MWCNTs the water flux decreased. The increase in permeability of the membrane was related with the hydrophilicity and adsorption effects of MWCNTs. However, increased viscosity of the casting solution at higher MWCNTs loading (>0.4wt %), retarded the exchange between solvent and non-solvent during phase inversion process, and smoother membrane surfaces and smaller pores appeared as a result. Hence, water flux was severely limited.

425520 Evaluating Techniques for Incorporating Metal-Organic Frameworks into Mixed Matrix Membranes for Organic Solvent Nanofiltration

Abstract:

Polymeric membranes have dominated the membrane market because of their low cost, ease of manufacture and mechanical flexibility, the latter of which allows for facile inclusion into industrial membrane modules. However, it is widely known that there is a trade-off between permeability and selectivity for traditional polymeric membranes. Moreover, there is still a need for more chemically and thermally stable membranes for organic solvent nanofiltration (OSN) applications. Inorganic membranes have such enhanced performance characteristics. Membranes based on metal-organic frameworks (MOFs) are particularly interesting because they have tuneable chemistries and well-defined pore sizes, which allow for customizable separations based on molecular sieving and adsorptive affinity. Though inorganic membranes offer significant improvements over polymeric membranes, they are costly to manufacture and difficult to consistently produce without selectivity-reducing defects. Mixed matrix membranes (MMMs) provide a promising alternative to pure polymeric and inorganic membranes. MMMs incorporate inorganic materials into a polymeric matrix. They are capable of achieving some of the superior performance characteristics of inorganic membranes while still maintaining the manufacturability and flexibility of polymeric membranes.

To date, our group has successfully synthesized MMMs based on MOFs for OSN using four distinct techniques. In the first technique, HKUST-1 was synthesized and incorporated into a cross-linked P84 membrane by mixing the MOF in the dope solution prior to phase inversion. While this is the most typical way of forming MMMs, it means that all permeation paths will go through both the polymer and MOF, which limits the influence of the MOF on the membrane performance. To create a potentially continuous MOF flow path, HKUST-1 was incorporated into an existing cross-linked P84 membrane via in situ growth (ISG) [1]. Furthermore, interfacial synthesis was also used to form a thin film of HKUST-1 on a cross-linked P84 support with the goal of enhancing flux relative to that of the ISG membrane [2]. Morphological and chemical composition are explored with FT-IR spectroscopy, SEM and EDX. This data is compared with flux and rejection data, evaluated using acetone-polystyrene solutions, to understand the underlying mechanisms behind membrane performances.

The final method of synthesis involved interfacial polymerization of a polyamide thin-film nanocomposite (TFN) membrane on a cross-linked P84 support using nano-scale ZIF-8, MIL-53(AL), NH2-MIL-53(Al) and MIL-101(Cr) [3]. The MOF's presence is inferred by a combination of contact angle measurements, FT-IR spectroscopy, SEM, EDX, XPS and TEM. TFN membranes show dramatically increased methanol and tetrahydrofuran permeance when compared to thin film composite controls, without sacrificing rejection.

SRP Advances in Packing Characterization

Micah Perry[1], Chao Wang[2] and Frank Seibert[3]
[1]Skyonic
[2]FRI
[3]UT-SRP

Distillation Honors: Dr. Bruce Eldridge
AIChE Fall Meeting
Session #283
Salt Lake City, Utah
November 10, 2015

Jim's research group, part of the Separations Research Program, in March 1985. Front row: Jimmy Humphrey, Antonio Rocha, Mike Prado, Rich Lahiere, Frank Seibert, Bruce Eldridge, Narut Ruchirote, and JRF. Back row: Paul Rathkamp, John Buchanan (visiting professor), José Bravo, Joan Schork, Sudhir Joshi, Chris Martin, and Gerald McGlammery. The group is in front of the Schoch Laboratories (Ch. E. Building).

5

Overview

Background

a_e, k_L and k_G Absorption/Stripping Data

Mass Transfer Modelling

Application to Distillation

James R. Fair Pilot Plant
SRP

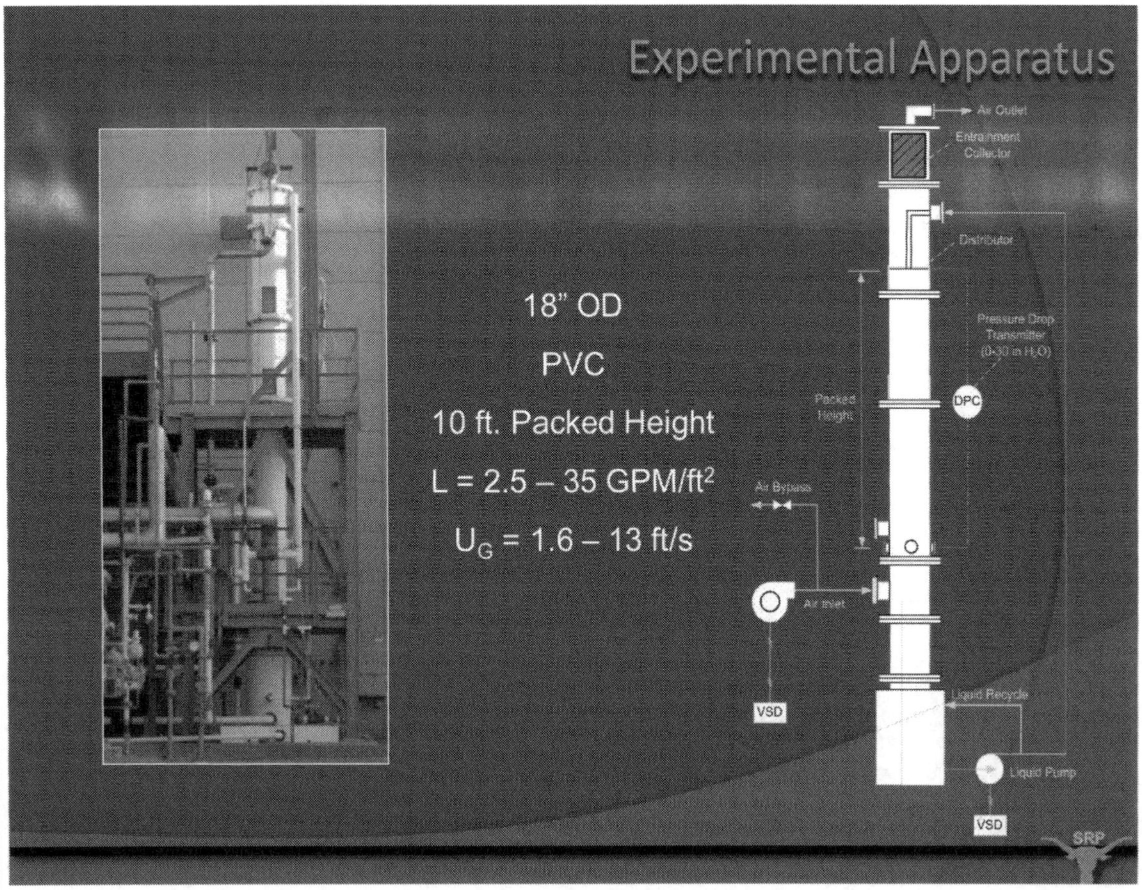
Experimental Apparatus
18" OD
PVC
10 ft. Packed Height
L = 2.5 – 35 GPM/ft²
U_G = 1.6 – 13 ft/s
Air Outlet
Entrainment Collector
Distributor
Pressure Drop Transmitter (0-30 in H2O)
DPC
Packed Height
Air Bypass
Air Inlet
VSD
Liquid Recycle
Liquid Pump
VSD
SRP

Volumetric Mass Transfer Coefficient

Ka

K ?   a ?

Example Packings

Structured Packings

Structured Packings	a_P (m^2/m^3)	ε	Corrugation Angle
RSP250	250	98%	"45°"
MP2X	205	99%	60°
GTPAK™350Z	350	91%	70°
MP250X	250	95%	60°
MP250Y	250	95%	45°
A350Y	350	95%	45°
B350X	350	95%	60°
RSP200X	200	98%	60°
GTPAK™500Y	500	90%	45°
GTPAK™350Y	350	91%	45°
GT-OPTIM PAK™ 250Y	250	98%	45°
MellaGrid 64Y	64	99%	45°

SRP

Effective Gas-Liquid Contact Area

- Absorption of Ambient CO_2 into 0.1 N NaOH
- Pseudo first order reaction
- Mass transfer resistance mainly in reaction film

$$a_e = \frac{u_G \ln(\frac{y_{CO_2in}}{y_{CO_2out}})}{ZK_G RT} \approx \frac{u_G \ln(\frac{y_{CO_2in}}{y_{CO_2out}})}{Zk_G^{'} RT}$$

$$k_g^{'} = \frac{\sqrt{k_{OH^-}[OH^-]D_{CO2,L}}}{H_{CO2}}$$

SRP 12

Contact Area Example

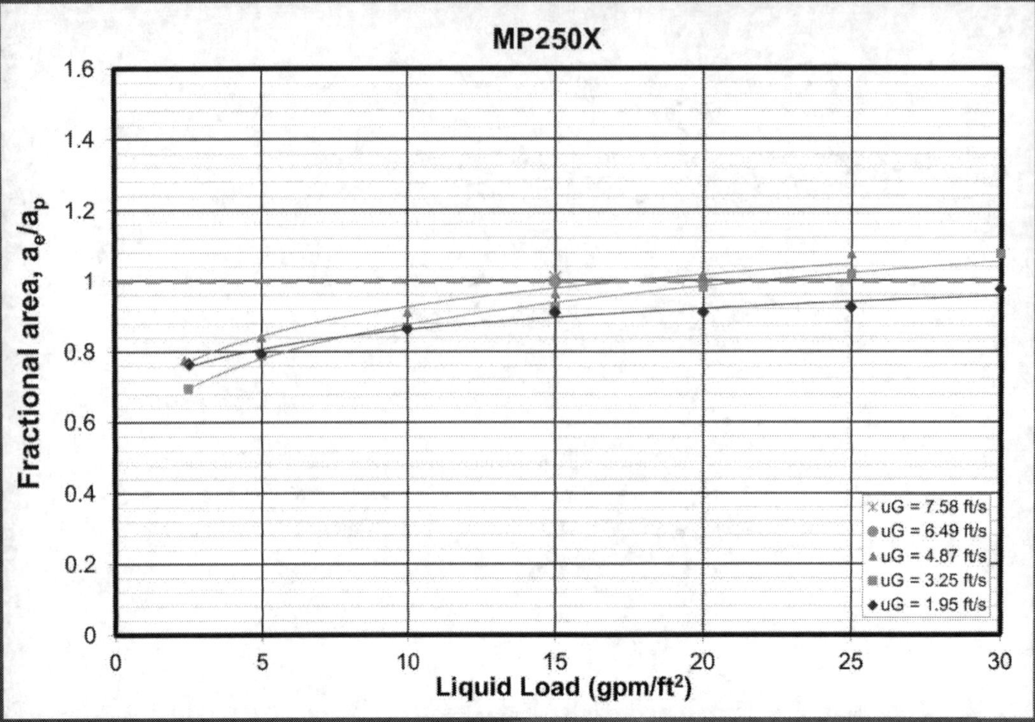

SRP

Effective Area Model

- a_e is a function of liquid flow rate per wetted perimeter Q/L_P, and surface tension σ (Tsai, 2010, modified by Wang 2014)

$$\frac{a_e}{a_P} = 1.41\left[\left(\frac{\rho_L}{\sigma}\right) g^{1/3} \left(\frac{U_L}{a_P}\right)^{4/3}\right]^{0.116}$$

- Q: Volumetric liquid flow rate, m^3/s
- Q/L_P can be expressed by U_L/a_P
- ρ_L: kg/m^3; σ: N/m

SRP

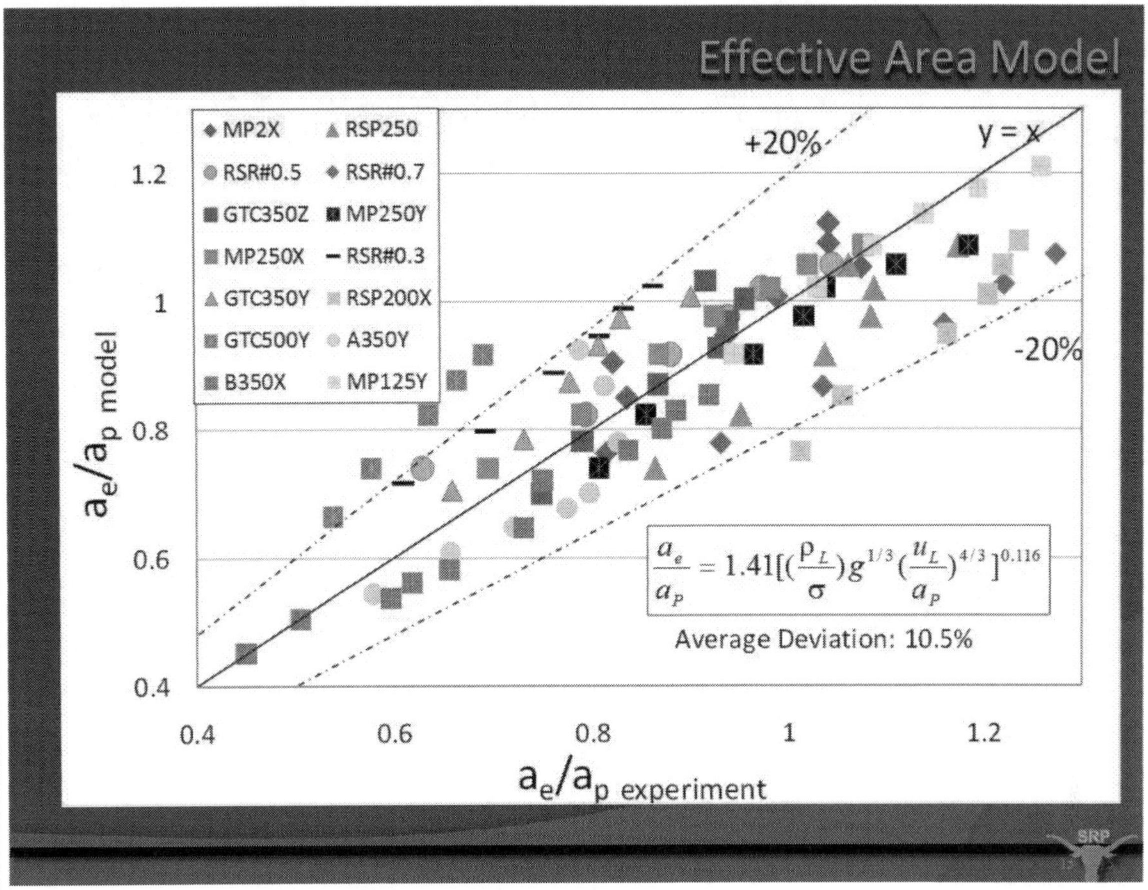

Effective Area Model
MP2X
RSP250
RSR#0.5
RSR#0.7
GTC350Z
MP250Y
MP250X
RSR#0.3
GTC350Y
RSP200X
GTC500Y
A350Y
B350X
MP125Y
+20%
y = x
-20%
$\frac{a_e}{a_p} = 1.41[(\frac{\rho_L}{\sigma})g^{1/3}(\frac{u_L}{a_p})^{4/3}]^{0.116}$
Average Deviation: 10.5%
a_e/a_p model
a_e/a_p experiment
SRP

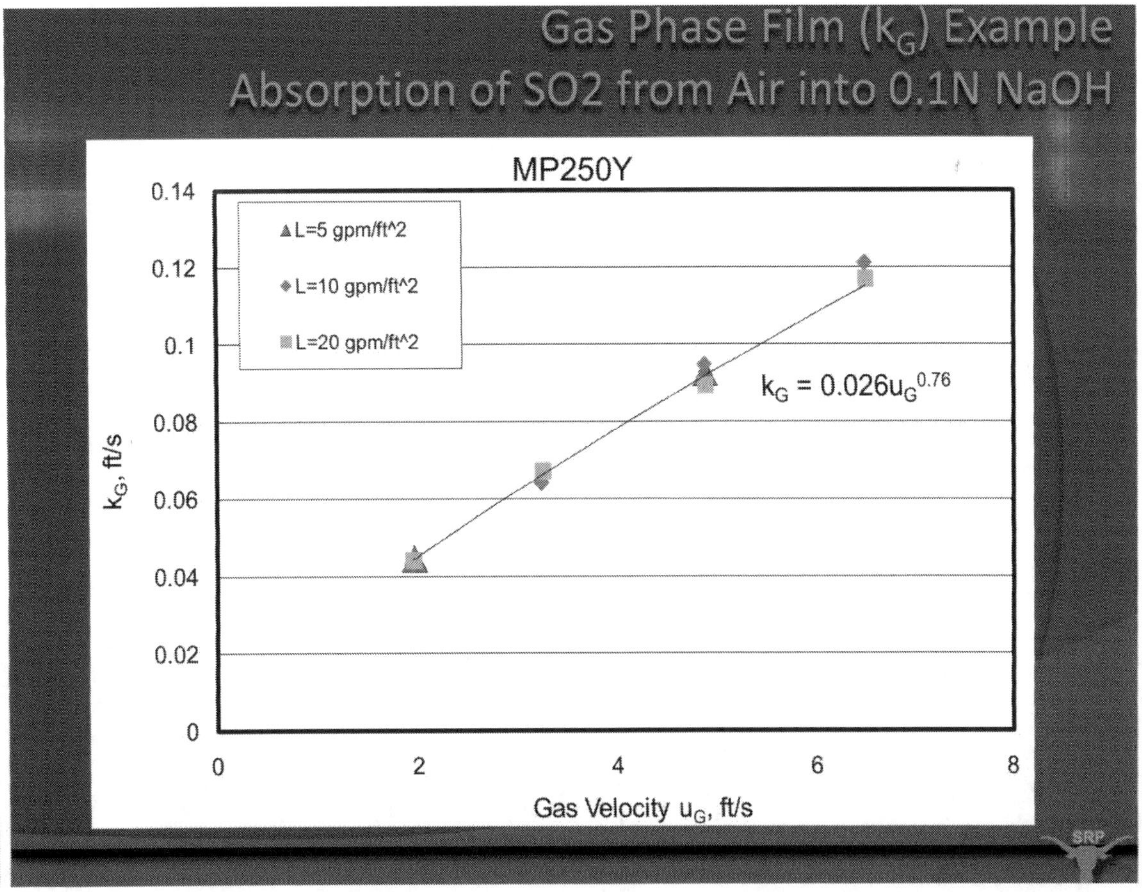

Gas Phase Film (k_G) Example
Absorption of SO2 from Air into 0.1N NaOH
MP250Y
L=5 gpm/ft^2
L=10 gpm/ft^2
L=20 gpm/ft^2
$k_G = 0.026u_G^{0.76}$
k_G, ft/s
Gas Velocity u_G, ft/s
SRP

Liquid Phase Film (k_L) Example
Air Stripping of Toluene from Water

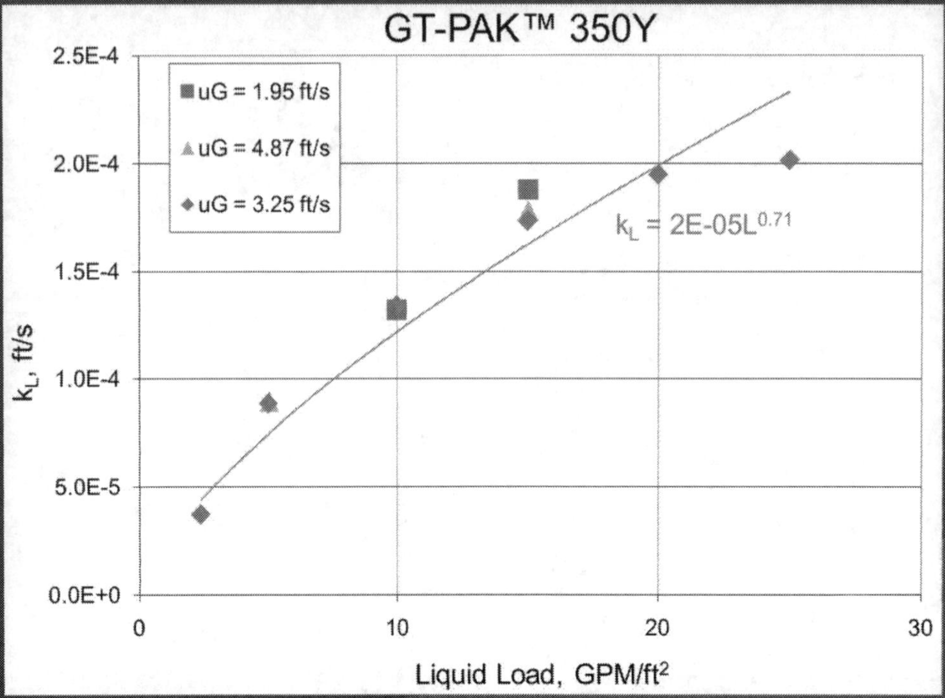

k_L Dependence on Corrugation Angle & Total Packing Area

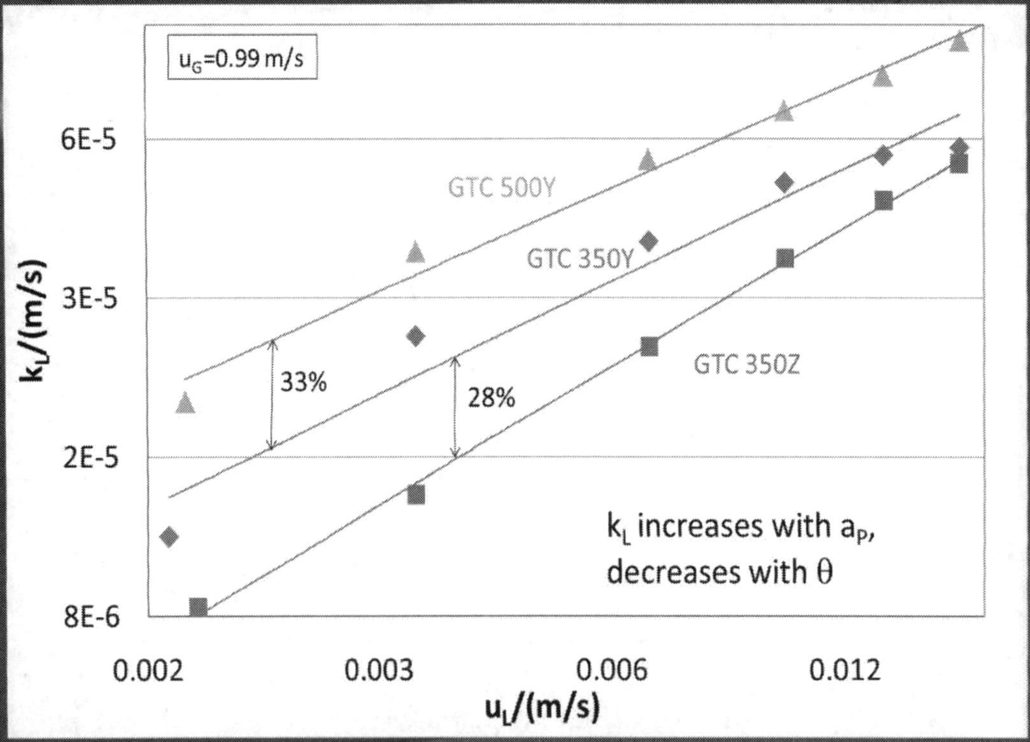

Packing Geometry (side view)

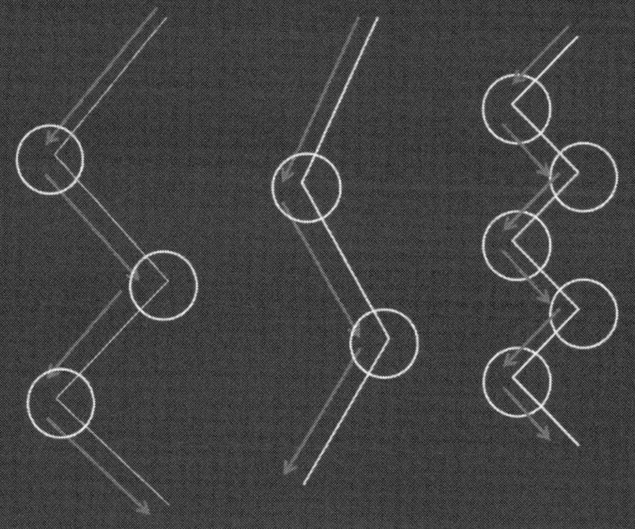

Low angle High angle Large area

- Lower corrugation angle or larger surface area
- More mixing points
- More liquid and gas interactions

SRP

General Model

$$Sh = A * Re^{b} * Sc^{0.5} * M_i^{c}$$

$$M_i = \frac{3}{16} \sin(\theta) \cos(\theta) \left[\cos\left(\frac{\alpha}{2}\right)\right]^2 \sin\left(\frac{\alpha}{2}\right)$$

θ = corrugation ∡ ; α = channel opening ∡ = 90

Resistance	A	b	c
Gas	0.83	0.58	0.30
Liquid	1.79	0.74	0.42

Characteristic Length (L) = a_p^{-1}

Reference: Wang, PhD Dissertation, The University of Texas at Austin, 2014

SRP

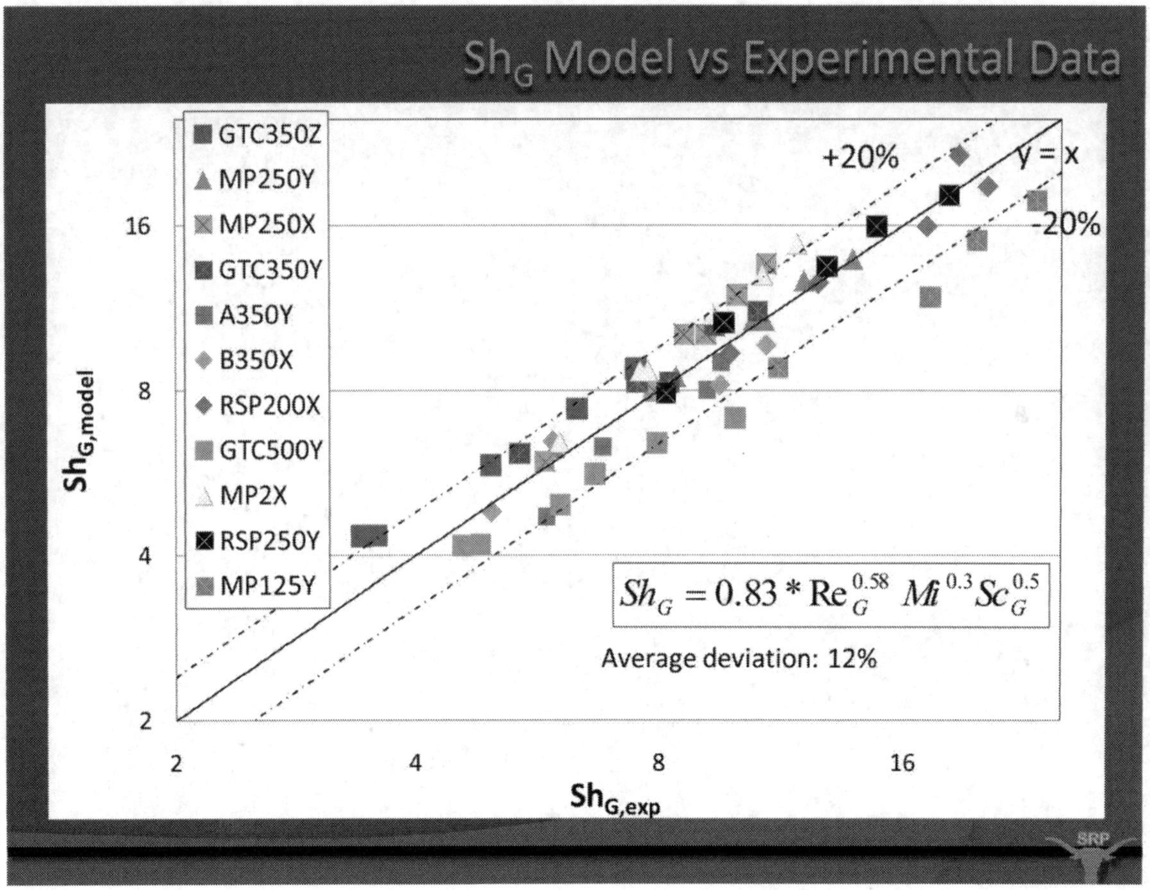

Sh_G Model vs Experimental Data
GTC350Z
MP250Y
MP250X
GTC350Y
A350Y
B350X
RSP200X
GTC500Y
MP2X
RSP250Y
MP125Y
+20%
y = x
-20%
$Sh_G = 0.83 * Re_G^{0.58} Mi^{0.3} Sc_G^{0.5}$
Average deviation: 12%
$Sh_{G,model}$
$Sh_{G,exp}$
2
4
8
16
SRP

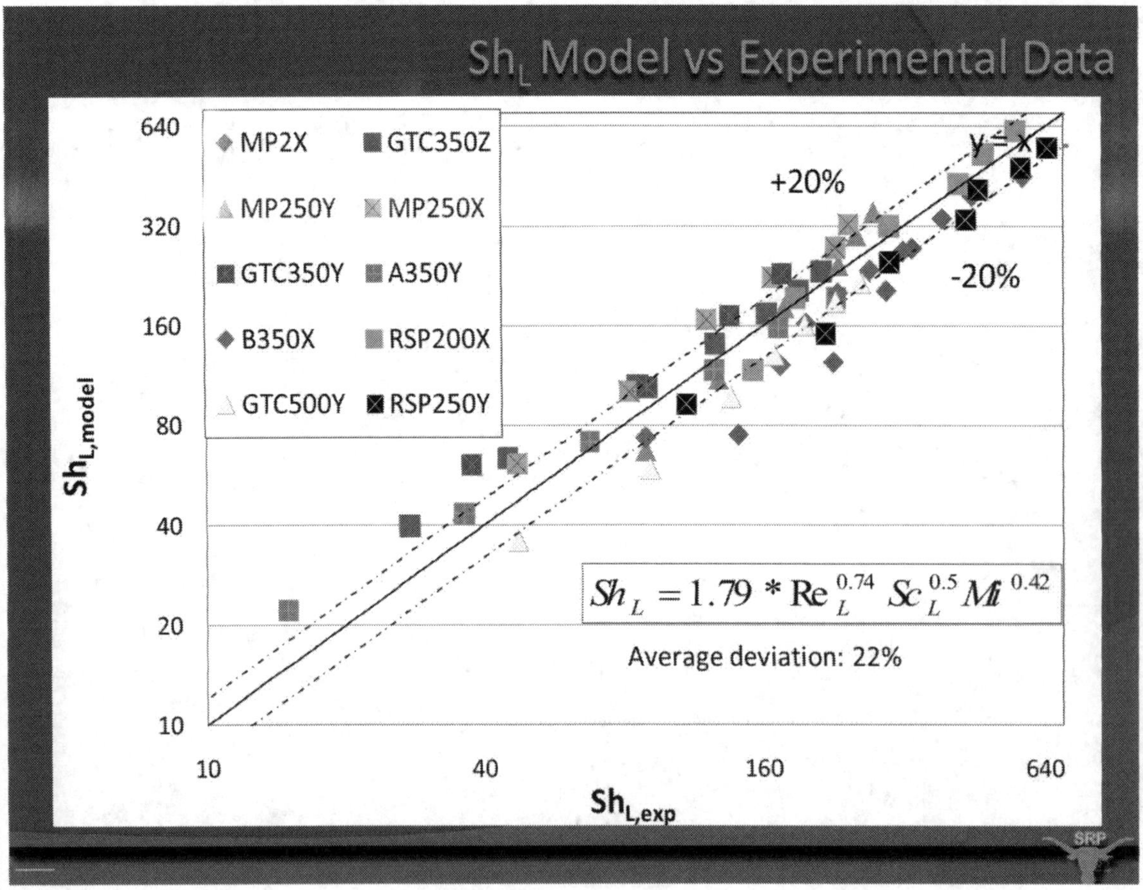

Sh_L Model vs Experimental Data
MP2X
GTC350Z
MP250Y
MP250X
GTC350Y
A350Y
B350X
RSP200X
GTC500Y
RSP250Y
+20%
y = x
-20%
$Sh_L = 1.79 * Re_L^{0.74} Sc_L^{0.5} Mi^{0.42}$
Average deviation: 22%
$Sh_{L,model}$
$Sh_{L,exp}$
10
20
40
80
160
320
640
SRP

Physical Properties of Air/Water System

Temperature	60 ˚F	80 ˚F	100 ˚F
Liquid Density, kg/m^3	1,000	1,000	1,000
Liquid Viscosity, cP	1.29	1.01	0.89
Diffusivity of Toluene in Water, cm^2/s	6.89E-6	9.14E-6	9.48E-6
Vapor Density, kg/m^3	1.22	1.18	1.14
Vapor Viscosity, cP	0.0182	0.0187	0.0193
Diffusivity of SO_2 in Air, cm^2/s	1.19E-1	1.27E-1	1.35E-1

Small Changes in Physical Properties → Re, Sc, σ Variations?

SRP

Physical Properties of Cyclohexane/n-Heptane System

Pressure, bar	0.165	0.33	1.65	4.14
Avg. Temperature	49 ˚C	61 ˚C	114 ˚C	152 ˚C
Liquid Density, kg/m^3	667	656	608	560
Liquid Viscosity, cP	0.32	0.29	0.19	0.15
Liquid Diffusivity, cm^2/s	4.5E-05	5.2E-05	9.0E-05	12.7E-05
Vapor Density, kg/m^3	0.65	1.18	5.44	13.1
Vapor Viscosity, cP	6.1E-3	7.7E-3	8.3E-3	9.1E-3
Vapor Diffusivity, cm^2/s	2.22E-1	1.19E-1	0.31E-1	0.144E-1
Surface Tension, dynes/cm	18	17	14	8
Relative Volatility	1.96	1.84	1.57	1.42
Slope of Equilibrium Line	0.89	0.91	0.95	0.97

SRP

Application to Distillation?

$$HTU_{OV} = HTU_V + \lambda\, HTU_L$$

λ = stripping factor

For stripping factor ~ 1,

$$HETP = HTU_{OV} = HTU_V + HTU_L$$

25

Packings Evaluated

Structured Packings	a_P (m^2/m^3)	Corrugation Angle
MP500Y	500	45°
MP250Y	250	45°
RSP250	250	"45°"
RSP350	350	"45"
GT-OPTIM PAK™ 250Y	250	45°
GT-OPTIM PAK™ 350Y	350	45°
A3-500	500	60°
RJ MaxPack	250	45°
RJ MaxPack 90	250	90°

SRP

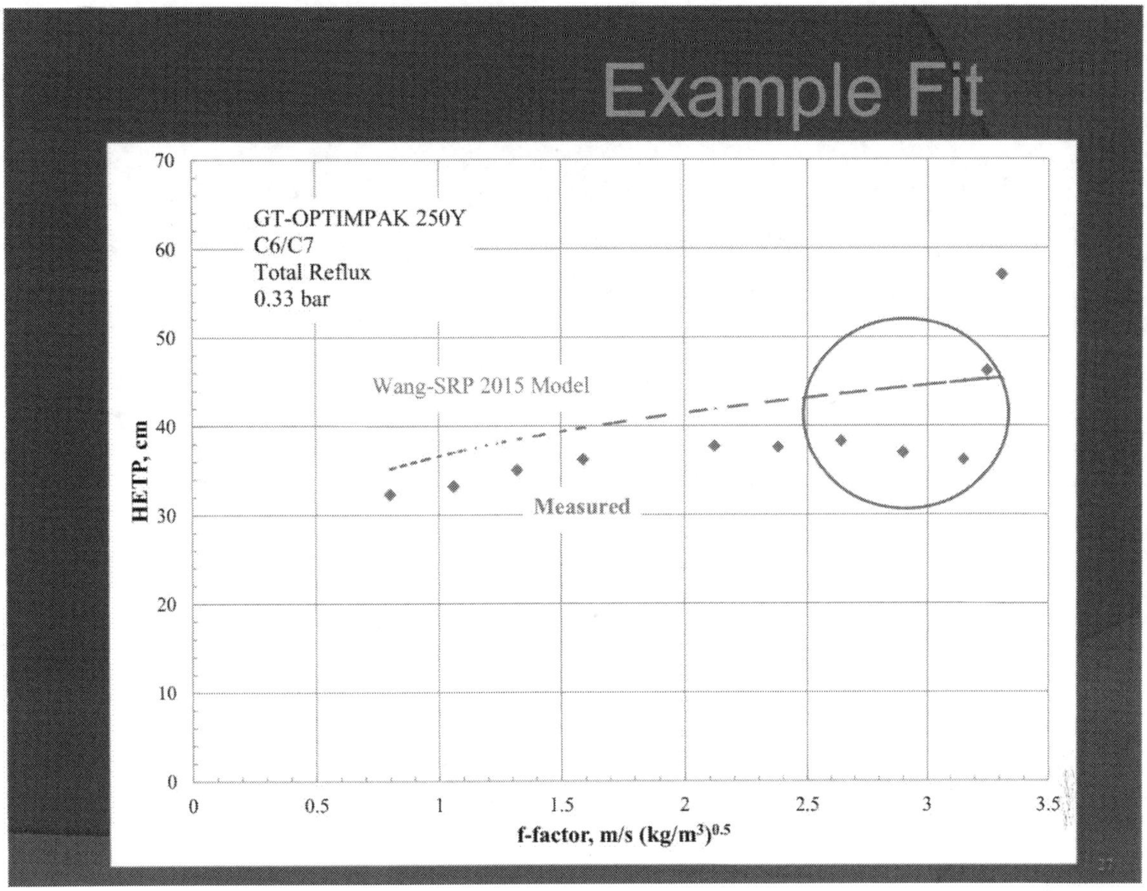
Example Fit
GT-OPTIMPAK 250Y
C6/C7
Total Reflux
0.33 bar
Wang-SRP 2015 Model
Measured
HETP, cm
f-factor, m/s (kg/m³)^0.5
70
60
50
40
30
20
10
0
0
0.5
1
1.5
2
2.5
3
3.5

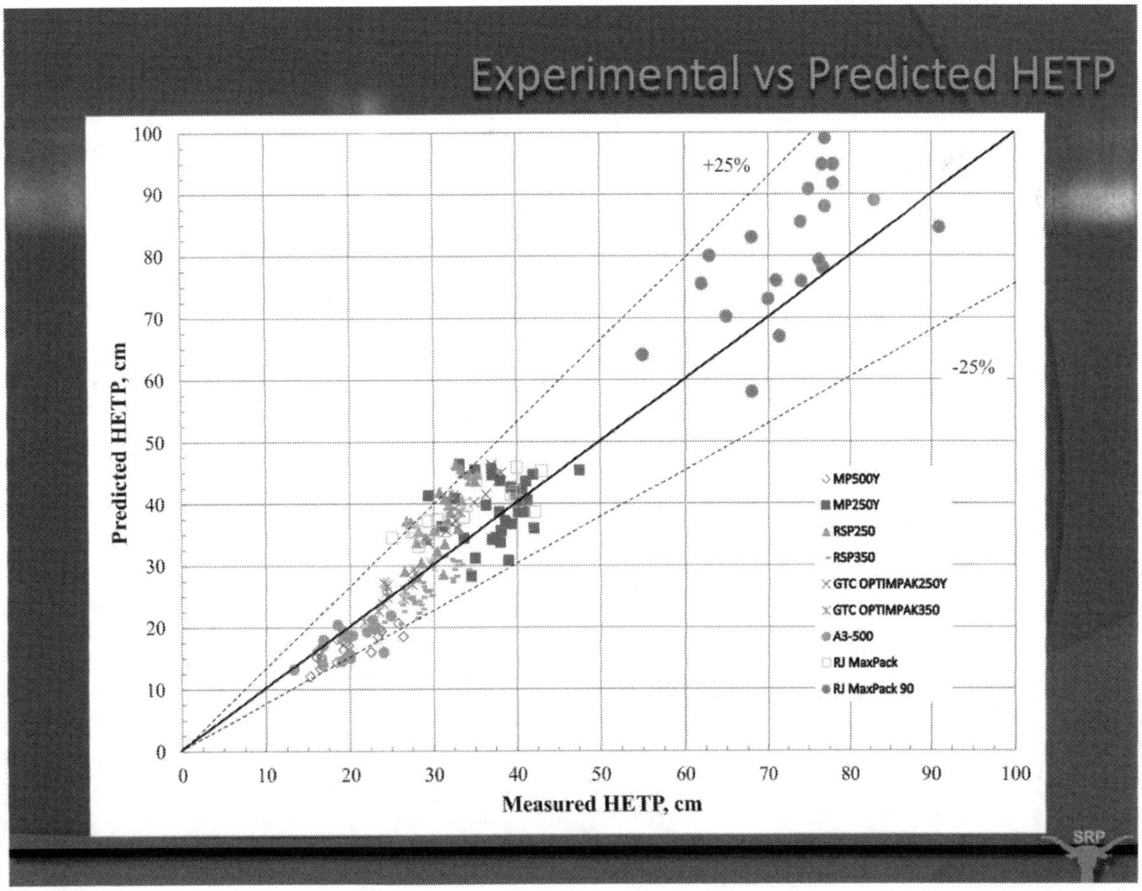
Experimental vs Predicted HETP
+25%
-25%
Predicted HETP, cm
Measured HETP, cm
MP500Y
MP250Y
RSP250
RSP350
GTC OPTIMPAK250Y
GTC OPTIMPAK350
A3-500
RJ MaxPack
RJ MaxPack 90
0
10
20
30
40
50
60
70
80
90
100
SRP

Average Error of Model (<80% of flood)

Structured Packings	Avg. Relative Error %
MP500Y	14.1
MP250Y	12.3
RSP250	23.1
RSP350	10.9
GT-OPTIM PAK™ 250Y	17.8
GT-OPTIM PAK™ 350Y	4.6
A3-500	12.9
RJ MaxPack	17.5
RJ MaxPack 90	15.1*

* Angle of 87° Used in Calculations

SRP

Summary

- Variety of structured packings studied
- Direct measurements of a_e, k_L and k_G
- Novel modelling – MTC ~ mixing point density
- Models fit SRP absorption/stripping data and SRP C6/C7 HETP data < 80% of flood
- Future refinement in contact area and k_L modelling

30

Acknowledgements

Process Science and Technology Center

Texas Carbon Management Program

Raschig GmbH

Sulzer Chemtech

Montz GmbH

GTC

Chris Lewis

Gary Rochelle

Michael Basden

SRP

430701 Designing CO2 Absorption Columns with Activated Amine Solutions

Abstract:

CO2 absorption into chemical solvents is a growing industrial market since decades. Nowadays activated Methyldiethanolamines (aMDEA) in aqueous solutions are quite often used as solvents. There are several amine compositions available on the market promoted by companies like BASF, DOW, Evonik, Hitachi, Ineos, Mitsubishi Heavy Industries or Siemens to name only a few of them. All solvents from these companies are activated and piperazine is the most popular activator to enhance the absorption efficiency in CO2 chemisorption towers. The reaction kinetics in the solvent plays an important rule for binding the CO2 efficiently into the liquid as CO2 has regularly a very low solubility in aqueous solvents.

The prediction of a chemisorption process is quite different to physical diffusion processes like rectification, absorption or stripping towers. For designing a CO2 activated amine absorption tower the chemical reaction in the liquid phase have to be understood beside the gas and liquid phase mass transfer diffusivities. The reactivity of the solvent is changing significantly along the tower. At the column top the lean amine is entering the column with all its reactivity whereby at the column bottom the solvent is almost inactive. The highly exothermal CO2 reaction with the activated amine is generating heat in the solvent leading to temperature peaks along the tower.

Last not least tower internals like structured or random packings or mass transfer trays are applied in industrial towers to handle the reaction and temperature effects in the gas and liquid phase. The paper will discuss especially structured and random packings used in industrial CO2 chemisorption columns. It will disclose information about packing characteristics and parameter to look at for selecting the preferred tower internal. The paper will also discuss prediction models recommended to be used for such processes.

Process Science and Technology Center (PSTC) in Austin TX has done a significant amount of research to understand the mass transfer conditions in CO2 amine absorbers. The paper will provide information how the PSTC research is supporting the industry for providing an optimized column design.

430818 Challenges for Separations and … By Definition … Challenges for the Chemical Industry

Abstract:

The Chemical Process Industries are very strongly identified with the large scale separations that are required to make our modern processes possible. However, the success and perceived maturity of the separations endeavor has been a two-edged sword for those seeking to advance the state of the art. There has been little if any success in obtaining research funding in an area that is perceived to be the exclusive purview of industrial practitioners. Ironically, industrial practitioners are hesitant to adopt or test new processes because the cost of testing is high and the results of failure can be catastrophic. There are still promising areas that can be investigated, and there is still merit in searching for improved separations concepts and methodologies.

415887 Modeling and Experimental Study of a Dividing Wall Column

Abstract:

Originally patented in 1949, the dividing wall distillation column (DWC) is gaining renewed attention in the United States. [1] The benefits of a DWC are significant, including an estimated energy savings of thirty percent over traditional separation technologies. Despite these advantages, there are minimal open literature experimental studies of DWC's.

The University of Texas and industrial partners have constructed a 6 inch diameter pilot scale DWC. The UT research program utilized a ternary alcohol feed with experiments conducted at various compositions and operating conditions. The experimental results were compared to literature data for a 2.4 inch diameter column. [2]

In addition to the mass transfer studies, air-water experiments were conducted to fundamentally characterize the column hydraulics. These studies included the quantification of the vapor split using pressure drop and liquid load across the dividing wall section.

Both study results were incorporated into a steady state AspenPlus model.

1. Wright, O. R. Fractionation Apparatus. U.S. Patent 2.471.134, 1949.
2. Niggemann, Gerit, Christoph Hiller, and Georg Fieg. "Experimental and Theoretical Studies of a Dividing-Wall Column Used for the Recovery of High-Purity Products." Industrial & Engineering Chemistry Research 49.14 (2010): 6566-577.

411839 The Influence of Consortium Research on Industrial Mass Transfer Technology

Abstract:

The Separations Reserach Program was estabilished in the 1980's by Dr. James Fair to provide research capabilities through a joint university/industry effort. This paper will dsicuss several of the research efforts conducted within SRP (Now PSTC which is lead by Dr. Bruce Eldridge) with an emphasis on column packing research and how that research has been applied successfully in commercial technology.

421772 Optimal Operation of Simulated Moving Bed Reactor: Model Correction and Parameter Estimation

Abstract:

Simulated moving bed reactor (SMBR) offers a better economic and environmental alternative to conventional sequential integration of batch reactor and separator operations. Specifically, SMBR improves separation resolution, increases productivity, and reduces solvent consumption [1]. Yet despite these benefits, the complexity in SMBR modeling and design is extremely challenging for industrial implementation. Within the past decade, work has been done on the optimal design and operation of SMBR; however, these investigations were focused on single-objective optimization or used heuristic based algorithms [2-4].

This work demonstrates a practical and deterministic model-based approach to optimize an industrially relevant SMBR process and to identify the optimal SMBR operating parameters. The SMBR operation is applied to an industry case study for the continuous production of a solvent, propylene glycol methyl ether acetate (DOWANOL^{TM}PMA) through an acid-catalyzed esterification reaction of 1-methoxy-2-propanol and acetic acid.

The model-based approach is demonstrated by lab-scale SMBR experiments. The model is capable of identifying the operating strategy to satisfy productivity and conversion objectives and constraints for product purity. The initial model containing kinetic and adsorption parameters was obtained from batch reaction and single-column reactive chromatography experiments [4]. From these experiments, a mathematical model to predict the performance of the SMBR is developed. With this model, a multiple objective optimization problem was formulated [5] and several optimal operations, particularly conversion targets from 70% to 95% were experimentally validated. We will show that the initial model of SMBR performance is predictive of SMBR experimental results. Moreover, this model's accuracy can be improved when the experimental SMBR data is used to recalculate the parameter values of the SMBR model.

References:

1. Seidel-Morgenstern, A., L.C. Keßler, and M. Kaspereit, *New Developments in Simulated Moving Bed Chromatography*. Chemical Engineering & Technology, 2008. **31**(6): p. 826-837.

2. Dünnebier, G., Fricke, J., Klatt, K-U., *Optimal Design and Operation of Simulated Moving Bed Chromatographic Reactors*. Industrial & Engineering Chemistry Research, 2000. **39**(7): p. 2290-2304.

3. Lode, F., Francesconi, G., Mazzotti, M., Morbidelli, M., *Synthesis of methylacetate in a simulated moving-bed reactor: Experiments and modeling*. AIChE Journal, 2003. **49**(6): p. 1516-1524.

4. Oh, J., Agrawal, G., Sreedhar, B., Donaldson, M. E., Schultz, A. K., Frank, T. C., Bommarius, A. S., Kawajiri, Y., *Conversion improvement for catalytic synthesis of propylene glycol methyl ether acetate by reactive chromatography: Experiments and parameter estimation*. Chemical Engineering Journal, 2015. **259**(0): p. 397-409.

5. Agrawal, G., Oh, J., Sreedhar, B., Tie, S., Donaldson, M. E., Frank, T. C., Schultz, A. K., Bommarius, A. S., Kawajiri, Y., *Optimization of reactive simulated moving bed systems with modulation of feed concentration for production of glycol ether ester*. Journal of Chromatography A, 2014. **1360**(0): p. 196-208.

413214 Membrane Cascade with Solvent Recovery – Process Innovation for the Purification of Molecules

Abstract:

For the purification of a target compound from impurities, an ideal membrane should completely retain the former while allowing the latter to permeate through or vice versa. In real life, the purification process is often a trade-off between the yield and purity of the target compound due to the non-ideality of the membranes used. For example, the separation of of polyethylene glycol (PEG) 400 from PEG 2000 by constant volume diafiltration (CVD) in a single-stage membrane system resulted in a 41% yield loss of PEG 2000 in order to achieve 98% purity, although the rejection of PEG 2000 was substantially higher than that of PEG 400 (96% and 51% respectively)[1]. By installing a second-stage membrane to recover the PEG 2000 that permeated through the first-stage membrane, the yield loss was greatly reduced to only 6% with the same final purity. Similarly, the modelling of the Membrane Enhanced Peptide Synthesis (MEPS) in a two-stage membrane cascade shows that a significantly higher yield of final peptide (98 % versus 71 % in a single-stage process) can be achieved with a similar purity (93%). These examples suggest that the use of membrane cascades can overcome the problem of non-absolute rejection of the desired product and maintain high yields for a similar purity of the target compound at the expense of a larger consumption of fresh solvent than a single-stage process. To reduce this solvent consumption and the associated material cost and environmental impact, solvent recovery can be performed downstream. In a case study of the separation of an active pharmaceutical ingredient (API) (roxithromycin macrolide antibiotic (Roxi)) from genotoxic impurities (GTIs) (4-dimethylaminopyridine (DMAP) and ethyl tosylate (EtTS)), adsorptive solvent recovery by activated carbon exhibited lower energy consumption and a smaller carbon footprint than distillation (96% and 70% lower respectively)[2]. Alternatively, solvent recovery through a tight membrane is also technically feasible as demonstrated in two case studies where the requirement of fresh solvent (tetrahydrofuran (THF)) was reduced by 90% in the first case[3] and where membrane-based solvent recovery outperformed adsorptive- and distillation-based solvent recovery in terms of operability and carbon footprint in the second case[4]. In short, diafiltration in membrane cascade coupled with downstream solvent recovery is a promising process for the purification of compounds in the chemical industry.

Reference:

1. Kim, J. F., da Silva, A. M. F., Valtcheva, I. B., & Livingston, A. G. (2013). When the membrane is not enough: A simplified membrane cascade using Organic Solvent Nanofiltration (OSN). Separation and Purification Technology, 116, 277-286.

2. Kim, J. F., Székely, G., Valtcheva, I. B., & Livingston, A. G. (2014). Increasing the sustainability of membrane processes through cascade approach and solvent recovery—pharmaceutical purification case study. Green Chemistry, 16(1), 133-145.

3. Sereewatthanawut, I., Lim, F. W., Bhole, Y. S., Ormerod, D., Horvath, A., Boam, A. T., & Livingston, A. G. (2010). Demonstration of molecular purification in polar aprotic solvents by organic solvent nanofiltration. Organic Process Research & Development, 14(3), 600-611.

4. Kim, J. F., Szekely, G., Schaepertoens, M., Valtcheva, I. B., Jimenez-Solomon, M. F., & Livingston, A. G. (2014). In Situ Solvent Recovery by Organic Solvent Nanofiltration. ACS Sustainable Chemistry & Engineering,2(10), 2371-2379.

431940 Hybrid Distillation – Membrane Process to Increase Capacity and Reduce Costs for Production of Propylene and Ethylene

Abstract:

Ethylene and Propylene are two primary feedstocks for manufacture of petrochemicals. Therefore, there is significant economic benefit in increasing capacity of existing distillation operations while simultaneously reducing costs. Existing distillation for separation of olefins from paraffins (e.g., propylene from propane) is very difficult requiring large distillation columns and significant energy consumption. In principle membranes are small and should be easily retrofitable into distillation columns.

A hybrid distillation-membrane process that can efficiently separate and recover the olefin from the paraffin would provide substantial economic benefit to these polymerization processes. Membrane processes have been previously evaluated for separating ethylene/ethane or propylene/propane. In many cases, silver salt facilitating agents have been incorporated into membranes to preferentially transport ethylene or propylene. Historically with earlier membranes, while good separations have been initially demonstrated in the laboratory, membrane stability problems have prevented development of commercial systems.

Using customized amorphous fluoropolymer (CAF) membranes, Compact Membrane Systems has developed stable olefin-paraffin separation membranes with very good propylene flux and propylene/propane selectivity. The permeance and selectivity combination of the membranes are above the industry Robeson curve. In addition results were stable when exposed to high levels of conventional foulants (e.g., sulfur containing gases) in the laboratory. Using basic data from these laboratory results payback times for hybrid distillation-membrane processes were calculated to be less than 1-2 years. Laboratory experimental data and hybrid distillation –membrane system modeling results will be discussed more extensively at the presentation.

435448 Bech-Scale Tests of Isopropyl Alcohol Dehydration with Zeolite Membrane

Abstract:

In FY2009-2013, we conducted a project to develop fundamental technologies of zeolite membranes for application to important chemical processes such as dehydration of isopropyl alcohol (IPA). The dehydration of IPA by distillation requires a large amount of energy because IPA forms azeotrope with water at about 14 %(wt) of water. We, 6 companies and 6 universities, concurrently carried out the development of element technologies. In particular, in FY2012-2013, we performed the bench-scale tests at the IPA production plant site in the Kawasaki Plant of JX Nippon Oil & Energy. Y-type zeolite (Hitachi-zosen) and SSZ-13 (Mitsubishi Chem.) tubular membranes have been developed, and the permselectivities of these membranes overwhelmingly exceeded targets. Process simulation suggested that in comparison with current distillation for the IPA dehydration, hybridization of distillation and membrane decreases >50% of the energy demand. The payout time would be < 1 year on a glass-roots basis. The bench-scale test performed in FY2012-2013 was performed using the real stream in a petrochemical plant. Operation exceeding 200 h was successfully done.

Fig. 1 IPA production plant at the KAWASAKI plant of JX-Nippon Oil and Energy, and the bench-scale zeolite membrane test plant set at the KAWASAKI Plant.

410697 Use of the Attainable Region method to specify optimal milling and leaching conditions for a joint process

Abstract:

Abstract

Particle size reduction is a process carried out either on dry materials or on slurries where a material is gradually reduced in size to produce material of the desired size class. It is an energy intensive process concerned with the relationship between energy input and the particle size made from the given feed size. It is of paramount importance to mill to the right size while minimising the energy costs. In this study, the specific rate of breakage and primary breakage distribution functions for gold ore sample taken from Carletonville plant, Johannesburg, were determined in a laboratory size ball mill. Grinding at about U= 1 gave the maximum amount of size class for leaching (-150+75 μm), and is a good powder-ball loading ratio to give efficient breakage in a ball mill for the gold sample. The AR analysis is then used to optimise leaching and milling time in terms of costs.

422165 Separation of Diethyl Citrate and Phase Equilibria in Mixtures with Ethanol and Water

Abstract:

The potential harmful effects caused by petrochemical plasticizers, specifically phthalates, are a continuous concern for consumers and producers. Plasticizer market is growing mainly in the biobased sector (based on natural resources) and stagnation is observed in the traditional phthalates. Among the variety of biobased plasticizers, the esters from citric acid (with or without acetylation) are preferred as ingredient in the fabrication of polymers and resins used in medical devices, food packaging, cosmetics, and many other applications. This preference is because citrates are generally recognized as safe (GRAS) by different regulatory agencies worldwide.

Current citrates production is carried out in semibatch processes and large excess of alcohol is required to overcome chemical equilibrium limitations of the reaction. The need for the excess alcohol removal and product purification make the process energy intensive, inefficient and costly. In this regard, intensification by using reactive distillation (RD) technologies is envisioned as an alternative to improve process sustainability.

For a complete RD feasibility study, thermodynamic and kinetic data associated to the reactive system are required. This include properties of intermediate esters that in the case of the esterification with ethanol correspond to monoethyl citrate and diethyl citrate. These are low-volatile, water-soluble acidic esters, and differ only in an ethyl radical, which makes difficult their separation and further purification. Despite molecular distillation and preparative chromatography used in the past were not efficient enough, a simple pH-controlled solvent extraction was successful to accomplish intermediate ester separation.

A mixture of ethyl citrates rich in dyethyl citrate (DEC) was prepared by incomplete esterification of citric acid and anhydrous ethanol. Reaction was carried out in a 2 L stainless steel reactor coupled with a reflux condenser and a mechanical stirrer. An initial citric acid/ethanol molar ratio of (1:5) and 5 wt.% Amberlyst 15 catalyst were used during reaction. The temperature was set at 73 °C and the reaction was carried out for 5 hours. After the reaction, the excess ethanol was removed by vacuum distillation, leaving behind a yellowish liquid mixture of citric acids and citrates.

By performing a pH controlled separation, trietyl citrate (TEC) and DEC were isolated from the citrates mixture, and 97 wt% DEC was obtained with citric acid as major impurity. Even if TEC and citric acid are commercially available to be used as standards in HPLC calibration, intermediate esters are not. In this sense, calibration of HPLC diode array detector was performed to improve quantitative analysis in further experiments.

Once isolated, diethyl citrate was characterized by NMR, DSC, and vapor pressure tests. Binary vapor liquid equilibria at constant temperature in mixtures with ethanol and water were obtained, and parameters for activity-based models were regressed. Good agreement of predicted and experimental phase equilibrium data were observed, and the model can be used in further reactive distillation feasibility studies.

430245 Development of an in-Line Membrane Process for Bioreactor Harvest

Abstract:

This talk will describe the development of a membrane application for reducing harvest volume to enable process fit within an existing facility. Selection of operating modes and process system component selection will be discussed.

431343 Design and Evaluation of Advanced Amsphere(tm) A3 Protein A chromatography Resin: The relationships and considerations of bead structure, pore structure, surface chemistries and ligand design on affinity resin performance targets

Abstract:

Industry needs, patent expirations of ligand design and market dynamics have accelerated the development and availability of many protein affinity resins. Technology providers must consider a range of design parameters such as resin backbone, bead size, pore size, surface chemistry modification, ligand construct and how they relate to performance criteria such as DBC, life time, caustic stability, and column packing among others. Approach to designing advanced Amsphere A3(tm) affinity resin will be discussed. The trade-offs between resin design and resin performance will be illustrated through a case study. Data comparing four commercially available affinity resins with multiple monoclonal antibodies (MAb) will be discussed.

432919 Recovery of Autologous Plasmin from Serum with Affinity Membranes

Abstract:

Plasmin is a protein of the fibrinolityc system with activity against fibronectin and laminin, components of the vitreoretinal interface. It can be used for treatment of diabetic rethinopaty, macular pukers and vitreomacular adhesion since it works by dissolving the proteins that link the vitreous to the macula, resulting in posterior detachment of the vitreous from the retina. Plasmin can be obtained by enzymatic activation of blood plasminogen using urokinase or streptokinase.

The purification of plasminogen from blood is performed with packed column chromatography, by exploiting the affinity between plasminogen and L-lysine. However, due to its low concentration, which is about 0.2 g/L in human blood, with a single step affinity purification from serum or plasma it is not possible to obtain sufficient amount of plasminogen for the subsequent enzymatic activation .

In this work we describe several strategies that have been investigated in order to improve the membrane chromatography step which is performed with L-lysine affinity membranes. The affinity membranes were obtained by immobilization of L-lysine on regenerated cellulose membrane supports and characterized in terms of ligand density, binding capacity and selectivity. The effect of operating parameters like flow rate, type of blood source (serum, plasma or Cohn fractions) and elution conditions on the activity of plasmin conversion have been investigated in detail. The comparison with a chromatography column packed with a commercial lysine affinity beads demonstrated the superior performance of the affinity membranes and the feasibility of the proposed process.

Implementation and Characterization of Solvent Detergent Viral Inactivation in Single-Use Bags

2015 AIChE Annual Meeting, Salt Lake City, UT

November 10, 2015

Michael A. Cunningham, Ph.D.

Presentation Outline

1	Background and Project Description
2	Experimental Methods and Results
3	Best Practices Guidance at Scale
4	Summary/Discussion

2

Virus Contamination – Risk Mitigation

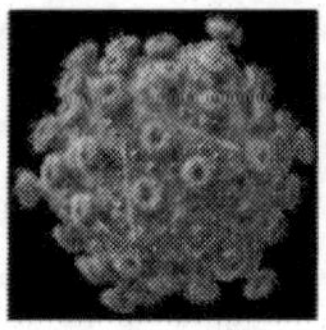

Plasma Product Contamination

- 1980s:**HIV** contaminated Coagulation Factor VIII
- 1994: **Hepatitis C** contaminated IgG
- 1995: **vCJD** contamination
- Other threats…**Hepatitis E**, **West Nile Virus**, …

Three Pillared Approach to Virus Safety

- Virus Safety of Raw Materials
- Implementation of Viral Clearance Technologies
- In-process and Final Product Testing

→ Solvent Detergent Viral Inactivation

Solvent Detergent Viral Inactivation Project Description

Application Space:

- Plasma-based processes where virus filtration not feasible
- Well-characterized pathogen safety risk mitigation process
- Inactivates enveloped viruses (HBV, HIV)
- Increasing single-use trend in plasma processes

Study Objective:

Demonstrate Feasibility of Conducting SD VI in Mobius® Single-Use Process Containers

- Chemical Compatibility with Single-Use Bags
- Efficacy of SD VI Procedure
- Characterization of Leachables Profile
- Guidance for Best Practices

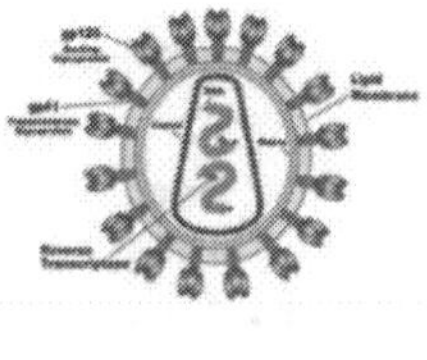

4

Presentation Outline

1	Background and Project Description
2	Experimental Methods and Results
3	Best Practices Guidance at Scale
4	Summary/Discussion

5

Chemical Compatibility of Pureflex™ Film

Residuals Testing:

Solution	Non-Specific Binding (%)
0.3% TnBP/1% Triton® X100	2
0.3% TnBP/1% Tween® 80	1

Minimal non-specific binding to film

Pureflex ™ Film Tensile Strength Testing:

Mechanical Property	PBS	TnBP Triton X100	TnBP Tween 80
% Elongation at Break	●	●	●
Tensile Strength	●	●	●
Secant Modulus	●	●	●
Toughness	●	●	●

No impact of chemical exposure to Pureflex ™ film mechanical properties

Environmental Stress Crack Resistance Test:

Reagent	Post-Exposure Cracking
PBS	●
TnBP/Triton® X100	●
TnBP/Tween® 80	●

Mobius® bag components integrity is not impacted by chemical exposure

6

Virus Spiking Studies – SD VI in Single-Use Bags

Materials:

Blood Fractions	Protein Conc. (g/L)	Solvent/Detergent	Virus
Crude Human Plasma	57	1%(v/v) TnBP + 1% (v/v) Triton® X-100	Xenotropic Murine Leukemia virus (XMuLV)
			Bovine Viral Diarrhea Virus (BVDV)
Human IgG	15	0.3%(v/v) TnBP + 1%(v/v) Triton® X-100	XMuLV & BVDV
		0.3%(v/v) TnBP + 1%(v/v) Tween® 80	

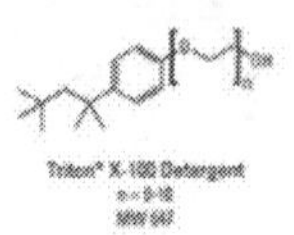

7

S/D VI Study Design

Virus → **Plasma / IgG N=1** → T=0'

Buffer → **Plasma / IgG N=1**

Virus → **Plasma / IgG N=2** → T=0'

S/D → **Plasma / IgG N=2**

Incubate for 6 hrs at 31°C
Sample at T=5', 30', 60', 360'

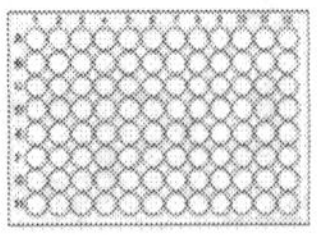

- **Add virus to matrix**
- **Remove T=0 sample for titer**
- **Add S/D**
- **Incubate & sample**
- **Titer immediately, LVP for increased sensitivity**

Human Plasma: VI by 1%TnBP+1%Triton® X100

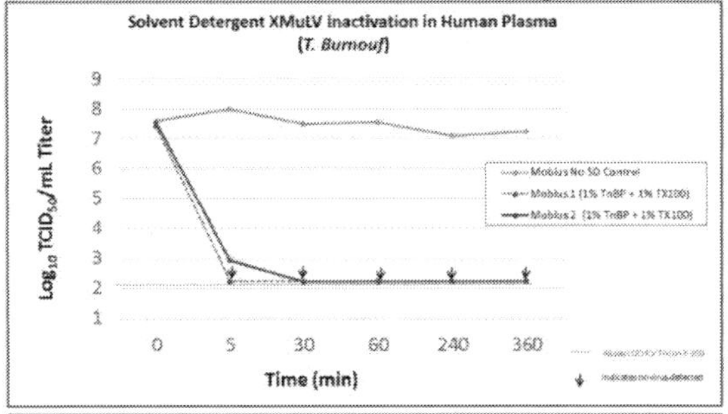

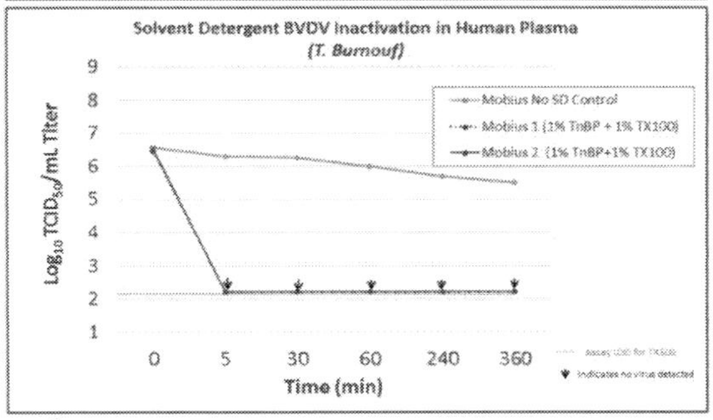

- **Controls:** No change in XMuLV titer & 1 log reduction of BVDV over 6 hrs
- **1% TnBP + 1% TX100® Treated:** Rapid virus inactivation

Human Plasma: 1% TnBP + 1% TX100 LRV Results						
Virus	Device	LRV at Incubation Time (min)				
		5	30	60	240	360
XMuLV	Mobius® 1	≥5.8	≥5.3	≥5.3	≥4.9	≥5.0
	Mobius® 2	5.1	≥5.3	≥5.3	≥4.9	≥5.0
BVDV	Mobius® 1	≥4.2	≥4.2	≥4.2	≥4.2	≥4.2
	Mobius® 2	≥4.3	≥4.3	≥4.3	≥4.3	≥4.3

BVDV LRV calculated from T0 of Control treatment

9

Human IgG: VI by 0.3%TnBP+1%Triton® X100

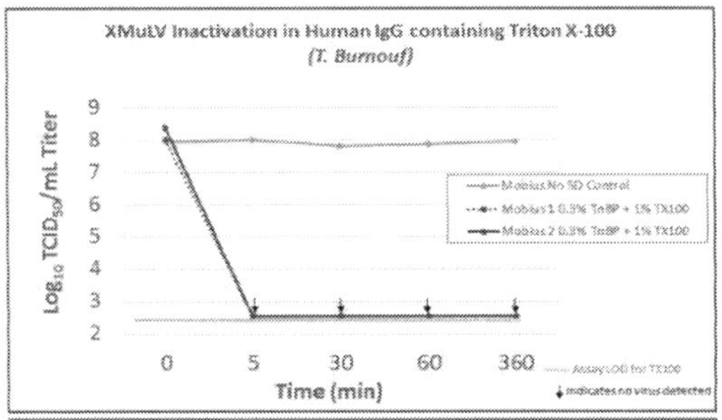

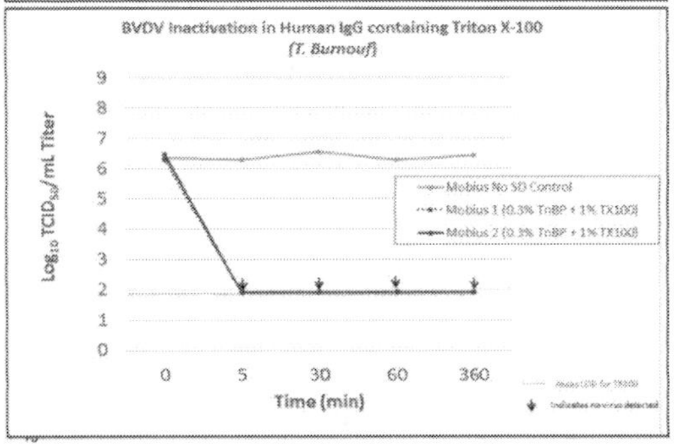

- **Controls:** No change in XMuLV or BVDV titer over 6 hrs
- **0.3% TnBP + 1% TX100® Treated:** Rapid virus inactivation

Human IgG : 0.3% TnBP + 1% TX100 LRV Results					
Virus	Device	LRV at Incubation Time (min)			
		5	30	60	360
XMuLV	Mobius® 1	≥5.5	≥5.3	≥5.3	≥5.4
	Mobius® 2	≥5.5	≥5.3	≥5.3	≥5.5
BVDV	Mobius® 1	≥4.5	≥4.4	≥4.6	≥4.5
	Mobius® 2	≥4.4	≥4.6	≥4.4	≥4.5

Human IgG: VI by 0.3%TnBP+1%Tween® 80

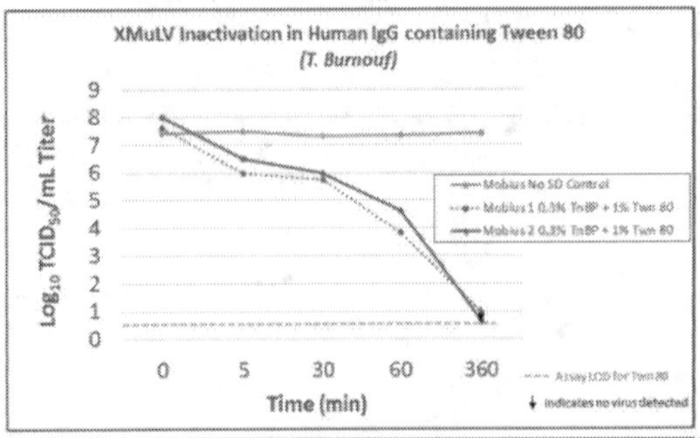

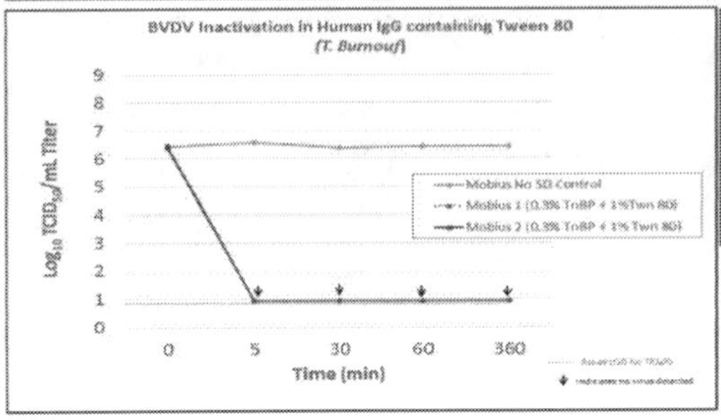

- **Controls:** No change in XMuLV or BVDV titer over 6 hrs
- **0.3% TnBP + 1% Twn 80® Treated:** Slow decrease in XMuLV titer, but BVDV inactivation was rapid

Human IgG : 0.3% TnBP + 1% Twn 80® LRV Results					
Virus	Device	LRV at Incubation Time (min)			
		5	30	60	360
XMuLV	Mobius® 1	1.5	1.6	3.5	≥6.5
	Mobius® 2	1.0	1.3	2.8	6.8
BVDV	Mobius® 1	≥5.6	≥5.5	≥5.5	≥5.6
	Mobius® 2	≥5.6	≥5.5	≥5.5	≥5.6

11

S/D VI Does Not Impact Plasma Composition, Activity

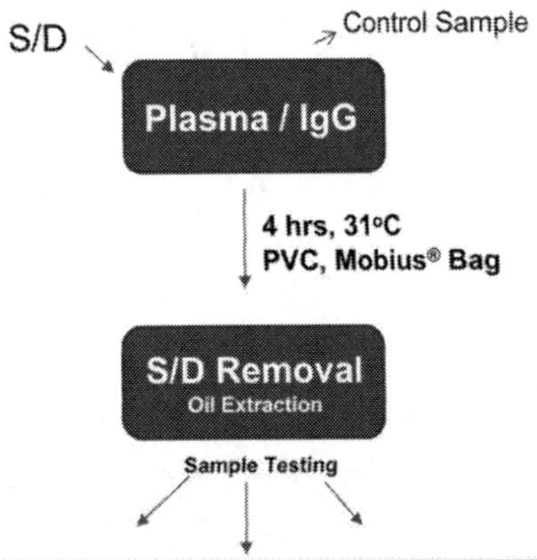

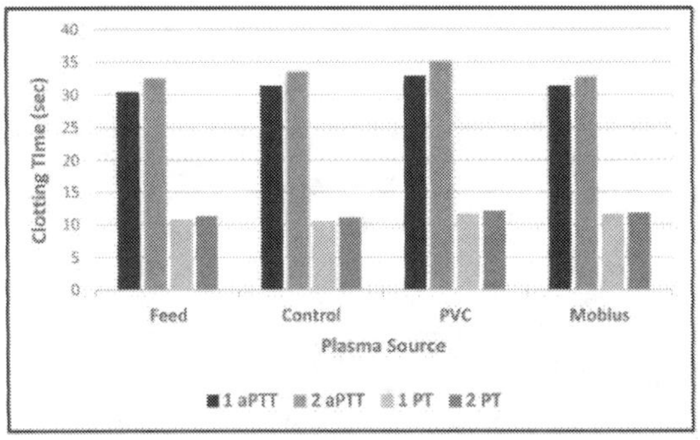

- **Plasma tests completed:**
  - **Total protein – no change**
  - **IgG/IgA/IgM levels – no change**
  - **FVIII and FVIII activity – no change**
  - **Clotting times – no change**

Normal Clotting Times:

aPTT: 25 – 35 sec
PT: 10 – 14 sec

Leachables Profiles of Pureflex™ Film

1L Bag Study (Worst Case):

Solution	Total Leachables Level		
	VOCs (ppm)	sVOCs (ppm)	Total (ppm)
0.3% TnBP 1% Triton® X100	1.71	0.14	1.85
0.3% TnBP 1% Tween® 80	0.45	1.05	1.50

Leachables identified:

- 10 total leachables > 0.1 ppm, only 1 compound > 1 ppm

- No genotoxic compounds identified, based on toxicological assessment
- In context of holistic bioprocess, patient safety risk to leachables exposure in BDS is likely low
- Ongoing regulatory focus to determine best practices

13

Presentation Outline

1	Background and Project Description
2	Experimental Methods and Results
3	Best Practices Guidance at Scale
4	Summary/Discussion

14

Best Practices Recommendations at Scale

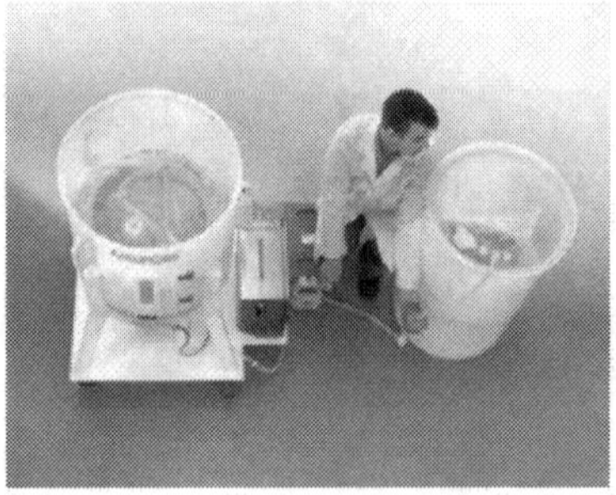

- Bottom drain port addition of solvent detergent solution minimizes foaming
- Attach clamp on drain line as close as possible to vessel (top right) to minimize dead leg
- Choose minimal speed that facilitates mixing (mitigate foaming or product quality risks)

15

Solvent Detergent Addition Through Drain Line Minimizes Foaming

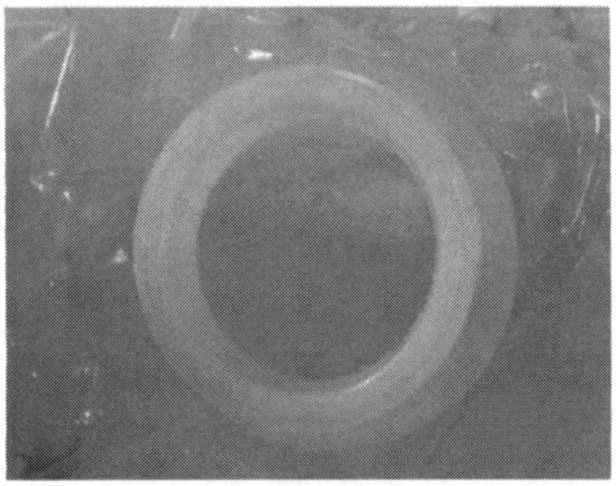

TnBP/Triton® X-100 in PBS

TnBP/Triton® X-100 in 5% BSA

16

Efficient Mixing at 50 and 500L Scales

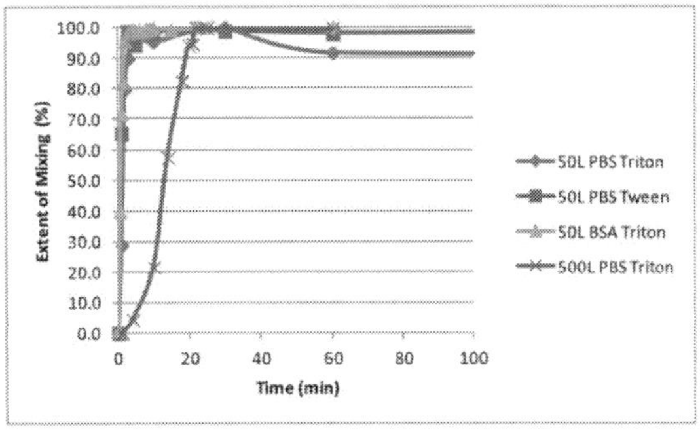

- Mixing speeds chosen to avoid vortex generation (50L-200rpm; 500L-100rpm)
- Samples collected at top and bottom of vessel (similar results)

- Mixing essentially completed by 5 min at 50L; 22 min at 500L
- 50L mixing efficiency similar in PBS and 5% (w/v) BSA

17

Presentation Outline

1	Background and Project Description
2	Experimental Methods and Results
3	Best Practices Guidance at Scale
4	Summary/Discussion

18

Summary

- Solvent detergent viral inactivation is the method of choice to inactivate enveloped viruses in plasma-based bioprocesses
- Implementation of single-use technologies in the SD VI application can create operational flexibility and reduce process validation requirements
- Results of this study demonstrate single-use technology implementation is feasible from compatibility and effectiveness perspectives
- Initial characterization of plasma post-SD VI demonstrates maintenance of coagulation activity, total Ig levels (vs. untreated control)
- Additional studies ongoing to further characterize leachables profile in the context of holistic plasma-based bioprocesses

19

Acknowledgements

- Lori Mullin
- Patricia Greenhalgh, Ph.D.
- Venkata Raman, Ph.D.
- Nitin Sood
- Jessica Shea
- Eric Youssef
- Pedro Almeida
- Elizabeth Goodrich

- Joe Cianciolo
- Paul Killian, Ph.D.
- Scott Forsberg

- Thierry Burnouf, Ph.D.

20

Questions

21

427695 Selection Support Method of Single-Use Vs. Multi-Use Technologies in Parenterals Manufacturing

Abstract:

The objective of this study is to create a decision supporting method for the choice between single-use and multi-use technologies in Parenterals manufacturing. There are two different manufacturing technologies for Parenterals, i.e., multi-use system (MU) and single-use system (SU). MU is a conventional technology using stainless steel, the equipment of which is cleaned with water for injection and sterilized with pure steam before and after batch production. In contrast, SU is a new technology using plastic-resin, the components of which are assembled and disposed by every batch. The choice of these two technologies is critical regarding quality, cost-effectiveness or environmental impacts, and there have been various case studies reported mainly for the drug substance manufacturing. However, a general decision-support tool is still in its infancy, and Parenterals manufacturing, a step after drug substance manufacturing to produce injectables, has not been investigated extensively.

This work presents a selection support method of SU vs. MU technologies in Parenterals manufacturing. The methodology consists of three steps. First, the manufacturing process is modeled and evaluated in terms of operating cost and life cycle CO_2 emissions as monetary and environmental objectives, respectively. Evaluation models were created that can reflect main features of each technology, such as preparation time, number of required operators or the area of the isolator. Second, in order to simulate the effects of input parameters, what-if analysis is performed. Among various parameters to be analyzed, pump speed can also influence the filling accuracy of the product, one of the most important quality attribute of Parenterals. This finding is based on an experiment, where coefficient of variations (CV) of the filling volume were measured with varying liquid viscosity, inner tube diameter, and pump speed. The result showed that the pump speed was the only parameter that caused significant difference on the CV, i.e., what-if analysis of pump speed can also be an evaluation on the product quality. Lastly, the results derived from above two processes are interpreted in terms of economy, environmental impact, and quality.

The methodology was demonstrated with considering the following cases: (a) different batch size in one batch production, (b) different production patterns under a fixed scale. As for case (a), SU is more cost-effective than MU when the batch size is small, and as the batch size increases, this tendency changes at certain point. Regarding CO_2 emissions, SU was found to be preferable regardless of the batch size. By performing what-if analysis on pump speed, the interval of batch size was obtained where the choice will differ depending on the priority of the filling accuracy. As for case (b), the operating cost of SU is smaller than that of MU in the pattern of "small-scale production of numerous products", and the other way round in the pattern of "large-scale production of small number of products". Similar to case (a), SU was found to be environmentally friendlier regardless of the production pattern, and the what-if analysis indicated the interval where the priority of the quality makes difference in the choice of the technology.

430192 On Adsorption Hysteresis in Closed-End Pores: Isotherm Reconstruction and Free Energy Analysis Via Flat-Histogram Monte Carlo Simulation

Abstract:

In studies of the thermodynamics of fluids confined in porous materials, a particularly interesting observed effect is that of adsorption-desorption hysteresis at subcritical temperature. This effect was observed and remarked upon nearly a century ago[1] and has been a point of discussion and, occasionally, contention ever since. An early explanation of hysteresis based on macroscopic thermodynamics, the Kelvin-Cohan relationship, proposed that the effect resulted from the formation of different gas-liquid menisci during the respective adsorption and desorption processes[2]. Later investigations linked hysteresis to fluid metastability and confirmed this argument via statistical mechanical means.

Based on the Kelvin-Cohan relationship, it was long assumed that hysteresis would not occur in a closed-end pore because the meniscus would be identically structured in the adsorption and desorption processes[2]. Some molecular simulations suggested the existence of hysteresis in a closed-end pore, but the uncertainty in the data precluded a firm conclusion[3]. More recent work by Do and coworkers has more confidently revealed the existence of hysteresis in closed-end pores of various shapes[4] via molecular simulations. Despite these results, disagreement over the existence of hysteresis in closed-end pores still persists in the field.

In the present work, we examine adsorption-desorption hysteresis in closed-end pores using flat-histogram Monte Carlo methods, in particular hybrid Wang-Landau/Transition-matrix Monte Carlo simulation. As we demonstrated recently[5], TMMC can easily and accurately compute adsorption isotherms exhibiting hysteresis including identification of the metastable regions, with high confidence. We now apply this method to adsorption in closed-end pores to revisit the results of Do and coworkers, as well as examining the purported metastable regions via ensemble macrostate probability distributions and free energy analyses.

[1] F. Rouquerol, J. Rouquerol, and K. Singh, Adsorption by Powders and Porous Solids, Academic Press: London, 1999.
[2] L. H. Cohan, J. Am. Chem. Soc., 60:433, (1938).
[3] L. D. Gelb, Mol. Phys, 100:2049, (2002).
[4] C. Fan, D. D. Do, and D. Nicholson, J. Colloid Int. Sci., 405:201, (2013).; Y. Zeng et al., J. Phys. Chem. C, 118:25496, (2014).
[5] D. W. Siderius and V. K. Shen, J. Phys. Chem. C, 117:5861, (2013).

434491 Multi-Component Gas Separation of Sour Natural Gas Mixtures Using All-Silica Zeolites

Abstract:

Raw natural gas is a complex mixture comprising of mainly methane but also other impurities such as ethane, propane, butane, C_{4+}, hydrogen sulfide, carbon dioxide, nitrogen, and water vapor. For sour gas fields, selective and energy-efficient removal of H_2S and CO_2 is one of the key challenges facing the natural gas industry. Separation using nanoporous materials, such as zeolites, can be an alternative to using highly energy-intensive amine-based absorption processes. In this work, adsorption of binary mixtures of H_2S/CH_4 and H_2S/C_2H_6 in the all-silica forms of the zeolite frameworks contained in the IZASC database is investigated using Gibbs ensemble Monte Carlo simulations. The recently developed 4-site TraPPE model for H_2S, that yields high accuracy for the H_2S/CH_4 and H_2S/CO_2 binary vapor–liquid equilibria, is used for the purpose of this study. The top-performing zeolites from this initial screening are further used to simulate multi-component gas mixtures that are more representative of actual natural gas compositions. All-silica zeolites are well-known for their hydrophobic character; adsorption selectivity of hydrogen sulfide over water in select frameworks is also computed as a means to quantify this hydrophobicity in presence of H_2S.

434548 Application of Classical DFT for Screening of MOFs for CO2 Capture

Abstract:

Application of Classical DFT for Screening of MOFs for CO_2 Capture

Yun Tian, Jia Fu, Shijie Sheng and Jianzhong Wu

Department of Chemical and Environmental Engineering, University of California, Riverside, California 92521, USA

ABSTRACT

Due to the widespread public concern of increasing CO_2 emission in the atmosphere, CO_2 capture has been one of the most urgent issues. With the large gas adsorption capacity and tunable structural properties of MOFs, numerous computational efforts have been dedicated in recent years to examine the performance of MOFs for carbon capture and sequestration from different gas mixtures. Compared with adsorption of LJ fluids (e.g. noble gases, hydrogen and methane) with no charge interactions with MOFs, the long-range Coulombic interactions are crucial for CO_2. Despite various methods for extracting atomic charges from the results of quantum mechanical (QM) calculations, performing the GCMC explicitly dealing with long-range Coulombic interactions at each step of simulation for thousands or even millions of MOFs may not be feasible.

To achieve the high-throughput screening of MOFs for CO_2 adsorption, we apply the classical density functional theory (DFT) combined with MHNC-RISM to efficiently solve for the site direct correlation functions as the input for DFT calculation. DFT methods have been proven to be much more efficient than simulations in the material screening for hydrogen and methane storage in MOFs with comparable accuracy with GCMC. Within the framework of site-DFT calculations developed for solvation, the long-range electrostatic interaction with full Ewald treatment will only need to be considered once for each material at zero loading, which will be much more efficient than the conventional GCMC calculations. In combination of semi-empirical schemes for the assignment of partial charges and molecular force fields, the computational efficiency of classical DFT makes it a promising tool for design of new MOFs in application to CO_2 capture.

431726 Insights into the Coordination Polymer Ligand Deformation upon CO_2 Adsorption through Grand Canonical Monte Carlo Simulations

Abstract:

Coordination Polymer Ligands (CPLs) are nanoporous-crystalline materials consisting of layers formed by copper and ionic pyrazine-2,3-dicarboxylate that are separated by organic pillar ligands. CPLs are promising materials for carbon capture applications because they have higher adsorption affinity for CO_2 than for other small molecules, such as O_2 and N_2. Previous works, however, have reported that the structure of these materials change upon CO_2 adsorption, and while changes in the crystallographic unit cell dimensions have been estimated to quantify those changes,[1-2] the actual structural transformation mechanism is still unclear.

In this work, we use grand canonical Monte Carlo (GCMC) simulations to predict CO_2 adsorption loadings and get insights into the structural transformation of CPL-2, CPL-4, and CPL-5 during adsorption. First, we calculated CO_2 adsorption isotherms based on the crystallographic structures and unit cell dimensions reported by Matsuda et al.[1] and Riascos et al.[2], and compared them with experimental data available in the literature. However, the calculated isotherms do not correctly describe the experimental data at high pressures even if unit cell expansion as reported by Matsuda et al.[1] and Riascos et al.[2] is considered. Interestingly, when we consider the potential rotation of the organic pillar ligands as the pressure increases the agreement with experimental data is improved significantly. Indeed, we were able to reproduce experimental adsorption isotherms at high pressures through a superposition of adsorption isotherms in which different degrees of rotation of the organic ligands were induced on the CPLs structures, strongly suggesting that this kind of transformation occurs in the studied CPLs as CO_2 loadings increase.

REFERENCES

[1] R. Matsuda, R. Kitaura, Y. Kubota, T. C. Kobayashi, M. Takata, and S. Kitagawa, "Incommensurate guest adsorption in bellows-shaped one-dimensional channels of porous coordination polymers," *Microporous Mesoporous Mater.*, vol. 129, no. 3, pp. 296–303, 2010.

[2] K. Riascos-Rodríguez, A. J. Schroeder, M. R. Arend, P. G. Evans, and A. J. Hernández-Maldonado, "Hysteretic adsorption of CO2 onto a Cu2(pzdc)2(bpy) porous coordination polymer and concomitant framework distortion," *Dalton Trans.*, vol. 43, no. 28, pp. 10877–10884, Jul. 2014.

419570 Understanding Gas Adsorption in MOF-5/Graphene Oxide Composite Materials: A Computational Study

Abstract:

Recently, composite materials have garnered a great deal of interest within the scientific community due to their unique structural and electronic properties. From a computational perspective, it is very difficult to model these materials as the self-assembly pattern and the position of the atoms remain uncertain. In this talk, we seek to enhance the understanding of the composite materials by investigating the MOF-5/graphene oxide (GO) structure, which was first synthesized by Petit et al[1]. Unlike other MOF/GO structures, it has been experimentally demonstrated that MOF-5/GO possesses significant degree of order, making the material relatively simple to model. Our study focuses on the CO2 adsorption properties at infinite dilution condition and in particular, the synergestic effects of MOF-GO interfaces on the adsorption strength are evaluated by making comparisons between the composite and the individual host materials. Furthermore, using the experimental adsorption isotherm data as reference values, a simple Monte Carlo simulation that shuffles the MOF and the GO layers is performed to identify the self-assembly pattern. Finally, additional functional groups are grafted to the MOF-5 layers in order to determine their effect on the adsorption properties with the aim of achieving better adsorbents for carbon capture application.

[1] Petit, C., & Bandosz, T. J. (2009). MOF–graphite oxide composites: combining the uniqueness of graphene layers and metal–organic frameworks. *Advanced Materials*, *21*(46), 4753-4757.

CO_2 Adsorption Properties of DMOF
A Computational Study

Jorge Rosa-Raices, Paul Meza-Morales, Maria Curet-Arana

Department of Chemical Engineering, University of Puerto Rico-Mayaguez
AIChE Annual Meeting 2015

DMOFs

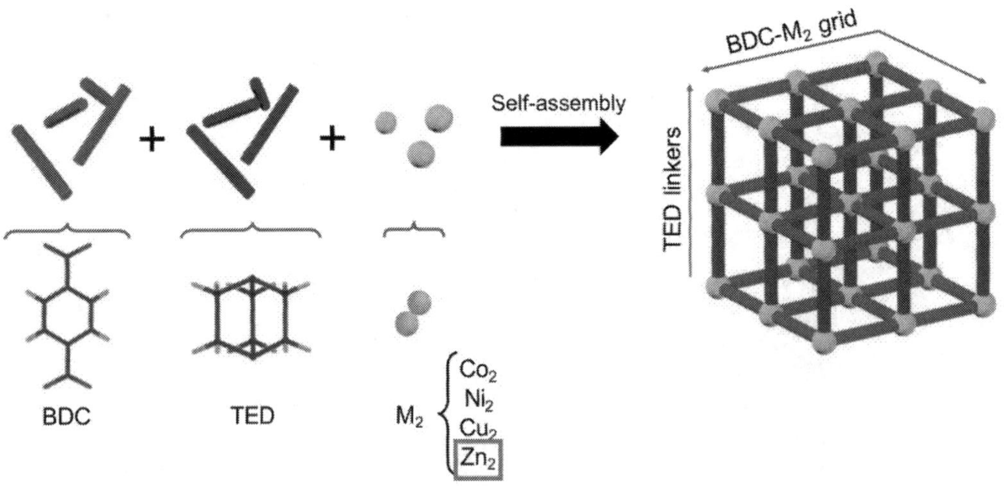

Zn-DMOF

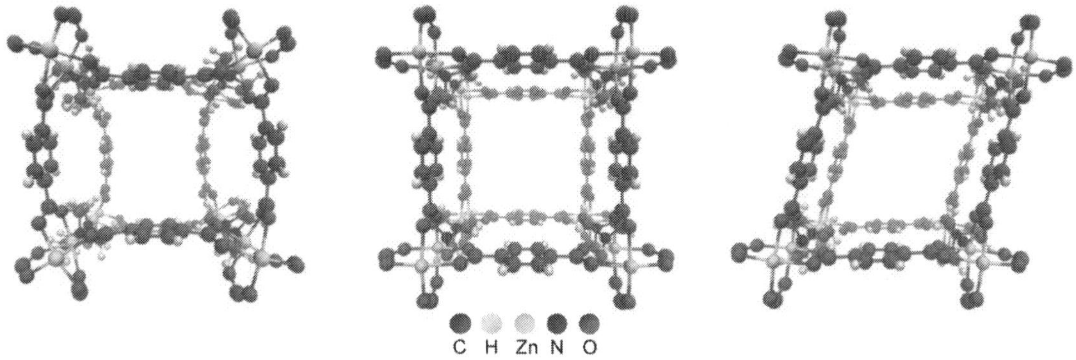

b

a

As synthesized[1]
(bent)

Upon activation[1]
(straight)

Upon benzene adsorption[1]
(skewed)

[1] Dybtsev, D. N., Chun, H., & Kim, K. (2004). *Angewandte Chemie*, *116*(38), 5143–5146.

Zn-DMOF

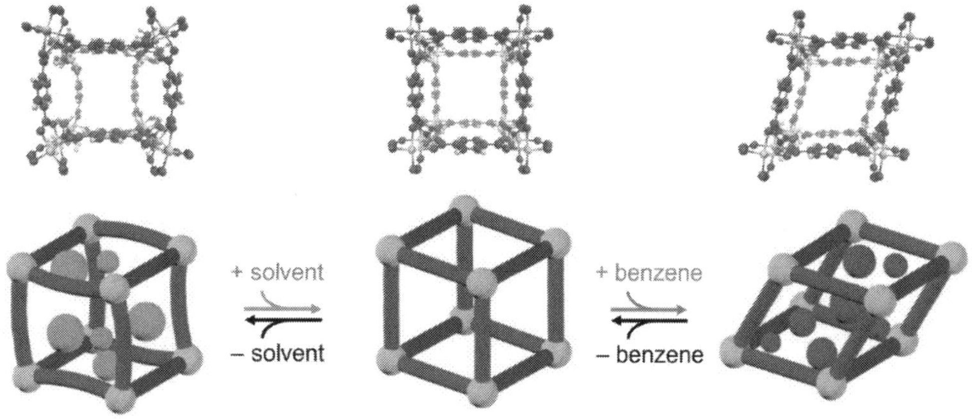

b

a

As synthesized[1]
(bent)

Upon activation[1]
(straight)

Upon benzene adsorption[1]
(skewed)

[1] Dybtsev, D. N., Chun, H., & Kim, K. (2004). *Angewandte Chemie*, *116*(38), 5143–5146.

BACKGROUND METHODS RESULTS CLOSURE 5

CO_2 adsorption onto Zn-DMOF

Experimental work

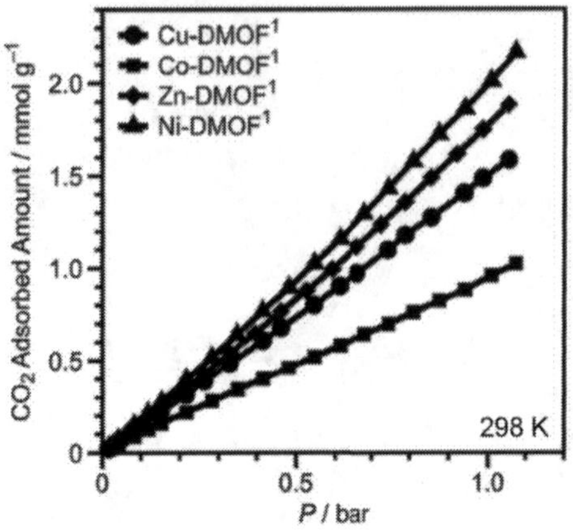

[1] S. Chaemchuen et al. (2015). *Microporous Mesoporous Mat.*, 201, 277–285.

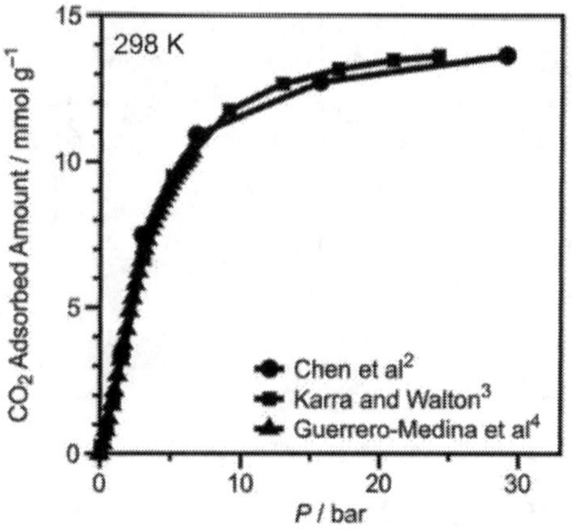

[2] Chen, Y. F., et al. (2010). *J. Phys. Chem. C*, 114, 6602–6609.
[3] Karra, J. R., & Walton, K. S. (2010). *J. Phys. Chem. C*, 114, 15735–15740.
[4] Guerrero-Medina, J. et al. (2015). *Microporous Mesoporous Mat.* 212, 8–17.

BACKGROUND METHODS RESULTS CLOSURE 6

Aims of this work

- Simulate CO_2 adsorption isotherms for Zn-DMOF using the Grand Canonical Monte Carlo method
- Provide insight into the interactions between CO_2 and Zn-DMOF through quantum mechanics calculations

Grand Canonical Monte Carlo

Adsorption simulation with RASPA[1]

INPUT
Thermodynamic variables *T, P*
Sorbate and sorbent models
Force field definitions: **UFF, TraPPE**

→ **SIMULATION**
Grand-canonical ensemble
Monte Carlo integration

→ **OUTPUT**
Adsorption isotherm

MC MOVES
Insertion
Deletion
Translation/Rotation

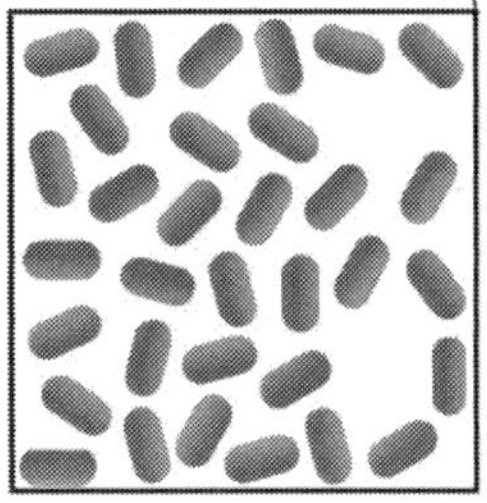

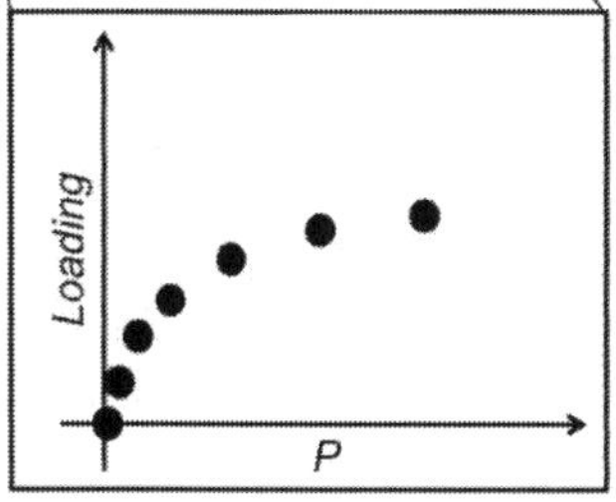

[1] Dubbeldam, D. et al. (2015). *Molecular Simulation*, (May), 1–21.

Interaction potential

$$U_{ij} = \underbrace{4\epsilon_{ij}\left[\left(\frac{\sigma_{ij}}{r_{ij}}\right)^{12} - \left(\frac{\sigma_{ij}}{r_{ij}}\right)^{6}\right]}_{\text{Lennard-Jones (LJ) Van der Waals}} + \underbrace{\frac{1}{4\pi\epsilon_0}\frac{q_i q_j}{r_{ij}}}_{\text{Coulombic (C) Electrostatic}}$$

Lennard-Jones (LJ)
Van der Waals

Coulombic (C)
Electrostatic

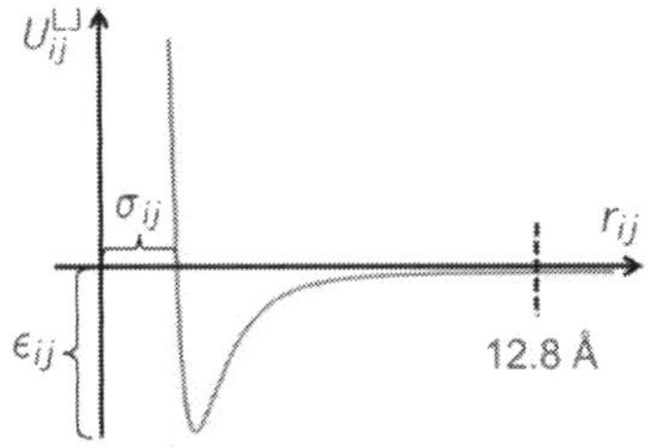

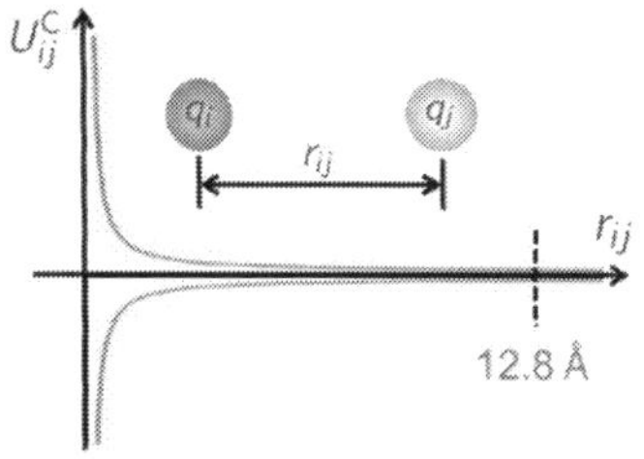

Partial atomic charges from *ab initio*[1]

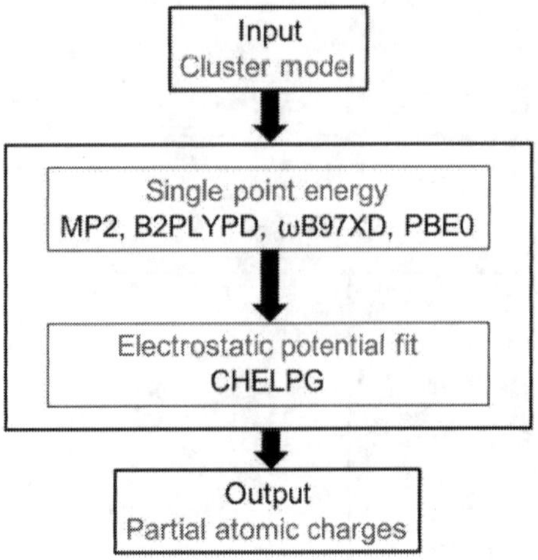

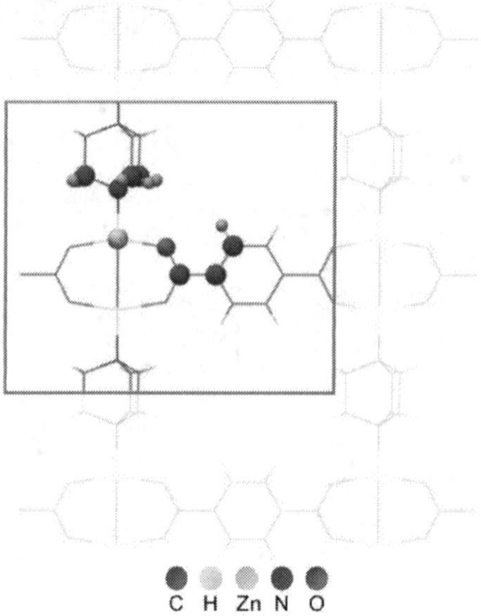

[1] Gaussian 09, Revision A.02, Frisch et al.

Partial atomic charges

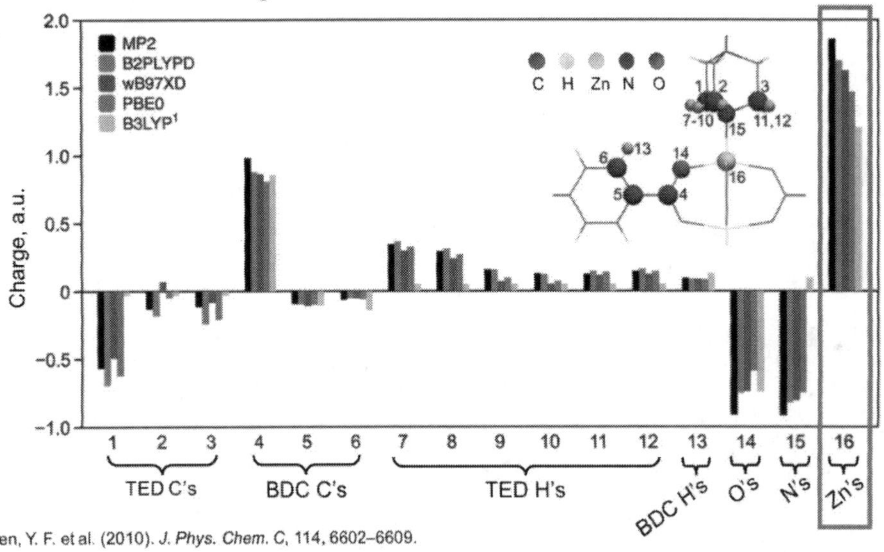

[1] Chen, Y. F. et al. (2010). *J. Phys. Chem. C*, 114, 6602–6609.

CO_2 adsorption isotherms

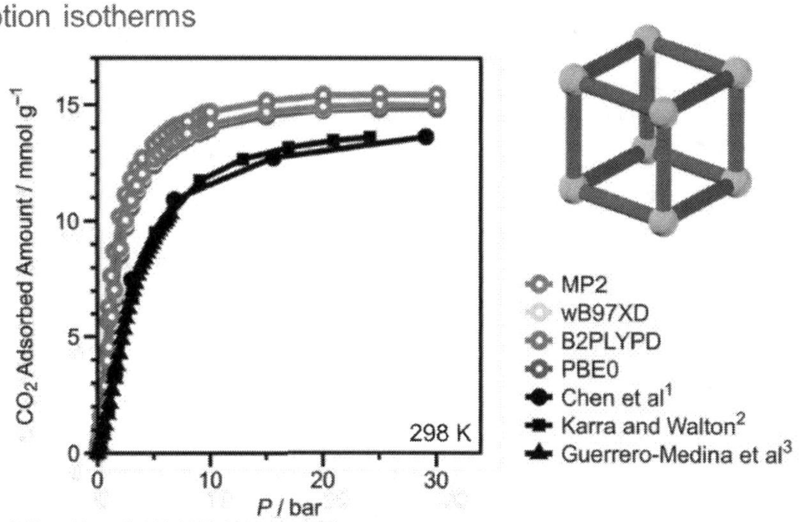

[1] Chen, Y. F., et al. (2010). *J. Phys. Chem. C*, 114, 6602–6609.
[2] Karra, J. R., & Walton, K. S. (2010). *J. Phys. Chem. C*, 114, 15735–15740.
[3] Guerrero-Medina, J. et al. (2015). *Microporous Mesoporous Mat.* 212, 8–17.

N_2 adsorption isotherms

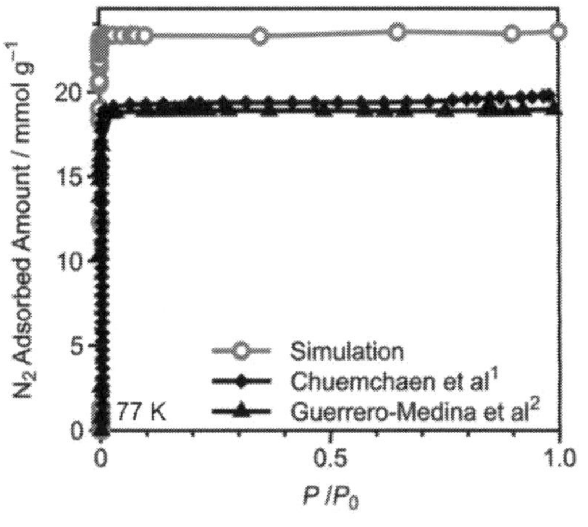

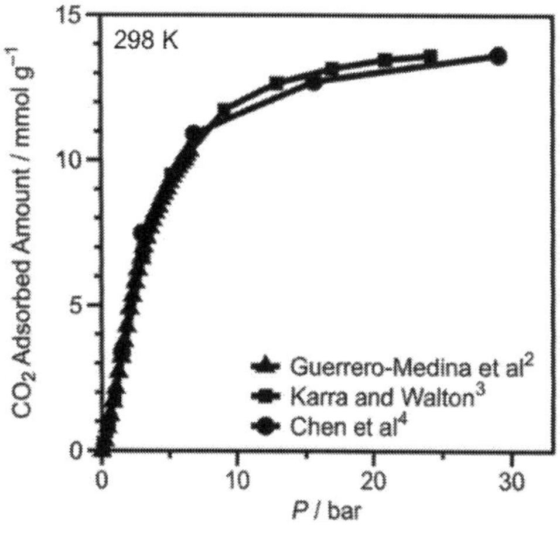

[1] S. Chaemchuen et al. (2015). *Microporous Mesoporous Mat.*, 201, 277.
[2] Guerrero-Medina, J. et al. (2015). *Microporous Mesoporous Mat.* 212, 8.
[3] Karra, J. R., & Walton, K. S. (2010). *J. Phys. Chem. C*, 114, 15735.
[4] Chen, Y. F., et al. (2010). *J. Phys. Chem. C*, 114, 6602.

Zn-DMOF

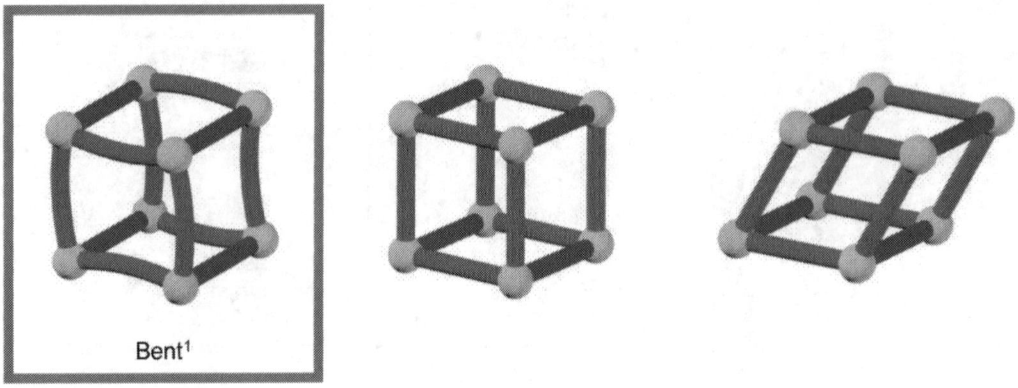

[1] Dybtsev, D. N., Chun, H., & Kim, K. (2004). *Angewandte Chemie*, *116*(38), 5143–5146.

N_2 adsorption isotherms

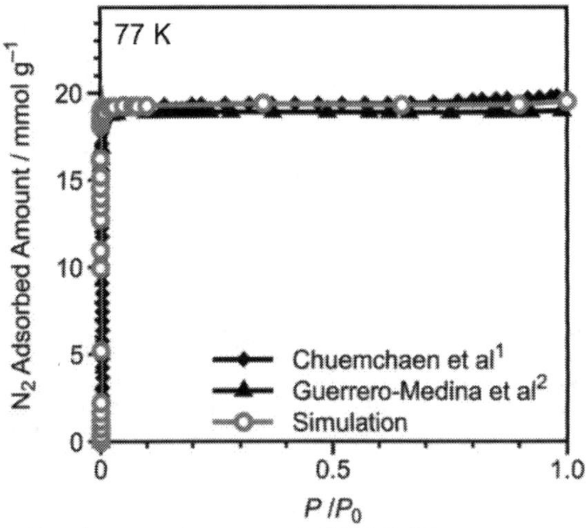

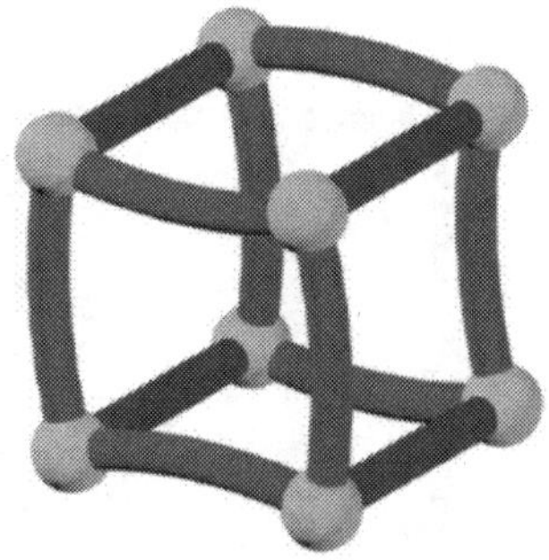

[1] S. Chaemchuen et al. (2015). *Microporous Mesoporous Mat.*, 201, 277–285.
[2] Guerrero-Medina, J. et al. (2015). *Microporous Mesoporous Mat.* 212, 8–17.

Structural changes upon CO_2 adsorption in Zn-DMOF

Experimental work[1]

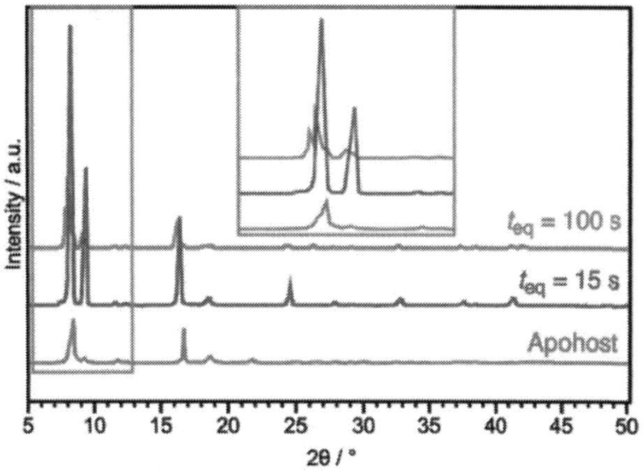

[1] Guerrero-Medina, J. et al. (2015). *Microporous Mesoporous Mat.* 212, 8–17.

Zn-DMOF

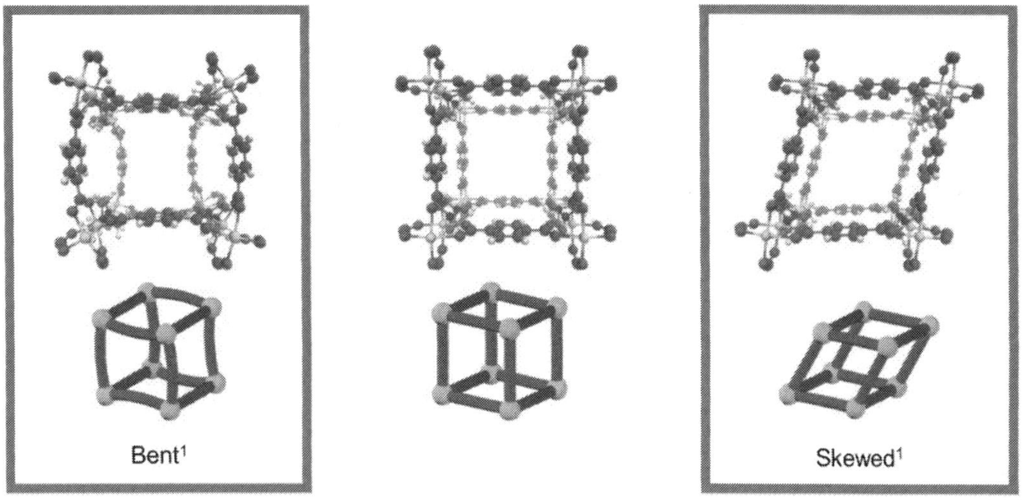

[1] Dybtsev, D. N., Chun, H., & Kim, K. (2004). *Angewandte Chemie*, *116*(38), 5143–5146.

CO_2 adsorption isotherms

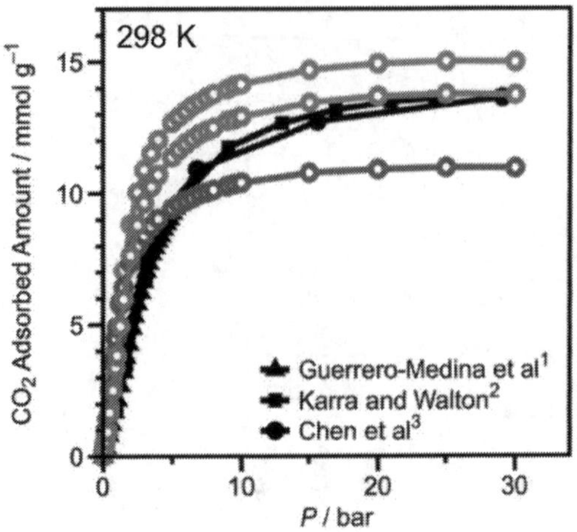

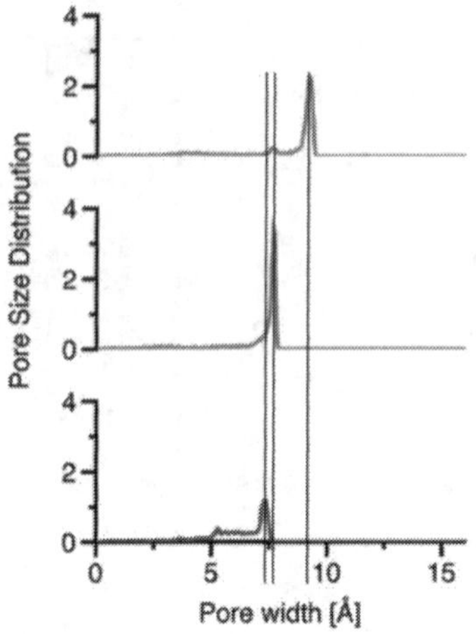

[1] Guerrero-Medina, J. et al. (2015). *Microporous Mesoporous Mat.* 212, 8–17.
[2] Karra, J. R., & Walton, K. S. (2010). *J. Phys. Chem. C*, 114, 15735–15740.
[3] Chen, Y. F., et al. (2010). *J. Phys. Chem. C*, 114, 6602–6609.

CO_2 adsorption isotherms

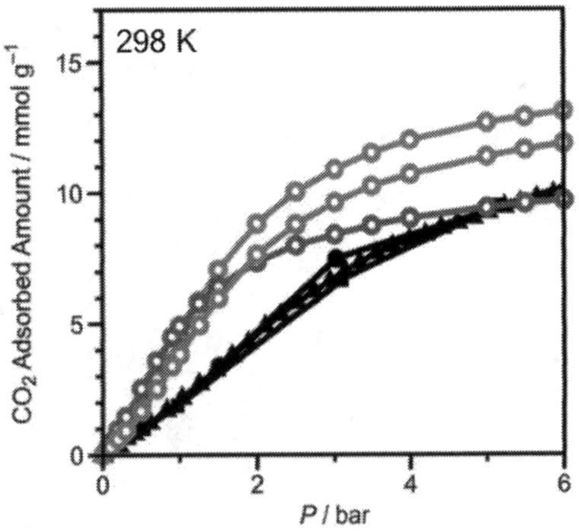

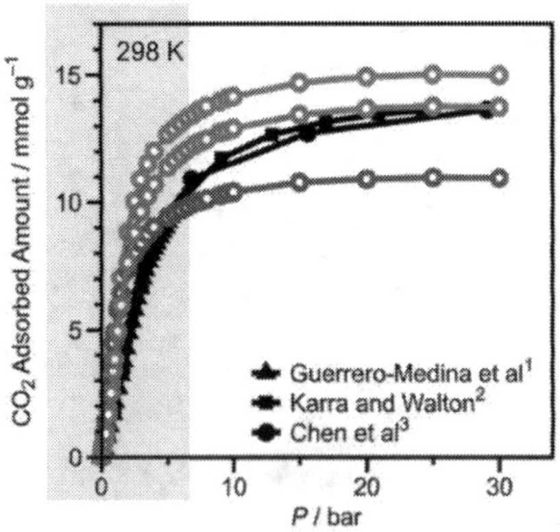

[1] Guerrero-Medina, J. et al. (2015). *Microporous Mesoporous Mat.* 212, 8–17.
[2] Karra, J. R., & Walton, K. S. (2010). *J. Phys. Chem. C*, 114, 15735–15740.
[3] Chen, Y. F., et al. (2010). *J. Phys. Chem. C*, 114, 6602–6609.

CO_2 / Zn-DMOF interacting sites

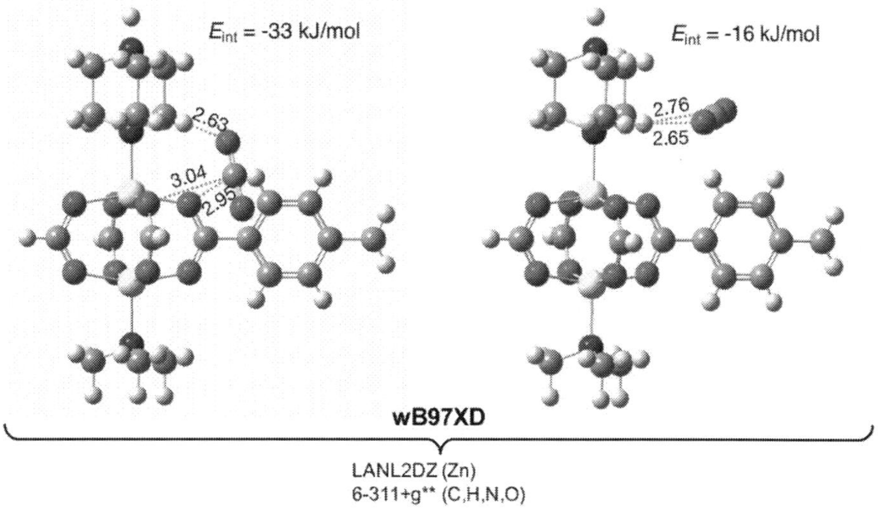

Concluding remarks

- From our simulations with the crystallographic data for the activated structure, the data is not able to describe the CO_2 or N_2 adsorption isotherms obtained by experiment
- Results using the skewed structure suggest that we should look further into the effect of CO_2–induced structural changes on the loading capacity of Zn-DMOF
- We have obtained minimal energy configurations of the CO_2–MOF fragment complex that show the crucial role of the paddle wheel unit in stabilizing the CO_2 interactions with the MOF

BACKGROUND METHODS RESULTS CLOSURE

Acknowledgements

People

- María Curet-Arana, PhD
- Paul Meza-Morales, mentoring graduate student
- Diego Gomez-Gualdron, PhD (Northwestern University)

Computing resources

Funding

PR NASA EPSCoR

NASA–EPSCoR (Grant number NNX13AD38A)
NSF (Grant number: EPS-1002410)

THANK YOU FOR YOUR ATTENTION

APPENDIX

CO_2 adsorption isotherms

Previous theoretical work

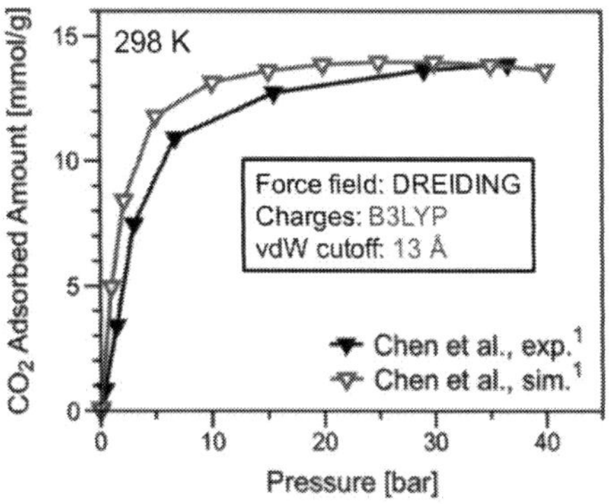

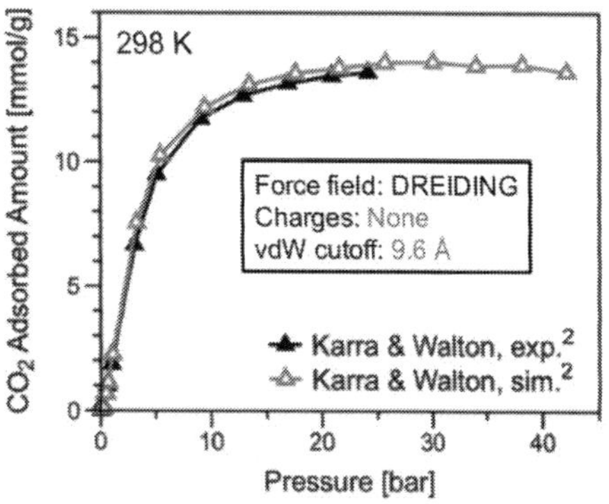

[1] Chen, Y. F., et al. (2010). *J. Phys. Chem. C*, 114, 6602–6609.
[2] Karra, J. R., & Walton, K. S. (2010). *J. Phys. Chem. C*, 114, 15735–15740.

CO_2 adsorption isotherms

Effect of reducing cutoff

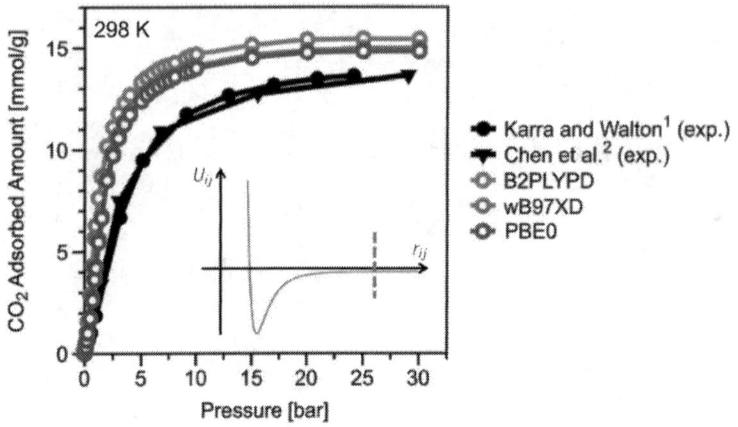

CO_2 adsorption isotherms

Effect of reducing cutoff

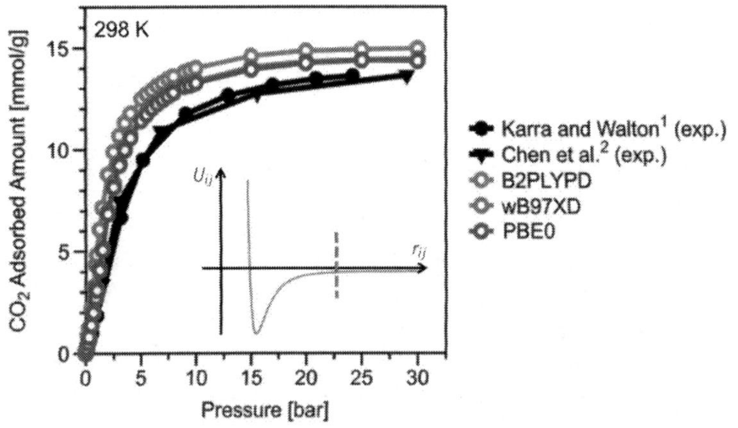

BACKGROUND METHODS RESULTS CLOSURE 27

Electrostatic potential maps

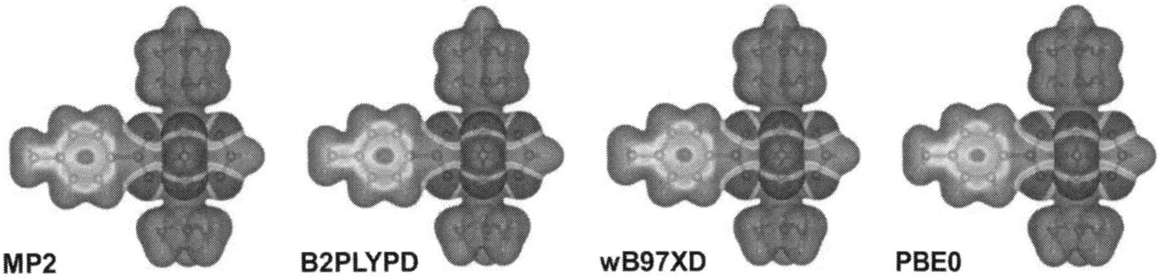

MP2	B2PLYPD	wB97XD	PBE0
aug-cc-pVDZ (Zn) D95V+* (C,H,N,O)	aug-cc-pVQZ (Zn) 6-311+g** (C,H,N,O)	LANL2DZ (Zn) 6-311+g** (C,H,N,O)	aug-cc-pVQZ (Zn) 6-311+g** (C,H,N,O)

–0.025 ISO = 0.01 +0.025

412423 Stimuli-Responsive Smart Membranes for Controlled Release and Separations

Abstract:

As emerging artificial biomimetic membranes, smart membranes that are able to respond to environmental stimuli are attracting ever-increasing interest from various fields. Their surface characteristics and/or permeation properties including hydraulic permeability (pressure-driven convective flow of solvents) and diffusional permeability (concentration-driven molecular diffusion of solutes) can be dramatically controlled or adjusted self-regulatively in response to small chemical and/or physical stimuli in their environments, such as temperature, pH, glucose concentration, and chemical or biological species. Such environmental stimuli-responsive smart membranes could find myriad applications in numerous fields ranging from controlled drug delivery, to separation, water treatment, chemical sensors, chemical valves, and tissue engineering, *etc.*[1] In this presentation, the trans-membrane mass-transfer and membrane separation is introduced as the beginning to initiate the requirement of smart membranes, and bio-inspired design of environmental stimuli-responsive smart membranes are introduced and discussed. Then, smart membrane types and their applications as smart tools for controllable mass-transfer in controlled release and separations are introduced. The research topics in the near future are also discussed.

Reference

1. L.Y. Chu. *Smart Membrane Materials and Systems - From Flat Membranes to Microcapsule Membranes*. Springer-Verlag Berlin Heidelberg (2011).

427894 Fouling Resistant Temperature and Magnetic Field Responsive Nanofiltration Membranes

Abstract:

Nanofiltration membranes are finding increasing industrial applications. Advantages over reverse osmosis membranes include much lower operating pressures yet significant rejection of multivalent ions. However membrane fouling is a significant concern that frequently limits the commercial viability of nanofiltration membranes. Here we have developed fouling resistant, self-cleaning nanofiltration membranes. A method aimed at fouling mitigation and control, using stimuli-responsive polymer brushes attached to nanofiltration membranes is proposed and examined.

This study examines the effect of poly-N-isopropylacrylamide (PNIPAM) brushes attached to NF-270 membrane surface. PNIPAM is a thermo-responsive polymer that has a lower critical solution temperature (LCST) of 32°C. Below the LCST PNIPAM brushes are in an expanded coil state and the intermolecular forces between the brushes and water are dominant - the polymer is hydrophilic. Above this temperature these grafted brushes form a collapsed dehydrated globular state as interactions between the brushes are more dominant. By changing the temperature of the feed the grafted layer can switch between a more swollen (hydrophilic) and collapsed (hydrophobic) state leading to detachment of adsorbed foulants.

Changing the temperature of the bulk feed can be energy intensive. We have developed a method to attach superparamagnetic nanoparticles to the ends of the grafted PNIPAM chains. In an oscillating magnetic field heating of the superparamagnetic particles attached directly to the PNIPAM chains leads to collapse of the chains when the temperature is increased above the LCST of the grafted chains. Thus we can change the conformation of the grafted chains without heating the bulk feed. The effects of polymer grafting degree varying both the chain length and chain density are studied. Our results indicate the importance of carefully tailoring the architecture of the grafted polymer chains in order to minimize membrane fouling.

430449 Responsive N-Isopropylacrylamide (NIPAAm) Ultrafiltration Membranes

Abstract:

Membrane fouling occurs when there is reversible or irreversible attachment of macrosolutes present in the water to the membrane surface. Membrane replacements due to fouling can add to the operating costs of membrane systems. Reversible fouling can be minimized by crossflow operation, backflushing and/or by chemical pretreatment. On the other hand, irreversible fouling cannot be minimized by backflushing or crossflow filtration, and it results in permanent flux decline. Among irreversible foulants, natural organic matter (NOM) is considered to be a major contributor. NOM is composed of a wide range of hydrophilic and hydrophobic components; hence, any static hydrophobic or hydrophilic membrane can get fouled. Therefore, a dynamic membrane able to alternate between being hydrophobic and hydrophilic is hypothesized to decrease fouling. Stimuli-sensitive polymers can change their conformation from a coiled and hydrophobic state to an uncoiled (or globular) and hydrophilic state in the presence of a stimulus. By using such polymers to make a membrane and then alternating between the phases, the membrane surface can be made dynamic.

The purpose of this study was to develop stimuli responsive membranes to control fouling using membrane casting dopes made of cellulose acetate with N-isopropylacrylamide (NIPAAm). NIPAAm is a stimuli-responsive polymer, which offers the potential to reversibly collapse or expand the membrane as a function of changes in temperature. By continuously activating the film, it is hypothesized that a dynamic stimuli-responsive surface can prevent foulants from attaching to the membrane surface.

Membranes were cast using phase inversion, were characterized chemically, morphologically, and were used in filtration experiments using bovine serum albumin (BSA), lipase and humic solutions. Flux studies were conducted at alternating cold and hot temperature cycles. CA-NIPAAm membranes offered an average higher flux during operation and lower protein and humic accumulation on the membrane surface as compared to regular CA membranes. CA-NIPAAm membranes showed higher flux recovery than CA membranes.

429303 Novel Photoresponsive Water Filtration Membranes

Abstract:

Materials with reversibly switchable properties hold great interest in the development of membranes with new functions such as fouling resistance and self-cleaning, adjustable pore size, and tunable flux. Photoresponsive membranes can show these changes without the need for changes to the feed composition. They can also show orthogonal response when combined with pH- or ionic strength responsive functionalities. This research focuses on light-responsive membranes that can self-clean to remove foulants upon exposure to light. We show the synthesis of new comb-shaped polymers with photo-regulated zwitterion formation and their use as responsive membrane selective layers. Zwitterions are charged molecules that comprise equal numbers of covalently bound cations and anions. The unique co-existence of opposite charges bridged through alkylene spacers provides zwitterions extremely high polarity. This feature leads to the exceptional anti-fouling properties for zwitterionic materials. This has exciting applications in the membrane field. We synthesized novel comb-shaped graft copolymers at different side-chain lengths with polyacrylonitrile (PAN) backbones and photoreactive sidechains using atom transfer radical polymerization (ATRP). The side-chains undergo a light induced transition between a hydrophobic neutral state and a zwitterionic state, allowing photoregulated control over membrane features. We used these graft copolymers to modify a commercially available UF membrane by coating a thin layer of the copolymer solution followed by precipitation in a non-solvent. Prior to any photo-treatment, coated membranes consist mainly of hydrophobic groups that dominate the membrane surface, and they increase upon visible light irradiation. We postulate that the hydrophobic groups promote the adsorption of hydrophobic solutes on channel walls, reducing flow rate. Upon reversing the photochemical response by irradiation with UV light, the hydrophobic groups are converted to zwitterionic groups that release the adsorbed molecules and permit water passage once again. This "self-cleaning" behavior has been has been shown by measuring pre- and post-UV water permeability after fouling with bovine serum albumin. We have found that a copolymer with shorter side chain length imparts better self-cleaning capability to the membrane. Changes in fingerprint IR peaks pertinent to neutral and zwitterionic forms upon light treatment have further confirmed the structural difference between the two forms at a molecular level. Additional work focuses on determining optimal phototreatment regimes for self-cleaning behavior with minimal energy use, and developing a better understanding of how polymer self-organization controls this responsive behavior.

434932 Visible-Light-Responsive Materials for Separation of Oil-Water Mixtures

Abstract:

Photocatalytic materials such as TiO_2 have attracted significant attention in altering the wettability of surfaces. These materials degrade any organic deposits and induce superhydrophilic properties to the surface upon irradiation with UV light. The excitation of TiO_2 by light with energy greater than the bandgap is the primary process underlying its vast applications in self-cleaning, antifogging, treatment of wastewater and most recently separation of oil-water mixtures. However, the large bandgap of TiO_2 (~ 3.2 eV for anatase and ~ 3.0 eV for rutile) limits its ability to absorb visible light or sunlight effectively. Given that only ~ 5 % of solar flux incident at the earth's surface lies in this spectral regime (λ < 390 nm), utilization of a large fraction of sunlight for excitation process can thus be enhanced by tuning the bandgap of TiO_2 to the visible spectral regime. In this work, we have developed novel approaches to tune the bandgap of TiO_2 to be responsive in the visible spectral regime by surface modification. Under visible light irradiation, we show that these modified TiO_2 surfaces undergo hydrophobic-to-hydrophilic conversion in both air and oil environments. We explain the mechanism behind the change in wettability in terms of the surface regeneration processes. Such unique visible light-induced hydrophobic-to-hydrophilic conversion of our surfaces enables on-demand separation of oil-water mixtures. We anticipate that the materials that have such tunable light-responsive wettability can find broad applications in oil-water separation, wastewater treatment, self-cleaning, and phase transitions.

428774 Stimuli-Responsive Charge Mosaic Membranes Prepared Using Layer-By-Layer Assembly

Abstract:

Charge mosaic membranes (CMMs) are capable of separating dissolved ions from neutral molecules of comparable size by facilitating diffusion of charged species through oppositely charged domains. We have developed a new class of the CMMs by utilizing layer-by-layer assembly to form positively and negatively charged polymeric nanotubes. Selecting different polyelectrolyte multilayer systems determines the nanotube functionality. These nanotubes are vertically aligned and traverse the membrane thickness with no overlap of adjacent charged domains. Initial studies revealed these CMMs have hydraulic permeabilites of 8 L m^{-2} hr^{-1} bar^{-1}. Since the charge mosaic membranes are fabricated from polyelectrolyte multilayer systems, they exhibit greater functionality and stimuli-responsiveness than CMMs fabricated from other materials (e.g., ion exchange resins). In particular, our CMMs demonstrate a pH-induced hysteresis in both pure water flux and salt rejection. Using profilometry and Fourier transform infrared spectroscopy, we identify that the hysteresis is a result of the polymer nanotubes swelling and deswelling as the functional groups are protonated and deprotonated. Challenging the membrane with feed solutions of varying ionic strength and feed component chemistries illuminated some of the underlying transport mechanisms of the CMMs. These properties demonstrate the increased functionality of these novel CMMs as a result of fabrication from layer-by-layer assembly of polyelectrolytes.

425536 Responsive Membranes for Hydrophobic Interaction Chromatography: The Effect of Salt Ions

Tuesday, November 10, 2015: 10:40 AM
155C (Salt Palace Convention Center)

S. Ranil Wickramasinghe[1], Zizhao Liu[1] and Xianghong Qian[2], (1)Department of Chemical Engineering, University of Arkansas, Fayetteville, AR, (2)Department of Biomedical Engineering, University of Arkansas, Fayetteville, AR

Responsive membranes for hydrophobic interaction chromatography: The effect of salt ions

Ranil Wickramasinghe, Zizhao Liu, Xianghong Qian

The use of environmentally responsive membranes for hydrophobic interaction chromatography (HIC) enables high recovery and high resolution during protein purification. Further the possibility of protein fractionation also exists. Here poly N-vinylcaprolactam (PVCL) has been grafted from the surface of commercially available regenerated cellulose membranes (0.45 μm pore size) via atom transfer radical polymerization. PVCL displays a lower critical solution temperature (LCST) at around 32 °C in DI water. The observed LCST of PVCL is a strong function of solution ionic strength and type of salt ion present. Increasing ionic strength leads to a decrease in the LCST. Above the LCST the grafted polymer chains collapse, dehydrate and appear more hydrophobic promoting protein adsorption. Below the LCST the grafted chains swell, hydrate and promote protein desorption.

Optimization of this novel system requires a systematic investigation of different salt types and concentrations. Here dynamic binding capacities of our PVCL modified membranes were determined at different pH, salt type and concentration at room temperature with two model proteins, a human immunoglobulin$_4$, IgG_4, and bovine serum albumin (BSA). In addition, BSA adsorption isotherms in the presence of monovalent (Na^+, NH_4^+), divalent (Mg^{2+}, Zn^{2+}) and trivalent (Al^{3+}) sulfate solutions were compared in order to investigate the effect of the cation). Selected BSA binding capacity data (left hand side y-axis) and recovery data (right hand side y-axis) are given in the figure below. As can be seen the ionic strength and type of salt not only affect the binding capacity, but also the recovery during elution. Finally, the responsiveness of the PVCL ligand synthesized in solution as well as grafted on silicon wafers was investigated. Molecular dynamics simulations were also conducted to elucidate the salt effect on the PVCL ligand.

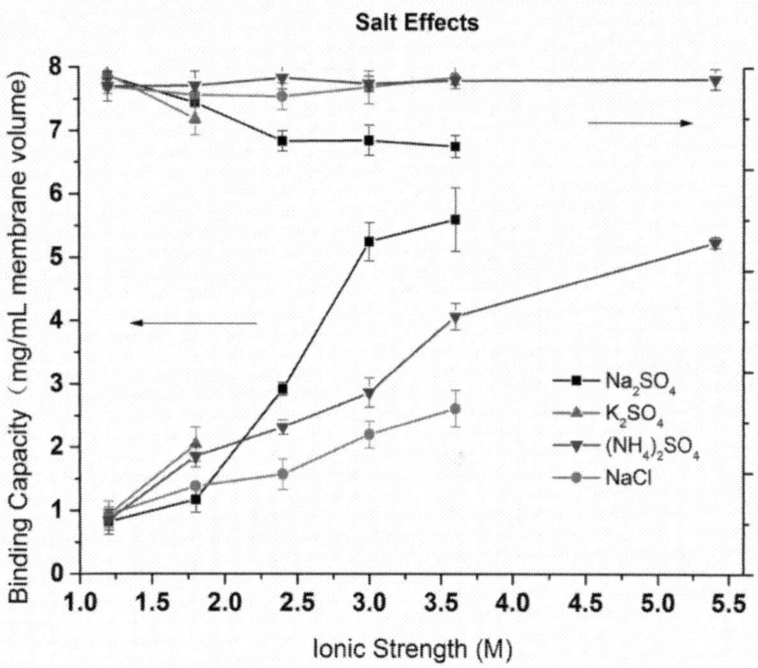
Salt Effects
Binding Capacity（mg/mL membrane volume)
Recovery (%)
Ionic Strength (M)
Na2SO4
K2SO4
(NH4)2SO4
NaCl
0
1
2
3
4
5
6
7
8
20
40
60
80
100
1.0
1.5
2.0
2.5
3.0
3.5
4.0
4.5
5.0
5.5

429166 Charged Nanopore Membranes By Random Copolymer Micelle Assembly

Abstract:

Membranes that can separate solutes by factors other than size (e.g. charge, hydrophobicity, chemical functionality) would enable the use of this energy-efficient, green technology in new applications. However, achieving membrane selectivity based on parameters other than size has been a challenge, especially when the molecules to be separated are small. Biological pores such as ion channels achieve exceptional selectivity by confining flow into very small pores lined with functional groups. Therefore, incorporating nanostructures that confine permeation into membrane selective layers can significantly enhance selectivity. In this study, we created membrane selective layers with nanopores lined with carboxyl groups using the deposition of micelles formed in methanol onto porous membranes. To prepare these membranes, we prepared an amphiphilic random copolymer that combines highly hydrophobic, fluorinated repeat units of 2,2,2-trifluroethyl methacrylate (TFEMA) with repeat units of methacrylic acid (MAA), PTFEMA-*r*-PMAA, by free radical polymerization. The copolymer contained 55 wt% TFEMA, and was insoluble in water. We found that this copolymer forms micelles in methanol with a multimodal size distribution, as shown by dynamic light scattering (DLS). Using cupric acetate in membrane casting solution was found to change the average size and the size distribution of the micelles. When these micelles are coated onto the surface of a porous support membrane whose pores are smaller than the micelles, a selective layer of micelles packed together is formed. If the membrane pore size is too large, micelles go inside the membrane, clogging inner pores and resulting in a membrane with very low flux. Using this method, we achieved membranes with pure water permeabilities up to ~8 $L/m^2.h.bar$. The resultant membranes exhibited rejection properties expected from negatively charged membranes with ~2 nm pore size, as indicated by the relative rejections of monovalent and divalent salts. The membrane showed charge-based selectivity between organic dye molecules, with a negatively charged dye being retained by >85% whereas a positively charged dye with a similar molecular size was retained by <20%. This is the first demonstration of a new approach to forming membranes from easily synthesized copolymers by micelle assembly. We believe this approach can be adapted to address various pore sizes and pore surface chemistries, leading to novel membranes with promising selectivity.

434171 Ion Transport Structure/Property Relationships in Charged Polymer Membranes

Abstract:

Global need for sustainable water and energy sources presents an opportunity for polymer science as many technologies for water purification and some emerging power generation technologies rely on polymeric membranes to control rates of water and ion transport. Ion exchange membranes are used in many of these technologies, and the presence of charged groups on the polymer backbone of such materials influences ion sorption and, subsequently, both rates of ion transport and the selective nature of ion transport. Membranes are often characterized using sodium chloride, which is widely prevalent in nature, but several applications expose membranes to a variety of different ions. For electric field-driven processes, ion transport properties, such as ionic resistance and electrochemical permselectivity, are sensitive to the specific nature of the ions present in the system. This presentation will discuss the influence of ion properties and ion-polymer interactions on the ion transport properties of polymer membrane materials.

Alkali Metal Chloride Sorption in Cross-linked Polymer Membranes

Eui Soung Jang, Donald R. Paul and Benny D. Freeman

Department of Chemical Engineering and the Texas Materials Institute

The University of Texas at Austin

10100 Burnet Road, Building 133, Austin, TX 78758, USA

2015 AIChE Annual Meeting

November 10th, 2015

Polymer Membrane-Based Technologies to Address Water and Energy Needs

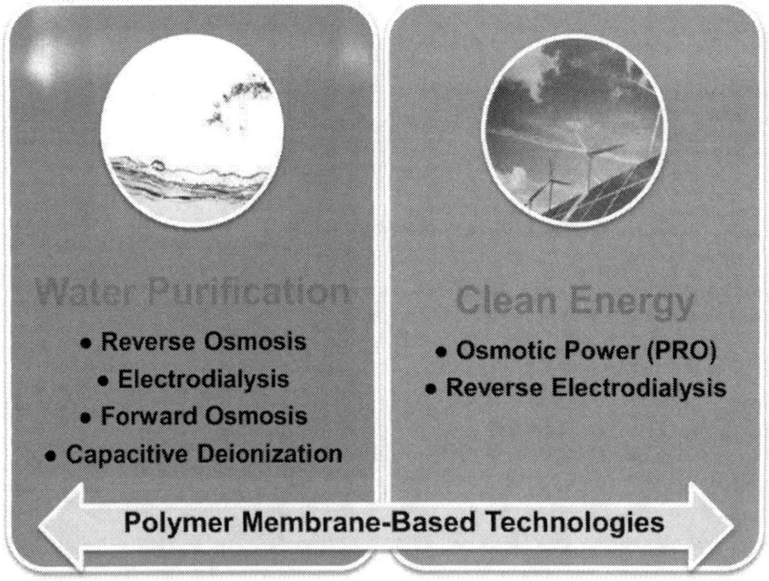

The fundamental understanding of structure-property relationship leads to rationally designing new membranes.

Geise, Lee, Miller, Freeman, McGrath, Paul, *J. Polym. Sci. Part B*, **48**,1685 (2010) .

2

Approach : Systematic Studies of Structure-Property Relationships

Structure-property relationships

- **Effect of fixed charge group**
- **Transport of various electrolytes**
- **Thermodynamic non-ideality**

Polymer systems

Uncharged Polymers
Cross-linked poly(ethylene glycol) diacrylate (XLPEGDA)

Charged Polymers
XL(AMPS-DEGDMA)

AMPS-DEGDMA: poly(2-acrylamido-2-methylpropane sulfonic acid-r-diethylene glycol dimathacrylate)

3

Solution-Diffusion Mechanism

Salt Transport

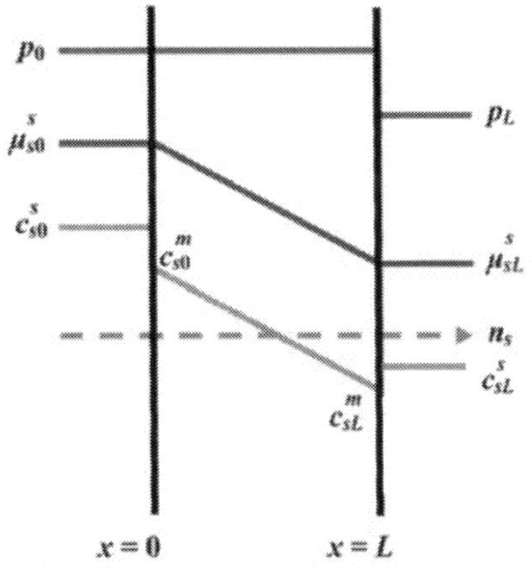

$$n_s = \frac{P_s}{L} \Delta c_s^s$$

$$\langle P_S \rangle = \langle D_S \rangle \times K_S$$

$$< P_i > = \text{Salt Permeability}$$

$$< D_i > = \text{Salt Diffusivity}$$

$$K_i = \text{Salt Solubility}$$

< > = average value

D. Paul, J Membrane Sci, 241 (2004) 371-386.

4

Uncharged Polymer Used in This Study

Cross-linked poly(ethylene glycol diacrylate) [XLPEGDA](M_n = 700)

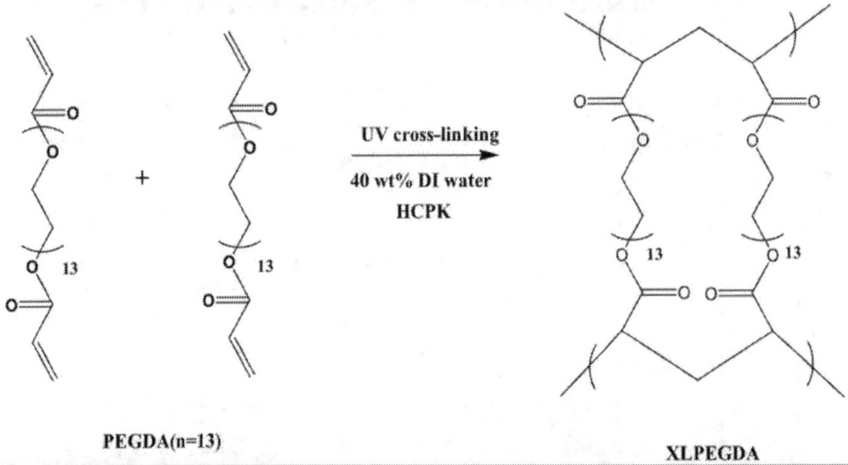

	Pre-polymerization water content [wt%]	Equilibrium water uptake [g(water)/g(dry polymer)]	Equilibrium water volume fraction
XLPEGDA	40	0.93	0.52

HCPK: 1-Hydroxy-Cyclohexyl-Phenyl-Ketone

5

Charged Polymer Used in This Study

Cross-linked poly(AMPS-DEGDMA)

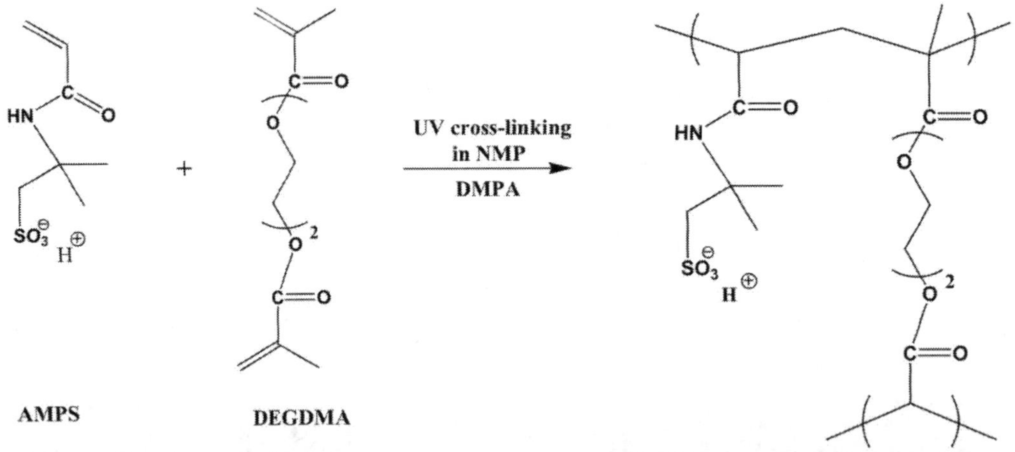

	IEC (meq/g dry polymer)	Equilibrium water uptake [g(water)/g(dry polymer)] (Na^+ form)	Equilibrium water volume fraction
CEM	1.45	0.89	0.52

AMPS: 2-acrylamideo-2-methylpropane sulfonic acid; DEGDMA: diethylene glycol dimethacrylate; DMPA: 2,2-Dimethoxy-2-phenylacetophenone

6

Uncharged vs. Charged Polymers

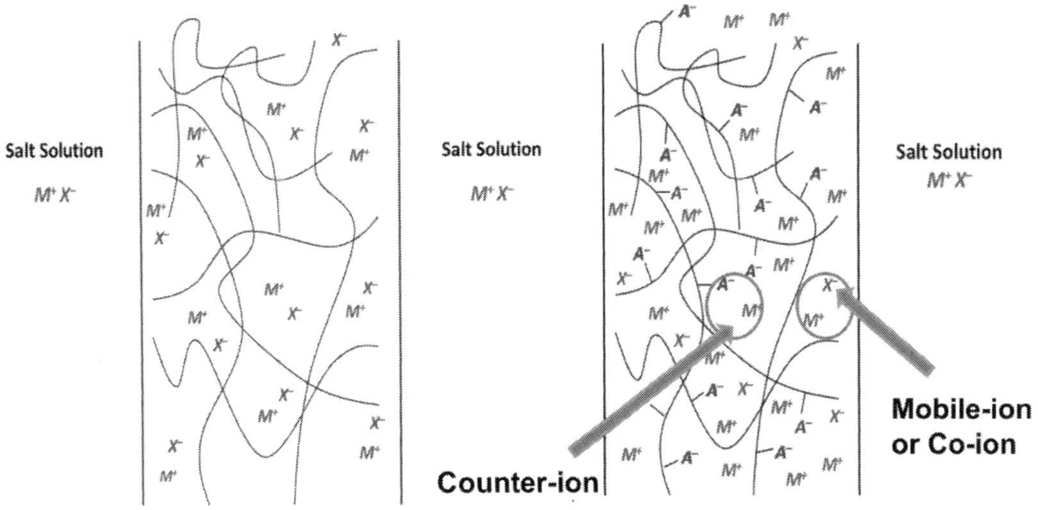

For sulfonated polymers:
$A^{-} = -SO_3^{-}$
(bound to the polymer backbone)

G.M. Geise, L.P. Falcon, B.D. Freeman, D.R. Paul, J Membrane Sci, 423 (2012) 195-208.

7

Ion Concentration: Uncharged vs. Charged Polymers

Uncharged polymer XLPEGDA

Charged polymer XL(AMPS-DEGDMA)

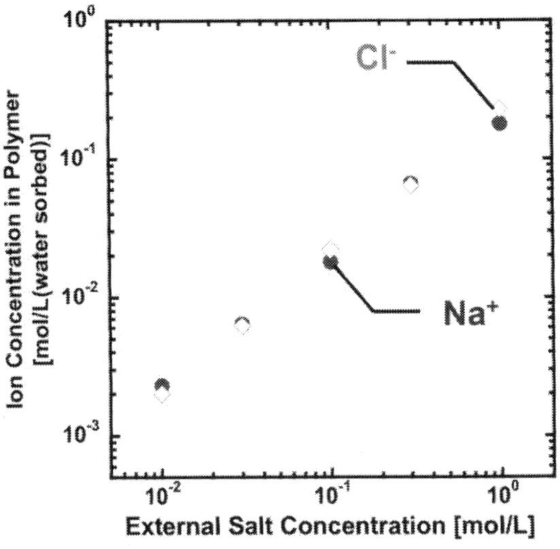

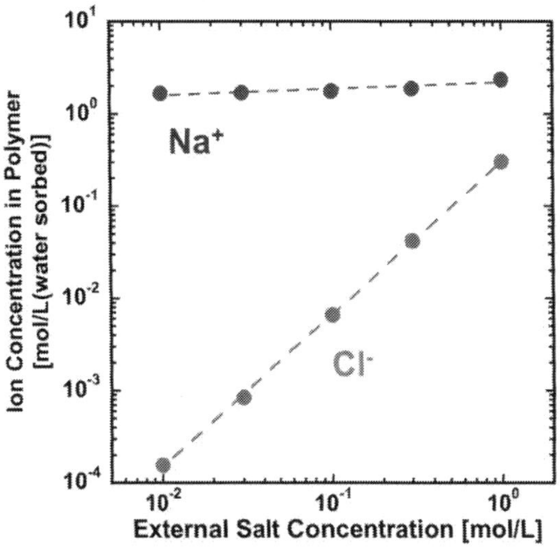

8

Ion Solvation in Water

Crystal Radius*

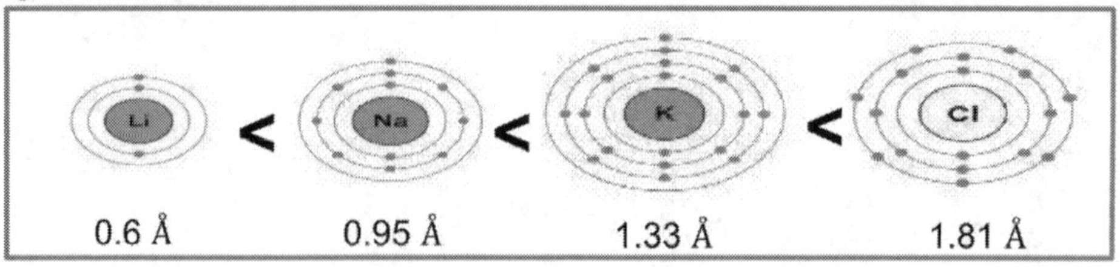

- Hydrated radius is inversely proportional to crystal radius because charge density decreases as size of atom increases.

Hydrated Radius*

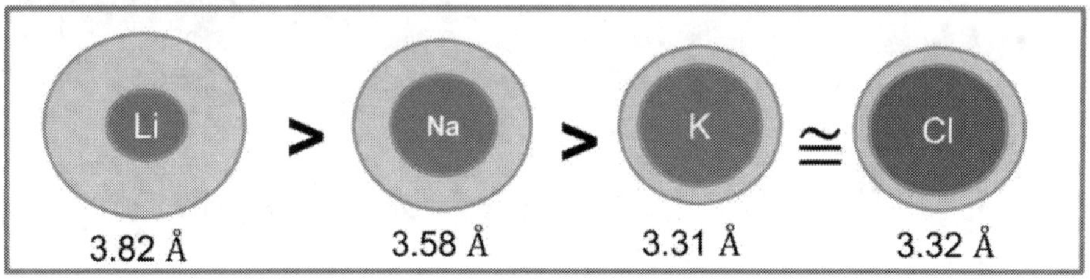

*E.R. Nightingale, The Journal of Physical Chemistry, 63 (1959) 1381-1387.

9

Ion Interactions with PEO

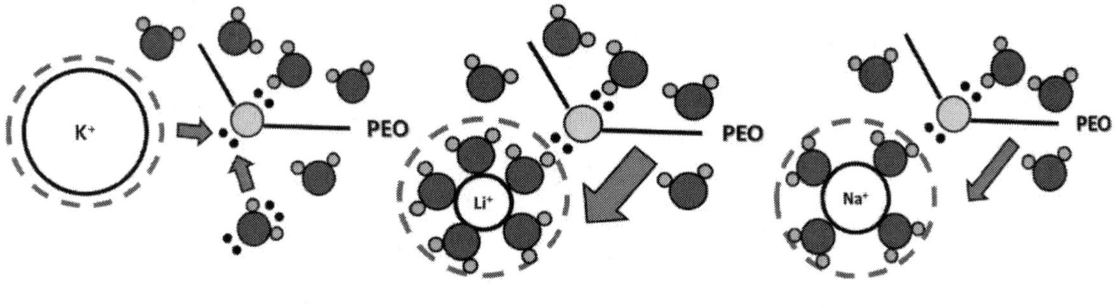

- **K^+** ion interacts directly with ethylene oxide (EO) in the polymer chain.
- **Li^+ and Na^+** ions interact with EO through water molecules.
- Ion binding to EO units follows:

$K^+ > Li^+ > Na^+$

R. Sartori, L. Sepulveda, F. Quina, E. Lissi, E. Abuin, Macromolecules, 23 (1990) 3878-3881.
M. Takano, K. Ogata, S. Kawauchi, M. Satoh, J. Komiyama, Polymer Gels and Networks, 6 (1998) 217-232.
Y. Okazaki, K. Ishizuki, S. Kawauchi, M. Satoh, J. Komiyama, Macromolecules, 29 (1996) 8391-8397.

10

Ion Sorption Coefficient in Uncharged Polymer

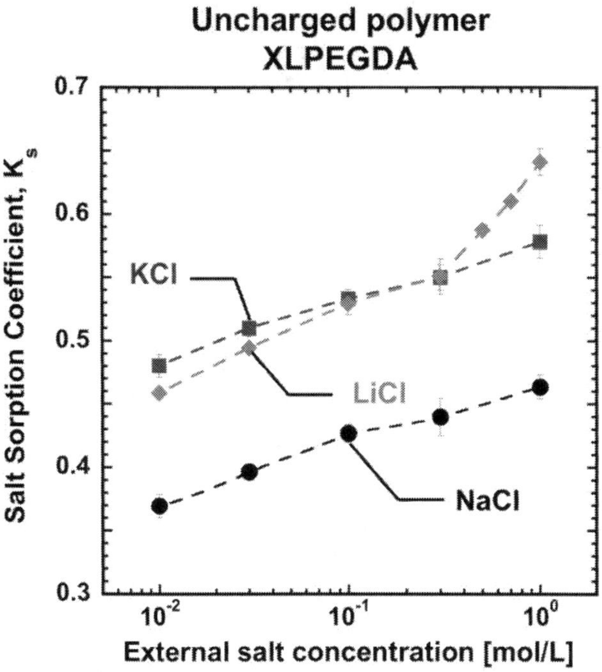

$$K_s = \frac{C_s^m}{C_s^s}$$

Hydrated Radius Order:

$$R_{Li^+} > R_{Na^+} > R_{K^+}$$

Ion-polymer interaction should be considered.

H. Yasuda, C.E. Lamaze, L.D. Ikenberry, Die Makromolekulare Chemie, 118 (1968) 19-35.

11

Free Water and Hydration Water in Charged Polymers

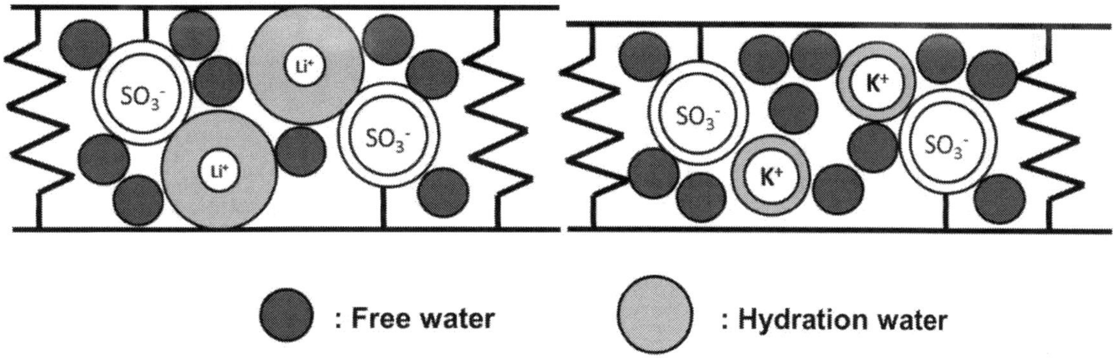

- Li^+ form CEM has less free water because of strong hydration of Li^+ ion.
- K^+ form CEM has more free water because of weak hydration of K^+ ion.
- Ion sorption coefficients in CEM follow free water content:

$K^+ > Na^+ > Li^+$

Helfferich, F. G. (1995). Ion Exchange, Courier Dover Publications.

12

Water Uptake in Alkali Chloride Solutions

Uncharged polymer
XLPEGDA

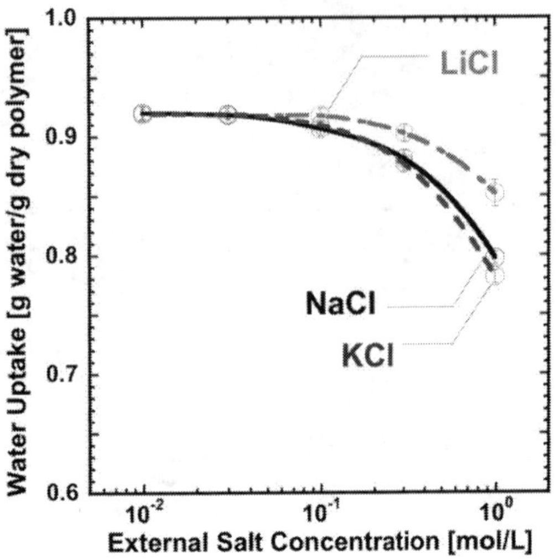

Charged polymer
XL(AMPS-DEGDMA)

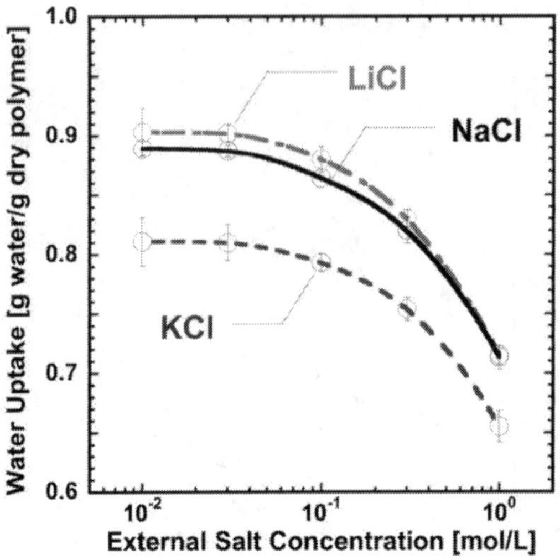

Ion Sorption Coefficient in Charged Polymer

Charged polymer

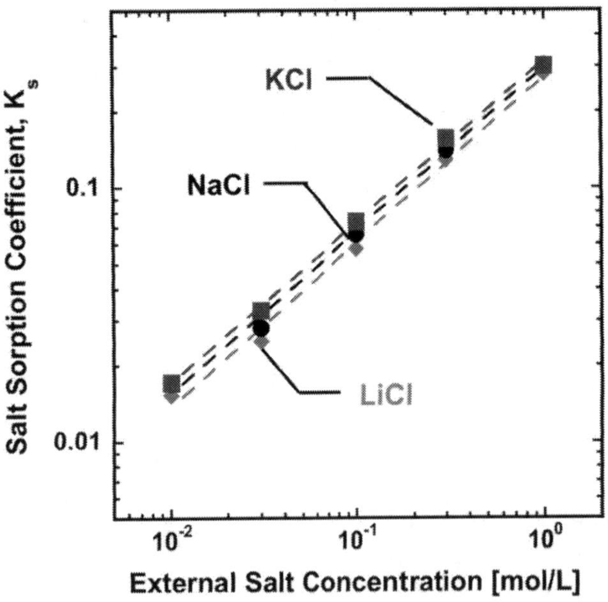

$$K_s = \frac{C_s^m}{C_s^s}$$

Hydrated Radius Order:

$$R_{Li^+} > R_{Na^+} > R_{K^+}$$

Mobile ion sorption depends on the counter-ion species and free water.

Activity Coefficient of Alkali Chlorides in Aqueous Solutions

Uncharged polymer XLPEGDA

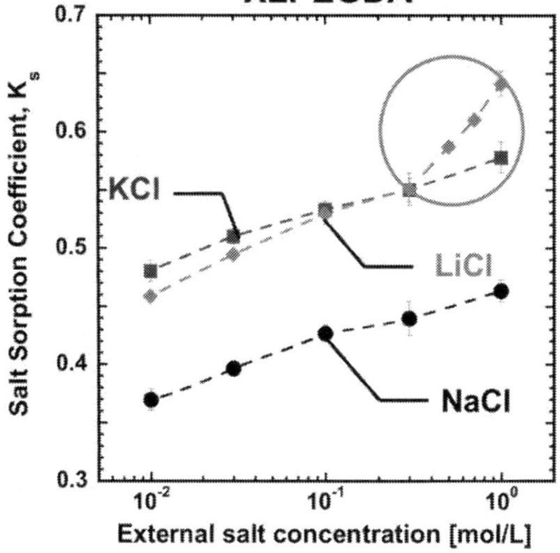

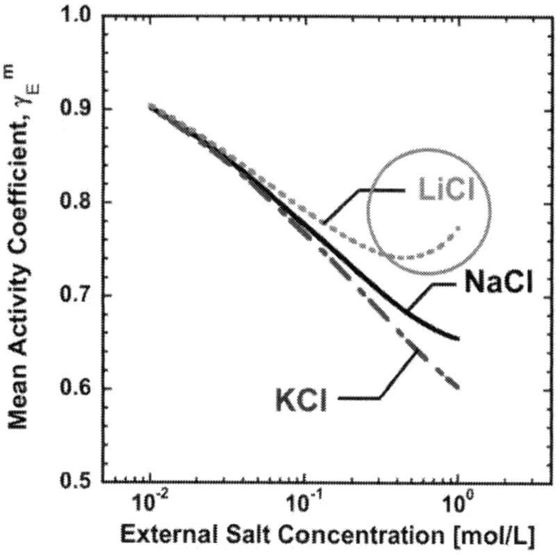

Activity coefficient affects salt sorption in the uncharged polymer.

K.S. Pitzer, G. Mayorga, The Journal of Physical Chemistry, 77 (1973) 2300-2308.
R.A. Robinson, R.H. Stokes, Electrolyte solutions, Courier Dover Publications, 2002.

Activity Coefficient of Alkali Chlorides in Uncharged Polymer

Uncharged polymer XLPEGDA

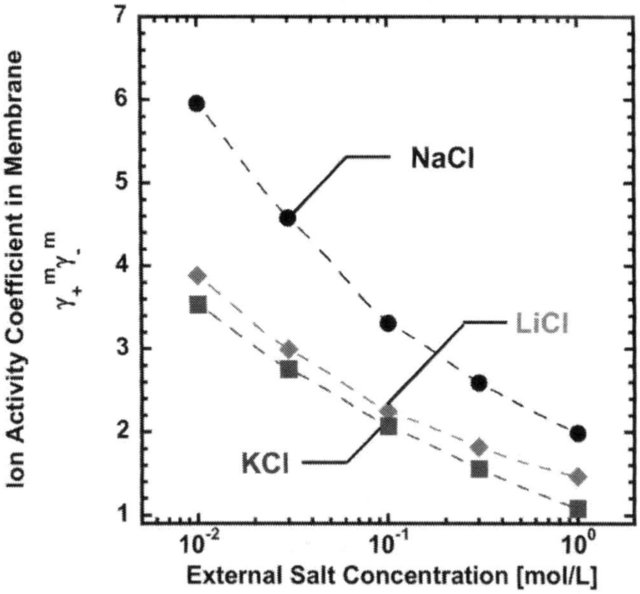

$$a_+^m a_-^m = a_+^s a_-^s$$

$$(\gamma_+^m \gamma_-^m) C_+^m C_-^m = (\gamma_+^s \gamma_-^s) C_+^s C_-^s$$

$$\gamma_+^m \gamma_-^m = \left(\frac{\gamma_\pm^s C_s^s}{C_s^m} \right)^2$$

Activity coefficient > 1 : extensive hydration of ions unfavorable for mixing.

R.A. Robinson, R.H. Stokes, Electrolyte solutions, Courier Dover Publications, 2002.

Activity Coefficient of Alkali Chlorides in Charged Polymer

Charged polymer XL(AMPS-DEGDMA)

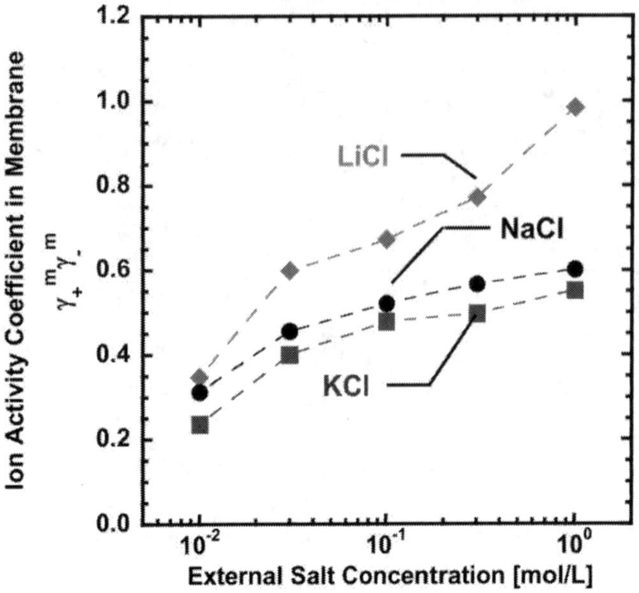

$$a_+^m a_-^m = a_+^s a_-^s$$

$$(\gamma_+^m \gamma_-^m) C_+^m C_-^m = (\gamma_+^s \gamma_-^s) C_+^s C_-^s$$

$$\gamma_+^m \gamma_-^m = \frac{(\gamma_\pm^S)^2 (C_S^S)^2}{C_+^m C_-^m}$$

Activity coefficients < 1 : ion pair formation or favorable mixing of ions with polymer.

Helfferich, F. G. (1995). Ion Exchange, Courier Dover Publications.
R.A. Robinson, R.H. Stokes, Electrolyte solutions, Courier Dover Publications, 2002.

Ion Activity Coefficients in Cross-linked Polymers

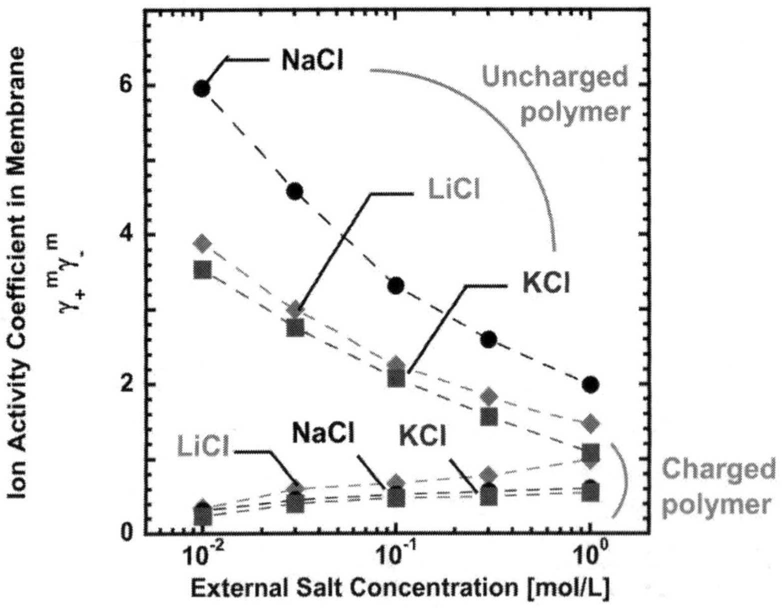

Polymer chemistry affects ion activity coefficients in the polymers with the same water content.

Conclusions

- Alkali chloride sorption in uncharged XLPEGDA polymer is affected by ion specific interaction with ethylene oxide units in the polymer chain.
- Mobile ion (i.e., chloride ion) sorption in negatively charged XL(AMPS-DEGDMA) polymer is influenced by counter-ion species.
- Ion activity coefficient in the membranes suggests that non-ideal behavior of ions significantly varies depending on polymer matrix chemical structure.

19

Acknowledgements

THE UNIVERSITY OF TEXAS AT AUSTIN

Prof. Benny D. Freeman
Prof. Donald R. Paul
Prof. Ho Bum Park
Jovan Kamcev
Ni Yan
Michele Galizia
Freeman Group Members

NSF

CBET 1160128

20

Thank you for listening

21

428807 Predicting Ion Sorption and Transport in Ion-Exchange Membranes

Abstract:

Predicting Ion Sorption and Transport in Ion-Exchange Membranes

Jovan Kamcev, Benny D. Freeman*, Donald R. Paul

McKetta Department of Chemical Engineering, The University of Texas at Austin, USA

**freeman@che.utexas.edu*

Fundamental understanding of ion sorption and transport in ion-exchange polymers is critical for development of high performance membranes. Despite the long history of literature on the subject, this topic is currently not well understood, making it difficult to build the appropriate connections between polymer structure and transport properties. Experimental techniques for characterizing ion sorption and transport in charged membranes have been established, however, a simple theoretical framework to describe the experimental findings is currently lacking. In this work, a model to predict ion sorption and permeability in ion-exchange membranes has been formulated and tested against experimental data. The model requires basic membrane information (chemical structure, fixed charge density, and water content). For the membranes used in this study, the model was completely predictive and no adjustable parameters were required.

Quantitative description of ion sorption in charged membranes equilibrated in salt solutions typically relies on a rigorous thermodynamic formalism (i.e., Donnan-type analysis). Previous studies have often overlooked the importance of including ion activity coefficients in such treatments, mainly because no reliable models or experimental techniques were available for obtaining these values in the polymer phase. In this work, Manning's counter-ion condensation theory, originally developed for polyelectrolytes, was extended to ion-exchange membranes and used to estimate membrane ion activity coefficients. Solution activity coefficients were obtained by the Pitzer model. The results of the model were compared to ion sorption data obtained experimentally for various commercially available ion-exchange membranes. Good agreement was found between the predicted values for membrane co-ion concentration and those obtained experimentally when solution and membrane non-idealities were included in the model. The generality and shortcomings of the model, as well as the applicability of polyelectrolyte theories to charged membrane systems, are discussed.

The model was extended to predict salt permeability coefficients in charged membranes within the framework of the solution-diffusion mechanism. According to this model, the permeability of a penetrant across dense polymers is comprised of the product between partition and diffusion coefficients. Ion diffusion coefficients in the membrane were estimated using the Mackie and Meares model. This model has no adjustable parameters and requires knowledge of polymer volume fraction and solution diffusion coefficient of the penetrant. Ion sorption coefficients were estimated by the model presented in this work. Thus, predicted values for permeability coefficients were obtained, and these values were compared to those determined experimentally. Good agreement between model and experiment was found without use of adjustable parameters. General applicability and limitation of this treatment are discussed.

415986 Evaluation of the Applicability of Polyelectrolyte-Based Membranes to Treating Real Wastewater Effluents

Abstract:

This study aims at evaluating the possibility of using polyelectrolyte multilayer (PEM) membranes to treating a real wastewater effluent, which in this case is an electrocoagulation-treated, anaerobic digestion (AD) effluent. A number of researchers have shown PEM membranes as highly permeable membranes with equivalent or higher salt rejection than some commercial nanofiltration (NF)/reverse osmosis (RO) membranes. However, the lab-based simple ion-containing solutions that are used to test these membranes fail to emulate the complexities involved in real effluents. Our study thereby focused on two typical polyelectrolyte-based systems and comparison of their performance (both flux and COD removal) with some commercial NF/RO membranes. It was found that irrespective of the pH of the effluent, the membranes showed higher permeability than commercial brackish water RO membranes with equivalent COD reduction. One major concern related to such wastewater effluents is the presence of natural organic matters (NOM) which tend to foul the membrane surfaces and lead to a decline in flux over a period of time. In our study, the PEM membranes demonstrated better anti-fouling properties than their commercial counterparts. In an effort to further reduce the fouling propensity of PEM membranes, some known anti-fouling polymers like poly ethylene glycol (PEG) or polydopamine were coated on their surfaces to prevent the flux decay over time. On being coated with these fouling resistant polymers using appropriate deposition conditions, the PEM membranes present a viable option for treating a wide variety of effluents.

427645 Electrochemical Characterization of an Ion-Exchange System Consisting of a Single Ion-Exchange Particle

Abstract:

Ion-exchange particles (resins) usually consist of highly cross-linked styrene and divinylbenzene polymers functionalized with functional groups that provide a so called fixed charge when the particles are immersed in an electrolyte. Depending on the type of the functional groups, these ion-exchange particles are divided into anion-exchange and cation-exchange particles. The former bear positive fixed charge and allow exchange and transport of anions, the later possess negative fixed charge and allow exchange or transport of cations. These ion-exchange particles find various applications, e. g. separation, purification or water decontamination. One specific area, where ion exchange particles play an irreplaceable role, is electromembrane separation processes as electrodialysis or electrodeionization used in removal of ions from water by means of an electrical field. They are a functional component of a so called heterogeneous ion-exchange membranes which besides the ion exchange particles contain a polymeric binder and polymeric fibers or they directly fill intermembrane space of the desalination units.

Because the ion exchange particles are the main functional component of these systems, it is of utmost importance to characterize their electrochemical properties as well as to describe their behavior in the electrical field. We developed a simple system that allows to study and electrochemically characterize a single ion-exchange particle by performing standard electrochemical measurements including current-voltage characteristics, chronoamperometry and potentiometry. At the same time the system allows observation of both electrolyte-particle interfaces, e. g. by using fluorescent microscopy, that is an invaluable tool in identification the fundamental phenomena that govern ion exchange, ion concentration polarization and associated electrokinetics. These phenomena can include electrohydrodynamic vortices or water splitting reaction occurring on the depletion side of the ion-exchange systems. Our electrochemical analysis and behavior is completed with carrying out Micro X-ray CT (computed tomography) scans of fully swollen particles being electrochemically characterized. The tomography of the particles adds important information on the structure and geometry of the measured system. In this paper we will present our experimental results for both anion- and cation-exchange particles and discuss the effect of the charge on the behavior of the ion-exchange systems.

436465 Integrating Tunable Anion Exchange with Reverse Osmosis for Inland Brackish Water Desalination

Abstract:

Ion Exchange (IX) and Reverse Osmosis (RO) are universally recognized as two fundamentally different processes for treating water. While RO uses semi-permeable membranes to separate dissolved electrolytes from water, IX uses functionalized polymeric materials, mostly in the form of beads or granules, to remove dissolved ions from water. Although seemingly unrelated, we present here the results of an investigation that acts as a bridge between the RO and IX processes to mitigate the problems associated with brackish water desalination. Most brackish water desalination plants are located inland and must resort to expensive concentrate disposal methods like deep well injection or evaporation ponds. Increasing the recovery of RO process would obviously reduce the volume of concentrate to be disposed of but cannot be implemented due to scaling of sulfate salts ($CaSO_4$, $BaSO_4$, etc.) and consequent fouling of RO membranes. Similar fouling is likely to occur in the presence of phosphate and aggravated by concentration polarization at the membrane-water interface. We have developed a hybrid Ione Exchange-Reverse Osmosis (HIX-RO) process where tunable anion exchange resins can eliminate sulfate precipitation without addition of external regenerants or anti-scaling chemicals. At the same time permeate recovery is enhanced and the volume of the concentrate can be cut in half or even lower. Results of the experimental study from a five-year long investigation will be presented.

CH2MHILL.

Atmospheric Preflash Drums

Andrew W. Sloley – CH2M HILL, Bellingham, WA

Prepared for Presentation at the 2015 AIChE Annual Meeting
Salt Lake City, UT, 8-13 November, 2015

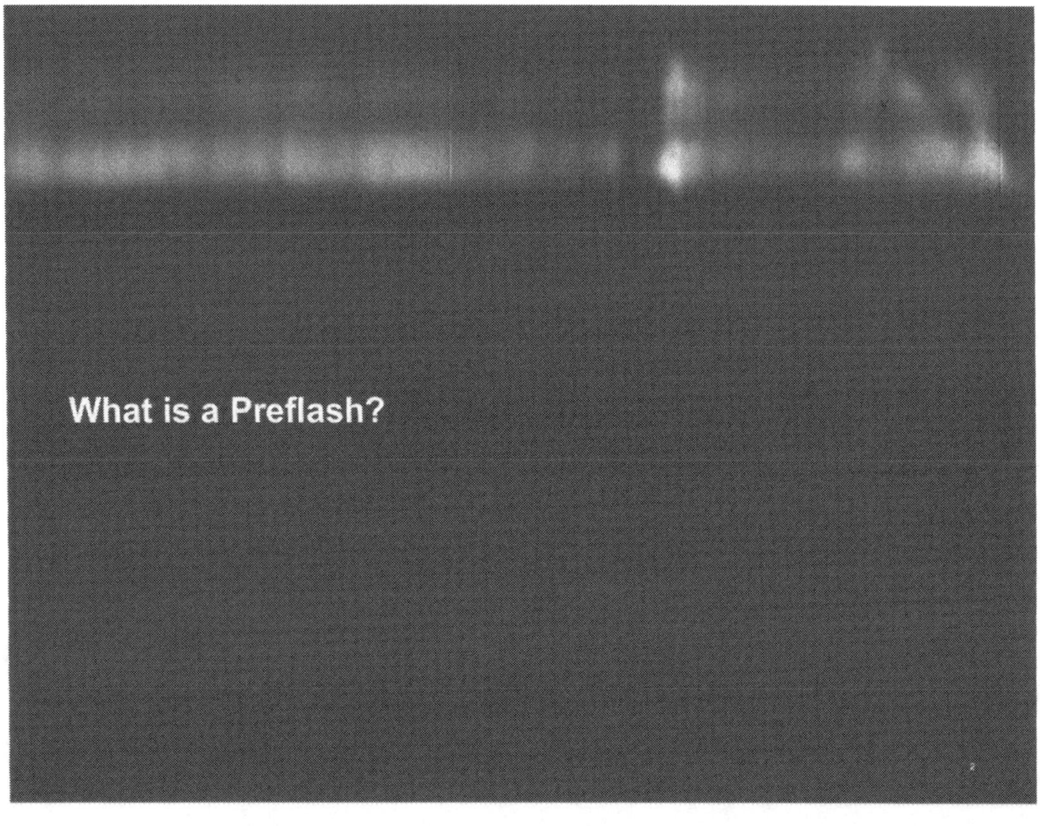

Atmospheric Preflash Definition

- A drum or tower upstream of the crude unit fired heater
- Downstream of the desalter
- Vaporizes lighter material and water
- Many different configurations
- This presentation covers drums
- Preflash drums
  - Connected to the atmospheric column
  - No reflux

Preflash Drums

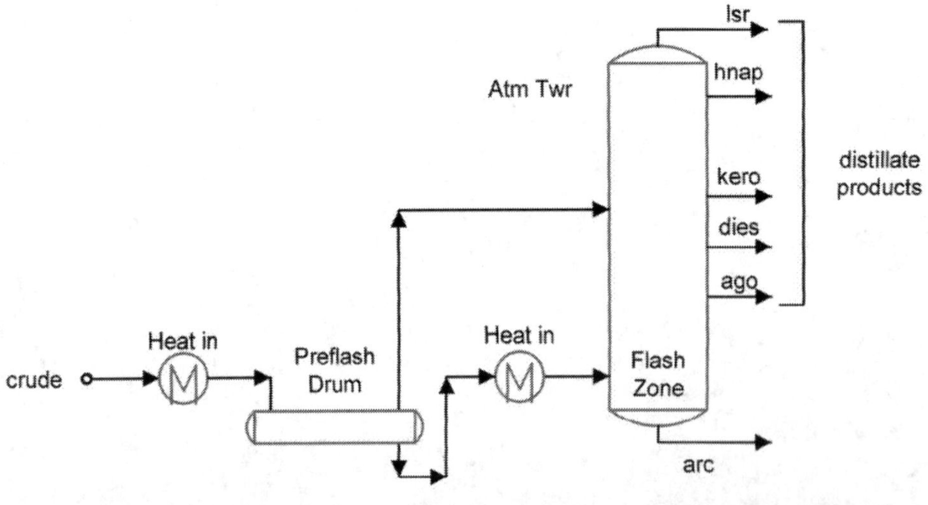

Crude Unit Survey – 50 Units

Configuration	Units
No preflash	23
Preflash drum	11
Preflash tower	16

Drum Type	11 Units
HP Drum	9
LP Drum	2
Tower connection (Bypass) (Endpoint – 99%) (Enthalpy match)	2
FZ connection	8
Transfer line	1

CH2MHILL.

Why Use a Preflash?

Technical Reasons to Use a Preflash

- Increases temperature before vaporization before fired heater
  - Allows more heat input upstream of heater pass-balancing valves
  - Light-ends vaporization
  - Water vaporization
- Reduces pressure drop in system
  - Reduced crude rate through exchangers
  - Reduced vapor formation in atmospheric heater
- Bypasses lower section of atmospheric column (sometimes)
  - Increases capacity by reducing vapor load in bottom of column
- Bypasses atmospheric tower completely (sometimes)
  - Preflash towers may add overhead systems
- Improve phase stability
  - Removes light paraffins from crude
  - Increases asphaltene solubility

CH2MHILL. 7

Configuration Options

8

Configuration Options

- High pressure preflash drum
  - Pressure control in the outlet vapor
  - Drum pressure independent of downstream tower pressure
  - High-pressure preflash
- Low pressure preflash drum
  - (No) pressure control in the outlet vapor
  - Floats on atmospheric tower pressure
  - Drum pressure close to atmospheric tower pressure

CH2MHILL.

High-Pressure Preflash Drum

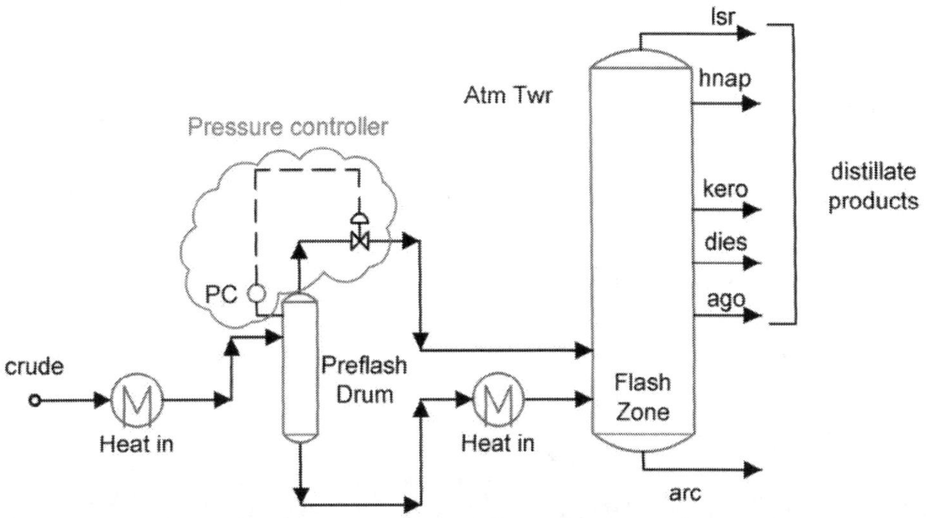

CH2MHILL.

Low-Pressure Preflash Drum

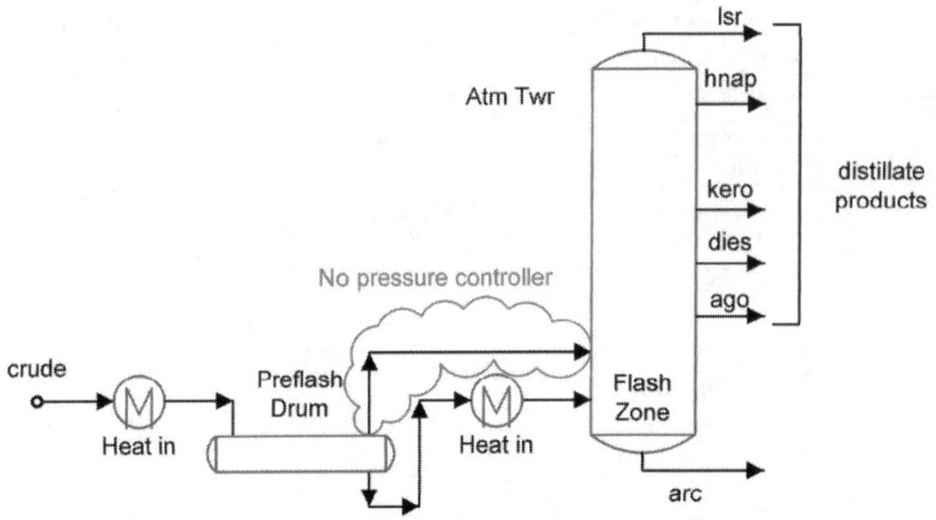

CH2MHILL.

11

Basis

12

Basis

- 100,000 bpd (11,920 cmpd)
- Temperature limit on crude at 690°F (366°C)
- High-pressure preflash
  - 80 psig (653 kPaa)
  - Temperature set to get 50% vaporization C5 minus
- Low-pressure preflash
  - 40 psig (377 kPaa)
  - Temperature set in the range seen in industrial units
- Three crudes evaluated at reasonable conditions for that crude

CH2MHILL.

Crudes

Crude	°API Gravity	Desalter Temperature, °F	Desalter Water Carryover, wt%
BCF 24	23.3	290	0.3
Tia Juana Light	31.6	270	0.2
Bakken	42.1	250	0.1

Crude	Specific Gravity	Desalter Temperature, °C	Desalter Water Carryover, wt%
BCF 24	0.9141	143	0.3
Tia Juana Light	0.8676	132	0.2
Bakken	0.8151	121	0.1

CH2MHILL.

Constraints Matter

- Product specifications constant
- Overflash from de-entrainment in crude tower constant

- Previous work
  - Constant yields, both temperature limit and duty limit on feed – 2013 and 2001
  - Constant separation, both temperature limit and duty limit on feed – 2014

CH2MHILL.

Results – Water

Water Changes Vaporization Temperature

- Significant changes in initial vaporization temperature depending upon water content

CH2MHILL.

The Importance of Water

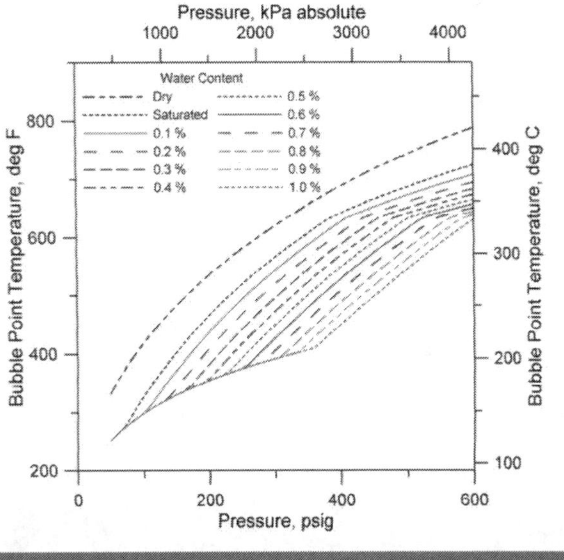

Tia Juana Light

CH2MHILL.

Comparing Crudes

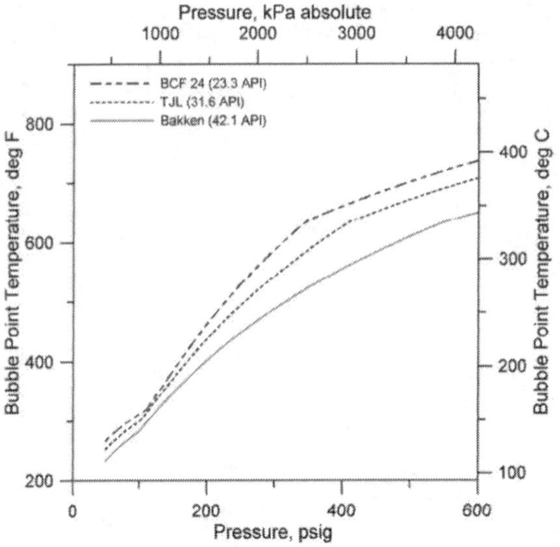

CH2MHILL.

19

Drum Connection Options and Configurations

20

Drum Connection to Atmospheric Tower

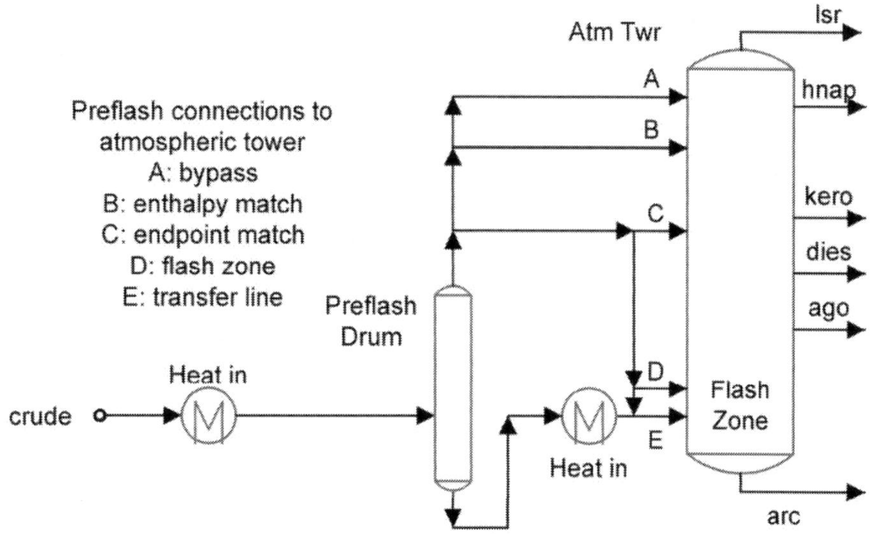

CH2MHILL. 21

Connection Options – 1

- Bypass
  - Moves vapor around a section of the tower
  - Hydraulic debottleneck
- Endpoint
  - Match distillation endpoint of vapor from preflash drum to endpoint of vapor inside atmospheric column
- Enthalpy
  - Match enthalpy of vapor from preflash drum to enthalpy inside atmospheric column
  - Most thermodynamically efficient

CH2MHILL. 22

Connection Options – 2

- Flash zone (segregated)
  - Feed vapor into flash zone, segregate vapor from heater outlet liquid
  - Reduces risk from foam carryover
  - Any residue in the foam goes into feed area
- Transfer line (or mixed flash zone)
  - Eliminates nozzle on the tower
  - Same as flash zone with heater outlet mixing with the preflash vapor

CH2MHILL.

Start in the Middle

- Tia Juana Light very close to 'average' US crude feed
- Then extend analysis to lighter and heavier crudes

CH2MHILL.

TJL – High-Pressure Preflash

Complicated, but close

	Units	TJL	TJL	TJL	TJL	TJL	TJL
Connection		No PF	Bypass	99%	Enthalpy	FZ	TL
LNap	Bbl/day	21,992	22,386	22,341	22,344	22,312	22,264
HNap		4,328	4,104	3,881	3,887	3,945	4,073
Kero		14,564	14,325	14,624	14,690	14,646	14,648
Dies		9,241	8,931	8,892	8,800	8,716	8,718
AGO		9,970	8,810	8,817	8,835	8,921	8,880
ARC		39,904	41,444	41,444	41,444	41,459	41,417
Cond	Mbtu/hr	130.0	133.9	134.7	134.8	136.5	139.8
Nap PA		36.0	36.0	36.0	36.0	36.0	36.0
Kero PA		36.0	36.0	36.0	36.0	36.0	36.0
Dies PA		42.0	28.3	28.3	28.3	28.3	28.3
Cut Point	°F	747	730	730	730	730	731

Goes up, bad

Close

CH2MHILL.

TJL – Low-Pressure Preflash

Complicated, but close

	Units	TJL	TJL	TJL	TJL	TJL	TJL
Connection		No PF	Bypass	99%	Enthalpy	FZ	TL
LNap	Bbl/day	21,992	22,539	22,396	22,473	22,317	22,424
HNap		4,328	4,160	3,845	3,960	3,970	3,658
Kero		14,564	13,721	14,724	13,923	14,514	14,600
Dies		9,241	8,758	8,078	8,832	7,896	8,346
AGO		9,970	7,173	7,308	7,161	7,621	7,197
ARC		39,904	43,649	43,649	43,650	43,681	43,774
Cond	Mbtu/hr	130.0	133.0	135.1	133.2	138.8	130.2
Nap PA		36.0	36.0	36.0	36.0	36.0	36.0
Kero PA		36.0	36.0	36.0	36.0	36.0	36.0
Dies PA		42.0	9.5	9.5	9.5	9.5	9.5
Cut Point	°F	747	706	706	706	706	705

Goes up, bad

Close **Goes up, bad**

CH2MHILL.

First Observations

- Available high temperature pumparound duty drops
- Less feed vaporization to atmospheric tower
- Energy efficiency benefits???
  - More preheat possible before feed vaporization
  - Less high temperature heat available
  - Energy efficiency changes highly dependent upon starting point
- Connection location only a small effect
  - Bypass used when column vapor limited – no choice
  - 99% versus enthalpy match – small yield effects – plant economics not clear
  - 99% (endpoint) most often used for design
  - Flash zone and transfer line locations are worst choices
  - Flash zone versus transfer line looks odd

CH2MHILL.

Low-Pressure Preflash – 99% Match (Endpoint Match)

	Units	BCF24	BCF24	TJL	TJL	Bakken	Bakken
Connection		No PF	99%	No PF	99%	No PF	99%
LNap	Bbl/day	12,312	12,421	21,992	22,396	35,004	36,008
HNap		1,773	1,706	4,328	3,845	9,020	8,726
Kero		9,961	10,006	14,564	14,724	15,157	13,899
Dies		9,871	8,752	9,241	8,078	11,118	10,582
AGO		7,369	5,554	9,970	7,308	12,357	8,835
ARC		58,713	61,561	39,904	43,649	17,345	21,951
Cond	Mbtu/hr	101.4	105.1	130.0	135.1	145.3	153.1
Nap PA		34.6	34.6	36.0	36.0	77.4	77.4
Kero PA		0.5	0.5	36.0	36.0	40.7	40.0
Dies PA		52.2	33.1	42.0	9.5	54.7	0.0
Cut Point	°F	714	682	747	706	799	731

CH2MHILL.

Adding an LP Preflash – What Changes?

	Units	BCF 24	TJL	Bakken	
ARC change	Bbl/day	+2,848	+3,745	+4,606	
HT duty change	Mbtu/hr	-19.1	-32.5	-55.4	

- Basis: same preflash drum temperature and pressure

CH2MHILL.

Further Observations

- Light crudes change more than heavy crudes
- Significant loss in high temperature duty available

- Why do many perceive preflash drums as an energy savings step?
- It depends where you start
- Look at what's been published…

CH2MHILL.

Previous Work – 1

- Limited fundamental analysis in most published reviews
- Energy recovery economics highly dependent on energy prices
  - Feintuch et al. 1985 most complete published evaluation
  - Compared energy recovery of 'no-preflash' versus 'preflash'
  - Only showed one energy cost
- Observations on Feintuch
  - Excellent technical content
  - Unit had only two pumparounds – OK for heavy crude
  - Heavy crude (20°API) – reduces available heat impact
  - Severe hydraulic limits
  - Capital costs too low – on both options
  - Low capital costs lead to heat integration target well past economic optimum
- Likely result – Optimistic (low) payback period for all cases

CH2MHILL.

Updating Feintuch (Sloley 2014, 2015)

- Assumed capital cost and energy savings accurate
- Capital escalated using Nelson cost factors
- No cost factors used for location differences on capital
- Energy sensitivity added for three locations
- Payback times still optimistic (low)

Basis	US $ / Mbtu	Energy Source	No Preflash	Preflash
			Payback Months	
Nymex 2014 08	3.82	Natural Gas	126	80
Europe 2014 08	7.27	Natural Gas	66	42
Japan 2014 06	15.23	LPG	28	20

CH2MHILL.

Previous Work – 2 – Increased Complexity

- Heat integration with a vacuum unit makes preflash options more attractive
  - Ji and Bagajewicz 2002
  - Author's analysis agrees with this (unpublished)

CH2MHILL.

TJL – Flash Zone versus Transfer Line

	Units	TJL	TJL	TJL	TJL	TJL
Connection		No PF	HP - FZ	HP - TL	LP - FZ	LP - TL
LNap	Bbl/day	21,992	22,312	22,264	22,317	22,424
HNap		4,328	3,945	4,073	3,970	3,658
Kero		14,564	14,646	14,648	14,514	14,600
Dies		9,241	8,716	8,718	7,896	8,346
AGO		9,970	8,921	8,880	7,621	7,197
ARC		39,904	41,459	41,417	43,681	43,774
Cond	Mbtu/hr	130.0	136.5	139.8	138.8	130.2
Nap PA		36.0	36.0	36.0	36.0	36.0
Kero PA		36.0	36.0	36.0	36.0	36.0
Dies PA		42.0	28.3	28.3	9.5	9.5
Cut Point	°F	747	730	731	706	705

Goes down, good

Goes up, bad

CH2MHILL.

Bakken – Low-Pressure Preflash

	Units	Bakken	Bakken	Bakken	Bakken	Bakken	Bakken
Connection		No PF	Bypass	99%	Enthalpy	FZ	TL
LNap	Bbl/day	35,004	36,008	35,996	36,003	36,015	36038
HNap		9,020	8,726	7,811	8,279	7,762	7,839
Kero		15,157	13,899	15,399	14,237	15,202	15,216
Dies		11,118	10,582	9,858	10,716	9,317	9,192
AGO		12,357	8,835	8,984	8,816	8,750	7,670
ARC		17,345	21,951	21,951	21,949	22,954	24,045
Cond	Mbtu/hr	145.3	150.1	153.1	150.4	160.4	168.6
Nap PA		77.4	77.4	77.4	77.4	77.4	77.4
Kero PA		40.7	40.0	40.0	40.0	40.0	40.0
Dies PA		54.7	0.0	0.0	0.0	0.0	0.0
Cut Point	°F	799	731	731	731	719	706

Goes up a lot, very bad

CH2MHILL.

More Observations – FZ versus TL?

- Mixing steam with the heater outlet liquid reduces the hydrocarbon partial pressures
- Mixing the hydrocarbon vapors with the heater outlet quenches the outlet liquid
- At low preflash vaporization – partial pressure dominates
- At high preflash vaporization – quenching dominates
- Heavy crudes – either partial pressure or quenching may dominate
  - More water carryover
  - Less vaporization
  - Transfer line versus flash zone may not be important
- Light crudes – quenching dominates results
  - Less water carryover
  - More vaporization
  - Can strongly favor flash zone over transfer line

CH2MHILL.

Most Preflash Drums Connect at the Flash Zone

Configuration	Units
No preflash	23
Preflash drum	11
Preflash tower	16

Drum Type	11 Units
HP Drum	9
LP Drum	2
Tower connection	2
FZ connection	8
Transfer line	1

If it's bad, why does this get done?

CH2MHILL.

Why is the Flash Zone Connection Used So Often?

- Preflash services are highly foaming
- Preflash drums and towers need *large* cross-section and holdup
- Many units undersized
- Undersized preflash leads to entrainment

CH2MHILL.

Results of Foaming

- Residue in feed from preflash column to atmospheric column
- If residue enters above the normal flash zone feed, residue ends up in distillate products – naphtha, kerosene, diesel, ago
  - Very expensive, shuts unit down
- Flash zone and transfer line connection removes this risk or solves problem
- Residue now enters the atmospheric column where it would go anyway
- Critical issue in preflash design and integration

CH2MHILL.

Summary

High-Pressure Preflash Drums

- Pressure controlled independently of atmospheric tower
- Pressure controller in vapor line from preflash drum to atmospheric tower
- Low vapor volumes
- Lower likelihood of foam carryover
- Majority of preflash drums
- Constraints relaxed
  - Liquid hydraulics downstream of preflash
  - Vaporization upstream of fired heater
  - Pressure drop in fired heater
  - Vapor load in atmospheric tower – depends upon connection
- Capacity increase limited by vaporization

Low-Pressure Preflash Drums

- Pressure floats on atmospheric tower pressure
- Direct vapor line from preflash drum to atmospheric tower
- High vapor volumes
- Higher likelihood of foam carryover
- Constraints relaxed
  - Liquid hydraulics downstream of preflash
  - Vaporization upstream of fired heater
  - Pressure drop in fired heater
  - Vapor load in atmospheric tower – depends upon connection
- Large capacity increases possible

Conclusions

- Bypass
  - If a bypass configuration is necessary to meet capacity requirements, no other choice
- Endpoint
  - Best overall yields
- Enthalpy
  - Most thermodynamically efficient – usually a yield cost
- Flash zone (segregated)
  - Most common option
  - Reduces risk of foam degrading product quality
  - Significant yield and heat recovery costs
- Transfer line (or mixed flash zone)
  - Avoid for light crudes
  - Avoid for high preflash vaporization

CH2MHILL.

Factors That Favor a Preflash

- Capacity more important than yield
- Capacity more important than energy recovery
- Heater duty limits instead of outlet temperature limits
- Vaporization in preheat train constraint
- Pressure limitations in preheat train (equipment constrained)
  - Heat exchanger limits
  - Fired heater limits

CH2MHILL.

Questions ?

Revisions

- 2015-11-07: Rev 0

CH2MHILL.

421828 Coking Mechanism in Refinery Fractionator Wash Beds

Abstract:

This paper will discuss a novel theory of the coking mechanism in a refinery fractionator wash bed.

The wash bed of a refinery fractionator is prone to coking/fouling due to the low liquid rates along with high vapor rates and high temperatures. Generally, the wetting rate at the top of the wash bed is minimized to prevent high-value product loss, but lower wetting rates lead directly to coke formation in the bed. So the refinery works to balance the economics of product recovery versus unscheduled downtime to replace a coked bed. Marathon has used Tracerco ThruVision™ technology to help manage the wash bed life while maximizing product recovery during the normal turnaround cycle. The ThruVision™ technology provides a detailed density map at a specific vertical elevation that can pinpoint specific areas of liquid maldistribution or solids build-up.

During a turnaround cycle, several towers have experienced heater trips that have led to step changes in the wash bed density seen by the ThruVision™ technology and ΔP. A review of this data from several refinery fractionators has guided the creation of a new theory on how the coke is formed in a refinery fractionator wash bed.

416268 Investigation of Fluid Dynamic Parameters in Packed Distillation Columns Using X-Ray Computed Tomography

Abstract:

The flow behavior of one or more liquid phases in packed distillation columns is hardly understood. The final aim of this work is to develop a model for the design and process optimization of distillation processes, using reliable data for the flow behavior of two liquid phases on a packing.

In this work the X-ray computed tomography (CT) is used as non-invasive measuring technique. The experiments are made with a customized computer tomograph at our institute that enables measurements of flows in vertical objects. The detector and the X ray source are rotating around the column and the column is not moved, so that the flow inside is not influenced by centrifugal force. As a result of the high spatial (80 µm) and temporal (1000 projections per second) resolution, dynamic processes inside separation columns can be monitored with a high precision. Fluid dynamic parameters in distillation columns (like interfacial area, liquid hold-up and liquid film thickness on the packing or between two liquid phases) can be directly estimated from the recorded CT images of a packing cross section at different height positions along the column. For the evaluation several steps of image treatment were applied to finally obtain segmented images with specific gray values for each component. The accuracy of the developed segmentation algorithm was verified on a dry and wetted packing through a comparison of the mass of the (wetted) packing and the calculated values out of the segmented images.

The experiments were performed in a glass column with an inner diameter of 100 mm and a packing height of 1000 mm consisting of five packing segments. For the investigation of the fluid dynamics five different special designs of Mellapak 500.Y packing from Sulzer Chemtech AG were used, to see the influence of the surface modification on the fluid dynamics. The special designs are a smooth, fluted and gridded texture with holes and a smooth and gridded texture without holes. Additionally the influence of the rotation of the packing segments to each other by 0° and 90° on the liquid distribution along the packing height and the accumulation of the liquid between two packing segments were determined. For the investigation of the influence of the substance properties on the flow morphology and the fluid dynamics, different liquids with low and high viscosities were used. The liquids were fed with a baffle distributor and the liquid load was varied between 5 and 50 m^2 / m^3 h and the F-factor was increased to 3 $Pa^{0.5}$.

In a former study we showed that the packing is not evenly wetted and the liquid film thickness is not constant over the entire packing. Hence we developed a method to determine the liquid distribution and the flow morphology on a packing cross section and the packing height. Due to the fact, that the packing surface is not evenly wetted, we determined the interfacial area between the packing and the liquid as well as between the liquid and the surrounding out of the CT-images. Thereby we were able to calculate a packing surface efficiency for different liquid and gas loads.

Furthermore the flow morphology of the liquids was investigated using a three dimensional visualization of short packing sections along the column.

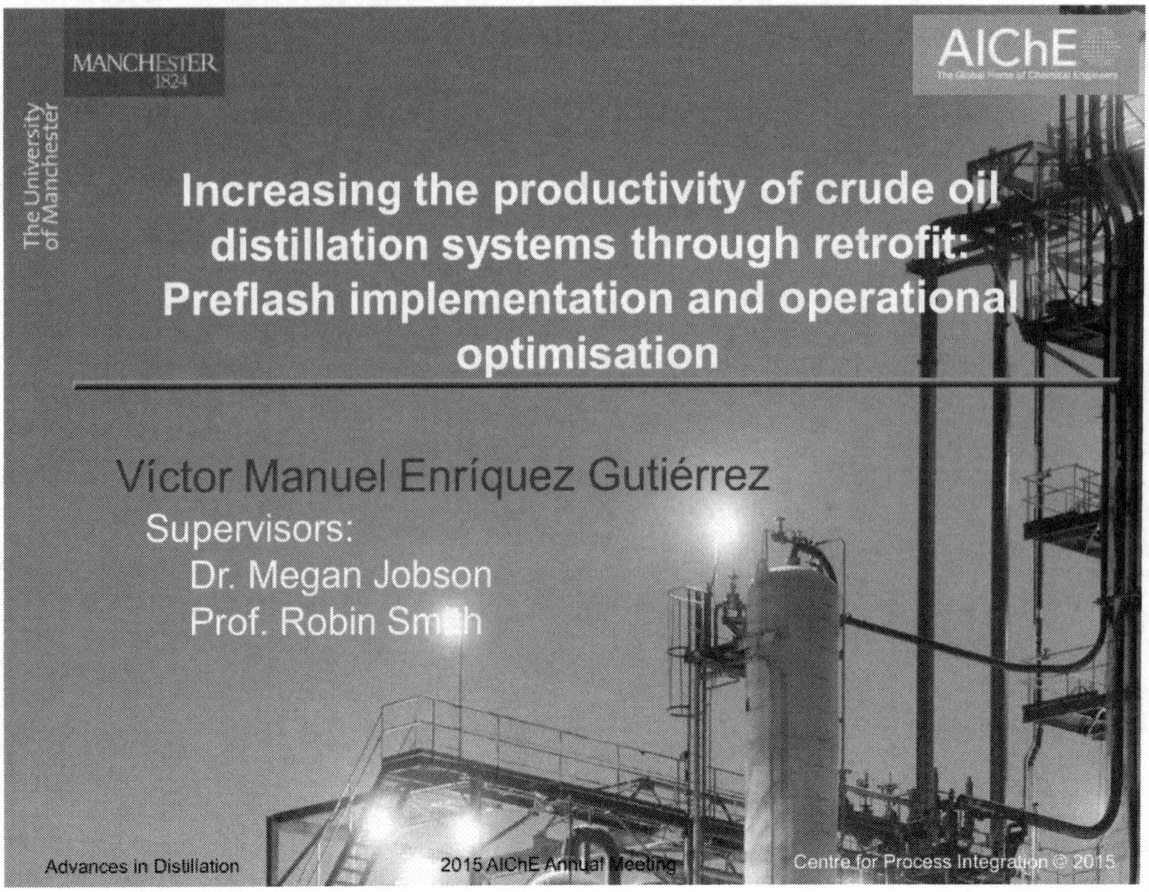

MANCHESTER 1824
The University of Manchester

AIChE
The Global Home of Chemical Engineers

Contents

1. Introduction
2. Objective
3. Proposed retrofit approach
    - 3.1 Distillation column simulation
    - 3.2 Hydraulic assessment
    - 3.3 HEN simulation and retrofit
    - 3.4 Objective function
4. Case study
5. Conclusions and future work

MANCHESTER 1824
The University of Manchester

AIChE The Global Home of Chemical Engineers

1. Introduction

MANCHESTER 1824
The University of Manchester

Introduction

AIChE The Global Home of Chemical Engineers

- What is retrofit?

  An engineering project that aims to increase the profitability of a process by maximizing the use of existing equipment

- This can be achieved by:
  - modifying the existing equipment
  - adding new equipment
  - changing equipment configuration
  - changing production/yield
- System constraints should be considered

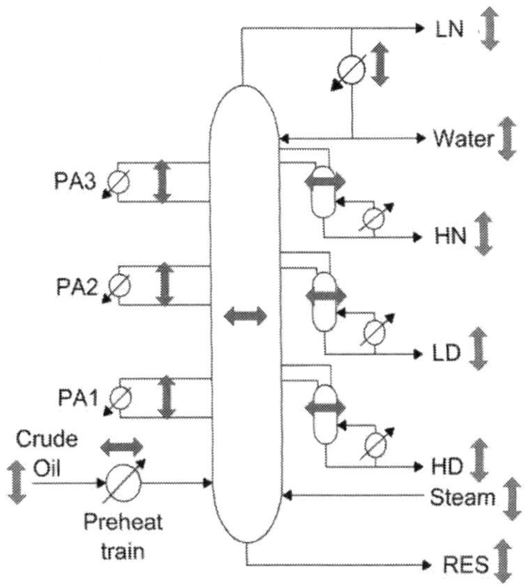

Introduction

- What is a crude oil distillation system?

Atmospheric distillation column

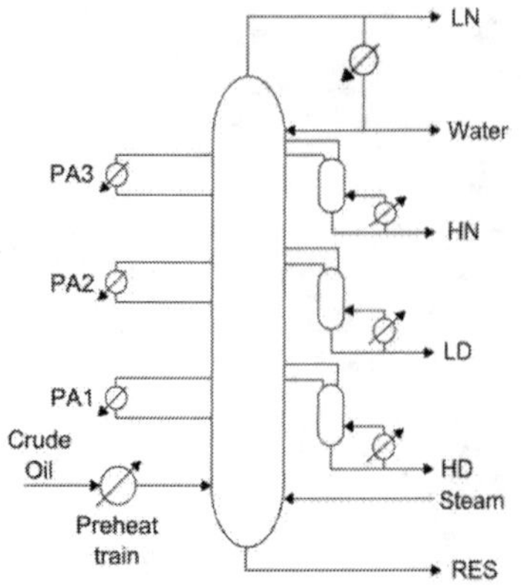

Heat exchanger network (HEN)

Maximise heat recovery == minimise operating costs

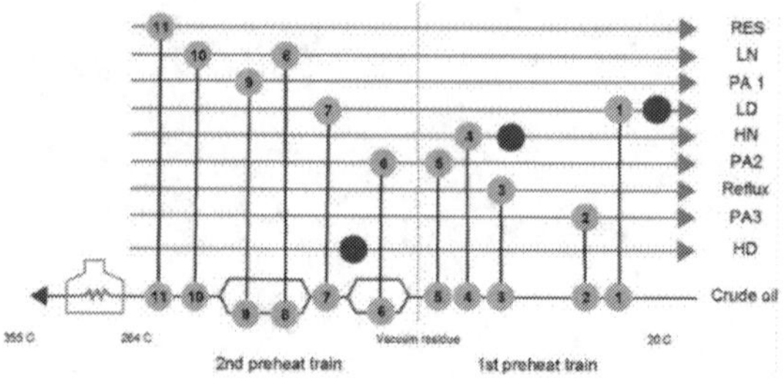

There is a complex interaction between the distillation column and the HEN

Introduction

- What is the problem?

The complex interaction between…

Distillation column + Heat exchanger network (HEN) =

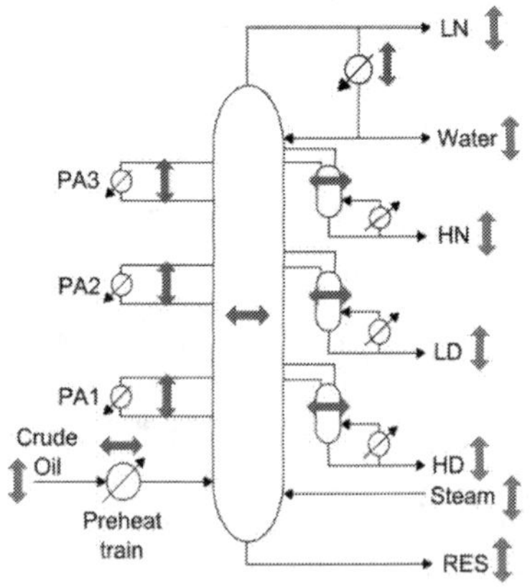

- ➢ Complex problem requiring detailed column hydraulics modelling
- ➢ Requires a method that accounts for the impacts on the HEN
- ➢ Many degrees of freedom and constraints
- ➢ Problem objectives non-trivial

MANCHESTER 1824
The University of Manchester

AIChE The Global Home of Chemical Engineers

2. Objective

MANCHESTER 1824
The University of Manchester

Objective

AIChE The Global Home of Chemical Engineers

To develop a retrofit approach:

- for increasing the productivity of crude oil distillation systems

that considers:

- operational and structural modifications
- system constraints.

Operational modifications	Structural modifications	System constraints
Preheat temperature	Adding preflash	Column hydraulics
Preflash temperature	Replacing column internals with high-capacity trays or structured packings	ΔT_{min} constraints
Pumparound flow rates and temperatures	Adding, deleting, repiping and resequencing heat exchangers	Product specifications
Stripping steam flow rates		

MANCHESTER 1824
The University of Manchester

AIChE The Global Home of Chemical Engineers

3. Proposed retrofit approach

MANCHESTER 1824
The University of Manchester

AIChE The Global Home of Chemical Engineers

Proposed retrofit approach

Optimisation variables (continuous)

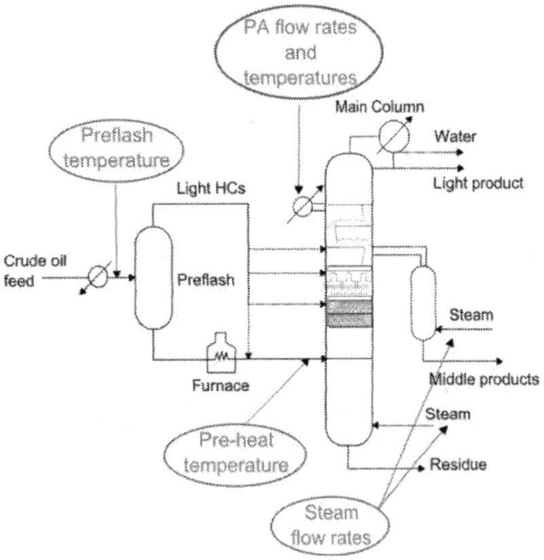

Structural design decisions (discrete)

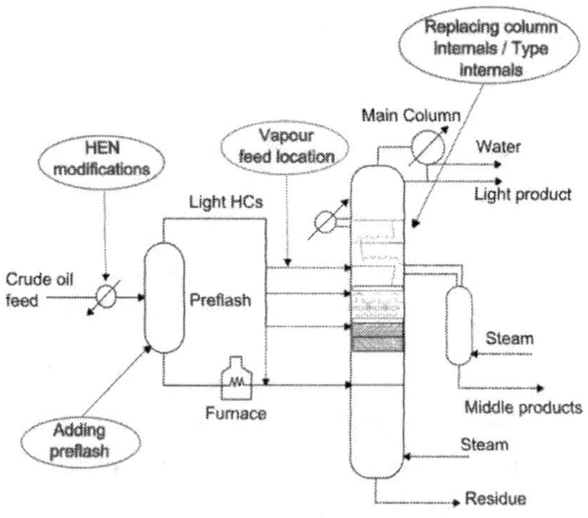

MANCHESTER 1824
The University of Manchester

Proposed retrofit approach

AIChE

Optimisation method

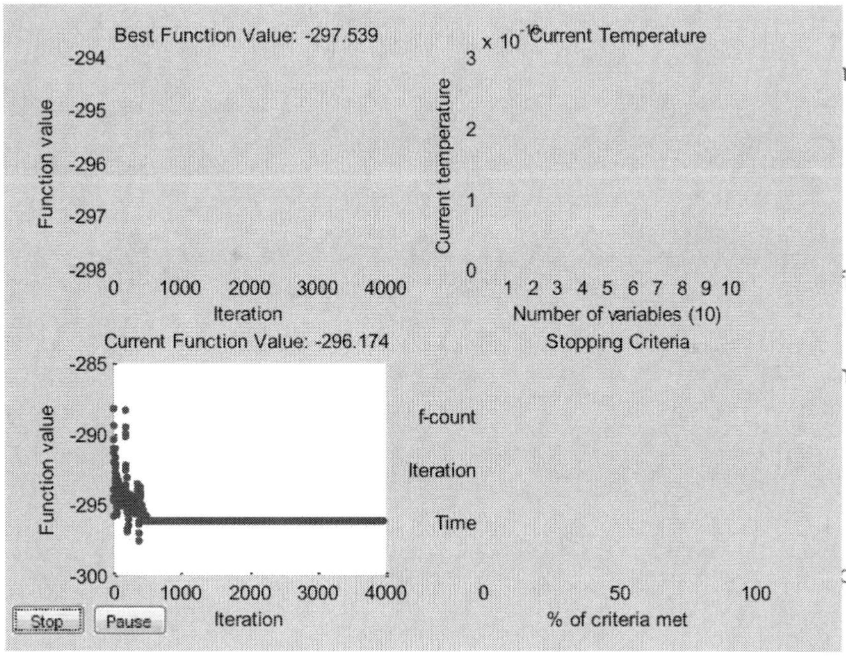

n MATLAB

ind the global
objective functions

nd on initial values

ptimal solution

MANCHESTER 1824
The University of Manchester

Proposed retrofit approach

AIChE

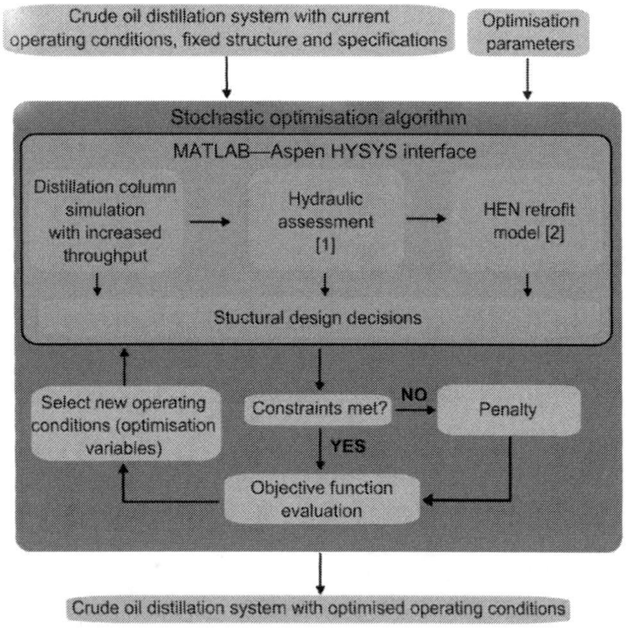

[1] Enríquez-Gutiérrez, V.M., M. Jobson, L.M. Ochoa-Estopier, and R. Smith, Retrofit of heat-integrated crude oil distillation columns. Chemical Engineering Research and Design, 2015.DOI: http://dx.doi.org/10.1016/j.cherd.2015.02.008.
[2] Ochoa-Estopier, L.M., M. Jobson, L. Chen, C. Rodríguez, and R. Smith, Optimisation of heat-integrated crude oil distillation systems. Part II: Heat exchanger network retrofit model. Industrial and Engineering Chemistry Research, 2015.DOI: http://dx.doi.org/10.1021/ie503804u.

MANCHESTER 1824
The University of Manchester

AIChE The Global Home of Chemical Engineers

3.1 Distillation column simulation

MANCHESTER 1824
The University of Manchester

Distillation column simulation

AIChE The Global Home of Chemical Engineers

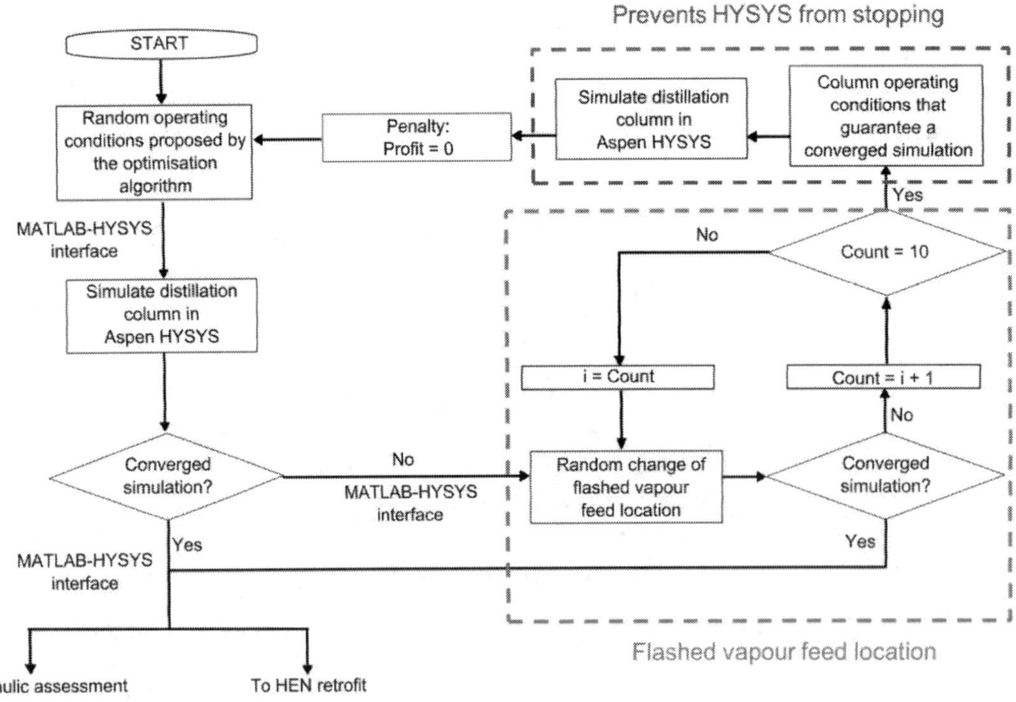

Distillation column simulation

Outputs from a converged rigorous simulation

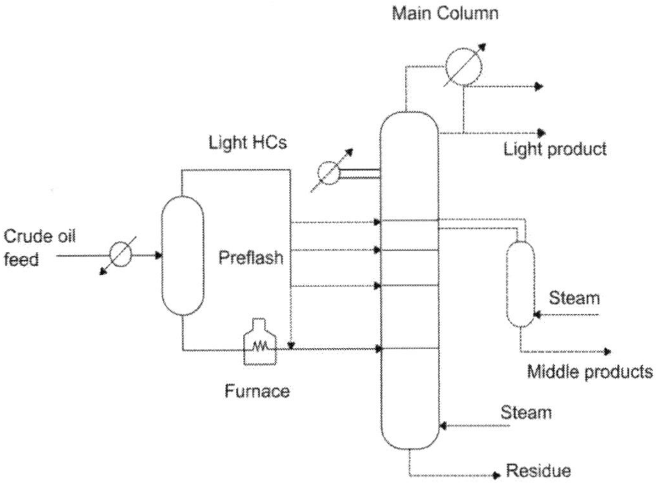

Hydraulic calculations:
For each stage
- Vapour and liquid phase flow rates
- Molar mass
- Density
- Viscosity
- Surface tension

HEN simulation:
Flow and temperatures
- Products
- Pump-arounds
- Condenser
- Reboilers
- Crude before preflash
- Crude after preflash

3.2 Hydraulic assessment

MANCHESTER 1824
The University of Manchester

Hydraulic assessment

AIChE

Trays — conventional & high-capacity

Structured packings

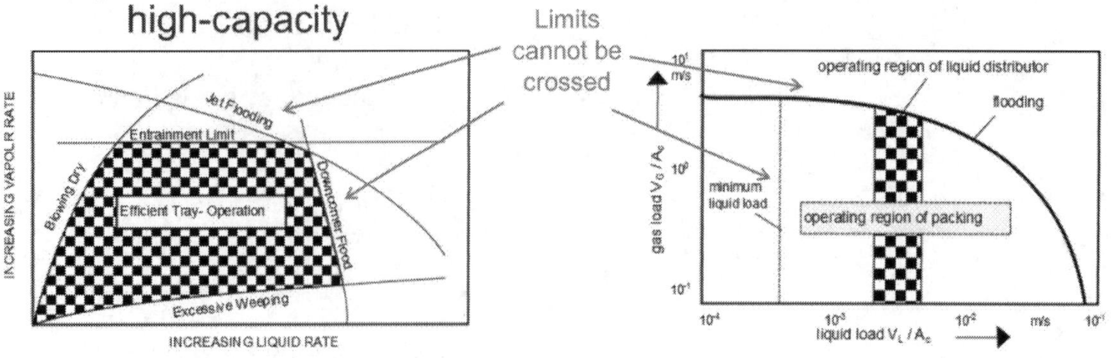

- Percent flooding[1,2] — 80%
- Liquid flow per weir length[1,2] — 100 m^3 m^{-1} h^{-1}
- Downcomer flooding[1,2] — 80%
- Downcomer exit velocity[1,2] — 0.46 m s^{-1}

- Flooding[3,4] — 80%

Apply regressed model[1,4] to predict pressure drop

[1] Branan, C.R., Rules of Thumb for Chemical Engineers. 4th ed. 2011, Burlington: Elsevier Science.
[2] Koch-Glitsch. Introduction to KG-Tower: Tray& Packed Tower Sizing Software Program, Version 2.0. 2006 www.koch-glitsch.com.
[3] Kister, H.Z., J. Scherffius, K. Afshar, and E. Abkar, Realistically predict capacity and pressure drop for packed columns. Chemical Engineering Progress, 2007. 103(7): p. 28-38.
[4] Enríquez-Gutiérrez, V.M., M. Jobson, and R. Smith, A design methodology for retrofit of crude oil distillation systems. Proceedings of the 24th European Symposium on Computer Aided Process Engineering-ESCAPE24, 2014. Part A: p. 1549-1554.

MANCHESTER 1824
The University of Manchester

Hydraulic assessment (in MATLAB)

AIChE

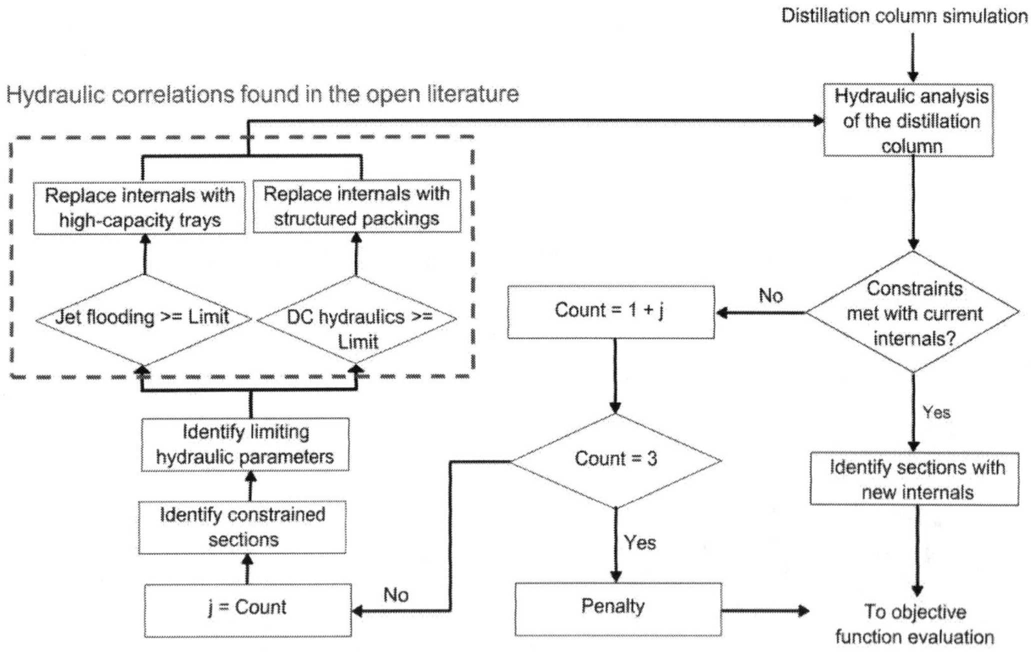

MANCHESTER 1824 The University of Manchester

AIChE The Global Home of Chemical Engineers

3.3 HEN simulation and retrofit

MANCHESTER 1824 The University of Manchester

HEN simulation and retrofit

AIChE The Global Home of Chemical Engineers

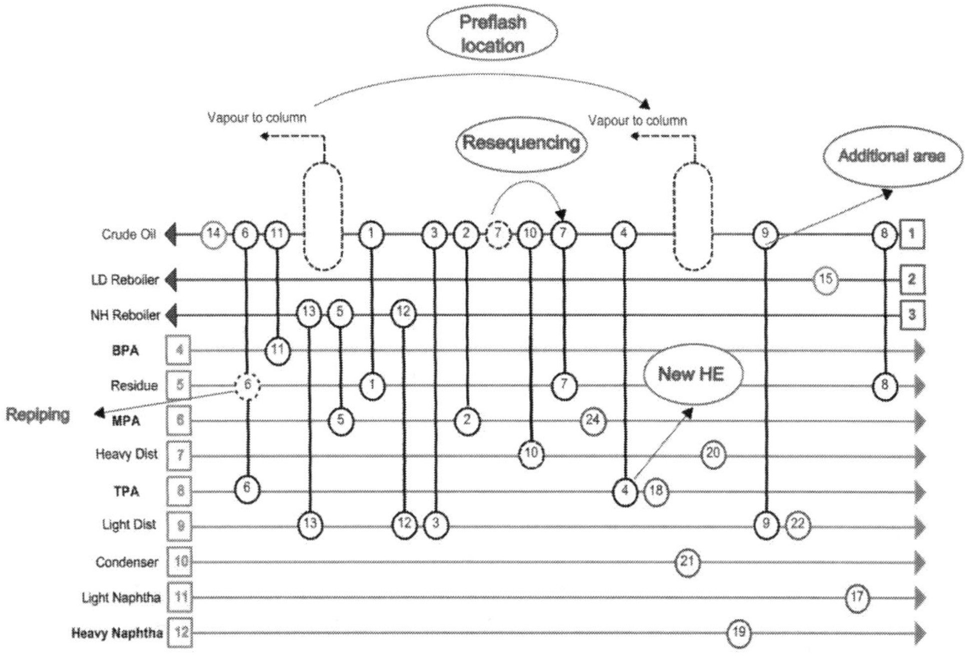

MANCHESTER 1824 The University of Manchester

HEN simulation and retrofit
(Ochoa-Estopier et al., 2015)

AIChE The Global Home of Chemical Engineers

HEN retrofit (using simulated annealing) – overview

1. Retrofit model:
- Add, delete, repipe, resequence a HE
- Add or delete a stream splitter
- Modify duty or split fraction

2. Feasibility solver:
- Minimum temperature approach constraints
- Stream enthalpy balance constraints

3. Simulation model:
- Linear equations formulated using principles of graph theory
- Temperature-dependent heat capacity considered

Ochoa-Estopier, L.M., M. Jobson, L. Chen, C. Rodríguez, and R. Smith, Optimisation of heat-integrated crude oil distillation systems. Part II: Heat exchanger network retrofit model. Industrial and Engineering Chemistry Research, 2015.DOI: http://dx.doi.org/10.1021/ie503804u.

MANCHESTER 1824 The University of Manchester

HEN simulation and retrofit

AIChE The Global Home of Chemical Engineers

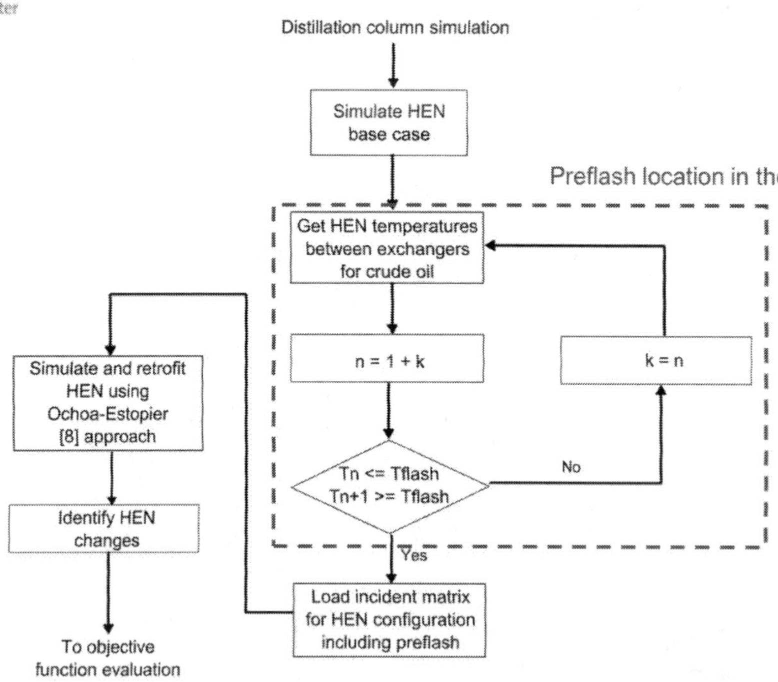

Ochoa-Estopier, L.M., M. Jobson, L. Chen, C. Rodriguez, and R. Smith, Optimisation of heat-integrated crude oil distillation systems. Part II: Heat exchanger network retrofit model. Industrial and Engineering Chemistry Research, 2015.DOI: http://dx.doi.org/10.1021/ie503804u.

MANCHESTER 1824
The University of Manchester

AIChE
The Global Home of Chemical Engineers

3.4 Objective function

MANCHESTER 1824
The University of Manchester

Objective function

AIChE
The Global Home of Chemical Engineers

Maximise net profit

Net Profit = Value of products − (Crude oil Cost + Utilities Cost + Stripping Steam Cost + Annualised Retrofit Cost) − Penalties

Can be easily found in literature

Transfer prices estimation

Cost models:
Column internals
Heat exchangers
Heat exchanger modifications
Preflash unit

$$ARC = RC\left[\frac{i(1+i)^n}{(1+i)^n - 1}\right]$$

$RC =$ Retrofit cost, $

$i =$ Fractional interest rate per year

$n =$ Number of years

MANCHESTER 1824 The University of Manchester

Penalties – reflect constraints

AIChE The Global Home of Chemical Engineers

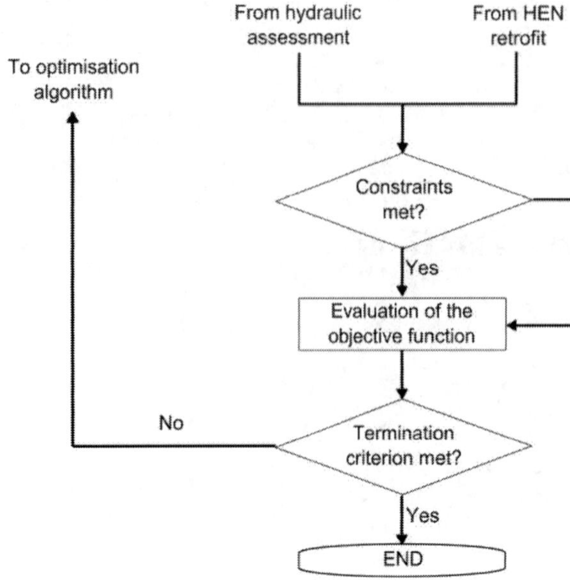

If constraints are not satisfied:

- product specifications
- hydraulic performance
- minimum temperature approach
- stream enthalpy balance

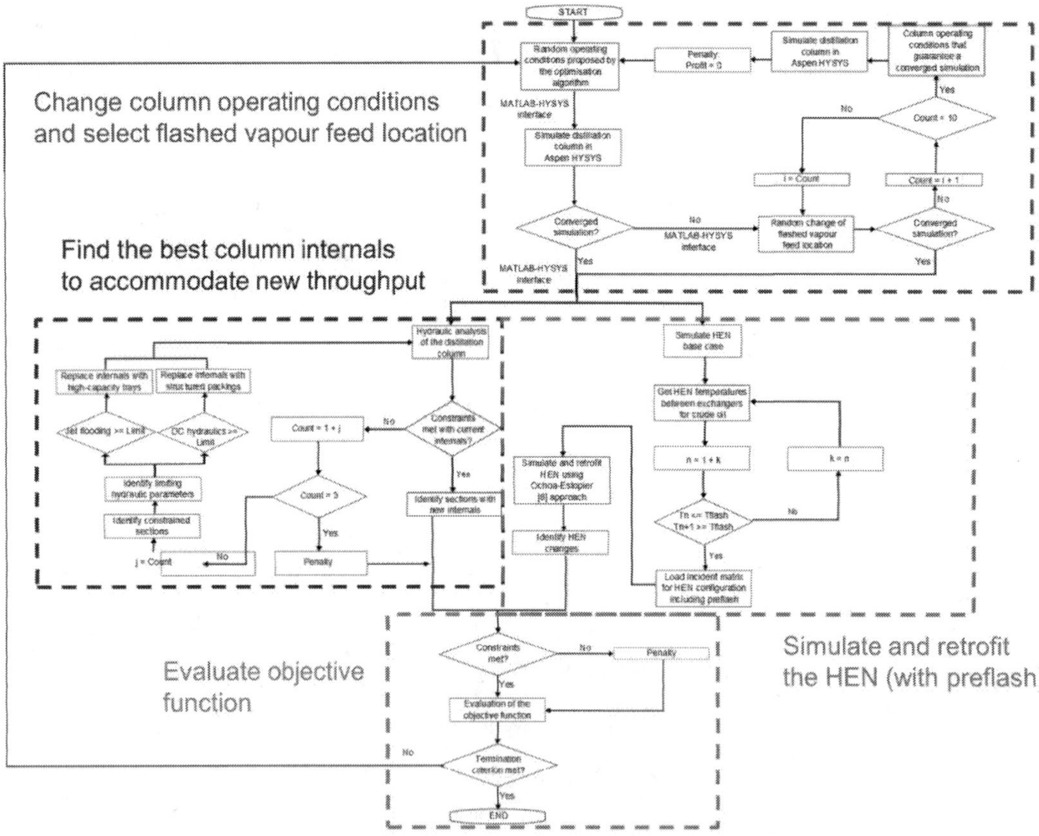

MANCHESTER 1824
The University of Manchester

AIChE
The Global Home of Chemical Engineers

4. Case Study

MANCHESTER 1824
The University of Manchester

Case Study

AIChE
The Global Home of Chemical Engineers

It is desired to increase the capacity of an existing crude oil distillation system by 30%.

Retrofit modifications	Scenario A*	Scenario B	Scenario C	Scenario D
Operational optimisation		X	X	X
Replace column internals in constrained sections **	X		X	X
Add preflash				X

* Operating parameters are increased pro-rata
** With 25° sloped downcomer high-capacity trays or Mellapak 452Y structured packing
All scenarios consider HEN retrofit

MANCHESTER 1824
The University of Manchester

Base case

AIChE
The Global Home of Chemical Engineers

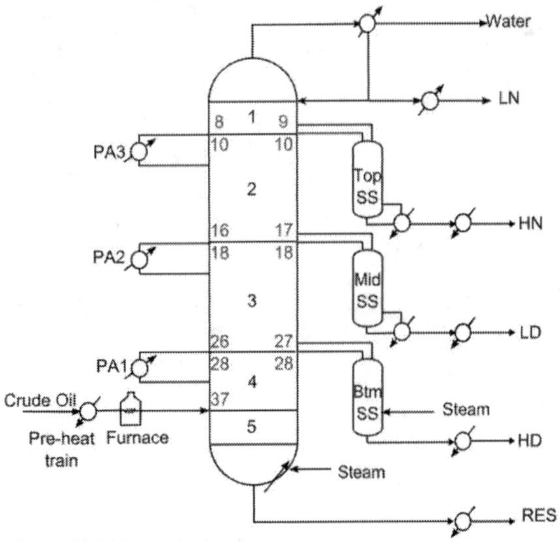

F = 100,000 barrels·day^{-1}
T = 25 °C
P = 3 bar
Operation = 8400 h·y^{-1}

Venezuela Tia Juana Light crude. Chen (2008)

	Number of stages	Stage numbers
Main fractionator	41	1 to 41
Top side stripper	6	42 to 47
Middle side stripper	7	48 to 54
Bottom side stripper	5	55 to 59

	Minimum diameter, m (HYSYS)	Actual column diameter, m
Section 1	3.5	5.5
Section 2	7	8
Section 3	7.5	8
Section 4	7	7.5
Section 5	7	7.5
Top SS	2	3.5
Mid SS	2.6	3.5
Bottom SS	2.2	3.5

Chen, L., Heat-integrated crude oil distillation system design. 2008, The University of Manchester: Manchester, UK.
Gadalla, M.A., Retrofit design of heat-integrated crude oil distillation systems. 2003, The University of Manchester: Manchester, UK.

MANCHESTER 1824
The University of Manchester

Base case

AIChE
The Global Home of Chemical Engineers

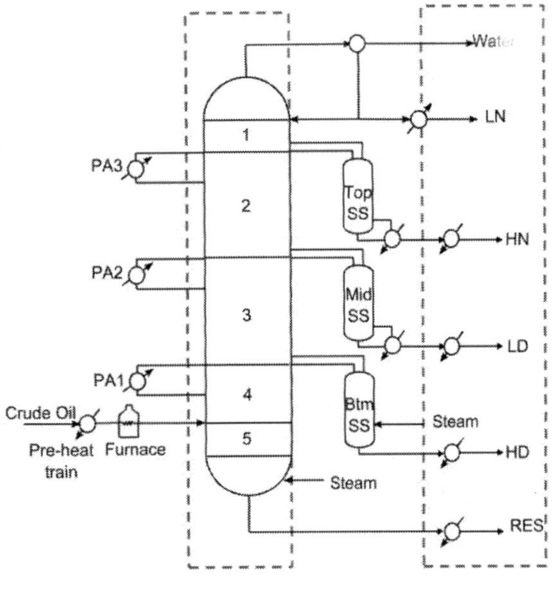

Column internals

Tower internals	Valve trays
Tray spacing, m	0.6069
% Active area	86 % — Sections 1 to 4 91% — Section 5 and SS
Downcomer clearance, m	0.06
Number of flow paths	4 — Sections 1 to 4 2 — Section 5 and SS

Product specifications

Products	T5 (°C, TBP)	T95 (°C, TBP)	Flow rate (kmol·h^{-1})
RES	353	798	635
HN	285	372	149
LD	190	317	653
HN	117	196	496
LN	3	118	678

Chen, L., Heat-integrated crude oil distillation system design. 2008, The University of Manchester: Manchester, UK.

MANCHESTER 1824
The University of Manchester

Base case

AIChE

Column hydraulics

Close to the limit

	Maximum approach to flooding, %	Maximum weir load, $m^3\ m^{-1}\ h^{-1}$	Maximum downcomer exit velocity, $m\ s^{-1}$	Maximum approach to downcomer flooding, %
Section 1	53.1	87.9	0.41	45.8
Section 2	52.6	67.7	0.31	35.2
Section 3	50.1	64.8	0.30	31.6
Section 4	43.9	22.2	0.10	10.9
Section 5	28.5	70.7	0.39	42.0
Top SS	21.0	36.5	0.20	34.1
Mid SS	29.2	55.6	0.31	51.8
Btm SS	12.8	20.5	0.11	19.1

Advances in Distillation   Increasing the productivity of crude oil distillation systems through retrofit   31

MANCHESTER 1824
The University of Manchester

Base case

AIChE

Based on heat exchanger network of Chen (2008)

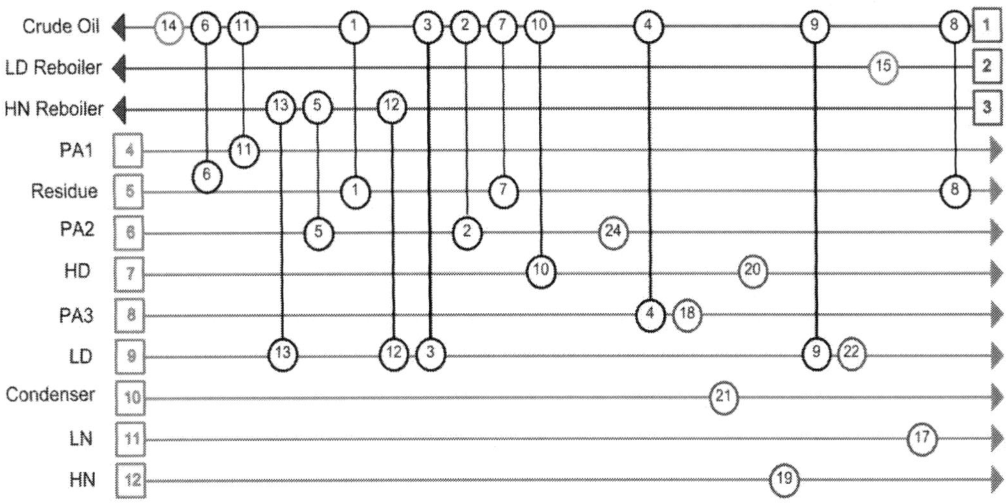

- 22 heat exchangers
- Available area: 5453 m^2
- Fired heating consumption: 62.9 MW

Chen, L., Heat-integrated crude oil distillation system design. 2008, The University of Manchester: Manchester, UK.

Advances in Distillation   Increasing the productivity of crude oil distillation systems through retrofit   32

Base case

Transfer prices

Product	Transfer price*,**
Light Naphtha	$74.0
Heavy Naphtha	$61.5
Light Distillate	$66.6
Heavy Distillate	$76.0
Residue	$66.5
Price crude oil*	**$46.0**

Cost of utilities

Utility	Price*	per unit
Steam, 3.5 bar	$6.8	t
Cooling water	$0.02	m^3
Fuel oil	$14	GJ

Process cash flows

Stream	Consumption*10^6	units	Million $
Light Naphtha	4.70	$bbl \cdot y^{-1}$	$343.31
Heavy Naphtha	4.71	$bbl \cdot y^{-1}$	$285.88
Light Distillate	6.65	$bbl \cdot d^{-1}$	$436.91
Heavy Distillate	2.94	$bbl \cdot y^{-1}$	$220.16
Residue	17.42	$bbl \cdot y^{-1}$	$1,142.96
Crude Oil	36.56	$bbl \cdot y^{-1}$	$1,659.24
Main Steam	189.37	$kg \cdot y^{-1}$	$1.26
HD Steam	39.45	$kg \cdot y^{-1}$	$0.26
Hot utility	1983.57	$MJ \cdot y^{-1}$	$26.48
Cold utility	19.67	$m^3 \cdot y^{-1}$	$0.31
HEN additional area	0	m^2	$0.00
Net profit			**$741.64**

* Estimated using Gary et al. (2007) operating cost for downstream operations.
** Crude oil and end product prices were taken from US Energy Information Administration data base.

Gary, J.H., *Petroleum refining : technology and economics*. 5th ed. 2007, Boca Raton, USA: Taylor & Francis.
US Energy Information Administration. *Petroleum Marketing Monthly*. 2015 01/07/15]; Available from: http://www.eia.gov/petroleum/marketing/monthly/pdf/pmmall.pdf.

Case study: results

MANCHESTER 1824 The University of Manchester

Case study: results

AIChE The Global Home of Chemical Engineers

Operating conditions

Parameter	Scenario A	Scenario B	Scenario C	Scenario D
Crude oil flow rate, kmol s^{-1}	3393.9 (+30%)	3393.9 (+30%)	3393.9 (+30%)	3393.9 (+30%)
Main steam, kmol s^{-1}	1560.0 (+30%)	1092.3 (-9%)	1108.0 (-8%)	1770.1 (+48%)
HD steam, kmol s^{-1}	325.0 (+30%)	412.9 (+65%)	407.2 (+63%)	269.4 (+8%)
Crude oil preheat temperature, °C	365.0 (0%)	350.8 (-4%)	351.3 (-4%)	352.7 (-4%)
PA1 duty, MW	14.7 (+30%)	12.6 (+13%)	12.7 (+14%)	13.8 (+23%)
PA2 duty, MW	23.4 (+30%)	16.1 (-10%)	27.3 (+53%)	16.3 (-9%)
PA3 duty, MW	16.3 (+30%)	18.3 (+43%)	11.9 (-7%)	16.5 (+28%)
PA1 ΔT, °C	20.0 (0%)	55.2 (+176%)	10.1 (-50%)	13.9 (-31%)
PA2 ΔT, °C	50.1 (0%)	42.5 (-15%)	87.7 (+75%)	93.2 (+86%)
PA3 ΔT, °C	30.0 (0%)	78.6 (+162%)	40.3 (+34%)	98.5 (+228%)
Preflash temperature, °C	—	—	—	252.6**
Optimisation time*, h	—	6.3	8.5	17.1

* On a computer with an Intel Core processor of 3.30 GHZ and 8.00 GB of installed RAM
** The flashed vapour is fed in Stage 26 (PA3 feed stage)

Scenario A — pro-rata increase, replace internals
Scenario B — operational optimisation
Scenario C — operational optimisation, replace internals
Scenario D — operational optimisation, replace internals, add preflash

Advances in Distillation Increasing the productivity of crude oil distillation systems through retrofit 35

MANCHESTER 1824 The University of Manchester

Case study: results

AIChE The Global Home of Chemical Engineers

Column retrofit

Parameter	Scenario A	Scenario B	Scenario C	Scenario D
Type of internal	Mellapak 452Y	—	Mellapak 452Y	Mellapak 452Y
Column section (s)	Section 1	—	Section 1	Section 1
Retrofit cost, $	79,139	—	79,139	79,139

HEN retrofit

Parameter	Scenario A	Scenario B	Scenario C	Scenario D
Fired heating (MW)	69.7 (+11.0%)	67.62 (+7.5%)	69.69 (+10.8%)	63.07 (+0.3%)
Required additional area, m^2	3609	2171	2076	1284
Number of HX needing additional area	20	18	17	12
Number of deleted HX	3	3	3	–
Retrofit cost, $	239,809	256,943	249,766	184,484

Scenario A — pro-rata increase, replace internals
Scenario B — operational optimisation
Scenario C — operational optimisation, replace internals
Scenario D — operational optimisation, replace internals, add preflash

Advances in Distillation Increasing the productivity of crude oil distillation systems through retrofit 36

MANCHESTER 1824 The University of Manchester

Case study: results

AIChE The Global Home of Chemical Engineers

Add preflash

Parameter	Scenario A	Scenario B	Scenario C	Scenario D
Retrofit cost, $	—	—	—	61,748

Net profit

Parameter	Scenario A	Scenario B	Scenario C	Scenario D
Total retrofit cost, $	318,948.	256,943	328,905	325,371
Net profit, Million $·y⁻¹	965.2 (+30.1%)	976.2 (+31.6%)	979.8 (+32.1%)	980.2 (+32.2%)
Payback, y	0.0014	0.0011	0.0014	0.0014

- In every scenario, the product specifications are satisfied.

Scenario A — pro-rata increase, replace internals
Scenario B — operational optimisation
Scenario C — operational optimisation, replace internals
Scenario D — operational optimisation, replace internals, add preflash

MANCHESTER 1824 The University of Manchester

Case study: conclusions

AIChE The Global Home of Chemical Engineers

- Introducing a preflash, together with column internals changes and optimal HEN retrofit can bring significant benefits
  - Fired heating demand, capital expenditure, ease of HEN retrofit
- In all cases, changing the PA temperature drops impacted significantly on system performance
- Section 1 of the main fractionator is the hydraulic bottleneck.
  - debottlenecking this section (operating variables and internals) can enhance energy efficiency of the system
- The optimisation time increases with the number of structural design decisions considered.
- Downtime is not considered in this analysis.

6. Conclusions and future work

Conclusions and future work

Conclusions

- The proposed retrofit approach is useful to analyse multiple retrofit scenarios with relatively little engineering effort.
- The high-level of detail used to analyse each unit allows one to exploit the interactions between the distillation column and the HEN while considering the system constraints.

- The novelty of the proposed approach is the incorporation of structural design decisions and a range of constraints into an optimisation environment.

Future work

- Include downtime considerations.
- Use more accurate hydraulic and cost correlations.
- Use rigorous simulation to develop metamodels in order to reduce the optimisation time.

MANCHESTER
1824
The University of Manchester

AIChE
The Global Home of Chemical Engineers

Thank you for your attention!!!
Questions?

431021 Elucidation of Anomalies in Reactive Distillation: The Node-Bumping Effect

Abstract:

The Residue Curve Map (RCM) technique is a tool for the synthesis and analysis of distillation columns. A Residue Curve is the locus of the compositions of the residual liquid during an isobaric batch distillation process, where the vapour removed is assumed to be in equilibrium with the liquid in the still. A Residue Curve Map is the family of all residue curves which could be obtained by using feed materials of different initial compositions in the still. A Reactive Residue Curve Map (RRCM) is similarly defined as the locus of compositions of the residual liquid during an isobaric batch process in which boiling and reaction occur simultaneously. The vapour that could be obtained can be obtained by mass balance. In this work, an equation is derived (this is the RRC equation) to model the liquid composition of a particular ideal three component system in a kinetically controlled simple reactive distillation process. From the RRC equation a dimensionless parameter Damkohler number (Da), is defined, which measures the amount of reaction relative to the product removal. A global RRCM viz. for both positive and negative compositions is then plotted for some arbitrarily chosen increasing Damkohler numbers. When Da is zero (separation alone) there are three nodes in the composition space. We notice that when Da is greater than zero the number of nodes increases from 3 to 5. Nodes that previously lay outside the Mass Balance Triangle (MBT) move inside the MBT at Da values greater than a critical value (0.5). Two singularities lie on top of each other at this critical value of Da. For this example, above this value of Da the saddle node that was at the corner moves into the composition space and the one that was outside at lower Da now becomes a fixed node on the corner of the MBT. We call this phenomenon node bumping. The composition space is thus divided into two distillation regions at Da values greater than this critical value. By monitoring changes in Da it is shown how to avoid multiple steady states where changes in composition of initial condition could result in profiles running in for what could be described as the incorrect direction (bifurcation regions). Tracking compositions outside the MBT also provides understanding as to why „reactive azeotropes can be formed.

INFLUENCE OF THE INLET DOWNCOMER IN THE MOMENTUM AND ENERGY TRANSFER ON A SIEVE TRAY USING CFD TECHNIQUES

*Gabriel Henrique Justi, Gabriela Cantarelli Lopes, José Antônio Silveira Gonçalves**.
Federal University of São Carlos, São Carlos, SP, Brazil.
**E-mail address: jasgon@ufscar.br*

Introduction

Separation processes are techniques which consume great quantity of energy. Among these processes, distillation columns are one of the most important and utilized techniques. Distillation is a physical process of separation based on the difference of volatility of the components in a mixture, which involves heat, mass, and momentum transfer. Prediction of tray hydrodynamics is important to avoid critical regions, as recirculation or stagnant regions, which directly affect the efficiency of the stages (Kister, 1992; Wankat, 2006).

Continuous improvements in computer technology and numerical methods have allowed to deal with physical and chemical phenomena from its fundamental equations and on a microscopic scale, through the Computational Fluid Dynamics (CFD) techniques. In this way, appreciable amounts of computational studies about sieve tray have been done in the last years (Tu et al., 2007; Anderson and Wendt, 1995).

In order to apply the CFD in distillation columns, it is important to understand the hydrodynamic in a distillation stage. Liquid falls through the downcomers by gravity from one tray to the another one below it. Downcomers play an important role in separation and hydrodynamic behavior of trays as to some extent, mass transfer occurs in the downcomer as well as on the tray deck. Thus, it is significant to focus on the simulation systems considering tray hydrodynamics in the presence of downcomers. Inlet downcomer conditions affect the froth height on the bulk of the tray as well as downstream conditions near the exit weir. As liquid flows through the downcomer clearance to the tray deck, a transition from a single phase to two-phases flow occurs which makes analysis in the region of the liquid entry quite difficult (Lockett, 1986).

Some of the studies involving were focused on the comprehension of hydrodynamics phenomena that occurs on sieve trays, making possible the analysis of important factors, such as: velocity distribution, liquid and gas volume fractions profiles along the tray, regions of intense turbulence and regions of liquid accumulation, etc. These kind of results allow the implementation of improvements in the design of the trays, in order to avoid critical regions that directly affect the mass and energy transfer occurring throughout the stages, such as retrograde flow and stagnant regions (Mehta et al., 1998; van Baten and Krishna, 2000, Liu et al., 2000, Gesit et al., 2003; Rahimi et al., 2006). In these studies, the authors did not model the inlet downcomer in the simulations, considering flat or parabolic liquid inlet velocity profile.

In order to verify the heat transfer of energy along the sieve trays, Noriler et al. (2008) presented a three-dimensional, transient and two-phase flow, in an Eulerian-Eulerian framework. Thus, the continuity equations and the energy and momentum conservation balances were applied for both fluids, for predictions of volume fractions and velocities, temperatures and pressure fields.

Rahimi et al. (2006) and Noriler et al. (2010) analyzed the mass and energy transfers along the sieve tray simultaneously. Both authors proposed a three-dimensional and multiphase flow, in an Eulerian-Eulerian framework. Rahimi et al. (2006) carried out their model in a stationary proposal, and Noriler et al. (2010) in a transient proposal. Thus, the continuity equations and the momentum, energy and chemical species conservation balances were applied for both fluids, for predictions of efficiency and temperature and concentration fields.

Recently, Rahimi et al. (2011) considered the complete geometry including sieve tray, inlet and outlet downcomers. These were modeled using CFD and based on experimental work of Solari and Bell (1986) for air-water system. The authors predicted only the tray hydrodynamics parameters, such as liquid velocity distribution, clear liquid height, average liquid volume fraction in froth and froth height.

In this context, the main purpose of this study is to investigate the extent of the effect of inlet downcomer in the momentum and energy transfer on a sieve tray using a two-phase, three-dimensional and transient CFD model in an Eulerian-Eulerian framework. A water-ethanol mixture at 1 atm was simulated.

Mathematical Modelling

The model was based on an Eulerian-Eulerian approach to model the liquid and vapor phases. Thus, each phase was treated as an interpenetrating continuum phase having separate transport equations. Therefore, the Reynolds averaged Navier-Stokes equations (RANS) for continuity, momentum, and energy transfer equations were numerically solved for each phase. Since the pressure and temperature gradients are small in a same distillation stage, all the system was considered at 1 atm. The balance equations for the phases are given below and The subscript "*k*" and "*j*" represents the liquid and gas phase.

Mass Continuity Equation

$$\frac{\partial}{\partial t}(f_k \rho_k) + \nabla \cdot (f_k \rho_k \mathbf{v}_k) = 0 \tag{1}$$

Momentum Equation

$$\frac{\partial}{\partial t}(f_k \rho_k \mathbf{v}_k) + \nabla \cdot (f_k \rho_k \mathbf{v}_k \mathbf{v}_k) = -f_k \nabla p_k + \nabla \cdot \left\{ f_k \mu_k^{\text{eff}} [\nabla \mathbf{v}_k + (\nabla \mathbf{v}_k)^{\text{T}}] \right\} + f_k \rho_k \mathbf{g} + M_k \tag{2}$$

Energy Equation

$$\frac{\partial}{\partial t}(f_k \rho_k h_k) + \nabla \cdot (f_k \rho_k \mathbf{v}_k h_k) = \nabla \cdot \left(f_k \lambda_k^{\text{eff}} \nabla T_k \right) + Q_k \tag{3}$$

Model Assumptions

The following assumptions were adopted on the model used in this study for a sieve tray: a two-phase system was modeled in the simulation:

- The gas phase is the dispersed phase and the liquid phase is the continuous phase;
- A mixture binary ethanol-water was used in this study;
- Due to the fact that there are no reactions and no other mass source, the terms related to them are null;
- the net rate of the mass transfer between the phases is considered to be small and also in this particular study, it not included in the modeling of the chemical species and thermodynamic equilibrium equations that it can be neglected; and
- No turbulence model was used for dispersed phase.

Closure Equations

In order to solve some terms of the Equations 2 and 3 additional equations are required. In this way, the terms that need to be modeled, such as effective viscosity, μ_k^{eff}, effective thermal conductivity, λ_k^{eff}, interphase momentum transfer, M_k, and interphase energy transfer, Q_k, are presented below.

The effective viscosity and thermal conductivity are related to the diffusion and the dispersion of the properties due to the turbulence effects. Thus, it can be written as a combination between diffusion flux and turbulent flux, where the "turb" superscript represents turbulent flux. Considering a Newtonian fluid and the eddy viscosity hypothesis, the effective viscosity and thermal conductivity can be written as:

$$\mu_k^{\text{eff}} = \mu_k + \mu_k^{\text{turb}} \tag{4}$$

$$\lambda_k^{\text{eff}} = \lambda_k + \lambda_k^{\text{turb}} \tag{5}$$

The μ_k^{turb} is then modeled using a turbulence model. The Shear Stress Transport (SST) turbulence model was used to describe the turbulence on the sieve tray. This turbulence model combines the advantages of the k-ε and k-ω models, with a possibility to describe accurately regions that are near the wall, the Equation 6 was used to calculate the eddy viscosity.

$$\mu_k^{\text{turb}} = \frac{\alpha_k k_k \rho_k}{\max(\alpha_1 \omega_k, SF_2)} \tag{6}$$

The SST model is well discussed in the literature (Menter, 1994; Wilcox, 2005).

From the prediction the turbulent viscosity and the concept of the turbulent Prandtl number, σh, it is possible to calculate The λ_k^{turb}, as shown in Equation 7 below,

$$\lambda_k^{\text{turb}} = \frac{\mu_k^{\text{turb}}}{\sigma_{\text{h}}} \tag{7}$$

The interphase momentum transfer term, M_k, was considered in this work only due to the drag force. The drag force per unit volume for gas-liquid flow can be written below, where the subscripts "l" and "g" represent the liquid and gas phase, respectively:

$$M_k = \sum_{k \neq j} M_{kj} = M_{gl} = -M_{lg} = \frac{3}{4}\frac{C_D}{d_B} f_g \rho_l (\mathbf{v}_g - \mathbf{v}_l)|\mathbf{v}_g - \mathbf{v}_l| \tag{8}$$

The drag coefficient value for the case of distillation is not well known. In this case, the drag coefficient has been estimated using the Krishna et al. correlation (Krishna et al., 1999), as showed below:

$$C_D = \frac{4}{3}\frac{\rho_l - \rho_g}{\rho_l} g d_B \left(\frac{f_g^{\text{ave}}}{U_S}\right)^2 \tag{9}$$

For average gas holdup fraction, f_g^{ave}, the Bennett et al. correlation (Bennett et al., 1983) was used as follows:

$$f_g^{\text{ave}} = 1 - \exp\left[-12.55\left(U_S\sqrt{\frac{\rho_g}{\rho_l - \rho_g}}\right)^{0.91}\right] \tag{10}$$

Note that when the drag correlation (Equation. 9) is substituted in Equation 8, the mean bubble diameter is canceled, i.e., eliminating the main problem in the prediction of gas-liquid bubbling flow.

Heat transfer between the phases must satisfy the local energy balance. Thus, the rate of energy transfer between phases can be written:

$$Q_k = \sum_{k \neq j} Q_{kj} = Q_{gl} = -Q_{lg} = h_{gl} A_{gl} (T_g - T_l) \tag{11}$$

The global heat transfer coefficient is defined as,

$$h_{gl} = \frac{\lambda_g \mathrm{Nu}_{gl}}{d_B} \tag{12}$$

An appropriate value of the global heat transfer coefficient can be obtained by using suitable correlations, such as Ranz-Marshall equation (Ranz and Marshall 1952):

$$\mathrm{Nu}_{gl} = 2 + 0.6(\mathrm{Re}_g)^{0.5}(\mathrm{Pr}_g)^{0.3} \tag{13}$$

$$0 \le \mathrm{Re}_g \le 200 \quad \text{and} \quad 0 \le \mathrm{Pr}_g \le 250$$

The dimensionless numbers are defined as,

$$\mathrm{Re}_g = \frac{\rho_l |\mathbf{v}_g - \mathbf{v}_l| d_B}{\mu_l} \tag{14}$$

$$\mathrm{Pr}_g = \frac{C_{p,g}\mu_g}{\lambda_g} \tag{15}$$

The interfacial area per unit volume between liquid and gas phases, known as the interfacial area density, A_{gl}, for the particle model is defined by:

$$A_{gl} = \frac{6 f_g}{d_B} \tag{16}$$

The mean bubble diameter, d_B, can be estimated relating it to the maximum bubble diameter, $d_{B,max}$, which is modeled by the correlation proposed by Davidson and Schuler (Davidson and Schuler, 1960) (Equation 18),

$$d_B = \frac{d_{B,max}}{2} \tag{17}$$

$$d_{B,max} = 1.23 \sqrt{\frac{p_t{}^4}{g}\left(U_S \frac{A_C}{A_B}\right)^2} \tag{18}$$

Numerical Methodology

Computational Domain and Numerical Mesh

The computational domains consists in a distillation sieve tray with diameter of 0.40 m. The three-dimension geometry was developed using the software ANSYS Design Modeler. Table 1 presents the specifications of the sieve tray geometry and Figure 1 shows the geometries used in the simulations. Both the geometry with and without the inlet downcomer were simulated.

The tray has a 75% of bubbling area, a weir height of 0.015 m, a downcomer clearance of 0.01 m, a 4.06% hole area with 0.005 m (length) rectangular holes arranged in a 0.025 m triangular pitch.

The grid used in the simulation was developed using the software ANSYS Meshing and consists in a hybrid grid with triangular prism and hexahedral elements. Figure 1 shows the details of the numerical mesh. Further ahead, the sensitivity tests to grid sizes different will be presents.

Table 1. Sieve tray details.

Tray diameter	0.40	m
Tray spacing	0.19	m
Weir henght	0.3091	m
Weir lenght	0.015	m
Donwcomer clearance	0.01	m
Downcomer lenght	0.15	m
Flow path lenght	0.2539	m
Hole lenght and pitch	0.005×0.025	m
Bubbling area (over total area)	75	%
Hole area (over bubbling area)	4.06	%

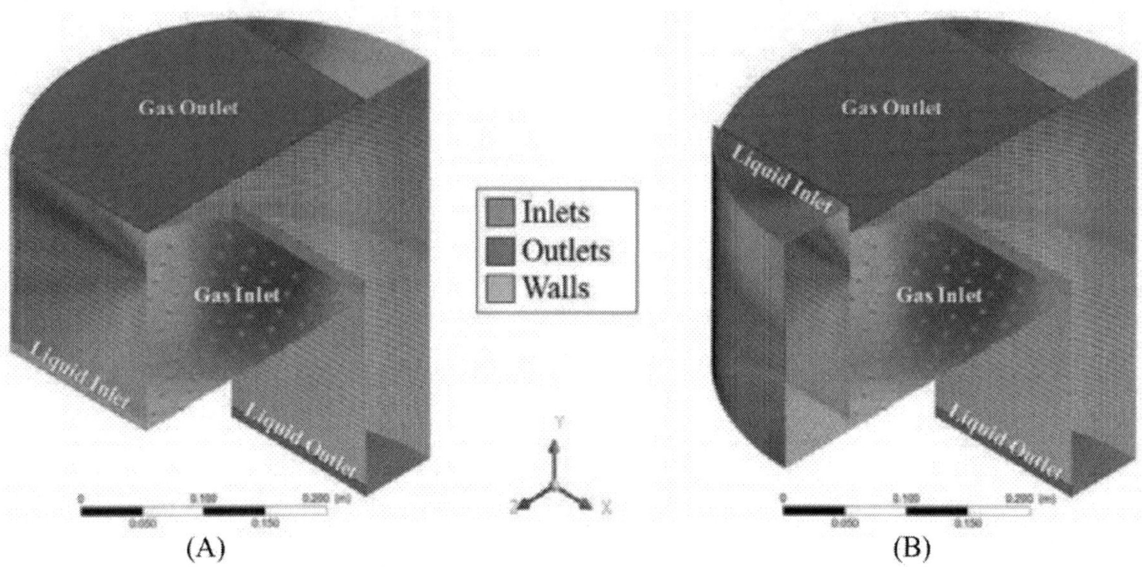

Figure 1. Computational domain, mesh and boundary conditions: (A) without inlet downcomer and (B) with inlet downcomer.

Boundary and Initial Conditions

The initial and boundary conditions are essential for closing the model and allow the numerical resolution of the model equations. Figure 1 illustrates the boundary regions of the sieve tray. Liquid enters the tray through the downcomer clearance area (labeled as Liquid Inlet), and leaves geometry through the downcomer clearance area that leads to the tray below (labeled as Liquid Outlet). Gas enters through holes at the bottom of the tray (labeled as Gas Inlet), and leaves the domain through holes at the top (labeled as Gas Outlet).

Gas Inlet: the gas mass flow rate entering through the holes on the tray is the same in each hole. Thus, the gas velocity that flows through the holes was calculated according to Equation 19. It was considered that only the gas phase enters this region, once the amount of liquid dragged by the gas is negligible. Additionally, it was specified that the mass fraction of ethanol in the gas phase was 0.7330.

$$V_{g,inlet} = \frac{U_S A_B}{N_H A_{H,i}} \quad (19)$$

Liquid Inlet: at the liquid inlet in the Figure 1, only liquid was considered entering through this region and was adopted a uniform velocity profile according to Equation 20. It was considered that only liquid phase enters this region, once the amount of dragged liquid by the gas is negligible. Additionally, it was specified that the mass fraction of ethanol in the liquid phase was 0.5402.

$$U_{l,inlet} = \frac{Q_l}{A_{cl}} \quad (20)$$

Gas and Liquid Outlets: at the liquid outlet region, a pressure condition was adopted and it was used in order to simulate a resistance in the liquid outlet, which actually exists due to the lower tray. This resistance involves the emergence of a buildup of fluid column in the downcomer outlet, which in this study was considered occupying of 50% of the downcomer length. At the gas outlet region a pressure condition was also used by specifying a relative pressure of zero.

Walls: a no-slip wall boundary condition was specified for both phases.

Domain Initialization: Good initial guesses of the flow variables are important not only to avoid a significantly longer computational time but also in some cases to avoid divergence. The initial condition applied in the simulation consisted of the domain filled only with stopped gas phase at a temperature of the 356.80 K. Thus, it was possible to observe the tray filling by the liquid phase.

Simulation Settings

The simulation settings was established using the ANSYS software CFX-Pre 14.5. The following strategy to minimize the computational effort in order to ensure the solution stability has been adopted. First of all, it was obtained the solution for the flow without the solution of energy equation reaching the quasi-steady condition. This result is used as initial condition for the next step which consists in the resolution of the model equations with energy equation.

The operation conditions used in the simulation were a liquid volumetric flow rate, Q_l, equal to 7.5×10^{-4} m^3/s, and a gas load factor F-factor, F_S, equal to 0.8 $m/s(kg/m^3)^{0.5}$. The F_S is defined as a function of the gas superficial velocity and the gas density, i.e.,

$$F_S = U_S\sqrt{\rho_g} \quad (21)$$

For continuity, momentum and energy transfer equations, the upwind interpolation scheme was used for the advective terms and First Order Backward Euler scheme was employed for the transient terms. For the equations related to turbulence, the High Resolution scheme was used. The coupled solution method was used to solve the model, with all the equations solved simultaneously as a single system.

The simulation was conducted using thirty-two AMD Opteron 2.40 GHz processors run (32×2.40 GHz) in parallel to simulate 10 seconds of multiphase flow on a distillation sieve tray using fixed time steps of 2.0×10^{-3} s. The Courant number used in the simulations was 0.5.

Physical properties of liquid and gas phases used in the simulation is presented in the Table 2.

Table 2. Physical properties of liquid and gas phases.

Ethanol-Water Mixture				Units
		Liquid	Gas	
Temperature	T_{inlet}	354.80	356.90	K
Ethanol mass fraction	$X_{A,inlet}$	0.5402	-	-
	$Y_{A,inlet}$	-	0.7330	-
Heat capacity	$C_{p,mix}$	3558.53	1692.59	J kg/K
Density	ρ_{mix}	839.28	1.292	kg/m^3
Dynamic viscosity	μ_{mix}	3.790×10^{-4}	1.070×10^{-4}	Pa s
Thermal conductivity	λ_{mix}	0.3955	0.0207	W/m K
Diffusivity	D_{AB}	6.709×10^{-9}	1.576×10^{-5}	m^2/s

Results and Discussion

A grid-size sensitivity test was performed using the GCI (Grid Convergence Index) method proposed by Roache (Roache, 1994) to quantify the numerical uncertainty. After the choice of the mesh, simulations the sieve trays with and without inlet downcomer were carried out to observe the effects of including this region at the computational domain in order to better understand the flow patterns and temperature profiles throughout the stage.

Grid-size Sensitivities

To evaluate the stability of the numerical grid, the clear liquid height (in an average volume) and the pressure (in an average ZX plane) were adopted as global and local variables, respectively. This kind of test is important to evaluate the influence of the grid size and the refinement of the grid cells in relation to the results of the simulations.

The tests were performed on three types of mesh: fine (grid 1), medium (grid 2), and coarse (grid 3), with elements ranging from 184 000 to 1 million of elements. The fine, medium and coarse grids have a characteristic length of 2.86, 3.68 and, 5.06 cm, respectively, resulting in a refinement ratio of 1.3 between the grids. Figure 2 and Table 3 show the details and the results of the grid independence tests.

The clear liquid height (global variable) showed a numerical stability from the grid 2 with a value of 1.8 cm, for which numerical uncertainty between the coarse and medium grids ($GCI_{fine,32}$) it was 29.83%, and between the medium and fine grids ($GCI_{fine,21}$) was 7.62%. For the local variable, pressure fields, the numerical stability was reached for a value of 19 Pa, for which the numerical uncertainty between the coarse and medium grid ($GCI_{fine,32}$) was 7.97%, and between the medium and fine grids ($GCI_{fine,21}$) was 3.43%. By the results acquired by the GCI method, it is concluded that the grid 2 (480000 elements) showed satisfactory performance, with a numerical uncertainty below 8% compared with the grid 1 (fine) for global and local variables, as shown Table 3.

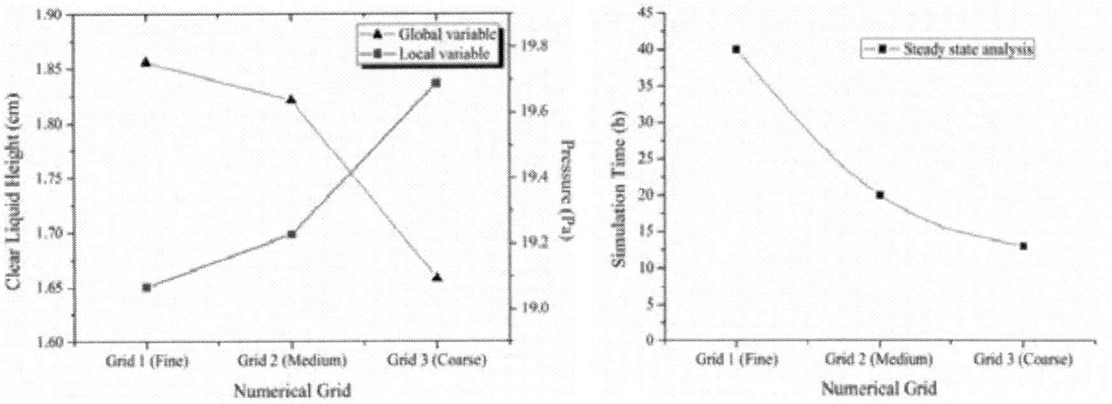

(A) (B)

Figure 2. Steady state simulation sensitivity of the simulation results by comparing the results for grids: (A) global and local variables and (B) simulation time as a function of the elements number.

Table 3. Numerical uncertainty calculation applied to grid under analysis.

Variables	Location	$GCI_{fine,32}$	$GCI_{fine,21}$
Clear liquid height	-	29.83%	7.62%
Pressure	YZ Plane (x = 0 mm)	7.97%	3.43%

In addition, as the computation time is of great importance in optimizing the simulation it is evident that grid 2 had a processing time (see Figure 2B) approximately two times lower than that grid 1 to obtain similar results.

Effect of Inlet Downcomer

The results presented in Figures 3 and 5 show the influence of the inlet downcomer on the sieve tray for liquid w velocity and liquid temperature, respectively. The results were analyzed for an instant of time equal to 10 seconds.

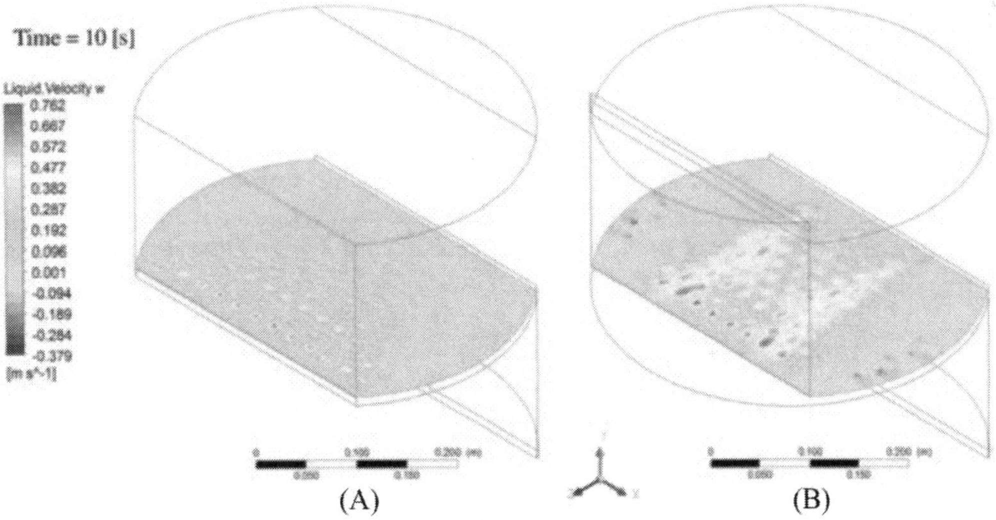

(A) (B)

Figure 3. Liquid w velocity situated in the ZX plane at 10 mm above the tray floor: (A) without inlet downcomer and (B) with inlet downcomer.

In Figure 3 are presented the results of the liquid w velocity, located on a ZX plane at 10 mm above the tray floor. The negative and positive values of the liquid w velocity refers to the axis direction of the computational domain, opposite to the flow direction. Liquid w velocity, in the geometry without inlet downcomer (Figure 3A), have presented a more uniform profile along the flow. This uniformity of liquid w velocity on the tray was due to the choice of the boundary condition type used at the liquid inlet. Based on this, the simulation without inlet downcomer did not present reverse fluid flow or recirculation zones. Liquid w velocity have more variations along the fluid flow in the geometry with inlet downcomer (Figure 3B). This fact, it is due to the natural formation of velocity profile from influence of the inlet donwcomer. Besides the velocity profile presents higher values at the center of the flow fluid (delimited by the color red, 0.762 m/s) and next to the wall of the sieve tray, presents stagnant zones (delimited by light blue) and regions of liquid reverse flow (delimited between blue color with values between -0.189 and -0.379 m/s).

The liquid w velocity profiles for the sieve tray with and without inlet downcomer can be shown in more detail in Figure 4.

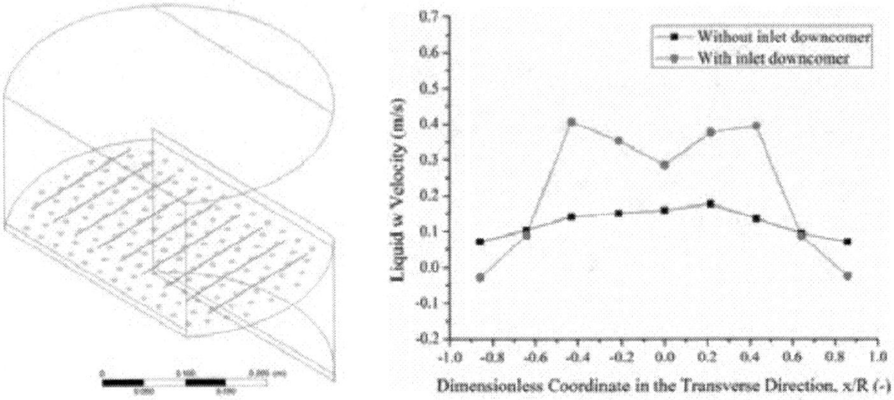

(A) (B)

Figure 4. Liquid w velocity profile: (A) lines positions of the CFD model and (B) average linear liquid w velocityalong the x axis for the geometries with inlet downcomer (red line) and without inlet downcomer (black line).

In Figure 4A the average linear liquid w velocity were measured along the lines perpendicular (z = -0.085 to z= 0.085 m) to the liquid flow direction on a ZX plane (y = 0.01 m) above the tray floor and were divided into nine lines along the x axis (x = -0.15, -0.1125, -0.075, -0.0375, 0, 0.0375, 0.075, 0.1125, and -0.15 m). The abscissa values represent the dimensionless coordinate in the transverse direction, where -1 and 1 represent wall region and 0 represents the center. In Figure 4B the results showed different liquid w velocity profiles for the two geometries. Therefore, if we analyze the liquid w velocity, the tray with inlet downcomer showed velocity gradients with high values (red line) at the central region of the sieve tray (see Figure 3B) and near the wall of the sieve tray the liquid velocity had a deceleration reaching values close to zero and negative, i.e., in regions of stagnant and retrograde liquid flow. Also in Figure 4B, the geometry without inlet downcomer presented a liquid w velocity profile more uniform (black line) in relation the x direction.

The liquid temperature profile on the ZX plane is shown in Figure 5, where temperatures ranged from 354.8 to 355.1 K. As already explained and discussed previously the influence and effects of the inlet downcomer in the velocity fields, is also observed for the liquid temperature. In general, the blue color indicate the colder liquid regions and the red color indicate the warmer liquid regions, as indicated in the legend. Average liquid temperature at the ZX plane analysis showed very close values for the two computational domains, i.e., 354.87 K for the geometry without inlet downcomer and 354.95 K for the geometry with inlet downcomer. However the temperature patterns are significantly different for the different geometries (Figures 5A and 5B). It highlights evident how the effects of flow reverse acting on the tray with inlet downcomer alter the liquid temperature profiles along the plate. This influence was due to the increased liquid residence time in the regions near the wall of the tray, allowing greater heat transfer between phases in the sieve tray with the inlet downcomer.

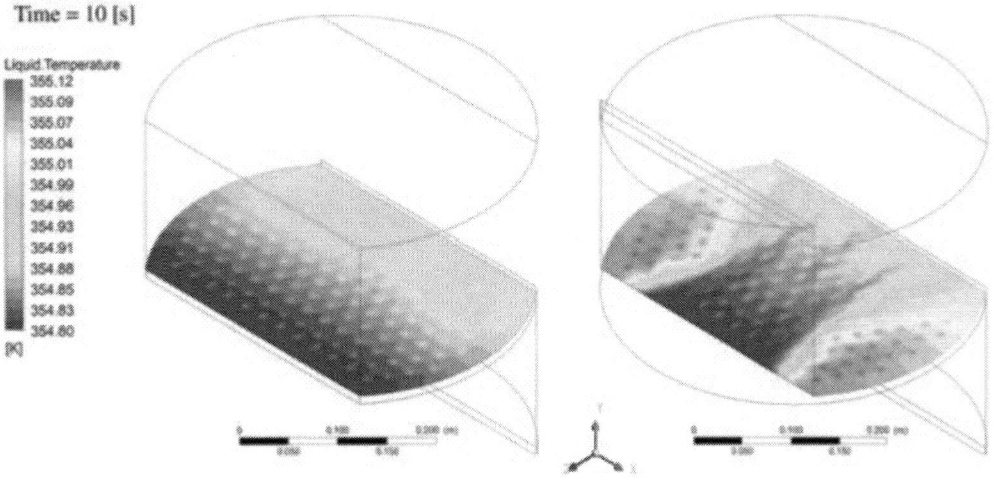

(A) (B)

Figure 5. Liquid temperature profiles situated in the ZX plane at 10 mm above the tray floor: (A) without inlet downcomer and (B) with inlet downcomer.

Based on the results obtained and the observations made for the liquid velocity and temperature profiles, it becomes evident the influence and importance of including the inlet downcomer in the computational model, to better understand the hydrodynamic and thermal behavior in a distillation column sieve tray.

Conclusions

A 3-D two-phase CFD model was developed in the Eulerian-Eulerian framework to compare the momentum, and energy transfer of sieve tray with and without inlet downcomer in the simulations. The CFD was used to predict w velocity, temperature and liquid volume fractions for a gas and liquid rates. The simulation results represented qualitatively and quantitatively the influence of the inlet downcomer in multiphase flow using CFD. These showed that probably the inlet downcomer has a great effect on mass transfer in distillation columns once nonuniform velocity and temperature patterns were observed in this case. Therefore, the CFD has proved a valuable tool for studying the flow in the internal distillation towers. The influence of the inlet downcomer on the mass transfer should be considered in further simulations. It can lead to better results and more realistic conditions compared to that obtained using an uniform inlet velocity.

Finally, the CFD can be used for a better understanding of the turbulent gas-liquid flow on the sieve trays and they can be used to optimize the design and operating conditions.

Notations

A_B	bubbling active area [m^2]	A_{gl}	interfacial area per unit volume [m^2 m^{-3}]
A_{cl}	liquid inlet area [m^2]		
A_C	column cross-section area [m^2]	$A_{H,i}$	area of each hole [m^2]
C_D	drag coefficient [-]	S_{MS}	mass source [kg m^{-3} s^{-1}]

C_p	specific heat [-]	S_M	momentum sources due to external body forces [kg m^{-3} s^{-1}]
d_B	maximum bubble diameter [m]		
$d_{B,max}$	mean bubble diameter [m]	T	temperature [K]
D_{AB}	diffusion coefficient [m^2 s^{-1}]	U_l	liquid inlet velocity [m s^{-1}]
f	volume fraction [-]	U_S	gas superficial velocity [m s^{-1}]
f_g^{ave}	average gas holdup fraction in the froth region [-]	**v**	velocity vector [m s^{-1}]
		V_g	gas hole velocity [m s^{-1}]
F_2	blending function [-]	V_{slip}	slip velocity [m s^{-1}]
F_S	F-factor [m s^{-1} (kg m^{-3})$^{0.5}$]	X_A	mass fraction of A (ethanol) in liquid phase [-]
g	gravity [m s^{-2}]		
h	enthalpy [J kg^{-1}]	Y_A	mass fraction of A (ethanol) in gas phase [-]
h_{gl}	global heat transfer coefficient [J m^{-2} s^{-1} K^{-1}]		
		Greek letters	
k	turbulent kinetic energy [m^2 s^{-2}]	α_1	turbulence model constant [-]
M	interphase momentum exchange term [kg m^{-3} s^{-1}]	λ	thermal conductivity [kg m^{-3} s^{-1} K^{-1}]
		μ	dynamic viscosity [kg m^{-1} s^{-1}]
Nu_{gl}	Nusselt number [-]	ρ	density [kg m^{-3}]
N_H	number of holes [-]	ω	turbulent frequency [s^{-1}]
p	pressure [Pa]	*Subscripts and superscripts*	
p_t	Triangular pitch [m]	eff	effective property
Pr_g	Prandtl number [-]	g, j	gas phase
Q	interphase energy exchange term [J m^{-3} s^{-1}]	$inlet$	at the inlet
		l, k	liquid phase
Q_l	liquid volumetric flow rate [m^3 s^{-1}]	mix	mixture property
Re_g	Reynolds number [-]	turb	turbulence property
S	invariant measure of the strain rate [-]	T	transpose of vector

Acknowledgments
The authors are grateful for the financial support of São Paulo Research Foundation (FAPESP), CNPq, and CAPES for this research.

References

1. Kister, H. Z. Distillation design. [S.l.]: McGraw-Hill New York, 1992.
2. Wankat, P. C. Separation process engineering. [S.l.]: Pearson Education, 2006.
3. Tu, J.; Yeoh, G. H. and Liu, C. Computational fluid dynamics: a practical approach. [S.l.]: Butterworth-Heinemann, 2007.
4. Anderson, J. D. and Wendt, J. Computational fluid dynamics. [S.l.]: Springer, 1995.
5. Lockett, M. J. Distillation tray fundamentals. [S.l.]: Cambridge University, New York, NY, 1986.
6. Mehta, B.; Chuang, K. and Nandakumar, K. Model for liquid phase flow on sieve trays. Chemical Engineering Research and Design, Elsevier, v. 76, n. 7, p. 843-848, 1998.
7. van Baten, J. M. and Krishna, R. Modelling sieve tray hydraulics using computational fluid dynamics. Chemical Engineering Journal, Elsevier, v. 77, n. 3, p. 143-151, 2000.
8. Liu, C.J.; Yuan, X.G.; Yu, K.T. and Zhu, X.J. A fluid–dynamic model for flow pattern on a distillation tray. Chemical Engineering Science, Elsevier, v. 55, n. 12, p. 2287-2294, 2000.

9. Gesit, G.; Nandakumar, K. and Chuang, K. T. Cfd modeling of flow patterns and hydraulics of commercial-scale sieve trays. AIChE journal, Wiley Online Library, v. 49, n. 4, p. 910-924, 2003.
10. Rahimi, R.; Rahimi, M. R. and Zivdar, M. Efficiencies of sieve tray distillation columns by cfd simulation. Chemical engineering & technology, Wiley Online Library, v. 29, n. 3, p. 326-335, 2006.
11. Noriler, D.; Meier, H.F.; Barros, A.A.C. and Wolf Maciel, M.R. Thermal fluid dynamics analysis of gas–liquid flow on a distillation sieve tray. Chemical Engineering Journal, Elsevier, v. 136, n. 2, p. 133-143, 2008.
12. Noriler, D.; Barros, A.A.C.; Wolf Maciel, M.R. and Meier, H.F. Simultaneous momentum, mass, and energy transfer analysis of a distillation sieve tray using cfd techniques: prediction of efficiencies. Industrial & Engineering Chemistry Research, ACS Publications, v. 49, n. 14, p. 6599-6611, 2010.
13. Rahimi, R.; Ameri, A. and Setooodeh, N. Effect of inlet downcomer on the hydrodynamic parameters of sieve trays using cfd analysis. Journal of Chemical and Petroleum Engineering, v. 45, n. 1, p. 27-38, 2011.
14. Solari, R. and Bell, R. Fluid flow patterns and velocity distribution on commercial-scale sieve trays. AIChE journal, Wiley Online Library, v. 32, n. 4, p. 640-649, 1986.
15. Menter, F. R. Two-equation eddy-viscosity turbulence models for engineering applications. AIAA journal, v. 32, n. 8, p. 1598-1605, 1994.
16. Wilcox, D. C. Turbulence modeling for CFD. [S.l.]: DCW Industries California, 2004.
17. Krishna, R.; Urseanu, M.I.; van Baten, J.M. and Ellenberger, J. Rise velocity of a swarm of large gas bubbles in liquids. Chemical Engineering Science, Elsevier, v. 54, n. 2, p. 171-183, 1999.
18. Bennett, D.; Agrawal, R. and Cook, P. New pressure drop correlation for sieve tray distillation columns. AIChE Journal, Wiley Online Library, v. 29, n. 3, p. 434-442, 1983.
19. Ranz, W. and Marshall, W. Evaporation from drops. Chem. Eng. Prog, v. 48, n. 3, p. 141-146, 1952.
20. Davidson, J. F. and Schuler, B. O. G. Bubble formation at an orifice in an inviscid liquid. Trans. Inst. Chem. Engrs., v. 38, p. 335, 1960.
21. Roache, P. J. Perspective: a method for uniform reporting of grid refinement studies. Journal of Fluids Engineering, American Society of Mechanical Engineers, v. 116, n. 3, p. 405-413, 1994.

431297 All Surface Areas Are Not Comparable.: BET Versus Rlrs

Abstract:

The expanding development of microporous solids such as zeolites and MOFs has driven researchers in these areas to employ conventional Brunauer-Emmett-Teller (BET) analyses to estimate surface areas and even to claim record-high surface areas. The assumptions of the BET model, however, assume the surface covers before pores fill, an assumption which is violated in microporous materials. As such, conventional BET analyses of adsorption data must be modified to take into account the different mechanisms (pore-filling vs. surface-covering) that exist when micropores are present.. The physics of sorption processes require that conventional "BET " analyses of adsorption must be modified to reflect the transition between initial physisorption to monolayer sorption to filling of the pores with condensing sorbate.

How should we interpret the resulting estimate of a "monolayer volume" (based on the standard volume adsorbed for nitrogen or argon) in terms of an "equivalent surface area" of a highly curved surface and/or of a three-dimensional pore network? Conventional BET analyses [1,2] of surface areas employ a calculation of the surface area covered by each molecule on a flat (two-dimensional) surface. For nitrogen, the area per molecule is estimated to be 0.162 nm^2/molecule, and for argon it is 0.138-0.182 nm^2/molecule, depending on the source of the estimate and the assumed reference state. Basic estimates of these areal densities come from liquid or solid densities and associated atomic/molecular volumes. When applied to truly three-dimensional zeolite and metal-organic framework (MOF) systems, the surface covered by each molecule is probably less: six molecules will surround a single molecule of similar size in a plane, but this does not make the surface of that molecule six times the flat covered surface of a single molecule! When applied to the truly three-dimensional metal-organic framework (MOF) systems, the surface covered by each molecule is probably far less: six molecules will surround a single molecule of similar size in a plane, but this does not make the surface of that molecule six times the flat covered surface of a single molecule!
We emphasize that the methods discussed in this section are *different* from the standard BET analysis, and emphasize that **using the "standard" BET range of 0.05 < P/P° < 0.3 for microporous materials should not be done.** Note that the C_{BET} values determined from the naïve application of the "standard" method are negative, and give ridiculous isotherms with asymptotes in them.

We review the work of Rouquerol and coworkers and its implementation and testing by Snurr and coworkers [4], a procedure we call the Rouquerol-Llewellyn-Rouquerol-Snurr modification of the Brunauer-Emmett-Teller method, or RLRS-BET method (pronounced "rollers"); the term "rollers-BET" is an allusion to the Snurr group's method of "rolling" a probe molecule along the surface to estimate the equivalent surface area. We emphasize that this method is different from the standard BET analysis, and that using the "standard" BET range of 0.05 < P/P0 < 0.3 for microporous materials **should not be done**---in many cases, the resulting C values are negative, yielding ridiculous isotherms with asymptotes in them. The RLRS-BET method, though in some cases yielding reasonable surface area estimates [4,5], is entirely empirical, and is not based on physically meaningful assumptions (as the original BET theory is for flat surfaces).

[1] P.H. Emmett, S. Brunauer, The Use of Low Temperature van der Waals Adsorption Isotherms in Determining the Surface Area of Iron Synthetic Ammonia Catalysts, J. Am. Chem. Soc., 59 (1937) 1553–1564.

[2] S. Brunauer, P.H. Emmett, E. Teller, Adsorption of Gases in Multimolecular Layers, J. Am. Chem. Soc., 60 (1938) 309–319.

[3] K. Hammond, and W.C. Conner, Analysis of Catalyst Surface Structure by Physical Sorption, Advances in Catalysis 56, 1-101 (2013)

[4] K.S. Walton, R.Q. Snurr, Applicability of the BET Method for Determining Surface

Areas of Microporous Metal–Organic Frameworks, J. Am. Chem. Soc., 129 (2007) 8552–8556.

[5] Y.-S. Bae, A.Ö. Yazaydín, R.Q. Snurr, Evaluation of the BET Method for Determining Surface Areas of MOFs and Zeolites that Contain Ultra-Micropores, Langmuir, 26 (2010) 5475–5483.

[6] J. Rouquerol, P. Llewellyn, F. Rouquerol, Is the BET Equation Applicable to Microporous Adsorbents, in: Stud. Surf. Sci. Catal., 2007, pp. 49–56.

432512 Characterization of Micro-Mesoporous Carbons By High-Pressure CO2 Adsorption with Hybrid QSDFT Methods

Abstract:

Conventionally used argon and nitrogen adsorption isotherms are of limited value for the characterization of very narrow carbon micropores due to the slow rate of adsorbate diffusion at cryogenic temperatures. In order to access the micropores of molecular dimensions, it is necessary to use an adsorbate such as CO_2, which may adsorb at elevated temperatures. Recent advances in sorption technology have enabled the accurate measurement of CO_2 isotherms at 273K over a wide range of pressures (up to 35 bar), allowing the resolution of both micropores (0.4 – 2 nm) and mesopores (2 - 50 nm) in a single adsorption experiment. Several such adsorption experiments were undertaken on characteristic non-porous carbons as well as hierarchical CMK-3 structures.

In this work, a novel method is developed for calculating the pore size distribution from high-pressure CO2 isotherms. For modeling adsorption, the method employs the Quenched Solid Density Functional Theory (QSDFT, [1]) with the roughness parameter that is kept constant in the mesopores range and is gradually reduced with the pore size in the range of micropores, converting QSDFT into Non-Local DFT (NLDFT) in the narrowest micropores of molecular dimensions (<0.5 nm). A cylindrical pore model is used for mesopores and a slit-shaped model for micropores. The pore size distributions obtained using the constructed hybrid kernel of QSDFT and NLDFT isotherms are shown to be in agreement with the conventional characterization methods based of N_2 and Ar isotherms. We demonstrate the capabilities of high pressure CO_2 characterization as an advanced characterization tool for micro-mesoporous carbons of hierarchical structure drawing on the example CMK-3 materials [2].

[1] Neimark, A.V. et. al., Carbon, 2009, V. 47, p.1617-1628.

[2] Jun et. al, J. Am. Chem. Soc., 2000, V. 122, p. 10712-10713.

432028 Neutron Scattering Characterization of CO2 Sorbent Materials : The Morphology of Polyethyleneimine in Mesoporous Silica and Its Performance Impacts

Abstract:

Composites of polyethyleneimine (PEI) and mesoporous silica are highly effective, reversible adsorbents for CO_2, both from flue gas and in direct air-capture applications. The morphology of the PEI within the silica pores can strongly impact the overall carbon capture efficiency and rate of saturation. Here, we directly probe the shape of the supported polymer as a function of weight loading through small-angle neutron scattering (SANS). Combined with textural characterization from physisorption analysis, the data indicates that PEI first forms a conformal coating on the pore walls, but at higher loadings the polymer aggregates and forms plugs that span the pore diameter. This model is consistent with the observation of reduced amine-efficiency (CO_2/N binding ratio) at low loadings and points to a tradeoff between achieving high chemical accessibility of the amine binding sites, which are inaccessible when they strongly interact with the silica, and high accessibility for mass transport, which can be hampered by diffusion through PEI plugs at high loading. We illustrate this design principle by demonstrating higher CO_2 capacity and uptake rate for PEI supported in a modified silica, which exhibits repulsive interactions with PEI.

425184 Extracting Information about Pore Sizes from Ultrasonic Experiments

Abstract:

The speed of sound in a medium provides information on its compressibility, and can therefore be used to investigate the properties of the fluid within a porous material. Ultrasonic experiments carried out during an adsorption measurement show that below saturation, the fluid compressibility is affected by Laplace pressure in the pores [1,2]. This effect has been recently confirmed by calculations based on macroscopic thermodynamics and classical density functional theory [3]. Additionally, these calculations predicted that compressibility of the fluid confined in the pore linearly decreases with the decrease of the pore size. Such a simple relation can provide a basis for calculation of pore sizes based on measurements of compressibility of the confined fluid. However, these results are based on macroscopic theories and therefore may not apply to pores below several nanometers in size.

Here we perform grand-canonical Monte Carlo (GCMC) simulations to calculate the compressibility of Lennard-Jones argon in silica pores from the fluctuations of number of particles. We confirm the linear relation from [3] for the pore sizes above 2.5 nm, but for smaller pores conventional GCMC simulations are inefficient, and fail to give a normal distribution for the number of particles. We therefore use transition-matrix Monte Carlo simulations [4] to calculate compressibility of the fluid in micropores. Our results provide a theoretical framework for determination of pore sizes from ultrasonic experiments on fluid saturated samples.

1. Page, J.; Liu, J.; Abeles, B.; Herbolzheimer, E.; Deckman, H. & Weitz, D. Adsorption and desorption of a wetting fluid in Vycor studied by acoustic and optical techniques Phys. Rev. E, 1995, 52, 2763
2. Schappert, K. & Pelster, R. Influence of the Laplace pressure on the elasticity of argon in nanopores EPL, 2014, 105, 56001
3. Gor, G. Y. Adsorption Stress Changes the Elasticity of Liquid Argon Confined in a Nanopore Langmuir, 2014, 30, 13564-13569
4. Siderius, D. W. & Shen, V. K. Use of the grand canonical transition-matrix monte carlo method to model gas adsorption in porous materials J. Phys. Chem. C, 2013, 117, 5861-5872

430143 The Impact of Activated Carbon's Pore Size Distribution on Heel Formation during Adsorption of Organic Vapors

Abstract:

This research investigates the effect of activated carbon pore size distribution (PSD) on heel formation during adsorption of organic vapors typically emitted from automotive painting operations. For this purpose five types of commercially available beaded activated carbons (BAC) were used. Micropore surface analysis, X-ray photoelectron spectroscopy, and derivative thermo-gravimetric (DTG) analysis were used to assess differences in the physical and chemical properties of these adsorbents. The samples have similar chemical characteristics (e.g. surface oxygen content) but different physical properties (e.g., surface area, total pore volume, micropore volume, and pore size distribution), allowing for isolating the contribution of carbon porosity on heel formation. The samples were tested for 5 consecutive adsorption/regeneration cycles using a mixture of nine organic compounds representing different organic groups. Mass balance cumulative heel was similar for BACs with similar micropore volume and it was 20% higher than heel for BAC with the lowest micropore volume, showing a strong correlation with the adsorbent micropore volume. Contrary to this, adsorption capacities and adsorption breakthrough times correlated with total pore volumes of the BACs not the micropore volumes. BET surface area, micropore volume, and total pore volume of the regenerated BACs decreased proportionally with the cumulative heel. PSD and pore volume reduction confirmed that the heel was mainly accumulated in narrow micropores, which can be occupied or blocked by some of the adsorbates. The results from this study help explain the heel formation mechanism and how it relates to adsorbent structure.

434616 Modified Nanofiltration Membranes for Hydraulic Fracturing Water Filtration

Abstract:

Water usage in hydraulic fracturing is growing. This water usage competes with agricultural and other needs. Further, the disposal of water used in hydraulic fracturing has been linked to earthquakes. We are working on developing techniques for recycling this water. One such technique is nanofiltration, which can selectively separate divalent ions from monovalent ions. However, selectivity espeically in real frack solutions remains a problem. We are working on developing methods for increasing the selectivity by modifying polysulfone based membranes.

432155 Functionalized Membranes: Evolution from Polyamino Acids to Designer Adsorbent Groups

Abstract:

Functionalization is an excellent way of adding new properties and capabilities to well-established and understood membranes. For example, sulfonation of a conventional polyethersulfone membrane will improve its hydrophilicity and inhibit protein binding. If the idea of surface functionalization is extended to the grafting of functional polymer chains, anti-fouling characteristics, surface charge, and permeability control can be enhanced by orders of magnitude. When this idea is applied to membrane adsorbers, very high sorption capacity can be achieved while minimizing the internal mass transfer resistance encountered in resin materials.

Professor Dibakar Bhattacharyya is a pioneer in the field of functionalized membranes. The work began with a challenge: Is it possible to create a sorbent material for heavy metals with a capacity of 1 g metal per g material? Building from his experience with immobilized enzymes, Prof. Bhattacharyya looked to biopolymers for a suitable candidate for the selective adsorption of lead and other heavy metals over calcium. Two candidates emerged – polyglutamic and polyaspartic acid – polyamino acids with carboxylate functionality and a natural selectivity for transition metals based ion acidity.

Since this early beginning, we have moved to fully synthetic polymer grafts based on polystyrene. Styrene is a highly versatile monomer, both for methods of polymerization and in diversity of side functional groups. We have focused on cationic polymerization of styrene and functional styrenes to add charged groups for catalysis and protein separation. We have also designed custom grafts based on Protein A chemistry for selective adsorption of antibodies. These membranes have also been scaled up for large scale production. Field trials have shown that functionalized membranes can give improved flux with no change in permeate quality for greater than 1 year of operation. These applications, as well as the future of functionalized membranes will be discussed.

408305 Biomimetic Membrane Systems Utilizing CNT Nanofluidics

Abstract:

Carbon nanotubes have three key attributes that make them of great interest for novel membrane applications 1) atomically flat graphite surface allows for ideal fluid slip boundary conditions and extremely fast flow rates 2) the cutting process to open CNTs inherently places functional chemistry at CNT core entrance for chemical selectivity and 3) CNT are electrically conductive allowing for electrochemical reactions and application of electric fields gradients at CNT tips. In general, the transport mechanisms through CNT membrane are a) ionic diffusion is near bulk expectation with no enhancement from CNT b) gas flow is enhanced by ~1-2 order of magnitude due to specular reflection off of flat graphitic surface c) and pressure driven flux of a variety of solvents (H2O, hexane, decane ethanol, methanol) are 4-5 orders of magnitude higher than conventional Newtonian flow [*Nature* **2005**, 438, 44] due to atomically flat graphite planes inducing nearly ideal slip conditions. Nearly all applications require chemical selectivity in what is allowed to pass across the membrane. However the act of placing selective functional chemistry at pore entrance or along the core of CNTs, dramatically/completely eliminates the enhanced flow effects by eliminating the near perfect slip boundary condition [*ACS Nano* **2011** 5 3867]. Needed is a mechanism to pump chemicals through the pore where the selective chemistry is. This is routinely achieved in protein channels where permeates are accelerated through regions of precise functionality. Also introduced here are biomimetic platforms using sequential enzymatic flow reactors, nanobubble valves for energy storage and monolayer catalytic reactors.

431666 Implementating Remediation Technology at a Superfund Site

Abstract:

Over the last 20 years, researchers at the University of Kentucky have been working to develop and implement membrane based remediation technologies for applications to TCE and PCB contamination at the Paducah Gaseous Diffusion Plant (PGDP) in western Kentucky. The PGDP was built in the early 1950s for the purpose of providing enriched uranium as part of the country's nuclear energy and weapons program. Following the initial discovery in 1988 of TCE and Technetium-99 contamination in nearby drinking water wells, US EPA placed the PDGP on the Superfund National Priorities (NPL) list on May 31 1994. Extensive PCB contamination was also discovered at the site. This paper will provide an overview of some of the challenges associated with implementing novel remediation technologies at a NPL Superfund site along with some of the lessons learned from this experience.

422144 Environmentally Relevant Separations and Db's Place

Abstract:

The Area of Environmentally Relevant Separations and DB's Place in it

Abstract

Several kinds of tough separation problems are encountered in situations of environmental interests. Such separations may be motivated either by purification of a stream or isolating a desirable species from the bulk. One type of tough problem, such as carbon capture from flue gases from power plants, is merely because of huge scale of operation in which economic considerations dominate. Another type is removing extremely low concentration of toxic components from a stream, such as arsenic from drinking water sources. Yet another type is recovering minute quantities of products from a much larger bulk of fluid, for example, isolating lipids for biodiesels from largely watery stream. Traditional low-cost separation methods of absorption, adsorption, membranes, and even distillation are the typical avenues with innovations to fit the purpose. DB, the researcher has had an impact in most of these approaches to devise methods that qualify as candidates for commercial considerations. The author has observed several innovations in DB's research that cleverly used membranes in various configurations where multi-functionality of ligands, sometimes aided by induced chemical reactions were employed to solve these tough problems. This presentation is about the author's eye witness account of that story.

422901 Adsorptive Membranes for Downstream Processing of Biologics

Abstract:

In this talk, I will describe our latest research efforts to develop best-in-class adsorptive membranes for purification of biologics. Using one or more examples, I will illustrate how membrane technologies can enhance load productivity and bed capacity compared to more conventional sorbent materials in wide use by the biopharmaceutical industry. I also will describe the formation, characterization, and performance of a new class of multimodal membranes that we have developed for protein purifications at high ionic strength. Incorporation of functional groups that provide orthogonal modes of interactions dramatically increases the range of ionic strength for operation of these membranes relative to conventional ion-exchange and hydrophobic interaction chromatography media, and makes them attractive candidates to face emerging challenges in downstream processing of biologics. I will end with an example that illustrates how our multimodal membranes can be used as the primary capture step for purification of monoclonal antibodies from Chinese hamster ovary cell culture supernatant. This ability to perform direct capture in place of Protein A chromatography is a major advancement to the field.

Finally, I would like to thank DB for his many contributions to membrane science and engineering research, the education and training of graduate and undergraduate students, and exemplary service to the separations community. His work is inspirational and his energy is infectious. I am truly grateful for the support and encouragement that he has given to me and my students over the years.

408147 A Novel Method of Continuous Crystallization to Avoid Filtration and Slurry Handling

Abstract:

Recently, development and elaboration of the unit operations necessary to implement continuous manufacturing of APIs and Drug Products have been being widely practiced successfully. However, fouling and clogging of lines in both continuous crystallization and flow chemistry still present a challenge in designing continuous processes. These persistent challenges are significant in handling slurries, traditional powder operations, and solid processing.

This work presents a novel unit operation designed to eliminate the need for slurry handling in the API manufacturing by utilization of a dynamic (falling) film based solution layer crystallization technique. Crystallization on an immobilized layer allows purification, solvent swapping and concentration of intermediate and final API streams to be conducted without the need for isolation or conveying of slurry or powder. In addition, in coupled with a powder free formulation operation this operation could facilitate the elimination of solids handling in drug manufacturing process altogether.

The results for different APIs and crystallization systems prove the performance of the technique for different process parameters, with high yield and level of purification, at high throughput and short processing (residence) time. In addition to the experimental results, a mathematical modeling study demonstrates the performance of the technique for different process parameters (temperature, flow-rate, supersaturation, growth-rate, and distribution coefficient) and the flexibility of the techniques for various process combinations, such as parallel, series, and scaled up/down.

409639 Indirect Ultrasonication in Continuous Slug-Flow Crystallization

Abstract:

In the pharmaceutical industry, the continuous generation of crystals of target size distribution has the potential to simplify and/or reduce post-crystallization operations [1]. The joining of two streams in dual impinging jet, coaxial, or radial micromixers has been demonstrated to produce seed crystals of uniform crystal size distribution (CSD) [2][3], but with a nucleation rate that depends on the liquid flow rate [2][3][4]. Ultrasonication is a technology for facilitating primary nucleation that has been useful in various industrial and academic studies [5]. The nucleation rate is a function of ultrasonication parameters rather than just liquid flow rate, which provides extra degrees of freedom for controlling the CSD. Ultrasonication-assisted crystallization seldom has clogging problems, and actually has the opposite effect and is commonly used to remove particles from containers for cleaning purposes [5]. Ultrasonication facilitates primary nucleation by inducing acoustic cavitation in the liquid solution using high-frequency mechanical vibrations from converters (connected sonication generator) [5].

In most studies of sonocrystallization, an ultrasonication probe (or horn) is in direct contact with the liquid solution, which may induce secondary nucleation from metal contact or contaminate the solution with metal after a long time of use [5]. This talk uses an ultrasonication horn and so has spatial localization of ultrasonication energy, but does not directly contact the horn with the liquid solution. This talk also describes the implementation of this indirect ultrasonication-assisted primary nucleation into a slug-flow cooling crystallizer to demonstrate rapid generation of a large amount of crystals of larger size than the nuclei. The main differences compared to high-quality past studies (e.g., [6]) is that this talk (i) uses an ultrasonic probe focused on a small spatial location instead of placing a long tube in a sonication bath, (ii) has order of magnitude lower time in which the solution is in contact with ultrasonication, and (iii) uses a simpler experimental system [7].

References

[1] Z. Sun, N. Ya, R. C. Adams, and F. S. Fang, "Particle size specifications for solid oral dosage forms: A regulatory perspective," *Am. Pharm. Rev.*, vol. 13, no. 4, pp. 68–73, 2010.

[2] M. Jiang, Z. Zhu, E. Jimenez, C. D. Papageorgiou, J. Waetzig, A. Hardy, M. Langston, and R. D. Braatz, "Continuous-flow tubular crystallization in slugs spontaneously induced by hydrodynamics," *Cryst. Growth Des.*, vol. 14, no. 2, pp. 851–860, 2014.

[3] M. Jiang, M. H. Wong, Z. Zhu, J. Zhang, L. Zhou, K. Wang, A. N. Ford Versypt, T. Si, L. M. Hasenberg, Y. E. Li, and R. D. Braatz, "Towards achieving a flattop crystal size distribution by continuous seeding and controlled growth," *Chem. Eng. Sci.*, vol. 77, pp. 2–9, 2012.

[4] A. J. Mahajan and D. J. Kirwan, "Micromixing effects in a two-impinging-jets precipitator," *AIChE J.*, vol. 42, no. 7, pp. 1801–1814, 1996.

[5] J. R. G. Sander, B. W. Zeiger, and K. S. Suslick, "Sonocrystallization and sonofragmentation," *Ultrason. Sonochem.*, vol. 21, no. 6, pp. 1908–1915, 2014.

[6] R. J. P. Eder, S. Schrank, M. O. Besenhard, E. Roblegg, H. Gruber-Woelfler, and J. G. Khinast, "Continuous sonocrystallization of acetylsalicylic acid (ASA): Control of crystal size," *Cryst. Growth Des.*, vol. 12, no. 10, pp. 4733–4738, 2012.

[7] M. Jiang, C. D. Papageorgiou, J. Waetzig, A. Hardy, M. Langston, and R. D. Braatz, "Indirect ultrasonication in continuous slug-flow crystallization," *Cryst. Growth Des.*, vol. 15, no. 5, pp. 2486–2492, 2015.

411833 Kinetic Modeling and Scaleup Assessment of Seeded Solvent Mediated Form Conversion

Abstract:

This presentation focusses on the challenges of developing a scalable form conversion process. The slurry conversion of solid forms involves unique challenges compared to typical cooling and anti-solvent crystallizations. Form conversion is often coupled with thermodynamics constraints involving polymorphs or other solid forms. This presentation will discuss the importance of phase diagrams, procedures for monitoring the kinetics of form conversion using Raman spectroscopy, developing a mathematical model of form conversion, examining the impact of process parameters and developing a scalable process. In particular, it was found that solubility is not the most significant driver for the kinetics of the process.

418206 Improving the Separation of Organic Small Molecules Purified Via Sublimation

Abstract:

Conjugated small molecule materials are currently being investigated as organic semiconductors for use in next generation display and solid-state lighting applications. Widespread commercial adoption of these technologies is limited by significant materials costs, partially due to an expensive purification step. This work investigates the current industrial purification technique, thermal gradient sublimation, in order to better understand the underlying mechanisms and guide process development. The purification of two common charge transport layer materials, N,N'-bis(naphthalen-1-yl)-N,N'-bis(phenyl)-benzidine (NPD) and 4,4',4"-tris(carbazol-9-yl) triphenylamine (TCTA), is found to be limited by vapor phase transport and deposition. Interestingly, mechanisms in the solid phase such as diffusion within the solid feed, reaction at the feed particle surface, and mass transfer within the bed of feed particles are not rate limiting. The vapor phase transport and deposition process is consistent with diffusion. These results provide a basis for improving existing thermal gradient sublimation processes and suggest guidelines for the design and operation of future large scale purifications of high value, low volume specialty chemicals.

Chemtec

Improvement of Design Assessment Methodologies for Melt Static Crystallization Process

2015 AIChE Annual Meeting, Salt Lake City, UT

L. Guillou

November 10th 2015
Prepared by: Loïc Guillou

POE-TMP-007-E

MSC: Design Assessment vs. Design & Modeling

fives

Design Assessment / Shortcut

- **Pre-evaluate**
- **Explore at low cost**
- **Comparatively rate (approx.)**

Design and Modeling

- **Secure your data**
- **Select ONE way to go and THE way to go**
- **Issue guarantees (or margins)**

2 approaches corresponding to 2 steps of a Project

Agenda

1. Who we are
2. Principle of MSC
3. Shortcut modelling for MSC processes
4. Shortcut modelling through statistic approach

Chemtec

Company Overview

Fives, an international industrial engineering Group

fives ultimate machines ultimate factory

Fives designs and supplies **machines, process equipment** and **production lines** for the **world's biggest industrial players.** Fives supplies **PROABD® MSC plants** since the 50's.

Fives, a pioneer which constantly innovates

Fives' strength is to develop

— **proprietary technologies**

— with **high added value for the customers**

— and **high energy and environmental performances**

Innovation data:

More than **33 million €** spent on R&D in 2014 (outside projects)

132 innovations patented in 3 years

532 patent families in force

1 859 patents in force

Chemtec

fives

Principle of MSC

Principle of Melt Static Crystallization operation

fives

- The Melt Static Crystallization (MSC) is a batch process for **separation and purification**
- The MSC is ruled by the **solid-liquid equilibrium** curve of the compounds, crystallization **kinetics** and **rheology**
- A MSC stage is divided into different steps, each managed through **heat load control**

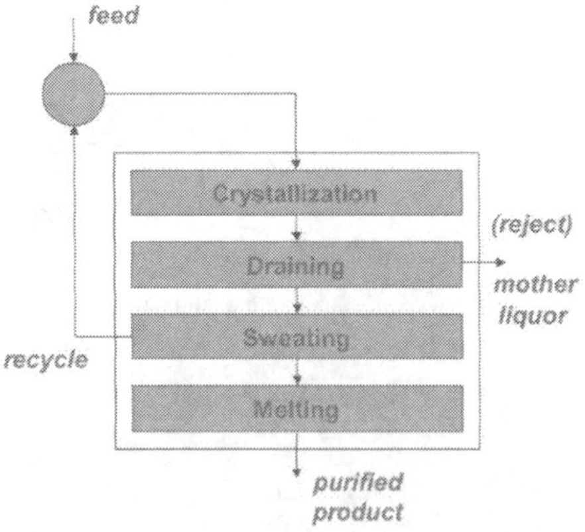

You want to know more? Visit our website and watch our MSC video

Multi-stages MSC = Fractional crystallization

fives

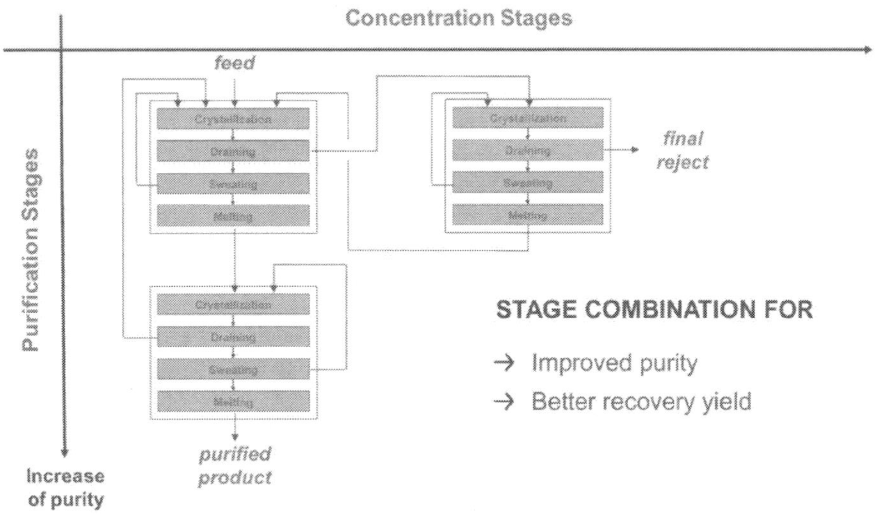

STAGE COMBINATION FOR

→ Improved purity

→ Better recovery yield

The PROABD® MSC unit

fives

A PROABD® MSC unit is composed of three parts synchronized by dedicated **advanced automation** :

- Product handling
- Crystallization group with **proprietary equipment**
- Utilities

Such a unit :

- is **fully integrated** to a continuous production line
- ranges in size from a **few liters to hundreds of m³** of crystallizer volume

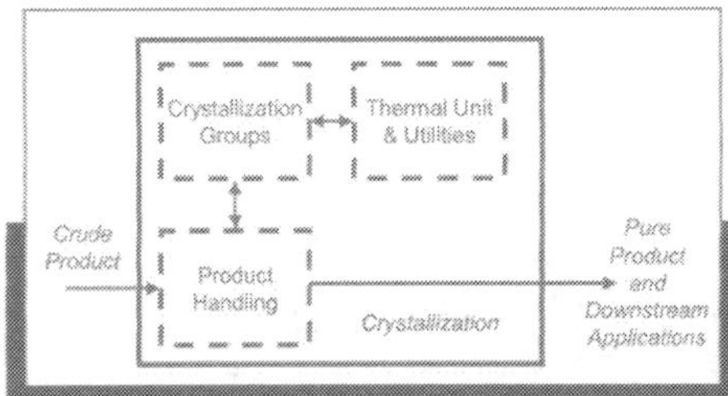

An unlimited list of compounds purified at lab & pilot scale

fives

- Carbochemicals & Petrochemical
  - naphthalene,
  - anthracene,
  - carbazole,
  - fluorene,
  - cresol
  - xylenes, …
- Polymer Industry
  - vinylformamide,
  - dicumyl peroxide,
  - benzoic acid
  - styrene,
  - methylene diphenyl diisocyanate (MDI),
  - toluene diisocyanate (TDI),
  - NCB (nitrochlorobenzene)
  - acrylic acid
  - bisphenol A,
  - L-lactid …
- Cosmetic Industry
  - levulinic acid
  - t-butylhydroquinone, …
- Food Industry
  - erythritol
  - menthol, …
- Fine Chemicals & Specialties
  - cyclododecanol
  - o-toluene sulfonyl chloride
  - dicyclopentadiene (DCPD)
  - chlorotoluene, …
- Agrochemicals
  - 2-methylimidazole
  - dichlorobenzene
  - o-phenylphenol
  - phosphoric acid, …
- Intermediates to Active Pharmaceutical & Ingredients
  - N-hydroxy-4-piperidinol
  - 2-picoline, …

Chemtec

fives

Shortcut modelling of MSC processes

Typical short cut modeling approach

fives

"Engineering judgment" [2] + SLE

$k_{eff} = C_s/C_l$ effective impurity distribution coefficient

C_s = concentration of impurities in solid
C_l = concentration of impurities in liquid

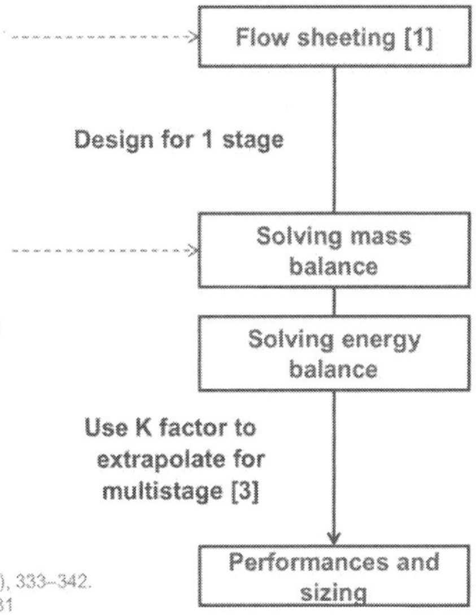

[1] Gilbert S.W., AIChE J., 37 (1991), 8, 1205-1218
[2] W. Marquardt, et al., Chin. J. Chem. Eng., 16 (2008), 333–342.
[3] Wellinghoff G. et al., Chem. Ind. Tech., 63 (1991) 881

Solving the mass balance

fives

The big deal with k_{eff}

1. Supposed to be constant…
2. …But is in fact compound dependant,…
3. …And process parameters dependant!

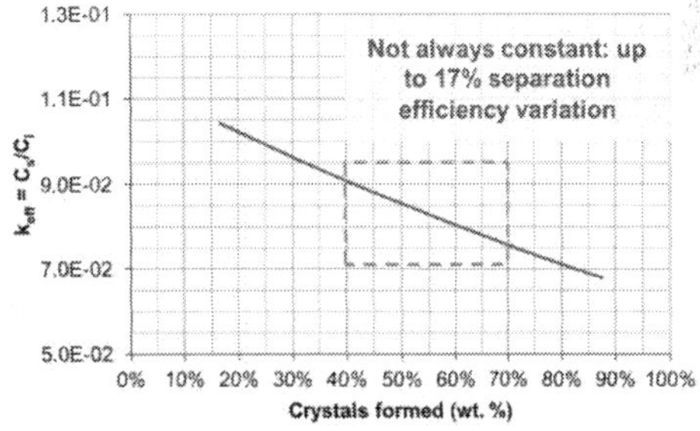

Lab rig for k_{eff} determination affects obtained values: ex. increased convection change growth behaviour vs. inclusions [5]

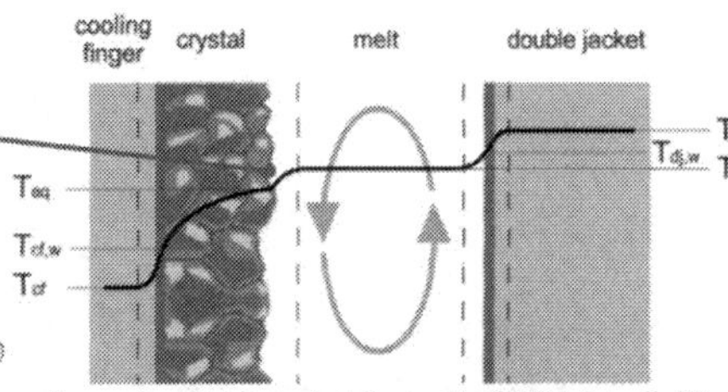

Example of triple jacket device for MSC lab tests [4]

[4] T. Beierling et al., Cryst. Growth Des. 2013, 13, 5229–5240
[5] A. Rich et al., Desal. & Water Treat., (2010) 13, 120–27

fives

Sizing

1. How many stages required – K factor determination?
2. MSC is a batch process but nothing is time related in the shortcut method ?
3. Which trust into growth rates values obtained in unscalable lab rig?

➢**Stage number and steps durations are missing but are a must have!**

➢**Sweating duration is key and often neglected!**

➢**Reported growth rate can be 10^{-6} to 10^{-4} m/s which is 10 to 100+ folds over estimated vs. real MSC growth rates!**

Data available is specific to few cases/compounds only…
… And amount of published data is not large enough…
… to allow supporting generalized MSC shortcut design!

fives

Chemtec

Shortcut modeling through statistic approach

Constituting a database – MSC methodology 101

fives

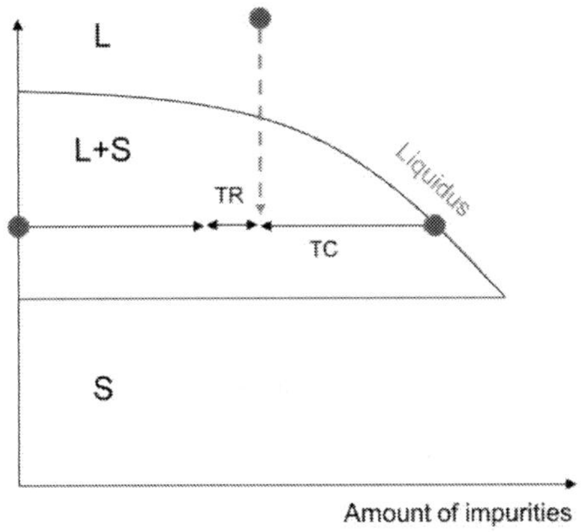

Instead of k_{eff}, use of mass ratio

1. TC mass of crystals formed
   TC is thermodynamic related

2. TR mass of mother liquor trapped
   TR is kinetics related

Constitution of a database on a specific lab rig with a fixed protocol

Constituting a database – MSC methodology 101

fives

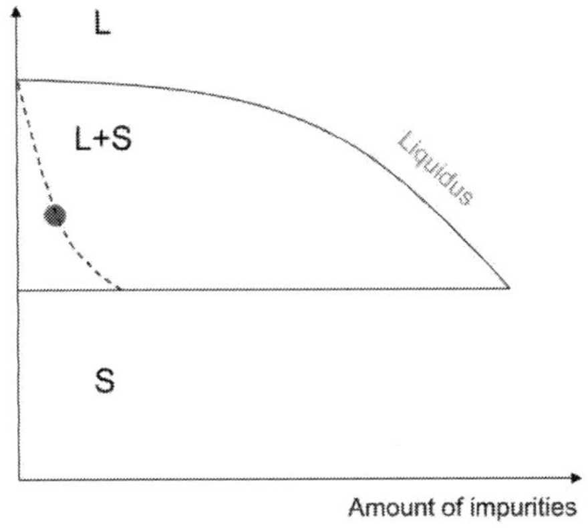

Instead of k_{eff}, use of mass ratio

1. TC mass of crystals formed
   TC is thermodynamic related

2. TR mass of mother liquor trapped
   TR is kinetics related

Final crystallizer content is calculated from TR and TC [6]

Method fully transposable to sweating

Constitution of a database on a specific lab rig with a fixed protocol

[6] Franke, M.B., et al., AIChE J. (2008), 54, 2925-2942

Constituting a database – turning k_{eff} pitfall

fives

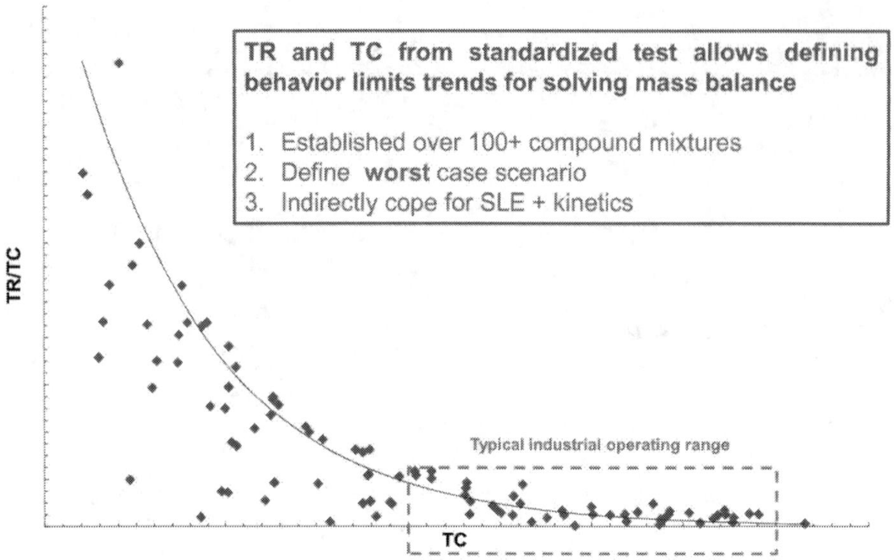

TR and TC from standardized test allows defining behavior limits trends for solving mass balance

1. Established over 100+ compound mixtures
2. Define **worst** case scenario
3. Indirectly cope for SLE + kinetics

Constituting a database – turning K factor pitfall

fives

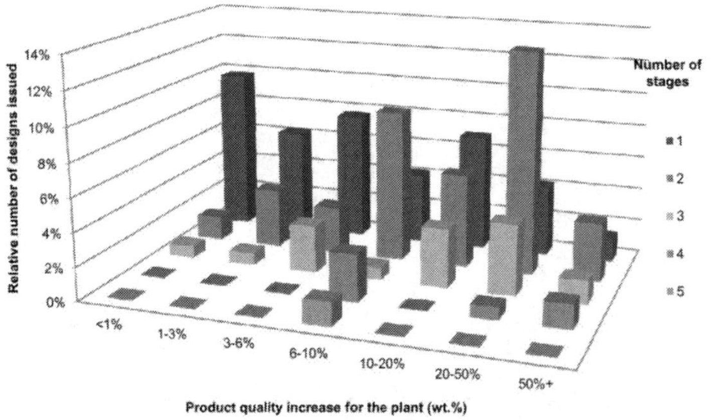

1. Large quality improvement ≠ large number of stages
2. Ultra purification ≠ large number of stages
3. **80% of designs = 1 or 2 stages only required**

K factor directly achievable through TR and TC

Constituting a database – turning step/stage duration pitfall

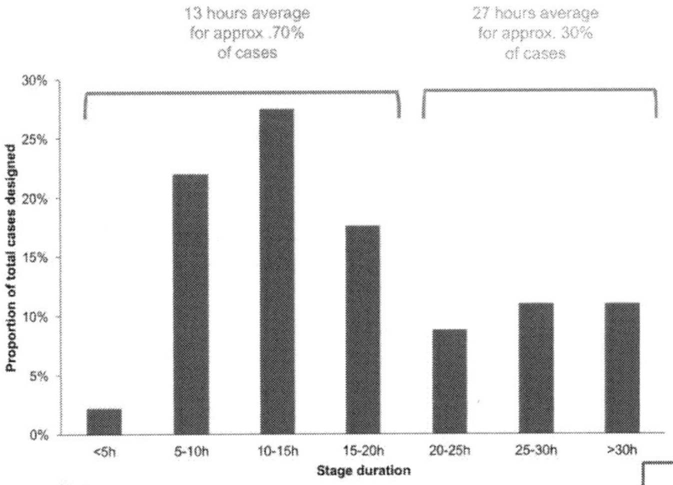

1. **13 hours duration for 70% of the cases**
2. Larger duration for lean feed or ultra-purifications

Time directly estimated from purification effort

MSC: Design Assessment vs. Design & Modelling

Design Assessment / Shortcut

- **Based on narrow number of flow schemes & using a few inputs …**
- **… Comparative design assessment can be sorted from trends & average metrics issued from large test database under normalized protocol**

This will be (latter) completed by adapted lab characterization but allows a rapid answer to the two most common questions:

- What may the process look like?
- How much may it cost?

fives ultimate machines ultimate factory

Loïc Guillou
Technology Manager
Chemtec

Loic.guillou@fivesgroup.com
T +33 3 89 36 32 00

Fives Proabd
CS 11437
20C rue de Chemnitz, Bât. B
Mulhouse Cedex , 68071 - **France**

435190 Continuous Separation of Enantiomers By Combined Crystallization-Distillation Systems: A Proof of Concept Study for Continuous Viedma Ripening

Abstract:

The separation of chiral molecules is of key importance in major industries, such as the pharmaceutical industry, where different enantiomers of a chiral compound can have markedly different biological effects. For instance, only one of the enantiomers of a drug might provide the therapeutic effect, while the other enantiomer shows little effect, no effect, or even harmful effects. Hence, the pharmaceutical industry is driven towards processes that allow separating enantiomers in a cost- and material-efficient way [1]. For systems of enantiomers that crystallize as conglomerates, the process of Viedma ripening has attracted considerable attention by the crystallization community [2,3]. In this process, a racemization reaction in solution (which may be catalyzed) and a complex interplay of crystal growth, dissolution, breakage and agglomeration lead to an enantiomerically pure product [4]. While the mechanism behind the process and how its performance can be improved were intensively investigated, a truly continuous process configuration of Viedma ripening has not been presented so far in the open literature.

In this work, we present such a continuous process configuration that relies on a combination of continuous crystallization, milling and distillation. Specifically, the process involves a mixed suspension mixed product removal crystallizer that is fed with a racemic solution of the substance. A suspension mill, operating in a loop around the crystallizer, serves as an attrition source. The outflow of the crystallizer is put forward to a continuous filtration unit where enantiopure crystals and the mother liquor are separated. The mother liquor is then subjected to a continuous distillation step in which part of the solvent is removed selectively, i.e., without the loss of chiral product or racemization catalyst. The remaining, concentrated mother liquor is recycled back to the crystallizer. Thus, the process allows obtaining enantiopure solids at a theoretical yield of 100%.

By using extensive process simulations based on population balance equation models [4], we show the start-up behavior (e.g., seeding requirements) and under which operating conditions (residence times in the crystallizer and the mills, feed concentration, milling intensity etc.) the process yields enantiopure solids in a continuous steady-state. We also present proof of concept experiments involving the amino acid asparagine in water as a model system including a racemization catalyst.

References

[1] H. Lorenz, A. Seidel-Morgenstern, *Processes to Separate Enantiomers*, Angew. Chem. Int. Ed, 53 (2014), 1218-1250.

[2] C. Viedma, *Chiral symmetry breaking during crystallization: Complete chiral purity induced by nonlinear autocatalysis and recycling*, Phys. Rev. Lett., 94 (2005), 065504.

[3] W.L. Noorduin, T. Izumi, A. Millemaggi, M. Leeman, H. Meekes, W.J.P. van Enckevort, R.M. Kellogg, B. Kaptein, E. Vlieg, D.G. Blackmond, *Emergence of a single solid chiral state from a nearly racemic amino acid derivative*, J. Am. Chem. Soc., 130 (2008), 1158-1159.

[4] M. Iggland, M. Mazzotti, *A Population Balance Model for Chiral Resolution via Viedma Ripening*, Cryst. Growth Des., 11 (2011), 4611-4622.

435733 Continuous Crystallization of L-Ascorbic Acid: Integration of Airlift Crystallizer with Membrane Distillation-Experiments and Results

Tuesday, November 10, 2015: 5:20 PM
155D (Salt Palace Convention Center)

Fatemeh Anisi, Process & Energy Department, Delft University of Technology, Delft, Netherlands and H.J.M. Kramer, P&E, Separation Technology, Delft University of Technology, Delft, Netherlands

Continuous Crystallization of L-ascorbic acid:

Integration of Airlift Crystallizer with Membrane Distillation-Experiments and Results

F. Anisi,[1] H.J.M Kramer[1] [1] *Department of Process and Energy, Faculty of Mechanical, Maritime and Materials Engineering, Delft University of Technology, Delft, the Netherlands*

Crystallization is one of the oldest important separation and purification processes in production of pharmaceutical compounds. The common problems in crystallisation processes are often include large variations in the product quality and inflexibility of the process. It is due to the entanglement of different phenomena such as primary nucleation, growth, secondary nucleation and so on. For a better control on the secondary nucleation and the growth of the crystals an airlift crystallizer has been developed, tested and analysed. Batch experiments with airlift crystallizer showed that rate of secondary nucleation is two orders of magnitude lower in such a crystallizer compared to conventional stirred crystallizer. Figure 1 shows how bimodal the CSD from stirred crystallizer is which confirms the higher rate of secondary nucleation in the stirred crystallizer. Therefore, it is a promising alternative for conventional ones.[1]

Membrane distillation technology can be used for supersaturation generation as an alternative to evaporation. Since it is not necessary to work with high temperatures, membranes are suitable options for thermo-labile compounds. During crystallization experiments supersaturation has been generated and controlled with continuous solvent removal and fed to the crystallizer. Seed crystals have grown by consuming the supersaturation.

Therefore, a continuous membrane-assisted crystallization using an airlift crystallizer is potentially believed to be more controllable, energy efficient and to produce a more uni-modal crystal size distribution.

In this contribution the airlift crystallizer and membrane-assisted unit are integrated in a skid. The set-up is capable of performing batch and continuous crystallization with airlift or stirred crystallizer. Continuous crystallization of L-ascorbic acid has been done in the airlift and stirred crystallizer while membrane controls the supersaturation during the process. The integrated system has been also modelled and various process conditions are tested.

Based on modelling results experiments are designed and show that generally use of membrane unit decreases the level of secondary nucleation but it is mainly suppressed in the airlift crystallizer rather than stirred one. The higher the residence time the larger the crystals grow. The results from the experiments are used to validate the model and furthermore to estimate the kinetic parameters and optimize the process via design of new experiments.

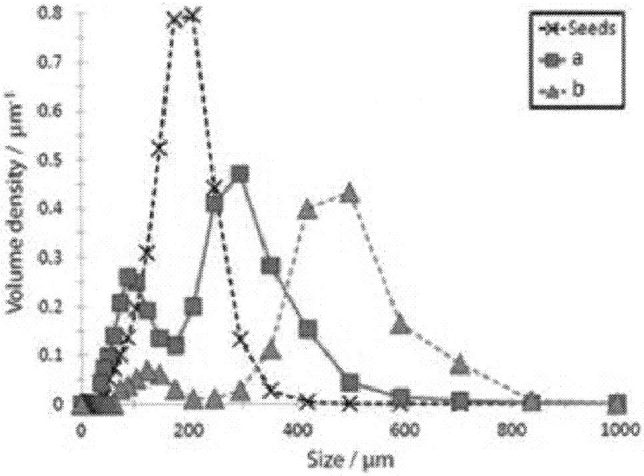

Figure 1:Crystal size distribution for different batch seeded cooling crystallization of L-ascorbic acid a) stirred crystallizer b) airlift crystallizer

USERPROPERTY * MERGEFORMAT [1] R. Lakerveld, J. Krochten, H. Kramer, An airlift crystallizer can suppress secondary nucleation at a higher supersaturation compared to a stirred crystallizer, Cryst. Growth Des. 14, 7 (2014)

420161 Steric Effects on Mass Transfer of C_{10} Hydrocarbons in BPL Activated Carbon

Abstract:

Adsorbent filters often use activated carbons to remove a wide range of toxic chemicals from air at high separation ratios. Many toxic compounds, such as pesticides, contain bulky functional groups or have significant branching. It is of interest to know if filter performance for these compounds could be diffusion limited due to steric hindrances in commercially available amorphous carbons. Most mass transfer studies in porous adsorbents have focused on highly ordered crystalline structures, such as zeolites, or carbon molecular sieves that have a very narrow pore size distribution, with the uniform structures allowing for shape selective separations. Few studies have attempted to find steric effects for amorphous adsorbents.

A concentration swing frequency response (CSFR) technique was used to measure mass transfer rates of various C_{10} hydrocarbons to explore steric hindrance effects on diffusion in amorphous BPL activated carbon. With the various C_{10} hydrocarbons, there are enough carbons to create different ring or branched shapes, but not so many that volatility is greatly reduced. By plotting amplitude ratios as a function of frequency and fitting to a mathematical model derived from transfer functions, this technique can easily distinguish among different mass transfer rate mechanisms and permits the accurate calculation of diffusion coefficients. The hydrocarbon adsorbates of interest are decane, α-pinene, limonene, and decalin. The fluorocarbon, perfluorodecalin, is also investigated. The CSFR experiment was performed at multiple gas-phase concentrations for each adsorbate, and a micropore diffusion model fits the data well. Adsorption isotherms were measured for each adsorbate on BPL activated carbon and described using a Toth model. Predicted isotherm slopes derived from the CSFR experiments compare well with the equilibrium isotherm slopes. For each adsorbate, the measured diffusion coefficients are found to be concentration dependent and increase as the adsorbate concentration increases in a manner consistent with Darken thermodynamic correction factors.

433702 Study of Adsorption, Desorption and Flow Phenomena of Methane -Ethane Binary Gas Mixtures in Shale Samples and Cores

Abstract:

Methane (CH_4) is the largest component of the natural gas produced from shale rocks. Although methane accounts for 87-96 mol% of the gas from these rocks, other components including hydrocarbons such as ethane (C_2H_6) and propane (C_3H_8), nitrogen (N_2) and helium (He) are also found in the gas produced. Methane and the various hydrocarbons are stored in the adsorbed state in the micro and mesopores of the shale rocks, and as free gas in the fracture networks. Though convective and diffusive transport accounts for the short-term behavior in gas production, desorption is thought to dominate the long-term dynamics of shale-gas generation.

The key objective of this study is the investigation of adsorption/desorption and diffusive and convective transport phenomena of methane/hydrocarbon mixtures in shale-gas rocks. In this talk, in particular, we focus our attention on the binary CH_4/C_2H_6 mixture, since ethane is typically the second largest component in shale gas, and is thought to compete for the same adsorption sites in shale-gas rocks as methane. We study the adsorption/desorption behavior of this mixture (and its individual components) in ground shale rock samples using thermogravimetric analysis (TGA), static volumetric and flow-chromatographic techniques. The sorption isotherms generated are important to predict the gas storage capacity of the shale samples, while the study of adsorption/desorption dynamics help us understand the role of desorption during the later times of gas production. The multicomponent potential theory of adsorption (MPTA) approach is used to model the experimental observations.

In this work, we also study the dynamics of shale-gas production from full-diameter shale-gas cores, under realistic field-scale pressure, temperature, and confining-pressure conditions. Depletion experiments from such cores storing CH_4/C_2H_6 mixtures at various compositions (simulating a model shale-gas) are carried-out, during which the pressure of the core, the flow of gas, and its composition are all continuously monitored and measured. These experiments, and related analysis and modeling, nicely compliment the studies with the ground shale samples, and help one to understand better the interplay between adsorption/desorption and transport phenomena that take place during shale-gas production from these rocks.

433857 Modeling Adsorption Rate of Metronidazole on Mesoporous Activated Carbon (CMK-3) and Activated Carbon (F400) from Aqueous Solution

Abstract:

The overall rate of adsorption of pharmaceutical compounds on activated carbon (AC) from aqueous solution was investigated as this information is required to design fixed-bed adsorbers. In a recent study, it was found that metronidazole (MNZ) can be adsorbed considerably on granular AC F400 and powder mesoporous AC (CMK-3). The latter was synthesized using a template mesoporous silica (SBA-15). In several studies, it has been reported that the rate of adsorption of organic compounds on mesoporous AC is much faster than on conventional AC; however, the rate of adsorption on granular AC (F400) was compared to that on powder mesoporous AC. The fast adsorption rate observed in the CMK-3 was attributed to its mesoporosity, but not to its particle size. The aim of this work is to compare the rate of adsorption of MNZ on granular AC with that on powder mesoporous AC. It is very well known that the rate of adsorption of MNZ on AC depends on particle size. Hence, the effect of particle size on the rate of adsorption on granular AC was studied in this work.

The concentration decay curves of MNZ during adsorption on CMK-3 and F400 carbons, were obtained in a batch stirred tank adsorber. The concentration decay curves of MNZ were interpreted using diffusional models assuming that the overall rate of adsorption of MNZ was due to external mass transfer and intraparticle diffusion. Moreover, the latter may occur by pore volume diffusion (Fick diffusion) and surface diffusion. Independent of the operating conditions, the results revealed that the surface diffusion model (MDS) successfully interpreted the experimental kinetic data and the surface diffusion controlled the intraparticle diffusion of MNZ in both ACs. The particle size of F400 carbon was reduced by grinding to study the effect of particle size on the rate of adsorption. The results revealed that the time to reach adsorption equilibrium decreased from 720 min to 20 min when the particle diameter was reduced from 1.02 to 0.081 mm. Independent of particle size, the MDS model fitted the experimental data quite well and the surface diffusion coefficient, Ds, of MNZ in F400 was 2.54×10^{-8} cm^2/s and not depended on particle size. As expected the values of the external mass transfer coefficient, k_L, increased when the particle size of F400 was reduced. The time to reach adsorption equilibrium in CMK-3 is very similar to that in powdered F400 carbon when the particle diameter of CMK-3 (d_p = 0.081 mm) is very similar to that of powered F400 (d_p = 0.081 mm). It was concluded that the adsorption rate of MNZ on CMK-3 was faster than that on F400 because the particle size of CMK-3 was very small and not to their larger mesoporosity.

417538 Understanding Unique Diffusion Behavior in Hierarchical Zeolites

Abstract:

Hierarchical zeolites are materials that contain both micro (<2 nm) and meso (2—50 nm) porosity and are proposed as potential solutions to alleviate transport limitations of conventional microporous zeolites. The introduction of mesopores adds considerable complexity to the adsorption and transport behavior exhibited by these hierarchical materials. Both experimental [1] and computational [2] studies demonstrate a crossover phenomenon where the sorbate loading in the conventional zeolite exceeds that in the hierarchical zeolite at low pressure, but the ordering is reversed at high pressure. Similarly, experimental measurements [3] point to unexpected transport behavior. In this talk, we present a molecular simulation study on the diffusion of linear and branched alkanes in MFI-type zeolites. Diffusion coefficients are computed at multiple temperatures from 363 – 543 K and as a function of loading. The comparison between the diffusional behaviors in conventional and hierarchical materials reveals surprising features in the latter systems, and we offer explanations in terms of free energy profiles in the two types of materials.

References:

[1] D. Xu, G. R. Swindlehurst, H. Wu, D. H. Olson, X. Zhang, M. Tsapatsis, *Adv. Func. Mater.* **2014**, *24*, 201 –208.

[2] P. Bai, D. H. Olson, M. Tsapatsis, J. I. Siepmann, *ChemPhysChem* **2014**, *15*, 2225–2229.

[3] C. C. Chang, A. R. Teixeira, C. Li, P. J. Dauenhauer, W. Fan, *Langmuir* **2013**, *29*, 13943–13950.

420576 The Effect of Crystal Diversity of Nanoporous Materials on Mass Transfer Studies

Abstract:

Nanoporous materials, like zeolites and metal-organic frameworks, find widespread applications in catalysis and separation technologies. Their proper design requires a detailed understanding of the mass transfer mechanisms with their surroundings. This is conventionally deduced from so-called uptake or release curves carried out with batches of nanoporous crystals. The analysis of these batch uptake curves is based on the assumption that all crystals are identical. However, in this presentation, we will show that this approximation may lead to inaccurate conclusions about the material properties. As a consequence of the crystal diversity, the real governing uptake mechanism may be obscured by mimicked mass transfer mechanism. Various experimental and theoretical cases will be presented. Results obtained from single-crystal and multiple-crystals uptake measurements, using different experimental approaches, including a liquid phase batch methodology[1] and the micro-imaging techniques[2], will be shown. These experiments were performed with the catalyst of excellence for the Methanol-To-Olefins (MTO) process, namely the zeolite SAPO-34[3]. Based on the experimental results, through a simple theoretical approach, several cases of batch uptake curves which are deformed by the crystal diversity will be shown and discussed in relation to the conventional analysis techniques.

Acknowledgements

J. Cousin Saint Remi is grateful to the Agency for Innovation by Science and Technology in Flanders (IWT) and The Research Foundation – Flanders (FWO) for the financial support.

References

1. Cousin Saint Remi, J., Baron, G. & Denayer, J. Nonuniform Chain-Length-Dependent Diffusion of Short 1-Alcohols in SAPO-34 in Liquid Phase. *J. Phys. Chem. C*, **2013**, *117*, 9758–9765.

2. Kärger, J. *et al.* Microimaging of transient guest profiles to monitor mass transfer in nanoporous materials. *Nat. Mater.* **2014**, *13*, 333–343.

3. Stocker, M. Methanol-to-Hydrocarbons: Catalytic Materials and Their Behavior. *Micropor. Mesopor. Mat.* **1999**, *29*, 3-48.

428863 Mechanistic Mass-Transfer Modeling for the Adsorption of Water Vapor By Molecular Sieves

Abstract:

Accurate modeling of adsorption kinetics is key to reactor design for engineered separation processes. Simple empirical models, such as the shrinking core or linear driving force model, are often adopted to describe adsorption kinetics for these processes. However, use of these empirical models introduces error to the macroscopic transport models for the overall system and can utterly fail to describe desorption. Additionally, these simple models occasionally involve many parameters that may vary with temperature, pressure, or other factors. These problems emphasize the need for a more fundamental way of modeling adsorption mass transfer processes at the microscopic level. Here, we will present an easy-to-use mass and energy transfer modeling framework. This framework is built using a high-resolution, conservative finite difference scheme and comes with built-in linear and non-linear solvers for extended flexibility. We will discuss briefly how this framework operates and then demonstrate the capabilities with real world problems. Our focus will be on accurate modeling of the mass-transfer mechanisms that govern the adsorption and desorption of water vapor in commercial adsorbents. This particular work will be applied to the separation and recovery of radioisotopic gases, such as tritium vapor (HTO), produced through the reprocessing of used nuclear fuel.

423930 Supported Liquid Membranes for Recovery of Cephalexin

Abstract:

I am very pleased to have this opportunity to speak in a session in honor of Professor DB Bhattacharyya. This presentation covers three configurations of supported liquid membranes (SLMs): (1) SLM with strip dispersion (SLM-SD), (2) SLM with feed dispersion (SLM-FD), and (3) SLM with organic dispersion (SLM-OD), for simultaneous removal and recovery of the antibiotic, Cephalexin, from aqueous feed solutions using a commercially available hollow-fiber module and the carrier Aliquat 336. In SLM-SD, the strip dispersion, formed by dispersing the strip solution in the organic membrane solution with a mixer, flowed through the shell side of the module, forming the SLM in the nanoporous hollow fibers. In SLM-FD and SLM-OD, the feed dispersion and the organic dispersion, formed by dispersing the feed solution in the organic membrane solution and a small amount of organic membrane solution in the feed solution, respectively, using a mixer, flowed through the shell side of the module, forming the SLM in the nanoporous hollow fibers. These three configurations can all ensure the stability of SLMs. For SLM-SD, Cephalexin has been extracted and concentrated with high recovery. Initial results from a 4-component feed mixture derived from its enzymatic synthesis have shown selective Cephalexin recovery possible. This allows Cephalexin recovery from the aqueous strip solution through complexation, without enzyme deactivation, which commonly occurs when complexation is conducted in enzymatic reaction mixtures. A mass transfer analysis has shown that the mass transfer resistance from the extraction reaction is dominant. To minimize this resistance, SLM-FD experiments were conducted, showing that the overall mass transfer coefficient obtained was significantly larger than (1.6 times) that attained by using SLM-SD. The mass transfer process of SLM-FD was elucidated, and a mathematical model was developed to describe the process. Based on this model, both the theoretical and experimental overall mass transfer coefficients were obtained as a function of Aliquat 336 concentration. These mass transfer coefficients were in reasonably good agreement. The mass transfer mechanism of the SLM-OD was also elucidated, and a mathematical model was developed to calculate the overall mass transfer coefficient. The overall mass transfer coefficient increased with an increase of organic-to-feed volume ratio or increase of Aliquat 336 concentration, but reduced with the increase of initial Cephalexin concentration. Among these SLMs, the reaction resistances were in the order: SLM-SD > SLM-OD > SLM-FD, and the overall mass transfer coefficients were: $K_{SD} : K_{OD} : K_{FD} = 1 : 1.3 : 1.6$, showing that SLM-FD and SLM-OD decreased the reaction resistances and thus had greater mass transfer than SLM-SD.

424338 REAL Time Monitoring of Membrane Mineral Scaling and Application to RO Plant Optimization

Abstract:

Reverse osmosis (RO) desalination is a relatively mature technology that is widely used in both seawater and inland brackish water desalination, as well as in water reuse application. In contrast with seawater desalination which is typically carried out at recovery below about 50%, inland desalination has to be carried out at a reasonably high recovery in order to reduce the challenge of concentrate management and avoid the loss of valuable water resources. However, RO deployment for treatment of brackish water is often impeded at high recovery operation by membrane mineral scaling due to precipitation of sparingly water soluble salts. At high recovery operation, the concentrations of mineral salts on the feed-side and near the RO membrane surface can increase to levels exceeding their solubility limits, which then leads to mineral scale formation on the membrane surface. Mineral surface scaling leads to permeate flux decline and can ultimately result in membrane damage. In order to effectively combat membrane scaling it is imperative to be able to deploy a capability for early detection and continuous real-time monitoring of mineral scaling. In order to address the above need, a novel direct membrane monitor (MeMo) was developed at UCLA as an ex-situ sensor for early detection and quantification of the evolution of mineral scaling in RO membrane plants. Mineral scale detection and monitoring with MeMo is accomplished via direct membrane surface optical imaging, under RO process flow conditions, along with specially developed real-time image analysis software. Previous work with an early MeMo prototype demonstrated effective detection of the onset of gypsum scaling in RO membrane operation. In the present work, the MeMo technology has been extended to enable, for the first time, detection of the onset and monitoring of calcium carbonate and silica scaling. Using advanced and automated high resolution real-time image analysis techniques, the MeMo quantifies the degree of mineral scaling in terms of surface area covered by scale and the surface number density of mineral growth crystals or deposited scale precipitate particles. Mineral scaling tests at different levels of solution saturation at the membrane surface were then carried out whereby the level of concentration polarization was adjusted via setting of the permeate flux and crossflow velocity. The rate of mineral scale nucleation was then quantified as a function of the level of solution saturation and shown to follow the classical nucleation theory. Finally, it is shown that with MeMo early scale detection RO plant operation can be optimized with respect to both operating conditions (flux and cross flow velocity) and antiscalant dose to ensure mineral scale-free operation. Results from field operation of a spiral-wound RO plant will be presented and discussed demonstrating the advantages of the MeMo approach for RO plant optimization.

424854 Polymeric Membranes Manufactured from Solutions in Ionic Liquid

Abstract:

The manufacture of asymmetric porous membranes by phase inversion and of multilayer composite membranes by dip coating is well established in the industry. These are versatile and simple processes, which lead to high performance membranes. Polymer dissolution is a mandatory step. We have been investigating ionic liquid as minor or major component in the casting solutions of polymers such as cellulose1, polyacrylamide, polytriazole and polystyrene-b-poly(4-vinyl pyridine)2 for the following reasons: (1) ionic liquids are among the few solvents dissolving cellulose and less aggressive than the other alternatives; (2) the membrane preparation by phase inversion in large scale involves in many cases partial solvent evaporation and the operator exposure to volatile components. Ionic liquids can substitute at least part of the volatile solvents, reducing problems related to the toxicity of organic solvent vapors; (3) ionic liquids can help to control the morphology, leading to structures different than those observed with other casting solutions; (4) in the case of block copolymers ionic liquids can induce or influence self-assembly and is a good additive to tailor morphology. Following the motivation supported by these aspects we have demonstrated the preparation of cellulose membranes with molecular weight cut-off in the range of 3 kg/mol and water permeance 14 L m-2 h-1 bar-1. These corresponds to the low range of ultrafiltration. The membranes are resistant to organic solvents. They can be used also as intermediate layer for the preparation of interfacial polymerized composite membranes. Polyacrylonitrile hollow fiber membranes with high porosity have been prepared from solutions containing ionic liquids. The membranes were tested for ultrafiltration and as supports for interfacial polymerization with final application in nanofltration and forward osmosis. Hydroxyl-functionalized polytriazoles have been synthesized in our lab for application in organophilic ultrafiltration3 and as support for interfacial polymerization4, using regular solvents like dimethyl formamide. More recently membranes have been also prepared using solutions in ionic liquid. Finally different ionic liquids have been used as additive for morphology control in the preparation of isoporous block copolymer membranes. References 1. S. Livazovic, Z. Li, A. R. Behzad, K.-V. Peinemann and S. P. Nunes, J. Membrane Sci., accepted. 2. P. Madhavan, R. Sougrat, A. R. Behzad, K.-V. Peinemann and S. P. Nunes, J. Membrane Sci., accepted. 3. S. Chisca, P. H. H. Duong, A.-H. Emwas, R. Sougrat, S. P. Nunes, Polymer Chemistry 2015, 6, 543-554. 4. P. H. H. Duong, S. Chisca, P. Y. Hong, S. P. Nunes, T.-S. Chung, ACS Applied Materials and Interfaces, accepted.

437498 Polymer Membranes for Gas and Liquid Separations

Abstract:

In the past 30+ years, polymer membranes have emerged as a viable and widely used technology to separate gas mixtures and purify water via desalination and filtration of, for example, wastewater. In desalination, for example, reverse osmosis membranes are the dominant technology for desalinating seawater, displacing thermally driven processes in much of the world. In gas separations, polymer membranes are widely used for air separation and hydrogen purification, and they are increasingly being used for natural gas separation.

In recent times, focus on grand challenges facing mankind, including development of energy-efficient methods to provide clean water in both the developed and developing world and mitigate carbon dioxide levels in the atmosphere, have driven research into new platforms of membranes in our laboratories. Additionally, the strong and often competing connections between water, energy, and food production, coupled with the shale gas and shale oil revolution currently under way in the United States, are providing new opportunities for low-energy separations.

This presentation will focus on two stories, each illustrating pathways beginning from fundamental molecular engineering of polymers to tune their properties for specific separations, followed by scale-up and commercial deployment. In one case, membrane materials were developed with a specific affinity for CO_2, resulting in membranes that are currently used for CO_2 removal from H_2 in H_2 production and in large-scale field tests for post-combustion carbon capture. In the second case, beginning from Office of Naval Research sponsored research in the 1990's, focused on separation of oil/water emulsions in bilgewater about Navy ships, porous membranes used for water filtration were surface modified to improve their lifetime and performance (i.e., resistance to fouling). Systematic tuning of the chemistry and application of these nanometer-thick surface modification coatings led to fouling-resistant, energy-efficient membranes now in commercial production for the oil and gas industry (purification of flowback water from hydraulic fracturing, desalination of water for fracking operations), municipal wastewater purification, and rural desalination. Finally, the presentation will provide a short section on a forward-looking view of the bright future emerging for fundamental science to drive innovation in polymer membranes for both new and existing applications.

437674 A Generic Method for Designing Potent Antimicrobial Peptides and Determining Their Mechanism of Activity: A TB Example

Abstract:

As pathogenic bacteria become resistant to traditional antibiotics, alternate approaches such as designing and testing new potent selective antimicrobial peptides (AMP) are increasingly attractive. Using a combination of "database filtering" bioinformatics, protein engineering and *de novo* design, four AMPs were designed for specific activity against *M. tuberculosis*. Here, we (i) describe the method, (ii) determine their mechanism of cell membrane disruption as a 4-stage barrel-stave pore formation using known mechanisms of natural AMPs, and (iii) demonstrate that the *in vitro* disruption of bacterial-membrane mimetic supported lipid bilayers correlates with AMP potency against *M. tuberculosis* cells using a Quartz Crystal Microbalance with Dissipation. This suggests a rapid *in vitro* screen for AMP efficacy. This approach could be useful as a generic design method and predictive assay for AMPs against other pathogenic bacteria.

432173 New Membrane Frontiers and Challenges

Abstract:

Membrane processes such as, Reverse Osmosis, Nanofiltration, Ultrafiltration, and Microfiltration are providing many successful, energy efficient applications ranging from high quality water production to water reuse to material recovery, but integration of nanotechnology and biological aspects in membranes should add immense value in the area of selective separations and catalysis. Some of the questions are: Can we make membranes that are tunable and responsive and that can be synthesized in a larger scale? Can we add functional moiety in membranes which can go through conformation changes and provide selectivity? Can we make membrane which is truly antifouling and long life (need to learn from nature), and what about graphene membranes? Can we integrate nanotechnology and membranes for efficient separations or catalysis? Transport of molecular species through protein channels are orders of magnitude enhanced over conventional nanoporous membrane materials, and enzymes allow the rapid cleavage of strong covalent bonds in aqueous environments. Thus nature provides lot of interesting clues in advancing membranes to new frontiers.

Sublimation as a Function of Diffusion

Gang Qian, Nathan T. Morgan, Russell J. Holmes, and E. L. Cussler
Department of Chemical Engineering and Materials Science, University of Minnesota
and
D. Wayne Blaylock and Robert D. Froese,
The Dow Chemical Company, Midland, MI

Abstract

In sublimation of large organic species, the dominant mechanism is volume diffusion, where experiments show a flux which varies with the sublimation tube diameter squared d^2, consistent with volume diffusion. These experiments are not dominated by laminar flow or Knudsen diffusion, where the flux varies with d^4 and d^3, respectively. The characteristics of this volume diffusion are discussed.

In the purification by sublimation of organic semiconductors, vapor is transported from high temperature to colder temperature. The pressure at high temperature is close to the equilibrium vapor pressure of the material being sublimed, and the vapor pressure in the colder region equals or exceeds the vapor pressure at the wall temperature. Results of some such purifications are consistent with sublimation controlled by diffusion. If diffusion is rate limiting, the reciprocal of the flux should vary linearly with the length. If diffusion is rate limiting, the total amount sublimed should vary with the square of the tube diameter. These expectations are supported by experiment. Predictions for convective flow and Knudsen diffusion are not supported by experiment. Diffusion does seem the key.

However, the success of these predictions does not mean that the sublimation mechanism results from intermolecular diffusion or even from self-diffusion. In intermolecular diffusion, there are two solutes;[2] because the solute being sublimed here is nearly pure, there is one solute here. In self-diffusion, there is one solute, but a portion of this solute can be separately identified by radioactivity, activation, or even optical rotation; here, no such identification is made. In diffusion in porous catalysts, the second species for diffusion is sometimes identified as the catalyst itself, in what is called a "dusty gas model".[3] Here, because the mean free path is much less than the tube diameter, this dusty gas model is inappropriate. While diffusion is a quick label for what is occurring, diffusion in the conventional sense is not involved.

We believe that these experiments exhibit "volume diffusion." This mechanism, rarely mentioned in the literature except in the analysis of multicomponent diffusion, was originally explored by Onsager[4] and studied experimentally in the context of ternary diffusion, especially by Dunlop.[5-6] More recently, the mechanism of volume diffusion was developed theoretically by Brenner,[7-8] whose approach is that used here.

In this approach, we treat volume as an additional property of the system which is not in general conserved. The flux of volume v_v, like any other property, includes a convective part, equal to the mass average velocity v_m, and a diffusive part j_v:

$$v_v = v_m + j_v \tag{1}$$

Note that each of these volumetric fluxes has dimensions of volume per area per time. The diffusive part j_v is given by:

$$j_v = \frac{j_D}{c} = \frac{D}{c}\frac{\Delta c}{L} \tag{2}$$

The diffusive part of the volume flux increases the volume flux beyond what would be expected from the mass flux alone, a result of the expanding fluid. The total flux j in sublimation should include viscous flow, volume diffusion and Knudsen diffusion as follows:

$$j = j_{\square} + \frac{1}{\frac{1}{j_D}+\frac{1}{j_{Kn}}} = \left[\frac{pd^2}{32\mu} + \frac{1}{\frac{1}{D}+\frac{1}{D_{Kn}}}\right]\frac{\Delta c}{L} = D'\frac{\Delta c}{L} \tag{3}$$

where the quantity in brackets D' is an overall transport coefficient including all three transport mechanisms. Equation 3 can be rearranged as shown to show the three zones of viscous flow, volume diffusion, and Knudsen diffusion, as shown in Figure 1.

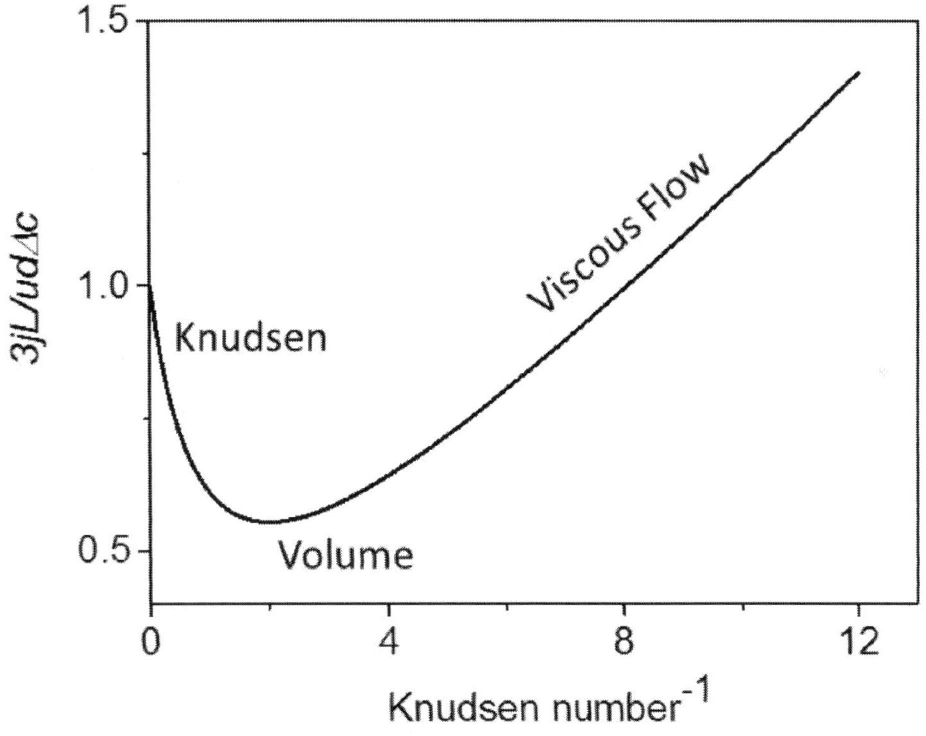

Figure 1. Dimensionless Flux *vs.* Knudsen Number. This graph illustrates the predictions of Knudsen diffusion, volume diffusion, and viscous flow.

While Equation 3 is physically clear, we will find it useful to rewrite this equation using results from the kinetic theory of a gas of hard spheres.[8] For Knudsen diffusion,

$$D = \frac{1}{3}du \tag{4}$$

where u is the molecular velocity. For hard spheres, u equals $[8k_BT/\pi m]^{1/2}$. For volume diffusion, the flux is:[7-9]

$$j_D = \frac{u}{6}\left(-2\lambda\frac{dc}{dz}\right) = -D\frac{dc}{dz} = -\left[\frac{u\lambda}{3}\right]\frac{dc}{dz} \quad (5)$$

where D equals to the quantity in square brackets. This diffusion coefficient does not arise from two species mixing without any net flow of volume, which is the usual form of diffusion.

We now neglect transport by Knudsen diffusion to look at two special cases. First, in experiments where the tube diameter varies, Equations 3-5 can be rewritten in terms of the total flux J $(= (\pi/4)d^2 j)$, in units of molecules per time:

$$J = \left[\frac{3\pi^2}{1016}\left(\frac{u\Delta c}{\lambda L}\right)\right]d^4 + \left[\frac{\pi}{12}\left(\frac{u\lambda\Delta c}{L}\right)\right]d^2 \quad (6)$$

Because the pressure is constant, the concentration c and the mean free path λ are also constant, so this equation gives the variation of J with d. When the first term on the right hand side of this equation dominates, the diameter d is large and the flux J varies with the fourth power of d. At smaller d but when the diameter is still larger than the mean free path λ, J varies with d^2. These expectations are supported by the results in Figure 2.[10]

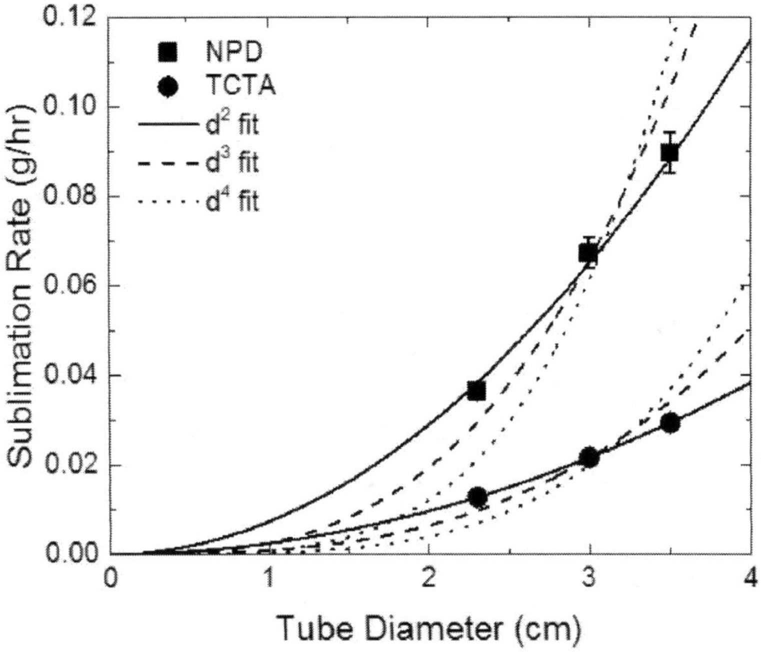

Figure 2. TCTA Sublimation Rate *vs.* Tube Diameter. The results at 310°C have a slope of 2.0, consistent with volume diffusion alone.

Thus the data seems consistent with volume diffusion, and not with other transport mechanisms. All these mechanisms are compared in Table 1.

Table I. Transport Mechanism for Sublimation While the laminar flow mechanism is consistent with experiment at high pressures, only the volume diffusion mechanism agrees with experiments at moderate pressures.

Phenomenon	**Number Solutes**	**Entropy Source**	***vs. d***	**mole flux**
Volume Diffusion[7-8]	1	Restrained Expansion	d^2	yes
Poiseuille Flow[10]	1	Momentum Conservation	d^4	yes
Slip Flow[11]	1	Momentum Conservation	d^3	yes
Diffusion[2]	2	Mixing	d^2	no
Knudsen Diffusion[8,12]	1	Pressure	d^3	yes
Self Diffusion[2]	2	Mixing	d^2	no
Constant Pressure Diffusion[2]	2	Momentum Conservation	d^2	yes
Collisionless Diffusion[13]	1	Free Expansion	d^2	yes

To explore the implications of this result, we rewrite Equation 3 in terms of the parameters of the kinetic theory of rigid spheres:

$$\frac{jL}{\Delta p} = ap + \frac{b}{1+cp} \tag{7}$$

The left hand side of this equation is the flux per area j per driving force $\Delta p/L$. As such, the left hand side is a transport coefficient. The right hand side of this equation gives the variation of this coefficient with pressure in terms of three parameters, which are:

$$a = \frac{3\pi^{\frac{3}{2}}d^2\sigma^2}{64\sqrt{m}(k_BT)^{\frac{3}{2}}} \tag{8}$$

$$b = \sqrt{\frac{8}{\pi}}\left(\frac{d}{3\sqrt{mk_BT}}\right) \tag{9}$$

$$c = \frac{\sqrt{2}\pi\sigma^2 d}{k_BT} \tag{10}$$

We can check Equation 7 with our experimental data for organic semiconductors and with other diffusion data, like those in Knudsen's original paper.[11] The data for TCTA, shown in Figure 3, do show the expected trends, but more definite conclusions are compromised by the scatter in the data. For Knudsen's carbon dioxide results like those in Figure 4, the fit is more reasonable. Impressively, the values of *a, b,* and *c* inferred from these figures agree surprisingly well with the parameters calculated from kinetic theory, as shown in Table II.

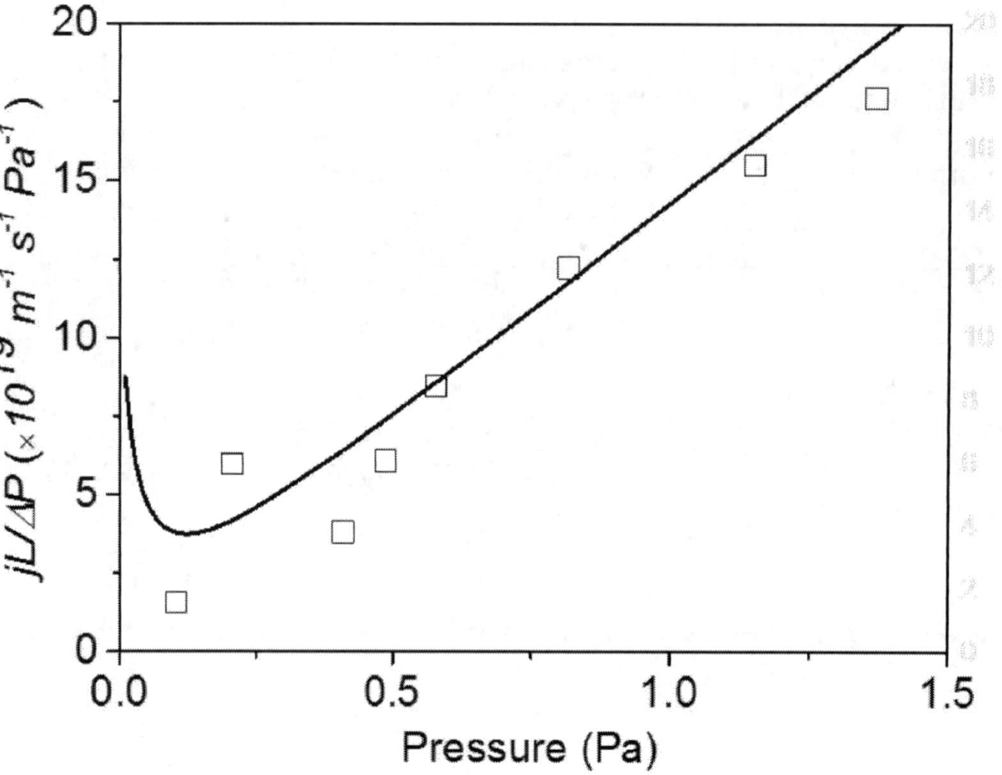

Figure 3. TCTA Transport Described with Volume Diffusion. The parameters found from this analysis are compared with those predicted in Table II.

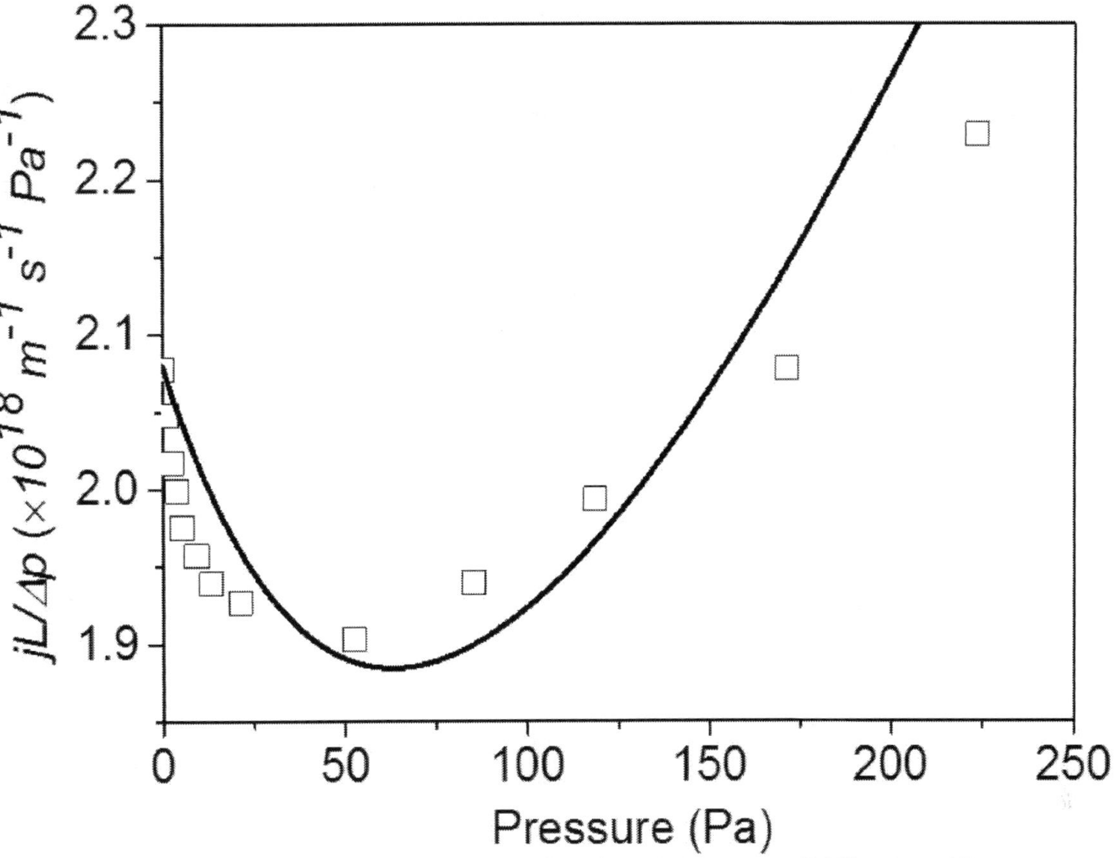

Figure 4. Carbon Dioxide Transport Described with Volume Diffusion. The parameters found from this analysis are compared with those predicted in Table II.

Table II: Measured vs. Predicted Values for TCTA Sublimation. The values given are for *a*, *b*, and *c* in Equations 25-27.

Parameters	Measured	Predicted
TCTA		
a	2.8×10^{20}	1.2×10^{20}
b	1.6×10^{20}	1.6×10^{20}
c	16	30
Carbon Dioxide		
a	7.0×10^{20}	2.4×10^{20}
b	2.1×10^{20}	2.1×10^{20}
c	0.2×10^{-2}	1.1×10^{-2}

We are also struck by the predictions of volume diffusion *vs.* those obtained by postulating slip flow.[12] The implication of volume diffusion is that we can get equivalent predictions without the empirical corrections like the first or second order slip and the varying viscosity. We look forward to making these tests in the future.

Acknowledgements

We have benefited from discussions with T. E. Schwartzentruber (Minnesota) and with Marc Donohue (Johns Hopkins). This work was largely supported by The Dow Chemical Company. Other support came from the State of Minnesota (LCCMR). RJH is a member of The Dow Chemical Company Technical Advisory Board.

References

01. McGhie AR, Garito AF, Heeger AJ. A gradient sublimer for purification and crystal growth of organic donor and acceptor molecules. *J Cryst Growth*. 1974;22(4):295-297. doi: 10.1016/0022-0248(74)90173-0.

02. Cussler EL. *Diffusion: Mass transfer in fluid systems.* Cambridge; New York: Cambridge University Press; 2009.

03. Mason EA, Malinauskas AP. *Gas transport in porous media: The dusty-gas model.* Amsterdam; New York: Elsevier; 1983.

04. Onsager L. Theories and problems of liquid diffusion. *Ann N Y Acad Sci.* 1945;46(5):241-265.

05. Dunlop PJ, Gosting LJ. Interacting flows in liquid diffusion: Expressions for the solute concentration curves in free diffusion, and their use in interpreting gouy diffusiometer data for aqueous three-component systems. *J Am Chem Soc.* 1955;77(20):5238-5249.

06. Cussler EL. *Multicomponent diffusion.* Amsterdam; New York: Elsevier Scientific Pub. Co.; 1976.

07. Brenner H. Beyond navier-stokes. *Int J Eng Sci.* 2012;54:67-98. doi: 10.1016/j.ijengsci.2012.01.006.

08. Brenner H. Kinematics of volume transport. *Physica A-Statistical Mechanics and its Applications.* 2005;349(1-2):11-59. doi: 10.1016/j.physa.2004.10.033.

09. Aranovich GL, Donohue MD. Exact solution to Einstein's evolution equation for fluid with exponential free paths distribution. In: Claes AL, ed. *Evolution equations.* Hauppauge, N.Y.: Nova Science Publishers; 2012.

10. Morgan NT, Zhang Y, Grandbois ML, Bell BM, Holmes RJ, Cussler EL. Mechanism for the separation of organic semiconductors via thermal gradient sublimation. *Organic Electronics.* 2015;24:212-218. doi: 10.1016/j.orgel.2015.05.029.

11. Knudsen M. Effusion and the molecular flow of gases through openings. *Ann.Phys.* 1909;29(179):179-193.

12. Dadzie SK, Brenner H. Predicting enhanced mass flow rates in gas microchannels using nonkinetic models. *Physical Review E.* 2012;86(3):036318. doi: 10.1103/PhysRevE.86.036318.

438771 Electrochemically-Mediated Adsorptive Processes for Chemical and Environmental Separation

Abstract:

Adsorptive separation processes are ubiquitous in chemical, biological, pharmaceutical and environmental processes, driven by specific interactions between the adsorbate molecules and the adsorbent surfaces. Reversal of the adsorption to recover the adsorbed species or to regenerate the adsorbent is accomplished by many means, including pH, ionic strength, solvent, temperature and pressure swings. In general, these methods require either the addition of chemicals, which may increase the environmental burden, or the heating/cooling of the bulk system, or significant changes in pressure, both of which are energy intensive. We have explored a variety of electrochemically-driven separation systems that are very attractive due to their flexibility of control through modulation of electrochemical potential, fast kinetics and possibility of energy recovery and integration with renewable sources. Our approach relies on heterogeneous redox-active species to mediate the specific interactions with target solutes, as well as take advantage of the large pseudocapacitive charge to provide faster kinetics, higher adsorption capacity and energy recovery performance.

We report the use of heterogeneously supported redox-organometallics, namely metallocenes, for the sorption and release of specific anions from the liquid-phase under oxidizing and reducing conditions, respectively, in both organic and aqueous media. Rather than rely solely on adsorbate charge or size, these redox-functionalized electrodes, through their electrochemical response to an applied potential, regulate the specific chemical interactions of the supported-organometallics with target functional groups of anions to achieve reversible and stoichiometric binding. We explore the mechanisms of this selective interaction through various spectroscopic methods, electrochemical characterization as well as electronic structure calculations. This technology is of scientific and economic interest for (1) sustainable water remediation and waste control, through the removal of highly toxic aqueous contaminants such as organic anions present in agricultural or industrial waste, and (2) improving chemical processes, as a powerful method for direct purification of charged products in chemical synthesis.

In a second example of redox-mediated separation processes, we describe a facile approach for the electrochemically-controlled capture and release of CO_2 from gas streams. In this instance, the metallocene electrode serves as an electron source for the reduction of a poly-quinone on the cathode side of the cell to foster the reaction of CO_2 with the reduced quinone moieties. The CO_2 is released during the stripping cycle on reversal of the inter-electrode potential. Again, the electrodes serve as a separation medium, while also allowing for some energy recovery during the swing operations.

The presentation will provide an overview of these new redox-mediated separation processes, including a description of the various redox-electrode preparation methods, and conclude with general observations on the strengths and limitations of such electrochemically-driven separation processes relative to more accepted separation operations.

437426 Calculation of Proteome Allocation Is Enabled By Genome-Scale Models

Abstract:

A fundamental problem of systems biology has been to understand how cellular proteome allocation strategies affect fitness under varying environments and genetic backgrounds. Recently, integrated genome-scale models of metabolism and gene expression (ME-Models) have been developed, which account for 80% of the *Escherichia coli* proteome by mass. These ME-Models enable calculation of proteome allocation and its phenotypic effects at the genome-scale. Using a ME-Model of *E. coli*, we identified a core proteome consisting of 356 gene products (44% of *E. coli* proteome mass), which was always expressed under 333 simulated growth conditions. The core proteome includes 212 genes not found in previous comparative genomics-based core proteome definitions, accounts for 65% of known essential genes in *E. coli*, and has 78% gene function overlap with minimal genomes (*Buchnera aphidicola* and *Mycoplasma genitalium*). Based on transcriptomics data across environmental and genetic backgrounds, the systems biology core proteome is significantly enriched in non-differentially expressed genes, and depleted in differentially expressed genes. Compared to the non-core, core gene expression levels are also similar across genetic backgrounds and exhibit significantly more complex transcriptional and post-transcriptional regulatory features. Future ME-Model developments include expanding model scope to the remaining 20% of proteome mass, and integration of ME with transcriptional regulation. These expansions will further enable systems-level investigation into the cellular control strategies that have evolved for proteome allocation under changing environments and stress conditions.

437766 Building "Body-on-a-Chip" Systems for Drug Development

Abstract:

Human surrogates for drug testing could reduce the cost of drug development by eliminating ineffective or toxic drugs earlier in the drug development cycle. Such surrogates can be created using tissue engineered constructs to represent key organs in a physical microscale model based on physiologically based pharmaceutic models (PBPK). We will describe our "pumpless" surrogate system and preliminary data using multiorgan models. We also will describe progress toward construction and operation of a 10+ organ human model.

438890 Catalytic Ionic Liquids for Biomass Conversion and Upgrading

Abstract:

Owing to its unique ability to dissolve lignocellulosic biomass under relatively mild conditions, 1-ethyl-3-methylimidazolium acetate, [C_2C_1im][OAc], has attracted rapidly growing interest as a solvent in biomass pretreatment. However, the catalysis role of [C_2C_1im][OAc] has been least studied for producing downstream platform chemicals in replacement of petro-derived chemicals. Herein, [C_2C_1im][OAc] is identified as an efficient catalyst for lignin depolymerization and sugar upgrading into platform chemicals under mild conditions. Specifically, [C_2C_1im][OAc] renders efficient lignin depolymerization into phenol derivatives (e.g., guaiacol and eugenol) in the presence of catalysts near 100 °C within 3 h; The carbene generated in the presence of [C_2C_1im][OAc] leads to quantitative self-condensation of 5-hydroxymethylfurfural (HMF) to a high-value biorefinery product, 5,5'-di(hydroxymethyl)furoin (DHMF)-a potential C_{12} kerosene/jet fuel intermediate through benzoin condensation. More catalytic reactions based on [C_2C_1im][OAc] have been proposed.

412702 Sustainable Product Separation from a Silylamine-Type Switchable Electrolyte in Electrochemical Reactions

Abstract:

Electrolyte separation from the products in an electrochemical reaction adds challenges to the overall manufacturing process [1], requiring additional separation units to be installed or causing downstream contamination of the products. While traditional salts serve as high conductivity electrolytes in aqueous systems, their poor solubility in organic solvents restrict electrochemical reaction performance in non-aqueous systems.

The ideal properties of an alternative electrolyte for electrochemical reactions are high conductivity, high solubility for various reactants and efficient separation. Ionic liquids (ILs) can be used as alternative electrolytes. ILs are defined as ionic salts with low melting points. The properties of ILs can be tuned by changing the cation or anion. They can have both good solubility in organic solvents and high conductivity, which leads to more promising organic-based electrochemical systems. In order to resolve both the product separation and electrolyte recovery concerns, switchable IL-like solvents can be used [2].

These switchable systems, known as reversible ionic liquids (RevILs), can be in an ionic state during the electrochemical processing and be changed to a molecular state for product separation and/or its recovery through use of an external stimulus. Silylamines can be used as a type of RevIL. They form a RevIL-state with addition of CO_2 and turn back to a molecular liquid-state with mild heating [3, 4].

We previously found that the conductivity of neat silylamine-type RevILs were insufficient to be used as an electrolyte when the model silylamine (3-aminopropyl) triethoxysilane (TEtoxySA) was examined electrochemically [5]. However, the RevIL conductivity drastically increased up to electrochemically-appropriate values in the presence of protic organic solvents [5], though it did not increase with aprotic organic solvents near as much.

Now that the concept of a switchable electrolyte has been proposed where solutions of TEtoxySA-RevIL and protic solvents have high conductivity, high solubility in organic solvents and a simple separation procedure with mild heat, they will be investigated in electrochemical applications. In this presentation, we report the electrochemical reaction and separation procedure using these new switchable electrolytes. Also, the effect of the switchable electrolytes on the reaction will be presented, comparing the system with traditional aqueous electrolytes. l Meeting)

References

[1] A.J. Fry, Electrochemical processing, organic, Kirk-Othmer Encyclopedia of Chemical Technology, John Wiley & Sons, Inc.2000, pp. 652-685.

[2] P.G. Jessop, S.M. Mercer, D.J. Heldebrant, CO_2-triggered switchable solvents, surfactants, and other materials, Energy & Environmental Science, 5 (2012) 7240-7253.

[3] A.L. Rohan, J.R. Switzer, K.M. Flack, R.J. Hart, S. Sivaswamy, E.J. Biddinger, M. Talreja, M. Verma, S. Faltermeier, P.T. Nielsen, P. Pollet, G.F. Schuette, C.A. Eckert, C.L. Liotta, The synthesis and the chemical and physical properties of non-aqueous silylamine solvents for carbon dioxide capture, ChemSusChem, 5 (2012) 2181-2187.

[4] J.R. Switzer, A.L. Ethier, E.C. Hart, K.M. Flack, A.C. Rumple, J.C. Donaldson, A.T. Bembry, O.M. Scott, E.J. Biddinger, M. Talreja, M.-G. Song, P. Pollet, C.A. Eckert, C.L. Liotta, Design, synthesis, and evaluation of nonaqueous silylamines for efficient CO_2 capture, ChemSusChem, 7 (2014) 299-307.

[5] J.D. Jimenez, S. Jung, E.J. Biddinger, Ionicity analysis of silylamine-type reversible ionic liquids as a model switchable electrolyte, Journal of The Electrochemical Society, 162 (2015) H460-H465.

418542 Precipitation of Cellulosic Biomass from Ionic Liquid/Cosolvent Mixtures By a Non-Reactive and Energy Efficient Gas Antisolvent Method

Abstract:

Select ionic liquids (ILs) demonstrate inherently large cellulose dissolution capacities making them highly touted solvents for biomass processing. Through an ionic liquid based dissolution and precipitation process cellulose crystallinity is largely reduced consequently enhancing subsequent chemical and biochemical reaction processes. Conventional antisolvents for biomass precipitation from ILs include polar protic liquids (i.e. H_2O, EtOH, etc.) which are highly effective at disrupting IL-cellulose interactions. Recently we have found that these liquid antisolvents are so effective that as little as 1 mass% residual water on an IL can reduce its cellulose capacity by as much as 55%. Therefore, ILs must be highly purified from liquid antisolvents prior to recycle. Preliminary analyses indicate that quantitative separation of an IL from a liquid antisolvent is highly energy intensive and could potentially impede large scale viability. Therefore, as an alternative to conventional liquid antisolvent processes we have developed a novel gas antisolvent method which precipitates cellulosic biomass from IL/cosolvent mixtures by compressed carbon dioxide at low to moderate pressures (i.e. 23 to 66 bar). The gas antisolvent separation process is especially unique as it is non-reactive and completely reversible. By simple depressurization of CO_2 to just a few bar pressure below the separation point, ionic liquid solvation power for cellulose is completely regenerated. This presentation will highlight the solid-liquid phase equilibrium effects of both conventional and novel separation processes. Additionally, the advantages of polar aprotic cosolvents will be highlighted as they apply to cellulose dissolution in IL/solvent mixtures and the gas antisolvent process. Spectroscopic techniques will identify key trends within the separation data. Finally an energy analysis will be presented to demonstrate the advantages of this novel CO_2 relative to liquid antisolvent separations.

419589 Nanoengineered Solid Acid Catalysts for Green Processes

Abstract:

Nanoengineered Solid Acid Catalysts for Green Processes

H. Liu, P.Shen, H.T. Zhang, S.J. Zhang*

Beijing Key Laboratory of Ionic Liquids Clean Process, Key Laboratory of Green Process and Engineering, Institute of Process Engineering, Chinese Academy of Sciences, Beijing, 100190, P. R. China

E-mail: htzhang@ipe.ac.cn

Improving catalytic activity and selectivity as well as reusability of catalysts is an unavoidable challenge because of stringent economical and ecological desire for sustainability on account of the principle of "green chemistry" and "green technology". The development of nanosciences and nanotechnologies make it possible to fabricate complicated artificial nanoarchitectures *via*the elf-assemble and reorganization of different anisotropic nanoscale building blocks (nanocrystals, nanosheets, nanowires, and polyhedrons of oxides, metal, sulfide, and nitrite).

Among different kinds of acid catalysts, solid acid catalysts are attracting more attention owing to their obviously environmental benign characteristics with respect to weak corrosiveness, safety, high efficiency, easiness of separation and recovery. Solid acid nanosheets are promising because of their exotic physicochemical properties and high specific surface areas and tunable acidity. They represent a diverse and largely untapped source of basic building blocks for rational creating complicated nanocatalysts with designable microstructures and programmed propertiesthe. In addition, many nanoparticles could be employ as basic building blocks with nanosheets to induce some synergistic effects that could improve catalytic performance. Our research study indicated that nanoengineered solid acid catalysts exhibited improved catalytic performance for Friedel-Crafts alkylation, acetalization and photocatalytic reduction reactions. Here we mainly present some progresses in retrosynthesizing complication layered nanohybrids by using solid acid nanosheets, and understanding the relationships between microstructures and improved catalytic performance.

427326 Synergistic Catalytic Performance in the Synthesis of Cyclic Carbonates Based on Ionic Liquids

Wednesday, November 11, 2015
Exhibit Hall 1 (Salt Palace Convention Center)
Suojiang Zhang, Institute of Process Engineering, Chinese Academy of Sciences, Beijing, China

Synergistic Catalytic Effect in the Synthesis of Cyclic Carbonates Based on Ionic Liquids

Suo-Jiang Zhang , Wei-Guo Cheng

Key Laboratory of Green Process and Engineering, Institute of Process Engineering, Chinese Academy of Sciences, P.O. Box 353, Beijing, 100190, China

**Corresponding author: sjzhang@home.ipe.ac.cn*

Carbon dioxide released annually to the atmosphere at ca. 30 000 Mt1 is also an attractive C1 feedstock as it is nontoxic, abundant, inexpensive. From the viewpoint of environmental protection and resource utilization, it is important to transform CO_2 into cyclic carbonates efficiently not only in academia but also in industrial applications(Scheme 1). Many scientists have made a concentrated effort to improve the performance of the ionic liquids (ILs) catalysts during the past decade due to their tunable structures. Synergistic catalytic effects of ILs catalytic system play an important role of promoting the cycloaddition reactions of epoxides. However, due to the diversity of different anion¨Ccation combinations it is difficult to make generalizations about synergistic effects.

THE GRAPHIC ON THIS PAGE WAS NOT AVAILABLE ON THE E-MEDIA VERSION OF THIS PAPER.

Scheme 1 Fixation CO_2 and epoxides into cyclic carbonates

In this work, we have investigated the synergistic effect in the catalytic system based on ILs through experimental and theoretical studies[1-5]. The results show that metal and ionic liquids, cations and anions of ionic liquids, the functional groups in cations or supports all have a synergistic effect on the cycloaddition reactions. Synergistic catalytic effects of ILs play an important role of remarkably reducing the activation energy and promoting ring-opening of epoxide. The design of catalytic system of ionic liquids can be possible as the synergistic effect on a molecular level is understood by using density functional theory (DFT) studies and in situ characterization. Significant developments in IL catalysts design and optimal process development for the synthesis of cyclic carbonates have been made with the possible prediction of the synergetic effect based on ILs.

Keywords: synergistic effect, ionic liquid, carbon dioxide, cyclic carbonate, process

References

1. Zhang SJ, Sun J, Zhang XC, Xin JY, Miao QQ, Wang JJ. Ionic liquid-based green processes for energy production. *Chemical Society Reviews*, 2014, 43: 7838-7869

2. Cheng WG, Chen X, Sun J, Wang JQ, Zhang SJ. SBA-15 supported triazolium-based ionic liquids as highly efficient and recyclable catalysts for fixation of CO_2 with epoxides. *Catalysis Today*, 2013,200: 117- 124

3. Sun J, Wang JQ, Cheng WG, Zhang JX, Li XH, Zhang SJ, She YB. Chitosan functionalized ionic liquid as a recyclable biopolymer-supported catalyst for cycloaddition of CO_2. *Green Chemistry*, 2012, 14: 654-660

4. Wang JQ, Sun J, Cheng WG, Dong K, Zhang XP, Zhang SJ. Experimental and theoretical studies on hydrogen bond-promoted fixation of carbon dioxide and epoxides in cyclic carbonates. *Phys. Chem. Chem. Phys.* 2012, 14: 11021- 11026.

5. Xu BH, Wang JQ, Sun J, Huang Y, Zhang JP, Zhang SJ. Fixation of CO_2 into cyclic carbonates catalyzed by ionic liquids: a multi-scale approach. *Green Chemistry*, 2015, 17: 108-122.

429678 Kinetics of Hydroformylation of Ethylene Using Homogeneous Rhodium-Phosphine Complexes

Abstract:

The hydroformylation reaction is a useful method to functionalize C-C bonds for providing aldehyde and alcohol compounds, and is also an important industrial catalytic process. In view of the importance, it is attracting continuous and increasing academic and industrial interest. Studies on the kinetics of hydroformylation of olefins not only are instructive for improvement of the catalytic system but also provide the basic information for design and scale-up of appropriate reactors.

In this research, the kinetics of hydroformylation of ethylene with triphenylphosphine modified rhodium catalyst was studied. The reaction was carried out batchwise in a flat-bottomed, magnetically stirred 150 mL steel autoclave equipped with a 120 mL teflon tube. Under the conditions of the Rh concentration of 0.0015mol/L, PPh_3/Rh=10, C_2H_4: CO: H_2 =1: 1: 1 (molar ratio), 2 Mpa and 100 °C, the selectivity for propionaldehyde was nearly 100%. The ethylene hydroformylation kinetics was studied as the function of the amount of catalyst and triphenylphosphine, partial pressure of ethylene, CO and H_2, and temperature. All these parameters were found to influence the rate of hydroformylation. On the range of conditions (catalyst, 1.024~2.2658 mmol·L^{-1}; PPh_3, 0.0058~0.3018 mmol·L^{-1}; P_{H2}, 0.26~1.27 MPa; P_{CO}, 0.11~1.22 MPa; P_{C2H4}, 0.18~1.13 MPa; T, 355.15~372.15 K), rates were observed to increase with the increase in the amount of the catalyst and triphenylphosphine; rates increased on increasing the C_2H_4 or H_2 pressure; rates increased on increasing the CO pressure up to certain values, and then decreased. The activation energy was found to be 61.52 kJ·mol^{-1}. A kinetic rate model based on the mechanism of hydroformylation of ethylene was found to fit with the experimental rate with ±5% deviation.

CH2MHILL.

Main Fractionator Control – Fundamental Concepts: Servo Control

Andrew W. Sloley – CH2M HILL, Houston, TX

Prepared for Presentation at the AIChE 2015 Annual Meeting, Salt Lake City, UT
08-13 November, 2015

Servo Versus Regulatory Control

- Servo
  - Response to setpoint changes
- Used to manipulate
  - Product rates
  - Product qualities
  - Heat recovery

- Regulatory
  - Response to environmental changes
- Used to
  - Maximize specific products
  - Respond to feed changes

Control in the Plant

- Manual action in response to changes
  - Servo or regulatory?
- Advanced control systems
  - Servo or regulatory?
- Operation in most plants makes distinction less clear

This Material

1. Servo: response to material balance
2. Servo: response to heat balance
3. Regulatory: composition and rate
4. Overhead systems
5. Side-draw
6. Wash oil rate
7. General control strategies

Atmospheric Column Yield Streams

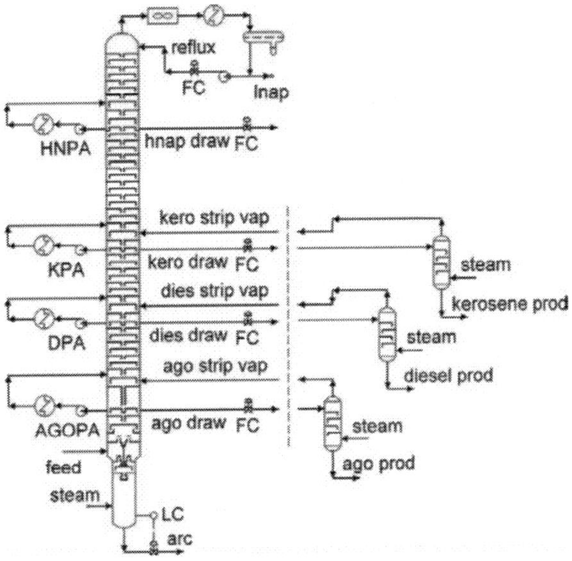

Atmospheric Column Simplified

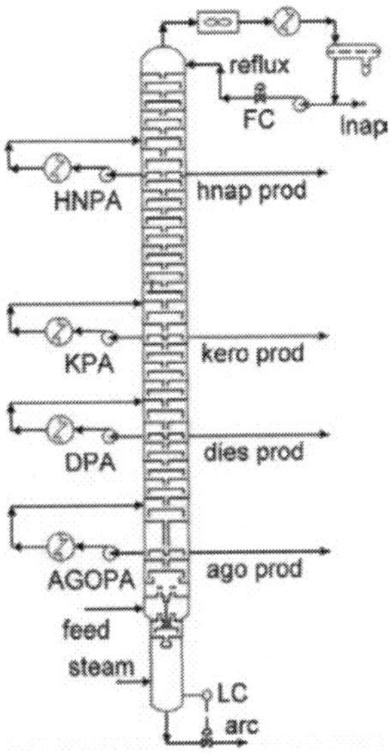

Base Operation

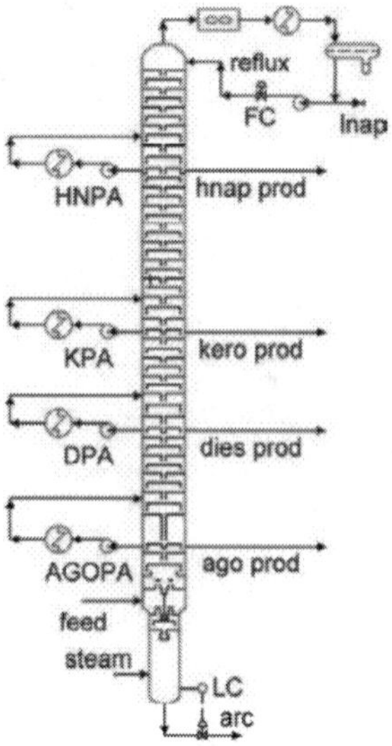

Stream	Rate	D86 5	D86 50	D86 95
Lnap	14,964	71.4	185.7	260.0
Hnap	7,299	262.0	313.9	350.0
Kero	19,091	368.4	445.7	534.7
Diesel	8,241	465.3	587.5	640.4
AGO	8,500	601.3	694.4	803.4
ARC	41,905	679.5		

Gap-Overlap	95-5		
	Base		
Lnap-Hnap	2.0		
Hnap-Kero	18.4		
Kero-Diesel	-69.4		
Diesel-AGO	-39.1		

Comparison Basis

- Flash on product streams
  - Difficult to modify downstream
  - Stripping steam rate controls
  - Always meet specification or better
- Overhead temperature
  - Controlled to meet water dew point
- Mass balance versus composition control
  - Mass balance (fixed product rates or volumes)
  - Composition – product specifications

Material (Volume) Balance Control

- Product draws must equal feed to column
- Quantity of any component fixed
- Product draw rate changes product composition
- Feed composition changes product draw rates

Response to Material Balance Changes

Constant Crude Rate and Crude Composition

- Increasing draw stream rate makes it heavier
  - Distillation curve temperatures rise
    - Initial, middle, and end points
  - Stream is heavier (higher molecular weight)
- Decreasing draw stream rate makes it lighter
  - Distillation curve temperatures drop
    - Initial, middle, and end
  - Stream is lighter
- Consequences cascade
- Representative examples

Blending and Reflux

- Changing product draw rates changes blending effects between streams
- Changing product draw rates changes reflux inside the tower
  - At a constant heat balance
- Blending effects typically stronger
- Examples use D86 distillations
- Other properties have stronger gain
  - Freeze point, cloud point

Yield Shift Example 1

Increase Draw, No Compensation, Product Rates Set

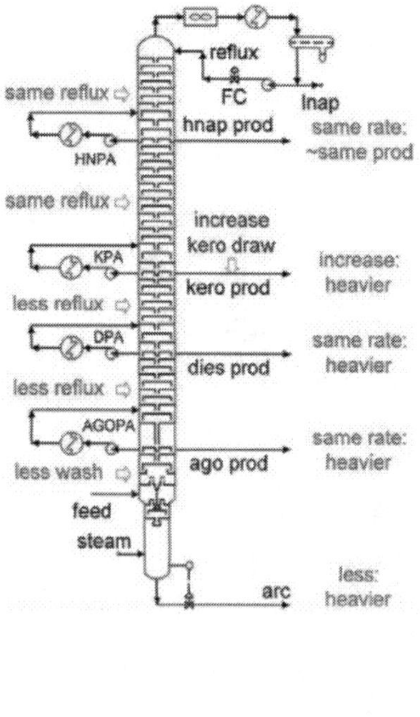

Yield Shift Example 1 – Results

Increase Draw, No Compensation, Product Rates Set

	Base				Example 1 +Kerosene Draw			
Stream	BPD	D86 5	D86 50	D86 95	BPD	D86 5	D86 50	D86 95
Lnap	14,964	71.4	185.7	260.0	14,964	71.5	185.8	260.3
Hnap	7,299	262.0	313.9	350.0	7,299	262.0	313.9	350.0
Kero	19,091	368.4	445.7	534.7	**21,089**	370.9	455.7	555.5
Diesel	8,241	465.3	587.5	640.4	8,240	476.5	609.9	667.3
AGO	8,500	601.3	694.4	803.4	8,502	614.4	721.6	842.1
ARC	41,905	679.5			**39,906**	682.3		

Yield Shift Example 2

Increase Draw, Compensate Above, Product Rates Set

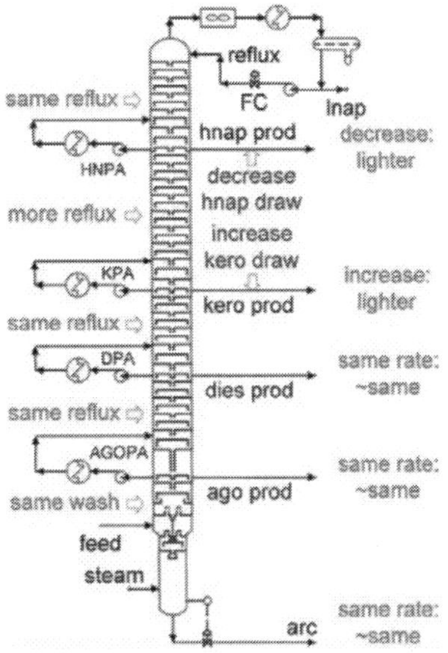

Yield Shift Example 2 – Results

Increase Draw, Compensate Above, Product Rates Set

	Base				Example 2 +Kerosene Draw -Heavy Naphtha Draw			
Stream	BPD	D86 5	D86 50	D86 95	BPD	D86 5	D86 50	D86 95
Lnap	14,964	71.4	185.7	260.0	14,964	71.5	185.2	260.3
Hnap	7,299	262.0	313.9	350.0	**5,299**	256.9	**302.1**	350.4
Kero	19,091	368.4	445.7	534.7	**21,090**	362.3	**435.1**	532.9
Diesel	8,241	465.3	587.5	640.4	8,240	465.3	587.4	640.4
AGO	8,500	601.3	694.4	803.4	8,502	601.2	694.0	803.6
ARC	41,905	679.5			41,906	679.5		

Yield Shift Example 3

Increase Draw, Compensate Below, Product Rates Set

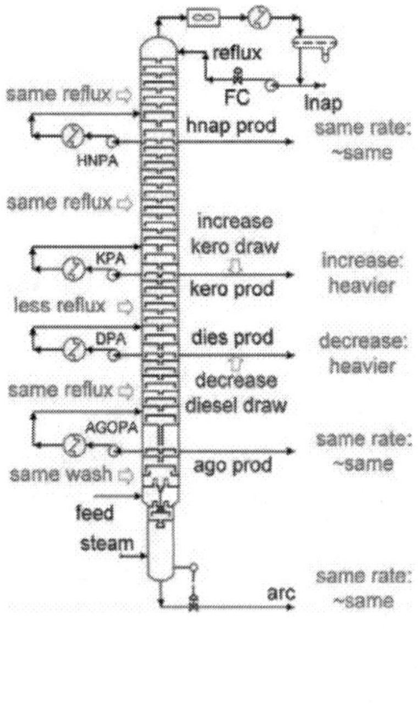

Yield Shift Example 3 – Results

Increase Draw, Compensate Below, Product Rates Set

	Base				Example 3 +Kerosene Draw -Diesel Draw			
Stream	BPD	D86 5	D86 50	D86 95	BPD	D86 5	D86 50	D86 95
Lnap	14,964	71.4	185.7	260.0	14,963	71.4	185.8	260.1
Hnap	7,299	262.0	313.9	350.0	7,299	262.4	314.3	350.2
Kero	19,091	368.4	445.7	534.7	**21,090**	370.6	**454.8**	553.4
Diesel	8,241	465.3	587.5	640.4	**6,241**	479.0	**599.0**	645.5
AGO	8,500	601.3	694.4	803.4	8,500	601.3	694.1	803.6
ARC	41,905	679.5			41,906	679.5		

Yield Shift Example 4

Increase Draw, Compensate Split, Product Rates Set

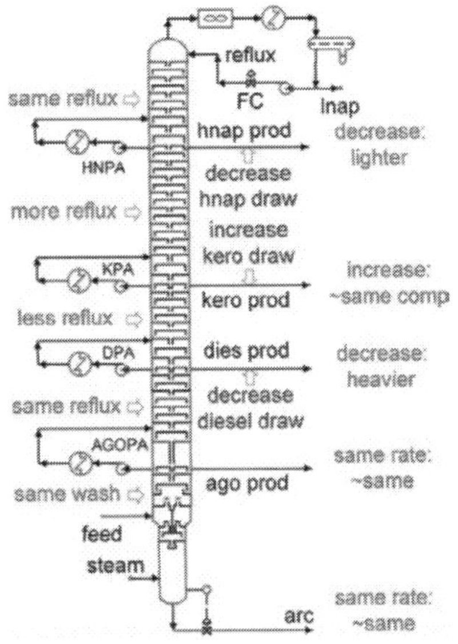

Yield Shift Example 4 – Results

Increase Draw, Compensate Split, Product Rates Set

	Base				Example 4 +Kerosene Draw -Hnap and -Diesel Draw			
Stream	BPD	D86 5	D86 50	D86 95	BPD	D86 5	D86 50	D86 95
Lnap	14,964	71.4	185.7	260.0	14,964	71.4	185.3	259.3
Hnap	7,299	262.0	313.9	350.0	**5,462**	256.7	**303.2**	332.9
Kero	19,091	368.4	445.7	534.7	**21,090**	362.9	436.7	**534.3**
Diesel	8,241	465.3	587.5	640.4	**8,078**	466.4	**588.4**	640.8
AGO	8,500	601.3	694.4	803.4	8,500	601.2	694.0	804.6
ARC	41,905	679.5			41,905	679.5		

Kerosene D86 Final Point same in both Base and Example 4

Decreasing Draw

- Decreasing draw has opposite changes

Internal Reflux Control

- Vary product draw rates
  - Changes amount of liquid inside the column
- Vary heat balance
  - Changes condensation of vapor to liquid
  - Changes vapor rate and liquid rate above duty change location
  - Changes liquid rate below duty change location
- Higher L:V ratio gives more separation

Internal Reflux Evaluation

- Use gap-overlap on distillation curves (D86 5/95)
- Gives consistent basis

$$gap\ or\ overlap_{5/95} =$$
$$ASTM\ D86_{5vol\%}\ Temp\ of\ Heavy\ Cut - ASTM\ D86_{95vol\%}\ Temp\ of\ Light\ Cut$$

- Positive = gap, negative=overlap
- More positive the value, the more separation
- In comparison to D86 5/95,
  - Flash more sensitive to light ends
  - Freeze and cloud more sensitive to heavy ends

Yield Shift Example 11

Increase Draw, Other Products' Specifications Set

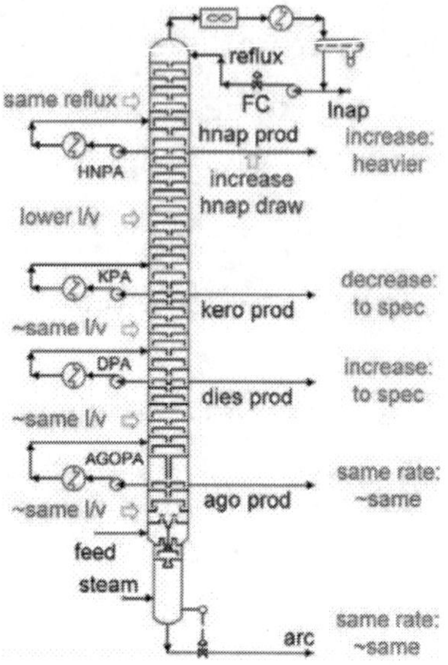

Yield Shift Example 11 – Results

Increase Draw, Other Products' Specifications Set

	Base		Example 11			Delta
Stream	BPD	D86 5-95	BPD	D86 5-95		
Lnap	14,964		14,964			
		2.0		7.1		+5.1
Hnap	7,299		9,298			
		18.4		6.0		-12.4
Kero	19,091		16,934			
		-69.4		-71.1		-1.7
Diesel	8,241		8,519			
		-39.1		-38.8		+0.3
AGO	8,500		8,500			

Reflux Control by Mass Balance Shift

- Higher rate of light products
  - Decreases fractionation effectiveness
- Products drawn to specifications
  - Blending effects result in alternating changes on column
  - Minimizes reflux changes
  - Mass balance dampens out the further away from the change
- Large flow rate changes required to change separation

Internal Reflux Control by Heat Shift

- Shift heat from one heat removal to another
- Reflux is heat removed at a lower temperature level beyond that required to condense the product
- More reflux – more heat removal at lower temperature
- Less reflux – less heat removal at lower temperature

To Improve	Reduce
Fractionation	Energy Efficiency
Energy Efficiency	Fractionation

Heat Shift Example 15

Shift Duty Up, Product Rates Set

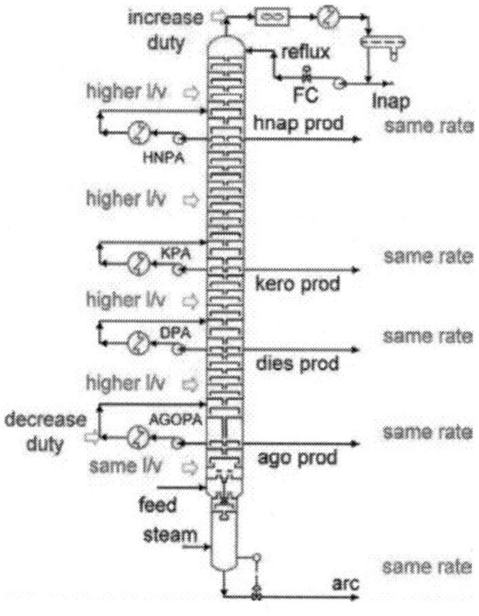

Heat Shift Example 15 – Results

Shift Duty Up, Product Rates Set

	Base		Example 15		Delta
Stream	BPD	D86 5-95	BPD	D86 5-95	
Lnap	14,964		14,964		
		2.0		5.0	+3.0
Hnap	7,299		7,299		
		18.4		20.2	+1.8
Kero	19,091		19,090		
		-69.4		-58.7	+10.7
Diesel	8,241		8,241		
		-39.1		-31.8	+7.3
AGO	8,500		8,500		

10M btu/hr shifted from AGO PA to Condenser

Heat Shift Example 19 – Results

Shift Duty Up, Product Specifications Set

	Base		Example 13			Delta
Stream	BPD	D86 5-95	BPD	D86 5-95		
Lnap	14,968		14,861			
		2.0		5.7		+3.7
Hnap	7,292		7,500			
		18.4		20.4		+2.0
Kero	19,142		19,210			
		-69.4		-59.4		+10.0
Diesel	8,396		8,183			
		-39.1		-33.1		+6.0
AGO	8,500		8,500			
ARC	41,905		41,746			

10M btu/hr shifted from AGO PA to Condenser

Heat Shift Example 20 – Results

Shift Duty Down, Product Specifications Set

	Base		Example 18			Delta
Stream	BPD	D86 5-95	BPD	D86 5-95		
Lnap	14,968		15,087			
		2.0		-1.8		-3.8
Hnap	7,292		7,067			
		18.4		16.3		-2.1
Kero	19,142		18,930			
		-69.4		-76.9		-7.5
Diesel	8,396		8,302			
		-39.1		-45.5		-6.4
AGO	8,500		8,500			
ARC	41,905		42,122			

10M btu/hr shifted to AGO PA from Condenser

Higher Products versus Lower Products

- Higher products have smaller changes in separation and rates than lower products
- Less liquid in the bottom of the column
- Same amount of duty shift changes the L:V ratio more in the bottom of the column

Duty Shift – L:V Ratios

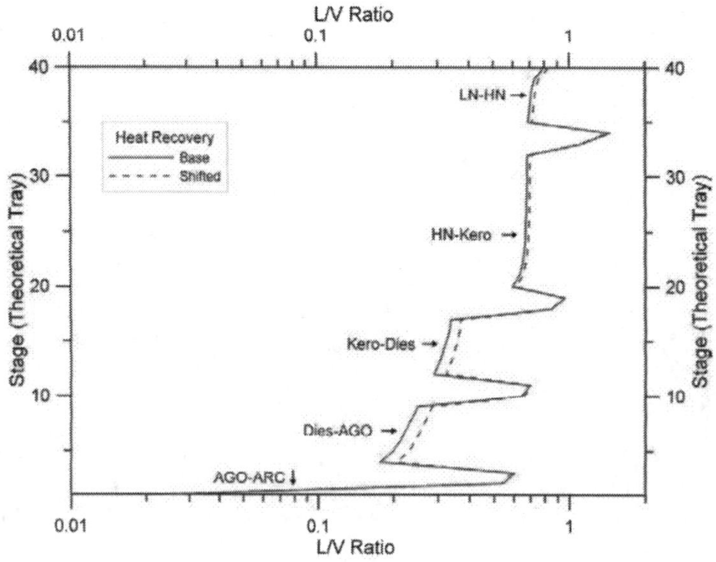

10M btu/hr shifted from AGO PA to Condenser

Heat Shift and Fractionation

- Change in fractionation strongly limited to zone of tower that duty shifts across
- Biggest change occurs where liquid rates are lowest
- Yields shift interactions
  - Product blending
  - Specifications

Summary

- Mass balance control linked to composition
- Blending effects important
- Mass balance control can shift separations
  - Large mass balance shifts required
- Lighter crudes will tend to have poorer separation
- Heat balance control can change separations
- Energy efficiency and separation efficiency conflict
- Energy balance shifts have larger results at bottom of column

Questions

418636 Optimal Control Theory for Batch Distillation with Missing Components

Abstract:

Batch distillation is commonly used unit operation in specialty chemicals. In order to maximize the profitability and improve separation using batch distillation column, optimal reflux operation can be used. The mathematical and numerical complexities of the optimal control problem get worse when there are missing components in the mixture whose properties could not be included in analysis. This paper presents a theory based on Ito's calculus and stochastic maximum principle for optimizing the column performance when some of the components are missing. This algorithm is implemented in the MultiBatchDS batch distillation process simulator. Finally, a numerical case-study derived from a real world batch distillation separation problem encountered at Firmenich is presented to show the scope and application of the proposed approach.

A model for flow through structured packings using the concept of effective surface

Sylvain Pasquier[1], Yohan Davit[1,2], Michel Quintard[1,2]

[1]Université de Toulouse ; INPT, UPS ; IMFT (Institut de Mécanique des Fluides de Toulouse) ; Allée Camille Soula, F-31400 Toulouse, France

[2]CNRS; IMFT ; F-31400 Toulouse, France

September 27, 2015

Modeling of the gas-liquid system in structured packings

1. Objective: packing-scale models for the liquid-gas phases in structured packings (distillation process, CO2 treatment, etc)

2. Proposition: construction of an effective surface at the liquid-gas interface

3. Upscaling of the effective system at the micro-scale for the gas

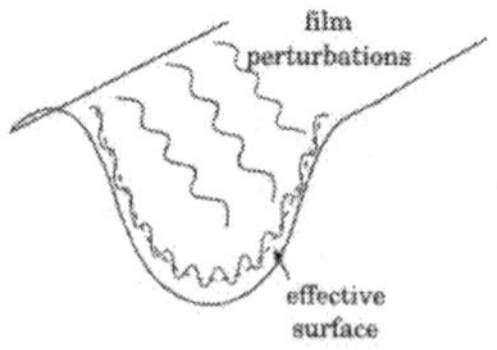

4. Numerical implementation of three-equation system at the packing-scale

Macro-scale form of laws issued from up-scaling

- Background: analysis of the macro-scale laws issued from up-scaling

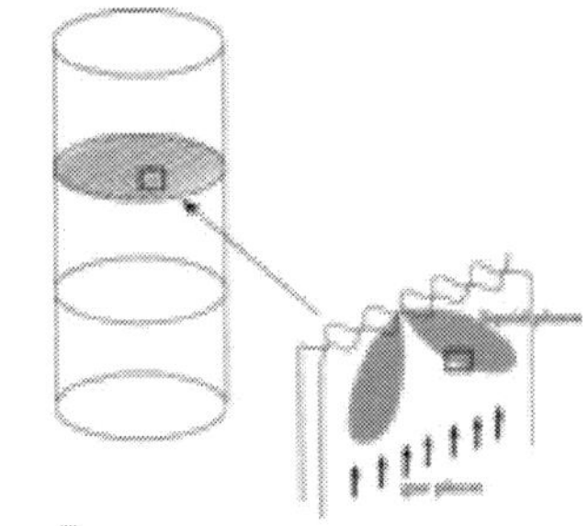

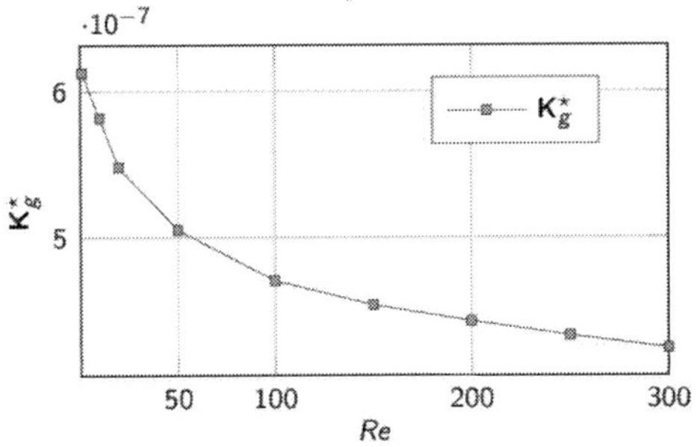

Figure 1 : Apparent permeability as a function of the Reynolds Number (Soulaine, Quintard 2014)

- Models at the macro-scale for a two-phase flow (with/without cross terms)

$$\langle \mathbf{u}_g \rangle = -\frac{\mathbf{K}_{gl}}{\mu_g} \cdot \left(\nabla \langle p_g \rangle^g - \rho_g \mathbf{g}\right) + \mathbf{K}_{lg} \cdot \langle \mathbf{u}_l \rangle$$

$$\langle \mathbf{u}_l \rangle = -\frac{\mathbf{K}_{lg}}{\mu_l} \cdot \left(\nabla \langle p_g \rangle^l - \rho_l \mathbf{g}\right) + \mathbf{K}_{gl} \cdot \langle \mathbf{u}_g \rangle$$

see Whitaker (1986), Lasseux (2006)

- Generalised Darcy-Forchheimer law for inertial flows (Whitaker 1996). See Soulaine, Quintard (2014) for the application to structured packings:

$$\langle \mathbf{u}_g \rangle = -\frac{\mathbf{K}_g}{\mu_g} \cdot \left(\nabla \langle p_g \rangle^g - \rho_g \mathbf{g}\right) - \mathbf{F}(\mathbf{u}_g) \cdot \langle \mathbf{u}_g \rangle$$

$$\langle \mathbf{u}_g \rangle = -\frac{\mathbf{K}_g^\star(\mathbf{u}_g)}{\mu_g} \cdot \left(\nabla \langle p_g \rangle^g - \rho_g \mathbf{g}\right)$$

A two-phase flow system : decoupling ?

- Mathematical problem in the **liquid** phase:

$$\nabla \cdot \mathbf{u}_l = 0$$

$$\rho_l \left(\frac{\partial \mathbf{u}_l}{\partial t} + \mathbf{u}_l \cdot \nabla \mathbf{u}_l\right) = -\nabla p_l + \mu_l \nabla^2 \mathbf{u}_l + \rho_l \mathbf{g}$$

$$\mathbf{u}_l = 0 \text{ at } \mathcal{A}_{ls}$$

- Liquid-gas boundary conditions

$$-p_g \mathbf{n}_{lg} + \mu_g \mathbf{n}_{lg} \cdot \left(\nabla \mathbf{u}_g + \nabla \mathbf{u}_g^T\right) = -p_l \mathbf{n}_{lg} + \mu_l \mathbf{n}_{lg} \cdot \left(\nabla \mathbf{u}_l + \nabla \mathbf{u}_l^T\right) + 2\sigma H \text{ at } \mathcal{A}_{lg}$$

$$\mathbf{u}_l = \mathbf{u}_g \text{ at } \mathcal{A}_{lg}$$

- Mathematical problem in the **gas** phase:

$$\nabla \cdot \mathbf{u}_g = 0$$

$$\rho_g \left(\frac{\partial \mathbf{u}_g}{\partial t} + \mathbf{u}_g \cdot \nabla \mathbf{u}_g\right) = -\nabla p_g + \mu_g \nabla^2 \mathbf{u}_g + \rho_g \mathbf{g}$$

$$\mathbf{u}_g = 0 \text{ at } \mathcal{A}_{gs}$$

Two one-phase flows ?

- Liquid film over the packing structure (Nusselt film)

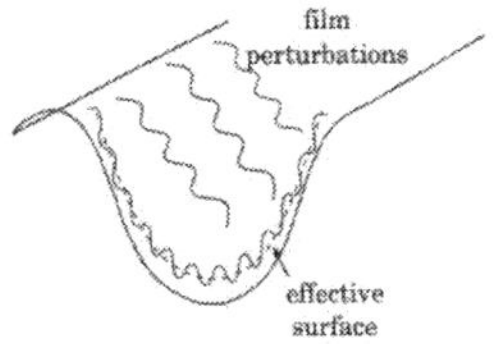

See Alekseenko (1995), Ruyer-Quil (1998), Trifonov (2008)

- Gas over a rough surface

$$\nabla \cdot \mathbf{u}_g = 0$$

$$\rho_g \left(\frac{\partial \mathbf{u}_g}{\partial t} + \mathbf{u}_g \cdot \nabla \mathbf{u}_g\right) = -\nabla p_g + \mu_g \nabla^2 \mathbf{u}_g + \rho_g \mathbf{g}$$

$$\mathbf{u}_g = 0 \text{ at } \mathcal{A}_{ls}$$

$$\mathbf{u}_l \ll \mathbf{u}_g \rightarrow \mathbf{u}_l \sim 0 \text{ at } \mathcal{A}_{lg}$$

Gas flow over a rough surface: effective concept surface

- An effective smooth surface is placed instead of the "rough" liquid surface
- Derivation of an effective condition on the smooth surface (slip condition)
- The decomposition domaine method is used, see Introïni and Quintard (2009), Achdou et al. (2002)
- For a 3-D rough surface, a "slip tensor" $\mathbf{M}|_{\Gamma_{eff}}$ characterises anisotropic effects

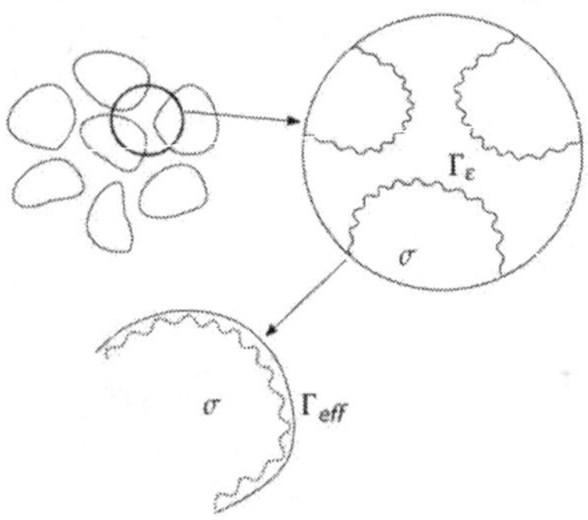

$$\mathbf{u}_g = \mathbf{M}|_{\Gamma_{eff}} \cdot \mathbf{n} \cdot \left(\nabla \mathbf{u}_g + \nabla^t \mathbf{u}_g\right) \cdot (\mathbf{I} - \mathbf{nn}) \quad \text{at} \quad \Gamma_{eff}$$

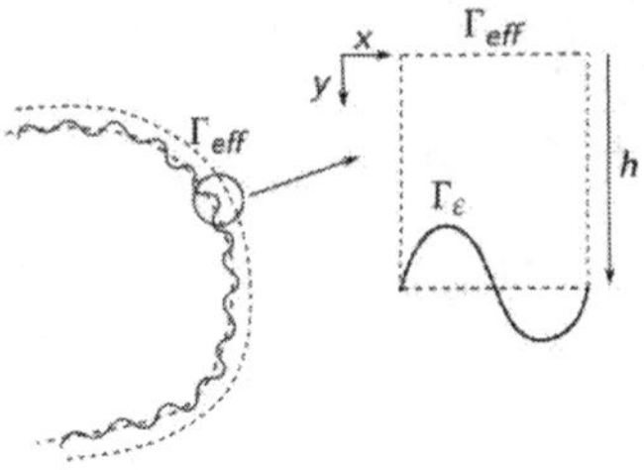

$$\nabla \cdot \mathbf{M} = 0$$
$$0 = -\nabla \mathbf{m} + \mu \nabla^2 \mathbf{M} \quad \text{in } \Omega_1$$
$$0 = \mathbf{n} \cdot \left[-\mathbf{mI} + \mu\left(\nabla \mathbf{M} + \nabla \mathbf{M}^T\right)\right] \quad \text{at } \Gamma_{eff}$$
$$\mathbf{M} = 0 \text{ at } \Gamma_\varepsilon$$
$$\mathbf{m}(\mathbf{x}) = \mathbf{m}(\mathbf{x} + \triangle \mathbf{x}) \text{ on } \Gamma_{per}$$
$$\mathbf{M}(\mathbf{x}) = \mathbf{M}(\mathbf{x} + \triangle \mathbf{x}) \text{ on } \Gamma_{per}$$

A closure problem is solved over a rough element to determine $\mathbf{M}|_{\Gamma_{eff}}$

Application to a sinusoïdal 3d pattern

- Application to a sinusoïdal profile: resolution of the closure problems over the unit-cell ($\mathbf{M}|_{\Gamma_{eff}}$)

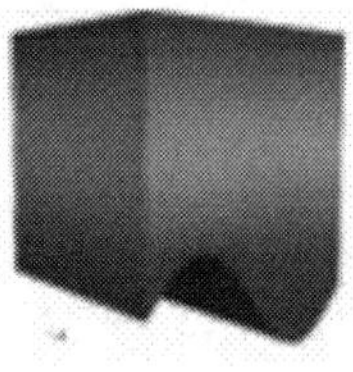

$$\mathbf{M}|_{\Gamma_{eff}} = \begin{pmatrix} M_{xx} & 0 & 0 \\ 0 & 0 & 0 \\ 0 & 0 & M_{zz} \end{pmatrix}$$

$$\mathbf{M}|_{\Gamma_{eff}} = \begin{pmatrix} 0.088 & 0 & 0 \\ 0 & 0 & 0 \\ 0 & 0 & 0.092 \end{pmatrix}$$

- M_{xx} and M_{zz} coefficients characterise anisotropic effects
- Resolution of the flow with the effective condition at Γ_{eff} and comparison to DNS (right)

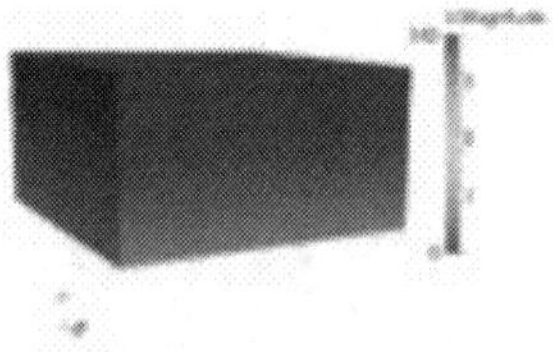

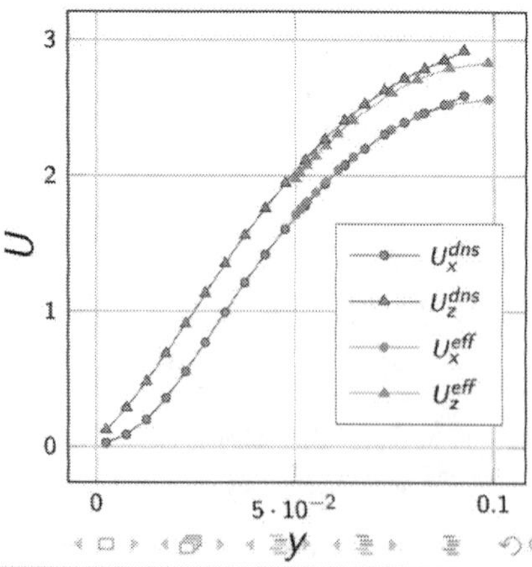

Upscaling : model at the macroscale

- Volume averaging of the problem at the microscale

$$\nabla \cdot \mathbf{u_g} = 0 \text{ in } \Omega_{eff}$$

$$\rho_g \left(\frac{\partial \mathbf{u_g}}{\partial t} + \mathbf{u_g} \cdot \nabla \mathbf{u_g} \right) = -\nabla p_g + \nabla \cdot \left(\mu_g \left(\nabla \mathbf{u_g} + \nabla \mathbf{u_g}^T \right) \right) + \rho_g \mathbf{g} \text{ in } \Omega_{eff}$$

$$\mathbf{u_g} = \mathbf{M}|_{\Gamma_{eff}} \cdot \mathbf{n} \cdot \left(\nabla \mathbf{u_g} + \nabla^t \mathbf{u_g} \right) \cdot (\mathbf{I} - \mathbf{nn}) \text{ at } \Gamma_{eff}$$

Upscaling method

$$\psi_i = \langle \psi_i \rangle^i + \tilde{\psi}_i$$

$$\langle \psi_i \rangle = \frac{1}{V} \int \psi_i dV$$

- Balance momentum mixing elements at micro and macro scales:

$$0 = -\nabla \langle p_g \rangle^g + \rho_g \mathbf{g} + \mu_g \nabla^2 \langle \mathbf{u_g} \rangle^g - \frac{\varepsilon^{-1}}{V} \int_{\mathcal{A}} \mathbf{n} \cdot \left[-\tilde{p}_g \mathbf{I} + \mu_g \left(\nabla \tilde{\mathbf{u}}_g + \nabla \tilde{\mathbf{u}}_g^T \right) \right] d\mathcal{A}$$

- Momentum balance law at the microscale:

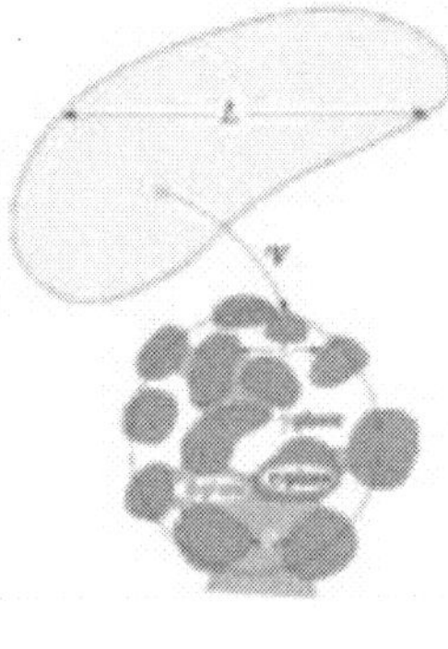

$$\langle \mathbf{u_g} \rangle = -\frac{\mathbf{K}_0 \cdot \left(\mathbf{I} - \mathbf{C}_0 \right)}{\mu_g} \cdot \left(\nabla \langle p_g \rangle^g - \rho_g \mathbf{g} \right) - \mathbf{F} \cdot \left(\mathbf{I} + \mathbf{C}_i \right) \cdot \langle \mathbf{u_g} \rangle \text{ in } \Omega_{eff}$$

  - $\mathbf{K}_0$ intrinsic permeability tensor
  - $\mathbf{C}_0$ correction to the intrinsic permeability
  - $\mathbf{F}$ inertial correction tensor
  - $\mathbf{C}_i$ correction to the inertial correction $\mathbf{F}$

Application to a generic porous media : Array of beads

- Unit cell : array of beads (2D)
- Resolution of the closure problem over the smooth unit-cell - DNS over the rough unit-cell

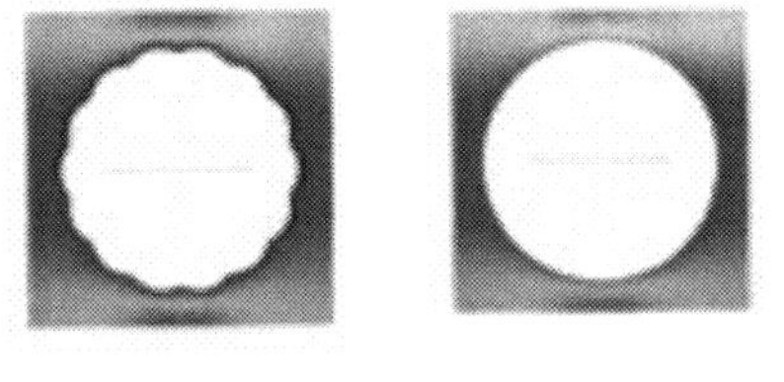

(a) rough unit-cell (b) smooth unit-cell

- Closure relations for deviations:

$$\tilde{\mathbf{u}}_g = \mathbf{B} \cdot \langle \mathbf{u_g} \rangle^g \quad ; \quad \tilde{p}_g = \mu_g \mathbf{b} \cdot \langle \mathbf{u_g} \rangle^g$$

- Variable change and closure problem:

$$\mathbf{B}^\star = (\mathbf{B} + \mathbf{I}) \cdot \mathbf{C}^{-1} ; \ \mathbf{C} = \frac{\varepsilon^{-1}}{V} \int_{\mathcal{A}} \mathbf{n} \cdot [-\mathbf{bI} + \mu \nabla \mathbf{B}] \, d\mathcal{A}$$

$$\nabla \cdot \mathbf{B}^\star = 0 \text{ in } \Omega \quad (6a)$$

$$\rho_g \mathbf{u_g} \cdot \nabla \mathbf{B}^\star = -\nabla \mathbf{b}^\star + \mu_g \nabla^2 \mathbf{B}^\star - \mathbf{I} \text{ in } \Omega_g \quad (6b)$$

$$\mathbf{B}^\star + \mathbf{M} \cdot \mathbf{n} \cdot \left(\nabla \mathbf{B}^\star + \nabla \mathbf{B}^{\star T} \right) \cdot (\mathbf{I} - \mathbf{nn}) = 0 \text{ at } \Gamma_{eff} \quad (6c)$$

$$\langle \mathbf{B}^\star \rangle^g = \mathbf{C}^{-1} \quad (6d)$$

$$\mathbf{B}^\star(\mathbf{r} + \mathbf{l}_i) = \mathbf{B}^\star(\mathbf{r}) \quad (6e)$$

$$\mathbf{b}^\star(\mathbf{r} + \mathbf{l}_i) = \mathbf{b}^\star(\mathbf{r}) \quad i = 1, 2, 3 \quad (6f)$$

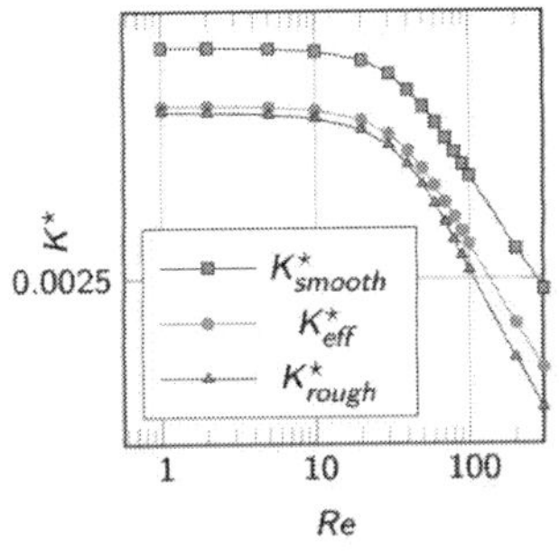

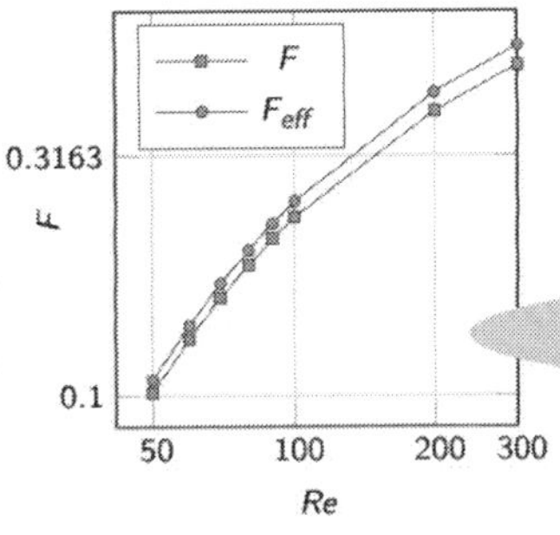

$$\langle \mathbf{u_g} \rangle = -\frac{\mathbf{K}_{eff}}{\mu_g} \left(\nabla \langle p_g \rangle^g - \rho_g \mathbf{g} \right) - \mathbf{F}_{eff} \cdot \langle \mathbf{u_g} \rangle$$

$$\mathbf{K}^\star_{eff} = \mathbf{K}_0 \cdot (\mathbf{I} - \mathbf{C}_0) \cdot [\mathbf{I} + \mathbf{F} \cdot (\mathbf{I} + \mathbf{C}_i)]^{-1}$$

Application to a packing unit-cell

- Unit cell : crossing of two packings
- A wavy profile is set up over the packing structure

$$\langle \mathbf{u}_g \rangle = -\frac{\mathbf{K}^{\star}_{eff}}{\mu_g} \cdot \left(\nabla \langle p_g \rangle^g - \rho_g \mathbf{g} \right) \quad ; \quad \mathbf{K}^{\star}_{eff} = \frac{\mathbf{K}_0 \cdot (\mathbf{I} - \mathbf{C}_0)}{\mathbf{I} + \mathbf{F} \cdot (\mathbf{I} + \mathbf{C}_i)}$$

$$\mathbf{K}^{\star}_{eff} = \mathbf{K}_0 \cdot (\mathbf{I} - \mathbf{C}_0) \cdot [\mathbf{I} + \mathbf{F} \cdot (\mathbf{I} + \mathbf{C}_i)]^{-1}$$

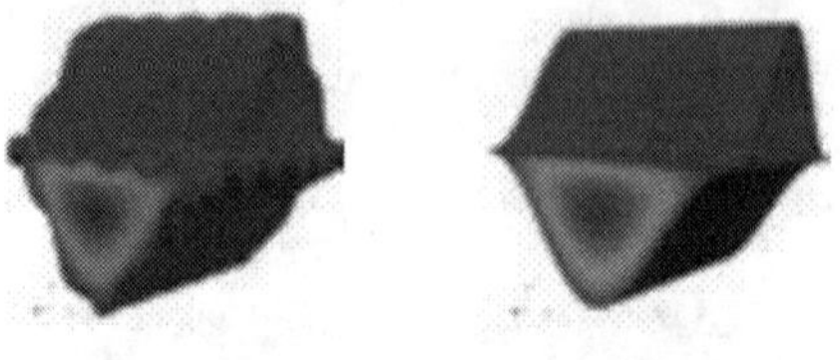

(c) rough unit-cell (d) smooth unit-cell

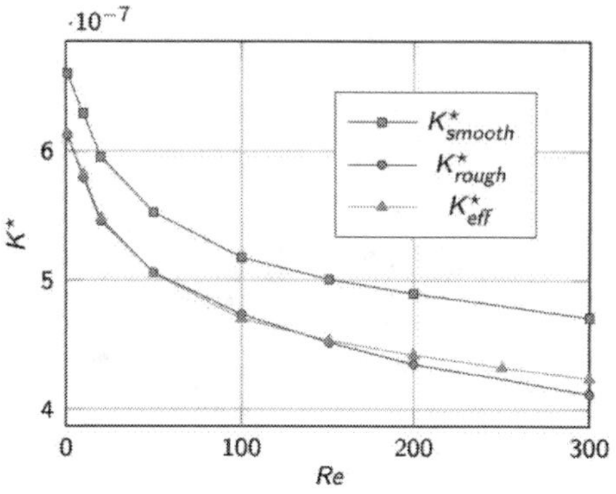

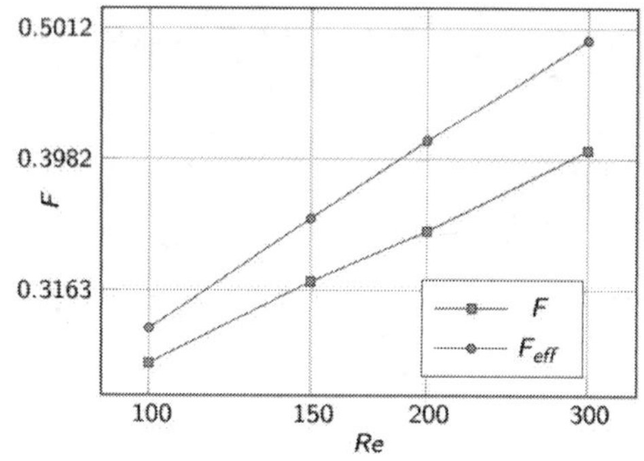

Conclusion on the "effective surface" method

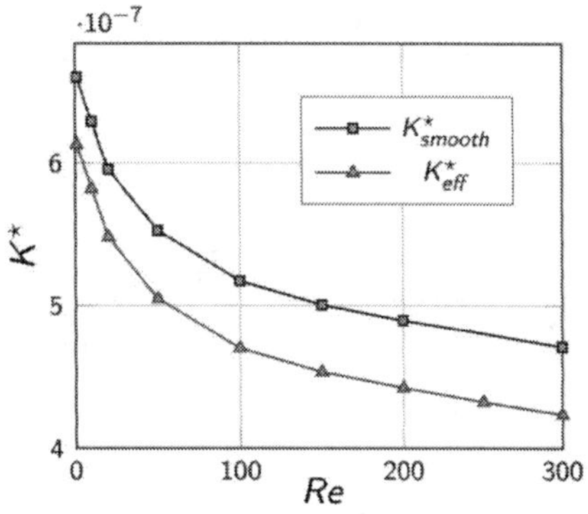

- The correction parameters permit to characterise the pressure drop induced by the rough liquid surface

$$\mathbf{K}^{\star}_{smooth} = \mathbf{K}_0 \cdot (\mathbf{I} + \mathbf{F})^{-1}$$

$$\mathbf{K}^{\star}_{eff} = \mathbf{K}_0 \cdot (\mathbf{I} - \mathbf{C}_0) \cdot [\mathbf{I} + \mathbf{F}\,(\mathbf{I} + \mathbf{C}_i)]^{-1}$$

- Anisotropic effects are taken into accound into the $\mathbf{M}|_{\Gamma_{eff}}$ tensor

- How to treat the anisotropic effects induced by the packing geometry ?

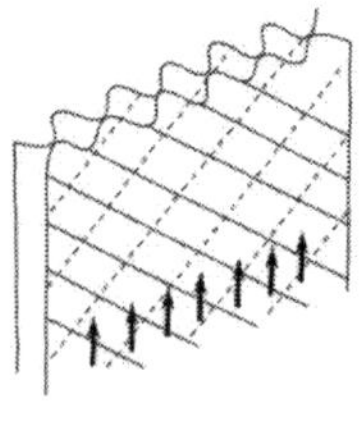

Anisotropic structure of the packings ?

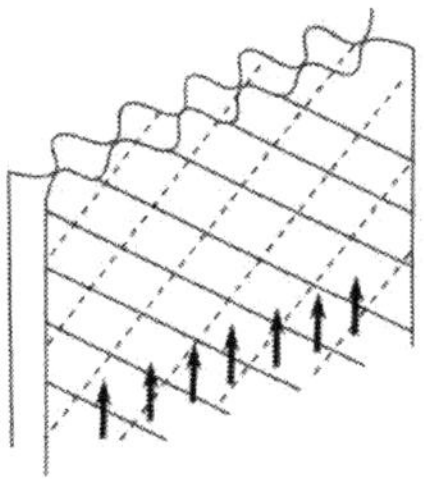

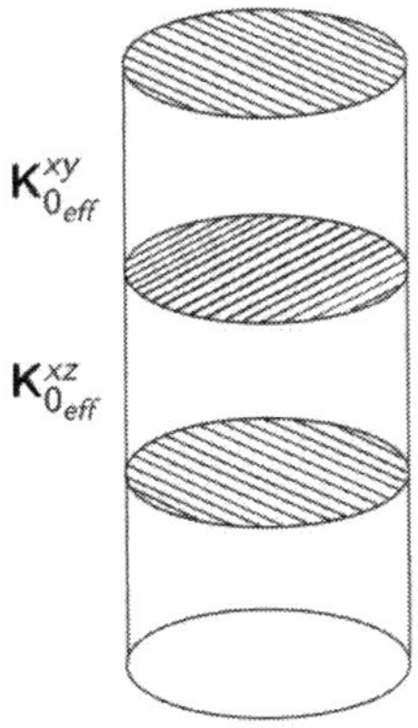

- Anisotropic form of the permeability tensor

$$\mathbf{K}_{0_{eff}}^{xy} = \begin{pmatrix} K_{xx} & K_{xy} & 0 \\ K_{yx} & K_{yy} & 0 \\ 0 & 0 & 0 \end{pmatrix} \quad \text{or} \quad \mathbf{K}_{0_{eff}}^{xz} = \begin{pmatrix} K_{xx} & 0 & K_{xz} \\ 0 & 0 & 0 \\ K_{zx} & 0 & K_{zz} \end{pmatrix}$$

depending on the pack orientation

$$\frac{\partial \varepsilon_g}{\partial t} + \nabla \cdot \langle \mathbf{u}_g \rangle = 0$$

$$\langle \mathbf{u}_g \rangle = -\frac{\mathbf{K}_{eff}^{\star}}{\mu_g} \cdot \left(\nabla \langle p_g \rangle^g - \rho_g \mathbf{g} \right) \quad ; \quad \mathbf{K}_{eff}^{\star} = \mathbf{K}_{0_{eff}}^{xy,z} \cdot (\mathbf{I} + \mathbf{F}_{eff})^{-1}$$

- High mass transfer $\dot{m}$ between the channels

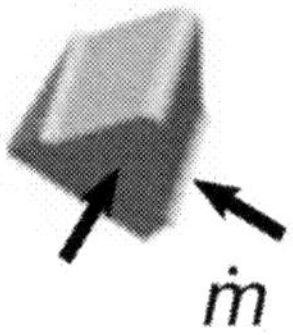

$$\mathbf{K}_{0_{eff}}^{xy} = \begin{pmatrix} K_{xx} & 0 & 0 \\ 0 & K_{yy} & 0 \\ 0 & 0 & 0 \end{pmatrix}$$

(See Soulaine and Quintard 2014)

A model for the liquid phase at the packing scale?

- For the liquid phase: separation at the macro-scale as two liquid films (Soulaine et al. 2014)

$$\frac{\partial \varepsilon_{l_1}}{\partial t} + \nabla \cdot \left\langle \mathbf{u}_{l_1} \right\rangle = 0 \quad ; \quad \frac{\partial \varepsilon_{l_2}}{\partial t} + \nabla \cdot \left\langle \mathbf{u}_{l_2} \right\rangle = 0$$

$$\left\langle \mathbf{u}_{l_1} \right\rangle = -\frac{\mathbf{K}_0}{\mu_l} \varepsilon_{l_1}^3 \cdot \left(\nabla \left\langle p_{l_1} \right\rangle^{l_1} - \rho_l \mathbf{g} \right)$$

$$\left\langle \mathbf{u}_{l_2} \right\rangle = -\frac{\mathbf{K}_0}{\mu_l} \varepsilon_{l_2}^3 \cdot \left(\nabla \left\langle p_{l_2} \right\rangle^{l_2} - \rho_l \mathbf{g} \right)$$

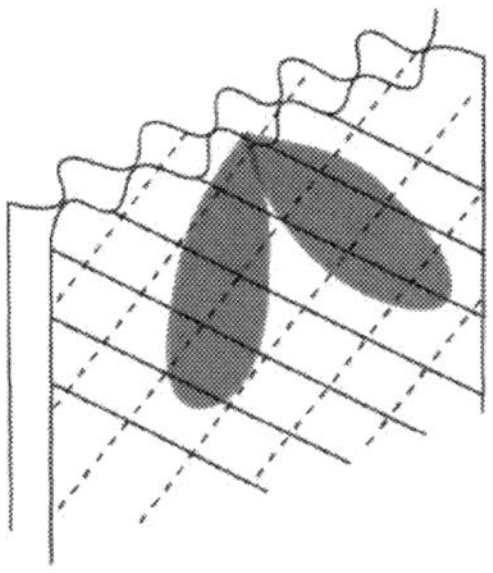

- The resulting three phases system is solved numerically using OpenFoam

Three-equation model at the packing scale

$$\left\langle \mathbf{u}_{l_1} \right\rangle = -\frac{\mathbf{K}_{l_1}}{\mu_l} \cdot \left(\nabla \left\langle p_{l_1} \right\rangle^{l_1} - \rho_l \mathbf{g} \right)$$

$$\left\langle \mathbf{u}_{l_2} \right\rangle = -\frac{\mathbf{K}_{l_2}}{\mu_l} \cdot \left(\nabla \left\langle p_{l_2} \right\rangle^{l_2} - \rho_l \mathbf{g} \right)$$

$$\langle \mathbf{u}_g \rangle = -\frac{\mathbf{K}_{eff}^{\star}}{\mu_g} \cdot \left(\nabla \langle p_g \rangle^g - \rho_g \mathbf{g} \right)$$

Conclusion

- Evaluation of the potential pressure drop in structured packings induced by the surface waves at the liquid-gas interface.

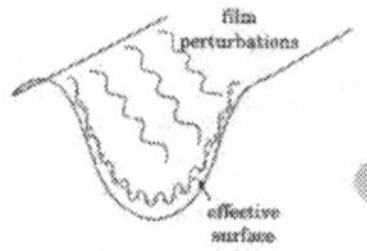

$$\langle \psi_i \rangle = \frac{1}{V} \int \psi_i dV$$

- Determination of the effective parameters in the macro-scale law for the gas phase.

$$\langle \mathbf{u}_g \rangle = -\frac{\mathbf{K}_{eff}}{\mu_g} \left(\nabla \langle p_g \rangle^g - \rho_g \mathbf{g} \right) - \mathbf{F}_{eff} \cdot \langle \mathbf{u}_g \rangle$$

- Numerical resolution of a three-equation system (gas-liquid-liquid) involving the effective parameters from the upscaling procedure.

Thank you for your attention

sylvain.pasquier@imft.fr

408463 Experimental and Theoretical Study of Multicomponent Batch Distillation

Abstract:

We present in this talk several important aspects of multi-component batch distillation.

These later include:

- Experimental implementation and results of a simple multi-component batch distillation,
- Theoretical prediction of temperature and composition evolution with time, and
- Liquid-vapor equilibrium calculations of ternary mixtures, including residue curves computations.

The selected multi-component mixture is composed of chloroform, acetone and methanol. This mixture is particularly interesting because it has several distillation boundaries and up to four azeotropes. During the simple batch distillation experiment, temperature of the vapor phase is measured and both the initial and final compositions are determined using Nuclear Magnetic Resonance (NMR). These experimental results are then compared to the theoretical calculations based on two approaches: (1) mass and energy balance approach (i.e, MESH equations) and (2) residue curve.

FUNDAMENTAL CHARACTERISTICS OF FILTRATION AND CLASSIFICATION OF FINE PARTICLES IN LIQUID BY ELECTROSPUN NANOFIBER MEMBRANE

Yasuhito Mukai, Nagoya University, Nagoya, Japan

Introduction

Recently, the nanofiber is attracting attention in the textile industry and being developed for various applications in such widespread fields as environment, power energy, electronics, medicine and biotechnology. One of the typical applications of the nanofiber material is a nanofiber membrane, and it is considered to be promising as an alternative to conventional membrane. The nanofiber membrane is expected to be a new filter for environmental purification with high separation performance and various separation functions since it has a nanosized effect, an extremely high specific surface area and a high porosity [1]. In this study, we focused on the application of an electrospun nanofiber membrane to 1. filtration and 2. classification of fine particles dispersed in liquid.

Application to Filtration

The nanofibers are applied to gas or air treatment such as an exhaust gas filter and a chemical and biological protective suit. For liquid treatment, however, the separation characteristics and mechanisms of the nanofiber membrane are poorly understood because few researches have been conducted [2, 3]. In this study, the nanofiber membranes with various fiber diameters and deposition weights were prepared, and the water permeation and particle filtration tests were performed to clarify such fundamental separation characteristics as fluid permeability, pore diameter and particle retention.

Preparation of Nanofiber Membrane

Figure 1 shows the photographs of nanofiber membranes used in this study by a scanning electron microscope (SEM). The nanofiber membranes were prepared by depositing nylon 6 nanofibers produced by electrospinning method on the surface of the base material. Nanofibers with different diameters were produced by controlling the concentration of the nylon 6 solution, and the mean fiber diameter D_f was estimated from the observed data of diameters of randomly-selected 100 fibers projected by SEM. The mean diameter D_f increased with increasing the nylon 6 solution concentration and ranged from 126 to 162 nm for the concentration of 12.5 to 25 wt%. This is attributed to an increase in viscosity under higher concentration. Also, the deposition weight w of nanofiber per unit base area was adjusted to 1.8, 2.6, 3.5 g/m^2 for each nanofiber membrane. These corresponded to the thicknesses of 29, 42, 56 μm. For reference, the mixed cellulose ester microfiltration (MF) membrane with nominal pore size of 1.0 μm was also used.

Permeation and Filtration Tests

A schematic layout of the experimental setup is shown in **Fig. 2**. The feed fluid was poured into the pressurized tank, and the membrane was set to the filter holder with a filtration area of 3.8 cm^2. The

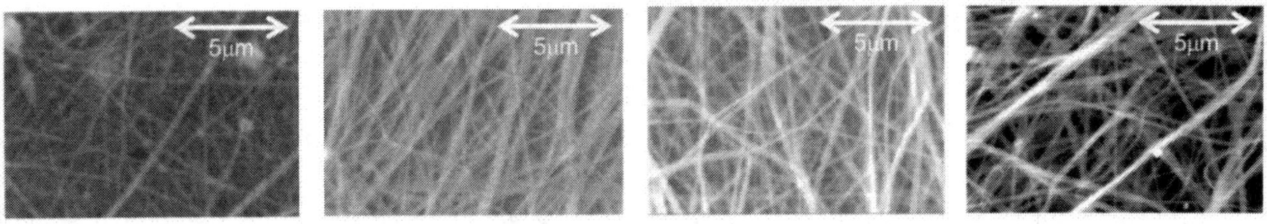

Figure 1. Photomicrographs of electrospun nanofiber membrane

water permeation and particle filtration tests for electrospun nanofiber membranes and conventional MF membranes were performed under the constant pressure p by applying compressed nitrogen gas.

In the permeation test, the ultrapure water passing through the membrane was collected in a flask and weighed in an electronic balance connected to a personal computer. Because the permeate volume v per unit membrane area was proportional to the permeation time t, the permeation rate q was obtained from the slope of the plot of v versus t.

In the filtration test, the mono-dispersed polystyrene latex (PSL) particles with uniform diameter d_p of 0.048, 0.100, 0.143 or 0.196 μm were employed. The test suspensions with the concentration s by weight of 2 ppm were prepared by dispersing PSL particles in the ultrapure water. The time variations of the PSL concentration s_f in the filtrate were measured by reading the absorbance to estimate the PSL rejection R.

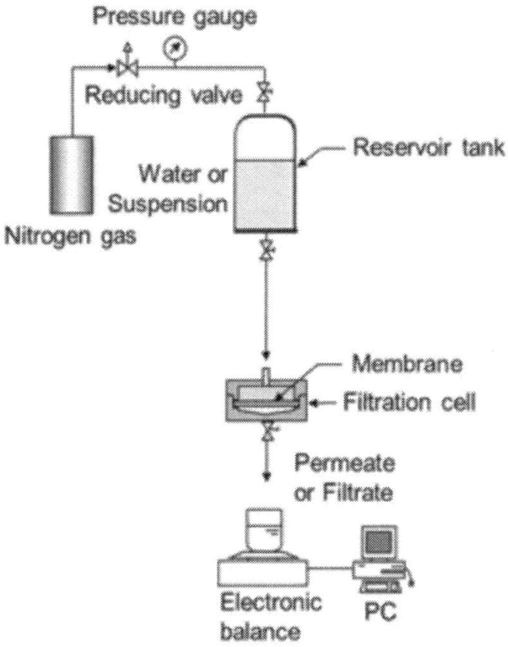

Figure 2. Experimental apparatus

Evaluation of Water Permeation Performance

The permeation tests using ultrapure water were carried out for the nanofiber membrane with D_f of 147 nm and w of 2.6 g/m^2 and the conventional MF membrane with d_m of 1.0 μm, and the permeation rates q were measured for various applied pressures p. The relationship between q and p is expressed by Darcy's equation (1).

$$q = K\frac{p}{\mu L} \tag{1}$$

where K is Darcy's permeability, μ is the viscosity of water, and L is the thickness of the membrane layer. Darcy's permeability K can be calculated from the slope of the plot of q versus p based on Eq. (1). Moreover, the pore size associated with the separation performance of membrane can be estimated as a hydraulic equivalent diameter d_e from the value of K using Eq. (2) [4].

$$d_e = 8.9\sqrt{\frac{K}{\varepsilon}} \tag{2}$$

where ε is the porosity of the membrane layer, and ε of the nanofiber layer can be estimated from the measured values of w and L.

In **Table 1**, the values of L, ε, q/p, K and d_e are summarized to compare the nanofiber membrane with the MF membrane. The distinctive characteristic of the nanofiber membrane is to be extremely thin and porous as is obvious from their values of L and ε. The permeation rate q of the nanofiber membrane is almost four times higher than that of the MF membrane although two membranes have almost the same hydraulic equivalent diameter d_e. This is also one of the major features of the nanofiber membrane, but the permeability K of the nanofiber membrane is not so large as we expected, which indicates that the main reason for the high permeation rate is the thinness of the nanofiber layer.

Table 1. Physical properties and water permeation properties of membranes

Membrane	Nanofiber membrane (D_f = 147 nm, w = 2.6 g/m^2)	Microfiltration membrane (d_m = 1.0 μm)
Material	Nylon 6	Mixed cellulose ester
L [μm]	42	150
ε [–]	0.95	0.80
q/p [m/(Pa·s)]	20.8×10^{-8}	5.1×10^{-8}
K [m^2]	8.8×10^{-15}	7.7×10^{-15}
d_e [μm]	0.86	0.87

Evaluation of Particle Retention Performance

The filtration tests for two membranes compared in the previous section were performed using the PSL suspensions under constant pressure p of 50 kPa. In **Fig. 3**, the rejections R of the PSL particles with different diameters of 0.100, 0.143 and 0.196 μm are plotted against the filtrate volume v per unit

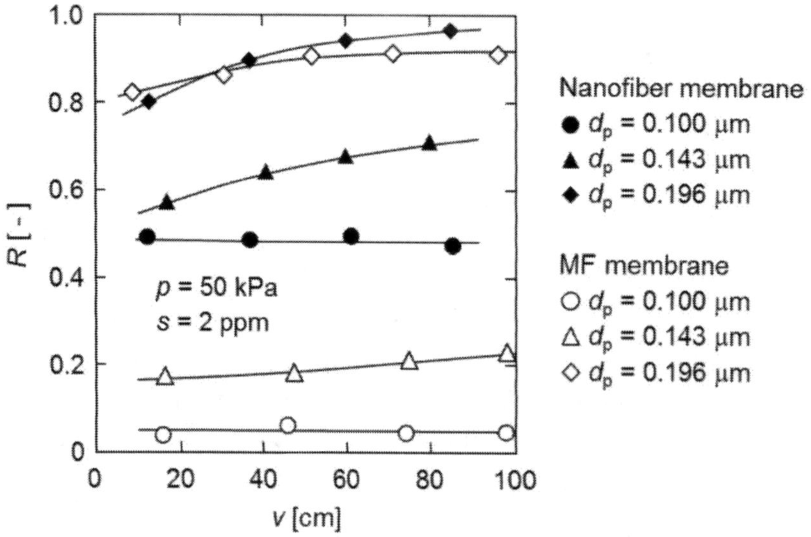

Figure 3. Time variations of PSL rejection by two membranes

membrane area. The rejection R was calculated from the PSL concentration s and s_f in the feed fluid and the filtrate by Eq. (3).

$$R = 1 - \frac{s_f}{s} \tag{3}$$

Both membranes are reasonably retentive for particles of 0.196 μm, but partially permeable for particles of 0.100 and 0.143 μm, which indicates that the critical particle diameter for these membranes can be estimated to be near 0.2 μm. This diameter is equivalent to one-fifth to one-quarter of their hydraulic equivalent diameter. In the SEM images of the nanofiber membrane during the filtration process shown in **Fig. 4**, it is observed that considerably smaller particles than the pore size are retained inside the pores. Thus, the particle retention by the nanofiber membrane progresses inside the nanofiber layer with the interruption by complex 3D network, the adhesion by extremely large surface area, and the cross-linking between particles.

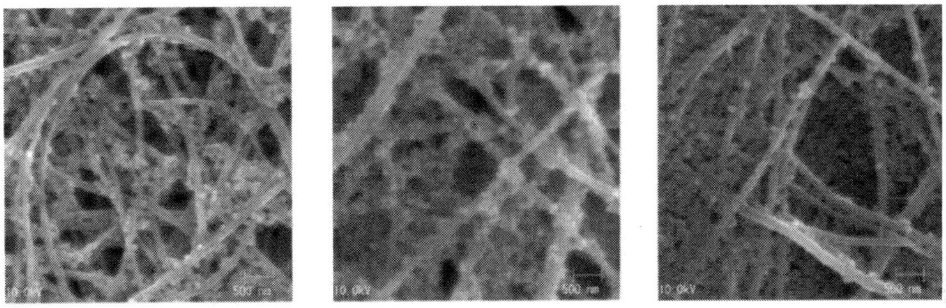

(a) v = 20 cm (b) v = 40 cm (c) v = 60 cm

Figure 4. SEM images of nanofiber membrane during filtration process

Effect of Preparation Conditions of Nanofiber Membrane

The water permeation tests were carried out using the nanofiber membranes with different fiber diameter D_f and deposition weight w. **Figure 5** shows the hydraulic equivalent diameter d_e calculated from Eq. (2) for various nanofiber membranes. The values of d_e in this figure are in the range of 0.7 to 1.2 μm and equal to about six times the fiber diameter D_f. Moreover, this figure clearly demonstrates that d_e has a tendency to decrease with decreasing D_f, which is due to an increase in the specific surface area. On the other hand, d_e is less affected by the deposition weight w of nanofiber.

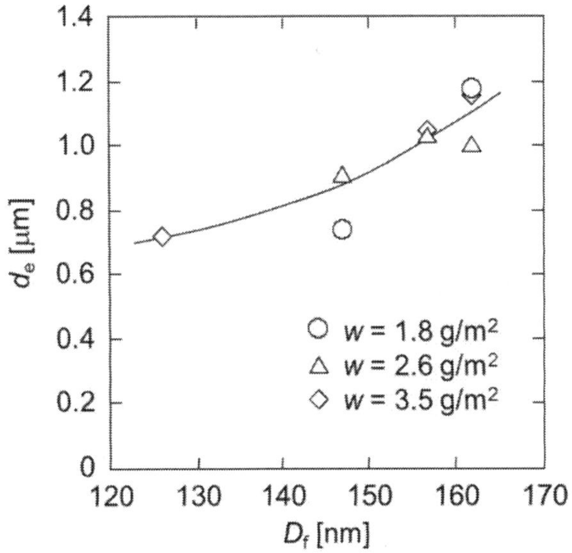

Figure 5. Relation between preparation conditions of nanofiber membrane and hydraulic equivalent diameter

In **Fig. 6**, the rejection R (the steady value at v of 100 cm) of PSL particles with d_p of 0.143 μm is plotted against the fiber diameter D_f. It is expected from Fig. 5 that the particle retention performance is improved by a reduction in the fiber diameter due to a decrease in the pore diameter. In fact, the PSL rejection tends to increase with decreasing D_f as shown in Fig. 6. On the other hand, the dependence of the particle rejection on the deposition weight w of nanofiber is not so significant as compared with the fiber diameter.

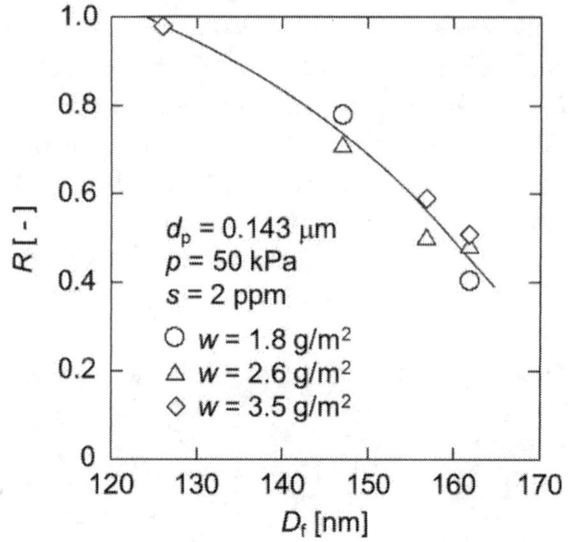

Figure 6. Relation between preparation conditions of nanofiber membrane and PSL rejection

Application to Classification

It is of importance in widely diversified fields to uniformize the particle size of particulate material, and therefore various studies on the classification methods have been conducted. One of promising methods is wet classification by using a porous membrane since the particles larger than the pore size are completely captured on the membrane and only smaller particles can pass through the membrane pores [5]. Actually, the polymer membrane is being applied to the classification of CMP slurry [6]. However, the excessive deposition of larger particles on the membrane retards the passage of smaller particles, resulting in a decline in classification efficiency. In this study, we focused on the nanofiber membrane which has complex 3D network structure and higher porosity as an alternative to conventional porous membrane. When the nanofiber membrane is applied to wet classification, it is expected that the larger particles are retained by the 3D nanofiber network and the smaller particles can pass smoothly through extremely porous space, resulting in high classification efficiency. In order to provide a fundamental understanding of the classification characteristics of fine particles through the nanofiber membrane, the fractionation experiments of the binary mixture of large and small particles were conducted under the various conditions of pH and particle concentrations.

Materials and membrane

Cross-linked polymethyl methacrylate (PMMA) particles with a diameter of 0.5 μm and 1.0 μm and an electrospun nanofiber membrane made of polyacrylonitrile with a fiber diameter of 400 nm were used for the fractionation experiment. The mean pore size of this nanofiber membrane was 1.4 μm, which was calculated from the permeation rate in the water permeation test. The feed suspension was prepared at the given small and large particle concentrations. The pH value of the feed suspension was not adjusted (pH 5.5) or adjusted to pH 11 by the addition of NaOH.

Experimental apparatus and procedure

The experimental apparatus shown in Fig. 2 was also used in this experiment. The experiment was performed under the constant pressure condition of 20 kPa. The time variations of small and large particle concentrations in the filtrate were measured.

Filtration characteristics of single component system

The time variation of the particle concentration in the filtrate obtained in filtration of each single component system was measured before the experiments of the binary mixture system. As a result, the large particles were completely retained from the beginning of filtration by the nanofiber membrane, and almost all small particles passed through the nanofiber membrane through the course of filtration.

Effect of pH on fractionation characteristics

The fractionation experiments of the binary mixture system were conducted at pH 5.5 and pH 11. **Figure 7** shows the transmittance T_p of each particle against the filtrate volume v per unit filtration area, respectively. The transmittance T_p of large or small particles is defined as the ratio of large or small particle concentration $s_{l,f}$ or $s_{s,f}$ in the filtrate to each particle concentration s_l or s_s in the feed suspension (i.e., $T_p = 1 - R$). The large particles were completely captured through the course of filtration at both pH conditions, while the transmittance of small particles at pH 5.5 tended to decrease with progression of filtration, but that at pH 11 kept almost one.

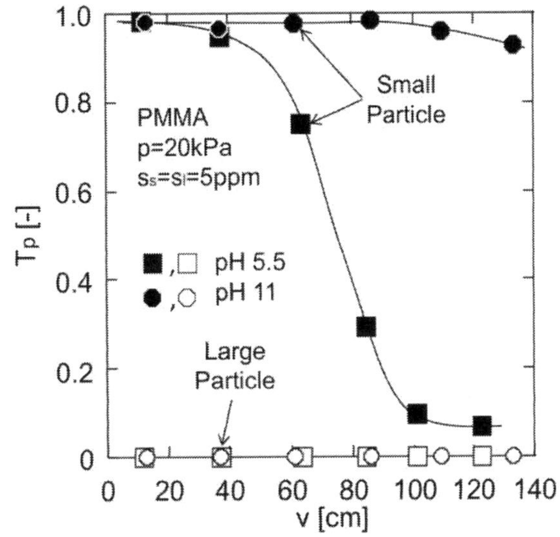

Figure 7. Transmittance of large and small particles

Figure 8 demonstrates the zeta potential of the nanofiber membrane and two kinds of PMMA particles. The zeta potential was measured by a streaming potential method for the nanofiber membrane and by an electrophoretic method for PMMA particles. Both the membrane and PMMA particles are negatively charged, and the absolute values of these zeta potentials tend to increase with an increase in pH. The difference of the fractionation performances between pH 5.5 and pH 11 can be explained based on the electrostatic interactions caused by these zeta potentials. The repulsive forces at pH 11 between similar or dissimilar particles and between the particles and the membrane become larger than those at pH 5.5, and consequently the retention of small particles by the large particle layer and the nanofiber membrane becomes less. In the following fractionation experiments, pH 11 was chosen because of better condition for fractionation.

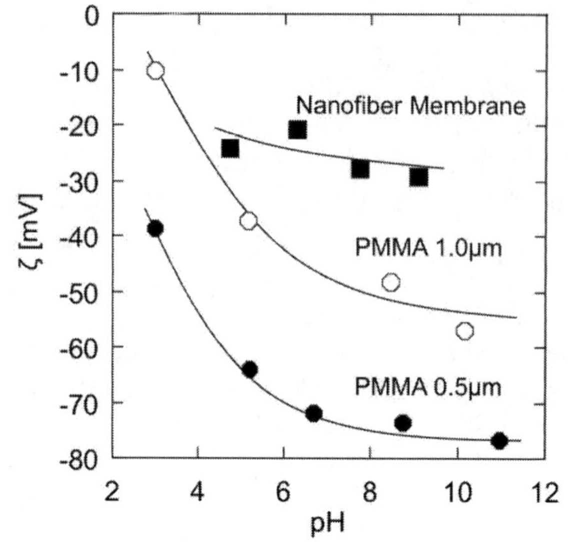

Figure 8. Zeta potential of nanofiber membrane and PMMA particles

Effect of concentration on fractionation characteristics

Figure 9 shows the time variations of the transmittance of small particles at different concentrations of large particles. The transmittance of small particles had a tendency to decline earlier with an increase in the large particle concentration. This indicates that the decline of the transmittance of smaller particles is associated with the amount of large particles trapped by nanofiber membrane.

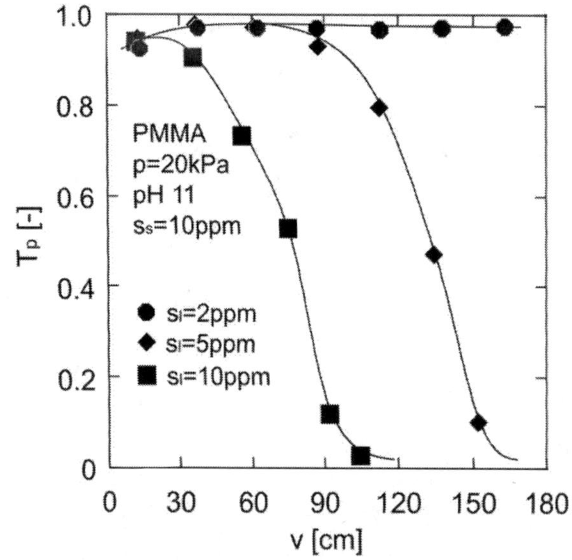

Figure 9. Transmittance of smaller particles at different concentrations of large particles

Conclusions

The electrospun nanofiber membranes were prepared and the water permeation and particle filtration tests were conducted to evaluate its separation performance. As a result, it was revealed that the nanofiber membrane has extremely high water permeation rate and is effective in the separation of the submicron-sized particles in liquid. In addition, the pore size of the nanofiber membrane decreased with a decrease in the fiber diameter, resulting in the improvement of the particle retention performance. Furthermore, it was also demonstrated that the classification performance of the fine particles by an electrospun nanofiber membrane was significantly affected by such conditions of the suspension as pH and particle concentration. It is of practical importance to exclude some factors interfering with the passage of smaller particles through the nanofiber membrane.

Acknowledgements

This work was supported in part by JSPS KAKENHI Grant Number 25420801. The author would like to express the sincere gratitude for the financial support.

References

1. Hongu, T. (2006), "Nanofiber Technology," Nikkan Kogyo Shinbun, Tokyo, pp. 26-27.
2. Yoon, K., B. S. Hsiao and B. Chu (2008), "Functional Nanofibers for Environmental Applications," *J. Mater. Chem.*, **18**, pp. 5326-5334.
3. Feng, C., K. C. Khulbe, T. Matsuura, S. Tabe and A. F. Ismail (2013), "Preparation and Characterization of Electro-spun Nanofiber Membranes and Their Possible Applications in Water Treatment," *Sep. Purif. Technol.*, **102**, pp. 118-135.
4. Furuta, M., Y. Mukai, E. Iritani and H. Nakakura (2004), "Estimation of Pore Size of Microfiltration Membrane Based on Pure Water Permeation Test," *Kagaku Kogaku Ronbunshu*, **30**, pp. 611-614.
5. Ma, W. and Y. Mukai (2012), "Basic Study on Wet Classification of Nanoparticles by Polymer Membrane," *Proc. 44th Autumn Meeting of the Society of Chemical Engineers*, Sendai, Japan, XA2P29.
6. Sumiya, M. (2008), "Improvement in Reliability of Polishing by Filtration of CMP Slurry," in *The Tribology*, Vol. 22, No. 11, Shinjusha, Tokyo, pp. 24-27.

414102 The Effect of Marangoni Currents and Gravitational Forces on the Separation of Two Different Sized Particles during Droplet Evaporation

Abstract:

When a homogenously mixed droplet with two different sized particles evaporates, the particles can separate by size into different rings with one ring for each particle size. This work shows that particle separation is significantly affected by Marangoni currents and by gravitation forces. The finite element method was used to create a transport model that tracks each particle concentration, fluid velocity, pressure, temperature, and vapor concentration. The model created is two dimensional with an axisymmetric assumption. The computational model shows that a majority of particles remain suspended in the droplet due to Marangoni currents. This is in contrast to previously published isothermal models which showed particles rapidly accumulating at the contact line. Experiments using 1 μm and 3 μm polystyrene particles support the computational results. Additional experiments with pendant and sessile droplets show that previous assumptions to neglect gravitational effects on 1 μm polystyrene spheres are incorrect. Finally, the importance of both thermal effects and gravitational forces on the separation of two different sized particles into rings is shown. This is achieved by comparing the depositions profiles for 1 μm and 3 μm particles with and without thermal and gravitational effects.

423355 Determination of Effective Diffusion Coefficient in Divergent Channel: Area Averaging Approach

Abstract:

In this contribution, the long term goal includes the use of the an area-averaging approach to obtain effective transport coefficients for the long-time regime, i..e the Taylor-Aris dispersion in Poiseuille flow in a divergent microchannel while the walls of channel are assumed to have an electrostatic potential. During the current work, electromigration coupled with diffusion and convection is study in a divergent system whereas the fluid in the geometry is assumed to be Newtonian, incompressible, and under steady state conditions. The system used here has been described by the convective-diffusive transport differential model coupled with the Laplace Equation for the electrostatic potential. Preliminary results will be presented and discussed.

413471 Characterizing the Scouring Efficiency of Fluidized Granular Activated Carbon (GAC) Particles in Membrane Fouling Mitigation Via Wavelet Decomposition of Accelerometer Signals

Abstract:

Particle fluidization, a promising unsteady-state shear means of mitigating membrane fouling in order to lower energy requirement and prolong membrane lifespan [1, 2], has garnered increasing attention in membrane-based water treatment processes. In particular, the fluidized particles facilitate back-transport in the polarization layer [3] and also play the role of mechanical scouring agents [4], the latter of which is dominant for millimeter-sized particles. To date, the hydrodynamics of the liquid-solid fluidization of the millimeter-sized GAC particles is not understood, although the understanding of the physical behavior of the liquid and solid phases is key to the fouling mitigation phenomenon. Hence, the current effort aims at bridging the gap in the knowledgebase of the dynamics of the solid particles in effecting the scouring mechanism on membranes to mitigate fouling.

In this effort, an accelerometer was successfully used to reveal the liquid-dolid fluidization dynamics of three GAC particle diameters (namely, 1.20 mm, 1.85 mm, and 2.18 mm) [5]. Specifically, because both liquid and solid phases contribute to the accelerometer signal, wavelet decomposition was used to extract information in the frequency ranges corresponding only to the solid phase [6]. The energy contained in the accelerometer signal corresponding to the solid phase was consistent with the extent of fouling mitigation in the filtration tests. Results indicate that the smallest particle diameter of 1.20 mm conferred the least scouring efficiency, while both the larger particle diameters of 1.85 mm and 2.18 mm provided similar scouring efficiency. The impact of particle diameter is due to the balance between particle inertia, and the difference between superficial liquid velocity and minimum fluidization velocity. Calculations of energy requirement indicate that, the energy requirement of all three particle diameters (d_p) were similar at lower scouring efficiency, whereas the energy requirement of the smallest $d_p$of 1.20 mm was the greatest at higher scouring efficiency.

References

1. Bixler, H.J. and G.C. Rappe, *Ultrafiltration Process.* United States Patent Office, 1970. **3,541,006**.

2. Zamani, F., et al., *Unsteady-state shear strategies to enhance mass-transfer for the implementation of ultrapermeable membranes in reverse osmosis: A review.* Desalination, 2015. **356**: p. 328-348.

3. Davis, R.H., *Modeling of Fouling of Cross-Flow Microfiltration Membranes.* Separation and Purification Methods, 1992. **21**(2): p. 75-126.

4. Meier, J., *Mechanical influence of PAC particles on membrane processes.* Journal of Membrane Science, 2010. **360**(1-2): p. 404-409.

5. Boyd, J.W.R. and J. Varley, *The uses of passive measurement of acoustic emissions from chemical engineering processes.* Chemical Engineering Science, 2001. **56**(5): p. 1749-1767.

6. Ren, J.Q., et al., *Wavelet analysis of dynamic behavior in fluidized beds.* Chemical Engineering Science, 2001. **56**(3): p. 981-988.

417946 Nonaqueous Lyotropic Ionic Liquid Crystals: Preparation, Characterization and Application in Extraction

Abstract:

NonAqueous Lyotropic ionic Liquid Crystals: Preparation, Characterization and Application in Extraction

Xianxian Liu, Qiwei Yang, Zongbi Bao, Baogen Su, Zhiguo Zhang, Qilong Ren, Yiwen Yang, and Huabin Xing*
Key Laboratory of Biomass Chemical Engineering of Ministry of Education,°°College of Chemical and Biological Engineering, Zhejiang University,°°Hangzhou 310027, P. R. China

Abstract

Ionic liquids (ILs) as unique materials have been used in various extraction processes and drawn much attention in the past decade because of their reduced environmental impact and enhanced separation selectivity. As we all know, ILs with their structural tunable property can be task-specifically designed and take more advantages than common solvents in the separation of H-bond donor solutes (such as tocopherols, oleic acid, naphthenic acid and phenol). However, the key drawback for IL-mediated extraction processes is the low extraction capacity for organic molecules due to the intrinsically strong polarity of ILs resulting from their charged structures, which seriously limits their application. In order to overcome this long-standing problem and expand the application of ILs, we designed a class of new non-aqueous lyotropic liquid crystals (LLCs) based on anionic surfactant carboxylate ILs (ASC-ILs) and developed an efficient IL-based LLC extraction process for some typical H-bond donor solutes, the distribution coefficients of these solutes, selectivity of targets solutes to typical impurities and recycle of ILs were performed through batch extraction experiments. The IL-based LLCs have self-assembled anisotropic nanostructures constituted by an ordered hydrogen-bond receptor interface and a nonpolar domain, which not only overcome the strong polarity of IL, but also provide outstanding cooperative effects between the hydrogen-bond and the van der Waals interactions. Particularly, comparisons of the extraction capacity between IL-based LLCs and other common extractants were also investigated. In addition, POM, infrared (IR) spectroscopy and X-ray diffraction (XRD) measurements were employed to investigate the interactions of ASC-IL-based LLCs and the microstructure of extraction phase. As expected, the distribution coefficients of IL-based LLCs to ¦Ä-tocopherol reached unprecedented values of 50-60 at very high feed concentrations of more than 100 mg/ml, which were 800-1000 times greater than those of common Meeting) ILs, traditional organic and polymeric extractants. Similarly, excellent extraction capacities for other H-bond donor molecules, such as oleic acid, naphthenic acid and phenol were also found with IL-based LLCs as extractants. The IL-based LLCs also demonstrate excellent separation selectivity for structure-related compounds and even homologues because of their great molecular recognition ability, which provides a new platform for the production of high-purity drugs and clean energies. In the cases of separation of ¦Ä-tocopherol to methyl linoleate, oleic acid to triolein and naphthenic acid to octane, surprisingly high selectivity were obtained, indicating the molecular recognition of the LLC extractant to these H-bond donor solutes is very accurate and exclusive. What's more, the IL-based LLC extraction process has the merit of easy recovery of solutes and simple regeneration of ILs through switching off the self-organized nanostructures of LLCs. Prospectively, IL-based LLCs, as new soft materials, successfully combine the unique properties of ILs and LCs open a new avenue for the development of high-performance extraction methods. And, we anticipate that this study will expand the application of ILs and LLCs to a wide variety of fields.

Figure 1. a) Distribution coefficient of d-tocopherol $D_{\text{¦Ä-toc}}$ in extractant¨Chexane biphasic systems at different feed concentrations C_f. The molar fraction of IL in the extractant was 0.1; b) Effect of the molar ratio of IL on $D_{\text{¦Ä-toc}}$ at a C_f of 100 mgmL_1; c) $D_{\text{¦Ä-toc}}$ values for different IL-based LLCs as extractants at C_f=100 mgmL_1. LLC1: [Ch][Lau]-acetonitrile (10:90, molar), LLC2: [Ch][Myr]-DMSO (10:90, molar), LLC3: [Ch][Pal]-DMSO (10:90, molar), LLC4: $[N_{2222}]$[Lau]-DMSO (50:50, molar), and LLC5: [Emim][Lau]-DMSO (40:60, molar).

References

1. Liu X. X.; Yang Q. W.; Bao Z. B.; Su B. G.; Zhang Z. G.; Ren Q. L.; Yang Y. W.; Xing H. B. *Chem. Eur. J.* DOI: 10.1002/chem.201500306.

2. Yang Q. W.; Xu D.; Zhang J. Z.; Zhu Y. M.; Zhang Z. G.; Qian C.; Ren Q. L.;Xing H. B. *ACS Sustainable Chem. Eng.* **2015**, *3*, 309-316.

Acknowledgements

The research was supported by the National Natural Science Foundation of China (21222601, 21476192, and 21436010), the Zhejiang Provincial Natural Science Foundation of China (LR13B060001) and the Program for New Century Excellent Talents in University of China (NCET-13-0524).

(464b) Direct Solvent Extraction with Cyanex 272 As Purifying Step for Nitric Acid Leach Solution of Nickel Laterite Ores

Ichlas, Z., Leaching and Separation Group, WA School of Mines, Curtin University
Ibana, D., Leaching and Separation Group, WA School of Mines, Curtin University
Originally presented on: 11/11/2015 8:55:00 - 9:20:00

The use of nitric acid in the leaching of nickel laterite ores by the so-called Direct Nickel process is attractive as it can be used to treat the full laterite ore profile and allows the reuse of the excess acid. It also allows the regeneration and reuse of the excess magnesia as its neutralisation agent. Clearly these are considerable distinct advantages over the conventional sulphuric acid leaching of these ores. As the use of nitric acid in leaching nickel ores is new, however, no method of separating nickel and cobalt from nitric acid solutions exists. This paper discusses the results of the work on developing a direct solvent extraction technique to separately recover cobalt and nickel directly from nitric acid solutions after the iron and aluminium precipitation steps.

The proposed technique involves two sequential solvent extraction circuits using bis(2,4,4-trimethylpentyl) phosphinic acid (Cyanex 272) in both; the first is to remove aluminium and zinc, and the second is to remove cobalt together with most of the remaining impurities. The bench-scale results showed that >95% aluminium and zinc removal can be achieved at optimised condition but significant amount of cobalt and manganese were also co-extracted. The cobalt and manganese, however, can be easily scrubbed out with acidified water and recycled back into the process. Aluminium and zinc can also be completely stripped from the loaded organic with only dilute sulphuric acid at pH 1. The bench-scale work showed that dilution of the first circuit raffinate, the addition of a modifier in the organic phase, or a combination of both is required in the cobalt and nickel separation circuit as, otherwise, significant precipitation occurs. At optimum conditions, devised as a result of this investigation, efficient cobalt and nickel separation was achieved with complete manganese co-extraction leaving only nickel, calcium, and magnesium in the raffinate. Subsequently, both are completely separated from their respective impurities with ease.

432724 Centrifugal Partition Chromatographic Process for Continuous Fractionation of Multicomponent Mixtures

Abstract:

Sequential Centrifugal Partition Chromatography (sCPC) is a new process in the portfolio of continuous liquid separation processes. sCPC is a cyclic liquid-liquid chromatographic process performed in two adjacent columns. Its particularity is that the feed mixture is introduced continuously between the columns, whereas the two product streams are collected sequentially at the opposite ends of the columns, in each step of the cycle. The components are separated based on their different partitioning between the phases of the used biphasic liquid system. One reason for the lack of industrial applications of this process is the limited number of published data on experimental demonstrations of its advantages.

In this work, sCPC is used to fractionate a multicomponent mixture of compounds with similar molecular structure into two fractions. A quaternary mixture of parabens (methyl-, ethyl-, propyl- and butylparaben) was selected as model feed mixture. The fractionation of the quaternary mixture was regarded as a separation of a pseudo-binary mixture. The approach for designing a sCPC process to separate of binary mixtures [1-3] was extended and used to select the unit operating parameters.

Three fractionation tasks were performed: (i) methylparaben was separated from the rest of parabens, (ii) methyl- and ethylparaben were separated from propyl- and butylparaben, and (iii) butylparaben was separated from the rest of parabens. Even though the separation factors of the parabens were relatively low (ca. 1.6), all desired fractions were recovered with purities higher than 99.9%.

The presented results demonstrate the potential of sCPC technology for fractionations of multicomponent mixtures containing components that are difficult to separate.

References

[1] J. Völkl, W. Arlt, M. Minceva, AIChE Journal 59 (2013) 241.

[2] E. Hopmann, J. Goll, M. Minceva, Chem. Eng. Technol. 35 (2012), 72.

[3] J. Goll, A. Frey, M. Minceva, J Chromatogr. A 1284 (2013), 59.

432993 Application of Deep Eutectic Solvents in Centrifugal Partition Chromatography

Abstract:

Centrifugal partition chromatography is a highly selective and versatile liquid-liquid chromatography technology for isolation, purification and separation of active compounds from natural resources (e.g. plants), biotech and biosynthetic products. The separation is achieved as a result of the different partitioning of the mixture compounds between two phases of a multi-solvent biphasic system. One of the phases is kept in place with the help of a centrifugal force field, while the second one is pumped through. Especially for the separation of hydrophobic mixtures there is a strong need of new biphasic solvent systems.

Deep Eutectic Solvents consist of two (or three) compounds capable of associating with each other to form a eutectic mixture with a melting point substantially lower than that of each individual compound.

In this work, the applicability of Deep Eutectic Solvents (DES) as solvents in liquid-liquid chromatography is evaluated. For this purpose, the partitioning coefficients of several natural compounds of different hydrophobicity in heptane/ethanol/DES biphasic systems are determined. The influence of the DES composition and the presence of water in the biphasic system on the partition coefficient is also examined. In addition, some process relevant biphasic system physical properties, as the density and viscosity of the phases, are evaluated.

Finally, the suitability of DESs as solvents in liquid-liquid chromatography is shown at pilot scale using a Centrifugal Partition Extractor column.

434895 Liquid-Liquid Equilibria in Aqueous Two Phase System (K2HPO4/1,3-propanediol/water/methanol) for 1,3-Propanediol Extraction

Abstract:

Biobased production of 1,3-propanediol is a good alternative to adding value to raw glycerin byproduct of biodiesel production process. Technologies used in downstream processing significantly influences the cost of capital and production. Aqueous two phase extraction (ATPE) has been reported as a suitable alternative for 1,3-propanediol extraction. This paper presents liquid-liquid equilibria (LLE) data for quaternary aqueous two phase system of water + 1,3-propanediol + methanol + K_2HPO_4 at T = 308.15 K . The experimental LLE data were correlated with NRTL-Electrolyte model. Distribution coefficients and selectivity were derived from the experimental LLE data to evaluate 1,3-propanediol extraction.

436276 Liquid-Liquid Extraction of Volatile Fatty Acids (VFAs) Using Surfactants

Abstract:

Liquid-Liquid Extraction of Volatile Fatty Acids (VFAs) using Surfactants

This research is focusing on the anaerobic digestion process, namely for VFA production. The acids of concern in this system, are acetic, propionic, and butyric acids. The main goal is to incorporate a liquid-liquid extraction stage during the digestion step to remove the produced VFAs from the resulting product. This will not only result in a VFA platform for specialty chemicals, but also prevent any product inhibition from the increase in acid concentration during the process. This requires ternary diagrams for each acid system separately, and then a final diagram for the three acid mixture system. Aspen Plus simulation is being used for theoretical modeling specifically using the UNIFAC and NRTL modeling tools. These two modeling tools are used for their interaction parameters based on the system that is being examined. As of now, the theoretical findings are promising with a rather large two phase envelope including mixtures of high water content, low acid content, and low surfactant content. Furthermore, the initial visual observations are almost identical to the theoretically modeled diagrams for the two phase envelope. The large two phase envelope will allow us to choose specific points that would fit in the process and even allow the desired in situ process. With an in situ process the system would be extracting the acids as they form, meaning that the composition of the system would be high in water and low in acids. While the initial visual observations will dictate the phase envelope line, phase compositional analysis is extremely important for tie line development making the diagram functional. Diagrams will be used to determine the surfactant amount needed to accomplish the desired extraction of the acids, based on the current system composition.

437432 Continuous Chromatography: Breaking the "Bottleneck" in Downstream Processing

Abstract:

Multi-column continuous chromatography has recently become an enabling technology that will "break the bottleneck" in downstream processing of bio-pharmaceuticals. This development is not only driven by the tremendous increases in expression levels in the upstream fermentations, especially for monoclonal antibodies, but also by the introduction of biosimilars/biobetters the market. Continuous chromatography has shown promises in reduction of manufacturing cost. However, up to today there has not been a reported case at the production scale. What are the remaining technical barriers when implementing the technology in the GMP environment? This presentation highlights how to overcome the major barriers when developing this technology platform.

418599 Design and Simulation of a Continuous Bioreactor

Abstract:

A novel continuous bioreactor consists of an outer shell cylinder and inner core cylinder, each with adjustable rotational velocities to control the mixing in the reactor and provide various shear environments. Inlet flows are controlled to provide additional medium and oxygen to the reactor volume. A spiroid is attached to the inside of the bioreactor to increase the gas-liquid contact area and improve oxygen transfer. This bioreactor has dramatic advantages in cell cultivation because of the high cell survival rate, enhanced growth rate, and overall high productivity.

Computational fluid dynamics (CFD) simulations were successfully performed to describe the range of flow patterns possible in this bioreactor. Simulations are conducted for multiphase fluids with various operating conditions to simulate the actual reaction environment in the bioreactor. Velocity profiles, shear distributions and turbulent energy dissipation were shown to predict the high performance of the bioreactor. In addition, various experiments along with mathematical models were used to calculate the volumetric mass transfer coefficients at different operating conditions for bioreactors with and without the spiroid. The comparison of the results supported the improved performance of the bioreactor with the spiroid. Cell lines were selected for cultivation in the bioreactor.

Key words: computational fluid dynamics, continuous bioreactor, oxygen transfer, cell cultivation

422745 New Synthetic Adsorptive Depth Filtration Media for the Primary and Secondary Clarification of Mab from CHO-S Harvests

Abstract:

We report the design and performance characteristics of a synthetic depth filtration media that demonstrates a greatly improved filtration capacity and decreased effluent turbidity for the clarification of mAb from CHO-S harvests. This new depth filtration media is constructed entirely from synthetic components, which include a silica-based filter aid, acrylic fibers, and a charged polymeric binder resin. When compared to commercial depth filtration media comprising mixtures of diatomaceous earth and cellulose, the synthetic depth filter media is shown to have greatly decreased organic extractibles and reduced WFI flushing requirements by TOC analysis. The synthetic depth filter media also presents a significantly increased filtration capacity versus the commercial benchmark due to the incorporation of a pre-filtration layer. As an added benefit, the silica filter aid is shown to impart a modest HCP clearance capability through an adsorption mechanism. The impact of HCP clearance on downstream process purification steps is briefly explored.

418751 Strategies for Implementation of a New CHO Platform Process

Abstract:

Platform processes are integral for the rapid development of new therapeutic molecules within the biopharmaceutical space. Since platform processes are inherently limited in their capacity for change, there are serious strategic questions related to timing and implementation of modifications to a platform. Under discussion are the implications of changing a platform process with a focus on mammalian cell culture development. This includes necessity and timing for the change, the method of exploration of new platform setpoints and conditions, qualification of the platform, and characterization of platform robustness.

A case study is presented regarding application and implementation of a new platform process. Exploratory studies were used to identify incremental improvements to an existing platform. Changes included culture feed and feed timing, gassing conditions, and seed train. When evidence was generated to support use of the new platform, a series of qualification studies on multiple established molecules were conducted to support the use of the new platform. Finally, after the platform was accepted a collection of process robustness studies were performed based on a platform risk analysis. These studies will provide the process knowledge to support a quality by design approach through both early and late stage development.

431560 The Impact of Cell Culture Media Variability on the Process Development of Biosimilars

Abstract:

Sourcing limitations and the efforts to minimize introduction of adventitious viral or prion agents in the production of biologics have over the years accelerated the adoption of serum-free cell culture media. The subsequent use of animal component free media containing plant derived hydrolysates in cell culture processes has come with observed variability in cell growth, productivity and product quality of biologics. In its purest and supposedly most consistent form, cell culture media has evolved to chemically defined media. As part of a media vendor qualification study, four proprietary media formulations manufactured by five vendors were tested for composition, particle size distribution and solubility characteristics. Media from two vendors were then used in functional cell culture testing. There was a direct correlation between the media powder particle volume mean diameter (VMD), time to solution and final media turbidity. Differences in media turbidities before and after filtration indicated particulate removal during filtration step, suggesting that some components were not going into solution and therefore unavailable to cells in the bioreactor. Differences between expected and measured levels of media components such as trace metals suggested that there were other extraneous sources of trace elements. Functional cell culture testing results showed that these variations in concentrations of media components had some product quality impact that can affect comparability. Media quality, among other process variables, needs to be controlled for consistency in cell culture process performance.

419187 Pervaporative Enrichment of 1,3-Propanediol from Model Fermentation Broths By Hydrophobic Specialty Polymers

Abstract:

While 1,3 PD can be easily separated from the other by products of bacterial metabolism, the real challenge lies in enriching it from dilute concentrations in aqueous mixtures which amounts to almost 50 – 70% of the total production cost in fermentative 1,3-PD production. The hydrophilic character of 1,3 PD compounds the difficulty of purification. Conventional and reported separation techniques have had their limitations in due cost and energy consumption and hence applicability. Membrane separation processes offer a cheap and energy efficient method of upgrading the value of specific components typically produced in dilute mixtures. Although pervaporation is an energetically advantageous process. 1,3-PD cannot be enriched by materials conventionally used for pervaporation. The development of novel materials with a good cost performance balance is essential for the establishment of industrial separation procedures for 1,3-PD enrichment from dilute broths. The present work describes a study on the pervaporation enrichment of 1,3-propanediol from model fermentation broths using three hydrophobic polymer families – functionalized siloxanes, cyclohexylamine base methacrylates and imidazolium ionic liquid based methacrylates with progressively increasing separation efficiency. While the highest 1,3-PD permeabilities were seen with the siloxane polymers, the imidazolium IL based methacrylates exhibited the best separation factors both with respect to water and glycerol. The effect of glycerol on transport coupling was evaluated. The permeability and separation factors from the pervaporation experiments with these polymers were siloxanes were collated to arrive at a Robeson's upper bound of pervaporation of 1,3-propanediol from aqueous solution by polymeric membranes. The development of novel materials with a good cost/performance balance is central to the establishment of industrial separation procedures for 1,3-PD enrichment from fermentation broths. The removal of this process bottleneck may increase the feasibility of using waste industrial glycerol as a sustainable resource for producing 1,3-PD as a biorenewable alternative to a petroleum platform.

IMPLEMENTATION OF A CUSTOMIZED GAS-SEPARATION MEMBRANE MODEL INTO COMMERCIAL FLOWSHEETING SOFTWARE TO SIMULATE A HYBRID CO_2 REMOVAL PROCESS FOR OXIDATIVE COUPLING OF METHANE

Alberto Penteado[1,2], Erik Esche[1], Günter Wozny[1]
(1) Process Dynamics and Operations Group, Technische Universität Berlin, Berlin, Germany
(2) CAPES Foundation, Ministry of Education of Brazil, Brasilia, DF, Brazil

Introduction

Ethene is one of the most produced basic chemical world-wide with a global production of 141 million tons in 2011 [1]. The oxidative coupling of methane (OCM) is a promising path for the conversion of methane into ethene, offering the possibility for utilizing natural gas, shale gas or biogas instead of oil as feedstock. It is a more direct route from methane to valuable chemicals compared to the Fischer-Tropsch process, which has synthesis gas as an intermediate [2]. The selective oxidation of methane to ethene can be seen below:

$$2CH_4 + O_2 \rightarrow C_2H_4 + 2H_2O$$

Due to the high temperatures required for catalyst activation in the OCM reaction (750 to 950 °C), products of non-selective oxidation such as carbon monoxide (CO) and carbon dioxide (CO_2) are also formed and should be removed in the separation section. The C_2 yield in the OCM reactor is not expected to exceed 28% [3], which highlights the need for an efficient downstream separation for an OCM process to become economically feasible.

The OCM process structure typically includes a reaction section, consisting of a reactor or reactor network; a compression section based on a multi-stage compressor with intermediate cooling and knock-out drums for water removal; a carbon dioxide removal section, commonly via an absorption/desorption system; and a distillation section, consisting of a demethanizer column and a C_2 splitter. This structure is reproduced in the block diagram in Figure 1 below.

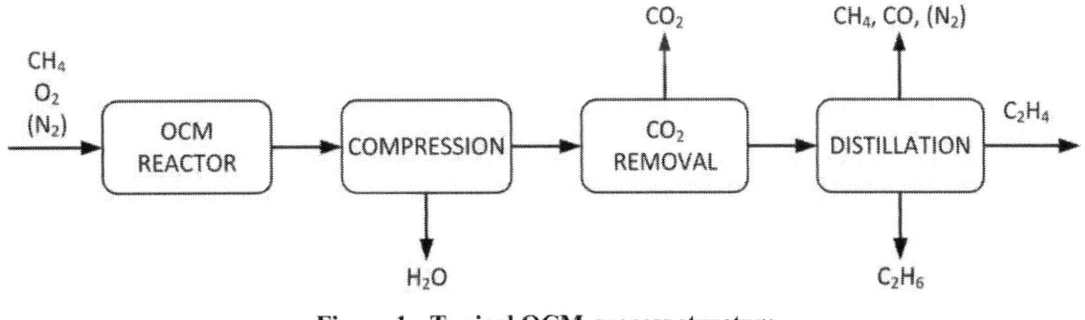

Figure 1 - Typical OCM process structure

The carbon dioxide removal is usually an energy-intensive step, since the amine solution used for the absorption needs to be regenerated in a stripping column by adding heat through the reboiler. Alternatively, a hybrid process can be employed. This consists of an upstream gas separation membrane section and a downstream absorption section. This concept has been successfully applied in a mini-plant in the Technische Universtiät Berlin, showing significant reduction in the specific energy required per kilogram of captured CO_2 [4]. In this contribution, the feasibility of applying this process at an industrial scale is assessed through simulations and preliminary cost estimations.

Process Concept

Amine solutions are often utilized for the sequestration of CO_2 in power plants or for natural gas sweetening. A similar approach is used for the OCM process. The OCM reactor outlet passes through the compression and water removal section and enters the CO_2 removal section, where it is fed to a packed column in counter-current flow with a 30 wt% Monoethanolamine (MEA) acqueous solution. The carbon dioxide is dissolved and chemically bound, producing a CO_2-free gas stream at the top and a CO_2-rich amine solution at the bottom. The rich amine solution is than flashed, pumped, and pre-heated before entering the stripper column, within which it is regenerated. A gas stream consisting mainly of CO_2 and water vapor is produced at the top of the stripper column, whilst a lean amine solution is produced at the bottom. The latter is then pumped, cooled, and recycled back to the absorption column.

In the proposed hybrid process, a gas separation membrane unit or unit cascade is added upstream of the absorption section. The carbon dioxide is partially removed in the membrane section, reducing the amine flow rate and reboiler duty in the absorption section for a fixed degree of CO_2 capture. This concept is represented in Figure 2, wherein a two-stage stripping cascade is employed.

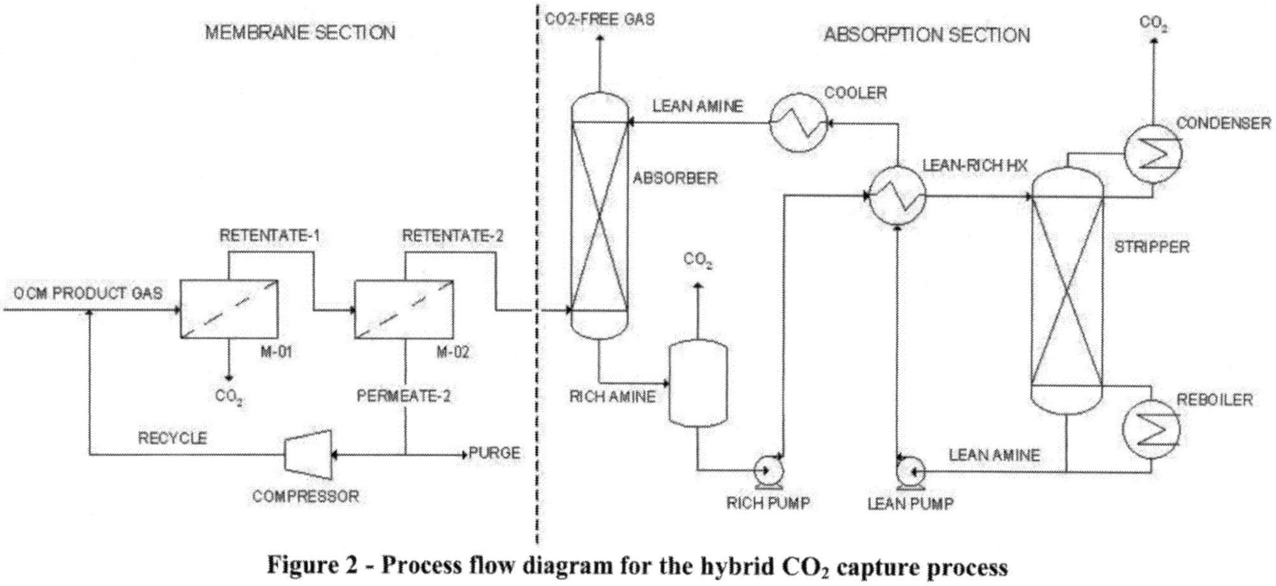

Figure 2 - Process flow diagram for the hybrid CO_2 capture process

Membrane units consist of a series of circular flat sheet membrane envelopes with segment cuts disposed in series inside a metal casting containing an inner tube for the permeate flow as seen in Figure 3. Both Polyimide (PI/Matrimid) and Poly-(ethylene oxide) (PEO) membranes are considered in this contribution. The first has higher selectivity to CO_2, whilst the second allows for a larger flow through the membrane. Membrane materials and modules are produced by coworkers at Helmholz-Zentrum Geesthacht Centre for Materials and Coastal Research, Germany, as described in [5].

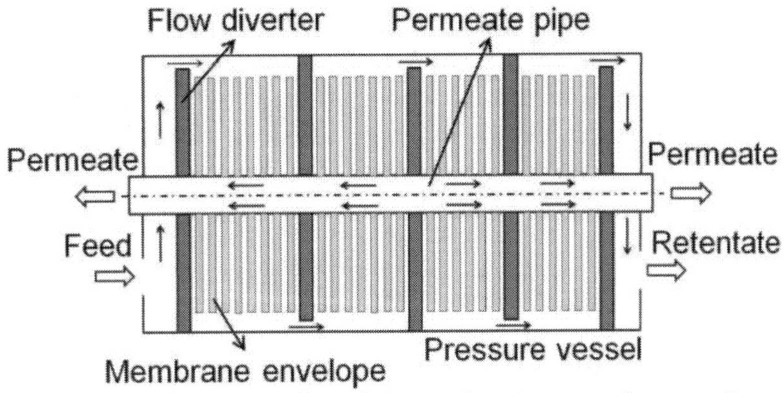

Figure 3 – Representation of the envelope type membrane unit

Modeling

The global model is divided into a membrane section and an absorption section. The model for the membrane section is implemented using the web-based modeling environment MOSAIC [6]. MOSAIC uses a document-based modeling approach to facilitate the documentation step and the exchange and reuse of models across platforms. It allows for automatic code generation for the model solution in different programming languages and environments. A code for the solution of the membrane model in Aspen Custom Modeler® (ACM) is generated and the ACM model is later exported to Aspen Plus® as a unit operation for the simulations. The absorption section is modeled in Aspen Plus® using available models from the library and from the example file [7].

Membrane Section

Equations for modeling gas permeation membranes are presented in [8]. A one-dimensional solution-diffusion model is applied and discretized using orthogonal collocation on finite elements with fourth order Lagrangian polynomials on Radau roots. Basic modeling equations are presented below: molar balances on permeate and retentate sides (Equation 1), permeance calculation for the PI membrane according to the Arrhenius law (Equation 2) and permeance calculation for the PEO membranes according to the free-volume theory (Equation 3). Further detail on modeling equations and implementation is given in [4].

$$\frac{dn_i^P}{dz} = -\frac{dn_i^R}{dz} = L_i \cdot \left(f_i^R - f_i^P\right) \cdot h \tag{1}$$

$$L_i^{PI} = L_i^{0,PI} \cdot exp\left(-\frac{Ea_i^{PI}}{R \cdot T}\right) \tag{2}$$

$$L_i^{PEO} = L_i^{0,PEO} \cdot exp\left[-\frac{Ea_i^{PEO}}{R \cdot T} + \sigma_i^2 \sum_{i=1}^{NC}\left(\frac{1}{\sigma_i^2} \cdot m_i^0 \cdot \frac{f_i^R + f_i^P}{2} \cdot exp\left(m_i^T \cdot T\right)\right)\right] \tag{3}$$

Permeabilities are obtained from co-workers at Helmholz-Zentrum Geesthacht Centre for Materials and Coastal Research, Germany [5]. Fugacitites are calculated using the Peng-Robinson equation of state on the retentate side, while ideal gas is assumed on permeate side.

Experiments carried within the mini-plant at the Technische Universität Berlin showed nearly isothermal conditions on both retentate and permeate sides, as well as negligible pressure drop and concentration polarization, therefore these effects are not taken into consideration. The model is initially used for the simulation and optimization of the mini-plant [4]. The model is then extended for the simulation of industrial size plants by using a scale-up by numbering-up approach. The inlet flow rate is divided by the design flow of the membrane module to obtain the required number of parallel modules. A single module is simulated and the resulting flows are then multiplied to the number of parallel modules to obtain the total retentate and permeate outlet flows.

Absorption Section

An unsymmetric Electrolyte NRTL model is used for the liquid phase and PC-SAFT equation of state is used for the vapor phase. Gaseous components, H_2, N_2, CH_4, C_2H_4, C_2H_6, CO, and CO_2 are selected as Henry components. The example file [7] is used as a starting point for the modeling of the absorption section. Implemented parameters from various sources are kept and the Henry constant for ethene in Monoethanolamine (MEA) is additionally estimated based on data from [9] to avoid an underestimation of ethene losses in the absorber. Columns are simulated using rate-based calculations with built-in correlations by [10] for mass transfer, by Chilton and Colburn [11] for heat-transfer and by [12] for the packing hold-ups. Kinetic parameters for the CO_2 absorption are taken from [13] and [14]. Plug flow is assumed for the vapor, while liquid is assumed to be well mixed. Column packing is Sulzer Mellapak 350X for the absorption column and 350Y for the stripping column. Other equipment (flash, pumps and exchangers) are simulated using the traditional MESH approach.

Simulations and Optimizations

Simulations are performed using Aspen Plus® version 8.6. All models are solved using the available numeric methods in the software. A sparse solver is used for the membrane modules, Broyden method is applied for converging tears in recycle calculations and SQP is used for sequential optimization of the membrane section.

Different reactor design and feeding policies for OCM are currently under investigation as described in [15] and [16]. As they impact the downstream separation processes drastically, an integrated approach is required to synthesize an efficient OCM process. In this contribution, multiple feed scenarios are considered to reflect possible reactor outlets of a plant producing 100 kton/year of ethene. All gas flows are assumed to be compressed, cooled, and dry, as if they were coming from the compression section. The feed scenario details are shown in Table 1 below.

Table 1 - Inlet gas conditions for the simulations

Feed Scenario	Temperature (°C)	Pressure (bar)	Molar Composition (%)			
			CO_2	C_2H_4	N_2	CH_4
I	35	10	27.4	15.4	34.4	22.8
II	35	10	9.0	9.0	16.7	65.3
III	35	10	24.5	4.5	8.0	63.0
IV	35	10	38.0	5.0	10.0	47.0

Scenario I reflects the use of nitrogen diluted flow in the reactor. Scenario II considers no dilution, but the use of a high methane to oxygen ratio instead. Scenarios III and IV consider carbon dioxide dilution, which has a positive impact on the C_2 selectivity and helps to control the high exothermicity of the OCM reaction network. For scenarios II, III, and IV, in which no nitrogen is present, other components (H_2, CO, and C_2H_6) are simulated as N_2, given that no permeances are available for those.

All simulations aim at the removal of 97% of the CO_2 content in the inlet gas. A base case consisting of a standalone absorption process is simulated for each scenario for the latter comparison with hybrid processes. Due to the size and complexity of the system, no actual optimization could be performed in the absorption section. Therefore, optimal design is carried in the form of extensive simulations and sensitivity analyses. A specific cost per kilogram of captured CO_2 is then calculated.

The membrane section is simulated separately. An objective function consisting of a simplified operational cost is used for its optimization. It comprises the compression cost, cost attributed to ethene losses and the cost for removing the remaining CO_2 through absorption at a fixed price calculated in the base case. Flowsheets containing a single PI membrane and a stripping cascade of PI and PEO membranes are considered. The decision variables are the membrane materials, the membrane areas and the recycle/purge ratio in the case of membrane cascades. Bad convergence for the recycle calculations limits the steps of the optimization solver. Many starting values are tested leading to different local optima. The downstream absorption section is then simulated only for the cases presenting a simplified operational cost lower than the base case.

Utility consumption and cost rates are calculated inside Aspen Plus®, allowing for the operational cost to be quickly assessed during the flowsheeting stage. Utility costs are set to their default values. Apart from the utilities, costs are also attributed to the ethene losses during the process, based on current market prices published on [17]. These costs are shown in Table 2 below.

Table 2 – Base costs for economic calculations

Type	Specific Cost	Unit
Cooling Water	2.12×10^{-7}	\$/kJ
Electricity	0.0775	\$/kWh
Steam	2.2×10^{-6}	\$/kJ
Ethene Loss	1.00	\$/kg

Cooling water is used in the condenser of the stripping column, amine solution cooler, and interstage coolers of compressors. Electricity is used in the amine circulation pumps and gas compressor. Steam is used in the reboiler of the stripping column. Ethene is lost via the permeate stream, the purge stream, and the absorber's bottom stream. The total operational cost (OPEX) is calculated as the sum of all applied utilities and the ethene losses.

Results

Table 3 shows the results of the cost estimations for the standalone absorption and for the hybrid processes for each feed scenario. It consists of the operational cost and the specific operational cost per ton of removed CO_2 and the membrane areas per module.

Table 3 - Operational costs for different feed scenarios

Scenario		OPEX (US$/h)	Specific OPEX (US$/ton$CO_2$)	PI area per module (m^2)	PEO area per module (m^2)
I	Standalone absorption	483.56	14.78		
	Single PI module	517.24	15.81	0.1	
	PI + PEO cascade	695.65	21.26	0.5	0.1
II	Standalone absorption	241.35	13.17		
	Single PI module	301.29	16.44	0.5	
	PI + PEO cascade	454.03	24.78	0.5	0.1
III	Standalone absorption	978.37	9.81		
	Single PI module	**924.75**	**9.27**	**1.0**	
	PI + PEO cascade	1132.32	11.35	2.3	0.1
IV	Standalone absorption	1456.66	10.46		
	Single PI module	**1353.27**	**9.72**	**2.5**	
	PI + PEO cascade	1548.33	11.12	2.0	0.1

For feed scenarios I and II, the cost for operating the hybrid process with a single PI module is higher than that of the standalone absorption. Even though the specific energy per kilogram of CO_2 removed and total utility costs are reduced, the total operational cost of the hybrid process is still higher because ethene losses are higher in the membrane process compared to the absorption. When utilizing CO_2 diluted streams, as in scenarios III and IV, the hybrid process with a single PI module becomes economically attractive. The higher CO_2 content in the feed stream increases the driving force for the separation in the membranes. Furthermore, the greater amount of CO_2 to be captured requires a larger amine recirculation rate, which also increases ethene loss in the absorption section up to the point that they become comparable to those in the membrane section. Economically feasible scenarios were found by applying a single PI module to remove carbon dioxide from feed scenarios III and IV.

The utilization of stripping membrane cascades intends to increase carbon dioxide removal for a fixed ethene loss, or reduce ethene losses for a fixed carbon dioxide removal. However, optimizations showed that the use of such membrane cascades is costlier than the use of single membrane modules or even the standalone absorption for the given application. This can be explained by the high cost of recompressing the permeate gas stream for recycle, or purging the gas stream containing ethene.

The diameters of the absorber and stripper do not change significantly between the standalone absorption and hybrid process with a single PI module for cases I and II. For cases III and IV the columns for the hybrid process have a smaller diameter as the total gas flow is significantly reduced by partially removing the CO_2 in the membrane section. Investment cost for the membrane material, which is to be replaced over time, is relatively small compared to the cost of the carbon steel casting. These are both relatively small if compared to the cost of the absorption and stripping columns with structured packing in stainless steel. Thus, the hybrid process also has the potential to offer a reduction in the capital investment costs.

Conclusions

MOSAIC is a convenient platform for documentation-based modeling of unit operations for subsequent export and integration into flowsheets. It allows the user to concentrate fully on the modeling activity, avoiding typographic mistakes and difficulties in model reuse, which are typical for coding in different environments. In this contribution, a gas separation membrane module is successfully exported from MOSAIC and integrated into Aspen for simulations and optimizations.

While very attractive for applications like natural gas sweetening or biogas enrichment, the utilization of gas permeation membranes for the OCM process is highly sensitive to product (ethene) losses. Hybrid processes employing a single PI membrane offer an economic advantage for OCM when CO_2 dilution is used in the reactor. Applying gas permeation membranes in this case leads to a significant reduction in the operational costs. For scenario IV, this reduction amounts to 103.39 US$/h. The use of CO_2 dilution is especially interesting for combining OCM with Dry Reforming of Methane as described in [18].

Outlook

A more effective and efficient approach for process optimization needs to be developed. The use of simplified models in an initial stage, especially in the absorption section, could ease the optimization and pave the way for the subsequent implementation of more rigorous models. Despite being a convenient platform for flowsheeting and preliminary cost analysis, Aspen Plus® is far from being a powerful optimization tool. The use of other tools needs to be considered for carrying a proper process optimization, notably simultaneous optimization, for the entire system. To this end, MOSAIC offers a built-in connection with the NEOS Server [19], which is a web-based service for optimization using state-of-the-art solvers.

Capital investment estimations need to be performed for the most promising scenarios in order to properly assess the advantages of the hybrid process. Cost of membrane units and membrane material replacement are yet to be estimated, and both negatively impact the hybrid process. More detailed economic analyses need to be performed, including the determination of the return on investment and payback times.

Finally, it is necessary to evaluate the operational and capital investment costs of the OCM process as a whole to assess its feasibility. Alternative unit operations like membrane reactors and pressure and temperature swing adsorption need to be considered in the OCM process superstructure. The integration of OCM with other processes as ethane cracking and methane dry reforming also need to be properly assessed.

Acknowledgement

Present work is developed with the support from CAPES, Coordination for the Improvement of Higher Education Personnel – Brazil (Grant Number BEX 11946/13-0). Financial support by the Cluster of Excellence "Unifying Concepts in Catalysis" (DFG EXC 314) is gratefully acknowledged.

Nomenclature

Latin letters

Ea	activation energy	kJ/mol
f	fugacity	Pa
h	membrane height	m
m^T	parameter in free-volume theory	K^{-1}
m^0	parameter in free-volume theory	Pa^{-1}
n	molar flow	kmol/s
z	membrane length	m
L	permeance	kmol/(m^2 s Pa)
R	universal gas constant	kJ/kmolK
T	temperature	K
y	molar fraction	kmol/kmol

Greek letters

σ	Lennard-Jones molecular diameter	Å

Superscripts

P	permeate
R	retentate
0	reference condition
PI	polyimid membrane
PEO	poly-(ethylene oxide) membrane

Subscripts

i	component index

References

[1] W. True, "Global ethylene capacity continues advance in 2011," *Oil & Gas Journal,* vol. 110, no. 7, 2012.

[2] A. T. Ashcroft, A. K. Cheetham, J. S. Foord, M. L. H. Green, C. P. Grey, A. J. Murrell and P. D. F. Vernon, "Selective oxidation of methane to synthesis gas using transition metal catalysts," *Nature,* vol. 344, pp. 319-321, 1990.

[3] Y. S. Su, J. Y. Ying and W. H. Green Jr., "Upper bound on the yield for oxidative coupling of methane," *Journal of Catalysis,* vol. 218, no. 2, p. 321–333, 2003.

[4] E. Esche, D. Müller, S. Song and G. Wozny, "Optimization during the process synthesis: enabling the oxidative coupling of methane by minimizing the energy required for the carbon dioxide removal," *Journal of Cleaner Production,* vol. 91, pp. 100-108, 2015.

[5] T. Brinkmann, C. Naderipour, J. Pohlmann, J. Wind, T. Wolff, E. Esche, D. Müller, G. Wozny and B. Hoting, "Pilot scale investigations of the removal of carbon dioxide from hydrocarbon gas streams using poly(ethyleneoxide)–poly(butyleneterephthalate) PolyActiveTM) thin film composite membranes," *Journal of Membrane Science,* vol. 489, p. 237–247, 2015.

[6] S. Kuntsche, H. Arellano-Garcia and G. Wozny, "MOSAIC, an environment for web-based modeling in the documentation level," *Computer Aided Chemical Engineering,* vol. 29, pp. 1140-1144, 2011.

[7] Aspen Technology, Inc., "Rate-Based Model of the CO2 Capture Process by MEA using Aspen Plus," Burlington, MA, USA, 2014.

[8] K. Ohlrogge und K. Ebert, Membranen: Grundlagen, Verfahren und industrielle Anwendungen, Weinheim, Germany: WILEY-VCH Verlag, 2006.

[9] J. J. Carroll and A. E. Mather, "A model for the solubility of light hydrocarbons in water and aqueous solutions of alkanolamines," *Chemical Engineering Science,* vol. 52, no. 4, pp. 545-552, 1997.

[10] B. Hanley and C.-C. Chen, "New Mass-Transfer Correlations for Packed Towers," *AIChE Journal,* vol. 58, no. 1, pp. 132-152, 2012.

[11] R. Taylor and R. Krishna, Multicomponent Mass Transfer, New York: Wiley, 1993.

[12] J. L. Bravo, J. A. Rocha and J. R. Fair, "A comprehensive model for the performance of columns containing structured packings," *I. ChemE. Symp,* vol. 129, no. A439, 1992.

[13] H. Hikita, S. Asai, H. Ishikawa and M. Honda, "The Kinetics of Reactions of Carbon Dioxide with Monoethanolamine, Diethanolamine, and Triethanolamine by a Rapid Mixing Method," *Chem. Eng. J.,* vol. 13, pp. 7-12, 1977.

[14] B. R. Pinsent, L. Pearson and F. J. W. Roughton, "The Kinetics of Combination of Carbon Dioxide with Hydroxide Ions," *Trans. Faraday Soc.,* vol. 52, pp. 1512-1520, 1956.

[15] E. Esche, H. Arellano-Garcia, G. Wozny and L. T. Biegler, "Optimal Operation of a Membrane Reactor Network," *Computer Aided Chemical Engineering,* vol. 31, p. 1321–1325, 2012.

[16] H. R. Godini, S. Jašo, S. Xiao, H. Arellano-Garcia, M. Omidkhah and G. Wozny, "Methane Oxidative Coupling: Synthesis of Membrane Reactor Networks," *Ind. Eng. Chem. Res.,* vol. 51, no. 22, pp. 7747-7761, 2012.

[17] A. Platts.com news & Analysis Feature, "www.platts.com," PLATTS McGraw Hill Financial, [Online]. Available: http://www.platts.com/news-feature/2014/petrochemicals/pgpi/ethylene. [Accessed 2015 07 07].

[18] H. R. Godini, S. Xiao, M. Kim, O. Görke, S. Song and G. Wozny, "Dual-membrane reactor for methane oxidative coupling and drymethane reforming: Reactor integration and process intensification," vol. 74, pp. 153-164, 2013.

[19] Wisconsin Institutes for Discovery at the University of Wisconsin in Madison, "NEOS Server," [Online]. Available: http://www.neos-server.org/neos/. [Accessed 17 09 2015].

434107 Membrane-Based Process for the Continuous Enzymatic Saccharification of Lignocellulosic Biomass

Abstract:

We are currently evaluating the feasibility of performing continuous enzymatic hydrolysis of lignocellulosic biomass to product sugars using a membrane-assisted reaction/separation process. The overarching technical goals are to continuously remove the sugars—this lowers product feedback inhibition—retain and recycle active enzyme, and continuously recover the co-product of lignin. Experimental data is being collected to support preliminary technoeconomic analysis of the speculative process designs versus the current state-of-the-art batch enzymatic hydrolysis. Prospective savings are anticipated due to higher time-averaged productivity from capital equipment investments, and lower enzyme usage. Our initial studies included characterizing cake-layer resistances for varying slurry solids loadings during crossflow filtration. These results highlighted the fact that the membrane's nominal permselective properties are less important than its module configuration and ease-of-cleaning because the cake formation controls the flux and fine tunes the selectivity. Thus, our current studies include two aspects. Firstly, we are coupling tubular ultrafiltration membranes to a continuous stirred-tank reactor and continuously removing the sugars and recycling the bulk of the water back to the reactor after concentrating them. Overall reaction kinetics and yield are being assessed and compared with classical batch hydrolysis results. Secondly, we are identifying sufficient protocols for in-situ membrane cleaning and duty cycling.

435012 Recovery of Nutrients from Swine Wastewater Using Ultrafiltration: Applications for Microalgae Cultivation in Photobioreactors

Abstract:

Algal biomass has been identified as a promising feedstock that could be used in a number of industrial applications, including biofuel production, aquaculture, and pharmaceutical production. Through the use of sunlight and CO_2, microalgae species are capable of producing biomass rich in lipids and carbohydrates, which can be extracted from the plant material and used in the production of commodities such as biofuels. Algae have been shown to have higher rates of productivity and higher lipid contents than traditional bioenergy crops, and do not require high-quality land for cultivation, and would not compete with current agricultural products for space.

While the high productivity and lipid content of microalgae make it a promising feedstock for a number of industrial applications, there are several challenges associated with their cultivation in large-scale production systems. The production of microalgae can require significant amounts of nitrogen and phosphorus fertilizer. In order to make the production of microalgae economical, it is important that low cost sources of nitrogen and phosphorus be available to producers. Considering the relatively high nitrogen content of microalgae (4 to 8 percent), the nitrogen inputs required for microalgae cultivation can be significant. In addition, the synthesis of nitrogen fertilizer produces around 2 kg of CO_2 kg^{-1}. The additional carbon production, if added to the overall lifecycle, undermines the favorable carbon balance for microalgae feedstocks.

Membrane separations technology constitutes an alternative method for the removal of biological contaminants from the nutrient-rich wastewater prior to use in microalgae cultivation. A membrane consists of a thin film that separates two phases, and can be used in liquid-liquid separations. Microfiltration and ultrafiltration involve the use of porous membranes to remove micro- or macro-particles from a solution. In this study we used ultrafiltration membranes for removing contaminants from swine wastewater for use in algal growth. Through several trials we have found that the algal wastewater is completely contaminant free and the growth rate of the algae is greater than optimal conditions.

433601 Use of Forward Osmosis in Treatment of Hyper-Saline Produced Water

Abstract:

Oilfield produced water forms the largest waste stream in the production of petroleum. The quality of produced water varies greatly, with some waters being close to potable while others are oily or extremely saline. Hyper-saline produced water is an exceptionally challenging problem as there are few treatment options available. Desalination of such hyper-saline water is extraordinarily expensive because we are limited to evaporative processes which have enormous energy requirements. Conventional desalination technologies, like reverse osmosis are unusable due to the high osmotic pressure which require exceptionally high hydraulic pressure to overcome.

Forward osmosis circumvents this issue by using osmotic pressure instead of hydraulic pressure. Using a high concentration draw solution, osmotic potential can force osmotic flow from even the highest salinity solutions. The draw solution can then be subsequently regenerated (or used) by another process chosen specifically for that draw solute. While forward osmosis will not use less energy than other technologies, it does have the capability of using lower quality energy (low temperature thermal) or operating with other advantages (low fouling propensity, for instance). FO achieves this while relying simply on an osmotic potential gradient generated by a concentrated draw solution across a semi-permeable membrane. In this work, the feasibility of using FO in treatment of produced water with TDS of about 240,000 ppm was studied. Both an asymmetric cellulose triacetate (CTA) and thin film composite (TFC) forward osmosis (FO) membrane from Hydration Technologies Innovation (HTI) were used for this study. Synthetic produced water was used as a feed solution to simulate the composition of the oilfield produced water. We evaluated the individual effect of the different salts in produced water (NaCl, $CaCl_2$, $MgCl_2$, $MgSO_4$, $NaHCO_3$, and $FeCl_3$) on the flux performance. Two types of draw solutions were used in this study: 6M ammonia-carbon dioxide draw solution and 4.8M magnesium chloride draw solution. Results show that although NH_3-CO_2 has the advantage of being economically separated and recycled to the FO process, it has high reverse solute flux across the membrane. This reverse solute flux can also have unforeseen effects such as the formation of low solubility salts that cause scale or a pH change (in this case, a rise), which can also impact the solubility of salts. While ammonia is a key solute that crosses the membrane, carbonate can also precipitate readily if it diffuses. The use of $MgCl_2$ showed promise as it gave higher water flux with less scaling, although the recovery method (membrane distillation or evaporation) would have greater energy demands and higher cost.

425116 Membrane Distillation for Separating Water from Ionic Liquid Solutions

Abstract:

Ionic liquids, salts molten at temperatures below 100 °C, have exciting potential as "green" recyclable solvents for industrial processes. Ionic liquids have been found that can dissolve the lignin from biomass at relatively low temperatures. The pretreated biomass is then ready for hydrolysis and fermentation to produce biofuels. Water is added to the ionic liquid solution to surround the ions so that the dissolved lignin precipitates out. This water must be removed prior to recycling the ionic liquid for pretreatment. Two ionic liquids known to be effective in biomass pretreatment are 1-ethyl-3-methylimidazolium acetate and 1-ethyl-3-methylimidazolium formate. These were diluted with water and their conductivities and densities measured at varying concentrations. Membrane distillation of the ionic liquid – water mixtures was performed using two different well-characterized PTFE membranes. The feed was maintained at 55 °C and the distillate at 20 °C. The feed and distillate were circulated counter-currently on opposite sides of the membrane at 1.5 L/min. Water flux obtained was ~10 $L/m^2 \cdot h$. Concentration of IL in the feed concentrate and distillate was determined by conductivity and density measurements. The membrane distillation process provided a concentration factor of 10 (from 5% IL to 50% IL) with near 100% rejection; as no ionic liquid was detected in the final distillate. Membrane distillation is a promising technique for re-concentrating ionic liquids for recycling.

441185 Investigation of water treatment alternatives in the removal of microcystin-LR

Abstract:

On August 2, 2014, the greater Toledo area woke up to a Do Not Drink or Boil Water Advisory. The advisory was due to the presence of a cyanotoxin (algal toxin) produced by cyanobacteria in Lake Erie called microcystin-LR in the drinking water supply that has a WHO provisional guideline of 1 mg/L. Upon entering the City of Toledo Collins WTP crib at the Lake Erie intake, potassium permanganate is added to the water for mussel control. While potassium permanganate is needed to control mussels, it lyses cyanobacteria cells, releasing algal toxins to the water. The water is then pumped nearly three miles to the Low Service Station, where powdered activated carbon (PAC) is added to the water for taste and odor control, and the water is transported approximately six miles to the WTP (High Service Station). PAC is generally effective for removal of algal toxins through adsorption onto its surface. At High Service, alum, lime and soda ash are added to the water for coagulation-flocculation, softening, and removal of metals. The water is then sand filtered, carbonated and chlorinated before being sent to the distribution system. However, traditional physicochemical water treatment processes, such as coagulation-sedimentation-filtration, have been shown to only be partially effective for the removal of whole algal cells and not effective for the removal of algal toxins. Furthermore, chlorination is effective for oxidizing algal toxins at relatively high free chlorine concentration as long as the pH is below 8; however, for corrosion control, Toledo water is kept at a pH above 9. Therefore, the treatment process was not enough of a barrier to prevent microcystin-LR from entering the drinking water supply. By August 4, the water treatment plant increased its PAC dosage by nearly four times, and while the toxin was removed, a significant amount of PAC sludge was produced and the cost of PAC addition was unsustainable. The objective of this project is to study alternative water treatment processes, namely membrane separations, biofiltration and ozonation, for the effective removal of algal toxins.

422768 Functionalized Carbon Nanotube Nanocomposite Membranes for Water Desalination

Abstract:

Cost-effective purification and desalination of water is a global challenge. Reverse osmosis (RO), the current method of choice, is a membrane-based semipermeable process that separates salts from water, and it is limited by the intrinsic properties of the membranes used in the process. Biofilm fouling on the surface of the membranes is another critical problem in RO applications. In this study, we utilized zwitterion functionalized carbon nanotubes (CNTs) and incorporated them in polyamide thin film composite membranes to improve membrane permselectivity and surface biofouling resistance for water desalination. The CNTs within the polyamide layer were partially aligned via a vacuum filtration step during membrane synthesis, and the fabrication of the zwitterionic CNT/polyamide nanocomposite membranes was optimized by varying the loadings of CNTs and the crosslink density of the polyamide. The permeation flux was found to increase by approximately three-fold as the fraction of CNTs was increased from 0 to 20 wt% for all the salt solutions tested in our study. The steric hindrance to transport due to the zwitterionic functional groups prevents the permeation of the majority of ions through the CNT pores resulting in high salt rejection. The charges of the zwitterionic groups do not appear to produce a significant Donnan potential to reject ions, and thus surface charge had no effect on membrane selectivity. The semi-aligned zwitterionic CNTs appeared to have their tips exposed at the membrane surface and interacted with the feed water to form a strong hydration layer, resulting in improved surface biofouling resistance. The adsorption rate of protein foulants on the nanocomposite membrane surface was significantly reduced compared to the control membrane without CNTs, and the adsorbed fouling layer could be easily removed by flushing with water. After washing, the nanocomposite membrane recovered 100% of the decreased water flux whereas the control membrane only recovered 10% of the decreased flux resulting in a permanent loss of 30% in water permeation.

442568 Influence of Nanoparticle Processing and Additives on PES Casting Solution Viscosity and Cast Membrane Characteristics

Abstract:

Nanoparticle incorporation into porous polymeric membranes has received continued attention with the concurrent continued development of reactive metallic and metal oxide nanoparticles for water treatment applications. Nanoparticles comprised of metals such as iron, nickel, palladium, silver, aluminum, and copper have been shown to remove both organic and inorganic water contaminants, and novel multi-metallic nanoparticle combinations and morphologies continue to be reported. Iron nanoparticles (Fe NPs) remain at the forefront of efforts focused on water remediation due to the low cost of the metal and the versatility of Fe NPs in addressing a wide variety of water contaminants. The incorporation of these nanoparticle materials into a support material such as a polymeric membrane immobilizes the nanoparticles, preventing movement within the treatment system, and may also provide enhanced contaminant removal. In particular, there is an interest in embedding nanoparticles in phase inversion polymeric membranes, where the nanoparticles can be incorporated into the membrane casting solution and retained in the solid membrane phase when the membrane polymer precipitates.

In this work, the incorporation of iron nanoparticles in a polyethersulfone membrane was studied. Nanoparticle stabilizer type, casting solution additives, and nanoparticle processing by centrifugal or magnetic separation were varied, and specific membrane characteristics were evaluated. Specifically, membrane casting solution viscosity, cast membrane thickness, pure water flux, and internal morphology were evaluated. While the addition of the additives ethanol and polyvinylpyrrolidone to a polyethersulfone-dimethylacetamide solution causes an expected increase in viscosity, the addition of nanoparticles can cause an increase or decrease in viscosity depending on the ligand stabilizer used during nanoparticle synthesis and casting solution additive concentration. Viscosity can also be affected by nanoparticle separation method, but again, changes are also dependent on casting solution composition. Varying changes in membrane thickness are observed and can be correlated to viscosity. Pure water flux decreases for all samples when nanoparticles are added, but the extent of the change is affected by casting solution composition. Internal morphology can partially explain the decrease in flux for nanoparticle-embedded membranes.

423873 New Fouling-Resistant High-Flux Membranes for Seawater Desalination

Abstract:

This presentation covers new high-flux and fouling-resistant reverse osmosis membranes synthesized through interfacial polymerization for seawater purification. We will review and discuss the state-of-the-art membranes in the thin-film-composite (TFC) structure prepared by interfacial polymerization. Recent advances have been in high flux TFC interfacially polymerized membranes for increasing water productivity and decreasing energy consumption. We have synthesized high flux membranes by incorporating a hydrophilic additive in the aqueous amine solution during interfacial polymerization, resulting in increasing water flux significantly through an additional pathway for water transport while maintaining high salt rejection via charge repulsion. For seawater desalination using synthetic 3.28 wt% NaCl solution at 800 psi (5.52 MPa) and 25°C, the membrane synthesized with the hydrophilic additive, *o*-aminobenzoic acid–trimethylamine salt, showed a very high flux of 1.81 m^3/m^2/day (44.4 gallons/ft^2/day (gfd)) and a salt rejection of 99.41%. This flux was 83.5% higher than the membrane synthesized without the hydrophilic additive (0.99 m^3/m^2/day or 24.2 gfd). The high-flux membrane was further tested using seawater from Port Hueneme, CA and exhibited a very good and stable desalination performance for 30 days. FTIR confirmed the good membrane stability. The membrane synthesized with hydrophilic additive showed significantly improved fouling resistance using sodium alginate, a common contaminant derived from seaweed, as the model foulant. In the presence of 25 ppm sodium alginate in the synthetic 3.28 wt% NaCl feed solution, the water flux reduction of the membrane synthesized with the hydrophilic additive was 11.0% compared to 18.2% for the membrane synthesized without the additive. With 50 ppm sodium alginate in the feed solution, the water flux reduction of the membrane synthesized with the hydrophilic additive was 19.8% compared to 31.8% for the membrane synthesized without the additive. Analysis of the fouling resistance using the resistance-in-series model indicated that the values of the ratio of the mass transfer resistance due to the foulant for the membrane synthesized with the hydrophilic additive to that for the membrane synthesized without the hydrophilic additive were 0.30 and 0.29 for 25 ppm and 50 ppm sodium alginate in the feed solutions, respectively. Analysis using atomic force microscopy (AFM) showed a smoother membrane surface for the membrane incorporated with the hydrophilic additive. In addition, the contact angle of the synthesized membrane reduced from 74° to 58° with the incorporation of the hydrophilic additive, which confirmed the increase of surface hydrophilicity.

424290 Synthesis, Characterization and Performance of Polyamide Membranes Structured with Hydrophilic Polymeric Brush Layers

Abstract:

Several modification methods have been used for surface nanostructuring of polyamide surfaces in order to increase surface hydrophilicity (i.e. decrease protein/bacteria fouling propensity), including physical modification (i.e. embedding nanoparticles, coating, etc.), chemical modification, UV or gamma irradiation, or plasma activation. In the present work, hydrophilic polymer chains were graft polymerized onto the surface of polyamide (PA) membrane via atmospheric pressure plasma-induced graft polymerization (APPIGP). The resulting polymeric brush layer lowers the free energy of hydration to a level suitable for reducing surface fouling. Prior to graft polymerization, the PA Surface is activated via atmospheric pressure plasma treatment to form free-radical peroxy initiation sites. The surface density of active sites depends on the type of plasma gas as illustrated in the study for helium and hydrogen plasma, plasma exposure time, and plasma power. Subsequently, graft polymerization was accomplished using an aqueous vinyl monomer solution resulting in a "brush" polymeric layer of chains that are terminally and covalently bonded to the polyamide surface. The rate of graft polymerization and thus the resulting chain size is affected by the initial monomer concentration, reaction temperature and time, and polymerization control agents. In the present work, the monomers acrylic acid, 2-hydroxyethyl methacrylate, and vinylsulfonic acid were selected to form surface brush layers of poly(acrylic acid) (PAA), poly(2-hydroxyethyl methacrylate) (PHEMA), and poly(vinylsulfonic acid) (PVSA). Surface topography was quantified in terms of feature height distribution determined by atomic force microscopy (AFM). Surface hydrophilicity/hydrophobicity was assessed via contact angle measurements with different solvents to determine the polar and dispersive components of the surface energy. It is shown that both surface roughness and free energy of hydration (i.e., wettability) can be tailored through the selected vinyl monomer chemistry, surface plasma exposure, and polymerization conditions. Studies on fouling resistance with model foulants correlated with surface hydrophilicity and topography thereby allowing tailor-design of fouling-resistant membrane surfaces for challenging water desalination applications.

439718 Biofouling Resistant Membranes for International Desalination Applications

Abstract:

With the rapid growth and development of countries around the world, and the inherent need of clean water, it is imperative to treat brackish and seawater sources. The two leading desalination technologies are thermal and membrane technologies, specifically reverse osmosis (RO). While nanofiltration (NF) is used for the removal of other substances from water source, it is also commonly used for the desalination of water. While both RO and NF can produce potable water, RO removes a large majority of solutes, which leaves the permeate water void of any essential nutrients (calcium, magnesium ions, etc.), placing the nutrient levels below that of the required World Health Organization standards. This means nutrients must be added back (i.e. reminerailization) to bring the water back to the standard levels for drinking water. NF methods, on the other hand, remove fewer solutes, and thus may avoid the need to remineralize the water. Thus, it is a more desirable technology for developing nations.

This project seeks to address a major technical challenge in membrane technology, mitigation of membrane biofouling, due to rejected chemicals and microbes, by impregnating the solutions with cost effective nanoparticles. Most of the research and development in the area of biofouling prevention has focused on pretreatment of the feed water, improved cleaning solutions and cleaning procedures. In this project, biofoulingresistant (BRN) nanocomposite membranes, loaded with green silver nanoparticles were studied from synthesis to macroscale production.

Filtration experiments were performed with synthetic brackish water containing biofoulant *Pseudomonas fluorescens Migula* to test regular cellulose acetate (CA) membranes and CA membranes loaded with silver nanoparticles (CA-AgNP). CAAgNP membranes resulted in lower flux declines when compared to CA membranes. CA membranes showed a flux decline of approximately 26% as compared to 20% for CA AgNP membranes during 7 hours of filtration. Though the differences are small in the decline of flux, the flux recovered after backwashing was higher for the CAAgNP membranes (93%) than for the CA membrane (84%). Therefore, filtration experiments using *p. fluorscens* showed that CAAgNP membranes were more effective at mitigating biofouling formation.

American Institute of Chemical Engineers Annual Meeting, Salt Lake City, UT, November 8-13, 2015

Optimization of Continuous Crystallization Processes to Maintain Content Uniformity

Christopher Burcham
Small Molecule Design and Development
Eli Lilly and Company

Niall Mitchell
Life Sciences, Consumer Goods and Specialty & Agrochemicals
Process Systems Enterprise Limited

PSE

Lilly

Desire End to End Understanding of Process Changes to Product Performance

solvent/additive, unit ops, recipes, impurities → **Drug substance manufacturing** → polymorph, morphology, PSD, purity

dosage form wet/dry gran. direct comp. → **Drug product manufacturing** → Content uniformity

dosage amount → **Drug delivery: pharmacokinetics** → Cmax, AUC

Drug delivery: pharmacodynamics → Efficacy, Safety

Drug substance manufacturing ↔ Drug product manufacturing ↔ Drug delivery: pharmacokinetics ↔ Drug delivery: pharmacodynamics

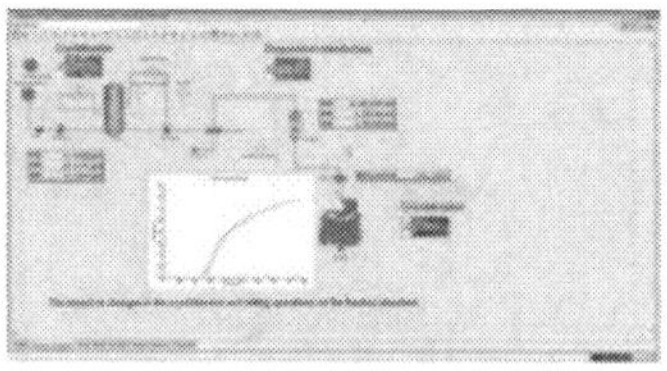

- Quantify effect of disturbances & decisions
  - uncertainty in process knowledge
  - common cause variability

  on Key Performance Indicators
  - Critical Quality Attributes
  - process economics, operability, safety

Desire End to End Understanding of Process Changes to Product Performance

solven
unit op
im

Drug s
manu

poly
morph
p

ery:
hamics

ty

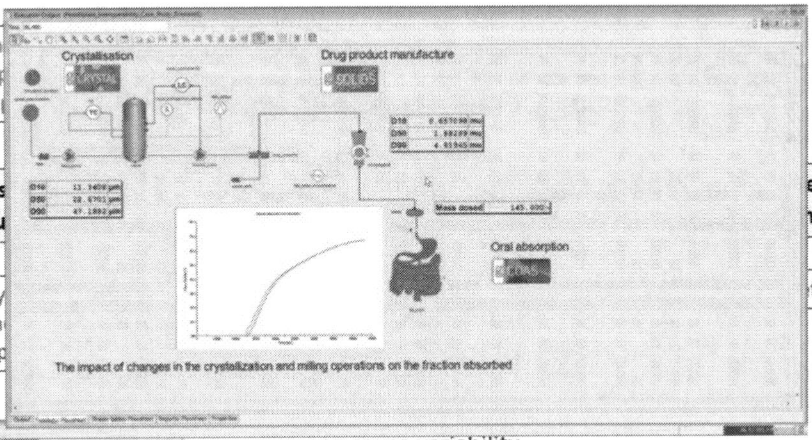

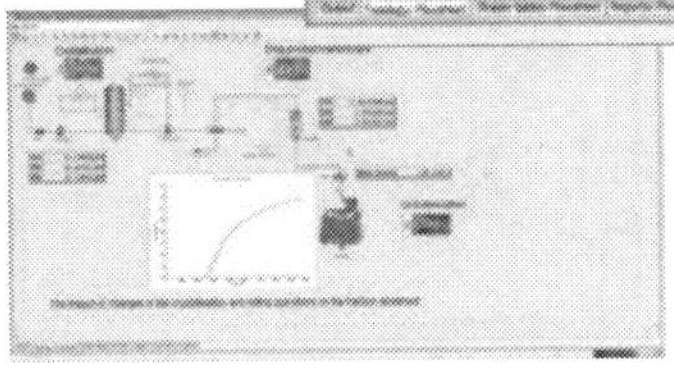

- common cause variability

on Key Performance Indicators

- Critical Quality Attributes
- process economics, operability, safety

Systems Based Pharmaceutics

- A holistic approach to the development and optimization of drug manufacture and drug delivery.
- A single model based framework for the modelling of:
  - drug substance manufacture
  - drug product manufacture
  - oral absorption and pharmacokinetics
  - *Beginning to end manufacturing process.*
- Considers the impact of changes to the crystallization process on downstream unit operations (ie drug product operations such as roller compaction, tablet compression), on quality indicators (ie dissolution testing) and performance expectations (ie absorption modeling).
- A framework to support optimization across the complete manufacturing system.

Objective

To optimize a continuous crystallization process such that:

1. Physical property control is achieved via crystallization conditions.
2. Drug product quality constraints (e.g. unit dose uniformity) are met.
3. Process meets manufacturing requirements for yield and productivity.

- Utilize a population balance model coupled with product constraints (unit dose uniformity in a drug product) and manufacturing constraints (filtration time).
  - Determine optimal processing conditions to maximize yield.
- Illustrative example:
  - Model low dose drug with crystallization kinetics for paracetamol.
    - High growth, low nucleation

Physical Property Control for Pharmaceuticals: Solid Oral Dosage Forms

- Drug substance physical properties, namely particle size can effect:
  - Content uniformity.

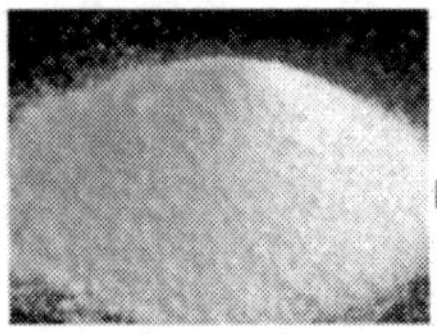
www.pallmannindustries.com

www.sweco.com

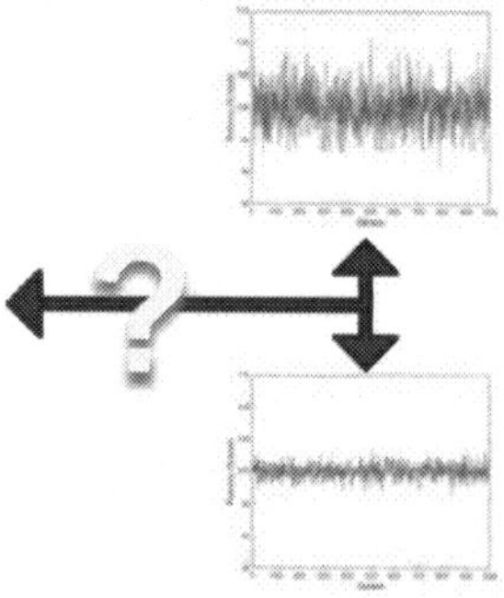

  - Bioavailability

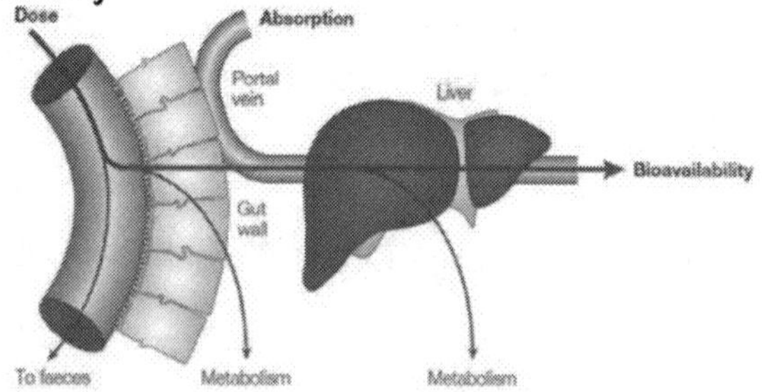

van de Waterbeemd, H, and Gifford, E. (2003). *ADMET In Silico Modelling: Towards Prediction Paradise?* Nat. Rev. Drug Disc. **2**, pp 192-204

Content Uniformity

- The USP guidance for acceptable Uniformity of Dosage Units
  - USP38-NF33 S1, <905>Uniformity of Dosage Units
  - Tier 1 (n= 10) testing
    - Provided the mean potency is within 1.5% of target, then a tablet RSD for potency not more than 6.25%
  - Tier 2 (n=30) testing
    - Provided the mean potency is within 1.5% of target, then a tablet RSD for potency not more than 12.5%
    - No tablet potency outside 75%-125% for up to n=30 tablets
- In brief, RSD < 6.25%

Content Uniformity Model

- Content uniformity models are used to predict volumetric diameter needed to meet USP Stage 1 criteria for tablets
- Model assumptions:
  - Random, dilute mixture (Poisson distribution)
  - Spherical particles
  - Representative PSD of drug substance in dosage unit
  - No segregation etc.

$$C_v = y\left(\frac{\pi\rho}{6G}\right)^{1/2}\left(\sum_i f_i D_i^3\right)^{1/2}$$

$f_i \equiv$ Mass fraction of drug existing as particles of diameter, d_i

$G \equiv$ Dose

$\rho \equiv$ True density

$y \equiv$ Mass fraction of excipient

Johnson MCR 1972. Particle size distribution of the active ingredient for solid dosage forms of low dosage. Pharm Act Helv 47A:546-559.

Content Uniformity Model

$$G = \sum_i np_i m_i \equiv \text{Mass of drug in a sample}$$

$$\operatorname{var}(G) = \sum_i np_i m_i^2$$

$$np_i m_i = p_i M$$

$$\operatorname{var}(G) = M \sum_i p_i m_i$$

$$m_i = \frac{\pi}{6} d_i^3 \rho$$

$$\sigma(G) = \left(\frac{\pi \rho M}{6}\right)^{1/2} \left(\sum_i p_i d_i^3\right)^{1/2}$$

$$p_i = f_i \frac{G}{M}$$

$$C_v = \left(\frac{\pi \rho}{6G}\right)^{1/2} \left(\sum_i f_i d_i^3\right)^{1/2}$$

- Spherical particles
- Ideal mixing
- Small drug fraction
- Sampling corresponds to a Poisson distribution

$n \equiv$ Total number of particles in the sample
$m_i \equiv$ Mass of drug in a particle of diameter, d_i
$p_i \equiv$ Number fraction of drug existing in the sample as particles of diameter, d_i
$f_i \equiv$ Mass fraction of drug existing as particles of diameter, d_i
$M \equiv$ Total sample mass

Johnson MCR 1972. Particle size distribution of the active ingredient for solid dosage forms of low dosage. Pharm Act Helv 47A:546-559.

Content Uniformity Models

USP38-NF33 S1, <905> Uniformity of Dosage Units.

- Rohr's et al model:
  - Assumes log-normal PSD
  - Design tool – target x_{90}.

$$d_{50} = \sqrt[3]{\left(\frac{6Dose}{\pi\rho}\right) \exp\left(-4.5 \ln^2 \sigma_g \left(\frac{RSD}{100}\right)^2\right)}\ 10^3$$

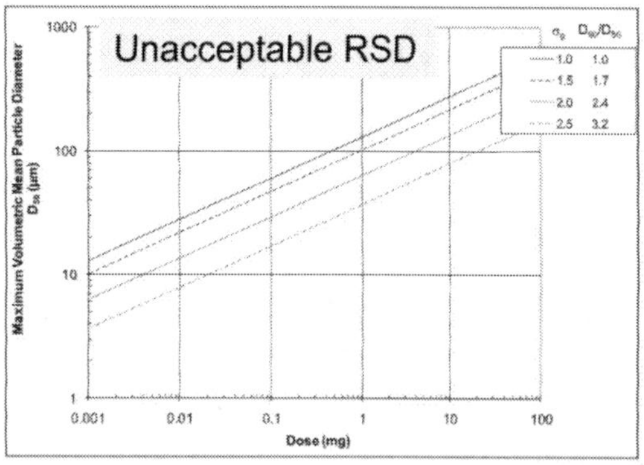

$C_v = 3.84\%$ $\rho = 1.3\ \text{g/cm}^3$

Rohr BR, *et al* (2006). *J of Pharm. Sci.* 95(5) pp 1049-1059.

- D[6,3] model:
  - No assumptions about PSD shape.

$$RSD = \sqrt{\frac{\pi}{6} \cdot D[6,3]^3 \cdot \frac{\rho}{Dose}}$$

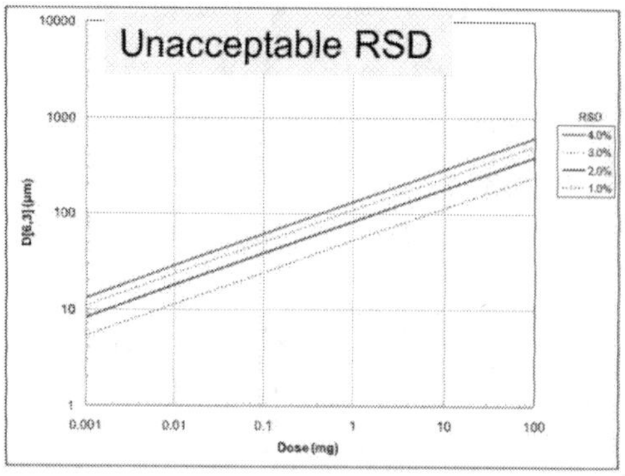

Hilden, J. *et al* (2012). *J Pharm. Sci.* 101(7) pp 2364-2371.

Absorption

- Dependent on solubility (thermodynamic), dissolution rate (kinetic) and permeability.

Class I • High solubility • High permeability • Generally well absorbed	Class II • Low solubility • High permeability • Solubility / dissolution limited absorption
Class III • High solubility • Low permeability • Permeability limited absorption	Class IV • Low solubility • Low permeability • Generally poorly absorbed

BCS I if entire dose dissolves in 250ml (pH 1 to pH 8)

Amidon, G. L. et al. (1995). *Pharm Res*, **12**, 3, pp 413-420.

Physiology of Drug Absorption

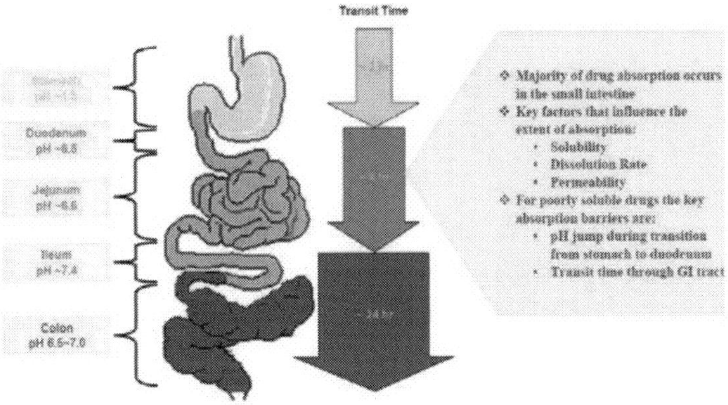

Martin, A. (1993). *Physical Pharmacy*, 4th Ed. Lea and Febinger, Philadelphia.
Martinez, M.N. and Amidon, G. J. (2002). *J. Clin. Pharmacol.*, **42**, pp 620-643.

Dissolution Rate Limited

- Drug substance physical properties, namely particle size effects:
  - Absorption/bioavailability of the drug substance can be impacted by the dissolution rate of the drug substance according to the Noyes Whitney Equation:

$$\frac{dm}{dt} = \frac{D_{AB}}{\lambda} A\left(C^* - C_b\right)$$

$$A = (Dose)SSA$$

$$SSA \propto \frac{1}{D}$$

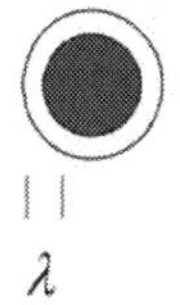

Impact of Particle Size on Content Uniformity as a Function of Dose

- True density of 1.33 g/cm³, log-normal distribution with a standard deviation of 1.72, $X_{90}/X_{50} = 2$

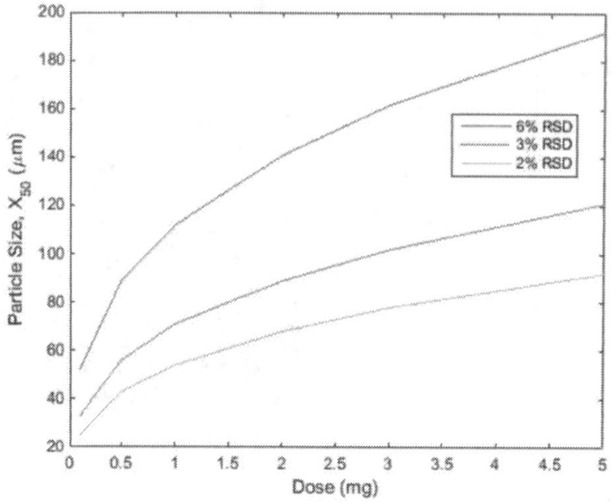

Allowable and Attainable Regions

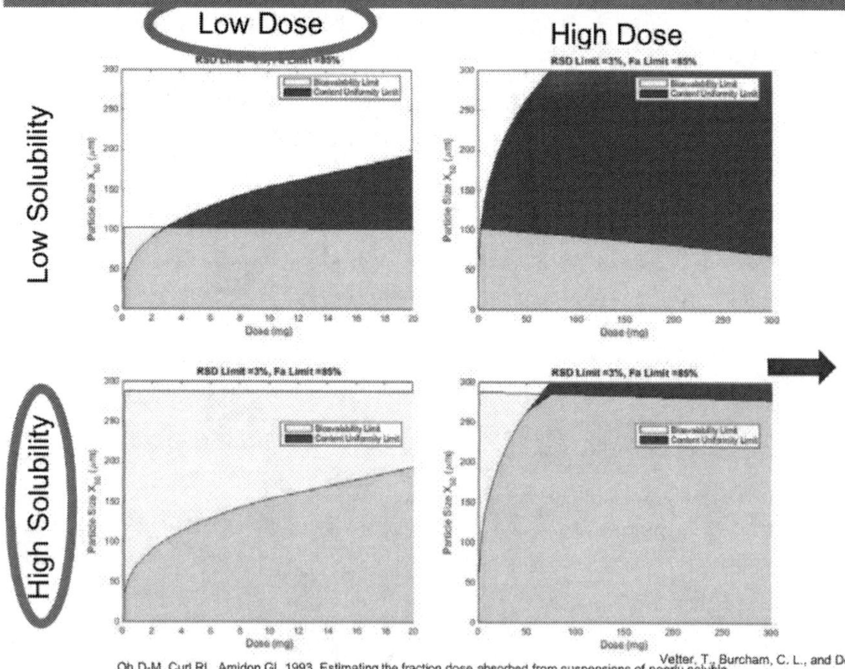

Oh D-M, Curl RL, Amidon GL 1993. Estimating the fraction dose absorbed from suspensions of poorly soluble compounds in humans: A mathematical model. Pharm Res 10(2):264-270.

Vetter, T., Burcham, C. L., and Doherty, M. F. (2014). "Regions of Attainable Particle Sizes in Continuous and Batch Crystallization Processes." *Chem. Eng. Sci.*, **106**, 167-180.

Allowable and Attainable Regions

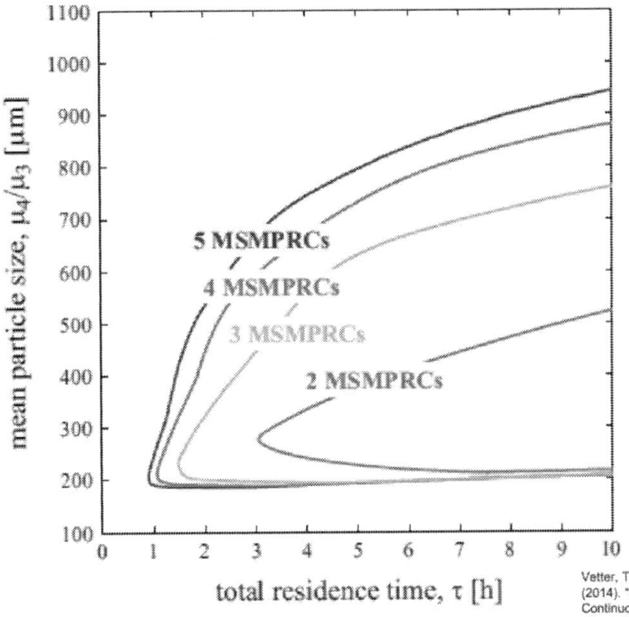

Vetter, T., Burcham, C. L., and Doherty, M. F. (2014). "Regions of Attainable Particle Sizes in Continuous and Batch Crystallization Processes." *Chem. Eng. Sci.*, **106**, 167-180.

Impact to Manufacturability – Filtration Time

- Filtration time[1]:

$$t=\left(\frac{\mu c\alpha}{2A^2\Delta p}\right)V\left(t\right)^2+\frac{\mu R_m}{A\Delta p}V\left(t\right)$$

$$c=\frac{c_s}{1-\left[m_F/m_c-1\right]c_s/\rho}$$

- Cake resistance dependence upon particle size[2]:

$$\alpha=\frac{18}{\rho}\frac{K}{d_{vg}^2\exp\left(4\ln^2\sigma_g\right)}\frac{\nu\left(\varepsilon\right)}{\varepsilon^2}$$

$$\nu\left(\varepsilon\right)=\frac{10\left(1-\varepsilon\right)}{\varepsilon}$$

t	time
V	Filtrate Volume
μ	Filtrate Viscosity
α	Cake Resistance
A	Filter Cross Sectional Area
Δp	Pressure Drop over the Filter
R_m	Filter Media Resistance
m_F	Wet Cake Mass
m_c	Dry Cake Mass
c_s	Concentration of Solids in the Slurry
$\rho_{filtrate}$	Density of the Filtrate
ε	Cake Porosity
ρ	True Density of the Solid Phase
d_{vg}	Volume equivalent diameter on a number basis
K	Dynamic Shape Factor
σ_g	Lognormal Standard Deviation
ν	Void Fraction

1. McCabe, W. L., Smith, J. C., and Harriott, P. (1985). *Unit Operations of Chemical Engineering*, 2nd ed., McGraw-Hill, New York, pp. 873-877.
2. Endo, Y. and Alonso, M. (2001) 'Physical meaning of specific cake resistance and effects of cake properties in compressible cake filtration', *Filtration & Separation*, pp. 42-46.

Modeling the Crystal Size

- Population Balance Model

$$\frac{\partial n(L,t)V}{\partial t} = F_{in}n_{in} - F_{out}n(L,t) - V\frac{\partial\left(G(L)n(L,t)\right)}{\partial L} + (B-D)V$$

- Birth (B)
  - Nucleation
    - Primary
      - Homogeneous (from solution)
      - Heterogeneous
    - Secondary
      - From existing crystals
        - Attrition
        - Activated
- Death (D)
  - Dissolution
- Birth and death
  - Attrition, breakage and agglomeration
- Growth (G)

Kinetic Models and Parameters

- Nucleation
  - Primary

$$B_{prim} = k'_n(\Delta C)^n$$

  - Secondary nucleation:

$$B_{sec} = k_N m_2 (\Delta C)^b$$

- Growth

$$G(L) = k_g \exp\left(-\frac{E_{A,g}}{RT}\right)(\Delta C)^g$$

$$\text{if } C - C^* > 0$$

Parameter	Value	Units
k'_n	$1.597*10^{10}$	(#/min)
n	2.276	-
k_g	9.979	(m/s)(m³/kmol)^g
$E_{A,g}$	40.56	kJ/mol
g	1.602	-

Agitation Rate (rpm)	k_N (#/s m^2)	b
200	$2.656*10^7$	2.232
250	$6.397*10^7$	2.339
300	$26.632*10^7$	2.314

- No agglomeration, attrition, breakage.

Frawley, P.J., Mitchell, N.A., O'Ciardha, C.T., and Hutton, K.W., (2012). "The effects of supersaturation, temperature, agitation and seed surface area on the secondary nucleation of paracetamol in ethanol solutions." *Chem Eng Sci*, **75**, pp. 183–197.

Flowsheet

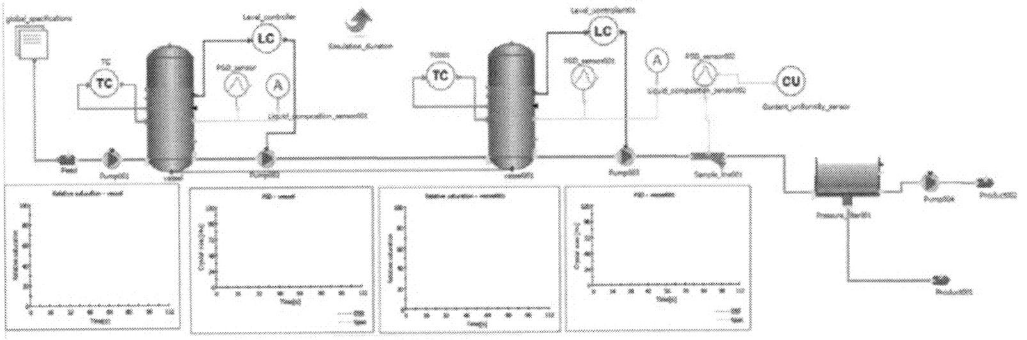

Flowsheet with Continuous Seeding

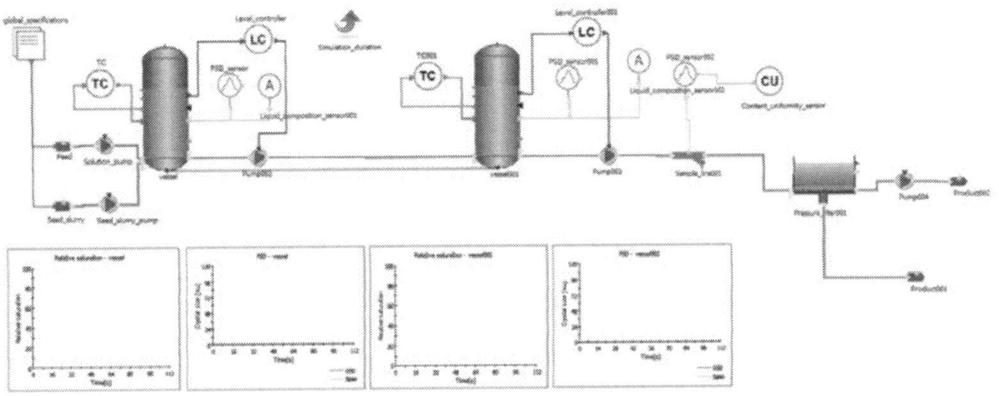

Optimization Problem

- Optimize: Maximize X_{50} of the product
- Constraints:
  - Tablet RSD < 3%
  - Cake resistance < 10^{10} m/kg
  - Relative supersaturation < 2.0
  - Yield > 90% theoretical yield
- Controls:
  - *Fraction of seed to first MSMPR (0.01 to 0.5 wt/wt)*
  - Temperature in the first MSMPR (30 to 50 °C)
  - Residence time in the first MSMPR (40 to 270 min)
  - Temperature in the second MSMPR (5 °C)
  - Residence time in the second MSMPR (40 to 270 min)
  - Solution concentration in the feed

11/9/2015 2015 AIChE Annual Meeting 21

Attainable Dose Range (Without Continuous Seeding)

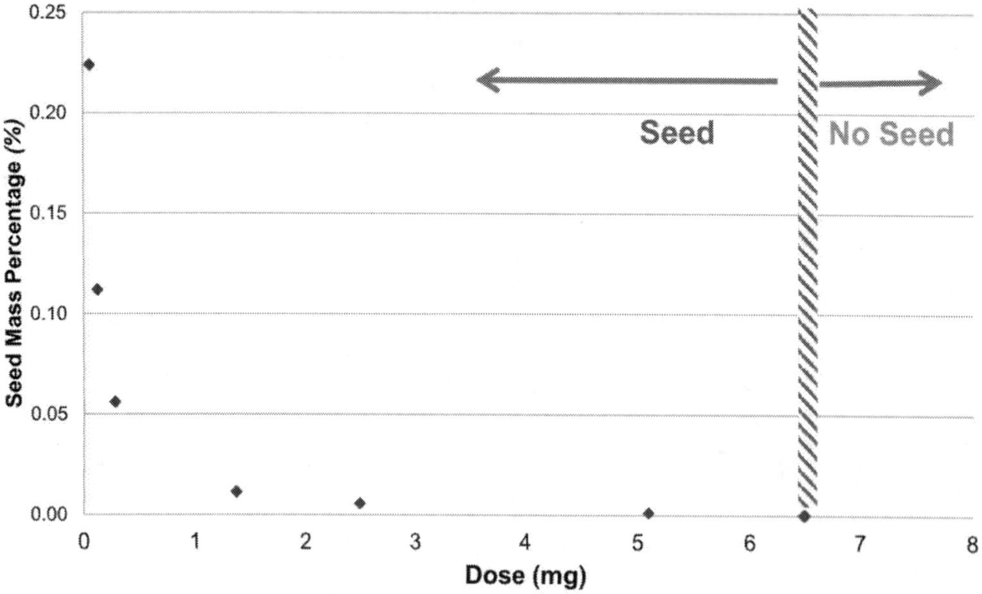

11/9/2015 2015 AIChE Annual Meeting 22

Increasing Secondary Nucleation

By Increasing Agitation Rate

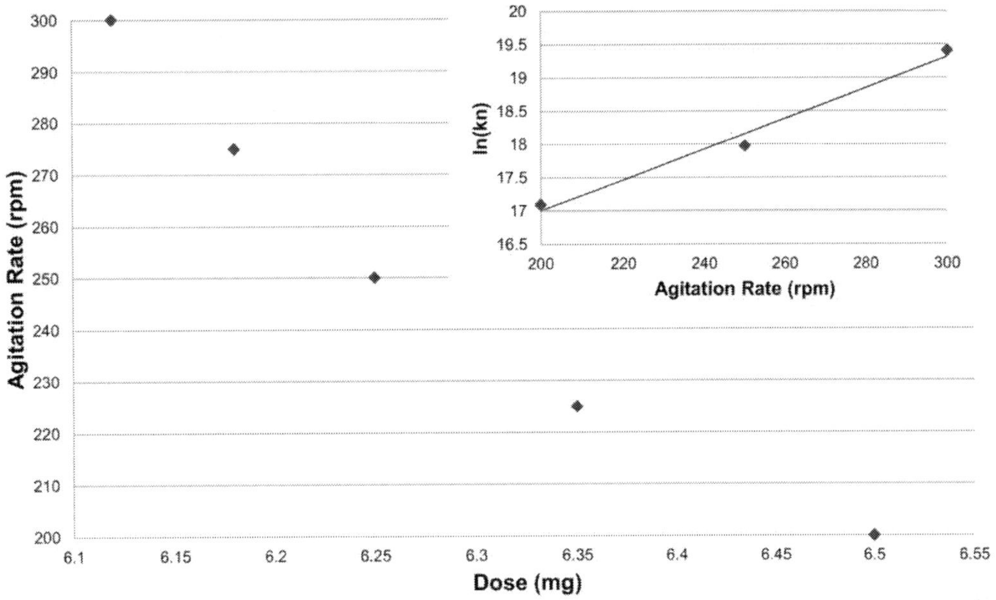

Continuously Seed MSMPR 1

0.000011 kg seed/kg feed, Fixed feed rate

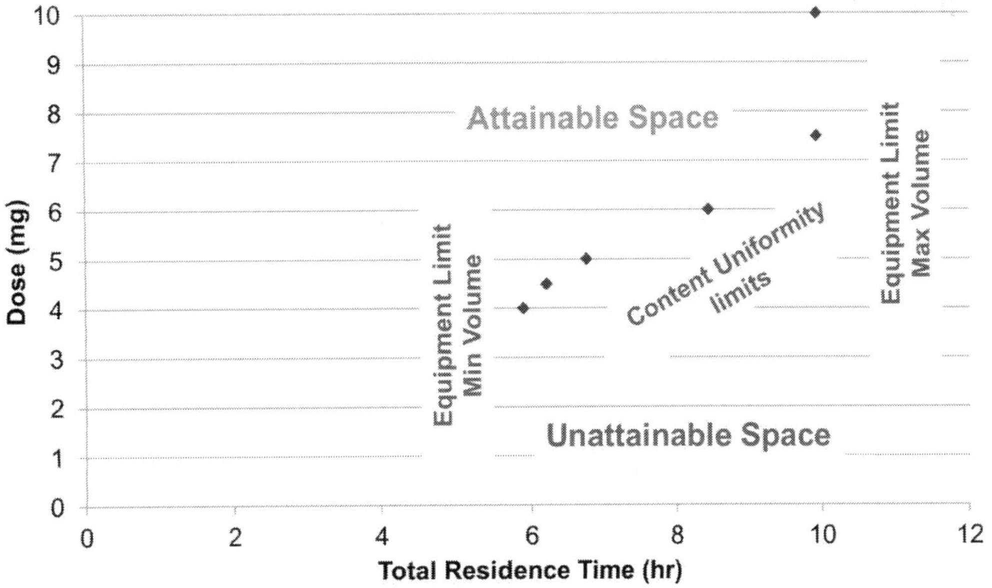

Conclusions

- Integrating product quality constraints (content uniformity) with manufacturing constraints (filtration time), economic constraints (yield) with a process model (operational parameters) allows for attainable regions to be probed.
  - Minimizes experimentation.
  - Allows for "what-if" scenarios to be considered.
- For this case, with the crystallization kinetics considered,
  - Seeding is needed to meet constraints on content uniformity RSD (3%) for doses below 7 mg due.
  - Increasing nucleation by an order of magnitude (via increasing the agitation level) is not sufficient to decrease the product particle size.

AUTHOR INDEX

AUTHOR INDEX

AUTHOR INDEX

AUTHOR INDEX

AUTHOR INDEX

AUTHOR INDEX

AUTHOR INDEX

AUTHOR INDEX

AUTHOR INDEX

AUTHOR INDEX

AUTHOR INDEX

AUTHOR INDEX

AUTHOR INDEX

AUTHOR INDEX

AIChE
120 Wall Street, FL 23
New York, NY 10005-4020

ISBN 978-1-5108-1873-6

CURRAN ASSOCIATES INC.
proceedings.com

Separations Division 2015

Core Programming Area at the 2015 AIChE Annual Meeting

Salt Lake City, Utah, USA
8-13 November 2015

Volume 2 of 2

AIChE
The Global Home of Chemical Engineers

Separations Division 2015

Core Programming Area at the 2015 AIChE Annual Meeting

Salt Lake City, Utah, USA
8-13 November 2015

Volume 2 of 2

ISBN: 978-1-5108-1873-6

Printed from e-media with permission by:

Curran Associates, Inc.
57 Morehouse Lane
Red Hook, NY 12571

Some format issues inherent in the e-media version may also appear in this print version.

Printed by Curran Associates, Inc. (2016)

For permission requests, please contact AIChE
at the address below.

AIChE
120 Wall Street, FL 23
New York, NY 10005-4020

Phone: (800) 242-4363
Fax: (203) 775-5177

www.aiche.org

Additional copies of this publication are available from:

Curran Associates, Inc.
57 Morehouse Lane
Red Hook, NY 12571 USA
Phone: 845-758-0400
Fax: 845-758-2634
Email: curran@proceedings.com
Web: www.proceedings.com

TABLE OF CONTENTS

VOLUME 1

VOLUME 2

2015 AIChE Annual Meeting
November 8-13, Salt Lake City, US

Application of Raman and ATR-UV/Vis Spectroscopy for a Model-Free and Model-Based Active Polymorphic Feedback Control of Crystallization Processes

E. Simone, A. Borsos, Z.K. Nagy

PURDUE UNIVERSITY ROYAL SOCIETY OF CHEMISTRY erc European Research Council Loughborough University

Overview

- Introduction (Polymorphism of OABA, Experimental rig)
- Model-free active polymorphic feedback control (APFC)
- Model-based APFC:
  - Model development
  - Parameters identification and validation
  - Optimization
- Conclusions

PURDUE UNIVERSITY ROYAL SOCIETY OF CHEMISTRY erc European Research Council Loughborough University

Polymorphism of OABA

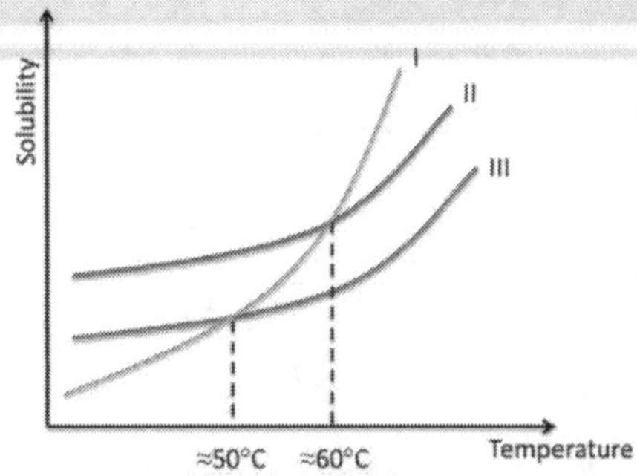

- Three known forms of OABA
- Form I and II are normally nucleated from solution
- Form III is very difficult to nucleate but can be obtained by melt crystallization or polymorphic transformation above 60° C

Form I: orthorhombic cell, **zwitterions** and neutral molecules (1:1 ratio)

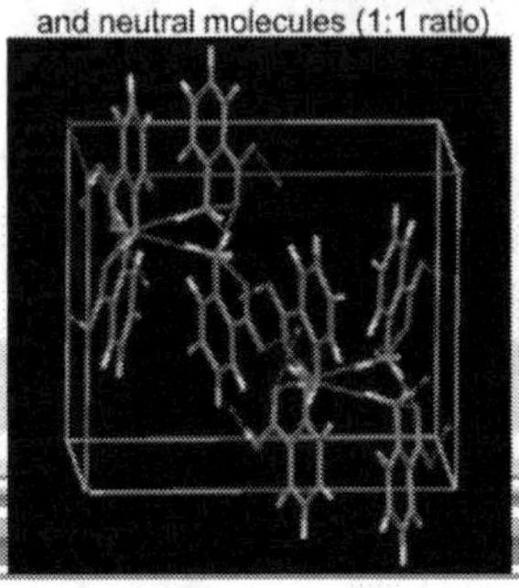

Form II: orthorhombic cell, **dimers** of neutral molecules

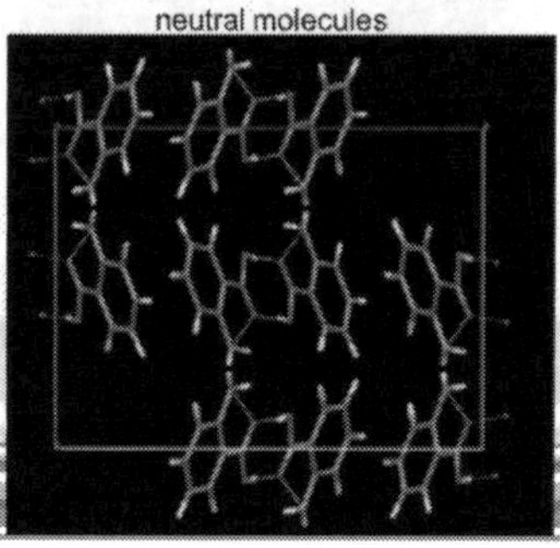

Form III: monoclinic cell, **dimers** of neutral molecules

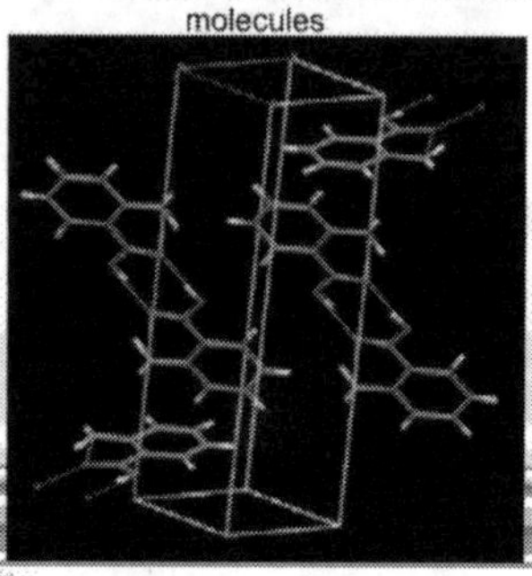

PURDUE UNIVERSITY — ROYAL SOCIETY OF CHEMISTRY — erc European Research Council — Loughborough University

Jiang et al., *Crystal Growth and Design*, 2010

Experimental rig

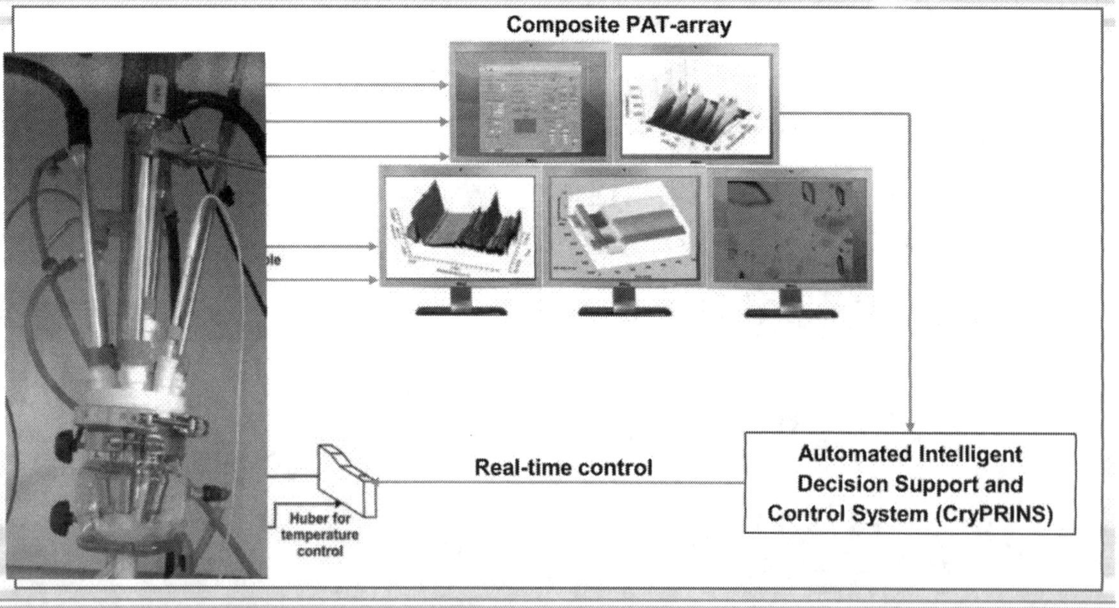

PURDUE UNIVERSITY — ROYAL SOCIETY OF CHEMISTRY — erc European Research Council — Loughborough University

Definition of Active Polymorphic Feedback Control

- Detects the formation of a **polymorphic mixture** and eliminates the metastable form by triggering a *controlled dissolution cycle*, and allowing the growth of the stable form using *supersaturation control*
- Uses a combination of **Raman spectroscopy** (calibration-free), **ATR-UV/Vis spectroscopy** and **CryPRINS**

PURDUE UNIVERSITY · Royal Society of Chemistry · erc European Research Council · Loughborough University

Active polymorphic feedback control (APFC) strategy

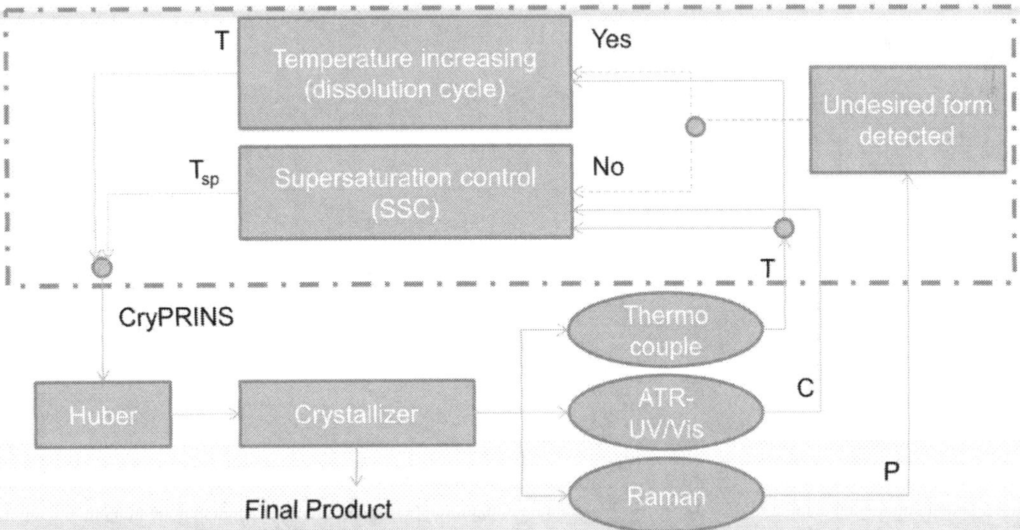

- 2° derivative Raman spectrum was used
- The Raman peak at 770 cm^{-1} was used to reveal the presence of the metastable form in the slurry

PURDUE UNIVERSITY · Royal Society of Chemistry · (Simone et al., *Crystal growth and design*, 2014) · erc European Research Council · Loughborough University

APFC using Raman and supersaturation control (1)

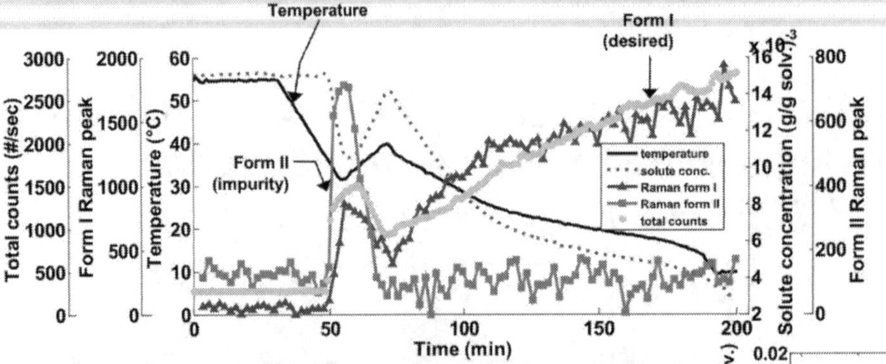

- Seeding of mixtures of OABA polymorphs
- PVM to check the purity of the final form

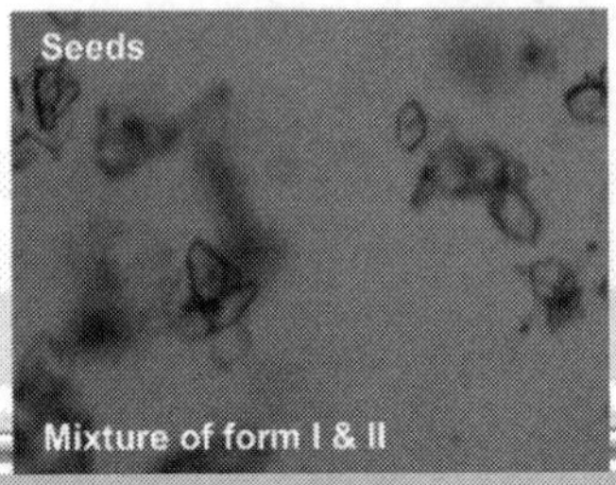

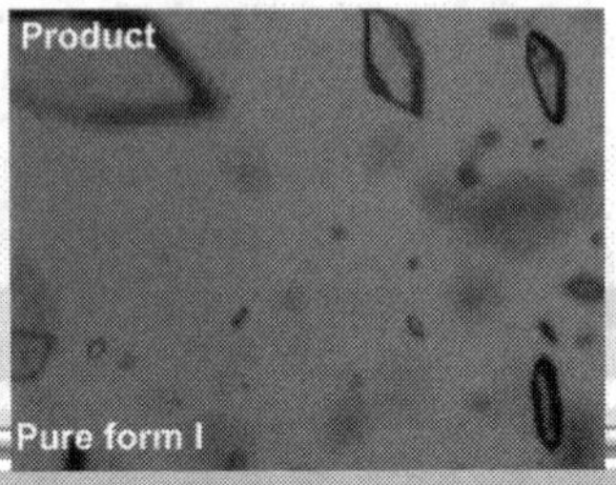

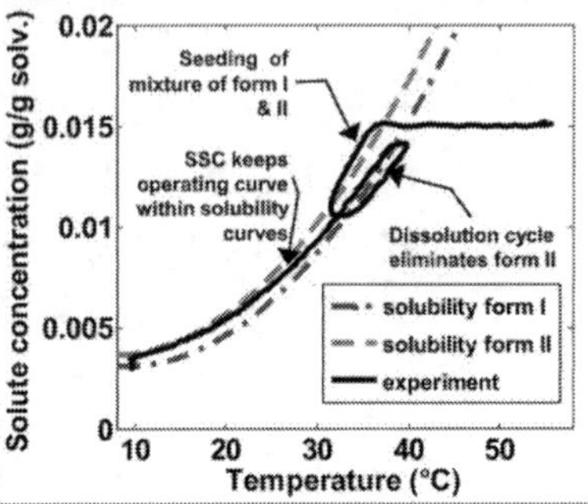

(Simone et al., *Crystal growth and design*, 2014)

APFC using Raman and supersaturation control (2)

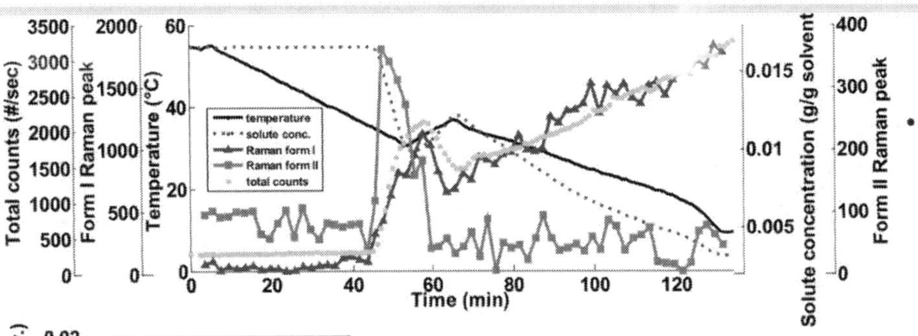

- Nucleation of mixtures of OABA polymorphs

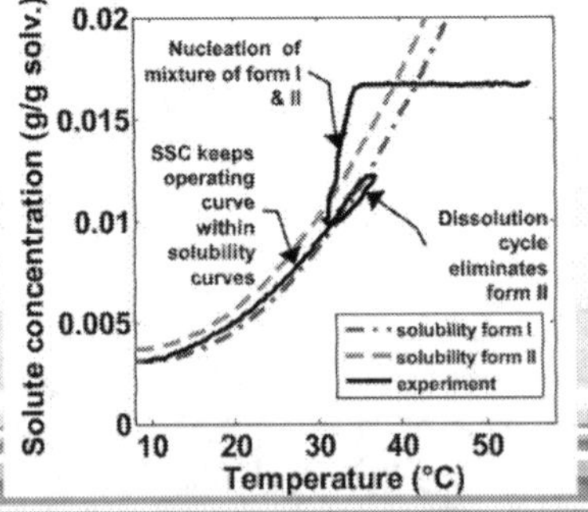

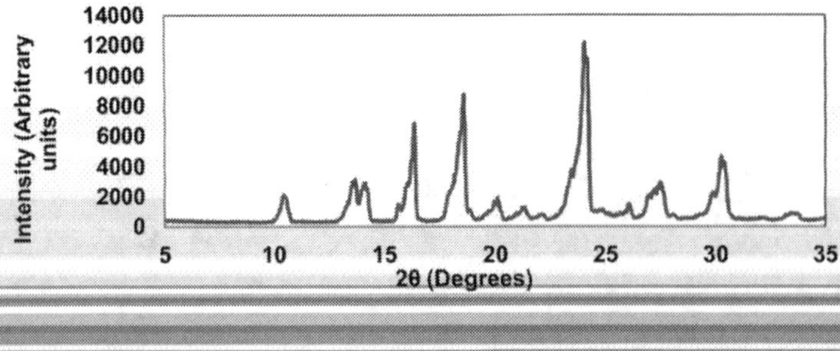

(Simone et al., *Crystal growth and design*, 2014)

Kinetic parameters for the model based APFC

- Dissolution of form II
- Dissolution of form I
- Growth of form II
- Growth of form I
- Nucleation of form I (primary and secondary)

PURDUE UNIVERSITY | ROYAL SOCIETY OF CHEMISTRY | erc European Research Council | Loughborough University

Population balance equations

- Two sets of equations, one for each polymorph
- Method of moments used to express the PBEs

$$\frac{d\mu_{0,I}}{dt} = B_I$$

$$\frac{d\mu_{i,I}}{dt} = i(G_I - D_I)\mu_{i-1,I} \qquad i = 1,2,3$$

$$if\ \mu_{1,I} \leq 0\ then\ D_I = 0$$

$$\frac{d\mu_{0,II}}{dt} = 0$$

$$\frac{d\mu_{i,II}}{dt} = i(G_{II} - D_{II})\mu_{i-1,II} \quad i = 1,2,3$$

$$if\ \mu_{1,II} \leq 0\ then\ D_{II} = 0$$

$$\frac{dC_{liq}}{dt} = 3k_v\rho_c\left(-G_{II}\mu_{2,II} - G_I\mu_{2,I} + D_{II}\mu_{2,II}\right)$$

$$D_{I,II} = A_{dI,II}exp\left(\frac{-B_{dI,II}}{T}\right)(1 - S_{I,II}) \quad G_{I,II} = A_{gI,II}exp\left(\frac{-B_{gI,II}}{T}\right)(S_{I,II} - 1)^{a_{I,II}} \quad S_{I,II} = \frac{C}{C_{satI,II}}$$

$$B_I = A_b S_I^{a_b} + K_{b1}S_I exp\left(\frac{-K_{b2}}{(ln\ S_I)^2}\right) \qquad B_{II} = 0$$

PURDUE UNIVERSITY | ROYAL SOCIETY OF CHEMISTRY | erc European Research Council | Loughborough University

(Hermanto et al., *AIChE Journal*, 2007)
(Wantha & Flood, *Chemical Engineering and Technology*, 2013)

Kinetics parameters estimation

- **Specific experimental design** to separate different mechanisms
- **Concentration only** fitted (using *fminsearch* and *fmincon* in Matlab 2013)
- Initial crystal size distribution from **Malvern Mastersizer**

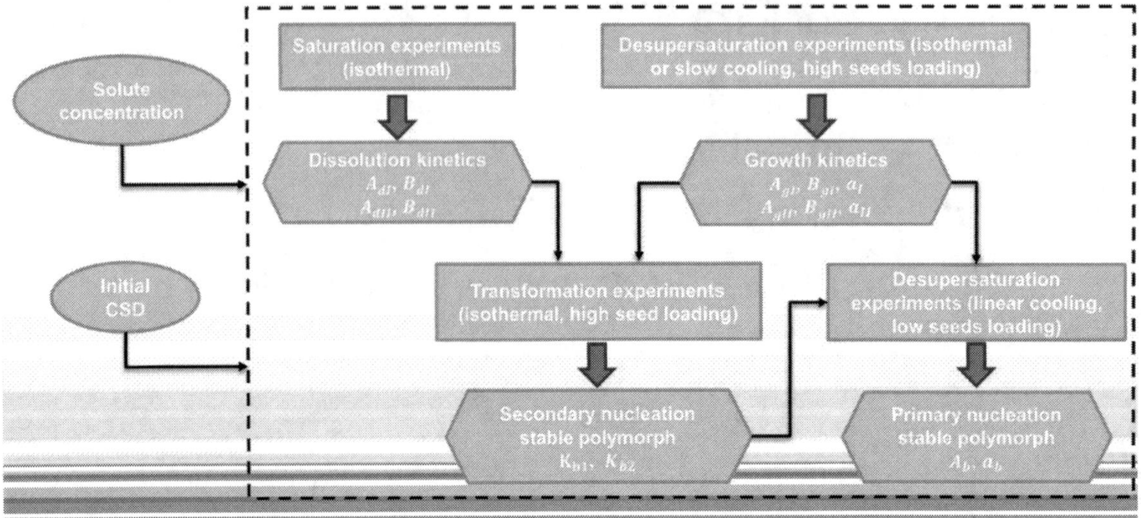

PURDUE UNIVERSITY | ROYAL SOCIETY OF CHEMISTRY | (Simone et al., *manuscript in preparation*) | erc European Research Council | Loughborough University

Estimated kinetic parameters

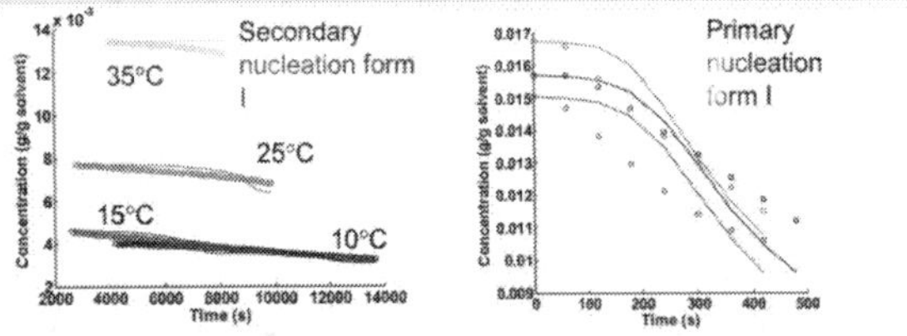

Kinetic parameter	Constant	Value	Upper limit of the interval of confidence	Lower limit of the interval of confidence
D_I	A_{dI}	4.3675	2.078	9.181
D_I	B_{dI}	4376.9	4160.1	4604.5
D_{II}	A_{dII}	1.148e-3	7.478e-4	1.764e-3
D_{II}	B_{dII}	1845.3	1719.7	1980.1
G_I	A_{gI}	2.385e8	1.149e8	4.950e8
G_I	B_{gI}	11033.6	10832.4	11238.5
G_I	a_I	1.131	1.018	1.245
G_{II}	A_{gII}	1.258e-4	6.15e-6	2.576e-3
G_{II}	B_{gII}	2032.9	1378.3	2998.3
G_{II}	a_{II}	1.003	0.766	1.239
B_I	K_{b1}	1.574e10	1.149e10	2.156e10
B_I	K_{b2}	1.483	0.380	0.409
B_I	A_b	1.656e8	7.385e6	3.710e9
B_I	a_b	2.36	-13.813	18.53

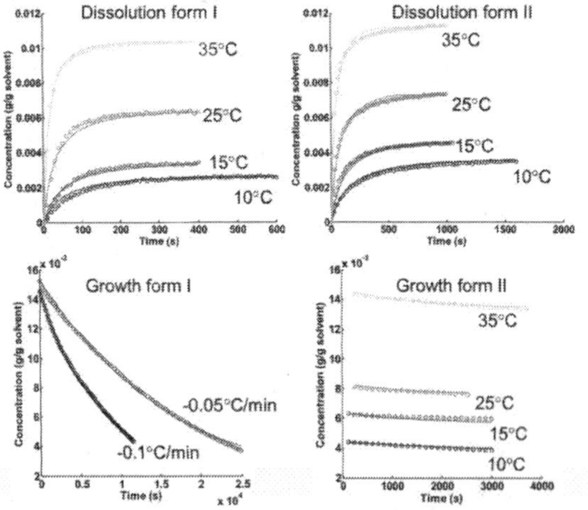

- ❑ Primary nucleation is the more uncertain parameter estimated
- ❑ The best fitting was obtained for growth and dissolution kinetics

PURDUE UNIVERSITY | ROYAL SOCIETY OF CHEMISTRY | (Simone et al., *manuscript in preparation*) | erc European Research Council | Loughborough University

Model validation (form II)

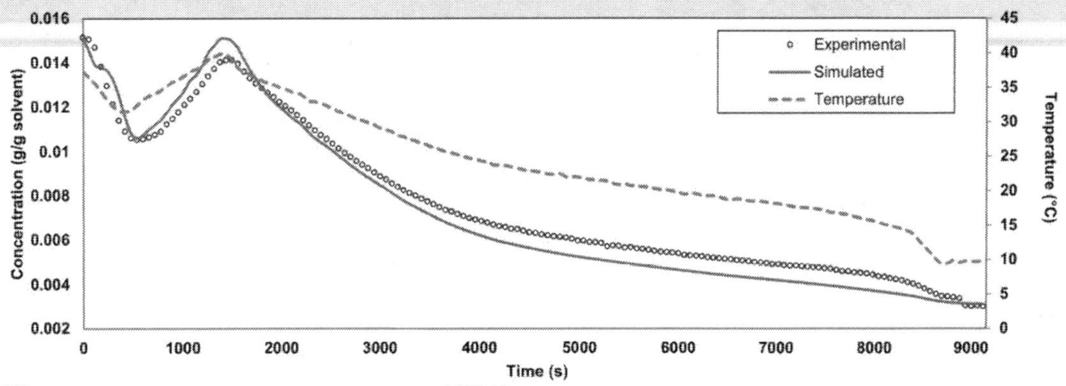

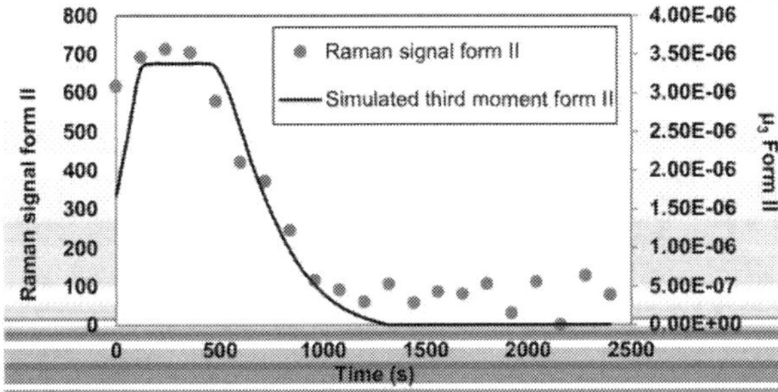

- The kinetics parameters estimated were used to calculate solute concentration during an APFC
- The simulated profile is very close to the experimental

PURDUE UNIVERSITY

(Simone et al., *manuscript in preparation*)

Loughborough University

Model validation (form I)

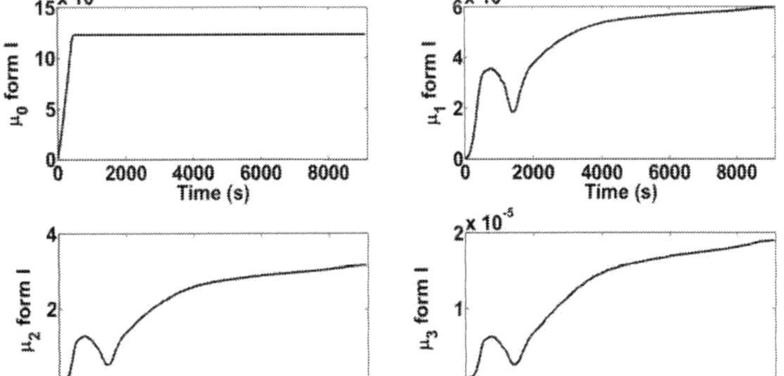

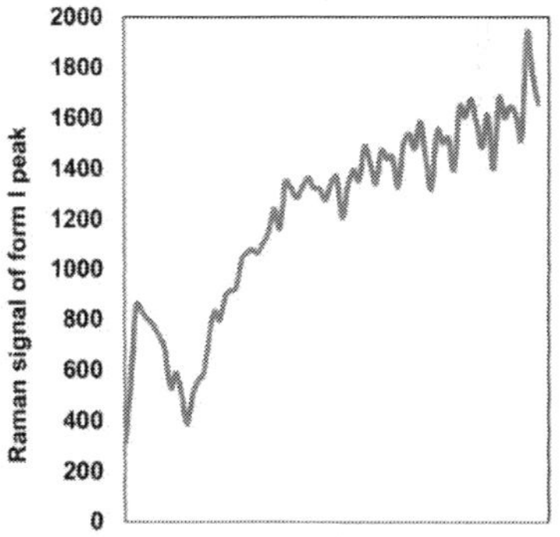

Initial conditions:

- Seeding temperature: 37.26°C
- Concentration of the solution: 0.015 g/g solvent
- Total mass of solid in the vessel: 0.6g
- Mass of solvent: 400 g

PURDUE UNIVERSITY

(Simone et al., *manuscript in preparation*)

Loughborough University

Optimization conditions

- Minimization performed using both the *genetic algorithm toolbox* and *fmincon* in Matlab 2013

$$\min_{T(k)}\left(-\bar{L}_{I,end}\right)\ with\ \bar{L}_{I,end}=\frac{\mu_{1,end}}{\mu_{0,end}}$$

Optimization problem | Objective function

$$-0.5\le\frac{dT}{dt}\le 0.5$$

$$11℃\le T\le 45℃$$

$$c_{end}\le c_{max,end}=0.005\text{g/g solvent}$$

$$\mu_{II1,end}=0$$

Constraints

PURDUE UNIVERSITY | ROYAL SOCIETY OF CHEMISTRY | erc European Research Council | Loughborough University

(Simone et al., *manuscript in preparation*)

Optimal temperature profile

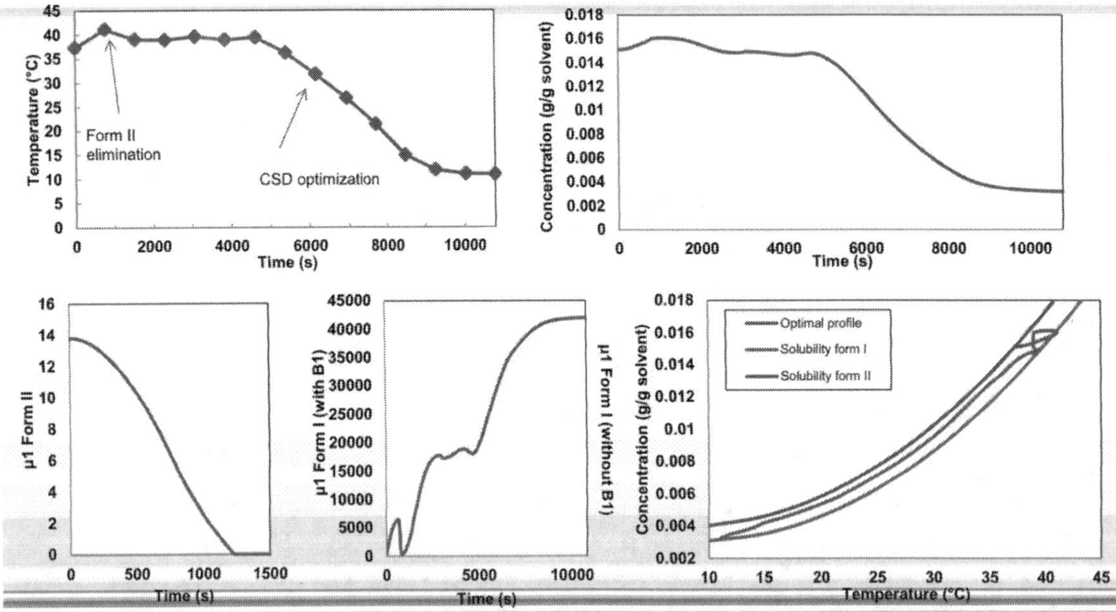

PURDUE UNIVERSITY | ROYAL SOCIETY OF CHEMISTRY | erc European Research Council | Loughborough University

(Seki et al., *www.aidic.it/isic18/webpapers/73**Seki**.pdf*)

(Majumder & Nagy, *AIChE Journal*, 2013)

Conclusions

- The **effectiveness** of the APFC in obtaining the pure desired polymorph of OABA was demonstrated **experimentally**
- The principal kinetic parameters of OABA were estimated using **sequential experiments to decouple mechanisms**
- A model based on the **method of moments** was validated and applied in a process optimization framework
- The optimal profile has two dissolution cycles to both eliminate form II and maximize the size of form I

Thanks for your attention

E. Simone, Z.K. Nagy

e.simone@lboro.ac.uk

Acknowledgements:

- European Research Council for the funding (grant no. **280106-CrySys**)
- Royal Society of Chemistry for the travel grant

Description of Technical Importance

Batch crystallization is commonly employed in the fine-chemical and pharmaceutical industries to obtain solid products of high purity. We are exploring the use of crystallization for a different application: the separation of non-radioactive salts from nuclear waste. If successful, such an operation could significantly reduce costs and expedite waste clean up efforts.

Regardless of the application, control of the mean crystal size is often critically important; yet, establishing such control has remained an outstanding issue. In this work we leverage process analytical technology to develop a feedback control method and demonstrate—for the first time—explicit control over the mean size of crystals produced by unseeded, batch cooling crystallization.

Mass-Count Framework for Crystal Size Control

Daniel J. Griffin, Martha A. Grover,
Yoshiaki Kawajiri, and Ronald W. Rousseau

Georgia Institute of Technology

Background and Introduction

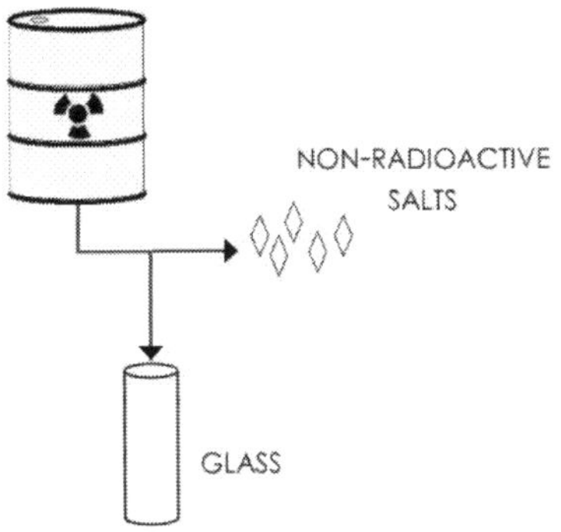

Legacy Waste Cleanup

56 MILLION GALLONS OF RADIOACTIVE WASTE STORED IN 177 TANKS

US DOE CLEANUP:

REMOVE, **VITRIFY (TURN INTO GLASS)**, AND ENCASE IN STEEL CANISTERS

RESEARCH ON SEPARATION OPERATIONS

4

Crystallization of Salts from Waste

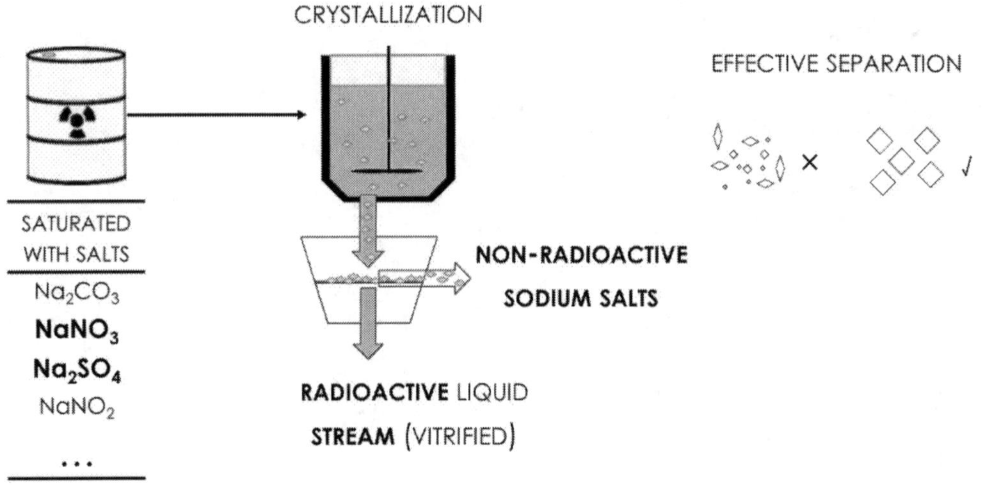

NEED TO **CONTROL CRYSTALLIZATION TO PRODUCE LARGE CRYSTALS THAT CAN BE EASILY FILTERED**

5

Research Objective and Outline

OBJECTIVE

establish a methodology for reliably controlling the

mean crystal size

unseeded, batch cooling crystallization

OUTLINE

1. EXPERIMENTAL SYSTEM AND ONLINE MEASUREMENT TOOLS
2. MASS-COUNT FRAMEWORK
3. DATA-DRIVEN MODELING AND DYNAMICS PROGRAMMING
4. RESULTS

6

1. Experimental System and Online Measurement Tools

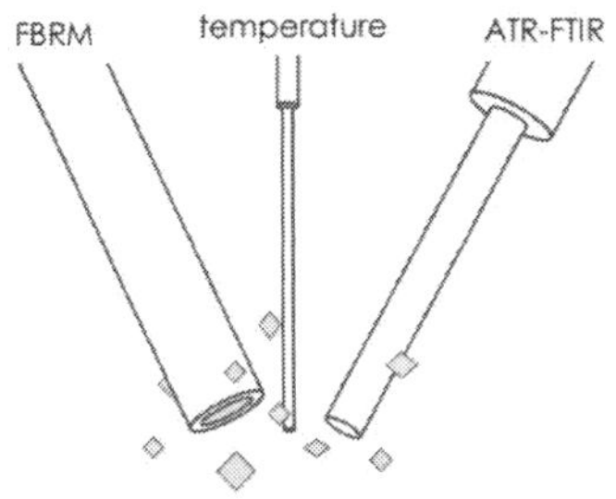

Experimental System

study crystallization from a two-component electrolytic solution

110 g **$NaNO_3$** and 7.25 g **Na_2SO_4** per 100 g **water**

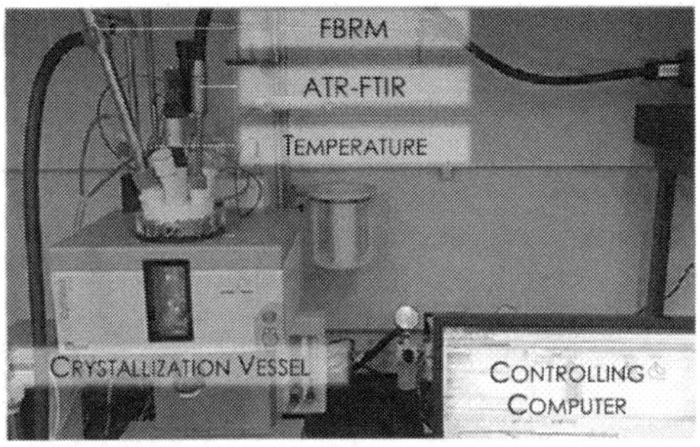

lab crystallization system

8

Online Measurement Tools

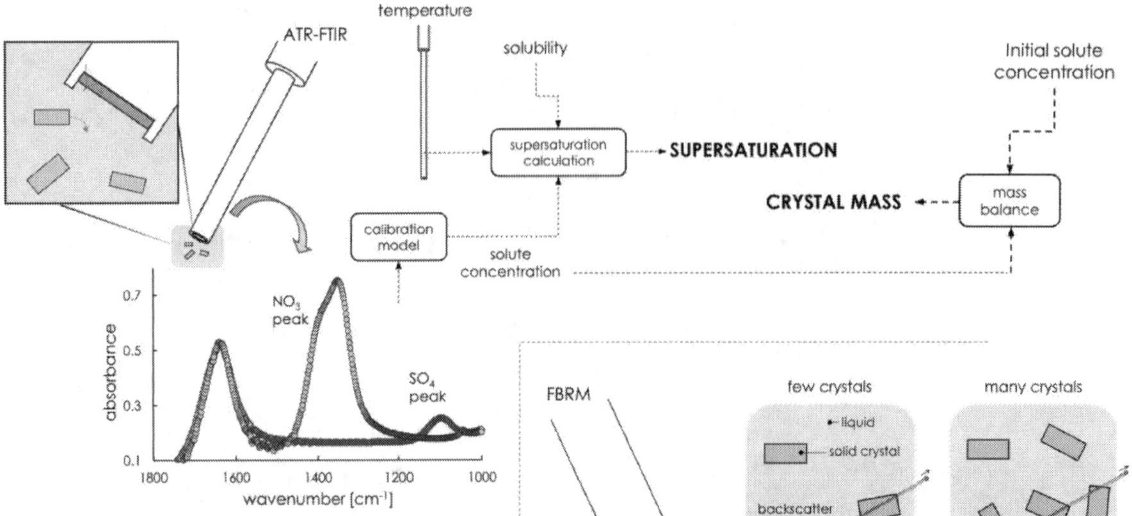

Dunuwila, D. D. and Berglund, K. A. 1997. **ATR FTIR Spectroscopy for In Situ Measurement of Supersaturation.** *Journal of Crystal Growth.*

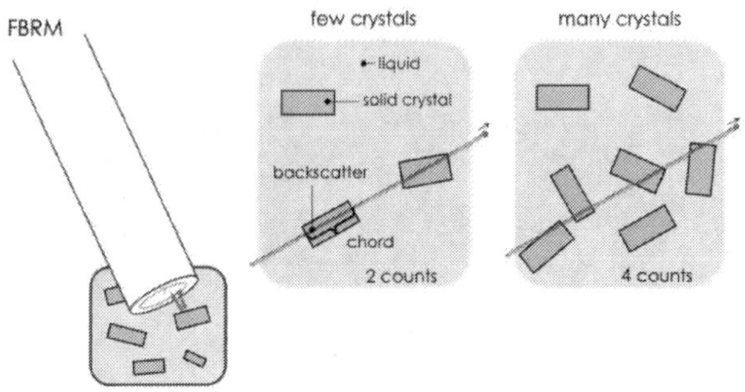

FBRM: **CHORD COUNT** ~ CRYSTAL NUMBER

Heinrich, J. and Ulrich, J. 2012. **Application of Laser-Backscattering Instruments for In Situ Monitoring of Crystallization Processes - A Review.** *Chemical Engineering & Technology.*

9

2. Crystallization Dynamics from a Different Perspective:

Mass-Count Framework

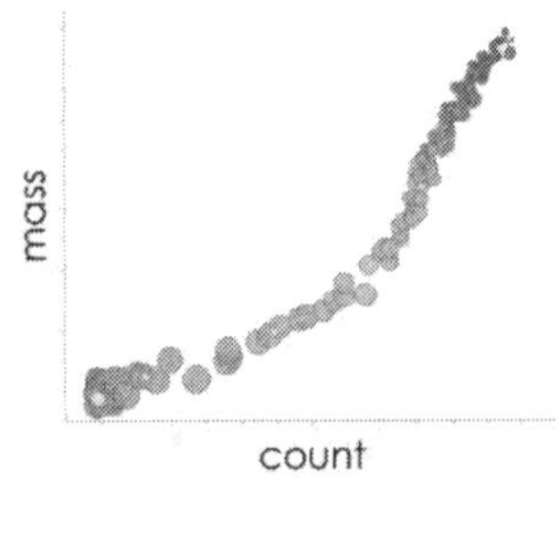

Mass-Count Plot

UNSEEDED, COOLING CRYSTALLIZATION

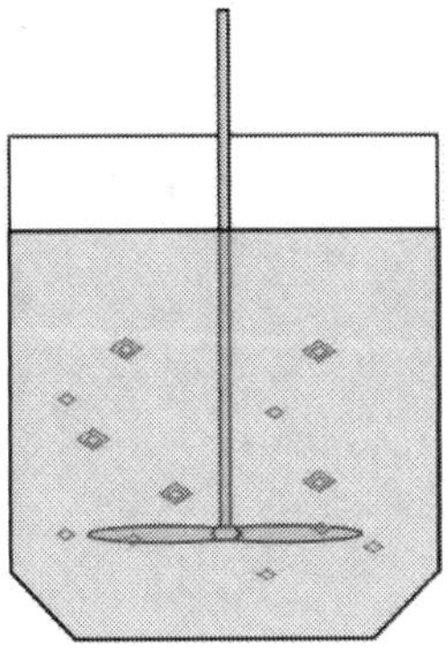

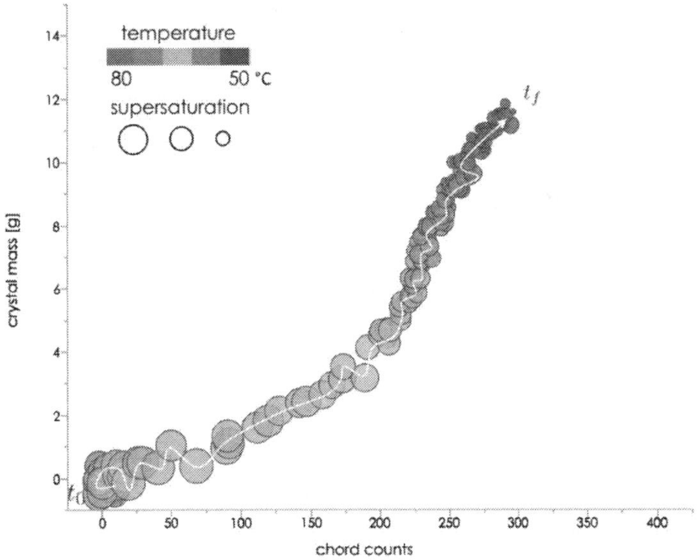

Griffin, D. J., Grover, M. A., Kawajiri, Y., and Rousseau, R. W. 2015. **Mass-Count Plots for Crystal Size Control.** *Chemical Engineering Science.*

11

Mass-per-Count Indicates Crystal Size

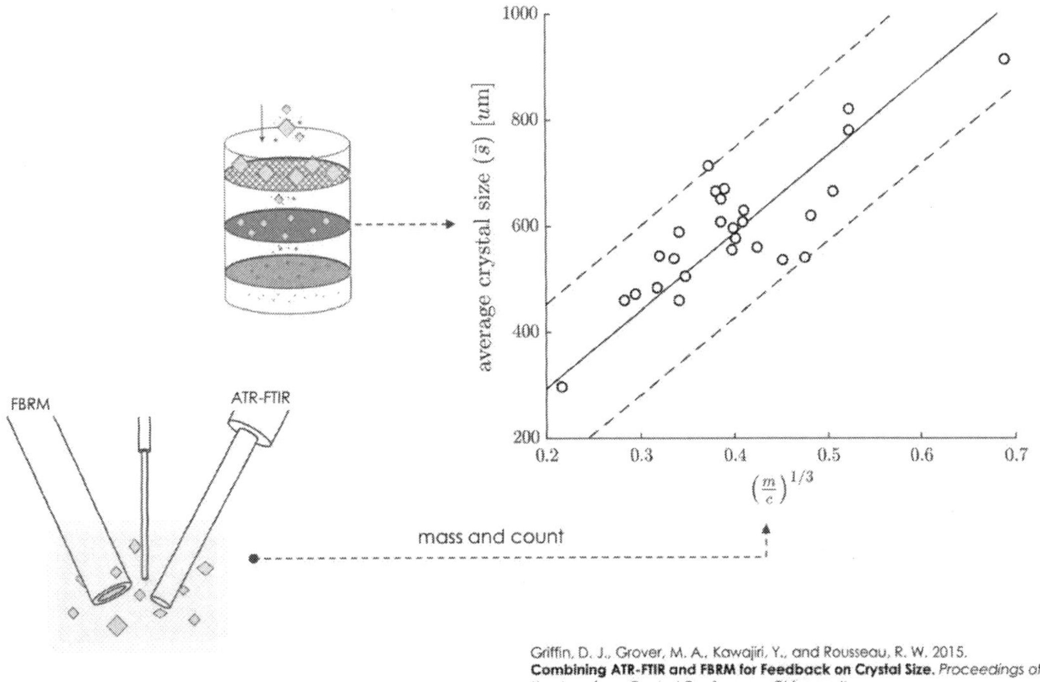

Griffin, D. J., Grover, M. A., Kawajiri, Y., and Rousseau, R. W. 2015. **Combining ATR-FTIR and FBRM for Feedback on Crystal Size.** *Proceedings of the American Control Conference*, Chicago, IL.

12

Clarified Control Objective

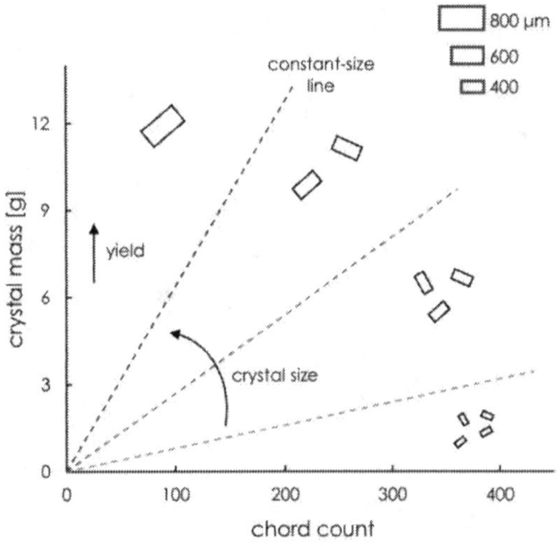

Clarified Control Objective: reach a target crystal mass and number

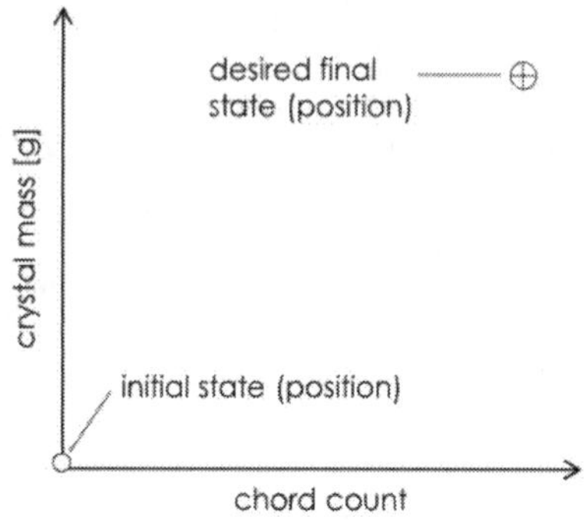

13

3. Data-Driven Modeling and Dynamic Programming

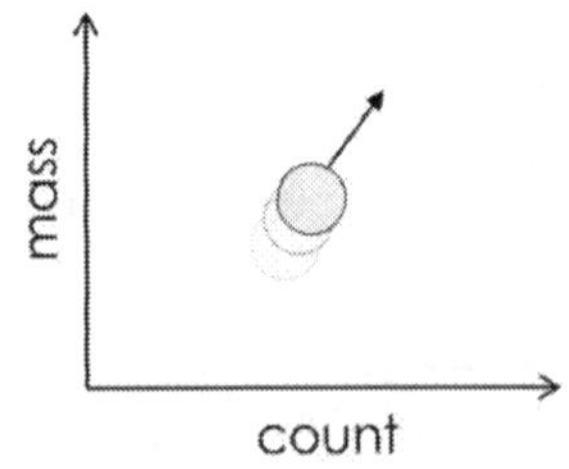

Data-Driven Modeling and Dynamic Programming

data-driven modeling

MEASUREMENT DATA →

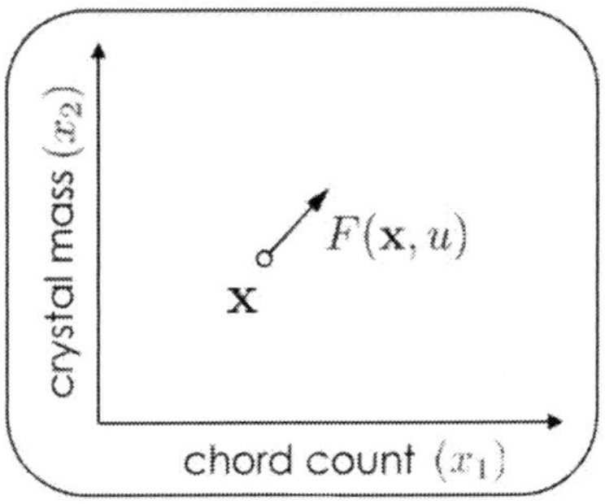

↓

dynamic programming

target size (final MC position) and batch time →

$$\underset{u_t}{\text{minimize}} \quad \Phi(\mathbf{x}_t, u_t)$$

subject to

$$\mathbf{x}_{t+1} = F(\mathbf{x}_t, u_t) + \mathbf{x}_t$$

→ CONTROL POLICY

15

Model From Data

MEASUREMENT DATA

'weighted-regression' algorithm

http://dgriffin36.github.io/Crystallization-Control-Feedback-Policy-from-Data/

↓

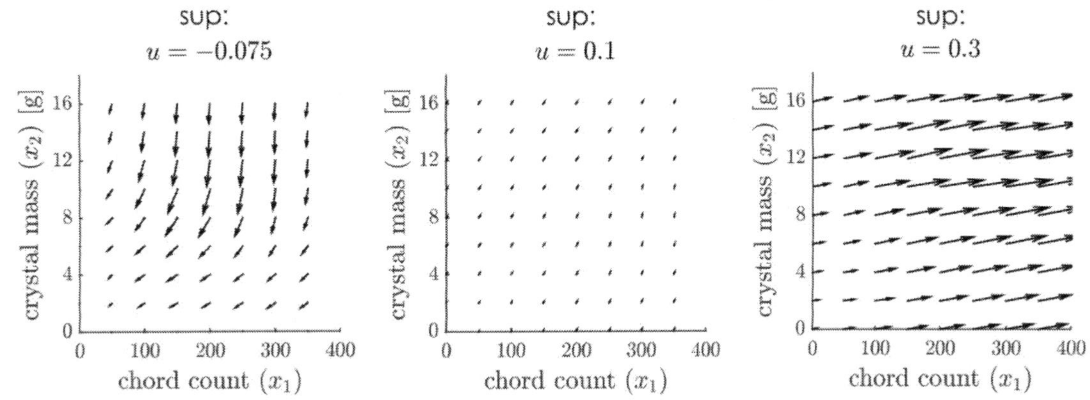

16

Dynamic Programming

Input-effort

$$\underset{\text{control policy } u_\tau = \pi(\mathbf{x}_\tau)}{\text{minimize}} \quad \left\{ \rho \sum_{\tau=0}^{N-1} u_\tau^2 + \sum_{\tau=1}^{N} (t_\tau / t_N)^\gamma d(\mathbf{x}_\tau, \mathbf{x}^\oplus) \right\}$$

batch time

distance-to-target

$$\text{subject to} \quad \mathbf{x}_{\tau+1} = F\big(\mathbf{x}_\tau, u_\tau = \pi(\mathbf{x}_\tau)\big)\Delta t + \mathbf{x}_\tau, \; \tau = 0, \ldots, N-1;$$

$$\mathbf{x}_0 = \mathbf{x}^{\text{init}}.$$

model of dynamics

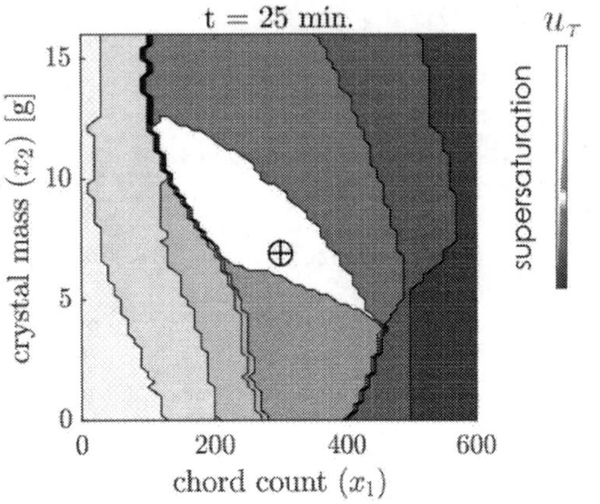

http://dgriffin36.github.io/Crystallization-Control-Feedback-Policy-from-Data/

17

Feedback Control Strategy

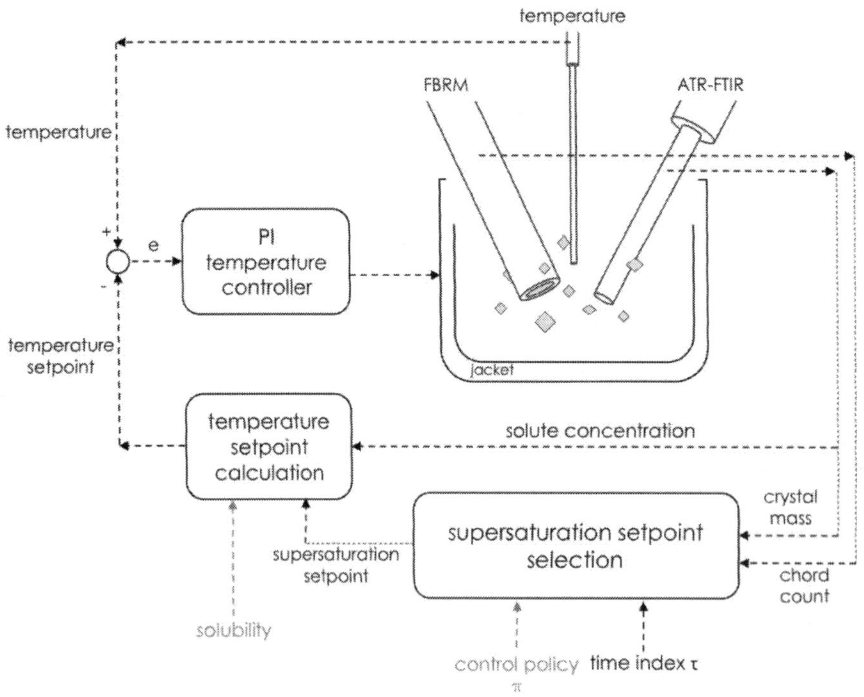

18

4. Results:

Application of strategy to control crystal size

Three Test Runs

Run number	Target final position: (count, mass (g))	Target crystal size (μm)	Set batch time (min.)
1	(300, 7)	420 (261–579)*	30
2	(200, 10)	540 (381–701)	60
3	(75, 11)	775 (612–937)	120

*95% confidence interval for the measured average crystal size if the mass-count target is hit exactly

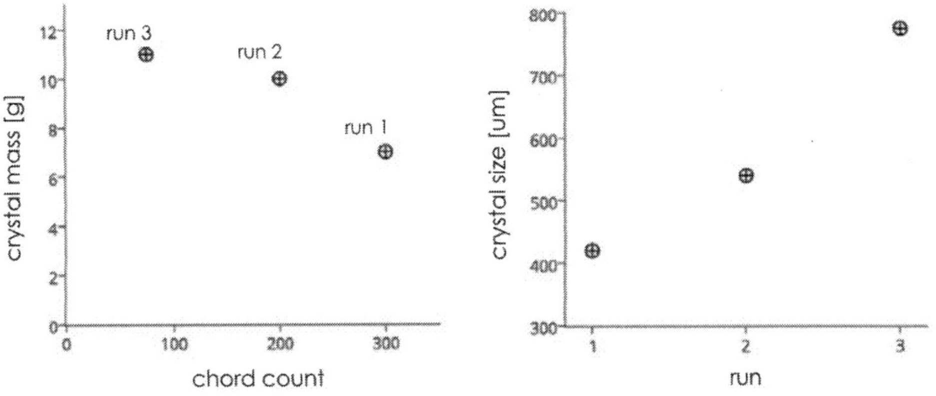

for each run we identify the appropriate control policy

$$u_\tau = \pi(\mathbf{x}_\tau)$$

20

Results

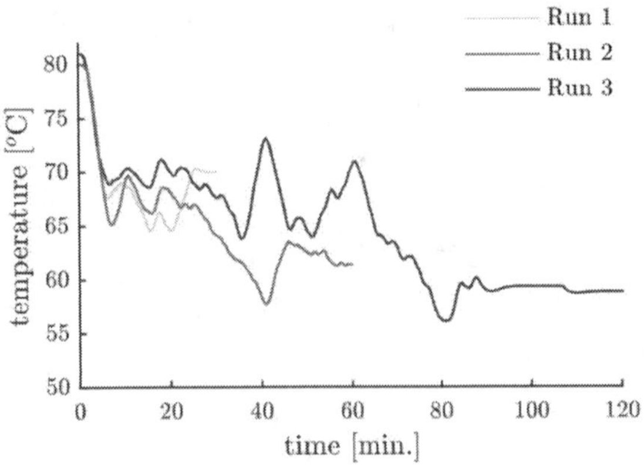

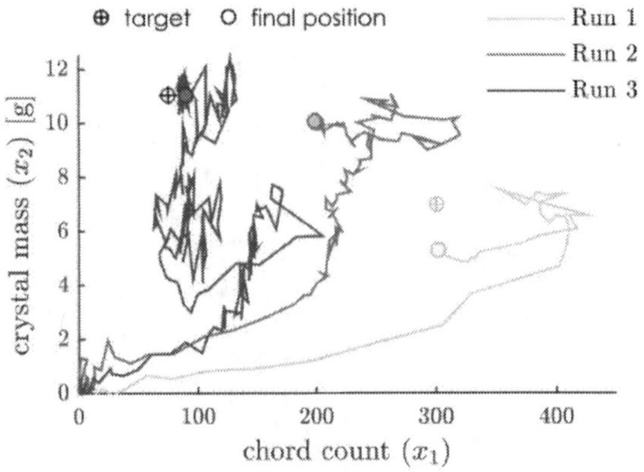

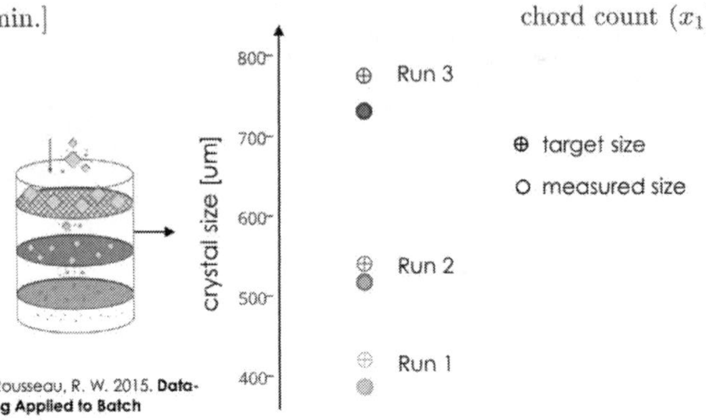

Griffin, D. J., Grover, M. A., Kawajiri, Y., and Rousseau, R. W. 2015. **Data-Driven Modeling and Dynamics Programming Applied to Batch Cooling Crystallization.** *Submitted.*

21

Summary and Conclusion

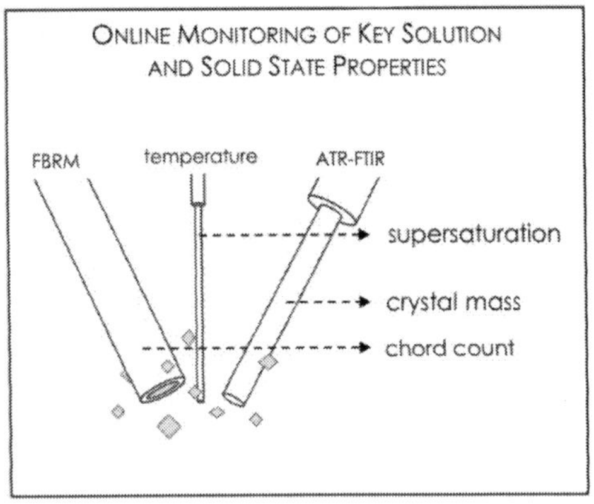

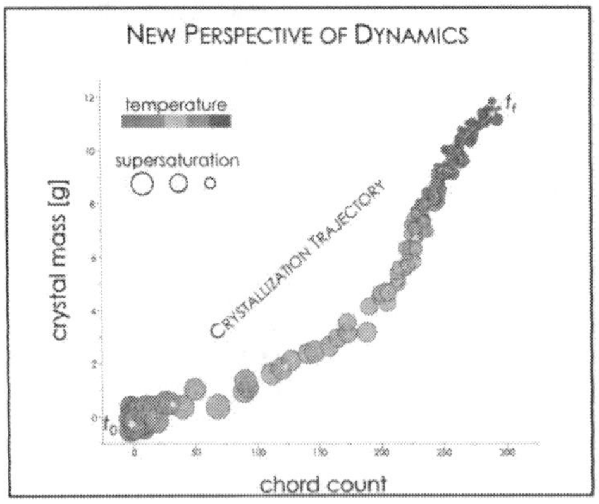

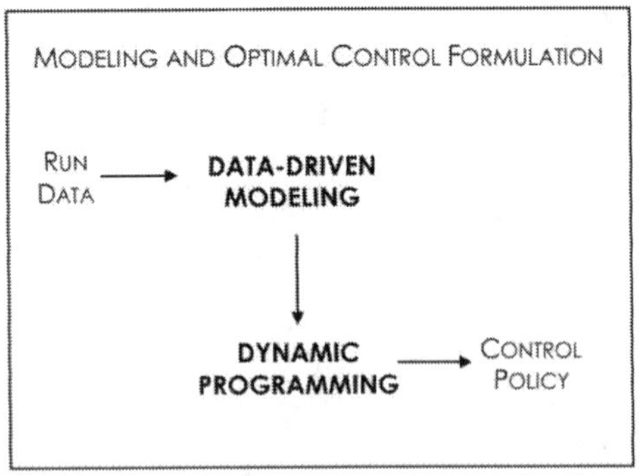

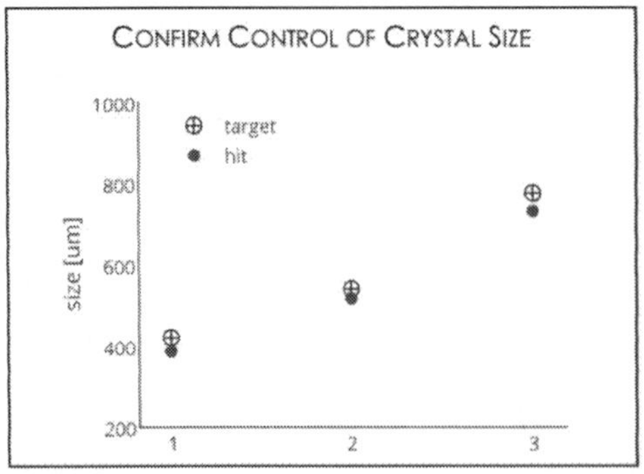

Acknowledgments

Grover, Kawajiri, and Rousseau Research Groups

Prof. Evangelos A. Theodorou (Aerospace Engineering GATECH)

- CONSORTIUM FOR RISK EVALUATION WITH STAKEHOLDER PARTICIPATION (CRESP)
- NUCLEAR ENERGY UNIVERSITY PROGRAMS (NEUP)
- INNOVATIONS IN FUEL CYCLE RESEARCH
- GEORGIA RESEARCH ALLIANCE, AND THE CECIL J. "PETE" SILAS ENDOWMENT

23

434276 Nonlinear Model Predictive Control of a Batch Crystallization Process

Abstract:

Batch crystallization is crucial in the pharmaceutical industry because more than 90% of the active pharmaceutical ingredients (API) are in the form of crystals. The crystal size and shape distribution is important for both downstream processing and product quality. The size and shape distribution can be controlled by manipulating the cooling profile of the reactor, which directly affects the supersaturation ratio. The dynamic evolution of the crystal size and shape distribution has been modeled as a system of differential algebraic equations derived using the method of moments [1].

Here, we present a nonlinear model predictive control (NMPC) formulation for batch crystallization using the JModelica [2] platform. This DAE-constrained optimization problem is solved by discretizing the system using Radau collocation and optimizing the resulting algebraic nonlinear problem with Ipopt. We analyze the performance of the NMPC in terms of set point change, model mismatch, and system noise. For all cases, the performance of the NMPC is much better than the open-loop optimal control strategy.

Nevertheless, many state variables cannot be measured accurately online and some model parameters change from batch to batch. This challenge drives the need for an optimization approach like nonlinear moving horizon estimation (NMHE) to estimate unknown states and parameters prior to solving the NMPC problem. Therefore, using the same JModelica framework, we also present a NMHE formulation for the batch crystallization process. Given the batch process, and combining these two formulations, we obtain an expanding horizon estimation problem and a receding horizon model predictive control problem. We test the performance of this control strategy using a case study of a 90-minute batch crystallization process with 90 control steps and sampling steps. The combined solution time for the NMHE and the NMPC formulations is well within the sampling interval, allowing for real world application of the control strategy. Furthermore, significant performance improvements are possible with this online, model-based control strategy over the existing open-loop optimal solution.

References:

[1] Acevedo, D., Tandy, Y., and Nagy, Z. K., Multiobjective Optimization of an Unseeded Batch Cooling Crystallizer for Shape and Size Manipulation, Industrial & Engineering Chemistry Research, 54(7), pp 2156-2166,2015.

[2] Åkesson, J., Årzén, K. E., Gäfvert, M., Bergdahl, T., & Tummescheit, H., Modeling and optimization with Optimica and JModelica. org—Languages and tools for solving large-scale dynamic optimization problems, Computers & Chemical Engineering, 34(11), pp1737-1749, 2010.

423850 Modeling of Crystallization of Melt-Based Oral Dosages in a Drop-on-Demand Manufacturing System

Abstract:

Under the NSF supported Engineering Research Center for Structured Organic Particulate Systems (NSF ERC-SOPS), a dropwise additive manufacturing process for solid oral drug production has been developed as an alternative to conventional pharmaceutical manufacturing methods [1]. This mini manufacturing process for the production of pharmaceuticals utilizes drop on demand printing technology for automated and controlled deposition of melt-based drug formulations onto edible substrates [2]. A supervisory control system, including on-line monitoring, automation and closed-loop control, is implemented on the process, in order to produce individual dosage forms with the desired dosage amount and crystal morphology [3].

The temperature of the deposited drops is controlled indirectly by controlling the temperature of the substrate using a Peltier device placed beneath the substrate on the xy-stage. This allows control of the rate of solidification of melts and thus control of nucleation and crystallization phenomena. Precise control of the drop solidification process occurring on the substrate is very important since the crystallization temperature profile has a strong effect on product solid-state characteristics, influencing the dissolution properties and hence the bioavailability of the drug [3]. Recently, Içten et al. [3] reported that applying a fast cooling rate of 10 °C/min to the deposits containing naproxen results in faster dissolution profiles compared to a slow cooling rate of 1 °C/min, indicating that the crystallization temperature profiles can be used to enhance the solubility of dissolution limited drugs such as naproxen.

This paper presents a model based on the non-isothermal Avrami kinetic equation [4] for the solidification and crystallization processes of the drug deposits produced using the dropwise additive manufacturing process. Temperature dependent crystallization kinetics is monitored under different cooling profiles via an optical microscopy with a hot-stage. Using the proposed model, the temperature profiles leading to the desired solid-state parameters such as the mean feature size can be optimized, which allows achievement of the desired product quality, in this case, dissolution rate.

1. L. Hirshfield, A. Giridhar, L. S. Taylor, M. T. Harris, G. V. Reklaitis, "Dropwise Additive Manufacturing of Pharmaceutical Products for Solvent-based Dosage Forms," *Journal of Pharmaceutical Sciences*, vol. 103, issue 2, pages 496-506, 2014.
2. E. Içten, A. Giridhar, L. Taylor, Z.K. Nagy, G.V. Reklaitis, 2014, Dropwise Additive Manufacturing of Pharmaceutical Products for Melt-based Dosage Forms, Journal of Pharmaceutical Sciences, DOI 10.1002/jps.24367.
3. E. Içten, Z. K. Nagy, G.V. Reklaitis, 2014, Supervisory Control of a Drop on Demand Mini-manufacturing System for Pharmaceuticals, Comp. Aided Process Eng., 33, 535-540.
4. M. Avrami, 1939, Kinetics of phase change I General Theory, Journal of Chemical Physics, 7, 12, 1103-1112.

421342 Towards Predictive Modeling of Crystallization Fouling: Taking into Account Fouling Layer Structures

Abstract:

Towards Predictive Modeling of Crystallization Fouling: Taking into Account Fouling Layer Structures

Jie Xiao, Feng Zhang, Jian Han, Xiao Dong Chen

Suzhou Key Laboratory of Green Chemical Engineering

School of Chemical and Environmental Engineering

College of Chemistry, Chemical Engineering and Materials Science

Soochow University, Suzhou, Jiangsu Province 215123, China

Abstract

The fouling layers formed on heat exchanger surfaces exhibit complicated structures, which essentially determine flow hydrodynamics, fouling kinetics and hence the heat transfer performance. In-depth understanding of deposit layer formation and its structural effects on the fouling process could lead to effective fouling mitigation or even prevention strategies. The exploration tasks, however, are not trivial due to a lack of knowledge of the crystallization process and the complicated deposit structure.

In this talk, I will summarize our past and on-going efforts on numerical simulation of different stages of a complex crystallization fouling process, focusing on deposit layer growth and its impact on the fouling process. The developed method incorporates a pseudo-dynamic scheme where the dynamic fouling process is approximated as a set of sequential steady-state processes taken place in a continuously varying geometric domain. This unique approach allows the characterization of mass, momentum and heat conservations of a continuous flow of liquid over a growing fouling layer. Dynamic evolution of the fouling layer surface and its intricate interactions with hydrodynamics and fouling kinetics can then be rigorously taken into account. The introduced model was validated using the experimental data for a calcium sulphate fouling system. The effects of the solution chemistry and operating conditions on fouling resistance evolution were quantified through a comprehensive parametric study. Furthermore, we are developing methods that can be used to quantitatively evaluate the effects of fouling layers' *interior* structures on fouling dynamics. We have systematically investigated four representative schemes for fouling layer characterization: i.e., a homogeneous porous medium that is impermeable to water (HoIm), a heterogeneous porous medium that is impermeable to water (HeIm), a homogeneous porous medium that is permeable to water (HoPe), and a heterogeneous porous medium that is permeable to water (HePe). Under the same operational conditions, four models offer significantly different prediction results on the fluid velocity, temperature distribution and fouling resistance. It can be concluded that numerical model development should take the fouling layer structure into account, and the scheme of HePe that best resembles a real fouling layer structure should be a promising option.

Keywords: Crystallization fouling; fouling inhibition; fouling layer structure; CFD model; pseudo-dynamic scheme

434926 Modeling Crystallization in Hydrogels Under Double-Diffusion Conditions

Abstract:

In nature, the deposition of biominerals usually takes place in gel-like extracellular matrix environments where the transport of ions to growth sites is regulated. These matrices play an important role in controlling the nucleation, growth, and the final morphology of biominerals. Hydrogels can serve as excellent matrix models for biomineralization. The hydrogel matrix provides a template for crystallization, while at the same time controls the diffusion of ions, regulates the nucleation and growth rates, and directs the hierarchical organization of the minerals. In this work, we investigate the crystallization of calcium oxalate, the primary mineral constituent of kidney stones, in hydrogels. We employ a double-diffusion experimental set-up in which a hydrogel (serving as the crystallization medium) is brought in contact with calcium and oxalate ion reservoirs that serve as sources for the reactants, thus allowing for counter diffusion of the reactant ions across the hydrogel column and precipitation of calcium oxalate in the gel. We study the nucleation, growth, polymorphic outcome and morphological evolution of the calcium oxalate crystals as affected by the gel density, reservoir concentrations, molar stock solution ratios, and additives. The experimentally observed precipitation patterns (time and location within the gel) in the calcium oxalate-gel systems are in very good agreement with theoretical predictions based on the relative diffusion coefficients of both ions. Such understanding of the diffusion and precipitation processes can aid in predicting the time scales of nucleation and growth of the crystals and can facilitate further use of gel matrices as templates for biomineral formation or for development of bio-inspired materials.

430770 Polymer adsorption in nanopores and nanoporous substrates

Abstract:

Polymer adsorption is a ubiquitous phenomenon occurring in a wide range of disciplines and finding numerous technological and healthcare applications. The mechanisms of polymer adsorption on surfaces and in pores are complex due to a competition of various entropic and enthalpic factors. Due to adsorption of monomers to the surface, the chains gain in enthalpy yet loose entropy due to confining effects. This competition leads to the existence of critical condition of adsorption when the enthalpy gain and entropy loss are in balance. The critical conditions are controlled by the confining geometry and effective adsorption energy, which depends on the solvent composition and temperature. We present a Monte Carlo simulation study of polymer adsorption in nanopores focusing on the explanation and quantification of the critical conditions. We show that the main contribution comes from the chains located at the external surface of porous substrates. These chains are adsorbed in pores only partially and attain so called "flower" conformations. The incremental gauge cell MC method allowed us to determine the free energies of adsorbed chains and to calculate the partition coefficient as a function of adsorption potential, pore size, and chain length. We found and justified the critical conditions for chain length independent separation in polymer chromatography. Drawing on a case study example of separation of a series of linear polystyrenes, we found quantitative agreement with the experiments.

425143 Adsorption-Induced Surface Stresses Based on Electronic Structure DFT Calculations

Abstract:

All porous materials deform during fluid adsorption. Although for most of the materials the experimentally observed strains are small, this phenomenon has a number of important applications, including sensors [1], actuators [2], CO2 sequestration [3]. In addition to those, adsorption-induced deformation has potential for characterization of porous materials because deformation is very sensitive to the pore sizes. In recent years there has been a major progress in developing theory of adsorption-induced deformation [4-6]. However, these studies were focused on calculating the pressure in the fluid phase, while the stresses in the solid phase were not considered explicitly. The current work fills this gap, presenting the electronic-structure density functional theory modeling of the dry and wet silica surfaces. In addition to the information on change of the surface energy with fluid adsorption, our calculations show how the components of surface stress tensor change. This change is the key parameter necessary for quantitative understanding of the adsorption-induced deformation phenomenon.

1. Baimpos, T.; Gora, L.; Nikolakis, V. & Kouzoudis, D. Selective detection of hazardous VOCs using zeolite/Metglas composite sensors *Sensors and Actuators A: Physical*, 2012, 186, 21-31

2. Bertinetti, L.; Fischer, F. & Fratzl, P. Physicochemical basis for water-actuated movement and stress generation in nonliving plant tissues *Physical Review Letters*, 2013, 111, 238001

3. Brochard, L.; Vandamme, M.; Pellenq, R. J.-M. & Fen-Chong, T. Adsorption-induced deformation of microporous materials: coal swelling induced by CO2-CH4 competitive adsorption *Langmuir*, 2012, 28, 2659-2670

4. Ustinov, E.; Do, D. Effect of adsorption deformation on thermodynamic characteristics of a fluid in slit pores at sub-critical conditions. *Carbon* 2006, 44, 2652−2663.

5. Gor, G. Y.; Neimark, A. V. Adsorption-induced deformation of mesoporous solids: Macroscopic approach and density functional theory. *Langmuir* 2011, 27, 6926−6931.

6. Long, Y.; Palmer, J. C.; Coasne, B.; Sliwinska-Bartkowiak, M.; Jackson, G.; Müller, E. A. & Gubbins, K. E. On the molecular origin of high-pressure effects in nanoconfinement: The role of surface chemistry and roughness *J. Chem. Phys.*, 2013, 139, 144701

431341 Large-Scale Topologically-Guided Design of Metal-Organic Frameworks and Application to Cryo-Adsorbed Hydrogen Storage

Abstract:

Metal-organic frameworks (MOFs) are porous crystalline materials constructed by combining molecular inorganic and organic building blocks into different topological nets. The combinatorial possibilities give access to an almost unlimited diversity of pore structures with different chemical and textural properties that can in principle be exploited in a wide variety of applications such as catalysis, separations, and gas storage. Major challenges are to efficiently identify what combinations of building blocks and topologies constitute MOFs that are both outstanding and synthesizable and to discern the performance boundaries of these materials. Here, we demonstrate an automated, topologically-guided approach to computationally construct MOFs and use it to generate a population of 13,512 MOFs encompassing 41 distinct topologies.

All structures were evaluated using grand canonical Monte Carlo simulations to determine the viability of using the MOFs for cryo-adsorbed hydrogen storage at 100 bar. We find a topological dependence of hydrogen storage performance, which is linked to the topological dependence of property-property relationships of the investigated MOFs. A large number of structures were found to have higher deliverable capacities than the incumbent hydrogen storage technology – which corresponds to a compressed hydrogen gas (CHG) tank at 700 bar and room temperature – and than cryo-compressed hydrogen gas at 100 bar and 77 K. The heats of adsorptions for the optimal structures were below 5 kJ/mol, which is beneficial for the thermal management of a potential cryo-adsorbed storage system.

Among the topologies that yielded structures for which we predicted cryo-adsorption loadings higher than the CHG tank, we focused on a rare (6,4)-connected MOF topology which had only been accessed synthetically using a "missing linker" strategy to generate 6-coordinated planar inorganic units [1]. Our automated generation approach produced MOFs of this topology based on organic 6-coordinated planar units instead. We selected some structures of this topology for more detailed study. We found that this topology is inherently non-catenating and that increasing the storage pressure beyond 100 bar (at 77 K) starts to negate the benefits of placing an adsorbent material in the tank. Two of the investigated structures were synthesized and activated successfully by our experimental collaborators.

[1] D. Feng, W.-C. Chung, Z. Wei, Z.Y. Gu, Y.P. Chen, D.J. Darensbourg, and H.-C. Zhou, *J. Am. Chem. Soc.*, **2013**, *135* (45), pp 17105–17110

413501 Adsorptive Separation of Xylene Isomers Using Zn2(BDC)2(DABCO) [MOF1] Metal Organic Framework

Abstract:

Adsorptive Separation of Xylene Isomers using $Zn_2(BDC)_2(DABCO)$ [MOF1] Metal Organic Framework

Arun G and Sudeep N Punnathanam

Isomeric forms of dimethylbenzene(xylene) especially the *ortho* and *para* isomers, find wide applications in the polymer, pharmaceutical and dye industries[1-3]. However, the separation of these isomers, especially the *ortho* isomer, remains a challenging task till date, due to the closeness of its melting and boiling points with the *meta* isomer[3]. The metal-organic framework MOF1, also known as $Zn_2(BDC)_2(DABCO)$, has been shown to selectively adsorb ortho-xylene from a mixture of xylenes[5]. In this work, Grand Canonical Monte Carlo(GCMC)[4] simulations are used to study the adsorption of xylenes on MOF1 in order to understand the reasons for selectivty towards o-Xylene. MOF1 shows guest dependent framework flexibility[6], i.e., the structure of the adsorbent changes with the chemical nature and loading of the adsorbate. In the case of MOF1, two distinct structures, namely a large pore and a narrow pore, are observed. Comparison of GCMC simulations with experiments reveal that at high loading of xylenes, the MOF1 undergoes a structural transition from the large pore to the narrow pore structure. An examination of density distributions inside the pore reveal two adsorption sites for ortho-xylene but only one for meta and para isomers. The presence of an additional adsorption site is the source of selectivity of MOF1 towards the ortho-isomer. The existence of this additional site is due to favorable alignment of the CH_3 groups in the ortho isomer towards the oxygen atoms of the benzene dicarboxylate linkers which make up the pore walls of the adsorbent. Such favorable alignment is not possible with the meta and the para isomers.

References:

1. Reynold, L.A, Quinizarin, Org. Synth. Coll., 1941. 1, p. 476.

2. Neurath, A. R.., Design of Microbicide for Prevention of Sexually Transmitted Diseases Using A Pharmaceutical Excipient. AIDS Patient Care ST, 1999, 27(1), p. 11-21

3. Fabri, J., Graeser, Ulrich, and Simo, Thomas A, Xylenes, Ullmann's Encyclopedia of Industrial Chemistry. 2002: Wiley-VCH, Weinheim.

4. Frenkel, D., Smit, B., Understanding Molecular Simulation 2e, Academic Press, 2002.

5.Nicolau, M., Marco.P.M., et al., Single- and Multicomponent Vapor-Phase Adsorption

of Xylene Isomers and Ethylbenzene in a Microporous Metal – Organic Framework., J. Phys. Chem. C 2009, 113, p.13173–13179.

6. Dybtsev, D.N., H. Chun, and K. Kim, Rigid and Flexible: A Highly Porous Metal-Organic Framework with Unusual Guest-Dependent Dynamic Behavior. Angew. Chem., 2004. 43: p. 5033-6.

422868 Temperature and Loading-Dependent Diffusion of Light Hydrocarbons in ZIF-8 As Predicted through Fully Flexible Molecular Simulations

Abstract:

Zeolitic imidazolate framework-8 (ZIF-8), a prototypical member of the ZIF class of metal-organic frameworks (MOFs), has been experimentally shown to exhibit interesting sieving properties with regard to increasing alkane chain length and morphology. The ability to predict diffusion characteristics of these systems using molecular simulations would create opportunities to find similar systems suitable for industrial separations. However, the challenge with simulating diffusion of light key hydrocarbons in ZIF-8 is that diffusion is too slow to observe with standard molecular dynamics (MD). Previous simulation studies have determined that adsorbate diffusion through cage-type nanoporous materials proceeds by an activated hopping process between low energy states and can be modeled using dynamically corrected transition state theory (dcTST). Umbrella Sampling (US), a free energy sampling method, has been applied to study small molecule and C1-C4 hydrocarbon diffusion in a fully flexible ZIF-8 model. Diffusion coefficients of methane, ethane, ethylene, propane, propylene, n-butane, and 1-butene in ZIF-8 are reported over a temperature range of 0 to 150°C and loadings from infinite dilution to saturation. The results demonstrate how the adsorbate influences the ZIF-8 window during the crossing event.

433850 Hydrogen Isotope Separation By Carbonaceous Materials

Abstract:

Deuterium is useful for developing many of functional materials, such as long-lived LSI, optical fibers with low light attenuation, and deuterium-modified drugs. Moreover, a compact fusion reactor, which uses deuterium and tritium as fuel, has attracted much attention because of the recent technical breakthrough made by Lockheed Martin. Though, separation of hydrogen isotopes are difficult and requires energy intensive processes, quantum molecular sieving, which separates lighter molecules from heavier ones because of the differences in zero-point energies of adsorbed species, is one of the candidates that may be used as the deuterium production process. The promising pore structure for the hydrogen isotope separation is a regular carbon cylindrical pore with a molecular hydrogen size (ca 0.3 nm). However, there are very few adsorbents with such a regular pore structure. We have therefore focused attention on deposition of carbons on the pore surface of zeolites (zeolite-carbon complex: ZCC), deposition of carbons on the surface of activated carbons (molecular sieve carbon: MSC), and mechanical deformation of single wall carbon nanotube (SWNT), to control the size of internal pore and pore entrance. ZCCs and MSCs were synthesized in silico using quench molecular dynamics simulation. The selectivities of D_2 over H_2 of the obtained structures (ZCC, MSC, and deformed SWNT) were evaluated using grand canonical Monte Carlo simulations with the Feynman-Hibbs effective potential (FH-GCMC) and transition state theory, and we have found that those structures are promising for kinetic quantum molecular sieving of hydrogen isotopes.

428637 Enhanced Ammonia Production with Combined Reaction-Separation: Kinetic of Reaction and Adsorption

Abstract:

Enhanced Ammonia Production with Combined Reaction-Separation: Kinetic of Reaction and Adsorption

Mohammadmahdi Malmali, Kevin Wagner, Alon V. McCormick, Edward J. Cussler

Chemical Engineering and Material Science, University of Minnesota, 421 Washington Ave. SE, Minneapolis, MN 55455-0132

According to US Department of Health ammonia is the most commonly produced industrial chemical in US. About 13 million metric tons of ammonia is consumed in US, which more than 80% of this amount was used in agriculture sector as different forms of fertilizer. Ammonia is produced via catalytic Haber-Bosch process. Despite a century of optimization and improvement in catalyst, single-pass ammonia production is reaching only up to 20% conversion. This process also requires large capital costs with large amount of sacrificial fossil fuels to obtain raw materials. We have developed a novel reaction-separation process that enables us to increase the conversion to above 95% by simultaneous and selective adsorption of ammonia with magnesium chloride. The feed for this process is obtained from totally renewable resources, in which hydrogen and nitrogen are obtained from electrolysis of water and membrane separation of air, respectively. This process has a great potential of being a local scale source of ammonia production for farmers in North America, Sub-Saharan Africa and elsewhere in the world.

Here, we present latest results for simultaneous reaction-separation process, including a detailed in-depth study of adsorption of ammonia in a fixed bed. We investigate altered MgCl2 particle sizes, bed pressure, feed composition, temperature, absorbent dilution and their effects on absorbent capacity and the sharpness of the breakthrough curve.

We also investigate these effects on the desorption curves, which will be important for an industrial scale continuous operation process. Our results indicate that diffusion of the ammonia in the bulk gas to the particle surface is fast, and the process is dominated by a recrystallization of the ammonia with the magnesium chloride. We also investigated the application of knowledge gained from adsorption studies to improve the kinetics of ammonia production in the reaction separation process. The improved adsorption characteristics enables us to gain faster reaction kinetics and optimize a viable process.

432952 Modeling and Kinetics of Phytate Adsorption from Corn Stillages on a Weak-Base Ion Exchange Resin

Abstract:

Dry-milling process accounts for over 80% of the total corn ethanol production in the United States and it generates at least 3 times the volume of ethanol produced as by-products, the stillages. These co-products have been processed and primarily utilized as animal feed. Phosphorus is predominantly found as phytate in corn, existing up to 80% in this form. Phytate is not fully degraded by the corn ethanol bioprocessing steps including saccharification and fermentation; and the downstream processes does not provide additional degradation of this material, which produces end-products, such as animal feed Dried Distillers Grains with Solubles (DDGS), rich in phytate phosphorus. It is not desirable to have the animal diet rich in phytate for monogastric animals, such as poultry and swine, since the digestive track of monogastrics cannot degrade this form of phosphorus, producing, alongside with other effects, a phosphorus-rich manure. This often may cause environmental concerns, especially when it is applied as fertilizer, leading to saturation of phosphorus in the soil, and runoff to waterbodies. Ion exchange has been studied to extract phosphorus compounds, and weak-base exchange resins have been utilized to extract and purify phytate from acid digestates from cereal materials, such as wheat and rice. Experimental results shows that, the weak-base resin used in this particular study has a high specificity for phytate, and does not have evident adsorption on nutritional ingredients of stillage, such as fat, fiber, proteins and residual starch. The purpose of this work is to evaluate the mass-transfer kinetics on the resin beads, understanding the breakthrough curves and Freundlich, Langmuir and Dubinin-Radushkevich adsorption models on the weak-base resin. The results for these adsorption phenomena will serve as basis for thermodynamic calculations on the resin, in order to establish enthalpies and entropies of the studied system under different conditions, such as dilution rate, pH, and temperature. Experimental set up using the weak-base resin has been able to achieve total phytate adsorption from stillage samples, and desorption values as high as 92% from the resin beads. Our studies indicate a mild-acid pH (on the range from 4 to 6) and temperature stable resin (from 4 °C to 75 °C). Understanding the kinetics and modeling is a key step in optimizing this system on a larger scale.

428118 Uranium Adsorption from Seawater By Amidoxime-Functionalized Polyethylene Fibers: Modeling and Experiments

Abstract:

At a concentration or 3.3 ppb, uranium in seawater can become a valuable resource for future nuclear energy. This goal can be realized by developing cost-effective methods for extracting uranium from seawater. Based on recent work in Japan, Europe, and the U.S., a cost-effective approach should include development of a highly selective adsorbent of relatively high uranium capacity, on the order of 4-10 g U/kg-adsorbent that can be recycled over six times during its lifetime. The adsorbent has to work in a passive mode, taking advantage of the flow of seawater currents. Progress toward the development of such adsorbent will be reported in this presentation. Hollow, high-surface-area polyethylene fibers grafted with amidoxime ligands have shown a uranium capacity over 4 g U/kg-adsorbent in experiments with natural seawater. Attempts to optimize the adsorbent performance during the synthesis and conditioning steps will be discussed. Efforts to understand the adsorption mechanisms include spectroscopic characterization and transport and reaction-kinetic modeling. FTIR and NMR spectra are obtained to determine the reaction mechanism, which is still not fully understood. In earlier modeling work, we have shown that, under some circumstances, uranium uptake from seawater by amidoxime-based adsorbent is reaction limited due to the presence of carbonate ions which compete with amidoxime. In other circumstances, transport effects become important. A current modeling effort includes both transport and kinetic mechanisms to describe uranium uptake by single-adsorbent fibers and adsorbent braids.

432703 Adsorption Studies of Carbon Dioxide, Methane, Ethane and Ethylene on 3A, 4A and 5A Zeolite Molecular Sieves

Abstract:

Light olefins from the petrochemical industry have a steady and growing demand because they are starting blocks for the synthesis of higher added value chemicals such as polymers, rubbers, plasticizers, gasoline additives, and others products. Olefins (e. g. ethene, propene) are generally obtained by steam or thermal cracking of light oil fractions in refineries. Taking into account the limited availability of light oils, the large reserves of natural gas have become a potential source for methane-derived chemicals such as olefins. In this direction, oxidative coupling of methane (OCM) has been developed to upgrade natural gas into valuable chemicals, and it is a promising alternative process to produce ethylene from methane.

In the OCM process, a fluidized bed reactor is used to react methane and oxygen at 750-850°C (exothermic reaction) using selective catalyst, producing a mixture of ethylene, ethane, carbon dioxide, and carbon monoxide. The conversion of reaction is ~30% and a large amount of methane is still present in the final product. Carbon dioxide is also present in the outlet stream (concentrations as high as 25%) and needs to be removed in order to reduce the gas flow for downstream processing (reducing costs and avoiding operational difficulties). This separation is generally accomplished by reactive absorption with amine solutions (MEA and MDEA). Once CO_2 is removed, the paraffin/olefin mixture separation should be addressed. This separation is very difficult because the C2s have similar physicochemical properties, and it is mostly accomplished by cryogenic distillation, which is very energy intensive and therefore expensive.

In order to reduce paraffin/olefin separation costs, pressure swing adsorption (PSA) has demonstrated to be a very efficient operation characterized by reduced energy consumption. For this reason research related with synthesis and evaluation of novel adsorbent materials, and the process modeling and simulation of PSA units have increased in recent years.

Taking into account that different zeolites have been used as adsorbent materials for different applications (ethanol dehydration, natural gas sweetening, N_2–O_2 separation, isomers separation, etc.), and that they are selective for olefin-paraffin separations, this work intends to evaluate the adsorption performance of zeolite 4A and 5A as adsorbent for the separation of OCM product gases.

In order to characterize the adsorbent material and its potential to accomplish the target separation, adsorption isotherms and dynamics experiments of carbon dioxide, methane, ethane and ethylene on 4A and 5A zeolites were done. Zeolites 4A and 5A were synthesized using sodium metasilicate, sodium aluminate and deionized water. The formed hydrogel was matured at room temperature and then crystallized in a stirred tank reactor. The product was filtered, washed and dried, and the final solid was characterized by X ray diffraction. Zeolites were agglomerated with attapulgite clay as binder, by using an rotary tilted pan to obtain 4mm mean diameter particles. Zeolite agglomerates were characterized by measuring crush strength, shape, and density (bulk and particle).

To carry out the adsorption isotherms experiments, a Belsorp-Mini equipment was employed and the measurement temperatures were 77, 298, 308 and 320K. From those experiments different isotherm models were fitted to determine the size pore distribution, surface area and parameters to obtain thermodynamic properties.

Further dynamical experiments were carried out in a fully controlled OCM miniplant. The experiments started with a purge stage of the adsorbent packing with helium at 300°C for 12 hours. After column conditioning, the columns was set to the operating pressure (1 or 5 barg) with nitrogen. Finally, and after stabilization, the column was fed with a mixture of nitrogen (80% v/v) and OCM gas (20% v/v) until the equilibrium condition was reached. The desorption process was done at 300°C and 1bar g. The outlet compositions of the outlet gas mixture were measured with gas chromatography and infrared gas analyzer. Results indicated that both zeolites 4A and 5A are suitable materials to be used as adsorbents in PSA columns for paraffin/olefin separations in a OCM process.

Ethane and Ethylene Adsorption on MIL-53(Al): Experiments and Molecular Simulations

Rui P. P. L. Ribeiro, Bárbara C. R. Camacho, Andriy Lyubchyk, Isabel A. A. C. Esteves, Fernando J. A. L. Cruz, José P. B. Mota
LAQV, REQUIMTE, Departamento de Química, Faculdade de Ciências e Tecnologia, Universidade Nova de Lisboa, 2829-516 Caparica, Portugal.
pmota@fct.unl.pt

I. Introduction

Adsorption on nanoporous materials is an interesting option for the separation of ethane/ethylene mixtures. The adsorption equilibria of ethane (C_2H_6) and ethylene (C_2H_4) has been studied over several adsorbents as activated carbons, zeolites, single-walled carbon nanotubes. More recently, the adsorption characteristics of the C_2H_6/C_2H_4 pair on different metal organic frameworks (MOFs) have been reported [1, 2].

The MIL-53 family is an interesting class of MOFs, consisting of a trivalent metal cation (Al, Cr, Fe, etc.) linked to the carboxylate group of a terephthalate moiety. The adsorption equilibria data of C_2H_6 and C_2H_4 on MIL-53 MOFs present in the literature are scarce, and in the present work we develop a self-consistent framework, combining experimental and molecular simulations, to determine the adsorptive characteristics of C_2H_6 and C_2H_4 onto a MIL-53(Al) MOF. The MIL-53(Al) is composed by corner-sharing $AlO_4(OH)_2$ octahedra interconnected with 1,4–benzenedicarboxylate ligands, possessing one dimensional diamond-shaped nanopores (Figure 1), whose large pore and narrow pore conformations exhibit free dimensions of (0.85×0.85) nm^2 and (0.26×1.36) nm^2 [3 - 5].

The adsorption equilibrium of the pure components over MIL-53(Al) was measured in a high-pressure magnetic-suspension microbalance from Rubotherm GmbH (Germany), spanning temperature and pressure ranges of 303 – 353 K and 0 – 8 bar for C_2H_6, and 323 – 373 K and 0 – 1.7 bar for C_2H_4. Additionally, simulations were performed using the Grand Canonical Monte Carlo (GCMC) algorithm with a parameterization of solid–fluid dispersive interactions based on the TraPPE-UA force field [6]. The GCMC results were checked against the experimental data to validate the molecular force-field employed and establish a thermodynamical analysis by calculating the isosteric heat of adsorption.

Binary adsorption data (C_2H_6/C_2H_4) were determined by simulation using the validated force-field; the Ideal Adsorbed Solution Theory (IAST) was employed to evaluate the ideal behaviour of the adsorbed mixture.

II. Results and Discussion

Single component Equilibria

The absolute adsorption isotherms of C_2H_6 at 303, 323, and 353 K between 0 and 8 bar, and C_2H_4 at 323, 353, and 373 K between 0 and 1.7 bar are plotted in Figure 2. The results present a slightly higher adsorption capacity of MIL-53(Al) towards C_2H_6 than to C_2H_4. The experimental data and the GCMC calculations show good agreement, although the latter slightly over predict C_2H_6 adsorption at 353 K and C_2H_4 at 323 K above 1.5 bar. The molecular simulations considered the modelled MIL-53(Al) in its large pore form, MIL-53lp(Al), and the results demonstrate that this is the most stable conformation in the whole (P, T) phase space studied; this is in accordance with our previous results regarding CH_4 adsorption [7].

Adsorption Energetics

The isosteric heat of adsorption, $Q_{st,}$ was obtained from molecular simulations considering the co-variance formulation of Nicholas and Parsonage [8]. Results show that the isosteric heat, within the studied regions, is independent of temperature for both adsorbates and exhibits a linear dependence with adsorbate loading, increasing essentially with pore filling as a consequence of molecular packing within the adsorbent channels. The energetical dissimilarity between ethylene and ethane is only minorly reflected in the energetics of adsorption, since their isosteric heat values are similar, even in the very low pressure domain, which at zero-coverage results in Q_{st} = 22.5 kJ/mol (C_2H_6) and Q_{st} = 20.5 kJ/mol (C_2H_4). In this region, where entropic considerations can be neglected, the small ΔQ_{st} = 2 kJ/mol difference can be rationalized as arising from enhanced dispersive interactions between the benzene rings and ethane molecules.

The Q_{st} was also calculated from the experimental data, employing the integrated form of the Clausius-Clapeyron equation, and the obtained data show a fairly good agreement with the GCMC results, although the experimental values are slightly higher (*ca.* 5 kJ/mol).

Binary Adsorption

Upon validation of the employed force-field, GCMC computational simulations were performed for C_2H_6/C_2H_4 binary mixtures. The binary adsorption equilibria GCMC results were probed using the IAST and the Toth isotherm multicomponent extension. Figure 3 presents the binary adsorption results obtained at 323 K and 1 bar. Both the IAST and the extended multicomponent Toth model predict the GCMC simulation data well, indicating that the mixture does have ideal behaviour in the studied conditions.

Acknowledgements: Financial support from FCT/MEC (Portugal) through projects EXCL/QEQ-PRS/0308/2012, PTDC/AAC-AMB/108849/2008, PTDC/CTM/104782/2008, PEst-C/EQB/LA0006/2013 and UID/QUI/50006/2013 is gratefully acknowledged.

References: [1] M. Jorge et al., *Ind. Eng. Chem. Res.*, **53** (2014) 15475. [2] C. Gücüyener et al., *J. Am. Chem. Soc.*, **132** (2010) 17704. [3] T. Loiseau et al., *Chem. Eur. J.*, **10** (2004) 1373. [4] Y. Liu, et al., *J. Am. Chem. Soc.*, **130** (2008) 11813. [5] C. Serre, *Adv. Mater.*, **19** (2007) 2246. [6] M.G. Martin, J.I. Siepmann, *J. Phys. Chem. B*, **102** (1998) 2569. [7] A. Lyubchyk et al., *J. Phys. Chem. C*, **115** (2011) 20628. [8] D. Nicholson, N.G. Parsonage, *Computer Simulation and the Statistical Mechanics of Adsorption*, Academic Press, New York, 1982.

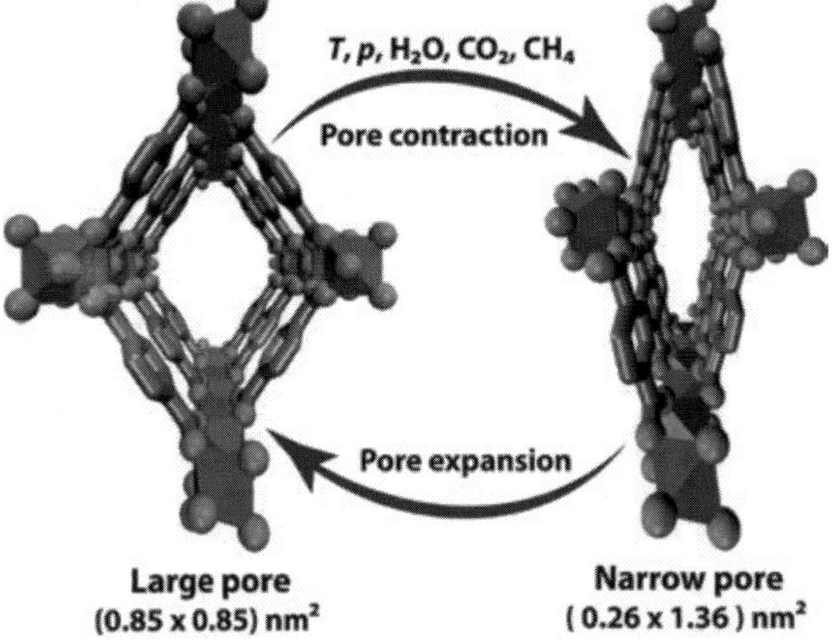

Figure 1. Structure of the two solid conformations of the MIL-53(Al) framework. The large pore (lp) form is orthorhombic [4] and belongs to the *Imma* symmetry space group whilst the narrow pore (np) analogue is monoclinic belonging to the *C2/c* category [5].

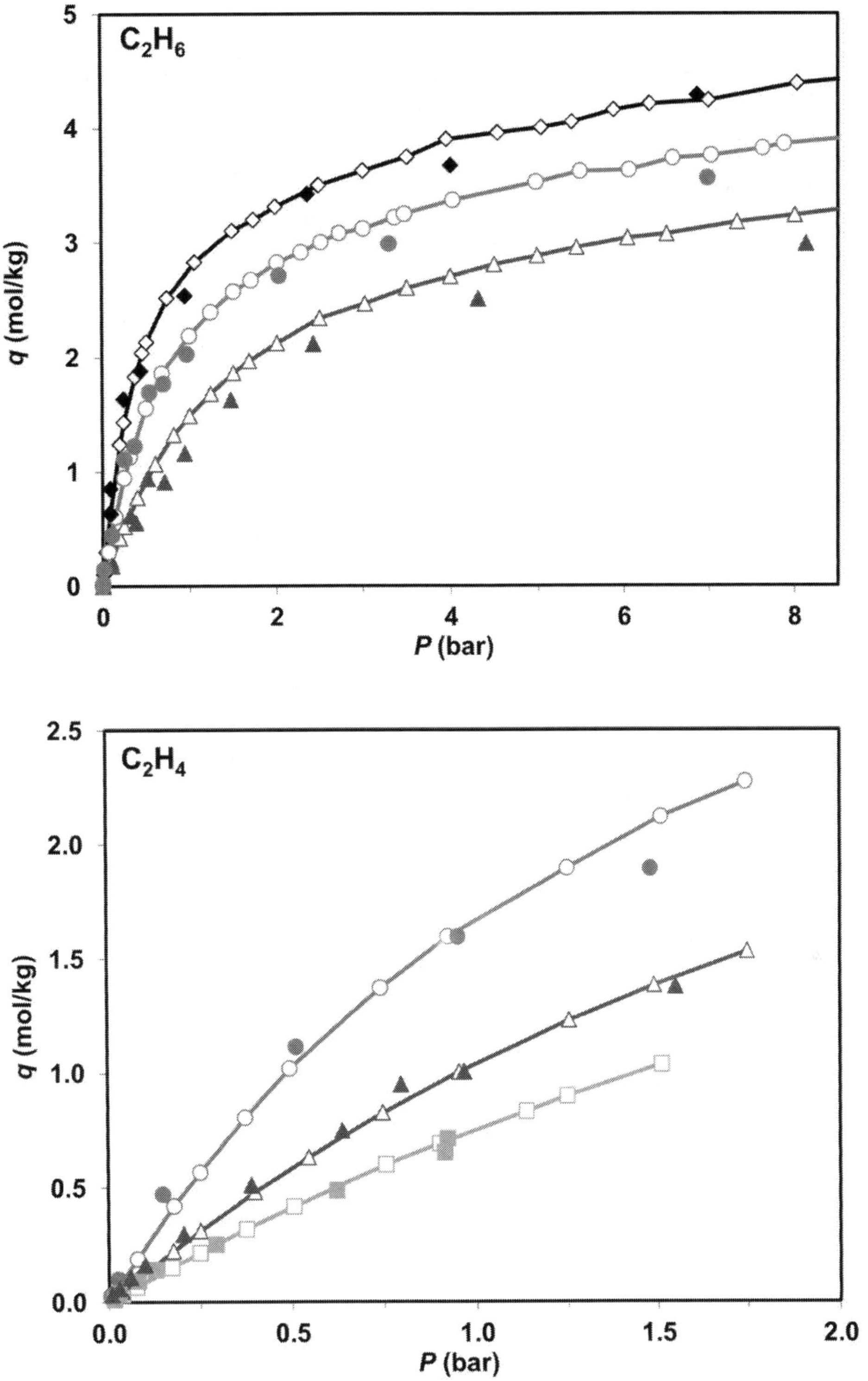

Figure 2. C_2H_6 and C_2H_4 absolute adsorption isotherms on MIL-53(Al). Comparison of experimental (filled symbols) and simulated by GCMC (open symbols) absolute adsorption equilibrium isotherms of C_2H_6 over MIL-53(Al) at 303 (♦), 323 (●) and 353 K (▲) and C_2H_4 at 323 (●), 353 K (▲) and 373 K (■). Solid lines are drawn as guide to the eye.

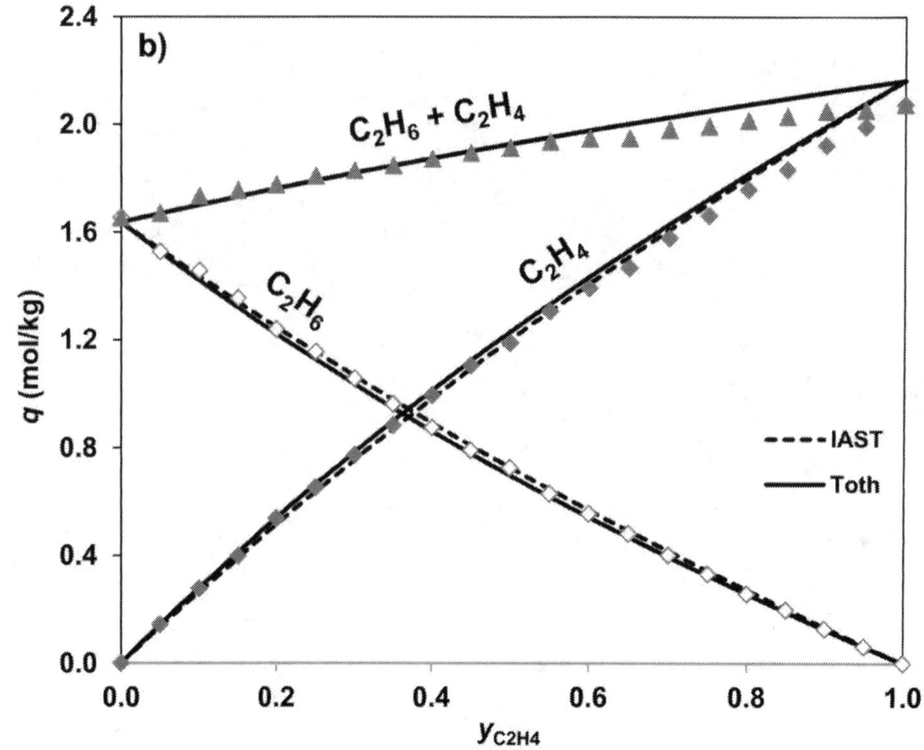

Figure 3. IAST and Toth model predictions and GCMC simulation adsorption equilibrium data for C2H6/C2H4 binary mixture on MIL-53(Al) at 323 K and 1 bar. Symbols – GCMC simulation results; Solid lines – IAST model predictions; Dashed lines – Toth model predictions

414806 Separation Techniques to Produce High Purity Hydrogen in LOHC Processes

Abstract:

Separation techniques to produce

high purity hydrogen in LOHC processes

Christoph Krieger, Karsten Müller, Wolfgang Arlt

Friedrich-Alexander-Universität Erlangen-Nürnberg, Germany

Institute of Separation Science and Technology

The storage of electrical energy with hydrogen has drawn rising attention in recent years. In this application hydrogen is produced in an electrolyzer, stored and in times of energy demand the chemical energy can be converted to electrical energy. However, hydrogen has a very low volumetric density. The reaction with a liquid organic hydrogen carrier (LOHC) in a hydrogenation reaction increases the volumetric density. However, after the storage the hydrogen has to be released in a dehydrogenation reaction.

Dibenzyltoluene as a LOHC has been examined for this study. Although dibenzyltoluene is a high boiling molecule, a certain proportion is evaporated and has to be separated from hydrogen after the dehydrogenation reactor. By cooling and separation of phases (demister) most of the dibenzyltoluene can be separated from hydrogen. However, certain power generation technologies, such as PEM fuel cells, requires for extremely high purified hydrogen: The purity requirements of hydrogen in PEM fuel cells according to ISO 14687 are 2 ppm for organic compounds. Due to the efficiency of the phase separator this purity cannot be fulfilled by just cooling and separation of phases.

The suitability of different separation technologies has been examined to determine the best separation technique to provide purified hydrogen. The focus is set on energetic assessment, size of the apparatuses and the practical feasibility. Within this study both modelling and experimental results will be compared.

Preliminary investigation shows that absorption and membranes as separation techniques cannot compete with adsorption, due to the energy demand of these separation processes. Therefore, the emphasis of this study is laid on adsorption processes.

410836 Separation of Complex Mixtures of Potential Lohc's

Abstract:

Separation of complex mixtures of potential LOHC's

Rabya Aslam, Karsten Müller, Wolfgang Arlt

Institute of Separation Science & Technology, Friedrich-Alexander-Universität Erlangen-Nuremberg

Liquid organic hydrogen carriers (LOHC) are potential compounds that can facilitate chemical energy storage of hydrogen, a pre-requisite for the storage of renewable energy, through reversible hydrogenation in a reasonable quantity. Among others, dibenzyltoulene (Marlotherm SH®, H_0-MSH) is a prominent LOHC which can store up to 6.2 mass % of hydrogen in the form of perhydro-dibenzyltoluene (H_{18}-MSH). High thermal stability even at harsh conditions (> 300°C) and reasonable hydrogen storage capacity also make it a favorable choice as a possible LOHC. For the application of MSH as LOHC, it is necessary to gather its thermo-physical data both in its lean (fully dehydrogenated) and rich forms (fully hydrogenated). However, MSH is not a pure compound but a mixture of various isomers of dibenzyltoluene. In the reversible cycle of hydrogenation and dehydrogenation reactions, it furthermore forms various stable intermediates with a different degree of hydrogenation. More than 24 stable isomers and intermediate compounds have been observed in partially hydrogenated mixtures formed in the reactions. These intermediates need to be separated into pure compounds and fractions (H_0-MSH, H_6-MSH, H_{12}-MSH, and H_{18}-MSH) not only for the measurement of thermo-physical data and toxicological analysis but also to study the detailed kinetics of hydrogenation and dehydrogenation reactions.

In this contribution, a process comprising distillation and chromatography, for the separation of a partially hydrogenated reaction mixture into fractions of isomers with the same degree of hydrogenation with more than 98% purity is presented. A pre-separation is done by a high vacuum distillation column with Sulzer packing. The developed HPLC method is successfully scaled up to a semi-preparative HPLC column (250 mm x 50 mm I.D.). The optimal method was developed on an analytical scale HPLC using phenylhexyl Silica as stationary phase and acetone/water mixture as eluent. Further purification of the separated fractions need different schemes. Fractions obtained from the semi preparative HPLC experiments have been used to measure various thermo-physical properties of the isomers like viscosity, density, refractive index, vapor pressure, heat capacity, and heat of vaporization.

435236 Synthesis of Azeotropic Multipressure Reactive Distillation Networks Involving Ternary Mixtures

Abstract:

The synthesis of a multi pressure reactive distillation network involving a ternary azeotropic mixture is investigated in this work. The network is comprised of vapor-liquid equilibrium trays, with liquid holdup that changes from tray to tray. Kinetically and/or equilibrium limited reactions are assumed to occur in the liquid phase.

The Infinite DimEnsionAl State-Space (IDEAS) conceptual framework, is first shown to be applicable to the problem, and is then employed to formulate an infinite linear program (ILP) that can rigorously assess the feasibility of the considered process synthesis problem. The formulated IDEAS ILP's infimum value is approximated by the sequence of optimal values of a number of finite dimensional linear programs (FLP) of ever-increasing size.

The method is demonstrated on a case study involving the multi pressure reactive distillation of methanol/isobutene/MTBE azeotropic mixture optimized for minimum holdup.

433268 A Heat & Mass Integration Approach to Reduce Capital and Operating Costs of a Distillation Configuration

Abstract:

We propose a general method to consolidate distillation columns of a distillation configuration using heat and mass integration. The proposed method encompasses all heat and mass integrations known till date, and includes many more. Each heat and mass integration eliminates a distillation column, a condenser, a reboiler and the heat duty associated with a reboiler. Thus, heat and mass integration can potentially offer significant capital and operating cost benefits. In this talk, we will study the various possible heat and mass integrations in detail, and demonstrate their benefits using case studies. This work will lay out a framework to synthesize an entire new class of useful configurations based on heat and mass integration of distillation columns.

Chemtec

Evaluation of the Environmental Impact of Coal Tar Distillations through LCA Methodology

2015 AIChE Annual Meeting, Salt Lake City, UT

L. Guillou, A. Gonzalez, K. Monatte

November 11th 2015
Prepared by: Loïc Guillou

POE-TMP-007-E

CTD: Shift the momentum from Feelings to Facts

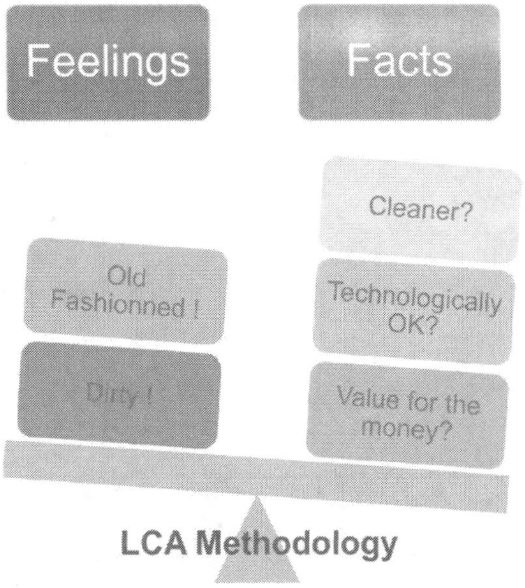

Agenda

1. Company Overview
2. Process evaluation through LCA
3. What is Coal Tar Distillation?
4. Hypotheses and scenarii
5. Results
6. LCA as a decision support tool

Chemtec

Company Overview

Fives, an international industrial engineering Group

Fives designs and supplies **machines, process equipment** and **production lines** for the **world's biggest industrial players.** Fives supplies PROABD® CTD plants since 1919.

Chemtec

Process evaluation through LCA

"Holistic technique to identify & quantify the potential impacts of a [...] process through its life cycle"[1]

[1]. M.A Curran, CEP, Oct. 2015, pp 26-35

Life Cycle Analysis

How is LCA working?

- LCA is a modeling approach based on a **multiscale and multicriteria** analysis methodology (ISO 14040 series)
- LCA delivers an **evaluation of the impact** through environmental KPI

What is the purpose of LCA?

- Identify **environmental "hot spots"**
- Identify and prioritize **opportunities for improvement** (e.g. Pareto type)
- **Compare** alternative processing/manufacturing routes and rate their preferability

What are the strengths of LCA?

- It is a **normalized** methodology
- It is **documented and proof based** demonstration

Methodology – ISO 14040 series

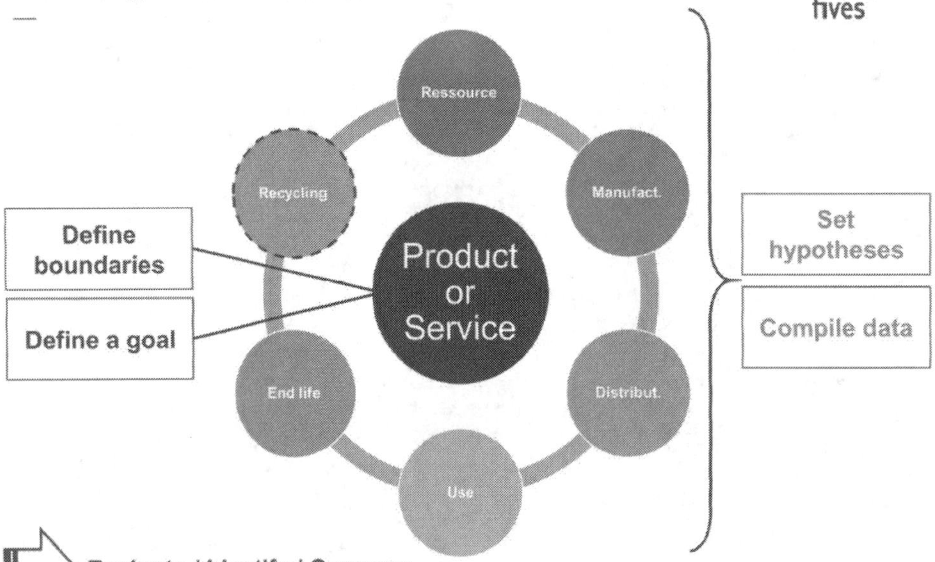

Evaluate / Identify / Compare

Quantis™ Suite 2 – LCA tool

1. **Modeling tool enabling**
   - Flow sheeting of the product/process in terms of:
   - Life cycle **steps**
   - Process **allocations** within each step
2. **Integrated databases (EcoInvent™ 2.0, Quantis™ water...)**
3. **Processing and reporting through 5 KPIs:**

- **Carbon footprint**
- **Water ressource**
- **Ecosystem quality**
- **Natural ressource stress**
- **Human health**

Highly correlated to ressource harvesting and product usage (Ecosystem quality, Natural ressource stress)

Chemtec

What is Coal Tar Distillation?

Carbochemistry

Valorization of Coke Oven Battery byproducts

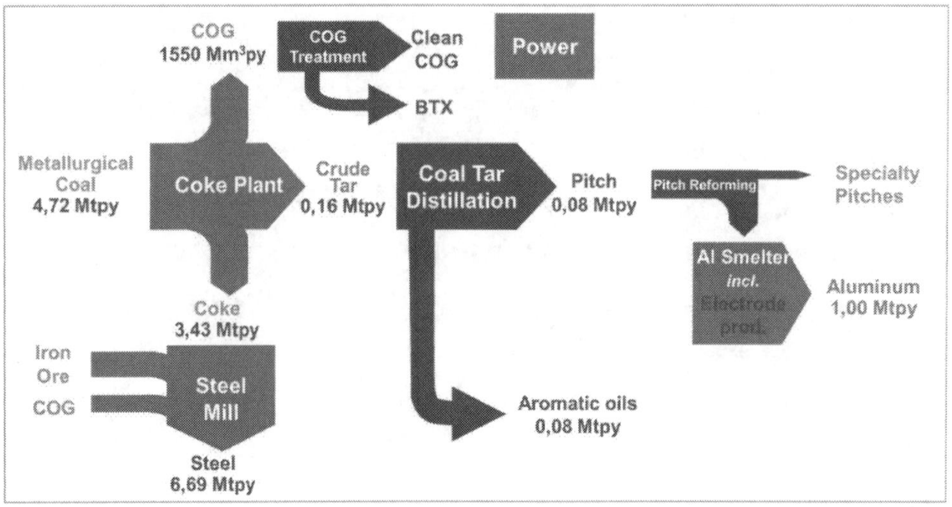

Coal Tar Typical Composition

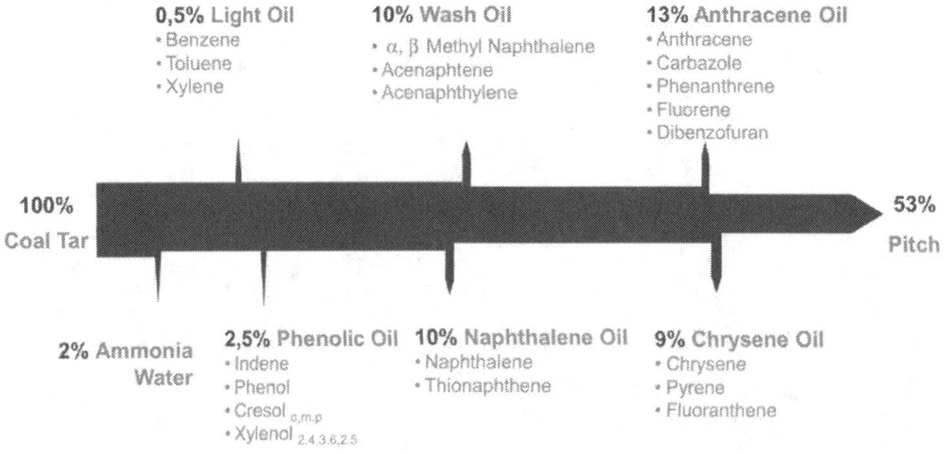

Coal tar industries' reputations issues

Poor image / reputation damages worldwide, because of :

- Association to popular **imagery** of steel mills/coal gasification plants
- Confusion between **compounds inherent toxicity** and plant operation practice
- **End life** of decades/century old assets and **heritage** of waste storage practice in pits
- **Environmental incidents**

⇨ **LCA and environmental KPI can help answering criticisms**

Bethlehem steel (PA-US), Congaree river (SC-US), HGD (Fr), Kingston coal gas plant (ON-CA), Kingston spillage (TN- US), Mulhouse water gas (FR), Sidney tar ponds (NS-CA), Air and soil pollution in Valasske Mezirici (CZ)….

Chemtec

Hypotheses and scenarii

Methodology applied to a CTD process – attributional LCA

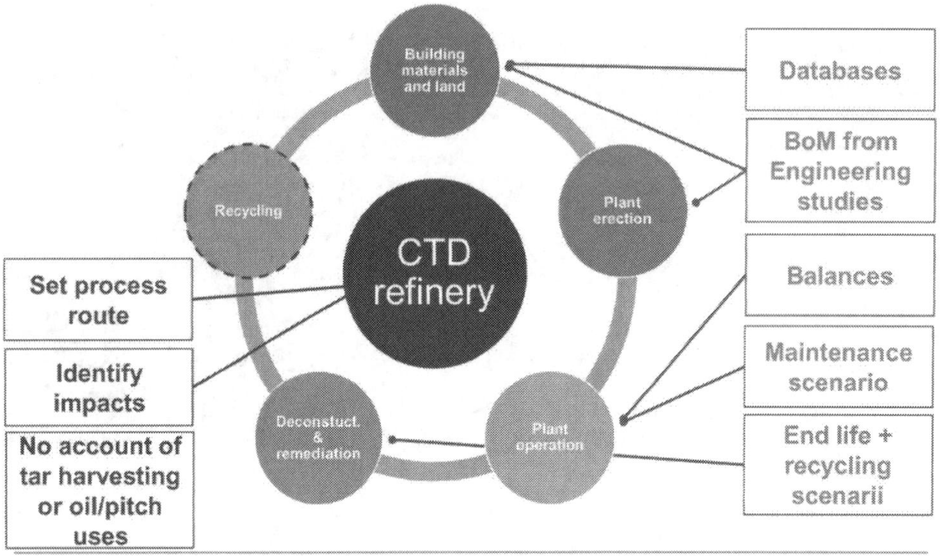

Detailed hypotheses – BFD

1. Process Basis of Design

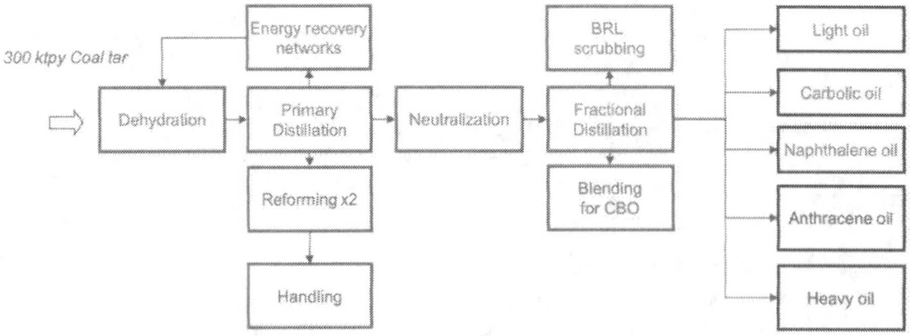

Detailed hypotheses

fives

2. **Material inventory + erection data**
   - FEED data (inc. Bill of material)
   - Mundane database information for raw material (concrete, structural steel...)

3. **Operations**
   - Energy balance
   - Mundane database information for available utilities (cooling water, N_2, IA...)
   - Composition model for fuel (COG) with **estimated** input burden based on coke manufacturing model
   - 30 years Time of Service (ToS)

Detailed hypotheses (Cont'd)

fives

4. **Maintenance**
   - Auditing of operation managers of different CTD plants with:
     - 3 to 30 years ToS
     - 10 countries (North America, Europe (western and eastern), Middle east, Asia)
   - Return of experience (Service and revamp projects handled by Fives *Pro*abd®)

- Definition of replacement frequency for itemized systems
- \+ Additionnal contribution (decontamination, waste managment...)

Detailed hypotheses (Cont'd)

5. **Deconstruction and decontamination (10 years)**

- Evaluation of the existing litterature for end life management of:

  Coal tar distillation plants
  Coke oven batteries
  Coal gasification and water gas plants
  Creosote impregnation plants
  Remediation of oil spills or chemical diffusion in underground waters (BTX, Quinoline…)

- Auditing of Site manager for on-going decontamination project of CTD plant (FR)

Difficult task as:

- Sites have been in operation over 50-90 years
- Operational practices were different
- Technologies & their reliability were different

⇨ **Definition of a worse case scenario:**

- Pollution diffusion in underground water
- Concentric radiuses of soil persistant pollution with Pitch or PAH
- Area restoration suitable for habitation & services

Chemtec

Results

Results

Relative impact of each step of the life cycle on KPIs

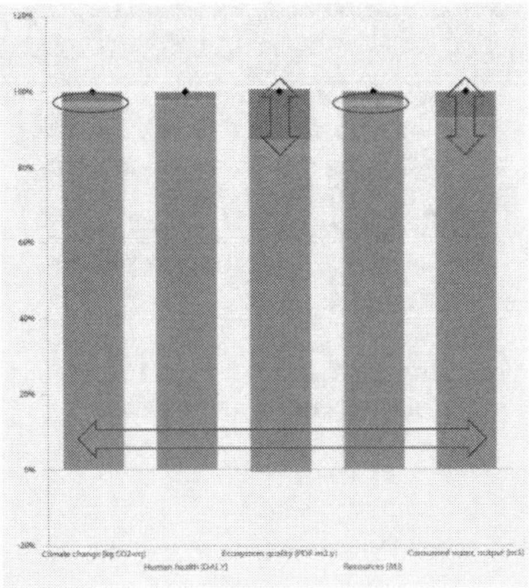

Construction / Erection

End-life

Operations + maintenance

1. **Operations overrule** all other KPIs
2. **End-life is not** as **impacting** as it can be believed despite toxicity of individual compounds
3. Construction impact is linked to raw material collection (minerals, metallurgy,…) and type (all inventory inc. « **electronics** »)

2015 AIChE Annual Meeting, Salt Lake City, UT Chemtec 21

Results (Cont'd)

Focus on Operations – Most impacting utilities

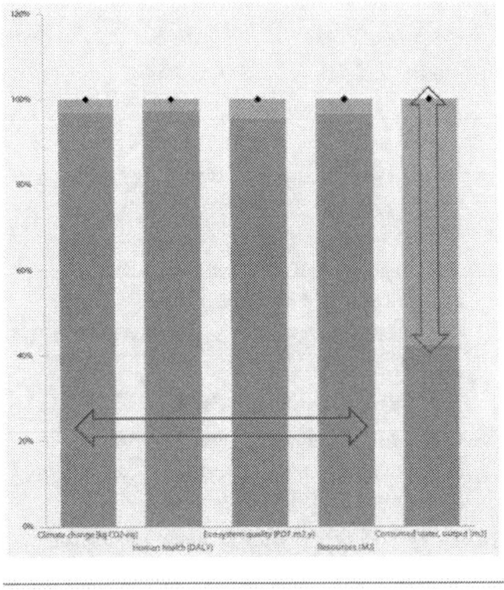

COG / Fuel

All others

1. **COG overrule** all other KPIs except stress on water ressource

2015 AIChE Annual Meeting, Salt Lake City, UT Chemtec 22

Results (Cont'd)

Focus on Operations – Most impacting utilities

fives

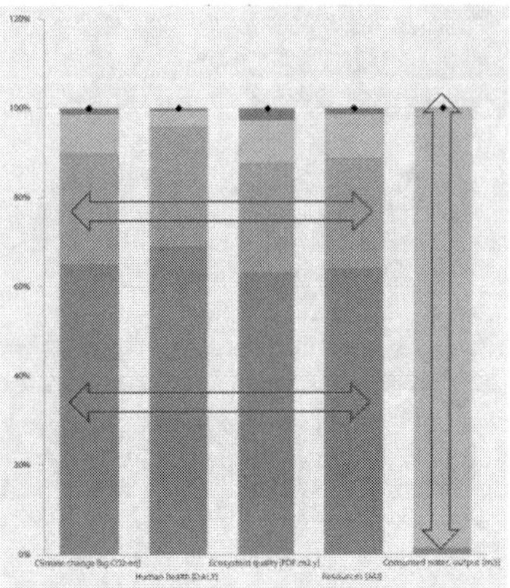

1. N_2 impact due to technology underlying **database assumptions** (ASU instead of PSA)
2. Despite use of chemicals DMW + **BFW is negligible**
3. Process steam impact results of **generation of waste waters**
4. Tracing steam impact results from balance with BFW and losses in winter time (-40°C)

Results (Cont'd)

Focus on End Life - Remediation

fives

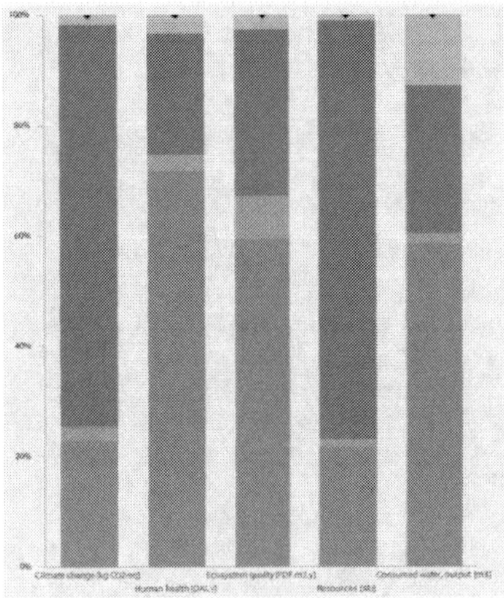

1. Incineration of soil and phytoremediations overrule
2. **Despite** soil incineration strictly **limited** to saturated soil (10% total contaminated area), impact is large
3. **Phytoremediation impact** = earthworks and plants used as fuel

Chemtec

fives

LCA as decision support tool

CTD improvement

Identification of key points to be worked out

fives

Hierarchy of hurdles and progress margins

1. **Combustion** systems are key
2. **Process steam** (steam stripping) shall be substituted
3. **Blending stations** are penalizing and are wasting energy
4. **Maintenance impact** is mainly occupationnal exposure and wastes destruction

1. **Quite common statement**
2. **Technology AND cultural change**
3. **Process scheme revision**
4. **Key equipment design improvement**

- **R&D program working on these 4 points**

From allocative LCA to comparative LCA

New process after LCA vs. other CTD processes (Operations step)

fives

Criteria	CTD Flex + ™	Multiflash	Bottom circulation	Improvement vs. best alternate technologies
Technology	**Vacuum**	Atmospheric	Atmospheric	
Fuel	**100**	100	116	**Eq.**
Elec. Power	**100**	152	> 110	**>10%**
CapEx (based on steel)	**100**	143	131	**Approx. 30%**
		Other performance criteria – product performance		
Naphthalene	**High yield , on specs**	Lower yield, off-specs	**High yield , on specs**	
CBO production	**Direct & adjustable**	Undirect / blend		
Pitch	**Controlled and adjustable characteristics sodium free**	No adjustment of characteristics uncontrolled ageing, sodium content		

From allocative LCA to comparative LCA

Is current CTD more polluting than equivalent industries?

fives

	Climate Change (kg CO2 eq.)	Human Health (DALY)	Ecosystem (PFD.m².y)	Resources (MJ)	Water stress (m3 eq.)
CTD refining	135	1.10^{-3}	16	2228	0.15
Crude refining	464	$0.5.10^{-3}$	[illegible] Inc. oil field	[illegible] Inc. oil field	0.19

1. *All results normalized for 1 ton of feed*
2. *Crude refining based on typical oil cuts distribution taken from "Training Manual Refining MB-107 – College of Managmenet & Economic studies, University of Petroleum & Energy Studies, Dehradun-India" and associated Ecoinvent database KPI*

1. Compare **orders of magnitude only**
2. **Globally equivalent**

*Differences (Ecosystem & Ressources) due to **database models assumptions for crude oil** (i.e impact of hydrocarbon harvesting at oil field, not accounted for coal mining for CTD LCA)*

CTD: Shifts the momentum from Feelings to Facts

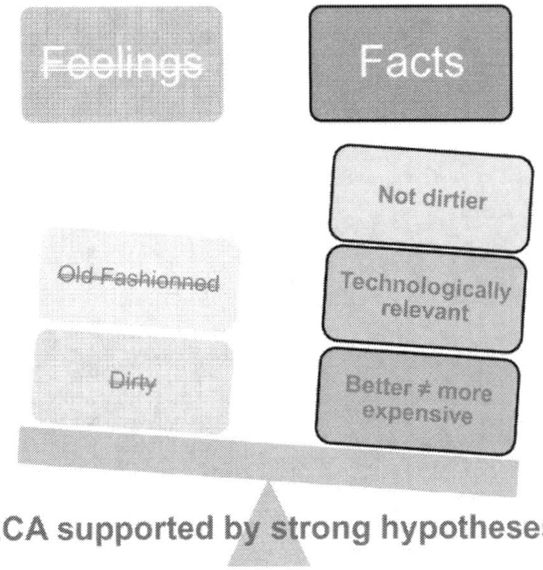

fives ultimate machines ultimate factory

Loic Guillou
Technology Manager
Chemtec

Loic.guillou@fivesgroup.com
T +33 3 89 36 32 00

Fives Proabd
CS 11437
20C rue de Chemnitz, Bât. B
Mulhouse Cedex , 68071 - **France**

technische universität dortmund

Fakultät Bio- und Chemieingenieurwesen

Continuous Enzymatic Reactive Distillation with Immobilized Enzyme Beads

fvt laboratory of fluid separations

Challenges for Chemical Processes

- New technologies for new sustainable products e.g. chiral compounds
- Active intermediates in pharmaceuticals
- Single enantiomers reaching 6.6 billion $ in 2001[1]

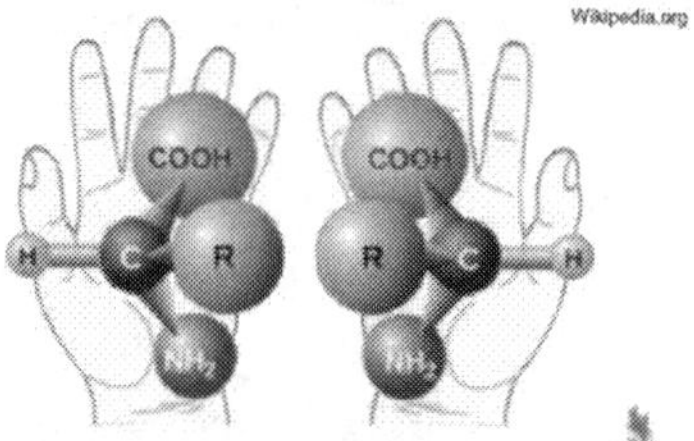

- Enzymatic reactions play important role in "green syntheses" [2]
  - Enzymes are enantioselective, green catalyst
  - Huge variety of different substrates and reactions
- Stirred Tank Reactor (Batch operation)
  - High productivity
  - Low space-time yield, equilibrium-limited reaction

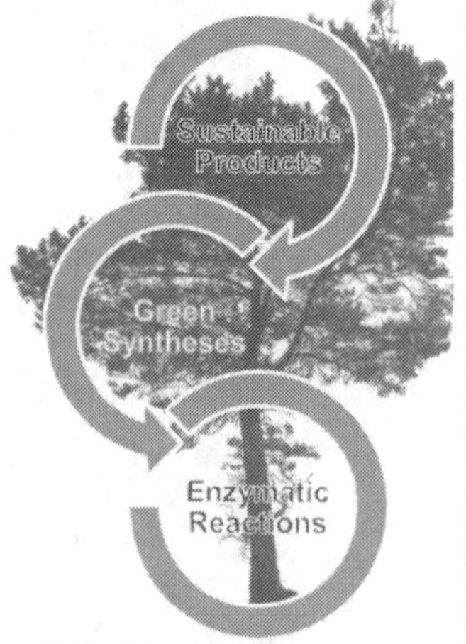

[1] Stinson, S.C: Chem. Eng. News. 79 (40), p. 79-97 (2001)
[2] Woodley, J. M. et al., Trends in Biotechnology., 26 (6), p. 321-327 (2008)

Challenges for Chemical Processes

- Process intensification

  ↓ A promising concept

- Enzymatic Reactive Distillation (**ERD**)

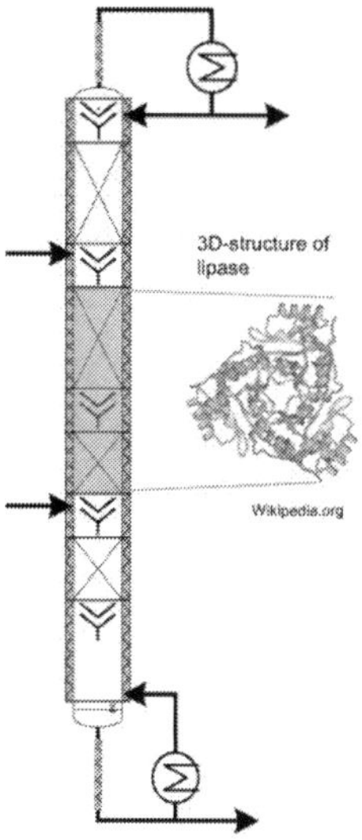

ERD

- RD integrates reaction and separation
- Reaction equilibrium shifted to the side of reaction product
- Immobilized enzyme

→ Continuous operation to be proven, technical realization and modeling

Concept - ERD

Challenges:

- How to provide enzymes inside distillation columns?
  - Structured packing
    - Katapak-SP
    - Filled with enzyme beads
  - Coating of wire gauze packing
    - Sulzer BX
    - Sol-gel method to entrap enzyme [1]
    - Stability sufficient for several batches [1]
- Experimental investigation and operating parameter influence

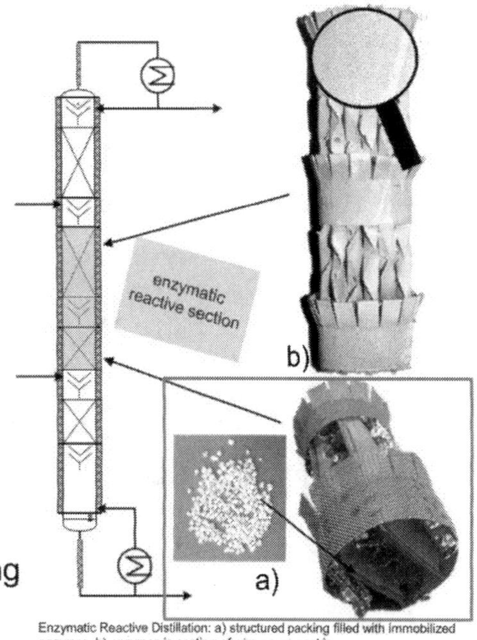

Enzymatic Reactive Distillation: a) structured packing filled with immobilized enzymes, b) enzymes in coating of wire gauze packing

[1] Heils, R. et al., Ind. Eng. Chem. Res., 53 (50), p. 19612–19619 (2014)

Experimental Set-up

- Column set-up
  - 6 packing sections made of glass
  - 7 liquid distributors
  - 12 m total heigth, 5.12 m packed column
  - 1 reactive section (1 m, Katapak-SP), 3 sections above, 2 below (Sulzer BX)
  - Vapor temperature measurement in distributors
  - Vacuum pump
- Analytics
  - Samples withdrawn in each distributor
  - GC analysis

packing section

Liquid distributor

Chemical system

- Chemical system
  - Transesterification of ethyl butyrate
  - Catalyzed by *Candida antarctica* lipase B (CalB)

ethyl butyrate (EtBu) | *n*-butanol (BuOH) | butyl butyrate (BuBu) | ethanol (EtOH)

- Boiling points

ethanol < *n*-butanol < ethyl butyrate < butyl butyrate

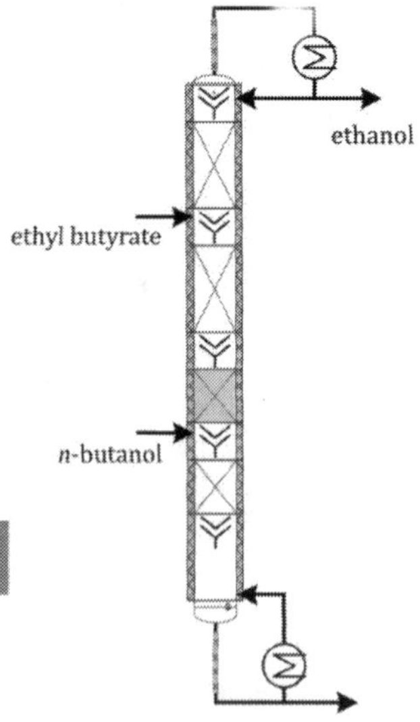

ERD experiments

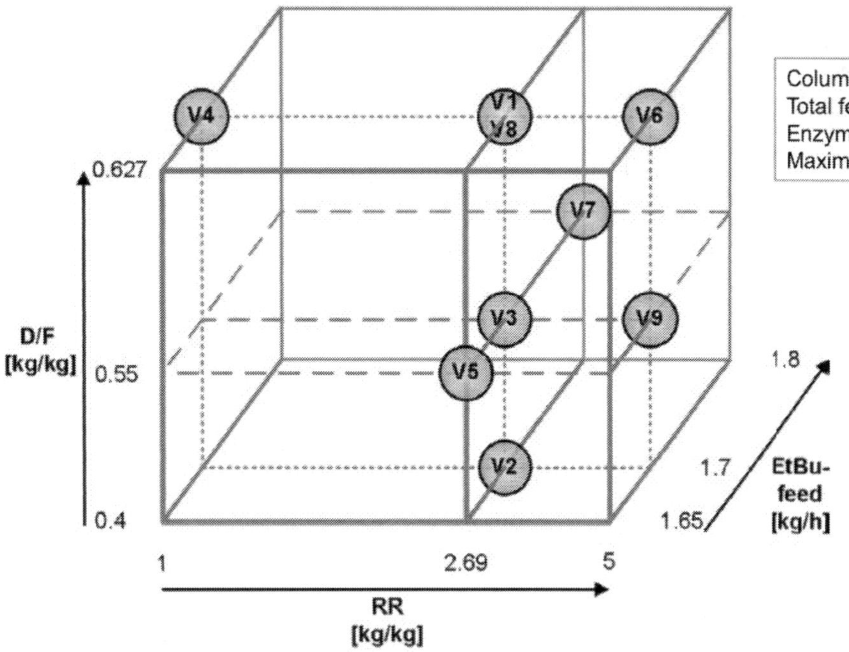

Column pressure: 145 mbar
Total feed mass flow: 2.0 kg/h
Enzyme bead mass: 56 g
Maximum temperature: 343 K

Experimental Results

- Variation of distillate-to-feed-ratio D/F

	V1	V2
D/F	0.60 kg/kg	0.39 kg/kg
RR	2.8 kg/kg	2.8 kg/kg
EtBu-feed	1.7 kg/h	1.7 kg/h
BuOH-feed	0.3 kg/h	0.3 kg/h
Pressure	146 mbar	147 mbar
Conversion BuOH	95.5 %	71.0 %

D/F ↑, Conversion ↑;
D/F ↑, mole fraction BuBu ↑

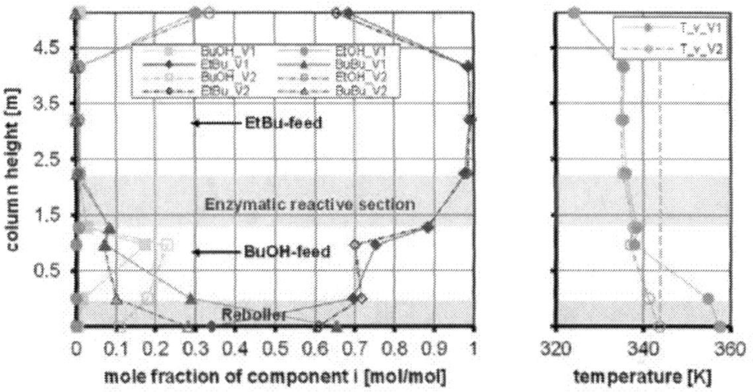

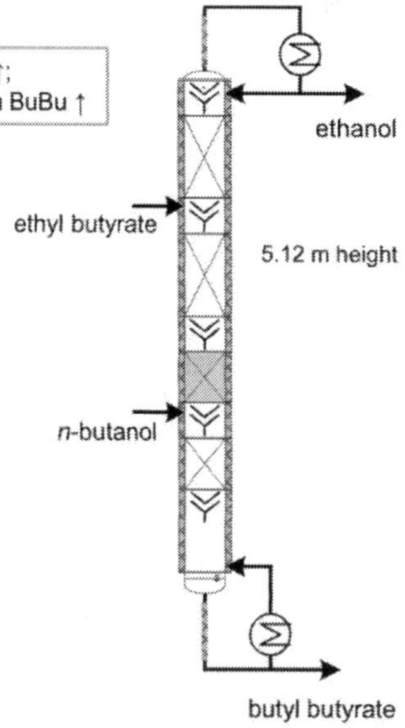

Experimental Results

- Variation of reflux ratio RR

	V4	V6
D/F	0.62 kg/kg	0.62 kg/kg
RR	1.0 kg/kg	5.1 kg/kg
EtBu-feed	1.7 kg/h	1.7 kg/h
BuOH-feed	0.3 kg/h	0.3 kg/h
Pressure	146 mbar	147 mbar
Conversion BuOH	73.6 %	95.2 %

RR ↑, Conversion ↑;
RR ↑, mole fraction BuBu ↑

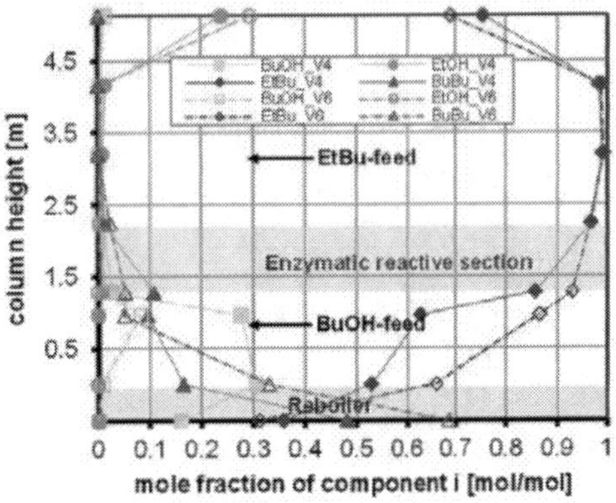

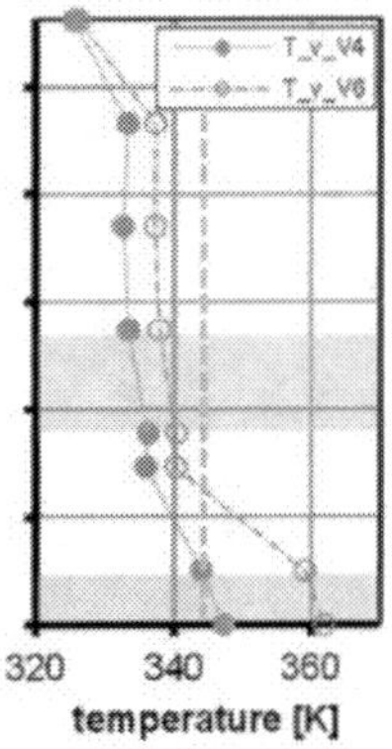

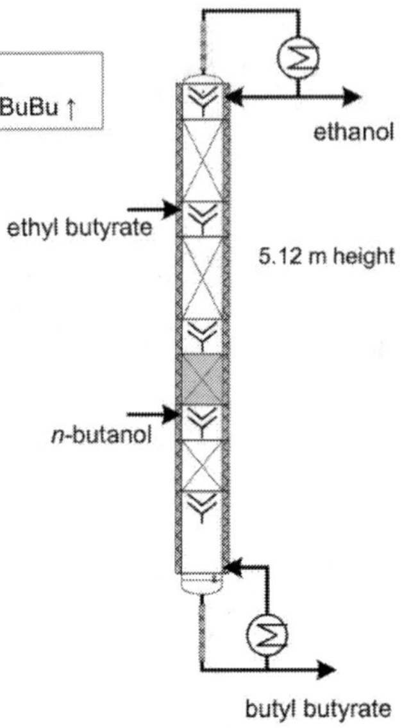

Experimental Results

- Variation of feed mass flow ratio (EtBu-feed/BuOH-feed)

	V5	V7
D/F	0.59 kg/kg	0.62 kg/kg
RR	2.8 kg/kg	2.7 kg/kg
EtBu-feed	1.65 kg/h	1.8 kg/h
BuOH-feed	0.35 kg/h	0.2 kg/h
Pressure	145 mbar	145 mbar
Conversion BuOH	96.0 %	95.2 %

Feed ratio ↑, Conversion =;
Feed ratio ↑, mole fraction BuBu ↓

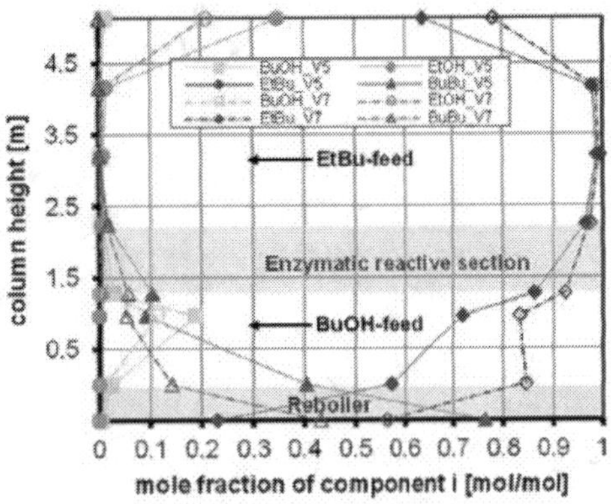

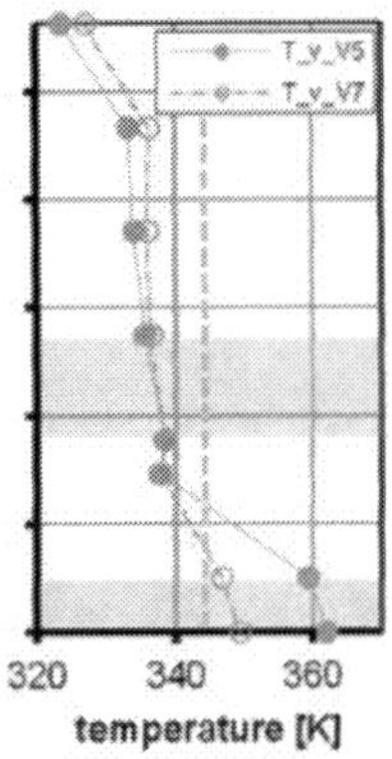

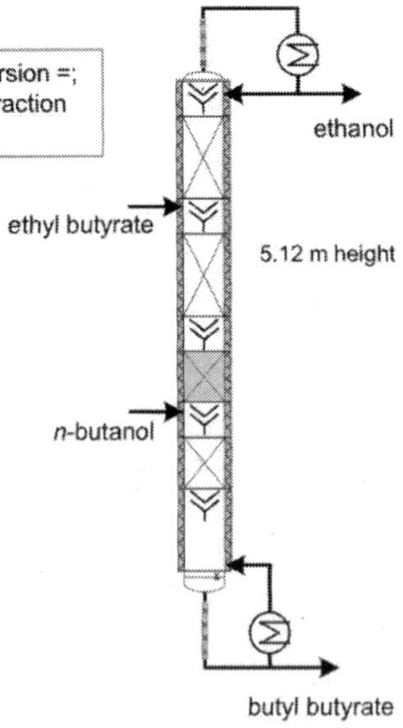

Experimental Results

- Reproduction

	V1	V8
D/F	0.60 kg/kg	0.62 kg/kg
RR	2.8 kg/kg	2.7 kg/kg
EtBu-feed	1.7 kg/h	1.7 kg/h
BuOH-feed	0.3 kg/h	0.3 kg/h
Pressure	146 mbar	145 mbar
Conversion BuOH	95.5 %	95.2 %

After ~70 h hot overflow
and 120 h cold overflow

→ Stable catalyst

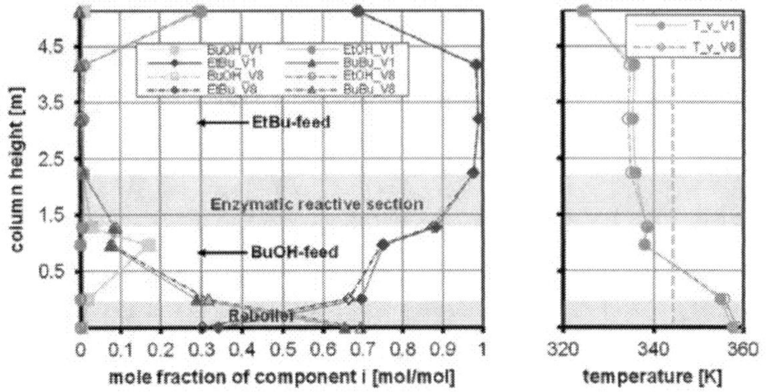

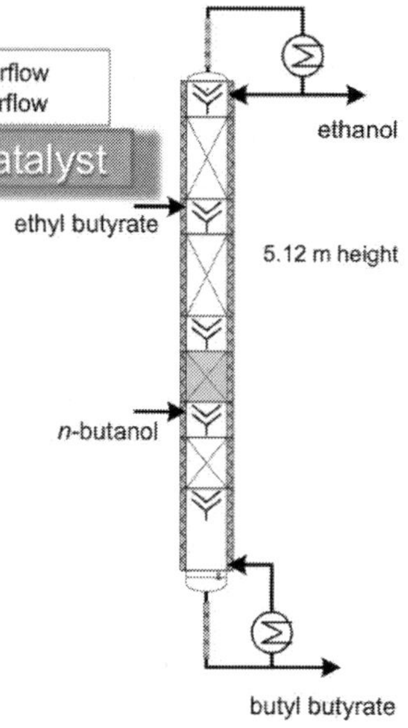

Modeling

- Simulation
  - Equation-oriented
  - Physical and thermodynamic data from Abrams UNIQUAC method [1]
- Equilibrium-stage modeling
  - Outgoing streams of a stage are in thermodynamic equilibrium
  - Hydrodynamics are implemented in the ERD model
- Reaction kinetics
  - Measured with enzyme beads
  - Kinetic model to be implemented into ERD model

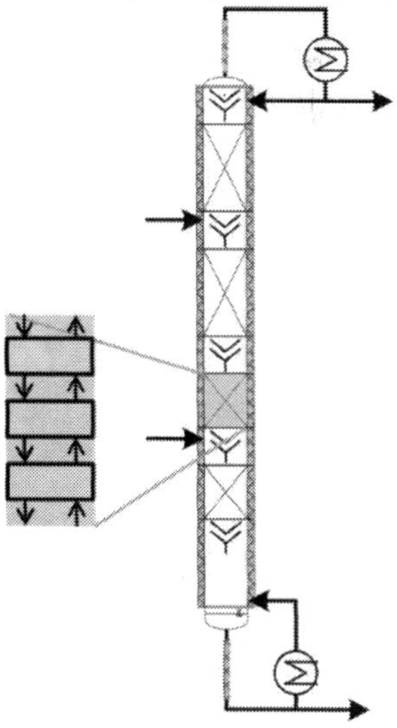

[1] Abrams, D. S. et al., AIChE J., 21 (1), p. 116–128 (1975)

Kinetic measurement

- Conditions
  - Closed system → no evaporation of the components
  - Mixing: Stirrer
- Investigated influences
  - Temperature
  - Substrate ratio

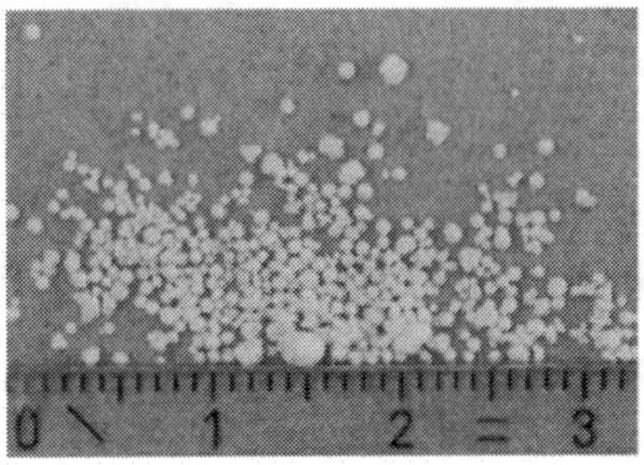

Enzyme beads

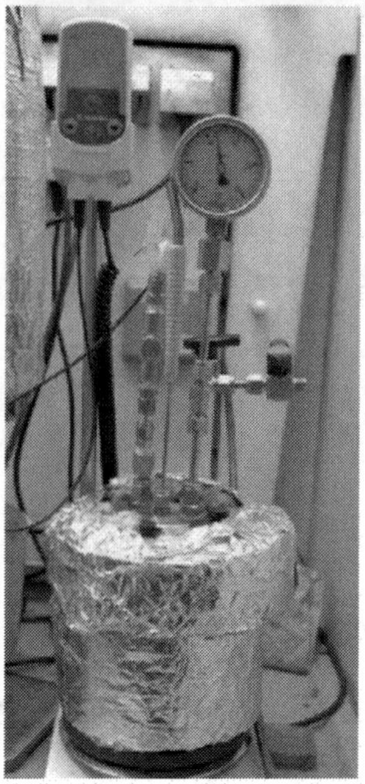

High pressure reactor

Determination of Kinetic Parameters

- Simplified Ping Pong Bi Bi mechanism [1]

$$\frac{d[BuBu]}{dt} = \frac{m_{cat}\, V_f\, V_r\left([EtBu][BuOH] - \frac{1}{K_{eq}}[BuBu][EtOH]\right)}{\frac{V_f}{K_{eq}}\left(K_{m,BuBu}[EtOH] + [BuBu][EtOH]\right) + V_r[EtBu][BuOH]}$$

- Genetic algorithm in MATLAB ®

Parameter	Value
V_f	$9.56 \cdot 10^{-4}\,\frac{mol}{l \cdot g \cdot s}$
V_r	$9.11 \cdot 10^{-5}\,\frac{mol}{l \cdot g \cdot s}$
K_{eq}	1.19
$K_{m,BuBu}$	$0.02\,\frac{mol}{l}$

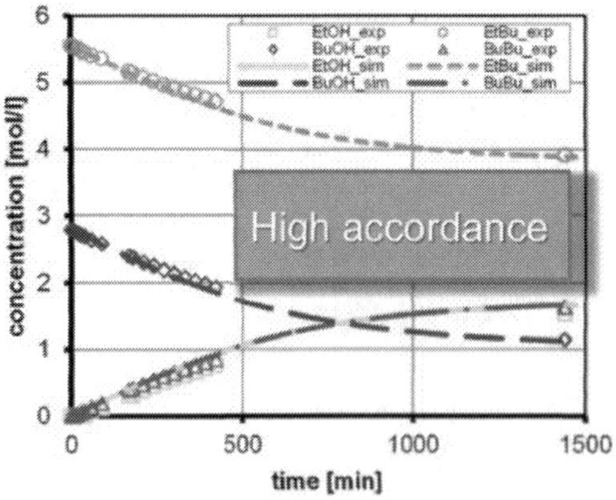

[1] Paiva, A. L. et al., Biotechnol. Prog., 19 (3), p. 750–754 (2003)

Conclusion & Outlook

- Feasibility of **Enzymatic Reactive Distillation** with enzyme beads was proven in pilot-scale
- Reproducibility was demonstrated

- Most promising operation variables
  - High distillate-to-feed-ratio (> 0.6 kg/kg)
  - High reflux ratio (> 5 kg/kg)
  - Reasonable high BuOH-feed (> 0.3 kg/h)

- Outlook
  - Modeling/model validation
  - Influence of ultrasound on enzyme kinetics and stability
  - Investigate synthesis of chiral compounds in **Enzymatic Reactive Distillation**

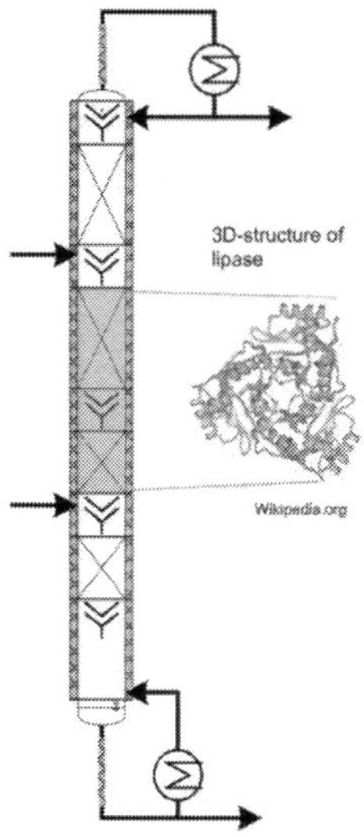

Acknowledgement

Thanks to:

- AlterEgo for funding
- CLIB-Graduate Cluster for funding
- Sulzer for packing provision

Any questions and remarks?

425085 Design Methods for the Heat-Integrated Distillation Column (HIDiC)

Abstract:

Major energy reductions can be obtained by heat integration in distillation columns. Potentially, this can lead to reductions in operation costs overcoming increased investment costs.

The types of heat integration for stand-alone distillation columns are external and internal, corresponding to vapor recompression and diabatic operation, respectively. Configurations, containing either or both types of heat integration, typically require compression. These configurations have a significantly higher number of design degrees of freedom, leading to complicated design procedures that often require models for simulation[1][2], though graphical methods also have been described[3],[4] for binary mixtures.

In this presentation, an iterative design algorithm for the conceptual design of internally or externally heat-integrated distillation configurations is presented. The algorithm provides means of selecting all central design variables such as the column pressures, the heat exchange area(s), and the thermal coupling of internally heat-integrated column stages.

The application of the design algorithm is illustrated for a number of separations of varying degree of difficulty, including the separation of methanol/water. The resulting design is compared against designs based on approaches suggested in literature. In addition, the proposed design algorithm is used for designing a separation of a multicomponent mixture. Simulations are carried out for outlining the techno-economic performance of all design.

References

[1] M. Gadalla, L. Jiménez, Z. Olujic, P.J. Jansens, 2007, A thermo-hydraulic approach to conceptual design of an internally heat-integrated distillation column (i-HIDiC), Computers and Chemical Engineering 31 (10):1346-1354

[2] A. Harwardt, W. Marquardt, 2012, Heat-integrated distillation columns: Vapor recompression or internal heat integration?, AIChE Journal, 58 (12):3740-3750.

[3] T. Ho, C. Huang, L. Lee, C. Chen, 2010, Extended Ponchon-Savarit Method for Graphically Analyzing and Designing Internally Heat-Integrated Distillation Columns, Industrial and Engineering Chemistry Research 49(1):305-358

[4] T. Wakabayashi, S. Hasebe, 2013, Design of heat integrated distillation column by using H-xy and T-xy diagrams, Computers and Chemical Engineering 56(13):174-183

417969 Influence of Ions and Fluid Dynamics on Coalescence in Liquid/Liquid Dispersions

Wednesday, November 11, 2015: 12:30 PM
155E (Salt Palace Convention Center)

Jörn Villwock[1], Johannes Kamp[1], Felix Gebauer[2], Hans-Jörg Bart[2] and Matthias Kraume[1], (1)Chair of Chemical and Process Engineering, Technische Universität Berlin, D-10587 Berlin, Germany, (2)Chair of Separation Science and Technology, Technische Universität Kaiserslautern, D-67663 Kaiserslautern, Germany

Influence of ions and fluid dynamics on coalescence in liquid/liquid dispersions

Jörn Villwock[1]; Johannes Kamp[1]; Felix Gebauer[2]; Hans-Jörg Bart[2]; Matthias Kraume[1]

[1] Chair of Chemical & Process Engineering, Technische Universität Berlin, Germany

[2] Chair of Separation Science and Technology, University of Kaiserslautern, Germany

Dispersions of two immiscible liquids are an integral part of several unit operations. The drop size distribution, which is the result of the competing phenomena drop breakage and coalescence, influences a decisive part of the overall process efficiency. Existing models for the coalescence rate include the physical properties, but also fluid dynamic characteristics like energy dissipation rate and relative velocity. However, these influencing parameters are implemented in existent models with different proportionalities and in some cases even contradictorily. To resolve this issue the system is reduced to the fundamental behaviour of single droplet coalescence and the impact of the influencing parameters is quantified separately.

With a fully automated test cell - based on the principle to observe the collision of a pendant drop and a second rising drop using high speed imaging (see Fig. 1) - serial examinations of binary droplet collisions under variable system conditions are possible. All experimental results were achieved using the standard test system for liquid/liquid extraction toluene/water. In this work the systematic analysis of influencing parameters on coalescence in liquid/liquid dispersions is presented. It focuses on the binary droplet coalescence and the following quantities will be discussed in detail: coalescence probability and coalescence time depending on the addition of different ion types in varying concentrations and on drop size and relative velocity. The experimental results are used to validate and refine existing models or to develop new models, which can be implemented in population balance equations to describe the drop size distribution in liquid/liquid dispersions.

Investigations with various ions (Cl^-, ClO_4^-, SO_4^{2-}, SCN^- and OH^-) in a concentration range of 10^{-5} - 1 mol/L showed a strong decrease of the coalescence probability with increasing concentration. However, general explanations failed as no clear influence on interfacial tension and zeta potential could be observed. The coalescence time increases up to an apparent critical ion concentration and subsequently decreases. Coalescence times also became longer with increasing drop size ratio as well as increasing collision velocity. With respect to coalescence probability a critical collision velocity was found. While below this critical value the coalescence probability mainly depends on the drop size, nearly no coalescence occurred above it. To test the transferability, the experiments are currently repeated with the system heptane/water having a higher interfacial tension.

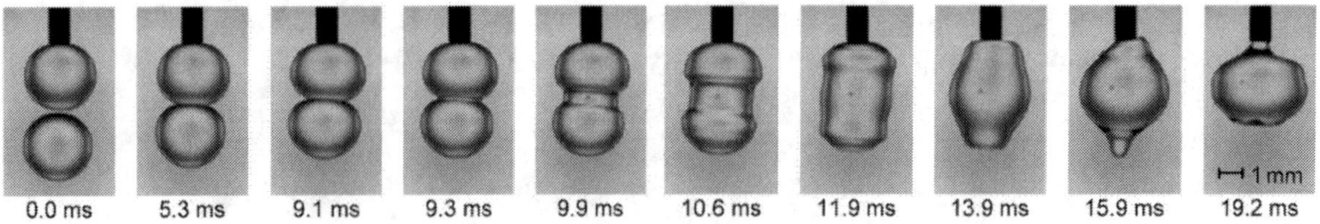

Fig. 1: Sample image sequence for coalescing droplets

Financially supported by DFG project KR 1639/19-2 "Coalescence efficiency in binary systems".

436138 Application of X-Ray Computed Tomography in Extraction Columns: A Feasibility Study

Abstract:

Computed tomography has become more important for the non-invasive investigation of technical devices. This measuring technique allows the observation of fluid dynamics and liquid distribution over the entire object without influencing the fluid flow. With suitable components it is even possible to determine the mass transfer of one liquid or gas phase into a liquid.

The experiments are performed with a customized computer tomograph at our institute that enables measurements of flows in vertical objects. The detector and the X ray source are rotating around the column and the column is not moved, so that the flow inside is not influenced by centrifugal force. As a result of the high spatial (80 μm) and temporal (1000 projections per second) resolution, dynamic processes inside separation columns can be monitored with a high precision. Fluid dynamic parameters in distillation columns (like interfacial area, liquid hold-up and liquid film thickness on the packing or between two liquid phases) can be directly estimated from the recorded CT images of a packing cross section at different height positions along the column. For the evaluation of the recorded CT-images several steps of image treatment were applied to finally obtain the relevant data directly from the images.

In chromatographic columns we are able to describe the breakthrough behavior of a liquid tracer and implement the experimental data in a CFD-model to predict a stationary phase for a specific separation problem. In packed distillation columns we can determine fluid dynamic parameters like hold-up, interfacial area and the liquid film thickness on the packing over the entire packing height and the entire cross section area. Three dimensional visualizations show that there are especially rivulet flows in the gaps between two plates and windings of the packing.

A decisive point in the design of extraction columns is the mass transfer between the phases. With computed tomography it is possible to determine the in-situ mass transfer in a suitable system. In extraction columns the droplet size is several millimeters and the velocity about 0.1 m / s. With computed tomography we are able to determine the shape of the droplets and by averaging the CT-images over the time it is possible to determine the temporal progress of the mass transfer from the surrounding into the droplet or the other way round. The change of the gray values in a CT image results in a concentration profile in the direction of the mass transfer. With this information we can determine a mass transfer profile into the droplet at several height positions of the column over the entire cross-sectional area.

412462 Investigation and Modeling of Polydisperse Drop Sedimentation in Liquid-Liquid Phase Separation

Abstract:

In the chemical industry, liquid-liquid extraction is a common separation technology. It is preferably applied when separation via distillation is inefficient or even impossible. Isolation of many bio-based chemicals, characterized by low volatility due to high oxygen content, may demand liquid-liquid extraction. The performance of extraction processes may suffer from poor phase separation properties.

In this work the phase separation has been studied in order to improve the design of settlers. Established design principles are based on separation experiments, carried out in laboratory scale settling cells. These experiments require optical monitoring of the separation process. Unlike academic test systems, industrial systems are often opaque, making optical monitoring impossible. Therefore, a new technique was used. With an ultrasonic scanning system the phase separation rate of opaque dispersions can be quantified. The speed of sound depends on the composition of the dispersion.

After phase separation, residual turbidity may limit the quality of the settler outflow. This is in particular true for high-viscosity and low interphase tension systems. As small droplets of turbid dispersions take much longer time to sediment and to coalesce they are a problem in industrial processes when they accumulate. With the ultrasonic scanning technique it is also possible to quantify residual turbidity of either phase and the rate of secondary sedimentation.

Current sedimentation models in the settler design are based on the assumption of monodisperse droplets. Application does poorly reflect hindered settling. Improvement is achieved by implementing a population balance. For a given drop size distribution the improved model can determine the change of the hold-up with height and time with high accuracy, since it takes into account the sedimentation rate based on each drop size, the hindered sedimentation through swarm effects and the countercurrent flow of the continuous phase. By comparing the error between model and experiment, the drop size distribution parameters can be fitted. The proposed scanning technique in combination with improved data interpretation and modeling will help to improve the design of settlers.

Turbidity control
Characterization, formation and separation of stable emulsions

R. Macher-Ambrosch, M. Siebenhofer

Motivation

- Turbidity is annoying (in liquid-liquid extraction)
- Water in oil: crude oil
- Oil in water: at the bottom of extraction columns
  → loss of solvent

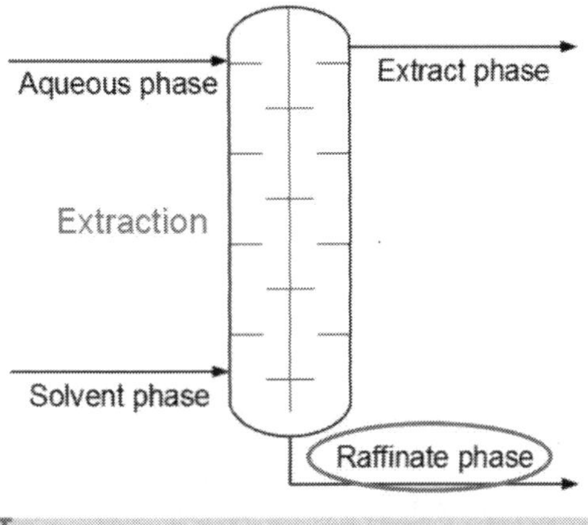

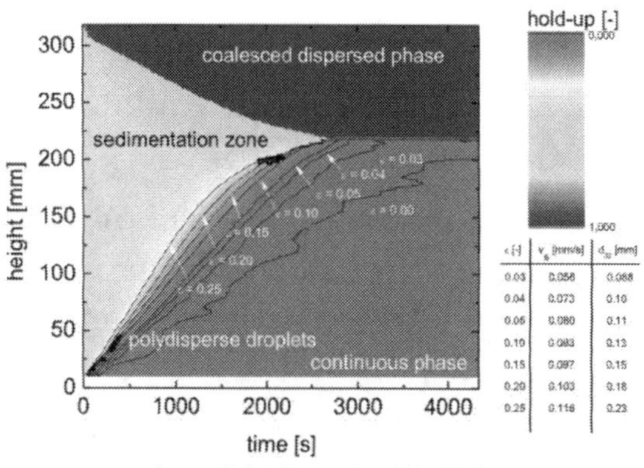

ε [-]	v_S [mm/s]	d_{32} [mm]
0.03	0.056	0.088
0.04	0.073	0.10
0.06	0.080	0.11
0.10	0.083	0.13
0.15	0.097	0.15
0.20	0.103	0.18
0.25	0.118	0.23

Source: P. Bol, Dissertation, 2015, TU Graz

Content

- Classification
- Background
- Stabilization
- Emulsion formation
- Emulsion splitting
- Splitting technologies
- Experimental basis
- Results
- Summary

Classification

- Microemulsions
  - Dropsize: 1 – 100nm
  - Thermodynamically stable!
    → Neither change in number nor size distribution

- Macroemulsions
  - Dropsize: 20nm – 10µm
  - Kinetically stable
  - Special case: Nanoemulsions
    - Dropsize: 20 – 500nm
    - Similar to microemulsions

Background

- Desired emulsions?
  - Food industry (eg. butter (w/o), milk (o/w))
  - Cosmetics (eg. creams)
  - Pharmaceutical industry (eg. injections)

- Annoying emulsions?
  - Liquid-liquid extraction
  - Heterogeneous azeotrope rectification

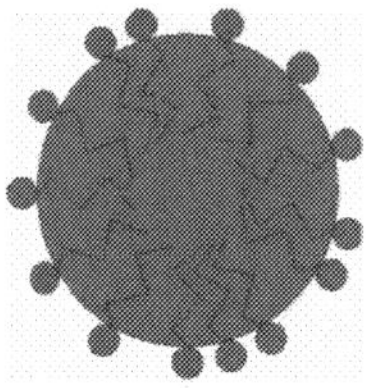

Stabilization

- DLVO-forces
  - Electric stabilization
  - Van der Waals forces
- Non DLVO-theory related effects
  - Surfactants
    - Membrane formation
    - Steric stabilization
- Brownian motion

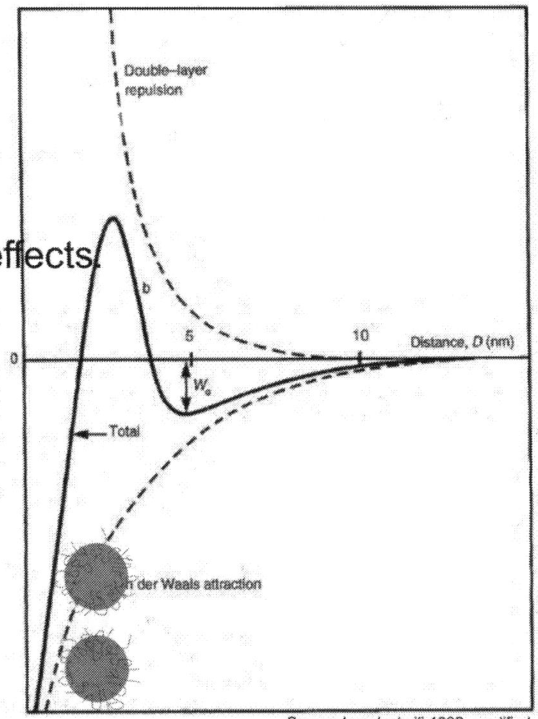

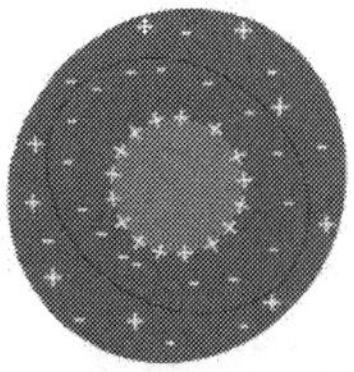

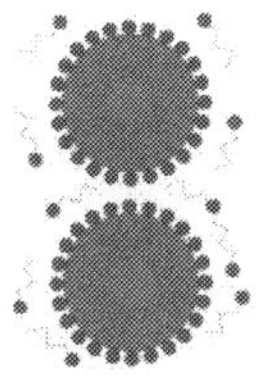

Source: Israelachvili 1992, modified

Emulsion formation

- High-energy emulsification
  - High shear stress in dispersing zone to deform droplets
    - Colloid mill or rotor-stator devices
    - High pressure homogenizer
    - Ultrasonic devices
- Intermediate-energy consumption
  - Membranes
    - Direct membrane emulsification
    - Modification of emulsions using membranes
- Low-energy emulsification
  - Phase inversion emulsification
    - Emulsion inversion point method (EIP)
    - Phase inversion temperature method (PIT)

Emulsion splitting

- **Coalescence:** droplets merge
- **Creaming/Sedimentation:** agglomeration
- **Flocculation:** aggregation without rupture of stabilizing layer
- **Ostwald ripening:** dispersed phase is transferred from small to big droplets
  - → avoided with narrow drop size distribution

Splitting technologies

- Mechanical splitting
  - Settler
    - Large footprint for slowly separating systems
  - Centrifugal devices
    - Smaller than settlers
    - Higher energy input
- Thermal splitting
  - Use of PIT:
    - Curvature in emulsion changes

    → Formation of bigger drops → separation

Splitting technologies

- Chemical splitting
  - Only for analytical purposes
  - Not splitting, but dissolution of dispersion
- Electrical splitting
  - W/O-emulsions:
    - Industrial application in crude oil processing for AC-devices
    - Low Power demand → Energy efficient way of separation
  - O/W-emulsions:
    - DC-applications in lab scale

Experimental Basis

- Turbidity: solvent (ShellSol-K) dispersed (0.5 wt%) in deionised water
- Surfactants (0.1wt% referred to solvent)
  - Anionic surfactant: 4-Dodecylbenzenesulfonic acid
- Equipment

Results: Mechanical splitting

- e.g. centrifugation (centrifuges: 100,000$)

Results: Thermal Splitting

- Energy consumption (1m³, Δ = 40 K: ~ 47kWh)

Results: Dissolution

- Chemical Splitting (e.g. dissolution with iso-propanol)
  - Costly and energy intensive

Results: DC-field

- Anionic surfactant: 4-Dodecylbenzenesulfonic acid

Anode  Cathode

Results: DC-field

- Cationic surfactant: Hexadecyltrimethylammonium bromide

Anode  Cathode

Results: DC-field

- Nonionic surfactant: Tween® 65

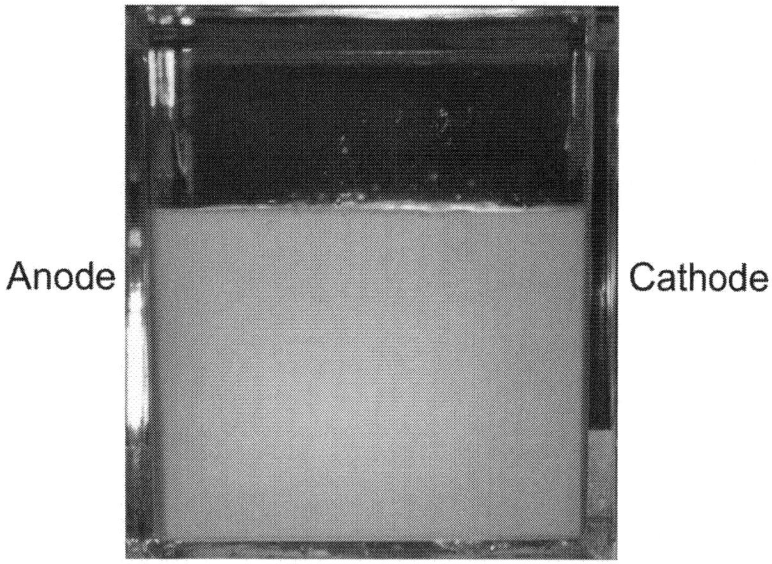

Field force supported sedimentation and coalescence

- Anionic surfactant

Summary

- Turbidity in the raffinate phase
  - Loss of solvent
  - Costly
  - Wastewater problem
- Reasons
  - Temperature swing
  - High shear
  - Surfactants
  - Combinations of
- Measures
  - Mechanical
  - Thermal
  - Chemical
  - Electrostatic

Turbidity control

Characterization, formation and cleavage of stable emulsions

R. Macher-Ambrosch, M. Siebenhofer
Institute of Chemical Engineering and Environmental Technology, NAWI Graz
+43 316 873 7475
r.macher-ambrosch@tugraz.at

412306 A Bionic System to Achieve Countercurrent Multi-Stage Micro-Extraction

Abstract:

Liquid-liquid extraction is an important operation unit in chemical processes that has been widely used in the petrochemical industry, nuclear reprocessing, and ore processing. However, conventional extraction systems, such as mixer-settlers, pulse columns, and centrifugal contactors, suffer from low efficiency, high energy consumption, and large equipment dimensions. Microextractoin provides a new option to solve these problems due to its high efficiency, safety, and small volume.

However, the existing microextraction systems are mostly cocurrent operated and the recovery efficiency can reach no higher than one theoretical separation stage. It is difficult to realize countercurrent flow at the micro-scale because viscosity and surface wetting dominate over gravity and inertial effects. Difficulty in achieving countercurrent operation has been a barrier to the application of micro-extraction.

The human cardiovascular system which transports blood to every part of the body gives us inspiration to solve the problem in the micro-fluidic system. In this work, a bionic system, analogous to the human cardiovascular system, was developed to realize countercurrent, multi-stage micro-extraction. The mass transfer performance and fluid dynamics of the system were studied. High mass transfer rates and high recovery efficiency were achieved. The system can be applied to any type of micro-extractor and may promote the application of micro-extraction in a wide variety of fields.

427131 Additive Manufactured Ceramic Internals for Small Scale Extraction Columns

Abstract:

Additive Manufactured Ceramic Internals for Small Scale Extraction Columns

S. Soboll, I.C. Stark, C. Schulze, N. Kockmann, TU Dortmund, Dortmund

K. Sauerzapfe, H. Wampers, Lapp Insulators Alumina GmbH, Redwitz

Continuous processes attain increasing interest in production of specialty chemicals and pharmaceutics, in which the amounts of produced goods are rather small. Additionally, small-scale equipment enables rapid process development in the lab with consistent scale-up capability. To improve thermal and chemical resistance and widen the operational windows, Lapp Insulators Alumina GmbH and TU Dortmund, Equipment Design have developed new concepts for small-scale plants with integrated key-components made of ceramics. Here, the development of a small scale extraction column [1] with ceramic components made of $Al_2O_3$and produced by 3D printing [2] shall be presented.

Different prototypes of stirrers, bearings, and plates were developed and their applicability in the column was evaluated. During this process, it took several iterations from design, manufacturing, and application of shaft-like geometries as stirrers for intensified dispersion in the column with internal diameter of 15 mm. Additionally, the proper fixation of the internals, which is usually done by connecting the plates and bearings by shafts, was a major challenge. The contribution will show some of the detailed solutions and achieved results already shown on the ACHEMA fair.

The column with an active height of 220 mm was investigated in terms of fluid dynamics and separation performance using the EFCE test system n-butyl acetate / acetone / water. The droplet size distribution was determined for different column loadings showing mean diameters ranging from 0.55 to 0.7 mm. Mass transfer and separation efficiency is in the same range as for the previously investigated column with metal internals [1] for direction disperse -> continuous and slightly lower in case of the opposite direction. Further investigations will focus on improved internals and further chemical systems.

[1] A. Holbach, E. Caliskan, H. Lee, N. Kockmann, Process intensification in small scale extraction columns for counter-current operations, Chemical Engineering and Processing 80 (2014), 21-28

[2] J. Homa, M. Schwentenwein, How can CIM benefit from additive manufacturing, Ceramic Applications Vol. 2/2014, 58

433229 The Research on a Polyamidoamine-Based Demulsifier for Oil/Water Separation

Abstract:

Oil-in-water emulsions widely exist in the waste water resulted from polymer flooding oil recovery process. Because there are large amounts of interfacial active substance and ultrafine oil droplets in this kind of emulsion, it is very hard to treat with traditional methods. To effectively deal with this kind of emulsion with ultrafine oil droplets and highly concentrated surfactants, one kind of chemical method, demulsification is adopted. In this thesis, the synthesized 3rd generation polyamidoamine (PAMAM) was studied as a typical demulsifier. FTIR and MNR were applied to analyze the molecular structure of the demulsifier. The influences of temperature, settling time, and concentration of the demulsifier used on the efficiency of the demulsifier were investigated in detail. At the optimal operating condition (demulsification time: 120 min; demulsification temperature: 30 °C; demulsifier concentration: 1500 mg L-1), the oil removal rate of the demulsifier could reach more than 90%. The surface tension and interfacial tension at the diesel–water interface were also measured to give a basic understanding of the demulsification mechanism.

436869 Three Types of Filter Media Produced from Electrospun Cellulose Acetate (Cac)-Polystyrene (PS) Composite Membrane for Separating Oil-Water Mixtures

Abstract:

Three Types of Filter Media Produced from

Electrospun Cellulose Acetate (Cac)-Polystyrene (Ps)

Composite Membrane for Separating Oil-Water Mixtures

L. Baghernejad[1] Chem., E. Iski[2] Chem.

Department of Chemistry and Biochemistry

O. Shoham[3], Department of Petroleum Engineering

R. Mohan[4], Department of Mechanical Engineering

S. Odueyungbo[5], Chevron Energy Technology Company

The University of Tulsa

Tulsa, Oklahoma 74104

(918) 812-7956

lida-baghernejad@utulsa.edu

ABSTRACT

(261 Words)

Oil-water mixtures are encountered in many domestic, industrial and environmental activities and are often undesi d and problematic. Particularly, oil-spills such as the sudden Mexican Gulf catastrophe have attracted tremendous interest towards the separation of oil-water mixtures. Membrane filtration is one of the most effective methods so far in separating oil-water mixtures. In this project, Cellulose Acetate (CAc) fibers were reinforced by Poly (styrene) (PS) fibers via co-electrospinning CAc and PS from one solution to produce a highly porous, nanofibrous and flexible composite membrane. Three types of filter media were developed from the electrospun CAc-PS membranes: 1) a hydrophobic-oleophilic filter medium (Filter #1), which separated oil from oil-water mixtures, 2) a hydrophilic-oleophobic filter medium (Filter #2), which separated water from oil-water mixtures and 3) a hydrophobic-oleophobic filter medium (Filter #3), which is expected to separate water and oil droplets from wet gas. These filter media were produced by heating and/or treating the electrospun CAc-PS membrane with two commercially available surfactants named PHOBOL-CP® and ULTRAPHIL® from the Huntsman-DuPont Textile Company. The filter media were characterized physically, mechanically and chemically. The filtration performances of the first two filter media were tested in a lab-scale batch filtration setup, while the performance of the third filter medium remains to be tested. These filter media are novel with respect to their material, structural topography and morphology, surface roughness and filtration performance. The uniqueness of these novel filter media can be attributed to their high reproducibility, scalability, low basis weight, flexibility, high separation efficiency (> 90%) and cost-effectiveness due to their abundant and cheap starting materials. In such terms, these filter media can compete with conventional filter media.

436152 Reducing Drilling Fluid Losses Using a Lab Scale Shale Shaker

Abstract:

The recovery of drilling fluid is essential in the oil and gas drilling under the environmental and economic point of view. These fluids are basically colloidal suspensions classified according to their base components in aqueous and non-aqueous fluids. The equipment typically utilized in solid-liquid separation units are divided into three groups sequentially arranged: shale shakers, hidrociclones (desander and desilter) and centrifugal decanter. Understanding the dynamic behavior of the shale shaker is very important to reduce the process load applied to the subsequent equipment. This paper presented a contribution on the experimental research on employment of g-force control loop dimensionless number applied to a vibrating screen prototype. The study was conducted based on the method traditionally employed in solids control units. The manipulated process variable is the rotation imposed on motor vibrators installed in the vibrating screen. The results obtained in the experiments indicate an appropriate and consistent operation of the proposed control loop. The dimensionless number g-force was maintained close to the set-point value, ensuring repeatability for the experiments.

The use of higher dimensionless values of g-force does not necessarily assured greater removing of residual liquid entrained with the solid material retained by the sieve screen. One possible reason for this behavior is the fact that the greater intensity of vibration provided by motor vibrators helped to increase the solid transport speed on the display screen, reducing the effect of elimination of retained fluid.

404743 Destabilization and Treatment of Produced Oil-Water Emulsions from EOR Application Using Polyacrylamides

Abstract:

Surfactant and polymer flooding technology can greatly enhance the oil recovery through the expansion of sweeping and displacing efficiency. The recovered oil from surfactant and polymer flooding emulsifies the residual chemical, which makes the separation of water from oil quite difficult, yet the impact of the enhanced oil recovery (EOR) chemicals on the produced water cycle is generally neglected in chemically-based EOR studies. This includes compatibility of EOR chemicals with the additives used to pre-treat the injected water or change reservoir wettability and result in producing oil/water emulsion after EOR breakthrough.

The largest waste produced in oil and gas industries is believed to be the produced water, as it contains different sort of organic and inorganic admixture. The discharging of produced water contaminates not only the water, both surface and underground, but also the soil. There are a number of treatment methods available for produced water. To separate water from oil in a much efficient manner and to reach the emission standard, a new class of water soluble polymer of polyacrylamides (PAMs) was used as destabilizing agents for water-oil emulsions, which have been stabilized by surfactant (Talowamine Acetate).

The impact of the surface charge form, the density of polyacrylamides in turbidity reduction , zeta potential, COD, viscosity of separated water were explored in this study. Different anionic polyacrylamide of different surface charge density and molecular weight were evaluated. Different anionic polyacrylamides were utilized, and at optimum dosage, anionic AN94 PAM at its optimum concentration was proved as the best way to reduce the residual turbidity compared with other PAMs mentioned in this research. The effects of electrolytes such as aluminum sulphates on produced water degree of flocculation in the existence of PAM were also investigated in terms of turbidity reduction and volume of separated water after jar test. The results showed that the volume of separated water increased more than 25% compared when only PAMs were used, and the turbidity of separated water improved significantly.

431325 Destabilization and Treatment of Produced Water-Oil Emulsions Using Anionic Polyacrylamide and Electrolyte of Aluminum Sulphate and Ferrous Sulphate

Abstract:

Surfactant and polymer flooding technology can greatly enhance the oil recovery through the expansion of sweeping and displacing efficiency. The recovered oil from surfactant and polymer flooding emulsifies the residual chemical, which makes the separation of water from oil quite difficult, yet the impact of the enhanced oil recovery (EOR) chemicals on the produced water cycle is generally neglected in chemically-based EOR studies. This includes compatibility of EOR chemicals with the additives used to pre-treat the injected water or change reservoir wettability and result in producing oil/water emulsion after EOR breakthrough.

The largest waste produced in oil and gas industries is believed to be the produced water, as it contains different sort of organic and inorganic admixture. There are a number of treatment methods available for produced water. To separate water from oil in a much efficient manner and to reach the emission standard, a new class of water soluble polymer of polyacrylamides (PAMs) was used as destabilizing agents for water-oil emulsions, which have been stabilized by surfactant.

The impact of the surface charge form, the density of polyacrylamides in turbidity reduction, COD, viscosity of volume separated water were explored in this study. The effects of electrolytes such as aluminum sulphates and ferrous sulphate on produced water degree of flocculation in the existence of anionic polyacrylamide were investigated in terms of turbidity reduction and volume of separated water after jar test. Different concentrations of both sulphates added into optimum concentration polyacrylamide selected from jar test were utilized, and at optimum dosage, anionic AN 934 PAM with aluminum sulphate at its optimum concentration was proved as the best way to reduce the residual turbidity compared with other additives mentioned in this research. The results showed that the volume of separated water increased more than 25% compared when only PAMs were used, and the turbidity and COD reduction of separated water improved significantly.

435174 Reliable Use of a Baghouse in High-SO3 Environment

Abstract:

Recent EPA regulations set very stringent particulate emission standards for coal-fired power plants, as well as limit mercury and acid gas emissions. This has led to an increased use of fabric filter baghouses, often in combination with sorbents for mercury and/or acid gas removal. EPRI and Southern Company have research collaboration around the use of a pilot-scale (20,000 ft3/min slip stream) baghouse on flue-gas from medium-to-high-sulfur fuel. This talk will review recent results around emission control, particulate filtration, pulse-jet cleaning, and baghouse O&M.

408584 TFC Membranes with Intrinsic Microporosity By Interfacial Polymerization for Organic Solvent Nanofiltration

Abstract:

Many of the conventional separation and purification processes in oil and gas, chemical, and pharmaceutical industries utilise large quantities of organic solvents and entail high energy consumption [1]. Such processes could be totally or partially replaced by membrane technology, with an order of magnitude lower energy consumption. To enable this, membranes need to be both chemically resistant to the solvents involved and provide high flux, in order to process large solvent volumes with a viable area and in a viable timeframe.

Existing state of the art polymeric Organic Solvent Nanofiltration (OSN) membranes are either integrally skinned asymmetric (ISA) or thin film composite (TFC) membranes [2]. However, ISA membranes age as the polymer chains relax towards equilibrium packing, losing permeance. As an alternative, one might consider using polymers of intrinsic microporosity (PIMs), in which equilibrium polymer packing results in significant free volume that will promote permeation. PIMs are a new class of polymers with great potential in separation due to their extremely high porosity. Non-network PIMs are soluble in certain solvents, making them suitable to cast a membrane by phase inversion, or fabricate a TFC membrane via dip-coating. However, their solubility in a range of solvents restricts their applications in organic solvent nanofiltration. Several efforts to make dip-coated TFC-PIMs membranes solvent resistant have been reported, including photo-crosslinking, blending with other thermally reactive polymers and chemical crosslinking. In this work, we show for the first time the formation of a network PIM-film in situ by interfacial polymerization to develop TFC-PIMs-like membranes for OSN applications.This novel preparation technique represents an advantage compared to conventional TFC-PIMs membranes prepared by coating, where further crosslinking is necessary. Moreover, forming the top layer by IP can produce thinner top layers than when prepared by dip-coating. Solvent stable crosslinked polyimide supports were used for the formation of these TFC-PIMs-like membranes.

To increase permeance we sought to design and control the nanostructure of the top layer in a better way and at a molecular level. This was achieved by incorporating rigid monomers during the IP reaction, resulting in highly porous polymer networks. These TFC-PIMs-like membranes exhibit comparable rejections and a significantly higher permeance for organic solvents, including methanol, acetone, THF, DMF and toluene when compared to commercial ISA OSN membranes and to TFC OSN IP membranes developed previously in our group, eliminating the need for an “activating solvent”. We believe that incorporating rigid monomers into the polymer structure provides higher free volume, which in combination with the polymeric network gives intrinsic microporosity to the top layer, which enables a great improvement of permeance without sacrificing selectivity. Such network TFC-PIMs-like membranes prepared by interfacial polymerization may lead to the next generation of high performance OSN membranes.

[1] G. Szekely, M.F. Jimenez-Solomon, P. Marchetti, J.F. Kim, and A.G. Livingston, “Sustainability assessment of organic solvent nanofiltration: from fabrication to application”, *Green Chem.*, **16**(2014) 4440-447.

[2] P. Marchetti, M.F. Jimenez Solomon, G. Szekely, and A.G. Livingston, "Molecular Separation with Organic Solvent Nanofiltration: A Critical Review", *Chem. Rev.*, **114** (21) (2014) 10735-10806.

412615 Sorption and Diffusion of Organic Vapors into PIM-1 and the Effects of Methanol Conditioning

Abstract:

Polymers of intrinsic microporosity (PIMs) are of great interest for fluid separation membranes due to their high permeability and moderate selectivity. To improve reproducibility, high free volume polymers such as PIMs are often soaked in methanol and subsequently dried before permeation and sorption analysis. This methanol preconditioning step is believed to remove residual solvent trapped within the microstructure of the polymer as well as erase the past thermal processing history. However, few studies have carefully investigated this process. Sorption data is analyzed to determine the diffusion coefficients of DMF and methanol in neat PIM-1 as a function of volume fraction. Repeated sorption isotherms of organics without methanol treatment show evidence of membrane conditioning and reduced solvent uptake over time. This work examines the efficacy of methanol treatments to remove high boiling solvents such as DMF from PIM-1. Surface area, pore volume, and thermogravimetric analysis are used to quantify the effectiveness of the methanol treatment at erasing the processing history of the polymer.

425009 Novel Composite Nanofiltration Membrane with Solvent Resistance Using Monomers of 1,2,4,5-Benzene Tetracarbonyl Chloride and Amines

Abstract:

Novel composite nanofiltration membrane with solvent resistance using monomers of 1,2,4,5-benzene tetracarbonyl chloride and amines

Yuyan Hai, Jinli Zhang, Wei Li*

(Key Laboratory for Green Chemical Technology MOE, School of Chemical Engineering & Technology, Tianjin University, Tianjin 300072, China. *liwei@tju.edu.cn)

Abstract

Nanofiltration (NF) has attracted more attention in the field of drinking water production and wastewater treatment, in particular, for removing pharmaceutical residues with low molecular weight in the range of 200-400 from organic solvents [1, 2]. However, it is still a challenge to manufacture solvent resistant NF membranes that can be substantially used to separate pharmaceutical residues from organic solvents.

1,2,4,5-benzene tetracarbonyl chloride (BTC), synthesized by pyromellitic dianhydride, is a cheap monomer to occur the interfacial polymerization with m-phenylene diamine (MPD) [3]. To prepare a NF membrane with high flux and good rejection towards small organic molecules, here we reported a method to prepare novel polyamide thin film composite NF membrane on polyetherimide supports via interfacial polymerization of BTC and a layer of amine-containing network.

Fig.1 (a) shows the separation performance of two NF membranes (M1 and M2) for the aqueous glucose solution. M1 is the NF membrane synthesized by BTC and MPD, and M2 is the novel NF membrane synthesized by BTC and a layer of amine-containing network. Both M1 and M2 have good resistance to acetone. At an operating pressure of 1.0 MPa, the M1 exhibits a glucose rejection of 90% at a flux of 25 L m^{-2} h^{-1}, while the M2 exhibits a glucose rejection of 98% at a flux of 52 L m^{-2} h^{-1}. Fig.1 (b) shows the rejection of M1 and M2 for the aqueous inorganic salts solution. The rejection of M2 is a little higher than that of M1, with the rejection decreasing in the order: Na_2SO_4 (86%) > $MgSO_4$ (82%) > $MgCl_2$ (80%) > NaCl (29%). The interaction mechanism between BTC and the layer of amine-containing network is studied deeply to disclose the reason that such NF membranes have good solvent resistance, in combination with the characterizations of attenuated total reflection Fourier transform infrared (ATR-FTIR), X-ray photoelectron spectroscopy (XPS), scanning electron microscopy (SEM), atomic force microscopy (AFM), contact angle, zeta potentials and solvent resistance assessment, etc.

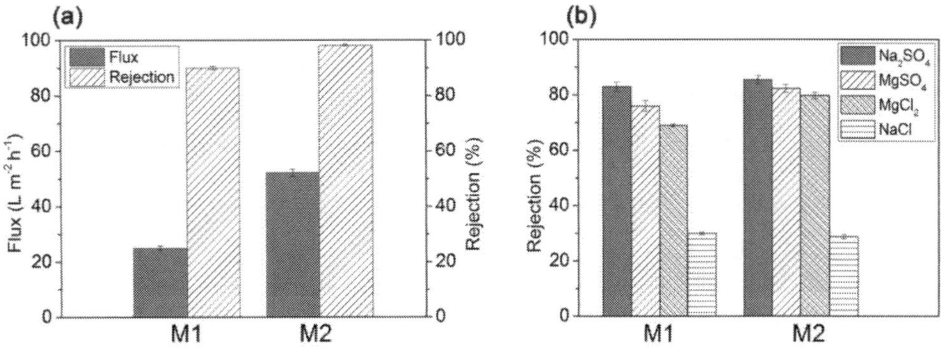

Fig.1 (a) Flux and rejection of M1 and M2 for the aqueous glucose solution at 1.0 MPa. (b) Rejection of M1 and M2 for the aqueous inorganic salts solution at 1.0 MPa.

Acknowledgement

This work was supported by National High-tech R&D Program of China (2012AA03A609).

References

[1] D. Rana, R. M. Narbaitz, A.-M. Garand-Sheridan, A. Westgate, T. Matsuura, S. Tabe and S. Y. Jasim, Development of novel charged surface modifying macromolecule blended PES membranes to remove EDCs and PPCPs from drinking water sources, *Journal of Materials Chemistry A*, 2014, 2, 10059.

[2] Ludmila Peeva, Joao da Silva Burgal, Irina Valtcheva, Andrew G. Livingston, Continuous purification of active pharmaceutical ingredients using multistage organic solvent nanofiltration membrane cascade, *Chemical Engineering Science*, 2014, 116, 183-194.

[3] Jinli Zhang, Yuyan Hai, Yi Zuo, Qian Jiang, Chang Shi, Wei Li, Novel diamine-modified composite nanofiltration membranes with chlorine resistance using monomers of 1,2,4,5-benzene tetracarbonyl chloride and m-phenylenediamine, *Journal of Materials Chemistry A*, 2015, 3, 8816-8824.

421416 Membrane Transport and Process Modelling in Osn: A Multiscale Approach

Abstract: **Membrane transport and process modelling in OSN: a multiscale approach**

Patrizia Marchetti [1] , Binchu Shi [1,2], Dimitar Peshev [3], Andrew G. Livingston [1,2] *

[1] *Department of Chemical Engineering, Imperial College London, Exhibition Road, London SW7 2AZ, United Kingdom*

[2] *Evonik Membrane Extraction Technology Limited, Unit 6 Greenford Park, Ockham Drive, Greenford, London, UB6 0AZ, United Kingdom*

[3] *Department of Chemical Engineering, University of Chemical Technology and Metallurgy, 8, Kl. Ohridsky Blvd, Sofia 1756, Bulgaria*

* a.livingston@imperial.ac.uk

In Organic Solvent Nanofiltration (OSN), a fundamental understanding of the basic separation mechanism is useful to advance the field and wisely tune new materials and processes. The development of membrane processes usually involves several stages, starting from feasibility tests at membrane scale, passing through fluid dynamics and mass transfer in membrane modules and finishing with large industrial scale process design (Figure 1) [1,2].

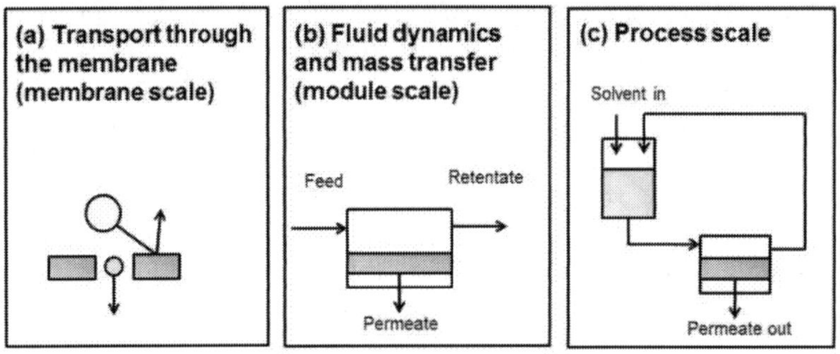

Figure 1. Modelling levels for the development of a membrane process [1].

Substantial research has focused on the description of the transport mechanism through membranes (Figure 1(a)) aiming at: *parameter estimation*, when experimental data for standard solute/solvent systems are available, and it is desired to estimate relevant parameters for as yet uncharacterized systems; and *prediction*, when the model parameters for the solvent-solute are available, and modelling can be applied to describe performance of a different solute/solvent system [1,3]. Most of these studies have used flat sheet membranes. On the other hand, there is lack of studies on the fluid dynamics and mass transfer characteristics in OSN spiral-wound modules (Figure 1(b)), and attempts to implement fundamental transport models on a process level (Figure 1(c)) are few, and are based on simple, non-predictive membrane transport models.

This paper reports a systematic comparison of different transport models using selected experimental data for various solutes, solvents and membranes [3]. Solution-diffusion, pore-flow and transient models were used to perform regression of experimental data, and prediction, based on the regressed model parameters. It was observed that solution-diffusion models described permeation through glassy membranes better than pore-flow models, against both positive and negative rejection data. Afterwards, the classical solution-diffusion model combined with the film theory was applied to model the performance of different rubber-coated spiral-wound modules (1.8"x12", 2.5"x40" and 4.0"x40") under various pressures and retentate flowrates. Correlations for the fluid dynamics and mass transfer in the modules were derived by regression of experimental data. Finally, OSN unit operations at membrane and module levels were made available in the Aspen Plus process modelling environment, to streamline process design using the so-called "OSN Designer" [2]. Matlab routines for the relevant models were interfaced to Aspen Plus by means of CAPE OPEN and custom OSN unit operations for common batch and steady-state membrane processes were developed. The suitability of this technique was demonstrated by comparing the model simulation to the experimental data for both ideal and non-ideal solutions.

References

[1] P. Marchetti, M.F. Jimenez Solomon, G. Szekely, A. Livingston, *Chem. Rev.* 114 (2014) 10735

[2] D. Peshev, A. Livingston, *Chem. Eng. Sci.* 104 (2013) 975

[3] P. Marchetti, A. Livingston, *J. Membr. Sci.* 476 (2015) 530

423328 Hybrid Membranes Based on Functionalized Metal-Organic Molecule Nanocages for Efficient Aromatic Hydrocarbons Recovery

Abstract:

Hybrid Membranes based on Functionalized Metal-Organic Molecule Nanocages for Efficient Aromatic Hydrocarbons Recovery

Cui Zhao, Naixin Wang, Lin Wang, Shunan Sheng, Shulan Ji*, Jian-Rong Li*

Beijing Key Laboratory for Green Catalysis and Separation and Department of Chemistry and Chemical Engineering, Beijing University of Technology, Beijing 100124, P. R. China.

The development of efficient and economic technology to separate aromatic hydrocarbons from their aliphatic mixtures has been of great concerns in chemical industry. Among them, the membrane-based separation has been recognized as one of the most promising approaches in aromatic hydrocarbons gathering.[1,2] However, the challenge for this membrane separation technique is the selection/preparation of high quality membranes with good selectivity and permeability.

Recent years, inorganic-organic hybrid membranes have gained urgent attention in the membrane separations.[3-6] Where, the unique nature of the inorganic fillers and their good dispersion and compatibility in/with the polymer are crucial for the separation performances of this type of membranes. Therefore, in order to solve the dispersion and uniformity problems, and at the same time to efficaciously control their loading in polymer, herein, our strategy is to use soluble metal-organic polyhedra (MOPs) molecules as fillers to prepare hybrid membranes, exemplified as below.

A nano-hybrid membrane with soluble MOP-*t*Bu, $[Cu_{24}(5\text{-}t\text{Bu-}1,3\text{-BDC})_{24}(S)_{24}]$ as fillers was initially fabricated by adopting the typical co-blending method, and used in the pervaporation separation of aromatic/aliphatic hydrocarbons.[7,8] It was found that this membrane is interfacial defect-free and homogeneous, and shows excellent performance for the toluene/*n*-heptane separation with a high total flux of 229.5 g/m^2 h and a separation factor of 19.0; while for the benzene/cyclohexane separation with a high total flux of 392.3 g/m^2 h and a separation factor of 15.39 at 40 °C. More remarkable, this MOP-based hybrid membrane had long-term stability in its performance in separating aromatic/aliphatic mixtures (see Figure. 1).

Furthermore, in order to improve the separation of aromatic/aliphatic hydrocarbons, a series of isostructural functionalized MOPs/hyperbranched polymer Boltorn W3000 membranes are molecularly designed and fabricated by a new method, which helps plugging the pores within the substrate.[9] Pervaporation tests employing these MOP-X (X = SO_3Na, OH, and *t*Bu) filled hybrid membranes demonstrated an significant improvement for the aromatic hydrocarbons separations. Particularly, the membrane with MOP-SO_3Na showed the highest permselectivity for a toluene/*n*-heptane mixture. Simultaneously, adsorption experiments and molecular simulations also confirmed that stronger polar functionalized groups have higher adsorption capacity towards the aromatic hydrocarbons, being in agreement with pervaparation experiments results.

Overall, in light of the versatile structures and customizable chemical functionalities of MOPs materials, a lot of hybrid membranes can thus be fabricated for different separation applications, after paving the road in this work.

Keywords: hybrid membrane, metal-organic polyhedra, pervaporation, aromatic/aliphatic hydrocarbons separation

References

[1] C. P. Ribeiro, B. D. Freeman, D. S. Kalika and S. Kalakkunnath, Aromatic polyimide and polybenzoxazole membranes for the fractionation of aromatic/aliphatic hydrocarbons by pervaporation, J. Membr. Sci., 390-391 (2012), 182-193.

[2] N. Wang, S. Ji, J. Li, R. Zhang and G. Zhang, Poly(vinylalcohol)-graphene oxide nanohybrid "pore-filling" membrane for pervaporation of toluene/n-heptane mixtures, J. Membr. Sci., 455 (2014), 113-120.

[3] T. Yang, G.-M. Shi and T. S. Chung, Symmetric and Asymmetric Zeolitic Imidazolate Frameworks (ZIFs)/Polybenzimidazole (PBI) Nanocomposite Membranes for Hydrogen Purification at High Temperatures, Adv. Energy. Mater., 2 (2012), 1358-1367.

[4] R. Zhang, S. Ji, N. Wang, L. Wang, G. Zhang and J.-R. Li, Coordination-Driven In Situ Self-Assembly Strategy for the Preparation of Metal-Organic Framework Hybrid Membranes, Angew. Chem. Int. Ed., 53 (2014), 9775-9779.

[5] B. Zornoza, C. Tellez, J. Coronas, J. Gascon and F. Kapteijn, Metal organic framework based mixed matrix membranes: An increasingly important field of research with a large application potential, Micropor Mesopor Mat., 166 (2013), 67-78.

[6] J. Campbell, G. Sz¨¦kely, R. P. Davies, D. C. Braddock and A. G. Livingston, Fabrication of Hybrid Polymer/Metal Organic Framework Membranes: Mixed Matrix Membranes versus In-Situ Growth, J. Mater. Chem. A, 2 (2014), 9260-9271.

[7] J.-R Li and H.-C. Zhou, Bridging-ligand-substitution strategy for the preparation of metal-organic polyhedral, Nat. Chem., 2 (2010), 893-898.

[8] C. Zhao, N. Wang, L. Wang, H. Huang, R. Zhang, F. Yang, Y. Xie, S. Ji and J.-R. Li, Hybrid membranes of metal-organic molecule nanocages for aromatic/aliphatic hydrocarbon separation by pervaporation, Chem. Commun., 50 (2014), 13921-13923.

[9] C. Zhao, N. Wang, L. Wang, S. Sheng, S. Ji and J.-R. Li, Functionalized Metal-Organic Molecule Nanocages Filled Hybrid Membranes for Effective Aromatic Recovery: Further Analysis by combining Adsorption Experiment and DFT Calculations, submitted.

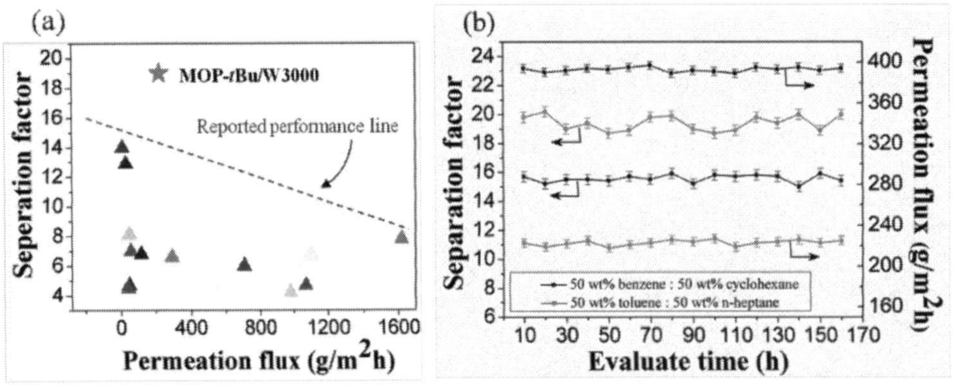

Figure 1. Pervaporation performances of MOP-*t*Bu/W3000 nanohybrid membrane and a comparison of pervaporation performances of some reported hybrid membranes towards the separation of toluene/*n*-heptane mixture.

420006 Organic Solvent Nanofiltration with Novel Polymeric Membranes

Abstract:

A very large variety of membranes polymeric or otherwise have been investigated over the last two decades for organic solvent nanofiltration (OSN). The materials and structures used to make OSN membranes include among others the following: poly (dimethylsiloxane) (PDMS); mixed matrix membranes (MMMs) of PDMS with zeolites and other fillers; asymmetric integrally skinned polyimide (PI) membrane crosslinked with aromatic or aliphatic diamines; polyaniline; polypyrrole; interfacially polymerized polyamide w/o carbon nanotubes; carbon-based membranes including graphene. One of the weaknesses of most polymeric membranes is their varying tendencies for swelling with demanding organic solvents aprotic or otherwise; this gets reflected in the solute rejection behaviors of such polymers. To this end we have started studying membranes from a particular class of polymers for OSN. We report preliminary results of solvent flux and solute rejection for such membranes. We also report results of molecular dynamic simulations of transport in one such polymeric membrane.

423339 Enhanced Rhamnolipids Production By Pseudomonas Aeruginosa BC1 in a Membrane Bioreactor with the Convenient Permeate Substances Separation

Wednesday, November 11, 2015: 12:30 PM
155C (Salt Palace Convention Center)

Yan Zhang[1], Zeyi Xiao[1], Chunyan Chen[2], Senqing Fan[1], Wenlei Liu[1] and Mingming Yang[1], (1)School of Chemical Engineering, Sichuan University, Chengdu, China, (2)College of Chemistry and Chemical Engineering, Southwest Petroleum University, Chengdu, China

Rhamnolipid biosurfactants are natural surfactants, emulsifiers, and a promising alternative to synthetic surfactants. A newly isolated Pseudomonas aeruginosa BC1 from oil-field wastewater was able to produce a glycolipid biosurfactant (i.e., rhamnolipid). Thin layer chromatotography (TLC), fourier transform infrared spectroscopy (FT IR) and high performance liquid chromatography-mass spectrometry (HPLC-MS) worked together to reveal that the biosurfactant consisted mainly of mono-rhamnolipid (RL1). The present work attempted to enhance RL1 production by Pseudom onas aeruginosa BC1 in a pervaporation membrane bioreactor with the convenient perme ate substances separation. The results showed that the enhanced fermentation could be ac achieved with high production of RL1 and variety of volatile substances separation in situ. Without volatile substances separation, RL1 production was 0.942g/l and 1.54g/l when gl glucose and glycerol as carbon source, respectively. However, in pervaporation membrane bioreactor, RL1 production reached 1.3g/l and 3.6g/l, an improvement of 38% and 134%, respectively. During the fermentation process in pervaporation membrane bioreactor, the convenient permeate substances separation in situ improved the ability of cell growth and metabolism obviously. Oil displacement experiments showed 76% heavy oil stripping from artificial quartz sands.

427064 Non-Conventional Analysis of Hydrogen Upgrading in Membrane Reactor By Process Intensification Metrics

Abstract:

Non-conventional analysis of hydrogen upgrading in membrane reactor by process intensification metrics

Giuseppe Barbieri[1,*], Enrico Drioli[1,2], Adele Brunetti[1]

[1] *National Research Council - Institute for Membrane Technology (ITM–CNR)*
c/o The University of Calabria, cubo 17C, Via Pietro BUCCI, 87036 Rende CS, Italy
www.itm.cnr.it; Tel. +39 0984 492029; Fax. +39 0984 402103; g.barbieri@itm.cnr.it

[2] *The University of Calabria - Department of Chemical Engineering and Materials, cubo 44A Via Pietro BUCCI, 87036 Rende CS, Italy*

The always increasing necessity arose in the last decennia in redesign the industrial processes with new unit operations more compact, efficient and, thus, able to address various environmental concerns, has led to the definition of new process indexes, so-called, metrics, that supply additional and important information to the conventional technical analysis for the selection of the type of technology and the identification of the operating condition windows for making a process more profitable [1].

In this work, some sustainability indexes, mass and energy intensities [2], volume and conversion indexes [3-4] were used in a no conventional evaluation of the up-grading stage in hydrogen production, i.e. the water gas shift reactor, by means of membrane reactors.

Defined as the ratio between the total inlet mass and total energy involved in the reactor, with respect to the hydrogen fed and produced by the reactor, mass and energy intensity provide useful information about the material exploitation and the energy efficiency of this new technology. On the other side, Volume and conversion index provide an indication on the gain offered by membrane reactor in terms of catalyst volume required and reactants conversion achieved, at the same operating conditions of a traditional unit. The comparative study of membrane reactor performance with respect to the conventional reactor was analysed as function of the main process variables, such as temperature, pressure, feed molar ratio and space velocity.

In the comparison between the two unit operations, the membrane reactor resulted always more material and energy intensive than a traditional reactor, particularly at the high feed pressure indicating that membrane reactor requires less material as feed and makes available more energy in producing the same amount of H_2. The advantage offered by a membrane reactor was also quantified in terms of ratios referred to the equilibrium condition of the traditional reactor. The membrane reactor resulted always more intensified than a traditional reactor operated in similar conditions and exceeded also the ideal performance achievable by a traditional reactor, at a temperature higher than 350°C. At the highest temperature (450°C) and gas hourly space velocity (40000 h^{-1}) the indexes for the membrane reactor were quite the same of the ideal values (the ones at equilibrium) of the traditional reactor; the membrane reactor can, thus, be operated with different combinations of operating conditions, achieving the same performance in terms of material exploitation and energy efficiency. An outlook of all the results in a phase diagram highlights a region where only MR has access. This is the most intensified in terms of both mass and energy intensity. Mass and energy intensities demonstrated, in line with the process intensification strategy, the assets of the membrane reactor technology also in terms of better exploitation of raw materials (reduction up to 40%) and higher energy efficiency (up to 35%). In addition, the catalyst volume required by membrane reactor was more than three times lower than the one of traditional unit, for achieving the same conversion.

[1] Drioli E.; Brunetti A.; Di Profio G.; Barbieri G.; "Process intensification strategies and membrane engineering", Green Chem. 2012, 14, 1561-1572

[2] Brunetti A.; Drioli E.; Barbieri G. "Energy and mass intensities in hydrogen upgrading by a membrane reactor", Fuel Processing Technology, 2014, 118, 278-286

[3] Brunetti A.; Caravella A.; Drioli E.; Barbieri G.; "Process intensification by membrane reactors: high temperature water gas shift reaction as single stage for syngas upgrading", *Chemical Engineering and Technology, 2012, 35, 1238-1248*

[4] Barbieri G.; Brunetti A.; Caravella A.; Drioli E.; "Pd-based Membrane Reactors for one-stage Process of Water Gas Shift", RSC Adv., 2011, 1 (4), 651-661

432020 Ultrathin Pd and Pd Alloy Membranes with Enhanced Sulfur Resistance Using Phosphorus Additive and Their Application in Membrane Reactors for Ethanol Steam Reforming

Abstract:

The conversion of fossil fuels, by gasification, reforming, or partial oxidation processes is important for the synthesis of hydrogen and syngas which are subsequently used in refining, and for production of chemicals and fuels. Gas-separation steps are needed in the implementation of these technologies and membranes are commonly used for this purpose. Membrane reactors are a promising application because they combine reaction and separation in one step, overcoming equilibrium limitations in some chemical transformations.

It is known that palladium-based membranes are effective materials for hydrogen separation with high permeance and selectivity and high mechanical strength. In the present work a novel membrane preparation technique was used employing an electric-field to uniformly deposit Pd nanoparticle seeds on a substrate, followed by deposition of ultrathin Pd or Pd-Cu layers on the activated surface by electroless plating (ELP). To increase the membranes' resistance to sulfur poisoning, phosphorus was incorporated in the structure and the peformance and sulfur resistance of the membranes were evaluated under exposure to a gas mixture of 100 ppm H_2S in H_2. The presence of phosphorus conferred structural integrity to the membranes, resulting in a more efficient regeneration under hydrogen flow after exposure to H_2S.

The prepared membranes were used in a membrane reactor and tested for ethanol steam reforming for hydrogen production, using catalysts such as Co-Na/ZnO. The reactions were also carried out in in a packed bed reactor to evaluate the benefits of employing membranes. Ethanol conversion and hydrogen product yields were significantly higher in the membrane reactors compared to the packed bed reactor.

433100 Preparation and Characterization of Supported Carbon Molecular Sieve Membranes, and Their Field-Testing for Process Intensification during Power Generation

Abstract:

Preparation and Characterization of Supported Carbon Molecular Sieve Membranes, and their Field-Testing for Process Intensification during Power Generation

Ashkan Garshasbi[1], Doug Parsley[2], Richard J. Ciora[2], Jr., Paul KT Liu[2], and Theodore T. Tsotsis[1]

[1]The Mork Family Department of Chemical Engineering and Materials Science, University of Southern California, 925 Bloomwalk, Los Angeles, CA 90089-1211, USA.

[2]Media and Process Technology Inc, 1155 William Pitt Way, Pittsburgh, PA 15238-1360.

High-temperature gas separations (GS) with inorganic membranes have attracted attention recently. In particular, the use of such membranes in membrane reactors (MR) has the potential to enhance process intensification and to increase energy savings and/or product yield. Though the potential benefits of high-temperature GS and/or MR processes are substantial, commercialization still remains elusive. A major technical barrier is the lack of robust inorganic membranes and full-scale modules which are suitable for use at the high-temperature and high-pressure conditions required. This team recently proposed and tested in the laboratory a technology, termed the "one-box" process, that employs carbon molecular sieve membranes (CMSM) in an MR for the water gas shift reaction of coal- and/or biomass-derived syngas. The process combines syngas contaminant removal and CO conversion into a single unit to produce high-purity hydrogen for power generation and/or chemical use.

Towards technology commercialization, full-scale multi-tubular CMSM modules have been constructed and characterized comprehensively and their effectiveness has been demonstrated. Our successful preparation of the full-scale multi-tubular inorganic membrane module overcomes some of the technical barriers which hamper the commercial implementation of inorganic membranes. Further multiple field-tests of the full-scale CMSM modules were conducted at the US National Carbon Capture Center (NCCC) coal gasification facility under conditions suitable for the "one-box" process, and this presentation will detail the results of these tests. During continuous use in raw coal-derived and/or biomass-derived syngas, the CMSM successfully rejected tar-like species present in the syngas without any evidence of fouling. Our current efforts focus on the lab-scale testing of reactive separations using these membranes in a broader range of pressure and temperatures akin to the real process, and on the field-testing of the "one-box" process by incorporating the catalyst used in the prior lab-scale study into the full-scale CMSM modules developed.

433330 Three-Phase Hydrogenation Reactions Using a Ruthenium Coated Polymeric Membrane Reactor

Abstract:

Three-phase Hydrogenation Reactions using a Ruthenium Coated Polymeric Membrane Reactor

John Stanford[1,2], Peter Pfromm[2], and Mary Rezac[2]*

[1] *IGERT Trainee in Biorefining*

[2] *Department of Chemical Engineering, Kansas State University*
Manhattan, Kansas, United States

*rezac@ksu.edu

Catalytic membrane reactors afford an alternative and potentially more efficient method for performing three-phase heterogeneous chemical reactions. Traditional three-phase reactors often present mass transfer limitations, namely relatively large diffusional distances to reach catalytic sites exacerbated by low gas solubility in the liquid phase. Membrane reactors can alleviate the inherent mass transfer limitations by directly and abundantly supplying gas to the catalytic sites located on the membrane surface, which acts as a gas/liquid phase contactor, and thus lessening the necessity for higher gas phase pressures. The reactions investigated in this work include the hydrogenation of 5-hydroxymethylfurfural (HMF) in an alcohol solvent and the hydrogenation of levulinic acid (LA) in an aqueous solvent.

Polyimide polymers are used in this work for membrane synthesis because of their good chemical and high temperature resistance. The polymer solutions are cast as asymmetric integrally-skinned flat sheet membranes and are then coated with a ruthenium catalyst on the non-porous surface. The completed membrane is positioned in a flow over configuration maintaining liquid contact on the metal coated surface while allowing hydrogen gas to permeate from the porous support side to the catalytic sites on the non-porous surface. The multi-functionality of the membrane reactor/contactor has allowed several areas to be investigated, including catalytic activity and reaction kinetics, membrane performance and characterization, and solvent/polymer interactions. Quantitative hydrogenation product formation with reaction kinetics similar to or better than more traditional three-phase reactors has been achieved. The membrane reactors are shown to be stable and catalytically active for several days of continuous operation. This work has also demonstrated that unless the dense separating layer of the polymeric membrane exhibits and maintains a high degree of 'defect free' quality, then the penetrating liquid reaction solvent reintroduces the mass transfer limitations for the gas phase that the membrane was intended to eliminate. Continued efforts to improve membrane reactor performance such as increasing hydrogen permeance, increasing catalytic site availability, and decreasing liquid phase permeance should yield an increasingly favorable comparison to traditional reactor systems.

Study of Mass Transport Regimes in Membrane Steam Reformers

M.A. Murmura, S. Cerbelli, M.C. Annesini, University of Rome "Sapienza", Italy

Abstract

The success of membrane reactors for hydrogen separation depends significantly on the development of membranes having high permeability and selectivity with respect to hydrogen. Pd-based membranes have been proposed and their performance is being continuously improved. In conditions of high permeating flux, external mass transfer resistances, such as those due to the transport of hydrogen from the bulk of the catalytic bed to the membrane surface, may limit the performance of the reactor. In the present work, a two-dimensional model has been developed and the influence of different operating conditions on the performance of a methane steam reforming reactor has been investigated. An isothermal reactor has been considered as a first step for a more complete analysis. The velocity profile has been evaluated through Darcy's law for flow through porous media, in which gas velocity is correlated to pressure drops and bed permeability. Both axial and radial components of velocity have been accounted for. Mass transport equations for all species present have been developed by considering consumption/production through the steam reforming reaction, and transport by convection and dispersion in the radial and axial directions. Variations of gas density due to pressure gradients and composition changes have also been accounted for. In these conditions, full coupling between mass and momentum transport has been obtained. The membrane has been considered to have infinite selectivity towards hydrogen and Sieverts' law has been used to describe hydrogen flux through the membrane. Dimensionless parameters which govern the system's behavior have been identified and, according to the value of these parameters, simplified models have been proposed. The equations of the general model have been solved using the finite elements solver COMSOL Multiphysics. Integral quantities, such as hydrogen production, permeation, and recovery, as well as concentration and mass flux profiles within the reactor have been evaluated. The most important results of this study are the definition of an optimal operating pressure, whose value depends on membrane permeability and reaction rate, and the identification of a reaction boundary layer, close to the membrane, in which all the hydrogen-consuming reaction occurs. These findings may serve as a basis for the optimization of reactor design.

Problem Description

The aim of this work is to describe transport regimes in a membrane reactor having a double-pipe configuration. The composition of the gaseous mixture to be treated has been set to be equal to the equilibrium composition obtained by considering that the methane steam reforming (MSR) reaction takes place at 773 K starting from a steam /methane mixture in 3:1 molar ratio. The reaction considered is

$$CH_4 + 2H_2O \rightleftharpoons 4H_2 + CO_2 \tag{1}$$

The mixture enters the annular gap between two co-axial cylinders. The wall of the outer cylinder (R_2) is impermeable to both components, whereas a membrane, with infinite selectivity towards hydrogen, is supported onto the outer wall of the inner cylinder (R_1). In the system considered, the aspect ratios, $L/R_1 = 10$ and $R_2/R_1 = 2$ have been adopted (Fig. 1). The subscripts "m", "w", "h", and "c" will be used to indicate methane, water, hydrogen, and carbon dioxide, respectively. In order to better understand the problem, the following procedure was adopted:

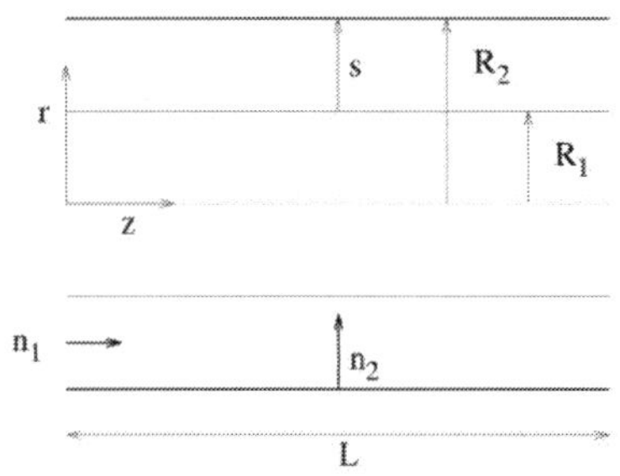

Figure 1: Reactor geometry and coordinate system

1. Study of the effect of operating conditions on a membrane separator for the treatment of a binary mixture
2. Study of the effect of operating conditions on a membrane reactor for MSR

Membrane Separator

The model of the membrane separator has been developed under the assumptions of uniform temperature and zero partial pressure of hydrogen in the permeate. The velocity field has been evalutaed by Darcy's law

$$\mathbf{v} = -\frac{\kappa}{\mu}\nabla P \tag{2}$$

where $\mathbf{v}$ if the mass average gas velocity, κ is the bed permeability, μ is the viscosity of the gas mixture, and P the pressure. In steady-state conditions, the overall mass balance is given by:

$$\nabla \cdot (\rho \mathbf{v}) = 0 \tag{3}$$

where ρ is the mixture's mass density, which may be expressed as

$$\rho = f(\omega_i)\frac{P}{RT} \quad \text{with} \quad \frac{1}{f(\omega_i)} = \sum_i \frac{\omega_i}{W_i} \tag{4}$$

where ω_i and W_i indicate mass fraction and molecular weight, respectively, of the $i-th$ component and f is the average molecular weight. The equation of continuity in isothermal conditions becomes

$$\nabla \cdot (-fP\nabla P) = 0 \tag{5}$$

Mass balance equations in terms of mass fractions are

$$\nabla \cdot \left(-\frac{\kappa}{\mu}fP\nabla P\omega_i - fP\mathbb{D}\nabla\omega_i\right) = 0 \tag{6}$$

Re-writing the above equations in terms of dimensionless variables allows the definition of characteristic groups. The following dimensionless variables have been chosen

$$\tilde{\mathbf{v}} = \frac{\mathbf{v}}{U} \;;\; \tilde{r} = \frac{r}{R_1} \;;\; \tilde{z} = \frac{z}{R_1} \;;\; \tilde{f} = \frac{f}{W_h} \;;\; \tilde{P} = \frac{P}{P_{atm}} \;;\; \tilde{\mathbb{D}} = \frac{\mathbb{D}}{\mathcal{D}} \tag{7}$$

where P_{atm} is the atmospheric pressure and $\mathcal{D}$ is the molecular diffusivity, supposed to be equal for all components of the system.

Substituting these variables in the continuity equation gives

$$\tilde{\nabla} \cdot \left(\tilde{f} \tilde{P} \tilde{\nabla} \tilde{P} \right) = 0 \tag{8}$$

where $\tilde{\nabla}$ is the dimensionless gradient operator.

The dimensionless form of the mass balance equation is instead

$$\tilde{\nabla} \cdot \left(\frac{U_c}{U} \tilde{f} \tilde{P} \omega_h \tilde{\nabla} \tilde{P} - \frac{1}{\mathrm{Pe}} \tilde{f} \tilde{P} \tilde{\mathbb{D}} \tilde{\nabla} \omega_h \right) = 0 \tag{9}$$

where

$$U_c = -\frac{\kappa}{\mu} \frac{P_{atm}}{R_1} \quad \text{and} \quad \mathrm{Pe} = \frac{U R_1}{\mathcal{D}} \tag{10}$$

are the characteristic permeation velocity through the packed bed and the Peclet number, respectively.

The continuity and mass balance equations presented above have been solved by assuming that Sieverts' law for hydrogen permeation applies, and hydrogen flux through the membrane is expressed by:

$$J_h = \mathbb{P}_m \left(\sqrt{P_h} - \sqrt{P_h^p} \right) \tag{11}$$

where $\mathbb{P}_m$ is the overall membrane permeability, P_h is the partial pressure of hydrogen in the retentate, and P_h^p is the partial pressure of hydrogen in the permeate, here set to zero. The main boundary conditions are reported

$$in \quad z = 0: \quad \left(-\frac{\kappa}{\mu} \nabla P \right) \cdot \mathbf{n_1} = U \quad ; \quad \omega_i = \omega_i^0 \tag{12}$$

$$in \quad z = L: \quad P = P_L \tag{13}$$

$$in \quad r = R_1: \quad \frac{1}{\mathcal{R}T} \left(-\frac{\kappa}{\mu} \omega_h f P \nabla P - f P \mathbb{D} \cdot \nabla \omega_h \right) \cdot \mathbf{n_2} = \mathbb{P}_m \left(\sqrt{\omega_h f P / W_h} \right) \tag{14}$$

$$\text{and} \quad \frac{1}{\mathcal{R}T} \left(-\frac{\kappa}{\mu} f P \nabla P \right) \cdot \mathbf{n_2} = \mathbb{P}_m \left(\sqrt{\omega_h f P / W_h} \right) \tag{15}$$

where $\mathbf{n_1}$ is a unit vector normal to the annular surface through which the gas flows, and directed inwards; and $\mathbf{n_2}$ is a unit vector normal to the cylindrical surface and directed towards the axis of the equipment ($r = 0$). It should be noted that, in what follows, inlet gas velocity has been maintained constant, meaning that increasing operating pressure leads to increasing inlet mass flow rate.

The dimensionless counterpart of the BCs in eqs.(12)-(14) writes

$$in \quad \tilde{z} = 0: \quad \left(-\frac{U_c}{U} \tilde{\nabla} \tilde{P} \right) \cdot \mathbf{n_1} = 1 \tag{16}$$

$$in \quad \tilde{z} = \frac{L}{R_1}: \quad \tilde{P} = \alpha \tag{17}$$

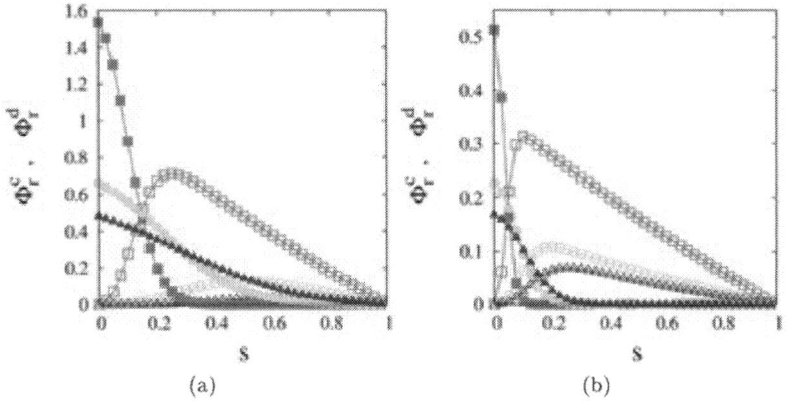

Figure 2: Radial mass flux of hydrogen by convection (empty points) and dispersion (filled points) vs. distance from the membrane, s, evaluated at $\tilde{z}$=1 (squares), 5 (circles), and 9 (triangles) for $\alpha = 20$, $\gamma = 5$, (a) $Pe = 10^2$, (b) $Pe = 10^3$ (figures are on different scales).

$$in \;\; \tilde{r} = 1: \quad \left((U_c/U)\, \tilde{f}\tilde{P}\omega_h \tilde{\nabla}\tilde{P} + (1/Pe)\, \mathbb{D}\tilde{f}\tilde{P} \cdot \tilde{\nabla}\omega_h \right) \cdot \mathbf{n_2} = -\gamma\sqrt{\omega_h \tilde{f}\tilde{P}} \tag{18}$$
$$(U_c/U)\, \tilde{\nabla}\tilde{P} \cdot \mathbf{n_2} = -\gamma\sqrt{\frac{\omega_h}{\tilde{f}\tilde{P}}}$$

where the following characteristic groups have been introduced

$$\alpha = \frac{P_L}{P_{atm}} \quad ; \quad \gamma = \frac{\mathbb{P}_m \mathcal{R} T P_{atm}^{-1/2}}{W_h} \frac{1}{U} \tag{19}$$

Figures 2(a) and (b) show the radial mass flux of hydrogen by convection and dispersion as a function of the dimensionless distance from the membrane, s. The curves have been evaluated for different axial positions at $\alpha = 20$, $\gamma = 5$, and at two different values of Pe. The figures highlight that radial convective transport is significant in the entire separator, particularly for high values of Pe, indicating that the competition between the different transport mechanisms must be accounted for to fully understand the behavior of the system. The main variables next used to describe the behavior of the system are the permeating flux, evaluated from the integration of Eq.(18)

$$\Pi_h = 2\pi\gamma \int_0^{\tilde{z}=L/R_1} \sqrt{\omega_h \tilde{f}\tilde{P}}\Big|_{\tilde{r}=1} d\tilde{z} \tag{20}$$

and recovery

$$\mathrm{R} = \frac{\text{permeate flow rate}}{\mathrm{H_2 \; inflow}} \tag{21}$$

Figures 3(a) and (b) show the permeate flow rate as a function of pressure at two different values of Pe for two values of γ. The black lines show the maximum permeate flow rate evaluated for Pe $\to 0$ in two limit-case conditions: the continuous line has a slope equal to 1 and refers to the case in which the membrane does not offer a significant resistance; the dashed line has slope equal to 1/2 and indicates the maximum permeate flow rate obtainable when the membrane permeability is the controlling resistance to mass transport.

For a separator operating with a given set of parameters, γ and Pe, a critical dimensionless pressure, which will be referred to as α_c, exists, marking the transition from the condition in which the controlling resistance is mass transport in the packed bed and to the one in which transport is limited by the membrane. The dimensionless pressure α_c indicates the optimal operating pressure. This may be seen from Figure ??, in which recovery as a function of pressure is shown to decrease when $\alpha > \alpha_c$. For increasing values of γ and Pe this transition occurs at increasing values of the dimensionless pressure.

A simplified analytical model is proposed for the prediction of the permeate flow rate as a function of the characteristic parameters α, γ, and Pe. For low pressures, transport in the packed bed is the limiting mechanism and Π_h has been evaluated by considering a simplified mass transport equation in which only axial convection and radial dispersion have been accounted for

$$\frac{\partial \omega_h}{\partial \tilde{z}} + \frac{\widetilde{D}_{rr}}{\mathrm{Pe}} \frac{\partial^2 \omega_h}{\partial \tilde{y}^2} = 0 \tag{22}$$

This equation, which has been solved with the boundary condition that the mass fraction of hydrogen in correspondence of the membrane surface is zero, is valid for $\alpha < \alpha_c$. The following expression for the permeate flow rate is obtained

$$\Pi_h \simeq 2\,\pi\,\omega_0\,\tilde{f}_{\mathrm{av}}\alpha \sum_{l=0}^{\infty} 2\,\frac{1-\exp\left(\lambda_l^2 \varepsilon L/R_1\right)}{\lambda_l^2} = 2\,\pi\,\omega_0\,\tilde{f}_{\mathrm{av}}\alpha\, g(\mathrm{Pe}) \tag{23}$$

In correspondence of $\alpha = \alpha_c$ the curve describing permeate flow rate changes slope, from 1 to 1/2. A comparison between the prediction of the simplified model and the results of the 2D fully coupled model is shown in Figure ??.

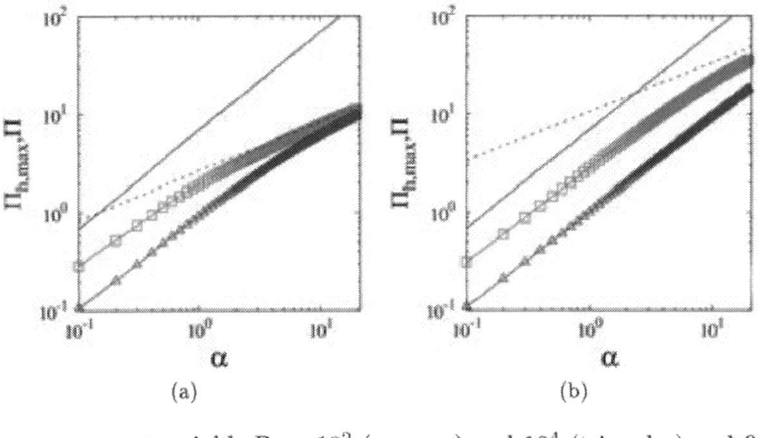

Figure 3: Hydrogen permeate vs.α at variable Pe $= 10^2$ (squares) and 10^4 (triangles) and fixed permeability (a)$\gamma = 5\times10^{-2}$ and (b) $\gamma = 2\times10^{-1}$. In the log-log scale of the picture, dotted and continuous black lines depict the slopes 1/2 and 1 respectively.

Membrane Reactor

When considering the presence of a reaction, mass balance equations are modified as:

$$\frac{1}{RT}\nabla\cdot\left(-\frac{\kappa}{\mu} f P \omega \nabla P - f P \mathbb{D}\cdot\nabla\omega_i\right) = r_i \tag{24}$$

The rate of methane consumption has been described by the expression:

$$R_m = k P_m (1-\eta) \tag{25}$$

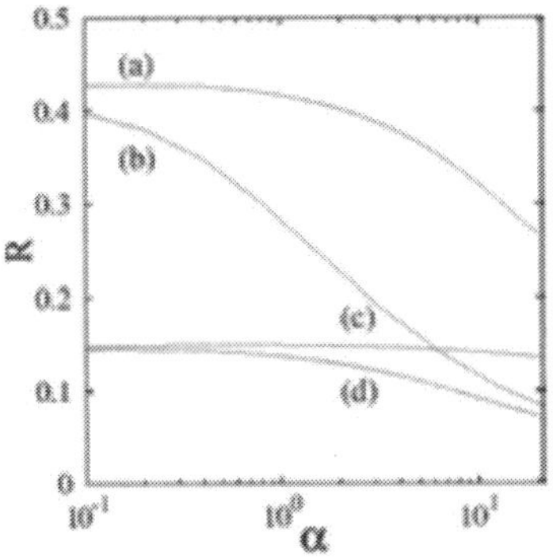

Figure 4: Recovery vs. operating pressure. Curves are labeled as: **a** : Pe $= 10^2$, $\gamma = 2 \times 10^{-1}$; **b**: Pe $= 10^2$, $\gamma = 5 \times 10^{-2}$; **c**: Pe $= 10^4$, $\gamma = 2 \times 10^{-1}$; **d**: Pe $= 10^4$, $\gamma = 5 \times 10^{-2}$.

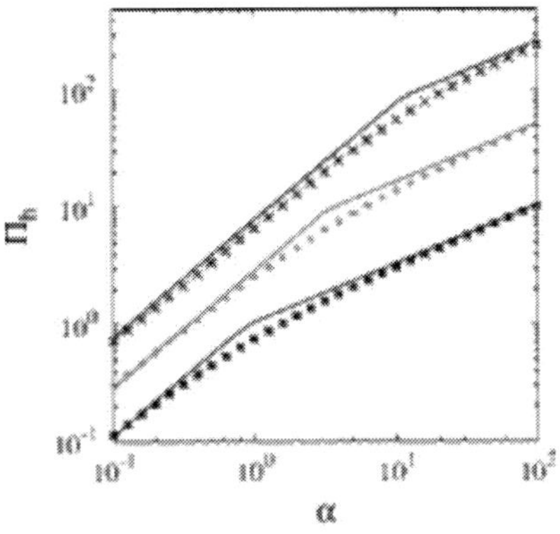

Figure 5: Comparison between model predictions (continuous lines) and results of the fully coupled transport model (symbols) at different γ and Pe values. (black) $\times$: $\gamma = 2 \times 10^{-2}$, Pe $= 10$; (red) $+$: $\gamma = 0.5$, Pe $= 10$; (black) $*$: $\gamma = 0.1$, Pe $= 10^2$.

where η represents the distance from equilibrium.

The mass balance equation takes on the following form in terms of dimensionless variables:

$$\widetilde{\nabla} \cdot \left(- \omega_i \tilde{f} \tilde{P} \widetilde{\nabla} \widetilde{P} - \frac{1}{\mathrm{Pe}} \tilde{f} \hat{P} \widetilde{\mathbb{D}} \cdot \widetilde{\nabla} \omega_i \right) = \nu_i \frac{W_i}{W_m} \mathrm{Da} \tilde{f} \tilde{P} \omega_m (1 - \eta) \tag{26}$$

with

$$\mathrm{Da} = \frac{RTkR_1}{U} \tag{27}$$

In this case the overall hydrogen balance must account even for hydrogen produced by the steam reforming reaction. This means that recovery, as defined by Eq. (21), increases for increasing values of Da because ω_h is kept high by the reaction, thereby maintaining a strong driving force for permeation. This may be seen from Figure 6, where the presence of a maximum in the curve of of recovery versus dimensionless operating pressure is evident. The maximum point moves to lower values of α and higher values of R for increasing values of Da. If recovery is defined as:

$$\mathrm{R}' = \frac{\text{permeate flow rate}}{\mathrm{H_2\ inflow} + \mathrm{H_2\ produced}} \tag{28}$$

to account for hydrogen produced by the reaction, the maximum in the curve is still present, even if less pronounced (Fig. 6).

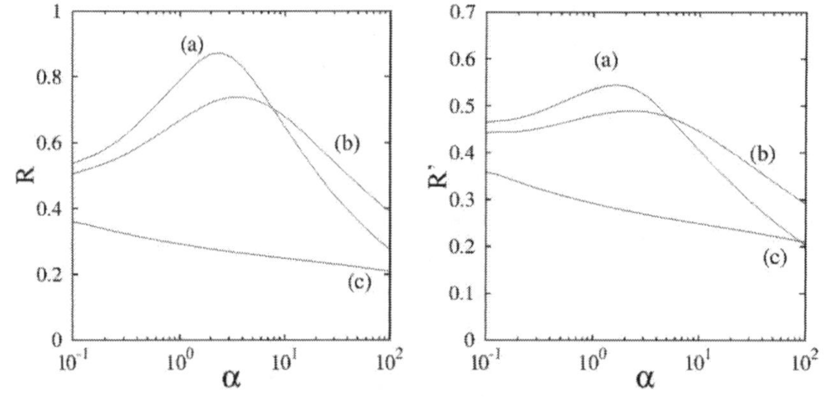

(a) (b)

Figure 6: Recovery, R (a) and modified recovery, R' (b) vs. operating pressure at $\gamma = 2 \times 10^{-1}$, Pe $= 10^2$. Curves are labeled as: **a**: Da $= 0$; **b**: Da $= 1$; **c**: Da $= 10$

(a) (b)

Figure 7: Reaction boundary layer evaluated for Da $= 5$, $\gamma = 1$ and Pe (a) 10^2 and (b) 10^3

The operating conditions also affect the catalyst volume that is actually active towards hydrogen production. The following figures show that a reaction "boundary layer" esixts close to the membrane, in which the distance from equilibrium, $1 - \eta$, is greater than zero. In the Figures shown, $1 - \eta$ varies between 0 (blue) and 1 (red). It was found that the thickness of the boundary layer decreases with increasing values of Pe (Fig. 7) and with increasing values of Da (Fig. 8). The effect of these parameters may be understood by considering the profile concentrations of methane, the limiting reactant in the conditions adopted. Hydrogen permeation causes an increase in methane concentration in proximity of the membrane, which in turn increases the rate of reaction. When Pe is high, concentration gradients are steep and ω_m drops rapidly as the distance from the membrane increases, leading to a small thickness of the boundary layer. Similarly, a high value of Da leads to a rapid consumption of methane, meaning that the active volume towards the reaction is reduced to a small portion of the reactor close to the membrane. Figure 9 shows that the value of γ influences not only the thickness of the boundary layer, but even the value of $1 - \eta$ in correspondence of the membrane. Low values of γ, i.e. low membrane permeabilities, lead to higher hydrogen

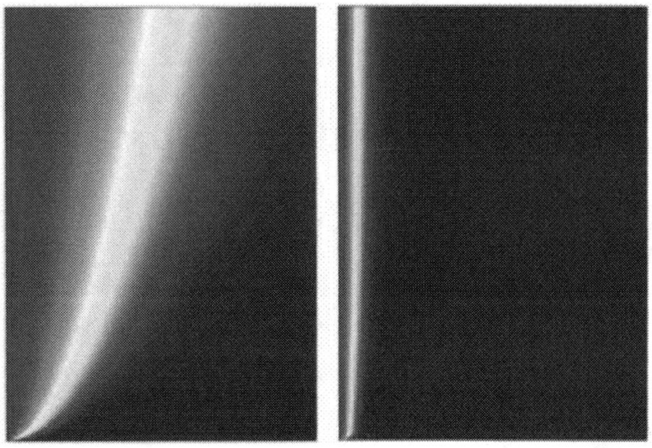

(a) (b)

Figure 8: Reaction boundary layer evaluated for $\gamma = 1$, $\mathrm{Pe} = 10^2$ and Da (a) 0 and (b) 5

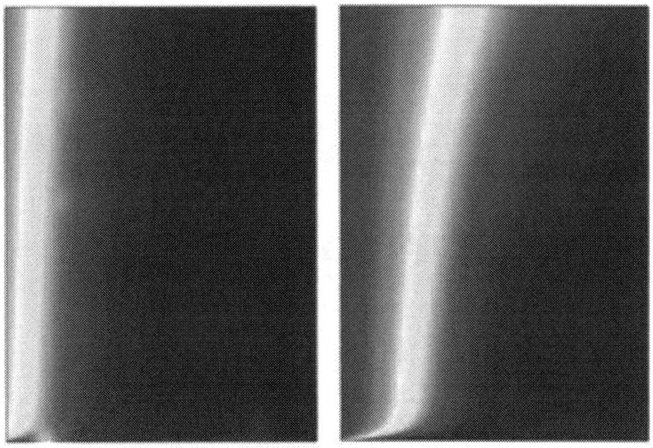

(a) (b)

Figure 9: Reaction boundary layer evaluated for $\mathrm{Da} = 1$, $\mathrm{Pe} = 10$ and γ (a) 10^{-1} and (b) 1

concentrations in correspondence of the membrane, therefore a smaller distance from equilibrium, shown by the fact that $(1 - \eta) < 1$.

Conclusions

A two-dimensional model has been developed to describe mass and momentum transport in a membrane reactor for methane steam reforming. The model accounts for axial and radial convective and dispersive mechanisms for mass transport, thus allowing the determination of concentration profiles within the reactor and the identification of the operating conditions in which concentration polarization becomes signficant. The preliminary analysis of the reactor led to the identification of two main transport regimes, in which hydrogen permeation is limited by transport within the packed bed and across the membrane, respectively. The pressure value for which this transition occurs conicides with the optimal operating pressure. By extending the model to a membrane reactor, the influence of the rate of reaction on the performance of the systema has been assessed. A boundary layer close to the membrane, in which hydrogen production occurs, has been identified and the influence of the operating parameters on the thickness of this layer has been studied.

426963 A CFD Study on the Impact of Convective Flow Characteristics and Species Distribution on the Performance of Full-Scale Membrane Reactors

Wednesday, November 11, 2015: 2:36 PM
155C (Salt Palace Convention Center)

Thomas Eppinger[1], **Nico Jurtz**[1] and Ravindra Aglave[2], (1)CD-adapco, Nuremberg, Germany, (2)CD-adapco, Houston, TX

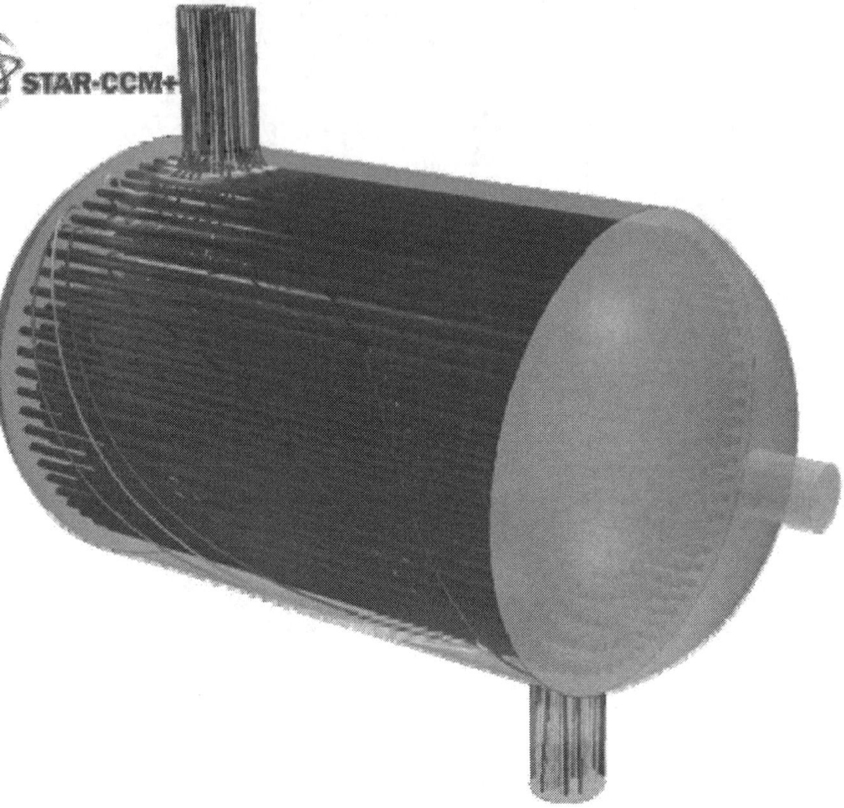

Membrane reactors are widely used within the industry to selectively separate particles or species from a stream. Well-known examples are the filtration of dust particles from flue gas or reverse-osmosis for producing drinking water from seawater or other pharmaceutical related separations. The main benefit of this reactor type is the modular structure and the low energy consumption compared to other separation methods.

At AIChE Spring Meeting 2015 an approach to simulate membrane processes by using the commercially available CFD code STAR-CCM+ was presented. The results showed very good agreement with experimental results and had the outcome that the performance and efficiency of a membrane reactors is not only a function of the molecular functionality of the membrane itself but also of the flow characteristics and species distribution in the reactor.

Based on this work the separation process in a full-scale multi-tubular membrane module consisting of more than 300 tubes like shown in the picture above is investigated. Two different approaches are compared: The first one uses a 1D-3D coupled approach that neglects radial profiles in the retentate side tubes while the permeate side gets spatially resolved. The second method uses advanced meshing technique to allow an efficient fully spatially resolved CFD simulation of the whole reactor. The results reveal the limits of one dimensional approaches and show how numerical simulation can help do develop more efficient reactors by doing slight design changes.

405378 Preparation and Characterization of Functional Polyamide RO Membrane Via Spin Assisted Layer-By-Layer Assembly

Abstract:

Layer by layer assembly is an attractive method for preparing ultrathin layer of composite membranes. In this work, spin assisted layer by layer (SA-LbL) assembly was used to prepare and modify polyamide membranes for water desalination. The polyamide membranes was prepared by two methods: 1) conventional Interfacial polymerization of MPD and TMC, 2) Spin assisted layer by layer of MPD and TMC. The prepared membranes were functionalized by two different sets of poly-electrolytes: Poly (ethylene imine) (PEI)/ poly (allylamine hydrochloride) (PAH), poly sodium styrene sulfonate (PSS)/Poly(acrylic acid) (PAA). Membrane characterization will be performed using atomic force microscope (AFM), Scanning Electron microscopy (SEM), Fourier transform infra-red (FTIR), and contact angle. It was found that the thickness of the membranes prepared by conventional interfacial polymerization is larger than that prepared by SA-LbL. On the other hand, functionalized membranes were found to have lower roughness and higher hydrophilicity. A model was developed to describe the effect of number of layers, polyelectrolyte concentration, and pH on contact angle, roughness, and membrane thickness via statistical models derived from the design of experiment.

Membrane performance is currently being investigated to assess the permeabilty, salt rejection, and fouling behavior under brackish water conditions.

415891 A Novel Hydrophilic Nanofiltration Membrane with Excellent Performance

Abstract:

A hydrophilic thin-film-composite (TFC) nanofiltration (NF) membrane with high flux and dual resistance to fouling and chlorine has been designed and synthesized through the interfacial polymerization (IP) of amino-functional polyethylene glycol (PEG) and trimesoyl chloride, which can be utilized for antibiotics separation. The higher hydrophilicity alongside the larger pore size of the PEG-based selective layer is the key to a high water flux of 66.0 L m^{-2} h^{-1} at 5.0 bar. With mean pore radius of 0.42 nm and narrow pore size distribution, the $MgSO_4$ rejections of the PEG based PA TFC NF membranes can reach up to 80.2 %. The hydrophilic PEG based membranes shows positive charged since the isoelectric points range from pH=8.9 to pH=9.1 and the rejection rates for different salts of the novel membranes are in the order of $R(MgCl_2)>R(MgSO_4)>R(NaCl)>R(Na_2SO_4)$. The pore sizes and water permeability of these membranes are tailored by varying the molecular weight and molecular architecture of amino-functional PEG. Due to the unique structure of the selective layer of the PEG based membranes consisting of saturated aliphatic construction unit (CH_2-CH_2-O), the membranes demonstrate dual resistance to fouling and chlorine. The membranes maintain good salt rejections and high water flux of PEG based membranes after treatment by 2000 ppm NaClO for 24 hours. Interestingly, the PEG based membranes exhibit excellent fouling resistance with a water flux recovery of 90.2 % using BSA as a model molecule. More importantly, the hydrophilic PEG based NF membranes have been exploited to separate several water soluble antibiotics (such as tobramycin, an aminoglycoside antibiotic applied in the treatment of various types of bacterial infections), showing excellent performance in concentration or removal of antibioics. By optimizing operating conditions, the advanced membranes show highly stable tobramycin solution flux of 37 L m^{-2} h^{-1} alongside tobramycin rejections up to 96 % with 50 ppm feed concentration under 8.0 bar. When the feed concentration increases up to 800 ppm, our membranes can still exhibit a high rejection over 92 %. Therefore, as a promising "green" technique capably declining the solvent emissions and energy-consuming, the designed PEG based NF membranes owning highly efficient antibiotic concentration ability together with dual resistance to fouling and chlorine have great potentials in substituting conventional separation techniques for concentration and purification of active molecules in pharmaceutical and fine chemicals industries.

416192 GO-Imbedded Nanocomposite Hollow Fiber Membranes with Improved Antifouling Properties for Oily Wastewater Treatment

Abstract:

On account of the increasingly generated oil-water emulsion in diverse industries, the development of anti-fouling membranes is one of the most important imperatives in in order to economically remove oil from water and meet the discharging requirements. Currently, membrane fouling is still an inevitable challenge in various membrane applications. Molecular design of the membrane materials with balanced physicochemical properties would be one of the major approaches to eliminate fouling. In this work, a group of GO-imbedded nanocomposite hollow fiber membranes was investigated for oily water treatment, with the objectives of improving GO-polymer interfacial interaction and membrane anti-fouling properties via the formation of a simultaneously covalent and ionic inter-network. 1-Methylnicotinamide chloride (MNA) was selected to bridge the two parties. The reaction scheme was proposed and evidenced by FTIR and XRD analyses. The resultant membranes were systematically studied with respect to membrane microstructure, ultrafiltration performance and fouling behaviors. The responses of the membranes to oil-water fouling were evaluated by the resistance-in-series model and cyclic UF experiments with periodic backwashing. Substantial reductions in reversible, irreversible and adsorption-induced resistances as well as flux drop were observed for the nanocomposite membranes. Generally, their superior anti-fouling properties benefit from the contributions of (1) the hydrophilic nature of GO and MNA, (2) appropriate GO:MNA ratio and (3) synergetic effects between GO and MNA to overcome the interfacial voids and produce a balanced membrane structure.

417734 A Surface Reaction Route To Increase the Loading of Antimicrobial Ag Nanoparticles in Forward Osmosis Membranes

Abstract:

Biofilm formation from bacterial growth and colonization is a known cause for membrane performance deterioration in water treatment. A broad-spectrum biocide such as silver nanoparticles (Ag NPs) is generally added to the membrane to impart biofilm resistance. Since biofilm formation begins with the membrane surface in contact with the feed water, a high loading of uniformly dispersed Ag nanoparticles (Ag NPs) on the membrane exterior surface is desirable for a good antimicrobial performance. This was accomplished in this work by a short alkaline treatment of a hydrolysable membrane (polyacrylonitrile, PAN) to introduce negative charge to the surface. A subsequent exposure to a Ag precursor solution followed by reduction under mild reducing condition (CO) limited the Ag NP deposition to within a narrow region of the membrane surface. The Ag content on the resultant membrane exterior surface in this study was also the highest in the literature. Different from previously reported methods, this method of Ag NP incorporation can be applied to thin film composite (TFC) membranes without any adverse effect on the water flux. The TFC membranes modified as such have shown good antibacterial resistance for 14 days under laboratory conditions optimized for the growth of *E-coli.*

424553 New Frontiers: Membrane Fouling Remediation for Enhanced Performance & Energy Saving for Sustainable Water & Industrial Wastewater Treatment Technologies

Abstract:

One of the key issues that keep coming up in current research trends is the sustainable water use and water re-usability. The low-pressure membrane applications are considered to be the most effective and sustainable methods of addressing environmental problems when it comes to treated water and wastewater that meets or exceed stringent environmental standards. Nevertheless, membrane fouling is one of the primary operational concerns that is currently hindering a more widespread application of ultrafiltration (UF) with a variety of contaminants. Membrane fouling leads to higher operating costs, excessive energy demand, reduced membrane life time, and increased cleaning frequency. As a consequence, an efficient and well-planned UF process is becoming a necessity for consistent and long-term monetary returns. Examining the source and mechanisms of foulant attachment to the membrane's surface is critical when it comes to the research of membrane fouling and its potential practical implementation. A mathematical model was developed for homogeneous and heterogeneous membranes in order to predict the amount of fouling and the increase in transmembrane pressure in UF of simulated latex effluent based on the analyses of particle attachments. A power consumption model, incorporating fouling attachments, as well as chemical and physical parameters involved in membrane fouling, was developed in order to ensure accurate prediction and the scale-up applications. Innovative remediation techniques were likewise developed and applied so as to minimize membrane fouling, enhance membrane performance, and save energy. Based on these achievements, it is essential to direct current research trends toward sustainable water and wastewater treatment technologies that can solve the existing industrial and environmental issues, especially when it comes to solutions that can be successfully commercialized on the global scale.

433036 Thin Film Composite (TFC) Membranes with Copper Nanoparticles: Preparation and Antibacterial Characteristics

Abstract:

Membrane biofouling has been a challenging problem that could severely restrict its applications in water desalination process. Copper is known for its antimicrobial properties and is abundantly available with low cost. In this paper, copper nanoparticles (Cu-NPs) with a mean diameter of 15nm were synthesized by the reduction of copper (II) chloride with sodium borohydride ($NaBH_4$), using cetyl trimethylammonium bromide ($(C_{16}H_{33})N(CH_3)_3Br$, CTAB) as a capping agent. After purification of CuNPs by dialysis, the particles were successfully immobilized onto the surface of thin film composite (TFC) membranes via either electrostatic interactions or by covalent bonding with cysteamine as a linker. The electrostatic method entailed simply dipping the newly-fabricated TFC membranes to the Cu-NPs suspension. The presence of CTAB created a cationic bilayer outside the Cu-NPs, which allowed the Cu-NPs be attached to the negatively-charged membrane surface through electrostatic interactions. The covalent bonding method utilized cysteamine ($C_4H_{12}N_2S_2$) to activate the thin film layer with thiol functional groups, which then allowed attachment of Cu-NPs onto the membrane via a stable covalent bonding. The resulting membranes by these two methods were labeled as TFC-CuNPs and TFC-S-CuNPs, respectively, in this study. Scanning electron microscopy (SEM) imaging and associated energy-dispersive X-ray spectroscopy (EDS) showed that a significant amount of Cu-NPs existed on both types of membranes. Surface hydrophilicity of the membranes was enhanced by the attachment of CuNPs, as indicated by the measured contact angle of 63.3 ± 0.8^o for TFC, 38.6 ± 0.2^o for TFC-CuNPs, and 58.0 ± 3.4^o for TFC-S-CuNPs. At 300psi trans-membrane pressure and feed NaCl concentration of 2000 mg/L, the corresponding water flux was increased from 47.1 $L/m^2/h$ for TFC, 49.4 $L/m^2/h$ for TFC-CuNPs, and 69.1 for TFC-S-CuNPs. The salt rejection was slightly decreased from 94.4% for TFC to 92.4% for TFC-CuNPs and 92.3 for TFC-S-CuNPs. The TFC membranes with CuNPs both exhibited excellent antibacterial properties against *P. aeruginosa* based on the disk incubation test and the SEM observation. Moreover, TFC-S-CuNPs with covalently-bound CuNPs were shown to be more stable and with better anti-bacterial properties than the TFC-CuNPs.

432374 Packing Characterization: Advanced Model Development and Absorber Optimization

Abstract:

Packing Characterization: advanced model development and absorber optimization

by Chao Wang[1,2], Micah Perry[2], Gary T. Rochelle[1], Frank Seibert[2]

[1]Department of Chemical Engineering, The University of Texas at Austin

[2]Separations Research Program, The University of Texas at Austin

Austin, TX 78712

Packing is widely used in post-combustion CO_2 absorption process because of its low pressure drop, good mass transfer efficiency, and ease of installation. In this paper, three dimensionless mass transfer models based on absorption system were developed. The effects of operating conditions and packing geometries were considered. The new concept, Mixing Number was proposed to represent the packing geometry influence on mass transfer. The three dimensionless mass transfer correlations developed in this work are:

$$\frac{a_e}{a_p} = 1.41[(\frac{\rho_L}{\sigma})g^{1/3}(\frac{u_L}{a_p})^{4/3}]^{0.116}$$

$$Sh_L = 1.79 * \mathrm{Re}_L^{0.74} Mi^{0.42} Sc_L^{0.5},\ k_L = Sh_L a_p D_L$$

$$Sh_G = 0.83 * \mathrm{Re}_G^{0.58} Mi^{0.3} Sc_G^{0.5},\ k_G = Sh_G a_p D_G$$

$$Mi = \frac{3\sin\theta\cos\theta}{16(\sin^2\theta + 1)^{3/2}}$$

The economic analysis of the absorber for a 250 MW coal-fired power plant was conducted, using the developed mass transfer models. The purpose is to determine the optimal design and operating conditions for the amine scrubbing post-combustion absorber. The Energy Cost (Energy) and the Annualized Capital Cost (CAPEX) for the absorber were calculated to determine the total processing cost as a function of the gas superficial velocity (u_G). For reaction film controlled system such as CO_2 absorption, the optimum operating percent of flood is between 50 and 80% which is lower than the normal distillation design (usually between 70 and 90% of flood). The optimal absorber design is to use packing with 200 to 250 m^2/m^3 surface area and high corrugation angle.

424127 Measuring Entrainment from a Spray for Industrial Applications

Abstract:

Liquid entrainment occurs between distillation trays, from spray nozzles, above mist eliminators, and in kettle reboilers. Information pertaining to the drop size distribution and total volume of entrainment is crucial information for the design and operation of such equipment. While many aspects of spray dynamics have been correlated and quantified, information regarding the entrainment rate is limited. This study of entrainment from a spray nozzle has been conducted in a spray column capable of reliably capturing entrained droplets and measuring the entrained droplet size distribution using Phase Doppler Interferometry (PDI). A method was developed to quantify the total volume of liquid entrained from a full cone Bete® nozzle using PDI droplet size data for a water/air and a 0.1 N sodium hydroxide (NaOH)/air system. PDI results showed that increasing the liquid rate results in a decrease in the entrained droplet Sauter mean diameter and an increase in total liquid entrained. Corresponding results showed that increasing the gas rate results in a decrease in entrained droplet Sauter mean diameter and an increase in the total liquid entrained.

424149 Interfacial and Surface Area Characterization of Dense Sprays with Application to Absorption

Abstract:

Despite the wide use of sprays, the fundamentals of spray absorption are poorly understood. There is a particular lack of understanding about the relationship between surface area and interfacial area for dense sprays. The available surface area has a strong impact on the overall efficiency of absorption using sprays. However, not all of the surface area of a spray is available for mass transfer. Reaction of the solute gas with the solvent drops can result in saturation of the solvent mass and a decrease in the available interfacial area for mass transfer. Comparison of sprays with packed columns and other contactors can conveniently be made on an interfacial area basis. Thus, there is a need to ascertain the effective interfacial area for spray absorption along with the surface area availability. This study presents droplet size measurements using a Phase Doppler Interferometer (PDI), surface area quantification, and interfacial area measurements with the CO_2-0.1 N NaOH system inside a laboratory spray column. The effect of L/G ratio and gas-liquid contact height on the surface area availability and interfacial area are elucidated. A methodology is presented to compare the surface area measurements to interfacial area measurements, and this work will ultimately help in providing fundamental insight into spray absorption and spray behavior.

424804 Dynamic Simulation of CO2 Absorption Using Different Solvents

Abstract:

The removal of CO_2 from gaseous streams is a very important industrial technology with applications in (i) reduction of greenhouse gas emissions from flue gases in coal-fired power plants and (ii) removal of radioactive C_{14} in the nuclear industry, among others. One of the well-known technological alternatives for CO_2 capture is absorption/stripping with aqueous solvents, such as alkanolamines and their mixtures. A dynamic model of the absorption process for dilute carbon dioxide mixtures in high alkaline solutions is proposed. A new reaction scheme that allows the use of different solvents has been added to increase the versatility of the proposed model. Chemical reactions between CO_2 and the high pH solution have been included in the model through the use of an enhancement factor for chemical absorption. Interfacial equilibrium constants have been calculated based upon Henry's constant values for the different gas-phase species calculated using fugacity ratios obtained by the Peng-Robinson equation of state. Mass and energy balances have been derived for all the species in the gas and liquid phases. Numerical techniques have been used to calculate the axial profiles of temperature and concentrations of the resulting species in both phases. The dynamic behavior has been compared to steady-state solutions. The model has been validated using experimental data available in the literature for different columns, packing materials, and operating conditions. Simulations under different operating variables and dynamic inputs have been conducted. Simulation results for different solvents have been compared and analyzed. These results show the potential of the proposed model to optimize the operation and control of these separation processes.

429180 Modeling of Absorption with Reversible Reactions with Nonlinear Kinetics

Abstract:

Design calculations based on mathematical models of absorption with chemical reaction require computationally efficient, accurate and robust algorithms. This is particularly challenging when there are multiple reversible reactions with nonlinear kinetics and disparate time scales. Approximate analytical techniques solutions have been developed for solving the governing ordinary differential equations (ODEs) in the context of steady-state film theory, One, based on an analysis developed by K.A. Smith, linearizes the ODEs based on the assumption of small perturbations from local reaction equilibrium states. It yields asymptotically exact absorption rate predictions when the reactions are very fast. The other, originally developed by Van Krevelen and Hoftijzer, does so based on the assumption that, in the kinetic expressions, the concentrations of nonvolatile reactive solutes may be fixed at their values at the gas-liquid interface. It yields asymptotically valid absorption rates in the limit of very slow reactions. Surprisingly, both approaches are generally quite accurate over wide ranges of the Hatta numbers. However, there are regimes where this is no longer true and numerical methods of solutions are required. The latter are the focus of this presentation.

426175 Determination of Porous Media Resistance Coefficients in Gas-Liquid CFD for Amine Absorber with Mellapak 250.X

Abstract:

Structured packings are commonly used for absorption and distillation processes due to its high mass transfer efficiency and low pressure drop. Columns with structured packing should be designed precisely because they are more sensitive to design error than other types of packing. Many experiments and numerical models have been established to investigate fluid behaviors in the structured packing columns. In this work, we performed computational fluid dynamics (CFD) simulation of an amine absorber packed with Mellapak 250.X(M250X) for CO_2 removal from natural gas. A porous media CFD model was applied to examine hydrodynamics and mass transfer on M250X in the CO_2-MEA (Monoethanolamine) system. Three momentum forces (dispersion, porous resistance and drag between gas and liquid), mass transfer of CO_2, and one chemical reaction were considered in this model. CFD results were compared to experimental data in order to validate model.

431217 Modeling of Mercury Capture By CuCl2-Impregnated Activated Carbon Sorbent in Fabric Filter

Abstract:

It is estimated that over 300 new fabric filters (FF) are going to be needed for the capture of mercury and fine particulate. We have found that cupric chloride-impregnated activated carbon ($CuCl_2$-AC) demonstrated excellent Hg(0) capture performance. Our previous studies show that mercury adsorption is a sequential process of Hg(0) oxidation and oxidized mercury adsorption over $CuCl_2$-AC. Therefore, it is essential to separately determine the chemical reaction and physical adsorption kinetics for the prediction of Hg(0) vapor removal by $CuCl_2$-AC for sorbent injection.

A model for the physical adsorption of mercuric chloride ($HgCl_2$) onto raw activated carbon (AC), 4% and 10% $CuCl_2$-impregnated activated carbon ($CuCl_2$-AC) sorbents in a fabric filter was studied based on the adsorption equilibrium and kinetics. Sorbent loading, particle size, filtration time, and $CuCl_2$ loading onto AC were found to be the major factors determining an $HgCl_2$ removal efficiency. Although high sorbent loading and small sorbent particle size result in high $HgCl_2$ removal efficiencies, inlet $HgCl_2$ concentration and superficial gas velocity were found to have little impact on $HgCl_2$ removal. Our operation case study result shows that sorbent injection for a short duration at a high loading followed by the discontinuation of the injection until a cleaning cycle can significantly save a total sorbent amount with high sorbent utilization while the same amount of $HgCl_2$ is removed. The adsorption of $HgCl_2$ onto raw AC and $CuCl_2$-AC in the filter cake was found to be primarily governed by the adsorption capacity difference in the Langmuir adsorption kinetic expression. The study results demonstrate how fundamental adsorption equilibrium and kinetics can be used to design a sorbent and predict its performance in a fabric filter. A modeling result for the chemical reaction followed by physical adsorption will also be presented in this talk.

435081 Mercury Removal in a Coal-Fired Power Plant Using SO3-Tolerant Sorbents

Abstract:

Recent EPA regulations limit mercury emissions from coal-fired power plants, leading to the widespread use of activated carbon sorbents to remove mercury emissions. However, the selectivity of these sorbents for mercury in the presence of acid gas/mist is poor, which limits their effectiveness in flue gas generated from combustion of medium- to high-sulfur coals, which have more acid gas/mist. This critical issue has led to proposal of several SO3-tolerant strategies for mercury sorbents. This talk will look at the proposed mechanisms for SO3 tolerant sorbents using recent* pilot scale (20,000 ft3/min) tests of multiple technologies, followed by full-scale power plant testing of the most promising approaches.

*Pilot-scale testing is completed. Full-scale testing is planned for Summer 2015 and therefore analysis will be available by Fall Meeting.

431374 Investigation of Sulfur Adsorption on Fly Ash for Oxy-Combustion

Abstract:

Oxy-combustion proposes to incorporate carbon capture and storage technologies into existing coal-fired power plants to lower carbon emissions. In an oxy-fired power plant, nitrogen is separated out prior to combustion, resulting in concentrated carbon dioxide as a final product that can be easily captured and stored. Sulfur oxide (SO_X) emissions can be lowered due to high concentrations of sulfur trioxide in the oxy-combustion flue gas, which results in sulfur retention on fly ash and ash deposits in the furnace. It has been shown that the extent of sulfur retention is heavily dependent on the alkali and alkaline earth metal (AAEM) species in coal such as magnesium (Mg) and calcium (Ca) and the sulfur retention increases as the AAEM:S molar ratio increases. While higher sulfur retention on the ash particles reduces the emission rates, it creates problems utilizing the fly ash for cement and concrete production.

Although there have been previous experimental studies, the mechanism of sulfur retention on fly ash is not well understood. In this study, a variety of computational and experimental methods were employed to examine the binding of SO_X species to the most prominent AAEM species, calcium and magnesium oxide. Firstly, density functional theory (DFT) calculations were performed using Vienna *ab initio* Simulation Package (VASP) to investigate the binding mechanism for SO_X on CaO and MgO surface. Calculations were also run with various other co-adsorbates to examine the effect their presence would have on any SO_X reactions on the surface. Secondly, combustion experiments were employed to investigate the AAEM oxides after exposure to a SO_2/SO_3 mixture plus a variety of other flue gas components. Samples were exposed to simulated flue gas of air and oxy-combustion at various temperatures with the resulting products being analyzed using x-ray diffraction (XRD) and diffuse reflectance infrared Fourier transform spectroscopy (DRIFTS). In addition to the investigation of individual oxides, samples of fly ash will also be characterized. Overall, the experimental results along with DFT calculations will be used to create a more complete picture of the SO_X binding mechanism on the fly ash metal oxides and compare air and oxy-combustion environments for sulfur capture.

431351 Effect of Operational Parameters on the Performance of a Multistage Fluidized Bed Adsorber

Abstract:

Fluidized bed adsorbers are particularly suited for treating large flow rates of dilute air streams such as VOC-laden air from automotive painting operations. This research investigates the effect of different operational parameters on the performance of a multistage fluidized bed adsorber for adsorption of a VOC (1,2,4-trimethylbenzene) on beaded activated carbon (BAC). A pilot-scale six-stage fluidized bed adsorber was tested to assess the effect of inlet VOC concentration, inlet gas flow rate, temperature, relative humidity, and adsorbent feed rate on VOC removal efficiency. The results showed that an increase in BAC feed rate could be effective up to a level beyond which there was no change in the removal efficiency of the adsorber. Increasing VOC concentration (100 to 300 ppm_V) resulted in a significant decrease (51%) in removal efficiency of the adsorber. Additionally, increasing inlet air flow rate (200 to 350 SLPM) reduced the removal efficiency by 20%. Increasing inlet stream temperature (20°C to 50°C) resulted in a small (<10%) decrease in removal efficiency. On the other hand, increasing the inlet stream humidity had negligible effect on removal efficiency. The obtained results help determining optimal conditions for operating a fluidised bed adsorber.

427667 Ammonia-Modified Activated Carbon Adsorbents for Selective CO2 Adsorption

Abstract:

Introduction

Carbon dioxide, which is a by-product of fuel processing and mainly released as one of the flue gases from fossil fuel combustion processes and industrial power plants, is the main reason for global warming and climate change. Adsorption is one of the most promising technologies in capture and storage of CO_2 in terms of its low cost, low energy demand, ease of applicability and reusability after many processes. Carbonaceous materials, such as activated carbon (AC) based adsorbents are ideal candidates for CO_2 adsorption due to their high surface area, big pore volume, high CO_2 selectivity, stable adsorption capacity, adequate adsorption/desorption kinetics and high mechanical strength after repeated adsorption/desorption cycles at ambient pressure and temperature. Thus, modification of AC through different thermal and/or chemical pretreatments has been a great interest for researchers aiming to enhance effectiveness of AC's specific properties in selective capture of CO_2.

The objective of this preliminary study is to increase the pure and selective CO_2 adsorption capacity of the AC-based adsorbents by introducing basic N-containing surface groups through liquid ammonia treatment methods. The effect of wet-NH_3 treatment methods, ammonia concentration and thermal pretreatments were investigated. The effect of gaseous ammonia (amination) and ammonia-oxygen (ammoxidation) treatment at high temperatures to produce highly

CO_2-selective AC-based adsorbents are currently studied.

Experimental

Commercial activated carbon supplied by NORIT ROX 0.8 was crushed and sieved into 355-250 μm (45-60 mesh) particle size and referred to as AC0. Then, approximately 20 gram of AC0 was placed in an extraction thimble and was washed with 200 ml of 2 N HCl acid solution for 12 hours in a Soxhlet apparatus under reflux to remove some ash content and sulfur accompanied with it. As the next step, the slurry was rinsed with 250 ml deionized (DI) water and washed again for 6 hours in the Soxhlet apparatus to remove HCl remaining on the support surface. Finally, the slurry was dried at 115 °C overnight. This support is referred to as AC1.

Current experimental procedure on wet NH_3 treatment methods were based on the former studies of ammonia-modified AC-based adsorption [1,3]. Samples were exposed to different thermal and/or chemical pretreatments prior to being used as adsorbent. 2 g of AC1 samples were placed in 150 mL of 10 wt.% and 25 wt.% NH_3 solutions (Merck-analytical reagent grade) and left for 48 h at room temperature. After this time, the solutions were washed with DI water and vacuum-filtered until the filtrate showed a pH value of 7. These supports were referred to as AC1-10NH3w and AC1-25NH3w. Finally, the samples were dried at 115 °C for 24 h. Another NH_3 treatment technique is incipient-to-wetness-impregnation method. 2 g of AC1 samples were placed in Büchner flasks and impregnated with 10 wt.% and 25 wt.% NH_3 solutions (2.1 ml DI water/g AC) and were dried at 115 °C for 24 h. Finally, the samples were calcined under 150 ml/min N_2flow at 250 °C for two hours. These supports were referred to as AC1-10NH3i-250 and AC1-25NH3i-250. AC1-25NH3w sample was subjected to thermal treatment under 50 ml/min helium flow for 2 h at 600 °C and this support is denoted as AC1-25NH3w-600He.

The adsorption/desorption tests of AC-based adsorbents were conducted in a Gravimetric Analyzer (IGA-Hiden Isochema) in the upstream for collecting pressure versus change in the adsorbent weight data and a Mass Spectra (Hiden Analytical) in the downstream for analyzing the gases leaving the adsorption unit, i.e., unadsorbed gas streams. The adsorption/desorption isotherms were obtained for 70-75 mg AC-based samples within 0-1000 mbar pressure range for each 100 mbar step, at room temperature (RT) and at 120 °C. Four different feed streams are used in the adsorption studies, which are namely; 50 ml/min CO_2 (pure CO_2), 50 ml/min CH_4 (pure CH_4), 5ml/min CO_2-45 ml/min CH_4 (10% CO_2-90% CH_4) and 25 ml/min CO_2-25 ml/min CH_4 (50% CO_2-50% CH_4). Prior to tests, samples were outgassed overnight at RT to eliminate humidity and trapped gasses. The performance of the adsorbents were tested and compared on the basis of their adsorption capacity (mg adsorbed/g adsorbent), mass uptake values (%) and selective adsorption ability of CO_2 over CH_4.

Results

The adsorption/desorption isotherms of pure and mixture gases were obtained in order to investigate the changes in CO2 adsorption behavior upon the modifications, heat/chemical treatments and/or impregnations applied to the AC-based adsorbents. When adsorption/desorption isotherms of pure gases at room temperature are compared, it was observed that mass uptake values of pure CO_2 isotherms (~10%) were six folds of the pure CH_4 isotherms (~1.7%) at 1000 mbar. When the adsorption temperature was increased from RT to 120 °C, almost 80% of the CO_2 adsorption capacity was lost, and at that temperature mass uptake values for pure CO_2 gas dropped down to 2%, while CH_4 adsorption capacity decreased down to 0.4%. The adsorbed amounts on all samples decreased with temperature. When pure CO_2 gas was sent to the gravimetric analyzer, it was seen that after adsorption process complete desorption occurred, i.e, both isotherms overlapped. However, as the composition of CH_4 in the feed stream increased, desorption has been more difficult and deviations from the adsorption data were observed. A maximum of 100 mg CO_2 and 17 mg CH_4 gases were adsorbed per gram of adsorbent at 1000 mbar and RT. When the adsorption capacity of the adsorbents for gas mixtures were compared, in case of the feed mixture of 50% CH_4-50% CO_2 (65 mg adsorbed/g adsorbent) the mass uptake values and adsorption capacities were almost two folds with respect to the feed mixture of 90% CH_4-10% CO_2(30 mg adsorbed/g adsorbent).

Pure and selective CO_2adsorption capacity of AC1-10NH3w sample was found to be slightly higher than AC1-25NH3w sample. High temperature treatment with helium (AC1-25NH3w-600He) increased the CO2 adsorption capacity both at RT and at 120 °C. It was also found that samples prepared through incipient-to-wetness-impregnation method (AC1-25NH3i-250) had higher adsorption capacities compared to samples prepared through solution and filtration method (AC1-25NH3w).

Pure CO_2 isotherms at room temperature and at 120 °C were fitted to Langmuir, Freundlich and Dubinin-Radushkevich (D-R) models to describe the adsorption mechanism, surface properties and affinity of the adsorbent. Langmuir model, which quantitatively describes the formation of a monolayer adsorbate on the outer surface of the adsorbent, gave as an average of 4.02 mmol/g adsorbent for theoretical monolayer saturation capacity and 0.00168 for Langmuir isotherm constant with correlation coefficients greater than 0.99. The increase in temperature decreased the two parameters of the Langmuir model, since the adsorption rate decreased. Fitting Freundlich model to experimental CO_2 adsorption data, which assumes a heterogeneous surface (multilayer adsorption) with a non-uniform distribution, resulted in an average of 0.07414 and 0.6354 for k and 1/n constants at RT (adsorption capacity and strength of adsorption) with correlation coefficients greater than 0.99, respectively. The increase in temperature resulted in higher 1/n values, which are very close to 1, indicating relatively uniform surfaces. The best fit was obtained in Dubinin-Radushkevich model with correlation coefficients greater than 0.999. This model was used to estimate the characteristic porosity of the adsorbent and the apparent energy of adsorption. The micropore capacity was found to be 0.221 and the obtained characteristic energy of the adsorbents with an average of 7.48 kJ/mol, which is less than 8 kJ/mol, confirmed the adsorption process to be a physical one in this study.

Based on the results of these preliminary tests, the effect of ammonia concentration on CO_2 adsorption by treating AC1 samples with 1wt.% and 5 wt.% NH_3solutions, enriching of AC surface with nitrogen by treatment with gaseous ammonia and a mixture of ammonia and oxygen at high temperatures are currently studied. The adsorbents will be characterized by Scanning Electron Microscopy/Electron dispersive X-Ray (SEM-EDX) for studying microstructural properties, High Resolution Transmitting Electron Microscopy (HRTEM) and HRTEM-EDX for detailed characterization of phases and dispersion of additives, and Fourier Transform Infrared Spectroscopy (FTIR) for studying type of AC surface groups.

Acknowledgement

Financial support provided by TÜBİTAK through project 113M263 and ROX 0.8 type activated carbon provided by NORIT are gratefully acknowledged.

References

[1] Zhang, Z., M. Xu, H. Wang and Z. Li, 2010, "Enhancement of CO_2 Adsorption on High Surface Area Activated Carbon Modified by N_2, H_2 and Ammonia", *Chemical Engineering Journal*, Vol. 160, pp. 571-577.

[2] Shaarani, F. W. and B. H. Hameed, 2011, "Ammonia-modified Activated Carbon for the Adsorption of 2,4-dichlorophenol", *Chemical Engineering Journal*, Vol. 169, pp. 180-185.

[3] Przepiorski, J., M. Skrodzewicz and A. W. Morawski, 2004, "High Temperature Ammonia Treatment of Activated Carbon for Enhancement of CO_2 Adsorption", *Applied Surface Science*, Vol. 225, No. 1-4, pp. 235-242.

423127 Multistage Countercurrent Liquid-Liquid Extraction Enabled By Membrane-Based Separations

Abstract:

AIChE 2015: Separation Divisions

Sessions: Developments in Extractive Separations

Title: Multistage Countercurrent Liquid-Liquid Extraction Enabled by Membrane-Based Separations

Author: Nopphon Weeranoppanant, Andrea Adamo, Galym Saparbaiuly, and Klavs F Jensen

Liquid-liquid extraction is often the most common workup strategy in the pharmaceutical production due to its high selectivity and large capacity at relatively small energy consumption. In many cases, multiple stages of extraction are required to achieve a high degree of separation. Industrially, mixer-settlers, centrifugal extractors, spray and tray columns are widely used. However, they have the disadvantages of excessively large footprint, limited flow capacity, and difficult scale-up procedures. Their moving parts and/or gravity-driven operation complicate, and in some cases preclude, their use on a laboratory scale. In this work, we construct a multistage countercurrent liquid-liquid extraction at the ml/min scale. The membrane-based separators are used to separate the two phases at each extraction stage, depending on their wettability nature and pressure differential across the membrane. By virtue of an integrated pressure control element that ensures successful separation and decouples the operation from downstream pressure, the setup is greatly simplified. The use of multichannel peristaltic pump as a means of interstage pumping significantly reduces the number of pumps required. We demonstrate two classical examples of solvent recovery: (1) extraction of acetone from an aqueous solution using toluene, (2) extraction of acetic acid from an aqueous solution with ethyl acetate, representing systems with relatively high and low interfacial tensions, respectively. Increasing number of stages is shown to significantly increase the degree of recovery. The extraction of the aqueous acetone solution with mass fraction of 0.50 using toluene, with a solvent-to-feed ratio of 1:2 yields 49.6%, 73.4%, 81.1%, and 94.9% recovery for 1, 3, 5, and 7 stages, respectively, and these are found to be in a close agreement with the simulation assuming 100% extraction efficiency. Similarly, the extraction of the aqueous acetic acid solution with mass fraction of 0.05 with a solvent-to-feed ratio of 4:1 yields 64.6%, 77.9%, 93.3%, and 99.6% recovery for 1, 3, 5, and 7 stages, respectively. Scale-up strategies, yielding high extraction efficiency at short residence time, are also presented.

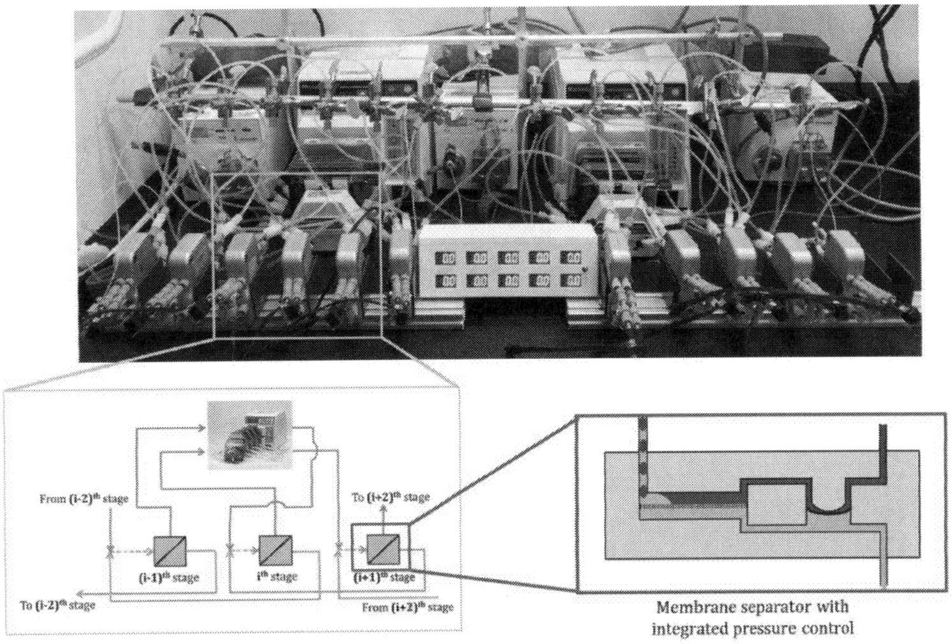

407856 On the Geometrical Features of the Optimal Membrane Contactor

Abstract:

Hollow fiber contactors have been successfully used for many decades in a wide range of applications. An important drawback of this contactor format is however the conflicting relation between throughput and kinetics. Fast kinetics can indeed be reached in smaller diameter fibers, but this occurs at the expense of lower axial velocities and a concomitant lower throughput when equal pressures are employed. An approach that is often used to increase the attainable flow rates consists of incorporating an array of fibers, but this is a tricky operation due to difficulties of fabrication, flow distribution, occurrence of fouling, regeneration, etc. An equivalent configuration, omitting the above mentioned problems, can be conceived by a using flat parallel channel plates that sandwich a membrane with spacers. For the application of liquid-liquid extraction, it is desirable that the interface is pinned at the membrane. The stability of this configuration is dictated by the Young-Laplace equation, which can result in stabilizing pressures of up to 5 bar, which is well in line with optimal flow conditions [1].

Our group has recently provided analytic expressions describing the extraction kinetics in function of operational and geometrical parameters [2] for both co- and counter-flow operation, which were validated using channels with a depth down to 100 µm (width 1 cm). In an attempt to enhance the extraction kinetics lower channel heights are required with a preferential counter-flow operation, for which increasing pressures are required to maintain a given linear velocity, resulting in excessively large pressure gradients across the membrane and a concomitant loss of liquid-liquid interface pinning at the membrane.

A fundamental solution for this stability problem in co-flow mode can in theory be provided by imposing equal in- and outlet pressures at both phases. We observed however that the corresponding flow rate ratio of both phases did not comply with the expected inverse viscosity ratio. The flow rate ratio decreased from 2.8 to 1.6 when inlet pressures increased form 0.1 to 1 bar. By estimating the corresponding dimensions of the channels using Poiseuille's law, the extent of bending of the flexible membranes (PTFE, polycarbonate) was estimated. The bending was subsequently confirmed and further studied using numerical simulation. Next, a microscopy-based measurement method was developed that enabled to follow bending of the membrane in function of the pressure. It appeared that bending decreased considerably from 95 µm to 10 µm (channel depth: 100 µm) with a decreasing distance between the membrane support structures from 3 mm to 400 µm. Also the shape of these support structures was vital, with radially elongated pillars as the most interesting structures due to a more uniform pillar spacing. Based on these findings, the optimal geometrical configuration can be conceived depending on the desired specifications of the membrane, the raffinate and the extractant phase. The attainable performance of membrane contactors is compared with emulsion-based approaches in large scale reactors as well as microreactors.

[1] Hereijgers, J., Breugelmans, T. & De Malsche, W. Breakthrough in a flat channel membrane microcontactor. *Chem. Eng. Res. Des.* **94,** 98–104 (2015).

[2] Hereijgers, J., Callewaert, M., Lin, X., Verelst, H., Breugelmans, T., Ottevaere, H., Desmet, G., De Malsche, W. A high aspect ratio membrane reactor for liquid-liquid extraction. *J. Memb. Sci.* **436,** 154–162 (2013).

424747 Study of the Extraction and Stripping Performance in a Flat Membrane Microcontactor

Abstract:

Solvent extraction (SX) is used throughout a range of different chemical industries: purification of pharmaceutical compounds, wastewater treatment, sample preparation, etc. In hydrometallurgy, SX even became one of the most important separation process [1]. Traditionally SX is performed in packed columns, centrifugal extractors or mixer-settler tanks. Although, these systems have been the workhorses for decades now, important problems like foaming, flooding, unloading, the need for a density difference and stable emulsion formation still present important disadvantages [2]. Non-dispersive contactors avoid such problems by contacting the fluids through a parallel flow profile. In order to maintain a stable interface and yet have a large operating range the interface needs to be stabilized either by placing coated guiding structures or equipping the contactor with a membrane [3]–[5]. This membrane can either be used in a flat sheet configuration or a hollow fiber configuration. Disadvantageous of the hollow fiber approach is that a bundle of fibers can lead to shell side by-passing and the potting adhesive is prone to attacks by the organic solvents. However, hollow fibers are self-supporting, resulting in inner diameters of typically 200-300 μm and walls of tens of micrometers thick [2]. The flat sheet configuration however does require a support, quickly leading to channels of hundreds of micrometer to millimeters deep. As flows are most of the time laminar inside such membrane contactors the mass transfer is purely diffusion based, leading then to slow extraction kinetics. With current microfabrication technology it is now possible to fabricate structured supports, resulting in depths of only hundred micrometer or less reaching equilibrium in matter of minutes [6].

In this work the extraction and stripping efficiency of cobalt from a cobalt/nickel mixture inside such a flat sheet membrane microcontactor was studied. The impact of different parameters on the mass transfer kinetics such as channel depth, concentration, wetting properties of the membrane, pH, saponification of the extractant (Cyanex 272) and upscaling effects was investigated. With channel depths of 100 μm complete extraction was reached after 3 min. Lowering the channel depth further to 50 μm, only 1 min was needed to extract all of the cobalt. It was also seen that buffering the aqueous feed instead of saponification of the organic extractant, increased the extraction kinetics by a factor of 2.5. This was in conflict with what we expected as extraction rate is reported to increase with saponified extractant [7], [8]. Due to the high partition coefficients it was also observed that the wetting properties of the membrane had a significant impact on the mass transfer kinetics. Using a hydrophilic membrane the stripping proceeded with a factor of 1.5 faster than with a hydrophobic membrane. By stripping cobalt from the loaded organic solution, the extractant was regenerated, allowing to operate continuously and recycle the organic phase between an extraction and stripping unit. Cobalt was successfully extracted for 6 hours leaving nickel behind in the raffinate. During this period no significant changes in extraction and stripping performance was observed and the interface remained stable. Only minimal breakthrough (1.87% of the total processed volume) at the stripping unit was observed, which could easily be removed from the buffering tank of the organic extractant, which can alternatively also be avoided by active pressure regulation.

Acknowledgements

J. Hereijgers greatly acknowledges the Agency for Innovation by Science and Technology (IWT) in the Flanders region for financial support. W. De Malsche is grateful to the Flemish Fund for Scientific Research (FWO) for financial support.

References

[1] D. S. Flett, "Solvent extraction in hydrometallurgy: The role of organophosphorus extractants," in *Journal of Organometallic Chemistry*, 2005, vol. 690, no. 10, pp. 2426–2438.

[2] A. Gabelman and S. Hwang, "Hollow fiber membrane contactors," *Journal of Membrane Science*, vol. 159, no. 1–2. pp. 61–106, 1999.

[3] D. Ciceri, J. M. Perera, and G. W. Stevens, "The use of microfluidic devices in solvent extraction," *Journal of Chemical Technology and Biotechnology*, vol. 89, no. 6. pp. 771–786, 2014.

[4] Z.-X. Cai, Q. Fang, H.-W. Chen, and Z.-L. Fang, "A microfluidic chip based liquid-liquid extraction system with microporous membrane.," *Anal. Chim. Acta*, vol. 556, no. 1, pp. 151–6, Jan. 2006.

[5] J. Hereijgers, T. Breugelmans, and W. De Malsche, "Breakthrough in a flat channel membrane microcontactor," *Chem. Eng. Res. Des.*, vol. 94, pp. 98–104, 2015.

[6] J. Hereijgers, M. Callewaert, X. Lin, H. Verelst, T. Breugelmans, H. Ottevaere, G. Desmet, and W. De Malsche, "A high aspect ratio membrane reactor for liquid-liquid extraction," *J. Memb. Sci.*, vol. 436, pp. 154–162, Feb. 2013.

[7] P. XING, C. WANG, S. XU, and Z. JU, "Kinetics of cobalt(II) extraction from sulfate aqueous solution by sodium salt of di-decylphosphinic acid (DDPA)," *Trans. Nonferrous Met. Soc. China*, vol. 23, no. 2, pp. 517–523, 2013.

[8] X. Fu and J. A. Golding, "EQUILIBRIUM AND MASS TRANSFER FOR THE EXTRACTION OF COBALT AND NICKEL FROM SULFATE SOLUTIONS INTO BIS (2,4,4-TRIMETHYLPENTYL) PHOSPHINIC ACID - 'CYANEX 272'.," *Solvent Extraction and Ion Exchange*, vol. 6, no. 5. Taylor & Francis Group, pp. 889–917, 12-Mar-1988.

Disc Supported Taylor-Couette Flow in Liquid-Liquid-Extraction

Enes Aksamija, Matthäus Siebenhofer

Disc supported Taylor-Couette flow in liquid extraction

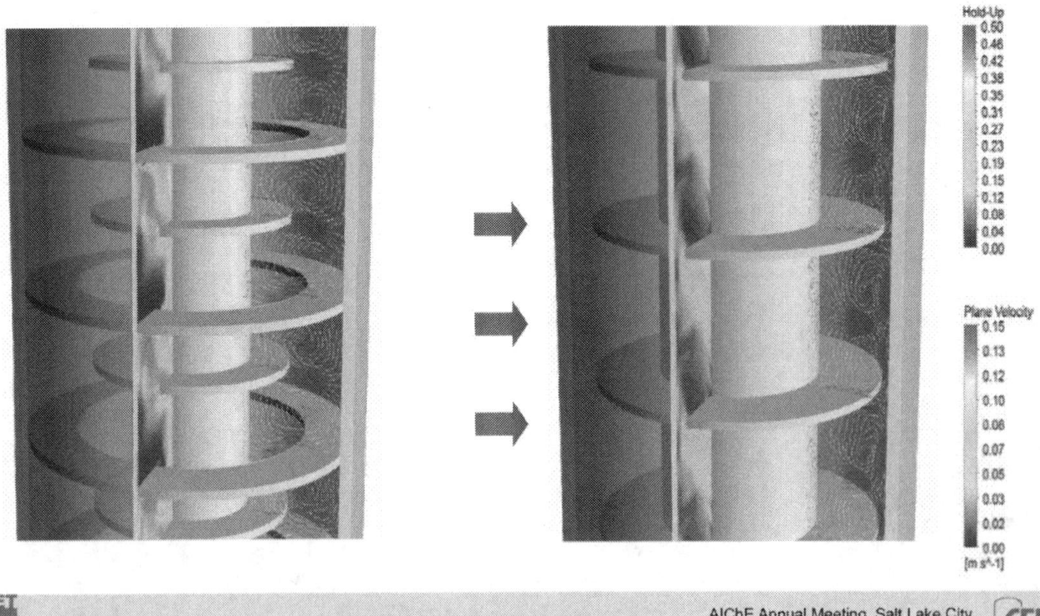

Disc supported Taylor-Couette flow

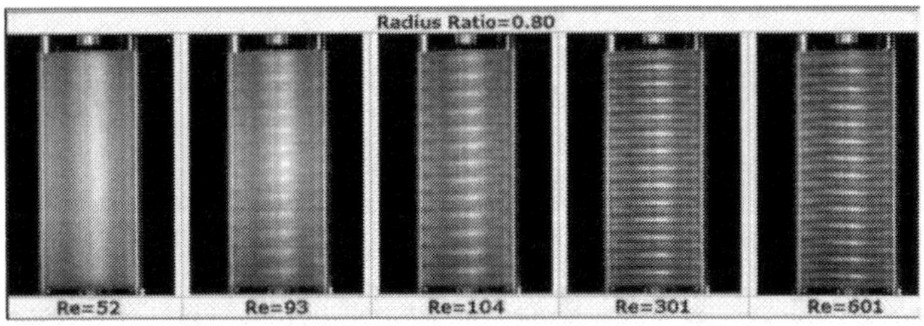

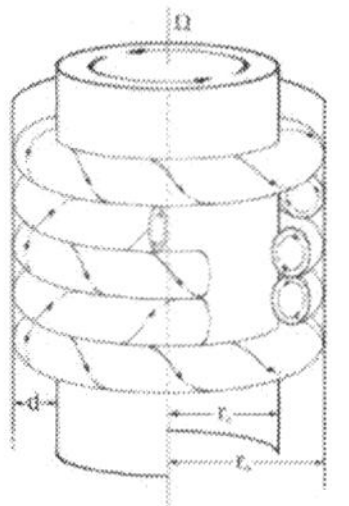

T. T. Lim, University of Melbourne, AUSTRALIA

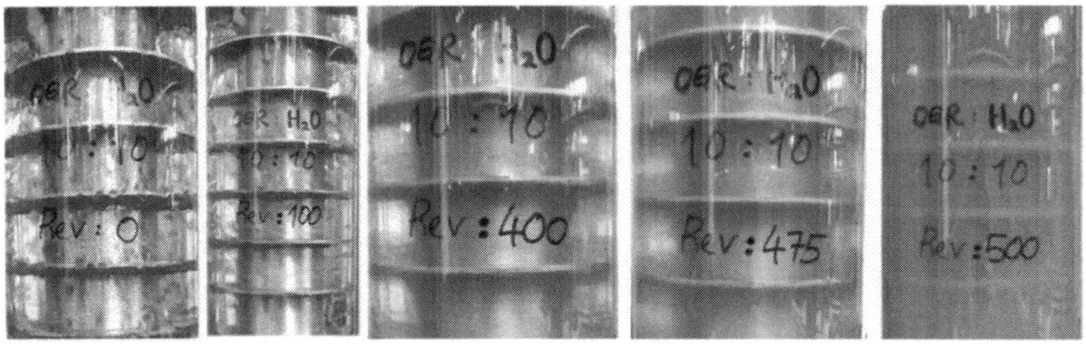

Aksamija E., Dissertation, TU Graz, 2015

Motivation

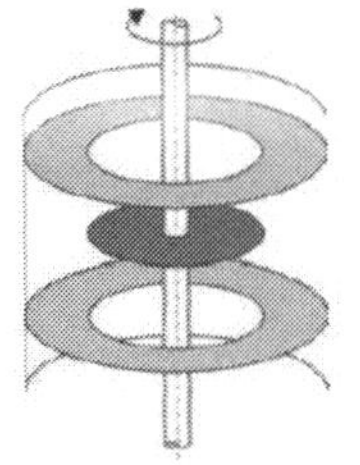

Figure 22. Rotating disk contactor (RDC)

Müller E. et al, Liquid-Liquid Extraction, Ullmann's Encyclopedia of Industrial Chemistry, Wiley-VCH Verlag GmbH & Co KGaA, Weinheim, 2012, Vol. 21, S. 271
Online ISBN: 9783527306732 DOI: 10.1002/14356007

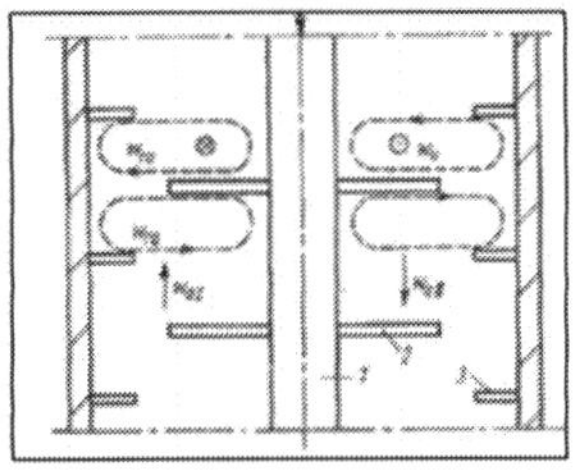

Weiß S. et al, Verfahrenstechnische Berechnungsmethoden, Teil 2, VCH Verlagsgesellschaft mbH, Weinheim, 1986, S. 257

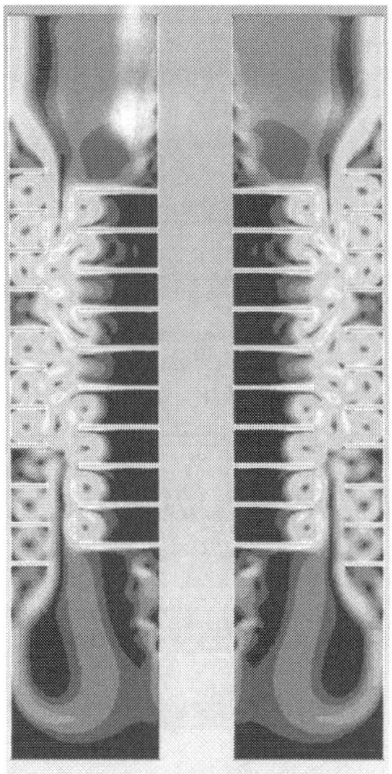

Rotating Disc Contactor (RDC)

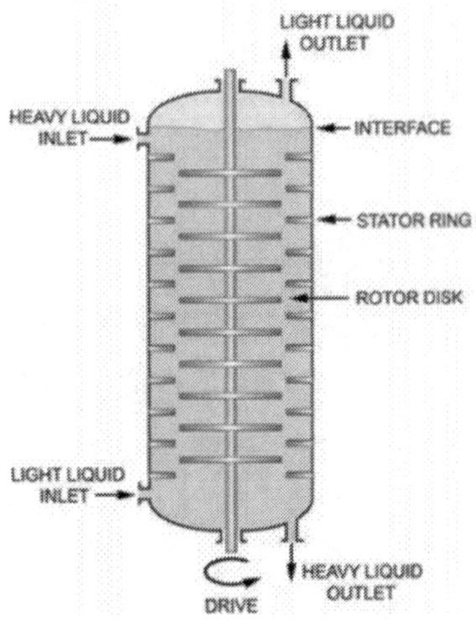

Empirical Design [KOSTERS (1983)]:

$$\frac{D_{Rotor}}{D_{Tower}} = 0.6 \quad \frac{D_{Stator}}{D_{Tower}} = 0.7 \quad \frac{H_{Comp.}}{D_{Tower}} = 0.08\ to\ 0.15 \quad D_{Shaft} = ?$$

- Simple, open design → **high throughput**
- No paddles → **viscous systems**
- Smooth energy input

Problems:

- Empirical design → problematic Scale-Up
- **High axial mixing** → limited separation efficiency

Outline

- RDC, design evaluation
- Taylor-Couette Disc Contactor (TCDC)
- TCDC
  - Onset of operation
  - Hold-Up
  - Flooding limit
  - Drop size distribution
  - Separation efficiency
- Design rules & Summary

Compartment height (PIV-Analysis)

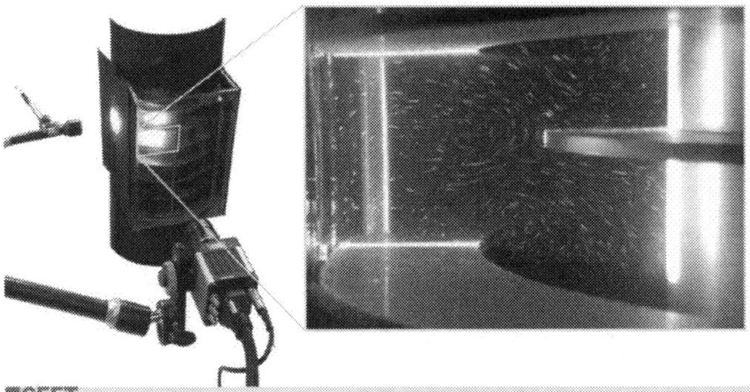

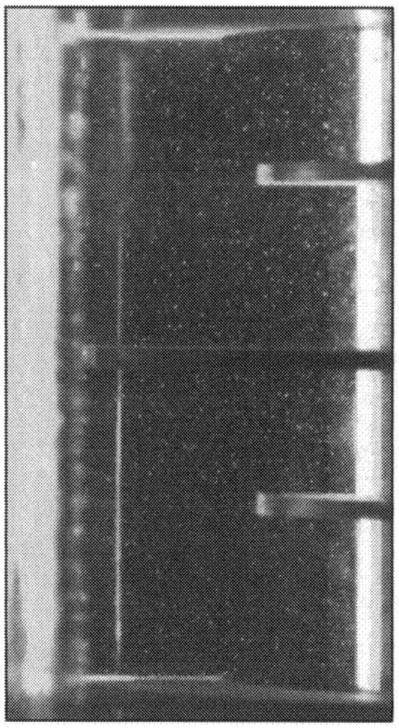

PIV - Results

H_c/D_c = 0.34: Flow pattern depends on load and circulation

0 [l/h], 400[rpm] 30 [l/h], 400[rpm] 130 [l/h], 400[rpm] 130 [l/h], 800[rpm]

Compartment height ($H_c/B_c \leq 1$)

Vortex inversion: inhomogeneous hold-up

coalescence below stator rings unstable

operation

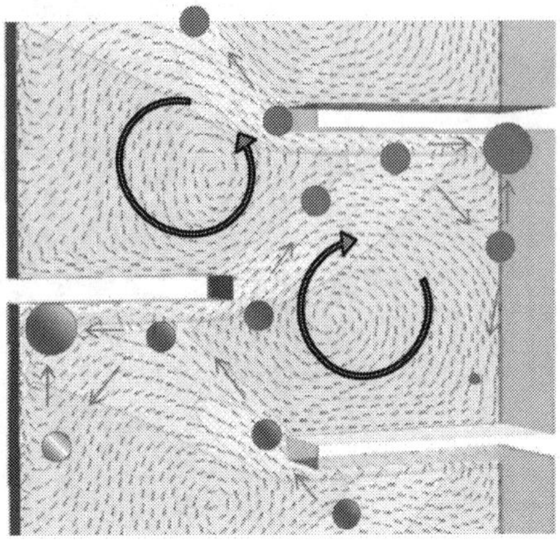

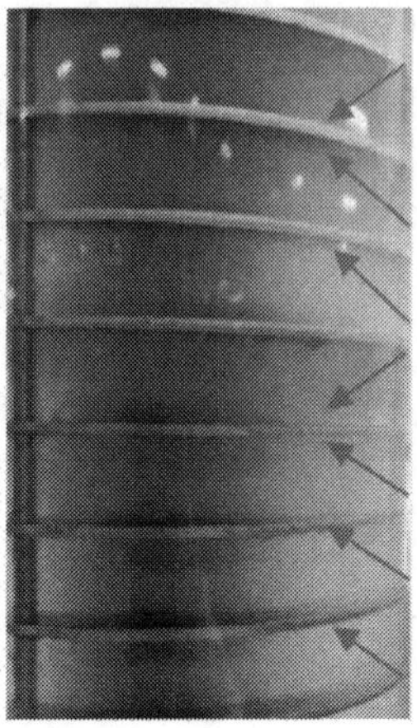

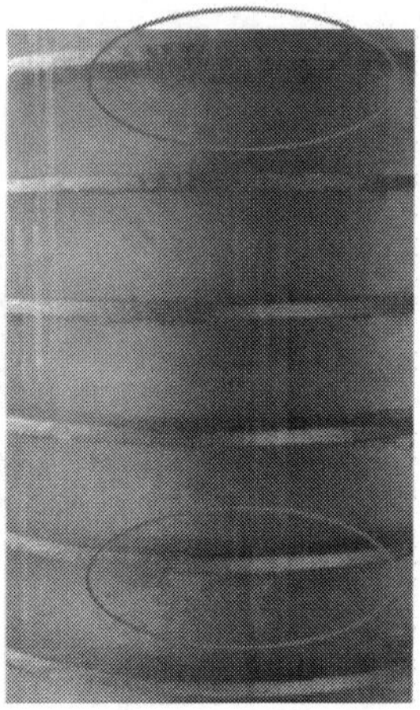

Design and Operation Characteristics of Liquid-Liquid Extractors with Disc Supported Taylor-Couette-Flow

RDC - CFD Optimisation

- Gap location

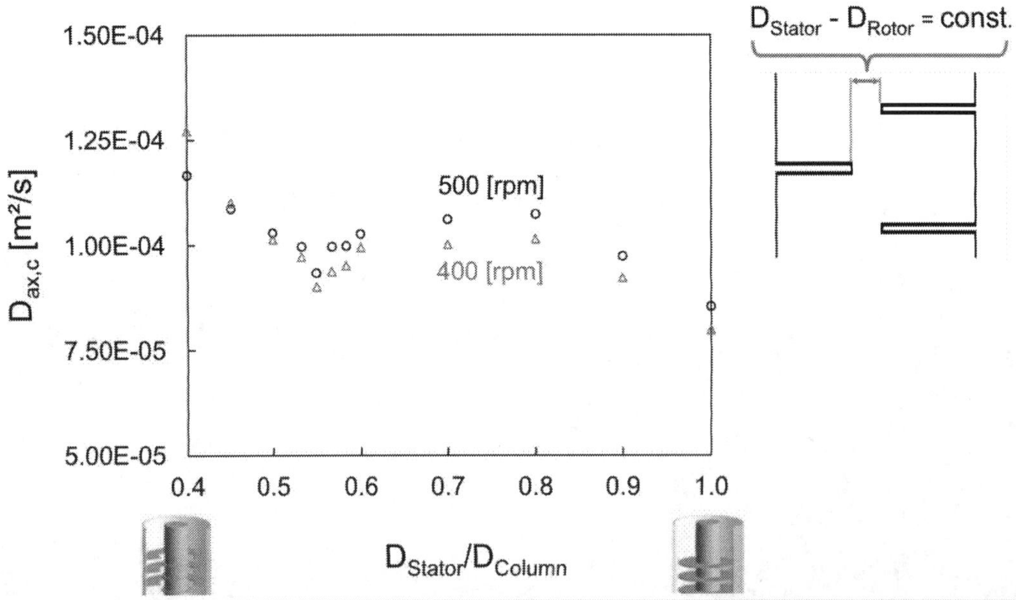

Shaft diameter & Compartment height

- Minimize dead zones
- Maximize vortex strength $\Gamma = \int_A \omega \, dA$

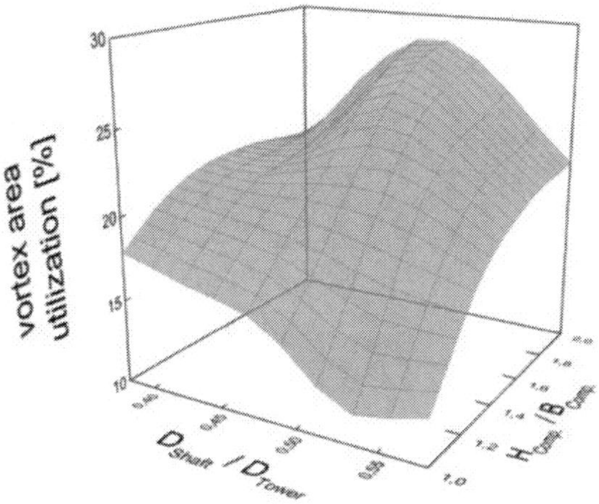

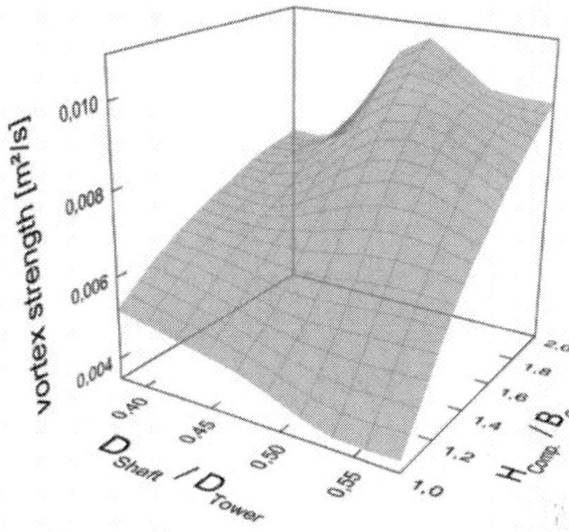

RDC100, 300 [1/min], $V_{cont.}$= 0 [l/h]

Design and Operation Characteristics of Liquid-Liquid Extractors with Disc Supported Taylor-Couette-Flow TU Graz

Shaft diameter & Compartment height

- Mean vorticity

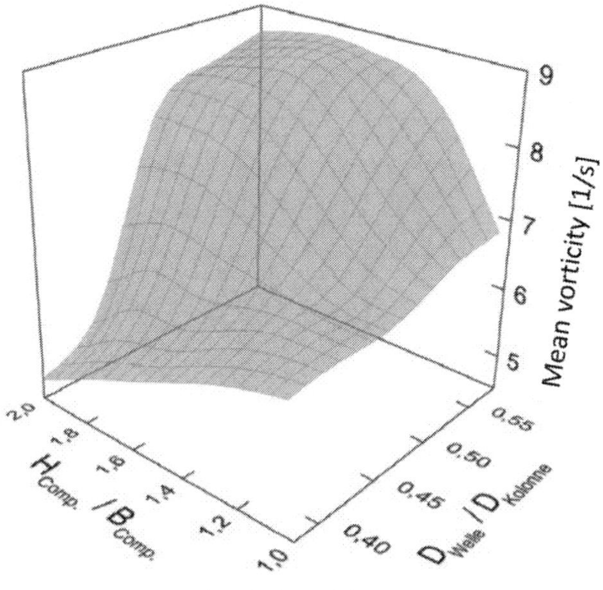

Onset of operation

- Sedimentation experiments with rigid spheres
- Polypropylene, 2mm & 2.5mm
- No axial flow
- Stepwise increase of agitation by 50 [1/min]

Onset of operation

- Residence time distribution, 2mm spheres

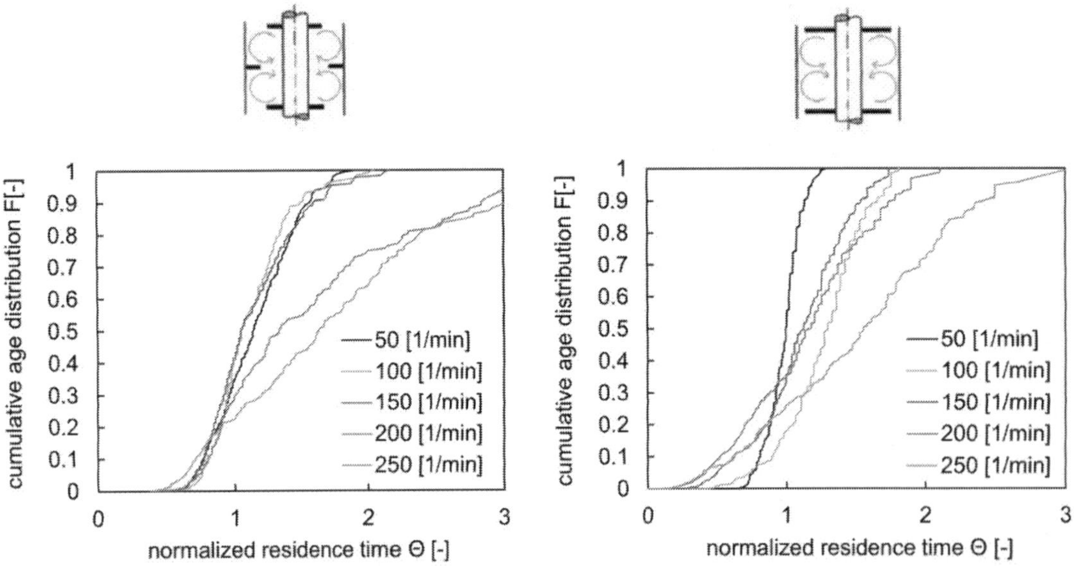

Drop size distribution (DSD)

- Endoscopic probe
- 1[m] active height, Shellsol-T / water
- Variation of B [m³/(m²h)] and n [1/min]

SOPAT
www.sopat.de

Drop size distribution (DSD)

- 450 [1/min]

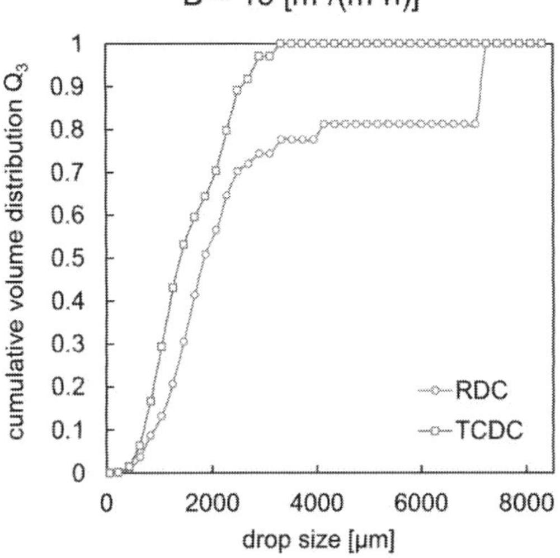

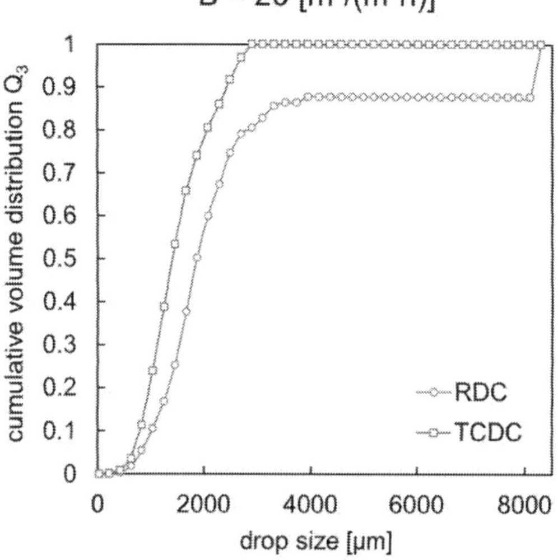

Drop size distribution (DSD)

- 20 [m³/(m²h)]

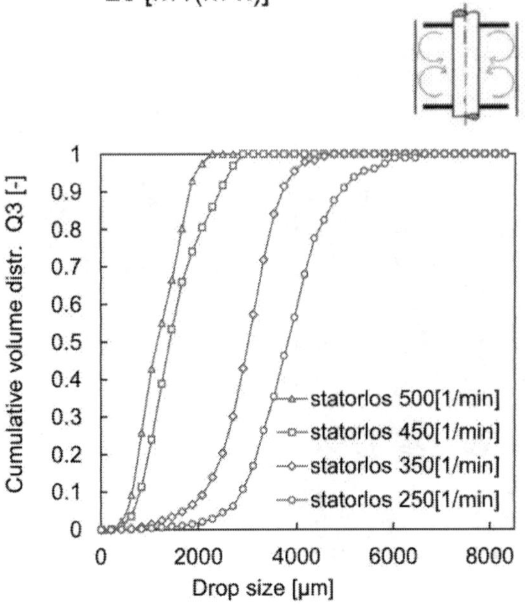

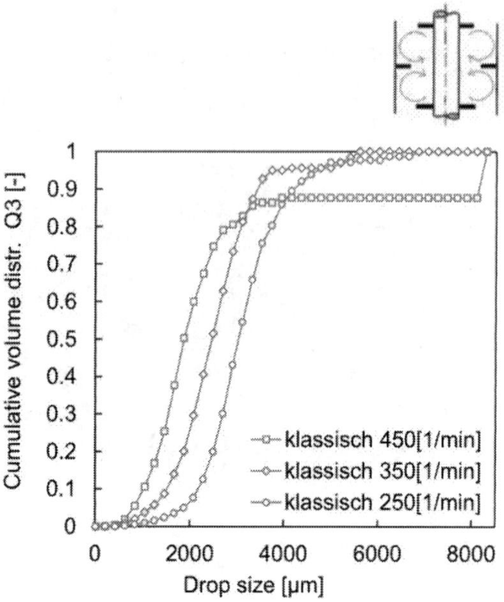

Design and Operation Characteristics of Liquid-Liquid Extractors with Disc Supported Taylor-Couette-Flow TU Graz

Drop size distribution (DSD)

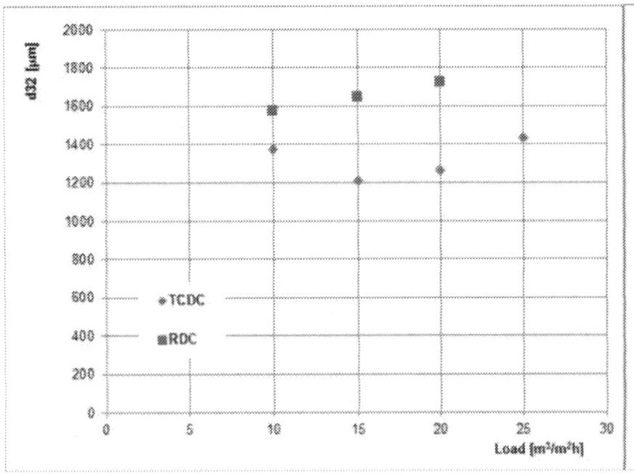

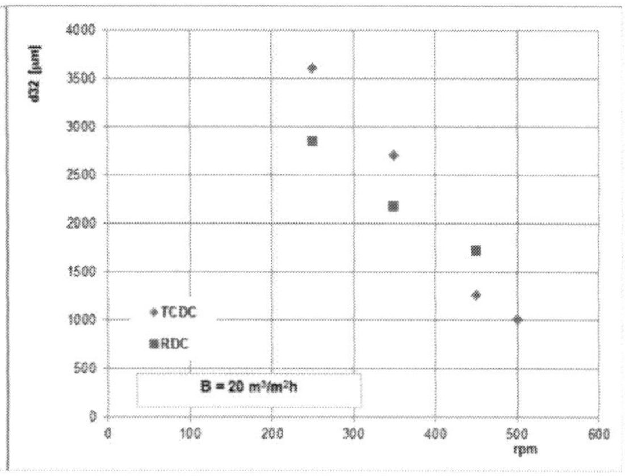

Drop size distribution (DSD)

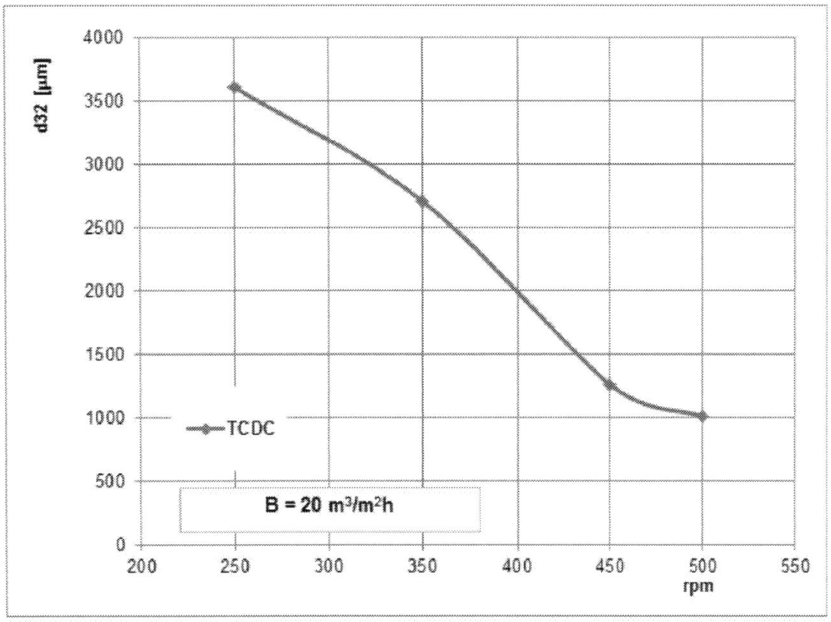

Hold-Up

- High hydraulic load, 30 [m³/(m²h)]
- Position of interface ≙ dynamic Hold-Up

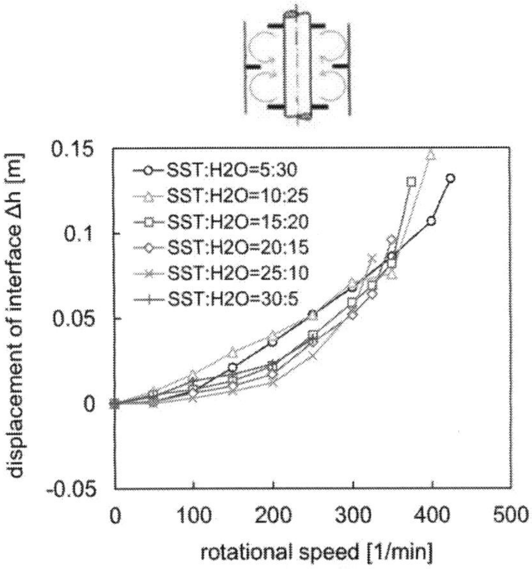

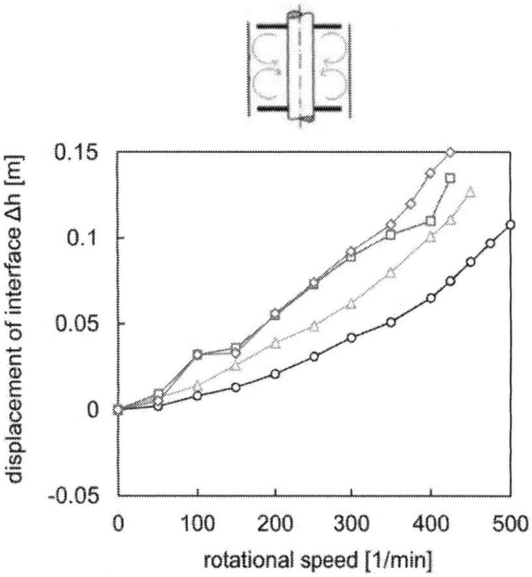

Hold-Up

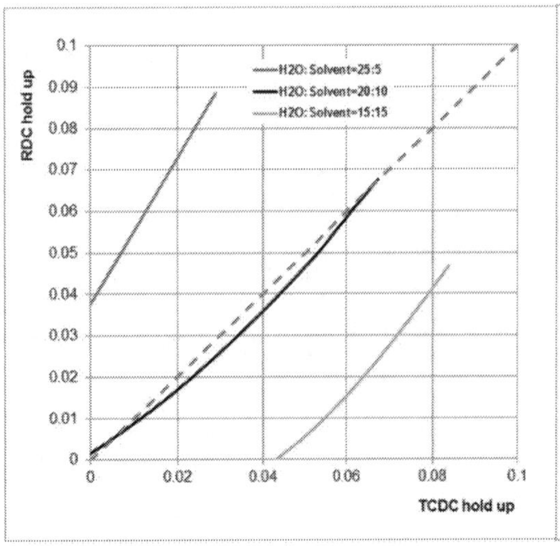

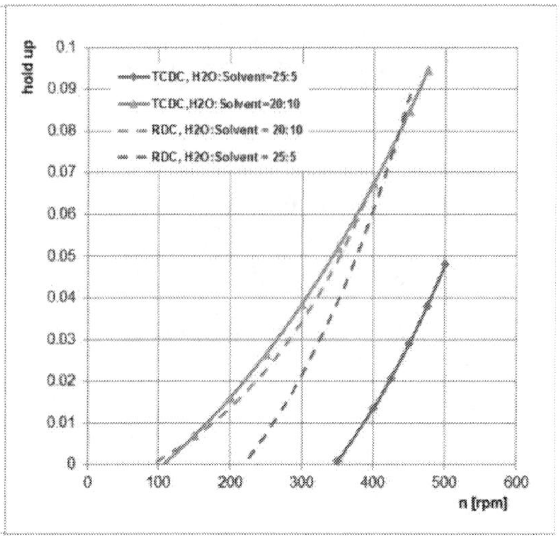

Flooding limits

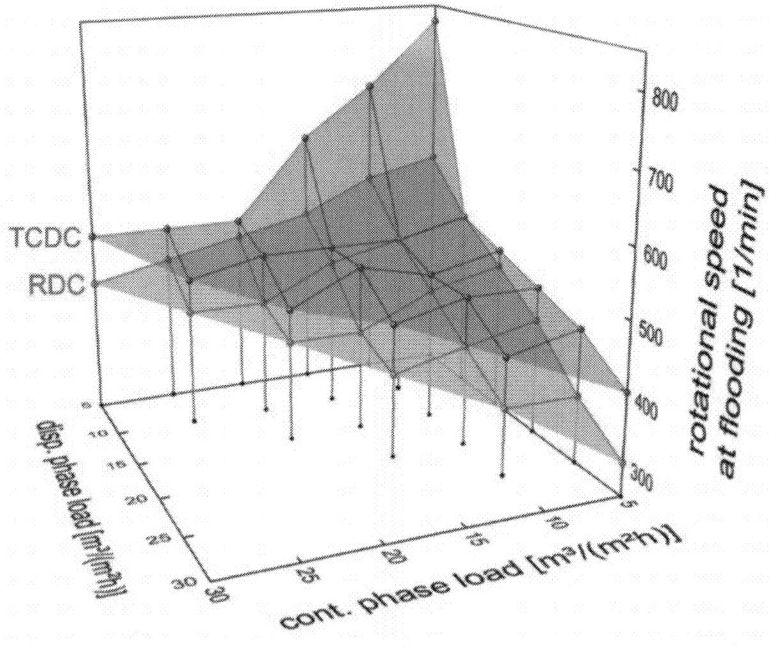

Flooding limits

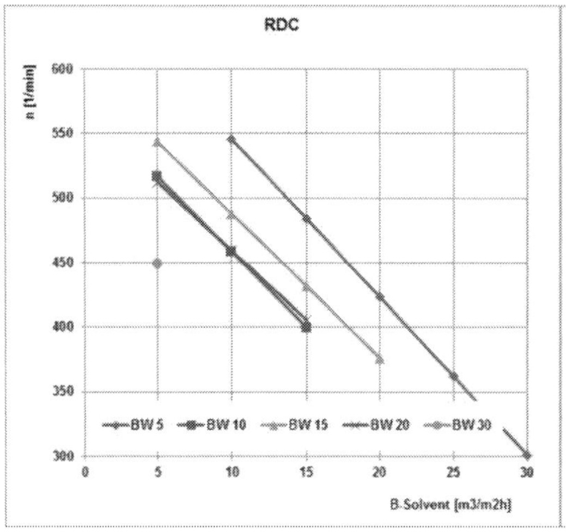

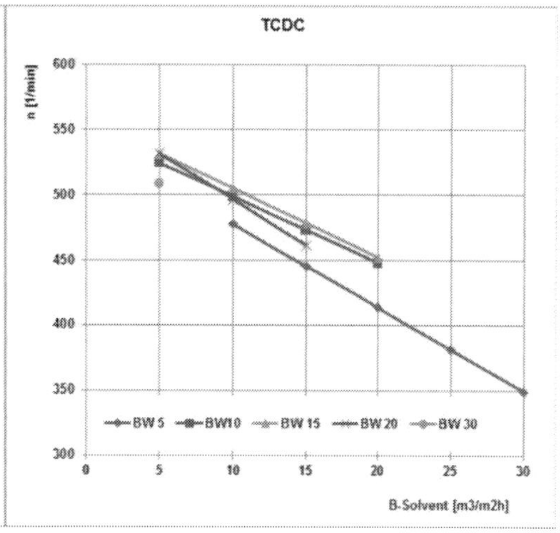

Operation range

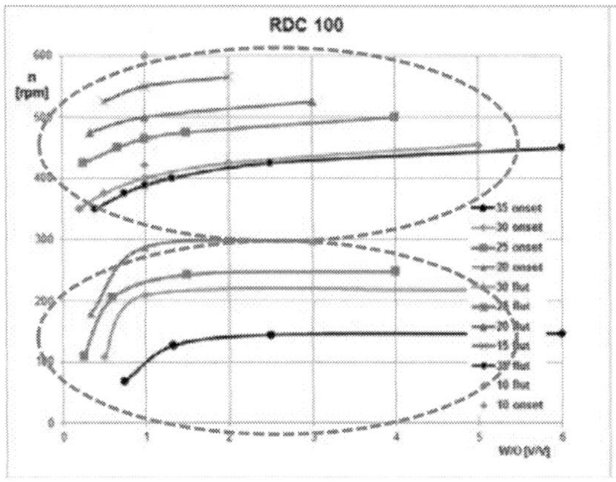

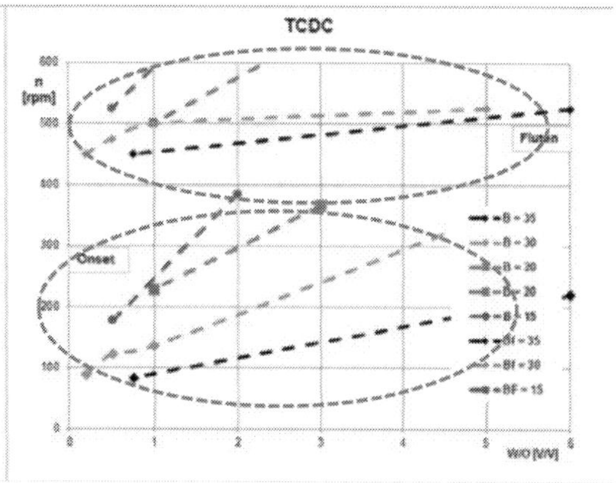

Validation

Test system: n-Butanol-SST-water; d ⟶ c

$$A_{solvent} \rightarrow A_{carrier}$$

$$-r_A'' = k'' * (c_A^* - c_A)$$

b) CSTR cascade

$$F_{A,0} - F_A = -r_A'' * V * a$$

$$c_{A,n} = \frac{c_{A,n-1} + k'' * \frac{V_R}{F_{V,0}} * a}{1 + k'' * \frac{V_R}{F_{V,0}} * a}$$

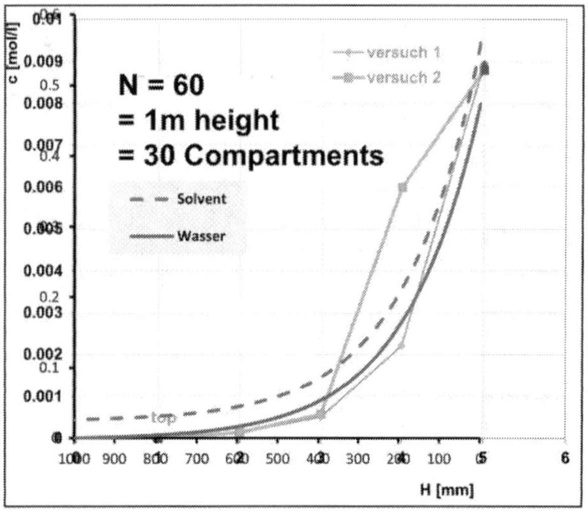

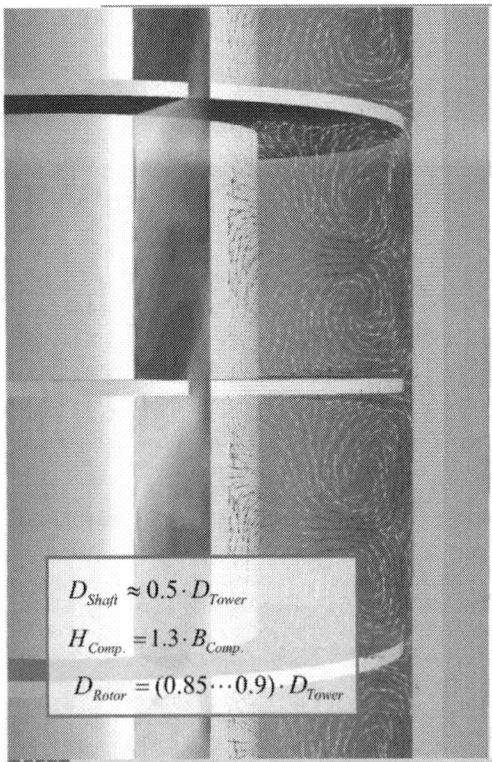

$$D_{Shaft} \approx 0.5 \cdot D_{Tower}$$

$$H_{Comp.} = 1.3 \cdot B_{Comp.}$$

$$D_{Rotor} = (0.85 \cdots 0.9) \cdot D_{Tower}$$

Summary

- Dynamic compartment separation → low axial mixing of both phases
- High flooding limits
- Narrow drop size distribution
- Mixer-Settler cascade
- No crud accumulation
- Low maintenance demand
- High separation efficiency
- Low investment cost

Disc Supported Taylor-Couette Flow in Liquid-Liquid-Extraction

Enes Aksamija, Matthäus Siebenhofer

434605 Numerical Simulation of Drops in a Disc and Doughnut Pulsed Column

Abstract:

The disc and doughnut pulsed column is one of the best choices in solvent extraction processes because of its much smaller footprints and low investment compared to the most commonly used mixer-settler type extraction units among various types of solvent extraction units [1-9]. The low mass transfer efficiency of commercial pulsed column is the only disadvantage, especially for systems with very low interfacial surface tension, and severe axial back-mixing [9]. A large interfacial area is needed for increased mass transfer efficiency. The optimal design of a pulsed column needs an intensive understanding of the multiphase flow dynamics.

Hydrodynamics of single and two phase fluid flow in a disc and doughnut pulsed column is studied in literature using experimental, modelling and simulation techniques [3-8]. The velocity profile, liquid holdup and the size distribution drops of in the column are investigated experimentally [3, 5, 6, 7]. The velocity contours and stream functions of the continuous phase and the hold-up distribution of both the phases in a single stage are determined using numerical simulations [8]; however, to the best of our knowledge, simulations reported in the literature, are performed only on a two dimensional geometry of a single stage.

The objective of the present work is to investigate the pulsatile flow of single and multi-phase fluids in a disc and doughnut column using the techniques of computational fluid dynamics, in order to understand the effect of the operating conditions on the drop dynamics and the geometric parameters on the distribution of the drop size of the dispersed phase and deformation, coalescence and breakage of the droplets.

A disc and doughnut pulsed column can be described as a circular column with an internal fitting that has several units stacked vertically in a repeated manner. Each unit comprises of a doughnut placed between two discs. The height between the disc and doughnut remains uniform throughout the column. All the geometric parameters such as diameter of the disc, height between the doughnut and the disc and aperture of the doughnut are non-dimensionalised by the column diameter. The model geometry for the present simulations is selected based on the dimensional analysis of the large scale columns.

The usual macroscopic mass and momentum conservation laws form the set of governing equations as all the phases are assumed to be interpenetrating continua. The low Reynolds number k-ε model is used to see the effect of turbulence in the continuous phase. The interaction between the phases is modelled using the "Schiller-Naumann" drag law. The effect of drop size distribution on the macroscopic hold up of the liquids is simulated considering multi group description of the dispersed phase. Two pairs of immiscible fluids i.e. water-toluene & water-butanol [4] are considered for the simulations.

In another set of simulations performed on a single stage of the column, the deformation, coalescence and breakage mechanism of the droplets is investigated using the volume of fluid (VOF) method. The effect of shear stress on the interface is neglected while tracking the surface. The boundary conditions at the inlet and outlet of the domain is selected based on the profiles at the central stage, obtained from the earlier simulation. The size distribution of droplets obtained from these simulations is used to form the multi-groups.

The governing equations along with the boundary conditions are solved using the finite volume method as implemented in the commercial software Ansys (Fluent) 14.0/14.5. Uniform structured mesh is created in the computational domain using the ICEM CFD 14.0. The first order upwind scheme is used to discretize the governing equations. The absolute convergence criteria are set at 1×10^{-9} (unit of field variables) in the analysis.

Single-phase simulations are carried out on a pulsed column with three stages, which is found to be the minimum number of stages to be simulated in order to ensure the periodic nature in the central unit. The velocity profiles and contours of the fluid in the central stage is compared with that reported in literature [3] and are found to be in good agreement with each other. Next, the flow of two immiscible liquids is simulated. The velocity contours and phase distribution obtained from the simulations are compared with the experimental observations [6].

Simulations based on the volume of fluid method show that the stable droplets form at the doughnut surface in the column after the flow reaches a stationary state. The drop is found to be sensitive to the flow intensity of the continuous phase. The fluctuations of the droplet surface increase with the flow intensity of continuous phase. The simulation result also indicates that probability of breakage and coalescence increases with the number density of the drops. Effect of the distribution of the drop size on the liquid hold-up in the column will be discussed in the presentation.

Key words: Disc and doughnut pulsed column, pulsating flow, volume of fluid.

References:

1. Angelov, G., Gourdon, C., 2012. Pressure drop in pulsed extraction columns with internals of discs and doughnuts. Chemical Engineering Research and Design, 90 (7), 877-883.
2. Angelov, G., Gourdon, C., 2009. Turbulent flow in pulsed extraction columns with internals of discs and rings: Turbulent kinetic energy and its dissipation rate during the pulsation. Chemical Engineering and Processing: Process Intensification 48 (2), 592 -599.
3. Bujalski, J., Yang, W., Nikolov, J., Solnordal, C., Schwarz, M., 2006. Measurement and CFD simulation of single-phase flowin solvent extraction pulsed column. Chemical Engineering Science 61, 2930 - 2938.
4. Mate, A., Masbernat, O., Gourdon, C., 2000. Detachment of a drop from an internal wall in a pulsed liquid-liquid column. Chemical Engineering Science 55 (11), 2073-2088.
5. Milot J, Duhamet J, Gourdon C, Casamatta G., 1990. Simulation of a pneumatically pulsed liquid-liquid extraction column. The Chemical Engineering Journal 45 (2), 111 -122.
6. Retieb,. S, Guiraud, P., Angelov, G., Gourdon, C., 2007. Hold-up within two-phase counter current pulsed columns via eulerian simulations. Chemical Engineering Science 62 (17), 4558-4572.
7. Kumar, R., Sivakumar, D., Kumar, S., Mudali, U. K., 2003. Modeling of hydrodynamics in a 25 mm Ø pulsed

disk and doughnut column. ISRN Chemical Engineering, 1-10.
8. Angelov G, Journe E, Line, A, Gourdon, C., 1990. Simulation of the flow patterns in a disc and doughnut column. The Chemical Engineering Journal 45 (2), 87 - 97.
9. Nabli, M. A., Guiraud, P., Gourdon, C., 1998. CFD contribution to a design procedure for discs and doughnuts extraction columns. Chemical Engineering Research and Design 76 (8), 951 - 960.

426766 Liquid-Liquid Extraction Performance of a New Single Stage Rotor-Stator Spinning Disc Extractor

Abstract:

Liquid-liquid extraction performance of a new single stage rotor-stator spinning disc extractor

Ling Lv, Bin Liang*,Changjun Liu, Siyang Tang,Hairong Yue

Multi-phases Mass Transfer and Reaction Engineering Laboratory, College of Chemical Engineering, Sichuan University, Chengdu 610065, China;

*Corresponding author: liangbin@scu.edu.cnAbstract:

Liquid-liquid extraction is one of the most important separation processes, which has been intensively used in chemical, pharmaceutical and hydrometallurgical industries due to its simplicity, high duty and low costs. Rotor-stator spinning disc contactor is supposed to have higher mass transfer efficiency than conventional column contactors. Here we designed and constructed a new countercurrent rotor-stator spinning disc extractor which has two settling chambers and one spinning chamber in between the shaft of the disc was placed horizontally. The performance of the extractor was evaluated with the water-ethanol-2-ethyl-hexanol system. The flow pattern, extraction efficiency, mass transfer coefficient were studied experimentally. A maximum of extraction efficiency of 88% and an overall mass transfer coefficient of $0.63 s^{-1}$was achieved at a spinning rate of 500 rpm and a liquid flow rate of $0.33\times10^{-6} m_{AQ}^3\ s^{-1}$ (water: 2-ethyl-hexanol = 1:1).The results indicated that overall mass transfer coefficient and extraction efficiency of rotor-stator spinning disc extractor were higher than rotating disc column extractor.

Keywords:

Mass transfer, ethanol, Counter-current flow, Spinning disc extractor, rotor-stator Reference

1. Hadjiev, D. and J. Paulo, Extraction separation in mixer–settlers based on phase inversion. Separation and Purification Technology, 2005. 43(3): p. 257-262.

2. Visscher, F., et al., Liquid–liquid mass transfer in a rotor–stator spinning disc reactor. Chemical Engineering Journal, 2012. 185-186: p. 267-273.

3. Visscher, F., et al., Rotating reactors – A review. Chemical Engineering Research and Design, 2013. 91(10): p. 1923-1940.

421683 Chemical Stability and Ion Transport in Polymerized Ionic Liquid Block Copolymer Anion Exchange Membranes with Various Cations

Abstract:

New highly hydroxide-conductive anion exchange membranes (AEMs) with high alkaline chemical stability are required in order to achieve long-lasting low-cost solid-state alkaline fuel cells (AFCs). In this study, we synthesized and investigated the chemical stability of methacrylate-based polymerized ionic liquids (PILs) consisting of various covalently attached cations: butylimidazolium, butylmethylimidazolium, trimethylammonium, pentamethylguanidinium, butylpyrrolidinium, trimethylphosphonium. The degradation pathways and extent of chemical stability of the PILs were determined using ^{1}H NMR spectroscopy after exposure to various pHs, temperatures, and times. Additionally, corresponding diblock copolymers consisting of a non-ionic component, methyl methacrylate (MMA), and a functional component, 2-bromoethyl methacrylate (BrEMA), were synthesized *via* the sequential reversible addition-fragmentation chain-transfer (RAFT) polymerization technique and subsequently functionalized with the same six cations to form a series of PIL block copolymers with the same composition and molecular weight. Ion transport, water content, and morphology were measured in each PIL block copolymer using electrochemical impedance spectroscopy, dynamic vapor sorption, small-angle X-ray scattering, and transmission electron microscopy. PIL block copolymer transport-morphology relationships at various humidities and temperatures with regard to cation type will be presented.

434811 Hydroxyl and Water Transport in Stable Anion Exchange Membranes

Abstract:

The potential of anion exchange membrane (AEM) fuel cells to provide inexpensive compact power from a wider variety of fuels than is possible with a proton exchange membrane (PEM) fuel cell, has continued to drive the research interest in this area. Alkaline catalysis in fuel cells has been demonstrated with non-precious metal catalysts, and with a variety of fuels beyond $H_2$and methanol. Alkaline fuel cells (AFCs), based on aqueous solutions of KOH, have serious drawbacks associated with system complexity and carbonate formation. Anion exchange membrane (AEMs) fuel cells have a number of advantages over both PEM fuel cells and traditional AFCs; however, although anionic conductivity in AEMs can be comparable to PEMs the chemical stability of membrane attached cations in hydroxide is still not always sufficient for practical applications. The real issue is water transport; water is both a product and a reactant in these systems and wet cations are much more stable than dry. Wetted cations are alos much more stable than dry cations. So an understanding of water in these membranes is essential.

Here we discuss water and anion transport in a series of thin mechanically robust state of the art experimental anion exchange membranes. The membranes are generally constructed from an isoprene block and a vinyl benzyl bromide block, either randomly or in di-, tri- or penta- block configurations. Post quaterniaztion leads to functionalized AEMs. We use electrochemical impedance spectroscopy to measure anion conductivity, multi-nuclear pulse field gradient spin echo NMR to measure self-diffusion, and broadband electric spectroscopy to measure the relaxation processes in these polymers. This information is coupled with microscopy and SAXS to explore the polymer morphology. Putting transport and morphology together allows us to describe a complete picture of water and anion transport in these systems. Recently, with the realization that AEMs derived from chlorinated polymer backbones are highly stable to hydroxide attack, we have in hand a series of films which are stable enough to study in the hydroxyl form. This is allowing us for the first time to compare experimental transport studies of hydroxyl with simultions based on chemically accurate theoretical models of hydrated hydroxide. One surprising outcome of this is that it is becoming apparent that many AEMs exsist in which hydroxyl transport is as facile as proton transport in stte of the art proton exchange membranes but with far less water. Of course the use of air or hydrocarbon fuels nessecitates understanding carbonate and bicarbonate transport as well, and we are just begining to formulate studies of these important anions in AEMs.

424859 Polybenzimidazole/Graphene Oxide Nano-Composites for Alkaline Direct Alcohol Fuel Cells

Abstract:

The objective of this study is to prepare Polybenzimidazole (PBI) composite with graphene oxide (GO) nano-sheets as a barrier layer to retard alcohol fuel permeabilities while retaining electrical conductivity. These films were doped with potassium hydroxide (KOH) solution to form hydroxide-conducting electrolytes in alkaline direct alcohol fuel cells. The alkaline fuel cell gave more than double power density than that with the H_3PO_4-doped PBI. The PBI/GO composite exhibited higher fuel cell performance than the PBI film by 30-60%. The peak power density of alkaline direct methanol fuel cell reached 200 mW cm^{-2} and that of alkaline direct ethanol fuel cell, 120 mW cm^{-2}. These results demonstrate that the PBI/GO composite is a potential electrolyte for high-performance alkaline fuel cell applications, especially for portable energy devices.

425768 "Smart MEA" Approach: Design of Catalyst Layer

Abstract:

It is well known that clean energy sources are currently in high demand. Different type of fuel cells, such as proton exchange membrane (PEMFCs) or anion-exchange membrane fuel cells (AEMFC) are a potential solution. At the moment, the technologies use platinum-based catalysts with a high loading of the catalyst, which are expensive and contribute to ~40% of the total cost of a PEMFC.[1] For the decades, considerable investment was involved into the examination of the non-platinum group metal (non-PGM) catalysts. One rapidly developing class of non-PGM catalysts for oxygen reduction reaction (ORR) are materials based on transition metal-carbon-nitrogen networks (M-N-C).[2] Those materials include Fe, Co or Mn precursors and nitrogen-containing organic precursors (either polymeric or low molecular weight ones). It is widely recognized that nitrogen species of M-N-Cs play a crucial role in the ORR mechanism. It has been suggested that the nitrogen stabilizes the attachment of the transition metals to the underlying carbon support creating the ORR active sites.[2-4] UNM group have examined a number of different organic precursors in our previous research. They include different combinations of transition metals with: tetramethoxy phenylporphyrin)[6], poly(ethyleneimine) [7], 4-aminoantipyrine[1], and carbendazim[8].

Composition of the catalyst from both chemical speciation and morphology (porosity, roughness, texture) plays crucial role in the interaction between those materials and ionomers (both proton and anion-exchange types). It also should be mentioned that the pore structure is critical to the transport of oxygen to active sites and removal of water (in order to prevent flooding). The variations of the pore-structure induced by the chemical changes introduced during fuel cell operation has to be understood in order to design non-PGM electrocatalysts with durability similar to DOE recommended. In present work, our group will discuss the focused ion beam/scanning electron microscopy (FIB-SEM) sectioning to obtain a visual 3D representation of catalyst layer morphology.

These 3D images will be used to generate detailed estimates of the evolution of structural parameters such as: specific surface area, total porosity, connectivity of pores, and others as a result of durability studies.

References

1. M. H. Robson *et al. Electrochim. Acta* **90**, 656-665 (2013).
2. F. Jaouen *et al. Energy Environ. Sci.* **4**, 114-130 (2011).
3. S. Kattel *et al. Phys Chem Chem Phys* **16**, 13800-13806 (2014).
4. U. I. Kramm *et al. J. Am. Chem. Soc.* **136**, 978-985 (2014).
5. U. Tylus *et al. J. Phys. Chem. C* **118**, 8999-9008 (2014).
6. K. Artyushkova *et al.*. *Journal of Power Sources* **226**, 112-121 (2013).
7. A. Serov *et al.*. *Applied Catalysis B: Environmental* **127**, 300-306 (2012).
8. A. Serov *et al.*. *Adv. Energy Mater.* **4**, 7 (2014).

429839 A Coarse-Grained Model of Polyphenylene Oxide – Trimethylamine Membrane for Alkaline Fuel Cells

Abstract:

Developing alkaline anion-exchange membrane fuel cells (AAEMFCs) remains challenging in part from difficulty synthesizing anion-exchange membranes with high OH^- ion conductivity, acceptable mechanical stabilities, and low chemical deterioration in high pH media for fuel cell operation. Molecular simulations provide a versatile tool to study the anion conductivity and stability of AEM materials in an effort to provide a fundamental understanding of the AEM design. However, available molecular models for membranes do not meet the quest of large spatial and temporal scales required to model the multiscale structure and transport processes in the polymer electrolyte membranes. Here, we present the development of coarse-grained models for hydrated polyphenylene oxide - trimethylamine (PPO-TMA) membranes. To our knowledge, this is the first coarse-grained model that includes water, ions, hydrophobic, and intramolecular interactions, all explicitly parameterized to reproduce multiple properties of interest for AEM, including ionic solvation and water-driven hydrophobic association. The uncertainty quantification (UQ) method is used in an iterative approach to parameterize the high-dimensional parameter space of the force field to reproduce multiple properties of the atomistic reference systems. Because of the reduced degrees of freedom in a coarse grained model, the structural, energetic, and dynamic properties cannot be simultaneously reproduced with arbitrary precision; however, the interdependence of these properties are considered collectively. The coarse-grained membrane model gives a reasonable description of the mobility of water and ions, and consequently, the solvation and the electro-osmotic drag, which are of utmost relevance for the operation of fuel cell membranes. We use the coarse-grained membrane model to explore how varying the alkyl amine pendant groups attached to the PPO backbone affects the morphology and dimensions of segregation in the membrane, and compare the simulation results with experiments. We anticipate that the large spatial and temporal simulations made possible by the coarse-grained model will advance the quest for anion-exchange membranes with improved transport and mechanical properties.

423363 Investigation of Anhydrous Proton Transport Mechanism on Functionalized Graphene

Abstract:

Proton transport at interfaces and surfaces is of tremendous importance in diverse areas such as proton exchange membranes in fuel cells, biological membranes and proton pump channels. We are specifically interested in the transport of protons at interfaces for a better design of new materials for proton exchange membrane (PEM) fuel cells. The practical engineering motivation for developing new materials is to conduct protons for improved performance and reduce costs of hydrogen fuel cells.

We focus specifically on functionalized graphene for studying surface transport of protons in anhydrous conditions using density functional theory. The hypothesis is that a facile anhydrous proton transport will occur on a surface that has a continuous network of hydrogen bonded OH groups showing "grotthuss-like" mechanism. A simple 1-D hydrogen bonded network of hydroxyl groups on graphane membrane with one and two excess proton concentrations are studied. In this work, we have computed activation energy barriers and investigated mechanisms for proton transport reactions on the surface of functionalized graphene with varying excess proton concentrations. Ab initio molecular dynamics (AIMD) simulations with various density functional approximations are used to determine equilibrum and dynamical properties of fully hydroxylated graphene surface with one excess proton concentration will be discussed. We have developed a simple lattice model for the 1-D hydrogen bonded network to explore the cross-over from normal to single-file diffusion. We have developed a simple lattice model for the 1-D hydrogen bonded network to explore the cross-over from normal to single-file diffusion.

434508 Investigating the Chemical and Morphological Structure of TR Polymers

Abstract:

Thermally rearranged (TR) polymers have shown outstanding separation performance for a number of current and emerging applications for membrane-based gas separations. In certain cases, these polymers surpass the best combinations of permeability and selectivity currently known for polymers. It is quite surprising, therefore, that a significant and outstanding question remaining in the field relates to the identification of their chemical structure.

TR polymers are traditionally formed from aromatic polyimides that contain nucleophilic chemical functionality ortho-position to their precursor diamine. Upon heating, these polyimides undergo a chemical reaction, whereby they become insoluble in all known solvents. This insolubility endows these polymers with outstanding chemical resistance to aggressive feed gas streams; however, it also precludes the use of traditional characterization techniques such as solution-state NMR for identifying their structure.

Herein, we seek to identify the structure of TR polymers by designing a partially soluble TR polymer. This polymer, prepared from a *polyamide*, which we will refer to as PA-TR, was characterized through a suite of 2-D solution-state NMR experiments to confirm its chemical structure, after which, we prepared a commonly studied polyimide-based TR polymer, PI-TR, with a nearly identical proposed chemical structure to that of PA-TR. Cross-polarization magic-angle spinning NMR reveals that both of these TR polymers are polybenzoxazoles.

Interestingly, despite having nearly identical chemical structure, these TR polymers have different gas transport property sets. Using positron annihilation lifetime spectroscopy to estimate the fractional free volume in these polymers, differences in transport properties are attributed to unique morphologies between each sample. Therefore, the outstanding transport properties for TR polymers are achieved not simply from their chemical functionality, but also their physical nature.

426994 Organic-Inorganic Composite Membranes for Energy and Environmental Applications

Abstract:

Organic-Inorganic Composite Membranes for Energy and Environmental Applications

Wanqin Jin*

Membrane Science & Technology Research Center, State Key Laboratory of Materials-Oriented Chemical Engineering, Nanjing Tech University, Nanjing, 210009, P.R.China.

Abstract

Membrane-based separation technology are currently considered as more environmentally friendly and energy-efficient than traditional separation methods. By combining advantages of organic and inorganic materials, organic-inorganic composite membranes are regarded as a hot research interest in fields of membrane technology and chemical engineering. This presentation will review our recent progress in organic-inorganic composite membranes, from the aspects of fundamental study and industrial implementation. It aims to give an overview of design, preparation and application of the organic-inorganic composite membranes that used for molecular separation. Various membrane types with different materials will be involved in the talk, including ceramic-supported polymer composite membranes, mixed matrix membranes, and graphene-based membranes. Physicochemical properties, morphologies, interfacial behaviors, transport characteristics and separation performance of the organic-inorganic composite membranes will be thoroughly discussed based on advanced characterization techniques. Meanwhile, the presentation will contain several typical applications of the membranes in fields such as bio-fuels production, organic compounds recovery, solvent dehydration, carbon dioxide capture and so on. In addition, large-scale production and industrial implementation of the organic-inorganic composite membranes will also be introduced. This work was financially supported by the National Basic Research Program of China (No. 2009CB623406), the Innovative Research Team Program by the Ministry of Education of China (No.IRT13070) and National Natural Science Foundation of China (Nos. 21406107, 2149580015, 21476107).

***Keywords*:** organic-inorganic composite membranes; fundamental study; industrial implementation

429313 Roles of Polymer Chain Entanglement in Determining Membrane Morphology

Abstract:

Though nonsolvent-induced phase separation has been widely used for a long time to prepare commercial polymer membranes, it still remains a challenge to get a clear insight into the mechanism associated with membrane formation. The main focus of the presentation is on how the polymer chain entanglement in casting solution affects its viscoelasticity and the resulting phase separation mechanism and membrane morphology.

We observed that the phase-separation mechanism of a polymer solution strongly depended on the degree of chain entanglement in the solution. A solution with higher degree of chain entanglement was more stable in the metastable region between binodal and spinodal, thus having higher tendency not to phase separate in the meta-stable region but to phase separate in the unstable region. So the solution inclined to phase separate via the mechanism of spinodal decomposition, resulting in membranes with interconnected pores. On the other hand, as the solution had lower degree of chain entanglement, the solution tended to phase separate via the nucleation and growth mechanism, forming membranes with cellular pores. It was also observed that chain entanglement had influences on polymer crystallization for a casing solution of semi-crystalline polymer, such as PVDF. A semi-crystalline polymer solution with higher degree of chain entanglement had slower crystallization rate; therefore, liquid-liquid phase separation dominated during membrane formation. And polymer crystallization became the dominant mechanism with a casting solution having lower degree of chain entanglement.

The chain entanglement in polymer solution was observed to be strongly related to the polymer-solvent interaction. The results we obtained indicated that lower quality solvent made solution with higher degree of chain entanglement. And the addition of nonsolvent in polymer solution could also enhance chain entanglement. Besides, the effect of molecular weight of polymer on chain entanglement and membrane morphology has been investigated. The results we obtained indicated that polymer chain entanglement, characterized by the viscoelastic property of casting solution, plays an important role in determining its phase separation mechanism and the resulting membrane morphology. Therefore, knowledge of the rheology of polymer solution is needed to clearly understand the formation mechanism of polymeric membranes.

431787 A Highly Permeable and Selective Aminofunctionalized MOF Cau-1 Membrane for Separation of Molecular Mixture

Wednesday, November 11, 2015: 4:25 PM
155D (Salt Palace Convention Center)

Jianhua Yang[1,2], Huimin Yin[2], Jinqu Wang[2], Jinming Lu[2], Yan Zhang[2], Y.S. Jerry Lin[3] and Michael Tsapatsis[4], (1)Chemical Engineering of Material Science, University of Minneapolis, Minneapolis, MN, (2)Chemical Engineering, Dalian University of Technology, Dalian, China, (3)School for Engineering of Matter, Arizona State University, Tempe, AZ, (4)Chemical Engineering & Materials Science, University of Minnesota, Minneapolis, MN

A highly permeable and selective aminofunctionalized

MOF CAU-1 membrane for separation of molecular mixture

Jianhua Yang*[a,b], huimin Yin [a], Jinqu Wang [a], Jinming Lu [a], Yan Zhang [a], Jerry Y. S. Lin [c] and Michael Tsapatsis*[b]

a:State Key Laboratory of Fine Chemicals, School of Chemical Engineering, Dalian University of Technology, Dalian, Liaoning, 116024, China. E-mail: yjianhua@dlut.edu.cn; b: Department of Chemical Engineering and Materials Science,College of Science and Technology, University of Minnesota, MN55455; c: School for Engineering of Matter, Transport and Energy, Arizona State University,

Tempe, AZ 85287, USA.

*corresponding authors:Jianhua Yang, E-mail add.: yjianhua@dlut.edu.cn; Michael Tsapatsis, Tsapa001@umn.edu

Abstract

Metal-organic frameworks are appealing candidates for molecular separation particularly for CO_2 and other molecular mixture separation because of the high porosity, structural flexibility, pore size and chemical tailorability as highlighted in several reviews.[1-5] In the form of powders, MOFs have exhibited an unprecedented CO_2 adsorption capacity/selectivity and high CO_2 adsorptive separation efficiency from N_2 and CH_4. The excellent preferential CO_2 adsorption ability are expected to lead to high CO_2 perm-selectivity over other gases for MOFs membranes. However, so far only the MOF-5 membrane reported by Zhao et al. showed efficient separation for CO_2/N_2 mixtures under high pressure and CO_2 content conditions.[6-7] Instead, other reported MOFs membranes including small and large pore size MOFs membranes have so far shown a poor CO_2/N_2 separation selectivity around 1 for either single gas or mixture system despite of their high adsorption selectivity.[8-11]There is a long way to explore MOF membranes with high performance for CO_2 capture.

In the present work, a continuous and thin amine-decorated microporous metal-organic framework CAU-1 membrane on asymmetric a-Al_2O3 support was synthesized and explored for CO2 separation. The obtained CAU-1 membrane was of 2¨C3μm thin and exhibited a high CO2 permeance of up to 1.34 °Á 10¯6 mol m^{-2} s^{-1} Pa^{-1} and a CO_2-N_2 selectivity of 17.4-22.8 for CO_2-N_2 mixtures, demonstrating for the first time that amino-functionalized MOF membranes can provide high CO2¨CN2 selectivity and possess the potential for CO2 capture from flue gas[12]. Separation of other molecular mixture will be studied as well.

SHAPE * MERGEFORMAT 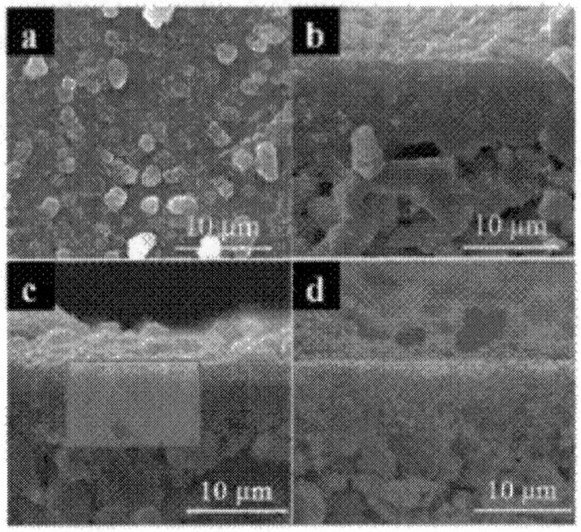

Fig.1 SEM image of the surface and cross-section of the CAU-1 membrane (a, b) and EDS mapping of the CAU-1 membrane: green C; red Al (c, d).

Reference:

1. J.R.Li,R.J Kuppler,H.C. Zhou,*Chem. Soc. Rev.* 2009; 38: 1477-1504.

2. D.Zacher,O.Shekhah,C.Wöll,R.A.Fischer,*Chem.Soc.Rev.* 2009; 38: 1418-1429.

3. M.Shah, M.C.McCarthy, S. Sachdeva, A.K.Lee, Hae-K.Jeong, *Ind. Eng. Chem. Res.* 2012; 51:2179¨C2199.

4. A.B¨¦tard and R.A.Fische, *Chem.Soc.Rev.* 2012; 112: 1055¨C1083.

5. K. Sumida, D.L.Rogow, J. A. Mason, T. M. McDonald, E.D.Bloch, Z.R.Herm,Tae-H.Bae and J. R. Long, *Chem.Soc.Rev.* 2012; 112:724¨C781.

6. Z.X.Zhao, X.L.Ma,A.Kasik,Z.Li,and Y. S. Lin, *Ind. Eng. Chem.Res.* 2013; 52: 1102¨C1108.

7. Y. Y. Liu, G. F. Zeng, Y. C. Pan and Z. P. Lai, J. Membr. Sci. 2011;379:46.

8. F.Zhang,X.Q. Zou,X.Gao,S.J. Fan,F.X. Sun,H. Ren,and G.S. Zhu, *Adv. Funct. Mater.* 2012; 22:3583-3590.

9. A.S.Huang,Y.F.Chen,N.Y.Wang,Z.Q.Hu,J.W.Jiang and J.Caro, *Chem. Commun.* 2012;48:10981¨C10983.

10. A.S.Huang,W.Dou and J.Caro, *J.Am.Chem.Soc.* 2010; *132* : 15562¨C15564.

11. X.L.Dong,K.Huang,S.N.Liu,R.F.Ren,W.Q.Jin and Y.S.Lin, *J.Mater. Chem.* 2012;22:19222-19227

12. Huimin Yin,a Jinqu Wang,a Zhong Xie,a Jianhua Yang,*a Ju Bai,a Jinming Lu,a Yan Zhang,a Dehong Yina and Jerry Y. S. Lin, *ChemCommu.* 2014; 50: 3699--3701

426673 Mild Detemplation of Sub-Micron Silicalite-1 Membranes for Xylene and Butane Isomer Separation

Abstract:

Conventional thermal detemplation of zeolite membranes with high temperatures at ~500 °C usually leads to defects or cracks due to the mismatched thermal expansions between zeolite membrane and substrate. Also, high temperature detemplation methods preclude the application of temperature-sensitive substrates such as polymer supports. In addition, the energy cost of thermal detemplation can limit industrial application. Here, we fabricated sub-micron silicalite-1 membranes supported on porous silica and used them as model membranes to evaluate two mild detemplation methods: (i) low temperature thermal and (ii) UV/ozone treatment. The silicalite-1 membranes were fabricated with exfoliated nanosheets as the seed layer followed by gel-free secondary growth, which led to sub-micron *b*-oriented silicalite-1 membranes on the porous silica supports made by quartz fibers. Xylene and butane isomer permeation tests demonstrated that thermal treatment at a nominal oven temperature as low as 280 °C for 8 hours or UV/ozone treatment for 6 hours is sufficient to establish high selectivity. The highest temperature during the UV/ozone treatment was around 80 °C. The implications of these findings to membranes supported on flat porous polymer sheets or polymer hollow fibers will be discussed.

15AIChE
Annual Meeting, Salt Lake City, UT

Nanodisperse Organic/Inorganic Hybrid Membranes for Efficient Bioalcohol Pervaporation

Hongwei Fan, Hao Yan, Naixin Wang, Jie Li, Shulan Ji, Guojun Zhang*

Center for Membrane Technology, College of Environmental and Energy Engineering,
Beijing University of Technology, Beijing100124, P. R.China

November 11, 2015 Salt Lake City

Outlines

Backgrounds

Bioalcohol recovery

Microbial fermentation

Recovery

- Azeotropic distillation
- Extractive distillation
- ....
- Pervaporation

Bioalcohol

- Energy-saving, cost-effective
- In situ and continuous removal of butanol

Backgrounds

Pervaporation (PV)

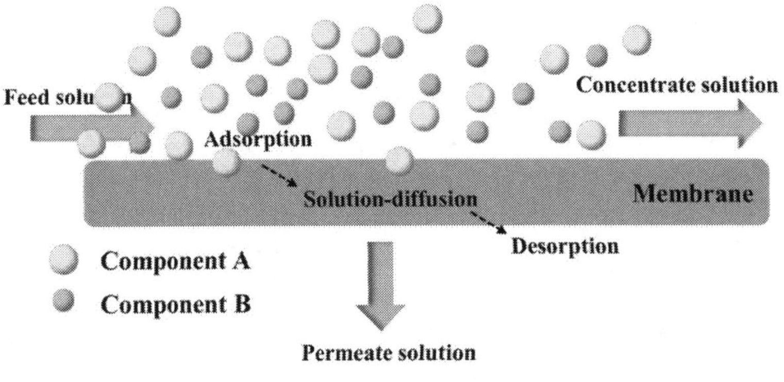

The key to successful application of pevaporation in bioalcohol recovery is developing an appropriate alcohol-permselective membrane with as high as possible flux and selectivity as well as operating stability.

Backgrounds

Membrane materials

- **Polymeric membranes:** swollen and unstable, poor performance
- **Inorganic membranes:** small-area, complex preparation process

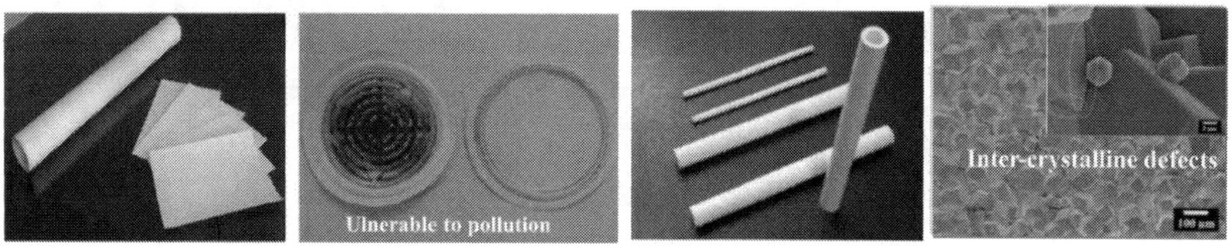

Polymeric membrane **Tubular-ceramic ZIF membrane**

- Hybrid membrane / Mixed matrix membrane (MMM)

- **High performance (anti trade-off)**
- **High stability**
- **Easily scale-up**

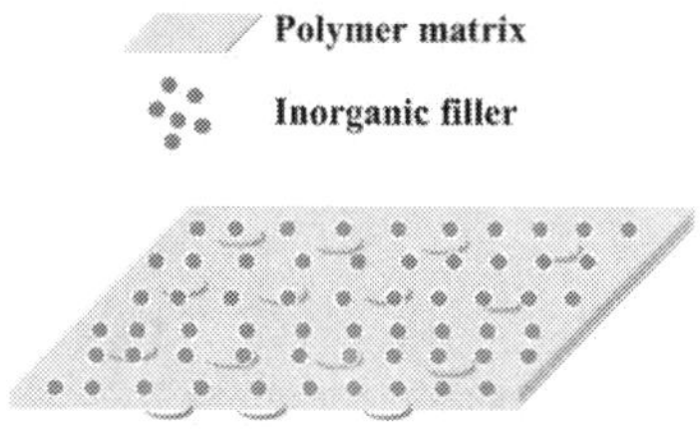

Hybrid membrane

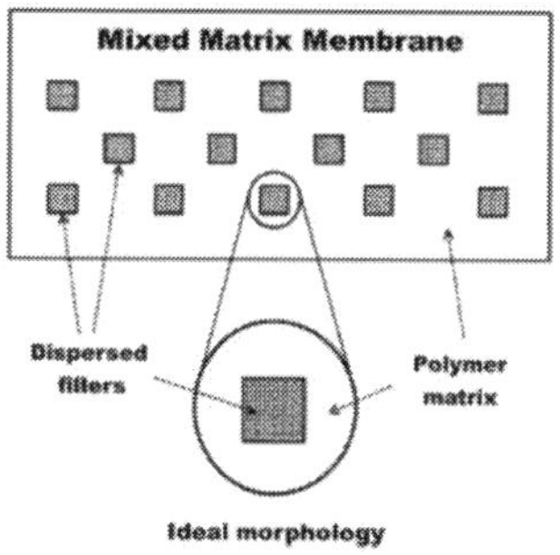

Backgrounds

General membrane-making process

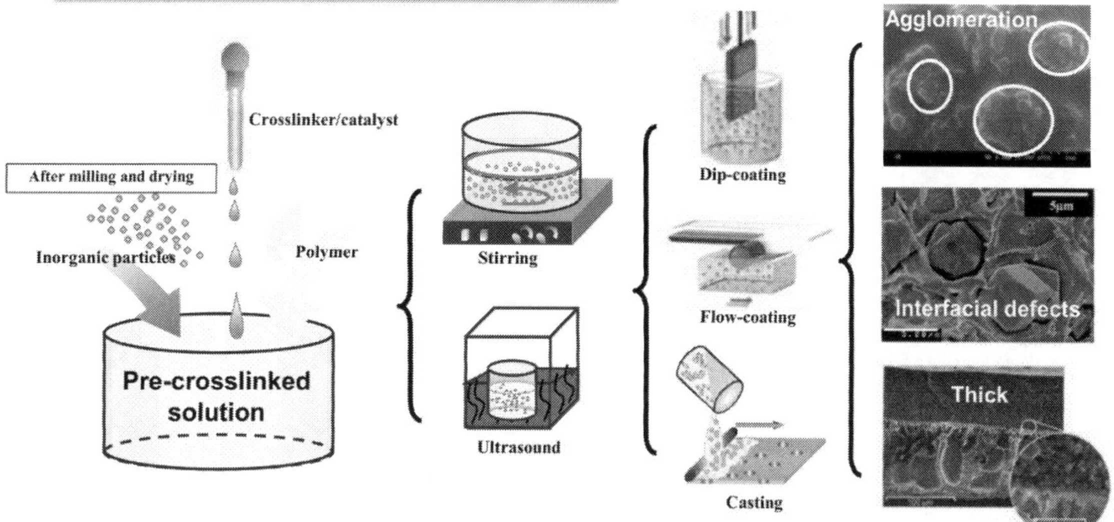

How to **effectively improve the compatibility and dispersibility** ? → Developing novel approaches

Research contents

- Preparation of nanodisperse ZIF-8/PDMS hybrid membrane by suspension-dispersed method and followed a defects-removal process

- Sonication-enhanced in situ assembly of well-dispersed SiO_2/PDMS hybrid membrane

Research contents

- Preparation of nanodisperse ZIF-8/PDMS hybrid membrane by suspension-dispersed method and followed a defects-removal process

- Sonication-enhanced in situ assembly of well-dispersed SiO_2/PDMS hybrid membrane

Experimental Section

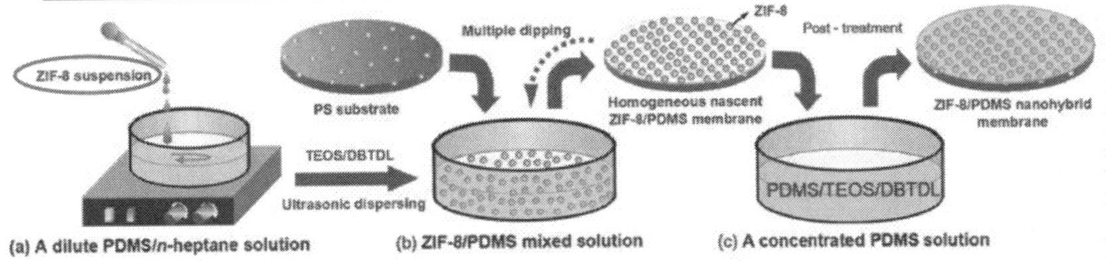

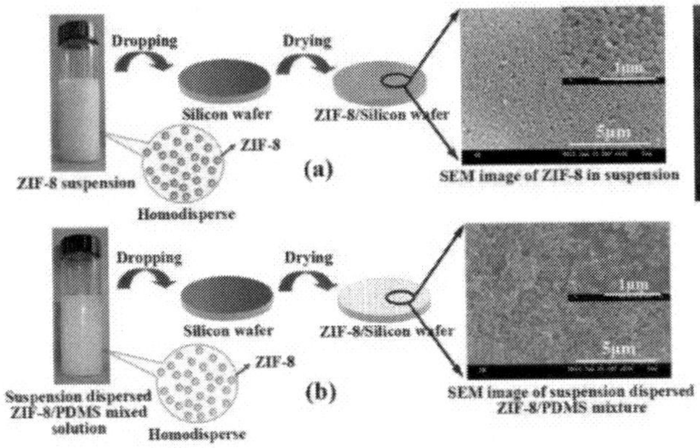

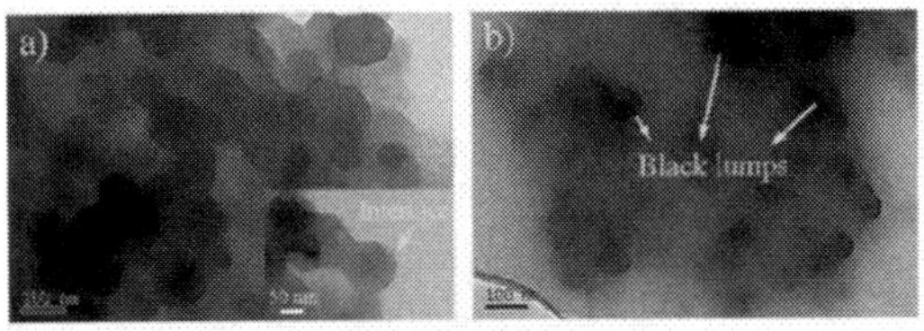

a) Suspension-dispersed ZIF-8/PDMS mixture
b) Powder-dispersed ZIF-8/PDMS mixture

- ZIF-8 suspension was mixed well with the PDMS solution;
- ZIF-8 nanoparticles were dispersed well in PDMS solution.

Hongwei. Fan, Naixin Wang, Shulan Ji, Hao Yan, Guojun. Zhang*. *J. Mater. Chem. A.*, 2014, 2, 20947–20957.

Characterization

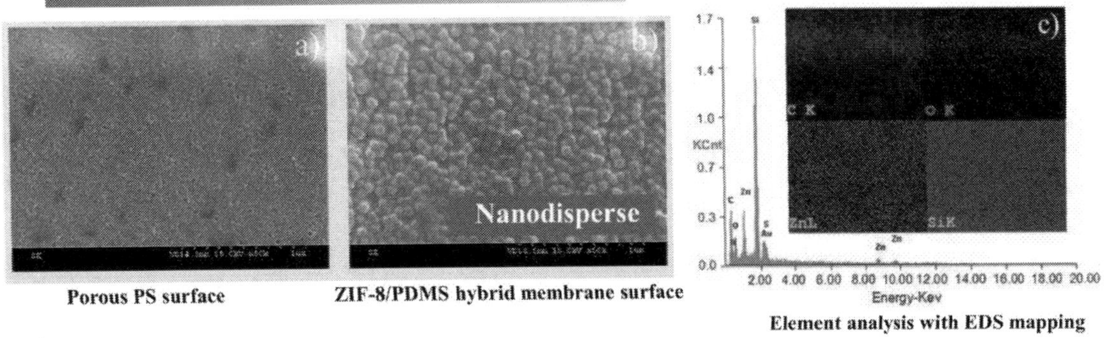

Porous PS surface　ZIF-8/PDMS hybrid membrane surface　Element analysis with EDS mapping

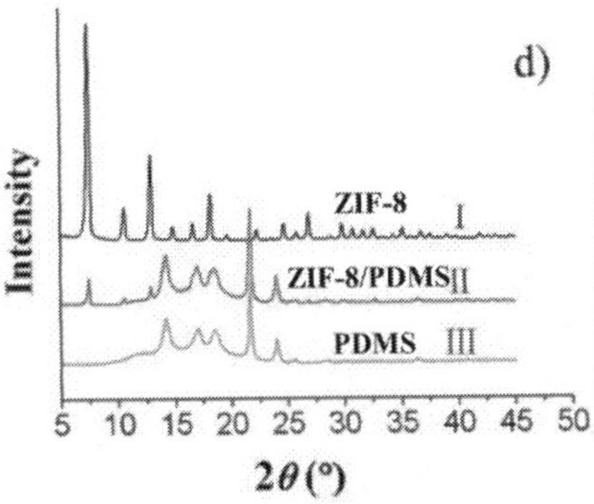

XRD

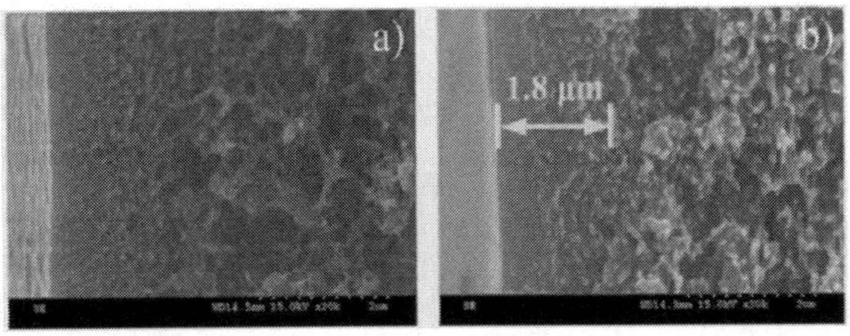

Porous PS cross-section　ZIF-8/PDMS hybrid membrane cross-section

Biobutanol Recovery

Feed solution (Feed: 5 wt.% butanol/water mixture at 80° C)

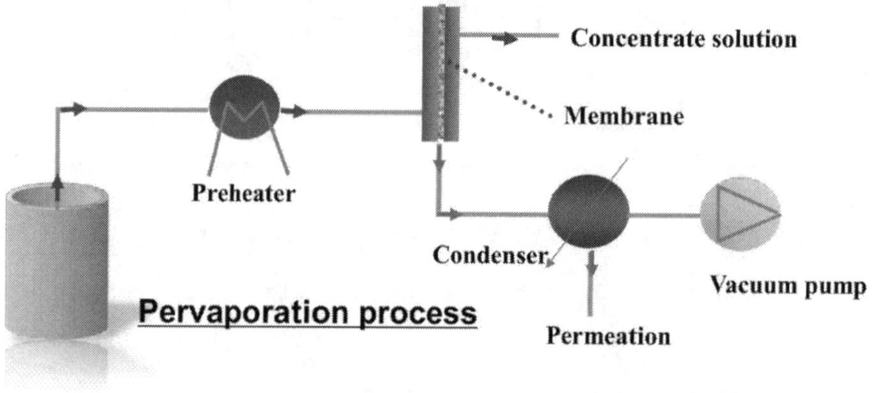

Separation factor $\alpha = \frac{Y_A / Y_B}{X_A / X_B}$

Permeate flux $J = \frac{W}{A \cdot \Delta t}$

Results and Discussion

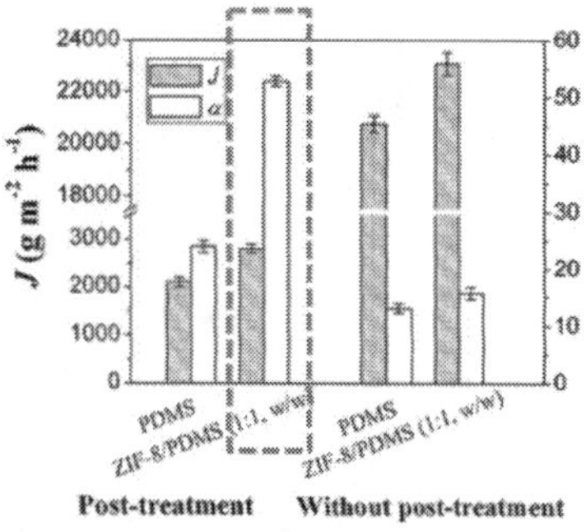

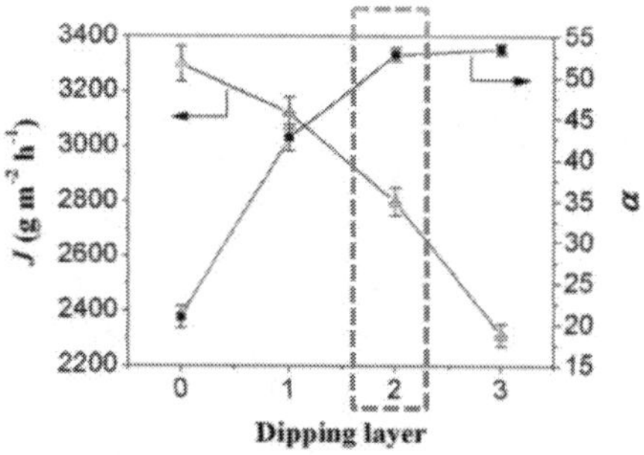

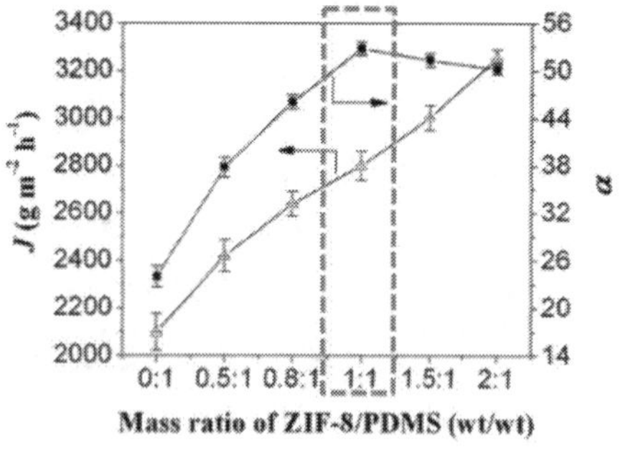

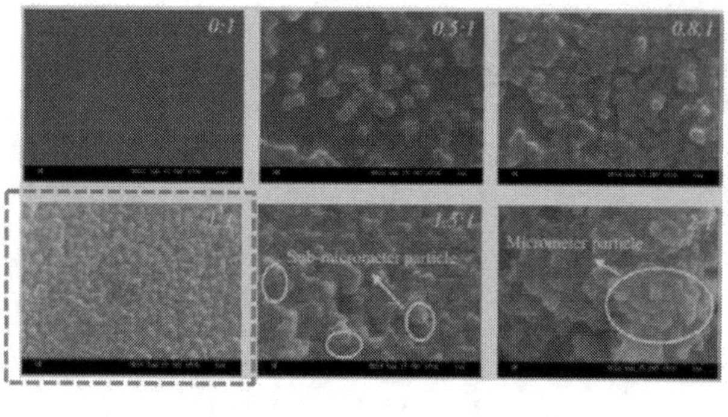

Results and Discussion

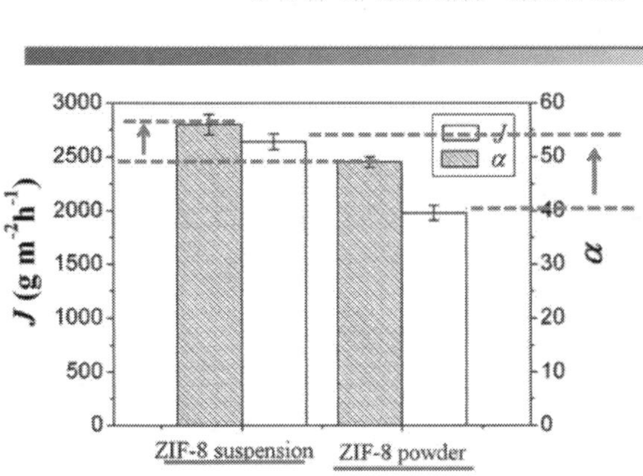

ZIF-8/PDMS nanohybrid membrane

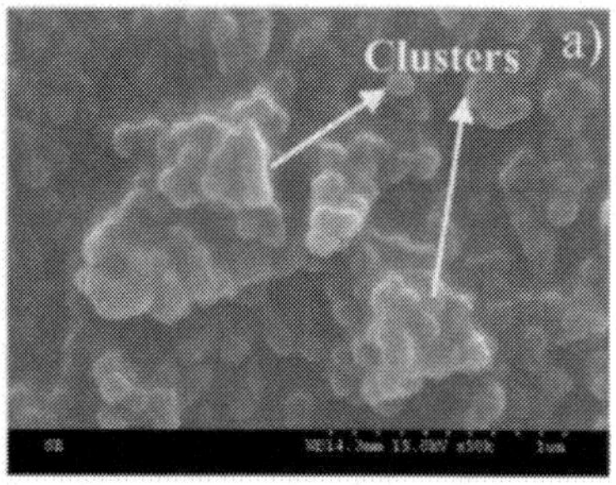

Powder-dispersed **ZIF-8/PDMS hybrid membrane**

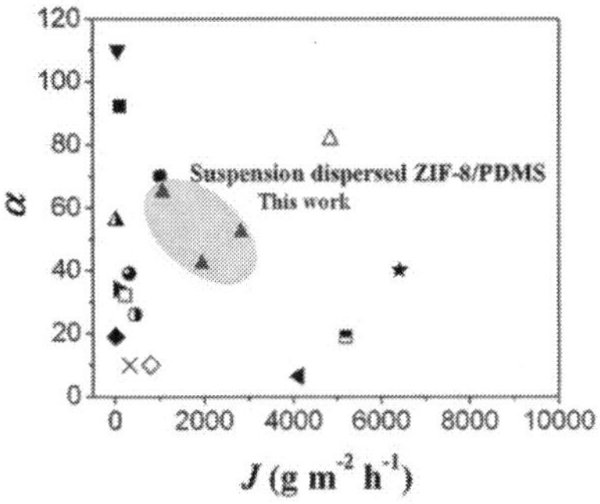

- **Nanodisperse ZIF-8/PDMS membrane exhibited an anti-tradeoff phenomenon**
- **Under the optimal conditions, the suspension-dispersed ZIF-8/PDMS membrane exhibted a high separation factor (52.81) and flux (2800.5 g m^{-2} h^{-1})**

Research contents

- Preparation of nanodisperse ZIF-8/PDMS hybrid membrane by suspension-dispersed method and followed a defects-removal process

- **Sonication-enhanced in situ assembly of well-dispersed SiO_2/PDMS hybrid membrane**

Experimental Section

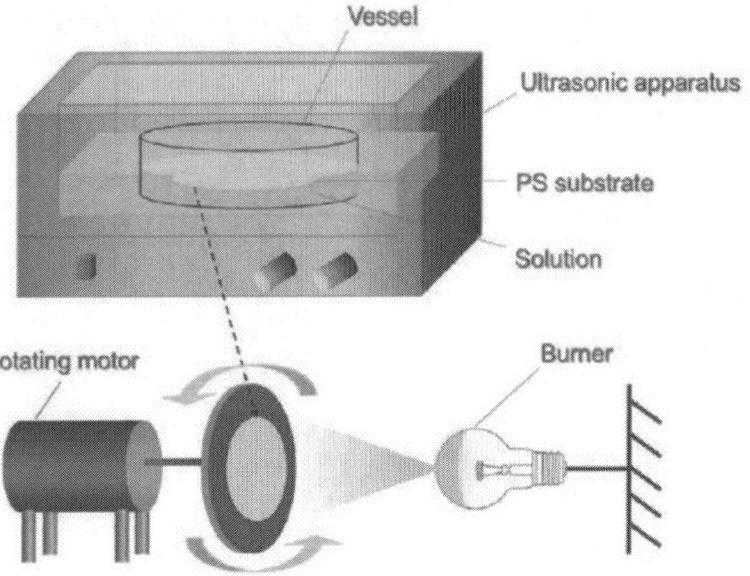

Sonication-enhanced in situ assembly **of the organic/inorganic hybrid membrane process**

Hao Yan, Jie Li, **Hongwei Fan**, Shulan Ji, Guojun Zhang*. ***J. Membr. Sci.*,** 2015, 481. 94–105

Experimental Section

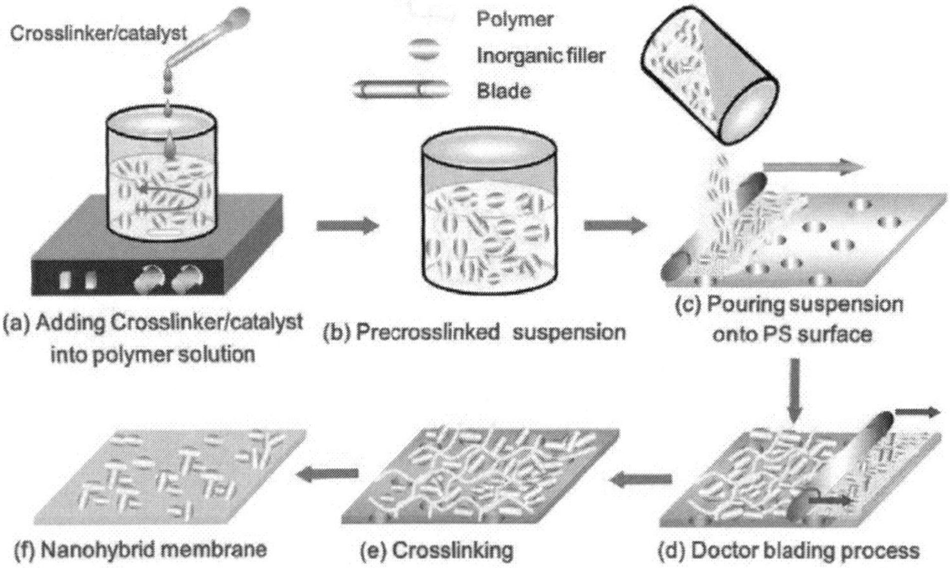

Schematic diagram of the hybrid membranes produced by the traditional method

Secondary aggregation of the particles

Characterization

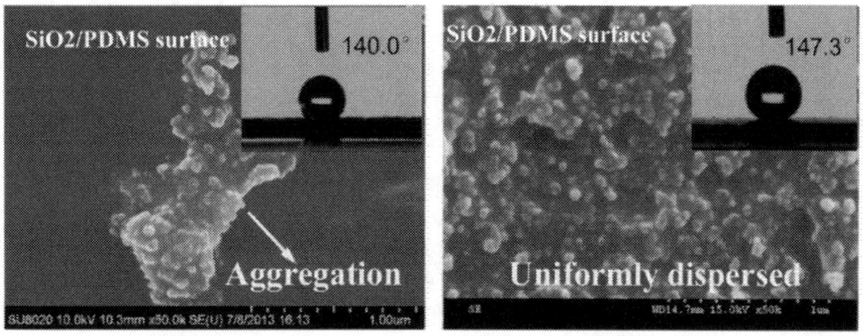

Traditional method **Sonication-enhanced in situ assembly**

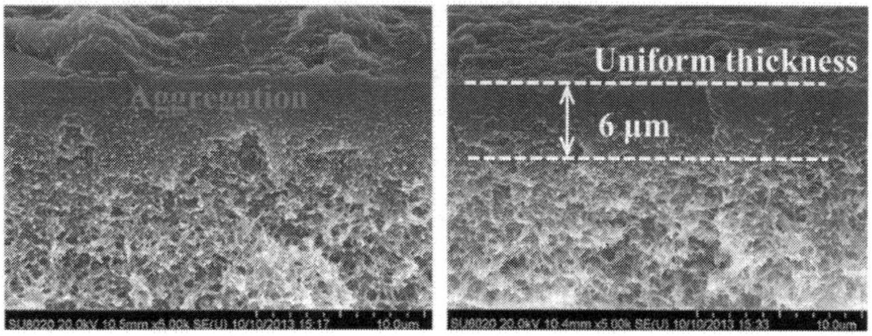

Traditional method **Sonication-enhanced in situ assembly**

Characterization

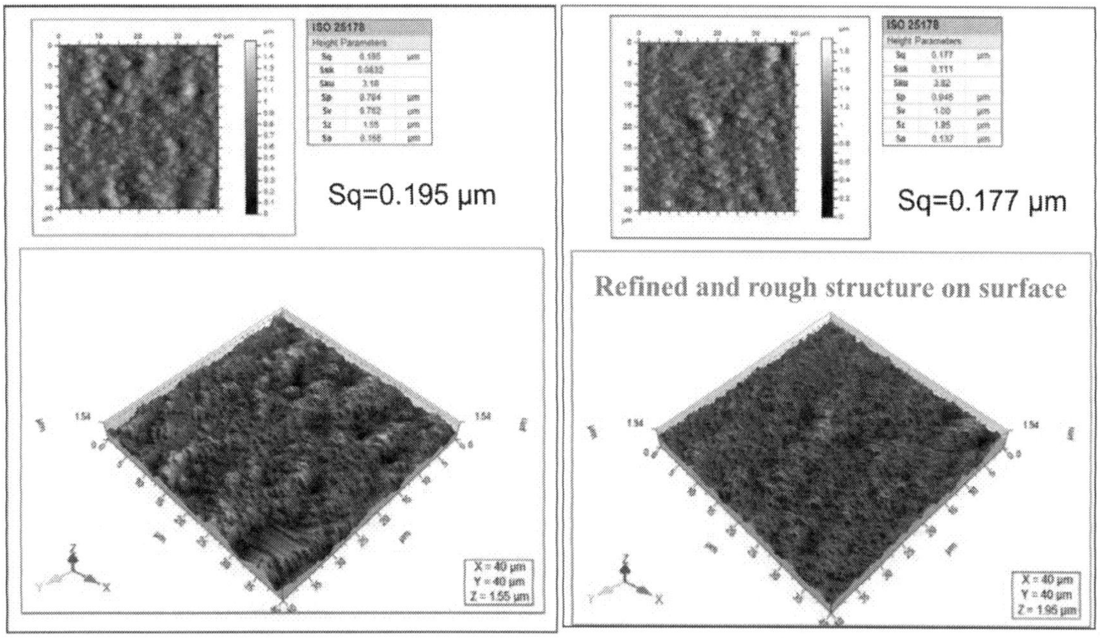

Traditional method **Sonication-enhanced in situ assembly**

Characterization

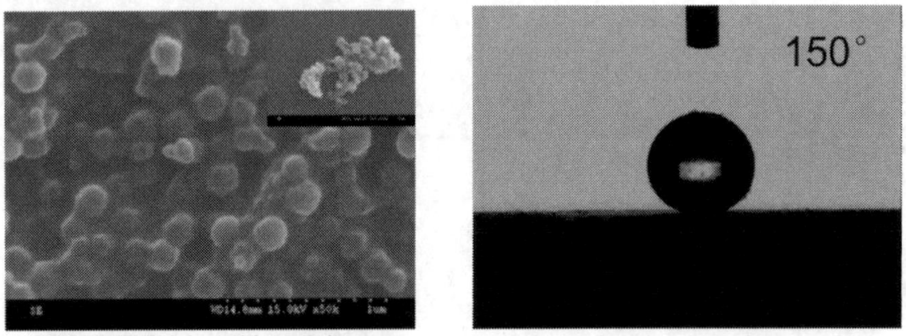

PDMS/ZIF-8 hybrid membrane

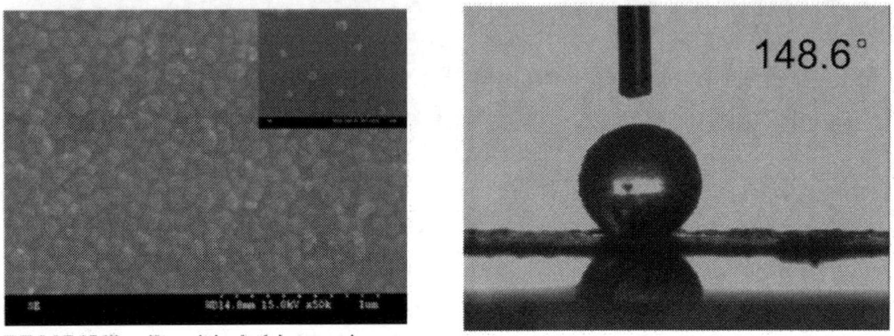

PDMS/Silicalite-1 hybrid membrane

Results and Discussion

What is the function of the ultrasonication in membrane-forming process?

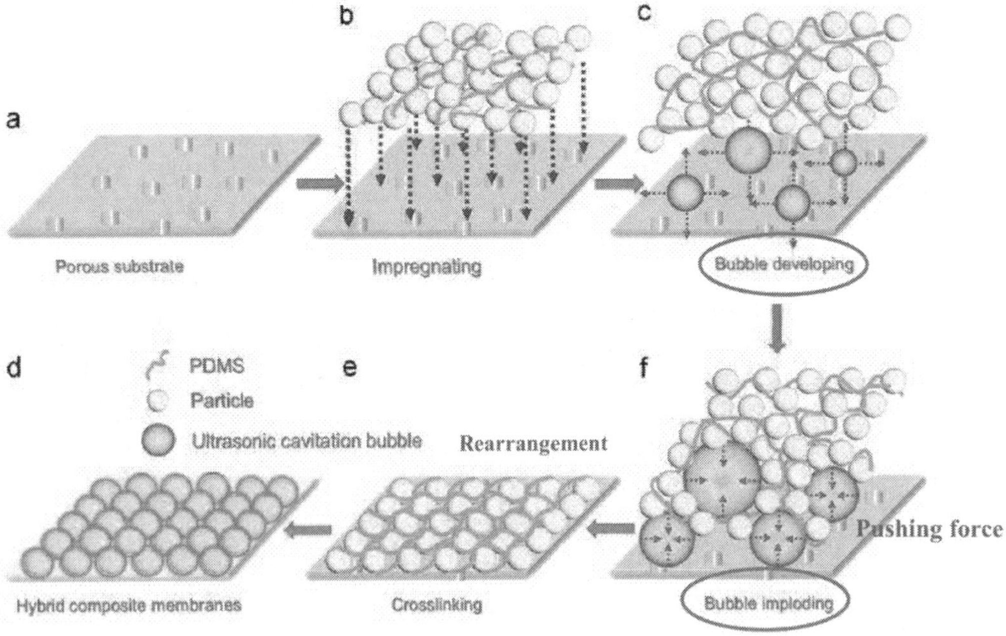

Results and Discussion

Separation of ethanol/water mixture by pervaporation process (Feed: 5 wt.% ethanol/water at 60°C)

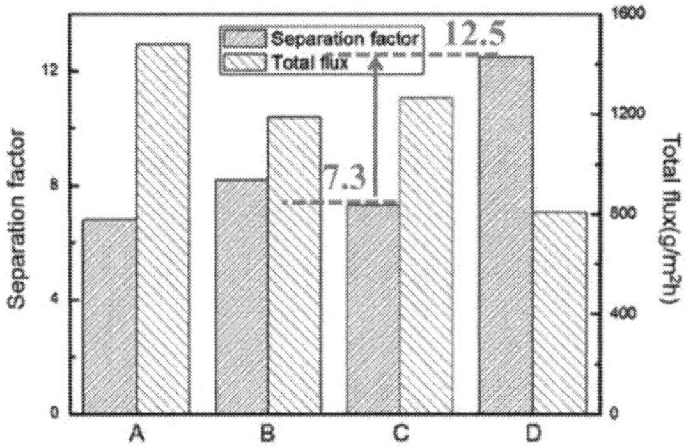

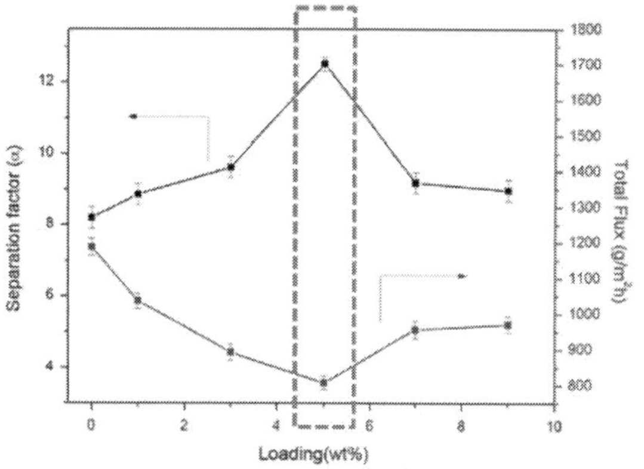

The effect of ultrasonic cavitation on membrane performanc

(C): PDMS/SiO2 hybrid membrane produced by traditional method;

(D): PDMS/SiO2 hybrid membrane produced by sonication in situ assembly.

The separation factor of SiO2/PDMS membrane for the pervaporation of 5% ethanol/water mixture increased from 7.3 to 12.5 after sonication

Results and Discussion

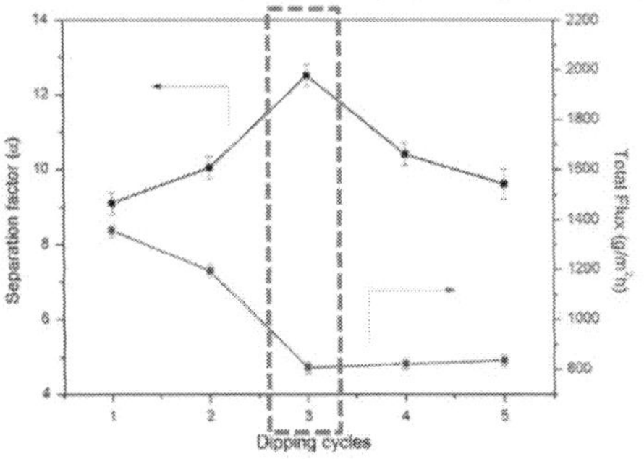

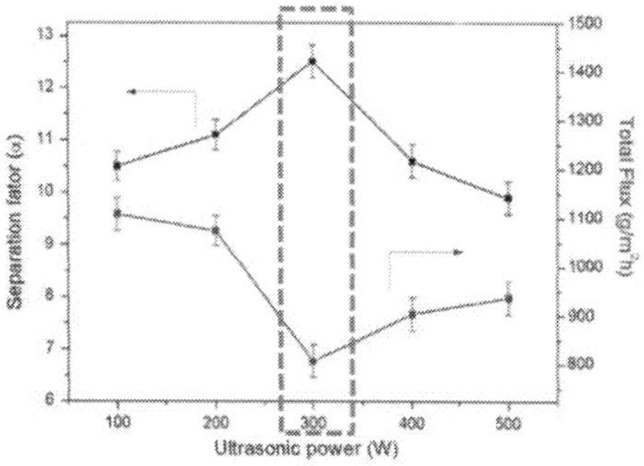

Prevaporation performance of different pervaporation membranes in ethanol/water mixtures.

Membrane	Feed concentration (wt%)	Temp (°C)	α	J (g/m² h)	Thickness (μm)	Test method of thickness	Water CA	Ref.
PDMS /Fumed silica	5	40	19	200	–	–	105	39
PDMS/Silicalite-1	6	50	20	145	150	Screw micrometer	116.5	40
PDMS/Nano-sized silicalite-1	6	35	14.3	330	100	SEM	–	41
PDMS/ZSM-5	5	50	16.7	85	70	SEM	–	42
PDMS/ZSM-5	5	40	14	408	10	SEM	138.7	43
PDMS/ZSM-5	5	50	12	144	100	–	–	44
PDMS/Carbon black	6	35	9	51	–	–	–	45
PDMS/Zeolite Y	6	35	4.5	750	100	–	–	46
PDMS /Fumed silica	5	60	12.5	807	6	SEM+EDX	147.3	This study

- **Optimal condition: SiO_2 loading, 5wt.%; 3 dipping cycles; Ultrasonic power, 300w.**
- **The PDMS/SiO2 exhibited a relatively high separation factor of 12.5 and flux of 807 g/m^2h**

Conclusions

- Directly dispersing a nascent ZIF-8 suspension in PDMS solution was confirmed an effective approach to overcome nanoparticle agglomeration during hybrid membrane formation;
- A homogeneous ZIF-8/PDMS nanohybrid was obtained by repeated immersion of the PS supporting membrane in a dilute ZIF-8/PDMS solution followed by removal of defects using a concentrated PDMS solution (post-treatment);
- Dispersion of the inorganic filler in the polymeric matrix in the nanoscale range could be markedly improved because secondary aggregation of nanoparticles on the supporting membrane could be avoided by sonication.

Acknowledgements

- **National High Technology Research and Development Program of China (No.2012AA03A607)**
- **Program for New Century Excellent Talents in University Ministry of Education, China(No. NCET-12-0604)**
- **Doctoral Fund of Innovation of Beijing University of Technology**

Prof. Guojun Zhang

Center for Membrane Technology

15AIChE
Annual Meeting, Salt Lake City, UT

Any Comments are welcome !

November 11, 2015 Salt Lake City

430437 Effects of MCM-41 Silica Nanoparticles on the Performance of Thin Film Nanocomposite Membranes for Reverse Osmosis

Abstract:

Reverse osmosis (RO) has been increasing applied for water desalination since the successful development of thin film composite membranes (TFC) by Cadotte three decades ago. In this work, a series of TFC membranes were prepared by reacting m-phenylenediamine (MPD) solution containing triethylamine/ camphorsulfonic acid salt (TEA/CSA salt) and calcium chloride with trimesoyl chloride (TMC) solution on a polysulfone (PSU) support sheet. The PSU support sheets were prepared by dissolving PSU in N,N-dimethylformamide (DMF) and then casting on a glass plate. In particular, isooctane with an intermediate boiling point was used for the first time as a solvent to dissolve TMC, which helped the membrane fabrication process. The TMC/isooctane solution contained an increasing amount of MCM-41 spherical silica nanoparticles (0.01%, 0.015%, 0.02%, 0.025%, 0.03% and 0.04%) during the fabrication of the thin-film nanocomposite. The performance testing demonstrated that the water flux was increased for all membranes with the silica nanoparticles though the salt rejection was often decreased slightly. With the addition of 0.015% of the silica nanoparticles, the water flux was increased from 44 to 61 L/m2hr while the rejection of NaCl was maintained essentially the same at 95%.

422417 Highly Porous Ti_4O_7 Reactive Electrode Membranes for Water Filtration

Abstract:

Previous work has demonstrated the effectiveness of Ti_4O_7 ceramic reactive electrochemical membranes (REMs) for contaminant oxidation and fouling prevention during water filtration, where contaminants/foulants are destroyed *via* electrochemical anodic oxidation at the surface of the membrane. However, these studies used low-porosity rigid ceramic REMs with limited configurations. To expand the potential applications of Ti_4O_7 REMS, in this study, highly porous and flexible REMs were fabricated *via* a simultaneous electrospinning/electrospraying (E/E) technique. The resulting membranes from this method produced a network of poly(sulfone) (PSU) nanofibers and Ti_4O_7 particles with intimate contact between both materials throughout the porous matrix, where changing the various electrospinning and electrospraying parameters and solution properties allows for membranes with various porosities, fiber sizes, and particle sizes. To date, few studies have explored the E/E method to develop membranes for water purification and to our knowledge previous studies have not utilized the E/E method to develop REMs. Characterization of the E/E REMs was performed using X-ray diffraction (XRD), scanning electron microscopy (SEM), and thermal gravimetric analysis (TGA). The electrochemical properties and water filtration performance of these E/E REMs were evaluated using electrical impedance spectroscopy (EIS), cyclic voltammetry (CV), and membrane filtration studies. Filtration results demonstrate that E/E REMs exhibited a significantly higher water flux than previously investigated rigid REMs along with an order of magnitude higher contaminant removal rate.

414870 Adsorptive Removal of Arsenic from Aqueous Solution By a Zirconium Ion Impregnated PVDF Membrane

Abstract:

Abstract:

Arsenic contamination in drinking water and groundwater has become an important environmental issue due to its acuteness and chronic toxicity. In this study, a novel zirconium ion impregnated PVDF membrane was developed for arsenic removal from aqueous solution. The zirconium ions were impregnated within the PVDF membrane by coordinated with ether and hydrogen bond which were produced by crosslinking of PVA (polyvinyl alcohol) and glutaraldehyde. The coating process was optimized by a Box-behnken experimental design approach. The kinetics study showed that the adsorption equilibrium was established within 24 h. The adsorption increased as the solution pH was decreased; the optimal pH for adsorption was 2.0. The experimental data were better described by Langmuir equation than Freundlich equation; the maximum adsorption capacity of 128 mg-As/g was achieved at pH of 2.0. The adsorption was not reduced in the presence of phosphate. The presence of humic acid, fluoride, carbonate and silicate had some negative effects on the adsorption. In the filtration study, 12.56 cm^2 membrane had 15.6 L arsenic solution to meet the maximum contaminate level of 10 ppb when treating a simulated arsenic contaminated water with an initial concentration of 97.6 ppb. The adsorption mechanism of adsorption was mainly associated with ion exchange between chloride and arsenic ions.

Key words: PVDF membrane, zirconium, PVA coating, arsenic, adsorption, filtration

415944 Mixed Matrix Membranes with Impermeable Water-Based Matrix for Water Recovery

Abstract:

Clean drinking water is necessary to human life and secures a reasonable standard of living for populations. Current fresh water supplies are under great pressure with increasing demand and decreasing supply. Oceans and wastewater are excellent candidates for new water supplies. Available osmotic processes have great potential for effective water recovery. However, current commercially available polyamide membranes degrade quickly with exposure to chlorine.

This presentation will focus on a new design of mixed matrix membranes with high chemical resistance and high rejection for small molecules [1]. In our mixed matrix membranes, a liquid-barrier chemical-resistant polymer matrix binds a layer of selective molecular sieve particles into a versatile composite film.

The main distinguishing characteristics of our new membrane design compared with current mixed matrix membranes include: (1) we use an impermeable polymer matrix enabling a broader range of chemical resistant polymers; (2) we use latex dispersions as the polymer matrix source. Latexes are environmentally friendly water based-solutions, sharing the qualities of low volatile organic compound (VOC), low cost, and non- toxicity.

We anchored 500–600 nm diameter zeolite particles on porous alumina supports by using silane covalent linkers to prevent forming an impermeable polymer layer between the molecular sieves particles and porous substrates. Then, we formed 400-500 μm thick PVDC films via dip-coating onto the substrate with the anchored molecular sieves. Next, we performed oxygen plasma to remove impermeable PVDC film from the surfaces of the molecular sieves. We report on flux and sodium chloride rejection of brackish water at 24 °C. We also will present a variety of characterizations of mixed matrix membrane morphology.

Reference

1. Nanocomposite membranes , Lind, M. L. Fink, S., Sasaki, K., Lin, J., USA, WO 2014159352 A1, 2014

422124 Brushing Off Salt Using a New Class of Synthetic Membranes

Abstract:

A new class of hydrophobic synthetic brush membranes are designed and tested. Tethered polymer chains or brushes have attracted enormous interest recently for both engineering and biomedical applications such as to stabilize colloids, to reduce friction between surfaces, to design novel fluidic devices, and to fractionate proteins in chromatographic columns. Here we tested them for desalination, which is an urgent challenge and an opportunity to meet future world water demand. The membranes were prepared by graft-induced polymerization of hydrophobic vinyl monomers onto light sensitive poly(ether sulfone) nanofiltration support membranes, without initiating agents, using atmospheric pressure plasma. Several hydrophobic vinyl monomers were tested to investigate the effect of chain length, the presence of linear versus branched chains, etc. Membranes were characterized by using several techniques, such as ATR-FTIR, Atomic Force Microscopy (AFM) and Quartz Crystal Microbalance with Dissipation (QCM-D) monitoring to characterize the brush layer in terms of graft density, length and viscoelastic properties. The modified membranes were finally tested in a high-pressure filtration unit against salt aqueous solution that reproduced seawater salinity. The main novelty here is the use of hydrophobic brush-like structures as a selective skin or dense layer attached to a non-selective support membrane. Possible advantages of these new desalination membranes include (i) ease and speed of synthesis as compared with interfacial polymerization and phase inversion, (ii) scalability, (iii) low cost production, and (iv) high fluxes due to hydrophobic resistance layer. The optimization of these easy-to-synthesize and scale-up membranes could guarantee an alternative to traditional reverse osmosis membranes, which still remains the major method of choice for desalination of sea water.

434647 Carbon Nanofiber Adsorbent Membranes: Applications in Treatment of Oily Wastewater

Abstract:

Carbon nanofibers nonwoven (CNFs) have been attracting increased attention for a wide variety of applications due to their high porosity and surface area. The low hydraulic resistance of nanofibers make them excellent candidate materials for water filtration, but CNFs have an added benefit of being a high performance sorbent, especially for oil, as a result of CNF hydrophobicity. We present our work on the performance of these unique materials for sorption of oil and removal of other dissolved subsances from oily water. We fabricated our CNFs by thermal treatments of electrospun polyacrylonitrile (PAN). Through a series of optimization studies, we identified how different fabrication procedures can change material properties that could impact material performance as a membrane or a sorbent. Mechanical strength, surface area, and fibers size are some of these properties. We performed batch dynamic sorption studies as well as normal flow filtration tests using emulsified crude oil. These materials show rapid adsorption kinetics of oils and very high sorption capacity relateive to other conventional sorbents.

435919 Novel Electrodialysis Metathesis (EDM) Relationship to Electrodialysis (ED) for Desalination and Recovery of Minerals from Impaired Water

Abstract:

Electrodialysis Metathesis (EDM) is a novel modification of the ED process consisting of quads as the repeating unit cell and uses the technique that applies double decomposition reaction for the interchange of cations and anions between the feed and a sacrificing solution. EDM has the potential to increase the desalination efficiency of reverse osmosis. It is also a promising technology to treat concentrate with scale formers like $CaSO_4$ and $CaCO_3$ that can further recovered as secondary products with marketable value. In doing this, augmenting water and reducing disposal cost can be possible through EDM technology. Current voltage relationship determines the resistance behavior in ED, however there is still knowledge gap to understand the current voltage relation in EDM. In this research we determined the current voltage relationship for the EDM process as a function of increasing number of quads, varying feed concentrations for a given superficial flow velocity as well as batch and continuous mode. We found that current density is directly related to the number of cell pairs and quads for ED and EDM, respectively, and the resistance increased with increasing cell pairs/quads. This current - voltage relationship study would help to provide the range of voltage for optimum energy usage in the EDM process.

412921 Propane and Propylene Separation By PSA Using a Novel Carbon Molecular Sieve

Abstract:

Cryogenic distillation is the current method for separating propylene/propane mixtures. It is a difficult separation requiring high reflux ratios, high energy input, and tall columns, because of the small difference in the volatilities of propylene and propane. Economical alternatives for the separation of propylene/propane mixtures are highly desirable, particularly for smaller streams where the contained propylene has significant value, but not enough to warrant additional distillation capacity. Adsorption based processes have been explored in recent years as potential replacement of or augment to the incumbent cryogenic distillation process to reduce the energy intensity or debottleneck existing assets. To make the adsorption processes economically advantageous, a highly selective adsorbent is the key since propylene is the most adsorbed component and is recovered in the effluent during adsorbent regeneration.

We have discovered new carbon molecular sieves (CMS) that show a high separation factor for olefin and paraffin, particularly for propane and propylene. The adsorption isotherms of propylene and propane on the CMS were measured at different temperatures that cover the commercial operating range. A temperature dependent, four-parameter Langmuir model was selected to fit the single component isotherm data and the model parameters were obtained by using the Aspen ADSORPTION^{TM} equilibrium parameter estimation tool. The heat of adsorption of each component was determined from the isotherm models according to the Clausius – Clapeyron equation. The adsorption kinetics of propylene and propane on the CMS were studied by fixed bed adsorption breakthrough experiments. Breakthrough curves were obtained at different feed flow rates. The mass transfer coefficients of propylene and propane were estimated using Aspen ADSORPTION^{TM} simulations.

The new CMS was experimentally investigated for its propane and propylene separation performance using a fully automated bench-scale PSA unit. The propylene enrichment, recovery and propane purity were measured for each PSA experiment. An Aspen ADSORPTION^{TM} model was developed for this PSA process and was validated against experimental data. The validated PSA model can then be used for future process optimization and scale-up.

Evaluation of Thermal Swing Adsorption for the Removal of Toxic Vapor Emissions from Condensate Storage Tanks

Oluwatosin A Oyelakin, Jesus Prado, Kaveh Farhadi, David Ramirez*

Department of Environmental Engineering, Frank H. Dotterweich College of Engineering, Texas A&M University - Kingsville, Kingsville Texas

Abstract

Oil and gas production from wellheads at shale formations generates toxic vapors during initial stabilization of flow between the producing wells and transportation to refinery or at processing facilities. Other important source of toxic vapors is during the temporary storage of oil in condensate storage tanks. Emission of toxic gases during tank venting is known to contribute significantly to toxic vapor emissions (TVEs) released into the environment as losses which are primarily classified as breathing, working and flash losses. This research focused on the design and testing of a thermal swing adsorption system to capture and recover vapors released into the environment through tank venting.

A bench scale thermal-swing adsorption (TSA) systems was designed, tested and operated at temperatures of 25°C and 43°C to capture and recover toxic vapor release from crude oil and condensate storage tanks. Benzene, toluene, ethylbenzene, and xylene isomers (BTEX) are examples of toxic vapor emitted from storage tanks which represent about 10% of the mixture of crude oil and condensate sample obtained at a well site (API is 54.2). BTEX adsorption onto a commercially available granular activated carbon was studied at different gas flow rate ranging from 1 sLpm to 6 sLpm, after which the bed was regenerated.

Experimental breakthrough curves for the vapor stream containing 1,000 ppmv BTEX resulted in adsorption capacities for toluene, benzene, m, p-xylene, o-xylene and ethylbenzene on the activated carbon of 215 ± 7 mg/g, 183 ± 5 mg/g, 56 ± 2 mg/g, 9 ± 2 mg/g and 3 ± 0.3 mg/g, respectively with overall liquid recoveries of 55-60% at 25°C. Different from existing TSA systems, the bench scale TSA system had one adsorber and operated for continuous treatment of the TVEs in four different phases namely; adsorption, desorption, regeneration with nitrogen and liquid recovery. Most of the BTEX vapor (>90%) was removed during the adsorption phase of the TSA cycle and before breakthrough started to occur which is an indication that the bench scale TSA setup was effective for the removal of TVEs.

Introduction

Unconventional energy exploration in United States is trending due to innovative drilling technology that provides the ability produce natural gas from shale in huge volume. The economic impact it offers has invariably change the discussion about the energy exploration across all tiers of government (Ladislaw et al., 2013). Currently, unconventional exploration provides 11.5% of the total energy consumed and 46% of the non-renewable energy sources in the U.S (McLeod et al., 2014). Evolving technologies such as downhole equipment and directional drilling that allow for more cost-effective drilling, hydraulic fracturing, and development practices are expected to keep natural gas production on the rise through the year 2035 (McLeod et al., 2014; Ladislaw et al., 2013; Koch, 2013; MIT Energy Initiative, 2011). While natural gas exploration continues to rise, the development of unconventional oil and gas resources may present environmental risks since air pollutants may be emitted during various stages of oil and natural gas production. Particulate matter, nitrogen oxides, sulfur dioxides, methane, volatile organic compounds, and various hazardous air pollutants are some of the key contaminants that may emanate throughout the oil and gas production process (Ratner and Tiemann, 2015; Lattanzio, 2013; Field et al., 2014; Alvarez and Paranhos, 2012). Furthermore, the EPA has stated that the industry is a major source of methane and VOC emissions, which react with nitrogen oxides to form ground-level ozone, or smog (Ratner and Tiemann, 2015). These compounds may impair local air quality and consequently impact the overall health of people – particularly those living in or around drilling sites.

As with other energy sources or fuel production activities, several studies on emission monitoring at wellheads of drilling sites have shown potential emissions release from oil and gas activities during the exploration process. Notable emission sources include pad, road, and pipeline construction; well drilling and completion, and flowback activities and natural gas processing, storage and venting, and transmission equipment (Ratner and Tiemann, 2015; Lattanzio, 2013; Field et al., 2014; Alvarez and Paranhos, 2012). Storage tanks are reported to be one of the largest sources of VOC emissions from an oil and gas production activities at drilling sites in Texas (TCEQ, 2012). Toxic vapors of interest in this study include benzene, toluene, ethylbenzene, and xylene isomers (BTEX).Toxic vapors may enter the atmosphere through tank venting. These emitted vapors comprised of lighter hydrocarbons and gasoline will cause a reduction in the gravity of the oil, thereby decreasing its value. Not only is this an economic disadvantage, but a safety hazard since this combustible mixture is formed near the tank battery.

The vapor recovery unit (VRU) is the TVE control technology most widely used in condensate storage tanks. VRUs can be simple to operate and achieve pollutant removal efficiencies over 90% (USEPA, 2006; USEPA 2010). A VRU system continuously maintains the pressure within the tank at safe operating parameters (USEPA, 2006; USEPA 2010; Richards, 2013).

Biofiltration, membrane technology, thermal oxidation, absorption, and adsorption are some technologies that have been explored for the control of vapor emissions (Ambrozek and Zwarycz-Makles, 2014; Cooper and Alley, 2011). Of these control technologies, incineration and biofiltration are categorized as destructive, while the others are labeled as recuperative methods (Lee et al., 2010). Currently, adsorption technology is the most widely used technology for the capture and recovery of volatile organic compound (VOC) emissions (Ambrozek and Zwarycz-Makles, 2014). The effectiveness of the adsorption process generally relies on the physical, chemical, and electrical properties of the adsorbent. Activated carbon is the most widely used adsorbent due to its unique physical properties that allow for the effective treatment of vapor and aqueous phase pollutants (Bansal and Goyal, 2005). The thermal-swing adsorption (TSA) system developed in this study can capture more than 95% of TVEs. The TSA system is also equipped with a condensation unit where the TVEs can be recovered as liquid. The development of the TSA system was based on a prior quantification and characterization study conducted from a sample containing a mixture of crude oil and a condensate (API 54.2) obtained from a well site at the Eagle Ford Shale. Table 1 shows the results of sample analysis that contained 9.93% of BTEX, a subgroup of VOCs, and provides a list of common organic vapors found from characterization experiments. The relative composition does not total 100 %t as some chemicals were not listed.

Table 1. Relative chemical composition of the crude oil sample used in this study.

No.	Compound	Relative Composition (%)
1	Benzene, 1,2,4-trimethyl-	0.184
2	o-Xylene	0.378
3	Pentane, 2-methyl-	1.29
4	Benzene	4.92
5	Benzene, 1,2,3-trimethyl-	0.085
6	Benzene, 1,3-diethyl-	0.024
7	Benzene, 1-methylethyl	1.14
8	Butane, 2,2-dimethyl	0.825
9	Butane, 2-methyl	0.966
10	Decane	0.076
11	Ethylbenzene	0.184
12	Heptane	20.7
13	Heptane, 3-methyl-	5.14
14	Hexane	1.55
15	Hexane, 2-methyl	1.14
16	Hexene	1.79
17	Nonane	0.528
18	Octane	2.18
19	Pentane, 3-methyl-	0.83
20	p-Xylene	0.184
21	Styrene	0.714
22	Toluene	4.27

Source: Oyelakin and Ramirez 2015

Despite developments of TSA systems, studies of the adsorption of multiple pollutants are still limited (Wang et al., 2011; Cortés et al., 2009; Sharma and Wankat, 2010). Nonetheless, few papers are available that addresses optimizations and mathematical models of the TSA process (Ambrozek and Zwarycz-Makles, 2014). This paper focused on the effect of the temperature on the equilibrium capacity of BTEX mixtures on granular activated carbon and the performance of a bench-scale TSA system.

Methodology

The operation of the bench scale TSA unit occurred in four different modes: initial stage as offline, followed by online, regeneration, and liquid recovery. A schematic of the bench-scale TSA system is shown in Figure 1. In the offline mode, the TSA unit generated a BTEX-laden gas stream at specified concentrations. In the online mode, the generated BTEX gas stream was conveyed into the adsorption column. The control system turned on the adsorption process. The first few effluent concentrations monitored were almost zero as most of the pollutants were adsorbed by the adsorbent. Over time, the adsorbent approached saturation and the effluent BTEX concentration downstream of the adsorption column began to rise.

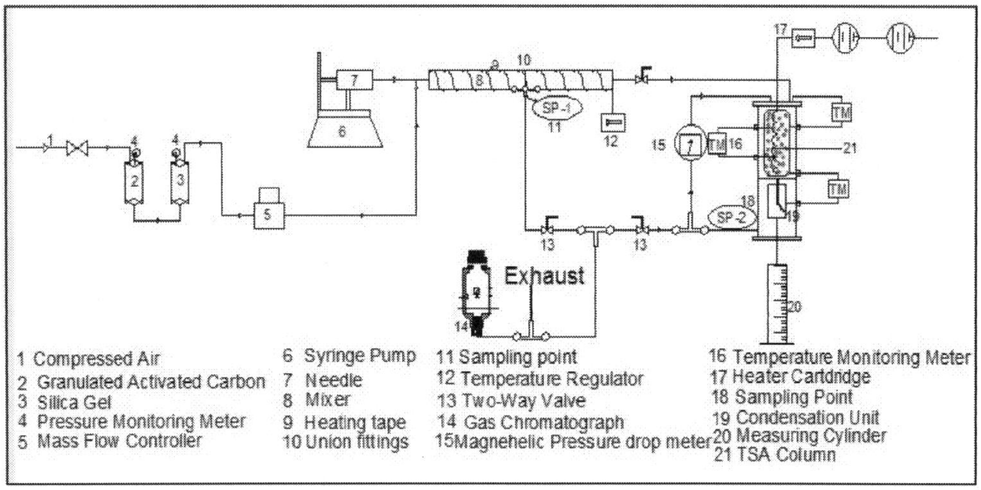

Figure 1: Bench Scale Schematics for the TSA System Developed

The regeneration mode involved passing nitrogen gas through the adsorption column upon saturation. The flow of nitrogen gas helped to purge the oxygen from the adsorption column to then desorb BTEX from the adsorbent. In the liquid recovery mode, the desorbed vapor formed a concentrated stream inside the vessel that condensed as liquid.

Adsorption capacity (q_e) of the HAPs adsorbed on the GAC was estimated using equation (1)

$$q_e = \sum \frac{Q(Co - Ct)\Delta t}{m} \tag{1}$$

Where Q is the gas flow rate for the pollutant (sLpm), C_o is the initial concentration of the pollutant (ppm), C_t is the outlet or effluent concentration of the pollutant during adsorption (ppm), m is the mass of the adsorbent (g) and Δt is the time step (min).

The GAC characterization involved the use of scanning electron microscope (SEM) to obtain the morphology of the GAC structure through an images produce. The SEM produces the image by focusing beams of electrons on the e surface of the GAC samples thereby producing a scan images. The sample preparation involved coating the GAC sample with alloy of gold metals in other to enhance conductivity which is done in the sputter coater before the sample is placed on the sample holder and later placed in the SEM upon which the vacuum pump is switched on. The SEM image is obtained by changing the X and Y direction so as to obtain a clearer image.

Results

Results: Granular activated carbon characterization using scanning electron microscopy

The morphology and the pores size distribution of the granular activated carbon (GAC, TIGGS 5CC) was investigated by using scanning electron microscope (SEM, Jeol Neoscope JCM 6000) and Fiji image J software for the pore size distribution analysis. Figure 2 illustrates the morphology of the granular activated carbon used before and after the experiment after which the carbon was regenerated.

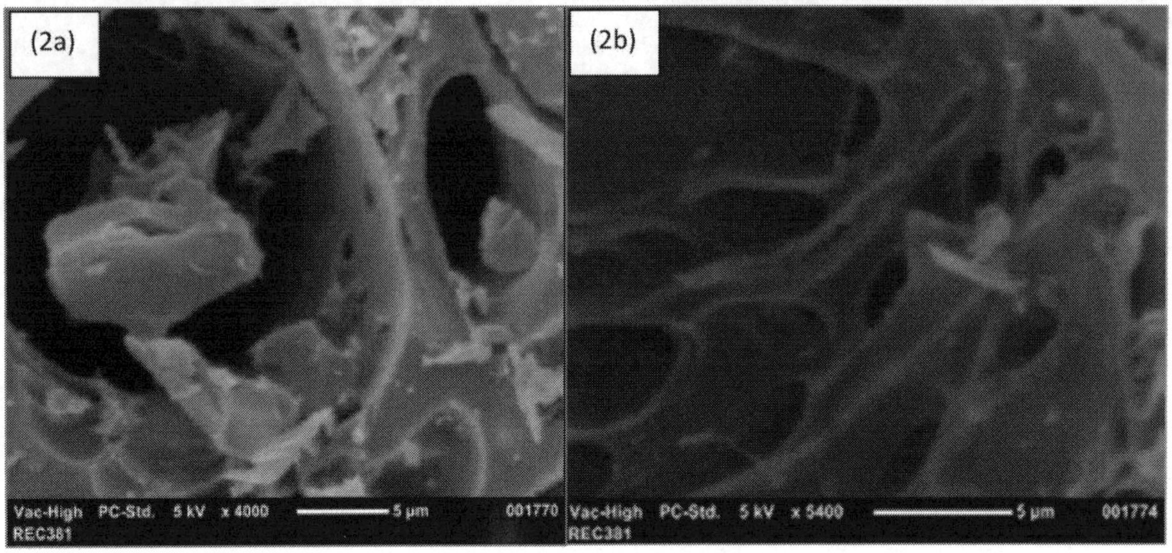

Figure 2: SEM images of a) filter fresh GAC, and b) filter regenerated GAC sample

Structural alteration in the morphology can be noted in Figure 2-b as compared to Figure 2-a. This change in morphology might due to effect of the regeneration temperature as there appears to be larger and more pores in the fresh sample (Figure2-a) compared to the regenerated GAC sample (Figure 2-b). Similar observation was noted when the samples were tested without filter , that is without passing the sample through a sieves to obtained a uniform surface (Figure 3).

Figure 3: SEM images of a) fresh GAC sample, and b) regenerated GAC sample without filter

Figure 4 shows the histogram for the particle size distribution of fresh and regenerated GAC samples when passed through 400 μ sieves. The samples had mean particle sizes of 0.178 μ for the fresh GAC sample and 0.144 μ after the GAC sample was regenerated.

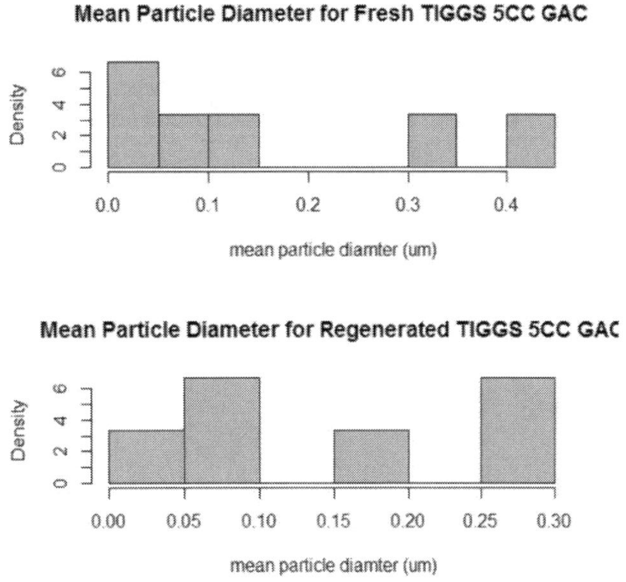

Figure 4 : Particle size distribution of the fresh and regenerated GAC samples

Sample filtering which implies passing the GAC sample to be analyzed through 400μm sieves screens was hypothesized to impact the particle size distribution. Figure 5 displays the histogram of two fresh GAC samples when either of the samples was filtered. The mean particle sizes of the filtered and unfiltered GAC samples were 0.163 μ and 0.157 μ, respectively.

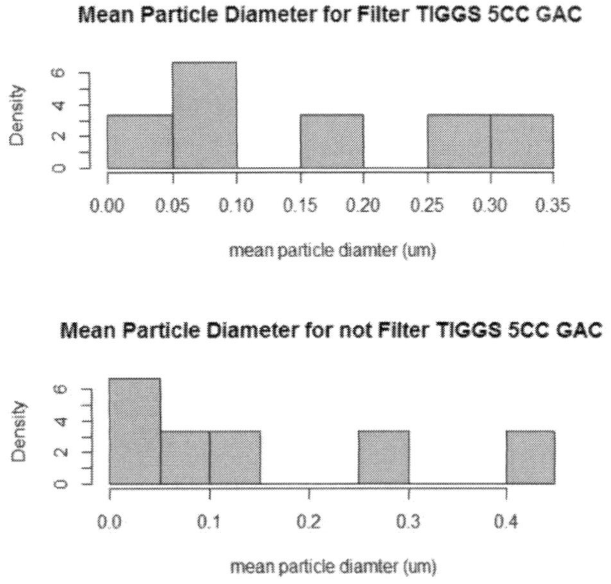

Figure 5: Effect of filtering on particle size distribution of GAC

The mean particle size of the GAC was reduced by 20 % after regeneration when the sample was filtered. About 4% reduction of particle size was measured when a fresh GAC sample was filtered. This implies filtering of GAC sample before testing with SEM has no significant impact on the morphological characterization. However, the 20 % reduction in mean pore size of the GAC is an indication that 80% of the sorbate were desorbed from the adsorption pore site.

Results from Figures 4 and 5indicated that the granular activated carbon utilized in this study had heterogeneous adsorption sites which may selectively adsorb adsorbate of similar molecular diameter in proximity with pore size on to the adsorption sites.

Effect of temperature on the equilibrium adsorption capacity

Adsorption capacity can be affected by factors such as gas temperature, adsorbate physical properties and concentration of the adsorbate in the gas stream. This study evaluated the impact of the gas temperature on the amount of adsorbate that can be adsorbed on a unit weight of the GAC used. Figure 6 illustrates the effect of temperature on the equilibrium adsorption capacities of BTEX on GAC.

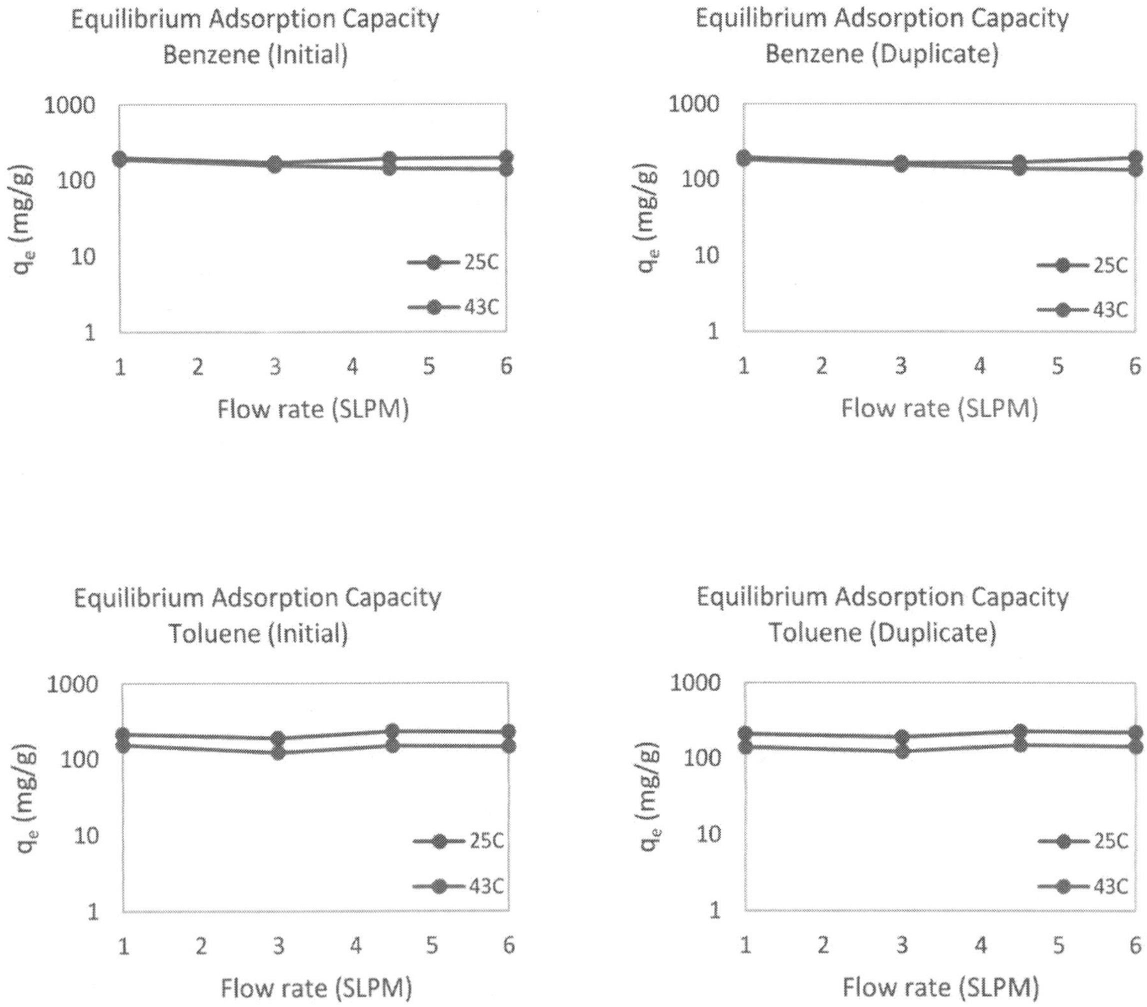

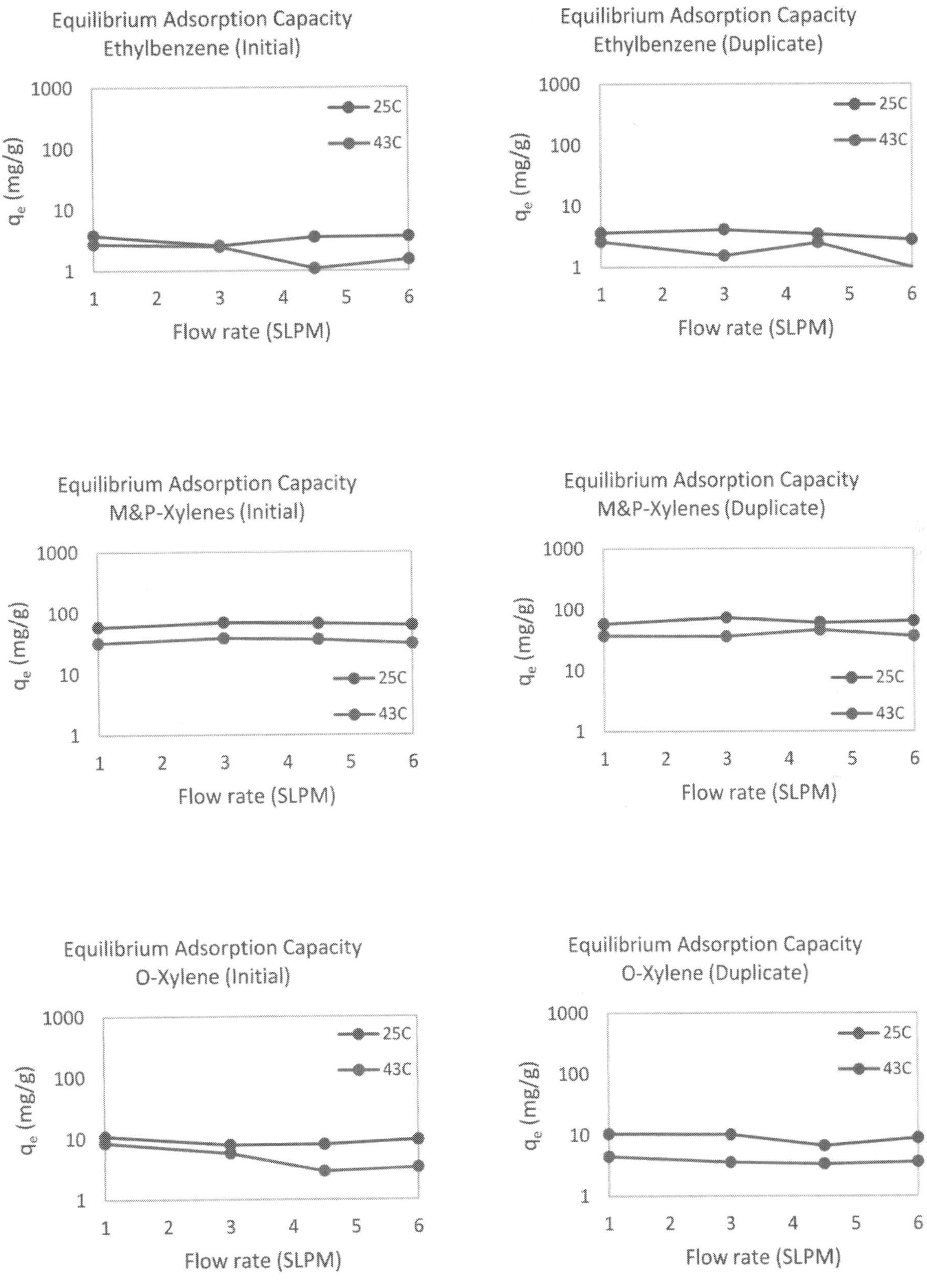

Figure 6. Effect of temperature on equilibrium adsorption capacity at various gas flow rates

At 1slpm of inlet gas flow rate, the equilibrium adsorption capacities for benzene was reduced by 5.05 % and 5.76% when the temperature increase from 25°C to 43°C., toluene percent reduction went from 28.14% to 32.89% , ethylbenzene change from 27.47% to 28.34, while m,p-xylene and o-xylenes were 45.23% to 35.96% and 22.04% to 57.22% respectively. Table 2 shows complete percent reduction equilibrium adsorption capacities at different two different temperature rates during the operation of the TSA systems.

Table 2.Percent reduction in equilibrium adsorption capacities from 25°C to 43°C

	1 slpm		**3 slpm**		**4.5 slpm**		**6 slpm**	
Components	Initial	Duplicate	Initial	Duplicate	Initial	Duplicate	Initial	Duplicate
BENZENE	5.05	5.76	8.39	6.93	25.38	17.16	29.77	30.03
TOLUENE	28.14	32.89	35.11	35.23	35.48	33.79	35.18	34.69
ETHYLBENZENE	27.47	28.34	3.91	62.59	69.41	27.75	58.17	64.52
M&P-XYLENES	45.23	35.96	44.47	50.76	45.27	24.34	50.25	43.58
O-XYLENE	22.04	57.26	27.31	64.91	64.01	49.33	65.03	59.26

Summary and Conclusion

The gas flow rate did not impact the equilibrium adsorption capacity of the carbon to effectively capture the BTEX emissions from the gas streams as previously published by (Oyelakin and Ramirez, 2015). However, the temperature affected the equilibrium adsorption capacity significantly as evident in Figure 6 and Table .2.

Sample filtering of the GAC before it is used for the treatment of TVEs may not be necessary as the particle size distribution did not show a significant change in mean particle size. However, reduction in the mean particle size of the regenerated sample as compared to fresh sample indicates that the BTEX adsorbed can be successfully desorbed at regeneration temperature between 80°C and 105°C and that 80% of the BTEX emission can be desorbed while carbon can further be used for another treatment cycle.

Future work includes comprehensive mass and energy balance analysis to account for the mass recovered in the liquid recovery to the mass desorbed as compared to the mass adsorbed.

Acknowledgement

Special thanks to Houston Advance Researched Center for the financial support for this project. Special acknowledgments to Nida Shaikh and Carlos Hinojosa for their technical assistance.

References

Alvarez, R. A., & Paranhos, E. (2012). Air pollution issues associated. *Air Pollution Issues Associated." EM Feature June.*

Ambrożek, B., & Zwarycz-Makles, K. (2014). Theoretical and experimental studies of the recovery of volatile organic compounds from waste air streams in the thermal swing adsorption system with closed-loop regeneration of adsorbent. *Energy Conversion and Management*, 85, 646-654.

Bansal, R. C., & Goyal, M. (2005). Activated carbon adsorption. CRC press.

Cooper, C. D., & Alley, F. C. (2011). *Air pollution control: A design approach*. Waveland Press.

Cortés, F. B., Chejne, F., Mejía, J. M., & Londoño, C. A. (2009). Mathematical model of the sorption phenomenon of methanol in activated coal. *Energy Conversion and Management*, *50*(5), 1295-1303.

Field, R. A., Soltis, J., & Murphy, S. (2014). Air quality concerns of unconventional oil and natural gas production. Environmental Science: Processes & Impacts, 16(5), 954-969.

Koch, Wendy. "U.S. forecasts natural gas boom through 2040." USA Today. USA Today, 16 Dec. 2013. Web. 22 July 2015 http://www.usatoday.com/story/news/nation/2013/12/16/doe-forecast-natural-gas-boom/4034723/

Ladislaw, S. O., Hyland, L. A., & Pumphrey, D. L. (2013). Realizing the Potential of US Unconventional Natural Gas. Center for Strategic & International Studies.

Lattanzio, R. K. (2013). Air Quality Issues in Natural Gas Systems. Congressional Research Service.

Lee, D. G., Kim, J. H., & Lee, C. H. (2011). Adsorption and thermal regeneration of acetone and toluene vapors in dealuminated Y-zeolite bed. *Separation and Purification Technology*, 77(3), 312-324.

MIT Energy Initiative. (2011). The future of natural gas: An interdisciplinary MIT Study. Massachusetts Institute of Technology, Cambridge, MA.

McLeod, J. D., Brinkman, G. L., & Milford, J. B. (2014). Emissions Implications of Future Natural Gas Production and Use in the US and in the Rocky Mountain Region. *Environmental science & technology*, *48*(22), 13036-13044.

Ratner, M., & Tiemann, M. (2014). An overview of unconventional oil and natural gas: resources and federal actions. *Congressional Research Service Report, 43148.*

Oyelakin O.A & Ramirez D. (2015). Control of Hazardous Air Pollutant from crude Oil and OCndensate Storgae Tanks using thermal Swing AdsorptionSystems. *108th Air& Waste Management Association Raleigh,NC*

Richards, Larry. 2013. *Fundamentals of Vapor Recovery*
http://www.beg.utexas.edu/pttc/archive/LarryRichards.pdf

Sharma, P. K., & Wankat, P. C. (2010). Solvent recovery by steamless temperature swing carbon adsorption processes. *Industrial & Engineering Chemistry Research*, *49*(22), 11602-11613.

TCEQ. 2006. *Technical Supplement 6: Aboveground Liquid Storage Tanks*
http://www.tceq.state.tx.us/assets/public/comm_exec/pubs/rg/rg360/rg-360-05/techsupp_6.pdf

TCEQ. 2012. *Calculating Volatile Organic Compounds (VOCs) Flash Emissions from Crude Oil and Condensate Tanks at Oil and Gas Production Sites.* Vol. 3. Austin.
https://www.tceq.gov/assets/public/permitting/air/Guidance/NewSourceReview/guidance)flashemission.pdf.

USEPA. 2006. *Lessons Learned Installing Vapor Recovery Units on Storage Tanks*
http://www.epa.gov/gasstar/documents/ll_final_vap.pdf

USEPA. 2010. *Reducing Methane Emission with Vapor Recovery on Storage Tanks*
http://www.epa.gov/gasstar/documents/workshops/vernal-2010/06_vrus.pdf

Wang, R. Z., Xia, Z. Z., Wang, L. W., Lu, Z. S., Li, S. L., Li, T. X., ... & He, S. (2011). Heat transfer design in adsorption refrigeration systems for efficient use of low-grade thermal energy. *Energy*, *36*(9), 5425-5439.

435104 New PSA Cycle for CO2 Removal during Closed-Loop Human Space Exploration Missions

Abstract:

The objective of this presentation is to provide modeling and/or experimental details about the development of a novel PSA cycle that can be used for enriching and recovering CO_2 from a spacecraft cabin that is suitable for use in a Sabatier reactor. This novel PSA cycle will utilize a combination of equalization, cocurrent depressurization and heavy reflux cycle steps to facilitate heavy component enrichment and recovery. Preliminary results obtained with the group's dynamic adsorption process simulator (DAPS) revealed that a novel PSA cycle design utilizing 13X zeolite is indeed capable of enriching a dry gas stream of N_2 containing 3500 ppm of CO_2 to over 95 vol% CO_2 at high recovery with a similar removal rate to that of NASA's current 5A zeolite system, which utilizes a temperature swing adsorption (TSA) cycle. Implications from these results potentially reach far beyond the needs of NASA and begin to hint at the extreme concentration of CO_2 from ambient air at relatively high recovery or the extreme concentration of any dilute species from a light gas stream using a simple PSA cycle. This presentation will discuss the most recent results obtained from DAPS and/or a multi-bed PSA system currently being modified for this unique application. This presentation will also discuss the most recent DAPS results obtained with a packed bed of commercial beaded adsorbent and/or a bed filled with a corrugated structured adsorbent that will facilitate higher mass transfer kinetics and lower flow resistance, thereby allowing for higher flow rates with significant bed size reduction.

418049 Evaluation of a Pressure Swing Adsorption Process for Oxygen Generation Utilizing a Rate Enhanced Adsorbent

Abstract:

The rise in usage of portable oxygen concentrators in the medical field as a viable alternative to oxygen tanks has increased the demand for rate enhanced adsorbents that allow for the operation of a rapid pressure swing adsorption process necessary for these devices. Rate enhanced adsorbents are developed specifically to lower the heat and mass transfer resistance of the adsorbent with the aim of reducing the cycle time at which a PSA process can efficiently operate. However, not all rate enhanced adsorbents exhibit the same rate characteristics due to differences in the size and formation of the adsorbent along with the amount and type of binder used to form the particle. Thus, determining how to efficiently utilize a specific rate enhanced adsorbent in a rapid pressure swing adsorption process requires an appropriate process evaluation that determines a range of suitable operating conditions with a particular focus on reducing cycle time (minimizing BSF) and identifying appropriate adsorption/desorption pressures. This study illustrates one method to evaluate a rapid PVSA process using a rate enhanced adsorbent while demonstrating a process that provides a high purity product (97% oxygen using argon free air) and extremely favorable recoveries (50-60% at pressure ratios between 4-5) with a BSF under 100 lbs. /TPD oxygen.

Using a commercial Li-X zeolite, a lab-scale two column PVSA system was used to assess the ability of the adsorbent for use in a rapid PVSA process. The evaluation included first examining the effect of cycle time on the oxygen recovery at a range of pressure ratios (varying from 2.5-7 by changing the desorption pressure). The adsorption/desorption pressure, amount of gas used for the feed and purge steps, and product purity were maintained as the cycle time was reduced. This provided a method in which any change in oxygen recovery was almost solely assessed to the reduction in cycle time and not variations in other process parameters. It also allowed for the determination of a kinetic limit for how fast the adsorbent could be used while still providing a high level of performance. Additionally, a range of adsorption pressures were examined to determine an ideal operating range using rapid PVSA for this adsorbent.

414950 Evaluation of Pilot Plant for High Temperature PSA-Oxygen Process Using Perovskite-Type Oxide Adsorbent

Abstract:

Evaluation of Pilot Plant for High Temperature PSA-Oxygen Process Using Perovskite-Type Oxide Adsorbent

Tomoya Fujimine, Tokyo Gas Co.

Kiyofumi Monma, Tokyo Gas Co.

Jun Izumi, ADSOTECH, Ltd.

Norio Miura, Kyushu University

In this study, we constructed a pilot plant for high-temperature PSA (HTPSA) oxygen production process having a production capacity of 5 $m^3N\text{-}O_2/h$ by using perovskite-type oxide of $La_{0.1}Sr_{0.9}Co_{0.9}Fe_{0.1}O_3$ (LSCF1991) as an oxygen sorbent. At high operational temperatures of 500-800 degree C for HTPSA-O_2, 1) the heating method for the adsorption tower, 2) the temperature control process, and 3) the selection of optimum operation temperature were examined in order to achieve reduction of electric power consumption rate by means of heat recovery technique. As a result, the rather lower electric power consumption rate of 1.5 $kWh/m^3N\text{-}O_2$ was estimated in this case.

Furthermore, in order to reduce the material cost of sorbent while increasing its oxygen adsorption ability, various perovskite oxides other than LSCF1991 were examined as a candidate of new oxygen sorbent. Among the oxides examined, we found that the Sr-Co-Fe-based perovskite oxide (SCF) that does not include expensive La element gave superior oxygen sorption behavior compared with LSCF1991.

To achieve a higher heat recovery ratio, the front and rear parts of the adsorption tower are filled with the heat storage material. In the adsorption step (Step 11) of the first cycle, the feed air is supplied from the front part to contact with the heat storage material for heat exchange between the feed air and the desorbed oxygen. The shortage of heat content is compensated for by an electric heater. Given that effluent nitrogen comes into contact with the heat storage material at the rear part of the adsorption tower, the heat in the nitrogen can be recovered. In the desorption step (Step 12), since the recovered oxygen comes into contact with the heat storage material in the front part, the heat in the desorbed oxygen can be recovered.

In the next (second) cycle, by supplying the feed air from the rear part of the tower in the adsorption step (Step 21), the feed air contacts the heat storage material in the rear part and heat exchange between the feed air and the heat storage material proceeds. Heat insufficiency is again supplemented by an electric heater. The heat in the effluent can be recovered with the heat storage material in the front part. Heat recovery from the desorbed oxygen can be undertaken with the heat storage material in the rear part in the desorption step (Step 22).

For each cycle, the directions of the feed air and the desorbed oxygen are switched to assure the increase in the heat recovery ratio between the feed air and the effluent nitrogen, and between the feed air and the desorbed oxygen.

During heat exchange with the heat storage material, 1) the electric power consumption rate of the rotatory equipment and heater and 2) the oxygen production rate are assessed on the parameter of a) the desorption pressure and b) cycle time.

As for a new oxygen sorbent, we examined the optimum composition of SCF with regard to oxygen sorption ability. As a result, it was found that $SrCo_{0.75}Fe_{0.25}O_3$ gave the largest amount of sorbed oxygen. Its amount of dynamic oxygen sorption was about 2.5 times greater than that of LSCF1991 in the evaluation by using a very small scale PSA apparatus. The performance of larger scale plant (such as 1,000 $m^3N/h\text{-}O_2$) charged with this new oxygen sorbent of $SrCo_{0.75}Fe_{0.25}O_3$ can be predicted in this study.

In the near future, we will mass-produce the new SCF oxygen sorbent for a pilot plant (the sorbent usage is 250 kg) and a prototype testing plant (the sorbent usage is 4 t). Then, the oxygen production ability and the electric power consumption rate of each plant will be evaluated.

Compared to conventional cryogenic separation process which gives the electric power consumption rate of 0.32 kWh/m^3N under the best condition, the present HTPSA-O_2 process using perovskite oxide sorbent would be capable of showing a much lower electric power consumption rate in the future.

428602 Total Reflux Dual-Piston Pressure Swing Adsorption System

Abstract:

The Dual-Piston Pressure Swing Adsorption (DP-PSA) is a closed system where pistons moving in cylinders at each end of the adsorbent column induce the cycling of fluid flows and pressure variations in the fixed bed. This system is uniquely suited to test adsorbent materials under a wide range of conditions: vacuum to 25 bar; 20-200 °C; cycle times as fast as 1 s. The fact that the movement of the pistons can be controlled independently allows also to run the system under both linear and non-linear frequency response conditions.

The development of the DP-PSA includes experiment design, measurement of leak rates, dead volumes for packed and unpacked column, friction, pressure drop across the adsorbing column and temperature gradient along the column. The system is controlled through LabVIEW running on a real-time computer. This allows the automated running of a series of experiment without direct supervision. During these experiments the absolute pressure and the pressure drop along the column as well as the temperature in the gas and solid phase at two positions along the column is recorded. The experimental runs are analysed with the general adsorption simulator CySim.

The system can be run in two modes: single pellet experiments and full packed column runs. Single pellet experiments, where only the pellets connected to the thermocouples are inside the system, were carried out to study the heat transfer between the zeolite pellets and gases (He, N2, CO2, and mixtures of N2 and CO2). The other mode is that the adsorbing column is packed fully with zeolite 13X pellets. Experiments with pure gases and mixtures of N2 and CO2 were run with different configurations of the system: cycle time; phase angle; stroke length ratio and initial temperature. The measured signals were compared based on the amplitude and time shifts and numerical simulations were used to compare the predictions of a dynamic model of the system with the experimental results. The DP-PSA has been shown to generate a very large set of experimental results, with varying conditions which allow to determine physical parameters of dynamic models. This is achieved without consuming gases given that several experiments are automatically carried out in a closed system.

The authors would like to acknowledge the support from the Engineering and Physical Sciences Research Council under grants EP/G062129/1 and EP/F034520/1.

Design of Reactive Chromatography for the Synthesis of Propylene Glycol Methyl Ether Acetate with Anion Exchange Resin

Jungmin Oh[1], Shan Tie[1], Gaurav Agrawal[1], Balamurali Sreedhar[2], Megan E. Donaldson[2], Alfred Schultz[2], Timothy C. Frank[2], Andreas S. Bommarius[1] ,Yoshiaki Kawajiri[1]

School of Chemical and Biomolecular Engineering,
Georgia Institute of Technology[1]
The Dow Chemical Company[2]

2

Motivation : DOWANOL^{TM} PMA

- Propylene glycol methyl ether acetate (DOWANOL^{TM} PMA)
- The U.S demand for PMA in 2007 was ~117million lbs.[1]

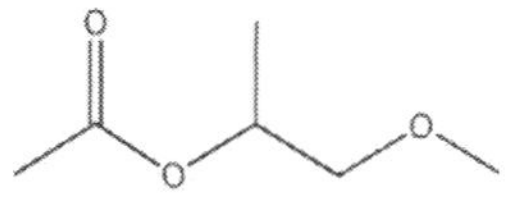

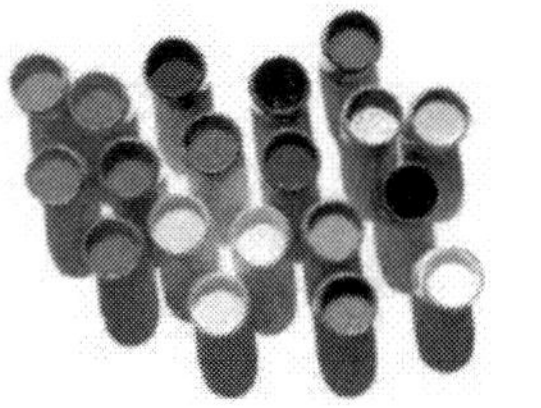

[1] The Dow Chemical Company, Product safety assessment **2008**
TM Trademark of The Dow Chemical Company "Dow" or an affiliated company of Dow

Potential options for production process of PMA

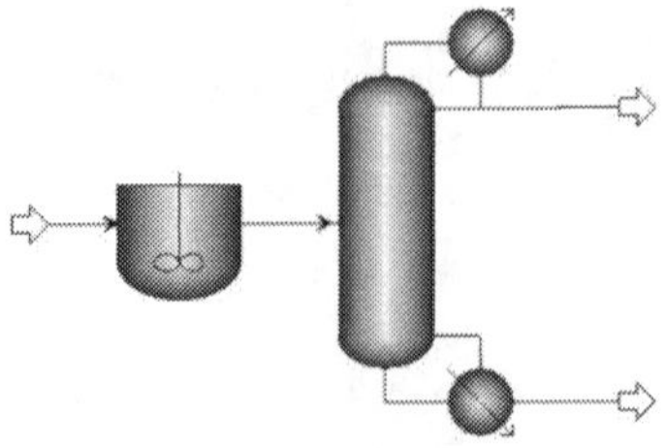

Reaction-then-separation multistep

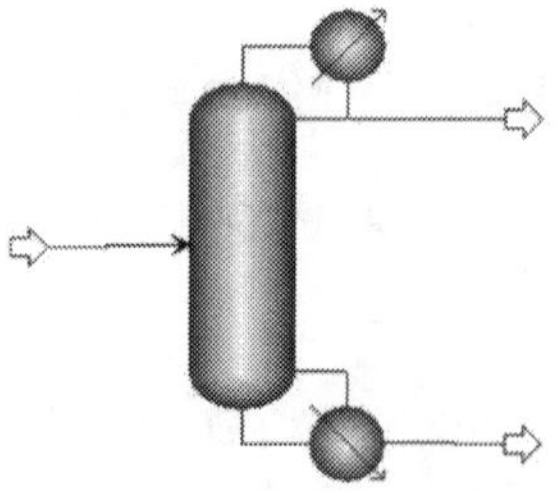

Reactive-distillation one-step

- Limitations
  - Separation : Azeotrope
  - Reaction : Thermodynamic equilibrium limit

Target : Production of PMA through Reactive Chromatography

$A + B \rightleftharpoons C + D$ low conversion

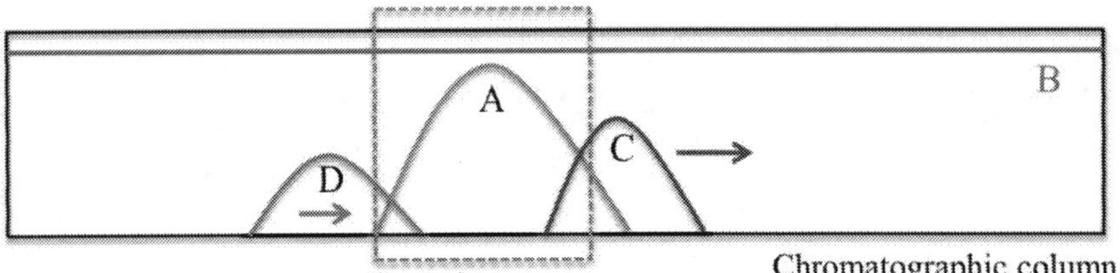

Reaction zone

Chromatographic column
Packed with catalyst (adsorbent)

Separation of products → Increased conversion

Motivation for Reactive Chromatography System

PMA Produced from equilibrium limited reaction

Transesterification:

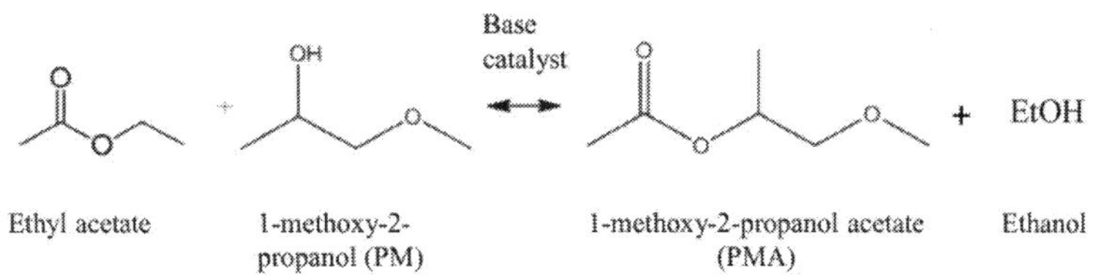

Esterification[1]:

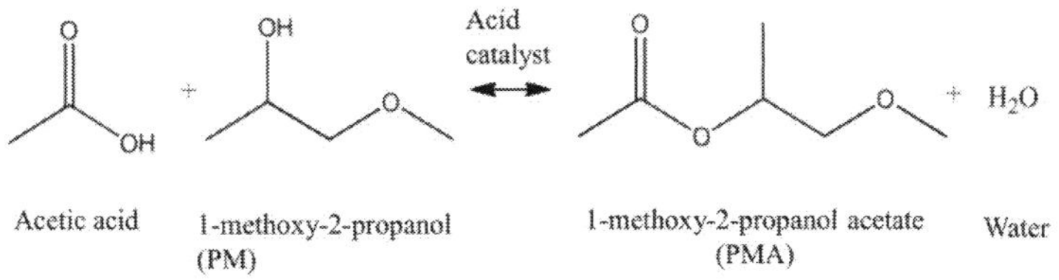

[1] J. Oh, G. Agrawal, B. Sreedhar, M.E. Donaldson, A.K. Schultz, T.C. Frank, A.S. Bommarius, Y. Kawajiri, *Conversion Improvement for Catalytic Synthesis of Propylene Glycol Methyl Ether Acetate by Reactive Chromatography: Experiments and Parameter Estimation*, **Chemical Engineering Journal** 2015 (397-409)

Motivation for Reactive Chromatography System

PMA Produced from equilibrium limited reaction

Transesterification:

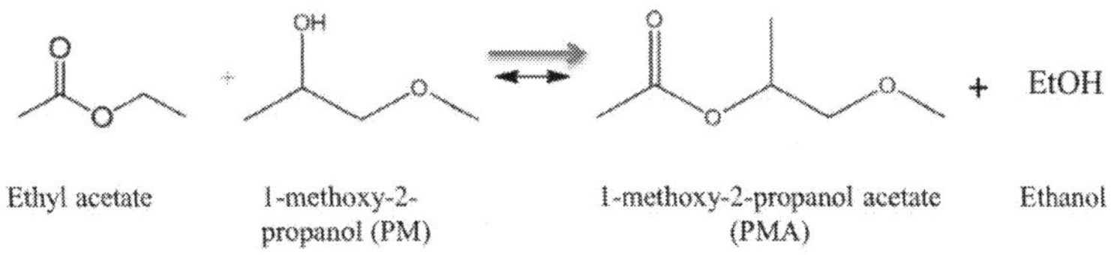

Esterification[1]:

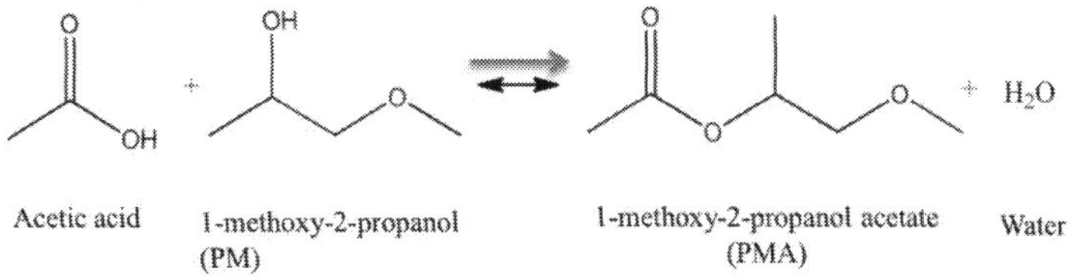

Separation of products increase final conversion

[1] J. Oh, G. Agrawal, B. Sreedhar, M.E. Donaldson, A.K. Schultz, T.C. Frank, A.S. Bommarius, Y. Kawajiri, *Conversion Improvement for Catalytic Synthesis of Propylene Glycol Methyl Ether Acetate by Reactive Chromatography: Experiments and Parameter Estimation*, **Chemical Engineering Journal** 2015 (397-409)

7

Motivation for Reactive Chromatography System

Transesterification:

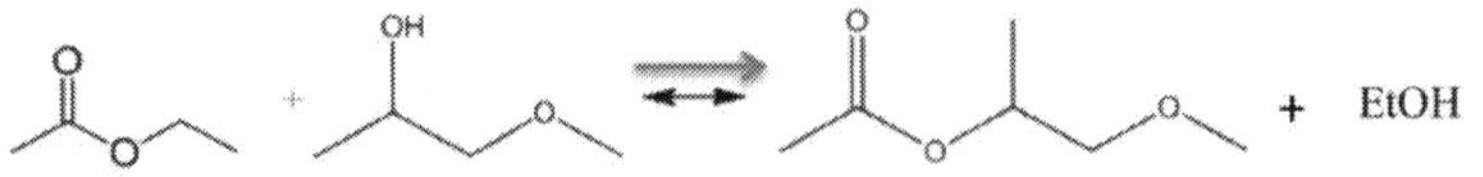

What are the advantages of Transesterification ?

[1] J. Oh, G. Agrawal, B. Sreedhar, M.E. Donaldson, A.K. Schultz, T.C. Frank, A.S. Bommarius, Y. Kawajiri, *Conversion Improvement for Catalytic Synthesis of Propylene Glycol Methyl Ether Acetate by Reactive Chromatography: Experiments and Parameter Estimation*, **Chemical Engineering Journal** 2015 (397-409)

Dow Georgia Tech Chemical & Biomolecular Engineering

8

Transesterification

Advantages[1]:

- Low operating temperature

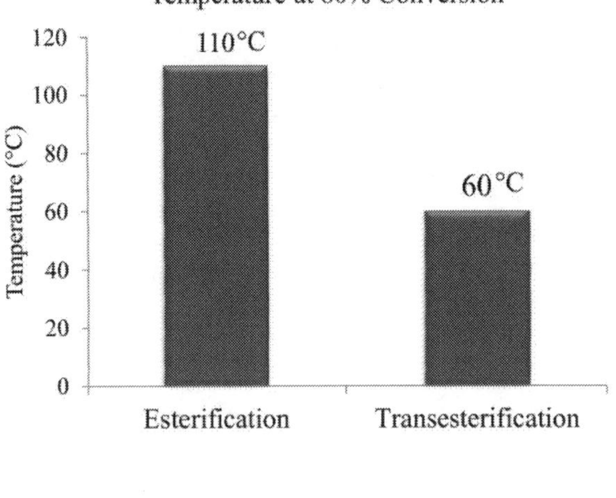

[1] J. Oh, G. Agrawal, B. Sreedhar, M.E. Donaldson, A.K. Schultz, T.C. Frank, A.S. Bommarius, Y. Kawajiri, *Conversion Improvement for Catalytic Synthesis of Propylene Glycol Methyl Ether Acetate by Reactive Chromatography: Experiments and Parameter Estimation*, **Chemical Engineering Journal** 2015 (397-409)

Dow Georgia Tech Chemical & Biomolecular Engineering

Transesterification

Advantages[1]:

- Low operating temperature
- Less desorbent consumption

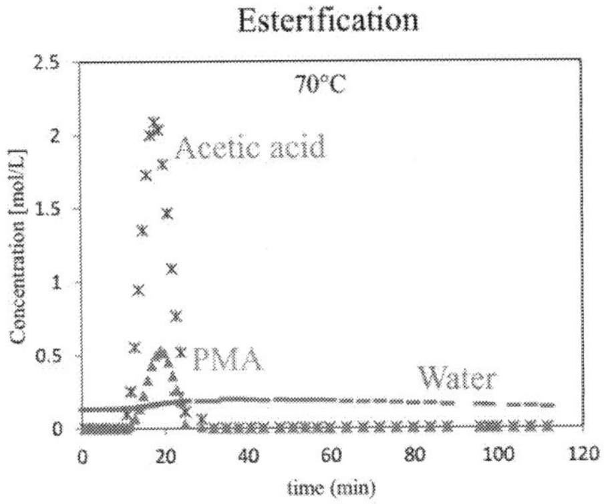

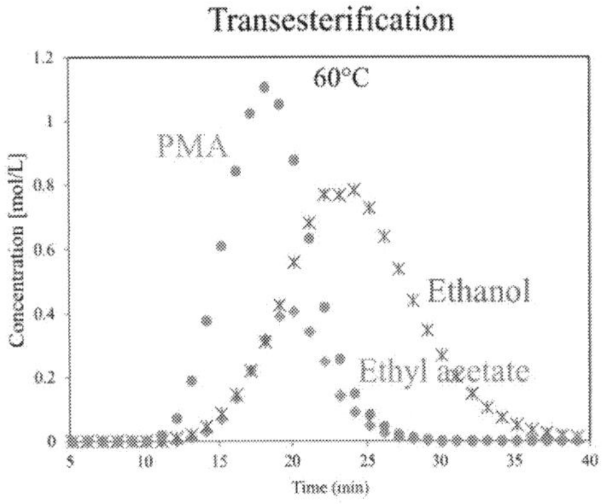

- ➢ High affinity of water
- ➢ More desorbent usage

[1] J. Oh, G. Agrawal, B. Sreedhar, M.E. Donaldson, A.K. Schultz, T.C. Frank, A.S. Bommarius, Y. Kawajiri, *Conversion Improvement for Catalytic Synthesis of Propylene Glycol Methyl Ether Acetate by Reactive Chromatography: Experiments and Parameter Estimation*, **Chemical Engineering Journal** 2015 (397-409)

DOW Georgia Tech Chemical & Biomolecular Engineering

Transesterification

Advantages[1]:

- Low operating temperature
- Less desorbent consumption

Challenges :

- Very few previous studies of transesterification in reactive chromatography
- Catalytic deactivation of anion exchange resin

[1] J. Oh, G. Agrawal, B. Sreedhar, M.E. Donaldson, A.K. Schultz, T.C. Frank, A.S. Bommarius, Y. Kawajiri, *Conversion Improvement for Catalytic Synthesis of Propylene Glycol Methyl Ether Acetate by Reactive Chromatography: Experiments and Parameter Estimation*, **Chemical Engineering Journal** 2015 (397-409)

DOW Georgia Tech Chemical & Biomolecular Engineering

Two approaches for reactive chromatography

Heterogeneous catalysis

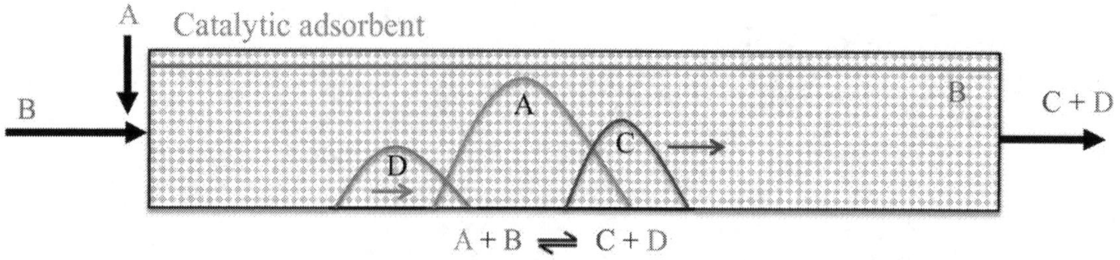

Homogeneous catalysis

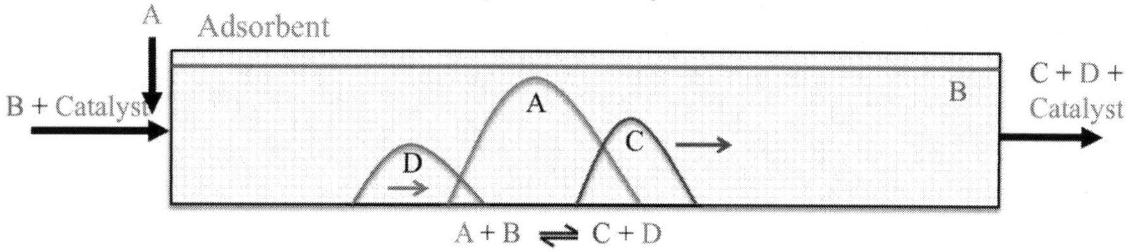

Dow

Georgia Tech | Chemical & Biomolecular Engineering

Homogeneous vs. Heterogeneous catalysis

Heterogeneous catalysis

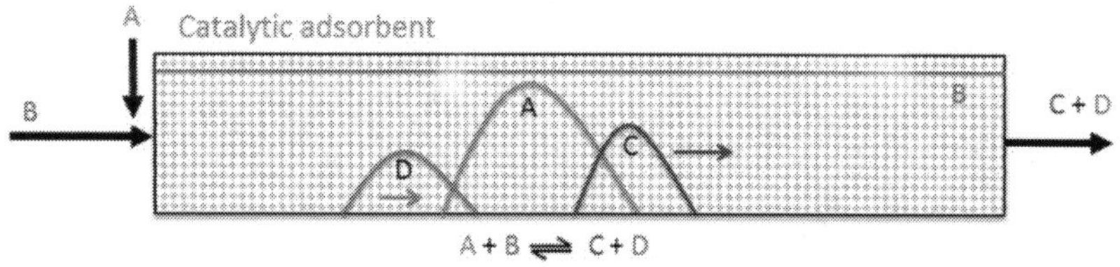

+ Catalyst remains in column: no separation needed
- Catalyst may deactivate
- Resin must perform catalysis and separation

Homogeneous catalysis

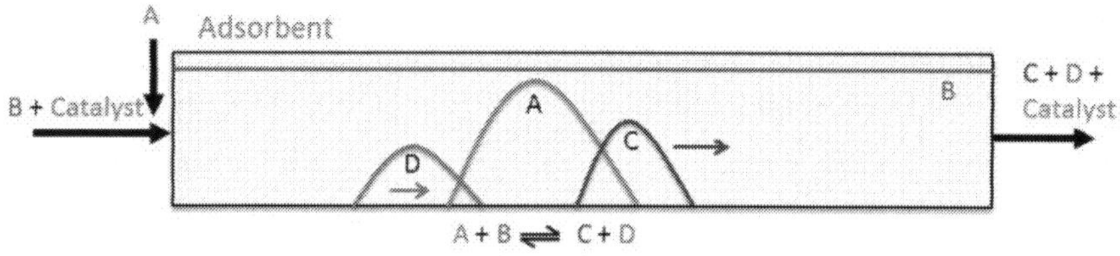

+ Adsorbent and catalyst can be optimized separately
- Catalyst must be separated from products

Dow

Georgia Tech | Chemical & Biomolecular Engineering

Overview

1. Heterogeneous Catalysts

2. Homogeneous Catalysts

3. Adsorbent Selection

4. Transesterification Experiment

Heterogeneous Catalyst Selection

- Small difference in swelling is desired for column packing

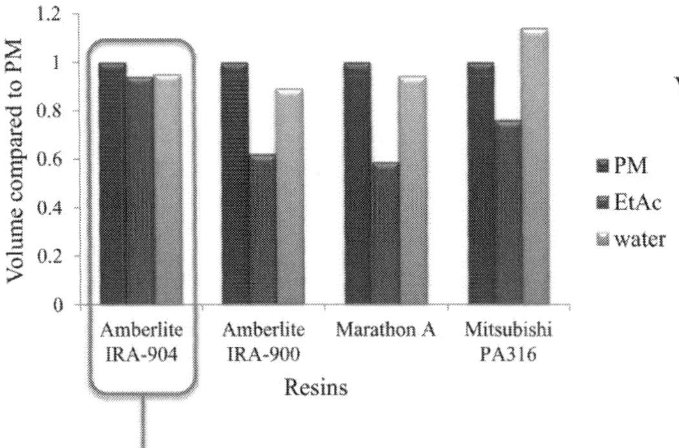

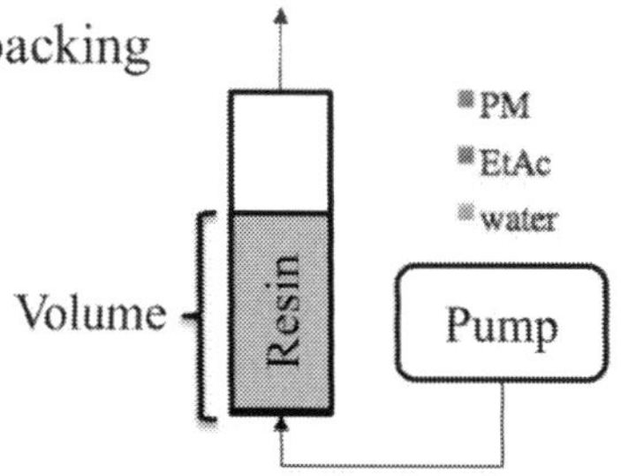

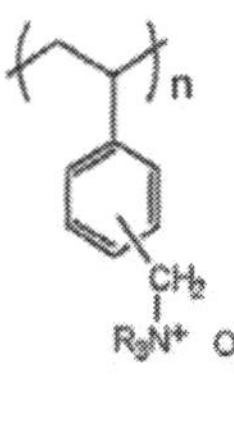

IRA-904

Minimum difference in swelling ratio

IRA904 Selected

IRA-904 Stability Test: Accelerated Stability Test

Continuous supply of reactants

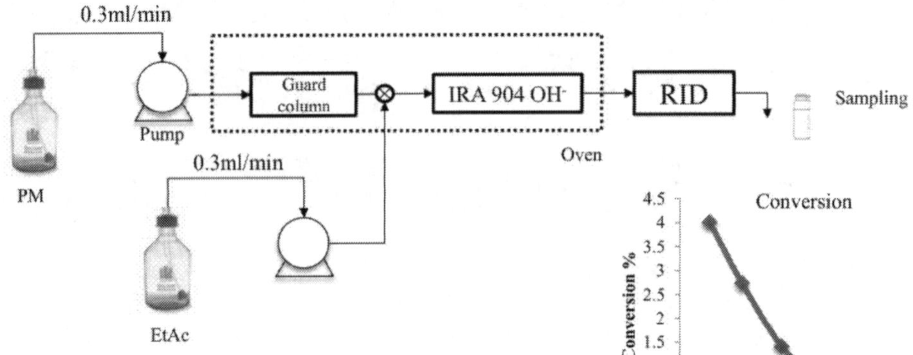

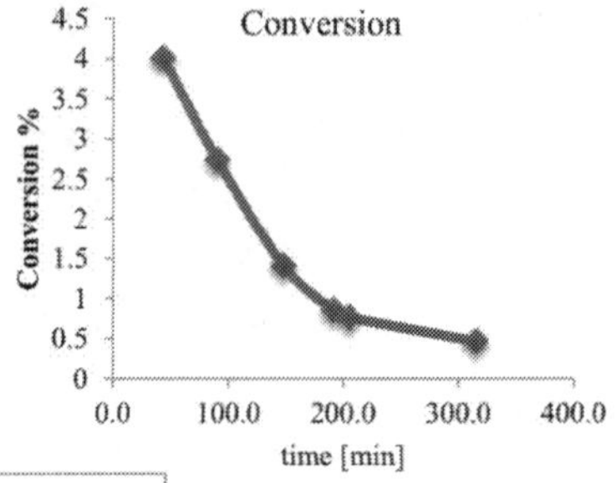

Deactivation of anion exchange resin, IRA-904

Hypothesis for Deactivation Mechanism

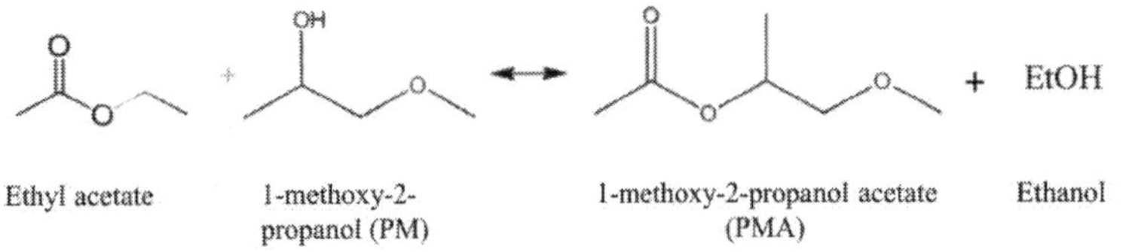

← Desired reaction

Side reactions

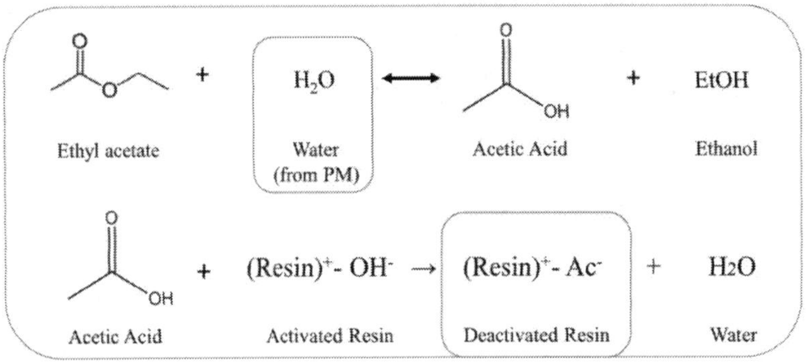

← Formation of acetic acid

← Catalyst deactivation by acetic acid

Even a small amount of water deactivates catalyst

Verification of Deactivation by Water

Side reactions

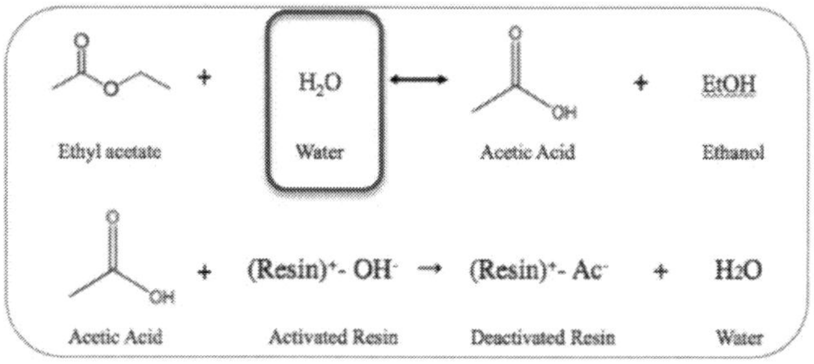

← Formation of acetic acid

← Catalyst deactivation by acetic acid

water

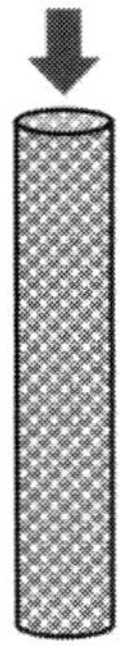

- Column packed with IRA904
- Initial wash with water

Verification of Deactivation by Water

Side reactions

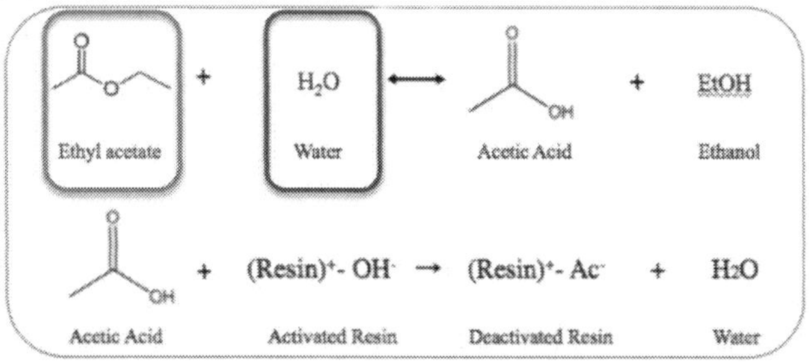

← Formation of acetic acid

← Catalyst deactivation by acetic acid

water EtAc

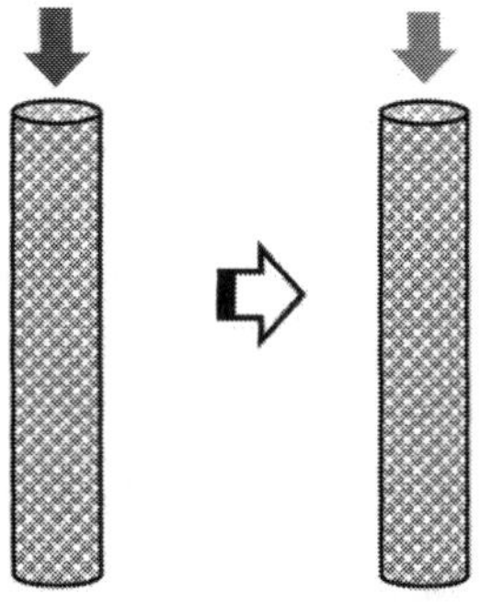

- Column packed with IRA904
- Initial wash with water

- Wash the resin with EtAc

Verification of Deactivation by Water

Side reactions

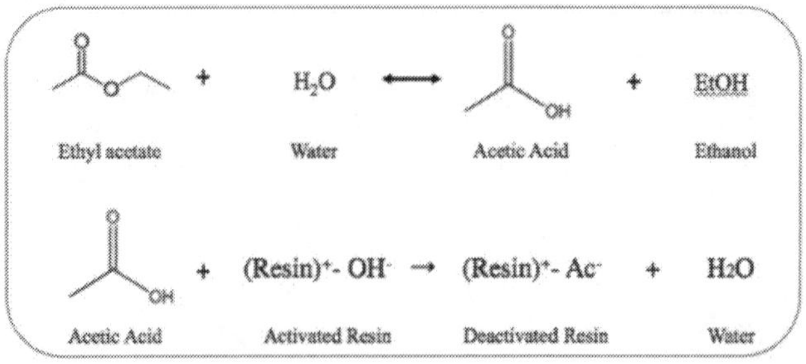

← Formation of acetic acid

← Catalyst deactivation by acetic acid

- Column packed with IRA904
- Initial wash with water

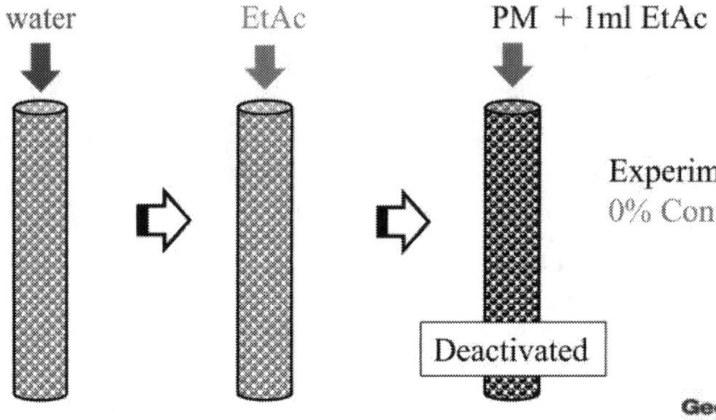

Experiment showed
0% Conversion

Verification of Deactivation by Water

Side reactions

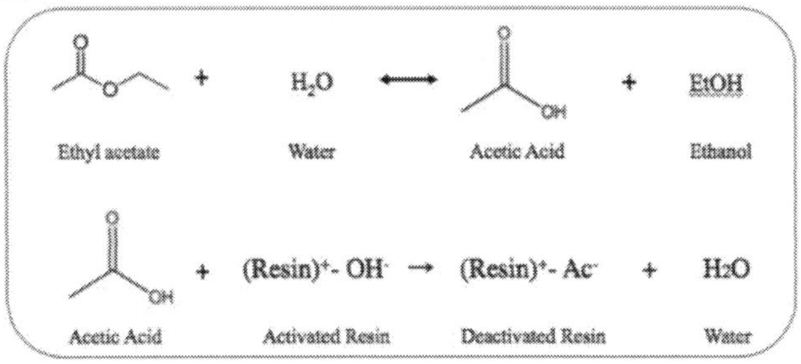

← Formation of acetic acid

← Catalyst deactivation by acetic acid

Complete deactivation verified by washing the column with water

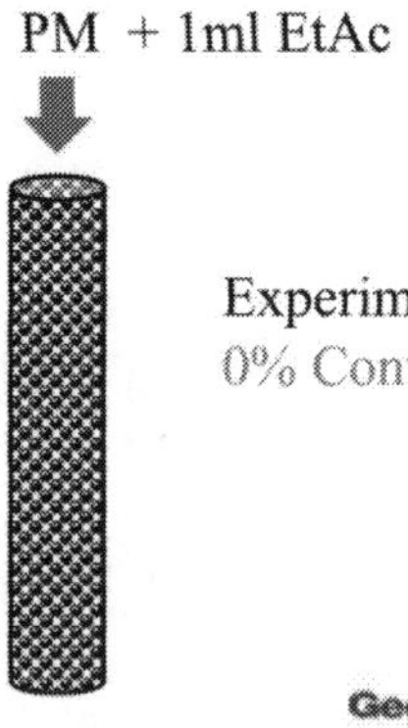

Experiment showed
0% Conversion

Overview

1. Heterogeneous Catalysts

Amberlite IRA904 tested, but deactivation occur due to water

2. Homogeneous Catalysts

1. NaOH

DOW

Georgia Tech | Chemical & Biomolecular Engineering

NaOH as Homogeneous Catalyst

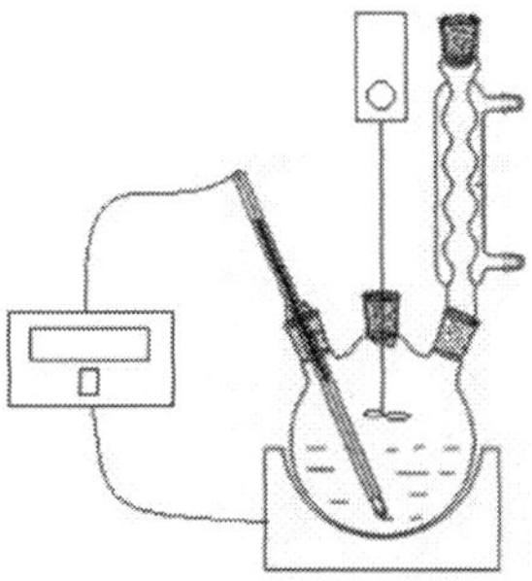

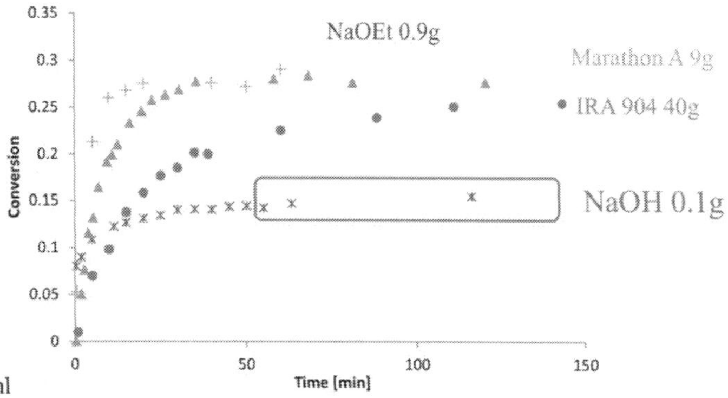

Batch experiment

- Temperature : 40°C
- Start with 100ml PM + 100ml EtAc

DOW

Georgia Tech | Chemical & Biomolecular Engineering

NaOH as Homogeneous Catalyst

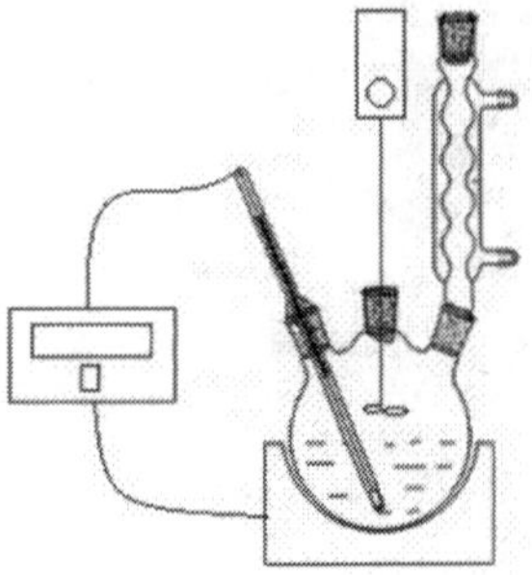

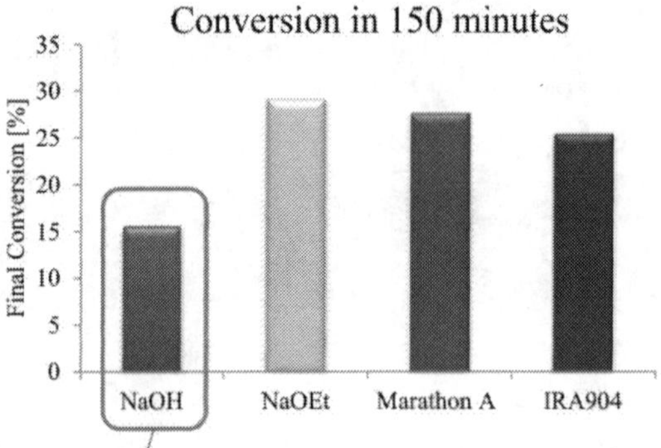

Batch experiment
- Temperature : 40°C
- Start with 100ml PM + 100ml EtAc

Due to deactivation of NaOH by water

Effect of Water Concentration on Homogeneous Catalysts

Experiment:
Batch experiments at 40°C with 0.1g of NaOH

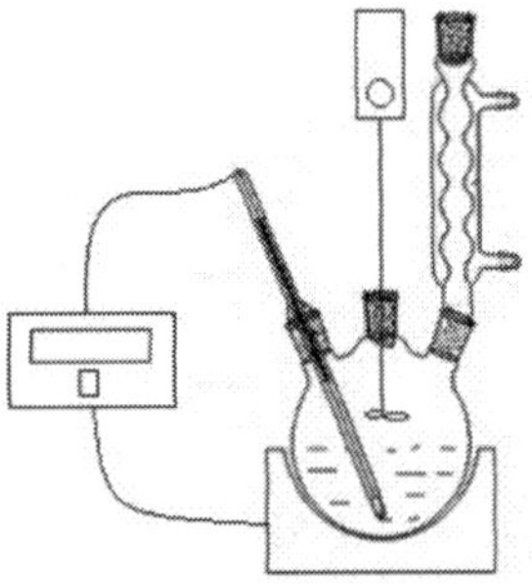

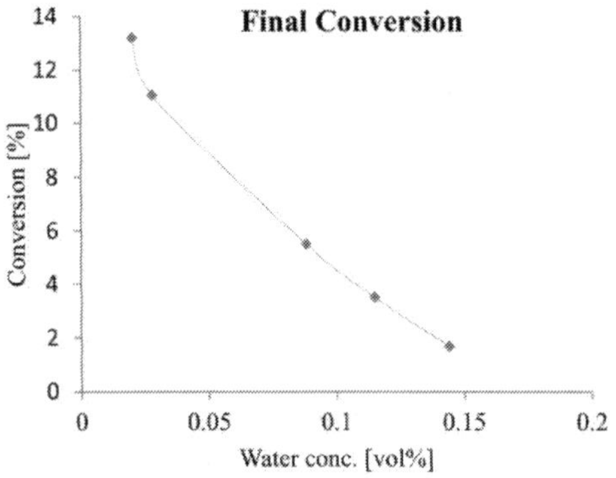

NaOH deactivates rapidly with increasing water concentration

Overview

1. Heterogeneous Catalysts
2. Homogeneous Catalysts
   1. NaOH : deactivates by water and have low solubility
   2. PM-Alkoxide

Alternative Homogenous Catalyst : PM-Alkoxide

Formation of homogeneous catalyst, PM-Alkoxide

$$C_4H_9O\text{-}OH + Na^{+}OH^{-} \leftrightarrow C_4H_9O_2^{(-)}Na^{(+)} + H_2O$$

PM — PM-Alkoxide

Remove by molecular sieve 3Å

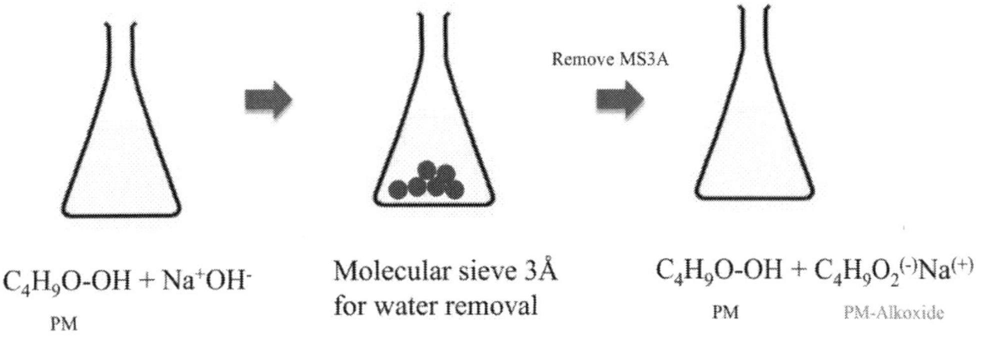

Advantages of PM-Alkoxide catalyst: high catalytic activity and non-hygroscopicity

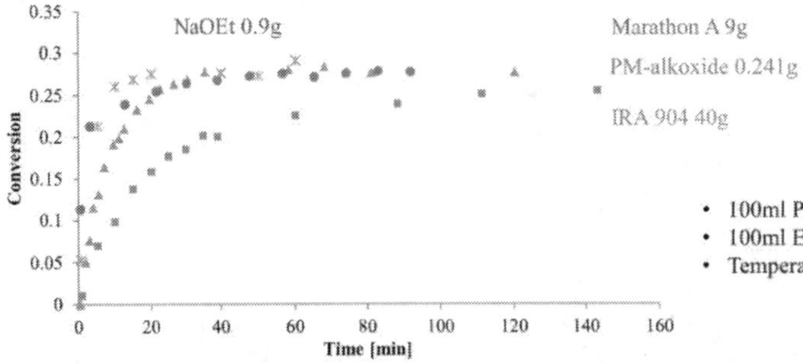

- 100ml PM + Catalyst
- 100ml EtAc
- Temperature : 40°C

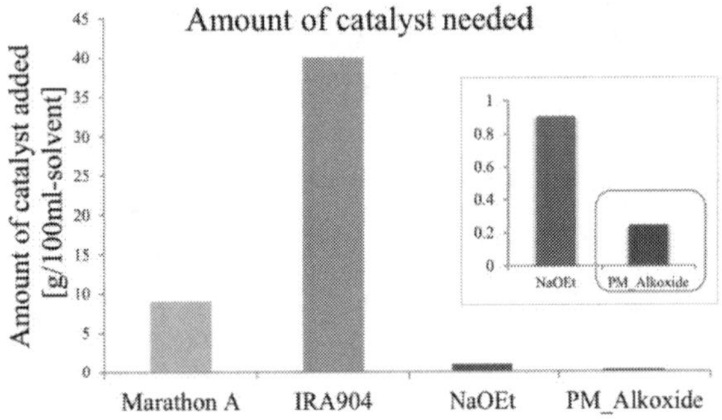

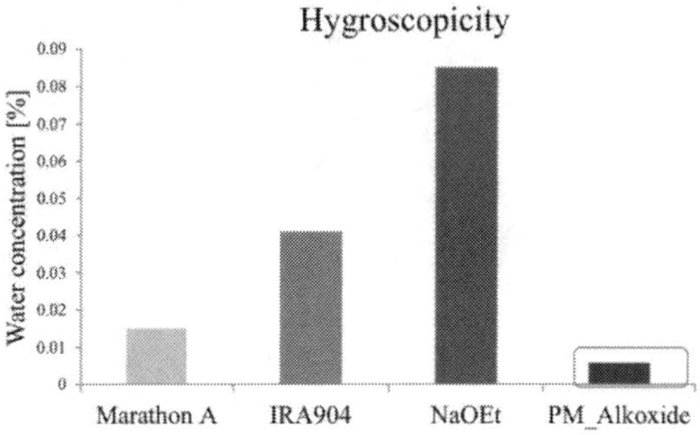

- Small amount of catalyst needed
- Water concentration remains low

Overview

1. Heterogeneous Catalysts
2. Homogeneous Catalysts
   1. NaOH : deactivates by water and have low solubility
   2. PM-Alkoxide
3. Adsorbent Selection

Screening of Separation Medium

5 ion exchange resins tested:

- ➢ Anion exchange resins: IRA904 OH^-, IRA904 Ac^-, IRA904Cl^-
- ➢ Cation exchange resin : IRA 200 Na
- ➢ Polymeric adsorbent : Optipore

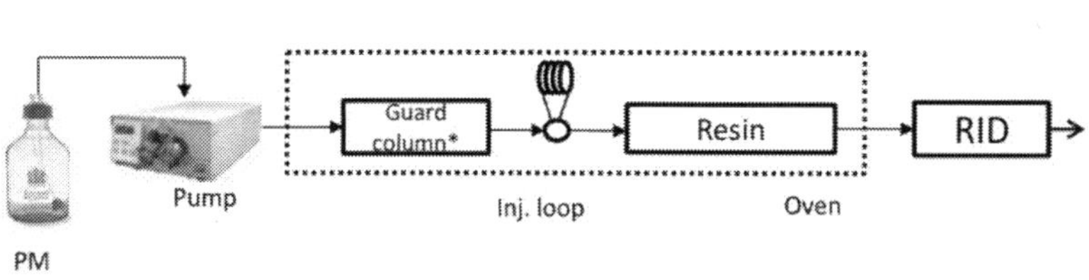

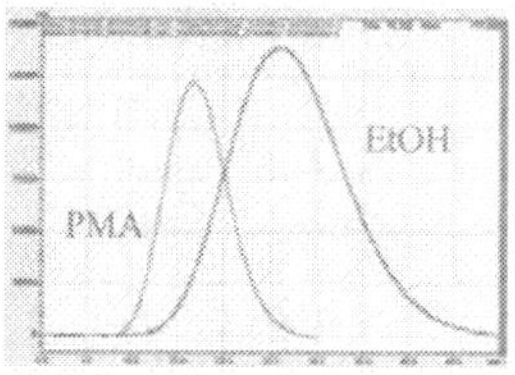

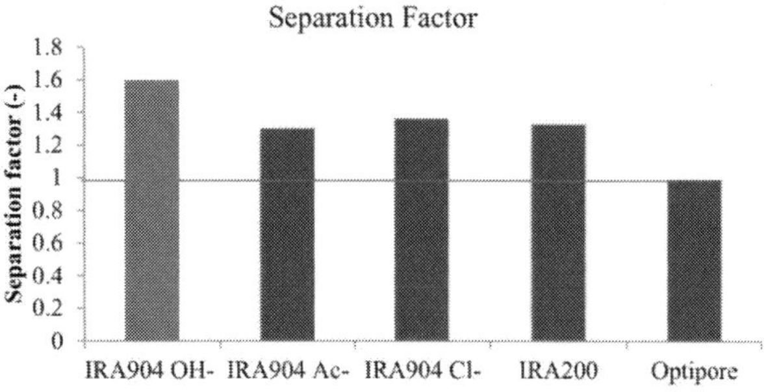

IRA904 OH^- best separation

Optipore no separation

Screening of Separation Medium : Conversion Test

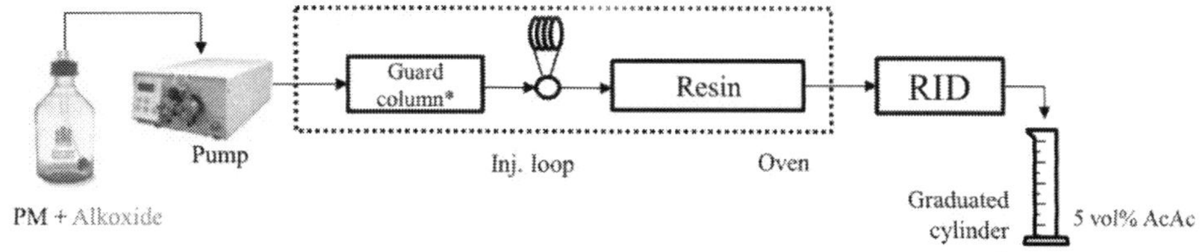

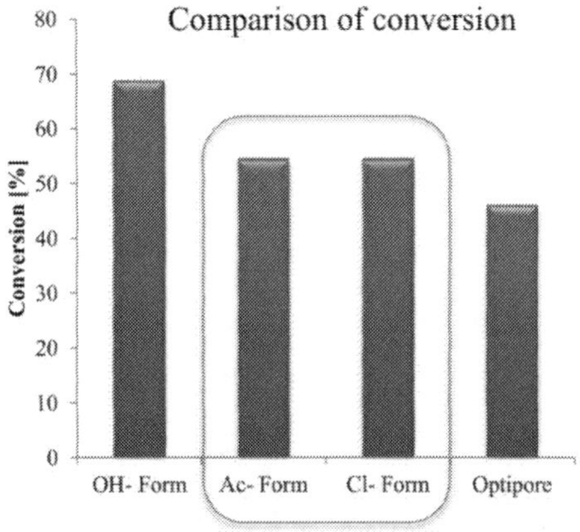

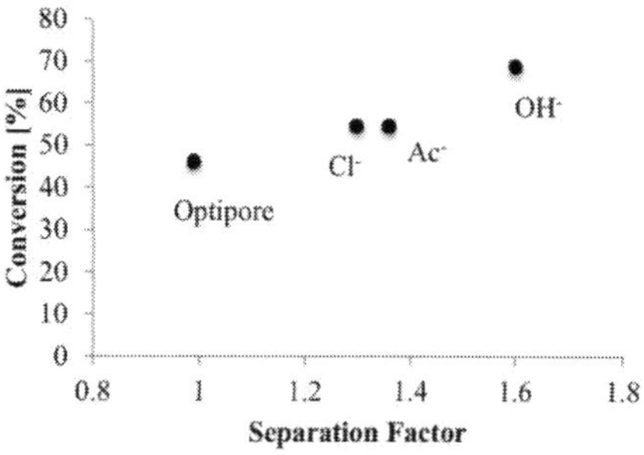

- Higher conversion with higher selectivity
- Both Ac^- and Cl^- form can be selected

Overview

1. Heterogeneous Catalysts

2. Homogeneous Catalysts

    1. NaOH : deactivates by water and have low solubility
    2. PM-Alkoxide

3. Adsorbent Selection

4. Transesterification Experiment

    With PM-Alkoxide as homogeneous catalyst and IRA904 Cl^- as adsorbent

Transesterification by Reactive Chromatography

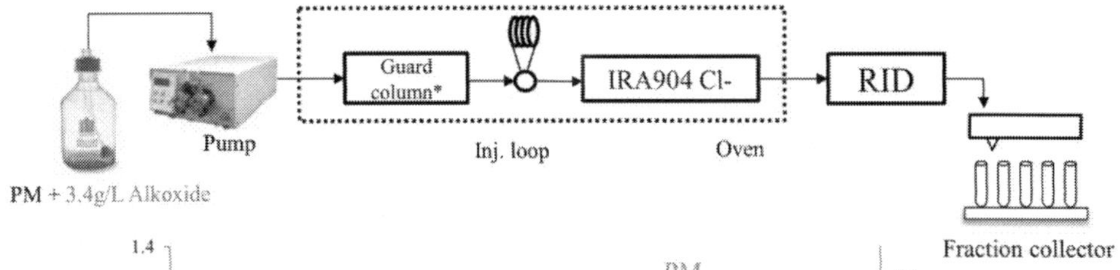

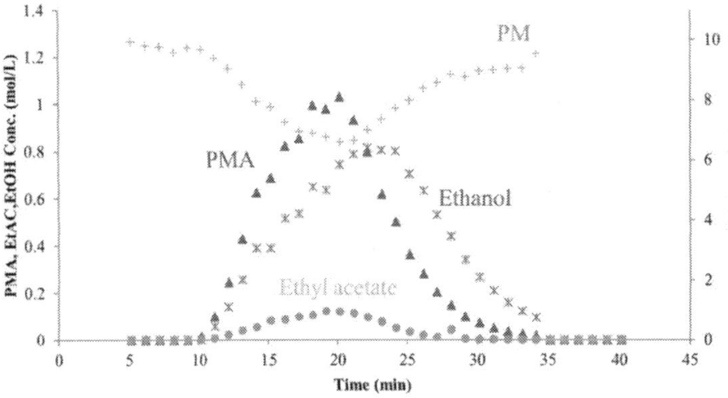

Final Conversion = 54 %
(Comparable to heterogeneously catalyzed reactive chromatography)

SMBR, reaction catalyzed only by homogeneous catalyst

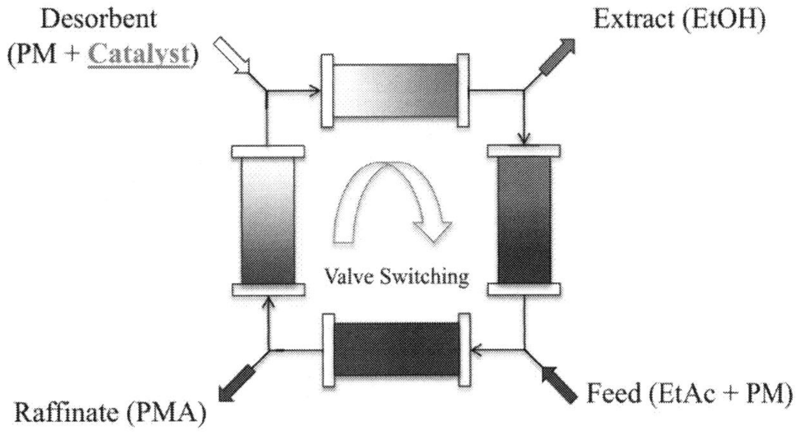

- Use resin as separation medium, but not as catalyst.
- Reaction is catalyzed only by homogeneous catalyst

Conclusion

- Novel process of applying transesterification to reactive chromatography for the synthesis of Dowanol™ PMA
- Screening of homogeneous and heterogeneous catalysts
- Study the deactivation mechanism of heterogeneous catalyst
- Adaptation of homogeneous catalyst in order to overcome the deactivation of anion exchange resin

Acknowledgments

- Dr. Kawajiri group
- Dr. Bommarius group
- The Dow Chemical Company
  - University Partnership Program

DOW®

Georgia Tech

428773 Size-Exclusion Simulated Moving Bed for Separating Organophosphorous Flame Retardants from a Polymer

Abstract:

Over 500,000 tonnes of flame retardants in electronic wastes are consigned to landfills each year. A room-temperature, size-exclusion simulated moving bed (SEC-SMB) was developed using the Standing Wave Design method. High purity (>99%) flame retardants were recovered with high yield (>99%) from a waste. The cost of such a process was estimated to be less than 10% of the purchase cost of the flame retardants. The SEC-SMB is over an order of magnitude more efficient than conventional SEC processes. Fast startup methods were developed to reduce the SMB start-up time by more than 18 fold. SEC-SMB can be an economical method for separating small molecules from polymers in wastes.

429007 A Continuum Approach to Chromatography Modeling

Abstract:

Some chromatographic applications are dealing with 'infinitely' complex feedstock composed of a large number of similar species with comparable properties. The feed mixture might be so complex that analytical methods fail to identify all the solutes present in the feedstock individually. In this case, the description of large multicomponent systems requires specific mathematical tools to reach an intelligible description of the complex solute mixture. Lumping methods are commonly used to reduce the number of solutes present in the feed mixture to a smaller number of pseudo-components with relatively homogeneous properties. This classical lumping approach is often of great help to model multicomponent systems but requires a tremendous amount of work to identify and characterize the different pseudo-components and their associated model parameters. There is often a trade-off between the precision of the lumping approach requiring a large number of pseudo-components and the experimental and computational efforts increase inherent to the modeling a multicomponent systems.

In this work, we propose to consider the solute mixture as a continuum. The continuum approach is based on the description of complex mixtures by continuous distributions instead of the classical discrete sampling. By doing so the precision of the results becomes independent of the number of considered components and of the sampling technique. Considering complex mixtures as a continuum considerably reduces the number of model parameters to the few parameters characterizing the solute distribution, and thus reduces the computational effort. The continuum modeling approach is first introduced in a general manner, and then applied to the modeling of three different chromatographic separations: polydisperse polymers by size exclusion chromatography, the mono-PEGylated lysozyme positional isomers by linear gradient elution on a strong cation-exchange media, and finally a mixture of polyclonal IgGs charge variants by IEC. The continuum model is shown to be well-suited for both isocratic and gradient elution chromatography modeling and very efficient when the number of solutes rises.

429144 Gas-PHASE Simulated Moving BED: LIGHT Olefins Paraffins Separation

Abstract:

Propylene is an important light olefin used in the manufacture of a wide variety of products. Apart from being a (petrochemical) raw material widely used in motor fuels for octane improvement, it is in the rubber and plastic industries where the largest of propylene share finds its application, as intermediate in production of polypropylene, acrylonitrile, propylene oxide, cumene/phenol, oxo alcohols, acrylic acid, isopropyl alcohol, and oligomers. This alkene is the second simplest unsaturated organic compound and its demand is ever increasing, mainly due to the increase of polypropylene production. The largest outlet, accounting for 66% of propylene demand globally, is polypropylene.

In the current study a 13X zeolite extrudates have been chosen as the adsorbent and isobutane as desorbent to separate a mixture containing 32% of propane and 68% of propylene. Adsorption equilibrium isotherms on the 13X zeolite extrudates has been determined gravimetrically at 50, 100 and 150 °C, up to 5 bar for propane and propylene; and up to 3.5 bar for isobutane. The Dual Site Langmuir (DSL) model was fitted to the adsorption equilibrium data and isosteric heat of adsorption as a function of adsorbate loading was calculated by the van't Hoff equation. Additionally, gas phase SMB cycles were designed and performed experimentally for production of polymer-grade propylene (> 99.95% eluent free bases) from 0.32/0.68 propane/propylene feed mixture, with a recovery of about 90% of the olefin in the extract.

For this purpose a senior LSRE team, making use of its own resources, designed and assembled a new pilot SMB unit able to separate gas mixtures. Due to the high complexity of the unit that was proposed to be assembled a detailed planning was required. Therefore, prior to the acquisition of the material and equipment, the unit was designed using the software "SolidWorks". This software allowed the determination of the exact amount of material required including the Swagelok fittings and tubing necessary for the assembly.

One should notice the high number of different items, more than eighty. Some of the items were required in very high numbers, such as 1/8"-Port Connectors: 252, 1/8" Union Elbow: 224, and ISO_BSP Parallel Thread Male Connectors: 130. Besides the number of required parts, this planning stage also allowed to study the best spatial distribution of the parts. In particular, the distribution of the columns inside the oven, and the layout of the gas distribution lines (inlets – feed and eluent, outlets – extract and raffinate), in order to reduce the dead volumes between columns, and between each column and the recycling streams.

The pilot gas-phase SMB unit contains eight columns in stainless steel, each with an internal diameter of 2.2 cm, and a length of 10 cm, with a metal sieve (0.5 micron, Swagelok, SS-2F-05) placed at the bottom end of each column. The gas-phase SMB unit can work between 1 and 7 bar of pressure, temperatures in the range of 303 – 523 K, and feed flow rates within 0.1-6 SLPM. The feed is obtained from the mixture of up to three individual gas streams, through the accurate control of three mass flow controllers. While, the eluent is obtained from a single gas stream controlled by an accurate mass flow control, with a maximum flow of 2 SLPM. The individual feed gas streams are mixed in a union cross before entering one of the columns. The eight columns are stainless steel with 2.12 cm internal diameter and 10 cm length, with a metal sieve placed at the bottom end to hold the adsorbent bed. The internal temperature of each column is acquired by type K thermocouples placed in the middle of the columns. The thermocouples are placed approximately at the centre (radial direction) of the adsorbent bed. The columns are placed inside of an air-forced convection oven (Termolab) equipped with resistances of 10 kW of power, and 2 PID controllers (Eurotherm), that allow the control of the temperature within 303 and 523 K with ±2 K accuracy. The inlet (feed, eluent, and 2 recycle entries) and outlet (raffinate, extract, and 2 recycle exits) streams are separated in 8 individual streams, by 2 manifolds 1/8", and each individual stream is connected to one column. Sixty-four solenoid valves are automatically actuated from the computer interface, to control the schedule of gas admission/exit of each column, accordingly with the SMB switch time. At the exit of each column a filter was placed to prevent damage to the check-valves, vacuum pumps, or other parts of the unit due to fines released as a consequence of adsorbent attrition. The direct connection between two consecutive columns is done via one lift check valve to prevent the back flow between the columns, and to ensure flow in only one direction. The system pressure is assessed by four pressure transducers with range of 0 – 7 bar. The pressure transducers are place at the top of every two columns, and the obtained data is acquired automatically by the computer interface. Two recycle loops (one before the feed stream and another before the eluent stream) are forced by means of two vacuum pumps with the purpose of re-pressurizing the exit of the columns before the feed or the eluent streams, to ensure perfect mixture. The flows of raffinate and extract are controlled by means of two Coriolis mass flow meters, up to a maximum of 2000 g/h. The internal composition profile of the SMB unit can be analysed through a sampling point that is located at the top of Column 7. Sampling at this point at different times of one cycle allows assessing the composition of different points in the four sections of the SMB. Additionally, the composition of the unit outlet streams, raffinate and extract, can also be analysed. This analysis at different times of one step allows assessing the composition of the raffinate and extract streams. These analyses are performed by gas chromatography using a capillary column, and a flame ionization detector (FID). The gas chromatograph is coupled with an automatic valve system and an 8-loops trapping selector with microelectric actuator. The 8-loops trapping selector with 50 μL loops allows to sample during the experiment at selected times. The GC is operated isothermally with an oven temperature of 373 K, a carrier gas flow rate of 5 ml min-1 of helium, and Split Ration of 60.

A LabView interface was developed and a relay controller from OMRON was programmed for switching the 64 valves automatically. The LabView interface allows the control of the feed, eluent, raffinate, and extract flow-rates by providing a certain set-point value to them. Additionally, the acquired signals of temperature (9 signals: 8 columns + oven), pressure (4 signals), and the actual flow-rates (5 signals) are presented by graphical and numerical displays in the interface, allowing the operator to have a clear knowledge of the operation conditions at any moment. The data acquisition is done by four Bus-Powered Multifunction Data Acquisition units. All the data, time, pressures, temperatures and flow rates, are automatically stored to a file.

Alternative membrane chromatography for water treatment

Chuanfang Yang*, Yanxiang Li, Lei Li and Lixia Cao

National Key Laboratory of Biochemical Engineering
Key Laboratory of Green Process and Engineering,
Institute of Process Engineering
Chinese Academy of Sciences, Beijing, 100190

2015 AIChE Annual Meeting, Salt Lake City, UT, Nov 8-13, 2015

Membrane chromatography allows faster mass transfer and lower flow resistance, but barely used for water treatment

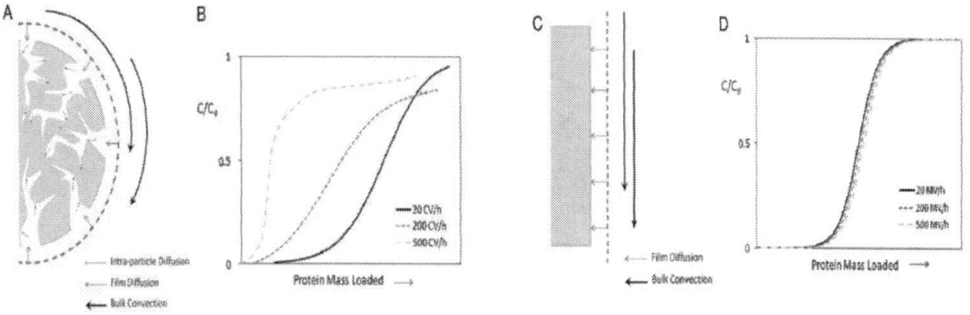

Packed bed column is mass transfer limited

Mass transfer is improved in membrane chromatography

Valerie. Orr et al., Recent advances in bioprocessing application of membrane chromatography, *Biotechnology Advances* 31 (2013) 450–465

Performance of membrane chromatography performance is governed by both membrane geometry and chemistry

- **Membrane geometry**
  - Small pore size for fast mass transfer
  - High porosity
  - Even flow distribution
  - Thickness requirement
  - Structural integrity
- **Ligand chemistry**
  - Kinetics
  - Isotherms
  - Selectivity
  - Regeneration
  - Density

Membrane is typically made by phase-inversion and has low loading capacity

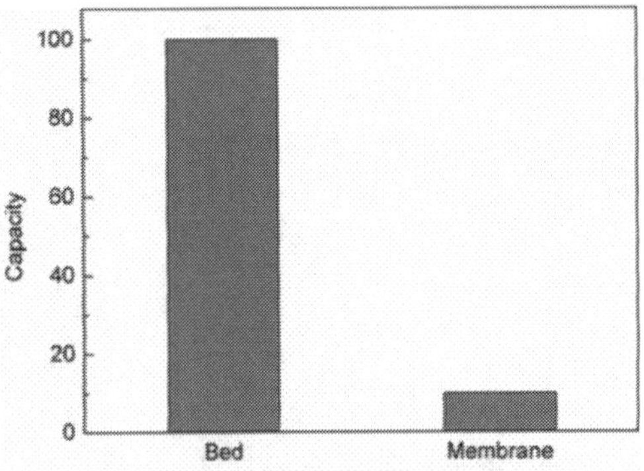

Up to an order of magnitude lower as compared with packed bed

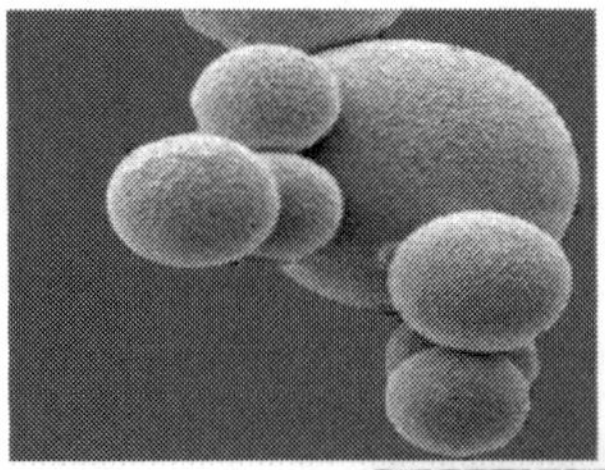

Resin

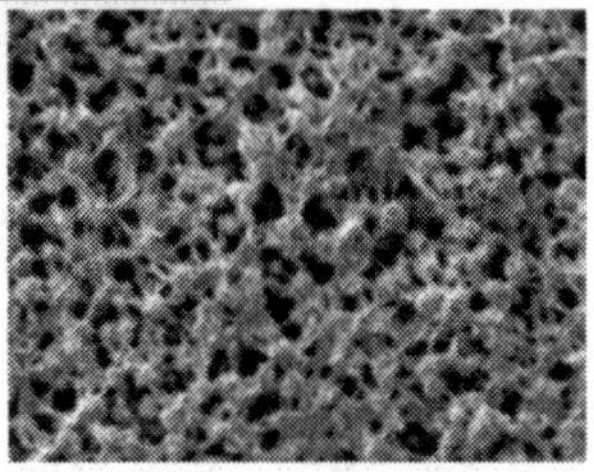

Membrane

Lower internal surface area

Questions

For water treatment:

1. Can we design alternative membrane chromatography?

2. What are the options?

Construct alternative membrane chromatography by using non-woven materials with large pores but added functionality

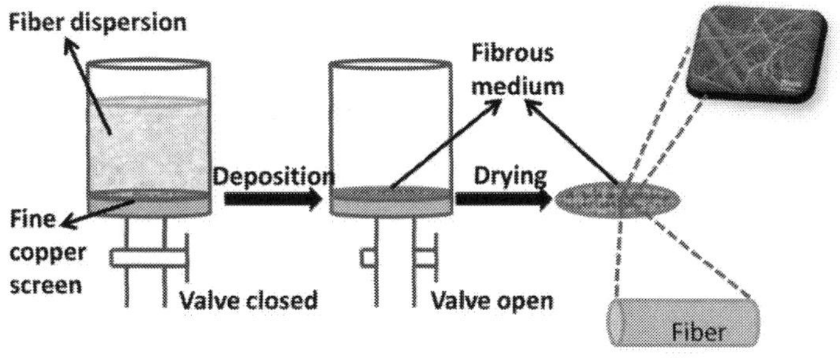

Wet-lay process of paper making

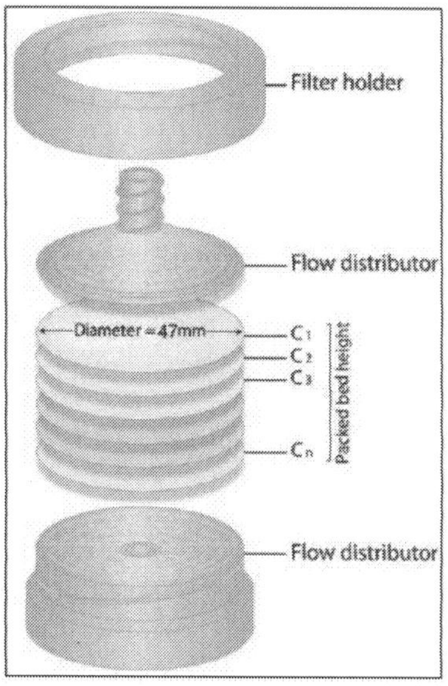

Stacked fibrous membranes

Construct alternative membrane chromatography by using non-woven materials with large surface area and natural functionality

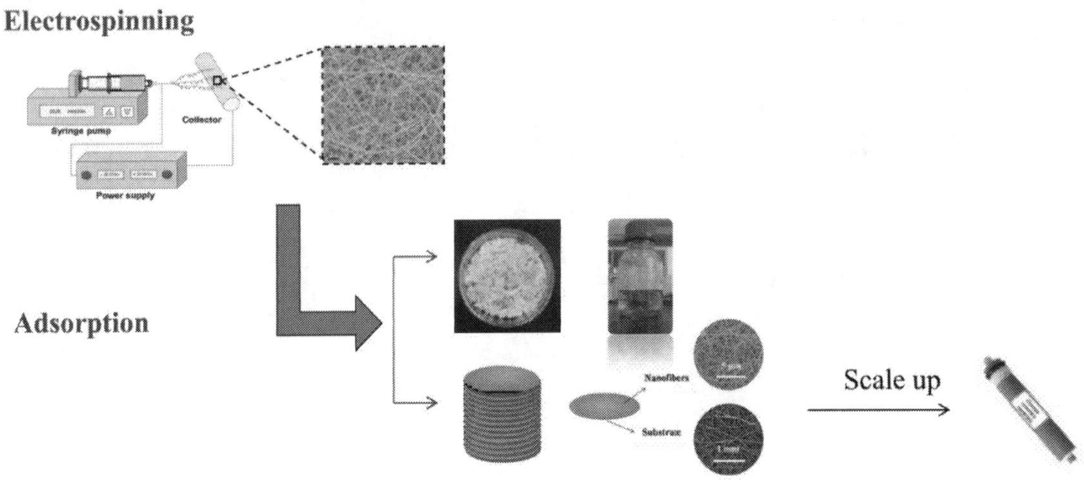

Making composite fibers using cellulose fibers and nano-TiO_2 particles

- Anchor TiO_2 on cellulose fibers
  - Control particles not to agglomerate
  - Control particles not to shed by taking advantage of the surface reaction
- Synergistic adsorption may be possible
- Green synthetic route is desired
- Easy fabrication of filter bed

Cellulose fibers

1. In-situ synthesized TiO_2/CF composite fibers

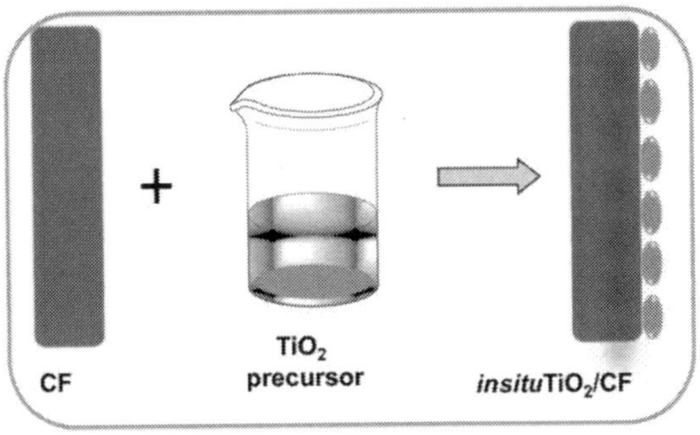

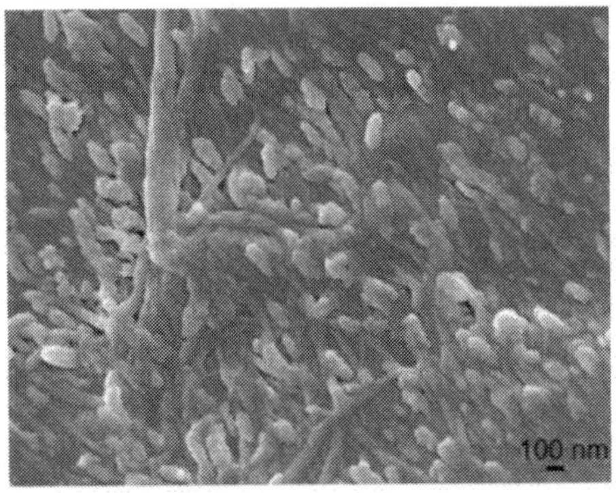

***insitu*TiO_2/CF (*in situ* synthesis)**

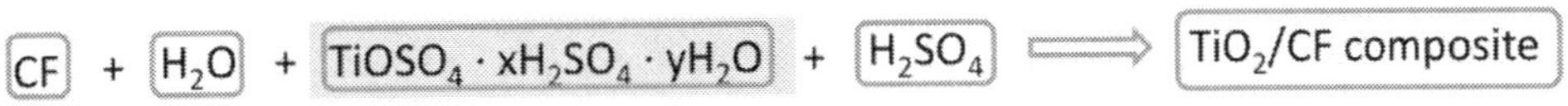

2: Nano-TiO_2 /cellulose composite made by self assembling

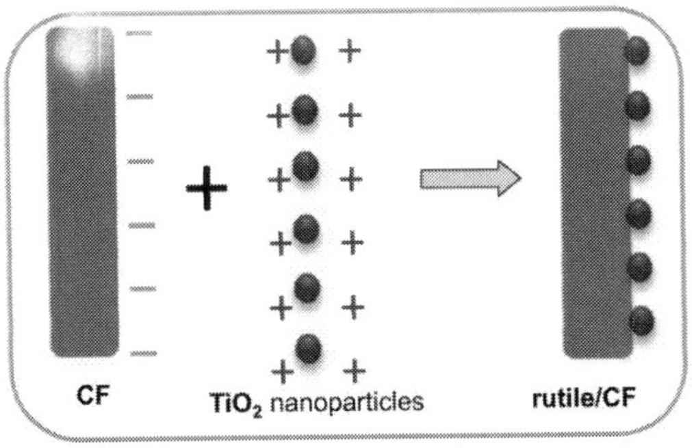

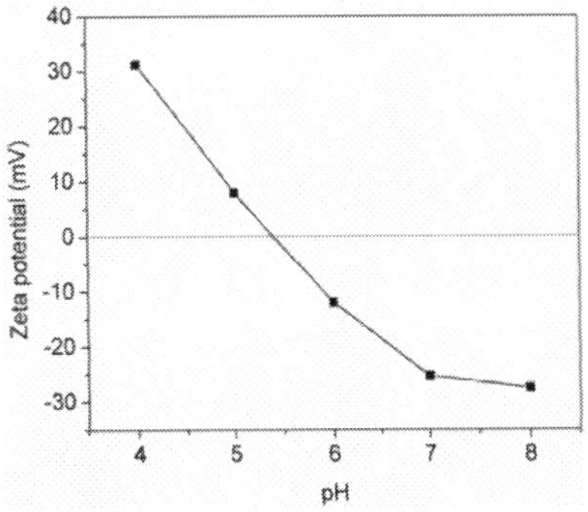

Zeta potential of TiO_2 vs. pH

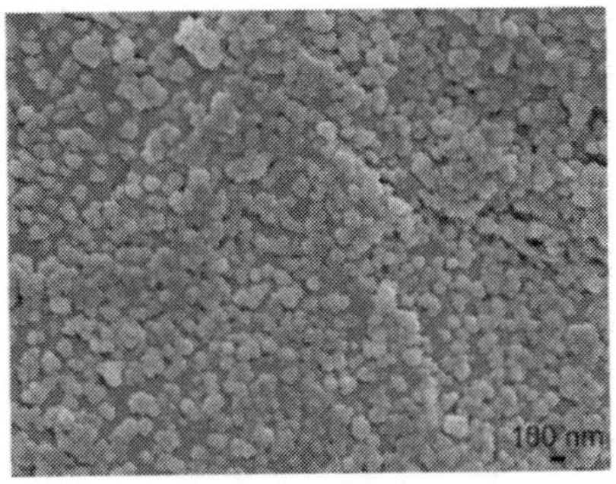

rutile/CF (electrostatic assembly)

CF and TiO_2/CF fibrous membrane

For in-situ synthesis, spindal TiO_2 forms progressively

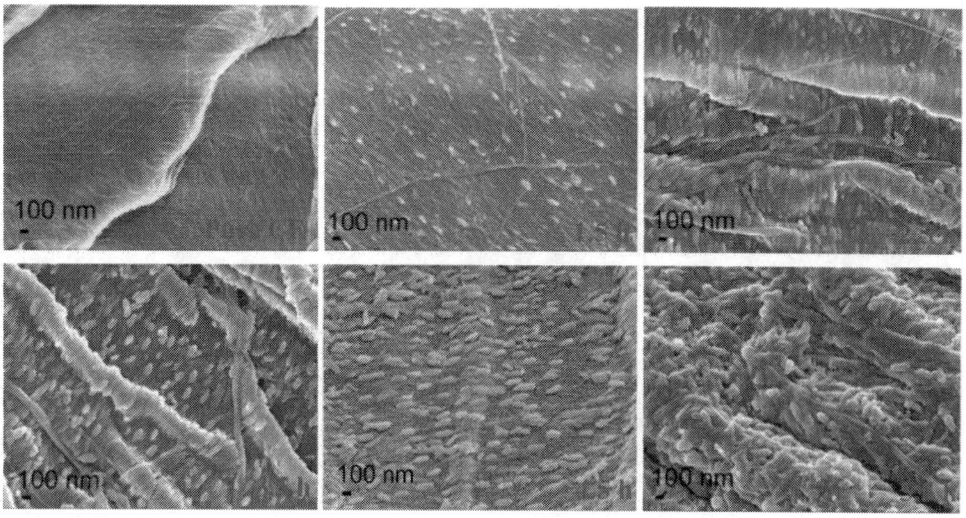

TiO_2 spindles: ~180 nm length
~ 50 nm diameter

TiO_2 particles do not cover the whole cellulose surface

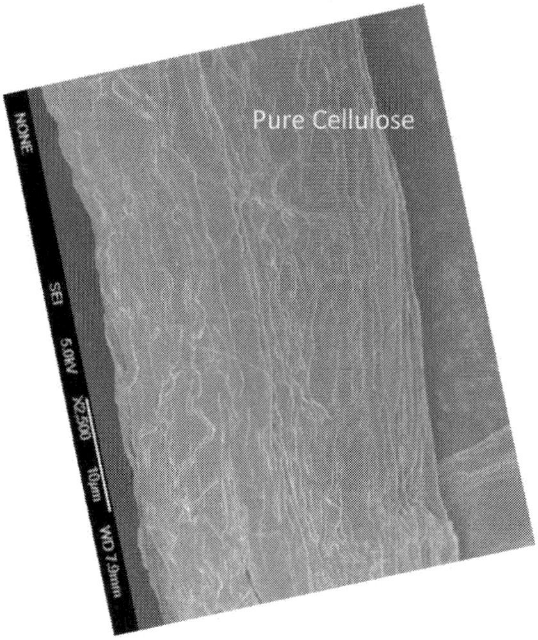

TEM and XRD indicate rutile TiO_2 nanocrystals

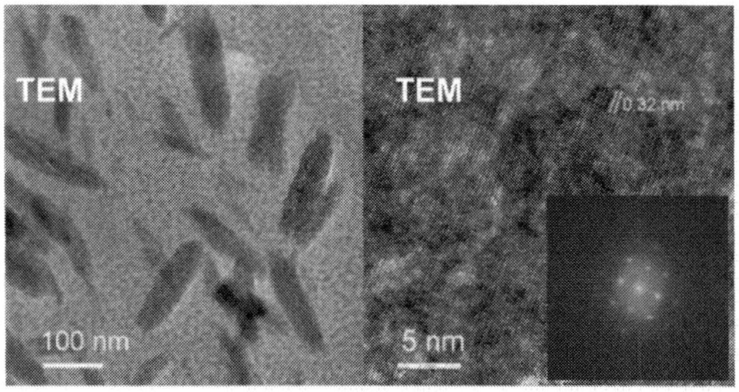

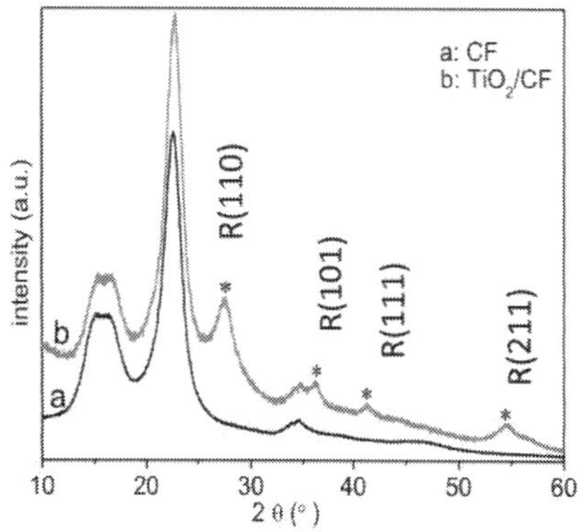

spindle crystal

Rutile
(0.32 nm fringe spacing)

Rutile
(27°, 36°, 41°, 54°)

Anatase

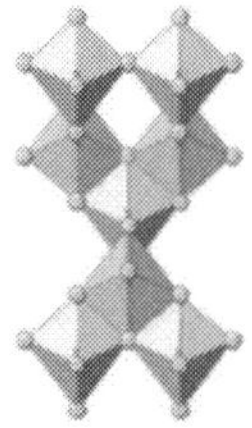

VS.

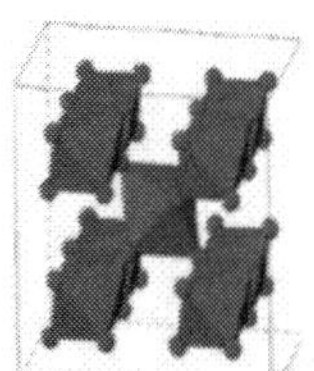

Rutile

All the adsorption fits pseudo-second order kinetics

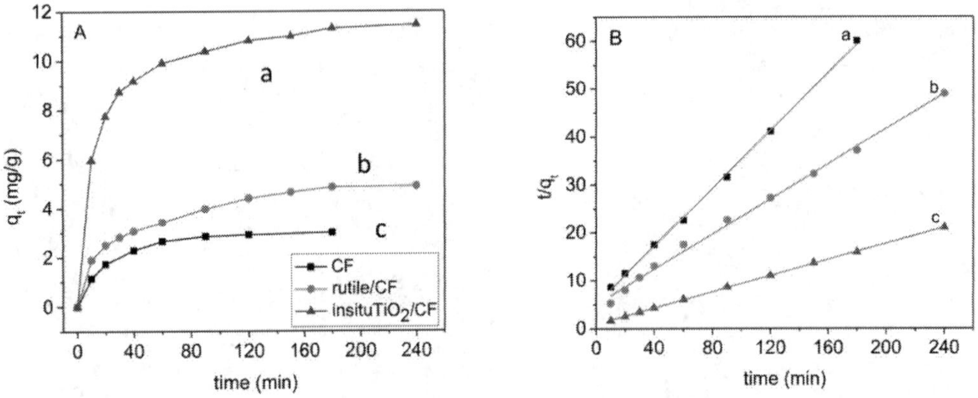

pseudo-second order kinetics

initial Pb^{2+} concentration = 10 ppm, pH = 6, adsorbent dose 0.4g/L

Breakthrough curves of different beds

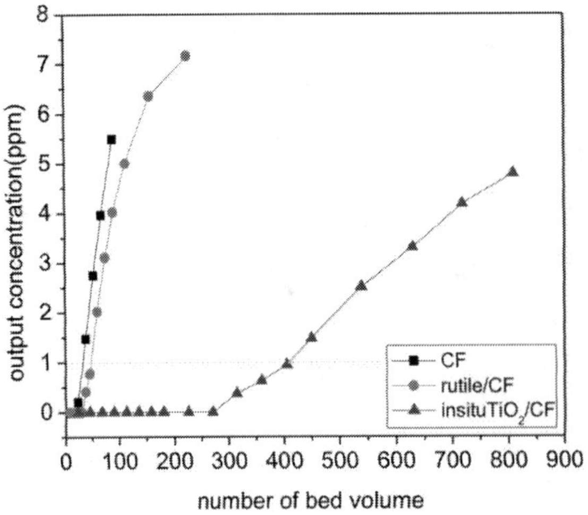

initial Pb^{2+} concentration = 10 ppm, pH 6, flow rate = 2.9 mL/min, two stacked membrane layers

Properties and adsorption performance of filter beds

Filter-bed	CF	rutile/CF	*insitu*TiO_2/CF
Flow rate (ml/min)	2.9	2.9	2.9
Static adsorption capacity (mg/g)	2.9	4.8	11.3
Number of BV at 10% breakthrough	31.5	48.5	409.3
Bed adsorption capacity (at 10% breakthrough) (mg/g)	0.5	0.8	6.8
Fiber specific surface area (m^2/g)	0.60	3.0	6.0
Porosity	0.51	0.51	0.50
TiO_2 content (wt%)	0	7.0	4.8

Effect of bed height

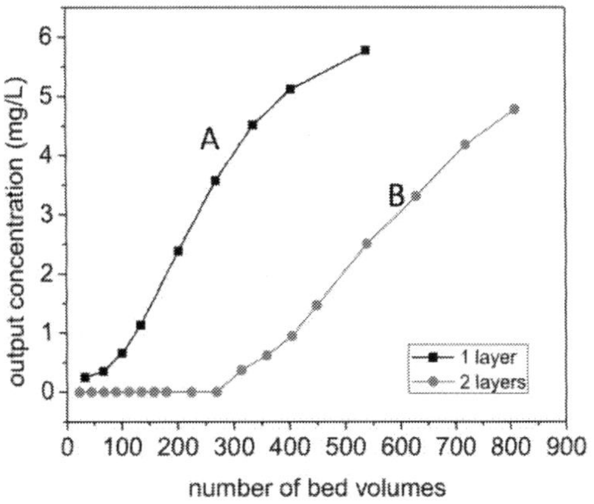

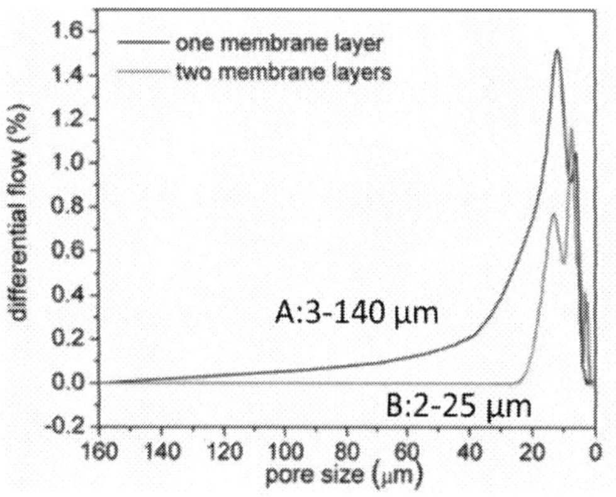

initial concentration = 10 ppm, pH = 6, flow rate = 2.9 mL/min

Yongming Wei, et al, More effective membrane chromatography, ***AIChE Journal***, 61(11): 3871-3878, 2015

Flow rate barely affects loading

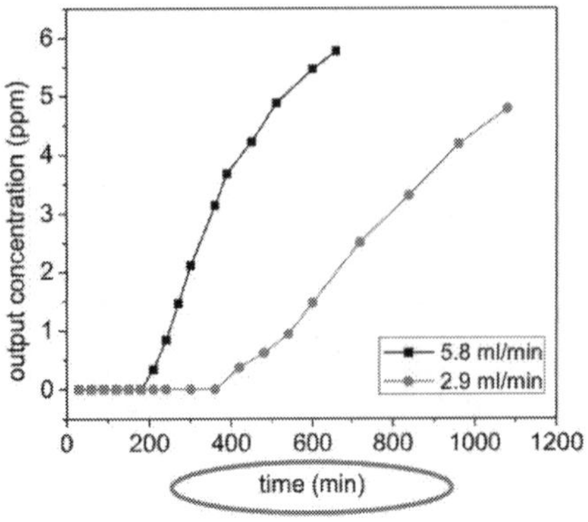

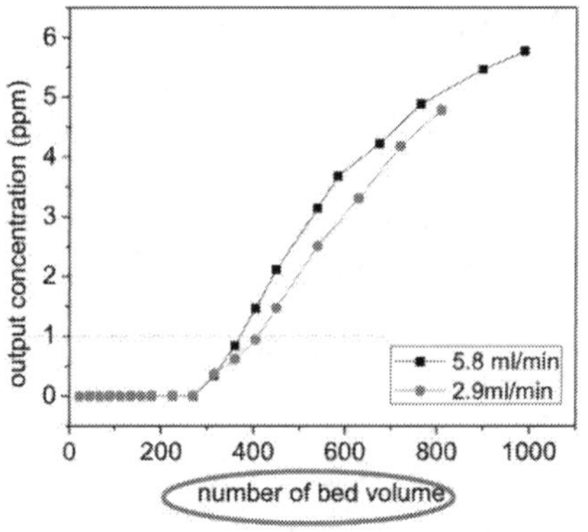

initial concentration = 10 ppm, pH = 6, two membrane layers

Selectivity

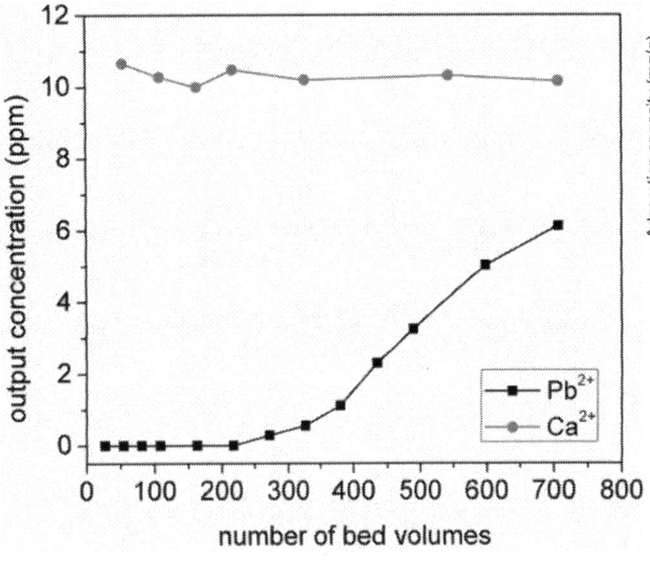

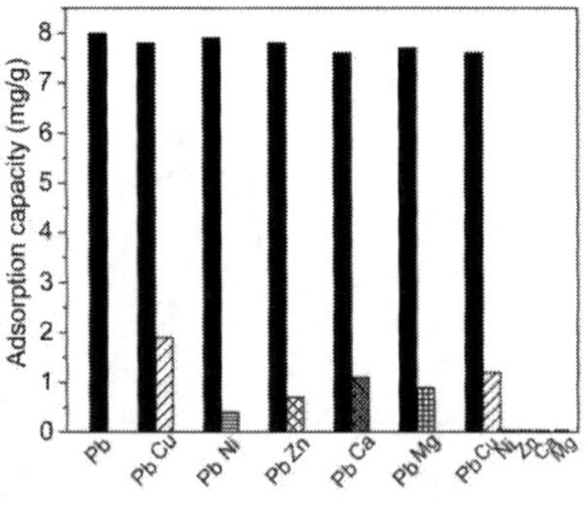

Re-generability

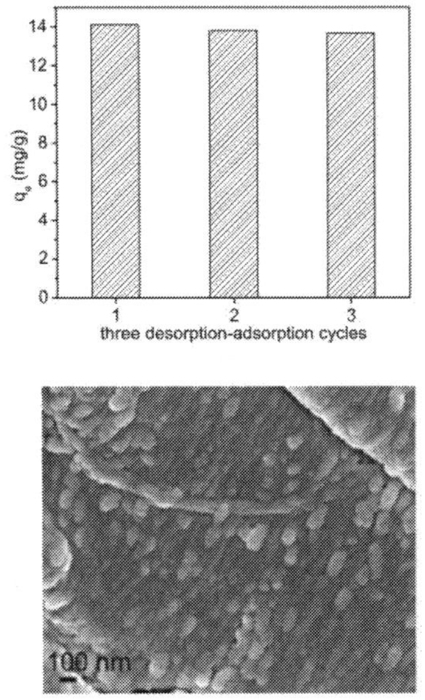

Fiber regeneration

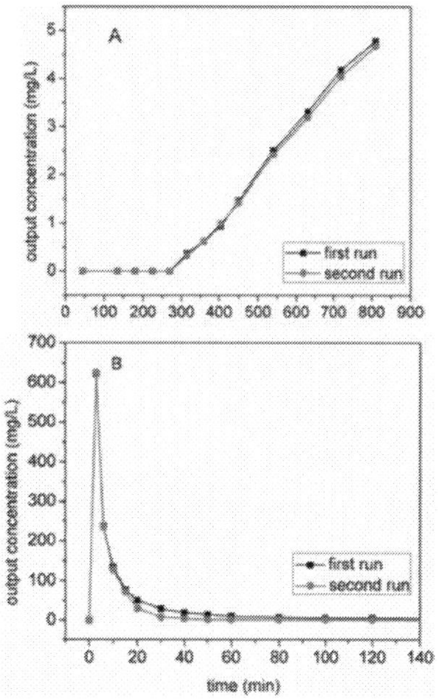

Bed regeneration

Breakthrough data fitted Dose-Response and A-B Models

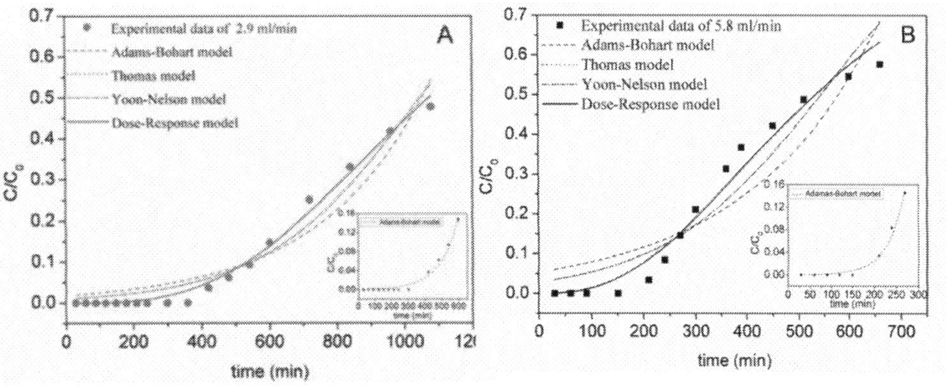

$$\frac{C}{C_0} = 1 - \frac{1}{1 + (\frac{C_0 V}{q_0 m})^a}$$

Dose-response model

$$\frac{C}{C_0} = e^{K_{AB} C_0 t - (K_{AB} N_0 Z/\nu)}$$

Adams-Bohart model

Conclusions

- Fibrous membrane chromatography can be easily made using wet-laid paper making technology or electrospinning
- The example shown here with TiO_2/cellulose fibers is simply a start
- Such alternative chromatography targets the removal of small amount of pollutants in water
- Designing functional fibers is key to success

413656 Biocatalytic Reactive Extraction in Micellar Two-Phase Systems: Implementation in a Continuous Process

Abstract:

Aqueous non-ionic surfactant solutions split into two phases above the so called cloud point temperature (CPT). Thereby an aqueous phase coexists with a surfactant-rich phase. This phenomenon is applied for separation processes like the cloud point extraction (CPE) for treatment of waste water or the purification of valuable compounds. Due to its mild operating conditions (low salt concentrations, low temperature) and the biocompatibility of many surfactants the cloud point extraction is generally considered as a promising method in biotechnological downstream processes. Recent studies describe the simultaneous biocatalysis combined with micellar extraction. However, this reactive extraction was not performed as a continuous process so far [1].

In this work, an aqueous micellar two-phase system containing the non-ionic surfactant Tergitol NP-7 is used to perform a biocatalytic reaction with a simultaneous separation of the products. The biocatalytic hydrolysis of penicillin G to phenylacetic acid (PAA) and 6-aminopenicillanic acid (APA) is used as a model reaction. Thereby, APA distributes predominantly into the aqueous phase while PAA accumulates in the surfactant-rich phase.

The liquid-liquid equilibrium of the aqueous micellar two-phase system changes during the reaction process since penicillin G elevates and PAA depresses the CPT. Moreover, the production of PAA decreases the pH which influences both, reaction kinetics and equilibrium composition as well as reactant distribution.

A pH-dependent study of the partition equilibrium of all compounds was performed to define the operating window for the reactive extraction process. Furthermore, the biocompatibility of Tergitol NP-7 with respect to the enzyme stability and activity was proven. Consequently, the biocatalytic reactive extraction was performed for the first time in a continuous countercurrent extraction column. The results prove the biocatalytic reactive extraction as a feasible alternative for biotechnological processes.

References: [1] Wang et al., Bioprocess Biosyst Eng (2006) 29: 157-162

417850 Applied Catastrophic Phase Inversion (ACPI) – a Continuous Non Centrifugal Phase Separation in Biphasic Whole-Cell Biocatalysis

Abstract:

As a reaction to decreasing fossil feedstock, the biotechnological production of fine chemicals has gained an increased academic and industrial interest over the last decades. Within this development biphasic whole-cell biotransformations have been shown to be reliable and efficient alternatives to common chemical synthesis routes, especially for the production of e.g. apolar enantiopure organic compounds. Biphasic whole-cell biotransformations provide high stereo selectivity as well as high product titers due to the presence of an organic phase serving as substrate reservoir and product sink. The organic phase thereby minimizes the toxification of the biocatalyst by the substrate and/or product. The main bottleneck for an industrial implementation is the formation of stable Pickering type emulsions due to the presence of whole cells. State of the art downstream processing of these emulsions includes the use of excessive centrifugation, chemical de-emulsifiers or thermal stress, leading to inefficient processes. In contrast, using the catastrophic phase inversion phenomenon applied within this work, which describes a sudden switch of emulsion type caused by addition of dispersed phase, Pickering type emulsions can be destabilized efficiently in a continuous setup. Four different reaction systems (bis(2-ethylhexyl) phthalate, ethyl oleate or dodecanol as organic phase in combination with *E.coli* JM101 or *Pseudomonas putida* VLB120) 2 were successfully separated.

Based on batch experiments to determine the influence of phase ratio and processing conditions, a simple prototype, consisting of a single mixer settler stage, capable of continuous emulsion separation was developed and constructed. In contrast to state of the art centrifugal separations, ACPI benefits from its simple setup as well as low operating costs and is therefore a simple applicable method that can generally be used for separation of stable Pickering type emulsions based on the knowledge of the point of inversion. This fact qualifies ACPI as a considerable alternative as first separation step in the DSP of biphasic whole-cell biotransformations.

426102 Protein Refolding in Aqueous Two-Phase Systems: Influence of System Design Parameters Using Invertase and Laccase As Model Proteins

Abstract:

Protein refolding represents the limiting step in the recovery of these biomolecules from inclusion bodies and from denaturing operations. Aqueous two-phase systems (ATPS) besides being a primary recovery operation provide good and affordable physicochemical conditions to refold these molecules. On their part, ATPS are formed when mixing two water-soluble substances (a salt and a polymer or two polymers) above a critical concentration giving as result two immiscible phases. When a biological sample is introduced in these systems, hydrophobic interactions, hydrogen bonding, Van der Waals and electrostatic forces between water, the phase forming chemicals and the sample molecules drive the latter ones to fractionate to either the top or bottom phase of the system once thermodynamic equilibrium is achieved. Traditionally, polyethylene glycol (PEG) is one of the most used polymers in developing ATPS strategies. This polymer additionally confers the system with protein refolding characteristics of importance when dealing with denaturized samples since it binds as an intermediate promoting appropriate folding with an analogous function to that of chaperones in *in vivo* folding. It therefore becomes important to characterize how the different system design parameters such as polymer molecular weight, tie-line length (TLL), volume ratio (V_R), and system pH influence protein partition and refolding from a denatured state assessed by measuring the partition coefficient (K_P, defined as the concentration ratio of the molecule of interest between the top and bottom phases) and the specific activity of the target biomolecules. In this work, 48 different PEG – potassium phosphate ATPS were tested for the partition and refolding of native and denatured invertase and laccase. The model enzymes in both forms have a partition preference for the bottom salt-rich phase where the polymer molecular weight and the tie-line length (TLL) selected for the system play a preponderant role in partition and refolding. Refolding percentages for invertase of up to 100% can be obtained in the polymer-rich phase and between 50% and 75% in the bottom phase. The achieved refolding percentages for laccase were significantly lower since only up to 35% of the loaded samples were correctly refolded in one of the tested systems. This work seeks to evaluate the partition behavior of native and denatured invertase and laccase in PEG – potassium phosphate ATPS and to characterize the influence of system design parameters upon refolding and enzymatic reactivation en route to establishing a potential strategy to define a simplified protocol for the *in situ* recovery and refolding of these enzyme from different sources.

420316 Co-Recovery of Lipids, Fermentable Sugars, and Protein from Bio-Oil Bearing Biomass Using Ionic Liquid Co-Solvents

Abstract:

This work will discuss the efficacy of an ionic liquid-methanol co-solvent system to extract lipids, fermentable sugars, and protein from bio-oil-bearing biomass. Application of this system for the recycle of fermentable sugars to grow bio-oil bearing yeast biomass and the recovery of phorbol esters from oil-seed biomass will also be presented. Finally, key issues for value as a commercial process will be discussed.

414223 Retaining and Quantifying Medium Chain Alcohols in Fermentation Broth

Abstract:

Primary alcohols such as 1-ethanol and 1-butanol can be produced through industrial fermentation processes using wild type microorganism strains. Today's recombinant DNA technology provides an opportunity for the production of longer chain primary alcohols through fermentation. Traditionally, the short chain alcohols, 1-ethanol and 1-butanol which have good solubility in fermentation broth, accumulate in the fermentation broth and can be recovered though distillation. However primary alcohols of longer chain length have deceasing solubility in water. Additionally, below their solubility limit, primary alcohols of increasing chain length have increasing vapor pressures, despite the opposite trend for pure primary alcohols. The increased volatility hinders quantification of alcohol titers in traditional aerobic fermentation systems, leading to difficulties in optimizing strains and fermentation conditions. This work evaluates various solutions to retain medium chain primary alcohols (1-pentanol to 1-octanol) in shake flasks, serum bottles and sparged bioreactors. The results and impacts of applying the solutions to actively producing cultures is presented.

Exploration into Liquid-Liquid Extraction Solvents to Remove *Saccharomyces cerevisiae* Inhibitors

Mahdieh Aghazadeh[1,2], *Abigail S. Engelberth*[1,2,3]

1. Department of Agricultural & Biological Engineering, 2. Laboratory of Renewable Resources Engineering, 3. Environmental & Ecological Engineering. Purdue University

Abstract

The process to convert lignocellulosic biomass bioethanol involves begins with pretreatment of the woody structure of the biomass. Pretreatment allows for better accessibility of the polymeric sugars for enzyme digestion, but also results in the release of detrimental compounds that can inhibit fermentation.

Laboratory results indicate that liquid-liquid extraction (LLE) is able to remove the common fermentation inhibitors and reduce the concentration in a pre-fermentation broth below an inhibitory threshold. Bench top studies were used in conjunction with process simulations to select an organic solvent for use in LLE. The goal was to identify an organic extraction solvent with the lowest miscibility with the biomass liquid hydrolysate while allowing the sugars and water to remain in the stream destined for fermentation.

Through careful study, an initial list of fifty solvents was narrowed to nine using an AspenPlus™ simulation. Laboratory experiments were then conducted to demonstrate that the affinity of each of the nine solvents to sugars is negligible. One solvent proved that it was also able to achieve complete extraction of acetic acid – the most ubiquitous inhibitor – from biomass liquid hydrolysate. Further simulation clarified the impact of LLE on the remaining known inhibitors. Lastly, a cost estimation analysis justified the economic feasibility of introducing an LLE process for inhibitor removal within a simulated biorefinery.

Introduction

The biochemical option for bioethanol formation involves: preparation of the biomass, pretreatment, enzymatic hydrolysis, fermentation, ethanol separation and recovery (Ladisch, Ximenes, Engelberth, & Mosier, 2014). Pretreatment is necessary to make the cellulose and hemicellulose accessible for enzymatic hydrolysis. Fermentation inhibitors are compounds that are inherently present in the solution after pretreatment of lignocellulose biomass. Examples of inhibitors include weak acids (e.g. acetic, formic, and levulinic), furans and phenols. All biomass is comprised of some combination of cellulose, hemicellulose, and lignin. When the weak ester bonds in the hemicellulose break, the acetyl groups are easily liberated during the biomass pretreatment, therefore acetic acid is present in all pretreated biomass (Grzenia, Schell, & Wickramasinghe, 2008).

Inhibition slows down the fermentation and reduces the yield of ethanol produced (Kim, Kreke, Hendrickson, Parenti, & Ladisch, 2013; Nilvebrant, Reimann, Larsson, & Jonsson, 2001). Casey *et al.* (2010) demonstrated that the yeast growth, substrate consumption, and ethanol volumetric productivity all decrease in presence of acetic acid.

Studies have been performed to determine the extent of removal of fermentation inhibitors prior to fermentation using ion exchange column, nano-filtration, and hollow fiber membranes (Grzenia et al., 2008; Nilvebrant et al., 2001; Weng et al., 2009; Weng et al., 2010). Liquid-liquid extraction (LLE) is a well-known process that has not been applied to this particular extraction issue. LLE uses the addition of an immiscible solvent to the liquid system that has higher affinity to the desired compounds. Solvent selection is a prominent part of this technology since purification of the desired compounds and recycle of the solvent are essential steps in this process.

Laboratory experiments have been performed to extract acetic acid via liquid-liquid extraction with trioctylphosphine oxide (TOPO) (Um, Friedman, & van Walsum, 2011). Zautsen *et al.* (2009)

measured the partition coefficients of inhibitors in many organic solvents and found that the better performing solvents may not be biocompatible with the fermentation microorganisms. Ahsan *et al.* (2013) found high liquid-liquid extraction efficiencies in their work to recover acetic acid from pre-hydrolysis liquor from a Kraft pulp process. Katikaneni and Cheryan (2002) conducted a comparative study to select the most effective method between LLE and esterification to recover acetic acid from an acetic acid fermentation process. They found that at higher acetic acid content and lower pH the extraction efficiency increases.

The research reported in this manuscript was conducted in an effort to identify the most efficient solvent to extract fermentation inhibitors using LLE and determine the subsequent economic impact on an existing biorefinery. This study examined multiple organic solvents using AspenPlus™ simulations, quantified the extraction performance on both a test solution and on biomass hydrolysate; and finally calculated the economic impact of this process on a biorefinery.

Materials and Methods

Simulation: AspenPlus™ version 7.3 (Aspen Technology Inc., Burlington, MA) was used to simulate the extraction of acetic acid, glucose, and xylose from a water solution using a 10 stage LLE column. The column was sized for economic purposes using simple flooding calculations.

Laboratory Experiment and Analysis: The LLE method was modified from the experimental method of Katikaneni and Cheyran (2002). LLE was performed in 50 mL vials with 10 mL of an aqueous feed and 10 ml of the solvent. Two types of solutions were created: a test solution containing only the compounds used in the simulation, and a corn stover hydrolysate solution with concentrations adjusted to match the simulation. The test solutions were made with 30 g/L glucose, 25 g/L xylose, and 10 g/L acetic acid. These concentrations were picked based on literature review on common acetic acid, glucose, and xylose concentrations in hydrolysate from lignocellulosic biomass (Casey et al., 2010; Garlock et al., 2011; Grzenia et al., 2008; Mao, Genco, van Heiningen, & Pendse, 2010). Biomass hydrolysate was the result of pretreating corn stover (40 mesh sieve) with 1% wt. sulfuric acid at 140 °C for 40 minutes. Pretreatment was conducted in 1" OD × 0.083 in. wall thickness, 316 stainless steel tubing capped with 1" Swagelock tubes and fittings (Swagelock, Indianpolis, IN). After adding the 10 mL of the organic solvent (solvent to feed ratio of 1:1), the vials were shaken for 5 minutes, left to equilibrate at room temperature for 12 hours, and then centrifuged for 20 minutes at 5000 rpm. All chemicals were purchased from Sigma (Sigma Aldrich, St. Louis, MO).

HPLC (Waters 2695 Separation Module, Waters Corporation, Milford, MA) was used to measure the glucose, xylose, and acetic acid contents of aqueous phase (Aminex HPX-87H Column, Bio-Rad, Hercules, CA – 300 x 7.8 mm pre-packed HPLC carbohydrate analysis column). The mobile phase was 5 mM sulfuric acid with a flow rate 0.6 mL/min, an internal temperature 35 °C and external temperature 65 °C.

Cost Estimation: The economic analysis and comparison was performed using Excel spreadsheet (Microsoft Excel 2010). Baseline costs for the biorefinery capacity, capital and operational cost were derived from NREL process design (David Humbird, National Renewable Energy, & Harris Group, 2011).

Results and Discussion

Removal of fermentation inhibitors, especially acetic acid, using liquid-liquid extraction (LLE) is the focus of this work. To establish the best performing solvent to extract acetic acid and other fermentation inhibitors in an LLE apparatus, a series of AspenPlus™ simulations and experiments were conducted. An initial list of fifty solvents were selected for screening based on the ternary phase diagrams of acetic acid, water and the solvent (Sorensen, 1980). To narrow down the fifty solvents to a more reasonable list, four benchmarks were developed. These benchmarks took into account both process and environmental constraints. The benchmarks are: 1) the consumption rate of the solvent should be as low as possible while achieving a high acetic acid yield, 2) water content in the extract stream should be as low as possible, 3) the sugar initially entering with the feed stream should leave the column in the raffinate stream and 4) the solvent must not be carcinogenic or toxic. Based on the fourth benchmark, eighteen solvents were immediately discarded due to their inherent toxic properties.

Simulation: Narrowing Down the Solvents

For the remaining thirty-two solvents, an AspenPlus™ simulation was developed to test the split fraction, $SF=\left(\frac{\text{acetic acid in the extract}}{\text{acetic acid in the feed}}\right)$, of the acetic acid between the water and the third solvent. Each of the solvents were run through the simulation varying the solvent volumetric flow rate until the split fraction of acetic acid was equal to a desired set point of 0.99. This meant that some solvents required a high solvent consumption rate to achieve the chosen SF and were thus eliminated, as they did not fit the first benchmark. The simulation allowed for the elimination of nine additional solvents; twenty-three of the solvents were able to achieve the desired SF within a ten-stage extraction column. Figure 1 displays the solvent flow rate and the water content in the extract stream for the twenty-three viable solvents. It is apparent in Figure 1 that there is a trade-off when selecting the ideal extraction solvent, as the solvent consumption rate decreases, the water content in the extract stream increases. The optimum range, based on Figure 1, includes solvents between ethyl acetate to methyl butyrate; these nine solvents consumed less than 150 kmol/hr for 500 kmol/hr of aqueous feed, while allowing the water content in the extract to remain low (less than 25 kmol/hr). The gray box in Figure 1 clearly displays the nine solvents that fit the first two benchmarks, which include: ethyl acetate, *n*-butyl acetate, cyclohexyl acetate, ethyl propionate, isobutyl acetate, *n*-pentyl acetate, isopentyl acetate, ethyl butyrate, and methyl butyrate.

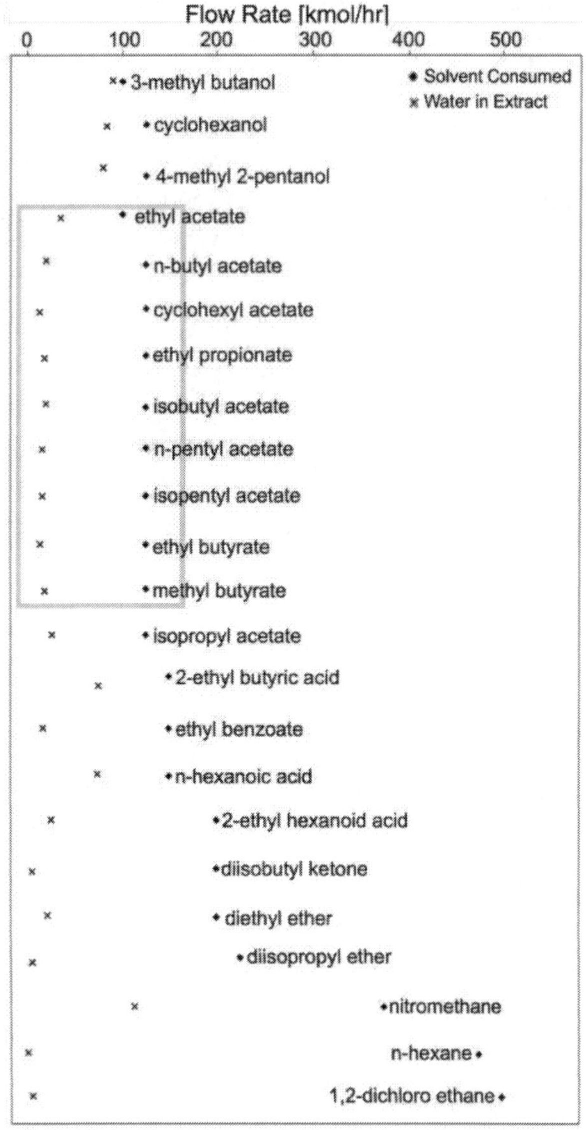

Figure 1: Solvent consumption rate and water content in the extract stream from the simulation results of the solvents extracting 99% acetic acid from a solution of water, acetic acid, glucose, and xylose. The compounds within the box were chosen as the best performing solvents based on their low solvent consumption while keeping the water in the extract as low as possible.

Experimental: Acetic Acid Extraction

The nine solvents that were selected via simulation were then subjected to a single-stage LLE experiment to determine their ability to remove acetic acid from the aqueous solution. The LLE was necessary to test the solvents with the aqueous solution because the destination stream of the sugars could be easily manipulated in the simulation based on the chosen property method. An initial LLE experiment with a test solution was conducted to evaluate the simulation results. Figure 2 shows the experimental results of the test solution for SF of acetic acid for each of the nine solvents tested along with the percent of xylose and glucose that remain in the aqueous phase (labeled as % Recovery). For all nine solvents, sugars remained predominately in the aqueous phase; 92%-95% for glucose, and 93%-99% for xylose.

To gain a more complete understanding of the performance of each of the nine solvents the LLE experiment was performed on a solution of corn stover hydrolysate. The procedure to make the hydrolysate was described in the material and methods section. A summary of the LLE experimental results comparing the test solution with actual corn stover liquid hydrolysate is shown in Figure 2. Note that the performance of the nine solvents is the same for the test solution as for the liquid hydrolysate. Using the data gathered from the LLE experiment, the nine solvents were ranked based on their ability to meet the benchmarks previously stated. The ranking of nine solvents, from best performing to unsatisfactory are: ethyl acetate, butyl acetate, isobutyl acetate, ethyl propionate, methyl butyrate, ethyl butyrate, cyclohexyl acetate, pentyl acetate, isopentyl acetate.

It was evident that ethyl acetate, butyl acetate, isobutyl acetate, and ethyl propionate were able to extract more acetic acid from the corn stover hydrolysate than from the test solution likely due to the presence of other molecules present in the hydrolysate such as sulfuric acid. The acid content, and thus lower pH, can enhance the LLE yield for acetic acid (i.e. produce a salting-in effect) (Cohn, 1943; Katikaneni & Cheryan, 2002; Wheelwright, 1991). The noteworthy result from Figure 2 is that ethyl acetate is able to extract 85% of the acetic acid from a real solution while allowing the sugars to remain with the raffinate.

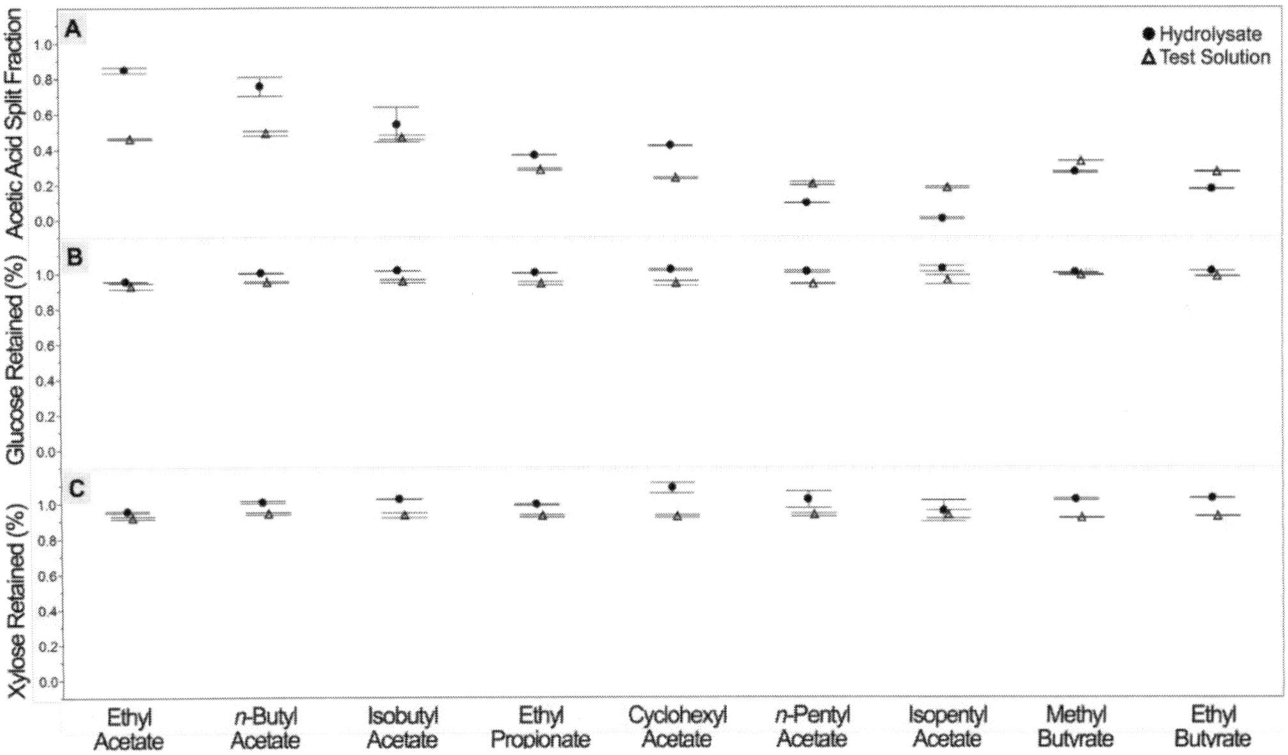

Figure 2: Comparison of the test solution and a real biomass hydrolysate to determine the performance of each of the nine organic extraction solvents. A) fraction of acetic acid extracted by each solvent – shown as split fraction, SF. A split fraction closer to one is more ideal as it indicates that the solvent removes more acid. B) Percent of glucose retained in aqueous phase, and C) percentage of xylose retained in aqueous phase. The ideal case is for the % retained to be as close as possible to 100% since the goal is for the sugars to not be extracted by the solvent.

Simulation: Extraction of Other Common Inhibitors

Based on the results gathered for the performance of the nine solvents to remove acetic acid from a solution of corn stover hydrolysate, the next question was to see how well the solvents were able to extract other common fermentation inhibitors from biomass hydrolysate. The common fermentation inhibitors were determined from existing literature and the higher end of the reported concentration was used in our simulation. Each of the inhibitors were added to the feed stream in the AspenPlus™ simulation. The inhibitors, along with the common concentration found in a post-pretreatment solution, are shown in Table 1.

Table 1: Concentration of additional fermentation inhibitors tested with each of the nine extraction solvents in the AspenPlus™ LLE simulation.

Inhibitor	Concentration (g/L)	Common Biomass Source	Reference
Acetic Acid	16.1	Hemicellulose acetyl bonds	(D. Humbird, Mohagheghi, Dowe, & Schell, 2010)
Furfural	2.9	Lignin degradation	(D. Humbird et al., 2010)
Formic Acid	3.5	Carbohydrate degradation	(Martin & Jonsson, 2003)
Levulinic Acid	2.6	Carbohydrate degradation	(Almeida et al., 2007)
Vanillin	0.43	Lignin degradation	(Almeida et al., 2007)
Cinnamic Acid	0.15	Lignin degradation	(Martin & Jonsson, 2003)
4-Hydroxybenzoic Acid	0.005	Lignin degradation	(Nichols et al., 2008)
Hydroxyacetaphenone	0.004	Lignin degradation	(Almeida et al., 2007)
Acetovanillone	0.008	Lignin degradation	(Almeida et al., 2007)
3,4 dihydroxybenzoic acid	0.000005	Lignin degradation	(Nichols et al., 2008)
Syringic Acid	0.44	Lignin degradation	(Almeida et al., 2007)
Hydroquinone	0.017	Lignin degradation	(Almeida et al., 2007)
Phenol	0.035	Lignin degradation	(Almeida et al., 2007)

The results of the simulation are shown in Figure 3. The solvents in Figure 3 are arranged along the x-axis in the order of how well they extract each of the inhibitors. Ethyl Acetate is shown to be able to fully extract eleven of the thirteen inhibitors. The order of the inhibitors in the legend has been arranged to demonstrate how likely each is to be extracted by the solvents tested. Furfural, cinnamic acid, hydroxyacetophenone, phenol, 4-hydroxybenzoic acid, and acetovanillone are all fully extracted by each solvent. The remaining seven inhibitors would require a higher solvent flow rate for better extraction. Based on the results shown in Figure 3, the ideal solvent is ethyl acetate due to its capacity to completely extract all inhibitors at a low solvent consumption (Flow rate = 100 kmol/hr) except acetic acid (SF =0.73) and formic acid (SF = 0.19). These findings are in agreement with the experiment results in Figure 2 and greatly simplify the solvent selection process as it predicts that the selected solvent from the experimental results will perform well with the other inhibitors as well.

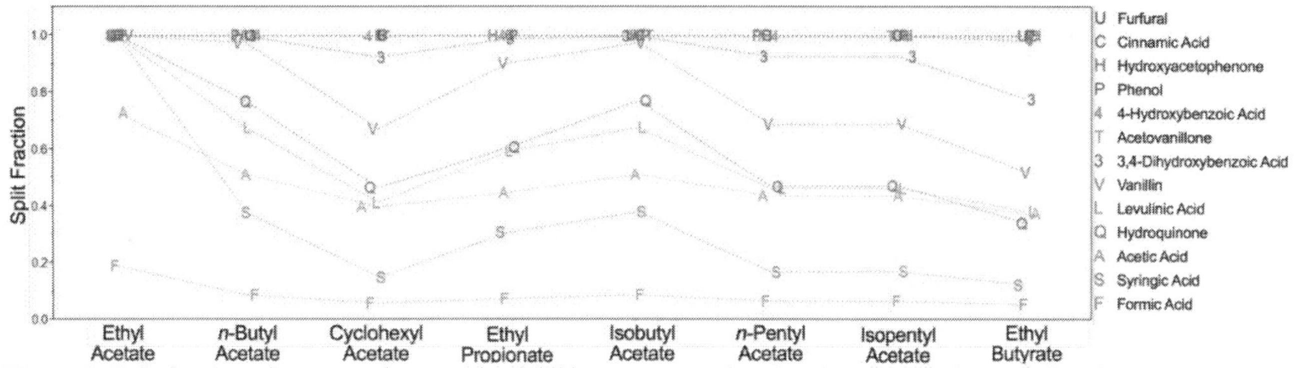

Figure 3: Split fraction of common fermentation inhibitors compounds using the selected nine organic solvents. Fraction of 1 indicates 100% extraction of the inhibitor. The legend explains the symbols on each line. The lines were included to more easily compare the performance of the inhibitors in each solvent.

Economic Analysis

The growth rate of *Z. mobilis* cells are affected by acetic acid concentrations (Franden, Pilath, Mohagheghi, Pienkos, & Zhang, 2013), and therefore the ethanol production rate changes based on the presence of acetic acid in the broth. Cell growth, and hence ethanol production, stops at high acetic acid content and reaches a maximum at zero acetic acid. Öhgren *et al.* (2006) demonstrated that the energy demand of the distillation column to purify the bioethanol is significantly influenced by the ethanol content at lower concentrations. When the amount of ethanol (% wt.) is 4 or less in the fermentation broth the energy – and therefore capital and manufacturing cost – increases significantly. However, at 5% wt. and above it is safe to assume that the energy requirement per flow-rate of the produced ethanol remains unchanged.

In an effort to estimate the impact of the LLE unit on the overall cost of the biorefinery, an itemized study of the capital and manufacturing costs was conducted. Total revenue of the biorefinery is a function of the sale of ethanol, fixed and manufacturing costs, as well as the loan payments. The feedstock and the labor expenses are not affected by implementing the LLE unit; as a result the manufacturing cost would be increased by the cost of the solvent. In addition, the flow-rates of the streams and the equipment sizing are not impacted; meaning that the capital investment would be only increase by the cost of the extraction column. Therefore it is reasonable to assume that incorporating liquid-liquid extraction in an existing corn stover biorefinery – modeled by Humbird *et al.* (2011) – to improve the final ethanol content of the fermentation broth will impact the fixed capital cost by the extraction column installation cost and the manufacturing cost through the price of the un-recycled solvent.

Engineering cost estimation equations were used to calculate the cost of the extraction column based on the flow-rate data of the NREL AspenPlus™ simulation. Flooding equations for a packed column were used to estimate the size of the extraction column and this was incorporated in the purchasing cost estimation of the tray and packed columns made of stainless steel (Seader, 1998; Turton, 2012). The total cost of this column (in 2011 dollars) is 4.4 million USD, which is 2 % of the total install cost of the plant (232 million USD) and only 1.1 % of the total fixed capital cost (423 million USD). Ethyl acetate F.O.B price is 850 Euro/tonne (Pitches, 2013). Assuming 90 % recovery at each run and solvent to feed molar ratio of 1 to 5, lead to an initial cost of 9.2 million USD of solvent cost for the first batch and only 38,000 USD of annual cost. This would contribute to 0.05 % of increase in the total annual manufacturing cost of the plant (76 million USD).

Conclusion

A group of twenty-three solvents were ranked on their ability to extract acetic acid from biomass hydrolysate using AspenPlus™ simulation. Laboratory results confirmed that ethyl acetate, butyl acetate, and isobutyl acetate have the capacity to reduce the acetic acid concentration below its inhibition threshold (less than 0.5 % wt.). Further simulation predicted that solvents that perform well to extract acetic acid also perform well to extract other common fermentation inhibitors. These non-toxic solvents have the capability to remove the fermentation inhibitors at low flow-rates, have no or negligible solubility with the pre-fermentation broth, and have minimum affinity to the sugars. Finally, a cost estimation analysis proved that incorporating liquid-liquid extraction using ethyl acetate as the extraction solvent within an existing corn stover biorefinery will have negligible impact on the manufacturing and capital costs of the plant – 0.05 and 2 %.

References

Ahsan, L., Jahan, M. S., & Ni, Y. (2013). Recovery of Acetic Acid from the Prehydrolysis Liquor of Kraft Based Dissolving Pulp Production Process: Sodium Hydroxide Back Extraction from the Trioctylamine/Octanol System. *Industrial & Engineering Chemistry Research, 52*(26), 9270-9275. doi:10.1021/ie401285v

Almeida, J. R. M., Modig, T., Petersson, A., Hahn-Hagerdal, B., Liden, G., & Gorwa-Grauslund, M. F. (2007). Increased tolerance and conversion of inhibitors in lignocellulosic hydrolysates by Saccharomyces cerevisiae. *Journal of Chemical Technology and Biotechnology, 82*(4), 340-349. doi:10.1002/jctb.1676

Casey, E., Sedlak, M., Ho, N. W. Y., & Mosier, N. S. (2010). Effect of acetic acid and pH on the cofermentation of glucose and xylose to ethanol by a genetically engineered strain of Saccharomyces cerevisiae. *Fems Yeast Research, 10*(4), 385-393. doi:10.1111/j.1567-1364.2010.00623.x

Cohn, E. J. (1943). *Proteins, amino acids and peptides as ions and dipolar ions*. New York: New York, Reinhold Publishing Corporation.

Franden, M. A., Pilath, H. M., Mohagheghi, A., Pienkos, P., & Zhang, M. (2013). Inhibition of growth of Zymomonas mobilis by model compounds found in lignocellulosic hydrolysates. *Biotechnology For Biofuels, 6*. doi:10.1186/1754-6834-6-99

Garlock, R. J., Balan, V., Dale, B. E., Pallapolu, V. R., Lee, Y. Y., Kim, Y., . . . Warner, R. E. (2011). Comparative material balances around pretreatment technologies for the conversion of switchgrass to soluble sugars. *Bioresource Technology, 102*(24), 11063-11071. doi:10.1016/j.biortech.2011.04.002

Grzenia, D. L., Schell, D. J., & Wickramasinghe, S. R. (2008). Membrane extraction for removal of acetic acid from biomass hydrolysates. *Journal of Membrane Science, 322*(1), 189-195. doi:10.1016/j.memsci.2008.05.030

Humbird, D., Mohagheghi, A., Dowe, N., & Schell, D. (2010). Economic Impact of Total Solids Loading on Enzymatic Hydrolysis of Dilute Acid Pretreated Corn Stover. *Biotechnol. Prog., 26*(5), 1245-1251. doi:10.1002/btpr.441

Humbird, D., National Renewable Energy, L., & Harris Group, I. (2011). *Process design and economics for biochemical conversion of lignocellulosic biomass to ethanol dilute-acid pretreatment and enzymatic hydrolysis of corn stover*. Golden, CO: Golden, CO : National Renewable Energy Laboratory.

Katikaneni, S. P. R., & Cheryan, M. (2002). Purification of fermentation-derived acetic acid by liquid-liquid extraction and esterification. *Industrial & Engineering Chemistry Research, 41*(11), 2745-2752. doi:10.1021/ie010825x

Kim, Y., Kreke, T., Hendrickson, R., Parenti, J., & Ladisch, M. R. (2013). Fractionation of cellulase and fermentation inhibitors from steam pretreated mixed hardwood. *Bioresource Technology, 135*, 30-38. doi:10.1016/j.biortech.2012.10.130

Ladisch, M., Ximenes, E., Engelberth, A., & Mosier, N. (2014). Biological Engineering and the Emerging Cellulose Ethanol Industry. *Chemical Engineering Progress, 110*(11), 59-62.

Mao, H., Genco, J. M., van Heiningen, A., & Pendse, H. (2010). KRAFT MILL BIOREFINERY TO PRODUCE ACETIC ACID AND ETHANOL: TECHNICAL ECONOMIC ANALYSIS. *Bioresources, 5*(2).

Martin, C., & Jonsson, L. J. (2003). Comparison of the resistance of industrial and laboratory strains of Saccharomyces and Zygosaccharomyces to lignocellulose-derived fermentation inhibitors. *Enzyme and Microbial Technology, 32*(3-4), 386-395. doi:10.1016/s0141-0229(02)00310-1

Nichols, N. N., Sharma, L. N., Mowery, R. A., Chambliss, C. K., van Walsum, G. P., Dien, B. S., & Iten, L. B. (2008). Fungal metabolism of fermentation inhibitors present in corn stover dilute acid hydrolysate. *Enzyme and Microbial Technology, 42*(7), 624-630. doi:10.1016/j.enzmictec.2008.02.008

Nilvebrant, N. O., Reimann, A., Larsson, S., & Jonsson, L. J. (2001). Detoxification of lignocellulose hydrolysates with ion-exchange resins. *Applied Biochemistry and Biotechnology, 91-3*, 35-49. doi:10.1385/abab:91-93:1-9:35

Öhgren, K., Rudolf, A., Galbe, M., & Zacchi, G. (2006). Fuel ethanol production from steam- pretreated corn stover using SSF at higher dry matter content. *Biomass and Bioenergy, 30*(10), 863-869. doi:10.1016/j.biombioe.2006.02.002

Pitches, J. (2013). Ethyl acetate. *ICIS Chemical Business, 284*(9), 39.

Seader, J. D. (1998). *Separation process principles*. New York: New York : Wiley.

Sorensen, J. M. a. W. A. (1980). *Liquid-liquid equilibrium data collection:Ternary and Quanternary Systerms*. Germany: Dechema.

Turton, R. (2012). *Analysis, synthesis, and design of chemical processes* (4th ed.. ed.). Upper Saddle River, NJ: Upper Saddle River, NJ : Prentice Hall.

Um, B. H., Friedman, B., & van Walsum, G. P. (2011). Conditioning hardwood-derived pre-pulping extracts for use in fermentation through removal and recovery of acetic acid using trioctylphosphine oxide (TOPO). *Holzforschung, 65*(1), 51-58. doi:Doi 10.1515/Hf.2010.115

Weng, Y.-H., Wei, H.-J., Tsai, T.-Y., Chen, W.-H., Wei, T.-Y., Hwang, W.-S., . . . Huang, C.-P. (2009). Separation of acetic acid from xylose by nanofiltration. *Separation and Purification Technology, 67*(1). doi:10.1016/j.seppur.2009.03.030

Weng, Y.-H., Wei, H.-J., Tsai, T.-Y., Lin, T.-H., Wei, T.-Y., Guo, G.-L., & Huang, C.-P. (2010). Separation of furans and carboxylic acids from sugars in dilute acid rice straw hydrolyzates by nanofiltration. *Bioresource Technology, 101*(13). doi:10.1016/j.biortech.2009.11.090

Wheelwright, S. M. (1991). *Protein purification : design and scale up of downstream processing*. Munich ; New York : New York: Munich ; New York : Hanser Publishers ; New York : Distributed in the U.S.A. and in Canada by Oxford University Press.

Zautsen, R. R. M., Maugeri-Filho, F., Vaz-Rossell, C. E., Straathof, A. J. J., van der Wielen, L. A. M., & de Bont, J. A. M. (2009). Liquid-liquid extraction of fermentation inhibiting compounds in lignocellulose hydrolysate. *BIOTECHNOL BIOENG, 102*(5), 1354-1360.

426353 Efficient CO2 Capture By Functionalized Graphene Oxide Nanosheets As Fillers to Fabricate Multi-Permselective Mixed Matrix Membranes

Abstract:

A novel multi-permselective mixed matrix membrane (MP-MMM) is developed by incorporating versatile fillers functionalized with ethylene oxide (EO) groups and amine carrier into polymer matrix. The as-prepared MP-MMMs can separate CO_2 efficiently owing to the simultaneous enhancement of diffusivity selectivity, solubility selectivity and reactivity selectivity. To be specific, MP-MMMs were fabricated by incorporating polyethylene glycol- and polyethylenimine-functionalized graphene oxide nanosheets (PEG-PEI-GO) into commercial low-cost Pebax® matrix. The PEG-PEI-GO plays multiple roles in elevating the membrane performances. Firstly, the high aspect ratio GO nanosheets in polymer matrix increase the tortuous path of gas diffusion and generate a rigidified interface between polymer matrix and fillers, enhancing the diffusivity selectivity. Secondly, PEG consisting of EO groups has excellent affinity for CO_2 to enhance the solubility selectivity. Thirdly, PEI with abundant primary, secondary and tertiary amine groups reacts reversibly with CO_2 to enhance the reactivity selectivity. Thus, the as-prepared MP-MMMs exhibit excellent CO_2 permeability and CO_2/gas selectivity. The MP-MMM doped with 10 wt% PEG-PEI-GO displays the optimum gas separation performance with a CO_2 permeability of 1330 Barrer, CO_2/CH_4 selectivity of 45 and CO_2/N_2 selectivity of 120, surpassing the 2008 Robeson upper bound lines.

435072 Gas Transport in Mixed-Matrix Membranes Formed from M_2(dobdc) Nanocrystals Dispersed in Polyimide-Based Copolymers

Abstract:

Metal-organic frameworks (MOFs) with coordinatively unsaturated metal sites have shown extraordinary performance for industrially relevant adsorption-based separations such as acid gas purification, hydrogen separations, and olefin/paraffin separations. Within this class of materials, one particularly promising candidate is M_2(dobdc), where M is a metal cation and dobdc (2,5-dioxido-1,4-benzenedicarboxylate) is an organic linker. Upon synthesis, these MOFs contain solvent molecules bound to the metal center; however, solvent exchange and heating can remove these bound molecules to expose unsaturated metal sites, thereby giving these materials exceptional adsorption selectivities. These strong adsorption sites, coupled with the ability to tune metal cations, organic linkers, and porosity make this family of MOFs of pressing interest for mixed-matrix membrane research.

Herein, we report the synthesis of mixed-matrix membranes formed from isostructural M_2(dobdc) MOF nanocrystals, where M = Ni and Mg, dispersed in a series of polyimides and polyimide-based random copolymers. The polyimide considered for this work was synthesized from 2,2'-bis-(3,4-dicarboxyphenyl) hexafluoropropane dianhydride (6FDA) and 2,3,5,6-tetramethyl-1,4-phenylenediamine (Durene). Poly(ethylene glycol) (PEG), poly(dimethylsiloxane) (PDMS), and hydroxyl-functional imide segments were copolymerized with 6FDA-Durene to specifically probe polymer/MOF interactions. To formulate structure-property relationships for this class of mixed-matrix membranes, several gas separations, including H_2/CH_4 and CO_2/N_2, were considered. In general, for diffusion-based separations, incorporation of flexible segments, such as PEG, into the polymer backbone negatively affects gas transport properties. However, flexible segments can effectively seal defects between the MOF and polymer, thereby permitting higher MOF loading and creating more mechanically robust, ductile mixed-matrix membranes. Transport properties are investigated for each of these systems to elucidate inherent tradeoffs between permeability, polymer composition, MOF loading, and MOF metal sites.

420319 Growth of High Quality, Thickness-Reduced Zeolite Mem-Branes Towards N_2/CH_4 Separation Using High-Aspect-Ratio Seeds

Abstract:

Greatly improved zeolite membranes were prepared by using high-aspect-ratio zeolite seeds. Slice-shaped seeds with large aspect ratio (AR) facilitated growth of thinner continuous SAPO-34 membranes with much higher quality. These membranes showed N_2 permeances as high as 2.87×10^{-7} mol/(m^2•s•Pa) at 22 °C while maintaining decent N_2/CH_4 selectivity (9~11.2 for equimolar mixture). On the basis of these thinner and high quality SAPO-34 membranes, fine tuning the local crystal structure by incorporating more silicon further increased N_2 permeance by 1.4 times without sacrificing the N_2/CH_4 selectivity. We expect employment of large AR zeolite seeds might be a viable strategy to grow thin and high quality zeolite membranes. In addition, fine tuning crystal structure by changing crystal composition might be a feasible way for further improving separating performance of high quality zeolite membranes.

421285 Silicon Carbide Sintered Supports and Silicon Carbide Membranes for Gas Separation Applications

Abstract:

Efficient separation of H_2 under high temperatures and pressures is fundamental to the development of clean energy industry, and has driven the research on inorganic membranes for two decades. Although several types of nanoporous membranes have been prepared, the fabrication of high-temperature and steam-stable inorganic membranes with high fluxes and large separation factors still remains a challenge. Of all the potential membrane materials, silicon carbide (SiC) is particularly promising due to its high temperature/corrosion resistance, high thermal conductivity, and excellent chemical/mechanical stability. In this work, we prepared highly permeable, mechanically strong porous SiC tubular supports for fabrication of nanoporous SiC membranes. We systematically investigated the effects of sintering temperature, the amount of sintering aids, the composition of the starting particles, and surface modifications on the transport characteristics of sintered SiC supports. The best SiC supports exhibit a He permeance of 5.82×10^{-5} $mol{\bullet}m^{-2}{\bullet}Pa^{-1}{\bullet}s^{-1}$, which is two orders of magnitude larger than the permeance of supports we previously prepared, and also 3~10 times larger than those of similar supports reported by other groups. These substrates are also mechanically strong, as manifested by their Vickers hardness tests. We have fabricated nanoporous SiC membranes on such supports by two different techniques: (i) the pyrolysis of thin pre-ceramic polymer films coated using slip-casting and dip-coating techniques, combined with a sacrificial polystyrene inter-layer method; (ii) the pyrolysis of pre-ceramic polymer films formed by the initiated chemical vapor deposition (iCVD) technique. These SiC membranes show good improvement on both permeance and selectivity compared to our previous membranes.

430882 Nanostructured Polymer-Based Membranes for Gas Separation and Other Applications

Abstract:

Nanostructured Polymer-Based Membranes for Gas Separation and Other Applications

Witopo Salim[1] and W.S. Winston Ho[1,2]

[1]William G. Lowrie Department of Chemical and Biomolecular Engineering, The Ohio State University, 151 West Woodruff Avenue, Columbus, OH 43210-1350, USA

[2]Department of Materials Science and Engineering, The Ohio State University, 2041 College Road, Columbus, OH 43210-1178, USA

Abstract

Advances in nanotechnology in recent years enhance the R&D in nanostructured polymer-based membranes. This work reports on the recent developments of the nanostructured polymer-based membranes, formed by incorporating nanomaterials and/or porous membranes with nanoscale pores, for gas separations, reverse osmosis, ultrafiltration, and other potential applications. Significant progress has been made on the development of nanostructured membranes for gas separations, particularly for CO_2 separations from H_2, N_2 and CH_4. A polymer/zeolite composite nanostructured membrane is presented as an example of the potential of nanostructured polymer-based membrane for the gas separation application of CO_2 capture from flue gas in power plants. Continued improvements on the nanostructured reverse osmosis polyamide membranes via interfacial polymerization have been made along with the efforts on developing novel nanostructured membranes for desalination. The modification of both the surface and bulk properties of the ultrafiltration membranes by nanomaterials can improve the membrane performance and open up the possibility of new applications in addition to those for the current ultrafiltration membrane processes. Amine-containing membranes composed of a selective layer on nanoporous polysulfone support are presented as an example of the application of ultrafiltration membrane for the CO_2 separation from a synthesis gas stream containing CO_2 and H_2S, i.e., for hydrogen purification for fuel cells or for pre-combustion CO_2 capture in IGCC (integrated gasification combined cycle) technology. Similarly, amine-containing membranes composed of a selective layer on nanoporous polyethersulfone support are another example of the application of ultrafiltration membrane for CO_2 capture from flue gas.

418041 Tailoring Free Volume Architecture in Polyimide Gas Separation Membranes Using Iptycene Structures

Abstract:

Development of robust polymeric membranes that can overcome the limits of permeability-selectivity tradeoff has been the main research focus in the field of membrane gas separation. Critical in the function of such polymeric membrane materials is the control of the free volume architecture defined by the chemistry and polymer chain dynamics to achieve excellent size sieving properties for gas separation. We have been working on using intriguing iptycene structure as the building block to construct hierarchically controlled "molecular cavity" architecture in polyimide membranes to simultaneously improve the permeability and selectivity. The rich structural hierarchy and versatile chemistry possibilities of iptycene structures offer unique opportunities for generating well-defined "molecular cavities" that are tunable to enable fast and selective molecular transport in these polyimide membranes. Our first generation of iptycene-polyimides prepared from triptycene-diamine and commercial 6FDA dianhydride showed very attractive gas separation performance because of the high fractional free volume introduced by the triptycene moieties. This talk will focus on two new generations of iptycene-containing polyimide membranes that were developed recently based on newly synthesized pentiptycene-diamine and triptycene-dianhydride, respectively. Specifically, replacing triptycene units with even bulkier pentiptycene moieties in the polyimide structure greatly improved the gas permeabilities with little loss in the selectivity due to larger and tunable internal free volume associated with pentiptycene units. To further manipulate the chain dynamics and free volume architecture, a triptycene-based dianhydride monomer was designed and synthesized to replace the commercial 6FDA dianhydride in the polyimide synthesis leading to highly rigid, fully iptycene-containing polyimides. These new series of iptycene-containing polyimide membranes will be discussed in depth regarding their synthesis, characterization, and gas transport properties. In particular, the fundamental correlations between the microscopic structures and macroscopic transport properties will be discussed for these new gas separation membranes.

433643 Highly Selective Layer-By-Layer Polymer Films for CO2/N2 Separation

Abstract:

The effects of global warming can be partially alleviated by the reduction of carbon dioxide emissions into the atmosphere. Flue gas from fossil fuel power generation plants is the largest contributor to CO_2 emissions. Currently, CO_2 is removed from flue gas (primarily in mixture with N_2) by technologies such as low temperature distillation, physical or chemical absorption, pressure swing adsorption and membrane separation. Cryogenic distillation, absorption technologies, and pressure swing adsorption are complex and energy intensive processes. Gas separation using membranes has garnered significant attention over the past two decades owing to the advantages of high energy efficiency, small footprint, continuous operation and lack of moving parts.

Polymeric gas separation membranes that combine low production costs with strong mechanical properties have been widely used in a variety of gas separation processes such as O_2 and N_2 separation from air and treatment of natural gas. The main obstacle encountered by polymer membranes is the inverse relationship between selectivity and permeability. A membrane with high selectivity is desired so that efficient separation occurs, but is important that the membrane is highly permeable so that a large amount of gas can be separated in a short amount of time and area.

Most polymeric membranes for CO_2/N_2 separation used the facilitated transport mechanism. In these membranes the targeted gas (*e.g.*, CO_2) reversibly bounds to a mobile/fixed reactive carrier that transports the gas across the membrane and releases it in the downstream side (low pressure). Despite the high permeabilities for CO_2 of most facilitated-transport membranes their selectivities are not high enough to be competitive with current CO_2 separation technologies. Depending on the gas purity required, it is expected that CO_2/N_2 selectivities in excess of 200 will be competitive with conventional absorption technology.

Poly(ethylene oxide) (PEO) have been widely studied for gas separation membrane since it offers an unique mechanism for CO_2 transport due to the dipole-quadrupole interaction between the polar ether group and the CO_2. However, PEO has a strong tendency to crystallize, which is detrimental for gas permeability.

Layer-by-Layer (LbL) technique, which was developed two decades ago, has been widely studied as a simple and versatile method to create multifunctional thin films. The ease of LBL assembly technique compared to the methods employed to fabricate the current gas separation films make it an attractive alternative. Extension of the LbL assembly technique to gas purification membranes promises the ability to tune overall gas permselectivity by varying the number of deposition or variables of the deposition process (*e.g.*, pH, deposition time, solution %wt) to suit a specific need. Kim et al. [1] reported a highly size-selective LbL light gas separation membrane. In their study, they demonstrated how gas transport properties of these assemblies can be tuned by modifying the number of layers deposited. There has been limited number of studies on LbL assembly for gas separation, most of them have focused on the effect of film thickness but not studied in depth the effect of deposition conditions.

In the present work, a PEO/PMAA LbL assembly was investigated for CO_2/N_2 separation using a constant-volume variable-pressure apparatus. Our aim is to investigate how the permeability and selectivity is affected by number of bi-layers (film thickness) and the deposition pH. Additionally, the temperature dependence of permeability and selectivity is presented. The activation energies for CO_2 and N_2 transport are reported for 10BL and 6BL films with deposition pH 2 and for a 10 BL with a deposition pH 4. The goal is to establish a processing-property relationship that will allow to obtained thin films with high selectivity and high permeability.

[1] Kim, Daejin, et al. "Highly Size Selective Ionically Crosslinked Multilayer Polymer Films for Light Gas Separation." Advanced Materials 26.5 (2014): 746-751.

442570 Development of Ultrafiltration Reactive Electrochemical Membranes for Contaminant Removal and Fouling Regeneration

Abstract:

The most significant issue affecting the utility of membrane filtration for water treatment is associated with membrane fouling. Membrane fouling is caused by particulate matter clogging pores, dissolved natural organic matter (NOM) adsorbing to the membrane surface, and biological growth causing biofilm formation. These various modes of membrane fouling can result in significant flux decline and the need for periodic off-line cleaning procedures. Physical and chemical cleaning procedures are necessary to maintain membrane performance, but result in increased costs associated with operation and disposal of spent chemical reagents. In this talk we will discuss our ongoing work focused on the development of novel reactive electrochemical membranes (REM) for water treatment applications. These membranes are synthesized from Ti_4O_7 materials and their high conductivity and ability to generate hydroxyl radicals via water oxidation allow them to be used for both contaminant removal and fouling regeneration. Model membrane foulants (polystyrene beads and NOM surrogates) were used as probes for elucidating the mechanisms associated with membrane fouling and the regeneration of membrane foulants at the REM surface. A new characterization technique based on electrochemical impedance spectroscopy for characterizing fouling at the REM surface will be discussed.

421605 Novel Nanofiltration Membrane with a Sulfonated Copolymer Selective Layer for the Removal of Heavy Metals

Abstract:

Facing increasing stringent regulations on wastewater discharge containing heavy metal ions, various industries are demanding more efficient and effective treatment methods. Among the various methods available, nanofiltration (NF) has been proven to be a feasible and promising option. However, it should be noted that the development of new membrane materials is constantly required for the advancement of this technology. To this aim, this study reports the first attempt to develop a composite NF membrane comprising of a molecularly designed pentablock copolymer selective layer for the removal of heavy metal ions. The resultant NF membrane has a reasonably high pure water permeability (*A*) of 2.4 LMH/bar with a mean effective pore diameter of 0.50 nm and molecular weight cut-off of 255 Da. In addition, the newly developed NF membrane can effectively remove heavy metal cations such as $Pb^{2}+$, Cd^{2+}, Zn^{2+} and Ni^{2+} with a rejection greater than 98.0%. Furthermore, the membrane also shows reasonably high rejections towards anions such as $HAsO_4^{2-}$ (99.9%) and $HCrO_4^-$ (92.3%). This observed performance can be attributed to (1) the incorporation of polyethyleneimine (PEI) as a gutter layer between the selective layer and the substrate and (2) the pentablock copolymer's unique ability to form a continuous water transport passageway with a defined pore size. The promising preliminary results achieved in the current study provide a useful platform for the development of new NF membranes for the removal of heavy metals.

427540 A Novel Reverse Osmosis Membrane with PAMAM/TMC for Effective Oily Saline Water Separation

Abstract:

Oily and saline wastewater is a significant waste stream generated in association with oil and gas production, the direct discharge of which will pollute the environment and impair people's health. Conventional polymer membranes suffer from low flux and serious fouling when used for treating oily and saline wastewater. Reported herein is the fabrication of a novel reverse osmosis membrane (PAMAM/TMC RO membrane) prepared with Ethylenediamine cored poly(amidoamine)(PAMAM) dendrimer and trimesoyl chloride(TMC) by interfacial polymerization on the PSf ultrafiltration membrane. A smooth and hydrophilic skin layer was constructed on the substrate PSf membrane and endows the composite membrane with an antifouling performance. In separation of the stable oily and saline water emulsion with an oil droplet size about 300nm, the rejection of oil droplet and NaCl was about 99% and 89.3% respectively, and the water flux was about 18.39 L/(m^2•h). After running for 24h, the flux only decreased about 3%, the rejection rate of oil and salt maintain 98% and 88%, exhibiting an excellent ability to resist oil pollution. Further investigation demonstrated its outstanding performance to resist chlorine, acid and alkali environment. Finally, it is believed that such membrane has great potential for practical applications.

430427 Role of Interparticle and Particle-Surface Interactions on Virus Removal By Ultrafiltration Membranes

Abstract:

Introduction

The reduction and inactivation of viral pathogens in natural waters is a major goal to achieve, due to the close relationship between these organisms and disease outbreaks. Available treatments based on bacteriological criteria are not always effective since viruses are more resistant and difficult to remove.

Membrane processes, like ultrafiltration, are increasingly used in potabilization to remove viruses and are a good barrier at nanometer level. The removal of virus particles is controlled by different mechanisms, such as electrostatic repulsion, attachment and size exclusion.

Ultrafiltration membranes are not expected to achieve significant retention of viral particles by size exclusion, since their pore size is similar or larger than the virus diameter; electrostatic repulsion and aggregation are therefore the predominant mechanisms during virus filtration.

Particles exhibiting the same charge as the membrane surface are generally less likely to be allowed through the membrane, as they are repelled on approach back to the bulk feed solution. Particle size increases due to aggregation, which occurs when electrostatic repulsive forces can be effectively overcome by van der Waals attraction forces, allowing attachment following inter-particle collision events. Electrostatic repulsion and attachment are strongly dependent on surface chemistry, pH, ionic strength, and chemical nature of the dissolved solids.

Non-pathogen viruses have been extensively used as models in water research. *Pseudomonas aeruginosa* bacteriophage PP7 is a good surrogate for poliovirus in water treatment processes, since both are icosahedral and have similar diameter (25 - 30 nm).

Previous work has shown that virus retention by polymeric ultrafiltration membranes was highly dependent on the water matrix. In particular, removal of PP7 was increased with high concentration of Ca^{2+}, low concentration of Mg^{2+}, simultaneous high or low concentrations of NO_3^- and HCO_3^-.

The filtrations were performed at cross flow in a custom made membrane filtration unit, connected to a feed tank through a peristaltic pump, a permeate tank on a scales, and control instruments (two pressure gauges and a flowmeter).

The objective of this work was to elucidate the mechanisms of virus filtration in relation to membrane surface - viral particle interaction and aggregation processes in order to be able to explain the experimental observations. Membrane surface and virus charges were measured, as well as particle size, for a variety of water matrixes. The inter-virus and virus-membrane interaction energy was modeled by Derjaguin, Landau, Verwey and Overbeek (DLVO) theory of colloidal stability.

Materials and methods

A flat sheet ultrafiltration membrane was employed in this work, made of a polyethersulfone dense skin and an amorphous support of cellulose. The average size of pore is 0.067 mm and MWCO is 50 kDa. *Pseudomonas aeruginosa* phage PP7 (ATCC 15692-B2) was used as model virus. Reagent grade NaCl, $NaNO_3$, $NaHCO_3$, $CaCl_2$, and $MgCl_2$ and Type I water was used in all experiments.

Preparation of the virus stock started with incubation of host bacteria in nutrient broth, and then PP7 was inoculated and incubated. Afterwards, the virus suspension was centrifuged and the supernatant filtered through a PVDF membrane. Purification of this stock consisted of dialysis through a 100 kDa MWCO membrane (SpectraPor Biotech CE, Spectrum Laboratories, USA) twice: first, against Type I water, and secondly, against the appropriate solution (NaCl, $NaNO_3$, $NaHCO_3$, $CaCl_2$, $MgCl_2$ of 1 mM, 10 mM, 100 mM ionic strength). The final suspension was again filtered and kept at 4°C overnight before measuring the zeta potential and hydrodynamic diameter.

The zeta potential of the bacteriophage and the membrane, as well as the hydrodynamic diameter of the bacteriophage were measured in a Zetasizer Nano ZS (Malvern, UK) at 25°C, using disposable cuvettes and the appropriate surface zeta potential accessory. Size of the bacteriophage was measured by Dynamic Light Scattering (DLS) and zeta potential by Laser Doppler Micro-electrophoresis. The membrane zeta potential was derived from electro-osmosis measurements.

For the DLVO modeling, PP7 was considered a sphere due to its icosahedral shape and the membrane was regarded as an infinite plate compared to the virus. Attractive van der Waals forces and electrical double layer repulsion were considered as the fundamental present interactions. Classical equations proposed by Hamaker (The London - van der Waals attraction between spherical particles, Phys. 4 (1937) 1058-1072) or collected in the existing literature were used. Moreover, non-DLVO forces such as Born repulsion, Lewis acid–base interactions and hydration repulsion were considered and their importance determined.

Results and discussion

Characterization: The average hydrodynamic diameter of the bacteriophage ranged between 44 and 84 nm, which differs from the diameter of the isolate virus of 25 - 27 nm.

PP7 had negative zeta potential for all the background solutions over the studied range of pH (5 to 8). For each salt, with increasing ionic strength, zeta potential of the bacteriophage increases (turning less negative).

For the flat membrane, the zeta potential showed always negative values, with only one exception: when the solution was $NaNO_3$ 10 mM for the whole range of pH considered.

The membrane is less negative than the virus for the case of NaCl 1 and 10 mM, $NaNO_3$ 1; 10 and 100 mM, $NaHCO_3$ 1 mM, $CaCl_2$ 1; 10 and 100 mM ionic strength, $MgCl_2$ 1 and 100 mM ionic strength. The membrane is more negative than the virus for the case of $NaHCO_3$ 10 mM ionic strength. Both the membrane and the virus have approximately the same charge for NaCl 100 mM, $NaHCO_3$ 100 mM and $MgCl_2$ 10 mM ionic strength.

DLVO modeling: The interaction potential energy for a virus like particle of PP7 and the flat sheet of polyethersulfone membrane in different background solutions was modeled. In the case of solutions formed by NaCl, $NaNO_3$, $NaHCO_3$, $CaCl_2$ and $MgCl_2$ 1 mM ionic strength, NaCl, $NaHCO_3$ 10 mM ionic strength an energy barrier can be predicted and a net repulsion potential between the primary aggregates and the membrane is expected. In the case of solutions formed by $CaCl_2$ and $MgCl_2$ 1 mM and 10 mM ionic strength, $NaNO_3$, NaCl, $NaHCO_3$ 100 mM ionic strength a secondary minimum is expected. And in the case of $NaNO_3$ 10 mM only a primary minimum is predicted, principally because of the positive charge of the membrane, and attachment to the membrane is more likely to happen.

When ionic strength was 10 mM, DLVO analysis was performed as function of pH; and it could be concluded that pH is not an important parameter when regarding attachment of the virus to the flat surface; which can be thought as the consequence of the almost constant zeta potentials for both the bacteriophage and the membrane. In all cases the total potential interaction are very close, and in particular it can be seen that the curve at pH=8 for NaCl, $NaNO_3$, $CaCl_2$ and $MgCl_2$ separates from the others, indicating extra repulsion for NaCl, $NaNO_3$ and $MgCl_2$ and more attraction for $CaCl_2$. For $NaHCO_3$ this situation is seen at pH=6 and less repulsion is expected.

Filtration: DLVO theory predicts that attachment of the bacteriophage to the membrane will increase when increasing the concentration of Ca^{2+} and Mg^{2+} respectively. However, net repulsion is only predicted at very low ionic strengths, when no secondary minimum would arise. This fact is in concordance with laboratory results since though low, removal was achieved.

DLVO theory predicts that attachment of the bacteriophage to the membrane will increase when increasing HCO_3^-; expecting net attraction at high ionic strength.

DLVO theory predicts that attachment of the bacteriophage to the membrane will increase when increasing NO_3^- from 1 mM to 10 mM and then decrease when increasing NO_3^- from 10 mM to 100 mM; expecting net attraction at 10 and 100 mM.

We can derive that when increasing both concentrations of HCO_3^- and NO_3^-, the effect of HCO_3^- outweighs the effect of NO_3^-. However, when one concentration increases and the other one decreases, the same asseveration no longer stands. These divergences between DVLO and experimental observations must be attributed to non-DLVO interactions.

Another limitation to the application of DLVO theory arises from the fact that viruses can be thought as soft particles (covered with a ion-penetrable surface layer of polyelectrolyte), as opposed to hard particles where the surface is clearly defined and no ion can penetrate it.

Conclusions

DLVO predicts the interactions studied between bacteriophage PP7 and polyethersulfone membrane, though further refinement should be made. It would be convenient to take into account that viruses are biological particles and therefore not perfect, rigid spheres with homogeneous surface; that commercial polyethersulfone membranes are not smooth homogenous surfaces; and when particles are approaching the supposition of constant charge no longer stands.

For the viral model, a partial removal is achieved. Ultrafiltration procedures to disinfect waters prove to be a technology to produce safe water, though its efficiency is affected by the chemicals present in the aqueous matrix.

431984 Combined Electrocoagulation and Membrane Distillation for Treatment of Hydraulic Fracturing Flow Back Waters

Abstract:

The reuse of wastewater has become an absolute necessity in recent years. Therefore, there is an urgent need to develop innovative and more effective technologies for treatment of wastewaters. Many of these wastewaters such as hydraulic fracturing flow back waters, contain very high dissolved solids (TDS). Membrane distillation has been proposed as a new unit operation for treatment of very high TDS wastewaters. One of the advantages of membrane distillation is the fact that low grade waste heat may be used. While membrane distillation could be used to concentrate the feed water to the solubility limit of the dissolved species present, leakage of feed water through the membrane pores into the distillate often occurs well before this level of water recovery. Leakage often occurs due the presence of oil and suspended solids in the feed which can adsorb on the membrane surface.

Here we have investigated the use of electrocoagulation as a pretreatment step for membrane distillation. Suspended solids and oil can be effectively coagulated by electrocoagulation followed by sedimentation prior to membrane distillation. Electrocoagulation has the potential to eliminate the disadvantages of classical pre-treatment technologies such as filtration and sedimentation. It is easy to operate and requires simple equipment. A laboratory scale electrocoagulation system containing aluminium electrodes was designed and built. Different ranges of current and voltage have been tested in order to optimize the process. In addition, effect of contact time, number of electrodes, settling time in the settler and operating temperature have been studied. Total dissolved and suspend solids and chemical oxygen demand have bene determined before and after electrocoagulation. We show that electrocoagulation can implemented as efficient and effective pretreatment step prior to membrane distillation. After pretreatment membrane distillation may be used to concentrate the dissolved specie in the feed to levels that are close to their solubility limit without leakage of water through the membrane pores. The high conductivity of high TDS wastewaters is an important factor that results in the efficiency of electrocoagulation as a pretreatment step for membrane distillation.

432820 Iron Functionalized PVDF-PAA Membrane for Treatment of Processed Water from Oil Industries

Abstract:

Membranes are finding wide applications in water reuse applications to meeting effluent discharge limits. Additional advantages can be obtained if membrane pores (for ex., microfiltration membranes) can be functionalized to provide catalytic activities for organic pollutant degradation. Polyvinylidene difluoride (PVDF) functionalized membranes immobilized with metal oxide nanoparticles provide such a robust system to catalyze the oxidation of a wide variety of organic compounds by using appropriate oxidants. An example application is, degradation of long chain organic acids that are often present in processed water from oil industries accompanied by high TDS.

In this study, commercial microfiltration membranes were functionalized with polyacrylic acid by in situ polymerization. Carboxylic acid groups present in the functionalized membrane was first ion exchanged with sodium followed by ferrous ions. Ferrous ions were eventually oxidized to iron oxide. In-situ generation of iron oxide particle prevented aggregation of this nanoparticles. Functionalized membrane was characterized and evaluated for its catalytic properties.

Oxidants like hydrogen peroxide, persulfate salts can be employed for the oxidation of the contaminants. Hydrogen peroxide is known to generate hydroxyl radicals with have an oxidation potential of 2.6 V and persulfate is known to generate sulfate radicals having an oxidation potential of 2.3 V. However hydrogen peroxide is relatively less stable and is known to be corrosive. Onsite generation of hydrogen peroxide can avoid the issues associated with transportations. Persulfate is relatively more stable, less corrosive and less expensive. In this study, both, Persulfate salts and hydrogen peroxide, were investigated for their effectiveness in catalyzing oxidation reactions.

Solution phase experiments in the presence of ferrous ions (1000 mg/l) and persulfate salt (5000 mg/l) was found to be effective in oxidizing 95% of the organic acids. Iron oxide functionalized membrane with an iron loading of 1.2 mg/lit was observed to oxidize 54% of organic acids in presence of 5000 mg/l potassium persulfate for the residence time of 6 seconds in membrane pores. As processed water from oil industries also contains high TDS of around 150 to 200 thousand mg/l, impact of TDS was also investigated in the study. Functionalized membranes was observed to be an effective platform for immobilizing iron oxide nanoparticle and catalyzing oxidation of organic acids. This project was supported by Nanostone Water Co, NSF KY EPSCoR program, and by NIH-NIEHS-SRC.

435026 Proper Accounting of Mass Transfer Resistances in Forward Osmosis: Improving the Accuracy of Model Predictions of Structural Parameter

Abstract:

Proper Accounting of Mass Transfer Resistances in Forward Osmosis:

Improving the Accuracy of Model Predictions of Structural Parameter

Ngoc Bui, *Lawrence Livermore National Laboratory, Livermore, CA*

Jason Arena, Jeffrey R. McCutcheon, *University of Connecticut, Chemical and Biomolecular Engineering Department, Storrs, CT*

Forward Osmosis (FO) and pressure retarded osmosis (PRO) have recently been revitalized as a sustainable and versatile membrane-based separation technology platform for water and power production, respectively. To design membranes for various FO processes, it is important to understand critical structure-performance relationships, especially with respect to mass transfer. This work demonstrates a more accurate method for calculating structural parameter (S) of asymmetric osmotic membranes using experimental data and a theoretical flux model which encapsulates all significant boundary layer phenomena. External boundary layer effects on the porous side of the membrane have been neglected in many current models. In these models, external concentration polarization (ECP) effects get combined with the internal concentration polarization (ICP), resulting in inflated S values. In this study, we proposed a new flux model in which ECP effects are accounted for so that S can be more accurately measured. This model considered the in-series resistances for solute transport based on intrinsic properties of the membrane as well as boundary layers at membrane surfaces and within the support layer. The results indicate that ICP is less severe than previously predicted and that cross-flow velocity, temperature and concentration of the draw and the feed solutions impact both external and internal concentration polarization. Our calculations also surprisingly show that changes in cross-flow velocity impact internal concentration polarization due to induced mixing within the support layer. Furthermore, new definitions of membrane reflection coefficient and total resistance to solute transport emerge from this work. Also, we suggest that it is critical to consider the "residence time" of solutes in the vicinity of the selective layer in determining the membrane selectivity.

423915 CFD Analysis of Transport in a Forward Osmosis System

Abstract:

A systematic approach to modeling and analyzing the performance data of a laboratory forward osmosis system was evaluated via the application of a detailed CFD transport model of the coupled fluid flow and solute mass transfer equations. The detailed modeling approach was compared with analysis based on the classical film model with the goal of assessing the impact of various model assumptions and associated implications regarding the significance of the FO membrane support layer to water and salt permeation. The analysis revealed that water permeability and solute transport parameter values that are determined from FO operation performance data in the RO operational mode (FO-RO) are of significantly higher magnitude relative to transport parameter values directly extracted from CFD analysis of FO experimental data. Moreover, the magnitude of the porous layer resistivity (K), when determined based on the FM model in which the transport parameters are extracted from FO-RO operational data, implies tortuosity/porosity ratio that is a factor ~2-5 higher than physically realizable with typical membrane porous supports. Comparative analysis of experimental FO data with the FM and CFD models showed that the former overestimates the permeation resistance of the FO membrane support layer by up to about an order of magnitude. Accordingly, the present study suggests that enhancement of FO membrane water permeability would in fact benefit from developing higher permeability membrane skin and improved channel hydrodynamics (to decrease external concentration polarization) as opposed to developing thinner ad higher permeability FO membrane support layers.

426892 Modeling Multicomponent Gaseous Diffusion for Membrane Vapor Extraction

Abstract:

A novel separation process, membrane vapor extraction (MVE), has recently been proposed to separate dilute biosolutes from broth exiting a fermenter [1]. In MVE, volatile aqueous biosolutes in the broth vaporize at the upstream side of a membrane, diffuse through the membrane pores, and subsequently dissolve into a nonpolar solvent highly favorable to the solutes but not to water. For successful MVE design and implementation, vapor phase transport of biosolutes and water through the membrane pores must be understood physically.

Previously, we modeled gaseous flow in membrane pores of an MVE separation unit for recovering biobutanol from water using the traditional binary convective diffusion model (CDM) assuming a completely nonvolatile, and aqueous insoluble solvent [1,2]. Neither assumption is rigorous. Unfortunately, CDM is inapplicable for ternary and higher component systems. Here we extend the binary analysis to a ternary system including finite solvent volatility and finite aqueous solubility using a modified dusty gas model (DGM) [3,4]. First, we establish that modified DGM for the binary system (i.e., nonvolatile solvent) is in accord with CDM for the butanol/water system. Next, we show that with finite solvent volatility butanol (and water) pore fluxes are impeded duty to collision with stagnant solvent vapor. Incorporating, in addition, solvent aqueous solubility in an MVE process leads to a countercurrent flux of solvent that further reduces butanol (and water) fluxes. For the solvent dodecane, little effect is found on the MVE unit performance. However, for strongly volatile and aqueous-soluble solvents, unit efficiency is compromised.

1. Liu, DE; Cerretani, C; Telles, R; Scheer, AP; Sciamanna, S; Bryan, PF; Radke, CJ; Prausnitz, JM. Analysis of Countercurrent Membrane Vapor Extraction of a Dilute Aqueous Biosolute. AIChE Journal, *Accepted, 2015.*
2. Bird, RB; Stewart, WE; Lightfoot, EN. Transport Phenomena. 2nd ed. New York, NY:Wiley, 2006; Chapter 17.
3. Veldsink, JW; van Damme, RMJ; Versteeg, GF; van Swaaij, WPM. The use of the dusty-gas model for the description of mass transport with chemical reaction in porous media. The Chemical Engineering Journal. 1995; 57:115-125.
4. Weber, AZ; Newman, J. Modeling gas-phase flow in porous media. International communications in heat and mass transfer. 2005; 32:855-860.

Numerical Simulations of Concentration Polarization in Reverse Osmosis Systems

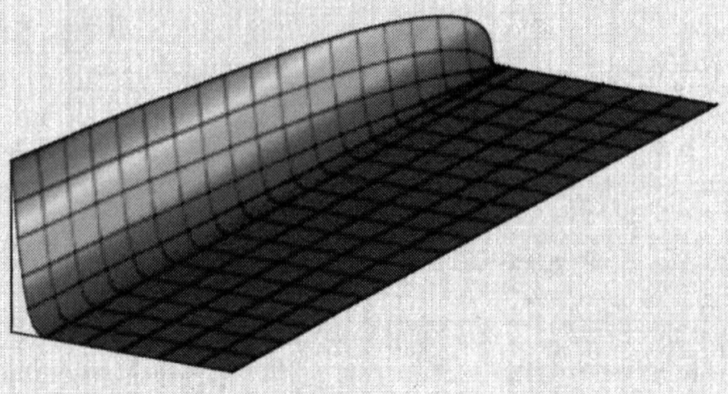

Claire Strebinger & Nils Tilton

12 November 2015

COLORADOSCHOOLOFMINES
EARTH • ENERGY • ENVIRONMENT

Concentration Polarization in RO Systems

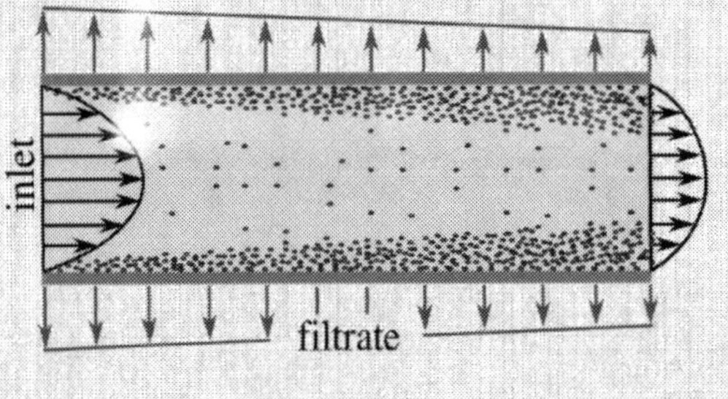

- Accumulation of solutes adjacent to membranes

Concentration Polarization in RO Systems

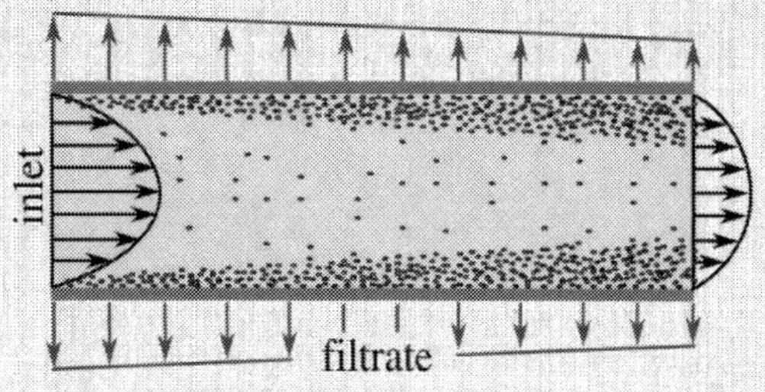

- Accumulation of solutes adjacent to membranes
  - Limits efficiency & recovery

    Leads to mineral scaling

    Reduces membrane life

    Increases maintenance costs

Concentration Polarization in RO Systems

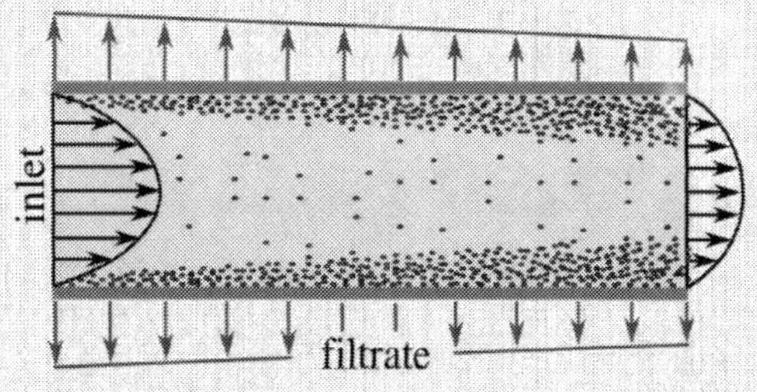

- Accumulation of solutes adjacent to membranes
  - Limits efficiency & recovery

    Leads to mineral scaling

    Reduces membrane life

    Increases maintenance costs
- Similar phenomena important in micro & ultrafiltration

  gel layers, cake layers, bio-fouling

Objective

- Dedicated methods for simulating mass transport in RO

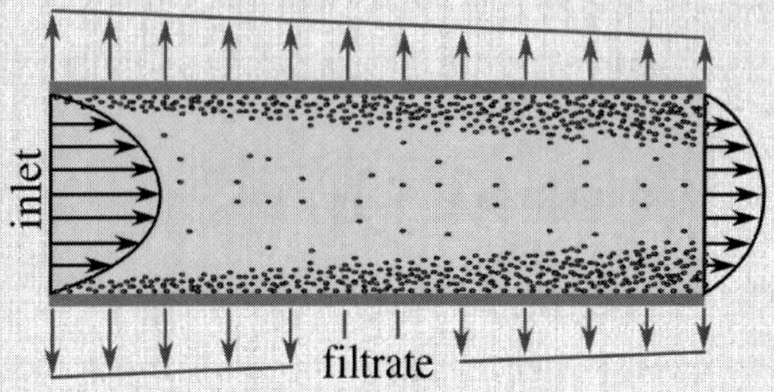

3/23

Objective

- Dedicated methods for simulating mass transport in RO
- Including effects of:
  1. Concentration polarization
  2. Competition between hydrodynamic and osmotic pressure
  3. Formation of mineral scale (long term goal)

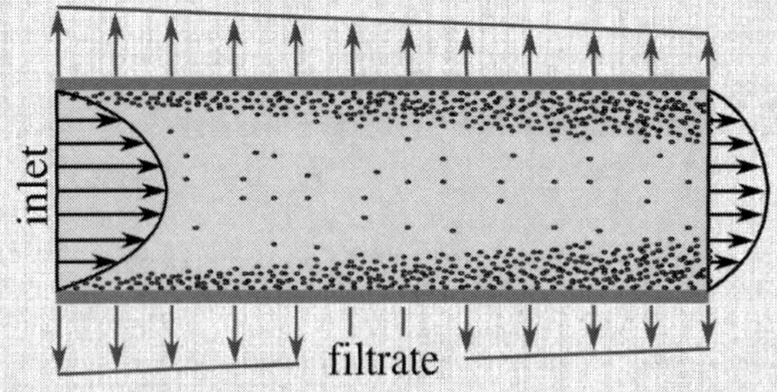

3/23

Objective

- Dedicated methods for simulating mass transport in RO
- Including effects of:
  1. Concentration polarization
  2. Competition between hydrodynamic and osmotic pressure
  3. Formation of mineral scale (long term goal)
- Computationally efficient for unsteady & 3-D simulations

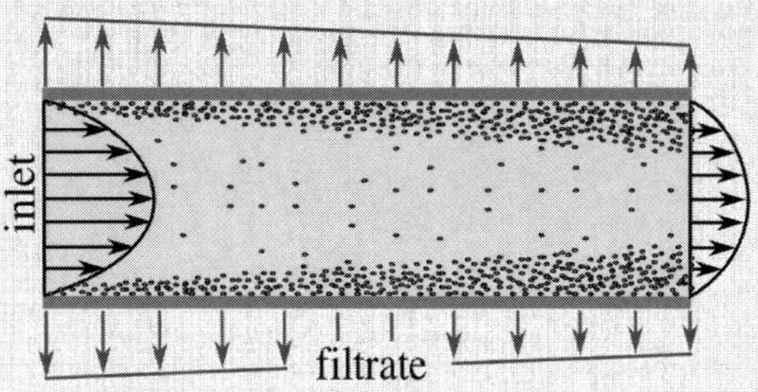

3/23

Objective

- Dedicated methods for simulating mass transport in RO
- Including effects of:
  1. Concentration polarization
  2. Competition between hydrodynamic and osmotic pressure
  3. Formation of mineral scale (long term goal)
- Computationally efficient for unsteady & 3-D simulations

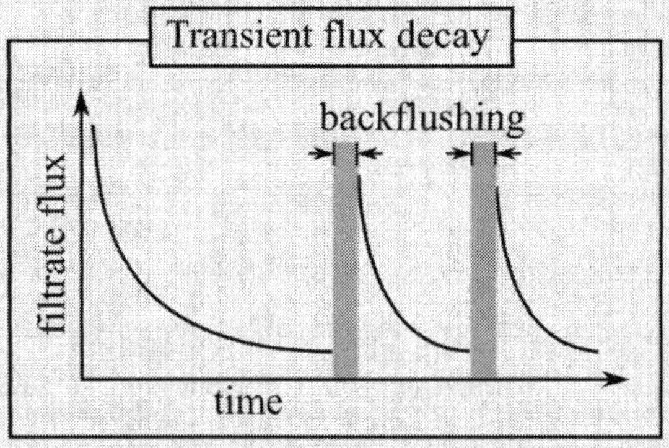

3/23

Objective

- Dedicated methods for simulating mass transport in RO
- Including effects of:
  1. Concentration polarization
  2. Competition between hydrodynamic and osmotic pressure
  3. Formation of mineral scale (long term goal)
- Computationally efficient for unsteady & 3-D simulations

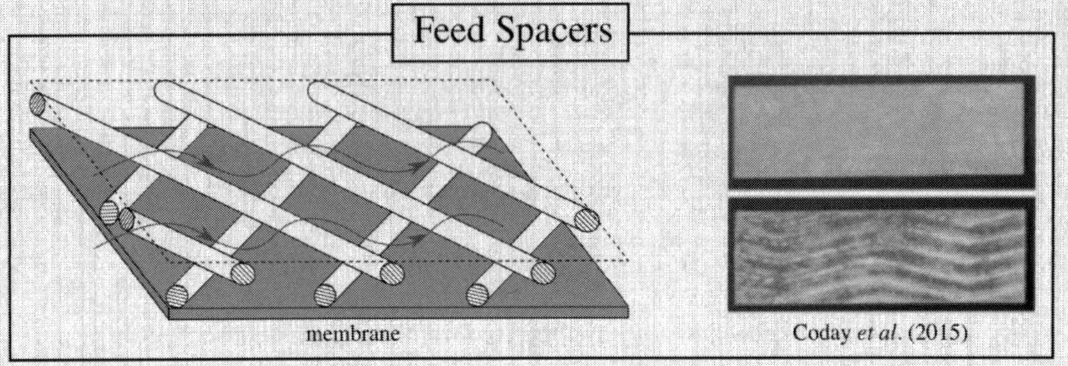

3/23

CFD of Membrane Filtration

- Challenging to simulate, due to velocity-pressure coupling

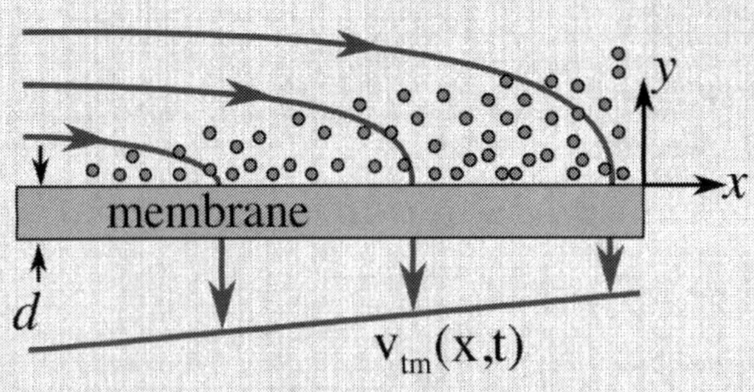

4/23

CFD of Membrane Filtration

- Challenging to simulate, due to velocity-pressure coupling

$$v_{tm}(x,t) = \frac{K}{\mu d}\left[p_{tm}(x,t) - p_{osm}(x,t)\right] \qquad \textbf{(Darcy's law)}$$

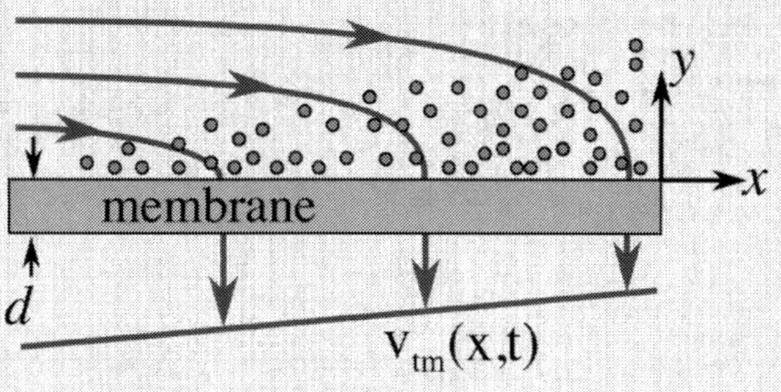

4/23

CFD of Membrane Filtration

- Challenging to simulate, due to velocity-pressure coupling

$$v_{tm}(x,t) = \frac{K}{\mu d}\left[p_{tm}(x,t) - p_{osm}(x,t)\right] \qquad \textbf{(Darcy's law)}$$

- Difficult to apply in transient & 3-D flows

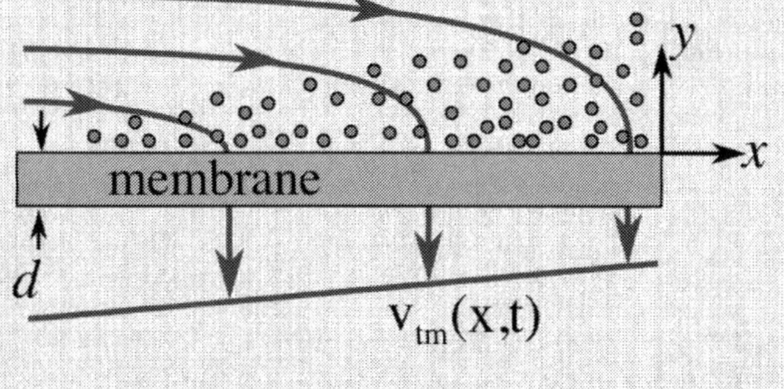

4/23

CFD of Membrane Filtration

- Challenging to simulate, due to velocity-pressure coupling

$$v_{\mathrm{tm}}(x,t) = \frac{K}{\mu d}\left[p_{\mathrm{tm}}(x,t) - p_{\mathrm{osm}}(x,t)\right] \qquad \textbf{(Darcy's law)}$$

- Difficult to apply in transient & 3-D flows
- Made worse by mineral scale: $K = K(x,t)$

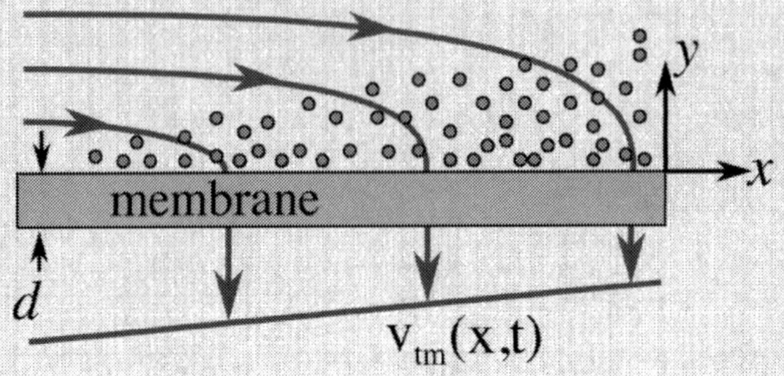

Do we need dedicated numerical methods?

- CFD increasingly common in filtration community
- But: velocity-pressure coupling remains significant hurdle

Cao *et al.* (2001)
Karode & Kumar (2001)
Li *et al.*
Koutsou*et al.*(2004)
Ahmad *et al.* (2005)
Dendukuri *et al.* (2005)
Koutsou *et al.* (2007)
Torras *et al.* (2009)
Koutsou *et al.* (2009)
Ahmad *et al.* (2011)
Bucs *et al.* (2014)
Gurreri *et al.* (2014)
Rosén & Trägårdh (1993)
Kotzev (1994)
Pellerin *et al.* (1995)
Geraldes *et al.* (1998)
Hansen *et al.* (1998)
Huang & Morrissey (1999)
Geraldes *et al.* (2001)
Miranda & Campos (2001)
Damak *et al.* (2004)
Lyster & Cohen (2007)
Hanspal *et al.* (2009)
Salcedo-Díaz *et al.* (2014)
Bhattacharyya *et al.* (1990)
Bouchard *et al.* (1994)
Lee & Clark (1997)
Wiley & Fletcher (2002)
Wiley & Fletcher (2003)
de Pinho *et al.* (2002)
Ahmad & Lau (2006)
Alexiadis *et al.* (2007)
Morão *et al.* (2008)
Lyster *et al.* (2009)
Li *et al.* (2011)
Ronen *et al.* (2015)

Do we need dedicated numerical methods?

- CFD increasingly common in filtration community
- But: velocity-pressure coupling remains significant hurdle
- Studies often
  1. Neglect transmembrane flow ($v_{tm} = 0$)

Cao *et al.* (2001)
Karode & Kumar (2001)
Li *et al.*
Koutsou*et al.*(2004)
Ahmad *et al.* (2005)
Dendukuri *et al.* (2005)
Koutsou *et al.* (2007)
Torras *et al.* (2009)
Koutsou *et al.* (2009)
Ahmad *et al.* (2011)
Bucs *et al.* (2014)
Gurreri *et al.* (2014)

Rosén & Trägårdh (1993)
Kotzev (1994)
Pellerin *et al.* (1995)
Geraldes *et al.* (1998)
Hansen *et al.* (1998)
Huang & Morrissey (1999)
Geraldes *et al.* (2001)
Miranda & Campos (2001)
Damak *et al.* (2004)
Lyster & Cohen (2007)
Hanspal *et al.* (2009)
Salcedo-Díaz *et al.* (2014)

Bhattacharyya *et al.* (1990)
Bouchard *et al.* (1994)
Lee & Clark (1997)
Wiley & Fletcher (2002)
Wiley & Fletcher (2003)
de Pinho *et al.* (2002)
Ahmad & Lau (2006)
Alexiadis *et al.* (2007)
Morão *et al.* (2008)
Lyster *et al.* (2009)
Li *et al.* (2011)
Ronen *et al.* (2015)

Do we need dedicated numerical methods?

- CFD increasingly common in filtration community
- But: velocity-pressure coupling remains significant hurdle
- Studies often
  1. Neglect transmembrane flow ($v_{tm} = 0$)
  2. Focus on 2-D steady flows

Cao *et al.* (2001)
Karode & Kumar (2001)
Li *et al.*
Koutsou*et al.*(2004)
Ahmad *et al.* (2005)
Dendukuri *et al.* (2005)
Koutsou *et al.* (2007)
Torras *et al.* (2009)
Koutsou *et al.* (2009)
Ahmad *et al.* (2011)
Bucs *et al.* (2014)
Gurreri *et al.* (2014)

Rosén & Trägårdh (1993)
Kotzev (1994)
Pellerin *et al.* (1995)
Geraldes *et al.* (1998)
Hansen *et al.* (1998)
Huang & Morrissey (1999)
Geraldes *et al.* (2001)
Miranda & Campos (2001)
Damak *et al.* (2004)
Lyster & Cohen (2007)
Hanspal *et al.* (2009)
Salcedo-Díaz *et al.* (2014)

Bhattacharyya *et al.* (1990)
Bouchard *et al.* (1994)
Lee & Clark (1997)
Wiley & Fletcher (2002)
Wiley & Fletcher (2003)
de Pinho *et al.* (2002)
Ahmad & Lau (2006)
Alexiadis *et al.* (2007)
Morão *et al.* (2008)
Lyster *et al.* (2009)
Li *et al.* (2011)
Ronen *et al.* (2015)

Do we need dedicated numerical methods?

- CFD increasingly common in filtration community
- But: velocity-pressure coupling remains significant hurdle
- Studies often
  1. Neglect transmembrane flow ($v_{\mathrm{tm}} = 0$)
  2. Focus on 2-D steady flows
  3. Reduce the coupling

Cao *et al.* (2001)
Karode & Kumar (2001)
Li *et al.*
Koutsou*et al.*(2004)
Ahmad *et al.* (2005)
Dendukuri *et al.* (2005)
Koutsou *et al.* (2007)
Torras *et al.* (2009)
Koutsou *et al.* (2009)
Ahmad *et al.* (2011)
Bucs *et al.* (2014)
Gurreri *et al.* (2014)

Rosén & Trägårdh (1993)
Kotzev (1994)
Pellerin *et al.* (1995)
Geraldes *et al.* (1998)
Hansen *et al.* (1998)
Huang & Morrissey (1999)
Geraldes *et al.* (2001)
Miranda & Campos (2001)
Damak *et al.* (2004)
Lyster & Cohen (2007)
Hanspal *et al.* (2009)
Salcedo-Díaz *et al.* (2014)

Bhattacharyya *et al.* (1990)
Bouchard *et al.* (1994)
Lee & Clark (1997)
Wiley & Fletcher (2002)
Wiley & Fletcher (2003)
de Pinho *et al.* (2002)
Ahmad & Lau (2006)
Alexiadis *et al.* (2007)
Morão *et al.* (2008)
Lyster *et al.* (2009)
Li *et al.* (2011)
Ronen *et al.* (2015)

5/23

Why is CFD of filtration difficult?

- Consider the governing equations:

$$\frac{\partial \mathbf{v}}{\partial t} + \mathbf{v}\cdot\nabla\mathbf{v} = \nabla p + \frac{1}{Re}\nabla^2\mathbf{v}, \qquad \nabla\cdot\mathbf{v} = 0$$

6/23

Why is CFD of filtration difficult?

- Consider the governing equations:

$$\frac{\partial \mathbf{v}}{\partial t} + \mathbf{v}\cdot\nabla\mathbf{v} = \underbrace{\nabla p}_{\text{problem}} + \frac{1}{Re}\nabla^2\mathbf{v}, \qquad \nabla\cdot\mathbf{v} = 0$$

6/23

Why is CFD of filtration difficult?

- Consider the governing equations:

$$\frac{\partial \mathbf{v}}{\partial t} + \mathbf{v}\cdot\nabla\mathbf{v} = \underbrace{\nabla p}_{\text{problem}} + \frac{1}{Re}\nabla^2\mathbf{v}, \qquad \nabla\cdot\mathbf{v} = 0$$

- No evolution equation for the pressure.

6/23

Why is CFD of filtration difficult?

- Consider the governing equations:

$$\frac{\partial \mathbf{v}}{\partial t} + \mathbf{v}\cdot\nabla\mathbf{v} = \underbrace{\nabla p}_{\text{problem}} + \frac{1}{Re}\nabla^2\mathbf{v}, \qquad \nabla\cdot\mathbf{v} = 0$$

- No evolution equation for the pressure.
- How to determine pressure without solving full coupled problem?

Why is CFD of filtration difficult?

- Consider the governing equations:

$$\frac{\partial \mathbf{v}}{\partial t} + \mathbf{v}\cdot\nabla\mathbf{v} = \underbrace{\nabla p}_{\text{problem}} + \frac{1}{Re}\nabla^2\mathbf{v}, \qquad \nabla\cdot\mathbf{v} = 0$$

- No evolution equation for the pressure.
- How to determine pressure without solving full coupled problem?
- **Common Solution:** Prediction-Correction Methods

Why is CFD of filtration difficult?

- Consider the governing equations:

$$\frac{\partial \mathbf{v}^n}{\partial t} + \mathbf{v}\cdot\nabla\mathbf{v} = \underbrace{\nabla p^{n-1}}_{\textbf{solution}} + \frac{1}{Re}\nabla^2\mathbf{v}, \qquad \nabla\cdot\mathbf{v} = 0$$

- No evolution equation for the pressure.
- How to determine pressure without solving full coupled problem?
- **Common Solution:** Prediction-Correction Methods
  - Use pressure from previous time step

Why is CFD of filtration difficult?

- Consider the governing equations:

$$\frac{\partial \mathbf{v}^n}{\partial t} + \mathbf{v}\cdot\nabla\mathbf{v} = \underbrace{\nabla p^{n-1}}_{\textbf{solution}} + \frac{1}{Re}\nabla^2\mathbf{v}, \qquad \nabla\cdot\mathbf{v} = 0$$

- No evolution equation for the pressure.
- How to determine pressure without solving full coupled problem?
- **Common Solution:** Prediction-Correction Methods
  - Use pressure from previous time step
- **Problem:** Concentrates errors on membrane surfaces[1]

[1] (Tilton *et al.* 2014, Comp. & Fluids)

Why is CFD of filtration difficult?

- Consider the governing equations:

$$\frac{\partial \mathbf{v}^n}{\partial t} + \mathbf{v}\cdot\nabla\mathbf{v} = \underbrace{\nabla p^{n-1}}_{\textbf{solution}} + \frac{1}{Re}\nabla^2\mathbf{v}, \qquad \nabla\cdot\mathbf{v} = 0$$

- No evolution equation for the pressure.
- How to determine pressure without solving full coupled problem?
- **Common Solution:** Prediction-Correction Methods
  - Use pressure from previous time step
- **Problem:** Concentrates errors on membrane surfaces[1]

  Difficulties satisfying Darcy's law

[1](Tilton *et al.* 2014, Comp. & Fluids)

6 / 23

Solution: improved prediction-correction scheme

3 step method:

7 / 23

Solution: improved prediction-correction scheme

3 step method:

1. Pressure prediction:

$$\nabla^2 p^* = -\nabla \cdot \left(\mathbf{v}^{n-1} \cdot \nabla \mathbf{v}^{n-1}\right) \qquad \text{(Raspo \textit{et al.} 2002)}$$

Solution: improved prediction-correction scheme

3 step method:

1. Pressure prediction:

$$\nabla^2 p^* = -\nabla \cdot \left(\mathbf{v}^{n-1} \cdot \nabla \mathbf{v}^{n-1}\right) \qquad \text{(Raspo \textit{et al.} 2002)}$$

2. Velocity prediction:

$$\frac{\partial \mathbf{v}^*}{\partial t} - \frac{1}{Re}\nabla^2 \mathbf{v}^* + \nabla p^* = -\mathbf{v}^{n-1} \cdot \nabla \mathbf{v}^{n-1}$$

Solution: improved prediction-correction scheme

3 step method:

1. Pressure prediction:

$$\nabla^2 p^* = -\nabla \cdot \left(\mathbf{v}^{n-1} \cdot \nabla \mathbf{v}^{n-1}\right) \quad \text{(Raspo } et\ al. \text{ 2002)}$$

2. Velocity prediction:

$$\frac{\partial \mathbf{v}^*}{\partial t} - \frac{1}{Re}\nabla^2 \mathbf{v}^* + \nabla p^* = -\mathbf{v}^{n-1} \cdot \nabla \mathbf{v}^{n-1}$$

3. Correction: Compute final p^n and $\mathbf{v}^n$

Solution: improved prediction-correction scheme

3 step method:

1. Pressure prediction:

$$\nabla^2 p^* = -\nabla \cdot \left(\mathbf{v}^{n-1} \cdot \nabla \mathbf{v}^{n-1}\right) \quad \text{(Raspo } et\ al. \text{ 2002)}$$

2. Velocity prediction:

$$\frac{\partial \mathbf{v}^*}{\partial t} - \frac{1}{Re}\nabla^2 \mathbf{v}^* + \nabla p^* = -\mathbf{v}^{n-1} \cdot \nabla \mathbf{v}^{n-1}$$

3. Correction: Compute final p^n and $\mathbf{v}^n$

- **Boundary conditions:** ensure Darcy's law satisfied exactly.

  (Tilton *et al.* Comp. Fluids 2014)

Benchmarking: Rotating Filtration (Solvent Flow)

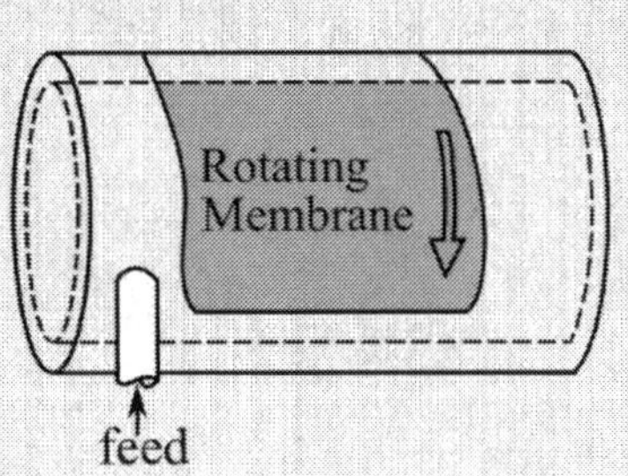

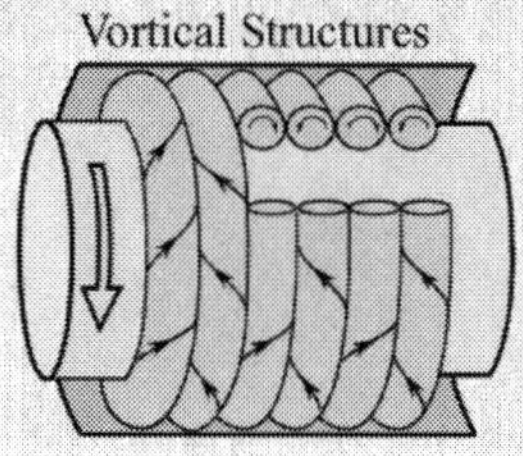

8/23

Benchmarking: Rotating Filtration (Solvent Flow)

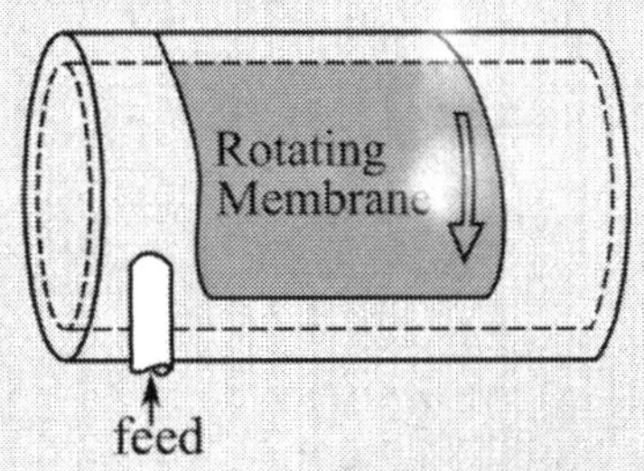

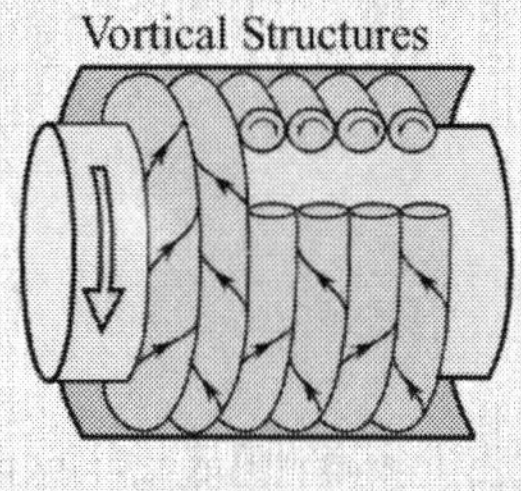

Numerical:

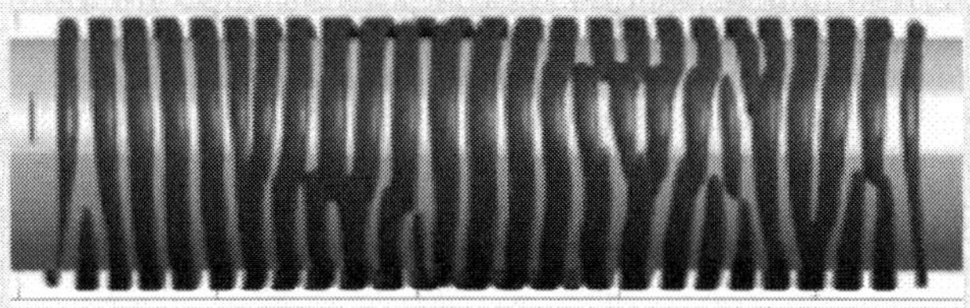

8/23

Benchmarking: Rotating Filtration (Solvent Flow)

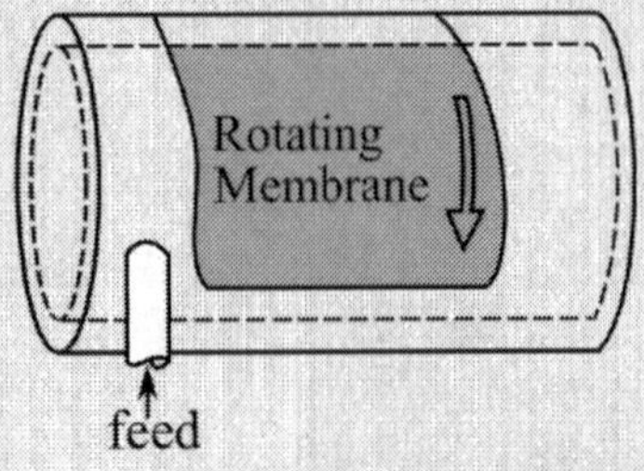

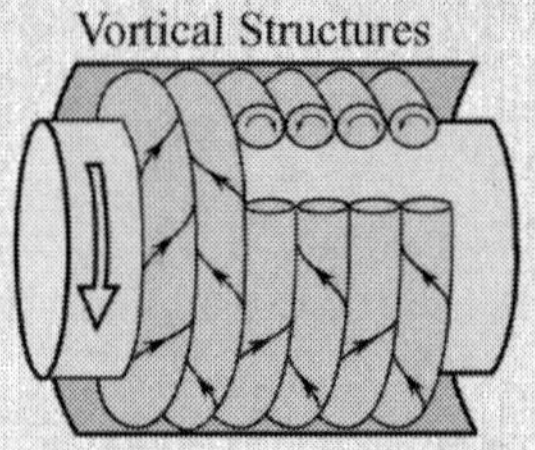

Numerical:

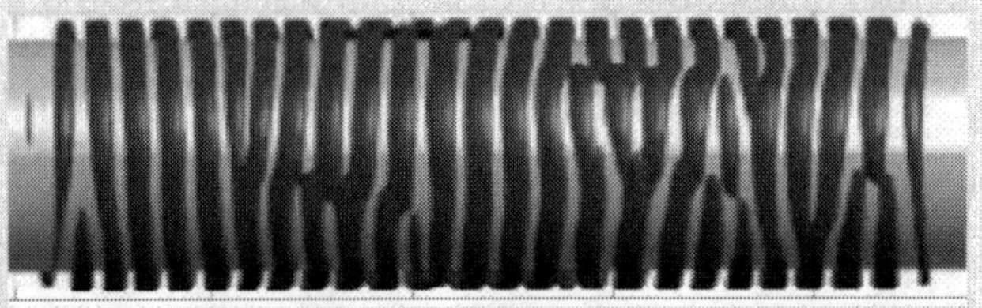

Experimental:
(Min & Lueptow 1994)

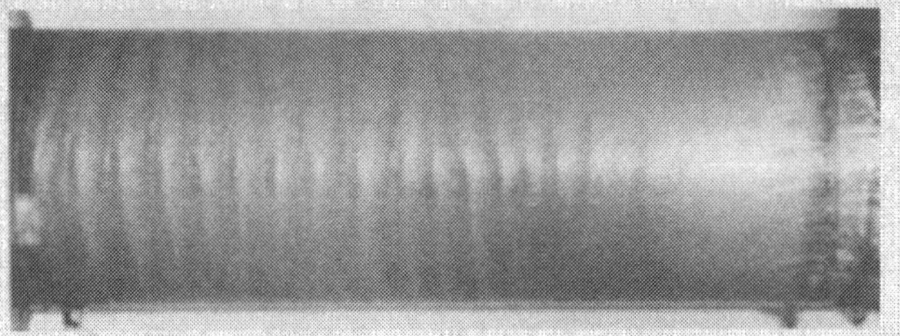

8/23

Concentration Polarization

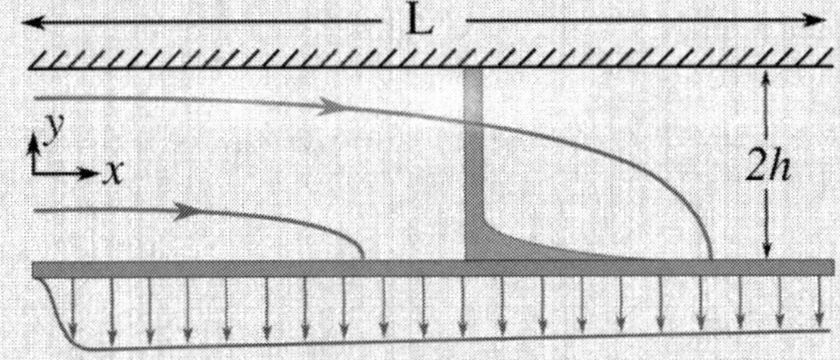

- **Geometry: 2-D Plate and Frame**

9/23

Concentration Polarization

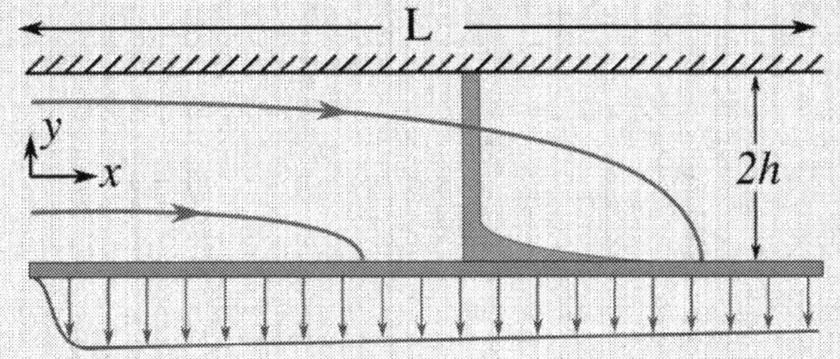

- **Geometry: 2-D Plate and Frame**
- **Unsteady Equations:**
  - Incompressible Navier-Stokes

    Continuity equation

    Advection-diffusion (ideal binary solution)

9/23

Concentration Polarization

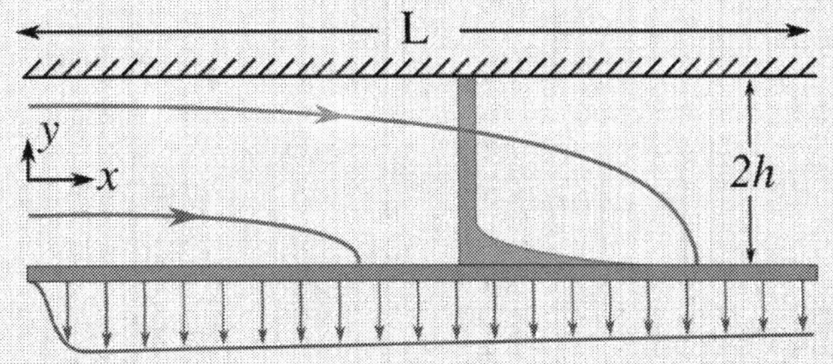

- **Membrane Conditions:**

Darcy: $v_{\text{tm}}(x,t) = -\frac{K}{\mu d}(p - Ac)$, K = permeability, d = memb. thickness

9/23

Concentration Polarization

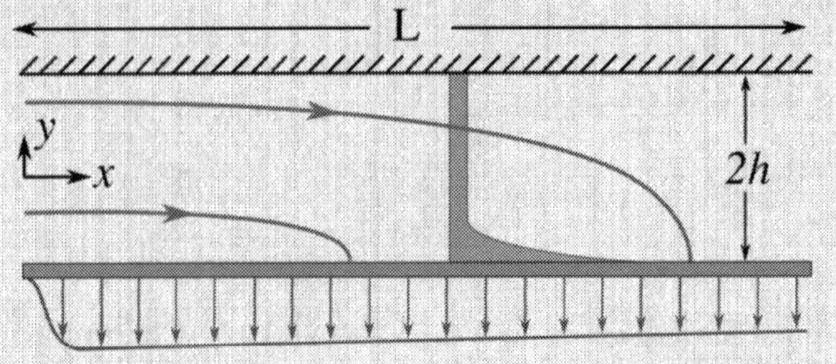

- **Membrane Conditions:**

Darcy: $v_{\text{tm}}(x,t) = -\dfrac{K}{\mu d}(p - Ac)$, K = permeability, d = memb. thickness

Rejection: $vc - \mathcal{D}\dfrac{\partial c}{\partial y} = 0,$ $\mathcal{D}$ = diffusivity

9/23

Concentration Polarization

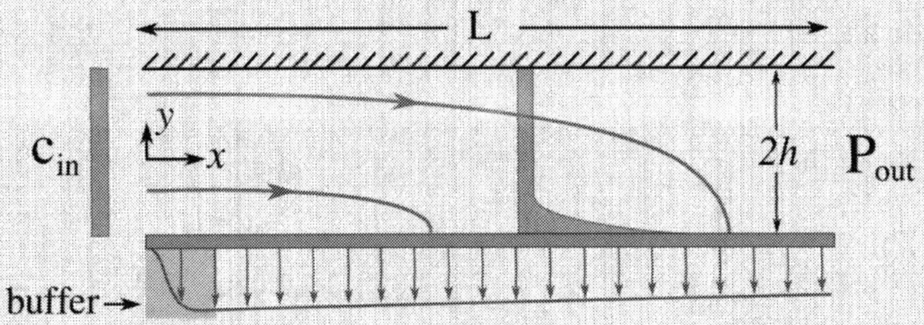

- **Inlet/Outlet Conditions:**

Inlet: $U = U_{\text{in}}(y)$, $c = C_{\text{in}}$, (applied using inlet buffer)

9/23

Concentration Polarization

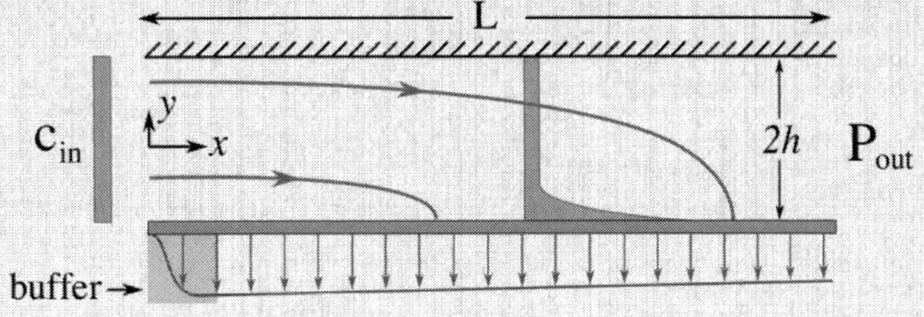

- **Inlet/Outlet Conditions:**

Inlet: $U = U_{in}(y), \qquad c = C_{in}, \quad \text{(applied using inlet buffer)}$

Outlet: $p = P_{out}, \qquad \frac{\partial u}{\partial t} + U_{in}\frac{\partial u}{\partial x} = 0,$

9/23

Concentration Polarization

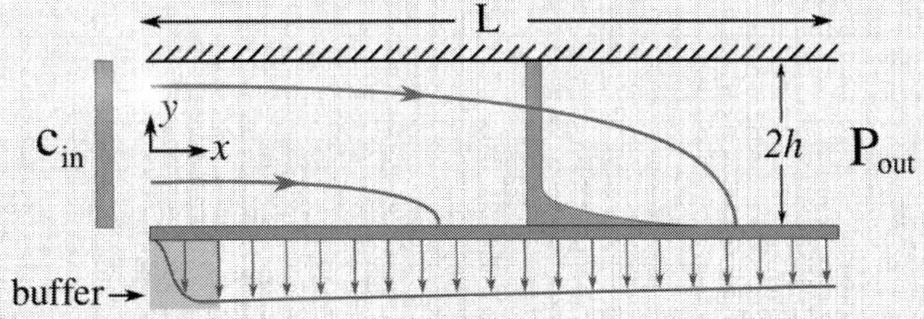

- **Parameters:**
  - Aspect ratio: $L/h = 500$ ($2h = 1$ mm)
  - $Re = U_{in}h/\nu = 250$
  - $Sc = \nu/\mathcal{D} = 700, 1000$
  - Permeability: $\sigma = K/(hd) = 10^{-10}, 10^{-11}$
  - Inlet osmotic pressure: $P_{in}^{osm} = AC_{in} \in [0, 7]$ bar
  - $P_{out} \in [10, 60]$ bar

9/23

Spatial Discretization

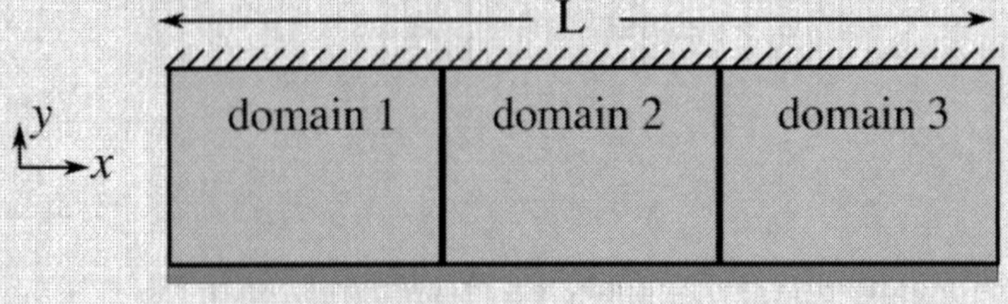

- **High-Accuracy Spectral Multi-Domain Method:**

10/23

Spatial Discretization

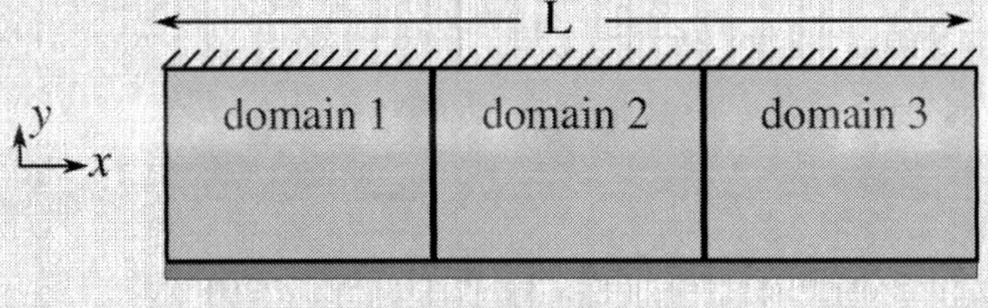

- **High-Accuracy Spectral Multi-Domain Method:**
- Chebyshev expansion:

$$\mathbf{u}(x, y, t) = \sum_{n=0}^{N} \sum_{m=0}^{M} a_{nm} T_n(y) T_m(x)$$

10/23

Spatial Discretization

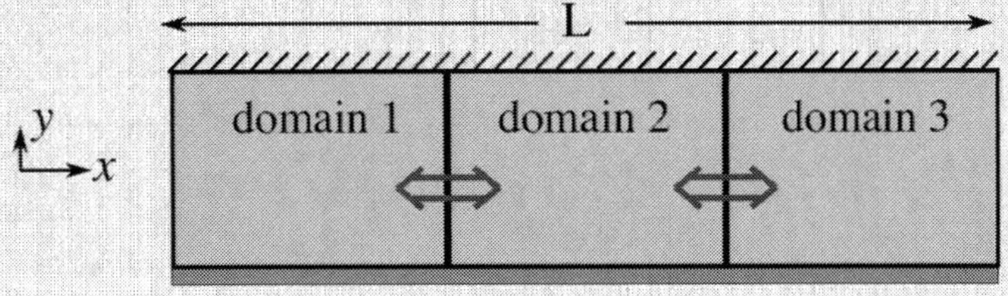

- **High-Accuracy Spectral Multi-Domain Method:**
- Chebyshev expansion:

$$\mathbf{u}(x,y,t) = \sum_{n=0}^{N}\sum_{m=0}^{M} a_{nm} T_n(y) T_m(x)$$

- C^1 continuity on interfaces

10/23

Spatial Discretization

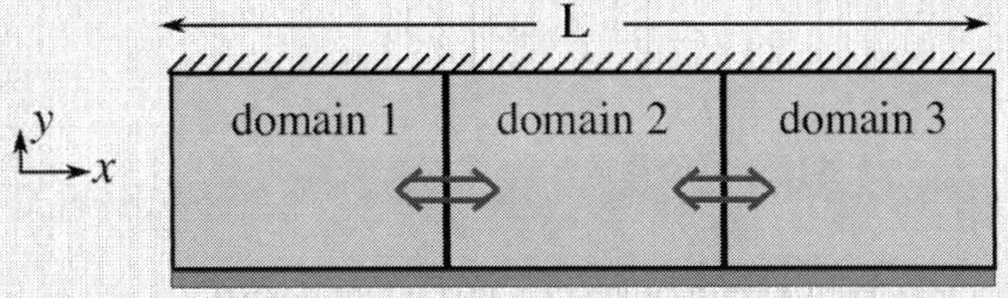

- **High-Accuracy Spectral Multi-Domain Method:**
- **Computational Efficiency:**

10/23

Spatial Discretization

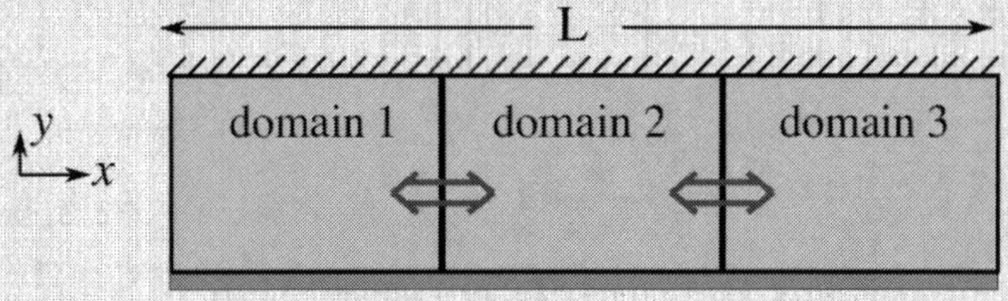

- **High-Accuracy Spectral Multi-Domain Method:**
- **Computational Efficiency:**
  - 50 % fewer grid points than comparable commercial softwares.

10/23

Results: Transient Solute Accumulation

time = 0.2 sec

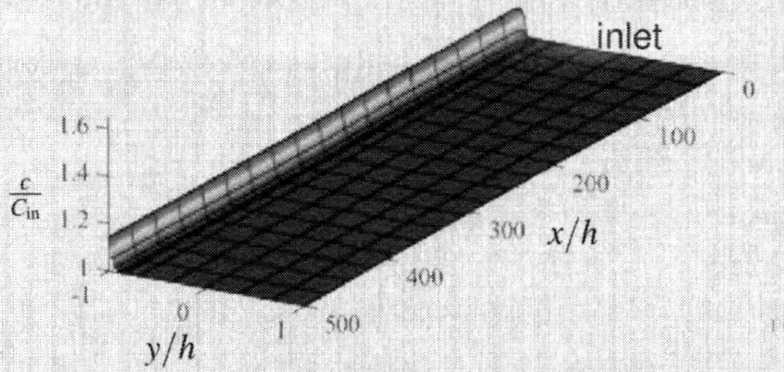

Re=250
Sc=700
$\sigma = 10^{-11}$
$P_{in}^{osm} = 3.4$ bar
$P_{out} = 24$ bar

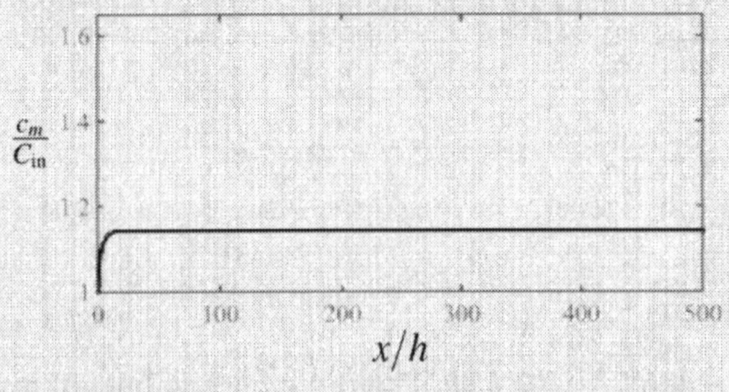

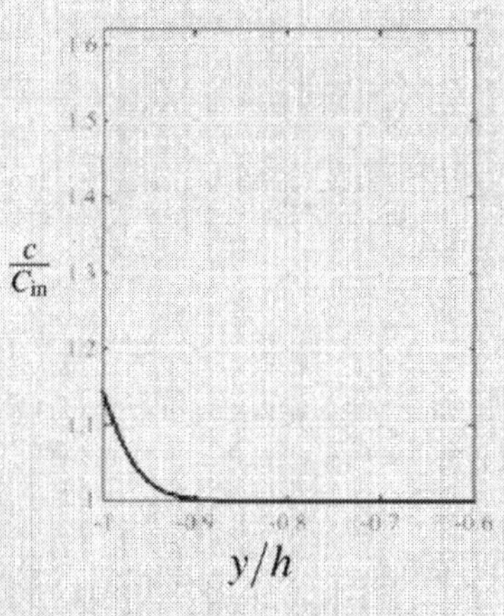

11/23

Results: Transient Solute Accumulation

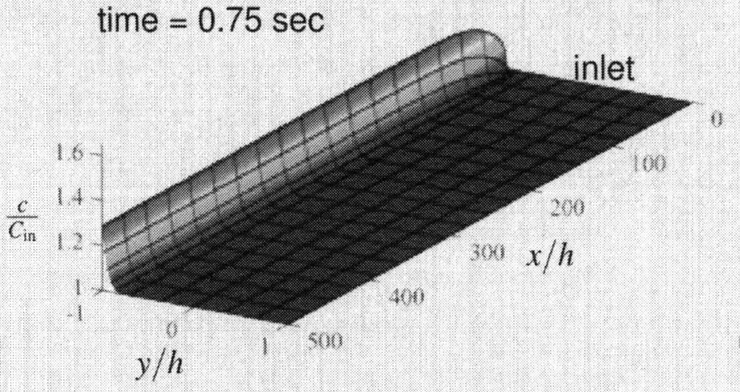

Re=250
Sc=700
$\sigma = 10^{-11}$
$P_{in}^{osm} = 3.4$ bar
$P_{out} = 24$ bar

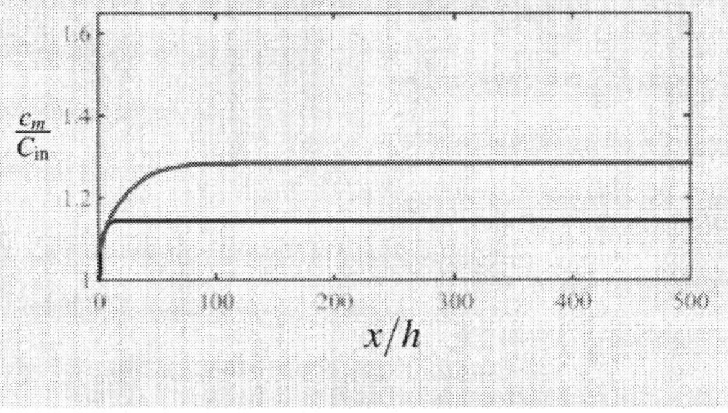

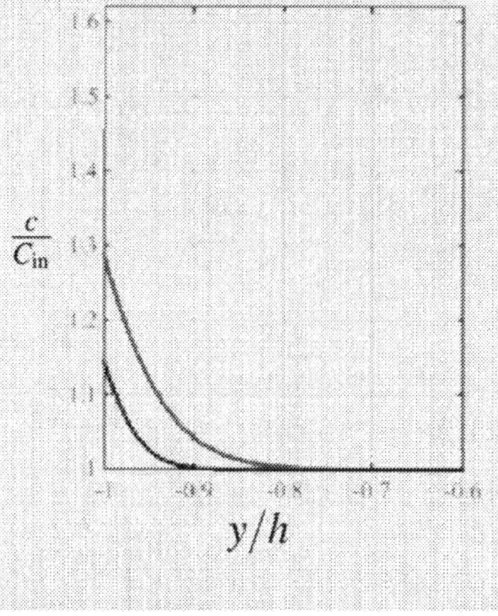

11/23

Results: Transient Solute Accumulation

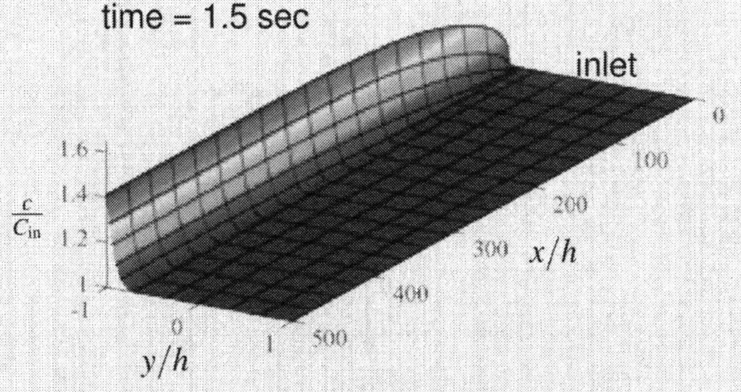

Re=250
Sc=700
$\sigma = 10^{-11}$
$P_{in}^{osm} = 3.4$ bar
$P_{out} = 24$ bar

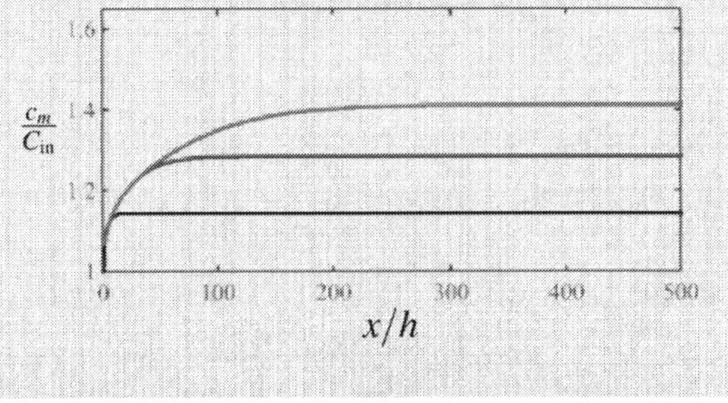

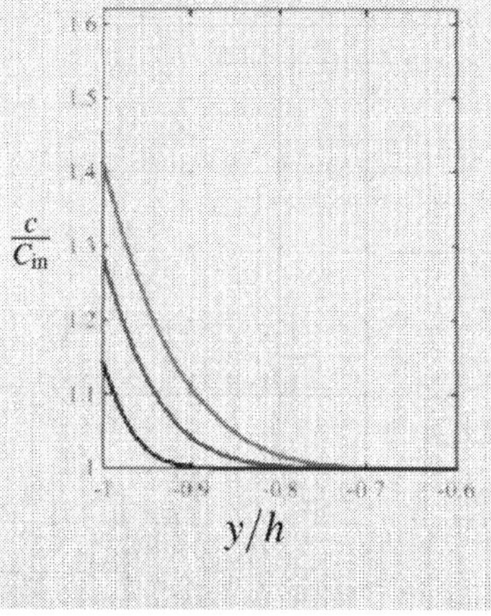

11/23

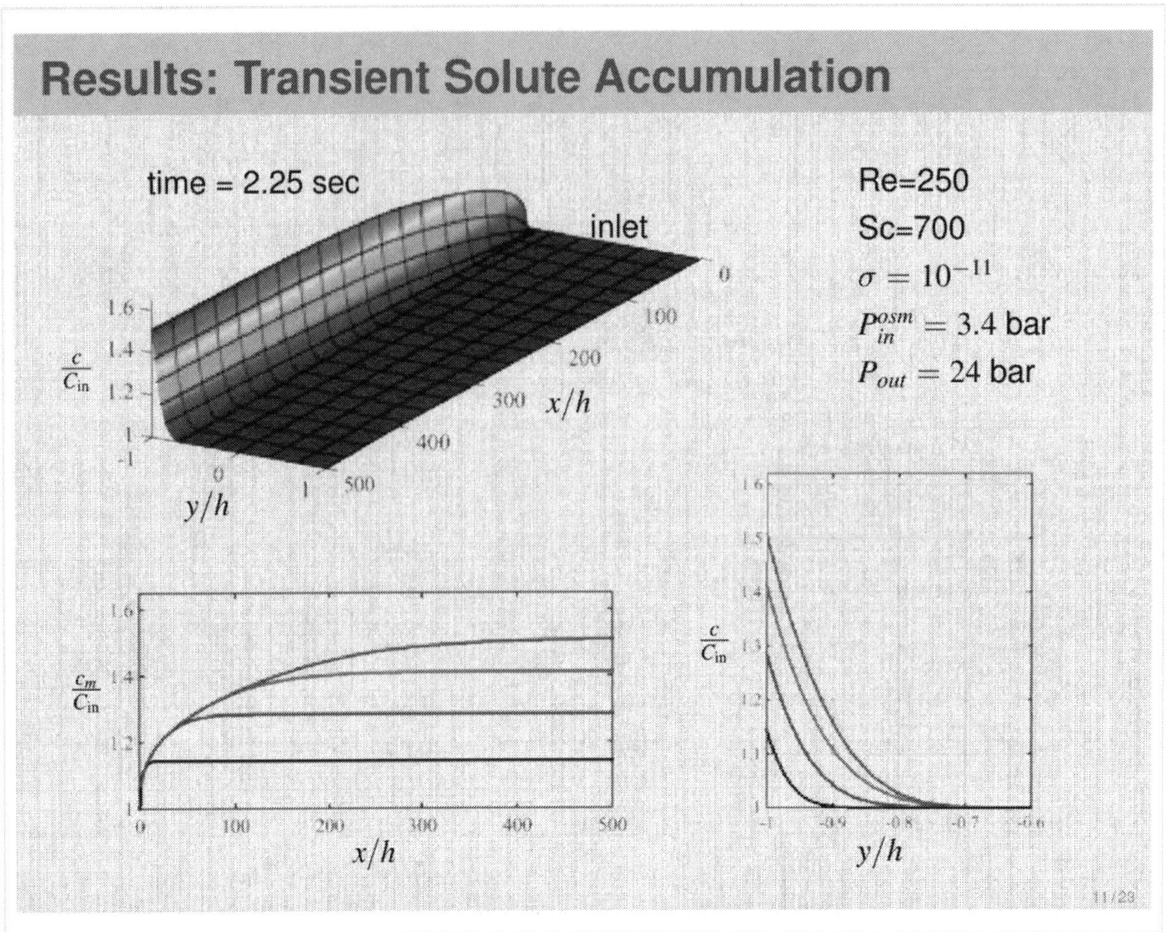
Results: Transient Solute Accumulation
time = 2.25 sec
inlet
Re=250
Sc=700
$\sigma = 10^{-11}$
$P_{in}^{osm} = 3.4$ bar
$P_{out} = 24$ bar
$\frac{c}{C_{in}}$
x/h
y/h
$\frac{c_m}{C_{in}}$
$\frac{c}{C_{in}}$
x/h
y/h
11/23

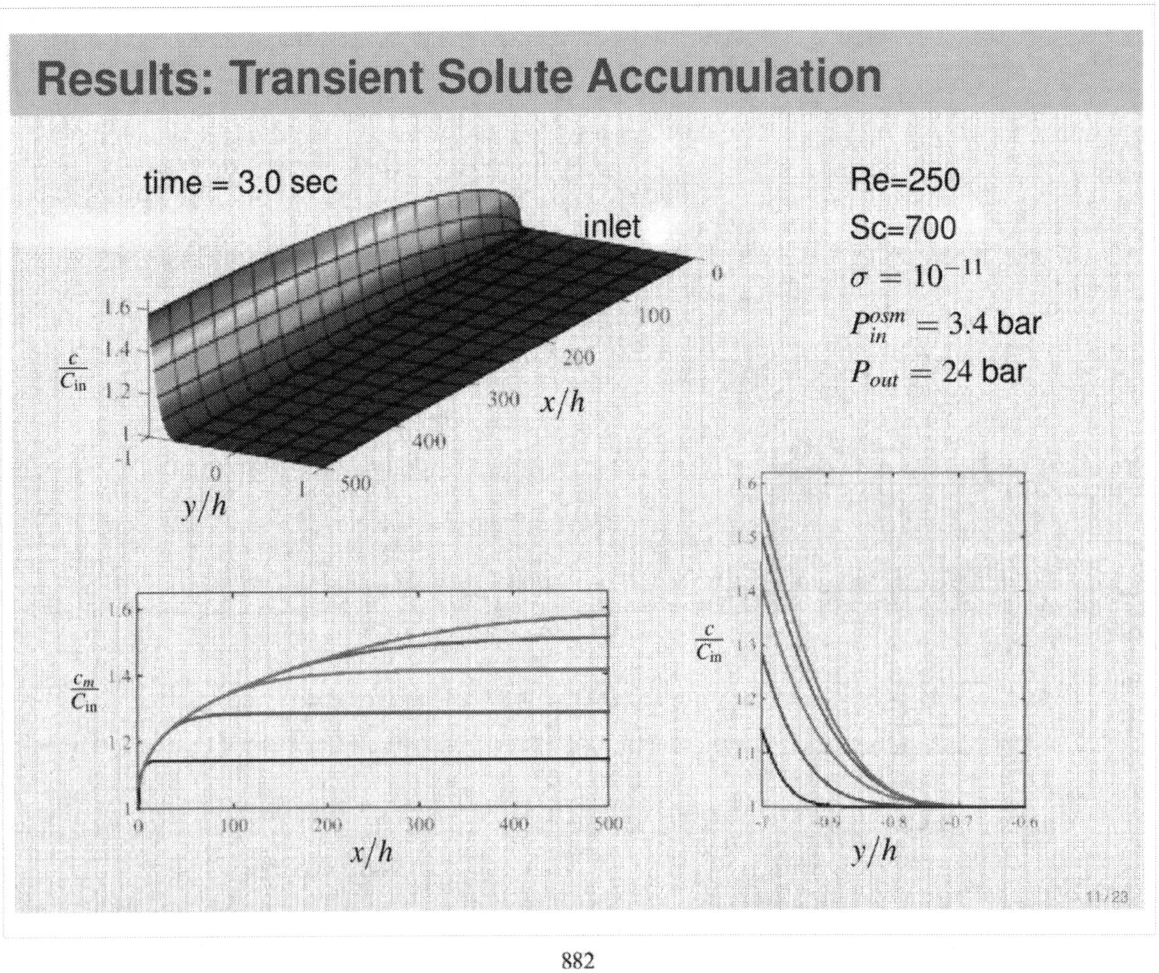
Results: Transient Solute Accumulation
time = 3.0 sec
inlet
Re=250
Sc=700
$\sigma = 10^{-11}$
$P_{in}^{osm} = 3.4$ bar
$P_{out} = 24$ bar
$\frac{c}{C_{in}}$
x/h
y/h
$\frac{c_m}{C_{in}}$
$\frac{c}{C_{in}}$
x/h
y/h
11/23

Results: Transient Solute Accumulation

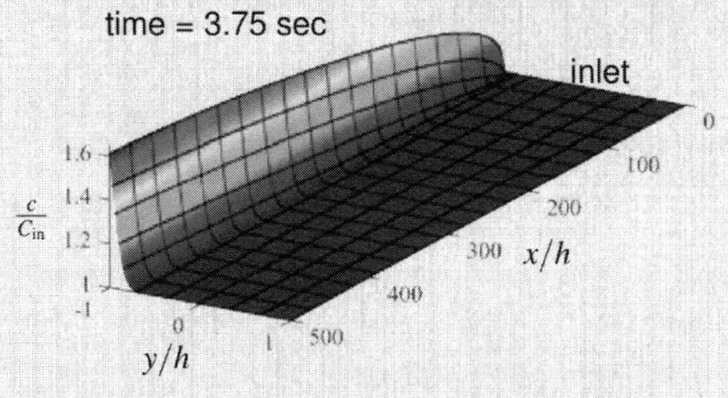

Re=250

Sc=700

$\sigma = 10^{-11}$

$P_{in}^{osm} = 3.4$ bar

$P_{out} = 24$ bar

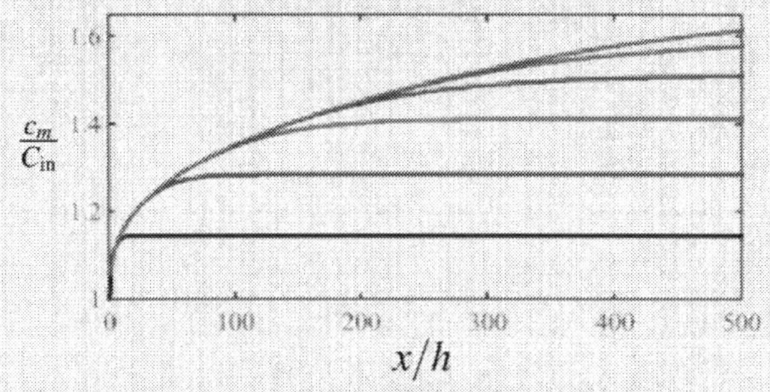

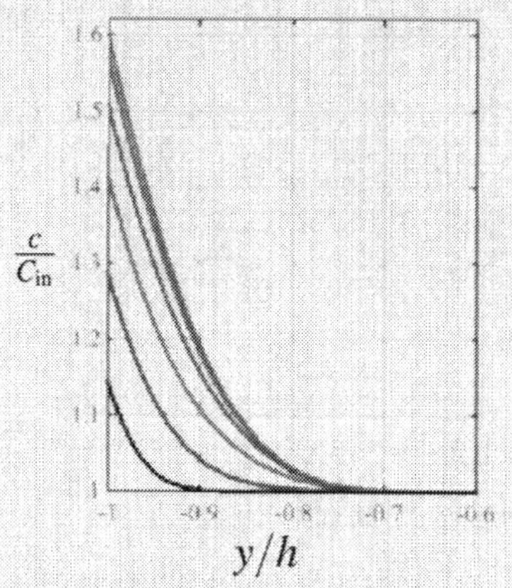

Results: Steady-State Polarization (Self-Similarity)

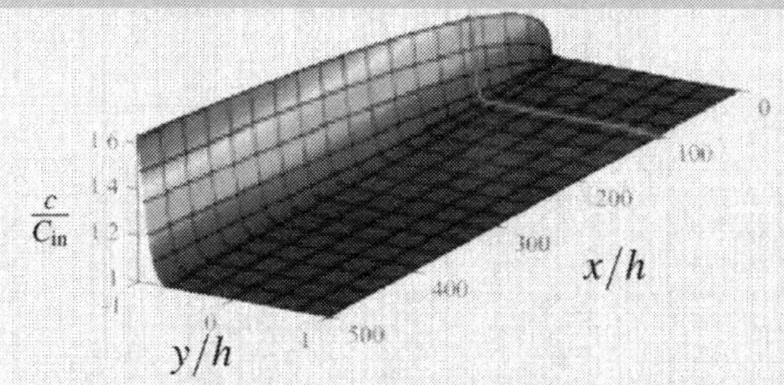

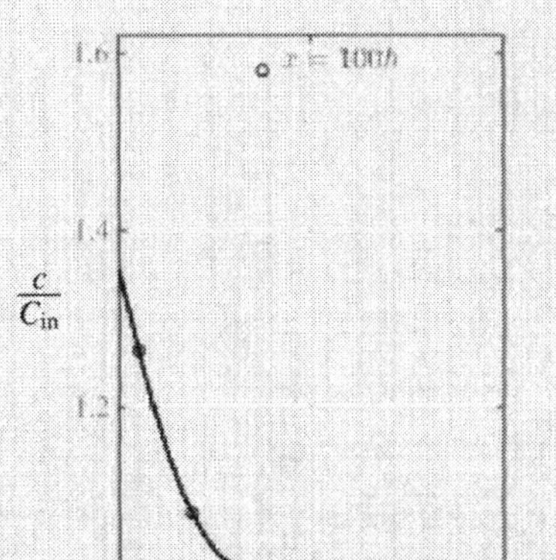

Results: Steady-State Polarization (Self-Similarity)

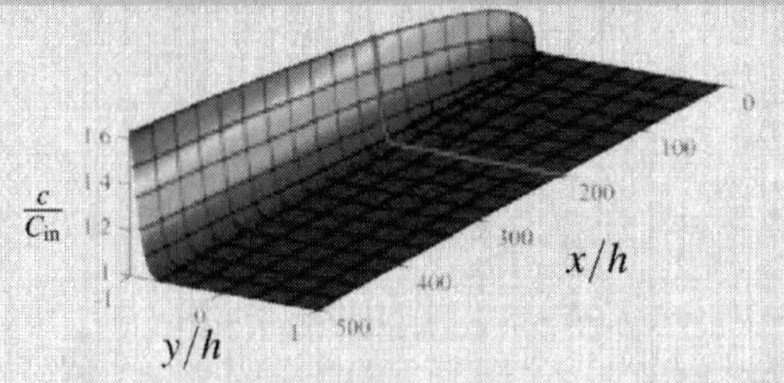

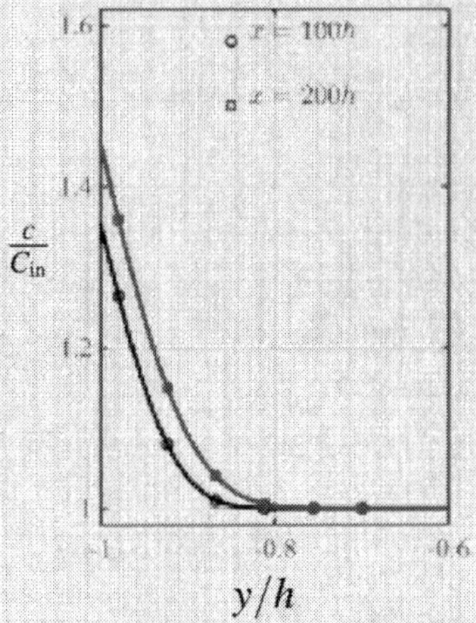

Results: Steady-State Polarization (Self-Similarity)

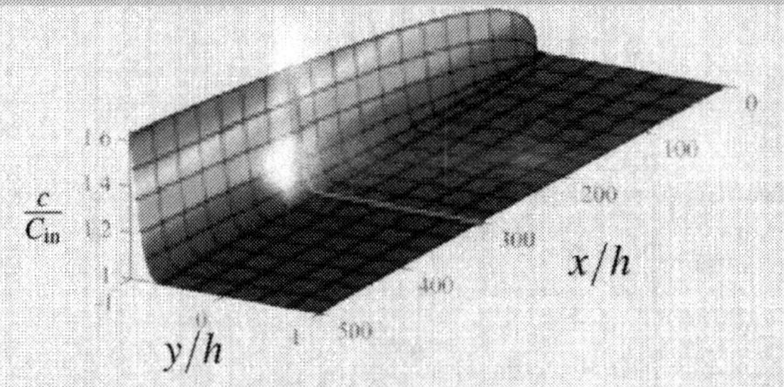

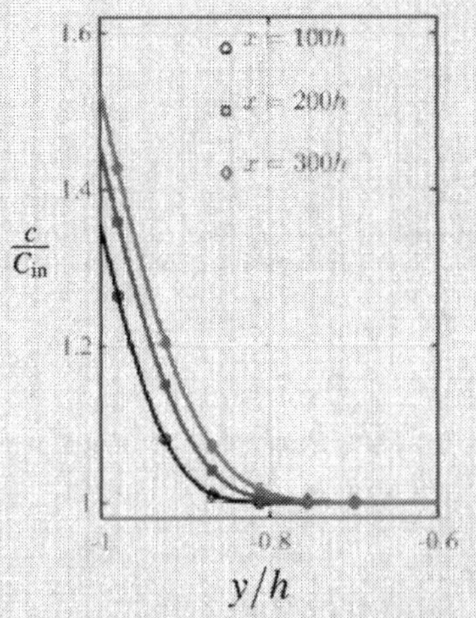

12/23

Results: Steady-State Polarization (Self-Similarity)

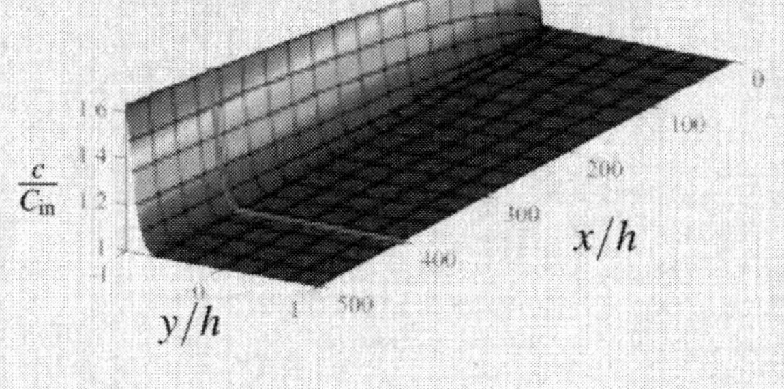

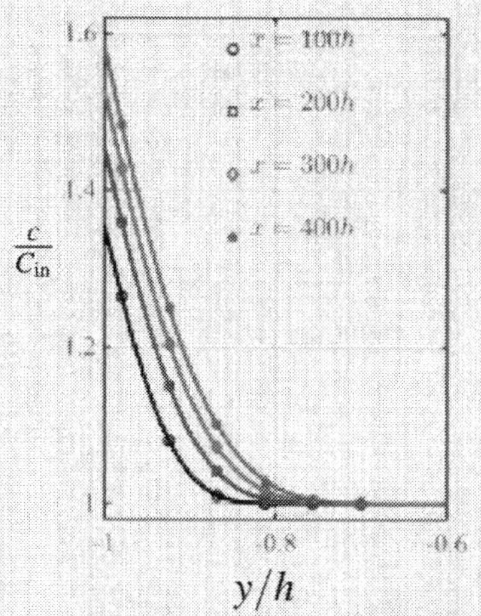

12/23

Results: Steady-State Polarization (Self-Similarity)

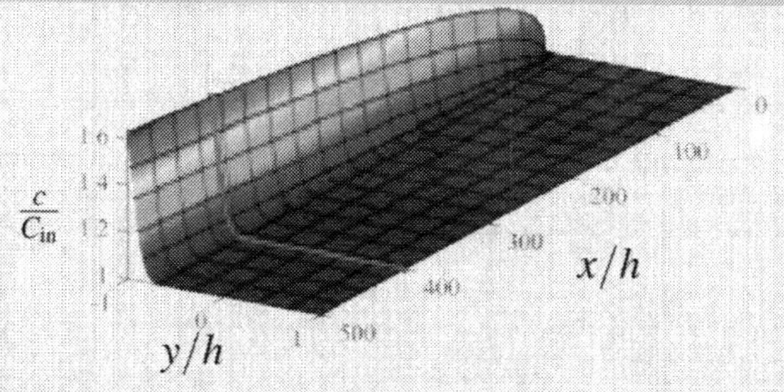

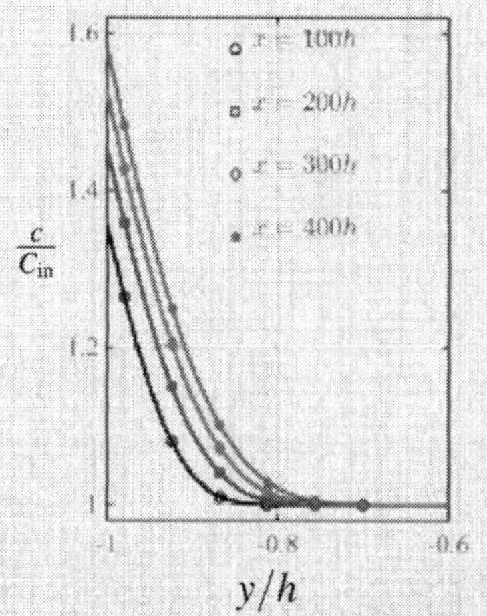

$$\widehat{c} = \frac{c - C_{\text{in}}}{c_m(x) - C_{\text{in}}}$$

$$\eta = \frac{y + h}{\delta(x)}$$

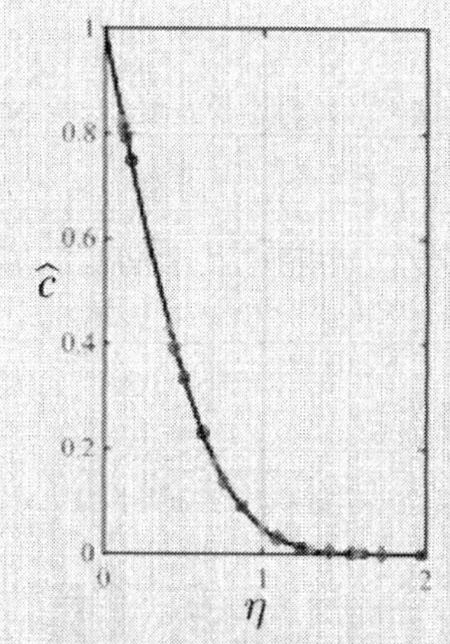

12/23

Steady-State Boundary Layer Depth

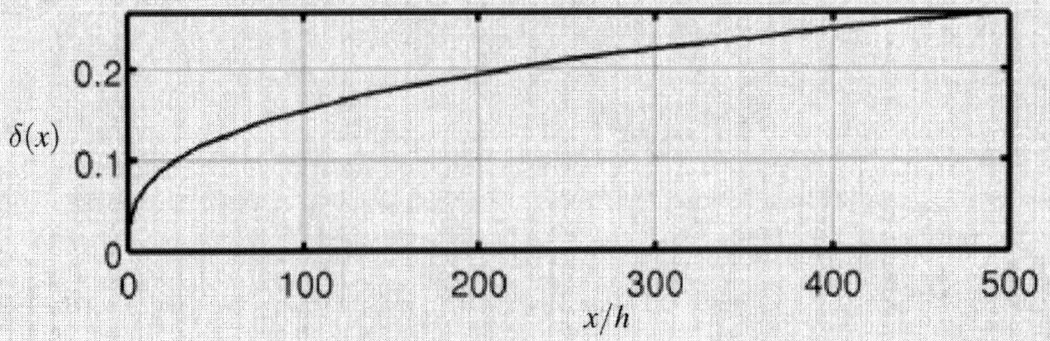

13/23

Steady-State Boundary Layer Depth

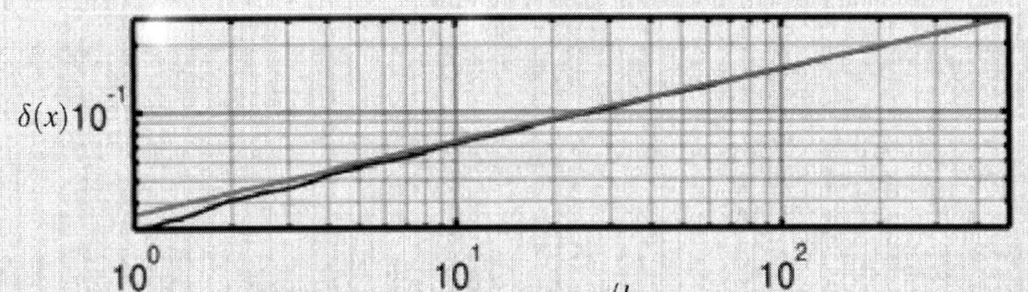

13/23

Steady-State Boundary Layer Depth

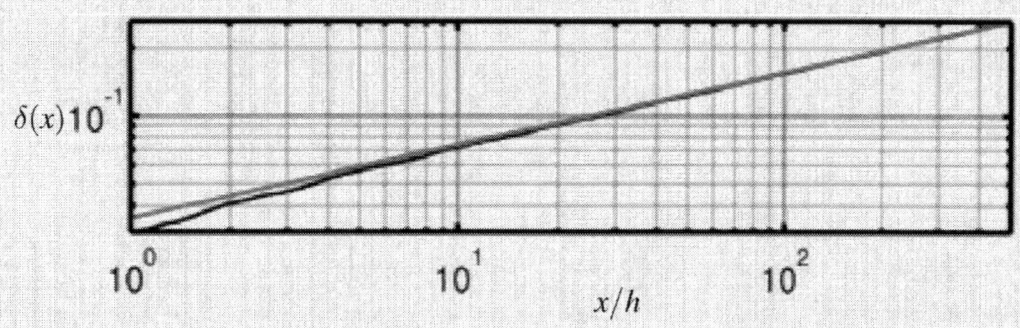

- We find that $\delta \sim x^{1/3}$

$$\delta(x) = 0.0353\, x^{0.323}$$

13/23

Steady-State Boundary Layer Depth

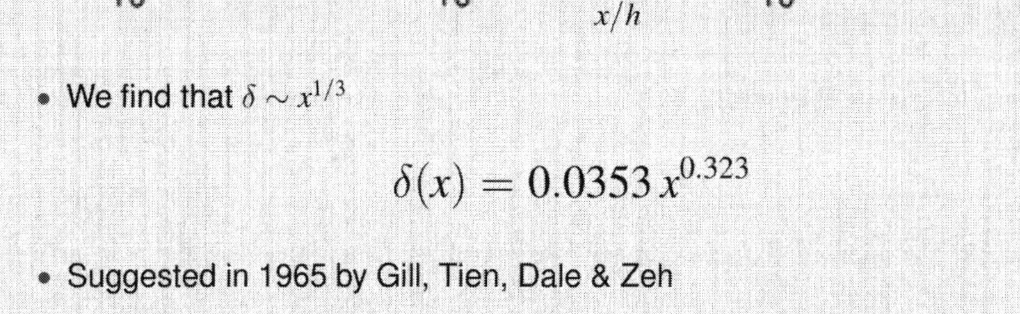

- We find that $\delta \sim x^{1/3}$

$$\delta(x) = 0.0353\, x^{0.323}$$

- Suggested in 1965 by Gill, Tien, Dale & Zeh

13/23

Potential Analytical Solution?

- A few attempts to find analytical solutions for CP

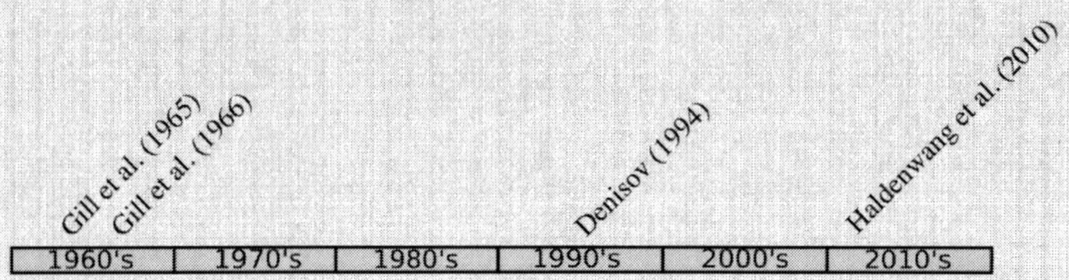

14/23

Potential Analytical Solution?

- A few attempts to find analytical solutions for CP

- Generally very complex

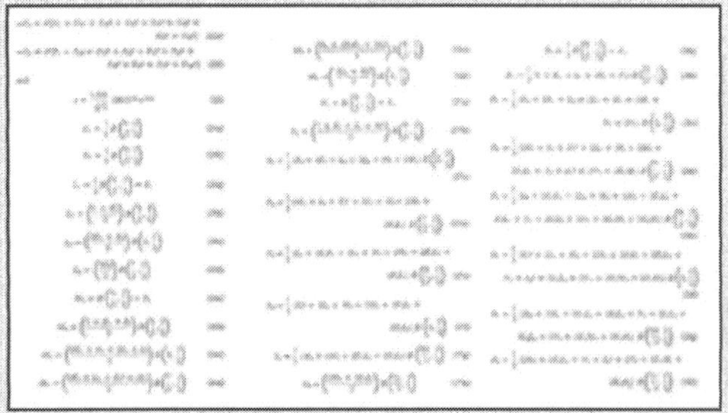

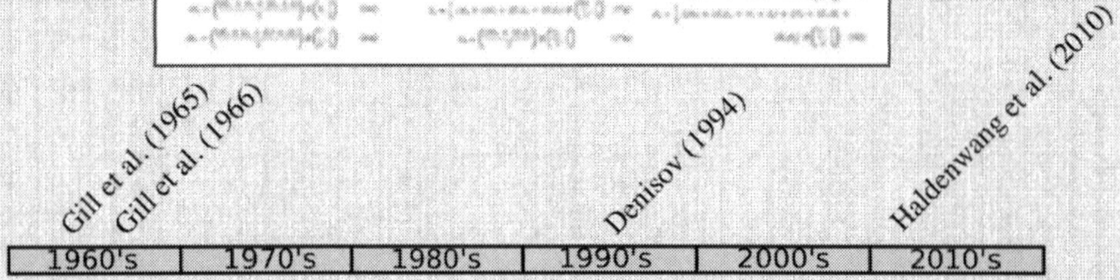

14/23

Potential Analytical Solution?

- A few attempts to find analytical solutions for CP
- Generally very complex
- Remain unverified with numerical results

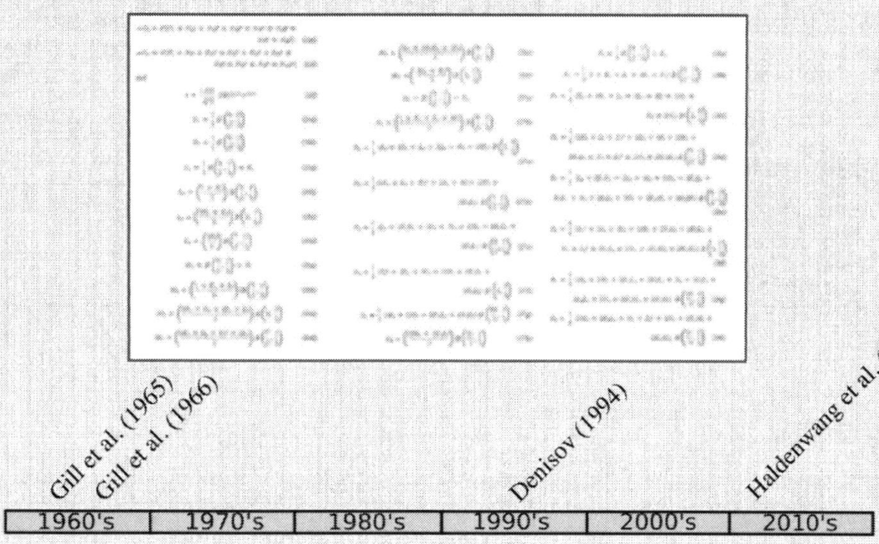

Steady State: Advection vs. Diffusion

$$u\frac{\partial c}{\partial x}+v\frac{\partial c}{\partial y} = \frac{1}{\mathrm{Pe}}\frac{\partial^2 c}{\partial x^2}+\frac{1}{\mathrm{Pe}}\frac{\partial^2 c}{\partial y^2}, \qquad Pe = \frac{U_{in}h}{\mathcal{D}}$$

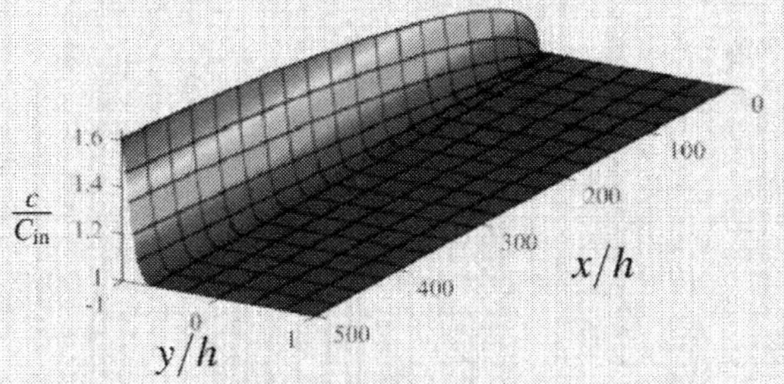

Steady State: Advection vs. Diffusion

$$u\frac{\partial c}{\partial x}+v\frac{\partial c}{\partial y}=\frac{1}{\mathrm{Pe}}\cancelto{\approx 0}{\frac{\partial^2 c}{\partial x^2}}+\frac{1}{\mathrm{Pe}}\frac{\partial^2 c}{\partial y^2},\qquad Pe=\frac{U_{in}h}{\mathcal{D}}$$

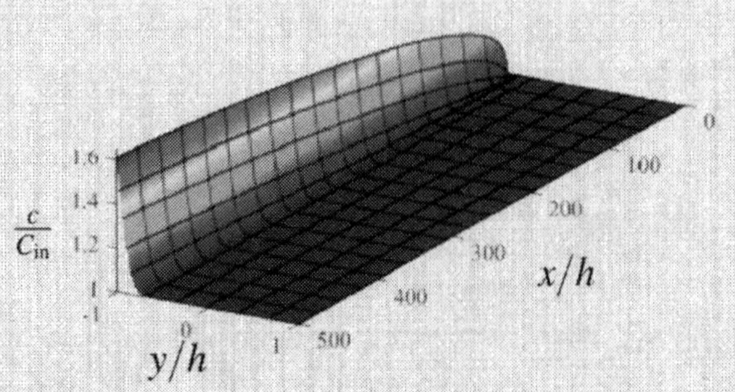

15/23

Steady State: Advection vs. Diffusion

$$u\frac{\partial c}{\partial x}+v\frac{\partial c}{\partial y}=\frac{1}{\mathrm{Pe}}\frac{\partial^2 c}{\partial y^2},\qquad Pe=\frac{U_{in}H}{\mathcal{D}}$$

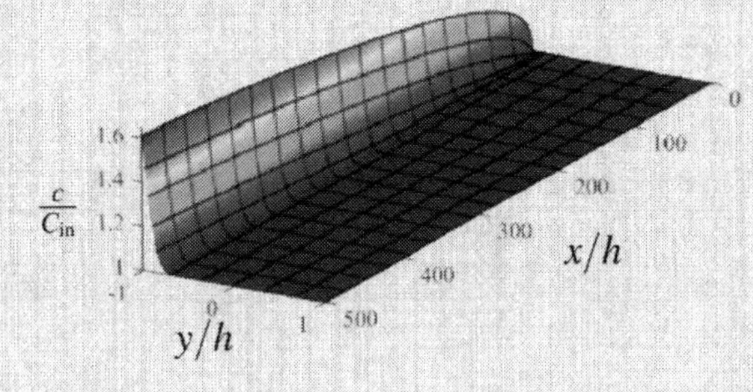

15/23

Steady State: Advection vs. Diffusion

$$\underbrace{u\frac{\partial c}{\partial x}} + \underbrace{v\frac{\partial c}{\partial y}} = \underbrace{\frac{1}{\mathrm{Pe}}\frac{\partial^2 c}{\partial y^2}}, \qquad Pe = \frac{U_{in}H}{\mathcal{D}}$$

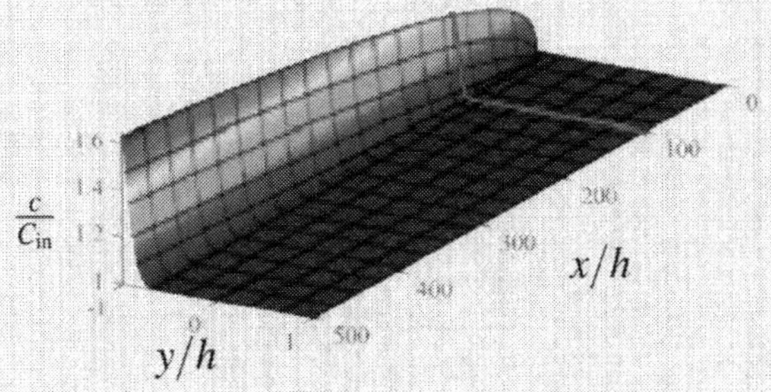

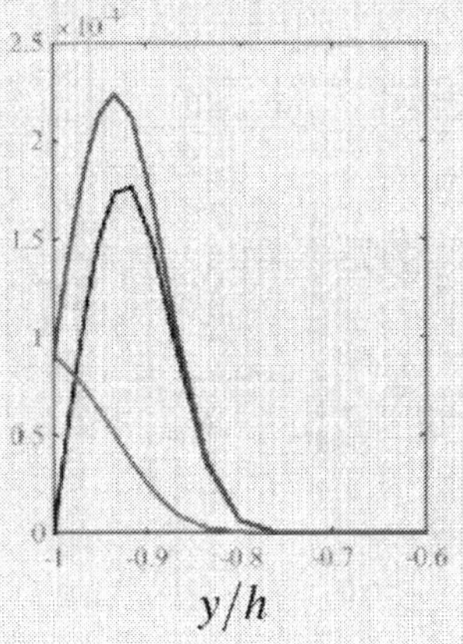

15/23

Steady State: Advection vs. Diffusion

$$\underbrace{u\frac{\partial c}{\partial x}} + \underbrace{v\frac{\partial c}{\partial y}} = \underbrace{\frac{1}{\mathrm{Pe}}\frac{\partial^2 c}{\partial y^2}}, \qquad Pe = \frac{U_{in}H}{\mathcal{D}}$$

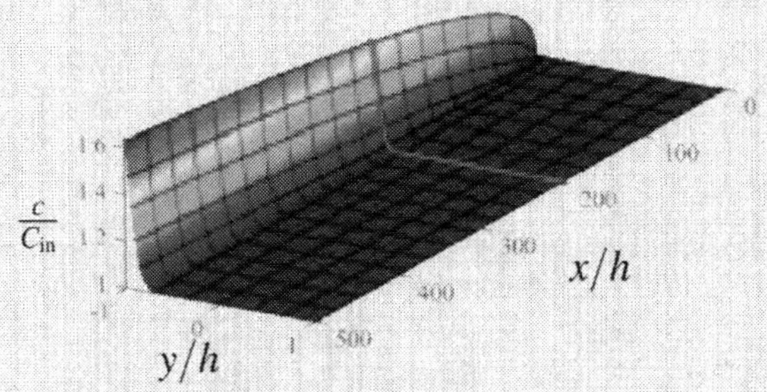

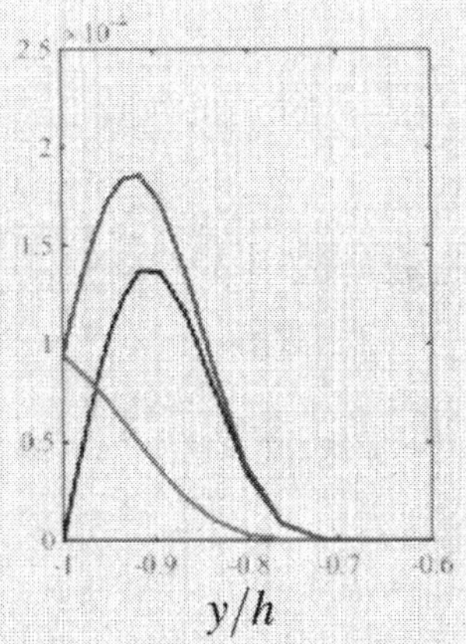

15/23

Steady State: Advection vs. Diffusion

$$\underline{u\frac{\partial c}{\partial x}} + \underline{v\frac{\partial c}{\partial y}} = \underline{\frac{1}{\text{Pe}}\frac{\partial^2 c}{\partial y^2}}, \qquad Pe = \frac{U_{in}H}{\mathcal{D}}$$

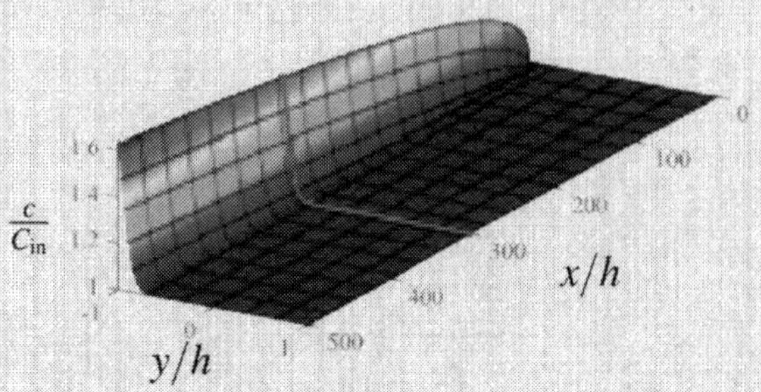

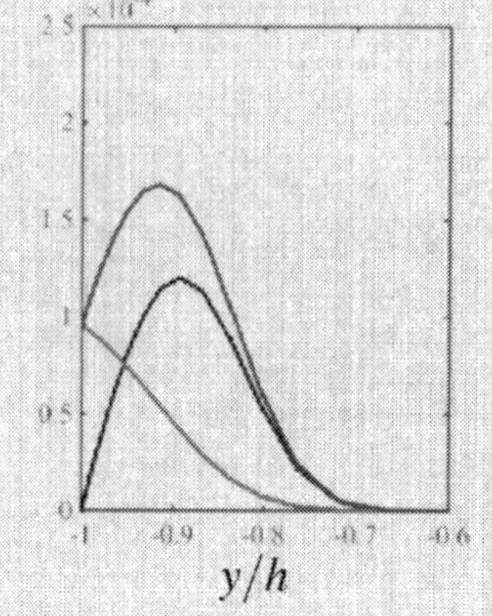

15/23

Steady State: Advection vs. Diffusion

$$\underline{u\frac{\partial c}{\partial x}} + \underline{v\frac{\partial c}{\partial y}} = \underline{\frac{1}{\text{Pe}}\frac{\partial^2 c}{\partial y^2}}, \qquad Pe = \frac{U_{in}H}{\mathcal{D}}$$

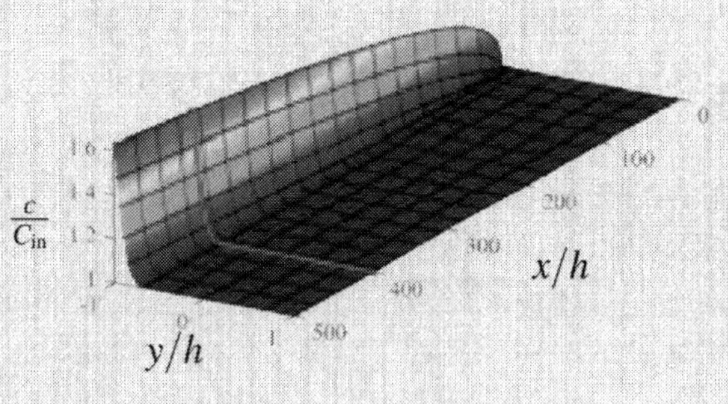

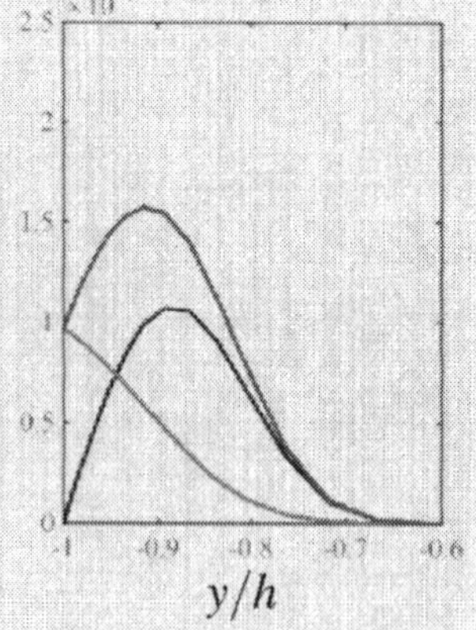

15/23

Steady State: Advection vs. Diffusion

$$u\frac{\partial c}{\partial x} + v\frac{\partial c}{\partial y} = \frac{1}{\mathrm{Pe}}\frac{\partial^2 c}{\partial y^2}, \qquad Pe = \frac{U_{in}H}{\mathcal{D}}$$

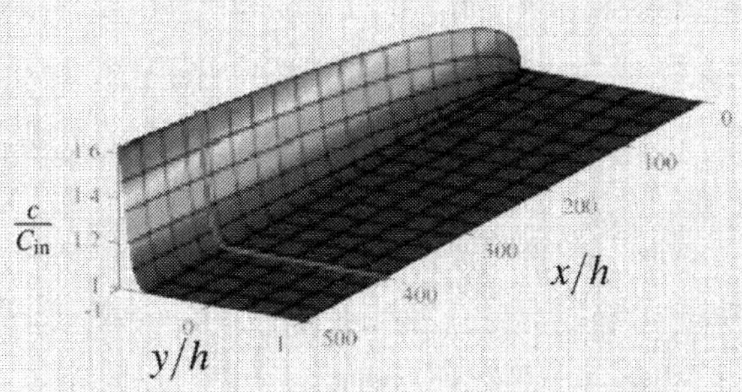

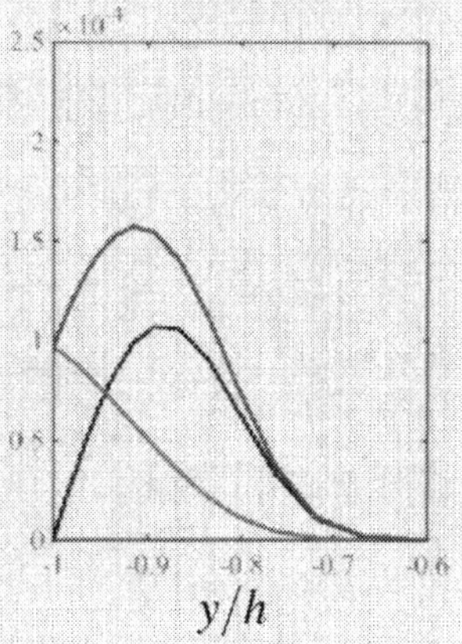

- Does not support the thin-film approximation

Steady State: Advection vs. Diffusion

$$u\frac{\partial c}{\partial x} + v\frac{\partial c}{\partial y} = \frac{1}{\mathrm{Pe}}\frac{\partial^2 c}{\partial y^2}, \qquad Pe = \frac{U_{in}H}{\mathcal{D}}$$

- But behavior changes at high pressures

Steady State: Advection vs. Diffusion

$$u\frac{\partial c}{\partial x} + v\frac{\partial c}{\partial y} = \frac{1}{\mathrm{Pe}}\frac{\partial^2 c}{\partial y^2}, \qquad Pe = \frac{U_{in}H}{\mathcal{D}}$$

- **But behavior changes at high pressures**

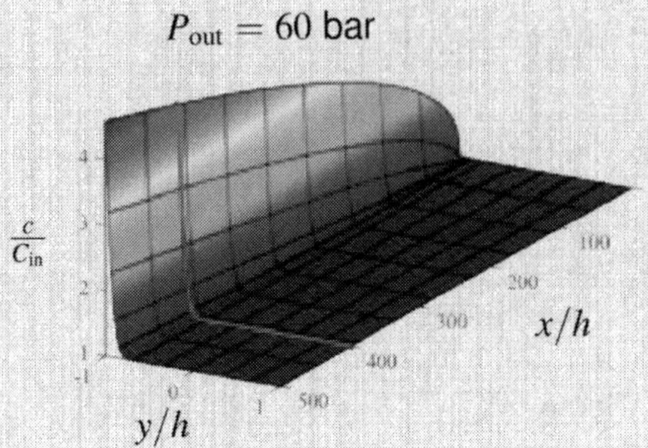

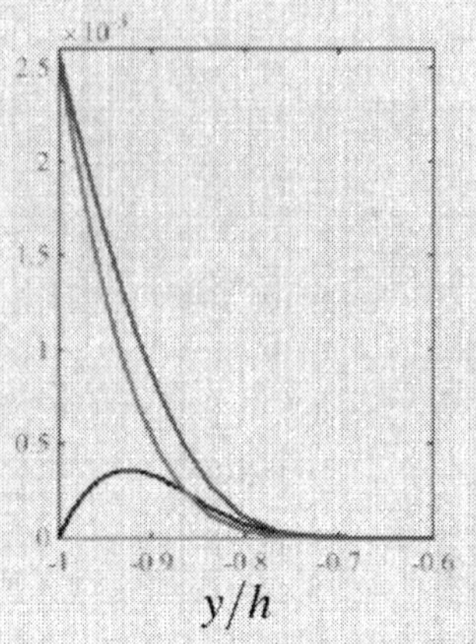

Filtrate Flux vs. Pressure

- $P_{out} = 10, 20, ..., 60$ bar, $Re = 250$, $Sc = 1000$, $\sigma = 10^{-11}$

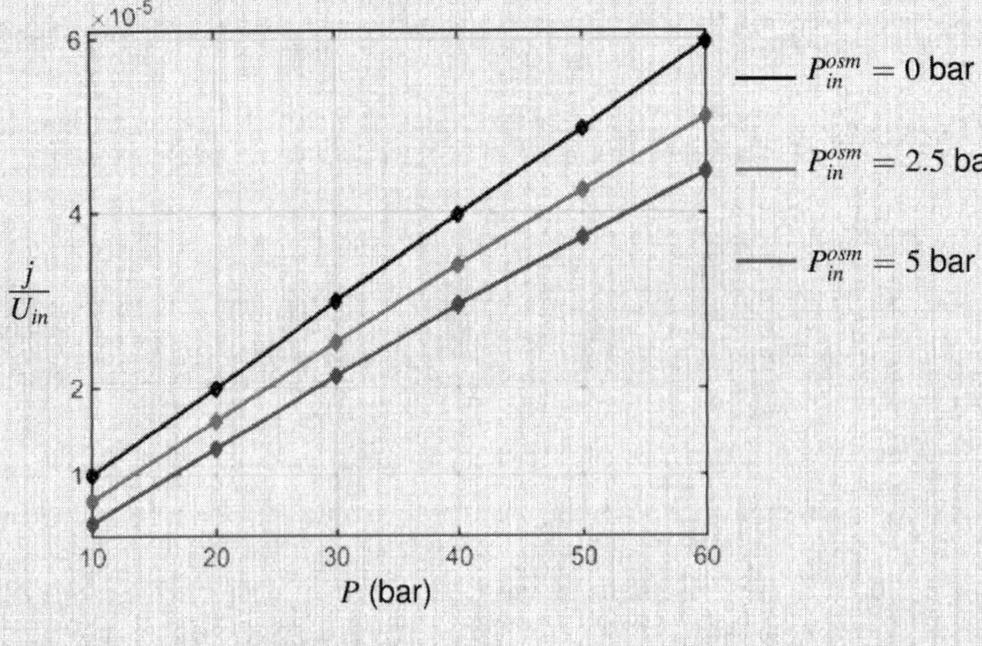

Filtrate Flux vs. Pressure

- $P_{out} = 10, 20, ..., 60$ bar, $Re = 250$, $Sc = 1000$, $P_{in}^{osm} = 2.5$ bar

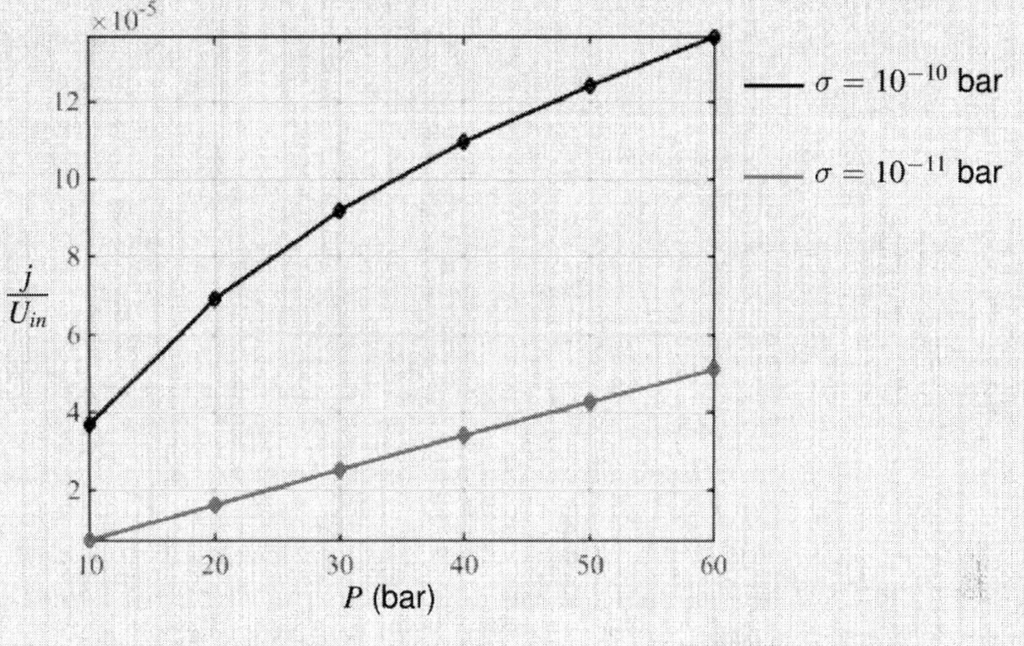

18/23

Future Work: Mineral Scale

- Model precipitation as a heterogeneous reaction on membrane ($y = -h$):

$$vc - \mathcal{D}\frac{\partial c}{\partial y} = -k_p(c)c,$$

- Where $k_p(c) = 0$ when $c < C_{sat}$

$$k_p(c) = A\left[1 + \text{erf}\left(B\frac{c - C_{\text{sat}}}{C_{\text{in}}} - D\right)\right],$$

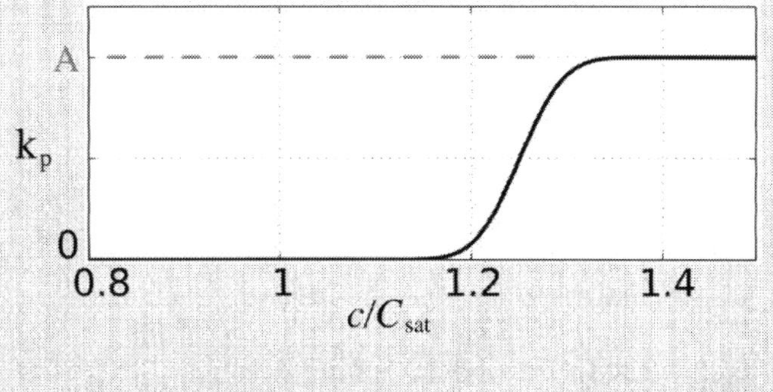

19/23

Future Work: Mineral Scale

- Make permeability a function of the local amount of scale (c_s):

$$\frac{\partial c_s}{\partial t} = k_p c,$$

$$K = K_0 \exp\left[-E \left(\frac{c_s}{hC_{in}} \right)^2 \right]$$

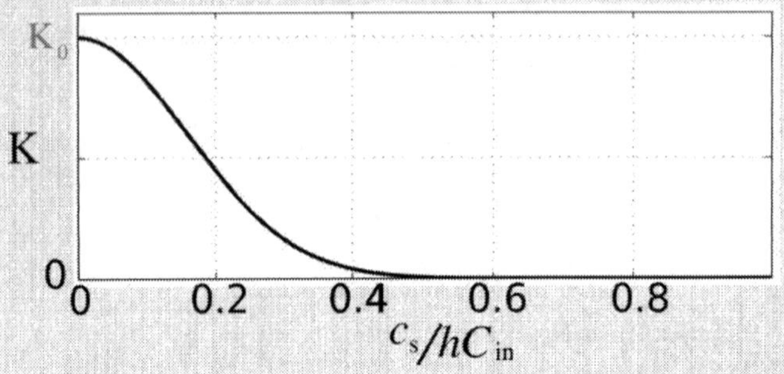

- **Blockage model:**
  - Supported by simulations/experiments (Lyster & Cohen 2009)

20/23

Future Work: Mineral Scale

- $L/h = 90$, $Re = 250$, $Sc = 1000$, $\sigma = 10^{-11}$, $P_{out} = 12.5$ bar, $P_{in}^{osm} = 3.4$ bar
- $C_{sat} = 1.1C_{in}$, $k_p(c)$, and $K(c)$ produce rapid precipitation

21/23

Future Work: Mineral Scale

- $L/h = 90$, $Re = 250$, $Sc = 1000$, $\sigma = 10^{-11}$, $P_{out} = 12.5$ bar, $P_{in}^{osm} = 3.4$ bar
- $C_{sat} = 1.1C_{in}$, $k_p(c)$, and $K(c)$ produce rapid precipitation

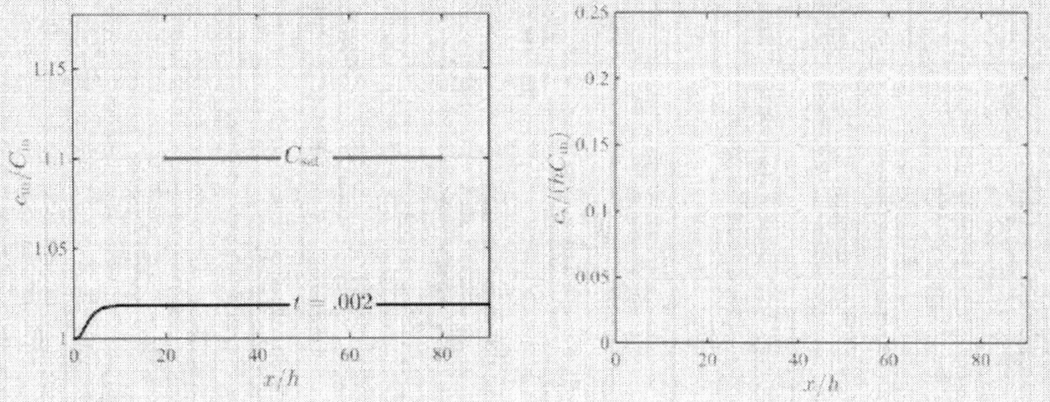

21/23

Future Work: Mineral Scale

- $L/h = 90$, $Re = 250$, $Sc = 1000$, $\sigma = 10^{-11}$, $P_{out} = 12.5$ bar, $P_{in}^{osm} = 3.4$ bar
- $C_{sat} = 1.1C_{in}$, $k_p(c)$, and $K(c)$ produce rapid precipitation

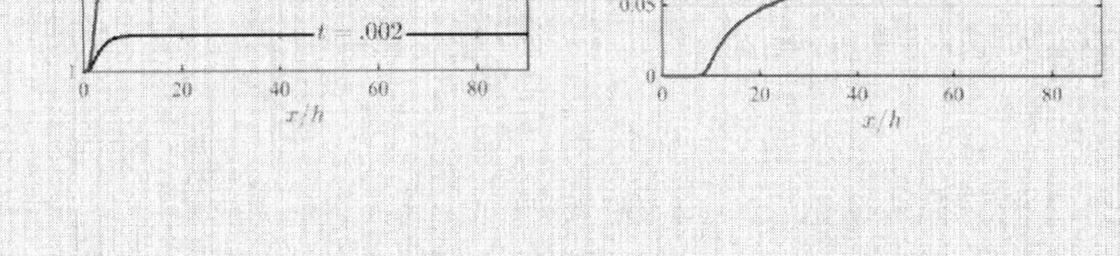

21/23

Future Work: Mineral Scale

- $L/h = 90$, $Re = 250$, $Sc = 1000$, $\sigma = 10^{-11}$, $P_{out} = 12.5$ bar, $P_{in}^{osm} = 3.4$ bar
- $C_{sat} = 1.1C_{in}$, $k_p(c)$, and $K(c)$ produce rapid precipitation

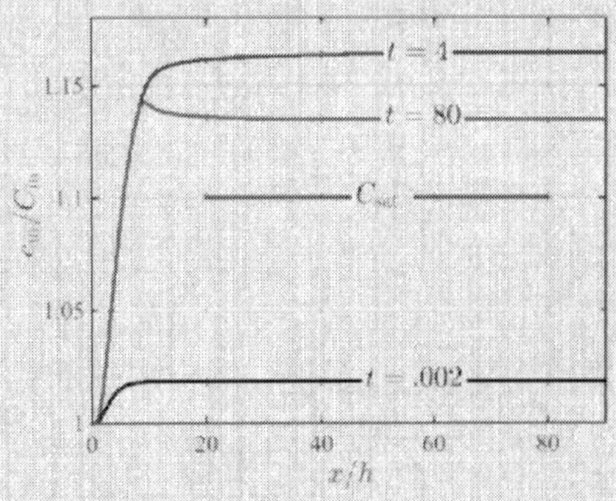

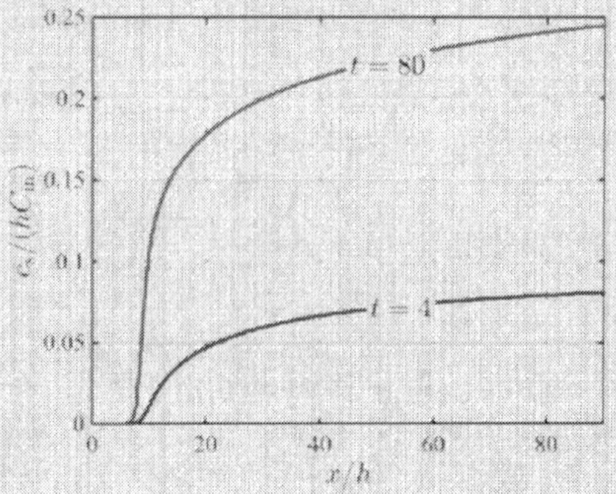

21/23

Future Work: Mineral Scale

- $L/h = 90$, $Re = 250$, $Sc = 1000$, $\sigma = 10^{-11}$, $P_{out} = 12.5$ bar, $P_{in}^{osm} = 3.4$ bar
- $C_{sat} = 1.1C_{in}$, $k_p(c)$, and $K(c)$ produce rapid precipitation

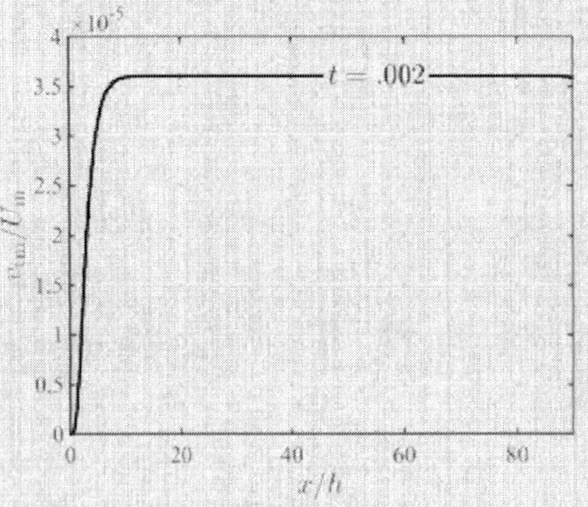

21/23

Future Work: Mineral Scale

- $L/h = 90$, $Re = 250$, $Sc = 1000$, $\sigma = 10^{-11}$, $P_{out} = 12.5$ bar, $P_{in}^{osm} = 3.4$ bar
- $C_{sat} = 1.1C_{in}$, $k_p(c)$, and $K(c)$ produce rapid precipitation

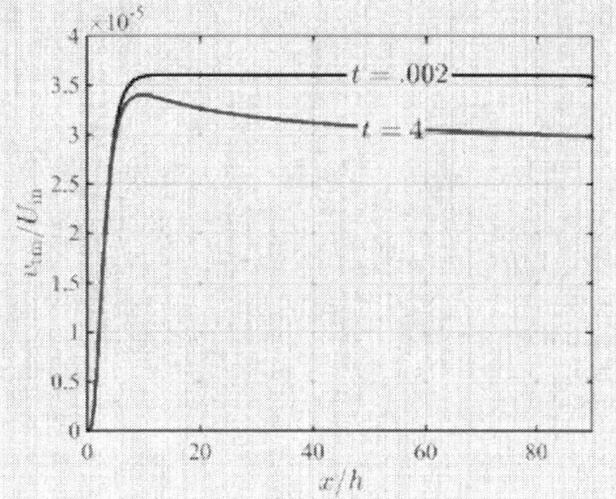

21 / 23

Future Work: Mineral Scale

- $L/h = 90$, $Re = 250$, $Sc = 1000$, $\sigma = 10^{-11}$, $P_{out} = 12.5$ bar, $P_{in}^{osm} = 3.4$ bar
- $C_{sat} = 1.1C_{in}$, $k_p(c)$, and $K(c)$ produce rapid precipitation

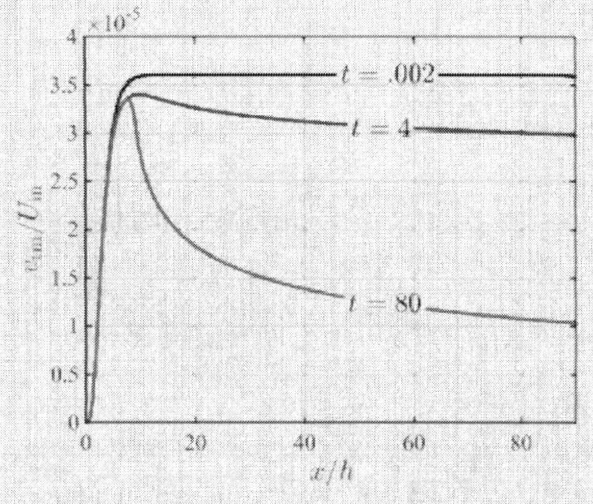

21 / 23

Summary

- Numerical method for transient concentration polarization
- Showed the existence of self-similarity
- Downstream growth as $x^{1/3}$
- Showed the thin-film approximation isn't always valid
- Investigated transmembrane flux vs. pressure
- Suggested potential avenues for modeling mineral scale.

Thanks!

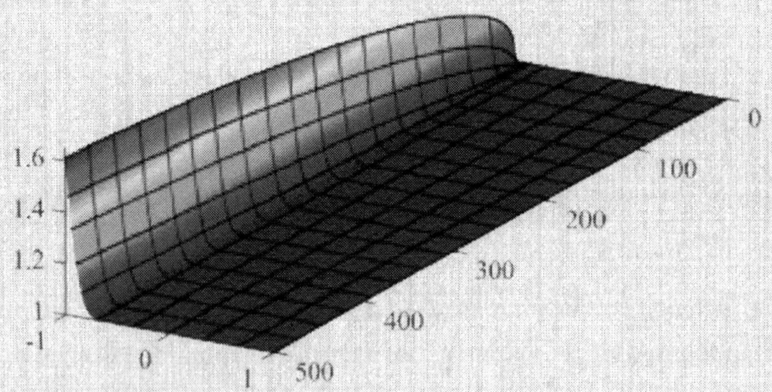

434719 Computational Fluid Dynamics Modeling of Internal Concentration Polarization in Forward Osmosis: Effects of Pore Geometry, Hydrodynamics, and Solute Diffusivity

Abstract:

Internal concentration polarization remains a primary limiting factor for water flux in forward osmosis. Modeling transport through the membrane support layer can provide new insights to guide design of membranes and membrane systems. Experimental results have definitively shown that membrane support layer structure and module hydrodynamics significantly influence the severity of concentration polarization. Previous numerical modeling efforts have confirmed the importance of the structural parameter in determining transport and have also characterized hydrodynamic conditions within support layer pores. We investigate here the interplay between pore geometry and cross-flow velocity in determining the extent of internal concentration polarization and its impact on water flux. In addition, we examine the potential for error to be introduced in transport models by the assumption of a constant solute diffusivity. A simplified pore geometry was adopted, with a single cylindrical pore of fixed diameter and length being considered. Diffusion of water and salt through the pore was modeled together with convection effects induced inside the pore by cross-flow velocity applied at the support layer – bulk solution interface. The bulk feed and draw solution concentrations were fixed at experimentally relevant values while cross-flow velocity, pore length, and pore diameter were varied. The draw solution concentration at the support layer – active layer interface and the resulting water flux were then determined via computational fluid dynamics simulations. Results obtained with concentration-dependent and concentration-independent solute diffusivities were compared. Implications of this work for future modeling efforts, for membrane support layer design, and for operating conditions will be discussed.

434277 Elucidating the Impact of Temperature Gradients Across Membranes during Osmosis

Abstract:

The forward osmosis (FO) process has long been touted as an emerging desalination technology that can operate off of a thermally regenerated draw solution. Using thermal energy to regenerate this solutions, however, can lead the draw solution being warmer than typical ambient conditions. Likewise, in industrial reuse or produced water treatment applications, the feed solution can be elevated in temperature. While each of these concepts has been discussed in the literature individually, lacking from the field is an understanding of how different solution temperatures can impact FO performance. We look at how various thermal gradients impact water flux, reverse salt flux and membrane intrinsic properties using both experiments and model based prediction to do this, we develop a model that effectively couples the effect of simultaneous heat and mass transfer across the membrane. The model is used to predict water flux and reverse salt flux in FO and PRO mode. We, then, compare these fluxes with experimentally obtained results. Experimental results suggest that the water flux increases substantially at higher system temperature (40°C), as has been determined before, while the reverse salt flux decreases at that temperature. In addition, operating at higher feed temperature (40°C) than the draw side (20°C), produces fairly comparable results to that operating at higher system temperature (40°C). The model prediction for water flux also matches well to the results obtained in these operating conditions. Reverse osmosis (RO) characterization reveals that membrane intrinsic properties such as pure water permeance increases with temperature while the salt permeability decreases. The decreasing trend in salt permeability is also seen in the reverse salt flux performance of the FO experiments where the membrane orientation is similar to RO. This suggests that operating at higher feed temperature results in better performance as compared to operating at higher system temperature where both feed and draw solutions are heated concurrently. Even though the predicted water flux by the coupled model compares well to the experimental value, the predicted reverse salt flux deviates as much as 30 % from experimental value. Overall, the coupled model performs well compared to similar models.

429642 Study of Permeation and Separation through Mesoporous Membranes Using Dynamic Mean Field Theory

Abstract:

Mesoporous membranes find application in multicomponent vapor phase separations such as removal of carbon dioxide from flue gas and recovery of volatile organic compounds (VOCs) from light gas streams through preferential adsorption and condensation of one of the species. However, design rules for mesoporous materials pertaining to specific separations are still unclear. Selectivity and permeance of membranes is determined by surface and capillary effects resulting from mesopore characteristics (geometry, surface chemistry) but truly predictive correlations for these effects based on mesoporous attributes are absent. In this work permeation through mesopores has been modeled using Dynamic Mean Field Theory (DMFT), a coarse grained lattice based density functional theory. DMFT has been applied to two types of membrane experiment: permporometry, which represents permeation of a light gas through a mesopore in presence of a condensable vapor at a controlled relative pressure, and VOC recovery, where the goal is to enrich condensable VOC, present in a stream of light gas at low concentration, through application of pressure gradient. DMFT predicts fluxes and selectivity, and provides deep insight on transport mechanisms. Interestingly, in simple slit or cylindrical pores, most of the light component flux was observed to be associated with the layer adjacent to the strongly adsorbed surface layer of the heavy component. Different pore geometries, and the influence of microporosity in the material structure (as in SBA-15), have also been studied.

416933 Accounting for Interfacial Solvent Effects within a Mechanistic Crystal Growth Model

Abstract:

The choice of solvent has a significant impact on crystal growth habit. It is therefore one of the most important design decisions in crystallization and its effect is one of the most desirable for a crystal growth model to predict.

Our group has developed a mechanistic crystal growth model, accounting for both spiral and two-dimensional nucleation & growth mechanisms [1,2]. Critical quantities in the model are kink and edge energies on the crystal surface, which affect the predicted relative face growth rates under these layered mechanisms. These energies are modified by solvation of the crystal surface, which must be effectively accounted for. However, the determination of this solvent-induced energetic modification must not compromise the ability to obtain fast growth shape predictions, in order for our model to remain a practical engineering tool.

Interfacial energy models based on the bulk interface approximation are suitable and their application to our mechanistic model is the focus of this work. The energy of an interface is expressed through a penalty term of each phase's cohesive energy and an adhesive reward. For the work of adhesion, the van Oss, Chaudhury & Good (vOCG) model [3] provides a functional form that appropriately separates interactions according to how they match across an interface. In applying this to our model, solid-state crystal interactions are determined using a force field and solvent interactions are determined using solubility parameters and appropriate correlations to interfacial energies. These cohesive energies can be used to determine the required parameters for the vOCG model and although there is some uncertainty in the required partitioning, shape predictions are often not sensitive to this. Heuristic rules are developed for the correct application of this interfacial energy model to different types of systems.

To test the success of this within our mechanistic growth model, predictions for vapor and solution grown crystal shapes of various systems have been compared to experimental results, which isolates the effect of the solvent and has indicated that this approach is successful.

References:

1. Snyder, R. C. and Doherty, M. F. Proceedings of the Royal Society A: Mathematical, Physical and Engineering Science 465(2104), 1145–1171 (2009).
2. Lovette, M. A. and Doherty, M. F. Crystal Growth & Design 12(2), 656–669 (2012).
3. Van Oss, C., Chaudhury, M., and Good, R. Advances in colloid and interface science 28, 35–64 (1987).

(659b) Influence of Diacylglycerol on the Physicochemical Properties and Crystallization Behavior of Palm Oil

Huang, X.,
Hao, H., Tianjin university, Tianjin, China
Wang, J., National Engineering Research Center for Industrial Crystallization Technology, School of Chemical Engineering and Technology, Tianjin University

Originally presented on: 11/12/2015 8:55:00 - 9:15:00

Abstract: In this study, diacylglycerol (DAG) oils were produced through enzymatic glycerolysis and chromatographic respectively and they were added to purified tri-glycerides of palm oil at different concentrations (2, 5, 10, 30 and, 50% w/w). The physical and chemical characteristics, i.e., iodine value, fatty acid composition and triglycerides composition were evaluated and only slight differences were observed. Then an exhaustive analysis of the crystallization behavior of the samples were per-formed by using Differential scanning calorimetry (DSC), Raman Spectroscopy, po-larized light microscopy, low-resolution magnetic pulsed nuclear resonance and X-ray diffraction over different ranges of crystallization temperatures. The aim of this study was to elucidate the mechanism changes of palm oil crystallization by the addition of two types DAG. The experiment results show that the influence of these DAGs on the crystallization behavior (e.g. nucleation rate, crystal growth rate, polymorph transform) of palm oil was depended on the DAG type, supercooling and diglycerides' concentration, e.g. the presence of a low concentration of enzymatic-DAG was found to significantly delay nucleation and crystal growth velocity of palm oil and induce the formation of β form at low degree of supercooling. Finally, we use Avrami model to interpret these influences and the results match the experiment successfully. This study will provide essential information for the palm oil industry for food application purposes.

427862 Investigation of the Formation of Nanoscale 2CL-20 · HMX Cocrystal

Abstract:

Energetic nanoscale 2CL-20·HMX, a cocrystal of CL-20 (2,4,6,8,10,12-hexanitro-2,4,6,8,10,12-hexaazaisowurtzitane) and HMX (1,3,5,7-tetranitro-1,3,5,7-tetrazocine) at a 2:1 molar ratio, was prepared by a novel method of bead milling an aqueous suspension of ε-CL-20 and β-HMX. Conversion of the coformers to the cocrystal form was monitored by powder X-ray diffraction (XRD) and scanning electron microscopy (SEM) analysis of specimens sampled at various milling times. Complete conversion to the cocrystal form was achieved. Rounded 2CL-20·HMX cocrystal particles with a mean size below 200 nm were produced. The conversion was further studied by varying the processing variables. A mechanism for the conversion of the coformers to the cocrystalline form is postulated based on the experimental results. As an inherently safe manufacturing method, the aqueous bead milling process has a great potential in advancing the cocrystal research and applications in the field of energetic materials.

427940 Kidney Stone Growth Modification: Insights from First Principles Calculations

Abstract:

Nearly 1 in 11 Americans will develop kidney stones during their lives[1]. Kidney stones are largely composed of calcium oxalate monohydrate (COM) crystals and develop as a result of heightened concentrations of oxalate in urine [2]. Previous work [2] has identified citric acid (CA) as an effective COM crystal growth modifier. Recently, Dr. Rimer's group at the University of Houston has shown that hydroxycitric acid (HCA) is a highly effective COM crystal growth modifier with notably different growth modification properties than CA. HCA is a molecule with the same structure as CA except for an additional hydroxyl functional group. In this work we use first-principles calculations to elucidate the growth modification mechanisms of CA and HCA at the molecular level. Ab-initio pKa calculations[3]demonstrate that HCA exhibits lower deprotonation constants than CA due to increased intramolecular hydrogen bonding. Accelerated Density Functional Theory calculations (accounting for solvent effects) revealed that HCA exhibits increased folding capability and binding affinity to both calcium ions and oxalate surfaces than CA. Most importantly this work introduces concepts that may be used in the rational design of effective kidney stone growth modifiers.

References

1. Scales, C. D., Smith, A. C., Hanley, J. M., and Saigal, C. S. European Urology 62, 160–165 (2012).
2. Qiu, S. R., Wierzbicki, a., Orme, C. a., Cody, a. M., Hoyer, J. R., Nancollas, G. H., Zepeda, S., and De Yoreo, J. J. Proceedings of the National Academy of Sciences of the United States of America 101(7), 1811–1815 (2004).
3. Feng, S., Bagia, C., and Mpourmpakis, G. Journal of Physical Chemistry A 117, 5211–5219 (2013).

429995 Inhibition of Calcium Oxalate Monohydrate Crystallization Using Organic Acids

Abstract:

Inhibition of Calcium Oxalate Monohydrate Crystallization using Organic Acids

Jihae Chung[1], Michael G. Taylor[2], Isadora Carnaval[2], Giannis Mpourmpakis[2], John R. Asplin[3], and Jeffrey D. Rimer[1]

[1]University of Houston, Department of Chemical and Biomolecular Engineering, Houston, TX 77204

[2]University of Pittsburgh, Department of Chemical and Petroleum Engineering, Pittsburgh, PA 15261

[3]Litholink Corporation and Renal Section, School of Medicine, University of Chicago, Chicago, IL 60637

Kidney stone disease is one of the most common disorders where the number of people afflicted is on the rise; however, there have been no significant advancements in therapeutic treatments for this disease over the past three decades. More than 80% of human kidney stones are calcium oxalate monohydrate (COM) polycrystalline aggregates. The current therapeutic for COM stones is citrate, an over-the-counter supplement with moderate inhibitory effect on COM crystal growth and relatively low compatibility with patients when consumed for an extended period of time. We have explored the effects of hydroxycitrate (HCA), which is a molecular analogue of citrate that is a more potent inhibitor of COM crystal growth, and is thus a possible replacement of citrate as a therapy. HCA differs from citrate by only one hydroxyl group, yet interestingly, this subtle structural difference leads to completely different modes of interaction with COM crystal surfaces. Here, we will compare the effects of citrate and hydroxycitrate on COM crystallization. Our studies focus on the effects of growth inhibitors at both macroscopic and microscopic length scales. In bulk crystallization assays, we quantified the effect of inhibitors on the bulk crystal habit as well as their impact on the kinetics of crystallization. We employ *in situ* atomic force microscopy (AFM) to probe the interfacial interactions between inhibitors and COM crystal surfaces in real time. Working in conjunction with collaborators, we are exploring the effects of HCA and other modifiers *in vivo* and *in silico* to establish a synergistic platform to rationally design crystal growth inhibitors, which can then be used to develop new drugs for pathological diseases involving mineralization.

432022 Continuous Production of Polymer Coated Drug Crystals, Particles and Nanoparticles

Abstract:

Polymeric coating can provide protection for fragile drugs from hydrolysis and degradation. With an appropriate polymer coating, nanoparticles are known to traverse the physiological human mucous barrier. Conventional approaches to such fabrication are quite cumbersome and often employ batch techniques. We have adapted the technique of solid hollow fiber cooling crystallization (SHFCC) as well as the technique of porous hollow fiber based anti-solvent crystallization (PHFAC) to continuously produce polymer-coated drug crystals and particles. Both submicron particles and nanoparticles have been coated. We have also continuously synthesized polymer-coated drug crystals by anti-solvent crystallization from a solution of the drug and the polymer. The synthesized polymer-coated drug crystals involve crystals of the drug Griseofulvin (GF) coated by a thin layer of the polymer Eudragit RL100. In the case of particles, submicron (550 nm) and nano-sized (12 nm) silica particles acted as the host particles; Eudragit RL 100 and poly (lactide-co-glycolide) were the coating polymers. For the polymer-coated drug crystals, the surface morphology, particle size distribution, and the polymer coating thickness were characterized by SEM, STEM, laser diffraction spectroscopy (LDS) and thermogravimetric analysis (TGA). To study the properties of the coated drug crystals, x-ray diffraction (XRD), Raman spectroscopy, and dissolution test were implemented. Coated silica particles were analyzed after vacuum drying using SEM, STEM, energy dispersive spectrometry, LDS and TGA. Scaleup has been demonstrated using hollow fiber modules of different dimensions. Extended duration runs have been conducted for two hours to demonstrate no variation in product quality.

432455 Mechanism of Hematin Crystallization and Inhibition in Biomimetic Solutions

Abstract:

Crystallization is governed by thermodynamic and kinetic events that result in molecule (solute) incorporation into crystal surfaces. These fundamental processes mediate both natural and synthetic crystallization pathways. Our aim is to understand the molecular pathways for the crystallization of hematin, which is a byproduct of heme detoxification in malaria parasites. Upon entering its human host, the parasite catabolizes hemoglobin in red blood cells and releases free heme (Fe^{2+}), which is rapidly oxidized to toxic hematin (Fe^{3+}) within its digestive vacuole (DV). The parasite utilizes this process to sequester the toxin into innocuous crystals within the complex DV environment. Hematin crystal growth ensures parasite survival within the human host and is thought to be suppressed by quinolone-class antimalarial drugs. These crystals are referred to as hemozoin (*in vivo*) or β-hematin (*in vitro*). Identification of antimalarial drug action could provide a foundation for drug design to overcome the continued resurgence of antimalarial drug resistance. Here, we will discuss the molecular mechanism of hematin crystallization and inhibition by quinolone-class antimalarial drugs in a biomimetic growth environment.

In situ atomic force microscopy (AFM) revealed that hematin crystallization occurs through a classical layer-by-layer mechanism involving the addition of individual molecules to the growth sites present on the crystal surface.[1] Our studies employed a two-phase medium which we designed to mimic the complex environment of parasite DVs, which is comprised of an acidic aqueous phase and lipids. We used citric buffer-saturated octanol (CBSO) as a biomimetic solution to grow hematin crystals.[2] During *in situ* AFM experiments, we directly observe the effects of hematin surface diffusion and molecular level fluctuations in the rates of attachment/detachment from step edges. Towards understanding the relationship of solvent interactions on molecule addition, we quantified the rate of layer generation and the velocities of anisotropic step advancement as a function of supersaturation in the absence and presence of antimalarial modifiers. These quantifications provide a foundation for determining the modes of action of antimalarial drugs (i.e. inhibition of layer formation and/or step propagation). We identify that current antimalarials act by unique modes of inhibition. Here, we will summarize these findings and discuss our current studies to elucidate the role of functional moieties in current antimalarials and how the organic solvent affects drug-crystal interaction. To this end, we use a combination of *in situ* AFM and chemical force microscopy (CFM) measurements wherein AFM tips are functionalized with moieties that mimic specific regions of antimalarial drug molecules. Our findings reveal interesting trends of molecular recognition between antimalarial drugs and crystal surfaces that when coupled with bulk crystallization assays is a potential platform for the design and screening of new drug compounds for malaria.

References

1. Olafson, K.N., M.A. Ketchum, J.D. Rimer, and P.G. Vekilov, *Mechanisms of Hematin Crystallization and Inhibition by the Antimalarial Drug Chloroquine*. Proceedings of the National Academy of Sciences U S A, 2015. **112**: p. 4946-4951.

2. Olafson, K.N., J.D. Rimer, and P.G. Vekilov, *Growth of Large Hematin Crystals in Biomimetic Solutions*. Crystal Growth & Design, 2014. **14**(5): p. 2123-2127.

431475 Process Intensification of Polyolefin Synthesis Using Novel Stable Perfluoro Based Membranes

Abstract:

Ethylene and propylene are the primary feedstocks for the manufacture of polyethylene and polypropylene, respectively. These olefins represent a major component of the polymer manufacturing cost. Consequently, there is significant economic benefit in minimizing losses of unreacted olefin from the process. Some olefin loss is a result of the need to remove paraffin from the polymerization reactor. Low concentrations of paraffin are present in the feedstock (ethane in the case of ethylene feed and propane in the case of propylene feed). The inert paraffin builds up in the reactor as the olefin is consumed by reaction and reactor effluent is recycled. This makes it necessary to vent some of the reaction mixture in order to limit the buildup of paraffin. The vent stream carries both paraffin and the more valuable olefin. A process intensification membrane that can efficiently separate and recover the olefin from the paraffin would provide substantial economic benefit to these polymerization processes. Membrane processes have been previously evaluated for separating ethylene/ethane or propylene/propane. In many cases, silver salt facilitating agents have been incorporated into membranes to preferentially transport ethylene or propylene. Historically with earlier membranes, while good separations have been initially demonstrated in the laboratory, membrane stability problems have prevented development of commercial systems.

Using perfluoro based membranes, Compact Membrane Systems has developed laboratory membranes which have shown stable (180+ days) and very good propylene flux and propylene/propane selectivity. These results are far above the industry Robeson curve. In addition results were stable when exposed to high levels of conventional foulants (e.g., sulfur containing gases). Using basic data from these laboratory results payback times for propylene plant process intensification were less than 1 year. Payback times for propylene/propane splitters were less than 1-2 years. These and other results will be discussed more extensively at the presentation.

434208 Highly Packed Artificial Water Channels in Lamellae Block Copolymer Film

Thursday, November 12, 2015: 12:50 PM
155C (Salt Palace Convention Center)

Yuexiao Shen, Department of Chemical Engineering, The Pennsylvania State University, University Park, PA and Manish Kumar, Department of Chemical Engineering, Pennsylvania State University, University Park, PA

Highly packed artificial water channels in lamellae block copolymer film

Yue-xiao Shen, Manish Kumar*

Department of Chemical Engineering, The Pennsylvania State University, University Park, PA, 16802

*Corresponding Author, Tel.: (814)-865-7519; email address: manish.kumar@psu.edu

The discovery of the high permeability of water channel proteins aquaporins (AQPs) and their analogs carbon nanotubes (CNTs) have created an explosion in materials where they are incorporated to develop novel filtration membranes.[1] AQPs can conduct single channel water transport with ~3 billion water molecules per second and are being extensively studied in biomimetic membranes for desalination. The bottlenecks of large-scale application are high cost of membrane protein production, questions regarding stability, and challenges in membrane fabrication.[1] CNTs based membranes have much better chemical stability, but the difficulty in fabricating vertically aligned and subnanometer diameter CNTs and low packing density in membranes also hold back commercial applications.[2]

Artificial water channels are alternative bioinspired analogues of AQPs created using synthetic chemistry.[1,3] They combine the advantages of biological channels and CNTs and improve upon them through their relatively simple synthesis and chemical stability. In the latest study, we have reported that the newly designed peptide-appended pillar[5]arene (PAP) artificial water channels (Figure 1a) have similar water conductance compared to AQPs and CNTs.[4] Furthermore, they are capable of aligning and forming highly packed two-dimensional arrays in lipid membranes.[4]

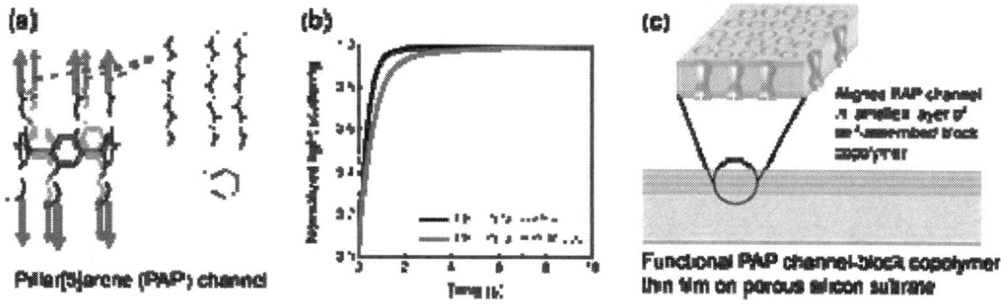

Figure 1. (a) The PAP channel forms a pentameric tubular structure through intramolecular hydrogen bonding between adjacent alternating D-L-D phenylalanine chains (D-Phe-L-Phe-D-Phe-COOH); (b) PAP channels are functional in PB-PEO block copolymers vesicles as demonstrated by stopped-flow light scattering measurement. mPCR means molar polymer to channel ratio; (c) Directed self-assembly of lamellar morphologies of block copolymer-aligned channels films on a chemically modified porous membranes.

In this article, we will present a new method to form porous and aligned PAP channels in the self-assembled lamellae block copolymer (BCP) films. We select polybutadiene-polyethylene oxide (PB-PEO) diblock copolymers, because PAP channels have been seen to be functional in the polymersomes (Figure 1b). PAP channels have tubular transmembrane structures and an outer surface which can favorably interact with the hydrophobic region of BCPs. Based on this theory, we expect that the PAP channels can be densely packed with orders in these lamellae films after annealing (Figure 1c).[5] We propose to utilize this functional porous film to fabricate composite desalination membranes.

Reference

1 Shen, Y.-x., Saboe, P. O., Sines, I. T., Erbakan, M. & Kumar, M. Biomimetic membranes: A review. *J. Membr. Sci.* **454**, 359-381 (2014).

2 De Volder, M. F. L., Tawfick, S. H., Baughman, R. H. & Hart, A. J. Carbon Nanotubes: Present and Future Commercial Applications. *Science* **339**, 535-539 (2013).

3 Barboiu, M. & Gilles, A. From Natural to Bioassisted and Biomimetic Artificial Water Channel Systems. *Acc. Chem. Res.* **46**, 2814-2823 (2013).

4 Shen, Y.-x. *et al.* Highly permeable artificial water channels that can self-assemble into two-dimensional arrays. *Submitted to PNAS.*

5 Li, M. & Ober, C. K. Block copolymer patterns and templates. *Mater. Today* **9**, 30-39 (2006).

413582 Application of Robust Aquaporin Incorporated Biomimetic Membrane in NF and FO

Abstract:

Aquaporin (AQP)-containing, planar, biomimetic membranes hold great potential in water purification and seawater desalination, due to the excellent permeability and the selectivity of AQP. However, there remain many challenges to production of robust and defect-free supported lipid bilayer (SLB) biomimetic membranes. In this work, we improved the stability of the AQP incorporated SLB membrane by layer-by-layer (LbL) method or covalent bonding method. The resulting biomimetic membranes exhibited excellent separation performances in nanofiltration (NF) and/or forward osmosis (FO) as well as robust structures that withstand a high operation pressure (in NF test) for a relatively long time. AQP show high permeability in SLB membranes. Furthermore, the AQP-containing SLB membranes possessed extra stability. When rinsed by a 0.24 mmol/L TritonX-100 (TX-100) solution for 5 minutes, the water flux and rejection (or reverse solute flux) of the biomimetic membrane only slightly changed. This research presents an effective method for preparation of biomimetic membrane with good separation performance and enhanced stability under harsh conditions.

416105 Aquaporin Based Biomimetic Membrane for Low Energy Water Purification

Abstract:

Current membrane technologies are important in many industrial separation processes, especially in water purification and desalination. However, the large operating pressure gradient that is required for conventional reverse osmosis filtration method makes water purification a very expensive process. Aquaporins are transmembrane water channel proteins that exist in biological plasma membranes. Aquaporins aid in biological water filtration processes by transporting water molecules at high speed, while selectively blocking out other kinds of solutes. Innovative aquaporin-reconstituted biomimetic membranes are envisaged to overcome the problem of high operating pressures needed, and hold great potential for the use in water purification processes, giving high flux while keeping the energy consumption low. It has been reported that the osmotic permeability of a single aquaporin is in the range from 6×10^{-14} to 24×10^{-14} $m^3\ s^{-1}$, and the permeability of an aquaporin-reconstituted biomimetic membrane is 167 mm/s/bar, which is by two orders of magnitude greater than commercial polymeric membranes. Therefore, aquaporin-based biomimetic membranes have attracted worldwide attention among those attempting to design a suitable biomimetic device for water treatment.

Three critical factors should be considered in designing an ideal separation membrane using aquaporins. The first one is that aquaporins must retain their water selectivity and transportation capability in the biomimetic membrane matrix. The second factor is that salt leakage should be minimum for the barrier layer. Last but not least, the mechanical strength of the membrane matrix must be sufficiently strong in the final application. Considering these three factors, we designed and fabricated novel aquaporin-based biomimetic membranes for forward osmosis and nanofiltration respectively. We functionally reconstituted E. coli protein aquaporin Z into vesicles that made from an amphiphilic triblock copolymer poly(2-methyloxazoline)-b-poly(dimethylsiloxane)-b-poly(2-methyloxazoline). Vesicles reconstituted with aquaporin Z were then coated onto a porous support membrane (cellulose acetate or alumina membrane). The structure and permeability of vesicles were confirmed by confocal laser scanning microscopy and a stop-flow apparatus. Membrane performance was characterized by nanofiltration and forward osmosis test. Both water flux and salt rejection were greatly increased for the Aquaporin Z embedded membrane compared with a membrane without Aquaporin Z, indicating that the water channel protein made a significant contribution to the water treatment performance.

428925 pH-Responsive Poly(ether sulfone) Composite Membranes Blended with Amphiphilic Polystyrene-Block-Poly(acrylic acid) Copolymers

Abstract:

Environmental stimuli-responsive membranes, or smart membranes, benefited from their environmental-sensitive permeability/selectivity, are attractive for applications in controlled drug delivery, biological/chemical separation, sensors, tissue engineering, and so on. The fabrication of environmental stimuli-responsive membranes mainly has two approaches: modification of existing porous membrane substrates by graft polymerization, and fabrication of membranes by modifying the membrane-forming polymers. Although the first approach could utilize most of existing membrane materials, it has such shortcomings as decrease of membrane pores and consequent flux decline, poor reproducibility, sophisticated membrane pretreatment or special equipment. Such shortcomings keep the fabrication of stimuli-responsive membranes by the first approach still far away from industrial implementation. The latter approach, imparting membrane materials with responsiveness before membrane fabrication with phase inversion method, is still hindered by the limited and yet uncontrollable modification methods (ozone and ray treatments). Controllable manufacture of stable stimuli-responsive membranes by this approach on large scale is still difficult.

pH-responsive poly(ether sulfone) (PES) composite membranes blended with amphiphilic polystyrene-*block*-poly(acrylic acid) (PS-*b*-PAA) diblock copolymers are fabricated through simple phase inversion in water. The anchor block PS has a lower surface tension than that of PES, which is beneficial to anchor the PS block near the membrane pore surfaces. Modification of membrane matrix surface including membrane surface and pore surface by PAA chains is achieved by surface-segregation of PS-*b*-PAA copolymers, and confirmed by water contact angle measurement on membrane surface and confocal laser scanning microscopy characterization of membrane cross-section. The effects of the blended PS-*b*-PAA copolymers on the microstructures and pH-responsive characteristics of the PES composite membranes are investigated systematically. The hydraulic permeation results show that the pH responsive characteristics of composite membranes blended with PS-*b*-PAA copolymers are proven to be significant, fast, reversible and durable with pH switching between 3 and 8. FITC-dextran sieving experiments of a PS-*b*-PAA blended composite membrane at ambient pH values of 3 and 8 show 28% and 90% rejection of 70 kDa solute, respectively. At pH =3 (< pKa), the membrane pores are in "open" state because of the shrunken conformation of the PAA chains on the pore surfaces; in contrast, at pH=8 (> pKa), the membrane pores are in "closed" state due to the swollen conformation of the PAA chains. The results provide valuable guidance for fabricating stimuli-responsive membranes by blending amphiphilic block copolymers.

430333 Polybenzimididizole Nanofiltration Membranes with Biomimetic Surfaces

Abstract:

Title:

Polybenzimidizole Nanofiltration Membranes with Biomimetic Surfaces.

Abstract:

Aquaporin is a bidirectional water channel protein present in cells, and it regulates the flow of water in and out of the cells while preventing the transport of ions, protons and hydroxyl ions. Biomimetic membranes attempt to artificially replicate this natural process of highly selective and efficient transport of different molecules across a semi-permeable membrane. Therefore, aquaporins have received worldwide attention because of their potential to form biomimetic membranes with high flux and selectivity for water reuse and desalination. However, challenges involved in the incorporation of aquaporin proteins in membranes limit their applicability. One of them is to attach aquaporins to the membranes without chemically altering or damaging the aquaporins during the binding to the membrane. The second challenge is to design and prepare an assembly that allows biomimetic membranes with aquaporins to sustain hydraulic water pressure gradients without losing their integrity and performance. The overarching objective of this project is to form a biomimetic membrane made of unaltered aquaporins dispersed in a polymeric membrane selective layer and capable of operation under high hydraulic pressure. To this end, polybenzimidazole (PBI) membranes were surface modified with treated aquaporins in order to achieve higher water flux and selectivity. Membranes modified with aquaporins displayed lower flux declines and higher flux recovery values after backwash as compared to unmodified PBI membranes. Also, modified membranes showed improved rejection values for both protein and salt solutions of different concentrations.

424450 Highly Water-Selective Hybrid Membrane By Incorporating Lamellar ZIF-8 into Polymer Matrix

Abstract:

Inspired by the structure and function of aquaporins in bio-membrane, novel water-selective hybrid membranes with superior water/ethanol separation performance were fabricated by incorporating lamellar ZIF-8 into sodium alginate (SA) matrix. The as-synthesized lamellar ZIF-8 exhibited a two-dimensional morphology and hydrophobic nanoporous structure. Due to the reinforcement effect of ZIF-8, the hybrid membranes exhibited higher mechanical strength, thermal stability, and swelling resistance. On one hand, the horizontally aligned structure of lamellar ZIF-8 within the hybrid membranes and could constructed ordered water channels; on the other hand, the selective and hydrophobic nanopores of lamellar ZIF-8 could render water channels with molecular sieving effect and fast water permeation. Moreover, the lamellar ZIF-8 could interfere the polymer chain packing and create additional interfacial channels. Accordingly, the hybrid membranes exhibited high water permeability and selectivity for water/ethanol mixture separation. Especially, for the membrane containing 4 wt% of lamellar ZIF-8, the separation factor was as high as 1840, while the permeation flux was 1218 g/m^2 h under 76 ^{o}C and 10 wt% feed water concentration. This study also revealed the lamellar ZIF-8 was more efficient than hexagonally shaped ZIF-8 as a filler of hybrid membrane for dehydration of aqueous solution.

430771 Rotary Wheel Adsorber for Carbon Capture

Abstract:

The key challenge in post combustion capture from gas fired power plants is related to the low CO_2 concentration in the flue gas (3 to 7% by volume). This means that conventional amine processes will result in a relatively high energy penalty while novel adsorbents and adsorption processes have the potential to improve the efficiency of separation. High-selectivity adsorbents are required to achieve relatively high CO_2 uptake at low partial pressures, which means that the separation process should be based on either very strong physisorption or chemisorption with thermal regeneration. From the process point of view, the main challenge is to develop efficient separation processes with rapid thermal cycles.

With this regard, adsorption processes based on rotary wheel technology represent a promising alternative to traditional Thermal Swing Adsorption (TSA) processes. They can treat large volume of gas with relatively low pressure drop and efficient heat integration allowing to perform rapid thermal swings with cycle times of few minutes, which is over one order of magnitude faster than tradition TSA fixed bed processes.

As part of the "Adsorption Materials and Processes for Gas fired power plants" (AMPGas) project an innovative bench scale Rotary Wheel Adsorber has been designed and it is being built at the University of Edinburgh. The apparatus is a 12-columns system able to capture CO_2 from dilute streams using rapid thermal swing adsorption cycles. Novel materials specifically designed for CO_2 capture from dilute streams have been developed and characterised as part of the project (Zeolites, amine-containing MOFs, amine-based Silicas, amine-based activated carbons and carbon nanotubes). By operating at conditions which are representative of the typical conditions of the flue gas from gas fired power plants, the system aims to produce experimental data needed to demonstrate the proposed capture process and validate the detailed model developed using CySim®, the adsorption simulator at the University of Edinburgh. The performance of the capture process are here evaluated by exploring the use of different types of adsorbents and cycle configurations, allowing to identify the optimal operating conditions for the different capture strategies.

431510 CO2 Capture from Flue Gas By PSA: Comparison of Bench to Full Scale Process Simulation When Utilizing a Structured Adsorbent

Abstract:

The primary objective of this work is to carry out a comprehensive modeling study to evaluate the performance of a structured adsorbent for capturing CO_2 from coal fired power plant flue gas at the process scale. This study is being carried out with the Dynamic Adsorption Process Simulator (DAPS) developed at USC and modified to account for pressure drop through a structured adsorbent with a one dimensional pressure drop correlation in place of the packed bed Ergun equation. The modeling results at the bench scale will be contrasted against those obtained from a 1-bed PSA system utilizing a structured adsorbent. The model will then be used to predict the performance of the PSA cycle at the process scale. This presentation will discuss the latest results obtained by Ritter and his group.

432394 Modeling and Economic Assessment of Rapid Thermal Vacuum Swing Adsorption of CO2 in Hollow Fiber Adsorbents

Abstract:

In recent years, there has been an increasing interest in capture and separation of carbon dioxide from flue gas due to concerns about rising CO_2 concentrations in atmosphere. The US Department of Energy investigated and concluded that the proposed CO_2 capture and sequestration (CCS) technology for retrofitting the existing coal fired power plants should not increase the cost of electricity beyond 35% [1]. Solid adsorbent based CO_2 separation process can potentially have lower energy requirements than liquid amine based absorption processes, which, in their current form, may increase the cost of electricity by 80% [2].

A novel process contactor in the form of hollow fiber loaded with amine sorbents was proposed and investigated for its application in a rapid thermal swing adsorption (RTSA) of CO_2 from flue gas [3]. Experimental studies of RTSA with hollow fiber amine sorbents demonstrated high breakthrough capacity and good stability during the rapid thermal swing. Also, mathematical modeling for the process was developed, rigorously validated and employed to study the economics of the process [4]. Although the process showed high promise in terms of energy recovery with its heat exchanger type configuration, the economic assessment of the process that meets the product purity and recovery constraints of 95% and 90% respectively could potentially have an increased cost of electricity beyond 35% [5]. The main reason for which is a high mass transfer diffusional resistance of high loaded silica supported amine sorbents, which results in curtailing the desorption step even before sweeping the bed clean of CO_2. This prohibits the utilization of the maximum swing capacity that can be achieved with the given thermal swing.

In this work an advance over the current process configuration of rapid thermal swing by introducing vacuum in desorption step is investigated. This can potentially increase the swing due to the additional driving force provided by the lower CO_2 partial pressure during desorption. As a penalty, this incurs an additional compression step for the low pressure CO_2 which is produced. The design and economics of the proposed rapid thermal swing vacuum adsorption that meets the constraints of CO_2 product purity and recovery will be presented along with an analysis of advanced cases that might further improve the parasitic loads by targeted materials and process improvements.

References

[1] Feeley. T.J., Fout, T.E., and Jones, A.P., DOE's Carbon Capture and Sequestration R&D Program, paper presented at the PowerGen conference at Orlando, FL, Dec 2007

[2] NETL, Cost and Performance Baseline for Fossil Energy Plants, Volume 1: Bituminous Cal and Natural Gas to Electricity, Rev 2, Nov 2010

[3] Fan, Y., Lively, R.P., Labreche, Y., Rezeai, F., Koros, W.J., and Jones, C.W., 'Evaluation of CO_2 adsorption dynamics of polymer/silica supported poly(ethylenimine) hollow fiber sorbents in rapid temperature swing adsorption', Int. J. Greenhouse Gas Control, 2014, vol. 21, pp-61-71.

[4] Kalyanaraman, J., Fan, Y., Lively, R.P., Koros, W.J., Jones, C.W., Realff, M.J., and Kawajiri, Y., Modeling and Experimental Validation of Carbon dioxide sorption on Hollow Fibers loaded with Silica-supported Poly(ethylenimine), Chem.Engg Journal, 2015, vol 259, 737-751.

[5] Swernath, S., Searcy, K., Rezeai, F., Labreche, Y., Lively R.P, Realff M.J., Kawajiri, Y., Optimization and Techno economic Analysis of Rapid Thermal Swing Adsorption of Carbon Capture from Coal Fired Power Plant, Computers Aided Chem.Engg, 2015, vol. 36, 253-278

2015 AIChE Annual Meeting

Investigation of Heat Transfer and Adsorption in Indirect Heated Adsorbent Beds

Gabriel Salazar Duarte[1,2],
Benedikt Schürer[1],
Christian Voss[1],
Dieter Bathen[2]
Salt Lake City, 12.11.2015

1. Linde AG, Engineering Division
2. Universität Duisburg-Essen, Lehrstuhl für Thermische Verfahrenstechnik

Contents

2

Introduction
Why use indirect heated adsorption processes?

Limitations of directly heated adsorption processes:

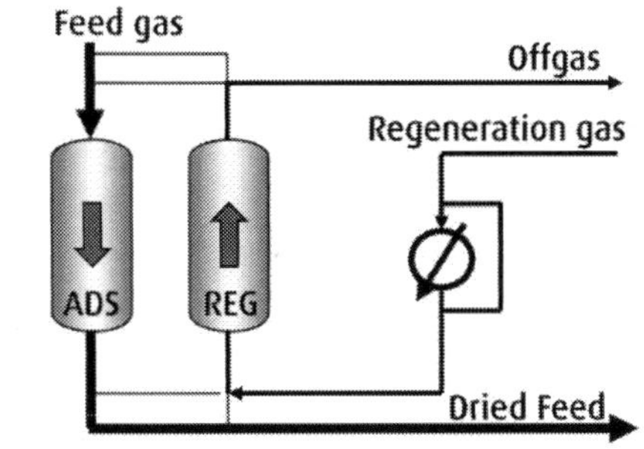

- Long heating and cooling times
- Regeneration is the limiting step
- Large adsorbent beds
- Dilution of desorbed component
  - ➢ Application is limited to removal of trace impurities

Introduction
Why use indirect heated adsorption processes? II

Advantages of indirectly heated adsorption processes:

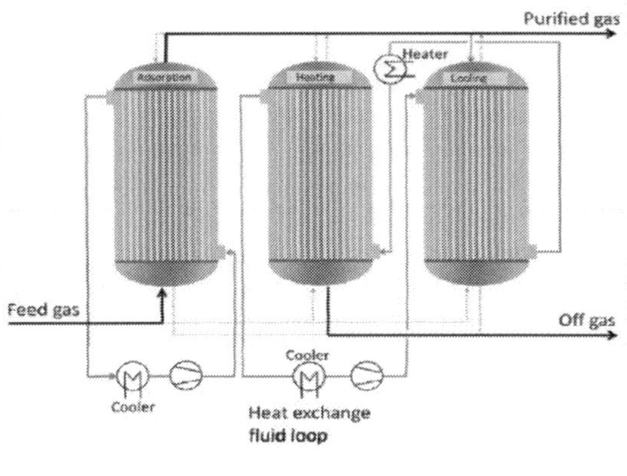

- Shorter heating and cooling times
- Less adsorbent required
- Separation of larger quantities (e.g. CO2 from flue gas)
- High concentrations of desorbed component

Introduction
Shell and tube heat exchanger type adsorber

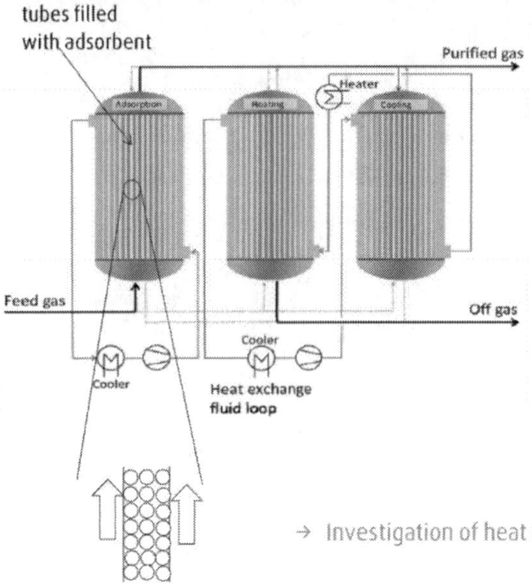

Concept:

- Two separated flow channels
- Heat exchange between channels
- Cooling/heating fluid shellside
- Adsorbent in the tubes
- Indirect heating with a utility fluid, e.g. water, possible

→ Investigation of heat transfer characteristics is very important

Contents

1. Introduction
2. **Experiments**
3. Modeling of an indirect heated adsorption process
4. Conclusion and Outlook

Experimental setup

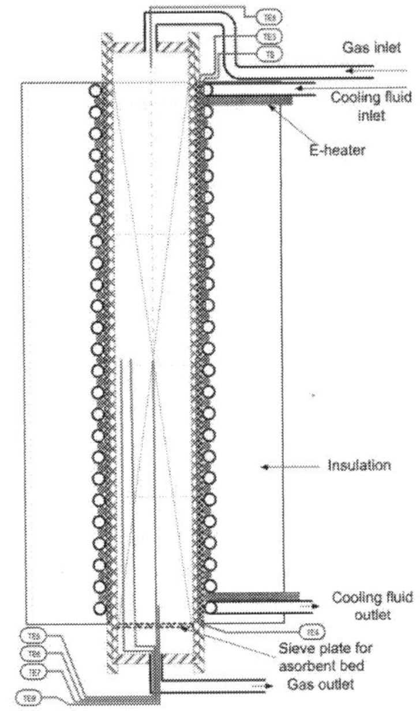

Theoretical model

α_w-Model (VDI-Wärmeatlas):

$$\overbrace{\left(\rho_g c_{pg}\epsilon + (1-\epsilon)\rho_s c_{ps}\right)\frac{\partial T}{\partial t}}^{Accumulation} = \overbrace{\Lambda_r \cdot \left[\frac{\partial^2 T}{\partial r^2} + \frac{1}{r}\cdot\frac{\partial T}{\partial r}\right] + \Lambda_{Ax}\cdot\frac{\partial^2 T}{\partial z^2}}^{Dispersion} - \overbrace{u_0\cdot\rho_g\cdot c_{p,g}\cdot\frac{\partial T}{\partial z}}^{Convection}$$

Boundary Condition:

$$\dot{q}\Big|_{r=r_i} = \alpha_w \Delta T \quad ?? \quad Nu_w = \frac{A}{\lambda_g}$$

Axial and radial dispersion:

$$\frac{\Lambda_{Ax}}{\lambda_g} = \frac{\lambda_{bed}}{\lambda_g} + \frac{Pe_0}{K_{Ax}}$$

$$\frac{\Lambda_r}{\lambda_g} = \frac{\lambda_{bed}}{\lambda_g} + \frac{Pe_0}{K_r}$$

Bed heat conduction (Zehner/Bauer/Schlünder Model):

$$\frac{\lambda_{bed}}{\lambda_g} = 1 - \sqrt{1-\psi} + \sqrt{1-\psi}\cdot k_c$$

$$k_c = f\left(\frac{\lambda_p}{\lambda_g}, N, B\right) \quad ??$$

Theoretical model
Parameter estimation

Minimization of the sum of least squares

Experimental Measurment

Estimation using Comsol® LiveLink for Matlab®

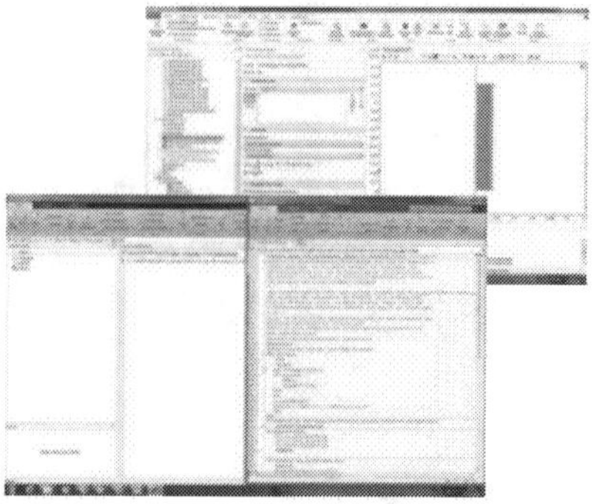

$$\min_{x} \sum (T_{model}(x) - T_{measurment})^2$$

Results
All experiments for the estimation of λ_p and A

Experiment number	Constant wall temperature in °C	Constant heat duty in W
1*	-	25
2*	-	37.5
3*	-	50
4*	-	62.5
5	-	75
6*	100	-
7*	120	-
8*	140	-
9*	160	-
10*	180	-
11*	200	-
12	220	

* Experiment was conducted twice with similar results

Results

Estimation of λ_p and A

Constant heat duty (25 W)

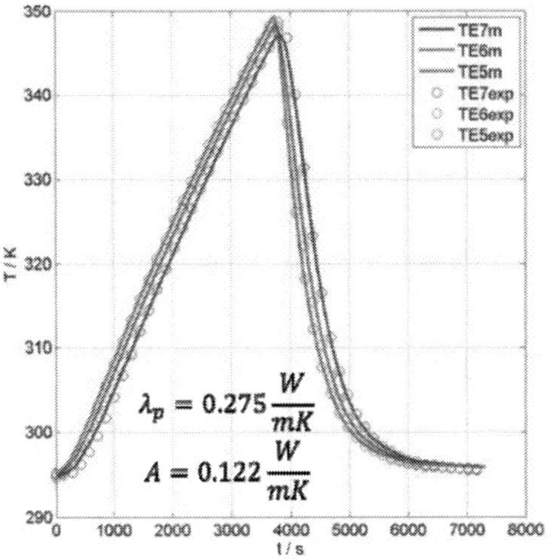

Constant wall temperature (100 °C)

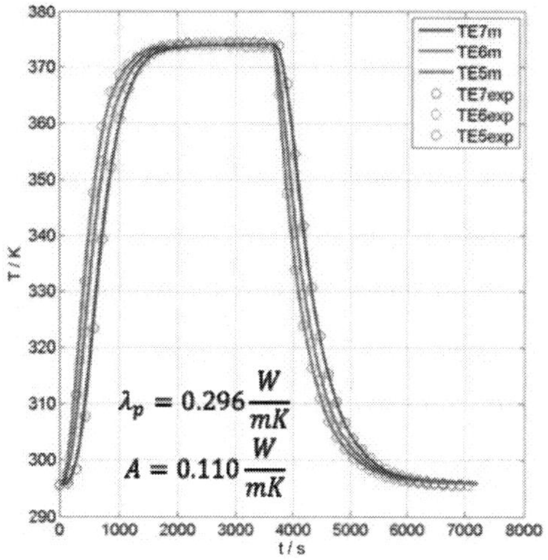

- Good agreement between experiments and simulations

Results

Estimation of λ_p and A : all experiments

λ_p

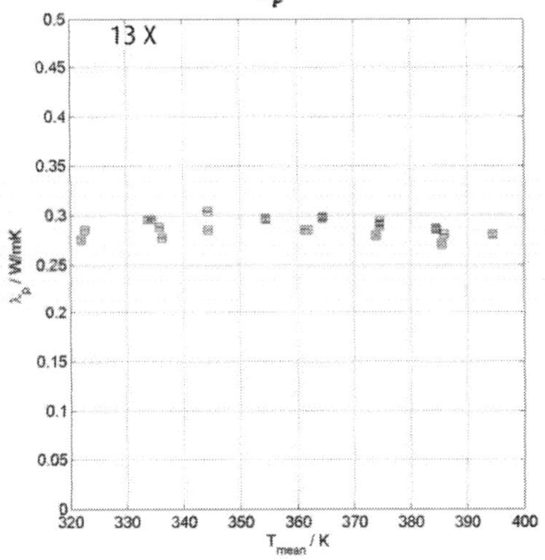

A

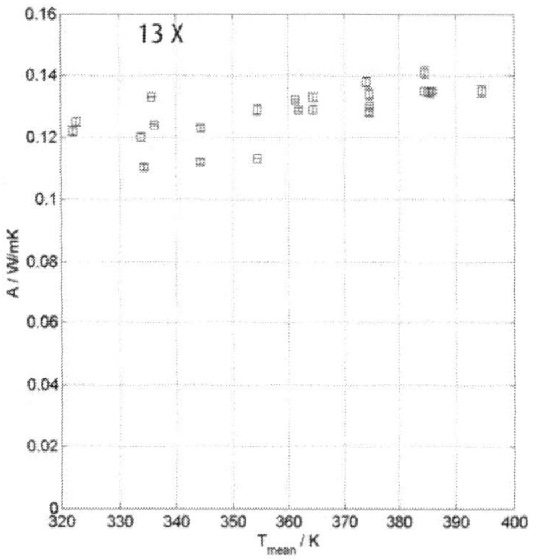

- Assumption of constant coefficients seems to be acceptable

Contents

Modeling of TSA process with indirect heat exchange

Three different phases only in thermal contact:

1. Adsorbent + gas
2. Wall
3. Heat transfer fluid

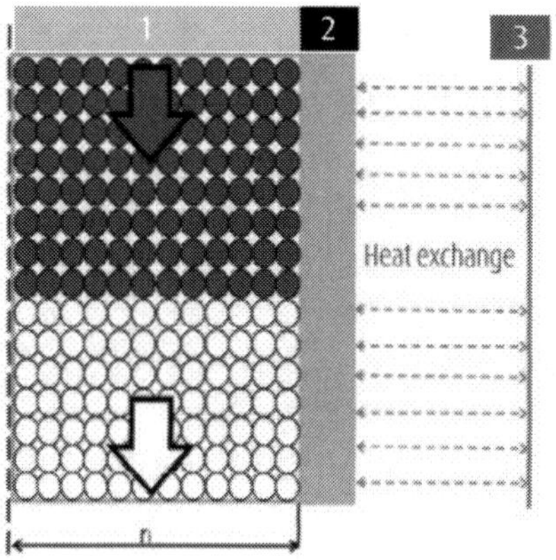

→ **Coupling by several partial differential equations**

Modeling of TSA process with indirect heat exchange II

Wall:

$$\frac{\partial T_{wall}}{\partial t} = \nabla(\frac{\lambda}{\rho_{wall} c_{p,wall}} \nabla T)$$

Adsorbent + gas (local thermal equilibrium and linear driving force):

$$\overbrace{\frac{\partial c_i}{\partial t}}^{Accumulation} = \overbrace{\nabla(D_i \nabla c_i)}^{Dispersion} - \overbrace{\nabla(u_g \cdot c_i)}^{Convection} - \rho_s \overbrace{\frac{1-\epsilon}{\epsilon}\frac{\partial q_i}{\partial t}}^{Adsorption}$$

α_w-Model (VDI-Wärmeatlas):

$$\frac{\Lambda_{Ax}}{\lambda_g} = \frac{\lambda_{bed}}{\lambda_g} + \frac{Pe_0}{K_{Ax}}$$

$$\frac{\Lambda_r}{\lambda_g} = \frac{\lambda_{bed}}{\lambda_g} + \frac{Pe_0}{K_r}$$

$$\frac{\lambda_{bed}}{\lambda_g} = 1 - \sqrt{1-\psi} + \sqrt{1-\psi} \cdot k_c$$

Heat transfer fluid:

$$\overbrace{(\rho_g c_{pg} \epsilon + (1-\epsilon)\rho_s c_{ps})\frac{\partial T}{\partial t}}^{Accumulation} = \overbrace{\Lambda_r \cdot \left[\frac{\partial^2 T}{\partial r^2} + \frac{1}{r} \cdot \frac{\partial T}{\partial r}\right] + \Lambda_{Ax} \cdot \frac{\partial^2 T}{\partial z^2}}^{Dispersion} - \overbrace{u_0 \cdot \rho_g \cdot c_{p,g} \cdot \frac{\partial T}{\partial z}}^{Convection}$$

Modeling of TSA process with indirect heat exchange
Investigated system

- One tube of the tube bundle system
- Numerical study using COMSOL Multiphysics®
- Simulated dry flue gas: 15% CO_2 and 85% N_2
- Geometry: Axial symmetrical 2D model coupled to a 1D model
- Surface to volume ratio ϕ estimated from unit cell

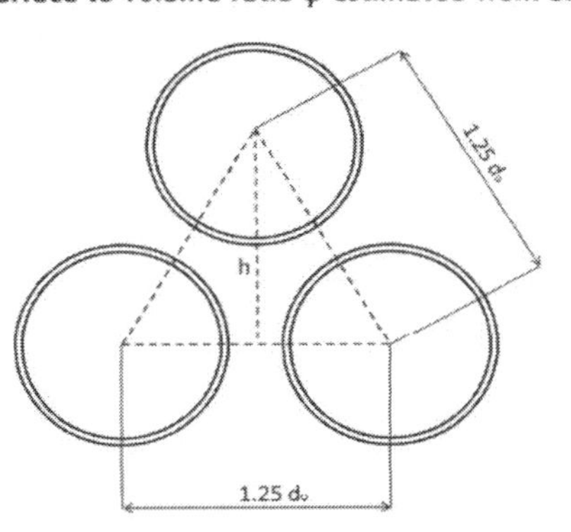

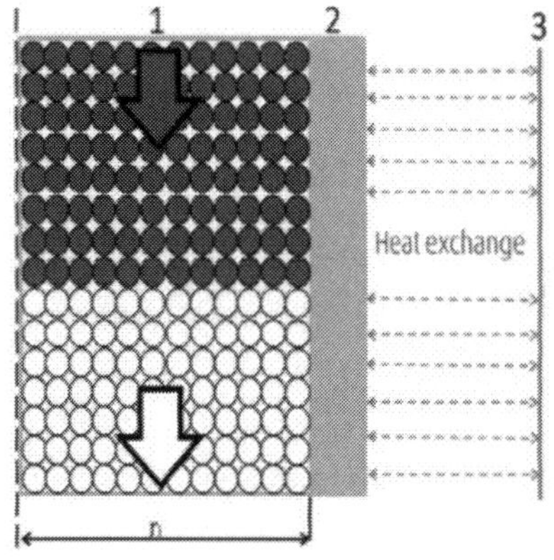

Modeling of TSA process with indirect heat exchange
Results

Parameter	Value
Values	
α_{Hex}	$500\,\text{W} / (\text{m}^2 \cdot \text{K})$
A	$0.128\,\text{W} / (\text{m} \cdot \text{K})$
λ_p	$0.288\,\text{W} / (\text{m} \cdot \text{K})$
d_i	32 mm
T_{Hot}	180 °C
Results	
Average purity	**98.75 %**
Recovery	**85.4 %**
Q_{specific}	$\mathbf{2.72\ MJ\ kg_{CO_2}^{-1}}$

Parametric study
Numerical study

— Parameters:
- heat conduction coefficient of the adsorbent
- heat transfer coefficient to the Hex-fluid
- Inner tube diameter
- Regeneration temperature

Parameter	Range
α_{Hex} in $\text{W}/(\text{m}^2 \cdot \text{K})$	150, 250, 350, 450, 550
λ_p in $\text{W}/(\text{m} \cdot \text{K})$	0.2, 0.4, 0.6, 0.8, 1
d_i in cm	2, 3.2, 4.4
T_{Hot} in °C	120, 140, 160, 180, 200

Results
Variation of α_{Hex}

Publication in progress at CIT

$\lambda_p = 0.288$ W/ (m K) $d_l = 32$ mm
$T_{Hot} = 180$ °C

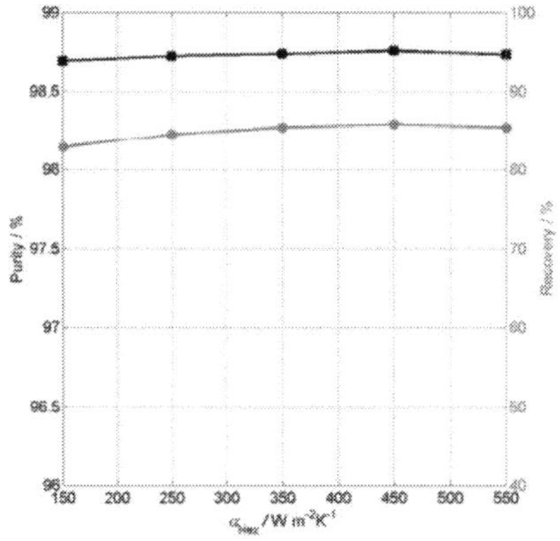

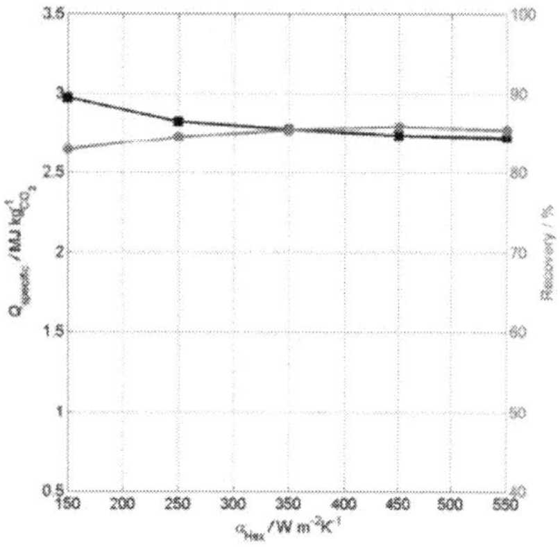

- No strong influence on the purity
- Small influence of the outer heat transfer coefficient on the recovery can be observed
- Energy requirement is reduced by increasing heat transfer coefficient

Results
Variation of λ_p

Publication in progress at CIT

$\alpha_{Hex} = 500$ W/ (m² K) $d_l = 32$ mm
$T_{Hot} = 180$ °C

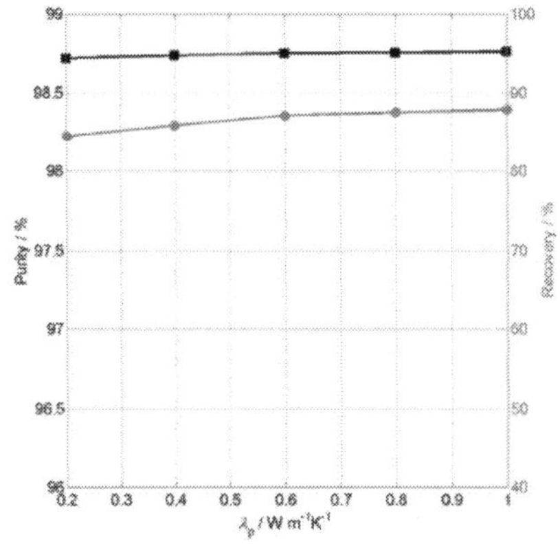

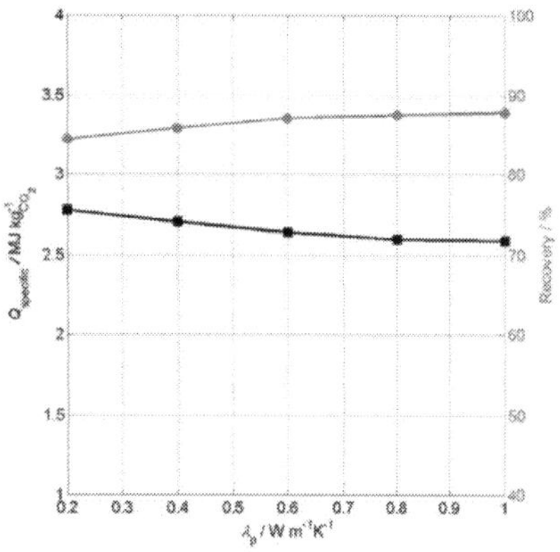

- No strong influence on the purity
- Recovery increases by increasing the heat conduction coefficient
- Energy requirement is reduced by increasing the heat conduction coefficient

Results
Variation of d_i

Publication in progress at CIT

– Ratio of incoming CO_2 to adsorbent mass and HEX velocity is held constant

$\lambda_p = 0.288\ \mathrm{W/(m\ K)}$ $\alpha_{Hex} = 500\ \mathrm{W/(m^2\ K)}$ $T_{Hot} = 180\ °\mathrm{C}$

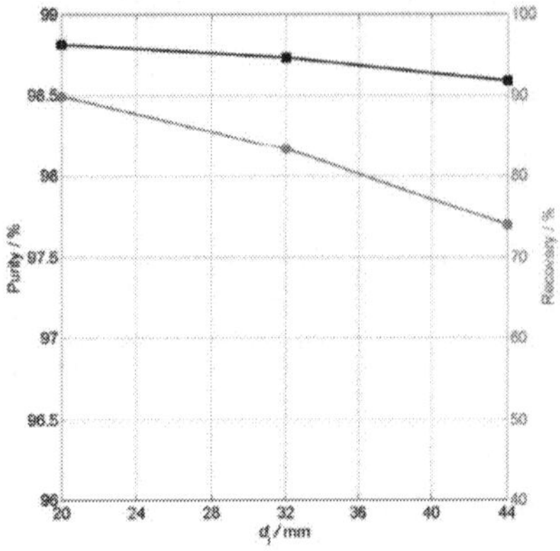

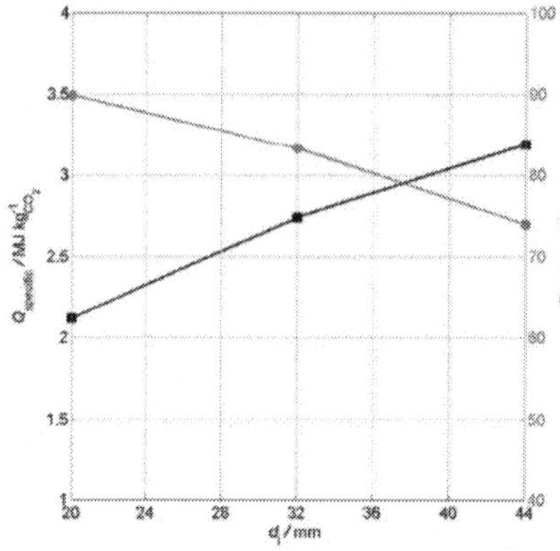

- No significant influence on the purity
- Recovery decreases by increasing the inner diameter
- Energy requirement increases by increasing the inner diameter

Results
Variation of T_{HOT}

Publication in progress at CIT

$\lambda_p = 0.288\ \mathrm{W/(m\ K)}$ $d_i = 32\ \mathrm{mm}$
$\alpha_{Hex} = 500\ \mathrm{W/(m^2\ K)}$

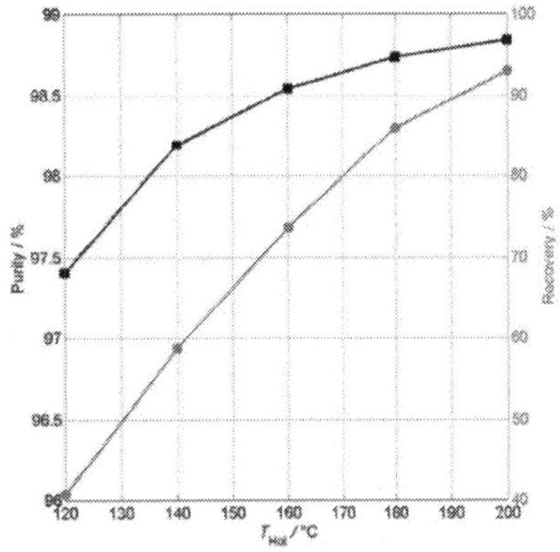

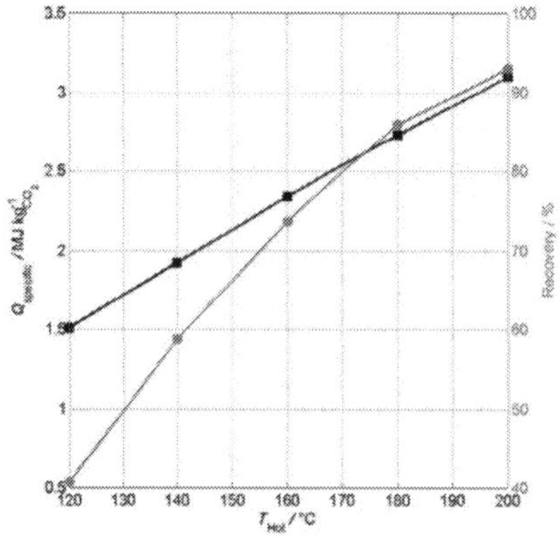

- Purity of the desorbed CO_2 decreases by decreasing the heating temperature
- Recovery increases by increasing the heating temperature
- Energy requirement specific increases by increasing the heating temperature

Results
Optimized values

Parameter	Value
Optimized values	
α_{Hex}	550 W / (m^2 · K)
λ_{p}	1 W / (m · K)
d_i	20 mm
T_{Hot}	180 °C
Results	
Average purity	**98.84 %**
Recovery	**91.26 %**
Q_{specific}	**2.01 MJ $kg_{CO_2}^{-1}$**

Contents

Conclusion and Outlook

Heat Transfer:

- Good agreement between simulation and experiments for non convective experiments
- Model parameters λ_p and A show no significant influence on boundary condition
- Very small influence of λ_p and A on temperature is observed
  - ➤ Assumption of constant coefficients is acceptable

Adsorption simulations:

- Better results at minimized radial thermal resistance
- High purity and recovery
- Energy requirement comparable to the benchmark process (amine wash)

Outlook

- Estimation of convective contribution of Λ_r and Nu_w
- Further detailed energetic estimations (compression work of the gas)
- Adsorption experiments

12.11.2015 2015 AIChE annual meeting 25

Thank you
for your attention.

434536 Adsorptive Processes for Purification of Carbon Dioxide Captured from Emission Gas

Abstract:

Recently, climate changes like flood and heat wave have become a big problem for the countries around the world. It is widely known that these climate changes have been caused by global warming from green-house gas (GHG). Thus, not only energy conservation and efficient use, but also carbon capture and sequestration or storage (CCS) technology are necessary to solve this issue.

According to the scenario of Department of Energy (DOE) in USA, the Carbon Sequestration Program aims at CO_2 capture over 90% purity and storage. In according to the EU program, the captured CO_2 with higher than 95% purity will be stored at a geological storage site. Some portion of captured CO2 can be used as a raw chemical material instead of storing it underground. However, in order to use the captured CO_2 as a source of other products, such as bio-diesel, plastic products, chemical substances, additional purification process is needed to supply 98% CO_2 in the industrial fields.

In the study, 90% CO_2 with N_2 balanced gas was used as feed gas. And 2-bed PSA and PVSA processes were experimentally and theoretically studied under 2.5 / 4.5 / 6.5 bar and 0.5bar vacuum. To improve purity and recovery, the simulation work was extended to 4-bed PSA and PVSA processes. CO_2 with higher than 98% purity and 90% recovery was successfully obtained from the developed adsorptive process.

427668 Optimum Adsorbent-Process Combination for Minimum Cost Carbon Capture Using a VSA Process

Abstract:

Vacuum Swing Adsorption (VSA) is one of the competing technologies being explored for carbon capture and concentration (CCC) from power plant flue gas. We have developed a simulation and optimization platform for CCC by VSA operated on different cycles. This platform has been used to evaluate energy-productivity performance of various adsorbents in order to identify the most favorable adsorbent-cycle combination. Energy and productivity are important determinants of the operating and capital costs of a VSA process. However, their relative share in the overall cost depends on the VSA cycle and its operating conditions in a complex way. Hence ranking adsorbents based on the energy-productivity performance may be very misleading.

Here, we have expanded our optimization platform to incorporate all the steps necessary to carry out detailed costing of a VSA process. The scope of the simulation and optimization platform is further expanded to include adsorbent isotherm characteristics, in addition to the process operating conditions, as decision variables. These expansions allow integrated adsorbent-process optimization including detailed costing. Thus we are able to find the minimum capture cost possible in a VSA process for CCC and at the same time determine the corresponding optimum adsorbent isotherm characteristics. We have answered these questions for two VSA cycles, namely a 4-step process with light product pressurization (LPP) and a 6-step process with light and heavy reflux in addition to LPP.

408475 Molecular Mobility As a Tool for Understanding the Impact of Polyvinylpyrrolidone (polymer) and Tpgs (surfactant) in Crystallization Kinetics of Amorphous Celecoxib

Abstract:

Formulating drugs (active pharmaceutical ingredients) as amorphous solid dispersions is an alternative for improving the solubility of poorly water soluble drugs. An amorphous solid dispersion ideally consists of the amorphous drug homogeneously distributed and thermodynamically stable within a polymer matrix. In addition to the polymer, the formulation may contain other materials such as surfactants to improve processability, dissolution and flow behavior, among others. Understanding the crystallization tendency of such compounds is crucial when assessing it's feasibility for a solid dispersion formulation, as re-crystallization of the drug is an undesired effect and may lead to potential market withdrawal of a commercial drug product. Celecoxib is a poorly water soluble (BCS class II) non-steroidal anti-inflammatory drug. When compared to other BCS class II compounds, it has relatively low molecular mobility (high viscosity) and crystal growth rate upon cooling from the undercooled melt state [1,2]. The presence of surfactants or polymers has demonstrated to significantly increase or inhibit the radial crystal growth rate of Celecoxib, respectively. Interestingly but not necessarily advantageous, the combination of both, results in a complex behavior where the polymer is able to reduce the growth rate acceleration caused by the surfactant resulting in an increase or decrease of the overall crystal growth rate depending on the miscibility of Celecoxib and the surfactant [3]. Even though, there is a large number of surfactants and polymers available for amorphous solid dispersions, few studies elucidate the impact of all of these in the crystallization kinetics of drugs. While the impact of surfactants and polymers in the crystallization growth rates of pure amorphous Celecoxib has been studied and reported, this impact analyzed in terms of the molecular diffusivity is a knowledge still not available. Crystallization of a drug is a diffusive process where the nucleation rate is dictated by the fluidity of the liquid (viscosity). Fundamental knowledge may be obtained when the crystallization tendency (or lack thereof) is analyzed in terms of the "glass forming ability" of the liquid, *i.e.*kinetic fragility. Here, the liquid's viscosity with temperature may follow Arrhenius or Vogel-Tamman-Fulcher (VTF) dependence, depending on the relaxation of the liquid molecules. It is commonly known that glassy liquids at high temperature follow Arrhenius dependence (linear), however upon lowering the temperature deviates from Arrhenius and follows VTF law (non-linear). Interestingly, some speculate that the VTF behavior transitions back to an Arrhenius dependence at low temperatures [4]. The viscosity is basically the resistance of a fluid to flow under an external applied stress, it is then related to the molecular relaxation and hence mobility of a material. We have characterized the rheological properties of Celecoxib with small amounts of PVP (polymer) and/or TPGS (surfactant) with the purpose of understanding the impact on the molecular diffusivity of the Celecoxib during crystallization from the undercooled melt state. We expect these results to complement the knowledge available of molecular mobility of Celecoxib with regards to thermodynamic quantities for solubility prediction [5]. Additionally, in general for when assessing a drug as a candidate for a solid dispersion formulation during developmental stages of a drug product. Amorphous samples where prepared using a cryogenic mill. Binary and ternary mixtures consisted of 10 wt.% TPGS or 10 wt.% PVP dispersed in a 90wt.% Celecoxib, and ternary mixtures of 10wt.% PVP, 10 wt.% TPGS and 80 wt.% Celecoxib. Samples were characterized in an Anton Paar MCR 301 series rheometer, using 20mm diameter parallel Peltier Plates. Temperature dependent Dynamic Viscosity was measured while cooling the sample from melting point at a cooling rate of 10°C/min. We measured the dynamic rheological properties of the Celecoxib samples. The dynamic viscosity data was analyzed in terms of molecular mobility, Activation Energy (E_a), and fragility (Strength Parameter) (D) of the drug near the glass transition (T_g) and melt temperature (T_m), this was possible by analyzing the viscosity data with Arrhenius and Vogel-Tamman-Fulcher (VTF) equations. The temperature dependent dynamic viscosity curves resulted in overall viscosity increase upon cooling from the melting temperature, which was expected due to the nature of the measurement. The rate at which the viscosity increased with temperature varied with sample. Generally speaking, pure Celecoxib and Celecoxib/TPGS resulted in a small degree of undercooling compared to Celecoxib/PVP and Celecoxib/PVP/TPGS, this due to sample solidification during measurement. Pure Celecoxib resulted in an increase in viscosity of approximately one decade at roughly 10°C degrees of undercooling; at this point the viscosity suddenly increased dramatically almost three orders in magnitude. This sample could not be measured more than 30°C degrees of undercooling. Celecoxib/TPGS resulted in a similar behavior as pure Celecoxib, with a reduced degree of undercooling, as the sample showed a marked increase in viscosity near the melt temperature. Celecoxib/PVP resulted in a significant increase in the degree of undercooling (~95°C) as the dynamic viscosity could be measured up to the glass transition temperature of the sample with an increase of roughly one decade of viscosity every 10°C in temperature. Interestingly, almost 10°C before the glass transition, the viscosity increase rate with temperature decreased significantly. Near the glass transition temperature, this sample showed roughly a viscosity increment of seven orders of magnitude. Similar behavior was observed for Celecoxib/PVP/TPGS, having similar viscosity values as the Celecoxib/PVP sample near melt temperature, as the glass transition approached during cooling, the Celecoxib/PVP/TPGS sample showed a decrease of one decade in viscosity compared to Celecoxib/PVP. Celecoxib/PVP/TPGS viscosity increase rate also was reduced near the glass transition. Analyzing the viscosity data in terms of diffusivity indicates that the molecular mobility of the Celecoxib is impeded by the presence of the PVP, this occurs by the hindering of local arrangements of the Celecoxib molecules within a crystal lattice (crystallization inhibition). This was evidenced by the significant increase in viscosity and degree of undercooling compared to pure Celecoxib. This induced hindering of Celecoxib molecules allowed for viscosity measurements near the glass transition temperature. On the other hand, the TPGS acted as a plasticizer where miscibility between TPGS and Celecoxib aided in the molecular mobility of Celecoxib molecules assisting therefore the crystallization near the melting temperature. The combination of Celecoxib with PVP and TPGS resulted in interesting results, generally a decrease of molecular mobility similar to the effect of PVP alone, however the plasticizing effect of TPGS was able to increase the mobility, hence lower viscosity values near the glass transition compared to the effect of PVP. The viscosity curves showed mostly linear dependences with regard to temperature, however some deviation from linearity were observed most significantly for the samples containing PVP and PVP/TPGS. Fitting the viscosity curves with the Arrhenius and VTF equations evidenced these dependence. The hindering of local arrangements of the Celecoxib molecules (crystallization inhibition) by PVP was evidenced by the increase in activation energy of Celecoxib/PVP (115.96kJ/mol) compared to pure Celecoxib (92.28 kJ/mol) near the melt temperatures, whereas TPGS decreased the activation energy necessary for crystallization to occur (60.52kJ/mol), evidencing thus the plasticizing effect of TPGS. Interestingly, the combination of PVP and TPGS was not capable of reducing the activation energy (98.89kJ/mol), as significantly as PVP alone complementing the similar viscosity values near the melt temperature. Deviations from the Arrhenius equation were observed as the degree of undercooling progressed (10°C-80°C), evidenced by a decrease in the R^2 value of up to 12% for the Celecoxib/PVP/TPGS sample, compared with R^2 = 10% of Celecoxib/PVP sample. These deviations could be successfully corrected with the VTF model, increasing the R^2 value to 99.9% for both samples within the same temperature range. A marked VTF-to-Arrhenius transition near the glass temperature was not observed for these two samples within the temperature range studied, however Arrhenius equation resulted in higher R^2values compared to the VTF model (10% difference between both). VTF fitting demonstrated that all samples had a calculated strength parameter (D) of less than 10, confirming the kinetic fragility (crystallinity) of the liquids. However, no correlation could be observed between the impact in strength parameter, and the crystallization tendency and molecular mobility. The impact on the crystallization of Celecoxib by PVP (polymer) and TPGS (surfactant) was studied in terms of molecular mobility. Over all it was demonstrated that the polymer decreases the mobility of the drug, whereas the surfactant acts as a plasticizer. PVP hinders the mobility of Celecoxib molecules inhibiting to some degree the crystallization of the drug, evidenced by the significant increase in measured degree of undercooling, dynamic viscosity and calculated activation energy. The surfactant acted as a plasticizer aiding the mobility of the drug allowing therefore the crystallization of the drug to occur at earlier, and lowers the activation energy necessary for crystallization to occur. The dynamic viscosity of the Celecoxib with polymer and/or surfactant follows Arrhenius temperature dependence near the melting point (10°C of undercooling), however the VTF model results in higher agreement as the undercooling degree increases.

413132 Process Development for a Co-Crystal of Telaprevir (VX-950) and 4-Hydroxybenzoic Acid

Abstract:

When development work was initiated, the existing process for generation of a co-crystal of telaprevir (VX-950) and 4-hydroxybenzoic acid (4-HBA) was uncontrolled precipitation from a solvent mixture. Anecdotally, there was the existence of a "magic" co-crystal batch that had superior dissolution performance and had been generated from the uncontrolled process. For a time, technical work was driven by this knowledge and there was an ill-fated attempt to reproduce this batch. However, when it was determined that the precipitation led to co-crystal with variable properties and performance, focus shifted to obtaining a more fundamental understanding of the crystallization. The objective for developing the process was then to design a controlled crystallization that would lead to co-crystal material with reproducible physical properties and dissolution behavior. There is a significant difference in solubility of the API and the coformer which led to difficulty in choosing an appropriate solvent system. This presentation will review the candidate solvent systems and modes of crystallization that were attempted. After choosing dichloromethane and methyl *tert*-butyl ether as the solvent system, an equilibrium phase diagram was generated for the solvent composition at the seed point. This phase diagram indicated that the process was at risk to produce co-crystal with excess VX-950. By slightly changing the parameters in the crystallization, it was found that only co-crystal would be formed. Further work involved optimization of the solvent composition at the end-point to improve yield and minimize agglomeration. Development and understanding of the phase diagram was essential to reliably producing VX-950/4-HBA co-crystal.

416357 Primary Nucleation Under High Shear Conditions

Abstract:

The influence of external energy on the kinetics of primary nucleation has been investigated under high shear conditions. It is shown in literature, that shear-induced molecular alignment and in particular agitation-enhanced cluster aggregation are mechanisms that appear to deserve further attention [1]. To study this phenomenon further, the induction time has been measured at different supersaturation, and levels of energy input, at various scales using magnetic stir bar and rotor-stator based IKA wet mills. Since there is a distribution of induction time possible, a statistical view is proposed for the value of induction time at a given condition [2]. There is an overall tendency in the experiments that the primary nucleation is promoted by increased input of shear imparted due to mixing. In small vials agitated by magnetic stir bars and in a larger scale reactor with high shear wet mill, the induction time was found to decrease with increasing wet mill tip speed to the power 0.66, the induction time is inversely related to the shear rate. By fitting the parameters of the classical nucleation theory to experimental data, it is shown that the results can be explained as an influence on the pre-exponential factor. Interfacial tension is estimated between API and solvent. An empirical equation is proposed to quantify the energy input from wet milling, and to predict the induction time.

References:

1. Liu, J.; Rasmuson, A.C. Cryst. Growth Des., 2013, 13 (10), pp 4385–4394
2. Kulkarni, K.A.; Kadam, S.S.; Meekes, H.; Stankiewicz, A.I.; ter Horst, J.H. Cryst. Growth Des., 2013, 13 (6), pp 2435–2440

AbbVie contributed to the design, research, and interpretation of data, writing, reviewing, and approving the publication. Huayu Li, Moiz Diwan and Fredrik Nordstrom are AbbVie employees at the time that they worked on this project.

417728 Microfluidic Spherical Crystallization and Co-Formulation of Drugs and Excipients – Controlling Morphology and Polymorphism

Abstract:

Microfluidic Spherical Crystallization and Co-formulation of drugs and excipients – controlling Morphology and Polymorphism

Eunice W. Q. Yeap, Ambika Somasundar, Denise Z. L. Ng, Yiji Wang, and Saif A. Khan*

Department of Chemical and Biomolecular Engineering,

National University of Singapore, SINGAPORE

There is considerable interest in the creation of "designer" pharmaceutical microparticles with controlled morphology and polymorphism, as drug product intermediates for facile formulations. Emulsion-based crystallization is an attractive route for controlling crystal size in downstream processing of active pharmaceutical intermediates (API), yet crystals formed in batch processes show limited control over key attributes like size and shape.[1] We circumvent these limitations, by coupling capillary microfluidics based emulsion generation, with evaporative crystallization for the production of monodispersed spherical API agglomerates. This platform has allowed us to produce exquisitely designed spherical agglomerates (of drug or drug mixtures), formed in uniform size, composition, morphology and polymorphism.[2-3] We study how precise control over key attributes of drug or drug-excipient microparticles can be achieved by a combination of process and formulation parameters.

In this work, we demonstrate the co-formulation of model hydrophobic API drugs – ROY and carbamazepine – with excipient, ethyl cellulose (EC). We generate monodisperse emulsion drops containing the API-excipient mixture, and subject these droplets to evaporative crystallization in thin films of continuous phase. We find that the co-formulation of API with EC not only shortens the API crystallization process, but also allows for the direct creation of highly monodisperse spherical microparticles containing API crystals and macromolecular excipient, with nearly complete and tunable control over both morphology and polymorphic outcome. Specifically, we control polymorphic selection of both ROY and CBZ in the API-excipient microparticles, by implementing different film thicknesses for the evaporative crystallization; the film thickness mediates the rate of solvent evaporation from the droplets, controlling the kinetics of nucleation and crystallization. We obtain two completely specific and different polymorphic outcomes for each drug when crystallized in the presence of EC under different film thickness. In summary, our methods pave the way for the synthesis and formulation of novel drug product intermediates that bridge the considerable gap between primary and secondary pharmaceutical manufacturing.

Word Count: 308

CONTACT

* S. A. Khan, Tel: +65-6516-5133; saifkhan@nus.edu.sg

[1] Chadwick, K.; Davey, R. J.; Mughal, R.; Marziano, I., Org. Process Res. Dev. 2009, 13, 1284 – 1290.

[2] Toldy, A. I.; Badruddoza, A. Z. M.; Zheng, L.; Hatton, T. A.; Gunawan, R.; Rajagopalan, R.; Khan, S. A., Cryst. Growth Des. 2012, 12, 3977 – 3982.

[3] Leon, R. A., Badruddoza, A. Z. M., Zheng, L., Yeap, E. W., Toldy, A. I., Wong, K. Y., Hatton, T. A. Khan, S. A., Cryst. Growth Des., 2015, 15 (1), pp 212–218

423568 Morphology Evolution of Quadrangular Dendrites of Erythritol

Abstract:

The change of the morphology of erythritol was investigated in this paper. Offline tools such as scanning electron microscope (SEM), XRD, DSC and optical microscope and online tools such as PVM and FBRM were used to test the effects of the factors including temperature, additives, solvents, cooling rate and so on. It was found that when the erythritol solution is added the same anti-solvent, its habit changes from dendritic structure to octahedron with the rising of the initial temperature, and the transition temperature changes with the kind of the anti-solvent. A possible explanation of this phenomenon has been proposed on the basis of experiments. The dendritic erythritol is the early stage forms of the octahedron. The morphology of erythritol experiences a transformation process of dendritic—snow-like with four corners—octahedron. The influences of other factors on this phenomenon have also been discussed in detail.

419326 Independent Determination of Solute-Solid Interactions in Phytoextraction

Abstract:

Independent Determination of Solute-Solid Interactions in Phytoextraction

Björn Dreisewerd, Juliane Merz, Gerhard Schembecker, TU Dortmund University, Dortmund, Germany

Phytoextraction is strongly influenced by the solute-solid interactions describing the adsorption/desorption equilibrium at the surface of the plant material. However, standard extraction experiments of plant material do not provide access to this equilibrium as various superimposing effects influence the concentration of the target component in the extract. The equilibrium at the surface is just one part of a sequence of different mass transfer steps during extraction. However, being able to differentiate between these effects is a necessary prerequisite to provide reliable predictive models for phytoextraction enabling to significantly reduce the experimental effort for designing a technical process.

Several dynamic and static procedures for the determination of the solute-solid interactions have been published, but none of them enable to determine the solute-solid interactions independently of other mass transfer steps and extraction phenomena. Therefore, misinterpretation of results is most likely.[1]

In this study, two experimental procedures are presented to overcome this gap. Both procedures consist of two steps: first, the production of a completely leached plant material to eliminate the interference of the extraction phenomena and afterwards, adsorption experiments using the completely leached plant material and plant extracts. The first experimental procedure is used for the determination of adsorption isotherms, while the second one allows the independent measurement of the adsorption kinetics. Both experimental procedures were applied to two different model systems, namely artemisinin from Artemisia annua using ethanol, hexane or hexane with 5 mol% of ethyl acetate (referred to as hexane/ethyl acetate in the following) as extraction solvents and steviol glycosides (rebaudioside A and stevioside) from Stevia rebaudiana Bertoni using water.

The results show that artemisinin in ethanolic extracts did not adsorb onto the plant material at any temperature investigated, while for hexane and hexane/ethyl acetate linear adsorption isotherms could be determined. These results fit the expectations based on the solubility of artemisinin in the solvent systems given. The solute-solid interactions determined were validated by predicting the concentrations and accumulated masses of artemisinin in different stages of leaching experiments. This validation showed a good agreement between the experimental and the theoretical data and, therefore, proved the applicability of the experimental procedure.

Regarding the solute-solid interactions of steviol glycosides, Langmuir adsorption isotherms were determined. Since rebaudioside A has an additional β-D-glucopyranosyl group compared to stevioside, the former one is more polar and, thus, adsorbed less onto the plant material's surface than stevioside in the presence of water. Moreover, adsorption of both rebaudioside A and stevioside strongly depended on temperature. Although adsorption usually is an exothermic process, adsorption of the steviol glycosides increased with increasing temperature. This behavior was most likely caused by solute-solvent interactions, particularly the strength of the hydrogen bond interactions between the steviol glycosides and water, which decrease with increasing temperature. Accordingly, the maximum loading of rebaudioside A was more strongly influenced by the temperature than the maximum loading of stevioside, since rebaudioside A has three additional hydroxyl groups.

However, validation of the adsorption isotherms determined was not successful, as the predicted concentrations and accumulated masses of both rebaudioside A and stevioside significantly differed from the experimental data of the leaching experiment compared with. Therefore, the surface of the plant material was analyzed by scanning electron microscopy, disclosing significant changes in the structure of the plant material during extraction. Hence, a changing adsorption behavior could be assumed. This hypothesis was proven by additional adsorption experiments after each stage of extraction, resulting in increasing loads with increasing number of stages, although the liquid concentration decreased.

The experiments performed prove that the experimental procedures proposed allow determining independently the solute-solid interactions in phytoextraction. However, the structure of the plant material has to be monitored in order to exclude changes in the adsorption characteristic due to structural changes of the extraction material.

[1] del Valle, J.M., and Urrego, F.A., 2012. Free Solute Content and Solute-Matrix Interactions Affect Apparent Solubility and Apparent Solute Content in Supercritical CO_2 Extractions. A Hypothesis Paper. The Journal of Supercritical Fluids 66, 157–175.

414804 Optimal Experimental Design in Plant-Material Extraction Considering the Process Costs

Abstract:

Plant-material extraction deals with the extraction of valuable components from natural raw materials. The kinetics and the achievable equilibrium of the extraction process can be improved be the application of pre-treatment methods like milling, drying or microwave treatment. Plant material contains a large number of different components of which often only a few are of interest. The selectivity and the yield of a process are strongly influenced by the polarity and the composition of the used solvent. Polar substances are in general extracted better by polar solvents and non-polar substances by non-polar solvents.

Due to the complexity of the plant material and the numerous influences on kinetics, equilibrium, a new approach to design plant-material extraction processes was developed. It is based on laboratory-scale experiments under different conditions. The process costs are determined considering fix (e.g. apparatus costs) and variable cost (e.g. price for plant material). Different experimental conditions will lead to different specific product costs. The optimal process is found, when the costs are minimized.

A systematic investigation of the pre-treatment methods and solvents requires a high number of laboratory experiments. Experiments are necessary for the modeling but costly and time-consuming. Therefore, an optimal experimental design method was used in this work. The Model-Based-Experimental Analysis (MEXA) is used to find the optimal conditions for the lab experiments. The tool MEXA fits extraction models to experimental data and determines the process conditions at which the specific process costs are minimal. Then those experimental conditions are found at which the error of the minimal product cost is smallest. This optimal condition will change, depending on the pre-treatment methods applied to the plant material. Thus the overall experimental effort is reduced while the specific product cost can be determined with a pre-defined accuracy.

417811 Process Concept for Reactive Extraction of Dicarboxylic Acids from Fermentation Broth

Abstract:

Dicarboxylic acids are experiencing an increased interest as intermediate chemical building-blocks within industrial biotechnology. Due to their two terminal carboxylic groups, as well as the high reactivity, they can be efficiently used e.g. for polymerization reactions, or precursors for classical chemical reactions. Due to recent advances within the upstream processing (metabolic engineering of microbial strains, optimization of fermentation conditions) product titers have now reached a level suitable for an industrial implementation. However, due to their high hydrophilicity, presence in a complex fermentation broth, as well as the common exclusion of thermal separation processes, the downstream processing still marks the bottleneck for industrial applications.

In order to overcome this bottleneck, reactive extraction using amines as extractants can serve as a viable tool to ensure a selective and efficient downstream processing [1]. Within this work a complete reactive extraction concept was developed (extraction and re-extraction) using ethyl oleate as biocompatible organic phase, *cis,cis*-muconic acid as model dicarboxylic acid from fermentation, and different water insoluble amines as extractants. Within this work, different parameters such as the influence of polar modifiers, the pH, and amine concentration were investigated. Based on the results it was possible to determine a best performing system for the reactive extraction. Yields of $94 \pm 4.0\%$ and higher were achieved for a system using a tri-amine as extractant and a biocompatible solvent mixture as organic phase. Several experiments were conducted to investigate the underlying mechanism during reactive extraction of the dicarboxylic acid. Reactive extraction was shown to be mainly based on complex formation of undissociated acid and the amine via H-bond. Furthermore, in order to show the applicability for the purification on a process scale, different re-extraction concepts were evaluated. These included the addition of water soluble amines or a pH shift in the re-extraction phase. Based on these results, a complete downstream processing concept for the biocompatible reactive extraction of *cis,cis*-muconic acid is presented. The results allow for a further development of

an in situ downstream processing concept for biocatalytic produced dicarboxylic acids.

[1] J. Gorden, T. Zeiner, C. Brandenbusch, Fluid Phase Equilibria, 393 (2015) 78-84

Recovery of Carboxylic Acids from Wastewater
Process Intensification by Reactive Extraction and Esterification

Susanne Lux
Andreas Toth
Daniela Painer
Matthäus Siebenhofer

Outline

- Motivation
- Reactive Separations – Process development
  - Conventional
  - Extraction and Esterification
- Experimental Setup and Operation
- Results
  - Extractant – Catalyst Interaction
  - Diluent Screening
  - Effects of Alcohol
- Summary

Outline

- Motivation
- Reactive Separations – Process development
  - Conventional
  - Extraction and Esterification
- Experimental Setup and Operation
- Results
  - Extractant – Catalyst Interaction
  - Diluent Screening
  - Effects of Alcohol
- Summary

Motivation

AUSTRIA - Styria

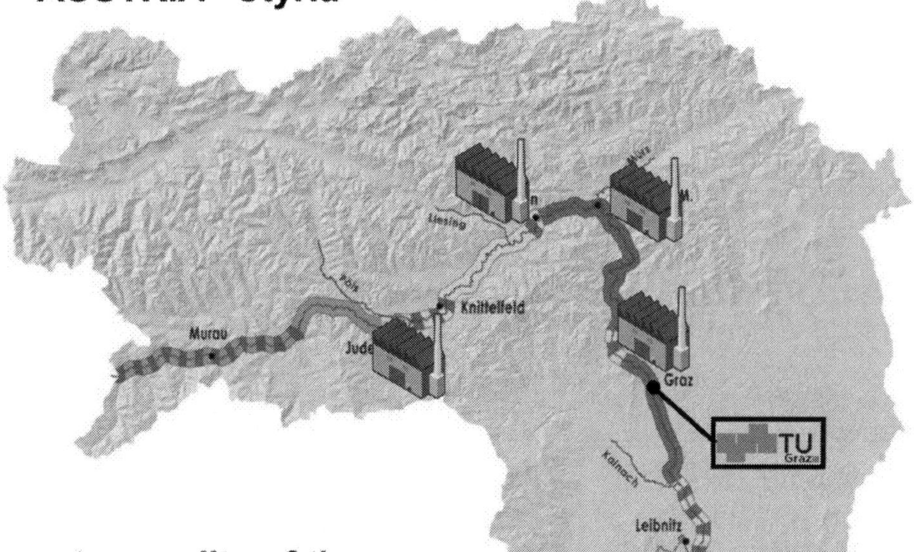

water quality of the Mur river in 1975

Source: http://www.umwelt.steiermark.at

Motivation

AUSTRIA - Styria

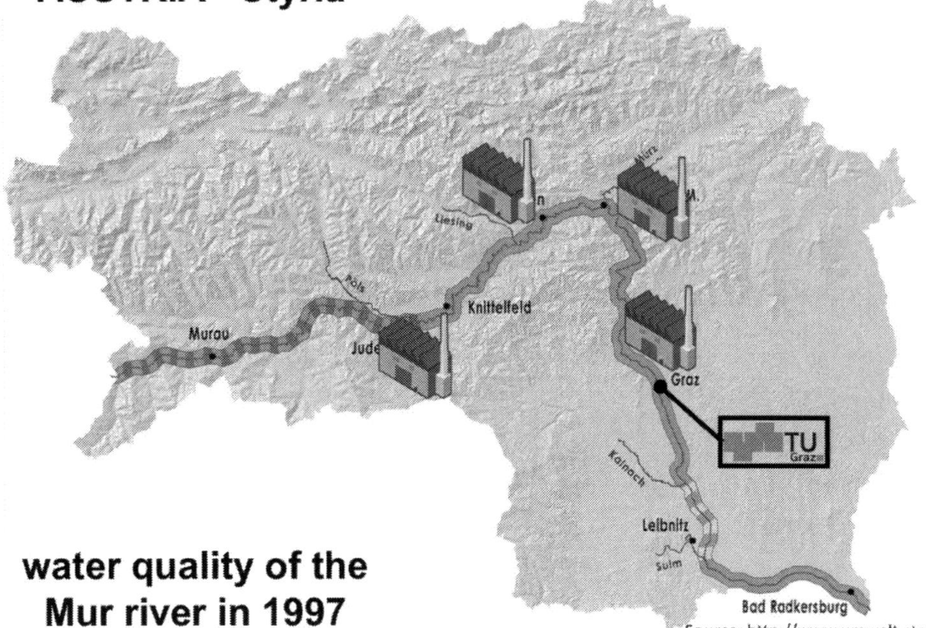

water quality of the Mur river in 1997

Source: http://www.umwelt.steiermark.at

Motivation – Facts

- 1,943 million tons/year of **primary pulp** in 2014 (Austria)
- 106 million m^3 of wastewater per year
- Acetic acid: about 60,000 tons/year

Source: http://www.austropapier.at/

Motivation

- Loss of valuable constituents
  - Carboxylic acids (e.g. acetic acid)
  - (Methanol)
- Search for efficient (economic) isolation processes
- **Esterification of Carboxylic Acids**
- Use of Esters
  - Solvents
  - Flavors and fragrances
  - Detergents
  - Pharmaceutics
  - etc.

Process Scheme – State of the Art

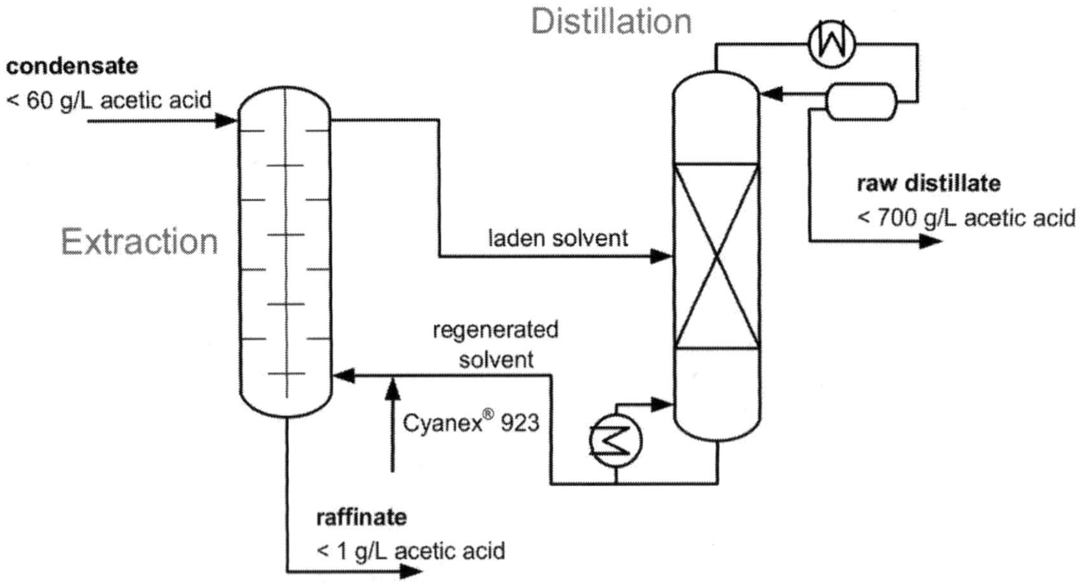

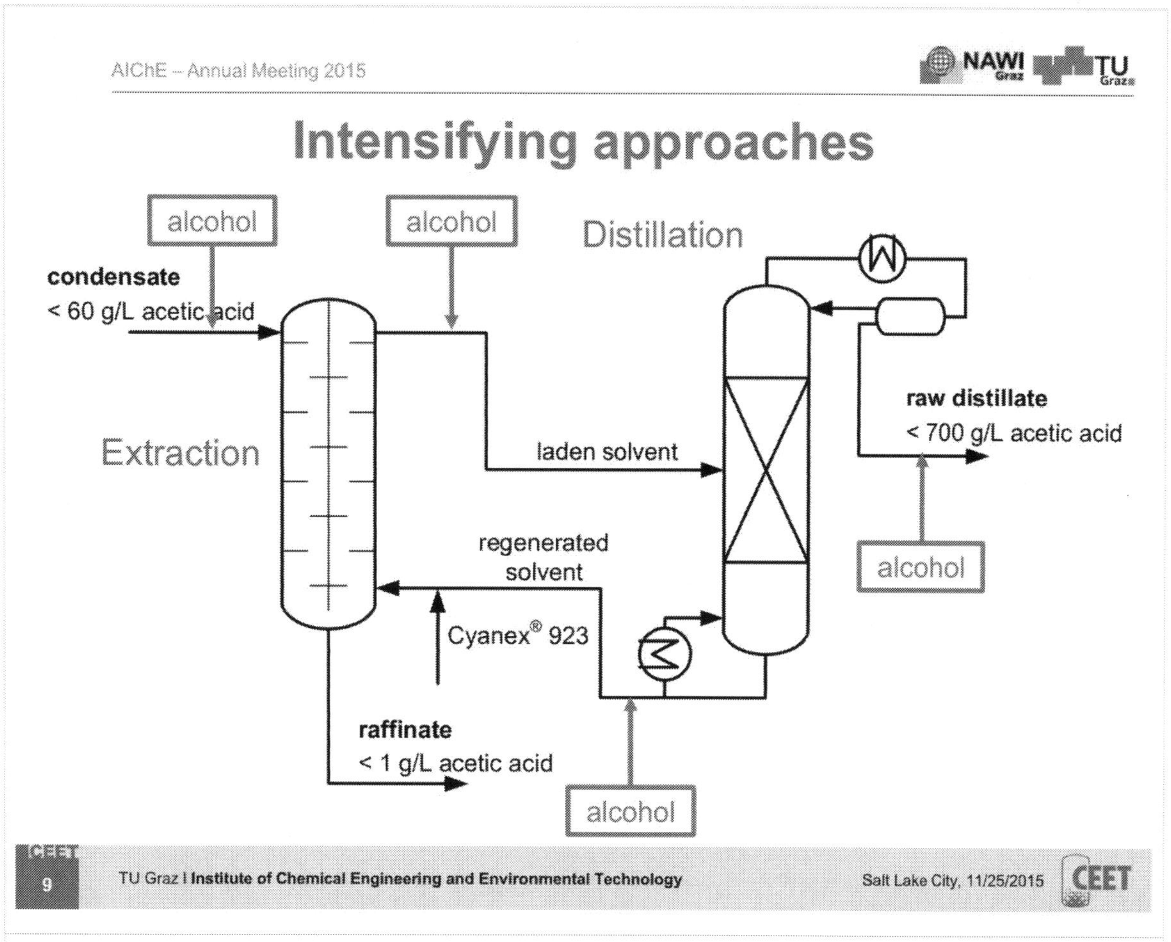

AIChE – Annual Meeting 2015
NAWI Graz
TU Graz
Intensifying approaches
alcohol
alcohol
Distillation
condensate
< 60 g/L acetic acid
Extraction
laden solvent
raw distillate
< 700 g/L acetic acid
regenerated solvent
alcohol
Cyanex® 923
raffinate
< 1 g/L acetic acid
alcohol
CEET
9
TU Graz I Institute of Chemical Engineering and Environmental Technology
Salt Lake City, 11/25/2015
CEET

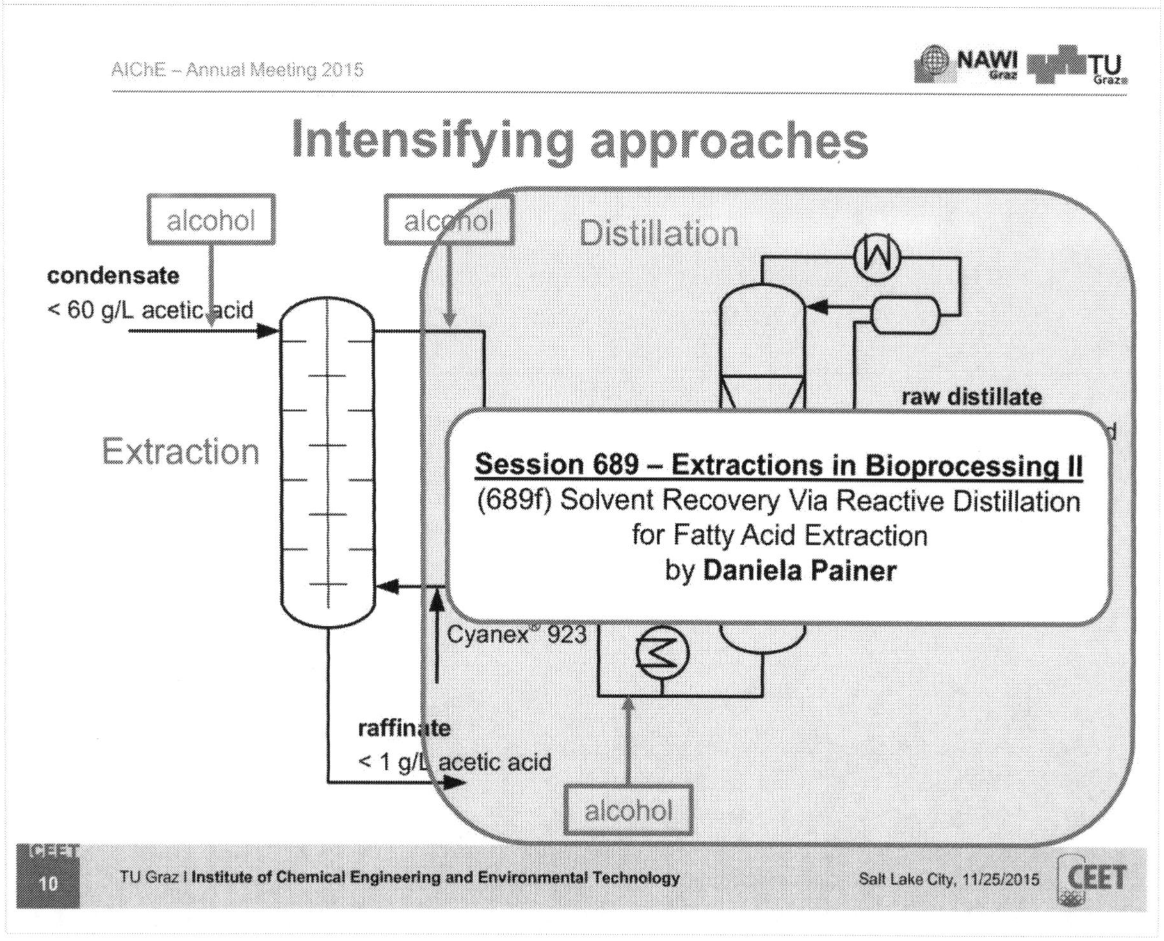

AIChE – Annual Meeting 2015
NAWI Graz
TU Graz
Intensifying approaches
alcohol
alcohol
Distillation
condensate
< 60 g/L acetic acid
raw distillate
Extraction
Session 689 – Extractions in Bioprocessing II
(689f) Solvent Recovery Via Reactive Distillation
for Fatty Acid Extraction
by Daniela Painer
Cyanex® 923
raffinate
< 1 g/L acetic acid
alcohol
CEET
10
TU Graz I Institute of Chemical Engineering and Environmental Technology
Salt Lake City, 11/25/2015
CEET

Intensifying approaches

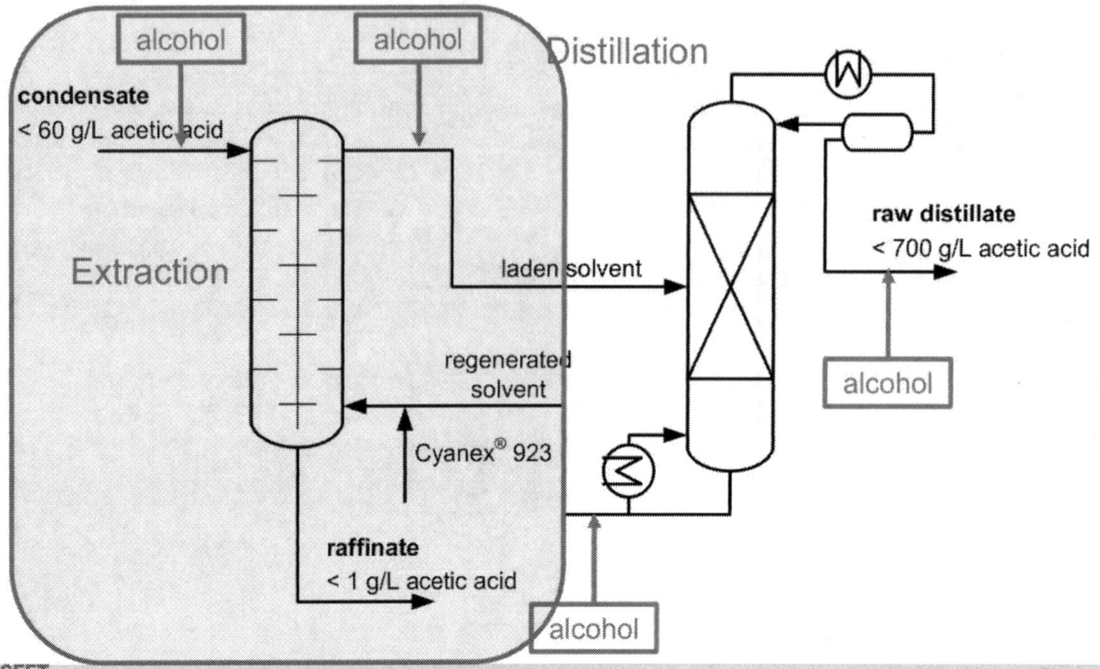

Outline

- Motivation
- Reactive Separations – Process development
  - Conventional
  - Extraction and Esterification
- Experimental Setup and Operation
- Results
  - Extractant – Catalyst Interaction
  - Diluent Screening
  - Effects of Alcohol
- Summary

"Conventional" Reactive Extraction

- Physical extraction with "chemical interaction" of the constituent and the solvent
- "Reaction" of acetic acid is basically the formation of Lewis acid/base complexes
- Product: constituent (e.g. acetic acid)
- Applied in
  - Extraction of carboxylic acids

Extraction and Esterification

- Physical extraction and chemical reaction
- Reaction: **Conversion of the constituent**
- Product: **Different chemical substance**
- Advantages
  - simplify separation
  - convert constituent into product

Extraction + Catalytic Esterification

- Catalysts
  - Sulfuric acid (homogeneous)
    - → difficult separation
    - → byproducts
    - → loss of catalyst
  - Alternative: heterogeneous catalysis
    Catalyst: e.g. **4-dodecylbenzenesulfonic acid**

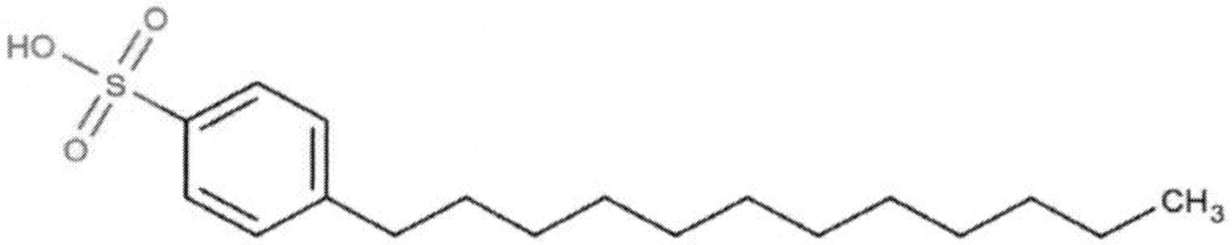

Heterogeneous Catalytic Esterification with Physical Extraction

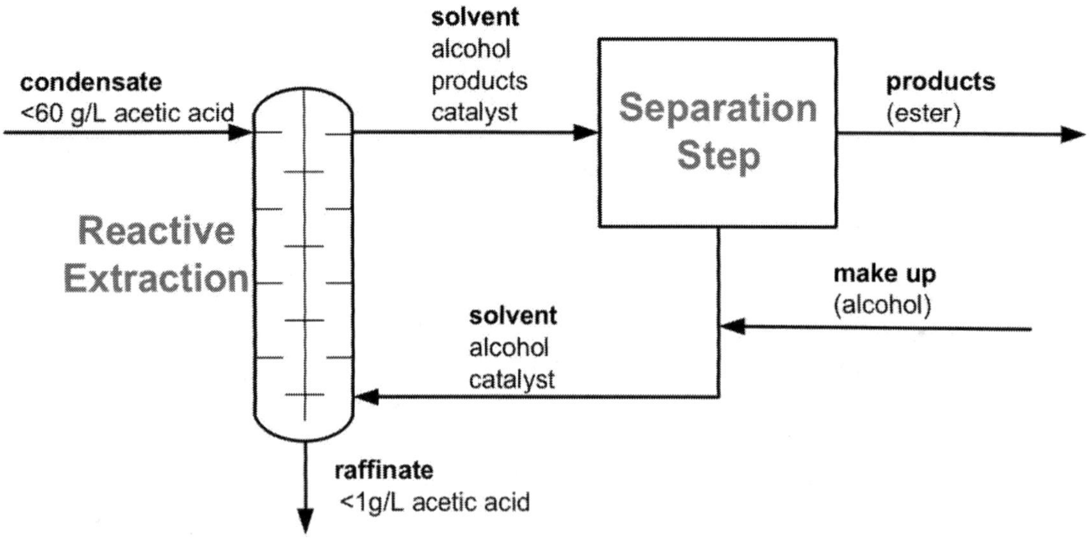

Outline

- Motivation
- Reactive Separations – Process development
  - Conventional
  - Extraction and Esterification
- Experimental Setup and Operation
- Results
  - Extractant – Catalyst Interaction
  - Diluent Screening
  - Effects of Alcohol
- Summary

Parameters

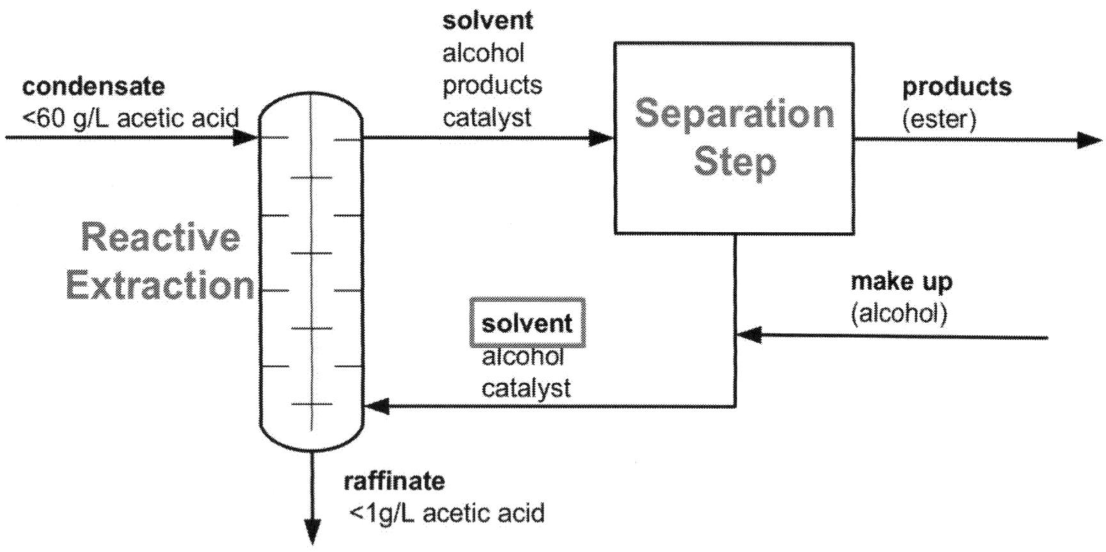

Parameters

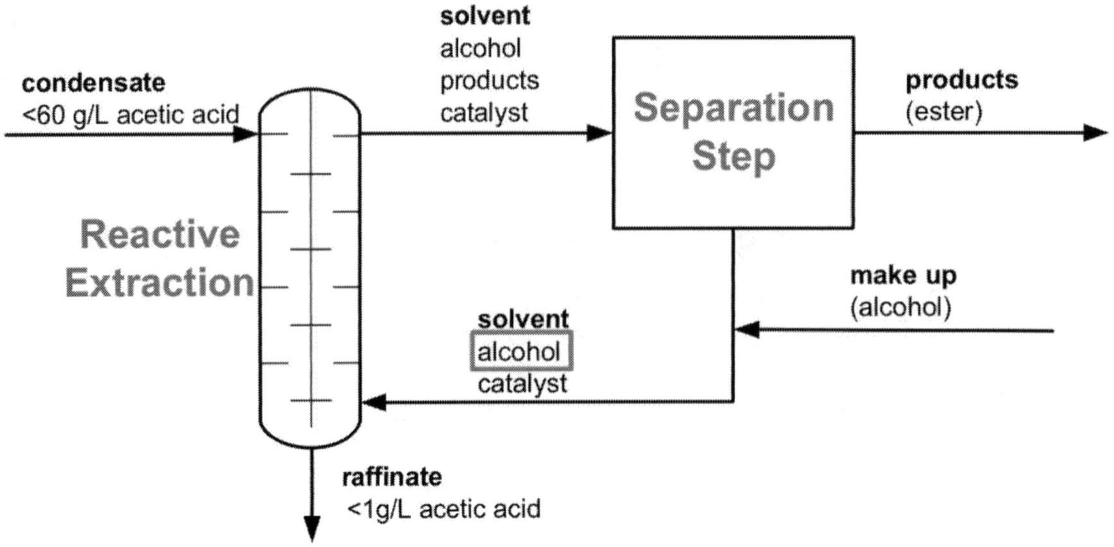

Experimental Setup and Operation

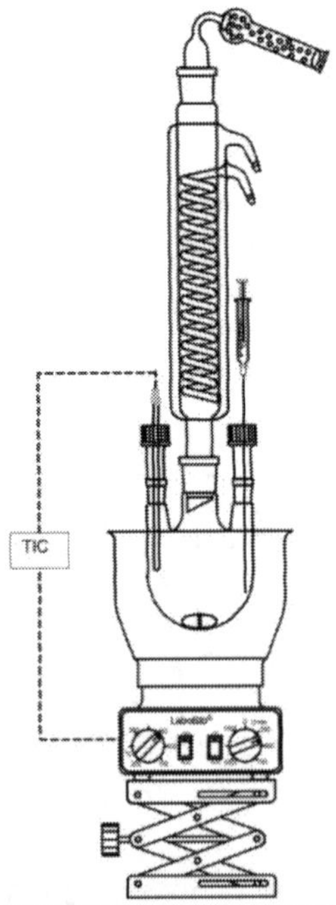

- Single stage batch extraction
- Model constituent: acetic acid
- Feed composition
  - Deionized water with acetic acid
- Solvent plus catalyst
  - Catalyst: 10 wt.-% of solvent phase
- Volumetric ratio of solvent phase to aqueous phase
  - 1:1
- Reaction time and temperature
  - 3h @ 50°C
- Molar ratio of alcohol to acetic acid
  - 3:1

Outline

- Motivation
- Reactive Separations – Process development
  - Conventional
  - Extraction and Esterification
- Experimental Setup and Operation
- Results
  - Extractant – Catalyst Interaction
  - Diluent Screening
  - Effects of Alcohol
- Summary

Solvent

- Extractant: **Cyanex® 923**
- Catalyst: **4-dodecylbenzenesulfonic acid**
- Diluents:
  - *n*-Heptane (C7)
  - *n*-Nonane (C9)
  - *n*-Undecane (C11)
  - ShellSol T (C11-C12)
  - ShellSol K (C11-C14)
- Solvent phase: a) Cyanex® 923 + diluent + catalyst
  b) diluent + catalyst
- Task
  - Influence of extractant – catalyst interaction on acetic acid conversion
  - Influence of diluent on conversion
  - Influence of alcohol on conversion and process

Extractant – Catalyst Interaction

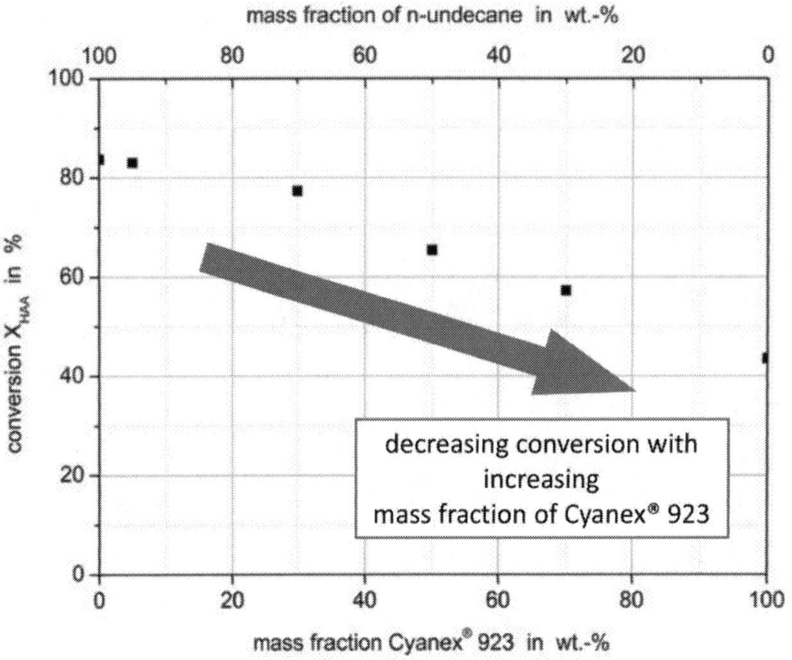

temperature
50°C

reaction time
3h

catalyst
10 wt.-%

alcohol
n-butanol

Influence of the Diluent on Conversion

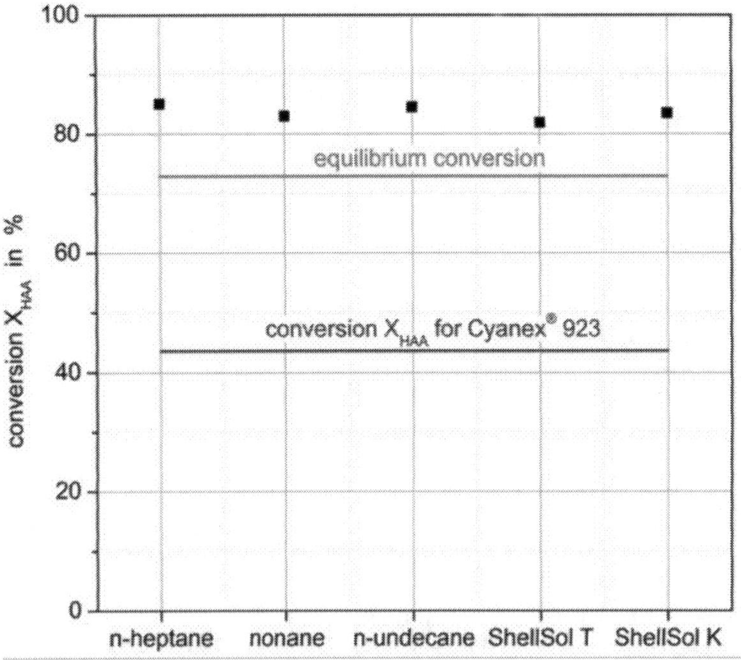

temperature
50°C

reaction time
3h

catalyst
10 wt.-%

alcohol
n-butanol

Effects of the Alcohol

- Effect on acetic acid conversion
- Effect on product isolation
  - Distribution of alcohol between organic and aqueous phase
  - Distribution of ester between organic and aqueous phase
  - Phase separation

alcohol (water solubility)	boiling point		ester (water solubility)	boiling point
methanol **(s)**	65 °C	↔	methyl acetate **(s)**	57 °C
n-butanol **(ls)**	118 °C	↔	butyl acetate **(ls)**	127 °C
(2-pentanol	119 °C		2-pentyl acetate	134 °C)
octanol **(is)**	195 °C	↔	octyl acetate **(is)**	211 °C

Effect of Alcohol on Conversion

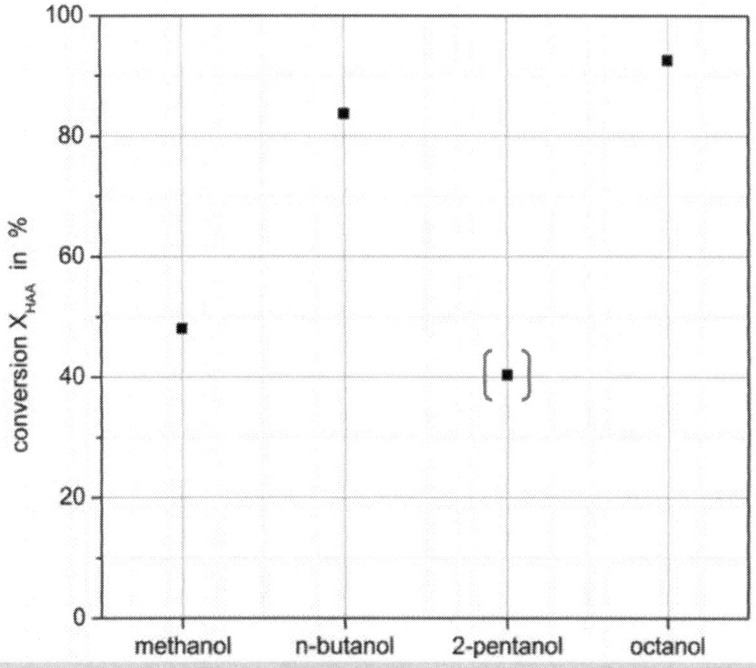

temperature
50°C

reaction time
3h

catalyst
10 wt.-%

Solvent
n-undecane

AIChE – Annual Meeting 2015

NAWI Graz TU Graz

Intensifying approaches

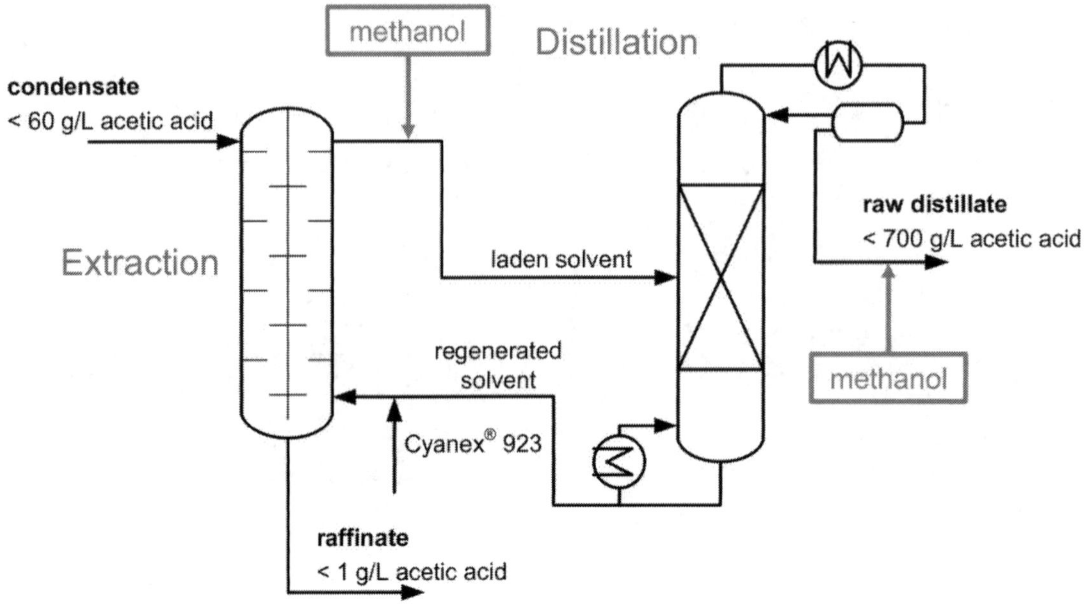

27 TU Graz I **Institute of Chemical Engineering and Environmental Technology** Salt Lake City, 11/25/2015 CEET

AIChE – Annual Meeting 2015

NAWI Graz TU Graz

Intensifying approaches

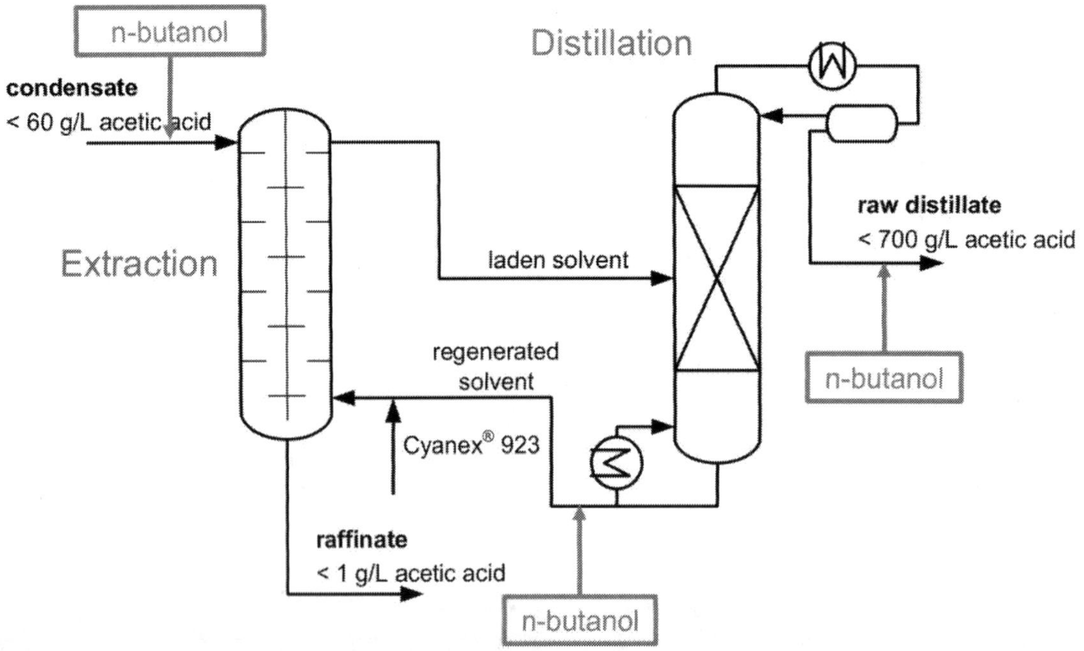

28 TU Graz I **Institute of Chemical Engineering and Environmental Technology** Salt Lake City, 11/25/2015 CEET

n-Butanol – Overall Balance

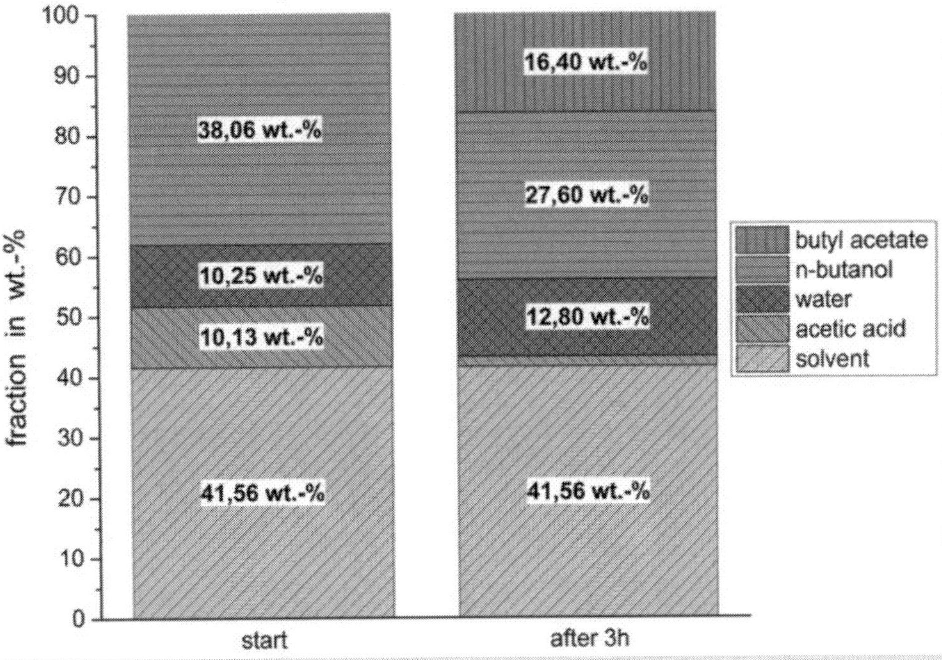

n-Butanol – Distribution of the Components after 3h

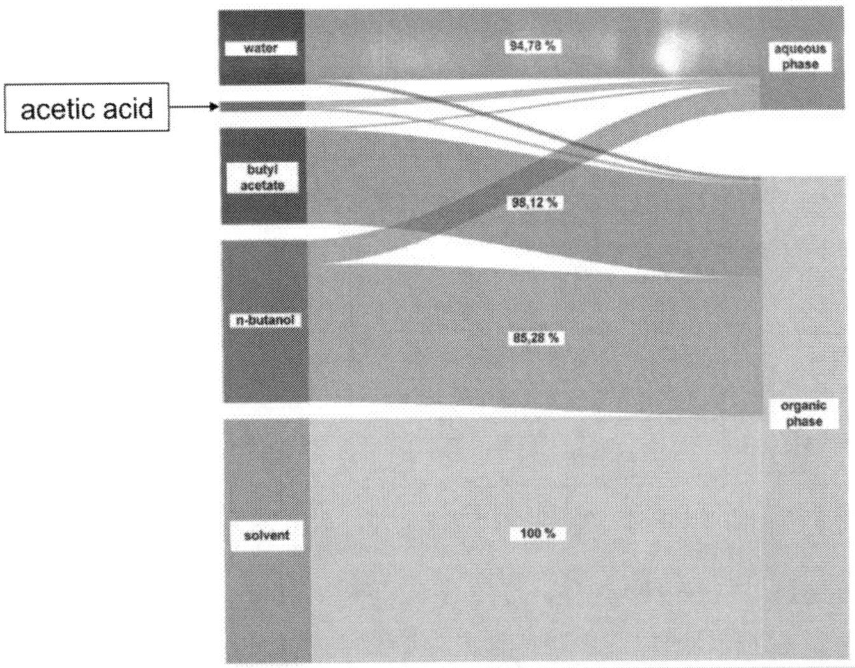

NAWI Graz TU Graz

Intensifying Approaches

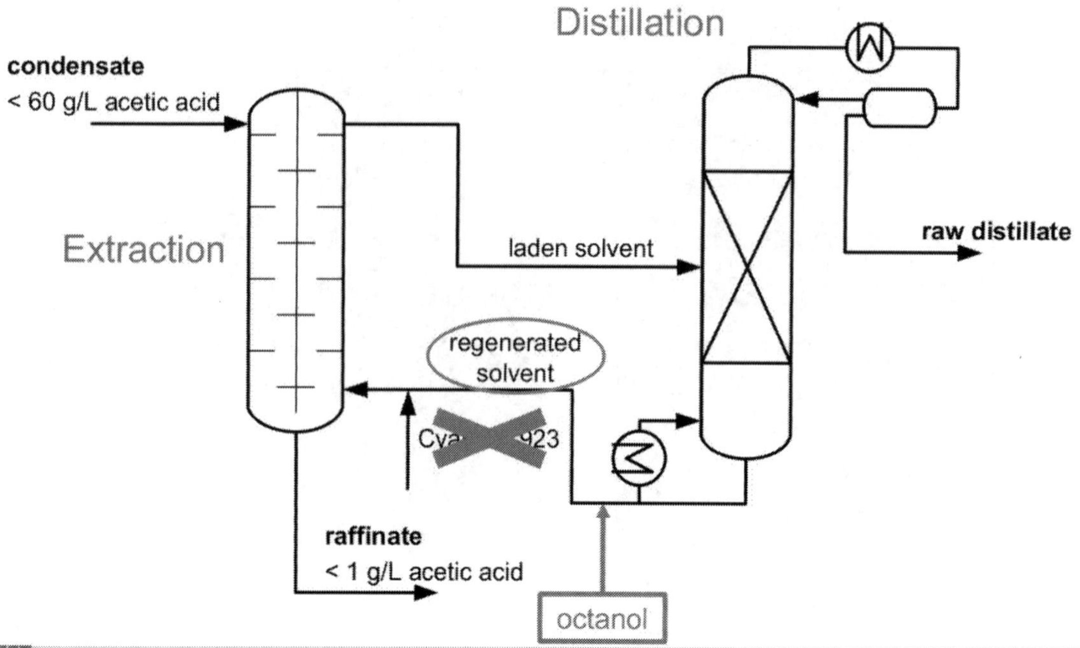

NAWI Graz TU Graz

Process: Physical Extraction with Heterogeneously Catalyzed Esterification

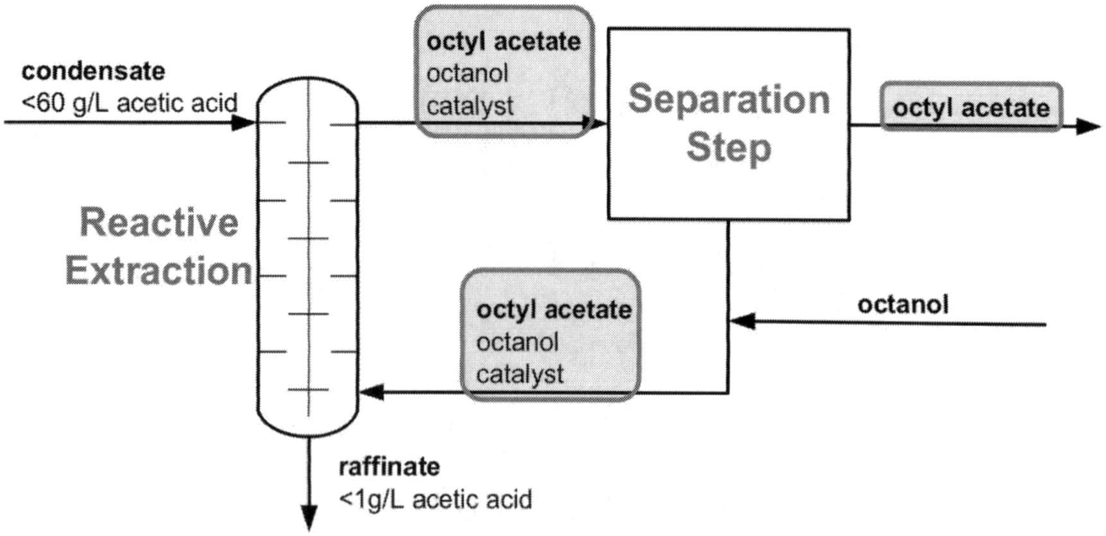

NAWI Graz TU Graz

Octanol – Overall Balance

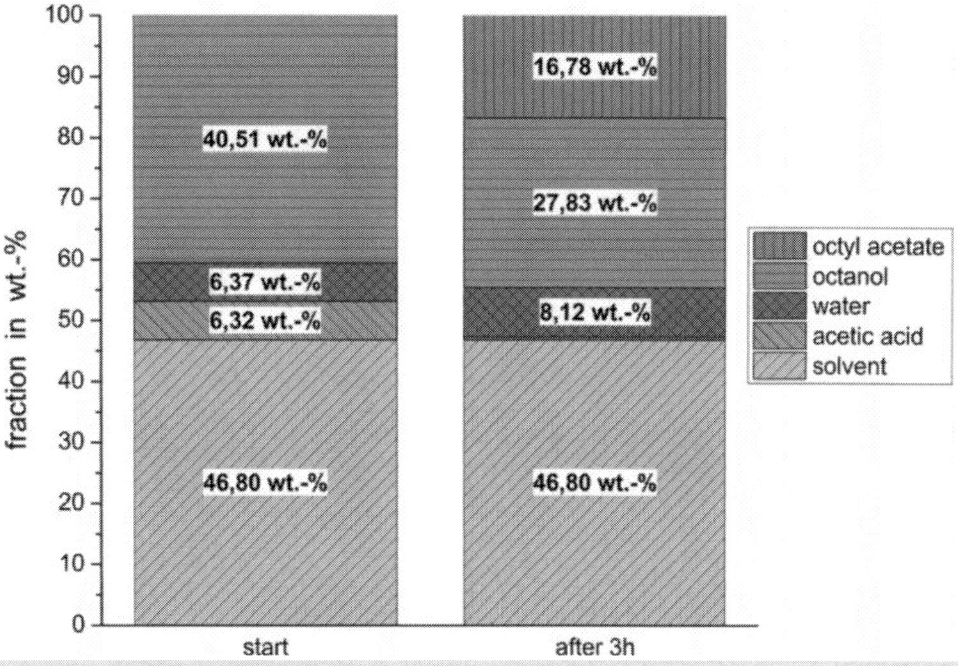

AIChE – Annual Meeting 2015

NAWI Graz TU Graz

Octanol – Distribution of the Components after 3h

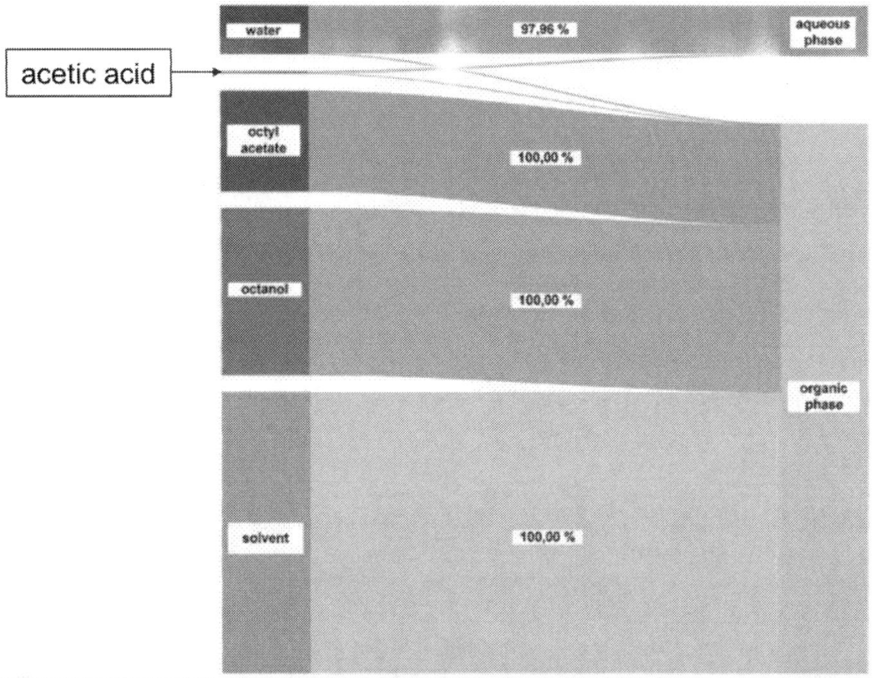

Outline

- Motivation
- Reactive Separations – Process development
  - Conventional
  - Extraction and Esterification
- Experimental Setup and Operation
- Results
  - Extractant – Catalyst Interaction
  - Diluent Screening
  - Effects of Alcohol
- Summary

Summary

- Several approaches possible and feasible
  - Reactants, solvents and catalysts determine process design
- Phase separation does improve reactive separation
  - About 10% higher conversion in single stage batch extraction
- Extractant – catalyst interaction may have a negative effect on conversion in reactive separations
- Process selection and design depend on the constituent and the product

Recovery of Carboxylic Acids from Wastewater

Process Intensification by Reactive Extraction with Esterification

Andreas Toth
Graz University of Technology
Institute of Chemical Engineering and Environmental Technology
Inffeldgasse 25C/II, A-8010 Graz
andreas.toth@tugraz.at

424674 Smart Polymer-Assisted Separation of Tocopherol Homologues and Fast Recovery of the Extractant

Abstract:

Due to the strong similarities in their structures and properties, the separation of natural homologues is quite challenging. Current separation methods of these products require large solvent consumption, have low separation efficiency, and result in significant energy expenditure. To address these issues, we have proposed a novel strategy to use ionic liquid polymers (PIL) or polyethyleneimine (PEI) as extractants for the liquid-liquid extraction of α-tocopherol from tocopherol homologues. The PIL was synthesized by RAFT copolymerization of N-isopropylacrylamide and ionic liquid, while the PEI is commercially available. The PIL was thermoresponsive in acetonitrile with upper critical solution temperatures (UCST) varying from 25.7 to 34.8 °C, and the PEI also had a stimuli-responsive property in acetonitrile triggered by CO_2. Both the two polymer extractants showed a good separation efficiency of tocopherol homologues. The distribution coefficient of δ-tocopherol between the extract (PIL as extractant) and raffinate phases (D_{δ}) and the selectivity coefficient of δ-tocopherol to α-tocopherol ($S_{\delta/\alpha}$) reached as high as 7.86 and 13.0, respectively, while D_{δ} (PEI as extractant) was 6.35 and $S_{\delta/\alpha}$ was 6.83. The PIL and PEI could be reused for multiple extraction cycles with a slight change in the tocopherol distribution and selectivity coefficients. The use of stimuli-responsive polymer extractants offers great benefits by reducing the back extraction steps that are normally required. This work demonstrates the potential of smart polymers for use in high performance separation of natural products.

Solvent Recovery via Reactive Distillation for Fatty Acid Extraction

D. Painer, S. Lux, M. Siebenhofer

Solvent Recovery via Reactive Distillation for Fatty Acid Extraction

Outline

- Motivation
- Liquid-liquid-equilibrium
  - Single acid extraction
  - Binary acid extraction
- Reactive distillation
  - Solvent recovery via reactive distillation
- Summary

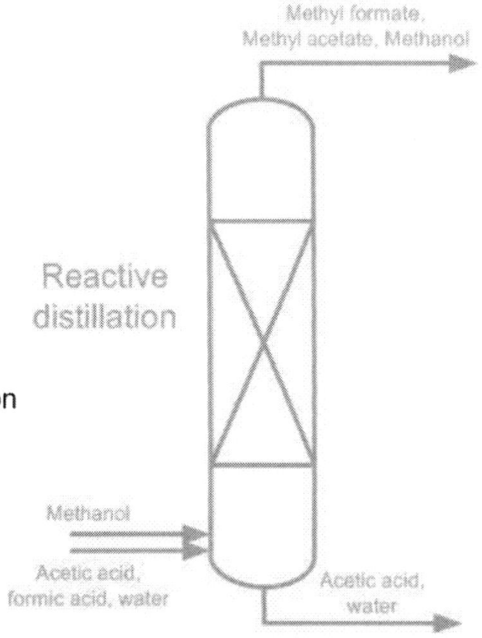

Motivation

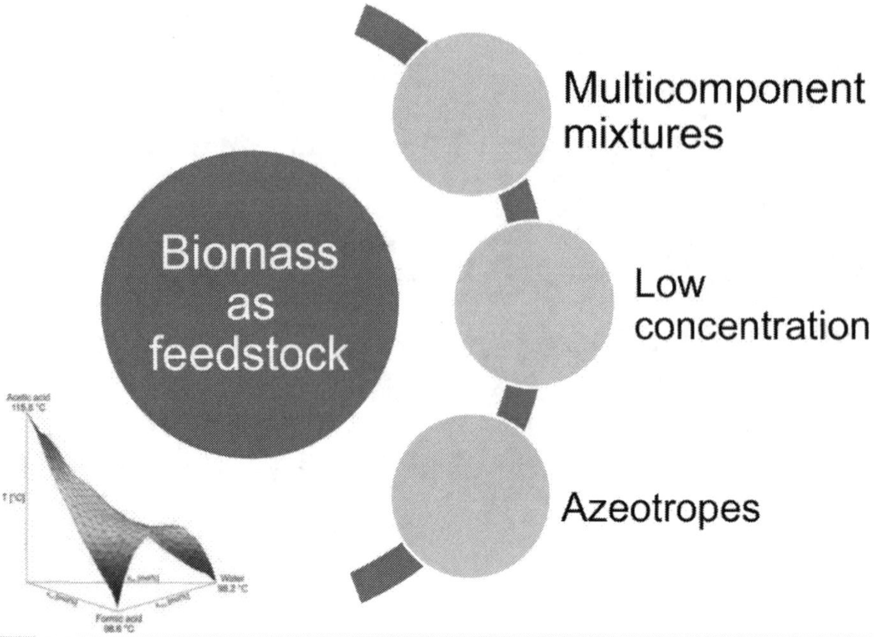

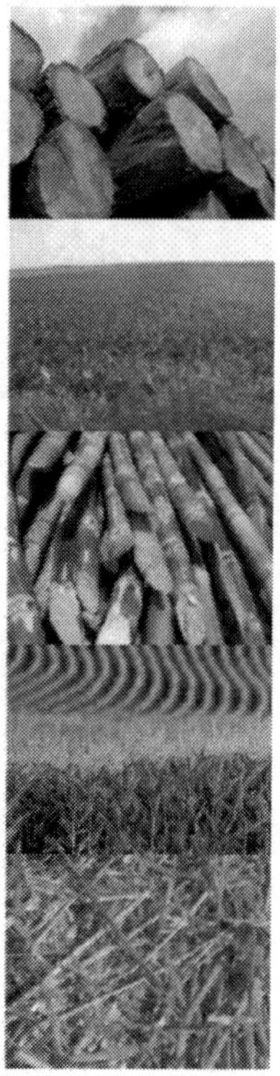

Extraction of carboxylic acids

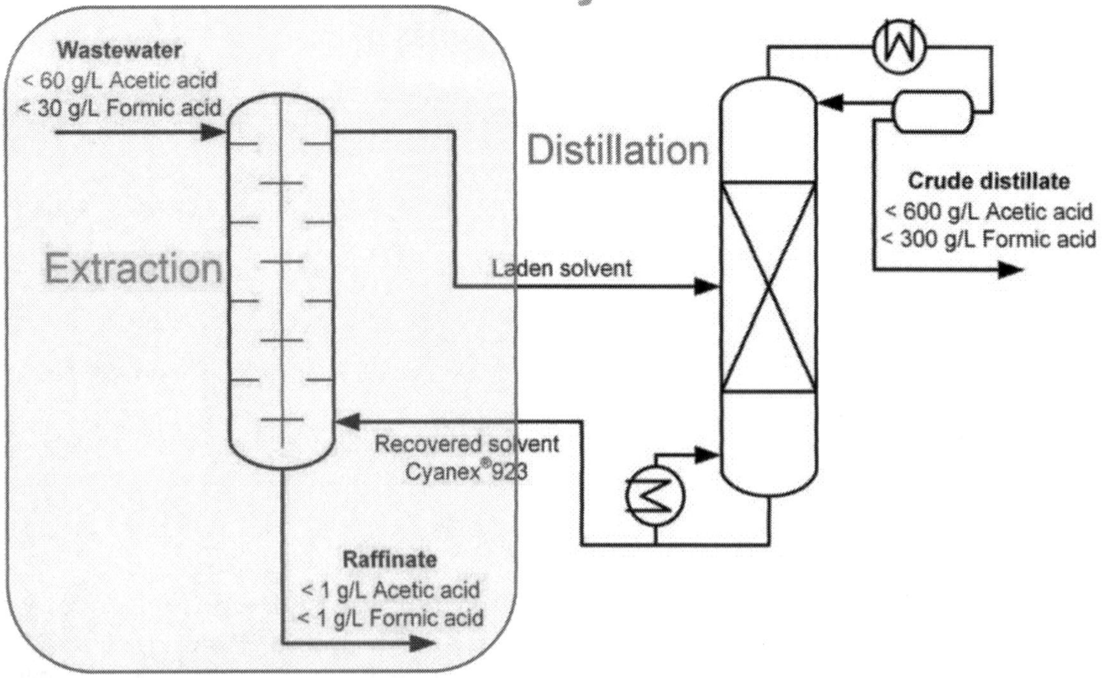

Solvent

- 70 wt.% Cyanex®923
- 30 wt.% Undecane

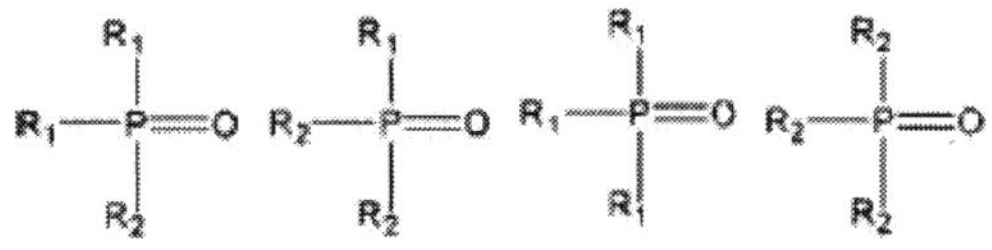

30.4% 37.4% 16.1% 8.5%

where: $R_1 - C_6H_{13}$, $R_2 - C_8H_{17}$

Wisniewski, M.; Pierzchalska, M.; J Chem Technol Biotechnol, 2005

$$n \cdot \mathrm{HAc} + \mathrm{S} \leftrightarrow \mathrm{HAc_n} \cdot \mathrm{S} \qquad K_{\mathrm{HAc},1} = \frac{[\mathrm{HAc_n} \cdot \mathrm{S}]_{\mathrm{org}}}{[\mathrm{HAc}]_{\mathrm{aq}}^{\mathrm{n}} \cdot [\mathrm{S}]_{\mathrm{org}}}$$

$$m \cdot \mathrm{HAc} + \mathrm{S} \leftrightarrow \mathrm{HAc_m} \cdot \mathrm{S} \qquad K_{\mathrm{HAc},2} = \frac{[\mathrm{HAc_m} \cdot \mathrm{S}]_{\mathrm{org}}}{[\mathrm{HAc}]_{\mathrm{aq}}^{\mathrm{m}} \cdot [\mathrm{S}]_{\mathrm{org}}}$$

Liquid-liquid-equilibrium T = 25 °C

$$a_{\mathrm{A,org}} = \frac{(n \cdot K_{\mathrm{A},1} \cdot a_{\mathrm{A,aq}}^{\mathrm{n}} + m \cdot K_{\mathrm{A},2} \cdot a_{\mathrm{A,aq}}^{\mathrm{m}}) \cdot a_{\mathrm{S,org}}}{1 + K_{\mathrm{A},1} \cdot a_{\mathrm{A,aq}}^{\mathrm{n}} + K_{\mathrm{A},2} \cdot a_{\mathrm{A,aq}}^{\mathrm{m}}}$$

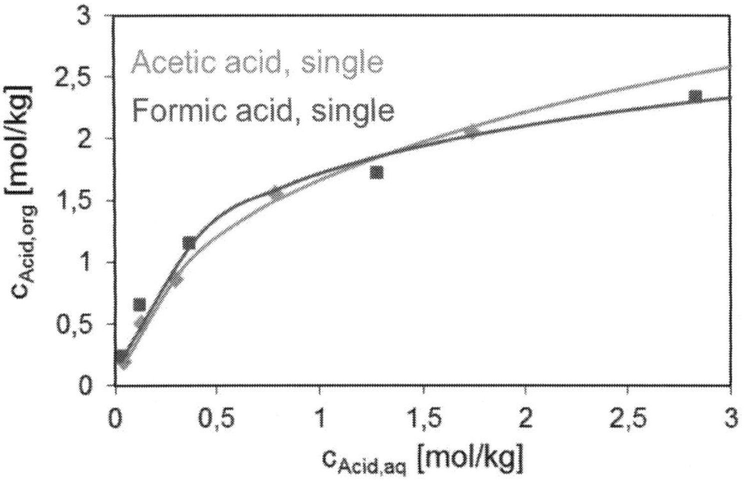

$c_{\mathrm{S,org}} = 1.89$ mol/kg

Acetic acid

$K_{\mathrm{HAc},1} = 2.64$

$n_{\mathrm{HAc}} = 1$

$K_{\mathrm{HAc},2} = 0.33$

$m_{\mathrm{HAc}} = 2.3$

Formic acid

$K_{\mathrm{HFo},1} = 3.59$

$n_{\mathrm{HFo}} = 1$

$K_{\mathrm{HFo},2} = 0.65$

$m_{\mathrm{HFo}} = 2$

Liquid-liquid-equilibrium T = 25 °C

$$a_{\mathrm{A,org}} = \frac{(n_{\mathrm{A}} \cdot K_{\mathrm{A,1}} \cdot a_{\mathrm{A,aq}}^{\mathrm{n_A}} + m_{\mathrm{A}} \cdot K_{\mathrm{A,2}} \cdot a_{\mathrm{A,aq}}^{\mathrm{m_A}}) \cdot a_{\mathrm{S,org}}}{1 + K_{\mathrm{A,1}} \cdot a_{\mathrm{A,aq}}^{\mathrm{n_A}} + K_{\mathrm{A,2}} \cdot a_{\mathrm{A,aq}}^{\mathrm{m_A}} + K_{\mathrm{B,1}} \cdot a_{\mathrm{B,aq}}^{\mathrm{n_B}} + K_{\mathrm{B,2}} \cdot a_{\mathrm{B,aq}}^{\mathrm{m_B}}}$$

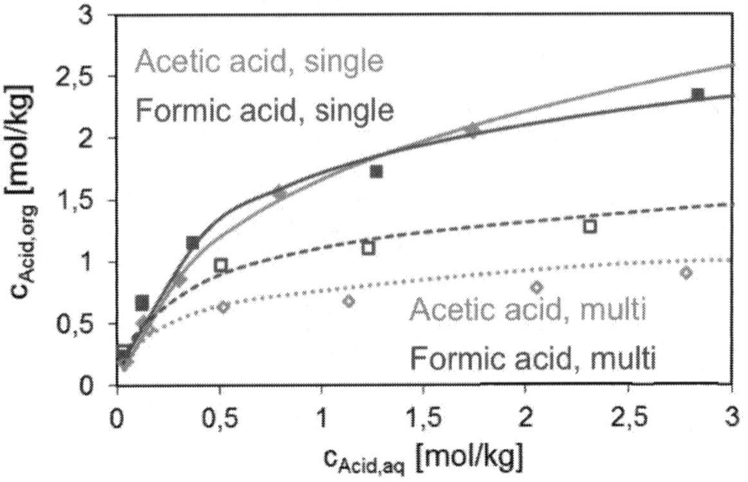

$c_{\mathrm{S,org}}$ = 1.89 mol/kg

Acetic acid

$K_{\mathrm{HAc,1}}$ = 2.64

n_{HAc} = 1

$K_{\mathrm{HAc,2}}$ = 0.33

m_{HAc} = 2.3

Formic acid

$K_{\mathrm{HFo,1}}$ = 3.59

n_{HFo} = 1

$K_{\mathrm{HFo,2}}$ = 0.65

m_{HFo} = 2

Extraction of carboxylic acids

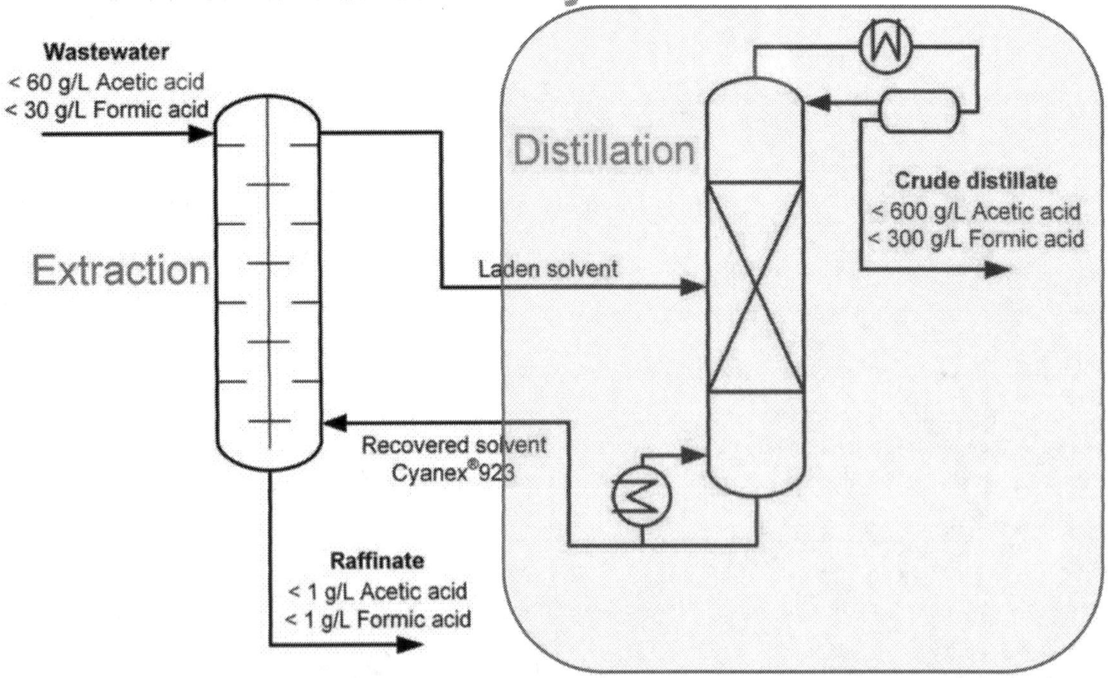

Boiling point surface

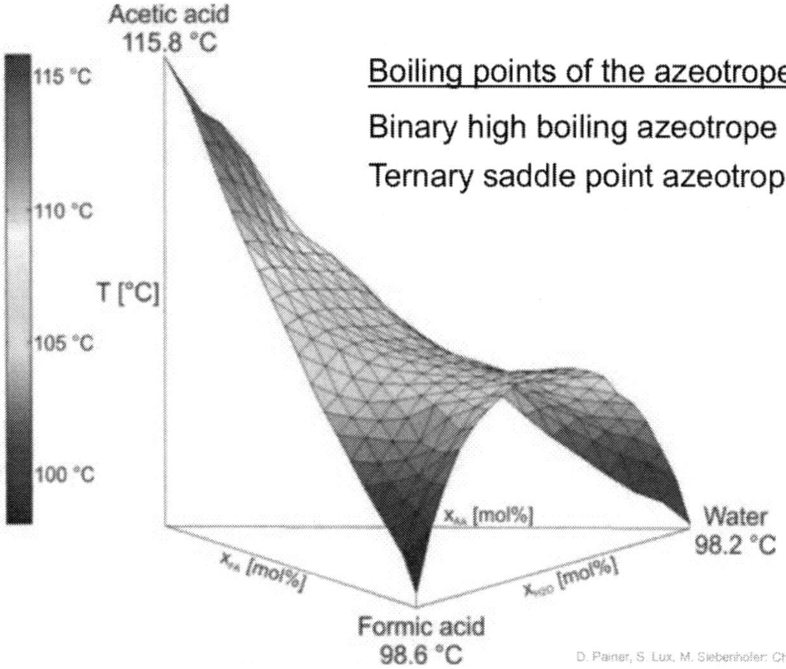

Boiling points of the azeotropes at 950 mbar:

Binary high boiling azeotrope	105.6 °C
Ternary saddle point azeotrope	105.2 °C

D. Painer, S. Lux, M. Siebenhofer: Chem. Ing. Tech. 2015, 87, No. 6, 843-847

Boiling point surface

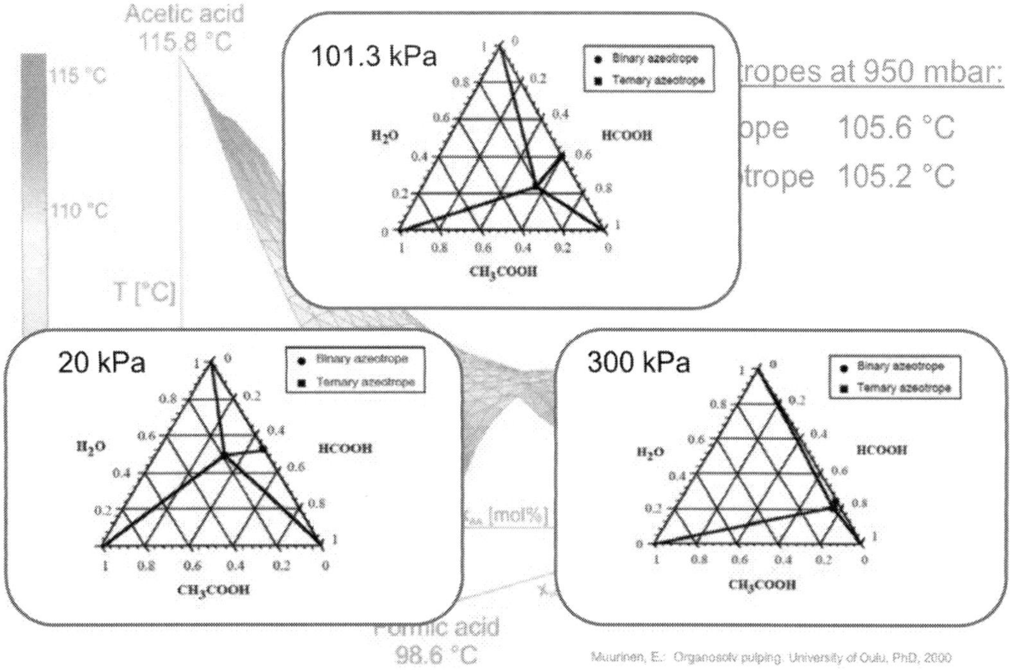

Muurinen, E.: Organosolv pulping. University of Oulu, PhD, 2000

Vapor-Liquid-Equilibrium

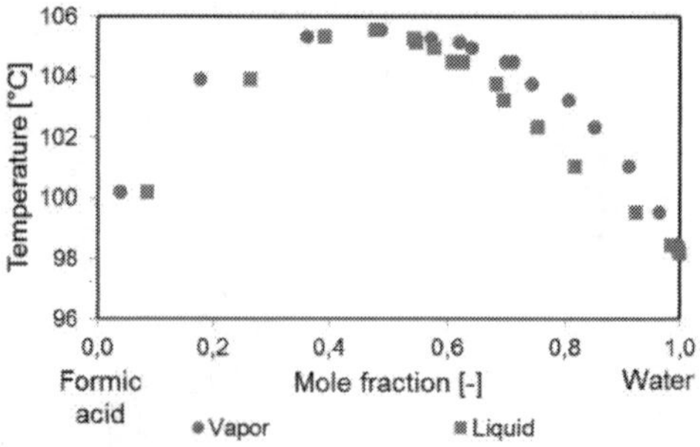

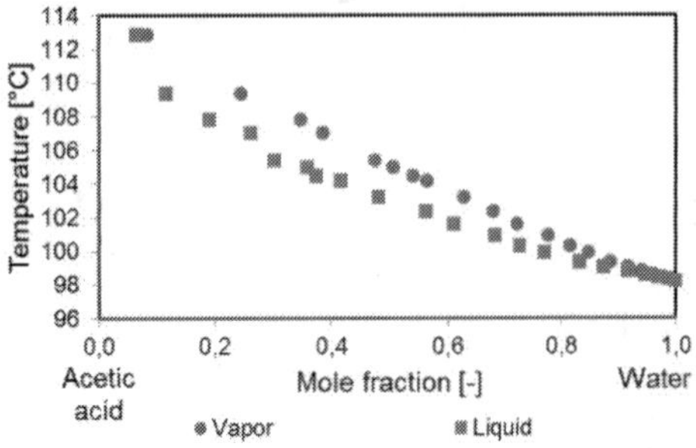

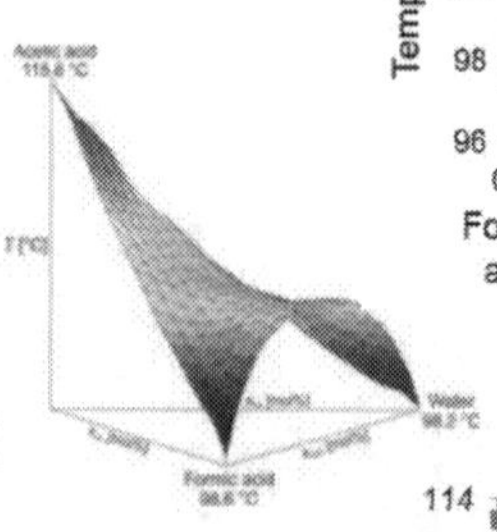

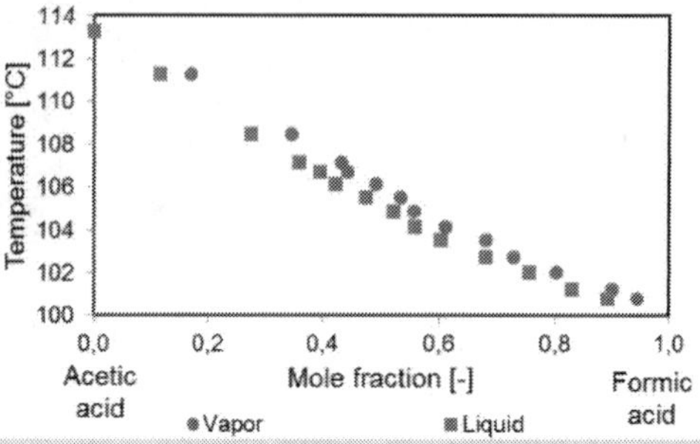

Process intensification solver strategy

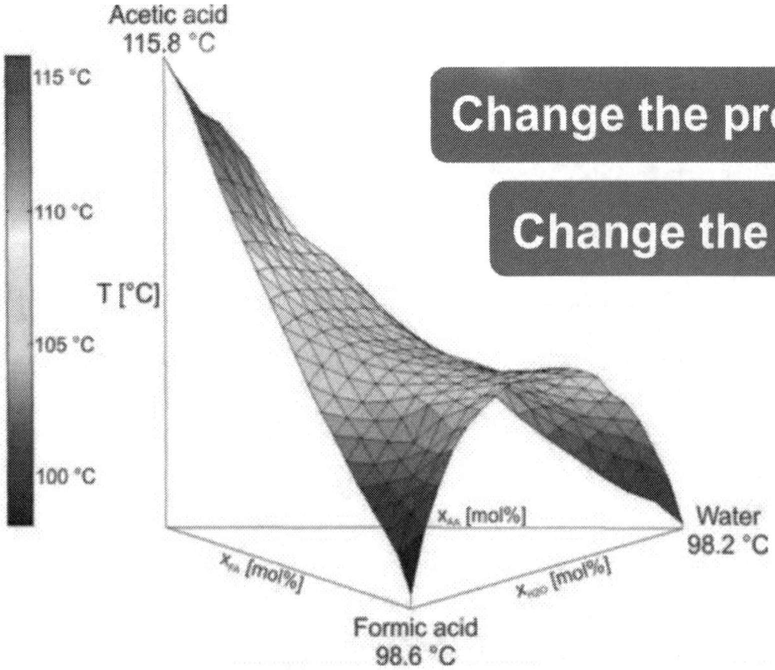

Change the properties!

Change the state!

Solvent Recovery via Reactive Distillation for Fatty Acid Extraction

Extraction of carboxylic acids

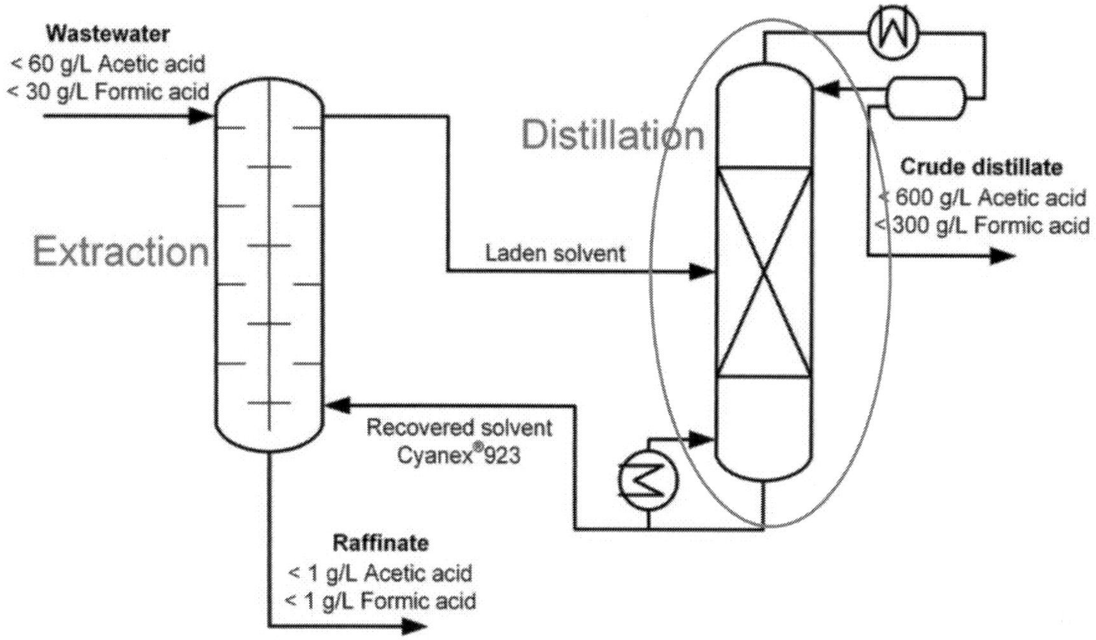

13 TU Graz | **Institute of Chemical Engineering and Environmental Technology** Salt Lake City, November 12, 2015

Solvent Recovery via Reactive Distillation for Fatty Acid Extraction

Extraction of carboxylic acids

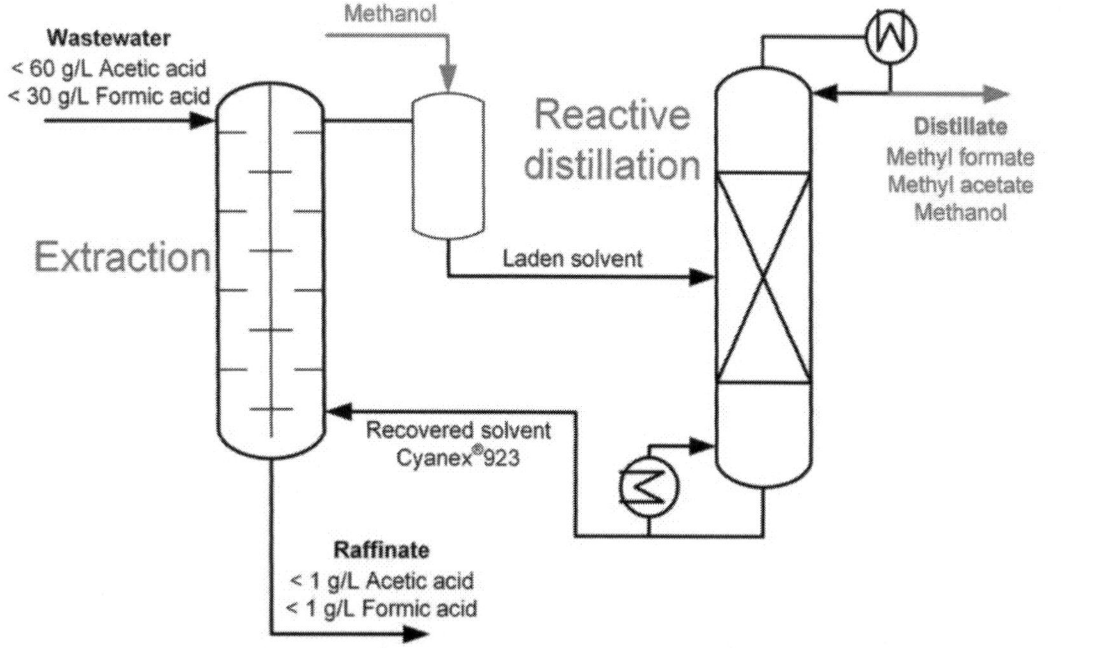

14 TU Graz | **Institute of Chemical Engineering and Environmental Technology** Salt Lake City, November 12, 2015

Change the properties

- Esterification with methanol

$$HCOOH + CH_3OH \overset{H^+}{\longleftrightarrow} HCOOCH_3 + H_2O$$
$$CH_3COOH + CH_3OH \overset{H^+}{\longleftrightarrow} CH_3COOCH_3 + H_2O$$

	Δhv [kJ/kg]	Boiling point [°C]
	Bp, 1.013 bar	1.013 bar
Water	2265	100
Formic acid	479	101
Acetic acid	398	117.9
Methanol	1097	64.6
Methyl formate	469	31.8
Methyl acetate	411	57

Esterification T = 40 °C, MeOH:Acid = 1

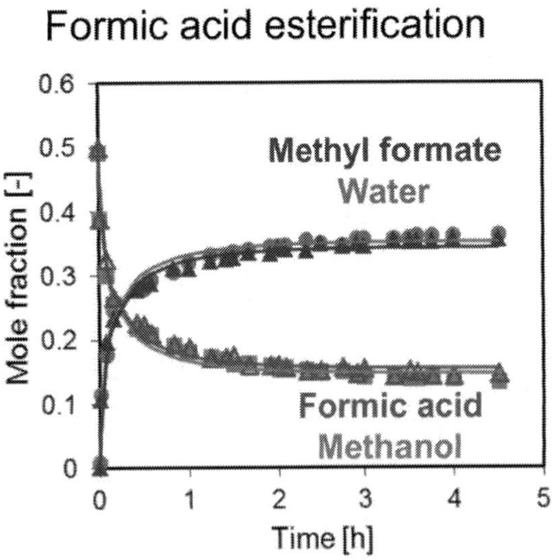

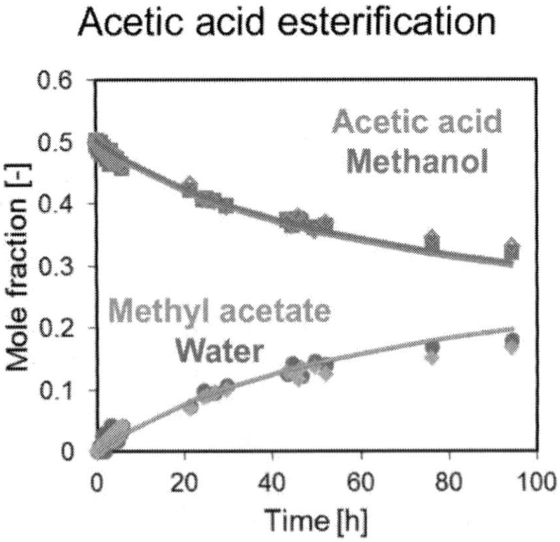

Esterification: Formic acid vs. Acetic acid

Forward reaction

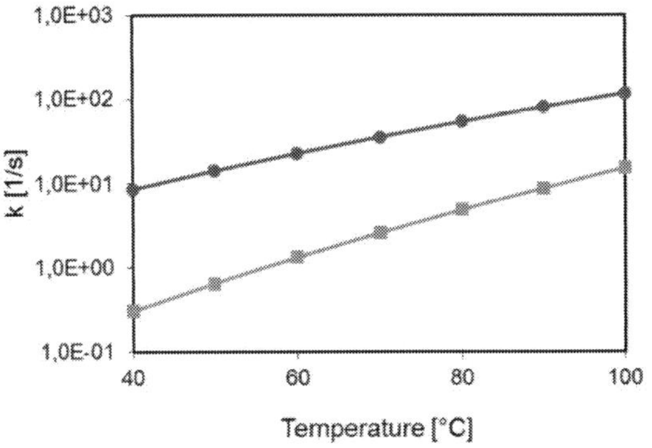

Back reaction

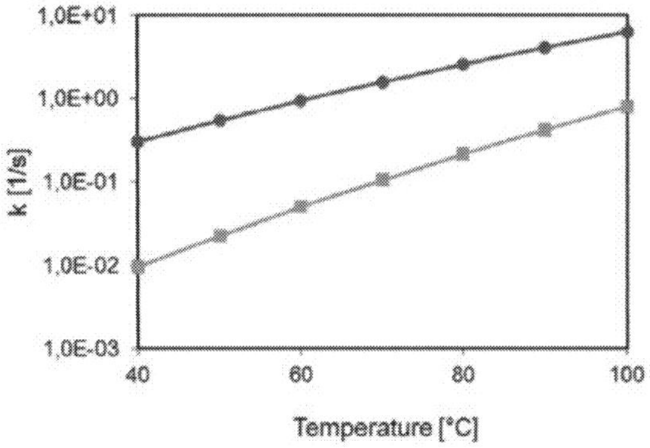

Formic acid esterification

Acetic acid esterification

Simultaneous esterification T = 50°C

$$-r_{\mathrm{A}} = \left(\frac{N_{\mathrm{ges}}}{V}\right)^3 \cdot \sum (a_{\mathrm{Acid}} \cdot K_{\mathrm{Diss}}) \cdot (k_{\mathrm{f}} \cdot a_{\mathrm{Acid}} \cdot a_{\mathrm{MeOH}} - k_{\mathrm{b}} \cdot a_{\mathrm{Ester}} \cdot a_{\mathrm{H_2O}})$$

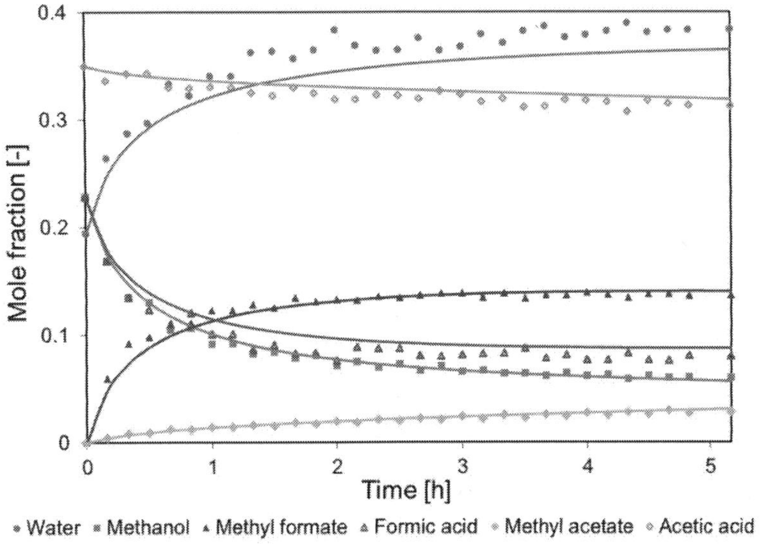

Esterification: Temperature effect

Formic acid esterification

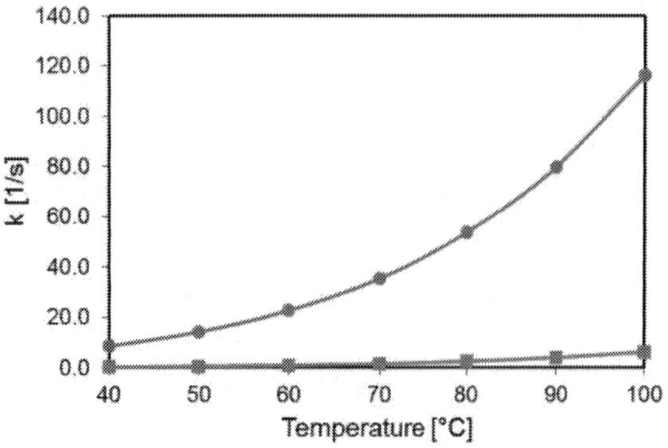

Acetic acid esterification

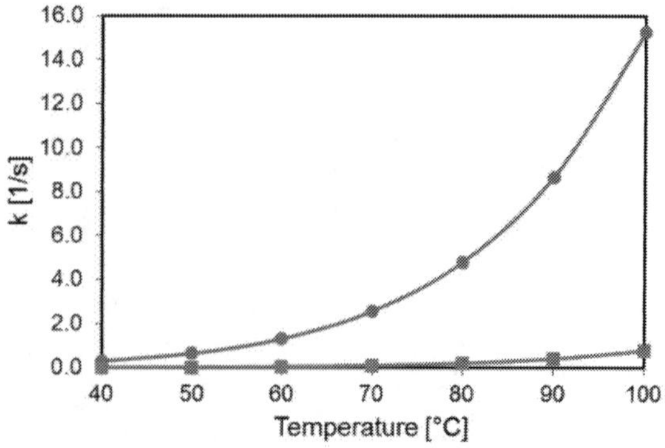

Forward reaction

Back reaction

Reactive Distillation

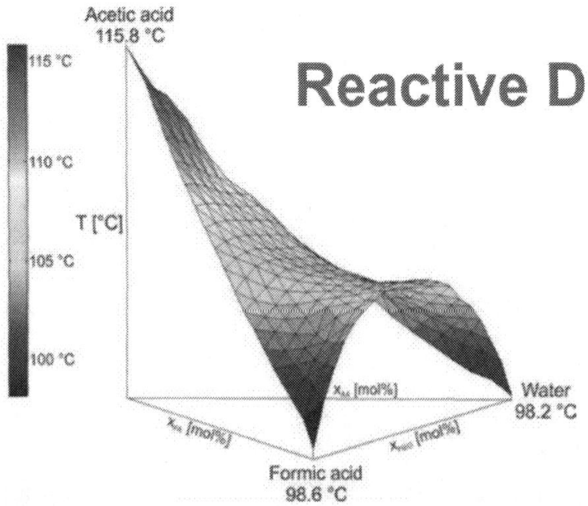

Continuous reactive distillation

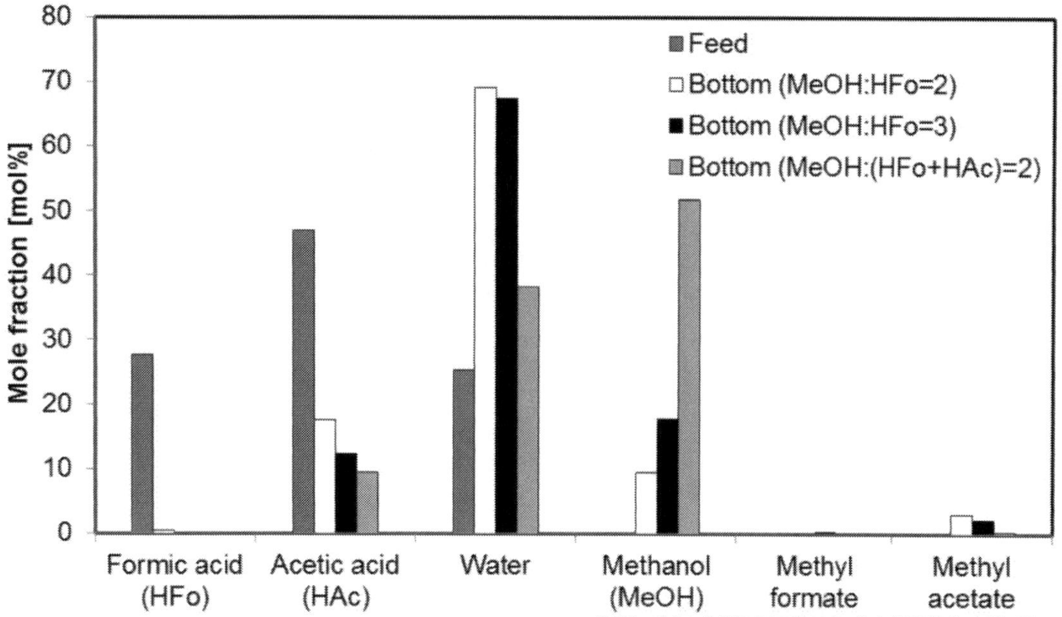

D. Painer, S. Lux, M. Siebenhofer: Chem. Ing. Tech. 2015, 87, No. 6, 843-847

Continuous reactive distillation

	MeOH:HFo=2	MeOH:HFo=3	MeOH:(HFo+HAc)=2
Distillate composition	**mol%**	**mol%**	**mol%**
Methanol	20.4	24.4	46.0
Methyl formate	41.6	33.5	27.6
Methyl acetate	38.0	42.1	26.2

Conversion	**%**	**%**	**%**
Formic acid conversion	98.0	99.3	98.7
Acetic acid conversion	60.8	71.5	64.4

D. Painer, S. Lux, M. Siebenhofer: Chem. Ing. Tech. 2015, 87, No. 6, 843-847

Solvent recovery via reactive distillation

PROOF OF CONCEPT

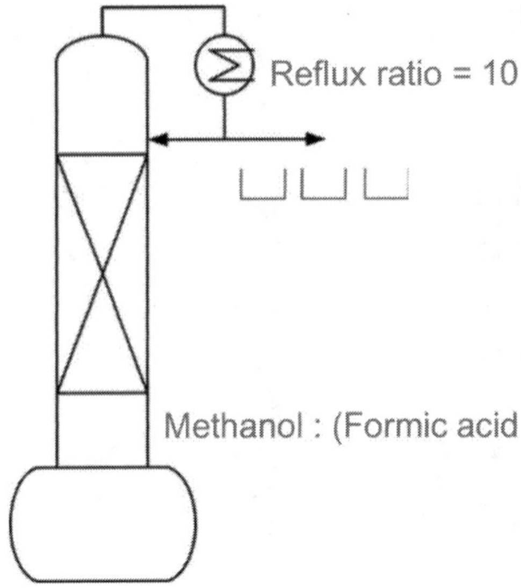

Solvent:

70 wt.% Cyanex®923
30 wt.% Undecane

Solvent Load:
13 wt.% Formic acid
24 wt.% Acetic acid
3 wt.% Water

Solvent recovery via reactive distillation

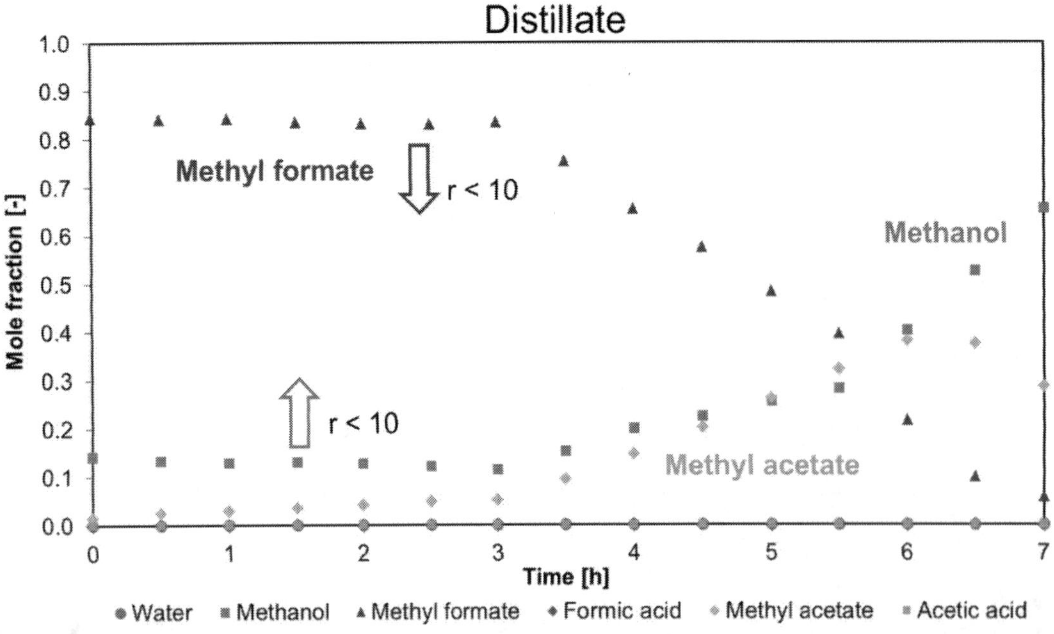

Solvent recovery via reactive distillation

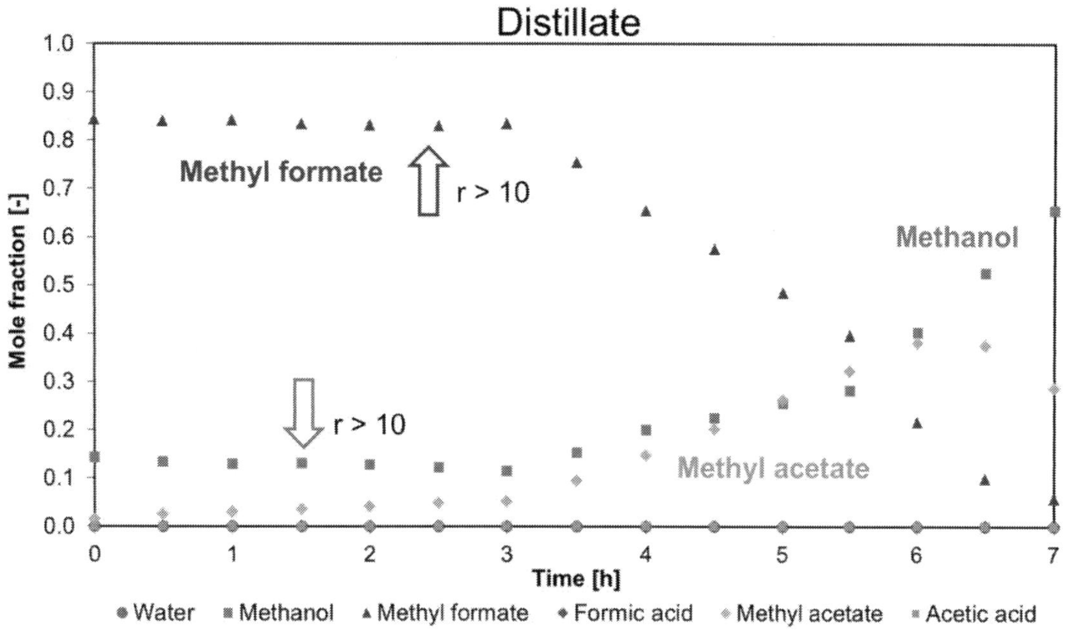

Continueous solvent recovery via reactive distillation

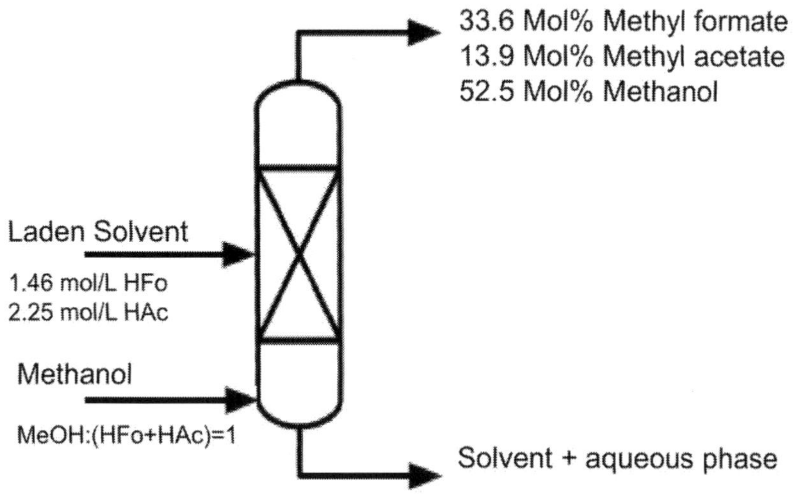

Continuous solvent recovery via reactive distillation

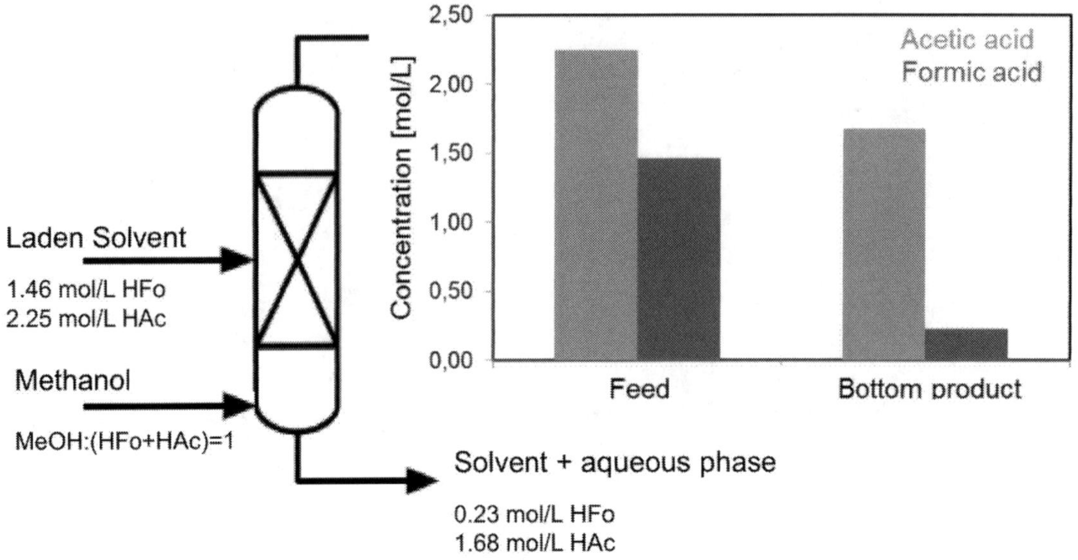

Continuous solvent recovery via reactive distillation

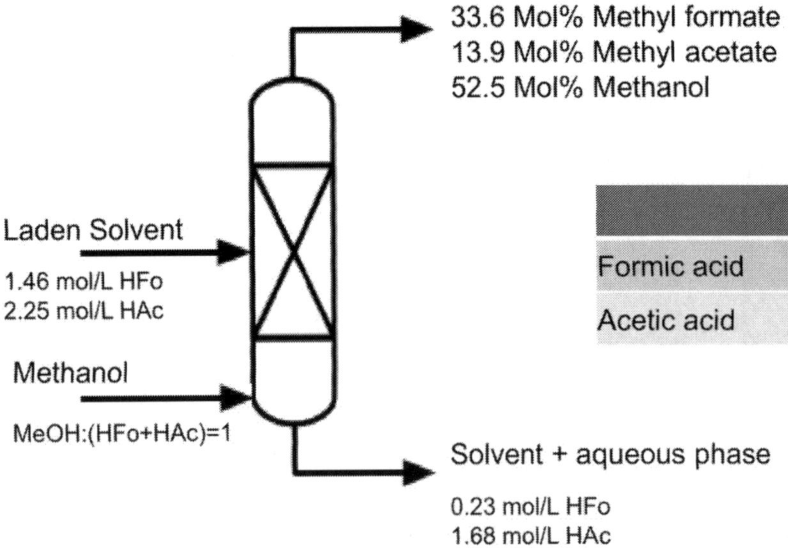

	Conversion
Formic acid	85.7 %
Acetic acid	31.2 %

Summary

- Challenge:
  - Dilute azeotropic mixture
- Process intensification solver strategy:
  - Change the properties
  - Change the state
- Reactive distillation
  - Conversion of formic acid = 99.3 %, Conversion of acetic acid = 71.5 %
  - Solvent recovery via reactive distillation is possible

Solvent Recovery via Reactive Distillation for Fatty Acid Extraction

D. Painer, S. Lux, M. Siebenhofer

Graz University of Technology
Institute of Chemical Engineering and Environmental Technology, NAWI Graz
Inffeldgasse 25c/I, A-8010 Graz
d.painer@tugraz.at

413820 Commercial Nanofiber Nonwoven As a Support for Thin Film Composite Membrane for Forward Osmosis

Abstract:

Product scale-up for any new emerging technology is key to seeing that technology succeed. Forward osmosis (FO) is one such technology that has been hampered by complications with scale-up. These challenges are largely limited to membrane manufacturing. Those companies that have seen some success with FO scale-up have developed FO membranes that resemble conventional RO membranes, though having certain chemical or structural differences. Many of the highly novel membranes, such as those widely discussed in the academic literature, will likely never see commercial reality because of the challenges associated with making those membranes in large quantity. Electrospun nanofiber supported thin film composite membranes are generally thought to belong to this group. Nanofibers have the properties sought after in FO TFC membrane support materials (high porosity, low tortuosity), and these membranes have shown great promise in the academic literature. However, the production of nanofiber mats at scale has been criticized. Only recently has industrial scale production of nanofibers from companies like DuPont, Donnaldson, and Hollingsworth & Vose brought nanofiber membranes into the commercial space. In this work, we demonstrate the use of a commercial nanofiber nonwoven based support media to manufacture thin film composite (TFC) membranes for forward osmosis. We evaluate the viability of this material by comparing it to laboratory grade nanofiber based TFC membranes and commercially available solution casted TFC membranes. The fiber diameter of the nanofiber mat ranges from 445± 135 µm. Scanning electron micrographs (SEM) unveils the flat and fused fibers due to calendaring with an overall porosity of 31%. ATR-FTIR spectrum confirms the successful formation of a polyamide selective layer by in-situ polymerization process and SEM imaging presents the uniformity of the polyamide layer. The TFC membrane has a very high Young's modulus, equivalent tensile strength, and low ductility as compared to a laboratory grade nanofiber based TFC membrane. The TFC membrane generates two times more water flux and ten times less salt flux as compared to a commercial HTI solution cast TFC membrane. These results proves that the commercial nanofiber support is capable of producing better membranes than commercial HTI TFC membrane while maintaining similar standards as the laboratory grade nanofiber supported TFC membrane.

419475 Highly Permeable Dual Selective Membranes for Promoted Use of Viscous Draw Solutes in Forward Osmosis

Abstract:

Recently, a wide range of large draw solutes have been successfully developed and employed in the forward osmosis (FO) process with the advantages of high water flux, minimum reverse flux and easy recovery. However, the high viscosity of these large draw solutes tends to suffer from more severe internal concentration polarization (ICP) effects and internal fouling of the draw solutes in the membrane sublayers. To solve this problem, a dual selective flat sheet membrane has been designed in this study in order to prevent the viscous draw solutes from entering the porous support layer and minimize the ICP effects. Dense and thin polyamide layers have been fabricated on both the top and bottom surfaces of highly porous cellulose ester support through interfacial polymerization (IP). The dual selective thin film composite (TFC) membrane demonstrated reasonably high water flux and low salt reverse flux using sodium chloride as draw solutions at the concentrations ranging from 0.25M to 1M. When tested with viscous draw solutes including sucrose, ferric citric acid complex (Fe-CA) and polyethylene glycol monolaurate (PEG 640ML), significantly mitigated ICP effects and reduced fouling propensity were observed compared to single selective TFC membranes. This study proves the great potential for the combined use of the newly developed dual selective membrane with viscous draw solutes for various FO applications.

423423 Novel Organic-Inorganic Thin Film Composite Membranes for Forward Osmosis Applications with Improved Water Flux and Anti-Fouling Performance

Abstract:

In this work, novel thin film composite (TFC) forward osmosis (FO) membranes are developed via interfacial polymerization on the polyacrylnitrile (PAN) substrate with blended diamines of m-phenylenediamine (MPD) and N-[3-(trimethoxysilyl) propyl]ethylenediamine (NPED) used. Effects of the PAN substrate hydrolyzed by sodium hydroxide for different time periods are studied on the membrane properties and performance of the resultant TFC membranes. By varying the monomer ratio of NPED and MPD in the aqueous phase, the membrane morphology and FO performance of the TFC membranes are also investigated. Both the PAN substrates and the TFC membranes are characterized by various techniques, including FTIR, SEM and water contact angle, etc. to examine changes in the chemical structures and the surface properties. The FO performance shows an improvement of the water flux and a sacrifice of the reverse salt flux when NPED/MPD mass ratio increases because of the enhanced hydrophilicity and larger free volume in the PA layer. In addition, the promising anti-fouling property of the TFC membranes with various MPD/NPED monomer ratios is also studied.

423451 Graphene Oxide Incorporated Thin-Film Composite Membranes for Forward Osmosis Applications

Abstract:

In this work, graphene oxide (GO) nanosheets with deep oxidation degree are incorporated into the polyamide selective layer to develop a novel thin-film composite (TFC) forward osmosis (FO) membrane with high water flux, good anti-biofouling property and reasonable draw solute rejection. The membrane morphology, surface hydrophilicity and FO performance of the GO embedded TFC membranes with different GO loading are investigated systematically. The influence of the draw solution concentration is also studied. The water flux of the GO incorporated TFC membrane exhibits nearly one time higher than that of the TFC membrane without embedded GO. The GO-embedded TFC membrane also possesses low fouling propensity in FO test. This study may provide important insights in developing novel TFC FO membranes in the future.

434669 Nanofibers in Osmotic Membranes for Sustainable Water and Energy

Abstract:

Nanofibers in Osmotic Membranes for Sustainable Water and Energy

Ngoc Bui, *Lawrence Livermore National Laboratory, Livermore, CA*

Jeffrey R. McCutcheon, *University of Connecticut, Chemical and Biomolecular Engineering Department, Storrs, CT*

To address global problems associated with water scarcity, new effective methods of producing water at low energetic and environmental costs are essential. Forward osmosis has recently gained increasing attention due to its various advantages in sustainable water, resource and energy production. Developing robust membranes with effective and controlled structures for osmotic transport is imperative for widespread adoption of this technology platform. Here we report a scalable approach to fabricate osmotic thin-film composite membranes with high permselectivity based on nanofibers and nanocomposite thereof. Nanofiber has recently emerged as an exciting platform for forward osmosis membranes. Possessing a high porosity, low tortuosity and an interconnected porous structure, nanofibrous mats exhibit exceptionally low structural parameters. This work highlights recent efforts to develop nanofiber-related materials for use in forward osmosis and pressure-retarded osmosis. Our best membranes showed an outstanding permselectivity, exceeding those reported in academic literature. Specifically, these membranes show a remarkable 7-fold and 3.5-fold enhancements in osmotic water permeability and water/sodium chloride selectivity, respectively, compared to standard commercial forward osmosis membranes. These new membranes, if tuned appropriately to further increase mechanical integrity, may help enable forward osmosis to effectively function in processes related to water treatment, desalination and power production.

411329 Enhance Dual Stage PRO Process for Power Generation

Abstract:

Abstract:

Enhanced Dual Stage PRO Process for Power Generation

Ali Altaee

Dual stage PRO process was evaluated for power generation using colossal salinity gradient resource. 2 M and Dead Sea brine were considered as the draw solution while fresh water and seawater of different salinities were the feed solutions of the PRO process. Closed-loop and open-loop PRO systems were investigated and power density was calculated in stage one and two of each PRO system. The impact of internal and external concentration polarization on the water flux was evaluated. As expected, the theoretical water flux was much lower than the actual water flux due to the effect of concentration polarization. Reverse salt diffusion was also calculated in open-loop PRO process which results in draw agent loss and lower process performance. The results show that dual stage PRO process is more efficient than conventional PRO process especially when the draw solution is concentrated brine. High efficiency Dual Stage PRO (DSPRO) process may replace the conventional PRO process in future because of its higher performance especially at high salinity gradient.

The present study investigates the application of dual stage PRO process for power generation using colossal salinity gradient resource. Impact of feed and draw solutions and operating parameters on the performance of dual stage PRO process was evaluated. Closed-loop and opened-loop dual stage PRO system were evaluated using 2 M NaCl and Dead Sea draw solutions respectively. The feed solution was either fresh water or seawater. The impact of concentration polarization and salt back diffusion was taken into account in the membrane flux calculation. The results showed that increasing feed salinity flow rates resulted in a trivial increase in the first stage permeate flow rate but there was a slight decrease of the second stage permeate flow rate. Whereas increasing the draw solution flow rate caused a tangible increase in the first and second stage permeate flow rates. Depending on the feed salinity and operating pressure, net power generation increased between 65% and 80% when the second stage was introduced. Increasing feed salinity from 32 g/L to 50 g/L didn't significantly affect the performance of dual stage PRO process because underperformance of the first stage was compensated in the stage of the PRO process. The net power generation was more than 90% of the gross power generation whereas power consumption was less than 10% of the gross power generation. Dual stage PRO process demonstrated high efficiency of power generation without compromising the net power consumption due to introducing the second PRO membrane process.

System-Level Analysis and Optimization of Pressure Retarded Osmosis for Power Generation

Mingheng Li

Department of Chemical and Materials Engineering
California State Polytechnic University, Pomona

minghengli@cpp.edu

Nov. 12, 2015

OUTLINE

1. Motivation
2. Simple Analysis in Literature
3. Analysis and Optimization Accounting for Flux Profile
   - Basic definitions
   - without Concentration Polarization
   - with Concentration Polarization
4. Concluding Remarks

MOTIVATION OF THIS WORK

- Pressure retarded osmosis is one of the promising "green" energy techniques that produces power from salinity gradient.
- There is a lack of computational framework to address scale-up issues in PRO
  - Optimal applied pressure much lower than $\Delta\pi/2$ is observed in some experiments (Xu et al., JMS, 2009; She et al., JMS, 2012; Sharif et al., Membranes, 2014).
  - Power generation in PRO is much lower than theoretical prediction.
- Mathematical models helps elucidate the understanding of energy issues in PRO.
  - Power density
  - Specific Energy Production (SEP)
  - Efficiency of osmotic to hydraulic energy

SIMPLE ANALYSIS BASED ON CONSTANT FLUX (see, e.g. Review by Helfer et al. JMS, 2014)

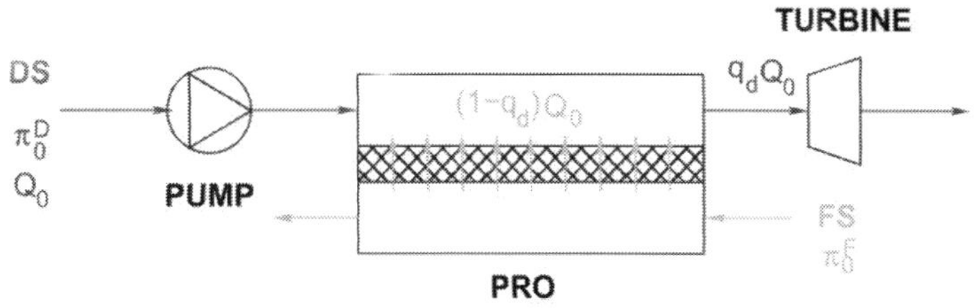

- Constant water flux along membrane

$$J_w = dQ/dA = L_p(\Delta\pi - \Delta P)$$

- Power Density (PD)

$$PD = J_w\Delta P = L_p(\Delta\pi - \Delta P)\Delta P$$

$$\Delta P_{opt} = \Delta\pi/2$$

This may be true for short membranes (i.e. under lab-scale experimental conditions)

BASIC DEFINITIONS IN PRO

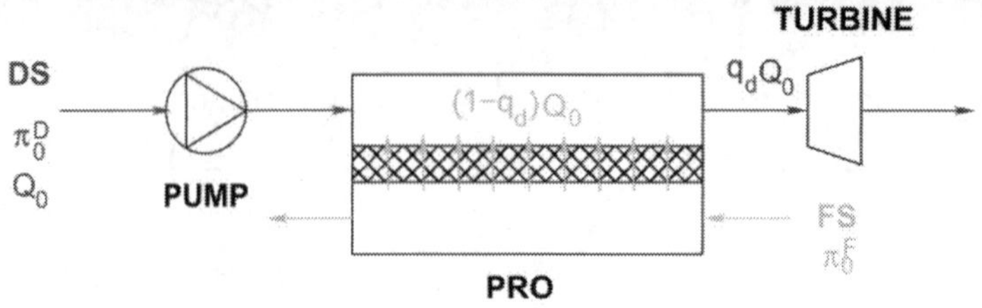

- Specific Energy Production (SEP)

$$SEP = Q_0(q_d - 1)\Delta P/Q_0 = (q_d - 1)\Delta P$$

q_d: dilution ratio at the end of the membrane
ΔP: applied pressure

- Normalized SEP or Osmotic to Hydraulic Efficiency

$$NSEP = SEP/\pi_0^D = (Q_0 SEP)/(Q_0\pi_0^D) = \eta_{O2H}$$

- Power Density (PD)

$$PD = (NSEP)(Q_0/A)\pi_0^D$$

1D MATHEMATICAL MODEL WITHOUT CP

- Local water flux (assuming $\pi^F = \pi_0^F$)

$$\frac{dQ}{dA} = L_p(\Delta\pi - \Delta P) = L_p(\pi_0^D\frac{Q_0}{Q} - \pi_0^F - \Delta P)$$

- Dimensionless form

$$\frac{dq}{dx} = \gamma\left(\frac{1}{q} - \theta\right)$$

$$q = Q/Q_0, \theta = (\Delta P + \pi_0^F)/\pi_0^D, \gamma = AL_p\pi_0^D/Q_0$$

- Solution (Li, AIChE J., 2015)

$$\gamma = \frac{1}{\theta}\left[1 - q_d + \frac{1}{\theta}\ln\frac{1-\theta}{1-q_d\theta}\right]$$

q_d: dilution ratio at the end of the membrane

SOLUTION TO q_d AND NSEP

- Profiles of q_d and NSEP

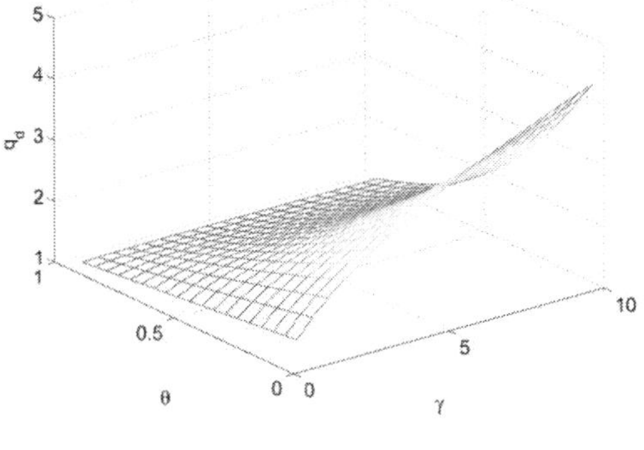

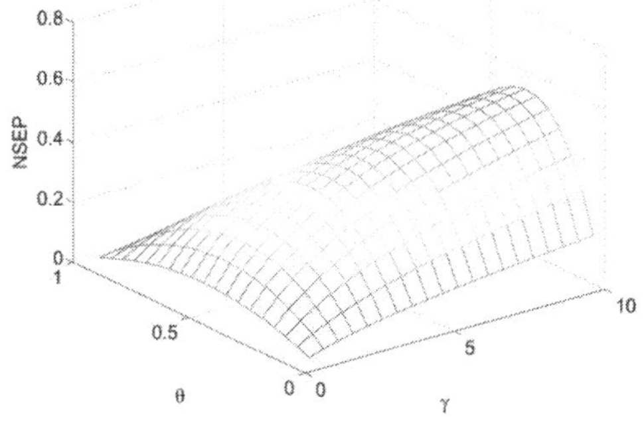

- Observations
  - q_d increases as ΔP reduces and/or γ increases.
  - At a fixed γ, there is an optimal ΔP corresponding the maximum NSEP.

CONSTRAINED NONLINEAR OPTIMIZATION

- Optimization model to determine optimal NSEP and ΔP

$$
\begin{aligned}
\max_{\alpha,z} NSEP &= (q_d - 1)\left(\frac{1}{\alpha} - r\right) \\
s.t. & \\
q_d &= \alpha - (\alpha - 1)e^{-z} \\
\gamma &= \alpha(1 - q_d + \alpha z) \\
1 - \alpha &\leq 0 \\
1 - q_d &\leq 0
\end{aligned}
$$

- Optimization of PD and optimization NSEP are essentially equivalent if Q_0/A and π_0^D are given, since $PD = (NSEP)(Q_0/A)\pi_0^D$.

CONSTRAINED NONLINEAR OPTIMIZATION OF NSEP

- Optimization results

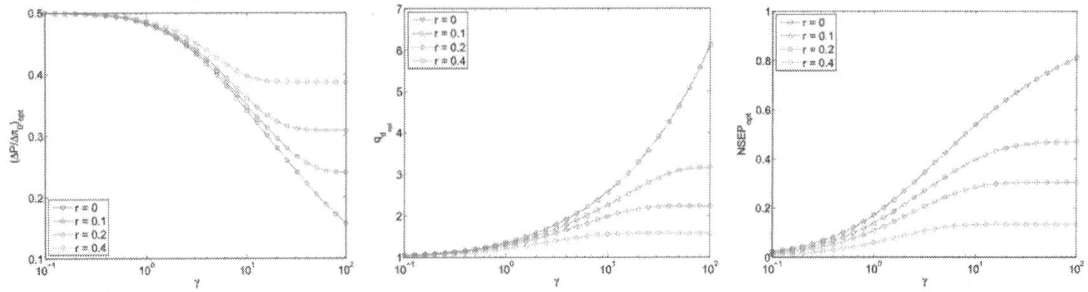

- Observations
  - The optimal ΔP shifts away from $\Delta\pi_0/2$ as $\gamma = AL_p\pi_0^D/Q_0$ increases.
  - An increase in r ($r = \pi_0^F/\pi_0^D$) *significantly* reduces q_d and NSEP.
  - When $r = 0$ (or fresh water is used as feed solution), the largest NSEP occurs at forward osmosis conditions, i.e., $\Delta P_{opt} = 0$.

OPTIMAL DRIVING FORCE IN PRO

- Optimal dimensionless driving force $\zeta = (\Delta\pi - \Delta P)/\pi_0^D$

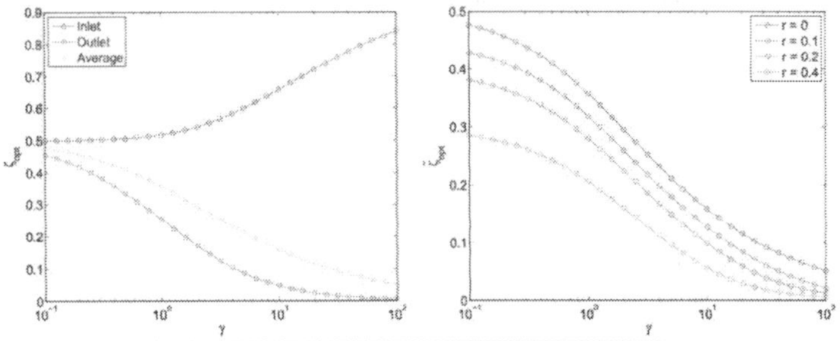

Inlet	Outlet	Average (ζ)
$1-\theta$	$1/q_d-\theta$	$(q_d-1)/\gamma$

- Similarity between RO and PRO
  - A larger γ allows the PRO to be operated closer to its thermodynamic limit, thus improving SEP.
  - A larger γ allows the RO to be operated closer to its thermodynamic limit, thus improving SEC (Li, IECR, 2010; Li, Desalination, 2012, Li, IECR, 2013).

NONLINEARITIES IN PRO

- Profiles of power efficiency and density when FS is fresh water

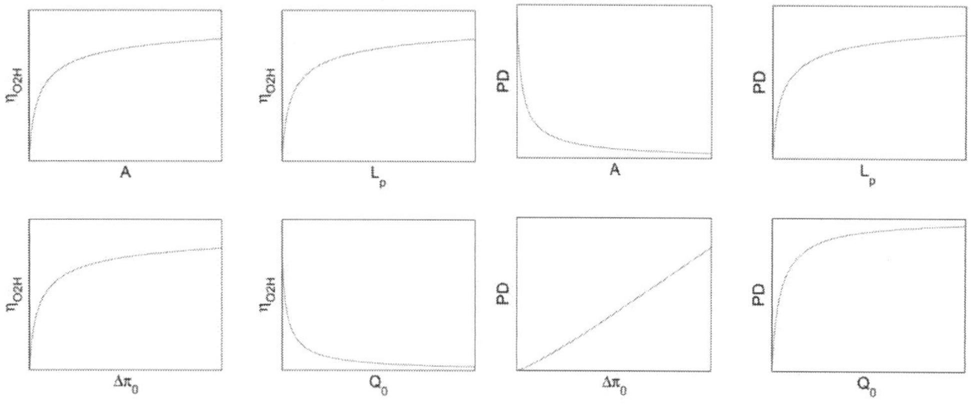

- Power density reduces as membrane areas increases. Therefore, its value in plant operation might be smaller than the one measured under lab-scale experimental conditions even if the same type of membrane is used.

PRO MODEL ACCOUNTING FOR CP

- Water flux accounting for concentration polarization (Achilli et al., JMS, 2009, Xu et al., JMS, 2010)

$$J_w = L_p \left[\pi_b^D \exp(-\frac{J_w}{k_m}) \frac{1 - \frac{\pi_b^F}{\pi_b^D} \exp(J_w K) \exp(\frac{J_w}{k_m})}{1 + \frac{B}{J_w}[\exp(J_w K) - 1]} - \Delta P \right]$$

$$\frac{dQ}{dA} = J_w$$

- Power Density

$$PD = \frac{\Delta P \int_0^A J_w dA}{A}$$

- J_w is in implicit form. Intensive computation is required to directly solve optimal ΔP.

FLUX IN EXPLICIT FORM UNDER TWO ASSUMPTIONS

- If $J_w/k_m << 1$ (Achilli et al., JMS, 2009) and $J_w K << 1$ (Lee et al., JMS, 1981), J_w may be approximated by (Li, AIChE J., 2015)

$$J_w \approx L_p'(\sigma\Delta\pi - \Delta P)$$

where $\sigma = 1/(1 + BK)$ and $L_p' = L_p/(1 + L_p\Delta\pi\sigma/k_m)$.

- Under such assumptions an analytical solution in PRO may be obtained (Li, AIChE J., 2015)

$$\gamma = \frac{1}{\theta}\left[(1 - q_d) + \frac{\sigma}{\theta}\ln\frac{\sigma - \theta}{\sigma - q_d\theta}\right]$$

where $\gamma = AL_p'Q_0/\pi_0^D$.

SHORT-CUT OPTIMIZATION ACCOUNTING FOR CP

- Optimization model

$$\begin{aligned} \max_{\alpha,z} NSEP &= (q_d - 1)\left(\frac{1}{\alpha} - r\right) \\ s.t. & \\ q_d &= \alpha\sigma - (\alpha\sigma - 1)e^{-z} \\ \gamma &= \alpha(1 - q_d + \alpha\sigma z) \\ 1/\sigma - \alpha &\leq 0 \\ 1 - q_d &\leq 0 \end{aligned}$$

- It is found that the model provides very accurate solution to ΔP_{opt}. However, flux profile, NSEP and PD may be better calculated using the original concentration polarization model and the derived ΔP_{opt}.

EFFECT OF ICP IN PRO

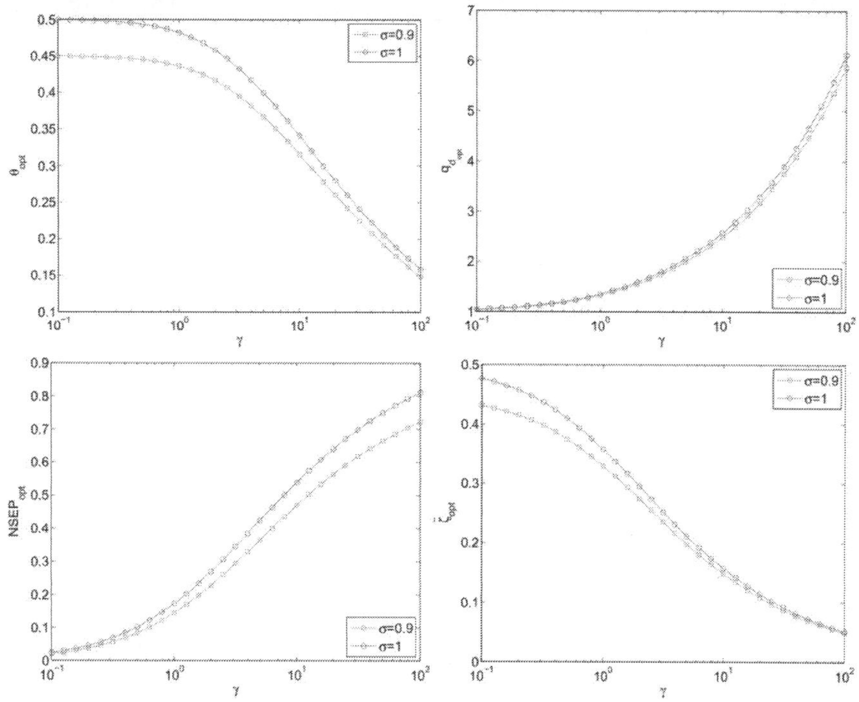

Mingheng Li AIChE Annual Meeting Nov. 12, 2015 15 / 21

CASE STUDIES

- Parameters used in the studies (Achilli et al., JMS, 2009)

Parameters	Value
k_m	8.48$\times 10^{-5}$ m/s
K	4.51$\times 10^{5}$ s/m
B	1.11$\times 10^{-7}$ m/s
L_p	1.87$\times 10^{-9}$ m/s/kPa
$\Delta\pi_0$	2763, 4882 kPa

- Four different Q_0/A (1, 2, 4, 8 $\times 10^{-6}$ m/s) are considered. The corresponding γ are between 0.5-10. These are comparable to the reverse process of industrial SWRO (Li, IECR, 2013).

Mingheng Li AIChE Annual Meeting Nov. 12, 2015 16 / 21

COMPARISON BETWEEN SHORT-CUT AND RIGOROUS OPTIMIZATION METHODS

Values obtained using the short-cut optimization method, if different from the rigorous method, are presented in parenthesis.

$\Delta\pi_0 = 2763$ kPa	ΔP, kPa	$\overline{J_w}$, μm/s	PD, W/m^2	γ	q_d	η_{O2H}, %
$\Delta\pi = \Delta\pi_0$	1330 (1316)	2.16 (2.18)	2.87	-	-	-
$Q_0/A = 8\mu$m/s	1300 (1295)	1.78 (1.79)	2.31	0.65	1.22	10.5
$Q_0/A = 4\mu$m/s	1260 (1259)	1.56 (1.56)	1.96	1.29	1.39	17.7
$Q_0/A = 2\mu$m/s	1190 (1186)	1.28 (1.28)	1.52	2.58	1.64	27.6
$Q_0/A = 1\mu$m/s	1070 (1067)	1.00 (1.01)	1.07	5.17	2.01	38.9
$\Delta\pi_0 = 4882$ kPa	ΔP, kPa	$\overline{J_w}$, μm/s	PD, W/m^2	γ	q_d	η_{O2H}, %
$\Delta\pi = \Delta\pi_0$	2390 (2325)	3.38 (3.47)	8.07	-	-	-
$Q_0/A = 8\mu$m/s	2260 (2244)	2.64 (2.66)	5.97	1.14	1.33	15.3
$Q_0/A = 4\mu$m/s	2140 (2133)	2.24 (2.25)	4.79	2.28	1.56	24.6
$Q_0/A = 2\mu$m/s	1950 (1943)	1.79 (1.80)	3.50	4.56	1.90	35.8
$Q_0/A = 1\mu$m/s	1680 (1685)	1.38 (1.38)	2.32	9.13	2.37	47.4

EFFECT OF DILUTION AND CP IN PRO

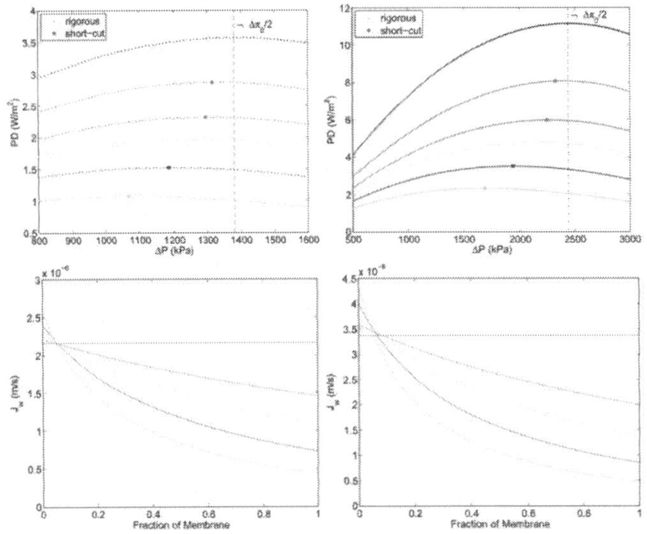

Black: $\Delta\pi = \Delta\pi_0$ and no concentration polarization.
Red: $\Delta\pi = \Delta\pi_0$ and with concentration polarization.
Magenta: $Q_0/A = 8 \times 10^{-6}$ m/s. Yellow: $Q_0/A = 4 \times 10^{-6}$ m/s.
Blue: $Q_0/A = 2 \times 10^{-6}$ m/s. Green: $Q_0/A = 1 \times 10^{-6}$ m/s.

CONCLUSIONS

- Characteristic equation in PRO to relate membrane properties, operating conditions and performance

$$\gamma = \frac{1}{\theta}\left[(1-q_d)+\frac{\sigma}{\theta}\ln\frac{\sigma-\theta}{\sigma-q_d\theta}\right]$$

where $\gamma = AL_p'Q_0/\pi_0^D$, $\theta = (\Delta P+\pi_0^F)/\pi_0^D$, $\sigma = 1/(1+BK)$.

- Short-cut optimization yields essentially the same solution as the rigorous solution. Moreover, it provides parameters to explain the effect of ICP, ECP and dilution in DS.
- Shift of optimal ΔP from $\Delta\pi_0/2$.
  - Dilution effect (i.e. γ is not zero).
  - Internal concentration polarization (i.e., $\sigma < 1$).
- Nonlinearities and conflicting power density and efficiency in process scale-up.

ACKNOWLEDGEMENT AND RELATED PUBLICATIONS

Financial Support from American Chemical Society Petroleum Research Fund (No. 55347-UR9) is gratefully acknowledged.

- Li, M. "Analysis and Optimization of Pressure Retarded Osmosis for Power Generation, " *AIChE Journal*, **61**, 1233-1241, 2015.
- Li, M. "Energy Consumption in Spiral Wound Seawater Reverse Osmosis at the Thermodynamic Limit," *Ind. Eng. Chem. Res.*, **53**, 3293-3299, 2014.
- Li, M. "A Unified Model-Based Analysis and Optimization of Specific Energy Consumption in BWRO and SWRO," *Ind. Eng. Chem. Res.*, **52**, 17241-17248, 2013.
- Li, M.; Noh, B. "Validation of Model-Based Optimization of Reverse Osmosis (RO) Plant Operation," *Desalination*, **304**, 20-24, 2012.
- Li, M. "Optimization of Multitrain Brackish Water Reverse Osmosis (BWRO) Desalination," *Ind. Eng. Chem. Res.*, **51**, 3732-3739, 2012.
- Li, M. "Optimal Plant Operation of Brackish Water Reverse Osmosis Water Desalination," *Desalination*, **293**, 61-68, 2012.
- Li, M. "Reducing Specific Energy Consumption in Reverse Osmosis Water Desalination: An Analysis from First Principles," *Desalination*, **276**, 128-135, 2011.
- Li, M. "Minimization of Energy in Reverse Osmosis Water Desalination using Constrained Nonlinear Optimization," *Ind. Eng. Chem. Res.*, **49**, 1822-1831, 2010.

421195 Molecular Design of High CO2 Reactivity and Low Viscosity Ionic Liquids for CO2 Separative Facilitated Transport Membrane

Abstract:

A facilitated CO_2 transport membrane is a membrane containing CO_2 carriers which react with CO_2 on the feed side, diffuse across the membrane and release CO_2 on the permeate side. This unique active transport by the CO_2 carriers allows it to show significantly higher CO_2 permselectivity compared to polymeric membranes. Ionic liquids, which are molten salts that exist as liquid around room temperature, have unique properties such as negligible vapor pressure, high thermal stability and easiness of designing the physical and chemical properties. In our group, facilitated transport membranes, which contain various CO_2 reactable ionic liquids, working as carriers and diffusion medium at the same time, have been studied and found to show promising performance. Although an ionic liquid is designable and there are many kinds of ionic liquid, however, it is challenging to find out the most suitable structure as a CO_2 carrier of facilitated transport membrane.

In this work, 10 ionic liquids composed of various combinations of anions (2-cyanopyrrolide ([2-CNpyr]$^-$), pyrrolide ([pyrr]$^-$), pyrazolide ([pyra]$^-$), 2-methoxypyrrolide ([2-Mtpyr]$^-$) and 3-methoxypyrrolide ([3-Mtpyr]$^-$)) and cations (tetraethylphosphonium ([P_{2222}]$^+$) and triethyl(methoxymethyl)phosphonium ([P_{222101}]$^+$)) were designed using molecular simulations as carrier candidates. Their properties before and after the reaction with CO_2 were calculated in order to find out the most suitable candidate for CO_2 separative facilitated transport membrane.

Molecular dynamics simulations were performed with LAMMPS in the isothermal-isobaric (*NpT*) and canonical (*NVT*) ensembles. Temperature was maintained at 350, 400, 450 and 500 K, respectively, by using the Nosé-Hoover thermostat. *NpT* simulations were performed for 2 ns to estimate the densities, after energy minimization. Subsequently, *NVT* simulations were conducted for 20 ns to calculate the diffusivities of ILs. In addition, 20 independent 3 ns *NVT* simulations (with initial configurations taken arbitrary from 20 ns *NVT* simulations) were run for viscosity calculation. In this study, 5 compositions of unreacted and reacted ionic liquids (unreacted, 25%, 50%, 75% and 100% reacted) were generated to express the extents of reaction in order to investigate the viscosity change after CO_2 absorption.

The viscosities at each temperature were calculated using the Green-Kubo relation

*fÅ=V,‹BT0‡Pabt0+t·Pabt0*dt

$$\eta = \frac{V}{k_B T}\int_0^{\infty} \langle P_{\alpha\beta}(t_0+t)\cdot P_{\alpha\beta}(t_0)\rangle \mathrm{dt} \quad (1)$$

where V is the volume of the system, k_B is the Boltzmann constant, T is the temperature and P_{ab} represents the off-diagonal components of the stress tensor. The angle brackets indicate the equilibrium average obtained by averaging over all the time origin t_0 and 6 components of P_{ab} (P_{xy}, P_{yz}, P_{xz}, $0.5(P_{xx}-P_{yy})$, $0.5(P_{yy}-P_{zz})$ and $0.5(P_{xx}-P_{zz})$). The viscosities at each time step were averaged over 20 simulation results to get converged integration of eq. (1). Then, viscosities between calculated temperatures were estimated using the Vogel-Fulcher-Tammann (VFT) equation

*fÅ=*A*T0.5expkT-T0*

$$\eta = AT^{0.5} exp\left(\frac{k}{T-T_0}\right) \quad (2)$$

where A, k and T_0 are fitting parameters. From the calculation, it was found that ionic liquids with P_{2221o1}^{+} cation showed lower viscosity than ionic liquids with P_{2222}^{+}, due to weaker electric interaction and larger free volume derived from electro donating and flexible methoxy group[1], allowing it to show higher mobility. [P_{2221o1}][pyrr] showed the lowest viscosity in this temperature range even after CO_2 absorption.

Furthermore, CO_2 reaction enthalpies of anions were calculated by using Gaussian09 (B3LYP/6-311+g(d,p)) after geometry optimization. The reaction enthalpies of 2-CNpyr$^-$,pyrr$^-$, pyra$^-$, 2-Mtpyr$^-$ and 3-Mtpyr$^-$ were -34.5, -98.8, -74.2, -62.7 and -93.8 kJ/mol, respectively. As the result, pyrr$^-$ was expected to have the highest reactivity, followed by 3-Mtpyr$^-$.

As a conclusion, [P_{2221o1}] [pyrr] can be expected as the most suitable ionic liquid for CO_2 separative facilitated transport membrane because of their high reactivity and low viscosity after CO_2 absorption.

***Reference*:**

[1] K. Tsunashima and M. Sugiya, *Electrochem. Commun.*, **9**, (2007), 2353-2358

Graphical Abstract

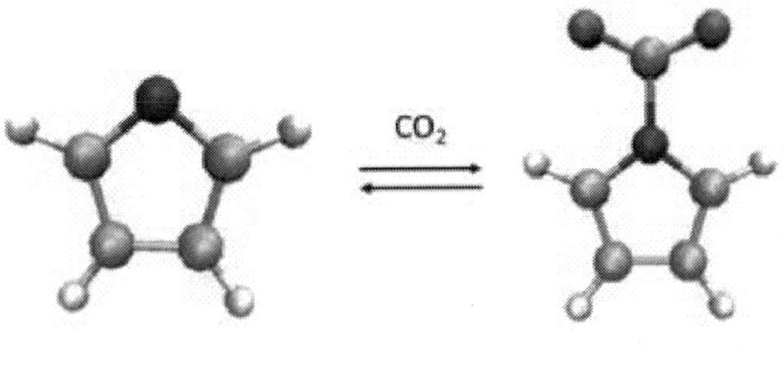

418604 Hydrogen Bond Acceptance Role in Designing Room Temperature Ionic Liquid (RTIL)-Membranes for the Treatment of Non-Polar, Non-Condensing Gasses

Abstract:

Prior research on the performance of RTIL-membranes theorized that membrane performance correlated with the RTIL properties of viscosity, molar volume, or the ability to chemically interact with the faster fluxing chemical species. Because RTIL-membranes are dominated by transport through the liquid phase, the lower the RTIL's viscosity the faster the transport by diffusivity. The other theories on membrane performance (molar volume and chemical interactions) relate to the RTIL-membrane selectivities being solubility dominated, instead of diffusion dominated. The literature reports on numerous attempts to increase the solubility of the faster fluxing species in order to increase the RTIL-membrane selectivities. However, it could also be fruitful to look at hindering the transport of the second species in the separation pair, for example nitrogen, air, or methane. Unfortunately, except for Carvalho and Coutinho's evaluation of only four ILs in 2011, this approach is the "road less traveled" in RTIL gas separations. In brief, Carvalho and Coutinho suggested that increasing the ionic liquid's hydrogen-bond acceptor ability would lead to lower methane solubility and, therefore, higher selectivities. To the best of our knowledge, the literature still does not contain any membrane research based on their proposal. Therefore, we tested of the hypothesis that rejection of methane will dominate in determining the water/methane selectivity of RTIL-membranes. We did find that RTILs that we fabricated with high hydrogen bond acceptance potential had significantly better performance (2700 GPU-water and water/methane selectivity = 100,000) compared to previously tested RTIL-membranes. However, their behavior and membrane fabrication is complicated. One hypothesis is that increasing the hydrogen-bond acceptance ability of the RTIL increases the ability of the RTIL to interact with and solubilize the polymers used in the liquid stabilizing porous matrix of the tested RTIL-membranes. We report a strong correlation ($R^2 = 0.98$) for an exponential fit between the water/methane selectivity and the Kamlet-Taft Parameter for hydrogen-bond accepting.

430321 Cellulose Triacetate Doped with Ionic Liquids for CO2/CH4 Separation

Abstract:

Polymeric membranes have been widely used for the removal of CO2 from natural gas. It is a leading technology for offshore platform applications and is gaining more competitive advantages over conventional amine absorption for onshore applications. The work horse membrane material for industrial CO2/CH4 separation is cellulose acetate/triacetate (CTA). The more widely use of CTA is, however, limited by its high crystallinity, leading to low CO_2 permeability. The goal of this work is to reduce the crystallinity in CTA by doping with ionic liquids such as [emim](BF_4) and [emim](DCA), to improve the CO2/CH4 separation properties. Both ionic liquids have CO_2/CH_4 solubility selectivity higher than CTA. In this project, we have prepared series of CTAs doped with various contents of [emim](BF_4) or [emim](DCA). The effect of ionic liquids on the crystal formation in CTAs was characterized using DSC and WAXD. Doping of ionic liquids can reduce or completely inhibit the crystal formation in CTA, resulting in an increase in gas solubility and permeability. The presentation will discuss the effect of the doping of ionic liquids on the physical properties of the blends, including morphology and pure-gas solubility, diffusivity and permeability.

430904 Heteroepitaxially-Grown Zeolitic Imidazolate Framework Membranes with Unprecedented Propylene/Propane Separation Performance

Abstract:

Heteroepitaxially-grown Zeolitic Imidazolate Framework Membranes with Unprecedented Propylene/Propane Separation Performance

Hyuk Taek Kwon[1], Panagiotis Krokidas,[3] Marcelo Castier,[3] Ioannis G. Economou,[3] Salvador Moncho Escriva,[4] Dusan Sredojevic,[4] Ed Brothers[4] and Hae-Kwon Jeong*[1,2]

[1]Artie McFerrin Department of Chemical Engineering and [2]Department of Materials Science and Engineering, Texas A&M University, College Station, TX 77843-3122

[3]Chemical Engineering Program and [4]Science Program, Texas A&M University at Qatar, PO Box 23874, Education City, Doha, Qatar

* Corresponding author: hjeong7@tamu.edu

Propylene/propane separation is one of the most challenging separations, currently achieved by energy-intensive cryogenic distillation. Despite the great potentials for energy-efficient membrane-based propylene/propane separation processes, no commercial membranes are available due to the limitations (i.e., low selectivity) of current polymeric materials.[1,2]

Zeolitic imidazolate frameworks (ZIFs) are promising membrane materials primarily due to their well-defined ultra-micropores with controllable surface chemistry along with their relatively high thermal/chemical stabilities.[3] In particular, ZIF-8 with the effective aperture size of ~ 4.0 A has been shown very promising for propylene/propane separation.[4] Despite the extensive research on ZIF-8 membranes, only a few of ZIF-8 membranes have displayed good propylene/propane separation performances[5-9] presumably due to the challenges of controlling the microstructures of polycrystalline membranes. Since the membrane microstructures are greatly influenced by processing techniques, it is critically important to develop new techniques.

Here, we report the first well-intergrown membranes of ZIF-67 (Co-substituted ZIF-8) by heteroepitaxially growing ZIF-67 on ZIF-8 seed layers. ZIF-67 membranes were sub-micron thick and well-intergrown, exhibiting impressively high propylene/propane separation capabilities. The presence of a methanol co-solvent in the secondary growth solution was critically important to achieve ZIF-67 membranes in a reproducible manner. Furthermore, when a tertiary growth of ZIF-8 layers was applied to heteroepitaxially-grown ZIF-67 membranes, the membranes exhibited unprecedentedly high propylene/propane separation factors of ~ 200 possibly due to enhanced grain boundary structure.

The new material was studied also computationally using molecular simulation. A new force field was generated for both ZIF-67 and ZIF-8 structures from ab initio calculations. Molecular Dynamics (MD) simulations of the diffusion of the two C_3 components show a great enhancement on the separation of propane/propylene in ZIF-67 compared with that in ZIF-8. A detailed study in the structural behavior shows that the presence of Co in the framework results in a shorter Me – linkage distance and "tighter" angles, which renders the pores an aperture with better sieving function. Further simulations using the Widom test particle insertion methodology shows that in both structures, the propane and propylene isosteric heat of adsorption are almost the same. Therefore, the separation of this mixture in ZIF-67 and ZIF-8 is primarily a diffusion-based process.

References and Notes

1. Baker RW. Future Directions of Membrane Gas Separation Technology. *Industrial & Engineering Chemistry Research.* 2002/03/01 2002;41(6):1393-1411.

2. Burns RL, Koros WJ. Defining the challenges for C3H6/C3H8 separation using polymeric membranes. *Journal of Membrane Science.* 1/31/ 2003;211(2):299-309.

3. Park KS, Ni Z, Côté AP, et al. Exceptional chemical and thermal stability of zeolitic imidazolate frameworks. *Proceedings of the National Academy of Sciences.* July 5, 2006 2006;103(27):10186-10191.

4. Zhang C, Lively RP, Zhang K, Johnson JR, Karvan O, Koros WJ. Unexpected Molecular Sieving Properties of Zeolitic Imidazolate Framework-8. *The Journal of Physical Chemistry Letters.* 2012/08/16 2012;3(16):2130-2134.

5. Brown AJ, Brunelli NA, Eum K, et al. Interfacial microfluidic processing of metal-organic framework hollow fiber membranes. *Science.* July 4, 2014 2014;345(6192):72-75.

6. Hara N, Yoshimune M, Negishi H, Haraya K, Hara S, Yamaguchi T. Diffusive separation of propylene/propane with ZIF-8 membranes. *Journal of Membrane Science.* 1/15/ 2014;450(0):215-223.

7. Kwon HT, Jeong H-K. In Situ Synthesis of Thin Zeolitic–Imidazolate Framework ZIF-8 Membranes Exhibiting Exceptionally High Propylene/Propane Separation. *Journal of the American Chemical Society.* 2013/07/24 2013;135(29):10763-10768.

8. Liu D, Ma X, Xi H, Lin YS. Gas transport properties and propylene/propane separation characteristics of ZIF-8 membranes. *Journal of Membrane Science.* 2/1/ 2014;451(0):85-93.

9. Pan Y, Li T, Lestari G, Lai Z. Effective separation of propylene/propane binary mixtures by ZIF-8 membranes. *Journal of Membrane Science.* 2/15/ 2012;390–391(0):93-98.

417738 Effect of Cross-Linker on 6FDA-Durene-ZIF71 Mmms for Hydrogen/Carbon Dioxide Separation

Abstract:

Interfacial void induced low selectivity is a critical drawback for mixed matrix membranes (MMMs) applied to gas separation. In this study, post-synthetic functionalization of fabricated MMMs by TAEA vapor-induced covalently interconnecting polymeric networks (CIPNs) technology has been employed to molecularly tune the 6FDA-Durene-ZIF-71 MMMs from CO_2-selective into H_2-selective via enhancement of the molecular-sieving properties. The new selective layer comprise of polyamide-based matrix and ZIF-71, interfacial void was refilled by vaporous TAEA treatment.

Both X-ray Photoelectron Spectroscopy (XPS) and Doppler Broadening Energy Spectroscopy (DBES) results reveal that non-selective interfacial voids between particle-particle and particle-polymer have been mitigated by construction of vapor-induced based CIPNs inside the ZIF-71 based MMMs. TGA-IR demonstrated the high thermal stability of resultant membranes and the feasibility of applying the fabricated membranes in a high temperature H_2/CO_2 separation application. Embedding 10 wt% of ZIF-71 nano-particles with size of less than 100 nm inside the TAEA-treated 6FDA-Durene membrane further enhances the pure H_2 permeability by 80% (from 300 Barrer to 540 Barrer) and the ideal H_2/CO_2 selectivity by 34% (from 46.8 to 62.5). The newly developed technology of tripodal amine-based vapor induced CIPNs may solve the most challenging issue in MMMs fabrication and may molecularly tune the MMMs to fit in a specific industrial gas separation application.

419840 Attainability and Energy Consumption Studies of Membrane Processes

Thursday, November 12, 2015: 2:15 PM
155B (Salt Palace Convention Center)

Zhiping Lai, Physical Science and Engieering, KAUST, Thuwal, Saudi Arabia and Ali Alshehri, Physcial Science and Engineering, KAUST, Thuwal, Saudi Arabia

Attainability and energy consumption studies of membrane processes

Ali Alshehri, Zhiping Lai*

Advanced Membrane and Porous Materials Center, Division of Physical Science and Engineering, King Abdullah University of Science and Technology, Saudi Arabia

After more than 50 years intensive studies in membrane materials using modern technologies, the performance of existing membranes has been dramatically improved. As a result, membrane technology has been successfully commercialized in many challenging separation processes such as seawater desalination, natural gas separation, CO_2 capture, and recovery of hydrocarbons, etc. With more and more experiences accumulated from these commercial membrane processes, conceptual design of a membrane process becomes feasible. We built a membrane module embedded in a commercial chemical engineering software, Aspen plus®. The membrane module included two membrane models. One is the simplest well mixed model where both feed and permeate sides are considered well-mixed. This model can be solved analytically, so it can be used to study the most important membrane process parameters. The other model is hollow fiber in counter flow configuration. This model requires complicated numerical solution but the results are more close to real membrane processes. Using this tool we can perform process simulation of not only membrane processes but also hybrid systems that combine membrane process with other separation processes such as distillation. Two important results shown in Figure 1 are obtained from the simulation results of a single-stage membrane process: (1) membrane selectivity S and pressure ratio g are the two control parameters and both exhibit a minimum value in order to meet a separation task that is defined by product purity and recovery ratio; (2) the minimum energy consumption is a monotonic function of the membrane selectivity. A minimum membrane selectivity is required in order to compete with a distillation process.

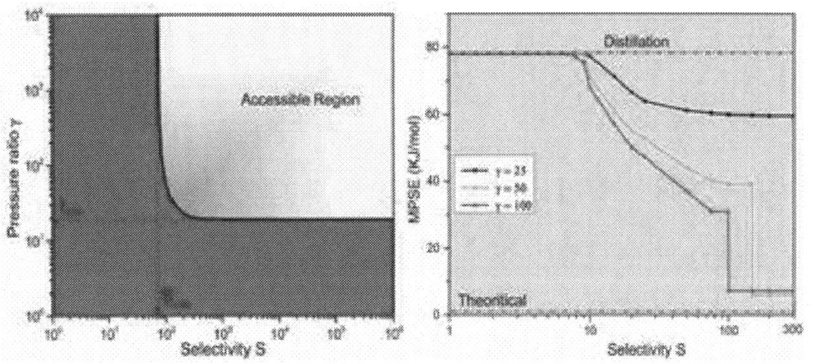

Figure 1: Left, the attainability of single stage membrane process is defined by selectivity and pressure ratio; right, the minimum energy assumption decreases monotonically with selectivity.

For multi-stage cascade membrane processes, the attainability is determined not only by selectivity and pressure ratio, but also by the recycle ratio which is defined as the ratio of flow rate of the stream recycled from one stage back to the previous stage to the flow rate of the feed. When the recycle ratio is 0, multi-stage membrane processes will decay to a single-stage membrane process; while if the recycle ratio increases to infinity, the minimum selectivity and pressure ratio of a n-stage membrane process have the following relationships with that of a single-stage membrane process: $S_n = \sqrt[n]{S_1}$, $\gamma_n = \sqrt[n]{\gamma_1}$. The minimum energy consumption of the multiple stage membrane processes also decreases monotonically with membrane selectivity.

425575 Modeling Gas Sorption in Glassy Hab-6FDA and in Its Thermally Rearranged Homologues

Abstract:

Modeling gas sorption in glassy HAB-6FDA and in its thermally rearranged homologues

Michele Galizia, Kevin A. Stevens, Zachary P. Smith, Donald R. Paul, Benny D. Freeman

J.McKetta Department of Chemical Engineering, 200 E. Dean Keeton Street, 78712 Austin,

and Center for Energy and Environmental Resources, 10100 Burnet Road, 78758 Austin,

The University of Texas at Austin, Texas (USA)

Due to their outstanding transport properties and unusual combination of high gas permeability and selectivity, thermally rearranged polymers have attracted a great interest in membrane science, especially for CO_2/CH_4 separation [1]. These polymers, first reported by Park et al. [2], are formed upon conversion of solvent-soluble polyimides, via thermal rearrangement, to insoluble polybenzoxazoles. The resulting materials show separation properties that, in some cases, surpass the upper bound defined by Robeson in 2008 [3].

Many research efforts have been devoted in the last decade to the fundamental understanding of gas transport properties of these polymers. In this work, for the first time, we attempted to analyze the sorption properties of TR-polymers from a theoretical point of view, by using the Non-equilibrium Lattice Fluid theory. Unfortunately, the characteristic lattice fluid parameters are not available in the literature for the pristine polyimide HAB-6FDA, nor for the corresponding thermally rearranged samples. These parameters are usually determined by fitting the polymer PVT data in the rubbery region to the Sanchez-Lacombe equation of state. However, thermodynamic data in the rubbery state are not accessible for these high performance polymers, thus the scaling parameters T^*, p^* and r^* were estimated by using an alternative approach first reported by Galizia et al. [4]. Interestingly, the model is able to describe with good precision the sorption isotherms of hydrogen, nitrogen, methane and carbon dioxide in both the pristine polyimide and corresponding thermally rearranged samples.

[1] D.F. Sanders et al., J. Membr. Sci. 2012 409-410 232-241

[2] H.B. Park, Science 2007 318 254-258

[3] L.M. Robeson, J. Membr. Sci. 2008 320 390-400

[4] M. Galizia et al., J. Membr. Sci. 2012 405-406 201-211

431138 Microfluidic Platform for Cocrystal Screening and Characterization

Abstract:

Microfluidic Platform for Cocrystal Screening and Characterization

Elizabeth Horstman,[1] Geoff G. Z. Zhang,[2] Yuchuan Gong,[2] and Paul J. A. Kenis[1]

[1]Department of Chemical & Biomolecular Engineering, University of Illinois at Urbana-Champaign, 600 South Mathews Avenue, Urbana, Illinois 61801,USA.

[2]Materials Science, Global Pharmaceutical R & D, AbbVie Laboratories, 1401 Sheridan Road, NCR13-317B, North Chicago, Illinois 60064, USA.

In pharmaceutical drug discovery, thousands of compounds are screened to determine if the active pharmaceutical ingredients (APIs) have appropriate physical and chemical properties. Most APIs on the market are crystalline so in depth solid form screening is done to determine the experimental conditions needed to produce the desired solid form (polymoph, cocrystal, or salt) of the API. Unfortunately, at the early stages of drug discovery only small quantities of APIs are available for testing, limiting the number of screening conditions. To address this challenge, we have developed microfluidic platforms for crystallization of pharmaceutical solid forms.[1-2] Utilizing a microfluidic platform allows the quantity of API needed per condition to be reduced from 5 mg to 5 µg per condition. Additionally, microfluidic platforms can be designed to be compatible with solid form analysis instruments used to differentiate solid forms such as Raman spectroscopy or X-ray diffraction.[3]

An increasing number of APIs in the pipeline have low aqueous solubility. To address this issue, APIs can be crystallized with other small molecules to produce a cocrystal. We developed a simple array based microfluidic platform comprised of single isolated microwells for high throughput screening. In this work, we present the discovery and characterization of cocrystals of piroxicam (PRX) and 2,5-dihydroxybenzoic acid (HBA). Two cocrystals were identified *via* solvent evaporation and subsequently characterized with X-ray diffraction. One was a 1:1 molar raito of PRX-HBA and the other was a solvated cocrystals containing acetone in a 1:1:1 molar ratio, PRX-HBA-ACT. The inclusion of acetone was unexpected since acetone was not used in the crystallization experiment and not identified until the structure of the cocrystal was determine. To further characterize the thermal properties of the cocrystals slurry crysalliztion was used to make bulk quantities (gram scale) of the cocrystals. Thermogravimetric analysis and differential scanning calorimetry were used to determine the desolvation temperature of PRX-HBA-ACT and the melting temperatures of the cocrystals. Powder X-ray diffraction data was taken as a function of temperature to determine the phase behavior of PRX-HBA-ACT during desolvation. Overall this work points to the importance of structure determination of pharmaceutical solid forms and the importance on impurities.[4]

References

1. M.R. Thorson, P.J.A. Kenis, et al. Lab Chip, 2011, 11, 3829-3837

2. M.R. Thorson, P.J.A. Kenis, et al. CrystEngComm, 2012, 14, 2404-2412

3. E.M. Horstman, P.J.A. Kenis, et al. Cryst. Growth Des., 2015, 15 , 1201–1209

4. E.M. Horstman, P.J.A. Kenis, et al. CrystEngComm, 2015, *DOI: 10.1039/c5ce00355e*

416468 Identification and Crystallization Development of Potential Conformational Polymorphs of an Active Pharmaceutical Ingredient

Abstract:

Solid form control is essential in the development of active pharmaceutical ingredient (API) as polymorphs may have different physical properties that could impact the drug product performance and processing. Only one polymorph (Form I) was initially reported in the early solid form screen of a development compound (Compound A). However, during the crystallization development and scale-up batches, very minor XRPD differences were observed between batches, which triggered a comprehensive solid form characterization using advanced solid state tools like ssNMR and Raman. It was found that a new polymorph (Form II) was identified in the crystallization development that had very similar XRPD pattern to Form I. Based on ssNMR and Raman data, Form I and Form II are likely conformational polymorphs though single crystal structure data have yet available for confirmation. Relative stability study between Form I and Form II showed that Form II is the most stable polymorph, which was later designated as the new development form.

Extensive efforts were made to develop a crystallization process that can produce the new Form II. A process was successfully developed in IPA/H_2O with seeding, extended H_2O addition and aging and in-process control (by SSNMR/Raman). This process was successfully applied to all the subsequent scale-up batches with good polymorph control, purification and recovery yield. Systematic process robustness studies on the crystallization and polymorph control, as well as efforts to modify the powder property for the drug product formulation, will also be discussed in the presentation.

436962 Depiction and Prediction of Practical Pathways for Crystallization

Abstract:

For many systems in which phase transformation is driven by supersaturation, the limits of viable processing conditions are conventionally displayed on a metastable zone width (MSZW) diagram. This type of diagram consists of a two-dimensional space defined by concentration and temperature axes, on which a metastable limit (line) and a solubility limit (line) are superimposed. The metastable limit corresponds to conditions for which nucleation is so rapid that it is seemingly instantaneous. The solubility limit corresponds to an equilibrium in which no nucleation occurs: there is no driving force and hence the time for transformation approaches infinity. The concentration difference between these limits at fixed temperature, or the temperature difference between these limits at fixed concentration, serves to define the MSZW. The limits are typically obtained from experimental observation (i.e. the diagrams are descriptive rather than predictive). Also, the MSZW is found to depend on the supersaturation *rate*. Although MSZW diagrams are useful, the absence of a time axis is a significant constraint to their application in process design. Kinetic data are intrinsically time-dependent, and this attribute is not conveyed in conventional MSZW diagrams.

We have recently [1] shown how the kinetics of supersaturation-driven phase transformations can be richly displayed on time-concentration-temperature-transformation (TCTT) diagrams, derived with reference to the time-temperature-transformation (TTT) diagrams that have long been used by metallurgists to summarize and develop processes used in steelmaking. The value of such diagrams lies in their ability to clearly suggest time-efficient process pathways to a final product, straightforwardly promoting desired structures and properties while also suppressing undesired characteristics and properties. Metallurgy and industrial pharmacology share a common goal in this regard, and transformation diagrams can avoid the trial-and-error aspects of the MSZW approach.

Another attractive aspect of TCTT diagrams is that (like TTT diagrams) they lend themselves to prediction, thus offering the possibility of adding further efficiency to process design. We have combined (i) classical modeling (Kolmogorov-Johnson-Mehl-Avrami) of nucleation and growth and (ii) modeling (Flory-Warner and Maier-Saupe) of the isotropic-to-nematic liquid crystal phase transformation for rod like mesogens as a function of concentration and temperature, to illustrate how specific TCTT diagrams for liquid crystalline systems can be developed from first principles. We propose that prediction of TCTT and similar diagrams can serve as a convenient guide for process design in polymer crystallization, pharmaceutical product crystallization, food processing, and many other phase transformation processes.

[1] J. Komadina, S.W. Watt, I.J. McEwen and C. Viney, Rate of lyotropic nematic phase formation: derivation and application of time-concentration-temperature-transformation diagrams, **DOI:** 10.1021/cg501459m, (2015).

AIChE Annual Meeting, 8th -13rd November 2015,
Salt Lake City, US

Tailoring Polymorphism of Ortho-Aminobenzoic Acid (OABA) Using Combinations of Several Solvents and a Structurally Related Additive

E. Simone[1], G. Steele[1,2], Z.K. Nagy[1,3]
e.simone@lboro.ac.uk

[1] Department of Chemical Engineering, Loughborough University, Loughborough, LE11 3TU, UK
[2] PharmaCryst Consulting Ltd, 13 Ashleigh Drive, Loughborough, Leicestershire, LE11 3HN, UK
[3] School of Chemical Engineering, Purdue University, West Lafayette, IN 47907-2100, USA

Loughborough University

Overview

- Introduction:
  - Polymorphism of OABA
  - OABA equilibrium in solution
  - Additives
- Methodology
  - Loughborough University rig
  - Experimental design
- Results and Discussion
  - Polymorphism
  - Shape and morphology
- Conclusions

Polymorphism of OABA (1)

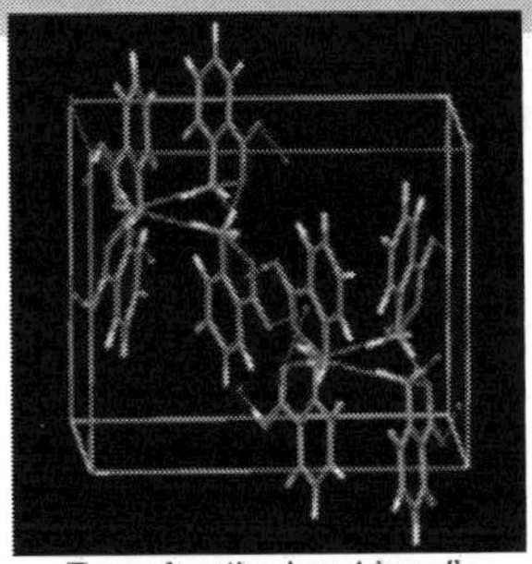

Form I: orthorhombic cell, **zwitterions** and neutral molecules (1:1 ratio)

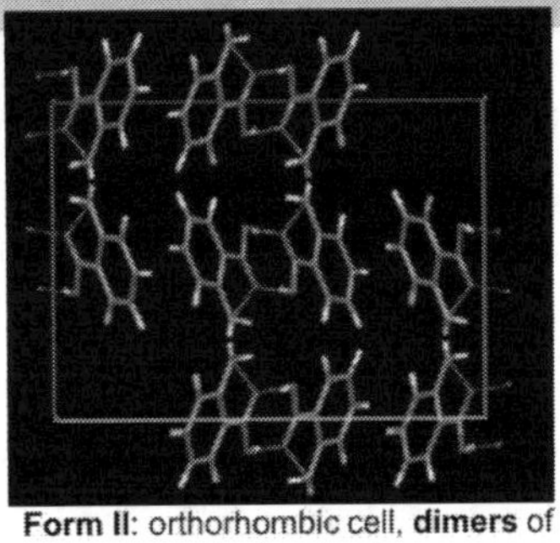

Form II: orthorhombic cell, **dimers** of neutral molecules

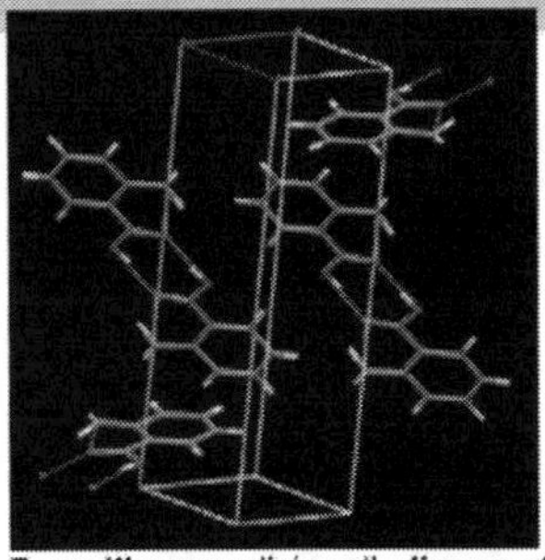

Form III: monoclinic cell, **dimers** of neutral molecules

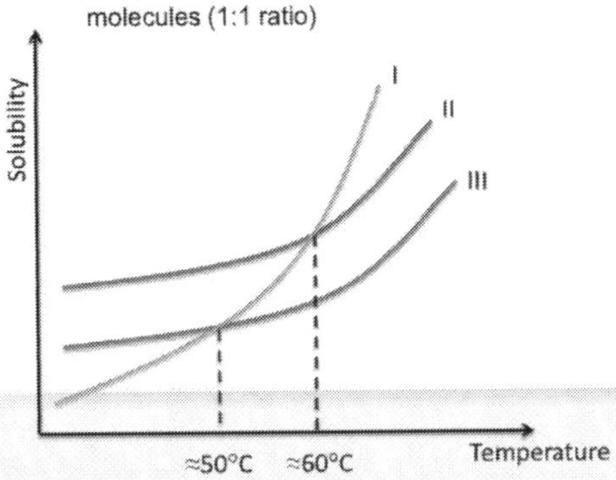

- Three known forms of OABA
- Form I and II are normally nucleated from solution
- Form III is very difficult to nucleate but can be obtained by melt crystallization or polymorphic transformation above 60° C

Jiang et al., *Crystal Growth and Design*, 2010

Loughborough University

Equilibrium of OABA in water

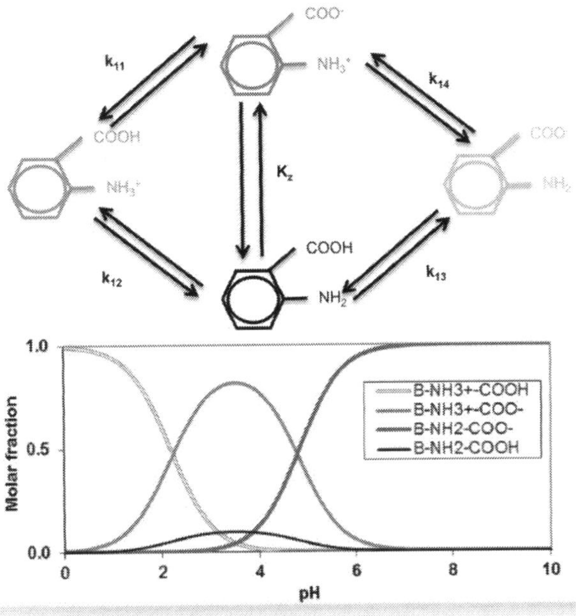

- OABA can exist in **solution** as a neutral molecule, with positive or negative charge or as a **zwitterion**
- Different forms of OABA presents different UV/Vis spectra

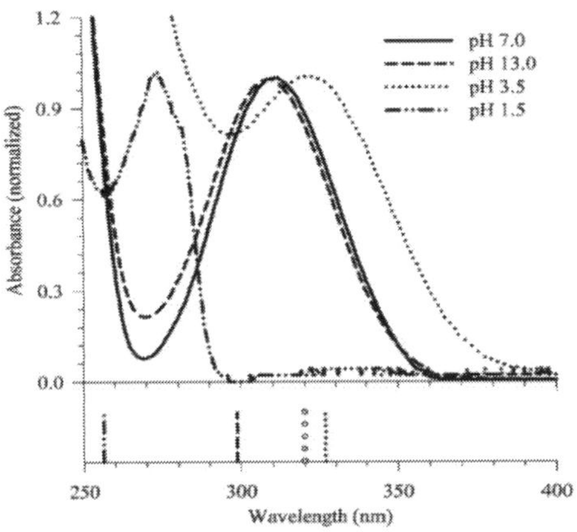

Zapala et al., *Biophysical Chemistry*, **2013**

Abou-Zied et al., *The Journal of physical Chemistry*, **2009**

Loughborough University

Equilibrium of OABA water and organic solvents mixtures

- Changes in the **UV/Vis spectrum** of OABA solutions can be associated with ***different amount of zwitterions in solution***

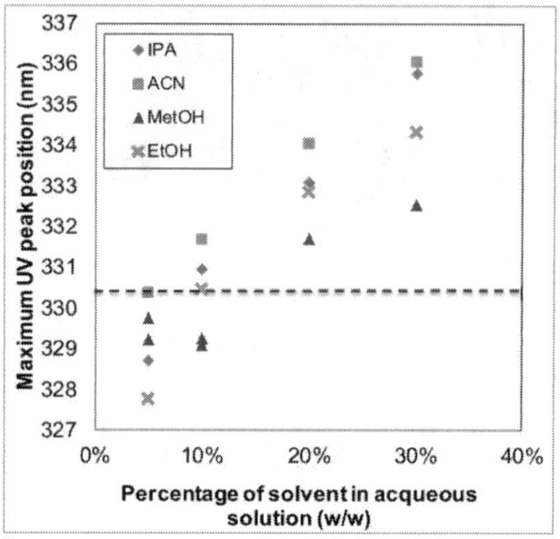

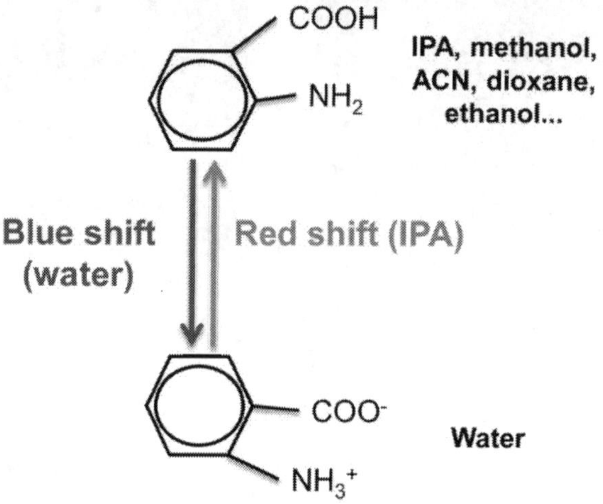

- Abou-Zied et al. concluded that the UV/Vis and fluorescence spectra are associated with the polarity of the solvent

Zapala et al., *Biophysical Chemistry*, **2009**
Abou-Zied et al., *The Journal of physical Chemistry*, **2013**
E. Simone and Z.K. Nagy, *CrystEngComm*, **2015**

Additives effect

- Additives can inhibit growth and/or nucleation of the crystals and modify their **morphology**
- They can also affect **polymorphism** (kinetic or thermodynamic effect)
- Three different categories of additives were tested:
  - Ionic salts
  - Structurally related compounds (**benzoic acid**)
  - Polymer (HPMC)

Garnier et al., *Journal of Crystal Growth*, **2002**
Mukuta et al., *Crystal Growth and Design*, **2005b**
Kubota et al., *Cryst. Res. Technol.*, **2001**
He et al. *Crystal Growth and Design*, **2001**

Experimental rig

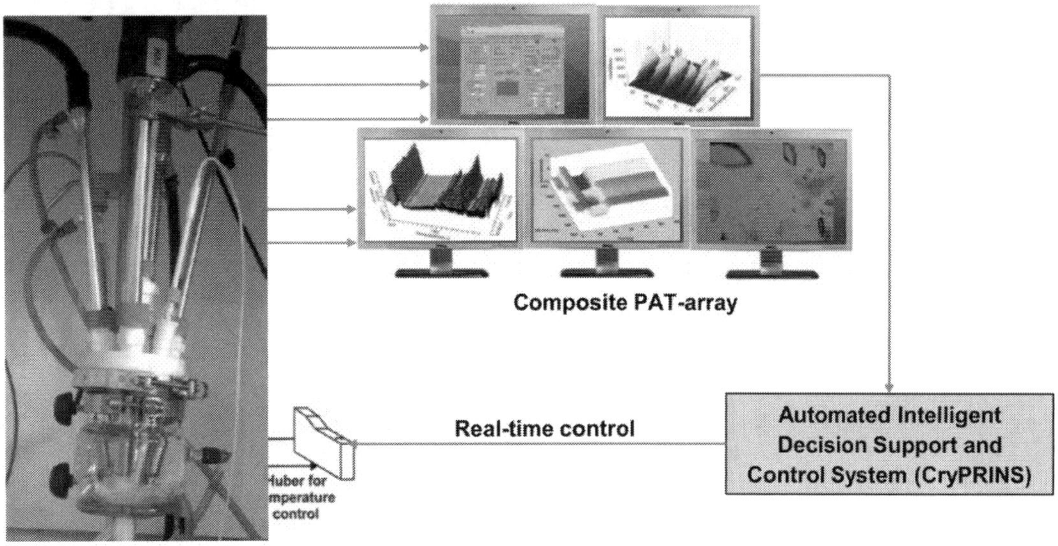

Integrated PAT array using Crystallization Process Informatics System (CryPRINS)

Loughborough University

Methodology

Solvent	OABA concentration (g/g solvent)	Maximum BA/OABA (w/w)
Water	0.0097	0.25
Water	0.0096	0.18
Water	0.0098	0.16
Water	0.0099	0.14
Water	0.0067	0.25
Water	0.0069	0.34
Water	0.0069	0.41
10% IPA	0.0089	0.29
10% IPA	0.0087	0.39
10% IPA	0.0088	0.40
10% IPA	0.0148	0.19
10% IPA	0.0148	0.21
10% IPA	0.0148	0.26
10% IPA	0.0148	0.34
20%IPA	0.0298	0.27
20%IPA	0.0299	0.27
20%IPA	0.0298	0.10
20%IPA	0.0163	0.35
20%IPA	0.0162	0.53
20%IPA	0.0162	0.64

Loughborough University

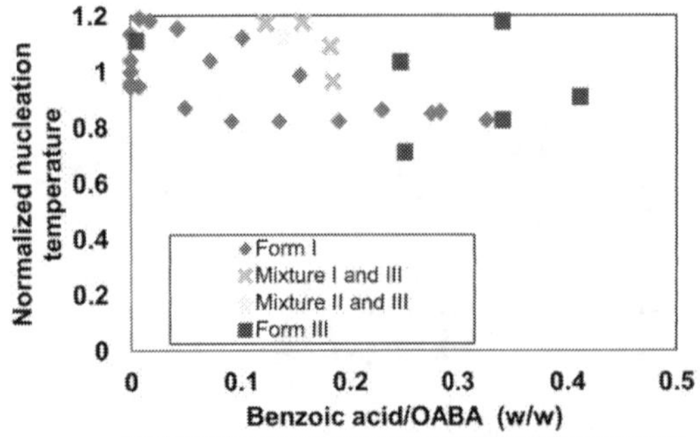

- In pure water (**high amount of OABA zwitterions**) form I nucleated
- Form II nucleated in mixtures IPA/water (mostly **non-ionized molecules** of OABA)
- **Form III** was nucleated in all tested solvents in presence of high concentrations of benzoic acid
- The **nucleation temperature** does not change significantly

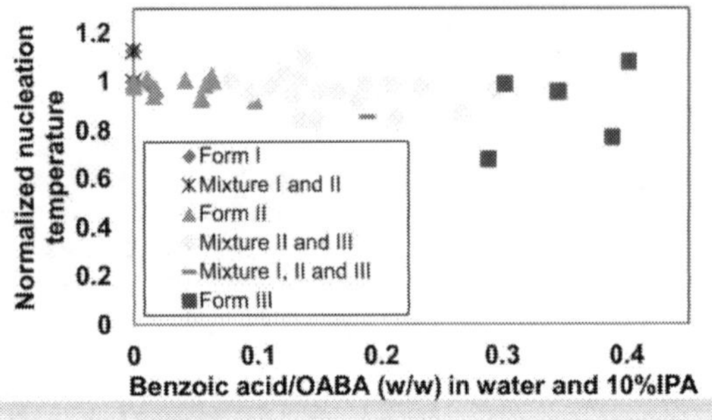

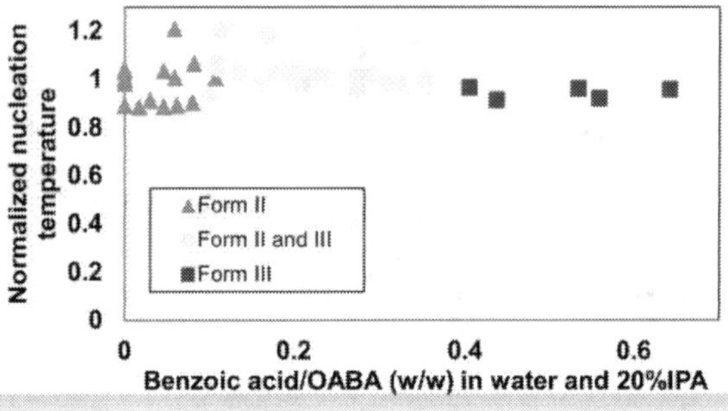

Simone et al., Submitted to CrystEngComm

Results: polymorphic outcome (2)

- ATR-FTIR spectra of the solid samples obtained in one of the experiments in 10%IPA and water

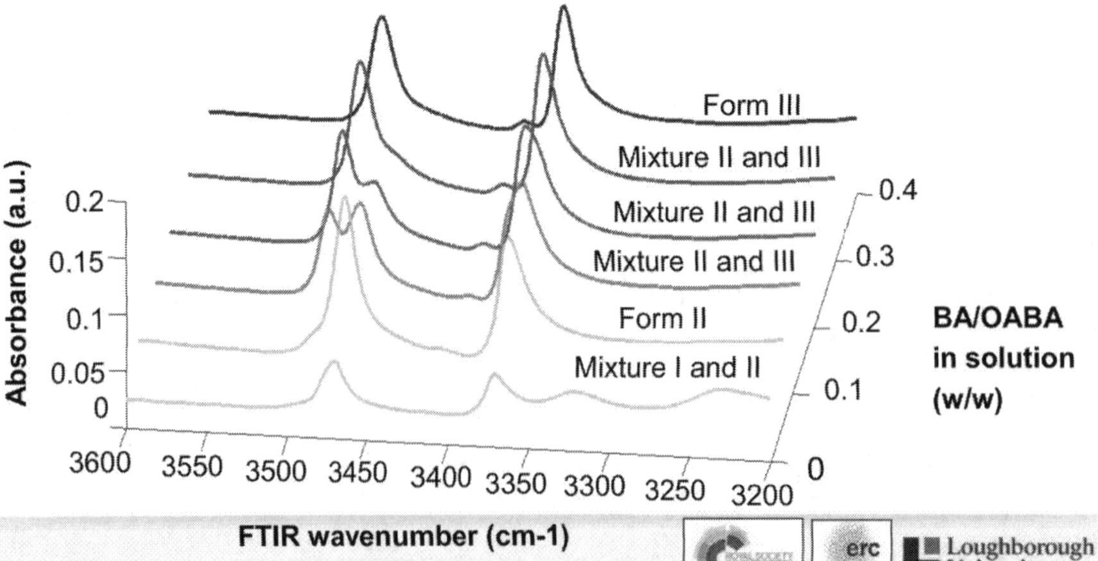

Simone et al., Submitted to CrystEngComm

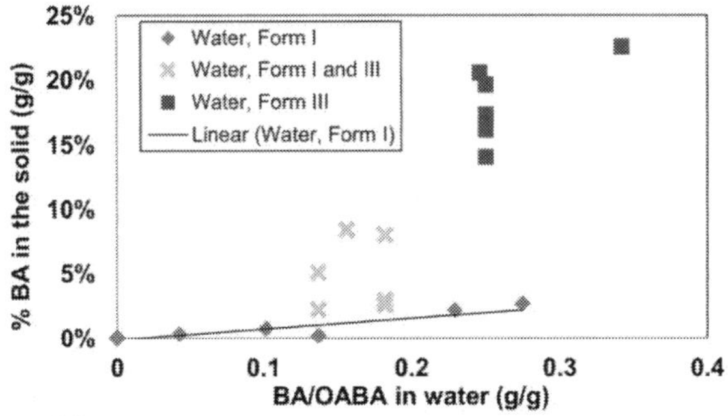

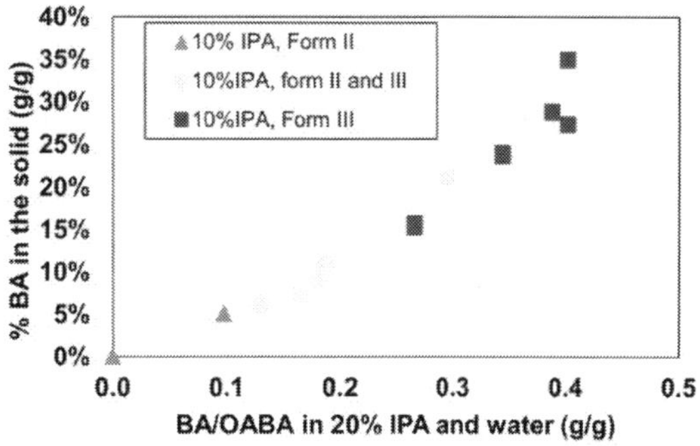

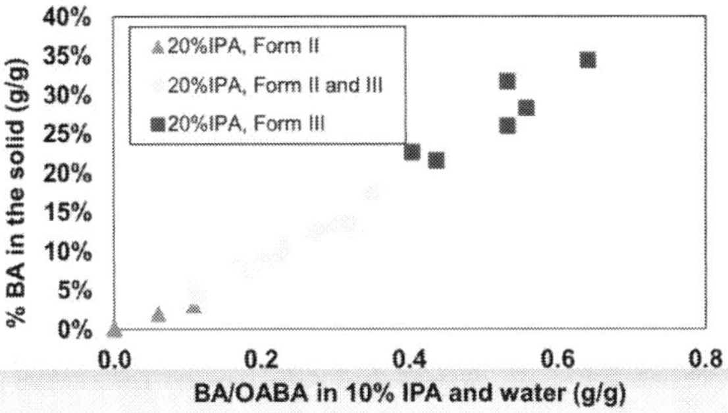

- Benzoic acid is easily incorporated in form II and III but not in form I (***SOLID SOLUTION***)

$$J = A \exp\left[-\frac{B\gamma^3}{T^3 S^2}\right]$$

Simone et al., Submitted to CrystEngComm

Results: polymorphic transformation

- Nucleation of a mixture of II and III
- Transformation II→III and then III→I

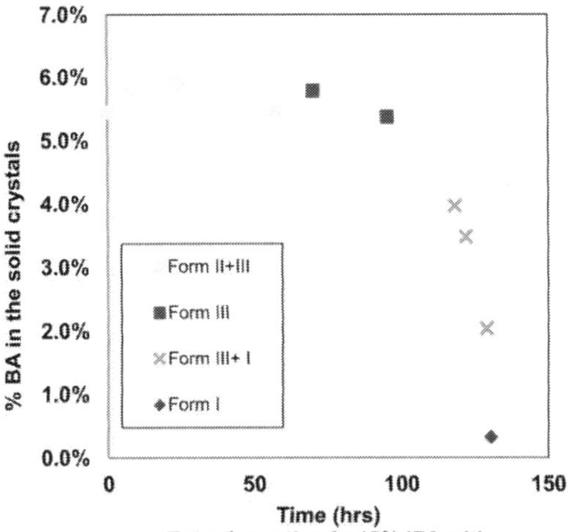

Transformation in 10% IPA with BA/OABA=0.12

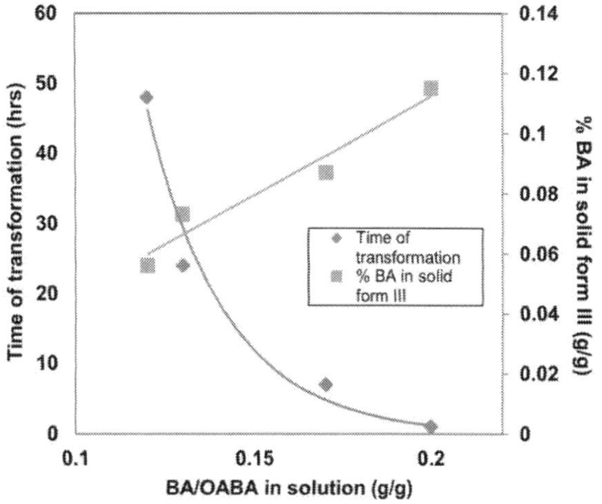

Simone et al., Submitted to CrystEngComm

Results: Morphology form I (1)

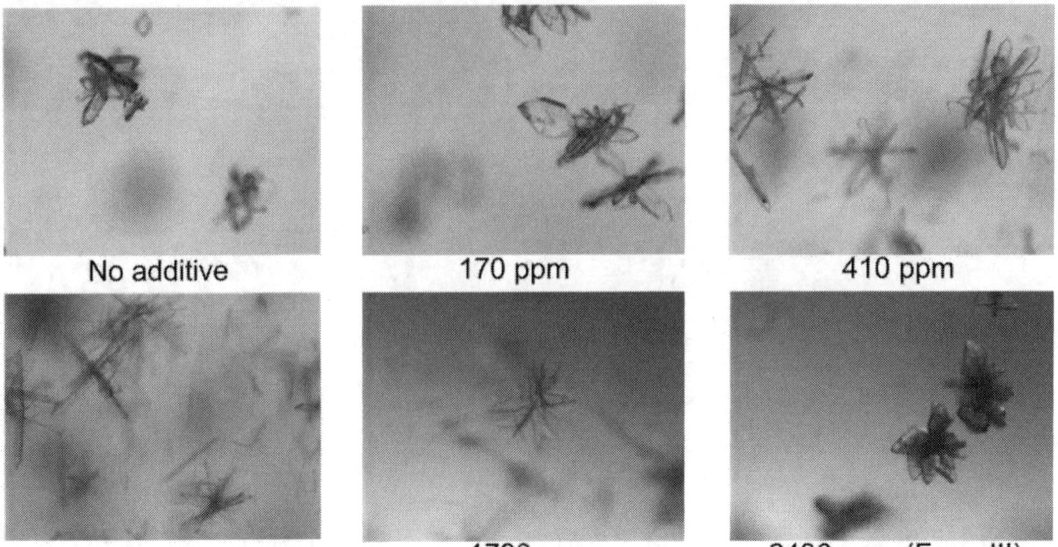

- Growth inhibition of a specific face, change in shape

Loughborough University

Results: Morphology form I (2)

- Form I is characterized by several intermolecular hydrogen bonds:
  - $NH_3^+ \cdots NH_2$
  - $O^- \cdots NH_2$
  - $O^- \cdots NH_3^+$
- BA seems to inhibit growth of form I in the b and a-axes

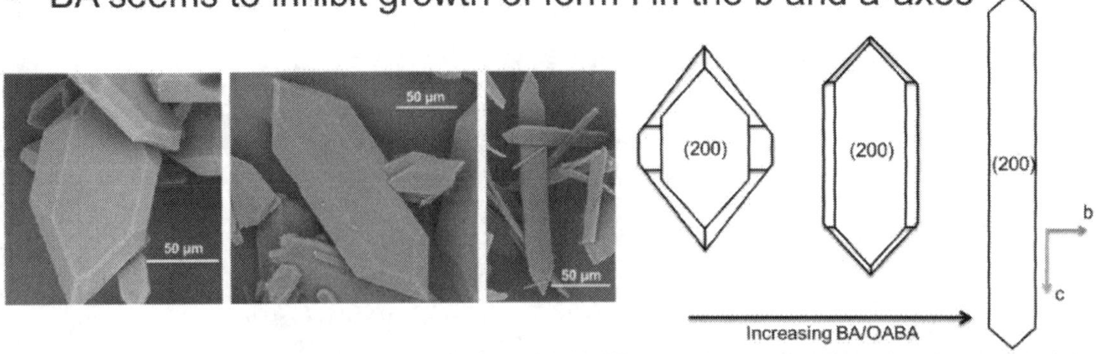

Loughborough University

Simone et al., Submitted to CrystEngComm

Results: Morphology form III

- **Solvent effect** on the morphology of form III
- The incorporation of large amounts of benzoic acid and the agglomeration of the crystals can **distort** the crystal lattice

Simone et al., Submitted to CrystEngComm

Loughborough University

Conclusions

- All the three known polymorphs of OABA were nucleated in solution using combinations of several solvent mixtures and benzoic acid
- Benzoic acid was found to be easily incorporated in form II and III (**dimers**) but not in form I (**zwitterions**)
- The growth of form I is inhibited by the additive (change in **shape**)

Thanks for your attention

Elena Simone, Zoltan K. Nagy
e.simone@lboro.ac.uk
zknagy@purdue.edu

Loughborough University

European Research Council (grant no. 280106-CrySys)
The Royal Society of Chemistry is acknowledged for the travel grant

(707e) Crystallization of Valnemulin Hydrochloride Heterosolvate: Crystal Structure, Desolvation Behavior, Solubility and Nucleation Mechanism

Ouyang, J. Sr.,
Hao, H., Tianjin university, Tianjin, China
Wang, J.,

Originally presented on: 11/12/2015 13:55:00 - 14:15:00

In this article, a new heterosolvate of valnemulin hydrochloride (HS-VLMHCl) was discovered and successfully prepared in ethanol solution of amorphous valnemulin hydrochloride (VLMHCl). Powder X-ray diffraction (PXRD) and thermal gravimetric analysis (TGA) were applied to characterize HS-VLMHCl. The crystal structure was determined by single-crystal X-ray diffraction for the first time, and it was found that both water and ethanol molecules participated in the forming crystal lattice with VLMHCl molecule ratio of 1:1:1. The result was also verified by thermal analysis and desolvation behavior successfully. To understand crystallization behavior of HS-VLMHCl, the solubility of HS-VLMHCl in water, methanol, ethanol, isopropanol, acetone, ethyl acetate were experimentally determined from 278.15 to 313.15 K using a static analytical method. In addition, the dissolution thermodynamic properties of HS-VLMHCl, including dissolution enthalpy and dissolution entropy in all test solvents, were obtained by using van't Hoff equation. Furthermore, the nucleation kinetic of HS-VLMHCl in ethanol system was investigated. The induction time was measured at 278.15 K, 283.15K, 288.15K at different supersaturation, and interfacial energies were calculated by using classical nucleation theory. The results reveal that nucleation process includes two steps: homogeneous nucleation and heterogeneous nucleation. The heterogeneous nucleation is the main mechanism at a low supersaturation, while the homogeneous nucleation predominates at a high supersaturation.

427227 Polymorphism in Dabigatran Etexilate Mesylate: Nucleation and Transformation

Abstract:

The active pharmaceutical ingredients (APIs) can exist as different solid-state forms including polymorphs, solvates, hydrates and amorphous form. Different solid-state forms of a drug usually exhibit different physical and chemical properties, such as stability, dissolution rate, bioavailability, etc. Therefore one of the primary goals of solution crystallization is to generate particles with desired crystal form in a reproducible manner. To accomplish this goal, one has to establish control over crystallization, which is a complex process including crystal nucleation and growth. The early stages of solution crystallization from the supersaturated solution to the formation of crystal nucleus play a decisive role in determining the crystal form. In this work, the solution-mediated polymorphic transformation (SMPT) process of dabigatran etexilate mesylate Form I occurring through dissolution of the metastable phase and crystallization of the stable phase was investigated by using several offline and online tools, such as PXRD, DSC, TGA, FBRM, Raman, and ATR-FTIR. The influence of solvent, temperature and supersaturated ratio were analyzed in detail. Because of the different solvent-solute interactions, SMPT process of Form I is prone to occur in alcohols rather than esters and ketones. Temperature is a key factor which will influence nucleation of different polymorphs and a new solvate was found. The supersaturated ratio could also have an effect on the nucleation of the new polymorphs which may result from the different conformations of the dabigatran etexilate mesylate molecular. It was confirmed by transformation experiments that the polymorphic transformation of Form I was controlled by the crystallization of the stable phase.

437890 Integrating CO_2 Capture with Steam-Reforming Reactions

Abstract:

CO_2 capture by carbonation is an exothermic reaction, while CO_2 generation by steam-reforming of hydrocarbons or oils is endothermic. Capturing CO_2 under the reforming conditions does not only shift the thermodynamic equilibrium toward completion but also mitigates the heat transfer issue. In this presentation, research results reported in the literature will be reviewed to delineate the progress and remaining challenges. The recent development of novel sorbents suitable for in-situ capture applications will be reported, and its application to hydrogen production from bio-oil will be discussed.

2015 AIChE Annual Meeting (Salt Lake City)

An Analysis of Extraction Mechanism and the Role of Grinding in Carbon Dioxide Mineralization Processes

Azadeh Hemmati[a,b], Jalal Shayegan[b], Tze Yuen Yeo[a], Hamidreza Kariminia, Paul Sharratt[a], Jie Bu[a*]

bu_jie@ices.a-star.edu.sg

[a] Institute of Chemical and Engineering Sciences (ICES)
A*STAR, Singapore

[b] Sharif University of Technology (SUT)
Tehran, Iran

Contents

- Why CO_2 Mineralization by Acid Extraction
- CO_2 Sequestration Efficiency and Energy Penalty of Grinding
- Mechanism of Mineral Acid Dissolution
- Conclusions

CO_2 Mineralization Acid Extraction Process: Theoretical Reactions

Mineralization →

Recycle (the NaCl from the last reaction is recycled to the first)

$$NaCl + H_2O \longrightarrow NaOH + HCl$$

$$2\,NaOH + CO_2 \longrightarrow Na_2CO_3 + H_2O$$

$$Mg_3Si_2O_5(OH)_4 + 6\,HCl \longrightarrow 3\,MgCl_2 + 2\,SiO_2 + 5\,H_2O$$

$$MgCl_2 + Na_2CO_3 \longrightarrow MgCO_3 + 2\,NaCl$$

Net reaction: $Mg_3Si_2O_5(OH)_4 + 3CO_2 \rightarrow 3\,MgCO_3 + 2SiO_2 + 2H_2O$

Purification →

$$Fe^{3+}(aq) + 3OH^-(aq) \leftrightarrow Fe(OH)_3(s) \qquad (P_1)$$

$$Fe^{2+}(aq) + 2OH^-(aq) \leftrightarrow Fe(OH)_2(s) \qquad (P_2)$$

Hemmati, A., et al., *Solid products characterization in a multi-step mineralization process.* Chemical Engineering Journal, 2014. **252**: p. 210-219.

3

Mineralization with pH-swing Process

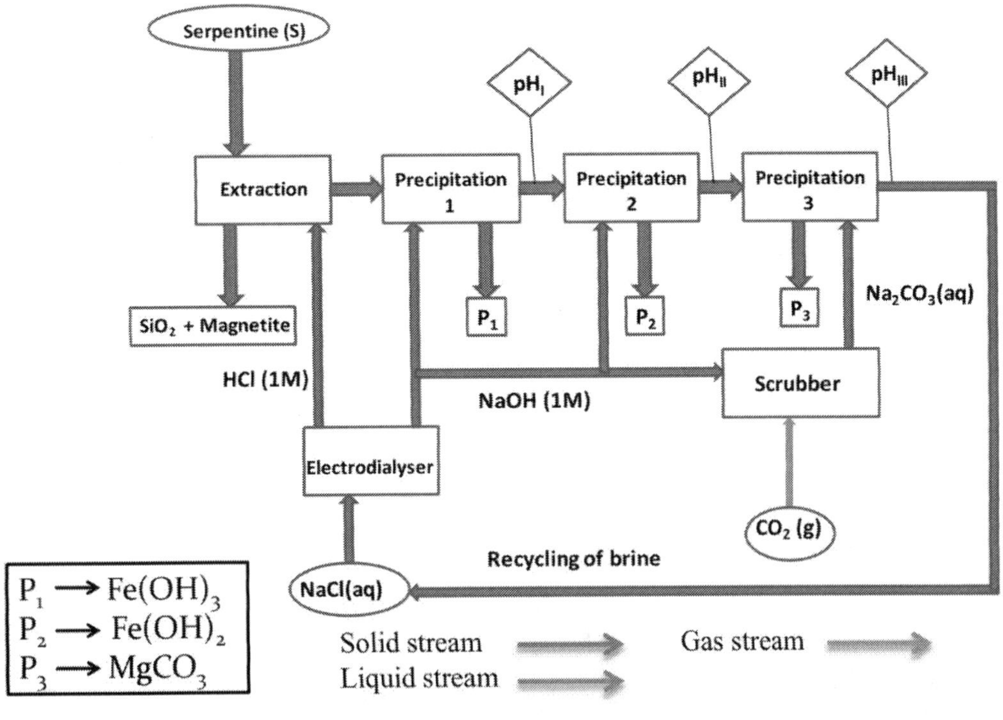

4

Materials

MgO	46.72%
SiO_2	42.70%
Fe_2O_3	8.40%

Raw materials
Serpentine
NaCl
CO_2
Water

Products
Silica
Magnetite
Iron hydroxide (II, III)
Hydrated $MgCO_3$

Indirect mineralization →

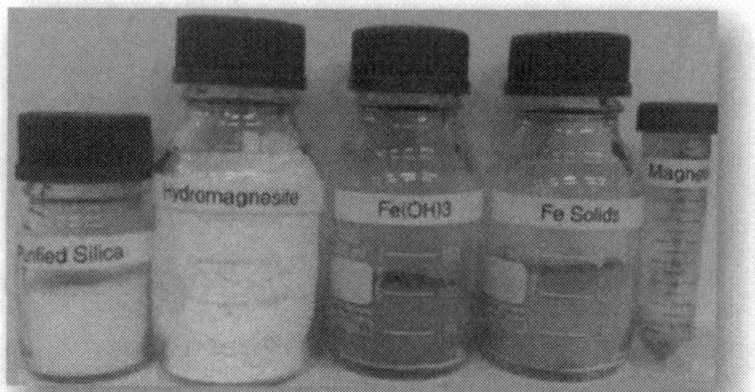

5

Magnesium Extraction

Serpentine crystals change to amorphous silica →

Silica composition after 6 hr extraction reaction

SiO_2	MgO	Fe_2O_3	Cr_2O_3	Al_2O_3	Others
82.87%	14.91%	1.33%	0.43%	0.43%	0.03%

6

Serpentine Solid Composition After Extraction

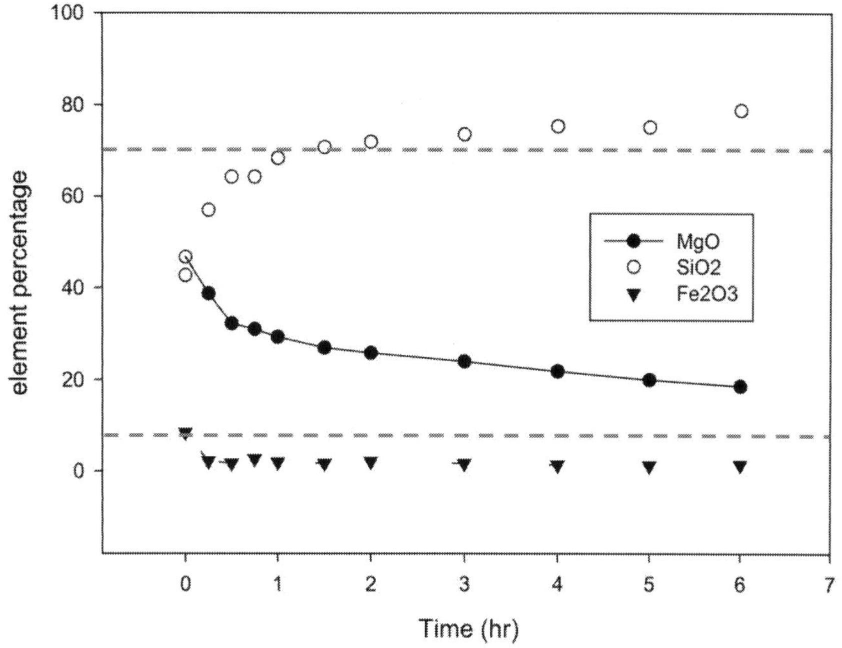

7

Particle Size Optimization

	D_{50} (μm) of mineral solid	Time	Stirring	Temp
Scenario 1	1,425	6 h	600rpm	8o °C
Scenario 2	625	6 h	600rpm	8o °C
Scenario 3	325	6 h	600rpm	8o °C
Scenario 4	187.5	6 h	600rpm	8o °C
Scenario 5	90.5	6 h	600rpm	8o °C
Scenario 6	30	6 h	600rpm	8o °C

8

Size Optimization for Serpentine

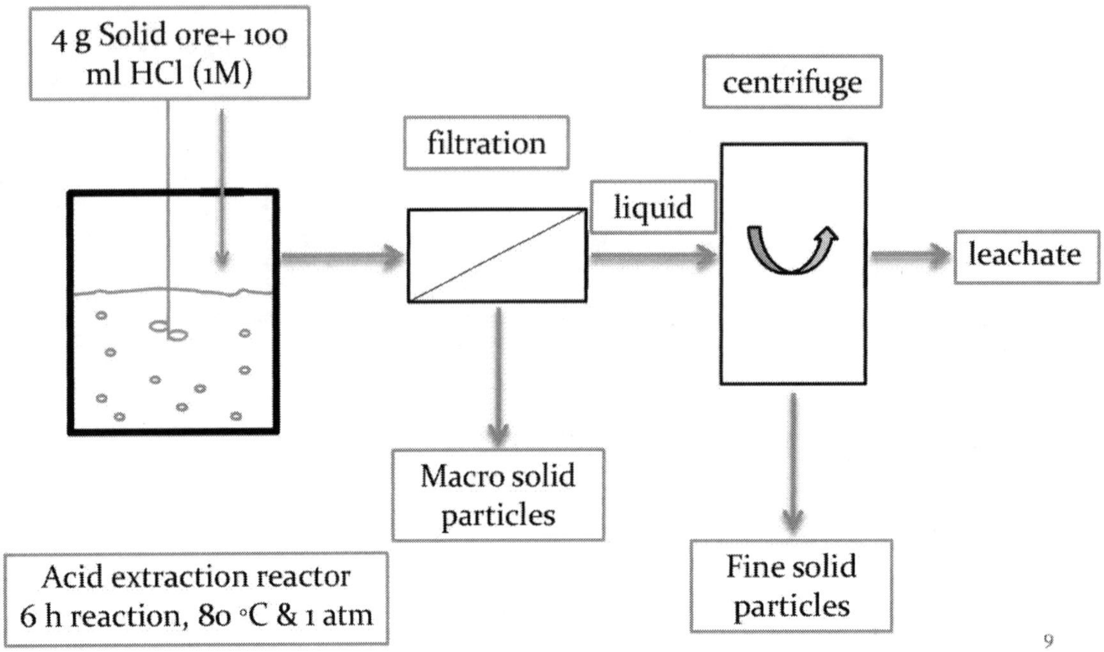

9

Mg-Extraction in Relation to Mineral Particle Sizes

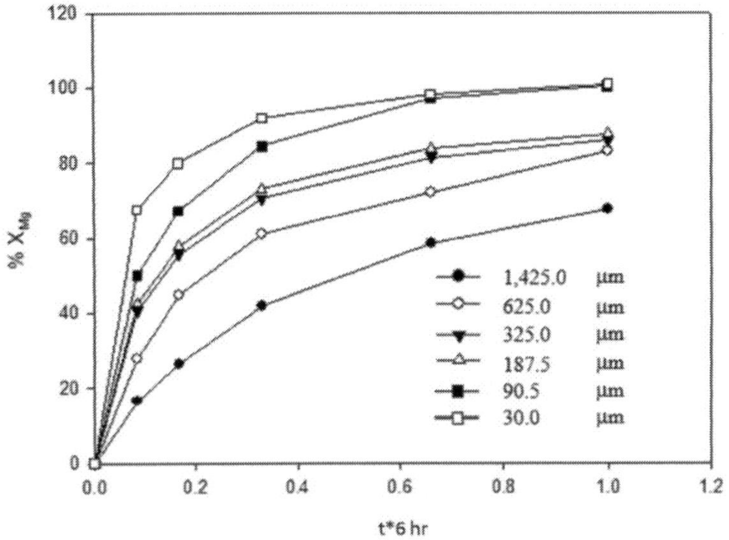

10

Mineral Grinding Energy

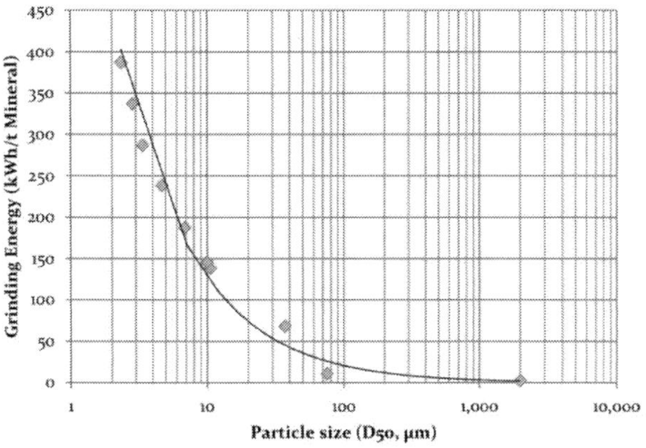

PD 50 (μm)	W_f (kWh/t mineral)
1425	4.2
625	6.55
325	8.54
187.5	12.47
90.5	19.1
30	33.74

•O'connor, W. et al. Aqueous mineral carbonation: Mineral availability, pretreatment, reaction parametrics, and process studies. A report produced by the Albany Research Centre for the US Department of Energy 2005.
•Gerdemann, S. et al. Environmental science & technology 2007, 41, (7), 2587-2593.
•Energy consumption for grinding based on bond's law (Particles size: 50mm > d > 0.05mm)

$$W_B = 10 \times C_B \left(\frac{1}{\sqrt{d_E}} - \frac{1}{\sqrt{d_A}}\right)$$

11

CO_2 Sequestration and Energy Penalty

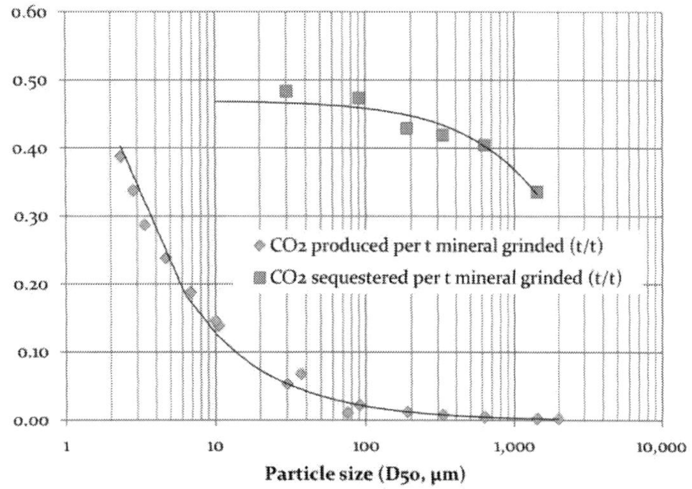

12

CO_2 Sequestration Efficiency and Energy Penalty

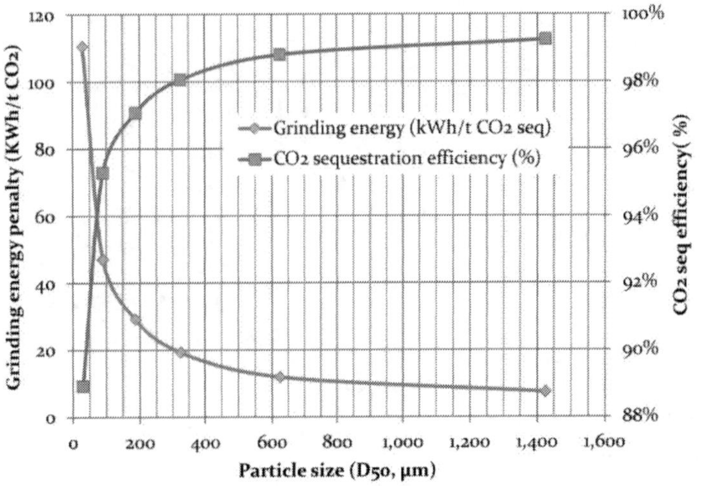

$$CO_2 Sequestration\ Efficiency = \frac{CO_2\ Sequestered - CO_2\ Emitted\ (Grinding)}{CO_2\ Sequestered}$$

13

Extraction Mechanism Investigation

Scenario	Stone Shape	Dimensions (cm)	Mass(g)	Surface removal	Agitation	Reaction time (h)
1	Sphere	R=0.694	3.6258	No	Yes (600 rpm)	90
2	Cube	2×1.2×0.8	6.9494	No	No	60
3	Cube	2×1.1×0.9	6.6307	Yes	No	60

Rate of Mass Decrease per Surface Area vs Time (Scenario 1)

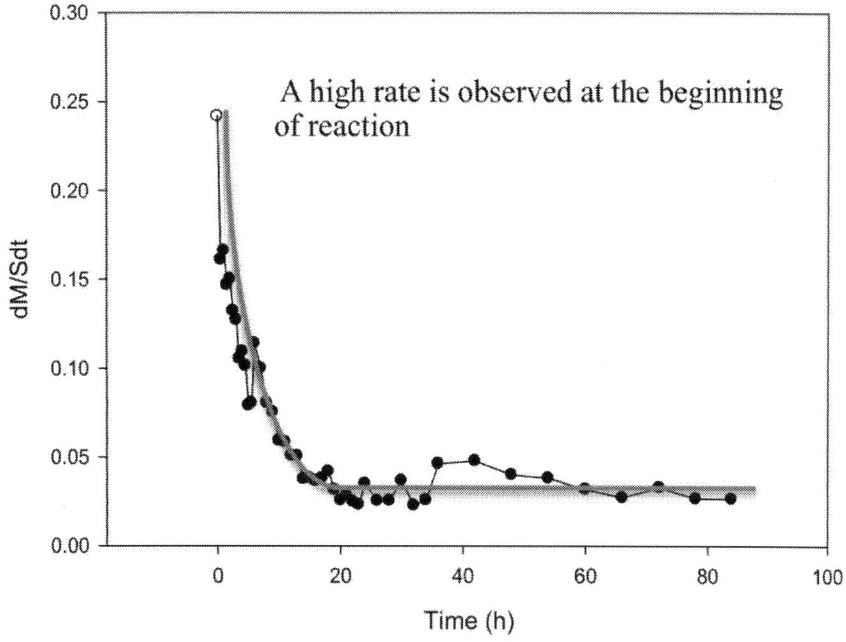

15

Extracted Mg/Extracted Fe (Scenario 1)

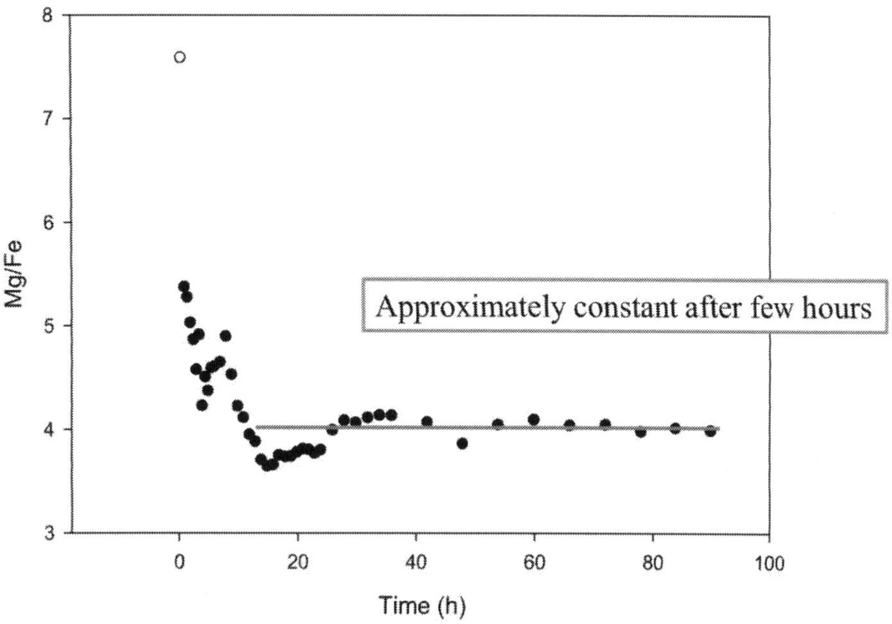

16

Study on Macro Particles: Scenario 2

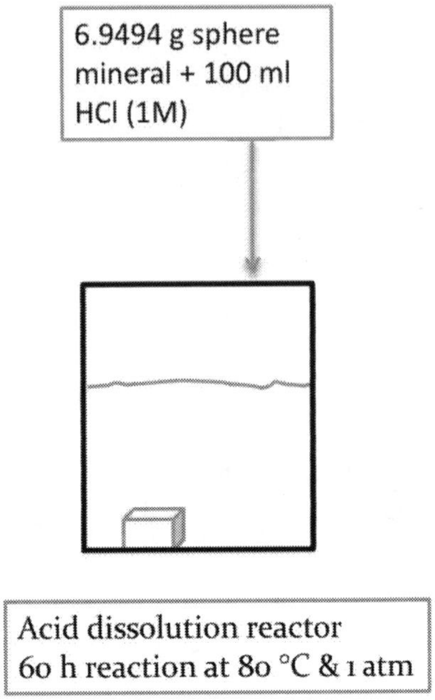

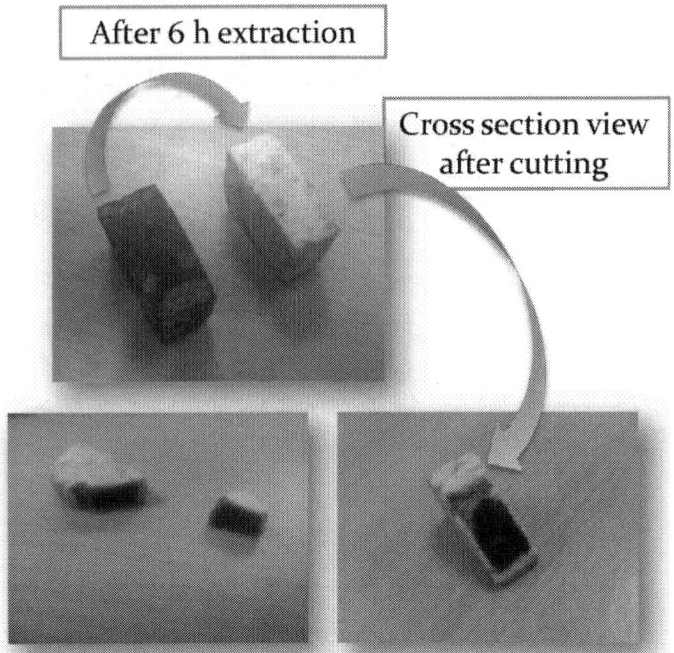

17

Reduction in Particle Mass (Scenario 2)

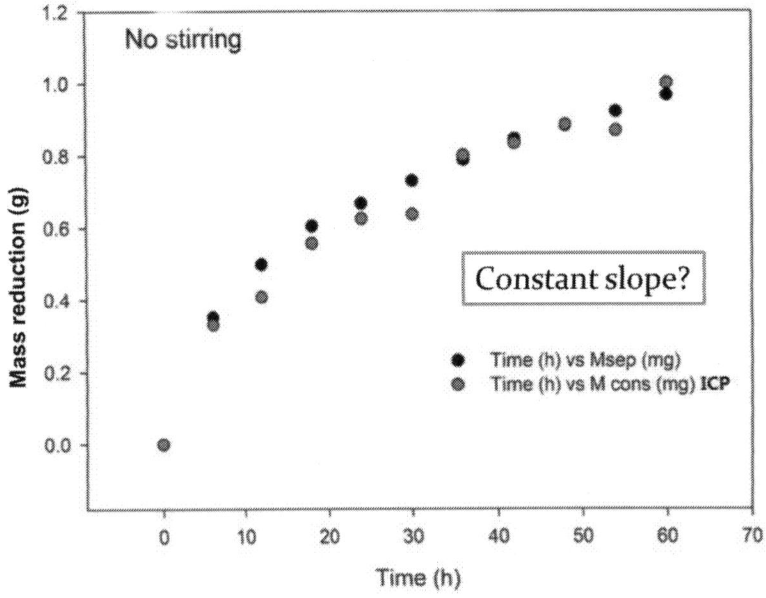

18

Study on Macro Particle: Scenario 3
Regular Removal of Surface Silica Layer

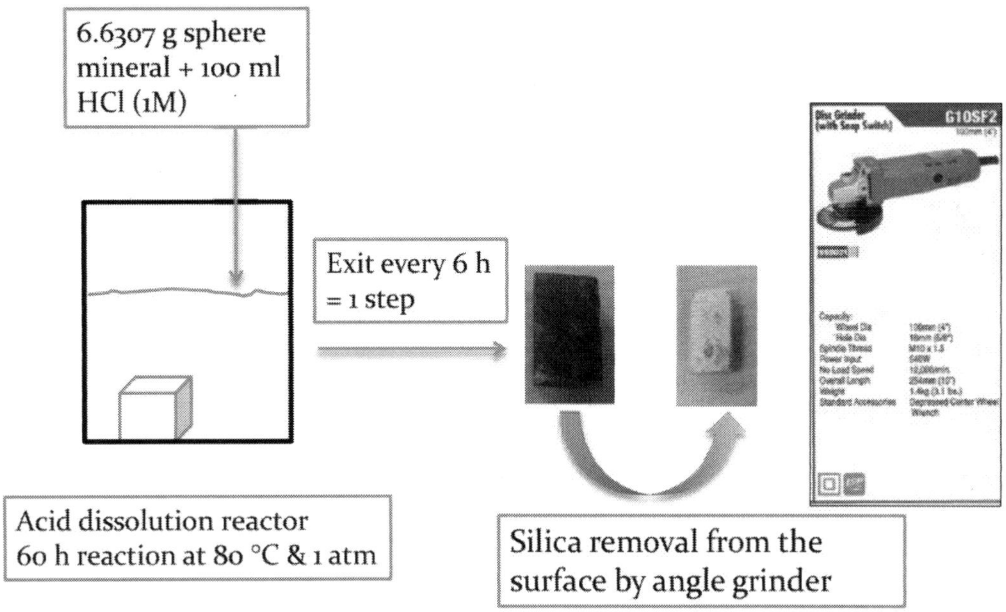

19

Reduction in Particle Mass (All Scenarios)

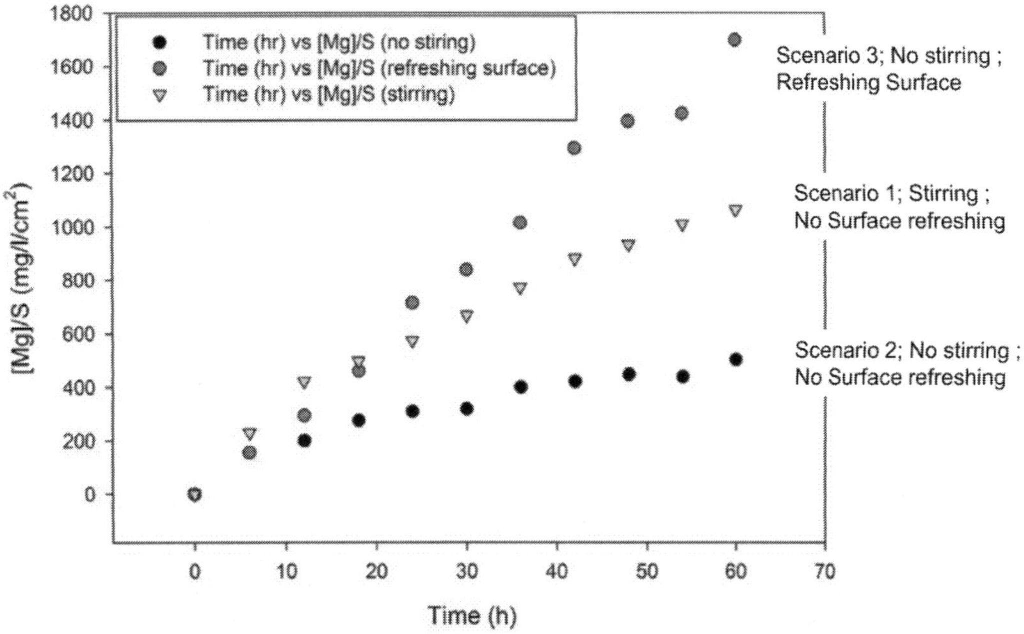

20

Scenario 1 – Shrinking Core Model (SCM)

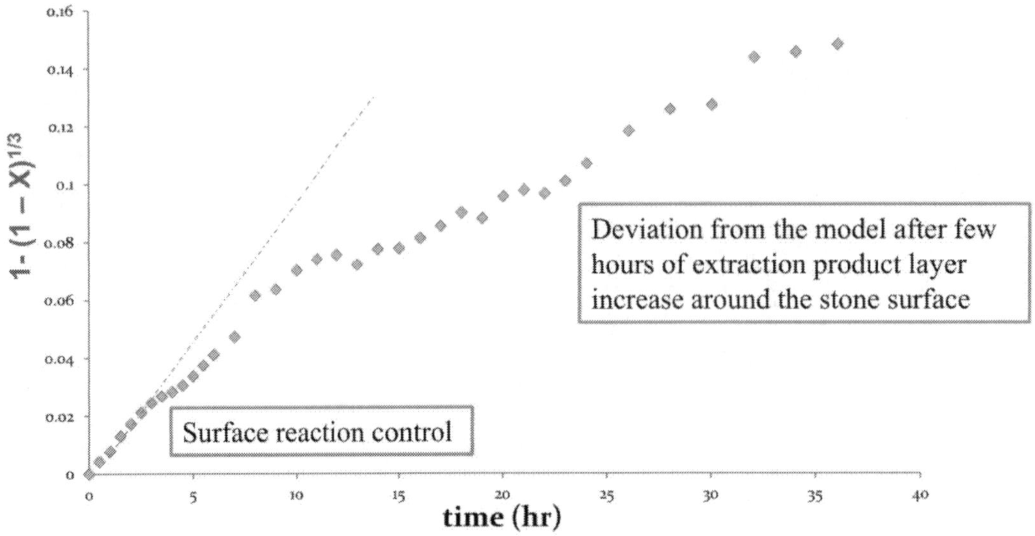

SCM Ref: 1.Yagi, S. and D. Kunii. *Studies on combustion of carbon particles in flames and fluidized beds*. in *Symposium (international) on Combustion*. 1955: Elsevier.
2.Levenspiel, O., *Chemical reaction engineering*. Vol. 2. 1972: Wiley New York

21

Fine Silica Particles Separated from the Surface of Large Particle

22

The Mechanism of Mineral Acid Dissolution
Crackling Core Model (CCM)

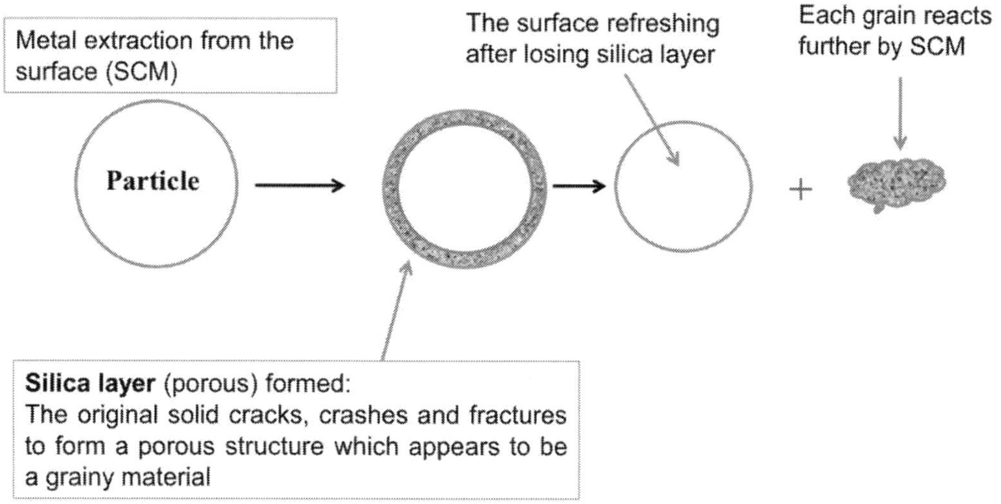

$$HCL(aq) + \frac{1}{6}Mg_3Si_2O_5(OH)_4(s) \rightarrow \frac{1}{2}Mg^{2+}(aq) + Cl^-(aq) + \frac{1}{3}SiO_2(s) + \frac{5}{6}H_2O(l)$$

23

Conclusions

- The grinding energy increases according to a power law as the mineral particle sizes get smaller. It is observed that the minerals particle could be grinded to ~500 μm with high CO_2 sequestration efficiency and low grinding energy penalty.

- The dissolution mechanisms follow the crackling core model for acid-leaching process. Chemical reaction control is important at the start of the dissolution, but the silica layer control dominates at the end.

- Acid-leaching processes do not require very small particle sizes, since the crackling core mechanism eventually results smaller particles and higher conversions.

24

Any question?

411061 Investigation of Carbon Content Analysis of Chlorella vulgaris and Nannochlropsis Oculata Using TOC Analyzer for CO2 Bioremediation and Biofuel Production

Abstract:

Global warming is become a current issue that world seek to find a solution to it through mitigation of the greenhouse gases. Carbon dioxide (CO_2) is one of the well-known greenhouse gases, which mainly came from human activities because of industrialization. Different CO_2 capture strategies have been suggested and the most attractive one is the combined carbon capture and wastewater treatment using microalgae. Microalgae can utilize CO_2 to increase its biomass which can be converted to a valuable biofuel products by conversion methods whether its thermochemical, biochemical, or chemical reaction based technology. Thus, characterization of the microalgae biomass for its carbon content is a decisive criterion for the selection suitable of conversion method. TOC analyzer is used to measure total organic carbon for *Chlorella vulgaris, Nannochloropsis oculta,* microalgae species. The effect of CO_2 concentration in the growth of microalgae and the amount of carbon produced is investigated. The increasing of the carbon content during microalgal biomass growth is well determined.
Acknowledgements: The authors would like to acknowledge the support provided by King Abdulaziz City for Science and Technology (KACST) for funding this work through project No. # KACST-ARP- A-T-23-26 (KACST-ARP#T-K-11-0431).

424086 Effect of B-Site Cu-Doping on $La_{0.75}Sr_{0.25}FeO_3$ Perovskites for the Conversion of Carbon Dioxide to Carbon Monoxide

Abstract:

The effect of Cu- doping on the conversion of CO_2 to CO was investigated through the Reverse Water Gas Shift - Chemical Looping process on $La_{0.75}Sr_{0.25}FeO_3$. Six perovskite powders were synthesized $La_{0.75}Sr_{0.25}Fe_{1-Y}Cu_YO_3$ (Y = 0, 0.1, 0.25, 0.5, 0.75 and 1) and characterized through X-ray diffraction, temperature-programmed oxygen vacancy formation, and temperature-programmed reduction (TPR). Incorporation of Cu facilitates the formation of oxygen vacancies at lower temperatures but also increases the instability of the perovskite. On the Y= 0.25, 0.1 and 0 samples, temperature programmed CO_2 conversion (TPO-CO_2) after isothermal H_2-reduction at 450 °C and post-reduction XRD were performed to evaluate the ability of the materials to convert CO_2 at the lowest temperature and to identify the crystalline phases active in the reaction. Conversion of CO_2 to CO was achieved 30 °C lower on the Y=0.1 sample versus the Y=0, but less CO was produced. An analysis of potential mass transfer limitations inferred that the reaction is not mass transport limited.

Experimental methods show that, at low Cu-dopings, the formation of oxygen vacancies under inerts exhibits a contrary trend to oxygen vacancies formed under H_2. DFT calculations on the Y=0, 0.1 and 0.25 samples suggested favorable generation of low amounts of oxygen vacancies by the Y=0.1 sample, and favored generation of high amounts of oxygen vacancies by the Y=0.25 sample

426761 CO2 Capture Using Highly Porous N-Doped Carbons Impregnated with Sodium

Abstract:

This talk presents the selective CO_2 uptake capacity of sodium-impregnated and N-doped carbon sorbents (SNSs). The SNSs were synthesized by a two-step process of KOH activation of polyacrylonitrile (PAN) and subsequent NaOH impregnation of the activated N-doped carbons. Especially, SNS2-20 synthesized with the ratio of KOH to PAN equal to 2 at 700 °C and 20 wt% NaOH based on the mass of PAN exhibited a maximum CO_2 uptake of 6.84 and 4.48 mmol g^{-1} at ambient pressure, and an exceptional CO_2 capacity of 3.03 and 1.90 mmol g^{-1} under flue gas conditions (0.15 bar CO_2) at 0 and 25 °C, respectively. These enhanced CO_2 capture performance is mainly due to the existence of narrow micropores (< 0.7 nm) and intrinsic N contents together with the strong basic sites generated by NaOH impregnation. In addition, it showed CO_2/N_2 selectivities of 59.5, 68.9 and 79.4 at 0, 25 and 50 °C for the gas mixture (CO_2:N_2=15:85). We will demonstrate the unaltered CO_2 uptake subject to even consecutive adsorption–desorption cycle experiments at 25 °C. Considering its simple preparation and superior CO_2 capture properties, SNS2-20 might be a promising sorbent for practical applications of CO_2 capture and storage.

432245 Recent Advancements in Metal-Organic Frameworks (MOFs) for CO2 Capture Applications

Abstract:

Recent advancements in metal-organic frameworks (MOFs) for CO_2 capture applications

Muhammad Ilyas*,

*Center for Environment and Water Research Institute, King Fahd University of Petroleum and Minerals, Dhahran, Saudi Arabia

ABSTRACT
As CO_2 is one of the major constituent of greenhouse gases responsible for global warming which is the current major environmental concern of the world. Major source responsible for CO_2 emission is power plants burning fossil fuel in the form of coal, oil and natural gas. Nanotechnology is one of the appropriate methods being explored worldwide for CO_2 capture process. Nano-materials are being developed to trap CO_2 molecules. Metal-organic frameworks (MOFs) have emerged as promising field for efficient and economical solution towards CO_2 capture. Metal-organic frameworks (MOFs) have been identified by US Department of Energy as most effective future technology for CO_2 capture. This research reviews the details of metal-organic frameworks (MOFs), their selectivity and properties which make them applicable for CO_2 capture applications.

Keywords: MOFs, CO_2 capture, nanomaterial's

432974 Investigation of Bifunctional Zeolites for the Adsorptive Desulfurization of Fuels

Thursday, November 12, 2015: 3:15 PM
255E (Salt Palace Convention Center)
Kevin X. Lee and Julia A. Valla, Chemical and Biomolecular Engineering, University of Connecticut, Storrs, CT

Investigation of bifunctional zeolites for the adsorptive desulfurization of fuels.

Kevin X. Lee and Julia A. Valla

Department of Chemical & Biomolecular Engineering, University of Connecticut, 191 Auditorium Road, Unit 3222, Storrs, CT 06269-3222, USA,

Phone: +1-860-486 4602, e-mail: ioulia.valla@.uconn.edu

Naturally occurring sulfur in transportation fuels has been considered a threat to environment and health safety. The combustion of sulfur produces toxic sulfur oxides (SO_x), which contributes to acid rain and catalyst deactivation. Furthermore, the slightest amount of sulfur can poison an entire fuel-cell system. Due to this global issue, many desulfurization methods have been developed to meet the stringent emission standards of 10ppm by 2017.[1] A promising method to achieve this goal is desulfurization by adsorption using zeolites because it is a cost effective and environment friendly process.[2] Zeolites are microporous aluminosilicates minerals that are widely used in catalysis, adsorption, and other energy applications due to their unique pore structure and large surface area. Unlike the energy intensive hydrodesulfurization method, zeolites can effectively remove refractory sulfur compounds such as alkyl-benzothiophenes and their derivatives at ambient conditions and without the consumption of hydrogen.

Further modification to the microporous zeolite can be done either by introducing mesoporosity and/or incorporating d-block metal cations.[3] Metal-incorporated zeolites, in combination with the ordered and open pore structures, were shown to be promising due to the sulfur-metal bond and the π-complexation interactions.[4,5] The crystallinity of the zeolite is also preserved to study the effect of acid sites and microposority on desulfurization.[6] Hierarchical zeolite Y and USY were prepared via desilication and surfactant methods.[7] The crystal structure and pore size were identified using x-ray diffraction (XRD) and Brunauer-Emmett-Teller (BET), respectively. Brönsted and Lewis acid sites were quantified using pyridine adsorption by diffuse reflectance infrared spectroscopy (DRIFT-FTIR). DRIFTS-FTIR experiments were also used to study diffusion limitations of thiophenic compounds through the zeolites. Ce, Ni and Cu metals were introduced onto the zeolites via either ion-exchange or wet incipient impregnation method. The metal loadings on the zeolite were determined using ICP.

The adsorption of a model fuel (eg. benzothiophene in octane) was studied via dynamic fixed-bed chromatography experiments. The sulfur concentrations in the zeolite and in the effluent after reaching equilibrium were evaluated using elemental analysis and the gas-chromatograph-sulfur chemiluminescence detector (GC-SCD), respectively. Breakthrough curves of final benzothiophene concentrations were generated and compared to evaluate the performance of each type of zeolite. The most promising zeolite for sulfur removal appears to be the Ni-impregnated on zeolite Y. Sigma bond connecting S in benzothiophene and Ni metal yields a strong S-M interaction and longer sorbent lifetime. This convincing result will allow for further investigation on other transition metals. Our results also demonstrate that zeolite Y is a more effective sorbent than USY, which suggests that the number of acid sites could play a more significant role than the pore size for benzothiophene. Precise and careful modification of zeolites is essential to optimize diffusivity, selectivity and adsorption rate.

The study has been extended to diesel fuels to examine the feasibility of sulfur adsorption with zeolites in industrial applications. The diesel fuel was spiked with one or several sulfur model compounds (eg. thiophene, benzothiophene and dibenzothiophene) as well as other aromatic hydrocarbons (eg. benzene, toluene, and pyridine). Dynamics experiments have been conducted to assess the capacity and the lifetime of each zeolite. This study explores wide applications of hierarchical pore structured zeolites as promising adsorbents of thiophenic compounds by investigating adsorption kinetics, selectivity, and capacity.

To simulate the adsorption behavior and predict sorbent performances, a mathematical model was developed based with the following assumptions: (1) isothermal conditions, (2) constant flow rate, (3) transport of diluted species, (4) constant particle and bed porosities, (5) internal mass transfer described by Linear Driving Force model, (6) adsorption based on Langmuir isotherm.[8] The overall differential-algebraic mass balance equation of the bulk (liquid) phase is given by:

$$\frac{dc}{dt} = D\frac{d^2c}{dz^2} - u\frac{dc}{dz} - \left(\frac{1-\varepsilon}{\varepsilon}\right)\frac{dq}{dt}$$

$$\frac{dq}{dt} = k(q_e - q)$$

$$q_e = \frac{K_L q_m c_e}{1 + K_L c_e}$$

Where c is the concentration of sulfur, D is the axial dispersion coefficient, u is the interstitial velocity, ε is the bed porosity, q ia the adsorbed sulfur concentration, k is the mass transfer coefficient, K_L is the Langmuir constant, and q_m is the sorbent capacity. The boundary conditions are given by:

$$z = 0, \quad c = c_0 - \frac{D}{u}\frac{dc}{dz}$$

$$z = L, \quad \frac{dc}{dz} = 0$$

$$t = 0, \quad c = 0$$

The study on temperature effects allowed for thermodynamic evaluation where the Gibbs free energy change is a function of the Langmuir constant.

$$\Delta G = -RT \ln K_L$$

The negative value of ΔG suggested that the adsorption of benzothiophene is a spontaneous process. The changes in entropy and enthalpy of the adsorption process were evaluated by the following equation:

$$\Delta G = \Delta H - T\Delta S$$

The analysis of thermodynamics indicated that the adsorption of benzothiophene is both spontaneous and exothermic. Metal-modified zeolite Y has shown to be a promising adsorbent for the desulfurization of fuels due to strong S-M or π-complexation interactions. Microporosity of the zeolite also plays an important role in providing acid sites for the sulfur compounds. For larger refractory thiophenic compounds where diffusion limitations may be an issue, the introduction of mesoporosity could enhance the adsorption of these molecules. The selectivity, capacity, lifetime, and kinetics of the adsorbents are dictated by the type of zeolite. The overall performance of each zeolite can be determined via the breakthrough curves. The mathematical model has shown to be a good representation of the breakthrough curves described by the non-linear Langmuir isotherm.

References

1. "EPA Proposes Tier 3 Motor Vehicle Emission and Fuel Standards," 2013.
2. Song, C.; Ma, X. New design approaches to ultra-clean diesel fuels by deep desulfurization and deep dearomatization. *Appl. Catal. B Environ* **2003**, 41, 207 – 238.
3. Tian, F.; Yang, X.; Shi, Y.; Jia, C.; Chen, Y. Adsorptive desulfurization over hierarchical beta zeolite by alkaline treatment. *J. Nat. Gas Chem.* **2012**, 56, 111 – 126.
4. Hernandez-Maldonado, A. J.; Yang, F. H.; Qi, G.S.; Yang, R. T. Desulfurization of transportation fuels by π-complexation sorbents: Cu(I)-, Ni(II)-, and Zn(II)-zeolites. *Appl. Catal. B Environ.* **2005**, 56, 111 – 126.
5. Velu, S.; Ma, X.; Song, C. Selective adsorption for removing sulfur jet fuel over zeolite-based adsorbents. *Ind. Eng. Chem. Res.* **2003**, 42, 5293 – 5304.
6. Sun, H.; Sun, L.; Li, F.; Zhang, L. Adsorption of benzothiophene from fuels on modified NaY zeolites. *Fuel Processing Technology* **2015**. 134, 284 – 289.
7. Li, K.; Valla, J.; Garcia-Martinez, J. Realizing the commercial potential of hierarchical zeolites: new opportunities in catalytic cracking. *Chem. Cat. Chem.* **2014**, 6, 46 – 66.
8. Raghavan, N. S.; Hassan, M. M.; Ruthven, D. M. Numerical simulation of a PSA system. Part I: Isothermal trace component system with linear equilibrium and finite mass transfer resistance. *AIChE J.* **1985**, 31, 385 – 392.

410328 Butanol Vapor Adsorption Behavior on Active Carbons and Zeolite Crystal

Abstract:

Butanol is considered to be a promising biofuel. However, the fermentation process for butanol production is restricted by its toxicity to the bacterial strains. Gas stripping technology can efficiently decrease butanol inhibition by removing it as it is produced. Adsorption can then be used to recover butanol from the vapor phase. Active carbon samples and zeolite were investigated for their butanol vapor adsorption capacities. Commercial activated carbon was hydrothermally modified with different H_2O_2 concentrations, and the specific surface area and oxygen-containing functional groups of activated carbon were tested before and after H_2O_2 modification. H_2O_2 hydrothermal modification increased the surface oxygen content, Brunauer-Emmett-Teller surface area, micropore volume, and total pore volume of active carbon. The adsorption capacities of these active carbon samples were almost three times that of zeolite. However, the un-modified active carbon had the highest adsorption capacity for butanol vapor (259.6 mg g^{-1}), compared to 222.4 mg g^{-1} after H_2O_2 hydrothermal treatment. Both modified and un-modified active carbon can be easily regenerated for repeatable adsorption by heating to 150 °C. Therefore, surface oxygen groups significantly reduced the adsorption capacity of active carbons for butanol vapor. This impact of surface oxygen functional group also enable engineers and scientists understand as well as predict loss of dynamic organic vapor adsorption capacity of carbon adsorbents during high temperature steam or air regeneration.

433258 Adsorption of Aqueous Rare Earth Elements Onto Pyrolyzed Recycled Tires

Abstract:

Extraction methods of Critical Materials[1] (e.g., rare earth elements, REE) is of interest to government initiatives and industry research. Currently all REEs are extracted through mining and mineral processing of REE ores. Established techniques such as precipitation and solvent extraction are used for metal recovery, especially useful for large-scale operations with high metal ion concentrations. In contrast, adsorption can be used to recover metal ions from low-concentration sources by means of relatively simple processes. Many adsorbents for REEs have been studied;[2-14] nevertheless, there is no adsorbent material in practical use as of date.

In this study we examine the feasibility of using carbon black derived from whole used tires obtained by pyrolysis as a sorbent material for REE. Other carbon materials were examined such as activated carbon, commercial carbon black (natural gas partial combustion), and functionalized forms of activated carbon and commercial carbon black. For the adsorption of light REEs (La, Ce, Nd, Sm, and Y), various operating variables were tested (i.e., temperature, pH, sorbent loading, interfering ions, and recyclability) as well as the kinetics of adsorption. Overall, the carbon black derived from recycled tires demonstrates better adsorption characteristics over the other carbon materials examined in this study and presents itself as a sustainable material for extraction of rare earth elements from low-temperature geothermal wells.

References:

1. Bauer, D.; Diamond, D.; Li, J.; Sandalow, D.; Telleen, P.; Wanner, B., US Department of Energy Critical Materials Strategy. **2010**.

2. Araki, K.; Yoshida, M.; Uezu, K.; Goto, M.; Furusaki, S., Lanthanide-imprinted resins prepared by surface template polymerization. *J. Chem. Eng. Jpn.* **2000,** *33*, 665-668.

3. Alakhras, F. A.; Dari, K. A.; Mubarak, M. S., Synthesis and chelating properties of some poly(amidoxime-hydroxamic acid) resins toward some trivalent lanthanide metal ions. *J. Appl. Polym. Sci.* **2005,** (691-696).

4. Bou-Maroun, E.; Goetz-Grandmont, G. J.; Boos, A., Sorption of europium(III) and copper(II) by a mesostructured silica doped with acyl-hydroxypyrazole derivatives — extraction, kinetic and capacity studies. *Colloids Surf. A Physicochem. Eng. Asp.* **2006,** *287*, 1-9.

5. Chen, Y. G.; Zhu, B. H.; Wu, D. B.; Wang, Q. G.; Yang, Y. H.; Ye, W. M.; Guo, J. F., Eu(III) adsorption using di(2-ethylhexyl) phosphoric acid-immobilized magnetic GMZ bentonite. *Chem. Eng. J.* **2012,** *181*, 387-396.

6. Choi, S. H.; Lee, K. P.; Sohn, S. H., Graft copolymer–lanthanide complexes obtained by radiation grafting on polyethylene film. *J. Appl. Polym. Sci.* **2003,** *87*, 328-336.

7. Das, N.; Das, D., Recovery of rare earth metals through biosorption: an overview. *J. Rare Earths* **2013,** *31*, 933-943.

8. Dupont, D.; Brullot, W.; Bloemen, M.; Verbiest, T.; Binnemans, K., Selective uptake of rare earths from aqueous solutions by EDTA-functionalized magnetic and nonmagnetic nanoparticles. *ACS Appl. Mater. Interfaces* **2014,** *6*(7), 4980-4988.

9. Jia, Q.; Wang, Z. H.; Li, D. Q.; Niu, C. J., Adsorption of heavy rare earth(III) with extraction resin containing bis(2,4,4-trimethylpentyl) monothiophosphinic acid. *J. Alloys Compd.* **2004,** *374*, 434-437.

10. Ramakrishnan, K.; Rao, T. P., Ion imprinted polymer solid phase extraction (IIP-SPE) for preconcentrative separation of erbium(III) from adjacent lanthanides and yttrium. *Sep. Sci. Technol.* **2006,** *41*, 233-246.

11. Shibata, J.; Matsumoto, S.; Yamamoto, H., A novel separation technology for a heavy rare earth residue using a solvent impregnated resin. *Solvent Extr. Res. Dev. Jpn.* **2000,** *7*, 167-175.

12. Wang, F.; Zhao, J.; Zhou, H.; Li, W.; Sui, N.; Liu, H., O-carboxymethyl chitosan entrapped by silica: preparation and adsorption behaviour toward neodymium (III) ions. *J. Chem. Technol. Biotechnol.* **2013,** *88*, 317-325.

13. Wu, D.; Sun, Y.; Wang, Q., Adsorption of lanthanum (III) from aqueous solution using 2-ethylhexyl phosphonic acid mono-2-ethylhexyl ester-grafted magnetic silica nanocomposites. *J. Hazard. Mater.* **2013,** *260*, 409-419.

14. Ogata, T.; Narita, H.; Tanaka, M., Adsorption behavior of rare earth elements on silica gel modified with diglycol amic acid. *Hydrometallurgy* **2015,** *152*, 178-182.

432355 Effect of Temperature and Heating Rate on Heel Build-up during Microwave Regeneration

Abstract:

This study investigates the effect of regeneration temperature and heating rate on irreversible adsorption of organic vapors typically emitted from automotive painting booths. Two types of beaded activated carbons (BAC), with different porosities (microporous versus mesoporous) were loaded with 1, 2, 4-trimethylbenzene and regenerated at two regeneration temperatures (288 and 400 °C) and at different heating rates (25, 50, 100, and 150 °C/min). The results showed that for the higher regeneration temperature (400 °C), increasing heating rate increased heel build-up by as much as 105% and 189% for the microporous and mesoporous BACs, respectively. The elevated heel formation at higher heating rates could be due to adsorbate coking as a result of exposure to high temperature. Conversely, for the lower regeneration temperature (288 °C), increasing the heating rate did not significantly affect the amount of heel build-up (< 15% and < 5% increase for microporous and mesoporous BACs, respectively). This could be because the lower regeneration temperature provided insufficient conditions for coking. The results from this study indicate that the heating rate during desorption could be optimized to allow fast desorption with minimal adsorbate decomposition.

429381 The Effect of Regeneration Purge Gas Impurity on Irreversible Adsorption of Organic Vapors

Abstract:

Irreversible adsorption or heel formation is a known challenge for organic vapors adsorption onto activated carbon as it decreases the adsorbent capacity and lifetime. The objective of this research is to investigate the effect of nitrogen purge gas purity on heel formation during thermal desorption of activated carbon. Kureha beaded activated carbon (BAC) was tested for 5 consecutive adsorption/regeneration cycles using a mixture of nine organic compounds typically emitted from automotive painting operations. Nitrogen, with different concentrations of oxygen (5, 625, 1250, 2500, 5000, and 10000 ppm) was used as purge gas during thermal desorption cycles. The mass balance cumulative heel was 9, 16, 24, 35, and 34% higher for samples exposed to 625 ppm, 1250 ppm, 2500 ppm, 5000 ppm, and 10000 ppm O_2, respectively, compared to sample exposed to 5 ppm O_2. Samples exposed to higher O_2 concentration (625 to 10000 ppm) during regeneration also lost their adsorption capacity at a faster rate (up to a 119% increase in capacity loss) compared to sample exposed to 5 ppm O_2. Derivative thermogravimetric (DTG) analysis showed that heel formation was due to strong physical adsorption for sample exposed to 5 ppm O_2 and a combination of strong physisorption and chemisorption for the other samples. In contrast to previously published results, less pore blockage occurred in samples exposed to higher O_2 concentrations most likely because of diffusion resistance due to chemisorbed adsorbates. BAC samples, which were exposed to 50 successive adsorption/regeneration cycles, were also characterized to provide further insight about the impacts of long term exposure to high O_2 concentration, showing trends consistent with short term exposure (5 cycles). The results from this study help explain the heel formation mechanism and how it relates to regeneration purge gas purity.

419668 Global Solution for Mercury Removal in Saturated Wet Gas Streams : From Rational Material Design to Innovative Process Issues

Abstract:

Elemental mercury is a natural contaminant of natural gas. In particular, mercury is a safety issue in LNG plants, as mercury forms amalgams with aluminum-based alloys, leading to corrosion issues on cryogenic exchangers required to liquefy natural gas.

Conventionally Mercury Removal Units (MRU) were located downstream Acid Gases Removal Units (AGRU) and dryers. However, in this configuration, mercury tends to contaminate all the processes streams and equipment located upstream of the MRU. Therefore, in recent gas plant design, MRU are always located at the most upstream possible location, i.e. generally just after condensate and water removal units. In such cases, natural gas is saturated with water and hydrocarbon condensate vapors. Moreover, in such a location, liquid carry-over in MRU beds is likely to occur. Liquid water has a very detrimental influence on conventional mercury guard bed performances: mercury capacity loss for sulfur-impregnated carbons and mechanical strength loss due to binder swelling induced by liquid water for bulk adsorbents. The use of water-resistant mercury adsorbent becomes thus an absolute necessity for MRU in gas plant after water and condensate removal units.

The way IFPEN has optimized MRU solutions dedicated to the total elimination of elemental mercury from saturated wet gas streams is based on a dual complementary approach : on one hand the development of water-resistant mercury adsorbent by deposition of CuS on optimized alumina and, on the other hand, the design of innovative cost-effective process solutions aiming at the reduction of relative humidity in MRUs.

The optimization approach of water-resistant mercury adsorbents will be presented. In particular, the effects of pore size and surface chemistry of dedicated alumina supports on water vapor capillary condensation mechanism will be discussed.

Concerning the process aspects, conventional solutions proposed to prevent any liquid presence in MRU beds rely on natural gas heating or dew pointing, which constitute energy-intensive operation units. Innovative energy-efficient process solutions aiming at the reduction of relative humidity of natural gas in MRU have been recently patented by IFPEN. For similar performances, heat consumptions for such operations have been reduced by an order of magnitude compared to existing solutions. Process modeling and economics calculation on the innovative solutions proposed by IFPEN will be presented.

AIChE Annual Meeting
Adsorbent Materials for Sustainable Energy and Chemicals

A Metal-Organic Framework with Immobilized Ag(I) for Highly Efficient Desulfurization of Liquid Fuels

Minhui Huang, Master Candidate

Key Laboratory of Biomass Chemical Engineering of Ministry of Education, Department of Chemical and Biological Engineering, Zhejiang University, Hangzhou, China

November 12, 2015

1 Motivation

Sulfur-contaminants

- Environmental pollution, such as acid rain, smog
- Poison catalytic converters for exhaust emissions

"Cleaner" fuels

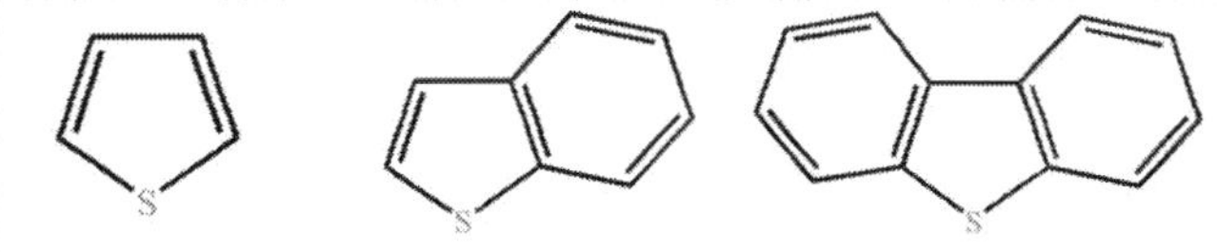

Thiophene (TP) Benzothiophene (BT) Dibenzothiophene (DBT)

Hydrodesulfurization (HDS)

- Eliminate Nonaromatic sulfur and thiophere efficiently;
- Fail for Sterically hindered organic sulfide: be difficult to be converted into H_2S

Adsorption separation

- A very promising energy-efficient alternative
- Adsorbents: metal-organic framework, porous organic polymers

2

Metal Organic Frameworks (MOFs)

Crystalline materials——combine metal organic complexes with organic linker species

Careful selection of MOF for separation

- Ultrahigh porosity with enormous surface areas
- Adjustable pore dimensions
- Chemical tunability

3

Ideal MOF materials for desulfurization

> - Porosity/pore size/specific adsorption sites
> - Metal ions: Cu(I), Ag(I), Pd(I), Pt(I) ions, superior ability for removing thiophenic compounds via π-complexation

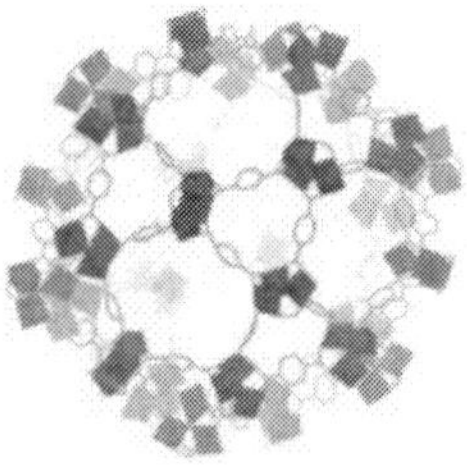

MIL-101

MIL-47

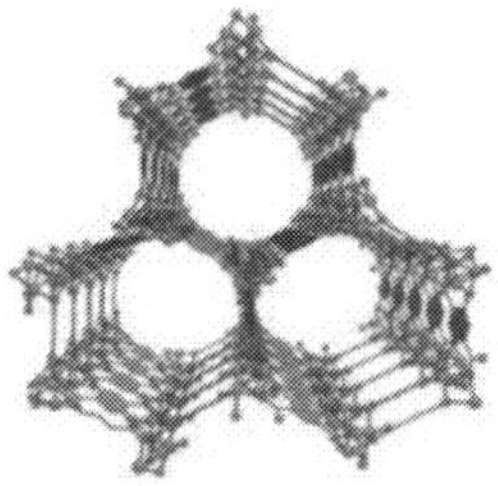

MOF-74

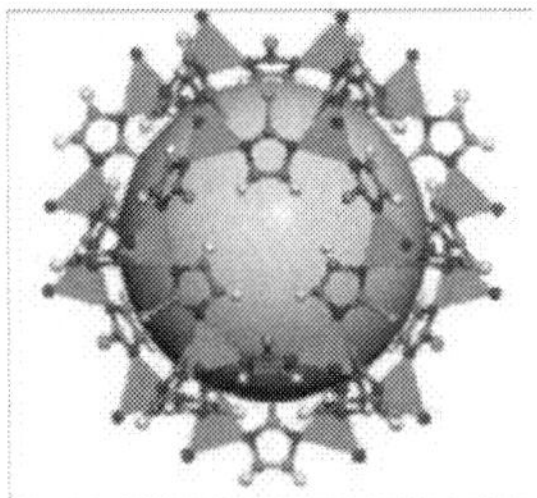

ZIF-8

- Focus: tuning the pores and porosities for no specific sites
- Be very difficult to remove thiophene derivatives to very low contents
- Metal ions incorporated zeolites, particularly Cu(I), Ag(I)

R. T. Yang, Science, 2003, 301, 79-81

4

1 Metal Organic Frameworks (MOFs)

Scheme:

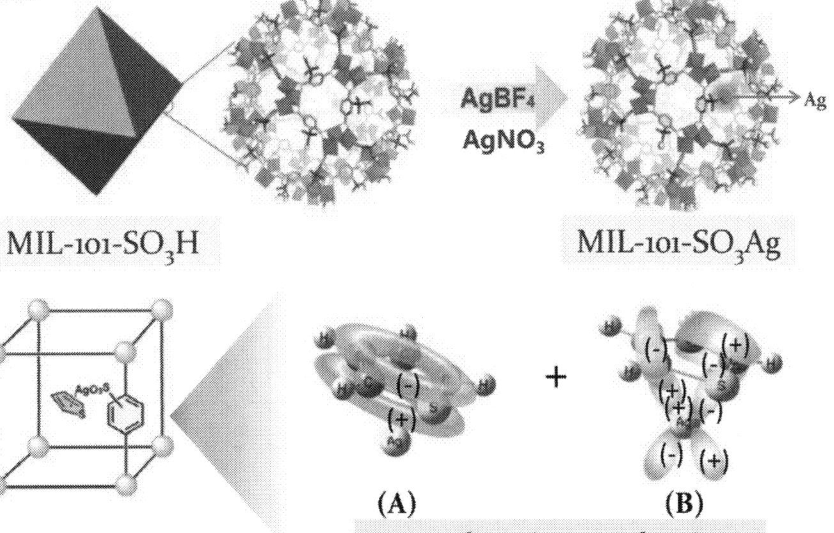

π-complexation mechanism

(A)——б-donation of π-electrons of thiophene to the 5s orbital of Ag(I)

(B)——d-π* back-donation of electrons from 4d orbitals of Ag(I) to π* orbitals of thiophene

R. T. Yang, Science, 2003, 301, 79-81

2 Synthesis

20.15 mmol Monosodium 2-sulfoterephthalic acid
520 μL, 37 wt%
10 mmol CrNO3·9H2O
60 mL ultrapure water

Solvothermal reaction
190 °C for 24 h

$AgNO_3$ (1 g) + 300 mg (Cr)-MIL-101-SO_3H

in 20 mL CH_3CN/H_2O (1:1,v/v)
Room temperature 24h
repeated one more time

Ion exchange

(Cr)-MIL-101-SO_3Ag

6

a. S. Kitagawa, *Adv. Mater.*, 2011, 23, 3294-3297; b. J. Juan-Alcaniz, J. Gascon, *Catal. Sci. Technol.*, 2013, 3, 2311.

2 Characterization

SEM & EDX

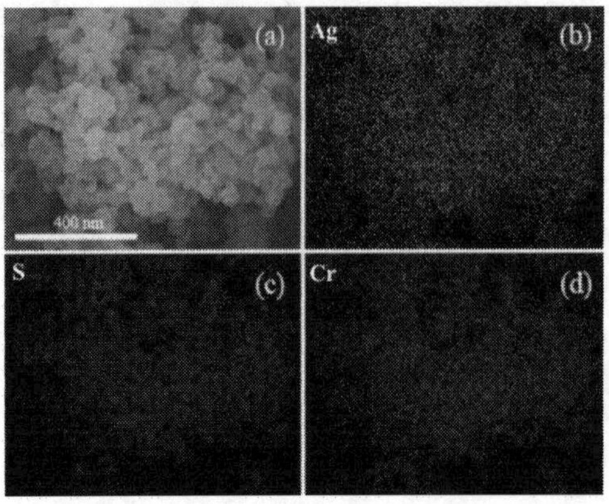

Element	Weight %	Atom %
C K	51.77	64.82
O K	32.63	30.67
S K	3.39	1.59
Cl K	0.16	0.07
Cr K	7.79	2.25
Ag L	4.26	0.59
Total	100.00	

- Ag(I) are evenly dispersed, ~37% $-SO_3H$ were transformed into $-SO_3Ag$
- Testify the binding of Ag(I) with the $-SO_3H$ groups

7

2 Characterization

N_2 adsorption& Pore Size

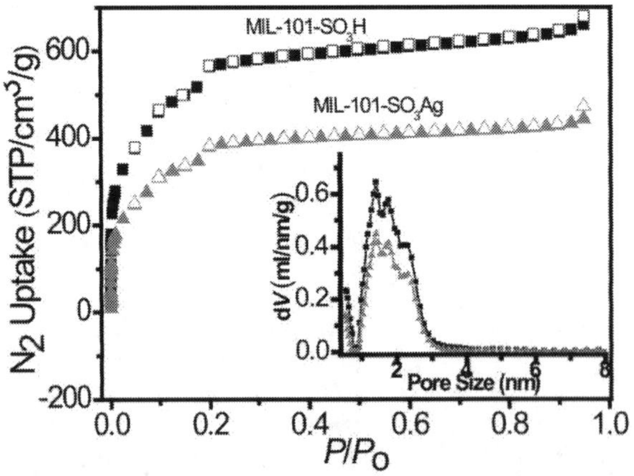

	Surface areas	Pore volume	Pore size
MIL-101-SO_3H	1856	1.35	15 ~18 Å
MIL-101-SO_3Ag	1253	1.00	15 ~18 Å

8

2 Characterization

XRD ● ---- ● XPS

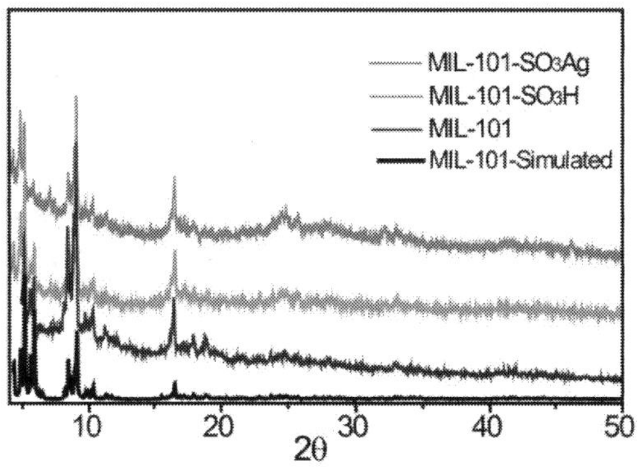

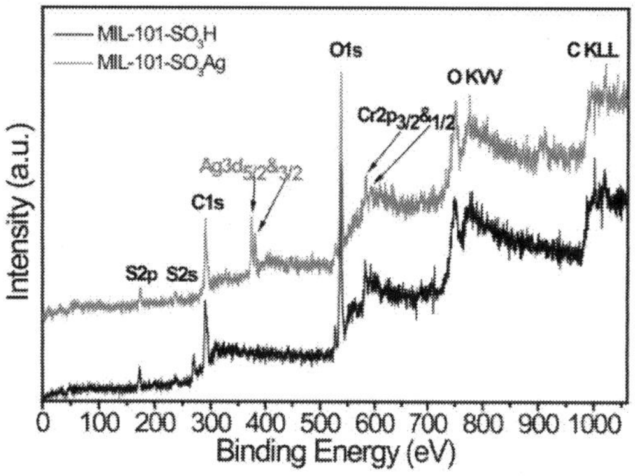

- MIL-101-SO_3Ag keeps the same structure as MIL-101-SO_3H
- XPS: a silver signal at binding energies of 373.5 and 379.5 eV, corresponding to the peaks of Ag $3d_{5/2}$ and Ag $3d_{3/2}$, respectively.

9

3 Adsorption Performance

Adsorption isotherms: DBT from *n*-octane solutions

30 °C over MIL-101-SO_3Ag (olive)/MIL-101-SO_3H (pink)

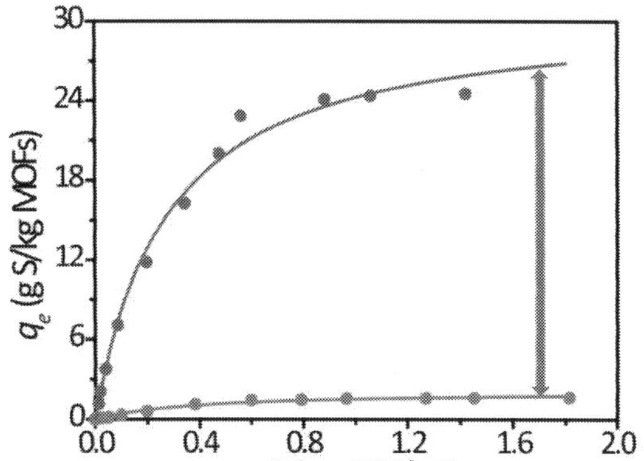

Curves——Langmuir equation

- 1253 m^2/g for MIL-101-SO_3Ag & 1856 m^2/g for MIL-101-SO_3H, such significant uptake increases are attributed to the Ag(I) sites

10

3 Adsorption Performance

Adsorption isotherms: BT from *n*-octane solutions

30 °C over MIL-101-SO_3Ag (olive)/MIL-101-SO_3H (pink)

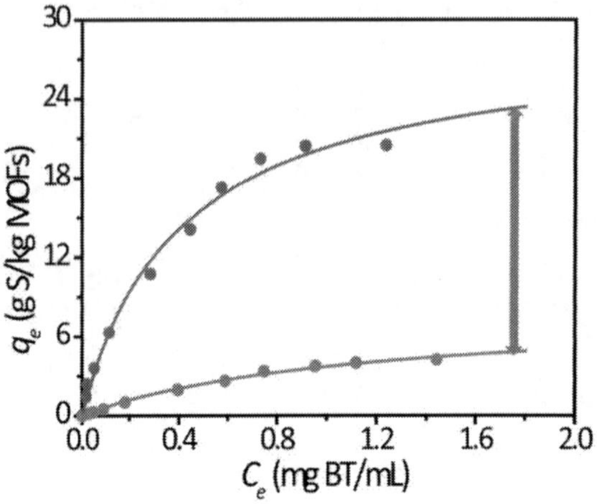

Curves——Langmuir equation

11

3 Adsorption Performance

Adsorption isotherms: TP from *n*-octane solutions

30 °C over MIL-101-SO_3Ag (olive)/MIL-101-SO_3H (pink)

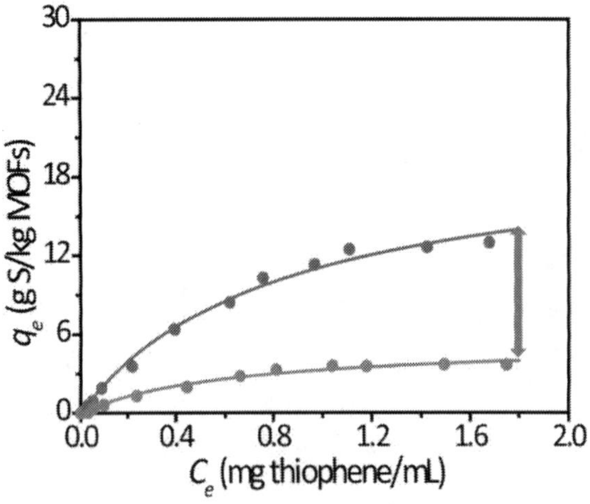

Curves——Langmuir equation

12

3 Adsorption Performance

Comparison of saturation uptake of S-compounds on the reported MOFs

Adsorbents	Solvent	TP (g S/kg MOFs)	BT (g S/kg MOFs)	DBT (g S/kg MOFs)
MIL-101(Cr)	*n*-octane/*iso*-octane	—	12.2[b]	17[c]
MIL-101-SO_3H	*n*-octane	5.47	7.95	2.14
MIL-101-SO_3Ag	*n*-octane	20.5(0.62*)	28.8(2.64*)	31.0(5.30*)
HKUST-1[a]	*iso*-octane	—	25(0.32*)	45(2.21*)
UMCM-150[a]	*iso*-octane	—	40(0.78*)	83(2.25*)

*Adsorption uptake at 10 ppm S concentration

- More delocalized π electrons in DBT and their effective interaction with Ag(I) ions
- 10 ppm S, (Cr)-MIL-101-SO_3Ag outperform HKUST-1 /UMCM-150 : Ag(I) sites play more important roles than Cu(II) sites

13

a. A. J. Matzger, *J. Am. Chem. Soc.*, 2008, 130, 6938-6939; b. S. H. Jhung, *Chem. Eur. J.* 2014, 20, 376-380; c. A. L. Nuzhdin, *Mendeleev Commun.*, 2010, 20, 57-58

3 Adsorption Performance

Regenerability

Immersed in the solvent and under sonication for 1 h under room-temperature conditions

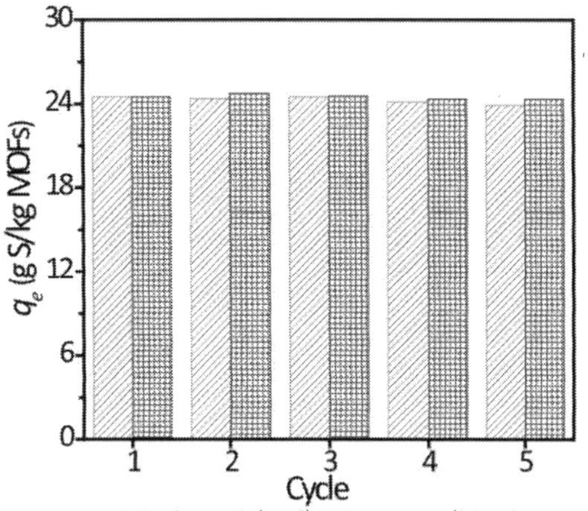

Methanol (red) /Acetone (blue)

- No obvious decay after five adsorption-regeneration cycles
- The regeneration is easily, used repeatedly

14

3 Adsorption Performance

Breakthrough curves

0.4 mg/mL DBT in *n*-octane for MIL-101-SO_3Ag (olive) and MIL-101-SO_3H (pink) at room temperature

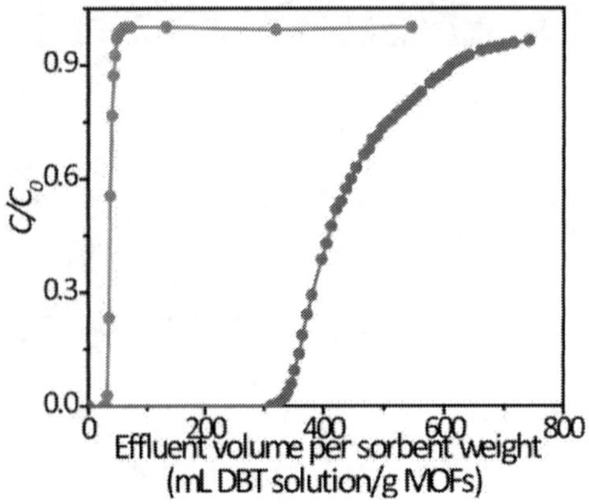

- Breakthrough point (defined as 1 ppm S), DBT——322 mL/g MOFs (22.5 g S/kg MOFs) for MIL-101-SO_3Ag, 30 mL/g MOFs for MIL-101-SO_3H
- Dynamic saturated adsorption capacities (integrating above the breakthrough curve), DBT——31.72 g S/kg MOFs

15

3 Adsorption Performance

Breakthrough curves

0.4 mg/mL BT in *n*-octane for MIL-101-SO_3Ag (olive) and MIL-101-SO_3H (pink) at room temperature

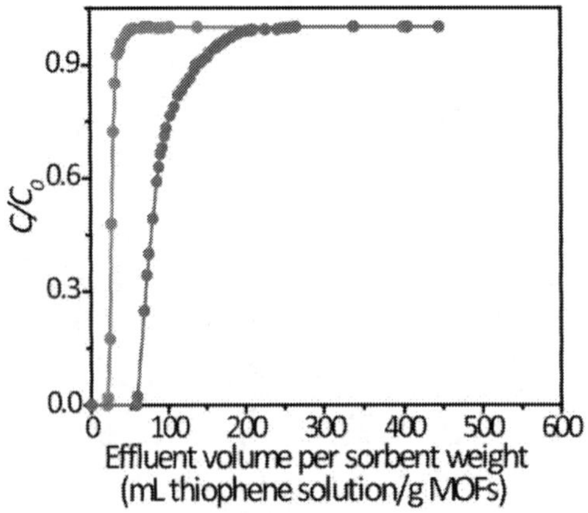

- Breakthrough point (defined as 1 ppm S), BT——156 mL/g MOFs (15.0 g S/kg MOFs) for MIL-101-SO_3Ag, 36 mL/g MOFs for MIL-101-SO_3H
- Dynamic saturated adsorption capacities (integrating above the breakthrough curve), BT——22.41 g S/kg MOFs

16

3 Adsorption Performance

Breakthrough curves

0.4 mg/mL TP in *n*-octane for MIL-101-SO_3Ag (olive) and MIL-101-SO_3H (pink) at room temperature

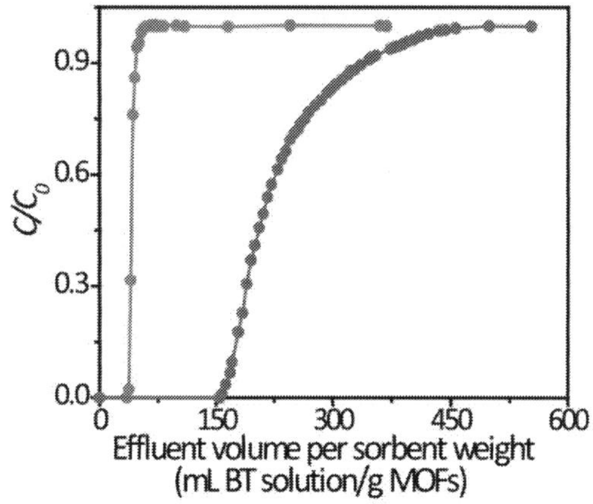

- Breakthrough point (defined as 1 ppm S), TP——58 mL/g MOFs (8.8 g S/kg MOFs) for MIL-101-SO3Ag, 18 mL/g MOFs for MIL-101-SO3H
- Dynamic saturated adsorption capacities (integrating above the breakthrough curve), TP——13.96 g S/kg MOFs

17

3 Adsorption Performance

Regeneration: elution curve

DBT from MIL-101-SO_3Ag
Acetone at flow-rate of 0.4 mL/min at ambient temperature for two hours

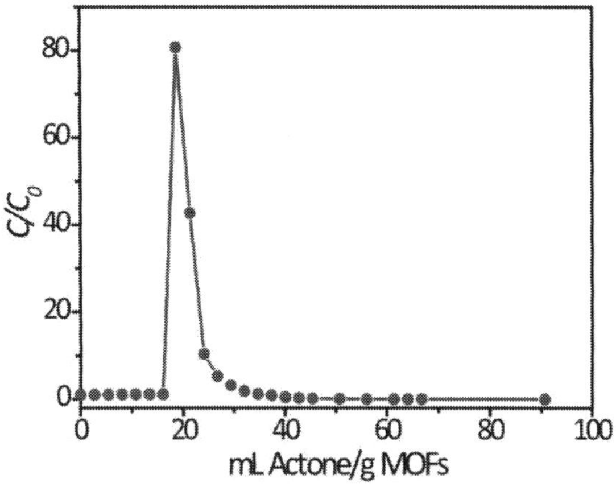

- The majority of thiophene derivatives can be eluted out within 30 minutes

18

3 Adsorption Performance

Breakthrough curves : performance of the recycled MOFs

0.4 mg/mL DBT in *n*-octane for MIL-101-SO_3Ag

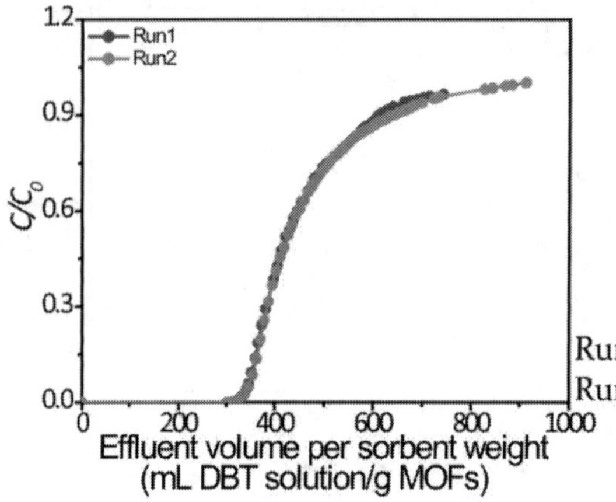

Run 1—— the initial
Run 2——after regeneration

- Run 1 and run 2 almost overlap——MIL-101-SO_3Ag is completely regenerated under these mild conditions

19

4 Conclusions

- Successfully prepared and characterized MIL-101-SO_3Ag
  - Ag(I) ions have preferentially strong interactions with thiophene derivatives through π-complexation
- Be stable under the moisture air, easily regenerated without losing its higher sorption capacity

Lead to some practical useful MOF materials for the ultradeep desulfurization!!

20

Acknowledgements

Fellows of the group:
Dr. Zongbi Bao
Dr. Baogen Su
Dr. Zhiguo Zhang
Prof. Huabin Xing
Prof. Yiwen Yang
Prof. Qilong Ren

National Natural Science Foundation of China

Thanks for your attention!

3 Adsorption Performance

The maximum adsorption uptakes
TP, BT, DBT are 20.5, 28.8, 31.0 g S/kg MOFs

Pseudo-second order kinetic constants (k_2) with correlation coefficients (R^2)

Adsorbate	q_e (g S/kg MOFs)	k_2 (kg/g S/min)	R_2
thiophene	12.11	0.080	0.994
BT	20.7	0.094	0.995
DBT	23.8	0.099	0.991

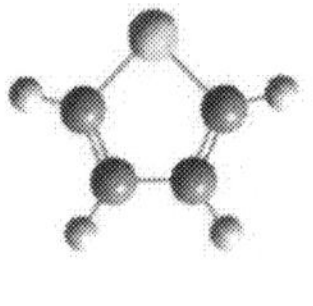
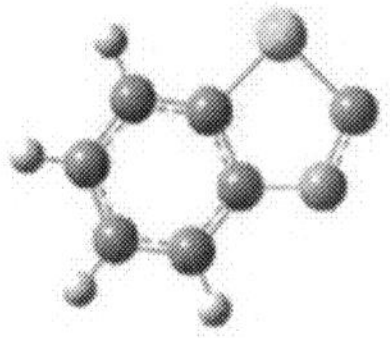
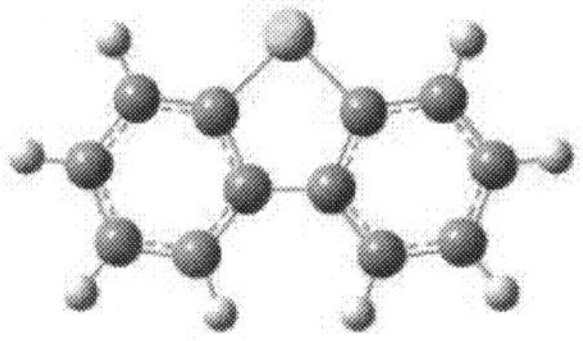

- More delocalized π electrons in DBT for their stronger interactions with Ag(I)
- The adsorption rates decrease when the molecule sizes increase and follow the order of TP > BT > DBT, as shown in their kinetic constants

23

429012 Ammonia Removal in Baitfish Retail Tanks and on-Farm Flow-through Grading/Holding Vats

Thursday, November 12, 2015: 3:15 PM
155F (Salt Palace Convention Center)

Pejman Ahmadiannamini[1], Satchithanandam Eswaranandam[1], Martin Christie[2], S. Ranil Wickramasinghe[2] and Xianghong Qian[1], (1)Department of Biomedical Engineering, University of Arkansas, Fayetteville, AR, (2)Department of Chemical Engineering, University of Arkansas, Fayetteville, AR

Ammonia Removal in Baitfish Retail Tanks and On-Farm Flow-through Grading/Holding Vats

Pejman Ahmadiannamini[1], Satchithanandam Eswaranandam[1], Martin Christie[2], Ranil Wickramasinghe[2], Xianghong Qian[1]

[1]*Biomedical Engineering, University of Arkansas, Fayetteville, AR*

[2]*Chemical Engineering, University of Arkansas, Fayetteville, AR*

Ammonia is the major metabolite of proteins for fish. Zeolite 13X is a promising adsorbent materials with high ion exchange capacity for ammonia removal in fishery water. However, the major concern is that zeolite particles will tend to degrade in water. Moreover, other ions and elements tend to leak into the solution preventing zeolite 13X as an effective adsorbent for ammonium removal in fishery water. In order to overcome the limitation of the direct adsorption approach, zeolite particles were incorporated within polymeric matrices in three different conformations, namely mixed matrix membranes; composite fibers; and pore-filled membranes.

Polymeric composite membranes were prepared via wet phase inversion processes by incorporating zeolite 13X particles into polysulfone matrices formed with and without PEG as a molecular porogen. With increasing zeolite content in the casting solution, the flux was decreased and the removal capacity was enhanced. Adding PEG in the casting solution resulted formation of more porous membranes with higher flux and lower of ammonium removal capacity. Similar solutions were also employed to spin composite fibers. It was found that employing PEG in spinning solutions resulted in lower ammonia removal capacities. Moreover, increasing zeolite and polymer contents resulted in increase and decrease of ammonia removal capacity of fibers, respectively.

Finally, zeolite pore-filled membranes were prepared via filtration of Zeolite 13X suspensions through flipped membranes in order to trap the zeolite powders within macrovoids of the membranes. To prevent particles shedding, the membranes were capped via interfacial polymerization method using an aqueous solution of piperazine and a solution of trimesoyle chloride in n-hexane. Higher zeolite particles were incorporated via pore-filled membranes compared to the mixed matrix membranes.Inductively coupled plasma mass spectrometry (ICP-MS) analysis showed no evidence of ion leakage into water treated by prepared zeolite 13X-polysulfone composite materials.

B

A

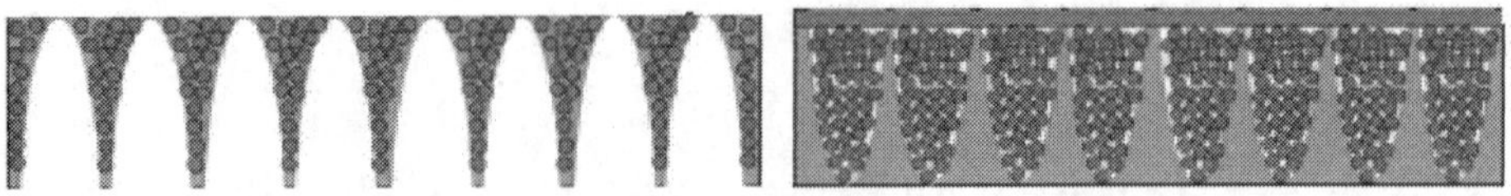

Schematic representation of zeolite 13X-polysulfone composite membranes: (A) mixed matrix membrane; (B) pore-filled membrane

432945 Cellulose Nanofiber Composite Membranes for Water Purification

Abstract:

Global water demand continues to increase because of population growth, urbanization, and climate change. Since there are limited natural water resources, water purification systems play a critical societal role. Membrane technologies are used widely to improve water quality for beneficial use. Therefore, the development of new membranes is an active area of research in the water purification field.

In this project, high-performance, nanofiber-based ion-exchange membranes were prepared for removal of heavy metals from impaired waters. Our hypothesis was that such membranes would have high metal ion capacities resulting from large surface area-to-volume ratios and high water permeability relative to barrier-type membranes such as reverse osmosis and nanofiltration. The nanofiber membranes were produced by electrospinning cellulose acetate nanofiber mats, which were strengthened by thermal-mechanical annealing and then converted to regenerated cellulose nanofiber mats. Subsequent surface modification transformed the membranes into ion-exchange membranes for removal of trace metal ions in water.

Cellulose acetate was selected as the starting material because it has been used widely to electrospin fibers due to simple processing. The electrospinning formulation and conditions yielded fiber mats having a fiber diameter in submicron range and a thickness range of 160 to 330 μm. The thickness range was reduced to 50 to 70 μm after annealing process. Following annealing, the nanofiber mats were converted to regenerated cellulose, which is a low-fouling material that can be modified using convenient surface chemistry to introduce the ion-exchange ligands.

Ion exchange using carboxylate anion ligands is a potential strategy to capture cadmium and nickel ions found in industrial wastewaters. Thus, the regenerated cellulose nanofiber mats were modified by grafting poly(itaconic acid) (dicarboxylic acid side groups) and poly(acrylic acid) (monocarboxylic acid side groups) to the surfaces of individual nanofibers that had been modified with polyglycidyl methacrylate (epoxide groups) by "Grafting to" method. The success of poly(itaconic acid) and poly(acrylic acid) grafting was confirmed by attenuated total reflectance Fourier-transform infrared spectroscopy.

Direct-flow filtration measurements showed that the polymer modified nanofiber membranes have water permeabilities of 1,500 L/m^2/bar/h, much higher than reverse osmosis or nanofiltration membranes. Results of metal ion static and dynamic ion-exchange capacities will be presented and compared to more traditional ion-exchange media.

424205 Surface Nano-Structuring of Polysulfone UF Membrane Surfaces

Abstract:

Membrane fouling remains as one of the major operational difficulties in both UF and RO membrane technologies. Various approaches have been proposed over the last four decades to modify membrane surface chemistry and topography in order to increase membrane fouling resistance. In the present work, surface nano-structuring via atmospheric pressure plasma-induced graft polymerization (APPIGP) has been developed to tailor-structure both polysulfone and polyamide UF and RO membrane surfaces, respectively, so as to increase surface hydrophilicity and thus also increase fouling resistance via the presence of a high density layer of terminally-anchored hydrophilic polymer chains. These nano-structured surfaces are created by generating high density of surface active sites on the base membrane surface, from which vinyl monomers are graft polymerized to create dense, terminally and covalently-bonded polymer brush layers. The above APPIGP approach was utilized to nano-structure polysulfone (PSf) membrane surfaces via graft polymerization of acrylic acid and 2-hydroxyethyl methacrylate monomers. The achieved level of surface hydrophilicity was evaluated with respect to surface activation via plasma treatment (i.e., reactive plasma gas species, radio frequency (RF) power, and plasma exposure time) and graft polymerization (i.e., initial monomer concentration, reaction temperature, and grafting time) conditions. The resulting modified polysulfone surface properties (chemistry and topography of polymer brush layers) were characterized by atomic force microscopy (AFM), contact angle measurement, and X-ray photoelectron spectroscopy (XPS). The results showed that significant reduction of surface wettability was achieved upon graft polymerization. Moreover, surface hydrophilicity and roughness of the modified polysulfone surface can tuned by varying plasma and grafting conditions, indicating that the APPIGP methodology could be beneficial for surface modification of commercial polysulfone UF membranes for improving the surface hydrophilicity and hence fouling mitigation.

415915 Mussel-Inspired Hybrid Coatings That Tailor Membrane Wettability for Hash Water Treatment

Abstract:

Mussel-inspired, hybrid coatings have be built firmly on hydrophobic polyvinylidene fluoride (PVDF) membranes in a facile manner via simultaneous polymerization of mussel-inspired dopamine and hydrolysis of commercially available and low-cost silane through a highly efficient one-step approach. The hydrophilicity of the hybrid coating can be tuned by adjusting silane concentration, which can be applied to rationally design optimized hybrid coatings to significantly improve the performance of PVDF ultrafiltration (UF) membrane and commercial PVDF microfiltration MF membrane, respectively. The hybrid coatings f are utilized to modify the PVDF UF membrane for treating protein-rich water with drastically enhanced antifouling performance and excellent dry storage ability, while the hybrid coatings can endow PVDF MF membrane with high hydrophilicity, underwater superoleophobicity and low oil adhesion for high-efficiency oil-in-water emulsion separation. More importantly, these hybrid coatings are extremely stable and can withstand rigorous washings or cryogenic bending. The simplicity and versatility of this novel mussel-inspired one-step strategy may bridge the material-induced technology gap between academia and industry, which makes it promising for eco-friendly applications.

432305 Functionalization of Silica Membranes with Lipid Bilayers for Sugar Recovery

Abstract:

Supported lipid bilayers on porous platforms offer biomimetic approaches to highly selective separations in dilute aqueous solute systems. Mesoporous silica materials (in the form of particles or thin films) with tunable pore morphology, chemical, thermal stability and ease of organic functionalization, are an ideal support for lipid bilayers. Our previous investigation of lipid bilayer interactions with mesoporous silica particles demonstrated both lipid bilayer enveloping and pore filling of the lipid bilayers as a function of pore size through vesicle fusion and evaporation deposition, respectively. Combining selective permeable lipid bilayers with high surface area silica materials suggests many potential applications for efficient, selective dilute aqueous solute separations. The recovery of aqueous fructose, which can be bound and released by boronic acid carriers, has been chosen as model separation, representing the recovery of carbohydrates from the depolymerization of biomass. The separation of fructose using lipid bilayer supported on mesoporous silica thin film composite membranes is examined, in which the carbohydrate carrier (boronic acid) is immobilized in the bilayer. The selective separation of carbohydrates from biomass hydrolysate mixtures will address significant challenges to the biochemical production of biofuels and chemicals from lignocellulose.

In this work, vertical oriented hexagonally close packed (HCP) silica thin films on macroporous support are used as the lipid layer supports. The composite silica thin film membrane was prepared by dip coating silica membranes on anodic alumina supports using cross-linked poly (ethylene glycol)-block-poly (propylene glycol)-block-poly (ethylene glycol) (P123) as a temporary pore blocking film. TEM study demonstrated that the silica membrane, synthesized using P123 as a pore template, has ordered pore structure. Furthermore, SEM imaging suggested that a continuous and uniform silica membrane was formed on the top of porous support. Solvent flux of ethanol and permeability of fluorescent molecules through the composite membrane were used to characterize the transport properties of the membrane. Through vesicle fusion, the lipid bilayers containing a lipophilic boronic acid carrier was formed on the single side of the platform. The diffusivity of fructose through the composite membrane before and after the immobilization of boronic acid into lipid bilayer is examined and used to identify the environmental factors to improve the transportation and selectivity for fructose. Specifically, the location of lipid bilayer (lipid filling and lipid enveloping), boronic acid composition and pH gradient are studied.

429254 Inverse Colloidal Crystal Membranes for Hydrophobic Interaction Chromatography

Abstract:

Membrane based hydrophobic interaction chromatography has gained interest due to its excellent performance in the purification humanized monoclonal antibodies. The membrane material used has typically been commercially available PVDF. Using a microporous membrane as a chromatographic support material is advantageous as one overcomes the adverse effects of hindered pore diffusion observed in resin based chromatography. However lower capacities for membrane adsorbers compared to packed beds remains a significant disadvantage.

We have developed high porosity uniform pore size inverse colloidal crystal (ICC) membranes. Inverse colloidal crystal structures have received extensive attention due to their highly periodic structures, high porosity and fully interconnected pores. Inverse colloidal crystals or inverse opals are produced from colloidal crystals which are long range ordered lattices assembled from polymeric colloids. Here we have investigated the use of ICC membranes as membrane adsorbers. In particular their high porosity as well as their highly interconnected and uniform pore structure will lead to low pressure drop and uniform flow through the membrane, both of which are highly desirable for membrane adsorbers. In addition, the presence of a high membrane porosity results in a high surface area for solute binding. The capacity of the ICC membranes developed here is up to 10 times greater than commercially available PVDF membranes with a similar pore size. This work highlights the importance of engineering the membrane pore structure in order to maximize capacity.

412377 Regenerable Amine-Functional Polymers for Carbon Dioxide Capture from Ambient Air

Abstract:

The continually increasing levels of atmospheric CO_2 resulting from the burning of fossil fuels has triggered an enhanced concern for developing efficient and economically plausible methods for capture and sequestration of this greenhouse gas. Commercially available processes for carbon dioxide capture mainly involve the use of amine-based liquid absorbents which require high energy consumption due to their relatively high heat capacities. Our work has focused on the synthesis of several amine-based polymers, namely, poly(vinyl amine), poly(allyl amine), poly(acrylamide), poly(vinyl formamide) and, poly(vinyl pyridine) and their fabrication as supported column packing materials for use in carbon dioxide separation from ambient atmosphere. The produced materials exhibit CO_2 adsorption capacities on the order of 50 mg/g at room temperature and atmospheric pressures. In addition, their regeneration is achieved at temperatures around 110 °C. Our results indicate that the ability of these materials to effectively separate CO_2 at the low concentrations typical of ambient conditions, offers great promise for their use in a number of environmental, industrial and scientific applications.

412796 MOF Supported Amine Materials for CO2 Capture from Air: An Investigation Under Dry and Humid Conditions

Abstract:

MOF Supported Amine Materials for CO_2 Capture from Air: An Investigation under Dry and Humid Conditions

Lalit A. Darunte, David S. Sholl, Krista S. Walton, & Christopher W. Jones

School of Chemical & Biomolecular Engineering, Georgia Institute of Technology, 311 Ferst Drive, Atlanta, GA 30332, USA

Abstract

Supported amine materials belong to popular class of solid adsorbent materials for CO_2 capture. While flue gas capture has been one of the most researched areas, air capture also offers an attractive and complementary option. Among commonly considered materials as a support for the amines, metal organic frameworks (MOFs) are promising materials with tunable pore structures and properties. We analyzed MIL-101 supported amine materials for CO_2 capture from air. Under constraints of the small pore size and relatively fixed availability of open metal sites to bind amines, we consider low and high molecular weight amine structures and the effects of amine structures on equilibrium CO_2 capacities. Under dry conditions and for similar nitrogen loadings, low molecular weight amines showed higher equilibrium capacities while high molecular weight amines showed superior cyclic performance. For practical air capture applications, the cyclic stability of materials and equilibrium capacity under humid conditions are important parameters. We have combined X-ray diffraction, gravimetric uptake measurements, a custom-built volumetric system, electron microscopy and infrared spectroscopy to develop an understanding of the effect of humidity on CO_2 capture and on the structure of the material. The kinetics of CO_2 capture, which have important implications on the working capacity of the sorbent, are also considered.

418506 Nitrogen-Functionalized Porous Carbons for Enhanced CO_2 Capture

Abstract:

Nitrogen-Functionalized Porous Carbons for Enhanced CO₂ Capture

Peter C. Psarras, Jiajun He, Jennifer Wilcox, Energy Resources Engineering, Stanford University, Stanford, CA

Separations Division

"CO2 Capture by Adsorption II: Adsorbents"

The global consequences of increased anthropogenic carbon dioxide emissions are well documented. To avoid the catastrophic long-term damages associated with climate change, much attention has been devoted to the development of CO₂ emission mitigation strategies. One of the more mature methods involves CO₂ absorption via amine-based solvents; however, this technology is notoriously energy intensive, as amine regeneration requires high regeneration temperatures. Additionally, this process is water-intensive, a factor that is becoming increasingly important in drought-ridden states like California. Solid sorbents represent an alternative method for CO₂ capture, with liquid solvents replaced by (typically) large surface area, carbon-based porous frameworks. These sorbents are generally low-cost, easily fabricated, and are far less energy intensive in terms of regeneration. Further, their design can be customized through the inclusion of surface-functionalized groups. Unfortunately, current sorbents display CO₂ uptakes that are deemed too low to be cost-competitive with traditional solvent-based scrubbing. Nitrogen-functionalization of these porous carbons could result in more efficient CO₂ capture, owing to the same chemistry exploited by basic solvents in capturing acidic CO₂. Similar studies have examined the effects of oxygenated functional groups on CO₂ uptake and selectivity over N₂.

This studt will employ grand canonical monte carlo methods to explore the effect of quaternary, pyridinic, pyrollic, and oxidized-N groups on CO₂ uptake in porous carbon sorbents. Doping amounts will be varied to ascertain the optimum coverage for CO₂ capture. Additionally, gaseous mixtures of CO₂/N₂ and CO₂/N₂/H₂O will be examined to assess CO₂ selectivity. Theoretical performance will be validated through comparison with experimentally obtained CO₂/N₂ isotherms over fabricated hierarchial micro/mesoporous N-doped carbon sorbents. The combination of these methods will help to inform on sorbent design by illustrating which groups are most important for enhanced CO₂ uptake. Design to include more of a particular functional group can be achieved through, for example, a change in carbonization temperature.

421498 Hierarchical LTA Zeolites with Amine-Functionalized Mesoporous Domain for Post-Combustion Carbon Dioxide Capture

Abstract:

Owing to global concern on climate change caused by rising atmospheric CO_2 concentration, the technologies for capturing CO_2 from anthropogenic sources, such as coal-fired power plants, have attracted a great amount of interests. Among various methods available, capturing CO_2 with solid adsorbents has been considered as an energy-efficient alternative to conventional amine scrubbing that consumes a large amount of energy for the regeneration of media. Efforts to develop solid adsorbents performing at conditions relevant to the flue gas released from coal-fired power plants lead to synthesis of various alkylamine-functionalized mesoporous materials including MCM-41 and SBA-15. Even though amine groups in such materials can effectively capture CO_2 molecules, the substrates typically have thick pores walls that cannot contribute to CO_2 uptake while increasing the mass of sorbent. Thus, optimizing the substrate structure can significantly improve the CO_2 uptake property of resultant materials.

In this work, we synthesized hierarchical LTA zeolites containing amine-functionalized mesoporous domain. Gas adsorption study revealed that CO_2 molecules are effectively captured by both alkylamine groups in the mesopore domain and active sites in the LTA zeolite. Hence, the amount of CO_2 uptake on this material was much greater than that for conventional aminosilicas at the conditions relevant to post-combustion CO_2 capture.

(728e) Enhancement of Carbon Dioxide Adsorption in Presence of Water in Zeolite Materials

Kim, J., KAIST
Jeong, W. S.,
Originally presented on: 11/12/2015 16:35:00 - 16:55:00

In the past decade, there has been significant amount of research devoted to discovering porous materials that possess large carbon dioxide uptake for carbon capture applications. However, less attention has been made on the effect of water in capture capabilities despite the fact that water comprises 5-8% of the total post-combustion flue gas composition. Unfortunately, water reduces the carbon dioxide capture capacity significantly in most materials as water and carbon dioxide molecules compete for the same adsorption sites with water molecules possessing stronger binding energies within the host frameworks. In this talk, we report a computational screening study on the effect of water in carbon dioxide adsorption for pure-silica zeolite structures. Using grand canonical Monte Carlo simulations, we have identified few zeolite structures that shows significant enhancement in the carbon dioxide uptake in presence of water. Material properties that are commony featured in these high performing zeolites are analyzed and quantified, making it easier to identify these structures prior to experimental synthesis.

434422 Fabrication of Modified-Amine Functionalized Adsorbents for Carbon Dioxide Capture

Abstract:

Amine functionalized adsorbents have been considered adequate for low temperature CO_2 adsorption process because of their high CO_2 adsorption capacity and selectivity. A large number of amine functionalized adsorbents had been developed and chemical grafting or physical impregnation was their main approach. Recently, the double-functionalized amine adsorbents fabricated by grafting aminosilanes on solid supports followed by impregnating amines were developed to compensate the low CO_2 adsorption capacity of amine-grafted adsorbents. Although they showed good adsorption properties including high amine efficiency, but exhibited low thermal stability in regeneration arising from the leaching of impregnated amine.

In this work, the double-functionalized amine adsorbents with chemical linker were fabricated to improve the performance in regeneration. Chemical linkers having the reactivity with amines were introduced into bare silica support and aminosilane-grafted solid adsorbents, respectively. The covalent bond between the linkers and the amines is expected to improve both the thermal stability by holding amines and the regeneration rate by lowering CO_2 binding energy of the amines.

As-prepared adsorbent showed more stable cyclic adsorption capacity and faster adsorption/regeneration rate compared to the double-functionalized amine adsorbents without chemical linker from the thermogravimetric anlysis (TGA). The improvement in the thermal stability was also observed from that results. The chemical bond between the chemical linkers and the amines is confirmed from IR spectra analysis. Adsorption/desorption cyclic testing was performed to assess the long term stability of the fabricated sorbent in the flue gas condition for potential application.

434465 Direct Evaluation of CO2 Adsorption Thermodynamics and Origins of Lower Amine Efficiency on Molecular Type I Adsorbents

Abstract:

The combination of a breakthrough reactor (BTR) and a commercial differential scanning calorimetry (DSC) is a versatile apparatus to study adsorption thermodynamic and kinetics of CO_2 adsorption on solid aminosilica materials under humid and dry conditions. The breakthrough reactor is utilized to measure CO_2 adsorption capacity and efficiency while the heat of adsorption and desorption are assessed in the DSC simultaneously.

In this study, a variety of linear molecular amines including diethylenetriamine (DETA), triethylenetetramine (TETA), tetraethylenepentamine (TEPA), pentaethylenehexamine (PEHA) on silica material have been utilized to synthesize impregnated Type I sorbents with various surface densities of amines on silica material have been used as models for polyethyleneimine (PEI). These amines contain two primary amines with varying primary:secondary amine ratios which increase in the order of DETA, TETA, TEPA, and PEHA. The experiments were carried out in a 10%CO_2/He gas flow at four different temperatures. The CO_2 adsorption capacity, heat of adsorption, and nitrogen efficiency of aminosilica materials for all loadings have been studied as a function of secondary amine increment.

The results indicate that at low surface densities, CO_2 adsorption decreases as temperature increases, while at high weight loadings, it increases with temperature. In addition, it was found that at low surface densities, lower secondary to primary amine ratios are in the favor of CO_2 adsorption while it behaves conversely at high weight loadings. Adsorption efficiency increases as surface density rises up to a moderate value (around 30 wt. % amine/SiO_2), and then decreases with further increases in the amine loading. It was also found that at low surface densities, the efficiency is directly proportional to the primary to secondary amine ratio while inversely proportional to temperature. At low surface densities, increasing amine weight loading leads to higher heat of adsorption even when normalized per mole of CO_2 adsorbed suggesting that a number of primary amines remain inaccessible to CO2 at lower temperatures. At high weight loadings, heat of adsorption passed through a maximum as temperature increases. Heats of adsorption are higher for lower secondary to primary amine ratios, which indicates that primary amines always have higher heats of adsorption.

These results demonstrate the adsorption thermodynamics in a real scenario (rather in a combined static adsorption equilibrium apparatus combined with calorimetry) and have recently been used to describe the spatiotemporal dynamics of an isothermal aminosilica adsorption bed.

435432 Life in Plastics

Abstract:

If you take the human existence as a black box. Then many inputs that go into this box and then there is also many outputs that come out of the box. There is also accumulation and growth within the box. The one thing that is clear is there is a large consumption of natural resources and energy that goes into this black box and there is also a large waste stream which is an output. If we take a step back and see how this human operation work, we can try to re-engineer the process to find more efficiencies.
I will start by telling the story of the history of waste.
A few of the earliest known occurrences of deal with waste can be found as early as 3000 B.C. Around this time first landfill was developed when Knossos, Crete digs large holes for refuse where garbage was dumped and filled with dirt at various levels. In 2000 B.C. China developed methods of Composting and Recycling, and recycling bronze for later use. And in 500 B.C. in Athens, Greece a new law claiming garbage must be dumped at least one mile from the city was implemented.

Now we are in 2014 and much has changed with the new innovations in technology, communication and infrastructure. We now have the necessary tools to re-think waste the old ways of doing things.
My current focus is on plastic waste, both industrial and consumer. Plastics have an interesting life-cycle that spans from fossil fuels from which they are made to landfills which they end up. If it is possible to recycle and reuse these resins in an energy efficient way then this is the best way to continue their life-cycle, however due to the mixed nature of plastics this is often difficult as there are many different types of resins as well as colors. The solution that I propose is to convert this mix waste stream into usable hydrocarbons which can be used as fuels or chemical feed-stocks.

In Salt Lake City, Utah we are doing just that. We have spent the last 4 years in collaboration with the department of chemical engineering at university of Utah developing a commercial process which converts mixed plastic waste streams into high quality low sulfur fuels. I will be presenting this journey with you.

423426 Investigation of Woody Biomass and Solid Residue Generated from Anaerobic Digestion and Their Allocation to Potential Energy Generation

Abstract:

In this work, a two-stage municipal solid wastes (MSWs) disposal process has been proposed and its feasibility study has been carried out. At first stage, anaerobic digestion of municipal solid waste, such as grass, leaves and animal manure, was conducted by mixing with anaerobic sludge. Cumulative biogas yield of grass and chicken manure through high-solid co-digestion was 373 mL/g volatile solid (VS) with initial total solid (TS) 20wt% at the temperature of 35°C within 40 days. After digestion, the solid residue was dried until the moisture decreased to 40 wt%. TG analysis results showed that 86wt% of dried residues were volatile solids which could be used as feedstock for gasification. At second stage, the dried solid residue was co-gasified with woody biomass for syngas generation in a fixed-bed downdraft gasifier. Co-gasification of woody biomass and sewage sludge has already been done in our preliminary study. The experimental results showed that 20 wt. % dried sewage sludge in the feedstock was effectively gasified to generate producer gas comprising over 30 vol. % of syngas with an average lower heating value of 4.5 MJ/Nm^3. Parameter studies were performed on the basis of optimizing syngas generation of this two-stage process. These included the effects of mixing ratio, heavy metal content and moisture content on the gasification performance. This two-stage process has also been compared with single one stage of anaerobic digestion and gasification process in terms of energy conversing efficiency. Through the two stage process, full utilization of MSWs can be realized to produce biogas and syngas through anaerobic digestion and gasification.

423466 Bio-Power Generation Characteristics of a Spark-Ignited Engine Operated on Syngas Derived from Biomass Downdraft Gasification

Abstract:

The power production from renewable sources must increase intensely to meet the growing demand of power across the globe. This study focusses on evaluating the operational performance of a 10 kW Spark Ignited (SI) engine operated on syngas produced from downdraft gasification of biomass. A commercial natural gas based SI engine is modified to accommodate biomass syngas and for the measurement and control of air-fuel ratio. The load variant from 1 to 5 kW was regulated by an electric load bank to evaluate the engine operational characteristics. The continuous operation of the gasifier and syngas cleaning system were monitored to analyze quality of the syngas that was injected into the engine. The variation of the biomass feedstock was observed to investigate its impact on syngas quality and engine operation. The major combustible elements of the syngas such as CO, H_2, and CH_4 were analyzed. Output power, engine efficiency, and exhaust emissions such as CO_2, CO, NOx, SO_2, H_2S, and hydrocarbons were also examined. The outcome of this study provides the engine performance operating with 100% syngas prior to the commercial scale-up for OSU's large bio-power generation system

436250 Recovery of Energy-Dense Hydrochar and Sugars from Sweet Sorghum Bagasse

Abstract:

Sweet sorghum, a C4 crop in the grass family has potential to produce ethanol biofuel due to its high carbohydrate content and high biomass productivity. It consists of 8% seed head, 19% leaf matter, 36% bagasse, and 37% juice and can be cultivated on marginal lands, has low input requirements, and is adapted to nearly all temperate climates. Sweet sorghum accumulates fermentable sugars in the stem which is extracted in a conventional roller press and fermented into ethanol using yeast and engineered microbes. A large amount of biomass residue (or bagasse) is left behind the extraction unit operation and its utilization will be critical to sustainability of ethanol production from sweet sorghum. Hydrothermal carbonization (HTC) is an effective technology for converting waste biomass and crop residues into a uniform, energy-dense solid fuel that has improved storage/handling characteristics. This study evaluates the HTC of sweet sorghum bagasse into an energy-dense hydrochar for subsequent energy use as fuel or further conversion by combustion/pyrolysis and a sugar rich water soluble product. HTC treatment of three varieties of sweet sorghum bagasse, namely, Theis, M81-E, and Dale, was performed in pressurized batch reactors at 175-275oC and 30 min residence time. HTC products were separated into hydrochar, water soluble phase, and gases, which were analyzed by standard laboratory methods. HTC resulted in 45-65% hydrochar that had 15-20% higher energy than the raw bagasse. The water soluble products showed about ~10,000 ug/mL sugars that consisted of glucose, sucrose, and fructose. Pelletization of hydrochar generated robust pellets that showed superior qualities than pellets from woody biomass feedstocks.

410260 Biofuel from Spent Coffee Grounds

Abstract:

As the fossil fuels reserves around the world are being depleted, the need for an alternative renewable source of fuel is evident. Currently, biodiesel serves as a promising source of renewable fuel. However, it is a fairly expensive process with its feedstock comprising the major cost. Therefore, the search for alternative feedstocks is important and spent coffee grounds (SCG) can be a large and economic source of vegetable oil. The present study involves the optimization of coffee oil extraction with hexane solvent and subsequent transesterification to biodiesel. This reaction was studied using inorganic base materials as catalysts, as well as the enzyme Lipase, both in solution and immobilized. The results so far indicate that an oil yield of 20 weight % from dry SCG can be obtained, and conversions to biodiesel of up to 50% can be achieved with both forms of the enzyme catalyst, with methanol or ethanol as reactants. The immobilized lipase is prefered because it can be used for extended times in a packed bed reactor (PBR) without production of undesired soap. The reaction kinetics were also studied in a lab scale PBR.

427591 Enhanced Electrochemical Performance of Waste Tyre Recycled Activated Carbon As an Anode in Li-Ion Battery

Abstract:

With the ever increasing global demand for energy supply, it has become imperative to address the issue of energy and environment sustainability. In this work, we present an effective synthetic procedure to obtain high purity activated carbon from pyrolitic tyre char and demonstrate its performance as a high capacity anode in Li-ion battery. Carbon obtained after pyrolysis of waste tire crumbs is first treated with concentrated HCl, followed by HF treatment which converts it to a high purity carbon black devoid of all metallic and sulphur impurities. The surface properties of the acid treated carbon is further tailored by activating it with KOH, resulting in a high specific surface area of 870 m^2/g with a high percentage of mesopores (avg pore size 3-5 nm). The morphology, structure and composition of the activated carbon have been studied by SEM, TEM, Raman, XRD, XPS and FT-IR techniques. Its electrochemical performance is analyzed by galvanostatic charge/discharge measurements at different current densities, cyclic voltammetry and impedance measurements. The activated carbon exhibits a first diacharge capacity of 880 mAh/g at 50 mA/ g current density, which is much greater than other (hard) carbon based anodes and more than twice that of commercial graphite anode. A high reversible capacity of ~ 620 mAh/g is maintained after 100 cycles with nearly 100% coloumbic efficiency. Our work opens a new avenue to tyre waste recycling as the carbon black processed by us exhibits enhanced electrochemical performance with potential application in energy storage devices.

433364 Catalytic Conversion of Automobile Shredder Residue to High Energy Fuels

Abstract:

Catalytic pyrolysis of automobile shredder residue to high energy fuels

FA Agblevor, K. Christian, S. Hatch, J. Long

USTAR Bioenergy Center, Biological Engineering, Utah State University, Logan UT 84322

Abstract

Automobile shredder residue (ASR) or auto-fluff is waste product from the extraction of metals from automobiles. The material is composed of plastics, vinyl, cellulose, wood, metals, leather, and other materials. ASR is currently disposed in public landfills. However because of the organic material content of this residue, it is potentially possible to convert the material to fuels using pyrolysis technology. ASR and red mud samples were characterized using several methods. The ASR samples showed large variability and had very high ash content. The ASR samples were pyrolyzed using sand and red mud. Pyrolysis oils obtained using sand as the pyrolysis medium were more viscous than those using red mud and they also contained polychlorinated biphenyls (PCB) at high levels. The yield of ASR pyrolysis liquids ranged from 2 to 32 wt%. Pyrolysis with red mud produced much lower viscosity oils and which also had very low PCB content. Whereas there were several types of PCBs in the sand pyrolysis oils, there was only one type PCB congener (trichlorobiphenyl) in the red mud ASR pyrolysis oils. Modification of the red mud with reduced the PCB content of the ASR red mud pyrolysis oils to 5 ppm compared to over 50 ppm in the sand ASR pyrolysis oils. The red mud ASR pyrolysis oils had close to neutral pH while the sand pyrolysis oils were acidic. The gas yield ranged from28 to 36 wt%. The red ASR pyrolysis gas was H2-rich while the sand catalyzed pyrolysis oil was CO2-rich. The char-yield was very high and ranged from 38 to 42 wt% and had ash content over 70%. The red mud ASR pyrolysis oil in its current state can be potentially used as heating fuel but needs upgrading to qualify as transportation fuel.

430405 Hydrothermal Carbonization for Nutrient Recovery

Abstract:

A variety of macro- and micronutrients are required for plant's healthy growth and maximum crop production. The most common macronutrients associated with commercial fertilizers are nitrogen, phosphorous, and potassium (NPK), each of which is commercially available in large quantities for modern farming. Until now, NPK containing fertilizers are produced by either accompanying chemical processes (e.g., Haber-Bosch process) or physical processing of mineral ores (e.g. potash). Although, most of the common waste streams (e.g. sewage sludge, dairy manure etc.) are very rich in macro-and micronutrients, the lack of economically and environmentally benign processes is hindering the use of nutrients from wastes.

Hydrothermal carbonization (HTC or wet torrefaction) is a pretreatment process, which makes a homogenized, carbon and nutrient rich solid product, often called hydrochar, from underutilized various wet feedstocks, including dairy manure and sewage sludge. Compared to raw biomass, hydrochar is more hydrophobic and more biologically inert, which favors better storage properties for further applications. In this pretreatment method, biomass is treated with hot compressed water in an inert environment in the temperature range of 180-260 °C. The ionic constant of water is highest in this temperature range. As a result, the oxygen containing volatiles are removing from the solid, leaving a stable carbon and nutrient rich solid hydrochar.

In this study, macronutrients especially NPK recoveries from various wastes were evaluated by HTC. Both dairy manure and sewage sludge were hydrothermally carbonized at 180-260 °C for 5-30 min. Physico-chemical properties of both solid and liquid products were determined by means of ICP-OES, Lachat, TOC/TN, and elemental CHONS analyses. Total nitrogen and NH_4-N in the HTC process liquid was increasing with HTC temperature and time. However, more than half of the starting nitrogen was recovered in the hydrochar. Phosphorus recovery in the solid hydrochar was also increasing with HTC severity but potassium was found entirely in the process liquid. In conclusion, HTC can be applied to extent for nutrient recovery especially for nitrogen and phosphorus in the solid product.

435161 Production of Bio-Diesel Additives from Dairy Manure in Supercritical Ethanol

Abstract:

Production of Bio-diesel Additives from Dairy Manure in Supercritical Ethanol

Xiaokun Yang[1], Mi Lu[1], Marcus Vinicius Pereira[1], Simon Poulson[3], Charles Coronella[1], and Hongfei Lin[1*]

[1]Chemical Engineering and Materials Science Department, [2]Agriculture, Biotechnology & Natural Resources Department, [3]Geological Sciences & Engineering Department,

University of Nevada, Reno, NV, U.S 89557

E-Mail: HongfeiL@unr.edu

Dairy manure is regarded as a waste material that needs substantial waste management. However, as the biomass feedstock it contains lipid, protein, lignin, hemicellulose, and cellulose, which can be used to produce value-added chemicals. Herein, dried dairy manure was converted in ethanol and ethanol-water solutions at near supercritical conditions. The highest conversion of manure reached to 72% (ash free) under supercritical ethanol at 245°C. In the presence of solid Lewis acid catalysts, Zr-KIT-6 or hydrated niobia, ethyl esters from cellulose, hemicellulose and lignin were the main products, which can be used as biodiesel additives. It was found that adding 5-10 % water in ethanol enhanced the decomposition of recalcitrant cellulose and lignin, and thus increased the manure conversion and the overall yields of ethyl esters. Various process conditions including temperature, pressure, biomass-to-catalyst ratio, water-to-ethanol ratio, etc. were examined and optimized. High-loading amounts of manure were used in this process which makes the catalytic process promising for future's scale-up to industrial scale. Lastly, the possible reaction mechanism based on Lewis acid catalysis in supercritical ethanol-water solvents was discussed.

423642 Carbothermal Zinc Recycling in a Packed-Bed Solar Reactor – Experimental Demonstration and Dynamic Modeling

Abstract:

Carbothermal Zinc Recycling in a Packed-Bed Solar Reactor – Experimental Demonstration and Dynamic Modeling

Waelz oxide (WOX) is produced in the dominating Zn recycling pyrometallurgical process, the so called Waelz process. The crude WOX generated contains 55–65% Zn in oxidic form, along with considerable amounts of chlorine, fluorine and lead compounds. A washing treatment for the production of dehalogenized WOX (washed WOX) is thus required before using WOX as a feedstock along with ZnS concentrates for the electrolytic production of primary Zn.

In the present study, an environmentally cleaner process route for the purification of crude WOX is experimentally investigated using concentrated solar process heat [1]. The solar-driven process comprises the thermal purification of WOX via volatilization of the Cl-, F- and Pb-compounds at temperatures between 1000 and 1285 °C to yield solar-clinkered WOX, which contains approximately 80 wt.% ZnO and can be used as feedstock for the subsequent carbothermal reduction.

Experimentation was carried out using a 10 kW_{th} packed-bed solar reactor [2]. It consists of two cavities in series separated by a thin ceramic plate, with the upper cavity serving as the solar radiative absorber and the lower cavity containing the reacting packed bed of solid feedstock. The laboratory scale reactor was subjected to solar flux concentrations of up to 2,500 suns (1 sun = 1 kW/m^2). Solar-driven purification of WOX at above 1265°C reduced the amount of Cl- and Pb-compounds to levels lower than those of washed WOX (< 0.1 wt.%) and led to ZnO contents of approximately 88 wt.% in the product.

The second process investigated was the solar-driven carbothermal reduction of crude, washed, and solar-clinkered WOX as well as commercial ZnO powder. Beech charcoal was used as reducing agent for a CO_2-neutral carbothermal process. Solar experimental runs were carried out with the 10 kW_{th} packed bed reactor at temperatures up to 1170 1320°C and ZnO:C molar ratios between 1:0.8 and 1:1.25. Condensed products contained 90 wt.% Zn.

A dynamic reactor model was developed to solve numerically the transient energy conservation equation coupling heat transfer with the reaction kinetics. Model validation was accomplished by comparison of packed-bed temperatures and Zn production rates with experimentally measured data obtained with the 10 kW_{th}packed-bed solar reactor. The validated model was further applied for optimizing the solar reactor configuration.

References

1. Wieckert C, Frommherz U., Kraeupl S., Guillot E., Olalde G., Epstein M., Santen S., Osinga T., Steinfeld A., "A 300 kW solar chemical pilot plant for the carbothermic production of zinc", *ASME Journal of Solar Energy Engineering*, Vol. 129, pp. 190-196, 2007.
2. Tzouganatos N., Matter R., Wieckert C., Antrekowitsch J., Gamroth M., Steinfeld A., "Thermal Recycling of Waelz Oxide using Concentrated Solar Energy", *Journal of Metals*, Vol. 65, pp. 1733-1743, 2013.

427565 Development of a Reactive Slurry-Slurry Approach to Improve Crystallization Process Performance and Scalability

Abstract:

The crystallization of a potent compound, hereafter refered to as compound A for the sake of confidentiality and currently in development at GlaxoSmithKline proved to be challenging in terms of obtaining acceptable yield, consistent and desirable API powder properties, and a good process robustness and scalability. In order to address these issues, the salt formation crystallization was redesigned and a relatively novel approach was developed that allowed significant improvement in yield (from ≈70% to ≈95%), consistent API powder attributes within the desired range and an improved process manufacturability. This presentation describes the approaches and the process design of the crystallization recipe which utilizes a seed bed of the base as mitigating factor to prevent spikes in supersaturation with respect to the salt. The new process design resulted in a relatively novel crystallization path, which is a direct result of the solubility behaviors of both the base and the salt. The process was scaled-up successfully in the pilot plant to manufacture material for clinical supply.

427844 Effect of Seed Type on the Resistance of API Agglomerates to Agitated Drying

Abstract:

Control of the Active Pharmaceutical Ingredient's (API) powder properties is often necessary to ensure successful drug product development and manufacture. This is, in part, accomplished by designing a crystallization process that can deliver consistent batch-to-batch powder properties. When an API forms agglomerates during the crystallization, the resulting powder often exhibits enhanced flow properties and bulk density, which are of benefit to the formulation. However, development of a crystallization to consistently produce agglomerates that can withstand downstream processing, particularly agitated drying, represents a significant challenge. Weak agglomerates that are prone to break during agitated drying increase the risk of batch-to-batch powder property variability and scale dependence.

This presentation describes the development of a crystallization process that consistently yields API agglomerates which are robust to agitated drying. During development, the impact of several factors on agglomerate formation and strength was evaluated, and a mechanism of agglomeration was proposed. These studies showed that seed type has a strong impact on agglomerate properties and, subsequently, on the extent of agglomerate breakage during drying. Seeds with varying properties were generated via wet milling, jet milling, and hammer milling, and evaluated in the crystallization on laboratory and pilot scale. Filter, conical, and helical dryers were used on pilot and manufacturing scale to investigate the propensity of agglomerates made from the different seed types to withstand agitation. Results from the studies enabled definition of a robust manufacturing process and powder property control.

429782 Continuous Nanoparticle Production Using Focused Acoustics

Abstract:

Continuous nanoparticle production using focused acoustics

Many drugs currently in development are either water insoluble or lipophilic, thus creating formulation challenges to administer these drugs at a high dose. In some situations, nanoparticles are an appropriate formulation strategy as a result of their intrinsic properties such as small shape and size (e.g., radius) and large surface area to volume ratios. Nanoparticles can be generated by grounds up crystallization, which is typically processed in batch mode. One basic approach of crystallization is to vary the solubility of the solute of interest with different methods, such as vary the ratio to non-solvent vehicle (anti-solvent).

Low frequency, unfocused ultrasound has previously been demonstrated for improving crystallization kinetics. However, unfocused ultrasound has efficiency limitations due the inability to focus the energy, creating issues in thermal management such as non-uniform mixing, and uncontrolled thermal fields. Additionally, these unfocused solutions are not able to provide sufficiently tight, uniform nucleation control at the chemical solvent boundary interface that is required for nanoparticle production. The Covaris Adaptive Focused Acoustics™ (AFA) technology is a highly controlled mechanical based process that provides a practical tool to precisely control both the nucleation formation and the rate of crystal growth. The high energy efficiency of AFA avoids excessive heat generation, reducing potential thermal damages to the sample and allowing greater control over thermally dependent chemical reactions. Moreover, the precisely controlled energy delivery improves the reproducibility of the sample processing for higher quality control.

As a model of nanoparticle formation, a Solvent:Anti-solvent crystallization method with an active AFA energy field was tested in different batch volumes as well as in a continuous flow format. Cinnarizine was used as a model

th

compound. Nanoparticles were measured using a Malvern Mastersizer 3000, and particle sizes were reported for 10 percentile, 50th percentile, and 90th percentile populations (d10, d50, and d90 populations respectively).

Initial studies were performed in triplicate on batch scale processing of 2 mL samples. Parameters evaluated include:

Solvent	Dimethylacetamide (DMA)
Drug solvent concentrations	50 mM, 125 mM, 250 mM
Stabilizer	0.2% PVP30 and 0.25 mM SLS
Stabilizer percentage	0.2% 1.5%, 3.0%
Solvent:Anti-solvent ratio	1:29, 1:14, 1:6.6

For comparison, a control was run at each setting. The control was mixing 2 mL samples via a magnetic stir bar in a 20 mL scintillation vial for the same time duration of 5 minutes.

In all cases, the particle sizes of the AFA-processed samples were lower and more reproducible than the control. A typical comparison is shown in the following tables:

S220x	Covaris Focused-ultrasonicator
300 W	Peak Incident Power (PIP)
50%	Duty Factor (DF)
1,000	Cycles Per Burst (CPB)
12x24	2 mL process vessel
300	processing time (seconds)
6°C	water bath temperature

Table 1. Results from the stir bar method (Control).

125mM-Control				
Ratio	**d10-Control**	**d50-Control**	**d90-Control**	**Span**
1:6.6	9.0	29.9	1730.0	57.5
1:6.6	8.0	26.8	66.5	2.1

1:6.6	8.0	26.5	64.2	2.1
Average	8.3	27.7	620.2	20.6
SD	0.45	1.54	784.70	26.1

All values are in microns

Table 2. Results from the AFA method.

125mM-AFA				
Ratio	**d10-AFA**	**d50-AFA**	**d90-AFA**	**Span**
1:6.6	0.0406	0.122	0.325	2.335
1:6.6	0.0406	0.122	0.333	2.387
1:6.6	0.0407	0.123	0.334	2.395
Average	0.041	0.122	0.331	2.372
SD	0.000	0.000	0.004	0.027

All values are in microns

d50 particle size populations for the control sample averaged 27.7 microns while the same method processed with AFA yielded average d50 particle size populations of 122 nanometers.

A continuous manufacturing process was evaluated with the following settings:

Drug solvent concentration 125 mM (46 mg/mL)

Solvent flow rate 800 µL/min

Anti-solvent flow rate 10 mL/min

Stabilizer 0.2% PVP30 and 0.25 mM SLS

Stabilizer percentage 0.2%

Solvent:Antisolvent ratios 1:12.5 (10 mL/min anti-solvent, 800 µL/min solvent)

With 125 mM at a 800 µL/min solvent flow rate, the production rate was 36.5 mg/min.

Higher total acoustic power input (Peak Incident Power) yielded both smaller particle sizes and repeatable results. At PIP values of 500 Watt, particle size distributions are listed in the following table.

Table 3. Particle size distributions at PIP = 500 W.

	Sample 9	Sample 10	Sample 11	Sample 12
d10	0.085	0.091	0.098	0.090
d50	0.180	0.191	0.202	0.187
d90	0.415	0.409	0.423	0.399
Span	1.834	1.663	1.611	1.654

All values are in microns

Control of particle size was demonstrated by varying PIP. Particles sizes at 300 PIP are listed in the following table.

Table 4. Particle size distributions at PIP = 300 W.

	Sample 5	Sample 6	Sample 7	Sample 8
d10	2.15	1.79	1.89	1.78
d50	3.48	3.23	3.03	2.86
d90	5.55	5.01	4.95	4.73
Span	0.98	1.32	1.01	1.03

All values are in microns

The current study demonstrates the technical feasibility of using Adaptive Focused Acoustics (AFA) to control nucleation and crystal growth for continuous flow Solvent:Anti-solvent crystallization. Nanoparticles were successfully and reliably produced in a continuous flow process with d50 particle size populations below 200 nm. By using the AFA system to control the crystallization process, a choice of nano- or micron-sized crystals can be easily generated with a simple change of PIP process parameter. In addition, the isothermal, reproducible AFA process enables control of fluid shear effects on secondary nucleation to minimize crystal size, accelerate the overall rate of crystal formation, and improve yield. The effect of AFA on the morphology of the crystal is currently under investigation.

430222 Crystal Growth Inhibition of a Poorly Water-Soluble Drug Using Polymeric Additives

Abstract:

A large percentage of active pharmaceutical ingredients (APIs) in the drug development pipeline demonstrate poor aqueous solubility as crystalline solid forms. This poses a problem for oral therapeutics due to decreased bioavailability. Utilizing the amorphous form of the drug rather than the crystalline form is an effective method to alleviate solubility issues, due to its greater overall free energy. Much emphasis has been placed on methods to stabilize the inherently unstable amorphous form and prevent crystallization. One such method utilizes polymers to act as a mechanical barrier, occupying growth sites. The effectiveness of this method, however, varies for each polymer-drug combination.

In this work, we study inhibition of crystal growth of the drug felodipine in solution using a variety of polymers. Using atomic force microscopy (AFM), we can determine the fractional surface coverage as well as the conformation of adsorbed polymers on crystalline felodipine. These results reveal insights into the mechanisms by which polymers can effectively inhibit crystal growth, as well as elucidate important polymer properties that impact their effectiveness

430469 Dynamic Modeling and Control of Multi-Stage Slug-Flow Crystallization

Thursday, November 12, 2015: 4:40 PM
Ballroom D (Salt Palace Convention Center)

Michael L. Rasche[1], Mo Jiang[2] and Richard D. Braatz[2], (1)Department of Chemical and Biomolecular Engineering, University of Illinois at Urbana-Champaign, Urbana, IL, (2)Department of Chemical Engineering, Massachusetts Institute of Technology, Cambridge, MA

The design of robust pharmaceutical crystallization processes has been of high interest in academia and industry for the last decade (Nagy and Braatz, 2012). Continuous-flow tubular crystallizers have shown potential for high reproducibility and process efficiency at low capital and production cost (e.g., Alvarez & Myerson, 2010; Lawton et al., 2009; Vacassy et al., 2000). A recent advance that combines the advantages of continuous and batch crystallizers is the air/liquid slug-flow crystallizer (Eder et al., 2010; 2011; 2012; Jiang et al., 2014; 2015) which has unique advantages including narrow residence time distribution, no stirrer generating secondary nucleation, and easy post-crystallization separation. The potential application of slug-flow crystallization in the final stage of pharmaceutical manufacturing is the focus of this design study.

Similar to batch crystallization, the design of a robust slug-flow crystallization process comes from a good understanding of how the supersaturation profile is affected by the design variables, such as the method and rate of supersaturation generation, and the corresponding implementation. For example, supersaturation should be minimized so as to avoid impurity incorporation and secondary nucleation (e.g., Jiang et al., 2012; 2014), which is important for slug-flow crystallization with its possibly short residence time (on the order of minutes) (Eder et al., 2012; Jiang et al., 2014) and with fast heat transfer associated with the large surface area-to-volume of tubular crystallizers.

Unlike past studies that involve mathematical modeling or analysis of similar continuous crystallizers (e.g., Eder et al., 2010; Kubo et al., 1998), this study only analyzes an experimental demonstration (Jiang et al., 2014) but also investigates the effect of and optimizes numerous design variables (e.g., number of heat exchangers, length of tubing in each heat bath/exchanger, mass flow rate through the heat exchangers) while minimizing the total equipment/material use in the design. This presentation describes the first dynamic simulation of slug-flow crystallization for a series of counterflow, concentric shell-and-tube heat exchangers.

The method of moments is combined with the numerical method of lines with a finite difference approximation to reduce the model described by four partial differential equations (balances on population, solute, and energy in both the shell and tube) to a system of ordinary differential equations. The system is to be designed so that, under normal operating conditions, the temperature drop in each heat exchanger produces an environment to minimize the maximum supersaturation, to promote molecular purity, while applying a constraint that ensures a nearly maximum yield.

Operating so that the exit concentration is near the minimum concentration provides the desired yield. The optimal crystal size is specified by the upstream nucleation rate. Excessive length of the tubes should be avoided, as that would introduce a time delay in the process dynamics that is undesirable for closed-loop control. The control objective is to minimize the maximum supersaturation in the system by manipulating the flowrate and inlet temperature of cooling water in each heat exchanger to suppress the effects of process disturbances.

Optimal steady-state spatial profiles for the four-stage slug-flow crystallizer are shown in Figure 1. The supersaturation reaches the minimized maximum value in less than 5 meters, and returns to that maximum once in each heat exchanger. The drop in supersaturation near the outlet also occurs in less than 5 meters.

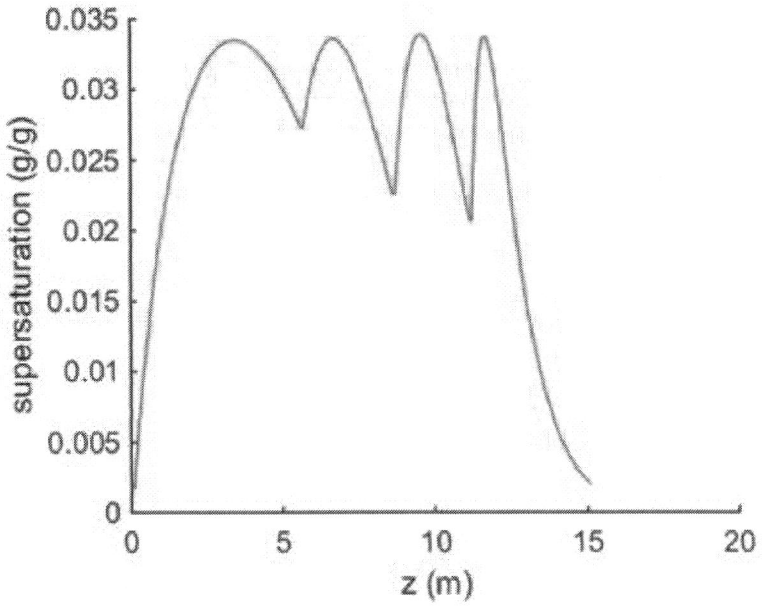

Figure 1: Supersaturation profile under steady, optimal conditions in the slug-flow crystallizer.

Investigating the dynamic response of the measured and product quality variables to disturbances, manipulations, and uncertainties provides a great deal of information and insight not available from steady-state models. The solute concentrations in the slug as it leaves each heat exchanger ("exit concentrations") are observed to be highly sensitive to cooling water temperatures and variation in the growth kinetics. The cooling water flowrates have relatively low effectiveness as manipulated variables for rejection of some disturbances. Increasing the cooling water flowrate or decreasing the cooling water temperature decreases the exit concentrations. The model shows that the effect of most input disturbances on concentration, yield, and average particle size are dampened by the process design automatically.

The response of maximum and average supersaturation is of vital importance as the driving force for growth as well as nucleation. While this system is designed to minimize secondary nucleation, supersaturation levels must be kept under control to avoid the incorporation of impurities into the crystalline matrix. Cooling water temperature and flowrate are seen to have a large effect locally. Inlet conditions for the slug stream have a decreasing effect with each subsequent heat exchanger. Persistent sources of relatively large deviation are again variation in the growth kinetics.

Outlet temperatures are also compared and evaluated in terms of value as measurements for feedback control, due to the relative ease and accuracy at which temperature can be measured. Not surprisingly, the greatest effect by far on the outlet slug stream temperature is the temperature of the cooling water.

Manipulating only the cooling water flowrates is shown to be insufficient to reject process disturbances that result from the deviation of growth kinetic parameters from nominal values. Additionally, without a tight control of the inlet temperature of cooling water, manipulation of the cooling water flowrates will be insufficient to suppress that resulting deviation as well. Changes in the inlet temperature and concentration for the slug stream are less problematic; the system design is inherently robust to disturbances which occur far upstream. The robustness allows the on-line measurement of particle size, a relatively expensive and inaccurate endeavor, to be eliminated from the design of a control algorithm; particle size is not observed to be affected greatly by upstream disturbances.

In an effort to increase the output response to changes in manipulated variables, the cooling water inlet temperature should be a manipulated variable rather than allowing it to be a disturbance. That is, a lower level regulatory control system should be implemented so that the cooling water inlet temperature is controlled, with the setpoint to the cooling water inlet temperature being a manipulated variable of the higher level control system. The extra degrees of freedom from controlling not only the cooling water flowrates, but also the cooling water temperatures, greatly increase the controllability of the slug-flow crystallization process. From a design standpoint, sensitivity analysis indicates the tubing length should be increased beyond nominal design conditions to serve as a buffer to disturbances.

Several control strategies motivated by the dynamic analysis are investigated for the multivariable control of the slug-flow crystallizer (e.g., Figure 2). Neglecting particle size measurements for the reasons mentioned above, there are two measurements of interest that can be taken at the outlet of each heat exchanger: temperature and solute concentration. There are also two manipulated variables for each heat exchanger: cooling water flowrate and cooling water temperature. A variety of factors are considered in selecting the best control strategy in terms of the number of measurements, number of controllers, and which controllers receive which measurements (and control which manipulated variables). Of the eight possible measurements, the four temperature measurements are inexpensive and accurate while four measurements of solute concentration would be relatively expensive and difficult. The design of this process does not provide evidence that intermediate solute concentration measurements would be useful process control elements, and it is unlikely that the final implemented control systems design would contain those measurements. With respect to the eight manipulated variables, the inlet temperature of cooling water could be altered by applying energy to a thermal bath or, perhaps more likely, combining multiple feeds from multiple baths at Y-mixers. The cooling water temperature displays a greater effect on process dynamics than the cooling water flowrate, but incompressible fluid flowrates have the advantage of being easily and quickly controlled. After comparing the pros and cons of different control strategies, an optimal control system design will be described.

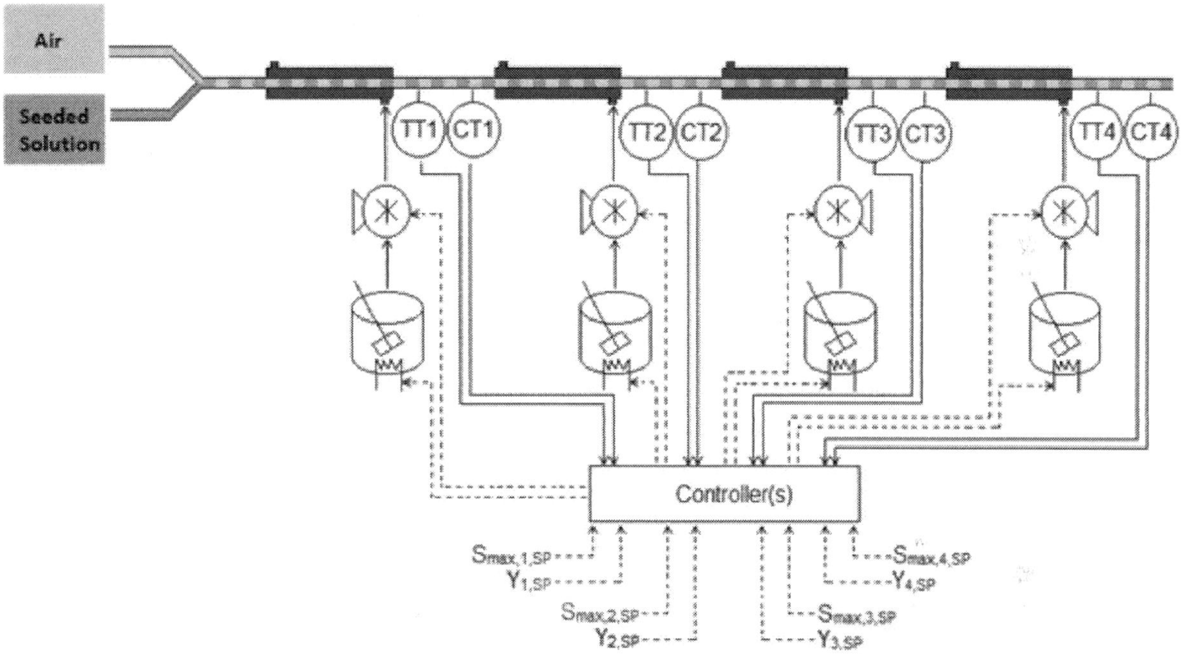

Figure 2: Example control strategy for the slug-flow crystallizer, which uses the complete set of eight measured and eight manipulated variables.

References:

Alvarez, A.J., Myerson, A.S., 2010. Crystal Growth & Design, 56, 349-369.

Eder, R.J.P., Radl, S., Schmitt, E., Innerhofer, S., Maier, M., Gruber-Woelfler, H., Khinast, J.G., 2010. Crystal Growth & Design, 10(5), 2247-2257.

Eder, R.J.P., Schmitt, E.K., Grill, J., Radl, S., Gruber-Woelfler, H., Khinast, J.G., 2011. Crystal Research and Technology, 46(3), 227-237.

Eder, R.J.P., Schrank, S., Besenhard, O., Roblegg, E., Gruber-Woelfler, H., Khinast, J.G., 2012. Crystal Growth & Design, 12(10), 4733-4738.

Jiang, M., Zhu, Z., Jimenez, E., Papageorgiou, C.D., Waetzig, J., Hardy, A., Langston, M., Braatz, R.D., 2014. Crystal Growth & Design, 14(2), 851-860.

Jiang, M., Papageorgiou, C.D., Waetzig, J., Hardy, A., Langston, M., and R. D. Braatz, R.D., 2015. Crystal Growth & Design, 15(5), 2486-2492.

Kubo, M., Kawakatsu, T., Yonemoto, T., 1998. Chemical Engineering Research & Design, 76(6), 669-676.

Lawton, S., Steele, G., Shering, P., Zhao, L., Laird, I., Ni, X.-Y., 2009. Organic Process Research & Development, 13(6), 1357-1363.

Nagy, Z.K., Braatz, R.D., 2012. Advances and new directions in crystallization control, Annual Review of Chemical and Biomolecular Engineering, 3, 55-75.

Vacassy, R., Lemaitre, J., Hofmann, H., Gerlings, J.H., 2000. AIChE Journal, 46(6), 1241-1252.

(730f) Polymorphism in the Development and Scale-up of Reactive-Crystallization of an Active Pharmaceutical Ingredient

Zhang, H., Sunovion Pharmaceuticals Inc.
Saranteas, K., Sunovion Pharmaceuticals Inc.
Originally presented on: 11/12/2015 17:00:00 - 17:20:00

Polymorphism in the Development and Scale-up of Reactive-crystallization of an Active Pharmaceutical Ingredient

Haitao Zhang, Kostas Saranteas

Chemical Process Research & Development, Sunovion Pharmaceuticals Inc., 84 Waterford Drive, Marlborough, Massachusetts 01752, U.S.A.

Abstract

Polymorphism is widely observed in pharmaceutical compounds and continues to be an important issue in drug development because of its impact on the physicochemical properties of drugs. It is well recognized that polymorphism and solvate formation affect the various pharmaceutically important physicochemical properties, such as stability, solubility, dissolution rate, crystal habit (shape), tableting behavior. Crystallization is a crucial step in the pharmaceutical industry in terms of the understanding and further control over the polymorphism during the manufacturing of active pharmaceutical ingredients (APIs).

Herein is described the observed polymorphic behavior of the API SEP-363856-01 in the reactive-crystallization process. This API is crystallized out from isopropanol (IPA) as the HCl salt through the addition of HCl/IPA to an IPA solution of the SEP-363856 freebase. As the API crystallizes it displays two distinct morphologies, the first a block like crystal (Form A) and the second a needle like crystal (Form B). Single crystal x-ray diffraction has shown Form A has a monoclinic crystal system while Form B has an orthorhombic crystal system. The experimental results indicate that the supersaturation control during the addition of HCl/IPA is critical in terms of the polymorphism control. Polymorphism transformation of kinetically preferred Form B to thermodynamically stable Form A is observed during a solvent-mediated phase transition in isopropanol. It has been shown that the transformation of Form B needles to Form A blocks is influenced by the quantity of Form B present and is accelerated with increasing temperature. A controlled HCl/IPA addition strategy has been proposed and successfully implemented in the lab development and scale-up manufacturing campaign. With this, the generation of unstable form B crystal can be avoided in the first place.

Solid phase polymorphism transformation is observed as well. The isolated samples of Form B needles held in the laboratory for a period of time yield a more rod like morphology, however XRPD analysis shows this morphology is consistent with the Form A blocks. Single crystal x-ray diffraction has shown Form A has a monoclinic crystal system while Form B has an orthorhombic crystal system.

433794 Extraction, Fractionation, and Purification of Lignin from Biomass By-Product Streams Via the ALPHA Process

Abstract:

Lignin is unique among biopolymers in having significant aromatic character, which makes it potentially useful for a wide range of applications, including foams, composites, resins, and carbon fibers. However, most of the commercial-grade lignins available today are recovered from the Kraft pulping process and thus have a relatively high metals and ash content, eliminating them from consideration for most high-value applications. Thus, less than 0.2% of the 50 million tons/yr of so-called "Kraft lignins" are being used for nonfuel products. Clearly, cost-effective separations processes for these lignins must be developed if lignin is to achieve its potential as a renewable biopolymer.

We have discovered that mixtures of acetic acid and water at elevated temperatures can be used to selectively extract Kraft lignins in terms of molecular weight. Of equal significance, these fractions are simultaneously purified by the solvent system, with the metals content in the lignin being reduced by almost an order of magnitude with each extraction step. Our process, which we call ALPHA (Aqueous Lignin Purification using Hot Acids), can be operated in two ways: (1) by using increasingly aggressive solvent mixtures to isolate metals-free (i.e., <100 ppm) lignin fractions of medium and high molecular weight, or (2) by "reversing" the process and using solvent mixtures of decreasing strength to isolate metals-free lignin fractions of low molecular weight.

435658 Extraction of Amines in Drilling Fluids By Solid-Phase Microextraction Coupled to GC/MS

Abstract:

Amine inhibitors are used in the petroleum industry to prevent or slow the corrosion reactions in downhole oil production. These inhibitors are pumped into the pipes with greases and lubricants for corrosion prevention and acid gas neutralization. A rapid method for amine determination in drilling fluids can be employed in this analysis. Solid-phase microextraction, SPME, involves the use of adsorption of analytes to a fiber coated with liquid phase chemistry. Once analytes have reached equilibrium with the liquid phase, the fiber is removed from the sample and presented to a gas chromatograph. The analytes are desorbed in the unit until a second equilibrium is obtained. This process can be automated and used with complex matrices as seen in the petroleum industry. Linearity and method detection limits will be discussed.

Challenges in Liquid-Liquid Extraction

D. Painer, S. Lux, E. Aksamija, M. Siebenhofer

Motivation

Feedstock biomass

- Complete utilization

Byproducts (e.g. pulp industry): multicomponent mixtures

- Low concentration of low value constituents; e.g. carboxylic acids
- Complex matrix

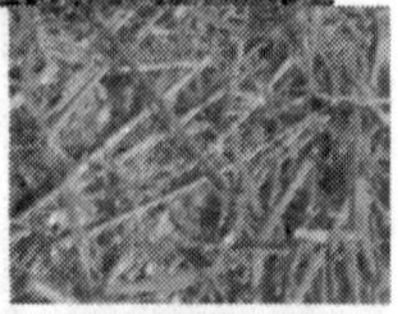

Outline

- The Task
- Equipment
- Processes
- Economics

Separation of acetic acid from spent cooking liquor condensates, case study

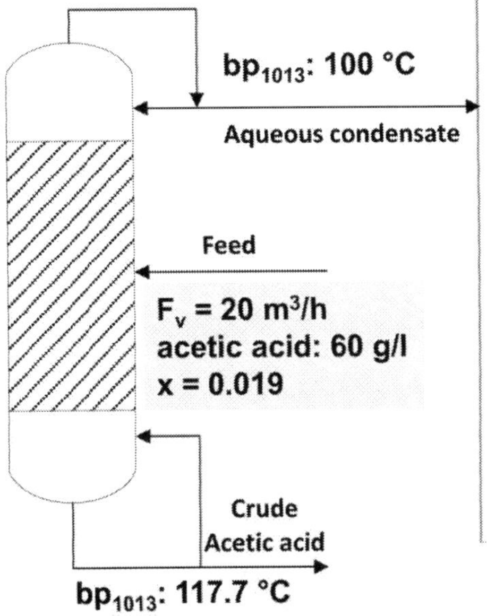

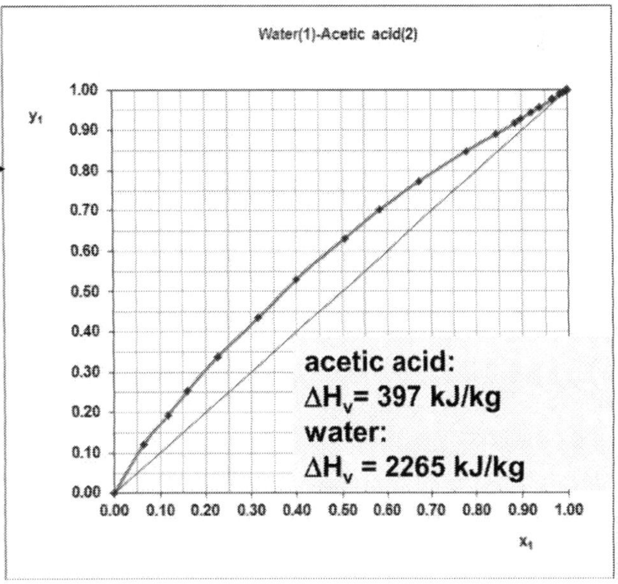

Separation of acetic acid from spent cooking liquor condensates, case study

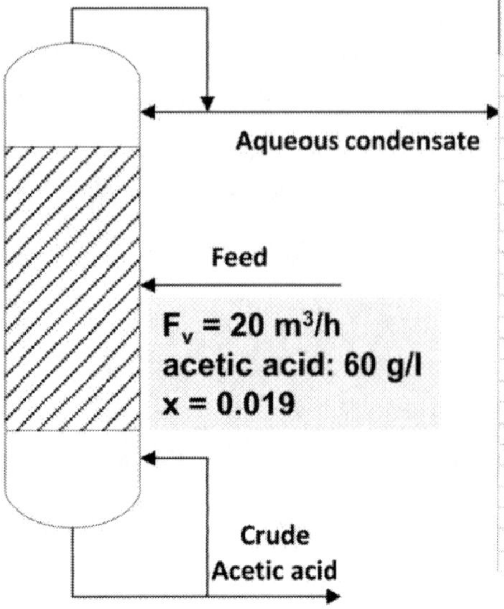

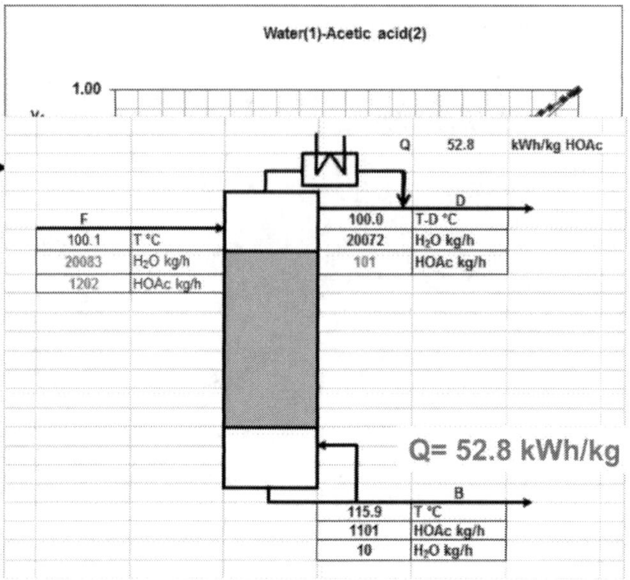

How to improve?

Can liquid-liquid extraction help?

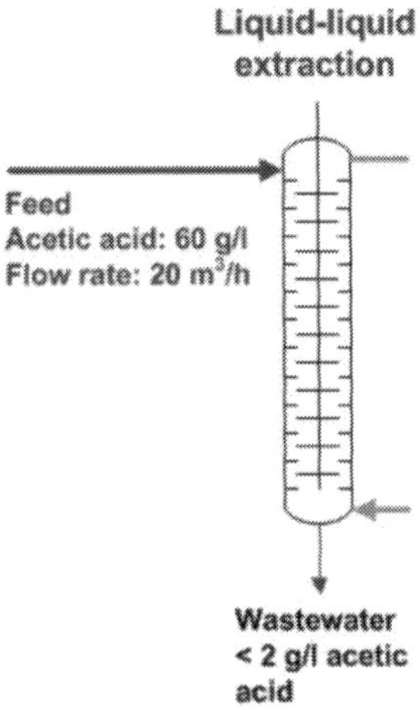

Can liquid-liquid extraction help?

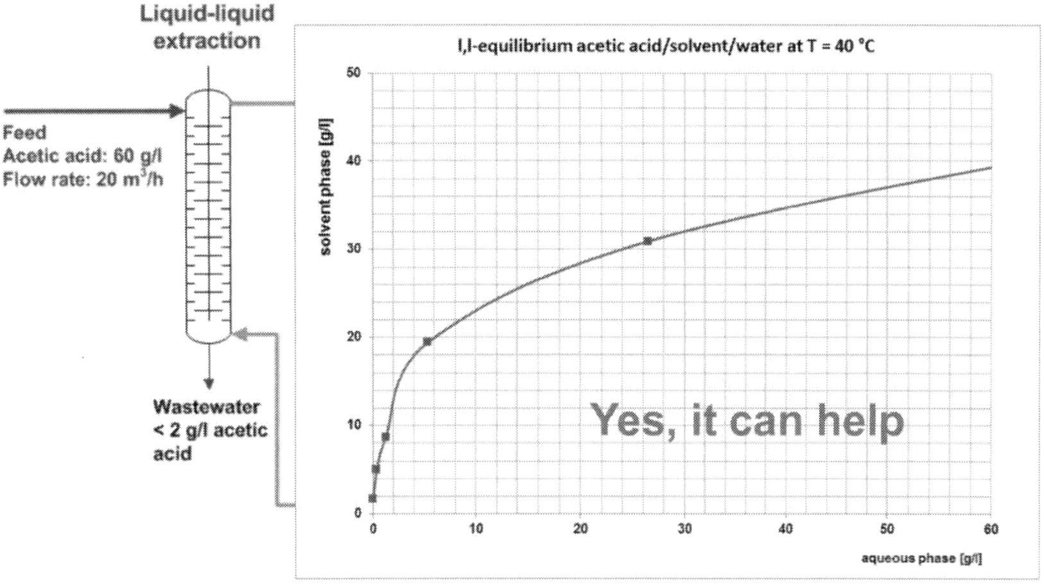

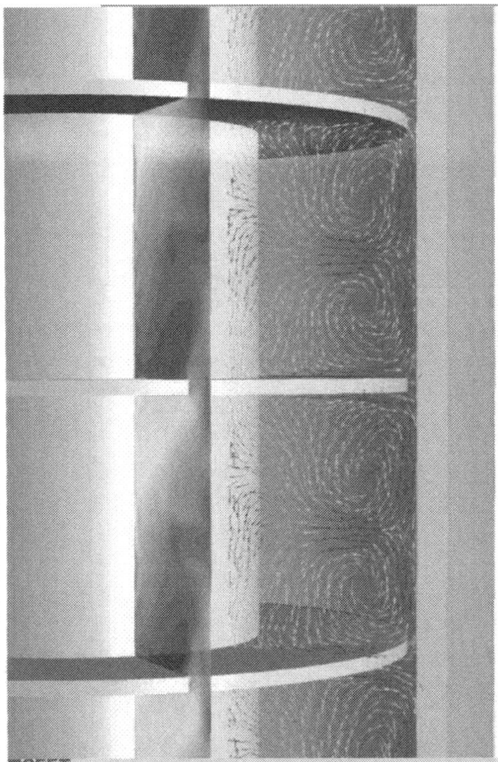

Equipment needs

- Complex matrix
- No dead zones
  - crud accumulation
  - flooding
- Smooth energy supply
- Sufficient separation efficiency
- Simple cleaning & maintenance
- Low investment costs

➡ **TCDC**

Onset of operation

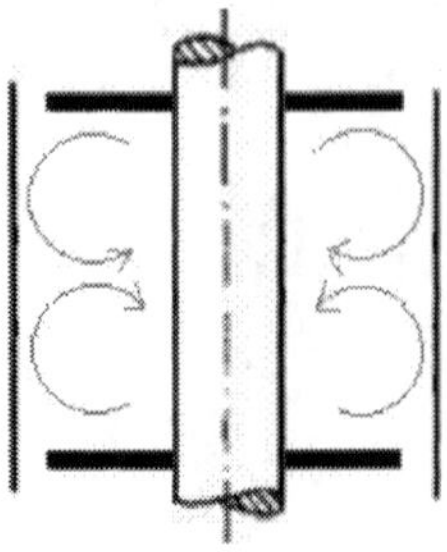

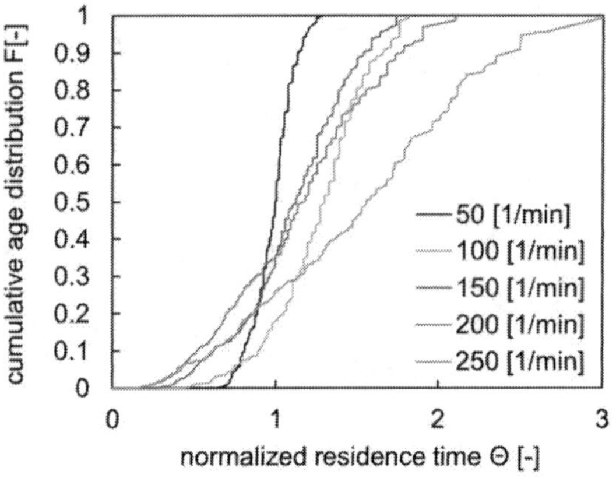

Onset of operation

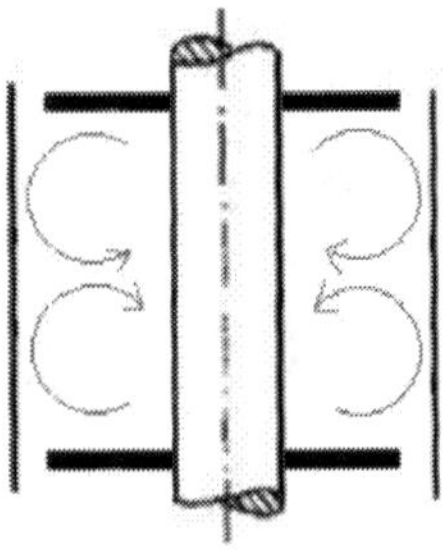

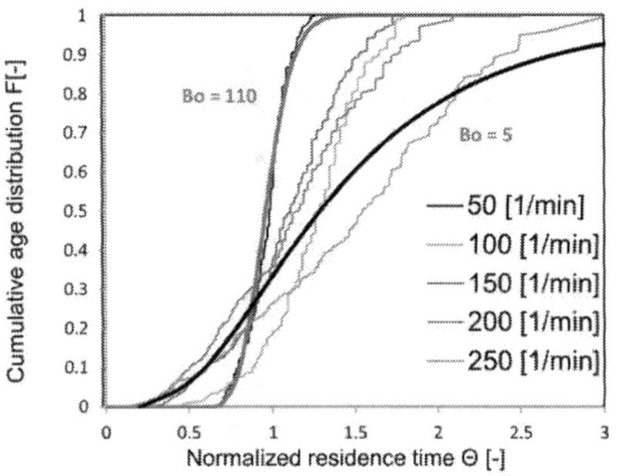

Hold-up

- High hydraulic load, 35 [m³/(m²h)]

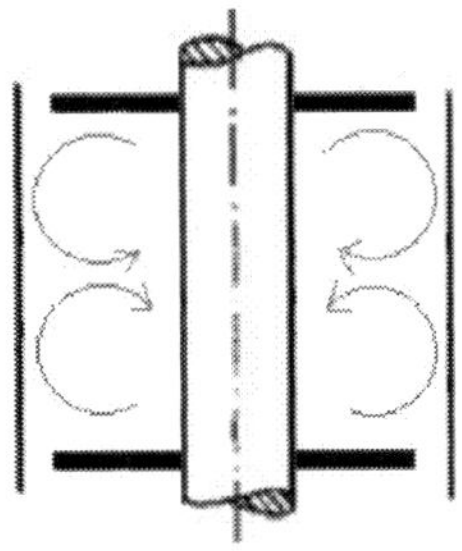

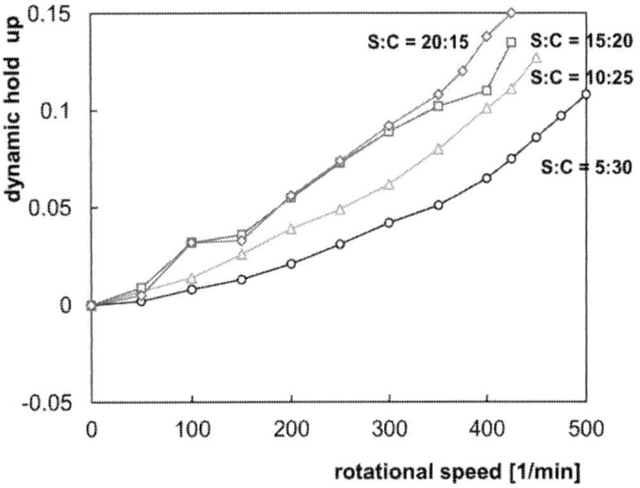

Dynamic hold-up

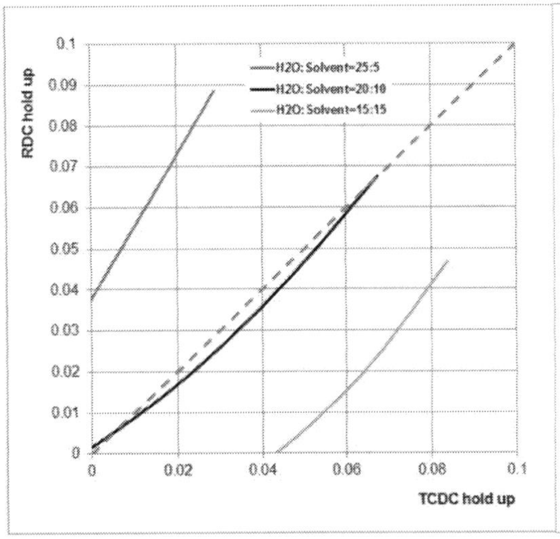

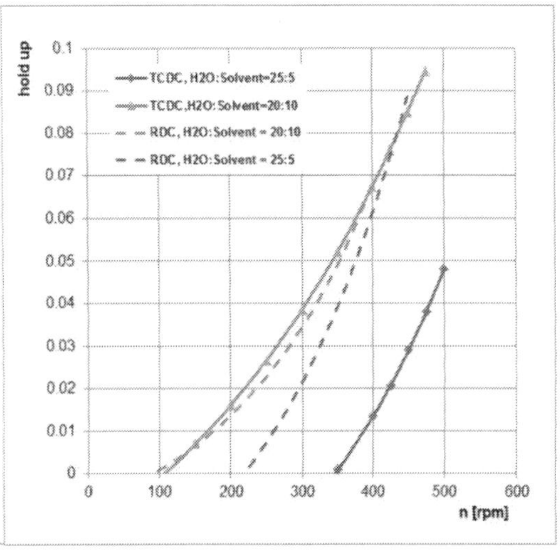

Drop size distribution (DSD)

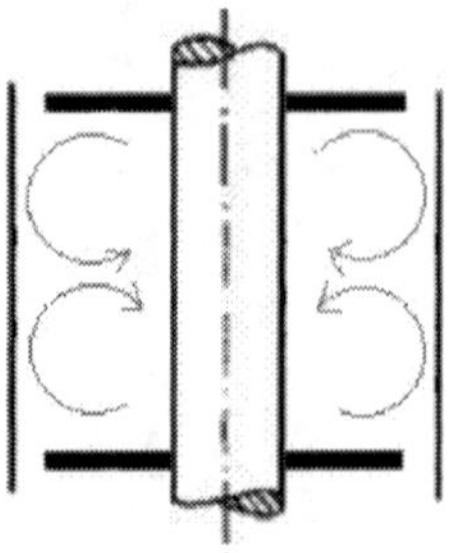

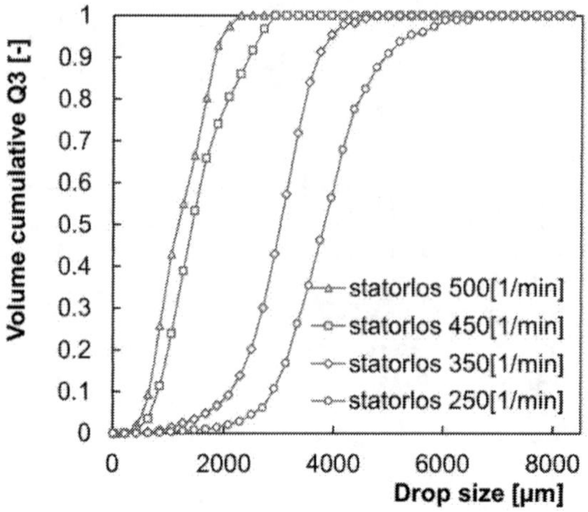

Challenges in Liquid-Liquid Extraction

Drop size distribution (DSD)

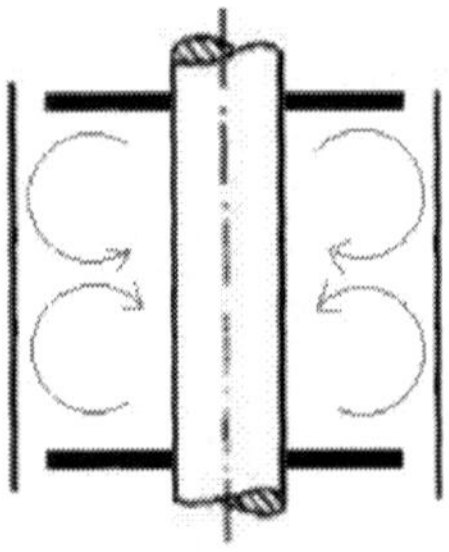

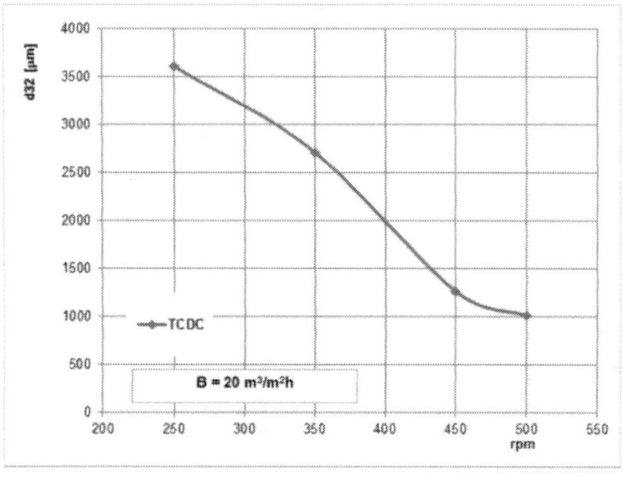

Operation range

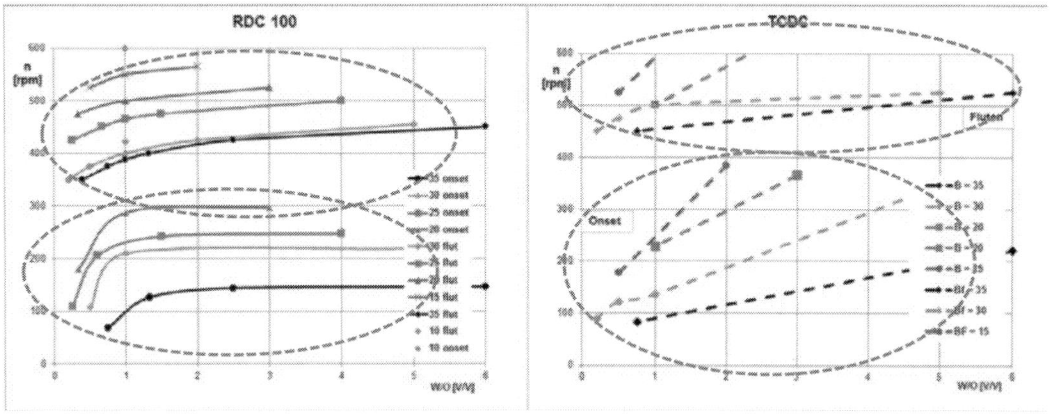

Two phase operation

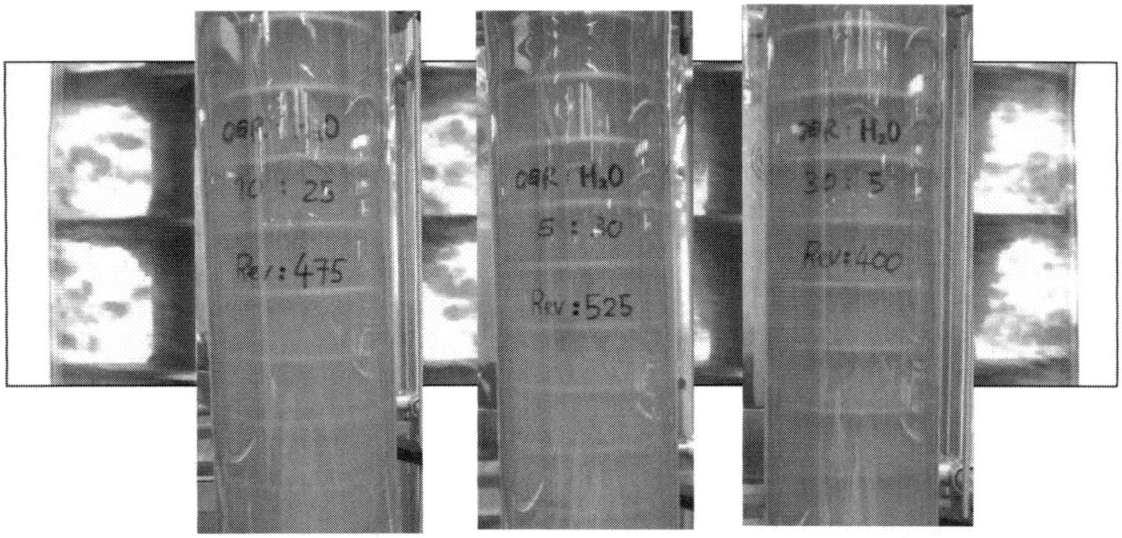

Modeling

Test system: n-Butanol-SST-water; d ⟶ c

$$A_{solvent} \rightarrow A_{carrier}$$

$$-r_A'' = k'' * (c_A^* - c_A)$$

b) CSTR cascade

$$F_{A,0} - F_A = -r_A'' * V * a$$

$$c_{A,n} = \frac{c_{A,n-1} + k'' * \frac{V_R}{F_{V,0}} * a}{1 + k'' * \frac{V_R}{F_{V,0}} * a}$$

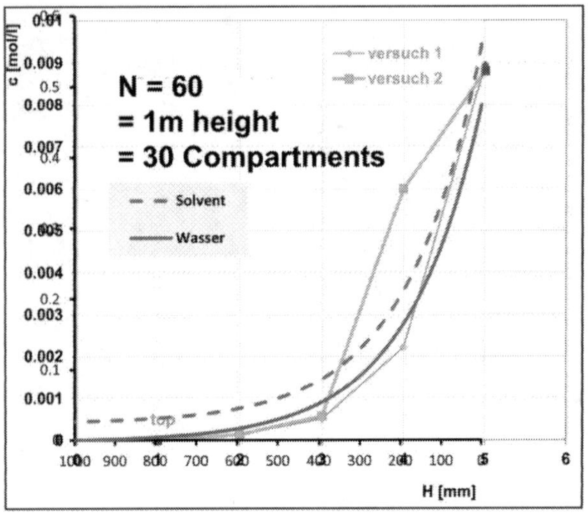

Liquid-liquid extraction can help.

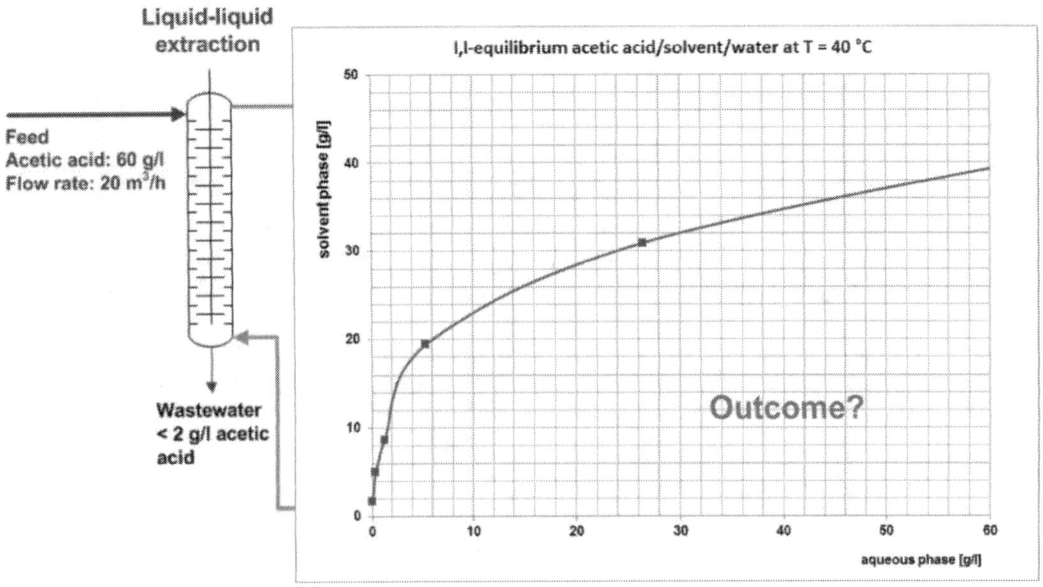

Liquid-liquid extraction can help.

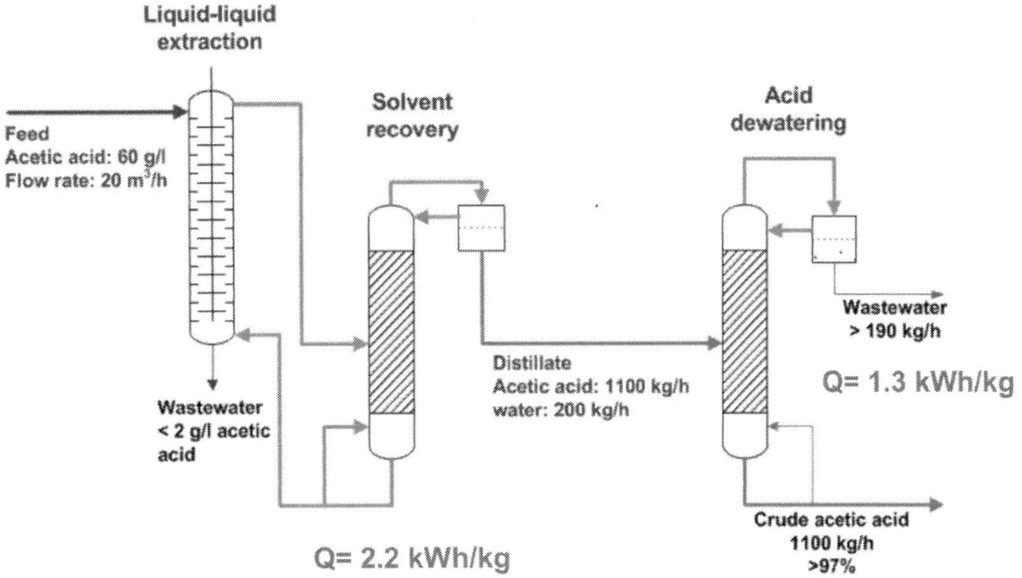

Challenges in Liquid-Liquid Extraction

Liquid-liquid extraction can help.

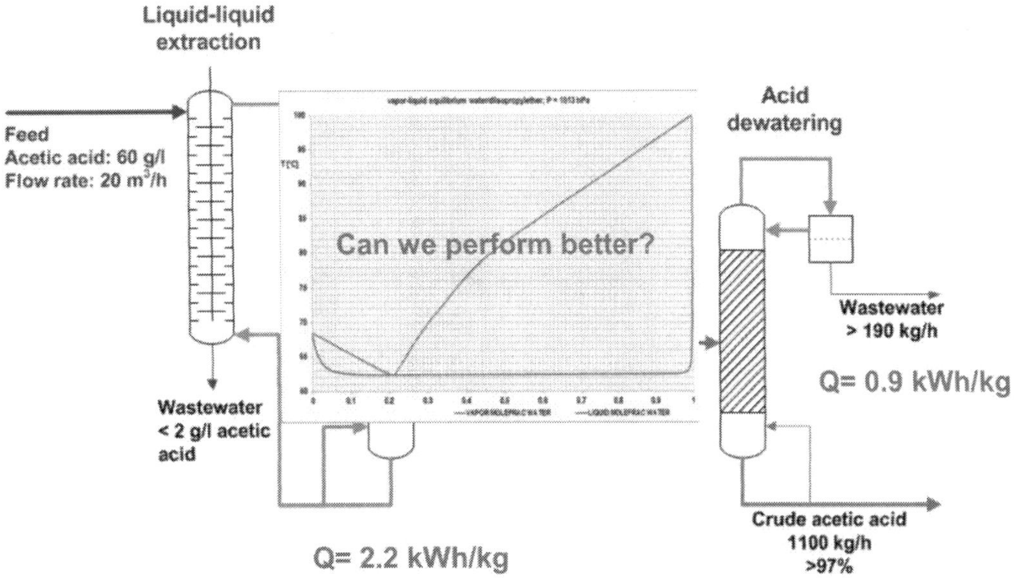

State of the art

Liquid-liquid extraction of carboxylic acids

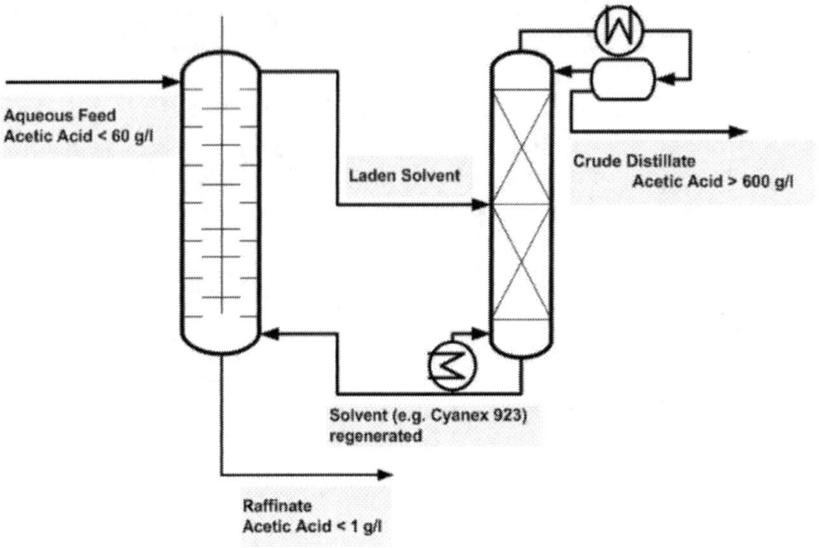

Liquid-liquid extraction plus esterification

$$CH_3COOH + CH_3OH \overset{H^+}{\longleftrightarrow} CH_3COOCH_3 + H_2O$$

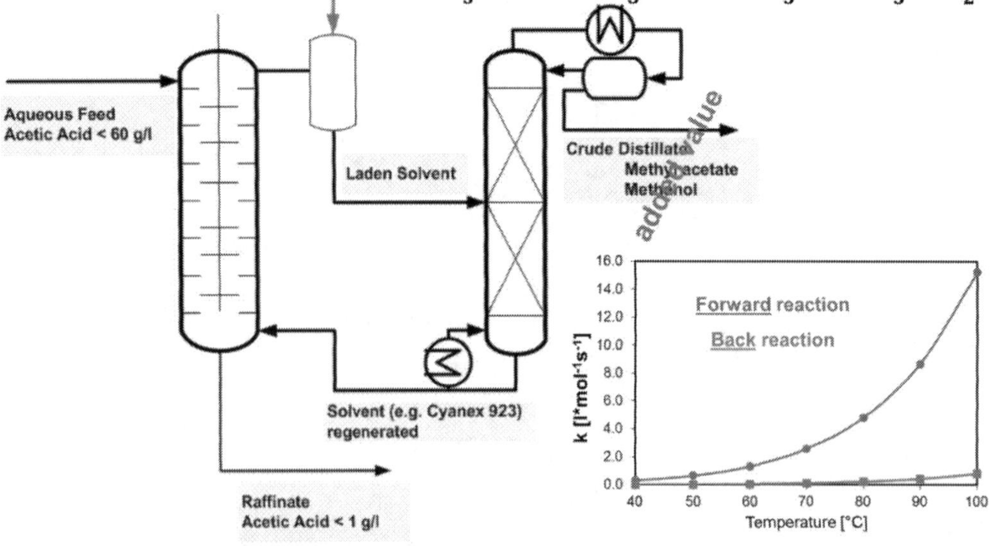

Liquid-liquid extraction plus esterification

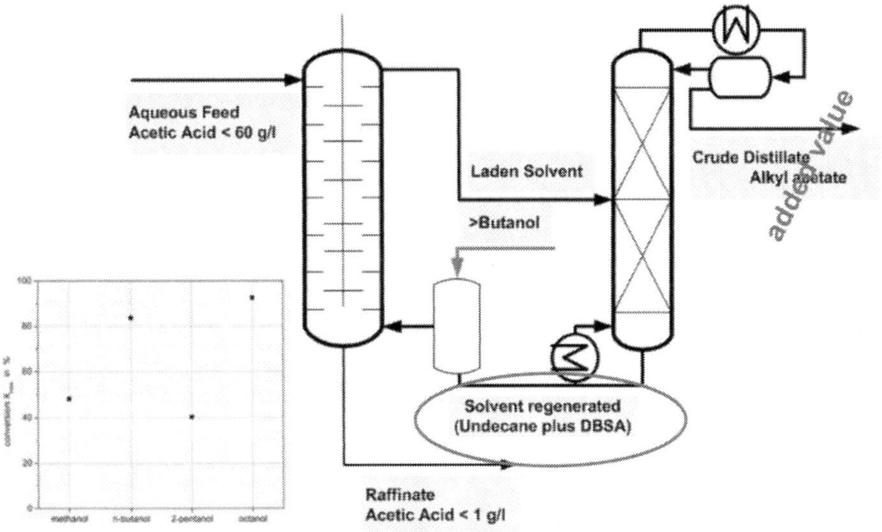

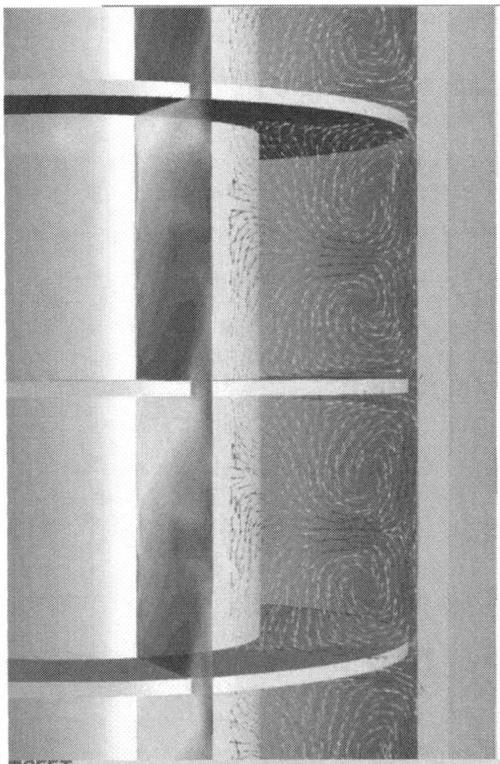

Summary

- Liquid-liquid extraction may contribute to economic feasibility of biobased bulk products by:
  - Adjusting equipment to the specific needs
  - Intensifying processes

421986 Selective Extraction of Rare Earth Elements from Scrap Magnets Via Membrane Solvent Extraction

Abstract:

The use of rare earth elements (REEs) is widely prevalent in various energy applications such as rechargeable batteries, wind turbines, photovoltaic thin films, and fluorescent lighting. It is critical to ensure sufficient supply of the REEs to meet the fast growing demand in many high-tech industries. One of the approaches to increase the REE supply is to recover REEs from end-of-life products containing REEs. In view of this, a membrane solvent extraction (MSX) system utilizing hollow fiber membrane modules was developed to selectively extract REEs from neodymium-based scrap magnets from several leading industrial manufacturers. Two extractants, TODGA (tetraoctyl digylcol amide) and Cyanex 923 (trialkyl phosphine oxides), were evaluated to determine the distribution coefficient and selectivity for Nd (neodymium) and Dy (dysprosium). The effects of several experimental variables on REE extraction in the MSX system such as REE concentration, membrane configuration, and molar concentration of acid were investigated. The results of selective extraction of Nd, Dy, and Pr (Praseodymium) without co-extraction of non-REEs from NdFeB scrap magnets will be discussed along with characterization of the recovered REEs with X-ray diffraction, energy-dispersive X-ray spectroscopy, and inductively coupled plasma-optical emission spectroscopy.

428137 Dependence of Colloidal Forces Between Bitumen and Silica on the Chain Lengths of Ammonium Surfactants in Solution

Abstract:

Oil sands are unconsolidated sand deposits impregnated with high-molar-mass viscous petroleum, which is an important unconventional oil resource and normally termed bitumen. To recover bitumen from oil sands, bitumen has to be liberated or detached from the sand grains. In hot-water extraction processes, surfactants play critical role in mediating interactions between bitumen and sand solids in the aqueous solution. Effect of ammonium chloride cationic surfactants with different hydrocarbon chain lengths (C8, C12, C16 and C18) on the colloidal forces between bitumen and silica surfaces were studied by atomic force microscopy (AFM). Investigation of the colloidal bitumen-silica interactions in various surfactants solutions allowed us to determine the dependence of the interaction behaviors on the surfactants chain lengths and the orientation and structure of the surfactants adsorbed on the bitumen and silica surfaces, as well as to have a deep understanding on the water-based bitumen extraction processes on a microscope.

The results showed that the hydrocarbon chain length of the cationic surfactants had a significant impact on both the long-range interactions and adhesion forces between bitumen and silica. With addition of C8 ammonium surfactant (TOAC) in the 1 mM KCl solution, the long-range repulsive force decreased to some extent and the corresponding adhesion force increased from about 3.3 mN/m to 4.6 mN/m. As the chain length of the surfactants increased to C12 (DTAC), the long-range repulsive force changed greatly to a strong attractive force and the adhesion further increased to about 6.6 mN/m. Interestingly, the long-range interaction force became repulsive again in CTAC surfactant solution, which has the chain length of C16. The adhesion force correspondingly decreased to 4.25 mN/m. Increasing the chain length to C18 (OTAC), the long-range repulsive force correspondingly enhanced and the adhesion decreased to 3.75 mN/m. It was further observed that addition of Ca^{2+} in various surfactants solutions would either depress or enhance the colloidal interactions according to the different chain lengths of the ammonium surfactants.
It was believed that electrostatic force and hydrophobic force between the bitumen and silica surfaces are the dominant factors of the colloidal interactions. Variation of the long-range interaction and the adhesion reflected that adsorption of the ammonium cationic surfactants molecules on the negative charged surfaces of bitumen and silica was in a manner of monolayer or bilayers according to the chain lengths of the surfactants, which was demonstrated by the Zeta potential measurements. Adsorption of the surfactants on the silica and bitumen surfaces significantly affected both the surface wettability and the surface charge characteristics. For surfactants with relative short chain lengths (C8 and C12), the adsorption of the surfactants molecules was mainly in a monolayer with the hydrocarbon chains toward in the solution, which enabled the silica and bitumen surfaces more hydrophobic and thus produced strong attractive hydrophobic force leading to the decrease of the long-range repulsive force or even a strong attractive interaction between the bitumen and silica. As the chain length of the surfactants increased to C16 and C18, the molecular order of the adsorbed surfactants increased and generated a bilayer adsorption with ammonium head group being on the most outside of the surface. Such a bilayer adsorption of the surfactants not only generated strong positive surface charges but also greatly decreased the hyrophobicity of the silica and bitumen surfaces. As a result, the silica-bitumen long-range interaction became repulsive force again and the adhesion considerably decreased as well. When the divalent metal ion of Ca^{2+} was added in the sufactants solution, the preferential adsorption of Ca^{2+} on silica and bitumen surfaces acted as a barrier to prevent the cationic surfactants from adsorbing and thus significantly affected the bitumen-silica interaction behaviors. In addition, the measured force profiles between bitumen and silica in various solutions could be well described by the extended DLVO theory considering an additional hydrophobic attractive force, which is a good evidence to support the variation of the surface wettability arising from the surfactants adsorption.

It is believed that the findings in this work deepen our understanding on the water-based bitumen extraction processes on a microscope. To have a high bitumen recovery and good froth quality, the surfactant type and concentration of the divalent metal ions in the processing slurry must be well considered.

(740a) Composite Organic-Inorganic Polymeric Membranes for CO2 Recovery Processes

Doghieri, F., University of Bologna
Miccio, F.,
Minelli, M., University of Bologna
Sarti, G. C., Alma Mater Studiorum - Università di Bologna
Medri, V.,
Landi, E.,
Originally presented on: 11/12/2015 15:15:00 - 15:36:00

A new process for CO_2 recovery from different gaseous streams is proposed, which is based on gas separation technology and takes advantage of peculiar characteristics of composite organic-inorganic polymeric membranes prepared combining rubbery polymers of good permselective properties for CO_2/permanent-gas mixtures with geopolymeric materials of suitable porosity and pore size distribution. In this work, PDMS and metakaolin-based geopolymers are specifically considered to produce composite membranes for the new gas separation process. Effective pure gas solubility and diffusivity isotherms for carbon dioxide, nitrogen and methane in the new polymeric materials as well as pertinent effective mixed gas permeance and selectivity of the membrane were measured by means of transient sorption and direct gas permeation experiments, respectively. Results indicates that significant improvements in productivity and separation factor are achieved with the new process, when compared to the case of conventional membrane separation processes.

ELECTROCHEMICAL CARBON DIOXIDE SEPARATION USING MEMBRANE-ELECTRODE ASSEMBLIES

Douglas S. Diez, Philip Cox, Ryan Reeves, Justin J. Hill, and Nicholas R. Schwartz *

Mainstream Engineering Corporation, Rockledge, FL

Corresponding Author: nschwartz@mainstream-engr.com

Abstract

A CO_2 removal system using membrane electrode assembly (MEA) technology to electrochemically separate CO_2 from an exhaled breathing within a breathing loop has been demonstrated. An anion exchange membrane (AEM) was used with catalyst coated electrodes to separate CO_2 from a humidified breathing gas mixture. At the MEA cathode, Pt-catalyzed reduction of CO_2 converts CO_2 into bicarbonate ions that are then electrochemically driven across the AEM. The reformation process of CO_2 at the anode to complete the gas separation was studied using oxidation catalysts, such as Ni, IrO_2, and RuO_2. Several membrane and electrocatalyst materials were characterized for electrochemical CO_2 separation. Electrode efficiency and CO_2 flux was characterized in asymmetric MEA with a Pt cathode and Ni anode using an AEM (Pt-Ni MEA) from 22 °C – 50 °C and CO_2 partial pressure at the cathode from 0.05 atm to 0.20 atm. Polarization response and electrochemical impedance spectroscopy was used to characterize performance in a environmentally controlled and metered atmosphere. Under optimal conditions Pt-Ni MEA device achieved a flux of 1.10 $L/m^2 \cdot min$ equivalent at 1.6 V with a constant current density of 10 mA/cm^2. The Pt cathode and RuO_2 anode MEA with an AEM (Pt-RuO_2 MEA) at 10 mA/cm^2 maintained an average voltage of 1.09 V with a CO_2 flux of 1.03 $L/m^2 \cdot min$.

Introduction

Chemical, biological, radiological, and nuclear (CBRN) environments pose a direct threat to soldiers, first responders, and rescuers. Self-contained breathing apparatus (SCBA) equipment is used for respiratory protection by isolating the user from the surrounding environment by providing contaminant-free breathing air directly to the face mask. This makes an uncomfortable and potentially hazardous environment. SCBA equipment is used to protect the user against contaminants by an isolated O_2 source, a CO_2 scrubber, and a positive pressure environment. Removal of CO_2 currently is done by adsorption, which generates heat and moisture, in addition to that added by the user. This heat increases the physiological burden on the user and decreases the performance of the protection equipment. Improper removal of heat and CO_2 poses many major physiological threats that require the user to discontinue operation.[1,2] Smart, reliable, and high-performance CO_2 removal is required to ensure protection without limiting the abilities of the operator.

Legacy CO_2 Separation Technologies

Common methods for capture and sequestration of CO_2 involve adsorption, membrane separation, and reactive systems. Commercial SCBA systems use heavy non-regenerable soda lime beds (weighing 6.3 lb) that generate over 140 W of heat at moderate breathing rates. Materials such as metal-organic frameworks (MOFs) and zeolites have high surface areas and adsorption capacity; however, these materials generally perform better at higher temperatures and pressures.[3] MOFs have been demonstrated as high performance and regenerable, but poor selectivity for gas mixtures.[3] Other amine-loaded mesoporous materials have a CO_2 adsorbent capacity higher than 1.25 g CO_2/g and is comparable to a metal-hydroxide adsorbent such as LiOH (0.919 g CO_2/g).[3, 4] Non-regenerative adsorbents have limited capacity in SCBA systems due to excessive weight, heat, and moisture generation.

Polymer membranes separate CO_2 from gas mixtures based on the selective diffusion and permeation through the membrane. One study demonstrated high CO_2 separation and selectivity over N_2 with a commercial nanofiltration composite polyamide membrane at high humidity and pressure.[5] Thin-film polydimethylsiloxane can be cross-linked at various temperatures to alter CO_2 permeation properties (selectivity of 6.7:1 and permeation rate of 3,970 Barrer at 400 kPa using polydimethylsiloxane crossed-linked at 75°C).[6] Membrane separation processes are impractical for SCBA equipment because high pressures are required, which introduces heavy, bulky, and high power components such as blowers and motors. Reactive CO_2 capturing systems with amine solutions such as monoethanolamine or 2-isopropylaminoethanol have a loading capacity near 100 g CO_2/L with an average enthalpy of 70 kJ/mol CO_2.[7,8] These systems are used primary for large-scale chemical processes and are not applicable for SCBA equipment.

Electrochemical Separation of CO_2 using Membrane-Electrode Assemblies

This work has focused on using a membrane-electrode assembly directed at CO_2 removal from humid, tidal flow, breathing gas streams for respiratory equipment. Separation of CO_2 from a mixed stream of humidified air can be accomplished using an AEM. Electrochemical CO_2 removal has been previously demonstrate; however, CO_2 separation can be improved by customizing MEAs for breathing air.[9, 10] These customizations include making a thinner AEM to reduce power requirements at identical flux and tuning the catalyst structure- and composition- properties to improve interfacial kinetics. An asymmetric MEA structure such as Pt/AEM/Ni improves separation by creating more facile kinetics involved with both bicarbonate formation at the Pt cathode and CO_2 reforming at the Ni anode. Figure 1 shows the mechanism for converting CO_2 to bicarbonate ions that are transported through the AEM by diffusion and applied potential to react and evolve gas on the anode side.

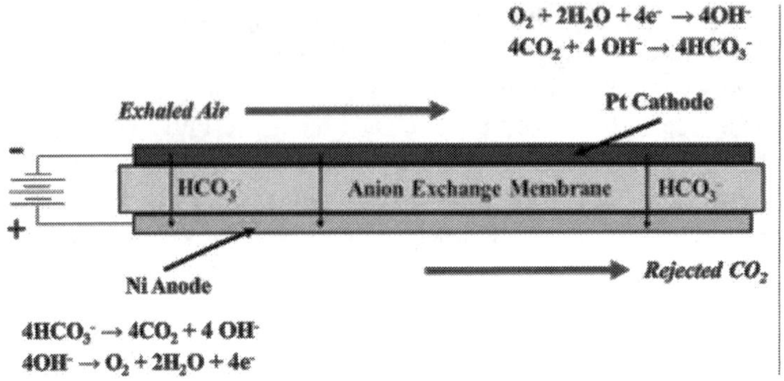

Figure 1. Electrochemical separation of CO_2 with an anion exchange membrane

Absorbed CO_2 reacts with hydroxide to form a bicarbonate ions and does not directly participate in the redox reaction at the membrane. The hydroxide group is generated from the reduction reaction with oxygen and water, which leads to water crossover and dehumidification. This bicarbonate ion is transported across the membrane, and then the reaction is reversed at the anode. The net reaction, shown as the reduction of CO_2, follows:

$$O_2 + 2\,H_2O + 4\,CO_2 + 4\,e^- \leftrightarrow 4\,HCO_3^- \qquad [1]$$

The reaction requires a 1 mole of O_2 for every 4 moles of CO_2 separated. CO_2 flux is proportional to the current density (charge passed). Thus CO_2 removal rate can be fixed exactly under galvanostatic control and indirectly with potentiostatic control but with more device protection. One study measured a current density of 6 mA/cm^2 at 1.2 V with a 1:1 CO_2:O_2 mixture supplied at 50 mL/min.[11] The measured flux was 0.2 mL/min·cm^2 and a CO_2:O_2 ratio above 3.56:1, which is 0.4% of the total flow.[11] At a respiratory rate of 95 L/min, nearly 4 L/min of O_2 is consumed and 3.4 L/min of CO_2 is generated. Thus an MEA with 1.7 m^2 of active surface area is required to remove 100% of the CO_2 with a asymmetric Pt cathode and Ni anode MEA using an AEM.[12] AEM thickness also increases the power requirement when operating under CO_2 removal rate control. The current density in this study and area required under CO_2 removal rate control uses 1.8 W of power at the maximum point during tidal gas flow. Negligible energy is dissipated as heat. An adsorbent would dissipate heat at a rate as high as 140 W to the breathing air. High temperature and humidity breathing air poses health hazards on SCBA users, while increasing respiration rate and O_2 consumption.

Methods

Several MEAs were fabricated with different catalysts and a single AEM material (Neosepta). Neosepta (A201) and fumasep® (FAB-PK-130) were subjected to humidity cycling to gauge physical stability. Manufacturer specified properties and values reported in the literature were used for comparison of electrochemical properties. Electrocatalysts for the reduction and oxidation reactions were characterized by cyclic voltammetry measurements using a rotating disk electrode (RDE) as the working electrode. Prior to testing, the catalyst loading and surface characteristics were inspected through scanning electron microscope (SEM) imaging.

Properties of Anion Exchange Membranes

Membranes were subjected to 28, 3 hour alternating dry/humid air flow cycles at 30°C. The resistivity before and after humidity cycling was calculated using the membrane thickness and electrochemical impedance spectroscopy measurements. Resistances increased for each of the tested membranes after the humidity cycling. Specific resistivity for each membrane before and after humidity cycling is shown in Table 1.

Table 1. Membrane Resistivity Calculated Before and After Wet-Dry Cycles

Membrane	Thickness (mm)	Resistivity Before (MΩ·cm)	Resistivity After (MΩ·cm)
Neosepta (A201)	0.042	0.021	0.103
Fumasep (FAB-PK-130)	0.140	2.576	17.50

Neosepta has an ion exchange capacity of 1.7 meq/g, which is greater than Fumasep with 0.79 meq/g.[13, 14] Specific resistance for Neosepta ranges from 2.4–4.1 Ω·cm^2 and has a conductivity between 1–40 mS/cm. Fumasep membrane material area resistance is reported as 10.6 Ω·cm^2 with a conductivity of 1.1 mS/cm.[13, 14] The humidity cycling revealed good physical strength for Neosepta and ultimately this membrane material was selected as the basis for the MEAs.

Catalyst Evaluation

The catalytic activity of the catalysts were measured for the oxidation and reduction reactions. Previous research has used Ni as the electrocatalysts for both the anode and cathode,[15, 16] while one study reported using Pt/C for the cathode electrocatalyst.[17] Lead (Pb), tin (Sn), Ni, palladium (Pd), Pt, copper (Cu), silver (Ag), and gold (Au) were examined in another study where it was determined that Cu, Ag, Pb, and Sn were unstable and were eliminated as potential catalysts.[15] Ni and Pt were also chosen for the catalyst evaluation along with palladium on activated carbon (Pd/C), iridium oxide (IrO_2), ruthenium oxide (RuO_2), nickel iron oxide ($NiFe_2O_4$), and cobalt (II,III) oxide (Co_3O_4).

The catalysts were evaluated with cyclic voltammetry (CV) using a rotating disk electrode (RDE) as the working electrode. The oxygen reduction and evolution overpotentials were measured for each working electrode catalyst drop-cast and dried on a 5 mm diameter glassy carbon electrode (GCE). An alkaline ionomer (5 wt% in 1-propanol) was added at a 10:1 ionomer-to-catalyst mass ratio to act as a binder for the catalyst to the GCE. The catalyst solution contained 1 mg/mL in 80 v/v% 1-propanol mixture dispersed by sonication. The catalyst loading on the GCE was calculated to be 48 mg/cm^2. Cyclic voltammetry was conducted in CO_2-saturated 0.5 M $NaHCO_3$ solution at 22°C with a scan rate of 50 mV/s for three cycles. A 3-electrode setup was used with the working RDE rotating at 1600 rpm, a Pt wire counter electrode, and a saturated calomel electrode (SCE) as the reference electrode.

CV measurements of cathode catalysts in CO_2-saturated 0.5 M $NaHCO_3$ with a sweep rate of 50 mV/s from 0 V to –2.0 V (vs. SCE) are shown in Figure 2, where the O_2 reduction overpotential of the CV curve. A single sweep from 0 V to –2.0 V is shown for clarity. These curves relate to the cathodic reactions:

$$O_2 + H_2O + 4e^- \rightarrow 4OH^- \qquad [2a]$$

$$4CO_2 + 4OH^- \rightarrow 4HCO_3^- \qquad [2b]$$

where only Rxn. 2a is captured by the CV scan because Rxn. 2b does not involve the transfer of electrons. As depicted in Figure 2, Pt displayed the lowest overpotential for Eqn. 1a followed closely by Co_3O_4. This means that for a Pt catalyst, a potential of –0.78 V (vs. SCE) is needed to activate Rxn 2a and additional energy through potentials in excess of –1.6 V (vs. SCE) is required to activate Rxn. 2a with the other catalyst. Pt was selected as the cathode electrocatalyst in the electrochemical cell to minimize the CO_2 ionization reaction energy demand.

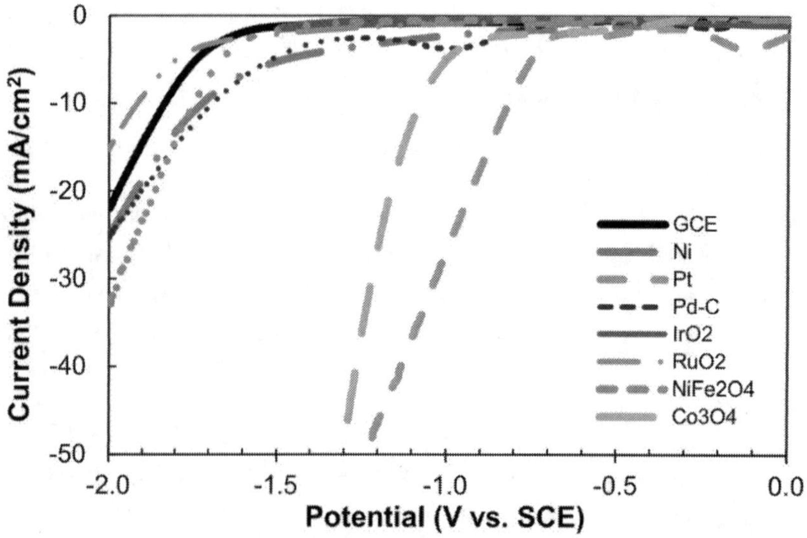

Figure 2. Cyclic voltammetry sweeps of potential cathode catalysts in CO_2-saturated 0.5M $NaHCO_3$

CV sweeps of anode catalysts in CO_2-saturated 0.5 M $NaHCO_3$ with a sweep rate of 50 mV/s from 0 V to 1.5 V (vs SCE) are shown in Figure 3 with the O_2 evolution overpotential included. A single sweep from 0 V to 1.5 V is shown for clarity. These curves relate to the anodic reactions:

$$4\ OH^- \rightarrow O_2 + H_2O + 4e^- \quad [3a]$$

$$4HCO_3^- \rightarrow 4CO_2 + 4\ OH^- \quad [3b]$$

where only Rxn. 3a is captured by the CV scan because Rxn. 3b does not involve the transfer of electrons. The O_2 evolution potentials were in the following order $RuO_2 < Co_3O_4 < IrO_2 <$ Pt < Pd/C < Ni < GCE. Based on the reported results, RuO_2, IrO_2, Co_3O_4, and Pt are all potentially more promising anode electrocatalysts. Ni has been used exclusively as the anode catalyst; however, RuO_2, IrO_2, Co_3O_4, and Pt warrant further testing in a planar electrochemical cell testing of CO_2 pumping.

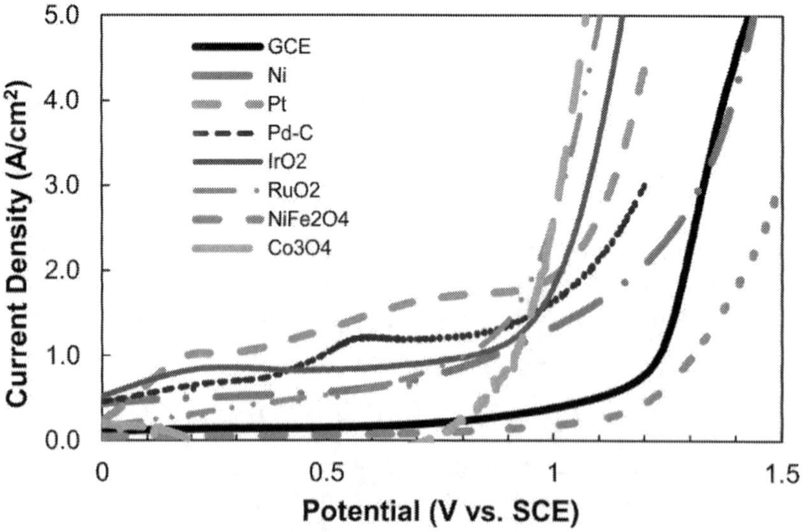

Figure 3. Cyclic voltammetry sweeps of potential anode catalysts in CO_2-saturated 0.5M $NaHCO_3$

The oxygen evolution and reduction potentials for the tested catalysts are listed in Table 2. These potentials only represent part of the overall reactions in CO_2 electrochemical pumping, relate to only Rxn. 2a and Rxn. 3a, and do not capture the catalyst activity of the CO_2 capture and evolution reactions. Testing the catalysts within a planar membrane electrochemical cell would provide a more thorough evaluation of the catalytic activity.

Table 2. Oxygen Reduction and Evolution Potentials of Alternative Catalyst Materials

Catalyst	Oxygen Reduction Potential (V vs. SCE)	Oxygen Evolution Potential (V vs. SCE)
GCE	−1.74	1.21
Ni	−1.63	1.09
Ni-SiO_2/Al_2O_3	−1.75	N/A
Pt	**−0.78**	0.97
Pd-C	−1.54	1.06
IrO_2	N/A	**0.95**
RuO_2	−1.75	**0.89**
$NiFe_2O_4$	−1.60	1.25
Co_3O_4	−1.24	**0.93**

Electrode Fabrication Methods

Pt-Ni MEAs were prepared by applying catalyst directly to the membrane to produce a catalyst-coated membrane (CCM) and by coating the microporous gas diffusion layer (GDL) electrode support and pressing the MEA together. The Pt catalyst ink was blended with a ratio of 1:10 Pt catalyst in 5 wt% ionomer solution. Additional water was added to wet the catalyst before adding ionomer at a mass ratio of 2:1 H_2O:Pt. After dispersion with probe sonication the ink was sprayed by air brush application. The Ni catalyst ink was prepared using a using a 1:6 Ni to ionomer ratio. The Ni ink was dispersed by stirring before application onto the membrane and GDL support. Each MEA used SGL Group (24BC GDL) electrode support and A201 membrane material. Pt and Ni loading for the 25 cm^2 membrane- and electrode-coated Pt-Ni MEA was calculated based on total catalyst ink casted to be 6 mg/cm^2 and 10 mg/cm^2, respectively. Each MEA was visually inspected and examined through SEM imaging before testing.

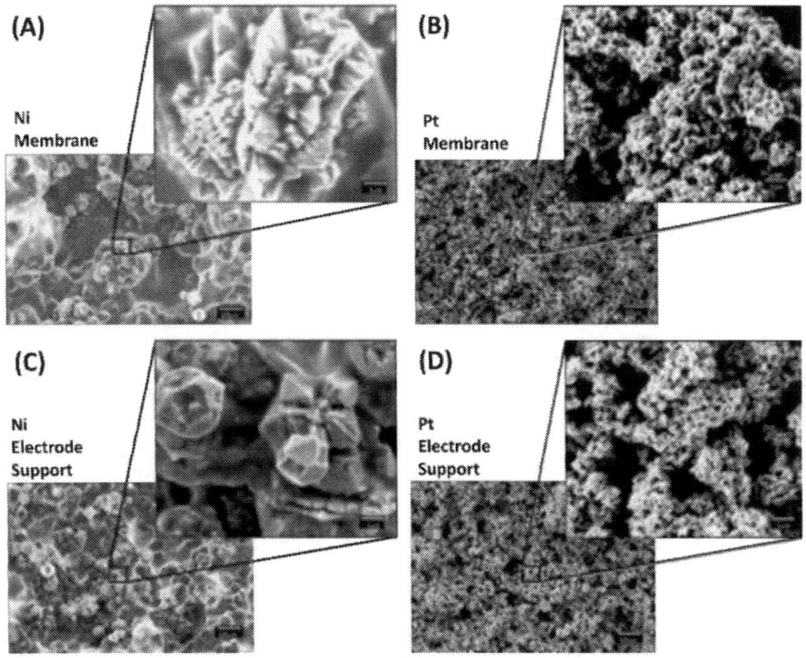

Figure 4. SEM image of (A) Ni catalyst spray casted on the membrane, (C) Ni applied to GDL electrode support, (B) Pt catalyst allied to the membrane, and (D) Pt applied to GDL electrode support

The Ni and Pt crystal structures are exposed and clearly seen throughout the electrode layer where the ionomer forms a matrix around the catalyst. These electrocatalyst coatings were highly uniform and had similar surface structures. In comparison to the Ni electrocatalyst layer, the Pt electrode layer was more uniform and microporous, resulting in a higher catalytic active area.

Results

Membrane- and Electrode Support-Coated Pt-Ni MEA

Impedance, linear sweep voltammetry (LSV), and steady-state current measurements were made for the membrane- and electrode support-coated MEAs after hydration. The equivalent series resistance (ESR) was evaluated before and after constant current CO_2 pumping. Before pumping both cells had a low ESR of 0.07 Ω, or 1.75 Ω·cm². After CO_2 pumping the resistance increased for the membrane- and electrode support-coated cell, to 9.00 Ω (225 Ω·cm²) and 2.46 Ω (61.5 Ω·cm²), respectively. A CO_2-rich gas mixture of 15 vol% CO_2, 15 vol% O_2, and 70 vol% N_2 was constantly fed to the Pt cathode side of the cell. Voltage sweeps from open circuit potential (OCP) to 1.2 V of the membrane- and electrode support-coated cell before CO_2 pumping measured a peak current at 0.71 A and 3.08 A, respectively. The CO_2 concentration of the N_2 sweep and the cell voltage were measured at constant currents of 10, 15, and 5 mA at 22°C.

The cell voltage and CO_2 concentration measurements for the membrane- and electrode support-coated MEAs maintained at 10 mA is depicted in Figure 5. After completion of the CO_2 pumping, a final impedance scan was completed to assess the effect of CO_2 pumping on the ESR.

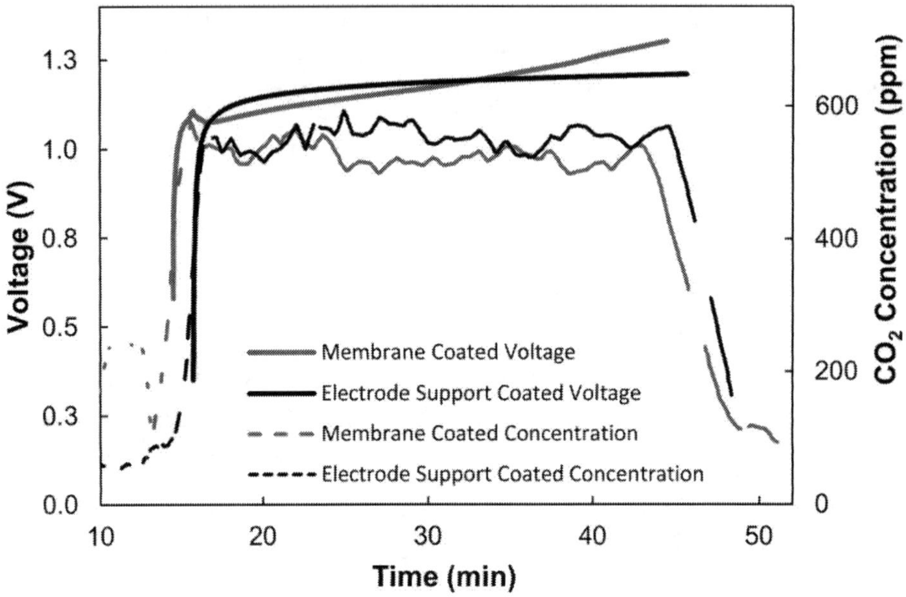

Figure 5. Cell voltage and CO_2 concentration for steady-state current at 10 mA for membrane- and electrode support-coated Pt-Ni MEA

The electrode support-coated cell maintained a lower, stable voltage and produced CO_2 concentrations closer to theoretical concentrations based on total charged passed for currents of 5 mA, 10 mA, and 15 mA. Baseline CO_2 concentrations from passive diffusion across the membrane was subtracted for the theoretical calculations. Average cell voltage, CO_2 flux, and calculated pumping efficiency (percentage of measured CO_2 to theoretical maximum amount of CO_2 separated) for the membrane- and electrode support-coated MEAs are summarized in Table 3. These measurements demonstrated improved CO_2 pumping efficiency through materials selection, formulation, and processing methods.

Table 3. Summary of Performance for CO_2 Pumping Pt-Ni MEAs

	Membrane Coating			*Electrode Support Coating*		
Pumping Current (mA/cm²)	Voltage (V)	CO_2 Flux (L/m²·min)	Pumping Efficiency (%)	Voltage (V)	CO_2 Flux (L/m²·min)	Pumping Efficiency (%)
0.2	1.38	0.027	89	1.09	0.030	100
0.4	1.17	0.059	97	1.17	0.061	107
0.6	1.62	0.082	90	1.34	0.091	100

Temperature & Concentration Effects on Electrode Support-Coated Pt-Ni MEA

The CO_2 separation performance of the 25 cm^2 electrode support-coated Pt-Ni MEA was measured at temperatures of 22°C, 30°C, 40°C and 50°C for each gas mixture with CO_2 concentrations of 0, 5, 10, and 15 vol% CO_2 with a constant oxygen concentration of 15 vol% O_2. Impedance, LSV, and steady-state constant current measurements were made to determine the dependence on temperature and CO_2 concentration. The measured CO_2 flux was determined by measuring the cell voltage and CO_2 concentration in the N_2 sweep with a Senseair K-33 BLG CO_2 sensor at a constant current of 50 mA for 20 minutes.

At each temperature and CO_2 concentration, impedance measurements determined the high frequency resistance. Due to the poor diffusivity of bicarbonate, the membrane resistance appeared to increase in the presence of CO_2. Higher temperatures would increase ion mobility within the membrane and lower the resistance; however, resistances increased slightly. As humidified gas was supplied to the cathode side only, non-uniform dehumidification was attributed to the change in resistance. Peak currents were measured from voltage sweeps from OCP to 1.2 V after the cell was at open circuit for 1 hour. The cell performance in air was dominate due to the ion mobility of hydroxide being greater than bicarbonate ion; however, no distinguishable trend was observed over the tested temperatures and gas concentrations.

The CO_2 flux was stable and the cell voltage was nearly 1.4 V for each run at 50 mA (2 mA/cm^2). The same electrode support-coated cell displayed signs of instability at lower currents of 5–15 mA at 1.4 V, indicating an optimal operating point. Production of water from the reaction with CO_2 and electro-osmotic drag maintains optimal hydration within the membrane and the testing conditions. In Table 4, the measured flux of CO_2 was used to calculate the pumping efficiency of the MEA by comparison to the maximum theoretical rate of separation using the CO_2 pumping reaction. The CO_2 pumping efficiency is the ratio between the actual and theoretical maximum amount of CO_2 separated through the MEA.

Table 4. Measured CO_2 Flux and Pt-Ni MEA Efficiency for Different Temperatures and CO_2 Concentrations

CO_2 Concentration (vol%)	Temperature (°C)	CO_2 (L/m^2·min)	Efficiency (%)
5	22	0.25	81.6%
	30	0.27	90.0%
	40	0.29	**95.8%**
	50	0.28	92.6%
10	22	0.27	90.9%
	30	0.29	95.3%
	40	0.30	**97.4%**
	50	0.31	95.5%
15	22	0.27	88.0%
	30	0.29	**99.7%**
	40	0.30	97.9%
	50	0.31	99.4%

The maximum measured CO_2 current efficiency for each concentration is highlighted in Table 4 with the respective temperature. The current efficiency of the Pt-Ni MEA is reduced at lower temperatures, near 22°C. Higher CO_2 concentrations tend to result in higher measured efficiencies. This data suggests that the cell is operated optimally at slightly elevated temperatures and CO_2 concentrations.

The Pt-Ni MEA tested at 50 mA (2 mA/cm^2) at 40°C was operated with an increased current of 250 mA (10 mA/cm^2) at 40°C and 40°C exposed to the 10% CO_2 gas mixture. The results are briefly summarized in Table 5. The CO_2 pumping capacity increased with the current although there was a decrease in efficiency at the higher loads. Higher efficiency was achieved with increasing the operational temperature; however, the drop in efficiency suggests the upper limit of operation was reached with the current cell configuration. For the final run at 50°C and 10 mA/cm^2, the MEA potential was stable over time at about 1.62 V.

Table 5. Electrode-coated Pt-Ni MEA Operation at Higher CO_2 Flux

Temperature (°C)	Current (mA/cm^2)	CO_2 (L/m^2·min)	Efficiency (%)
40	2	0.51	84.9%
40	10	1.01	67.8%
50	10	1.10	73.5%

We also observed that the voltage required to operate the cell depended strongly on the flow rate of the dry nitrogen sweep gas. Lower voltage, and therefore less power, was required when the flow was cut back. The strength and sensitivity of the cell to the dry gas flow on the anode were both related to water management in the MEA. Higher current would bring more water across the cell through the reaction and electro-osmotic drag, and the lower dry nitrogen sweep allowed the MEA to retain more of that water, ultimately increasing ion mobility.

Alternative Electrode Support-Coated Pt-RuO_2 MEA

A Pt-RuO_2 MEA was fabricated using RuO_2 catalyst (Alfa Aesar) blended with Tokuyama AS-4 solution at a ratio of 10.2 g of 5 vol% AS-4 solution per gram of catalyst for the anode catalyst forumation. The catalyst mixture was air brushed onto the surface of the PTFE-coated carbon electrode support (SGL 24BC) with a calculated loading of 6 mg RuO_2/cm^2. The Pt cathode electrode was produced using the same procedure previously discussed. The catalyst layer on the electrode supports were inspected through SEM imaging before MEA evaluation. A comparison of the Pt and RuO_2 catalyst microstructure is shown in Figure 6. The Pt formulation and coating process was optimized to produce a highly uniform and microporous electrode (Figure 6a). The RuO_2 layer (Figure 6b) appears to be uniform; however, the visible active catalyst layer on the surface is overloaded with ionomer.

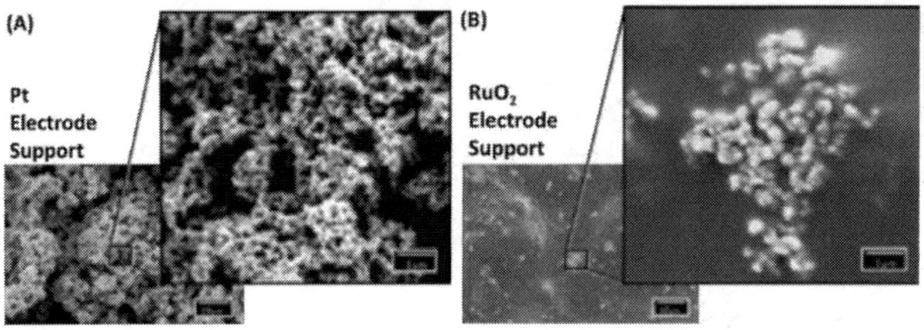

Figure 6. SEM images of MEA electrode layers showing (a) Pt and (b) RuO_2 catalysts on electrode support material

Electrochemical impedance spectroscopy, LSV, cell voltage, and CO_2 flux measurements during steady-state constant current operation were completed. The Pt-RuO_2 cell was fed a CO_2-enriched breathing gas mixture with 10 vol% CO_2 at 30°C for performance comparison to the previously tested Pt-Ni MEA. After the initial hydrolysis period, the MEA was exposed to the 10 vol% CO_2 breathing gas mixture, and the current was measured over a linear voltage sweep to 1.2 V. The Pt-RuO_2 peak current reached 18.0 mA/cm^2 at 1.2 V (compared to 0.98 mA/cm^2 at 1.2 V for the Pt-Ni MEA) at 30°C. The Pt-RuO_2 MEA was then held at a constant-current of 250 mA (10 mA/cm^2) and maintained a voltage of 1.09 V (summary of results for comparison are listed in Table 6).

Table 6. Comparison of Catalyst Performance at Constant Current of 10 mA/cm^2

Catalyst	Temperature (°C)	CO_2 (L/m^2·min)	Efficiency (%)	Voltage (V)
Pt-RuO_2	30	1.03	68.0	1.09
Pt-Ni	40	1.01	67.8	1.64

The results from the LSV and CO_2 pumping show an enhanced performance potential for the Pt-RuO_2 MEA over the Pt-Ni MEA, which would significantly reduce the cell area and power requirements. We expect to see large increases in the CO_2 removal capabilities as the voltage increases near 1.6–1.7 V with the Pt-RuO_2 catalysts. The RuO_2 catalyst formulation was not optimized and displayed erratic performance at higher voltages and current densities. It is expected that balancing catalyst and ionomer concentration and reducing ionomer wetting will improve electrode microstructure and performance. For a conservative design, the Pt-Ni MEA was chosen as the benchmark for the initial CO_2 separator for the SCBA system.

Conclusions

Electrochemical CO_2 removal from breathing air using advanced MEA technology was demonstrated as a viable component for enhanced SCBA systems. The Pt-Ni and Pt-RuO_2 MEAs produced a CO_2 flux of 1.03 $L/m^2 \cdot min$ and 1.01 $L/m^2 \cdot min$ at 10 mA/cm^2, respectively. An electrochemical separation of CO_2 system could eliminate heat generating absorbent packs by regulating humidity and heat within the breathing loop, reducing the burden on the cooling packs and increasing system capacity. Based on Pt-Ni MEA performance, the CO_2 removal cell would be 1.8 m^2 to remove 2 L/min of CO_2, or 44 MEAs of 17.7 cm x 23.0 cm. Stack configuration, fluid dynamics, current distribution, and life cycle performance are important aspects of cell design.[18] Separating exhaled CO_2 requires a unique design to accommodate for irregular flow and high-humidity gas. Electrode performance is dependent on catalyst size, microstructure, electrode support, ionomer, and fabrication methods, which influence mass transfer and water management. Absorption of CO_2 at the cathode, diffusion and migration in the membrane, and rejection of gas at the anode are the controlling steps for CO_2 transfer.[19] With Pt-Ni MEAs packaged in a stacked configuration, a 3D model of the CO_2 removal system indicates the system volume would be 23.1 cm x 18.0 cm x 16.5 cm (6.9 L) and weigh less than 5.7 lb (excluding power source), which is smaller than current absorbent cartridges (6.0 lb before use with weight added from CO_2 and moisture adsorption). Alternative anode catalyst materials such as RuO_2, IrO_2, and Co_3O_4 displayed higher catalyst activity, which could further enhance the rate of CO_2 removal. In addition to characterizing the MEA materials, we evaluated the micro-structure of the active catalyst area. The Pt-RuO_2 MEA has an enormous potential for performance enhancement, we expect the size of the CO_2 removal system to reduce dramatically.

References

1. Cunningham, S.; Burke, A.; Kelly, G., CFD Modeling of Breakthrough in Closed Circuit Rebreather Scrubbers. *Australian Journal of Basic and Applied Sciences* **2013,** *7* (5), 116-127.
2. Donnelly, M.; Putorti, A. *Exploratory Study of Airflow from SCBA Exposed to Elevated Temperatures*; NIST: 2013.
3. Yu, C.-H.; Huang, C.-H.; Tan, C.-S., A review of CO2 capture by absorption and adsorption. *Aerosol and Air Quality Research* **2012,** *12* (5), 745-769.
4. Nuckols, M. L.; Purer, A.; Deason, G. A., Design Guidelines for Carbon Dioxide Scrubbers. Naval Coastal Systems Center: 1985.
5. Andrew Lee, S.; Stevens, G. W.; Kentish, S. E., Facilitated transport behavior of humidified gases through thin-film composite polyamide membranes for carbon dioxide capture. *Journal of Membrane Science* **2013,** *429* (0), 349-354.
6. Berean, K.; Ou, J. Z.; Nour, M.; Latham, K.; McSweeney, C.; Paull, D.; Halim, A.; Kentish, S.; Doherty, C. M.; Hill, A. J.; Kalantar-zadeh, K., The effect of crosslinking temperature on the permeability of PDMS membranes: Evidence of extraordinary CO2 and CH4 gas permeation. *Separation and Purification Technology* **2014,** *122* (0), 96-104.
7. von Harbou, I.; Imle, M.; Hasse, H., Modeling and simulation of reactive absorption of CO2 with MEA: Results for four different packings on two different scales. *Chemical Engineering Science* **2014,** *105* (0), 179-190.
8. Goto, K.; Okabe, H.; Chowdhury, F. A.; Shimizu, S.; Fujioka, Y.; Onoda, M., Development of novel absorbents for CO2 capture from blast furnace gas. *International Journal of Greenhouse Gas Control* **2011,** *5* (5), 1214-1219.
9. Aeshala, L. M.; Rahman, S. U.; Verma, A. In *Effect of solid polymer electrolyte on electrochemical reduction of CO 2*, P.O. Box 211, Amsterdam, 1000 AE, Netherlands, Elsevier: P.O. Box 211, Amsterdam, 1000 AE, Netherlands, 2012; pp 131-137.
10. Narayanan, S. R.; Haines, B.; Soler, J.; Valdez, T. I., Electrochemical conversion of carbon dioxide to formate in alkaline polymer electrolyte membrane cells. *Journal of the Electrochemical Society* **2011,** *158* (2), A167-A173.
11. Landon, J.; Kitchin, J., Electrochemical Concentration of Carbon Dioxide from an Oxygen/Carbon Dioxide Containing Gas Stream. *Journal of Electrochemical Society* **2010,** *157* (8).
12. System Certification Procedures and Criteria Manual for Deep Submergence Systems. 1998.

13. Duan, Q.; Ge, S.; Wang, C.-Y., Water uptake, ionic conductivity and swelling properties of anion-exchange membrane. *Journal of Power Sources* **2013,** *243*, 773-778.

14. Yanagi, H.; Fukuta, K. In *Anion exchange membrane and ionomer for alkaline membrane fuel cells (AMFCs)*, Honolulu, HI, United states, Electrochemical Society Inc.: Honolulu, HI, United states, 2008; pp 257-262.

15. Pennline, H. W.; Granite, E. J.; Luebke, D. R.; Kitchin, J. R.; Landon, J.; Weiland, L. M., Separation of CO2 from flue gas using electrochemical cells. *Fuel* **2010,** *89* (6), 1307-1314.

16. Xiao, S.; Li, K., On the Use of an Electrochemical Membrane Module for Removal of COs2 from a Breathing Gas Mixture. *Chemical Engineering Research and Design* **1997,** *75* (4), 438-446.

17. Landon, J.; Kitchin, J. R., Electrochemical concentration of carbon dioxide from an oxygen/carbon dioxide containing gas stream. *Journal of the Electrochemical Society* **2010,** *157* (8), B1149-B1153.

18. Millet, P.; Ngameni, R.; Grigoriev, S. A.; Mbemba, N.; Brisset, F.; Ranjbari, A., PEM water electrolyzers: From electrocatalysis to stack development. *International Journal of Hydrogen Energy* **2010,** *35* (10), 5043-5052.

19. Xiao, S. Q.; Li, K., On the use of an electrochemical membrane module for removal of CO_2 from a breathing gas mixture. *Chemical Engineering Research and Design* **1997,** *75* (A4), 438-446.

409026 Single and Multi-Component Transport through Metal Supported Clioptilolite Membranes

Abstract:

Metal supported mixed-matrix membranes (MMMs), filled with natural zeolite particles, provide a mean to improve the gas separation performance of inorganic membranes. Microporous natural zeolite powders, clinoptilolite, in alumina silicate matrix were characterized for separation of hydrogen from light hydrocarbons. Both single and multi-component permeation test was used to evaluate the performance of the membrane. At room temperature and a feed pressure of 110 kPa large differences for the separation efficiency of hydrogen from lower n-alkanes were observed for the components under investigation.

The permeation of the weekly adsorbing molecules (e.g. hydrogen) was found to drop at room temperature and resulting in lower selectivities in presence of strongly adsorbing molecules in favor of hydrogen.

Experimental analysis showed that for mixtures of a fast, weakly adsorbing component and a slow, strongly/moderately adsorbing components, the maximum selectivity for hydrogen separation obtained with microporous is about a factor of 2 lower than the one obtained under ideal single gas permeations. Mixed matrix membranes coated on stainless steel supports were found to have as much as two times higher ideal selectivity for H2 over each light hydrocarbon than would be predicted by Knudsen diffusion.

(740e) Propylene/Propane Separation through FAU-Type Zeolite Membrane

Matsukata, M., Waseda University
Sakai, M., Waseda University
Sasaki, Y., Waseda University
Tomono, T., Waseda University
Originally presented on: 11/12/2015 16:39:00 - 17:00:00

Due to close boiling points of propane and propylene as well as their low boiling points, distillation separation of propane/propylene mixtures produced at the ethylene center has a large energy demands. Membrane gas separation would be a promising candidate to save separation energy dramatically. In this presentation, we would like to show our recent results on FAU-type zeolite membrane for such C3 mixture separation. FAU-type membrane was synthesized on the outer surface of porous alpha-alumina tube with 10 mm i.d. and 3 cm length (Noritake Co. Limited) by a secondary growth method consisting of seeding of FAU-seeds on the outer surface of support and its hydrothermal growth. The resultant membrane was treated in aqueous solutions containing with a variety of kinds of cation such as alkali and alkaline earth cation, and transition metal cation. Among them, we found that Ag-FAU membrane exceptionally showed high separation factor. For the equimolar C3 gas mixtures, the permeance of propane was 4.13×10^{-8} mol m^{-2} s^{-1} Pa^{-1} with the separation factor of 55.3 at 353 K.Figure 1 compares the permeances of propane and propylene in the unary and binary systems. Whereas the propylene permeances in the both systems were almost the same, the permeance of propane was markedly decreased in the presence of propylene, indicating that propylene strongly adsorbed in the membrane inhibited the permeation of propane. In addition, such a separation factor exceeding 50 will be able to enable us to produce polymer-grade propylene in one step of membrane separation.

□v

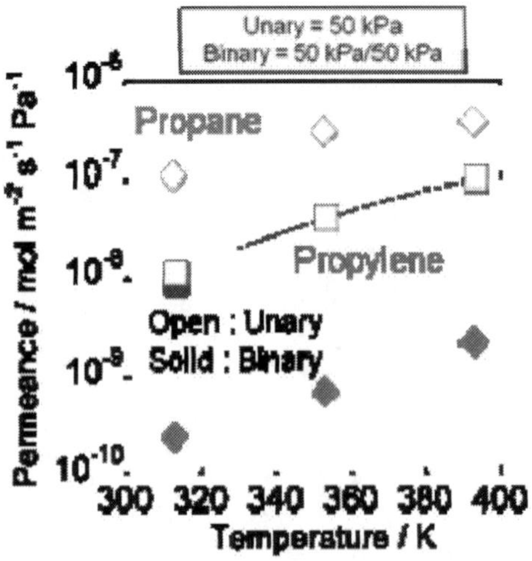

Figure 1 Comparison of the permeances of propane and propylene in the unary and binary systems.

428657 Amino Acid Ionic Liquids-Based Ion Gel Membranes with Superior Pressure Resistance for CO2 Capture Application

Thursday, November 12, 2015: 5:00 PM
155B (Salt Palace Convention Center)

Farhad Moghadam, Eiji Kamio, Ayumi Yoshizumi and Hideto Matsuyama, Center for Membrane and Film Technology, Department of Chemical Science and Engineering, Kobe University, Kobe, Japan

Amino Acid Ionic Liquids-based Ion Gel Membranes with Superior Pressure Resistance for CO_2 Capture Application

Farhad Moghadam□ð, Eiji Kamio□ð, Ayumi Yoshizumi□ð and Hideto Matsuyama□ð,*

□ð

* e-mail address: matuyama@kobe-u.ac.jp, tel.: +81-78-803-6180, fax: +81-78-803-6180

Abstract

Amino acid ionic liquids (AAILs)-based membranes have been recognized as an attractive alternative to conventional facilitated transport membranes (FTMs) due to unique properties of AAILs, such as the high CO_2 absorption capacity, selective reactivity with CO_2, negligible vapor pressure, high thermal stability, and tunable chemical structure. The AAILs-based gel membranes as well as supported ionic liquid membranes previously developed in our group exhibited good CO_2 permeability and CO_2/N_2 selectivity, but the poor stability under pressurized conditions is a major obstacle to their practical application [1,2].

Here we present a new class of AAILs-based ion gel membranes with not only superior CO_2 permeability and CO_2/N_2 selectivity but also excellent stability under pressurized conditions. The developed AAIL-based gel membranes consisted of two specific independent interpenetrating polymer networks, so-called doubel-network (DN) [3]. The DN was composed of two asymmetric polymer networks of which the first network was a rigid, brittle, and tightly cross-linked polyelectrolyte and the second network was a soft, ductile and loosely cross-linked polymer with good compatibility with AAILs. As the solvent of the DN gels, AAILs composed of phosphonium type cation and prolinate anion were used because of their excellent CO_2 transport properties [4]. The mechanical strength of the fabricated AAILs-based DN gel membranes were tested by compressive and tensile stress-strain measurements, and the DN gels composed of more than 80wt% AAILs showed more than 25 MPa of compressive and about 0.7 MPa of tensile fracture stresses.

According to the CO_2 permeation performances of the AAIL-based DN gel membranes, the DN gels membrane with 85 wt% of AAILs showed remarkable CO_2 permeability of more than 5000 barrer and high CO_2/N_2 selectivity of more than 170 (at 373 K and CO_2 partial pressure 10 kPa). From the trend of the CO_2 and N_2 permeabilities of the DN gel membranes at different CO_2 partial pressures, we confirmed that the permeation mechanism of CO_2 and N_2 were facilitated transport and solution–diffusion mechanisms, respectively. In addition, the pressure resistance of the DN ion gel membranes containing 85 wt% of AAILs was investigated by increasing the feed side pressure up to 500 kPa and the stable performance of membrane were confirmed (Fig.1). Moreover, the DN ion gel membranes exhibited good durability; the CO_2 permeation performance remained stable more than 5 days at 500 kPa.

Regarding the good mechanical strength of AAIL-based DN gels, it was possible to prepare thin membranes. We fabricated AAIL-based DN gel membranes with different thicknesses and evaluated the gas permeation mechanisms. From the obtained results, it was found that CO_2 and N_2 permeances showed reverse proportional relationship to the DN gel membrane thickness, which clearly demonstrated that the rate-controlling step of CO_2 and N_2 permeations across the DN gel membranes were intra-membrane diffusion. In the present stage, the AAIL-based DN gel membrane with the thinnest thickness (58 mm) showed the CO_2 permeance of ca. 120 GPU and CO_2/N_2 selectivity of ca. 100 under the conditions of T= 373 K and 10 kPa of the CO_2 partial pressure. However, as the rate-controlling step of the CO_2 permeation is intra-membrane diffusion, we could increase the CO_2 permeance by fabricating much thinner AAIL-based DN gel membranes.

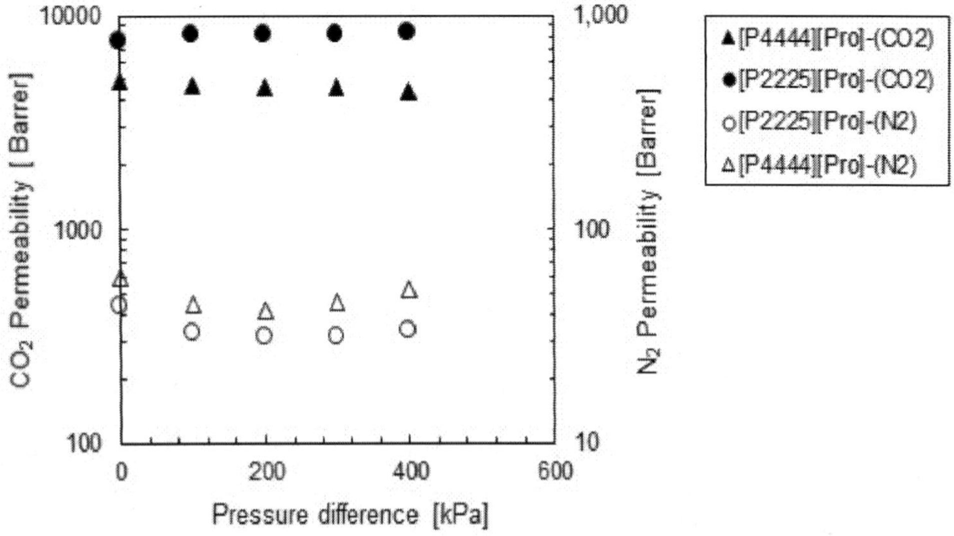

Fig.1. Pressure resistance behavior of AAILs-based PAMPS/PDMAAm DN ion gel membranes (Experimental conditions: T = 373 K, P_{sweep} = atmospheric pressure, $P_{CO2,feed}$ = 10 kPa)

References:

[1] S. Kasahara, E. Kamio, T. Ishigami and H. Matsuyama, ***Chem. Commun.***, 2012, **48**, 6903-6905

[2] S. Kasahara, E. Kamio, A. Yoshizumi and H. Matsuyama, *Chem. Commun.*, 2014, 50, 2996-2999

[3] J. P. Gong, Y. Katsuyama, T. Kurokawa and Y. Osada, *Adv. Mater.*, 2003, 15, 1155-1158

[4] S. Kasahara, E. Kamio, T. Ishigami and H. Matsuyama, *J. Membr. Sci.*, 2012, 415-416, 168-175

423667 Mixed-Conducting Multichannel Hollow Fiber Membranes for Oxygen Separation and Membrane Reaction

Abstract:

Mixed-conducting multichannel hollow fiber membranes for oxygen separation and membrane reaction

Jiawei Zhu, Wanqin Jin *

State Key Laboratory of Materials-Oriented Chemical Engineering

Membrane Science & Technology Research Center,

Nanjing Tech University, Nanjing 210009, P. R. China

* wqjin@njut.edu.cn

Due to the potential impacts on energy and environment, mixed-conducting dense membranes have attracted considerable interests in oxygen separation and membrane reaction process. Mixed- conducting hollow fiber membranes possess a thin wall and an asymmetric structure and can obtain high oxygen permeation flux. Therefore, the configuration of hollow fiber is considered as the promising one for the future industrial application. Nevertheless, the conventional single-channel mixed-conducting hollow fiber membrane may not meet the requirements of industrialization, suffering from low mechanical strength. Therefore, in our work, a novel mixed-conducting multichannel hollow fiber membrane with high oxygen permeation flux and excellent mechanical strength was firstly proposed by phase inversion and sintering technique to combine advantages of multichannel tubular membrane technology and conventional hollow fibers[1,2]. Due to excellent properties of the multichannel hollow fiber membrane configuration, the mixed-conducting multichannel hollow fiber membrane has been applied in oxygen separation in oxyfuel process for CO_2 capture[3]. In addition, to accelerate the commercial application of mixed-conducting membrane reactor for catalytic reaction processes, a robust mixed-conducting multichannel hollow fiber membrane reactor was constructed and characterized in our work[4]. Our work demonstrated that the mixed-conducting multichannel hollow fiber membrane can be a competitive candidate for industrial application in the field of energy and environmental.

Keywords: mixed-conducting; multichannel hollow fiber; membrane configuration; oxygen separation; oxyfuel process; membrane reactor

Acknowledgments

This work was supported by the Project of Priority Academic Program Development of Jiangsu Higher Education Institutions (PAPD) and the Innovative Research Team Program by the Ministry of Education of China (No. IRT13070).

Reference

1. Zhu J, Dong Z, Liu Z, Zhang K, Zhang G, Jin W. Multichannel mixed-conducting hollow fiber membranes for oxygen separation. AIChE Journal. 2014;60(6):1969-1976.

2. Zhu J, Liu Z, Guo S, Jin W. Influence of permeation modes on oxygen permeability of the multichannel mixed-conducting hollow fibre membrane. Chemical Engineering Science. 2015;122(0):614-621.

3. Zhu J, Guo S, Zhang Z, Jiang X, Liu Z, Jin W. CO_2-tolerant mixed-conducting multichannel hollow fiber membrane for efficient oxygen separation. Journal of Membrane Science. 2015;485(0):79-86.

4. Zhu J, Guo S, Liu G, Liu Z, Zhang Z, Jin W. A robust mixed-conducting multichannel hollow fiber membrane reactor. AIChE Journal. 2015: DOI 10.1002/aic.14835.

408894 A Theoretical Model for Accelerated Nucleation Due to Adsorbing Additives

Abstract:

Additives are used to control nucleation in many natural and industrial environments. However, the mechanisms by which additives inhibit or accelerate solute precipitate nucleation are not well understood. We modify classical nucleation theory (CNT) to predict reductions in surface tension and therefore nucleation barriers due to the adsorption of trace additives. Our model shows that nucleant efficacy depends primarily on additive concentration and adsorption equilibrium constant, suggesting that assays that screen for large equilibrium constants might facilitate nucleant design. We confirm this trend using a Potts-lattice gas model with surfactant-like additives in addition to solutes and solvents. We also show that the reaction coordinate is no longer nucleus size but is tilted toward the number of adsorbed additives coordinate. However, we show that rate corrections decay algebraically to zero as additive coverage approaches zero. Therefore, CNT can be simply modified by using a reduced surface tension due to quasi-equilibrated additive adsorption but only for weakly adsorbing and/or low concentration additives.

434267 Nucleation of Gas Hydrates in Interfacial Systems

Abstract:

Gas hydrates are solid crystalline structures formed when gases such as methane, ethane and carbon dioxide come in contact with water at low temperatures and high pressures. Hydrates remain a challenging flow assurance problem in the oil and gas industry. Recently, hydrates have also emerged as potential source of energy, route for carbon dioxide sequestration, alternative technology for water desalination and mode of gas transportation. The wide and promising applications in gas hydrates have renewed interest in understanding the molecular processes driving their formation, and stability. To this end, in our research, we use large-scale molecular dynamics simulations to probe the nucleation and growth of methane and tetrahydrofuran (THF) hydrates in interfacial environments. Specifically we investigate the effects of surface chemistry on hydrate nucleation by simulating solutions of methane/THF in water near hydrophobic and hydrophilic surfaces in hydrate forming conditions. We find that in case of methane at high concentration, the aggregation of methane molecules is promoted by the hydrophobic surface and inhibits hydrate formation due to the reduced concentration of methane in water. On the other hand no such effect is observed for the hydrophilic surfaces. In contrast, THF displays no preference to the surface and rapid nucleation is observed in both the systems. We conclude that for molecules with high solubility in water and for surfaces lacking any templating ability for hydrates, the surface chemistry does not affect hydrate nucleation.

434318 Is the Probability of Nucleation in a Metastable Zone Really a Function of the Rate of Nucleation?

Abstract:

Conventionally, in solution crystallization, the probability of nucleation at any supersaturation in a metastable zone width (MZW) experiment is determined using a survival probability analysis based on the functional dependence of the rate of nucleation on time-varying supersaturation. However, recent experimental results show that this approach leads to incorrect predictions of the probability of nucleation (and hence the mean supersaturation at which nucleation occurs) in an MZW experiment. In this work, we explore the root cause of this discrepancy. We distinguish the role of time in an MZW experiment from that in an induction time experiment by examining the characteristics of each type of experiment, and discuss the implicit assumptions involved in the traditional theoretical approaches used to model the metastable zone experiment. From this analysis, we show that the assumptions used in the development of the survival probability analysis based on the rate of nucleation are not valid for the case of the MZW experiment, and the concept of nucleation rate is irrelevant in analyzing the results of an MZW experiment. Instead, the experimental results are successfully explained by considering a survival probability analysis based solely on the energy barrier to nucleation. By developing a characterization of the lower boundary of metastability without reference to rate processes, we arrive at surprising conclusions that the lower limit of a metastable zone cannot be the solubility boundary and the practice of routinely extracting the kinetic parameters of nucleation from a measured metastable zone width needs to be reconsidered.

434713 Determination of Crystal Growth Rate of Racemic Ibuprofen Crystals from Enantiomerically Enriched Solutions

Abstract:

The influence of foreign molecules on crystallization kinetics has been widely investigated [1,2]. When growing crystals of pure enantiomers, the presence of the counter-enantiomer is known to typically reduce the crystallization kinetics of the preferred enantiomer [3,4]. For the case of racemic compounds, such investigations are rarely reported. However, if a retarding or promoting effect is determined in such an investigation it could lead to valuable insights about the underlying crystallization mechanisms. For instance, carefully investigating the effect of an enantiomeric enrichment in the solution on the crystallization kinetics of racemic crystals, could clarify whether interactions in the solution or interactions of the enantiomers on the crystalline surfaces are responsible for the change in kinetics.

In this work, we established the ternary phase diagram of the chiral molecule Ibuprofen in n-Heptane at various temperatures and described the thermodynamics using an activity coefficient model. Following this, batch desupersaturation experiments were carried out at different initial supersaturations, temperatures and enantiomeric excesses in the liquid phase with the aim of determining the growth kinetics of (R,S)-Ibuprofen crystals. We estimated the growth kinetics by fitting a population balance equation model to concentration profiles obtained by ATR-FTIR spectroscopy. The obtained growth kinetics provide insightful information about the crystal growth behavior of crystals of the racemic compound from enriched solutions.

References

[1] T. Vetter, M. Mazzotti, J. Brozio, *Slowing the Growth Rate of Ibuprofen Crystals Using the Polymeric Additive Pluronic F127,* Cryst. Growth Des., 7 (2011) 9, 1615-1617.

[2] K. Sangwal: *Additives and Crystallization Processes: From Fundamentals to Applications*, John Wiley & Sons, New York, 2007.

[3] A. Perlberg, H. Lorenz, A. Seidel-Morgenstern, *Crystal Growth Kinetics via Isothermal Seeded Batch Crystallization: Evaluation of Measurement Techniques and Application to Mandelic Acid in Water*, Ind. Eng. Chem. Res., 44 (2005), 1012-1020.

[4] L. Codan, C.F. Eckstein, M. Mazzotti, Growth Kinetics of S-Mandelic Acid in Aqueous Solutions in the Presence of R-Mandelic Acid, Cryst. Growth Des., 13 (2013), 652-663.

NREL
NATIONAL RENEWABLE ENERGY LABORATORY

Saturated Monoglyceride Polymorphism and Gel Formation of Biodiesel Blends

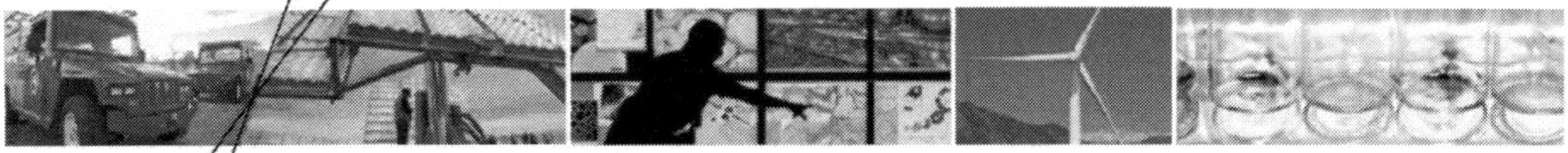

AIChE Annual Meeting

12 November 2015

Gina Chupka, Lisa Fouts, and Robert McCormick

National Renewable Energy Laboratory

Background: What is Biodiesel?

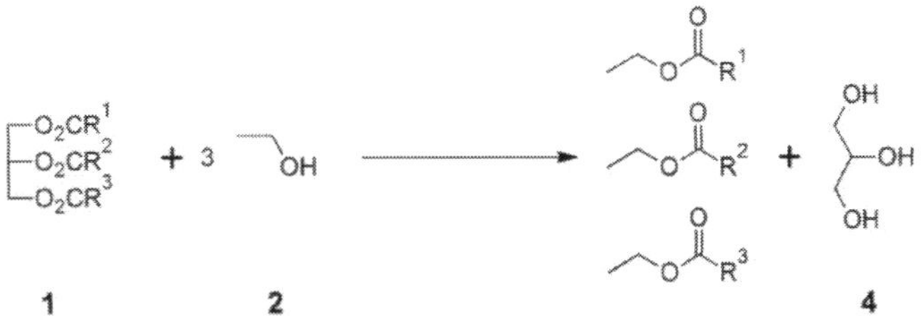

1 **2** **3** **4**

100 kg triglyceride fat/oil + 10 kg alcohol methanol = 10 kg glycerine (byproduct) + 100 kg mono-alkyl ester Biodiesel

Incomplete transesterification leads to mono, di, and triglycerides among other minor impurities such as unsaturated monoglycerides, glycerin, and water

Monostearin = saturated monoglyceride

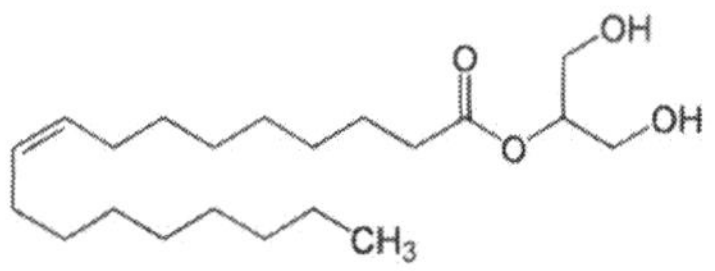

Monoolein = unsaturated monoglyceride

Background: Biodiesel Precipitates Above Cloud Point

- Cloud point (CP) is used as a measure of the lowest temperature at which petrodiesel or a biodiesel blend can be used without operational issues
- In some instances, unexpected fuel filter plugging is observed above the cloud point
- During some cloud point measurements, we observed an anomaly in the instrument signal that we could not explain

Background: Monostearin in Tallow-Derived B100

Initial crystals formed

Crystals after heating or standing at RT

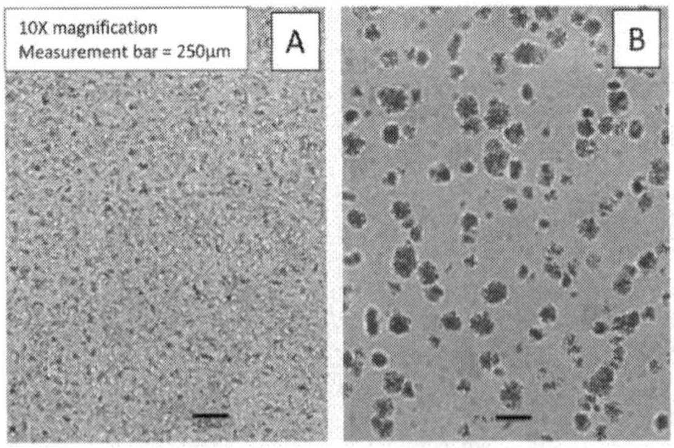

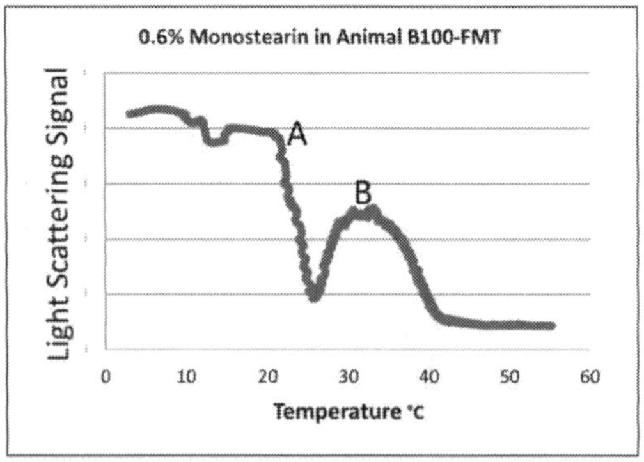

- One CP method utilizes light scattering to detect the CP
- Saturated monoglycerides (SMGs) are polymorphic and can undergo a solvent-mediated polymorphic phase transformation under short time scales
- Recovery of isolated crystals from biodiesel blends shows that they are the most stable beta crystal (Differential Scanning Calorimetry (DSC) and X-ray diffraction)

Chupka et al. *Energy & Fuels* 25 398–405 (2011) and *Energy Environ. Sci.* 5 8734-8742 (2012)

Background: Can SMGs Form Gels with Fuel Components?

- Previous gas chromatography-mass spectrometry results in our laboratory have found that precipitates are mainly composed of saturated and unsaturated monoglycerides, glycerin, and residual fuel
- Prior work showed that SMGs were the most significant component affecting the CP measurement: Chupka et al. *Fuel Process Technol* 118 302-309 (2014)
- Want to gain a fundamental understanding of how each impurity relates to filter plugging: What causes plugging in some cases but not others?

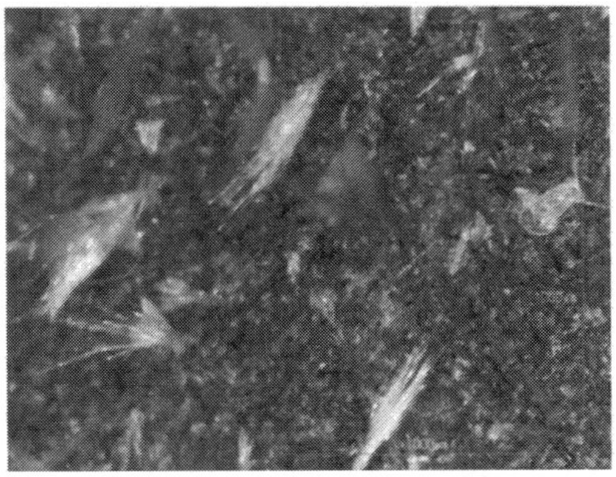

SMGs filtered from B20 soy biodiesel blend

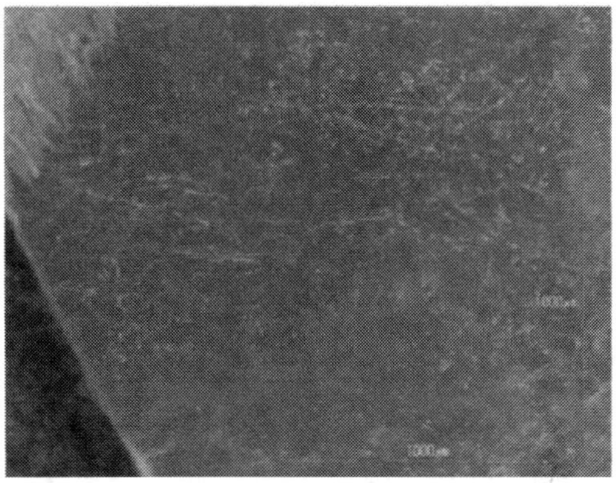

SMGs filtered from B20 canola biodiesel blend

Current Study Design

Questions to answer:

1. Because SMGs are most significant factor affecting CP, how does their solubility differ in various biodiesel and diesel type fuel components?
2. Can SMGs form co-crystals with other impurities in biodiesel?
3. Can SMGs form gels by trapping fuel within their crystal lattice?

- Monostearin solubility was determined in various pure components
- A simple model system with methyl oleate (biodiesel component) and dodecane (diesel component) was used as a "fuel" blend in which impurities were added to determine compounding effects
- A more complex diesel surrogate system consisting of decane, undecane, dodecane, tridecane, and tetradecane was also used to determine if chain length distribution was a factor
- Recovered precipitates were analyzed by several techniques

Solubility of Monostearin in Various Pure Components

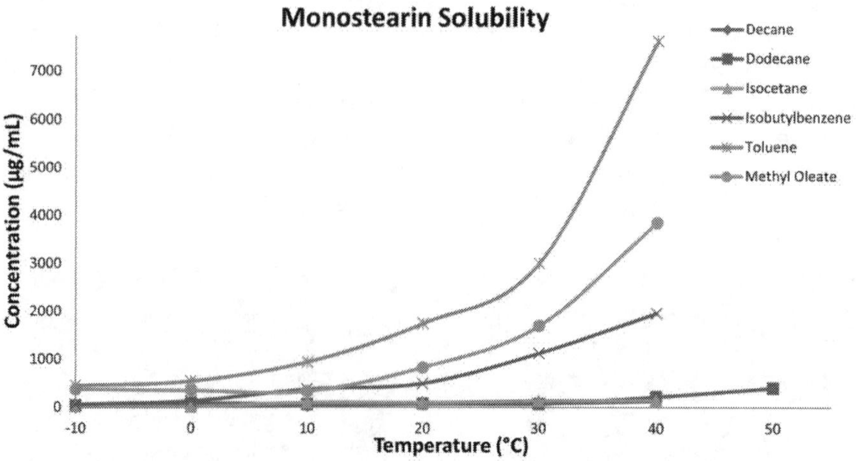

- Solubility of monostearin in diesel fuel hydrocarbon-type components (decane, dodecane, and isocetane) is much lower than that for toluene (aromatic diesel fuel component) and methyl oleate (biodiesel component), especially above 10°C
- These data suggest that the addition of aromatics or biodiesel can lower the CP of a blend by providing better solvency of the SMGs

Cloud Point and Time to Filter Results

Sample	CP (°C)	Time to Filter (Sec)
Methyl Oleate (MeO)/Dodecane	-15.9	7.4
0.025% Stearin MeO/Dodecane	12	44.8
0.025% Palmitin MeO/Dodecane	8	122.2
0.025% Palmitin/Stearin MeO/Dodecane	6.4	62.9
0.1% MO MeO/Dodecane	-14	67.0
0.025% Palm/Stear 0.1% MO MeO/Dodecane	10	231.1
MeO/Custom	-21.6	8.0
0.025% Stearin MeO/Custom	12	44.4
0.025% Palmitin MeO/Custom	9	95.2
0.025% Palm/Stear MeO/Custom	11	105.0
0.1% MO MeO/Custom	-21.6	116.0
0.025% Palm/Stear 0.1% MO MeO/Custom	6.5	142.6
Custom = Mixed hydrocarbon blend, MO = monoolein		

- As anticipated, SMGs increased the CP of the blends significantly
- The addition of both saturated and unsaturated monoglycerides increased the time to filter for all blends, even though the CP was not increased by the addition of MO
- The MO combined with SMGs lead to the highest filtration times
- Chain length distribution did not appear to have an affect on the CP or the time to filter

Co-Crystallization Results: Differential Scanning Calorimetry

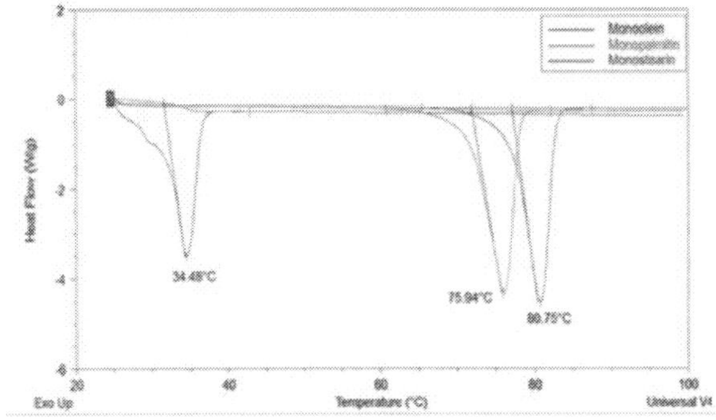

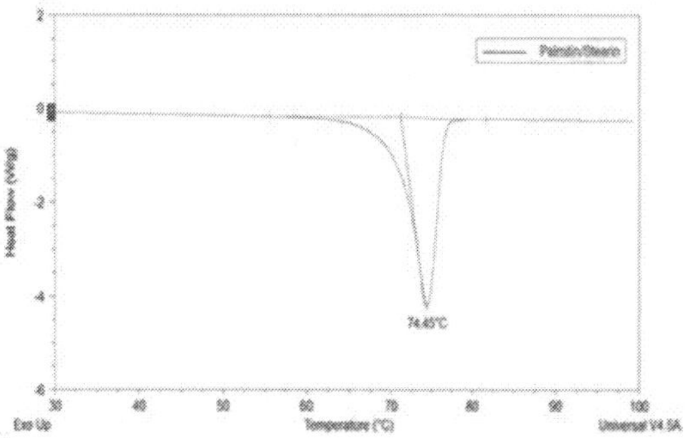

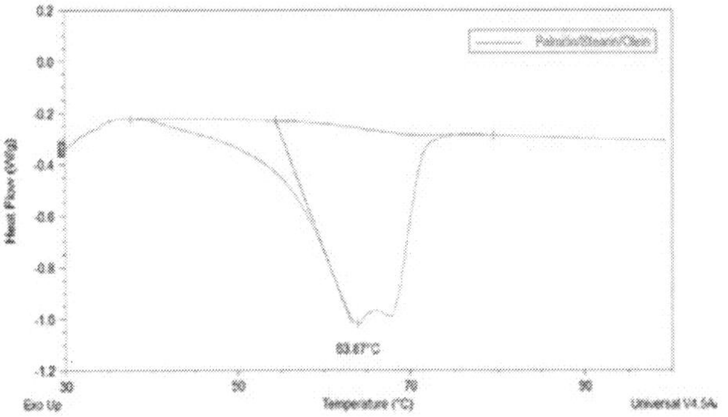

Crystals filtered from B5 Methyl oleate and dodecane surrogate

- From DSC thermograms, it can be seen that monopalmitin and monostearin co-crystallize which would be expected as their crystal structures are quite similar
- Surprisingly, we note that a co-crystal of monopalmitin, monostearin, and monoolein also forms which could indicate that the SMGs are able to provide a surface on which the monoolein can crystallize upon

Co-crystallization Results: X-ray Diffraction

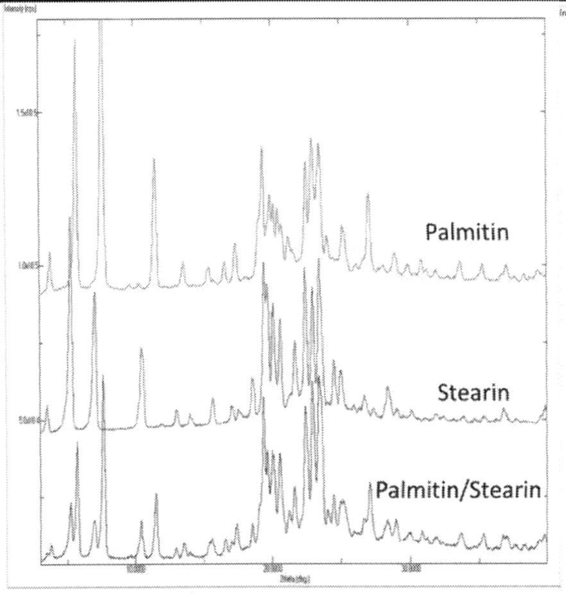

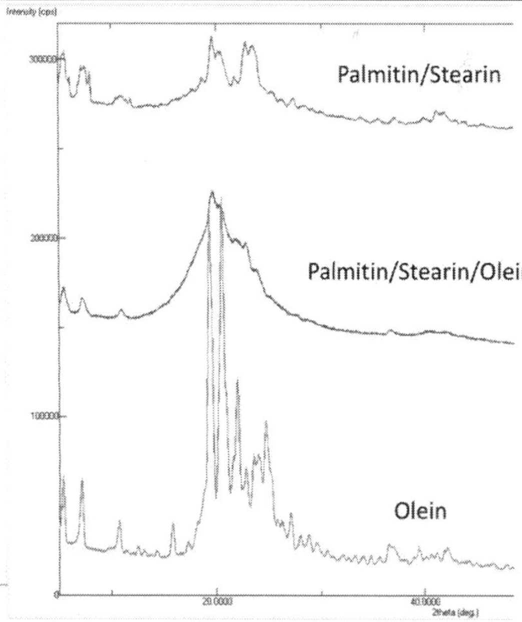

- Co-crystal of Palmitin/Stearin/Olein begins to take on amorphous character
- Could more amorphous crystal structure cause “coating” of filter?

Pure reference materials compared to filtered crystals from B5 Methyl oleate and dodecane surrogate

Rheology: Gel Temperature of Surrogate Blends

Sample	Gel Point, °C
B5 MO/Custom Surrogate	-2.3
Palmitin/Stearin	5.0
Palmitin/Stearin/Olein	6.4
Palmitin/Stearin/Olein/Glycerin	9.7
Palmitin/Stearin/Olein/Glycerin/Water	10.3

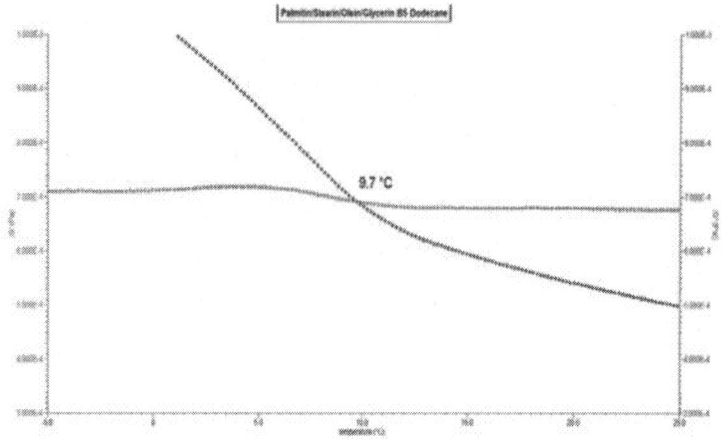

- For rheology samples, water and glycerin were also added as additional impurities to test their affect on gel temperature
- Gel point is defined as the cross over point of G' and G"
- Glycerin and water increased the gel temperature

Future Work

- **Measure gel strength of various blends by rheology**
- **Can we relate the measured gel strength to filter plugging?**
- **How do water and glycerin affect crystal structure of precipitates? Does more amorphous structure lead to "coating" and plugging of filter material?**

Conclusions

- SMG solubility showed that the addition of aromatics in diesel fuel and increased biodiesel content may help with solubility which could help keep SMGs in solution and decrease the CP of a blend
- SMGs increased the CP of blend samples greatly, while monoolein, an unsaturated monoglyceride did not
- SMGs can form co-crystals with other impurities in biodiesel and cause them to come out of solution before anticipated
- Loss of crystal structure to a more amorphous form is noted when monoolein is incorporated in the structure
- The addition of monoolein did increase filter times of the blend
- Additionally water and glycerin increased the gel temperature

Acknowledgments

We would like to thank the Vehicle Technologies Office, U.S. Department of Energy for funding for this study

Questions?

Monostearin Solubility Diagram-Hypothetical

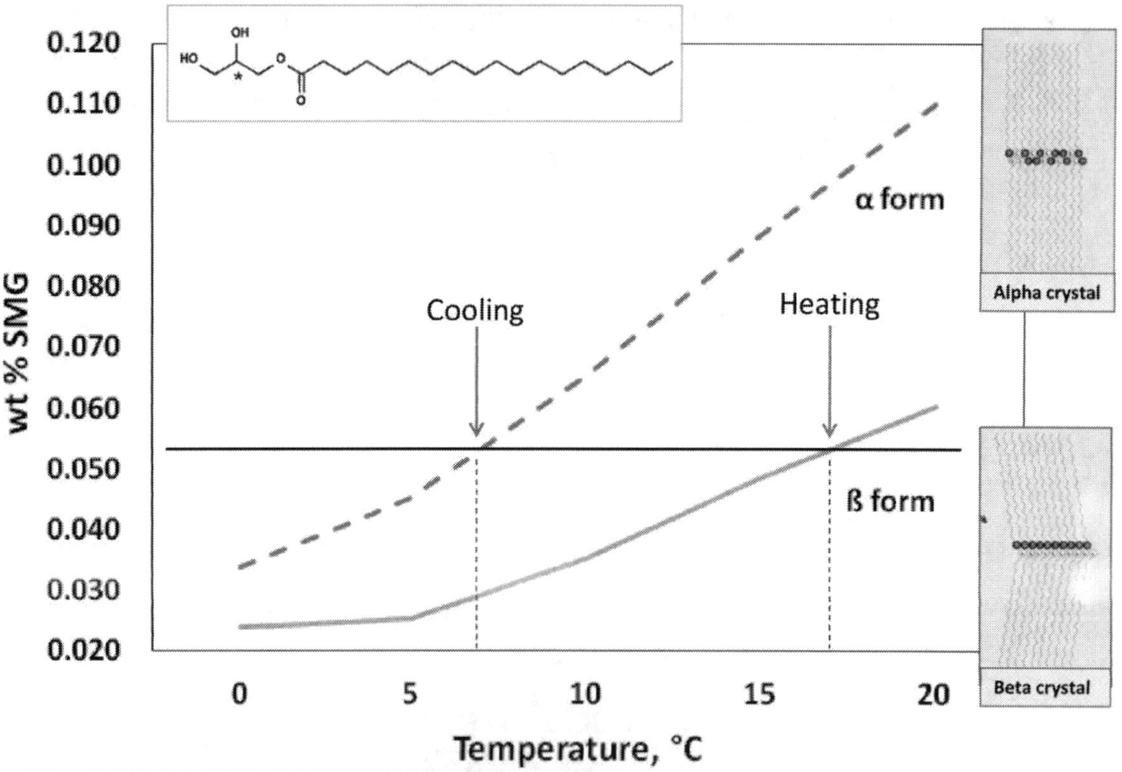

Structures from Chen, C.H., Terentjev, E.M., *Langmuir* 25 6717–6724 (2009)

429928 Ultrastable Core-Shell Structured Nanoparticles Directly Made from Zwitterionic Polymers

Abstract:
Colloid stability is crucial for nanoparticles (NPs) to remain nano-effects in various functioning media. Various methods had been utilized to modify hydrophobic core to render nanoparticle with colloid stability, especially in biorelated
environment. However, these methods were either involving complex chemical synthesis or low in efficiency due to unpredicted NP surface environment. Here, rather than modify the nanoparticle, for the first time we directly make selfassembled
core-shell nanoparticles from zwitterionic polymer using green chemistry. Zwitterionic polymers are novel generation biomaterials showing excellent nonfouling
properties. NPs protected by zwitterionic polymers have been found with long blood circulation characteristics and particularly suitable for drug delivery applications. Characterization confirmed the coreshell
structure and resulting zwitterionic nanoparticles had superior colloid stability in highly concentrated protein/PBS buffer solution and even freezedrying
condition. This work initiates a new concept of directly making stable nanoparticles through green chemistry and will inspire new strategy and green chemistry to be applied in nanotechnology.

431023 Can R&D be Standardized? Implementation of a Recipe Framework Supporting Lab to Patient Execution and Analytics

Abstract:

The benefits of a recipe framework have been recognized in manufacturing organizations for several years, but implementation in a development environment brings different challenges – and measurable benefits to drive appropriate scientific and engineering knowledge. To accomplish this task in a development environment a significant amount of structure needs to be implemented in the underlying laboratory systems, ensuring data can be captured with appropriate context. In this presentation, we will outline how we are implementing a flexible S88 recipe model across the organization to enable process definition, drive execution with context, and provide visualization of experiments/batches to compare this contextualized data across sites and scales. Detailed examples of the approach, resulting analysis tools, and challenges encountered, will be provided. This presentation will also outline how this framework can help a research and development organization easily analyze scale up and manufacturing processes, and enable transfer of process knowledge through development and commercialization.

414967 Enabling Rapid Elemental Analysis of Highly-Automated, Parallel Screening Studies for Pharmaceutical Process Development

Abstract:

In the pharmaceutical industry, utilization of lab automation for parallel, statistically designed experiments to optimize reagent and processing parameters allows for accelerated development of chemical processes. The resulting generation of comprehensive, high-fidelity data sets provides in-depth knowledge which ultimately informs a robust chemical manufacturing process. For a given chemical step that may require downstream removal of a metal catalyst, the use of X-ray fluorescence (XRF) enables rapid elemental analysis of samples for pharmaceutical process development. Together, automated parallel experimentation and XRF technology have proven to be a valuable partnership for the high-throughput analysis of large sample arrays for obtaining process knowledge and advancing pharmaceutical process development. Herein, we exhibit a case study where the combination of highly-automated design of experiments (DoE) studies and XRF allowed for rapid metals analysis and, subsequently, significant improvements in processing a late stage asset to remove trace metals.

423647 Pharmaceutical Process Development Using Computer Aided-Methods and Tools

Abstract:

A majority of pharmaceutical production systems involve batch processes, which have functioned well for pharmaceutical industries as well as the regulatory bodies. Batch vessels provide the flexibility of handling different unit operations in a multipurpose plant. However, batch processes are not very efficient for product quality assurance and have a number of drawbacks. Over the last decade the regulatory bodies required the pharmaceutical companies to demonstrate more process understanding. Moreover, financial drivers require profit expectations to be met due to increasing R&D cost and high competition [2]. Consequently, the pharmaceutical industry is expected to look for opportunities to rapidly evaluate the different alternatives for process improvement. To achieve this objective, Process Systems Engineering (PSE) methods and tools can play an important role, for example, (a) apply computer aided-methods and tools that are mature for other industries (such as chemical and petrochemical) also in solving problems in the pharmaceutical industries, (b) in the evaluation and implementation of alternative solutions and/or designs and (c) to evaluate opportunities for continuous manufacturing [1-2].

An integrated computer-aided framework for the development of pharmaceutical processes has been developed to tackle key research areas for pharmaceutical processes (such as continuous processing, separation and reaction processes, solvent selection/recycle/optimization and process intensification/integration [3]). The objective of the framework is to improve pharmaceutical process operation/design and to evaluate opportunities for continuous manufacturing through better process understanding of pharmaceutical processes by using systematic model-based methods. Computer aided-methods and tools are important for such a framework as they can assist in solving various problems such as property prediction, mixture separation (ternary and binary azeotropic separations, separation sequence and type), solvent selection and process modeling and in providing data through knowledge libraries. The developed framework consists of four main parts: reaction mechanism identification, reaction analysis, separation synthesis and design and finally process simulation, evaluation, optimization and operation.

To extend the applicability of the framework a solvent swap method has been developed and implemented in a computer-aided tool. The solvent swap method is based on the method proposed by Gani et al. 2006 [4] and is treated as a special solvent selection problem with additional criteria consideration. The objective of the method is the quick and reliable identification of swap solvents which are suitable for the following process step and it can increase the process efficiency (minimize solvent and product loss and reduce energy demands).

The application of the proposed integrated framework with the added features will be highlighted through the BHC (Boots and Hoechst-Celanese) patented synthesis pathway for ibuprofen and the production of region-someric products through Heck reactions [5]. The improved BHC synthesis of ibuprofen consists of three reaction steps [6] and it is used to verify the proposed integrated framework. Finally, the application of the solvent swap method using the computer-aided tool for both examples will be illustrated.

References:

[1] K. Plumb, 2005. Chem. Eng. Res. Des., 83, 730.

[2] K.V. Gernaey, A. E. Cervera- Pardell, J. M. Woodley, 2012. Comput. Chem. Eng., 42, 30

[3] C. Jiménez-González et al., 2011. Org. Process Res. Dev. 15(4), 900.

[4] R. Gani, M. Jones, I. Powell, J. H. Atherton, J. L. Cordiner, 2006. Chem. Eng. 30.

[5] R. L. Hartman, J. R.Naber, S. L. Buchwald, & K. F. Jensen, 2010. Angew. Chem. Int. Ed. Engl., 49(5), 899.

[6] Elango, V. Murphy, M., Smith, B.L., Davenport, K.G., Mott, G.N., Zey, E.G., Moss, G.L. (Hoechst Celanese Corp.), 1990. European Patent 4.981.995

Accuracy of Micro Powder Dosing and Feeding via a Vibratory Sieve-Chute System

M.O. Besenhard[1,2], E. Faulhammer[1,2], S. Fathollahi[1,2], G. Reif[2], V. Calzolari[3], S. Biserni[3], A. Ferrari[3], S.M. Lawrence[4], M. Llusa[1], A. Paudel, J.G. Khinast[1,2*]

1: Research Center Pharmaceutical Engineering (RCPE) GmbH, 8010 Graz, Austria
2: Graz University of Technology, Institute of Process and Particle Engineering, 8010 Graz, Austria
3: MG2, Via del Savena 18, I-40065, Pian di Macina di Pianoro, Bologna, Italy

4: GlaxoSmithKline, New Frontiers Science Park, Third Avenue, Harlow, Essex CM19 5AW, UK

* Corresponding author:

khinast@tugraz.at

Abstract:

This presentation describes our current studies on powder dosing [1] and feeding via a vibratory sieve-chute system, i.e., the MG2's Microdose®. Figure 1 shows the operating principle and parts of the set-up. The chute is tilted at a fixed angle of 5° and sieves with 5-10 holes of e.g. 0.7mm in diameter are fixed on its top. The powder was discharged from the sieve into the chute and the capsule body using gravity. Vertical vibration occurred with a broad range of frequencies and amplitudes. The fill weight (*fw* fw) during dosing events was recorded via a capacitance sensor, which had two parallel electrode plates encompassing the capsule body. The electrical field and the capacitance varied depending on of the powder quantity in the capsule. Due to this scale, fill weights could be recorded dynamically facilitating the analysis of filling characteristics, i.e., the fill rates and their robustness.

First, the range of frequencies and amplitudes was screened for settings that facilitated reasonable (no blocking of spilling in the sieve) fill rates for three lactose powders, serving as model substances. The filling characteristics were studied in detail within this operating space.

The filling rate was determined by averaging the slopes (*Δfw(t) Δt [mg s]* $\Delta fw(t)/\Delta t\ [mg/s]$) of *30* 30 filling profiles. The higher the probabilities of an unsteady filling behavior (as shown in Figure 5 b) or fluctuations in the filling rate are, the higher is the standard deviation ([mg/s]) of the evaluated slopes. Therefore, the latter can be used to characterize the robustness of the fill rate, which is of prime importance for selecting the settings for continuous (micro) feeding using such a system.

The results reveal similar operating spaces for all powders and the fill rate robustness varied distinctly in the operating space. Figure 2 depicts the results for one of the studied powders. Utilizing the gained knowledge on filling characteristics micro dosing studies have been performed, dosing capsules with 2.5 mg. These studies compared accuracy of dosing via a constant frequency and amplitude or a constant frequency but the amplitude manipulated by a PID controller. The results reveal that dosing was the most accurate when using constant settings, since the usage of PID control caused inhomogeneous powder beds on the chute, as shown in figure 3.

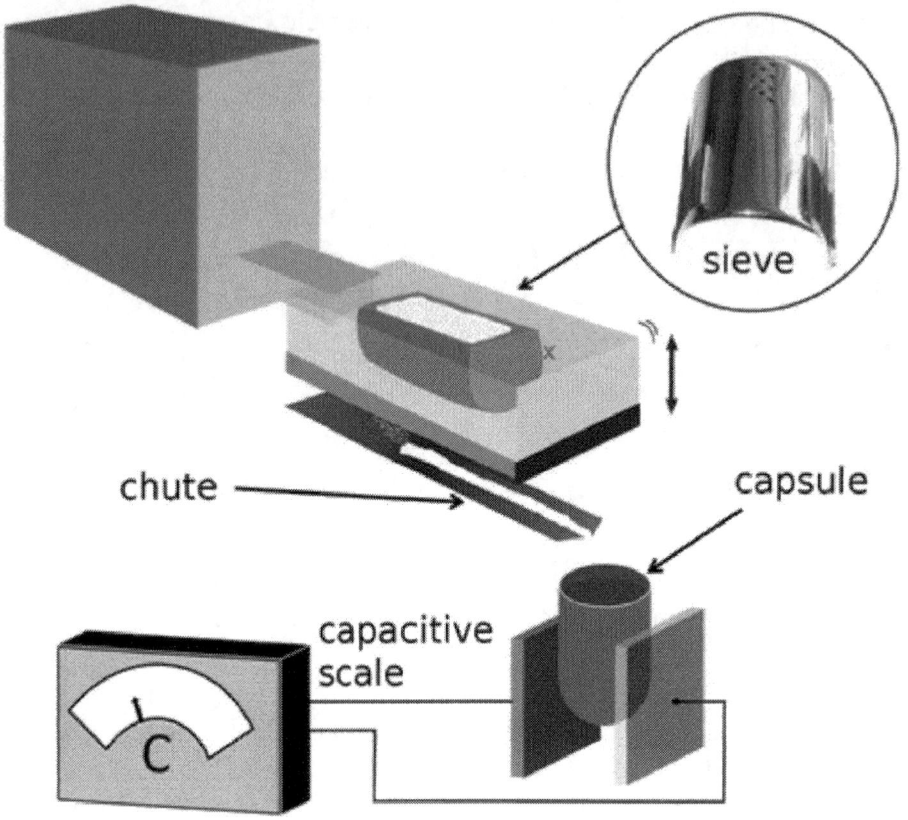

Figure SEQ figure * ARABIC 1: Principle of powder dosing via a vibratory sieve-chute system. The inlet shows a photo of the sieve used, and the red x depicts the position of an acceleration probe used to study the oscillation patterns.

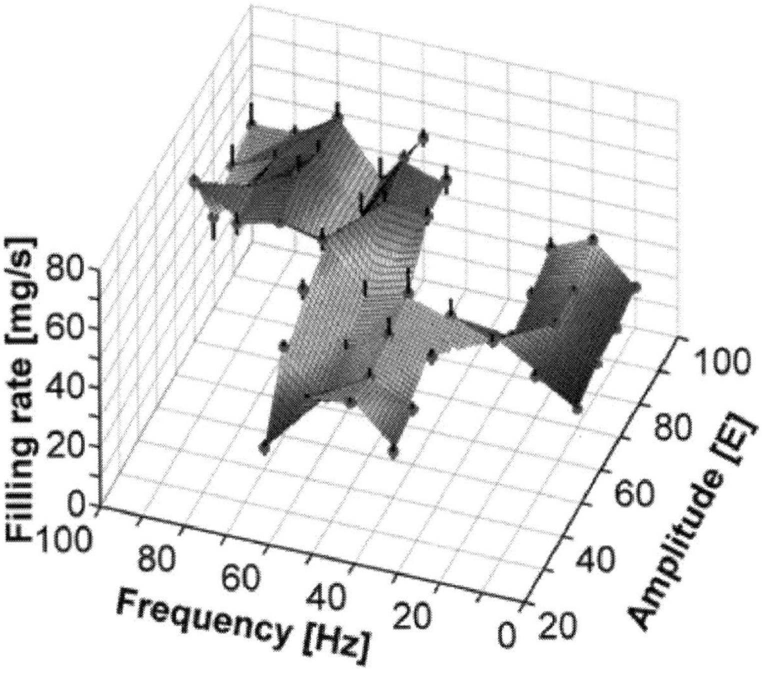

Figure 2: Filling rate and its standard deviations (vertical bars) for InhaLac 230 in the amplitude and frequency settings of the operating space.

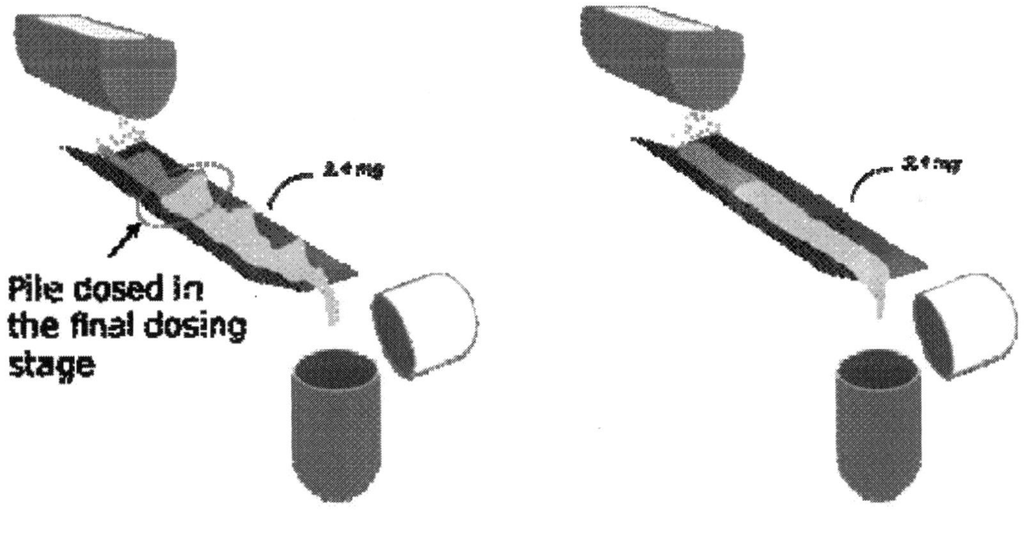

Figure 3: Powder bed in the chute during a PID controlled dosing (left) and dosing using a constant amplitude (right).

[1] Besenhard, M.O., Faulhammer, E., Fathollahi, S., Reif, G., Calzolari, V., Biserni, S., Ferrari, A., Lawrence, S.M., Llusa, M., Khinast, J.G. (2015). Accuracy of Micro Powder Dosing via a Vibratory Sieve-Chute System. *European Journal of Pharmaceutics and Biopharmaceutics,* accepted.

431162 Chromatographic Tools to Predict the Stability of MAbs for Faster Identification of Therapeutic Candidates

Abstract:

Chromatographic Tools to Predict the Stability of mAbs for Faster Identification of Therapeutic Candidates

S. Hedberg*, J.Y.Y Heng*, D.R. Williams* and J. Liddell**

*Surface and Particle Engineering Laboratory

Department of Chemical Engineering

Imperial College London, SW7 2BY,

United Kingdom

**Research and Development

Fujifilm Diosynth Biotechnologies

Billingham, TS23 1LH, United Kingdom

ABSTRACT

Protein-protein molecular interactions are known to be involved in protein solution aggregation behaviour and are a common issue for the manufacturing of therapeutic proteins such as mAbs. Much effort has been employed to gain a better understanding of aggregation, however the mechanisms leading to protein aggregation are still not fully understood. The osmotic second virial coefficient (B_{22}) is a fundamental physiochemical property that describes protein-protein interactions solution, which can be a useful tool to predict aggregation propensity of proteins.

One way of predicting aggregation propensity is self-interaction chromatography (SIC), which recently have shown to be a promising tool for better understanding of phase behaviour of proteins. Another technique, cross-interaction chromatography (CIC), has shown to be an even more high-throughput technique than its predecessor with possibly the same capabilities. This work consists of two experimental studies with therapeutic mAbs to improve SIC and CIC as a tool to predict protein aggregation.

The first part includes a 10 times scale-down study of therapeutic mAbs from laboratory scale macro-columns to micro-scale columns, which will enable the determination of B_{22} for the individual protein as well as the cross-virial coefficient, B_{23}, between two proteins. Micro SIC and CIC uses only a few milligrams of mAb in order to obtain a complete formulation study. The results from the first part of the study proved to give good comparable results between the micro and macro scales enabling the use of micro SIC for B_{22} determinations.

The second part of this work presents an extensive formulation study of mAbs, varying pH and salt, as well as the presence of different stabilisers as well as different external factors known to induce aggregation. The B_{22} and B_{23} values determined from the formulation study are then correlated with aggregation data obtained from size-exclusion chromatography. It was shown that over all test conditions, good correlations could especially be found between B_{22} and aggregation rate.

In summary this work hopes to lead to a better insight into the mechanisms of aggregation by looking at the extent of tailing of the peaks in the chromatograms and comparison with conformational studies.

REFERENCES

1. Dumetz, A.C., Snellinger-O'Brien, A.M., Kaler, E.W., and Lenhoff, A.M. (2007) *Patterns of Protein–Protein Interactions in Salt Solutions and Implications for Protein Crystallization.* Protein science. **16**(9): p. 1867-1877.

2. Quigley, A. (2012) *Protein Aggregation Behaviour and the Second Virial Coefficient,* Ph.D. Thesis, Imperial College London.

3. Tessier, P.M., Sandler, S.I., and Lenhoff, A.M. (2004) *Direct Measurement of Protein Osmotic Second Virial Cross Coefficients by Cross-Interaction Chromatography.* Protein science. **13**(5): p. 1379-1390.

4. Jacobs, S.A., Wu, S.-J., Feng, Y., Bethea, D., and O'Neil, K.T. (2010) *Cross-Interaction Chromatography: A Rapid Method to Identify Highly Soluble Monoclonal Antibody Candidates.* Pharmaceutical Research. **27**(1): p. 65-71.

Advancements of Accelerated Stability Studies and Modeling through Use of Nonlinear Kinetics and Bootstrapping

Don Clancy, Neil Hodnett, Rachel Orr, Martin Owen, and John Peterson

Primary corresponding author email: Donald.J.Clancy@gsk.com

What is Accelerated Stability? – 4 Steps

- Conduct a short term designed experiment at elevated temperature with varying humidity.

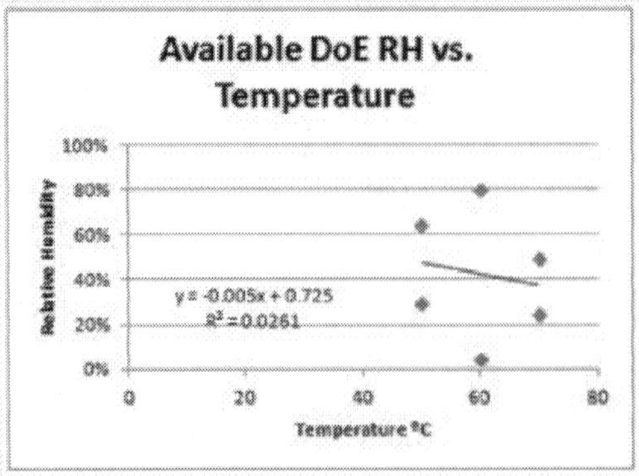

- Fit kinetic models for impurities and physical changes using short term data under accelerated conditions.
- Develop a long term model to predict container humidity and moisture content for a typical medicine bottle or blister pack.
- Extrapolate kinetic models from short term studies and integrate them with the long term humidity model to predict long term stability.

Background and Current Available Approach

- Currently, the most widely published approach for modeling drug product degradation based on accelerated stability data is that described by Waterman (Waterman, 2012), (Adami & Waterman, 2004). This is based on the modified Arrhenius equation shown in Equation (1):

$$ln(k) = ln(A) - \frac{Ea}{R \cdot T} + B \cdot (RH) \quad 1$$

Where "A", "Ea", and "B" are fitted terms, "R" is the gas constant, "T" is temperature in Kelvin, and "k" is defined as the change in impurity with respect to time.

Rearranging into a typical kinetic rate expression:

$$\frac{dy}{dt} = A \cdot exp\left(\frac{-Ea}{R \cdot T}\right) \cdot exp(B \cdot RH) \quad 2$$

Makes the implicit assumption the rates of reaction is an exponential function of RH, which implies moisture affects molecular mobility similar to temperature.

Background Continued - Isoconversion

- Equation 1 is further reduced to predict time to failure (t.alpha), or "isoconversion."

$$ln\left(t_{\alpha}\right) = \frac{Ea}{R \cdot T} - B \cdot RH - C \quad (3)$$

- This approach often requires significant extrapolation or interpolation of data to a failure limit to estimate the time till failure for each condition.

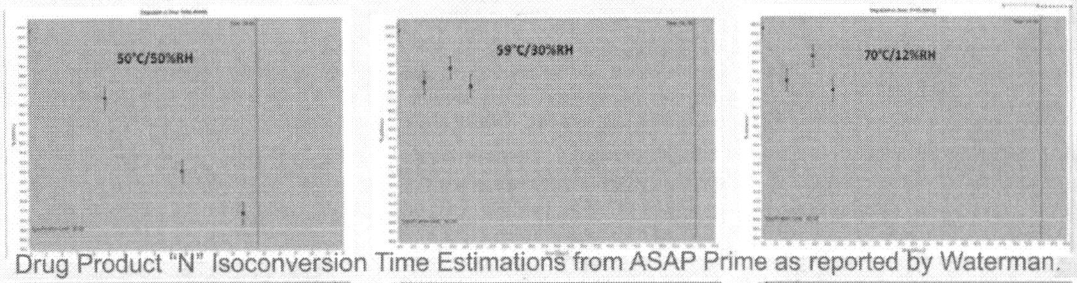

Drug Product "N" Isoconversion Time Estimations from ASAP Prime as reported by Waterman.

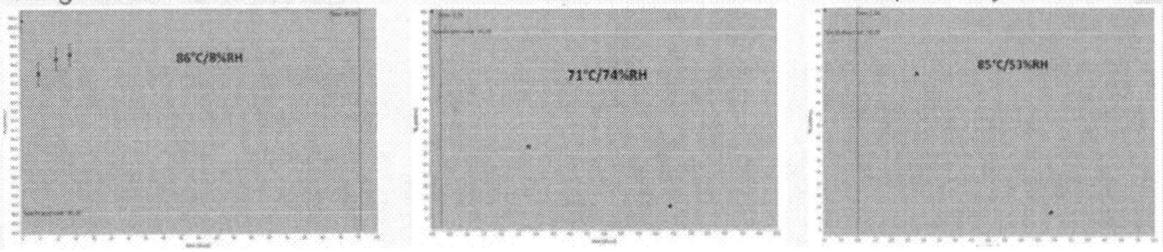

Kinetic Model Development With a Different Starting Assumption

Change in reaction conversion (called alpha in this analysis) as a solid API reacts with water is typically related to the "activity" of the reactants, raised to a power, as:

$$\frac{d\alpha}{dt} = k \cdot \left[H2O_{activity}\right]^{N1} \tag{4}$$

The "k" term in these equations is an Arrhenius factor relating an exponential relationship between temperature "T", the activation energy of the reaction "Ea" and a gas constant "R". Expanding on the Arrhenius term for equation (5) yields (7):

$$\frac{d\alpha}{dt} = K1 \cdot exp\left(\frac{-Ea}{R \cdot T}\right) \cdot RH^{N1} \tag{7}$$

Where, K1 is a pre-exponential term called the "frequency factor." For simplicity of nomenclature and units in the models described in this report, the activation energy divided by the gas constant (Ea/R) will be simplified to a fitted parameter "K2".

$$\frac{d\alpha}{dt} = K1 \cdot exp\left(\frac{-K2}{T}\right) \cdot RH^{N1} \tag{8}$$

Kinetic Model Development - Continued

Fitting Equation 8 can be challenging as K1, and K2 are correlated in an exponential relationship. A way of reducing this exponential correlation between K1 and K2 is to scale the exponent to a value of 1.0 at the average temperature in the study (Tref) by modifying equation 8 to create equation 9.

$$\frac{d\alpha}{dt} = K1 \cdot exp\left(\frac{-K2}{T} + \frac{K2}{T_{ref}}\right) \cdot RH^{N1} \tag{9}$$

What is actually fitted to the experimental data is the integrated form of the kinetic models, as the final impurity value at time "t" is what is observed. Integrating (9) with respect to time yields (10):

$$\alpha = K1 \cdot exp\left(\frac{-K2}{T+273.15} + \frac{K2}{Tref}\right) \cdot (H)^{N1} \cdot \Delta t + C \tag{10}$$

Where "C" is the value of alpha at time zero, and "H" is the humidity expressed as either RH or Vapor pressure. The type of humidity dependence will need to be tested within the experimental analysis as two alternate models.

Accounting for the Potential Curvature With Respect to Time

Model Type	Description	Typical Application	Nonlinear Model (Integrated Form) Used for Fitting:	Limitations	Terms Needed to Fit
A	Reaction slows exponentially with time	Reaction rate proportional to a limiting reagent that is close to being consumed	$\alpha = C + K3 \cdot \left(1 - exp\left(-K1 \cdot exp\left(\frac{-K2}{T+273.15} + \frac{K2}{Tref}\right) \cdot H^{N1} \cdot \Delta t\right)\right)$		4
B	Reaction slows down with time	Diffusion	$\alpha = K1 \cdot exp\left(\frac{-K2}{T+273.15} + \frac{K2}{Tref}\right) \cdot (H)^{N1} \cdot \Delta t^{N2} + C$	N2<1	4
C	Reaction is linear with time	Zero order reactions	$\alpha = K1 \cdot exp\left(\frac{-K2}{T+273.15} + \frac{K2}{Tref}\right) \cdot (H)^{N1} \cdot \Delta t + C$		3
D	Reaction speeds up with time	Power Law Induction Reactions	$\alpha = K1 \cdot exp\left(\frac{-K2}{T+273.15} + \frac{K2}{Tref}\right) \cdot (H)^{N1} \cdot \Delta t^{N2} + C$	N2>1	4
E	Reaction speeds up exponentially with time	Nucleation, or reaction rate dependent on product concentration	$\alpha = exp(C) \cdot exp\left(K1 \cdot exp\left(\frac{-K2}{T+273.15} + \frac{K2}{Tref}\right) \cdot (H)^{N1} \cdot \Delta t\right)$		3

Table 1 Kinetic Equations Available in the Accelerated Stability Dashboard

- Each of these 5 possible curvature types with respect to time is tested with either relative humidity or vapor pressure in the "H" term.
- All models are fitted and then compared.
- Data with large extents of conversion (>15%) may need an Avrami transform if exponential.

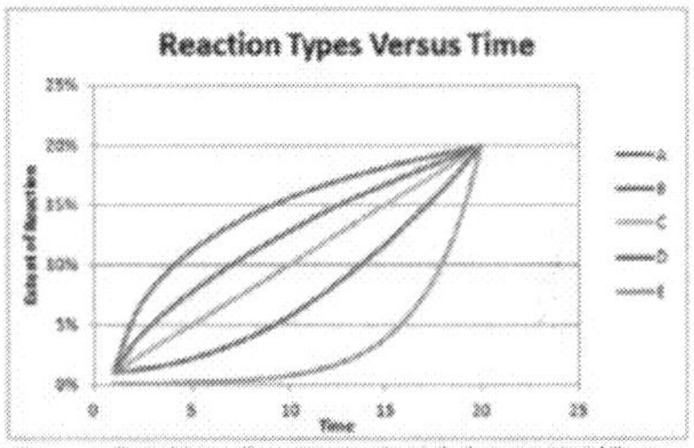

Figure 1 Typical Curve Shape Associated with Each Kinetic Model Type

Scientific Rational for Initial Model Selection

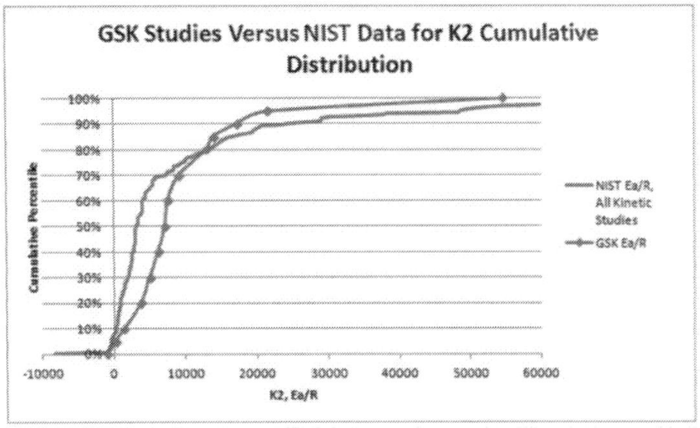

Figure 2 Cumulative Distribution of Ea/R for GSK studies Compared to All Organic Reactions in NIST Data Set

- To help decide if a potential kinetic model might be real, the activation energy K2 value is compared to the known potential distribution of activation energies to assess how unusual it is.
- Models with extremely unusual activation energy values are eliminated from further consideration.

Typical Power on Humidity Term (N1)

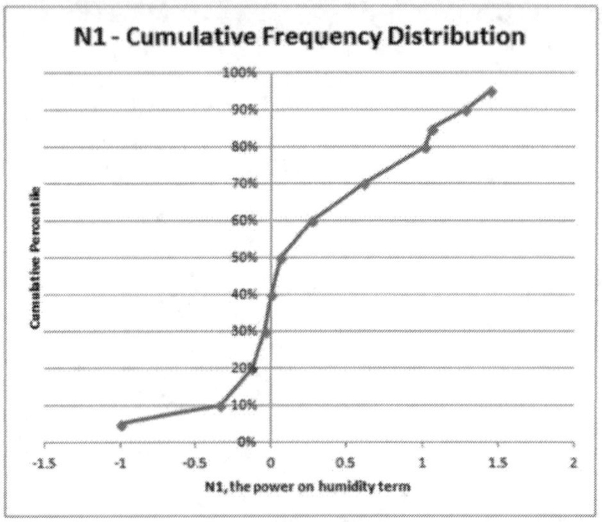

Figure 5 Cumulative Percentile Plot of N1 for GSK Accelerated Stability Studies

- From 120+ impurity analysis, the cumulative frequency distribution for the humidity power (N1) is found to typically fall between -1 and 1, which is the opposite of an exponential curve with respect to RH.
- There is a long tail to the right with N1 values as high as 6 being noted for some unusual hydrolysis reactions in the top 5% of the distribution.

Example Kinetic Model Selection in GSK Software within JMP

Potential Kinetic Models and Parameters

Model Type	K1	K2	K3	N1	N2	nParams	RMSE	Score	AICc	BIC
RH Linear Time Kinetic Model	0.69	8374	-	3.34	-	4	0.1115	-	-22.964	-21.742
AH Linear Time Kinetic Model	118	-8801	-	3.34	-	4	0.1115	-	-22.966	-21.744
RH Accelerating Kinetic Model	4.4	-17419	-	11.19	-	4	0.3514	-	25.237	26.459
AH Accelerating Kinetic Model	1	1000	-	1	-	4	1.3584	-	82.024	83.247
RH Decelerating Kinetic Model	0.33	12601	3.7	3.35	-	5	0.0221	-	-88.206	-87.939
AH Decelerating Kinetic Model	57.33	-4588	3.7	3.34	-	5	0.0221	-	-88.3	-88.032
RH Power Model	1.9	7500	-	3	0.5	5	0	-	-848.163	-847.896
AH Power Model	193.48	-7937	-	3	0.5	5	0.0001	-	-312.928	-312.66

Table 2 Fit of All Possible models to Data Set Created by an RH Model, with K2 Value of 7500, and N2 Value of 3.0, as applied to a typical Accelerated Stability study

- Potential kinetic models fit with highly unusual K2 and N1 parameters are first eliminated, as shown in red.
- Of the remaining models, the model with the best (lowest) Bayesian Information Criterion (BIC) is then chosen.

What Can Go Wrong in Accelerated Stability? Nitrazepam Stability (1977)

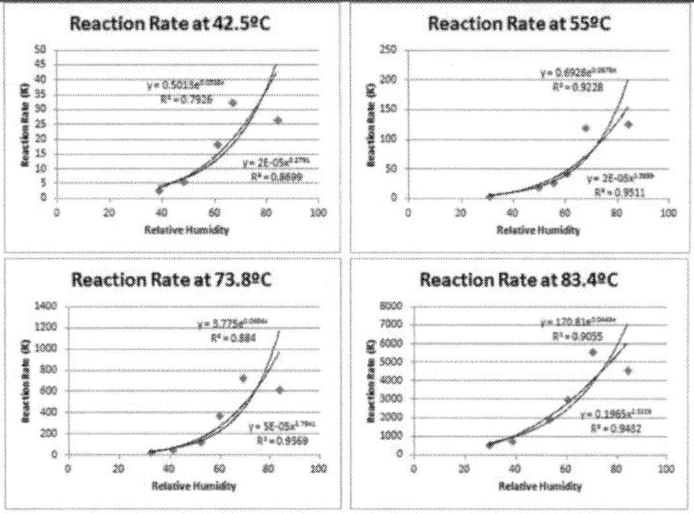

Figure 6 Data Visualization for Reaction Rate for Disappearance of nitrazepam at varying temperatures

The analysis of nitrazepam stability data from Kesselring's and Genton's 1977 paper (Kesselring & Genton, 1977) is of particular interest as it's an unusual case due to high level of humidity sensitivity and is a good example of what can go wrong in an Accelerated Stability study.

This study also has historical significance as it is referenced in Waterman's papers to support the hypothesis that the kinetics of degradation for drug products can best be modeled as an exponential function of relative humidity.

Analysis of Nitrazepam Stability

The data in the paper is expressed as reaction rate "K" which is the change in nitrazepam purity with respect to time. Our standard model given as equation 11 is simplified to equation (11) to predict K.

$$\frac{dy}{dt} = K = K1 \cdot exp\left(\frac{-K2}{T+273.15} + \frac{K2}{335.83}\right) \cdot (RH)^{N1} \quad (11)$$

Waterman's equation from equation 2, the "RH Exponential Model" is expressed as equation (12) using the standard nomenclature developed in this paper.

$$\frac{dy}{dt} = K = K1 \cdot exp\left(\frac{-K2}{T+273.15} + \frac{K2}{335.83}\right) \cdot exp(B \cdot RH) \quad (12)$$

- Since the Nitrazepam data set was the primary reference in prior work to the hypothesis that reaction rates are proportional to the exponential of RH, both equation forms will be tested.

Comparison of Model Fits

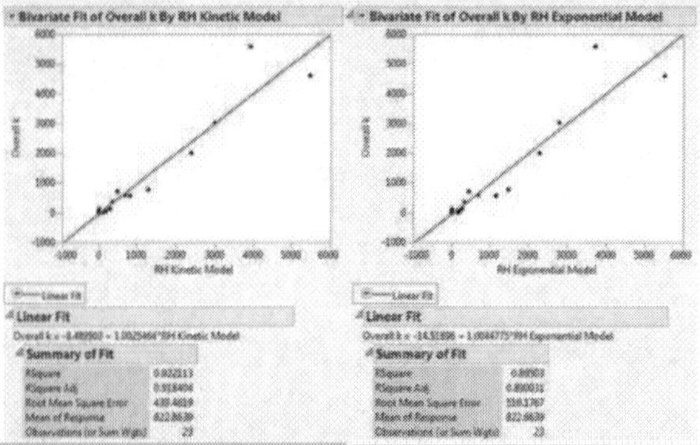

Solution (left)

SSE	DFE	MSE	RMSE
4056908.012	20	202845.4	450.38361

Parameter	Estimate	ApproxStdErr
K1	0.0167416035	0.02574611
K2	26757.744805	5939.75581
N1	1.8220013208	0.27078848
C	0	0

Solution (right)

SSE	DFE	MSE	RMSE
5469488.3438	20	273474.42	522.94781

Parameter	Estimate	ApproxStdErr
K1	4.8687759394	6.08294001
K2	26807.829412	6997.3303
C	1e-100	0
B	0.0285954924	0.00467036

Figure 7 Fits for Linear Kinetic Model as Derived versus Waterman's Exponential RH Model (Right)

- This data does not support the hypothesis that this reaction has an exponential rate with respect to humidity. A typical kinetic equation fits the data slightly better.

Dropping Highest Temperature Data Due to Change in Activation Energy

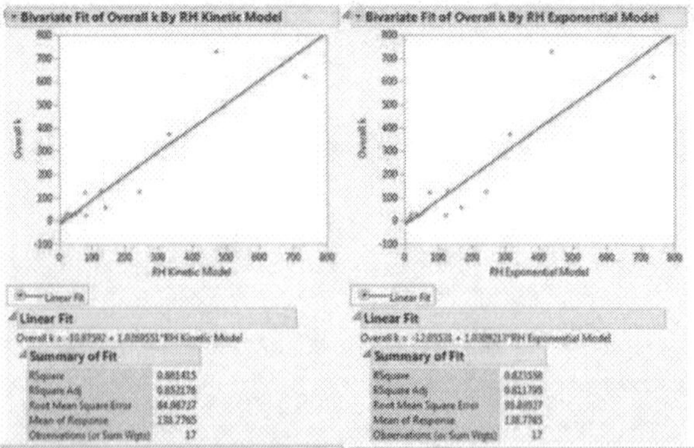

Solution (left)

SSE	DFE	MSE	RMSE
109627.14446	14	7830.5103	88.490171

Parameter	Estimate	ApproxStdErr
K1	0.0106708776	0.02166786
K2	10458.983857	2823.98606
N1	2.2911430108	0.4702406
C	0	0

Solution (right)

SSE	DFE	MSE	RMSE
139482.01185	14	9963.0008	99.814833

Parameter	Estimate	ApproxStdErr
K1	14.846054948	9.50787154
K2	10364.598576	3210.1629
C	1e-100	0
B	0.0348019883	0.00768587

Figure 8 Fits for Normal Kinetic Model by Equation 11 (Left) versus Waterman's Exponential Model (Right) for Overall Disappearance of nitrazapem, excluding data at highest temperature in study

- If the activation energy markedly changes as the highest temperature data is dropped, then likely there was a phase change, form change, or some other physical change impacting the results. Here the estimate changed from K2 value of 26,800 to 10,500.

- **RECOMMENDATION:** Get DSC and/or TGA data up to 90 degrees prior to the study.

Final Model Fit – Dropping Highest Humidity Data

Solution			
SSE	DFE	MSE	RMSE
3620.9308018	11	329.17553	18.143195

Parameter	Estimate	ApproxStdErr
K1	7.6732252e-8	8.45784e-8
K2	10836.303303	770.225387
N1	5.1846287097	0.26325205
C	0	0

Solution			
SSE	DFE	MSE	RMSE
4957.4557942	11	450.6778	21.229173

Parameter	Estimate	ApproxStdErr
K1	0.6966261868	0.22797375
K2	10723.079571	899.058984
C	1e-100	0
B	0.086179662	0.00482877

Figure 9 Parameter Fits for Normal Kinetic Model by Equation 11 (Left) versus Waterman's Exponential Model (Right) for Overall Disappearance of Nitrazapem, excluding data at highest temperature and highest humidity in study

- Dropping highest humidity in this study also results in significant changes in the estimate of the humidity sensitivity. Likely some change in physical behavior above 75% RH.
- Leads to conclusion that experimental designs need to have the flexibility to test for changes in humidity sensitivity at extremes in humidity.
- Finally, while the impact of RH may have looked exponential in this case, it fits standard kinetics consistently better. High values of N1 as high as 6 have been observed for some unusual hydrolysis reactions.

Initial Experimental Designs - Hexagon

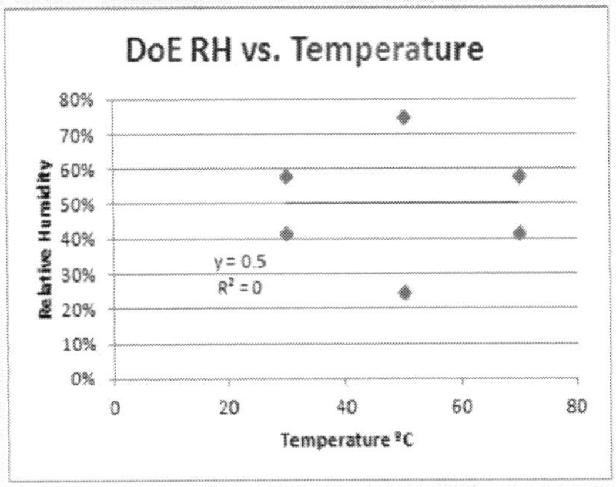

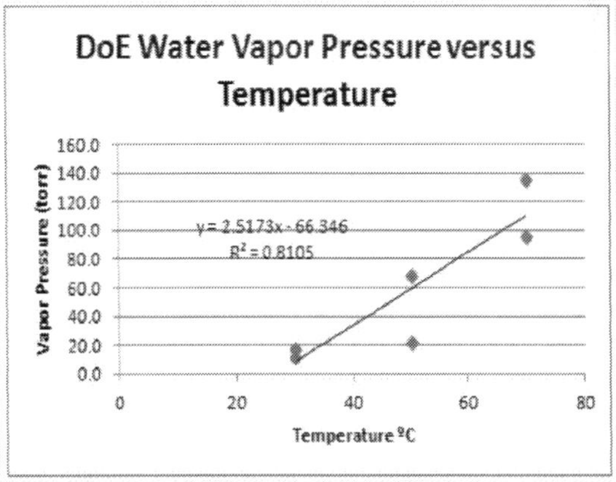

- A mathematically "optimized design" for fitting kinetic models looks like a square, and only tests a low and high humidity value in the design. This is a risky design due to the observed potential changes in kinetics at low or high humidity.
- A Hexagon is an attractive design, as it allows for 3 levels of temperature, and 3 levels of humidity within the design.
- Problem with this design is the potential correlation between temperature and vapor pressure makes fitting vapor pressure based models prone to error in parameter estimation.

Typical Initial Experimental Design - Offset Hexagon

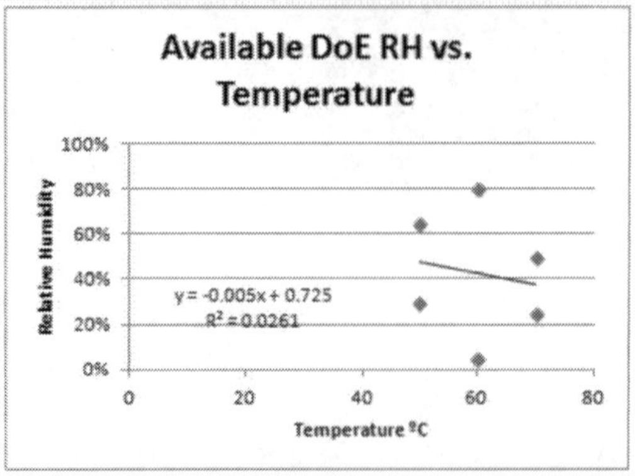

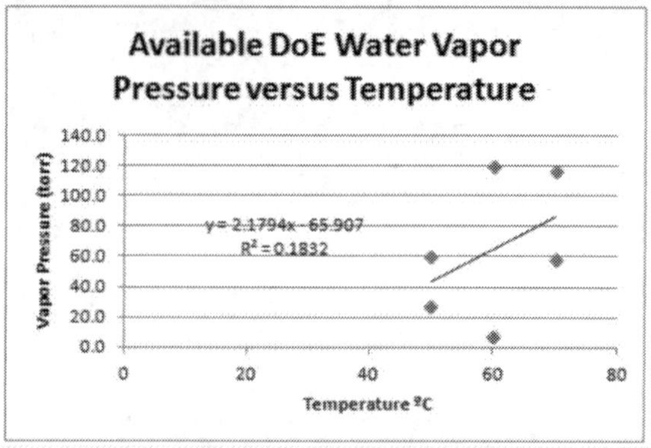

- Skewing the run points slightly reduces the resulting correlation between vapor pressure and temperature in the study without compromising the RH based design.

- Allows either RH or vapor pressure to be fit as the humidity term within the kinetic equations.

Estimation of Relative Humidity in Long Term Storage

- Moistures permeate into plastic storage bottles, following a standard permeation model, same process as polymer membranes:

$$Flux = P \cdot \left(\frac{SA}{thickness}\right) \cdot \left(VP_{exterior} - VP_{interior}\right) \cdot \Delta t \qquad (3)$$

Where

"Flux" is the water vapor transmission rate,
"P" is the permeability coefficient,
"SA" is the surface area,
"thickness" is the wall thickness,
"VP $_{exterior}$" is the water vapor pressure on the exterior of the bottle or capsule film,
"VP $_{interior}$" is the water vapor pressure in the interior of the bottle or capsule film and,
"Δt" is the change in time.

Permeability coefficients follow an Arrhenius relationship with temperature, and can be calculated at different temperatures using equation 4 (Abdel-Bary, 2003).

$$P = P0 \cdot exp\left(\left(\frac{Ea}{R}\right) \cdot \left(\frac{1}{273.15 + T_{ref}} - \frac{1}{273.15 + T}\right)\right) \qquad (4)$$

Solve for Equilibrium RH on Each Day

- Need to experimentally measure and define models for:
  - Moisture permeation of bottles at varying temperatures.
  - Moisture versus RH relationship of desiccant
  - Moisture versus RH relationship of tablet
- With known initial water content, and known permeation of water for a given day, four equations and four unknowns can be solved to find the equilibrium RH on each day.
  - Water Fraction in air
  - Water fraction in tablet
  - Water fraction in desiccant
  - Overall mass balance

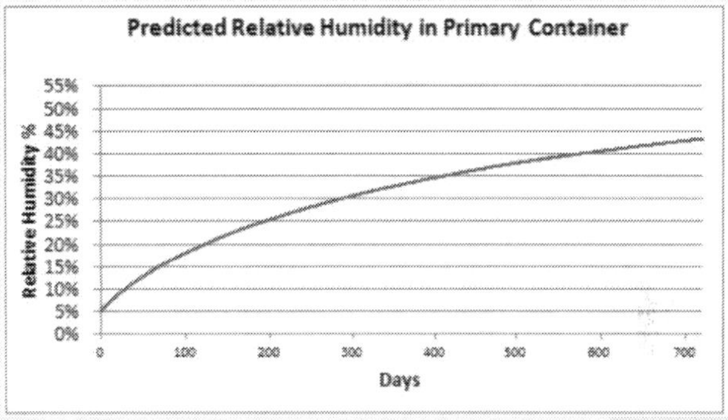

Estimation of Uncertainty in Prediction – Bootstrapping Approach

- The kinetic models can then be integrated over time with the estimated humidity curve to create a predicted line.
- To estimate the error of the prediction, the error in the kinetic model fit is assumed equivalent to the error in the experimental data.
- Each raw data point is perturbed with random noise equivalent to the error of the model. A new model is generated and plotted for the long term prediction.
- Process is repeated several thousand times to generate a distribution in the prediction.

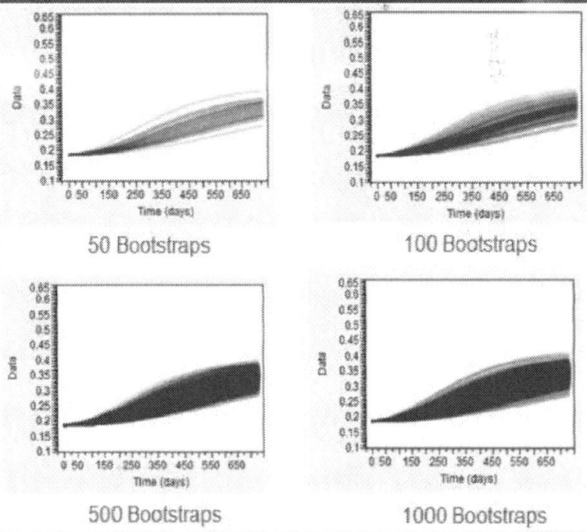

Example 1 from Drug Product 81:

	Specification Limit		35	35	35	34	60	80
	Assay/Imp		Impurity	Impurity	Impurity	Impurity	Impurity	Impurity
Temperature	Relative Humidity	Time	Imp101	Imp966	Imp545	Imp799	Imp283	Imp102
°C	%	days	parts per 10000	parts per 10000	parts per 10000	parts per 10000	parts per 10000	parts per 10000
71.3	77.6	4	6.7	14.8	9.3	7.7	38.2	13.2
79.3	27.7	4	8.3	13.5	9.5	6.9	36.8	14.0
61.8	39.9	7	7.4	6.0	9.4	6.1	34.1	13.1
71.5	51.3	7	7.5	12.9	9.7	7.2	35.6	13.7
71.3	77.6	7	7.2	13.2	8.8	8.8	37.8	14.1
79.3	27.7	7	9.8	16.3	9.9	7.5	37.7	14.0
61.9	6.6	10	9.2	5.7	9.4	6.0	34.9	13.8
61.8	39.9	10	8.1	6.9	9.4	6.4	34.6	14.1
71.5	51.3	10	8.2	14.6	9.7	7.8	36.1	13.4
50.4	74.8	14	6.8	6.0	9.5	6.6	34.4	13.7
61.9	6.6	14	10.1	5.5	9.6	6.5	34.6	14.5
50.4	74.8	10	6.9	5.4	9.4	6.6	34.3	14.3
61.8	39.9	0	5.2	2.0	9.5	5.8	29.8	13.1

Figure 13 Drug Product 81 Accelerated Stability Experimental Data

- Just eyeballing this accelerated data, surely Impurity 966 is the worst one?
- Eyeballing again, that impurity 102, can't possibly be a risk, it's nowhere near the specification limit.
- Lesson #1: Don't believe that eyeballing data is a good approach. Don't stop doing an analysis due to little response in the data.

Example #1: Comparisons of Model Predictions to Long Term Data for Drug Product 81

- Impurity 102 had a low activation energy, did not accelerate substantially at higher temperatures, but ended up the most critical impurity at lower temperatures.
- Impurity 966 was low risk with high activation energy, and showed little to no increase at low temperatures.
- Points out a problem with the isoconversion approach – low activation energy reactions.

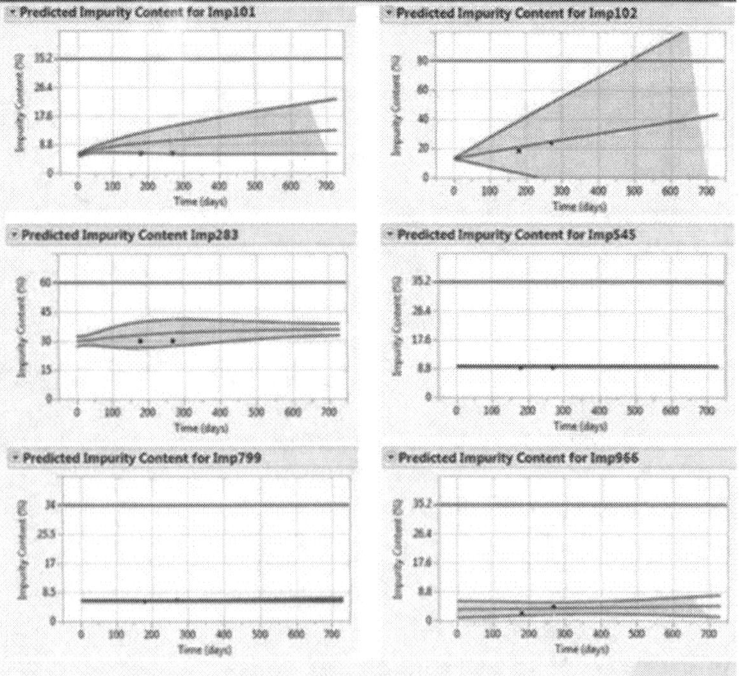

Drug Product 81 Impurity models overlaid with long term data at 30 deg C, 65% RH in HDPE Bottle with Desiccant

Example #2 – Drug Product "P"

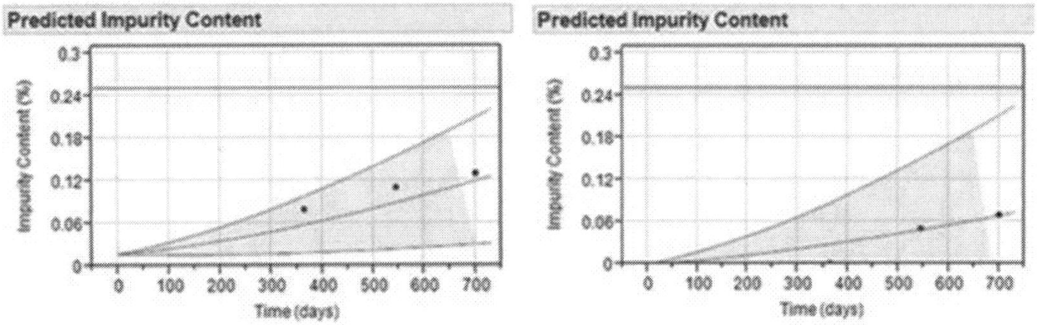

Figure 15 Drug Product P Impurities 0.94 (left) and 0.11 (right) models overlaid with long term data

•Predicted passing for both impurities of concern, and this proved true.

Example 3: Drug Product S

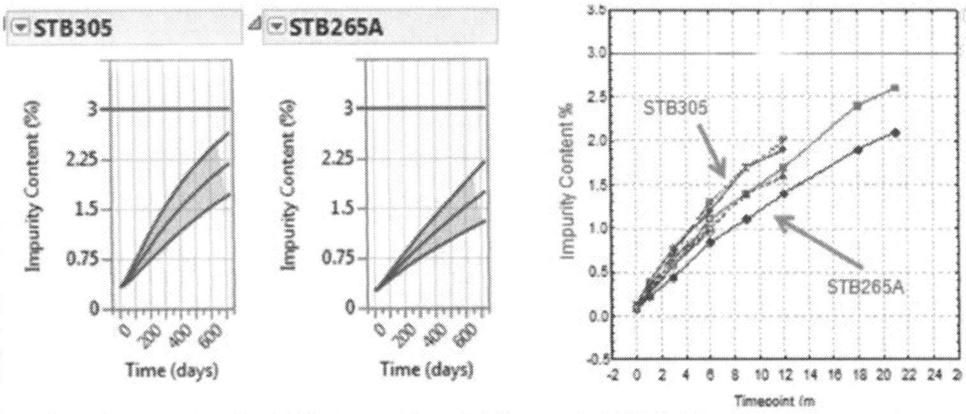

Figure 16 Drug Product S Batches 305 and 265A in Accelerated Stability Prediction (left and center) Compared to Long Term Observation (Right)

- Multiple formulations were compared in accelerated stability prediction (left), then run in long term stability (right).
- Accelerated stability predicted which formulation was going to have the most stable results.

Example 4: Desolvation of DMSO From an API

- Accelerated stability modeling can be used to predict some physical phenomenon, such as desolvation.
- Cannot predict other physical mechanisms, such as loss of free DMSO that became unbound from the API prior to the experiment start.
- The accelerated experiment must be run with similar DMSO vapor pressure as in final package in long term stability.

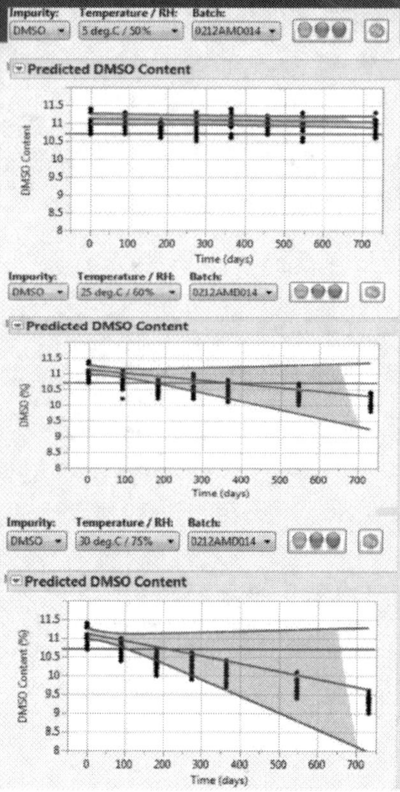

Example 5: Salt Disproportionation to Free Acid

- Can also predict salt disproportionation, and the subsequent nucleation and growth of free acid crystals.
- Note that this phenomenon was exponential in time, and followed an Avarami curve. What's predicted here is the Avarami transform of the data: Y*=ln(Y/(100-Y))

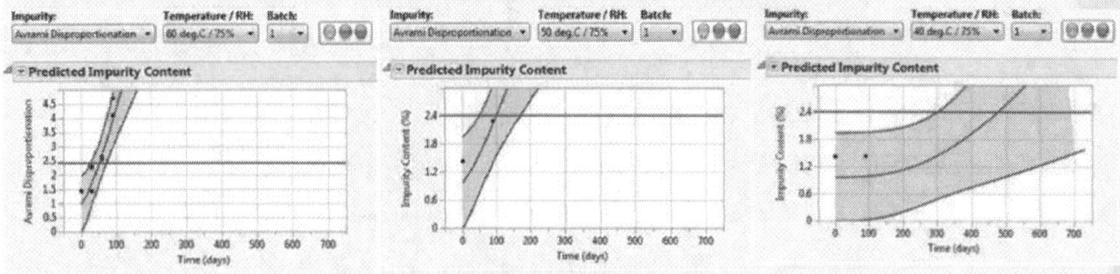

- The important prediction that the 30 degree 65% RH condition with desiccant will remain flat and undetectable, has proved true for the first year so far.

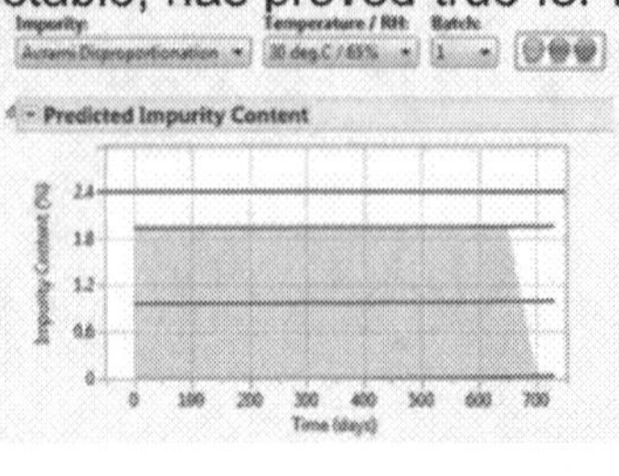

Example 6: Biological Product Main Isotherm Impurity

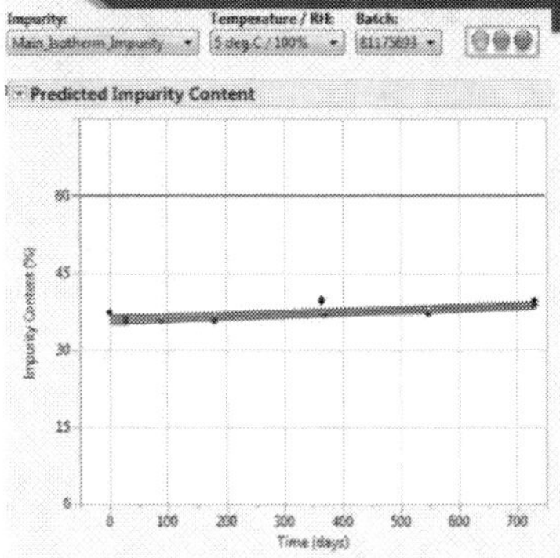

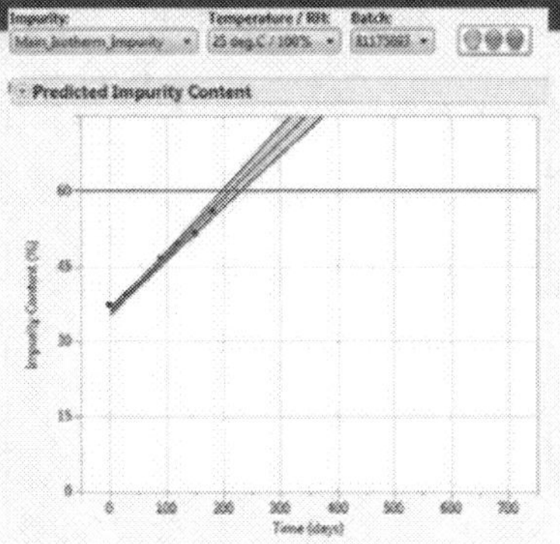

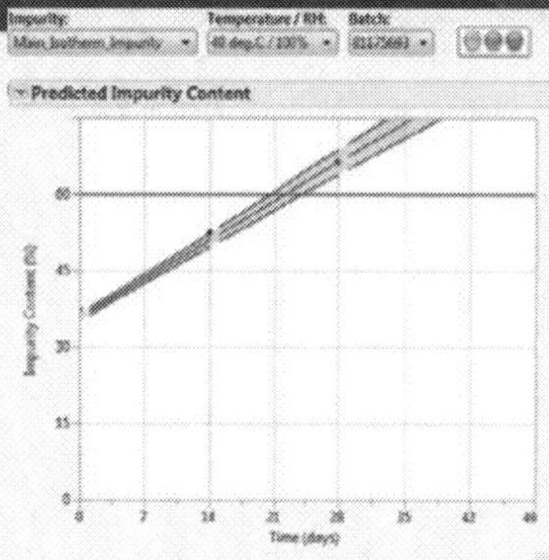

5 degrees 25 degrees 40 degrees

- Main isotherm impurities, and caliper reduced impurities are often well modeled as an Arrhenius model, linear with time, with no humidity impact since it's in liquid.
- Need lower temperatures in the study. Multiple years of shelf life at 5 degrees, reduces to 21 days at 40 degrees.
- Cannot predict below the freezing points generally around -15 degrees.

Example 7: Drug Product "N" Done in Both ASAP Prime and GSK Methodology

		Specification Limit	91.5	10	2.406846103
		RRT	impurity	Impurity	Impurity
Temperature	Relative Humidity	Time	Assay	Conversion	Avrami_Assay _Loss
°C	%	days	% a/a	% a/a	% a/a
50	50	0	100.3	0.7	-0.349720122
50	50	4.8	96.8	4.2	1.477558048
50	50	9.8	93	8	2.161962565
50	50	13.7	90.8	10.2	2.428848952
59	30	4.8	97.5	3.5	1.288031107
59	30	9.8	98.3	2.7	1.020348263
59	30	13.7	97.3	3.7	1.345654347
70	12	4.8	97.5	3.5	1.288031107
70	12	9.8	98.8	2.2	0.810480272
70	12	13.7	97	4	1.426703899
71	74	2.8	37.8	63.2	5.129115715
71	74	6.9	10.5	90.5	6.76909511
86	8	4.8	97.5	3.5	1.288031107
86	8	9.8	98.3	2.7	1.020348263
86	8	13.7	98.5	2.5	0.941354701
85	53	2.8	72.5	28.5	3.681438042
85	53	6.9	12.5	88.5	6.572394425

- Drug product "N" has an assay that declines with time.
- Since the extent of conversion is high (>15%) and exponential in time, an Avrami transform is used to linearize the data first.

Conversion = 101-Assay

Avrami Assay Loss = ln(100*(Conversion)/(101-Conversion))

Example 7: ASAP Prime Analysis

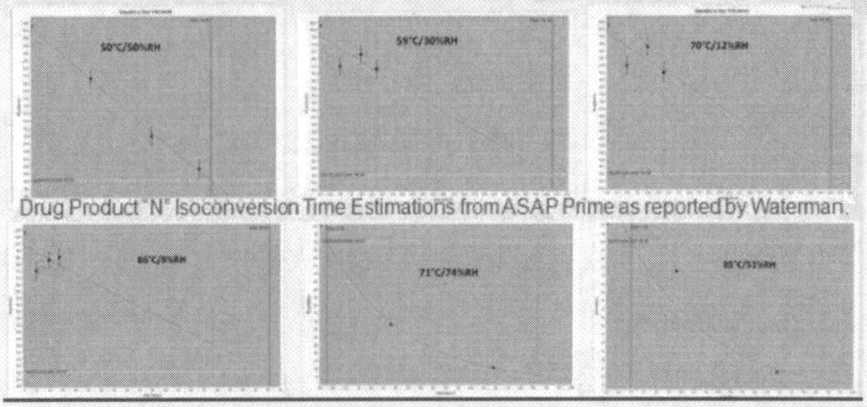

Drug Product "N" Isoconversion Time Estimations from ASAP Prime as reported by Waterman.

Temperature (°C)	%RH	Isoconversion Time (Days)	SD (Days)
50	50	14.6	1.2
59	30	56.4	18.4
70	12	55.0	17.5
71	74	0.3	0.1
85	53	1.0	0.0
86	8	95.6	44.8

Table 2 Calculated Isoconversion Time for Experimental Design Conditions

Example 7: ASAP Prime Predictions

ln A	18.7±3.1
E_a	15.6±2.0 kcal/mol (65.3±8.4 kJ/mol)
B	0.103±0.008
R^2	1.00

Table 3 Parameters for Moisture Modified Arrhenius Equation

Predictions were then made for shelf life in PVC blisters.

Condition	Measured Shelf-life (mos.)	ASAP*prime*™ Shelf-life (mos.) with PVdC
25°C/60%RH	17.9 (13.4-32.5)	11.2 (9.2-13.6)
30°C/65%RH	8.2 (5.7-11.3)	6.2 (5.3-7.3)
30°C/75%RH	4.8 (4.4-7.6)	5.2 (4.5-6.1)
40°C/75%RH	1.1 (0.8-1.3)	2.0 (1.8-2.2)

Table 4 Comparison between the measured long-term storage shelf-life and ASAP*prime*™ predicted shelf-life (to 91.5% potency)

- Due to the exponential assumption, an exponent cannot approach zero reaction rate at near zero humidity. The limit of an exponent of a small number is 1.
- As a result it tends to substantially overestimate reaction rates of humidity sensitive compounds extrapolated to low environmental humidity. These PVC blisters started with an RH of 6%.
- In this case the predicted failure time of the product in PVC blisters was 11 months, where the actual was approximately 18 months at 25 degrees C.

Example 7: GSK Methodology Prediction of RH in Blisters

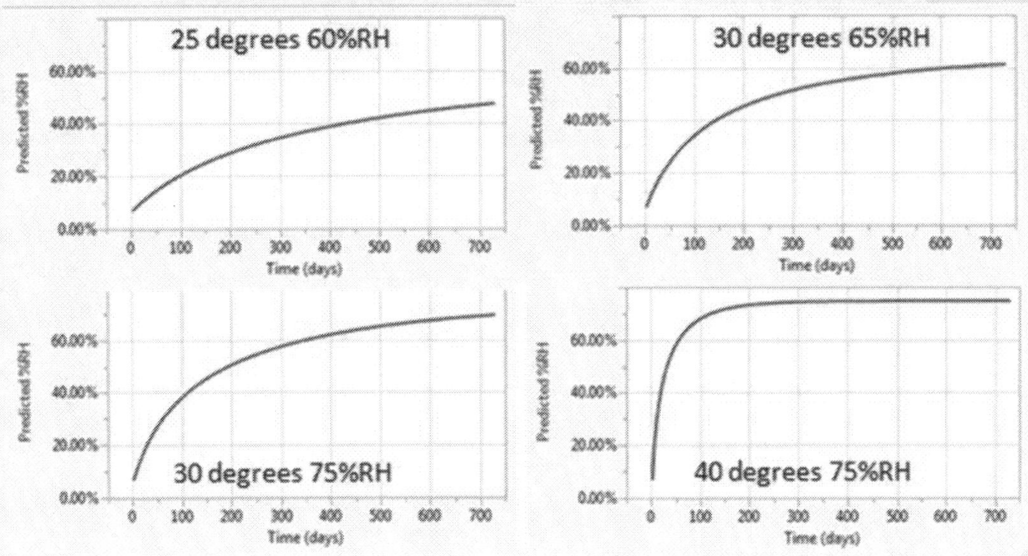

Example 7: Model Selection and Long Term Predictions versus Actual Long Term Data

Avrami_Assay_Loss Model Selection

Model Type	K1	K2	K3	N1	N2	nParams	RMSE	Score	AICc	BIC
RH Linear Time Kinetic Model	1.74	6562	-	2.61	-	4	0.6414	-	44.037	42.749
AH Linear Time Kinetic Model	45.74	-6632	-	2.59	-	4	0.6424	-	44.089	42.801
RH Accelerating Kinetic Model	-188386.9	15388	-	-3	-	4	3.0955	-	97.556	96.268
AH Accelerating Kinetic Model	-1	-7500	-	-3	-	4	3.197	-	98.654	97.365
RH Decelerating Kinetic Model	0.82	7344	6.82	3.67	-	5	0.4607	-	36.375	32.974
AH Decelerating Kinetic Model	80.08	-11170	6.82	3.63	-	5	0.4608	-	36.378	32.977
RH Power Model	3.77	4843	-	2	0.51	5	0.4789	-	37.688	34.287
AH Power Model	45.39	-5216	-	1.97	0.51	5	0.4786	-	37.665	34.264

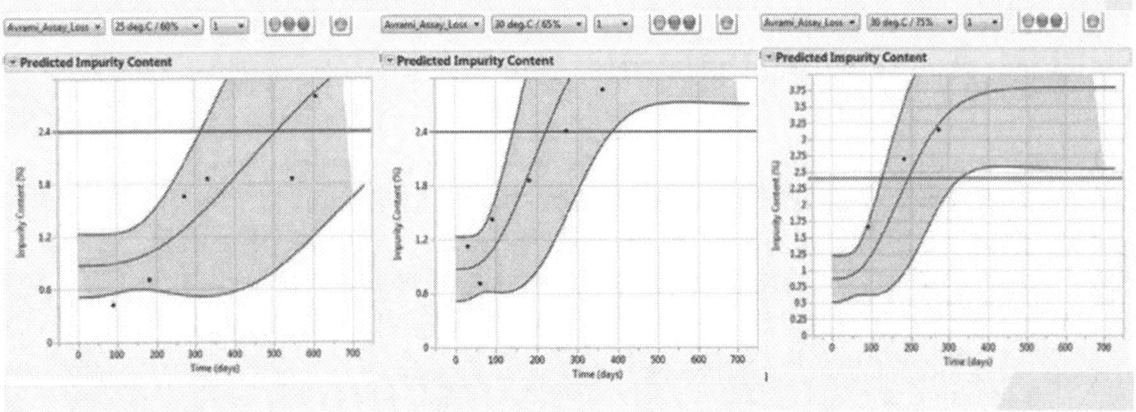

Example 7: Summary of Comparison of Results Between ASAP Prime and GSK's Methodology for Drug Product "N"

Storage Condition	Observed Time to Failure (months)	ASAP Prime Isoconversion method Prediction	GSK's Prediction
25 degrees 60% RH	17.9	11.2	16.8
30 degrees 65% RH	8.9	6.2	7.4
30 degrees 75% RH	5.3	5.2	6.3
40 degrees 75% RH	1.3	2	1.5
25 degrees 60% RH	percent error	37.4%	6.3%
30 degrees 65% RH	percent error	30.2%	16.7%
30 degrees 75% RH	percent error	1.2%	18.8%
40 degrees 75% RH	percent error	52.0%	12.5%
	Average percent error	30.2%	13.5%
	Slope actual vs. Predicted	1.84	1.10

- ASAP Prime's average error in prediction was 30% for this example.
- Methods described in this presentation resulted in an average of 13% error compared to actual long term data for this example.

Conclusions

- At GSK, we've observed that stability does not appear to be exponentially affected by increasing RH. Typically curvature is closer to cubed root or square root rather than exponential with power terms typically between -1 and 1, and the majority of data falling within -2 and 2 as would be typically expected for the order of reaction in reaction kinetics.

- We've chosen to pursue an approach using kinetics rather than focus on the isoconversion approach as it allows us to improve the logistics of doing a large number of accelerated studies.

- The aspiration of this model development would be to predict long term data ranging up to two years from a short 2 to 4 week accelerated stability study. The approach we've developed has thus far shown good fit with long term trends in available data. The number of studies is currently greater than 80, and impurities or physical phenomenon modeled greater than 300.

425091 Hydrodynamics and Mass Transfer Effects in Mini Vessel Dissolution Testing Systems

Abstract:

In the pharmaceutical industry, dissolution testing is routinely conducted to evaluate the *in vitro*performance of solid dosage forms in pharmaceutical development, determine the dissolution behavior of formulations and provide an optimization of drug release dosage forms, and as a quality control tool to insure that solid dosage forms have consistent dissolution properties. The USP Dissolution Testing Apparatus 2 is the device most commonly used for this purpose. In a typical test, a tablet is added to the vessel and samples are manually withdrawn from the vessels over time and analyzed for their drug content.

The USP 2 drug dissolution testing method is not always appropriate, especially for dissolution testing involving very small tablets or when not enough drug substance may be available for proper dissolution testing in the USP 2 Apparatus, especially during the initial stage of drug development. Small-volume dissolution testing apparatuses (mini vessel systems) can be advantageously used instead of the standard USP 2 system to overcome these limitations. The most common of these mini vessel systems is similar in shape and operation to the USP 2 apparatus, but it only requires a small volume vessel (200 mL) and a mini paddle impeller.

Before mini vessel systems can gain wide acceptance in industry, their hydrodynamics needs to be studied, and their operating conditions must be defined in order to compare the velocity distribution and mass transfer coefficient in a mini vessel system with those in the standard USP2 system.

Very few studies on mini vessel systems are available. Therefore, the objective of this work was to quantify the hydrodynamics and drug mass transfer rates in mini vessel systems, and determine the operating conditions under which mini vessel systems can be reliably used instead of USP 2 system for dissolution testing purposes.
The velocity distribution in a minivessel was obtained here using Computational Fluid Dynamics (CFD) at different agitation speeds. The predictions were compared to the velocity vector maps obtained using Particle Image Velocimetry (PIV). Four horizontal iso-surfaces were selected to be studied in detail. The CFC predicted and PIV measured flow patterns were found to be in good agreement. As expected, the 3D flow velocity magnitude increased as the agitation speed increased. However, the velocity in the inner core region below the impeller always remained very low.

The tablet-medium mass transfer process was additionally modeled, and the mass transfer coefficients at four different agitation speeds were obtained and compared to the standard system. The mini vessel system has a flow pattern similar to that of the standard USP 2 system. It was found that similar velocities at the tablet surface (on the iso-surface at y=7mm, i.e., at the vessel bottom) in the mini vessel system and in the USP 2 system were obtained when the mini vessel was operated a 100 mL, 125 rpm and the USP 2 system was operated at 900rpm and 100rpm, indicating that these two systems should result in similar dissolution profiles when operated under these conditions.

AUTHOR INDEX

AUTHOR INDEX

AUTHOR INDEX

AUTHOR INDEX

AUTHOR INDEX

AUTHOR INDEX

AUTHOR INDEX

AUTHOR INDEX

AUTHOR INDEX

AUTHOR INDEX

AUTHOR INDEX

AUTHOR INDEX

AUTHOR INDEX

AUTHOR INDEX